T0389011

NUMERICAL METHODS IN GEOTECHNICAL ENGINEERING IX

PROCEEDINGS OF THE 9TH EUROPEAN CONFERENCE ON NUMERICAL METHODS IN GEOTECHNICAL ENGINEERING (NUMGE 2018), 25–27 JUNE 2018, PORTO, PORTUGAL

Numerical Methods in Geotechnical Engineering IX

Editors

António S. Cardoso, José L. Borges, Pedro A. Costa, António T. Gomes, José C. Marques & Castorina S. Vieira

Faculty of Engineering, University of Porto, Porto, Portugal

VOLUME 2

CRC Press is an imprint of the Taylor & Francis Group, an **informa** business

A BALKEMA BOOK

CRC Press/Balkema is an imprint of the Taylor & Francis Group, an informa business

© 2018 Taylor & Francis Group, London, UK

Typeset by V Publishing Solutions Pvt Ltd., Chennai, India

All rights reserved. No part of this publication or the information contained herein may be reproduced, stored in a retrieval system, or transmitted in any form or by any means, electronic, mechanical, by photocopying, recording or otherwise, without written prior permission from the publisher.

Although all care is taken to ensure integrity and the quality of this publication and the information herein, no responsibility is assumed by the publishers nor the author for any damage to the property or persons as a result of operation or use of this publication and/or the information contained herein.

Published by: CRC Press/Balkema
Schipholweg 107C, 2316 XC Leiden, The Netherlands
e-mail: Pub.NL@taylorandfrancis.com
www.crcpress.com – www.taylorandfrancis.com

ISBN: 978-1-138-54446-8 (set of 2 volumes)
ISBN: 978-1-138-33198-3 (Vol 1)
ISBN: 978-1-138-33203-4 (Vol 2)
ISBN: 978-1-351-00362-9 (eBook set of 2 volumes)
ISBN: 978-0-429-44693-1 (eBook, Vol 1)
ISBN: 978-0-429-44692-4 (eBook, Vol 2)

Numerical Methods in Geotechnical Engineering IX – Cardoso et al. (Eds)
© 2018 Taylor & Francis Group, London, ISBN 978-1-138-33203-4

Table of contents

Finite element, discrete element and other numerical methods. Coupling of diverse methods

Artificial intelligence and neural networks

Ground flow, thermal and coupled analysis

VOLUME 2

Application of numerical methods in the context of the Eurocodes

Shallow and deep foundations

Embankments and dams

Tunnels and caverns (and pipelines)

Ground improvement and reinforcement

Offshore geotechnical engineering

Numerical Methods in Geotechnical Engineering IX – Cardoso et al. (Eds)
© 2018 Taylor & Francis Group, London, ISBN 978-1-138-33203-4

Preface

The European Regional Technical Committee (ERTC7) of the International Society for Soil Mechanics and Geotechnical Engineering (ISSMGE) and the Organizing Committee welcome all participants of the 9th European Conference on Numerical Methods in Geotechnical Engineering (NUMGE2018), held at the Faculty of Engineering of University of Porto (FEUP), in Porto, Portugal, from 25th to 27th June 2018.

This conference is the ninth in a series of conferences on Numerical Methods in Geotechnical Engineering organized by the ERTC7 under the auspices of the ISSMGE. The first conference was held in 1986 in Stuttgart, Germany, and the series continued every four years (1990 Santander, Spain; 1994 Manchester, United Kingdom; 1998 Udine, Italy; 2002 Paris, France; 2006 Graz, Austria; 2010 Trondheim, Norway; 2014 Delft, The Netherlands).

The conference provides a forum for exchange of ideas and discussion on topics related to numerical modelling in geotechnical engineering. Both senior and young researchers, as well as scientists and engineers from Europe and overseas, attend this conference to share and exchange their knowledge and experiences. Geotechnical engineering researchers and practical engineers submit their papers on scientific achievements, innovations and engineering applications related to or employing numerical methods.

The papers for NUMGE2018 cover topics from emerging research to engineering practice. For the proceedings the contributions are grouped under the following themes:

› Constitutive modelling and numerical implementation
› Finite element, discrete element and other numerical methods. Coupling of diverse methods
› Reliability and probability analysis
› Large deformation – large strain analysis
› Artificial intelligence and neural networks
› Ground flow, thermal and coupled analysis
› Earthquake engineering, soil dynamics and soil-structure interactions
› Rock mechanics
› Application of numerical methods in the context of the Eurocodes
› Shallow and deep foundations
› Slopes and cuts
› Supported excavations and retaining walls
› Embankments and dams
› Tunnels and caverns (and pipelines)
› Ground improvement and reinforcement
› Offshore geotechnical engineering
› Propagation of vibrations

Around 400 abstracts were submitted and the Authors of the approved abstracts were invited to submit full papers for peer review. A total of 204 papers were accepted for inclusion in the conference proceedings. The Editors would like to thank the Scientific and Reviewing Committees for their assistance in the review process.

The Editors are grateful for the support of the Chairman, Core Members and National Representatives of ERTC7, namely for promoting the conference on their respective home countries.

NUMGE2018 is jointly organized by SPG (Portuguese Geotechnical Society) and FEUP. These institutions and conference sponsors are gratefully acknowledged for their generous support.

The Editors want to express their particular thanks to the Authors, for their fundamental contribution to the success of the conference, and to the Participants, wishing that the 3 days of presentations and discussions would be fruitful for their future research and technical work.

On behalf of the Organising Committee and ERTC7, we welcome you to Porto hoping that you enjoy the scientific and technical aspects of the conference, as well as its social programme and the city of Porto.

ERTC7 (ISSMGE) –
Helmut Schweiger (Chairman) &
César Sagaseta (Past-Chairman)
NUMGE 2018 Organizing Committee –
António S. Cardoso, José L. Borges, Pedro A. Costa,
António T. Gomes, José C. Marques & Castorina S. Vieira
April 2018

Numerical Methods in Geotechnical Engineering IX – Cardoso et al. (Eds)
© 2018 Taylor & Francis Group, London, ISBN 978-1-138-33203-4

Committees

SCIENTIFIC COMMITTEE (ERTC7)

Core Members

Helmut Schweiger, *Austria (Chairman)*
César Sagaseta, *Spain (Past-Chairman)*
Philip Mestat, *France*
Steinar Nordal, *Norway*
Manuel Pastor, *Spain*
Juan Pestana, *USA*
David Potts, *United Kingdom*
Scott Sloan, *Australia*

National Representatives

Sergey Aleynikov, *Russia*
Katalin Bagi, *Hungary*
Ronald Brinkgreve, *The Netherlands*
Imre Bojtár, *Hungary*
Albert Bolle, *Belgium*
Harvey Burd, *United Kingdom*
Annamaria Cividini, *Italy*
Pedro Alves Costa, *Portugal*
George Dounias, *Greece*
Torbjörn Edstam, *Sweden*
Pit (Peter) Fritz, *Switzerland*
Maciej Gryczmański, *Poland*
Frands Haahr, *Denmark*
Ivo Herle, *Czech Republic*
Fritz Kopf, *Austria*
Tim Länsivaara, *Finland*
Tom Schanz, *Germany*
Herbert Walter, *Austria*

ORGANIZING COMMITTEE

António Silva Cardoso
José Leitão Borges
Pedro Alves Costa
António Topa Gomes
José Couto Marques
Castorina Vieira

REVIEWING COMMITTEE

Armando Antão, *Portugal*
Imre Bojtár, *Hungary*
Albert Bolle, *Belgium*
José Leitão Borges, *Portugal*
Harvey Burd, *United Kingdom*
António Silva Cardoso, *Portugal*
Annamaria Cividini, *Italy*
David Connolly, *United Kingdom*
Pedro Alves Costa, *Portugal*
Teresa Bodas Freitas, *Portugal*
António Topa Gomes, *Portugal*
Nuno Guerra, *Portugal*
Kianoosh Hatami, *USA*
José Vieira de Lemos, *Portugal*
Fernando Lopez-Caballero, *France*
José Couto Marques, *Portugal*
Arézou Modaressi, *France*
Paulo da Venda Oliveira, *Portugal*
Manuel Pastor, *Spain*
António Pedro, *Portugal*
David Potts, *United Kingdom*
César Sagaseta, *Spain*
Helmut Schweiger, *Austria*
Jorge Almeida e Sousa, *Portugal*
David Taborda, *United Kingdom*
Yiannis Tsompanakis, *Greece*
Ana Vieira, *Portugal*
Castorina Silva Vieira, *Portugal*
Helbert Walter, *Austria*
Lidija Zdravković, *United Kingdom*

Numerical Methods in Geotechnical Engineering IX – Cardoso et al. (Eds)
© 2018 Taylor & Francis Group, London, ISBN 978-1-138-33203-4

Institutional support

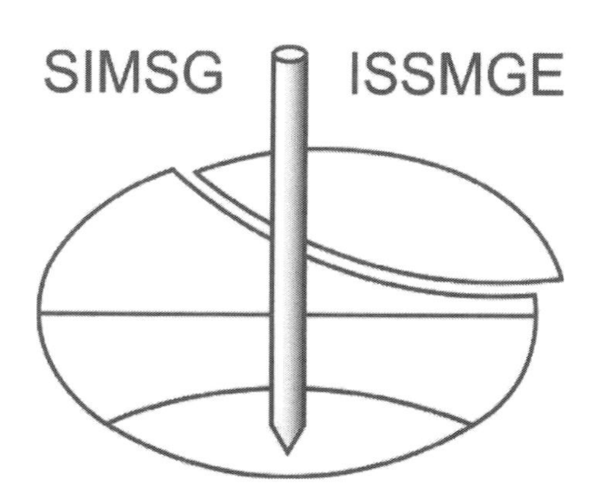

Sociedade Portuguesa de Geotecnia

PORTUGUESE GEOTECHNICAL SOCIETY

Numerical Methods in Geotechnical Engineering IX – Cardoso et al. (Eds)
© 2018 Taylor & Francis Group, London, ISBN 978-1-138-33203-4

Sponsors

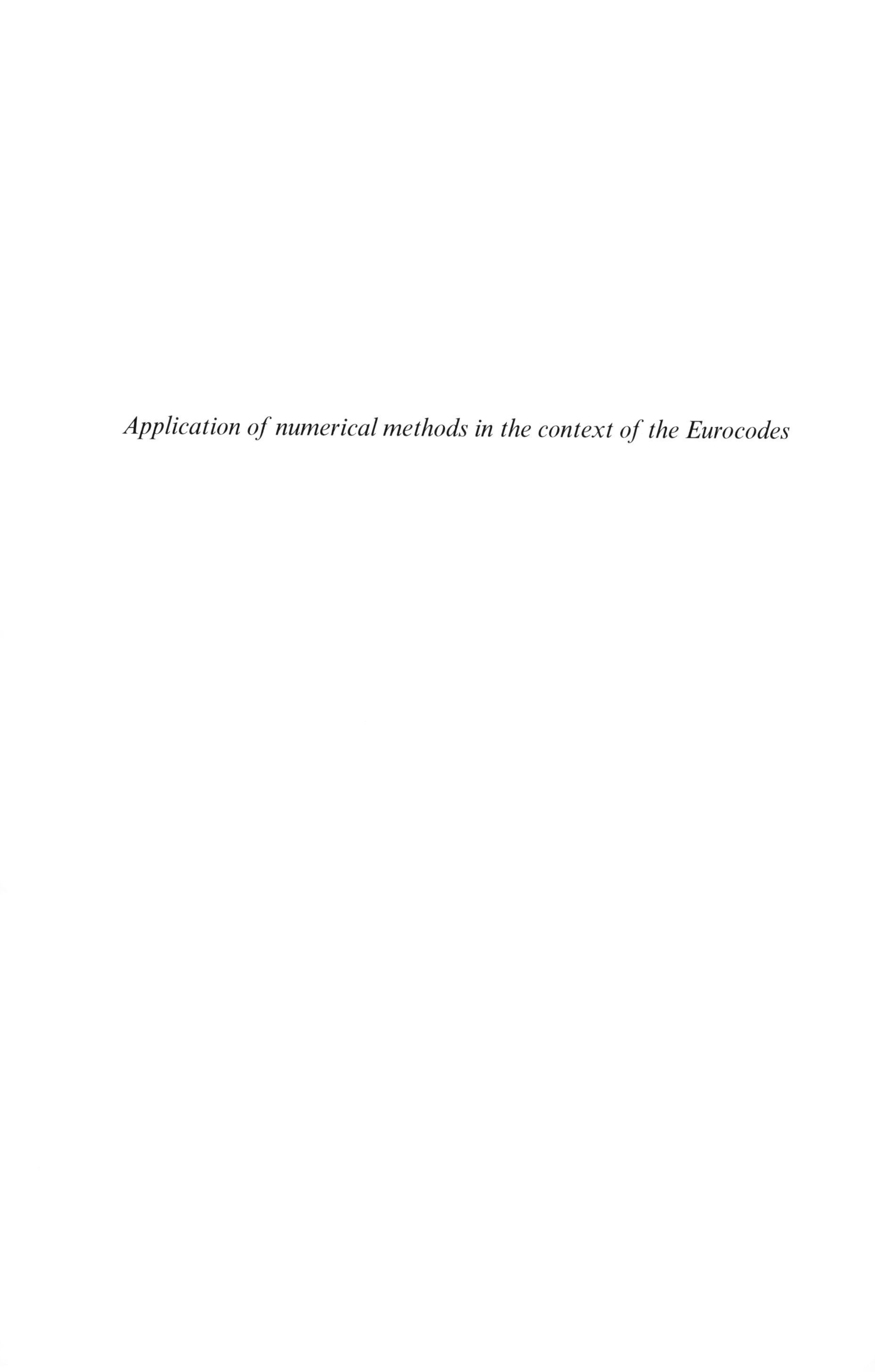

Application of numerical methods in the context of the Eurocodes

Numerical Methods in Geotechnical Engineering IX – Cardoso et al. (Eds)
© 2018 Taylor & Francis Group, London, ISBN 978-1-138-33203-4

Ultimate limit state design of retaining wall using finite element method and advanced soil models

Hoe C. Yeow
COWI UK Ltd., London, UK

ABSTRACT: The increasing use of the finite-element method in geotechnical design has raised the question of the compliance of this design approach with Eurocode requirements for the ultimate limit state conditions, especially when a more complex soil constitutive model has been used. In his papers on this topic, Yeow (2014, 2015) compared the design of simple and more complex geotechnical structures using advanced BRICK soil model in London Clay under Eurocode 7 (EC7) Design Approach 1 (DA1). In this paper, the analyses of the multi-propped retaining wall will be compared with another advanced constitutive soil model which is calibrated to model the London Clay in an attempt to compare and contrast the resulting wall design under DA1 of EC7 ultimate limit state.

1 INTRODUCTION

In the UK, the design of geotechnical structures needs to satisfy the requirement of DA1 under EC7.

DA1 combination 1 (DA1C1) introduces partial factors on actions (load) or the effects of actions; while

DA1 combination 2 (DA1C2) has the main partial factors on the material (strength).

Yeow (2014) published some typical geotechnical design analyses undertaken using the finite element method (FEM) and an advanced BRICK soil model (Simpson 1992) under Eurocode 7 (EC7) Design Approach 1 (DA1). The FEM analyses for the design were undertaken for both the combinations required in DA1 (DA1C1 and DA1C2) where considerations of initial stress, factoring of the strength and stiffness and making modelling excursions have been allowed.

To further assess the validity of the findings, Yeow (2015) reanalysed the retaining walls for Crossrail's Tottenham Court Road Western Ticket Hall (WTH) using the FE software *Oasys* SAFE and the *BRICK* advanced small strain soil model (Simpson 1992). In the assessment undertaken using the the *BRICK* soil model, the following observations were made:

- designing to DA1C1 is found to be more critical than the design under DA1C2 by factoring strength only;
- under DA1C2, factoring both the strength and stiffness produced a design more critical than DA1C1; and
- when taking modelling excursions under DA1C2, i.e. setting initial stress using characteristic parameters and applying factored strength only at critical stages of this multi-propped excavation, the design forces are found to be less critical than the design undertaken under DA1C1.

In this paper, the Hardening Small Strain Soil Model (HSS, Benz 2016) in Plaxis model is used to repeat the design analyses and confirm the findings of the assessments undertaken on the WTH using the *BRICK* soil model. The HSS parameters selected for these analyses have been based on the interpretation of the laboratory test data undertaken for the Crossrail project close to the WTH site. These parameters were also compared against a couple of London Clay advance triaxial tests undertaken by Gasparre (2001). The findings of this calibration will be the subject of another publication due to the length limitation of this conference paper.

2 GROUND CONDITIONS AND RETAINING WALL DESIGN

The ground conditions at the WTH site are typical of a central London site with superficial deposits overlying London Clay and Lambeth Beds, which in turn are underlain by Thanet Sand and Chalk, see Table 1. The site is under-drained and the porewater pressure profile is as shown in Figure 1.

The retaining wall design was undertaken using the *Oasys* FREW program, a pseudo finite element retaining wall design software. A 1 m thick

Table 1. Summary of ground conditions.

Stratum	Top level (mTD)
Made ground	+125.7
Terrace Gravel	+121.5
London Clay	+117.3
Lambeth Beds	+95
Thanet Sand	+75.5
Chalk	+72

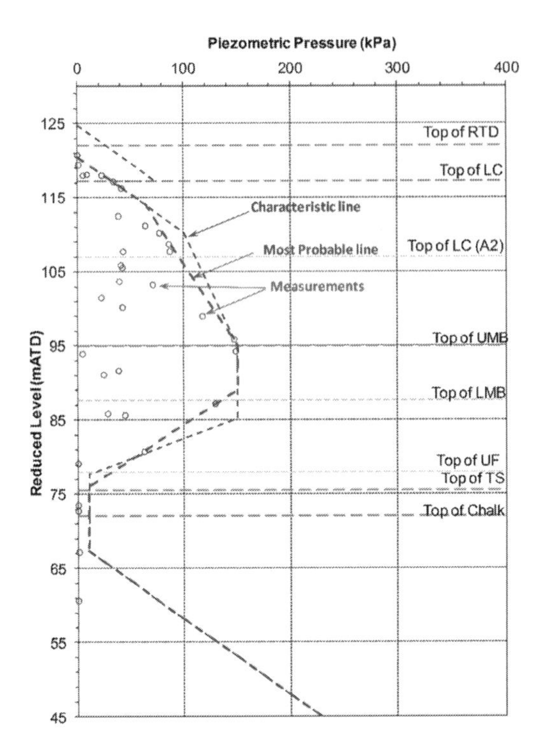

Figure 1. Groundwater profile at WTH.

diaphragm wall was provided for the construction of the WTH box using a bottom-up construction sequence with five temporary steel props. The excavation was about 30 m wide and 30 m deep and each excavation level was approximately 1 m below the temporary prop to provide the necessary clearance for installation. Figure 2 shows the layout of the temporary works for the construction of the underground station.

The characteristic soil properties of the soils other than those for London Clay are tabulated in Table 2.

The London Clay has been modelled using HSS and the parameters are included in Appendix A for reference. The remaining soils have

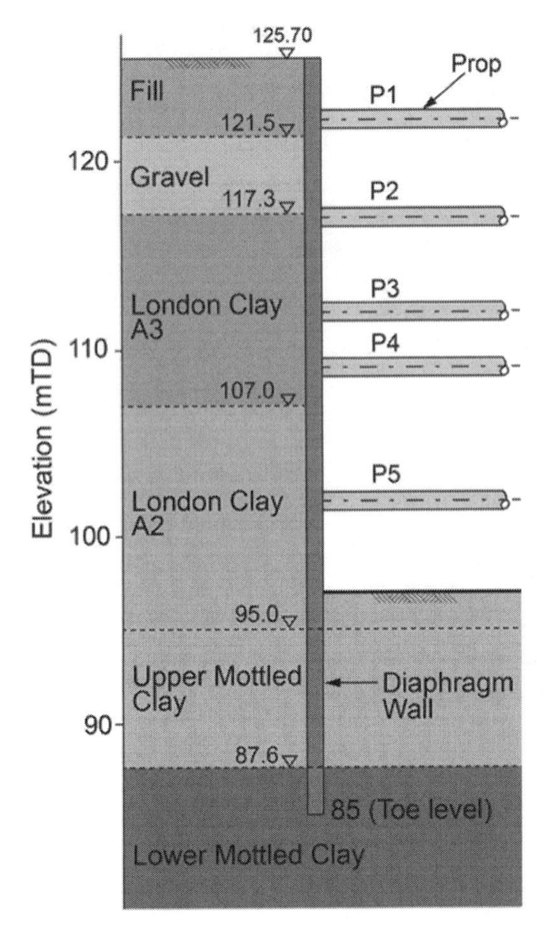

Figure 2. Cross section of the WTH box showing temporary props during construction.

Table 2. Characteristic properties of the soils.

Stratum	Young's modulus MPa	K_o	ϕ'	c'/c_u (kPa)
Made ground	10	0.6	25	-
Terrace Gravel	75	0.4	36	-
London Clay	Plaxis HSS	-	25	-
(FREW)			22	-
Lambeth Beds	148+3.8z	1–1.5	26	350
Thanet Sand	200	1.2	36	-

been modelled using a Mohr-Coulomb soil model. Further details of the design of the retaining wall are reported in Yeow *et. al.* (2014). The FE mesh used for the assessment is shown in Figure 3.

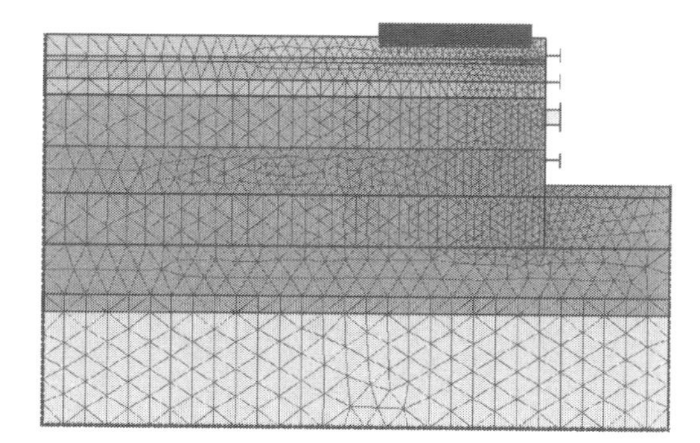

Figure 3. FE mesh.

3 RETAINING WALL DESIGN

In the UK, the design of geotechnical structures needs to satisfy both the combinations under the requirement of DA1 under EC7. There is, however, no recommendation in EC7 on any partial factor to be introduced for material stiffness. CIRIA 580 (Gaba et al., 2003) recommends that for the ultimate limit state (ULS) DA1C2 design case, the stiffness value should be half of the value used for the serviceability limit state (SLS) design.

In this paper the following partial factors have been used:

DA1C1:

- Actions, $\gamma_G = 1.35$ and $\gamma_Q = 1.5$
- Material properties, $\gamma_M = 1.0$

DA1C2:

- Actions, $\gamma_G = 1.0$ and $\gamma_Q = 1.3$
- Material properties, $\gamma_M = 1.25$ (effective stress strength) or 1.4 (total stress strength)

4 PRESENTATION OF THE DATA

In the paper the computed wall deflection and bending moments of the wall are compared by normalizing the parameters with the maximum values calculated from the conforming design which was performed using *Oasys* FREW program. This is to enable a direct comparison of the values computed relative to those in the conforming design. The normalized approach is also consistent with the approach adopted when the *BRICK* soil model was used in the studies (Yeow 2015); an agreement reached with Crossrail project for the publication of the detailed design information at that time.

5 METHODOLOGY FOR COMPARISON OF DESIGN OUTPUT

5.1 *DA1C1 by Pseudo—FE FREW and FE Plaxis*

For the retaining walls at WTH, DA1C1 was found to govern the ULS design. The comparison of the computed wall deflections and bending moment for both the conforming design using FREW model and the FE model using Plaxis HSS soil model are shown in Figures 4 and 5.

From these comparisons, although the maximum wall deflection is about 12% less in the FE assessment, the computed maximum bending moment from the FE model is similar magnitude to that from the FREW model. There are also some differences between the two bending moment profiles especially below the base slab level of +98 mTD. This is due to the different software used and the method used to model the base slab. It is also common for pseudo-FE software to predict slightly more lateral movement in the embedded section of a retaining structure, which is observed here. However, this is not considered critical to the comparison made for the governing area between the level of +100 and +106 mTD where maximum bending moment occurs. The maximum moment corresponds to the stage where the base slab is cast and the 5th level prop was removed and the subse-

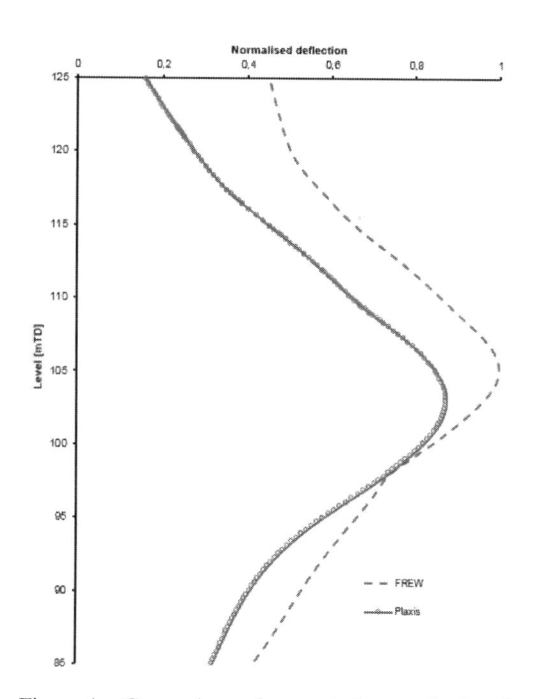

Figure 4. Comparison of computed normalized wall deflection.

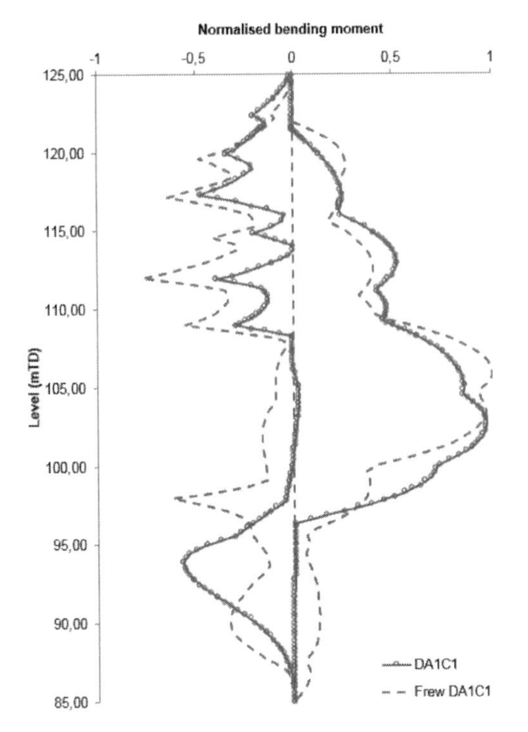

Figure 5. Comparison of computed normalized bending moments DA1C1 with $\gamma_F = 1.35$.

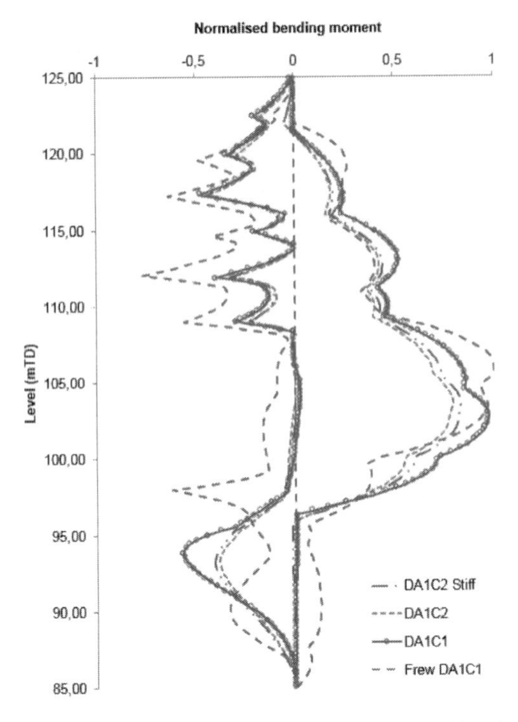

Figure 6. Comparison of computed normalized bending moment with partial factor on strength only and on both strength and stiffness.

quent stage where the 4th level of permanent slab was cast and the 4th temporary prop was removed.

5.2 *DA1C2 by factoring strength and stiffness*

Figure 6 shows the computed bending moments for DA1C2 by firstly factoring soil strength only and secondly factoring both the strength and stiffness.

Unlike the *BRICK* soil model (Yeow 2015), applying partial factors to the strength and stiffness parameters to undertake Combination 2 assessment in the HSS model is rather complicated. Therefore, in line with the same approach taken for the *BRICK* model, the following partial factors are applied:

- Partial factor on effective strength parameters, $\gamma_{\phi} = 1.25$ (equivalent of $\gamma_{cu} = 1.2$ at 0.01% strain to 1.33 at 0.1% strain);
- Partial factor on stiffness, $\gamma_{YM} = 2.0$ at 0.01% strain reduced to 1.3 at 0.1% strain.

As shown in Figure 6, the normalized bending moment computed from DA1C2 by factoring the strength alone is about 19% lower than that from the DA1C1 assessment. This confirms that DA1C1 governs the design and this is consistent with the finding of the conforming design using FREW. This is consistent with the findings of the previous papers by Yeow (2014, 2015).

Similarly, the computed normalized bending moments derived by factoring the strength and also stiffness is almost 15% lower than that computed from the DA1C1 assessment. This is not consistent with the findings from the previous paper (Yeow, 2015). This is probably due to the lower partial factor achieved on the stiffness, i.e. 2.0 reducing to 1.3, using the assumed set of the Plaxis HSS model parameters.

5.3 *DA1C2 by taking modelling excursion*

The concept of modelling excursions was initially proposed due to the concern that FE analysis does not capture the correct ULS mechanisms if the soil strength is modelled with factored strength from the start of the modelling sequence. The modelling excursion is therefore proposed as part of the analysis whereby factoring of strength is undertaken at any stages considered critical to the design.

For the WTH construction, as previously stated, the critical stage casting the base slab and removal of the 5th prop followed by the casting of the 4th level permanent slab and removal of the 4th level temporary prop. These two stages were therefore considered as part of the ULS design by taking modelling excursion after undertaking conventional DA1C1 analysis.

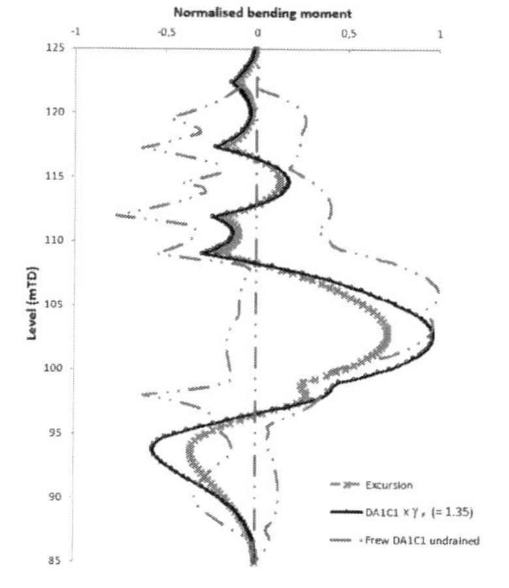

a) Cast base slab and remove 5th prop

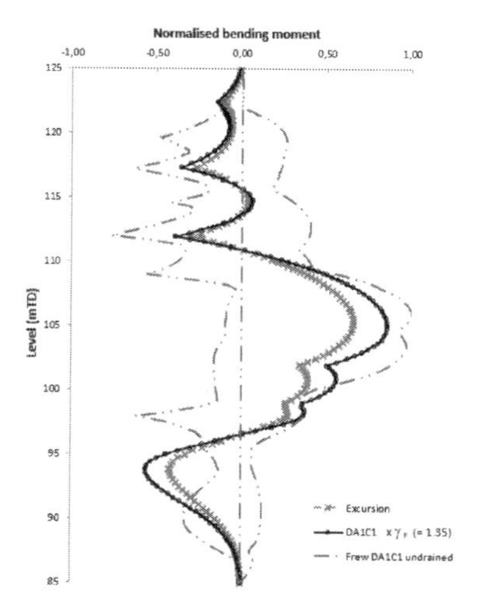

b) Cast 4th level permanent slab and remove 4th prop

Figure 7. Computed normalized bending moment for the two critical stages of the excavation.

Figure 7 shows the computed normalized bending moment for the two critical stages of the excavation. From this figure, by taking modelling excursions for both stages of the excavation did not produce forces greater than the bending moment computed from the DA1C1 design. This is not consistent with the findings previously derived from the simplified retaining structure models but consistent with the assessment undertaken using the *BRICK* soil model.

6 CONCLUSIONS

Factoring soil parameters for Combination 2 finite element assessments of retaining structure using Plaxis HSS model is complicated. Unlike the *BRICK* model, factoring the effective strength parameters in the HSS model did not produce an equivalent partial factor on the undrained strength when modelling undrained sequence of excavation. An attempt to factor the soil stiffness also did not achieve the required partial factor, especially in the range of strain normally observed in the soils adjacent to retaining walls.

When comparing the design assessment undertaken using the *BRICK* and Plaxis HSS soil models, the following conclusions could be made:

- DA1C1 produced the most critical design forces for both soil models;
- factoring soil strength did not produce more critical design in DA1C2 for both soil models;
- taking modelling excursion also did not produced more critical design for both soil models; and
- factoring both soil strength and stiffness in the HSS model did not lead to more critical design and this is not consistent to the findings when using *BRICK* soil model (Yeow 2015).

REFERENCES

BS EN 1997:1: 2004. Eurocode 7: Geotechnical design—Part 1: General rules.

BS EN 1997–2:2007 Eurocode 7 Geotechnical design, Part 2: Ground Investigation and testing.

UK National Annex to Eurocode 7: Geotechnical design—Part 1: General rules.

Benz, T 2006. Small-Strain Stiffness of Soils and its Numerical Consequences. *Ph.D. Thesis. Mitteilung des Instituts für Geotechnik der Universität. Heft 56.*

Gaba, A R, Simpson, B, Powrie, W, & Beadman, D R, 2003, Embedded retaining walls—guidance for economic design, *CIRIA 580*

Gasparre A. 2005. Advanced Laboratory Characterisation of London Clay. *PhD thesis, Imperial College London, London, UK.*

Simpson, B. 1992. 32nd Rankine Lecture: Retaining structures—displacement and design. *Geotechnique*, 42, 4, 539–576.

Simpson, B & Yazdchi, M. 2003. Limit State Design. *International workshop on Limit State Design in Geo Engineering practice, Cambridge, Massachussetts.*

Simpson, B & Hocombe, T. 2010. Implications of modern design codes for earth retaining structures

Yeow, H-C 2014. Ultimate limit state design using advanced BRICK model in finite element method, *Proceedings of Geotechnical Engineering, ICE, UK.*

Yeow, H-C 2015. Ultimate limit state design of retaining wall using finite element method and advanced BRICK model, *Proceedings of the XVI ECSMGE Geotechnical Engineering for Infrastructure and Development, UK.*

Yeow H, Nicholson D, Man C-L, Ringer A, Glass P & Black M. 2014. The application of observational method for the Crossrail Tottenham Court Road station, *Proceedings of Deep Basement and Underground Structures, ICE, UK.*

APPENDIX A

Plaxis HSS parameters assumed:

Constant for stiffness power law, m = 0.75

Reference stiffness modulus values:

$E_{oed}^{ref} = 10 \text{ to } 40 MPa$
$E_{50}^{ref} = 40 \text{ to } 95 MPa$
$E_{ur}^{ref} = 120 \text{ to } 425 MPa$

Poisson's ratio for unloading-reloading, = 0.2

Small strain shear modulus, G_o *= 100 to 300 MPa*

Shear strain when G_s *is 70% of* G_o*, = 0.002*

Reference stress, = 100 to 450 kPa

Angle of friction, ϕ' *= 25°*

Angle of dilation, ψ' *= 2°*

Numerical Methods in Geotechnical Engineering IX – Cardoso et al. (Eds)
© 2018 Taylor & Francis Group, London, ISBN 978-1-138-33203-4

Numerical analysis of a foundation of a cooling tower in difficult geotechnical conditions

W. Bogusz
Building Research Institute, Warsaw, Poland

M. Kociniak
BEROA Deutschland GmbH, Ratingen, Germany

ABSTRACT: Eurocode 7 offers general guidance on design of different foundation types and recognizes that foundations cannot be always considered in isolation from the structure. When a significant potential for stress redistribution exists, the exceedance of geotechnical ultimate limit state may become highly unlikely. However, it is often the geotechnical serviceability that guides the design as it may lead to a structural failure. The paper presents a case study with the results of a numerical analysis performed for a cooling tower of a conventional power plant, located in an area characterized by difficult geotechnical conditions. Firstly, a prediction of the settlement distribution has been conducted to assess the serviceability limit state and the behavior of the foundation. Then, a detailed distribution of moduli of subgrade reaction has been proposed for structural analysis. The measurements of vertical displacements of the foundation, conducted during the construction, validated the assumptions of the design.

1 INTRODUCTION

Increased complexity of structures and ground conditions is often a major challenge for designers; it requires them to make informed decisions regarding foundation solutions and efficiently managing geotechnical risk. Especially, structures susceptible to excessive differential settlements should be subjected to scrutinous analysis. Ensuring the proper reliability level and rational design of such structures justifies the use of advanced numerical methods. Nowadays, their application in geotechnical design is becoming a common practice, especially for structures with high potential consequences of failure and with significant uncertainties in relation to their soil-structure interaction. This involves not only the uncertainty due to inherent soil variability, but also the complexity of the structure itself. For spatial structures, such as cooling towers used at power plants, this issue is of utmost concern.

The paper presents a case-study analysis based on calculations performed for a cooling tower of a coal fired power plant located near Opole in Poland. The considered structure has been designed as a part of the power unit no. 5, which is one of two new power units recently constructed at the site. To ensure unlikelihood of reaching ultimate limit state (ULS), partial safety factors are commonly implemented in most recent codes (Simpson 2011), including Eurocode 7 (EC7), which requirements were followed in the design. With their use, reasonably likely characteristic values of strength parameters are transformed into ones that are very unlikely to occur for the ULS verification purposes. According to Simpson (2011), this approach, commonly used by structural engineers, allows to reach compatibility of structural and geotechnical design; however, the soil-structure interaction problems encountered in practice, especially in the case of unusual structures, are often far more complicated. Out of all limit states required by Eurocode 7 to be consider, it is the structural failure due to foundation movement that may guide the design in the case of spatial concrete structures, including cooling towers. Such structure has a reinforced concrete shell susceptible to excessive deformations, caused by significant differences in subsoil stiffness under its ring foundation. Although industry guidelines exist for the structural design (i.e. VGB PowerTech 2010) and sufficient knowledge is available to use non-linear constitutive laws for concrete (Harte & Wittek 2009), interaction of spatial structures with the subsoil has not been sufficiently addressed, yet. Majority of research papers involving FEM analysis of soil-structure interaction problems concentrate on accurate constitutive modeling, but usually with simplified stratigraphy, often represented as relatively simple geotechnical ground models, i.e. horizontal layering over relatively large area; this is often not the case in practical geotechnical design, as the strata distribution due to geological history may be one of the main factors guiding the design and the choice of a foundation method.

In the analyzed case, directly below the foundation level of the cooling tower, two types of subsoil were distinguished along its circumference: a soft rock with relatively high stiffness, and tertiary deposits of much lower stiffness. In this particular situation, the occurrence of the ULS in various structural elements may be a result of a geotechnical serviceability limit state (SLS), namely, excessive differential settlement. While all other relevant limit states were considered as well, the presented analysis focused on the displacement prediction for the ring foundation. Contrary to the ULS, for which the verification is frequently guided by extreme values of parameters, SLS may be reasonably based on their most probable, often average values; especially for soil strength and stiffness when advanced constitutive models are applied.

2 STRUCTURAL DESIGN AND CONSTRUCTION PROCESS

The new cooling tower has 185,0 m of height. The main structural elements of the shell, with exception of cooling tower's internal elements, are: ring foundation with external diameter of 113,5 m, 6,0 m width, and 1,5 m of height; 36 meridional columns supporting the shell; and the shell itself, 176,1 m of height, with varying thickness. In order to provide sufficient stiffness and stability of the shell, its lower part has significantly higher thickness (0,90 m) than the upper part (0,20 m). Due to varying thickness of the shell, the load increment with height has had a non-linear characteristic.

The execution of the ring foundation was the first stage of the construction. Precast concrete meridional columns (Fig. 1a) were installed afterwards, followed by the installation of precast lintel beams, increasing the stiffness of the lower part of the shell. Joints between structural elements were filled with cast-in-place concrete, thus enclosing the perimeter of the first ring of the shell. Then, it was used as the starting point for climbing formwork during casting of the shell (Fig. 1b). Following rings of the shell were cast-in-place (Fig. 1c), as so called "lifts", in the number of 118, each 1,5 of height, with exception of last two at the top, designed as a part of cooling tower's cornice.

The larger diameter of the shell at the bottom (100,5 m), compared to the upper part (73,0 m) and the throat (65,2 m) of the cooling tower (Fig. 2a), also contributed to the non-linear loading during construction. These geometrical characteristics and loading non-linearity were taken into account in the numerical analysis.

3 GEOTECHNICAL CONDITIONS

Usually, the design assumptions about geotechnical conditions at the site are based on the hypothesis about site geology interpreted from a number of observations (Baecher & Christian 2003). Furthermore, geotechnical investigation should be focused on layers governing the limit states under consideration to provide supplementary information, updating the established ground model, with consideration given to its uncertainties. One of the main uncertainties (Vardanega & Bolton 2015) concerning geotechnical conditions at every

a)

b)

c)

Figure 1. Construction of the cooling tower: a) meridional precast concrete columns; b) climbing formwork; c) continuous concreting of the shell.

site results from the heterogeneity of subsoil. This includes both, spatial variability of strata and the distribution of parameters for each and every significant strata that have to be taken under consideration. This is especially important in the presence of difficult ground conditions.

Primary geotechnical investigation, preceding design phase, included a large number of boreholes as well as extensive in-situ (CPTU, DMT, DPH, CSWS/SASW) and laboratory (TRX, BET) testing over the entire area allocated for the new power units. In the vicinity of the cooling tower no. 5, a total of 15 soil profiles were available from the original geotechnical investigation report, prepared at the preliminary design stage. Additional 24 CPTU tests were performed along the circumference of the foundation during the detailed design phase of the cooling tower, to refine the spatial distribution of strata assumed in the analysis and to decrease the uncertainty in the assumed geotechnical ground model. This risk-driven approach to geotechnical investigation proved to be an optimal solution to balance investigation cost and the confidence about spatial strata distribution.

Identified soils at the location of the power plant were mostly considered as load bearing strata for spread foundations, despite their variability. The cooling tower of the power unit no. 5 has been one of the structures located in the most difficult geotechnical conditions present in the area. Ground investigation revealed that the older strata denivelation varies from 5 to 15 m below ground level due

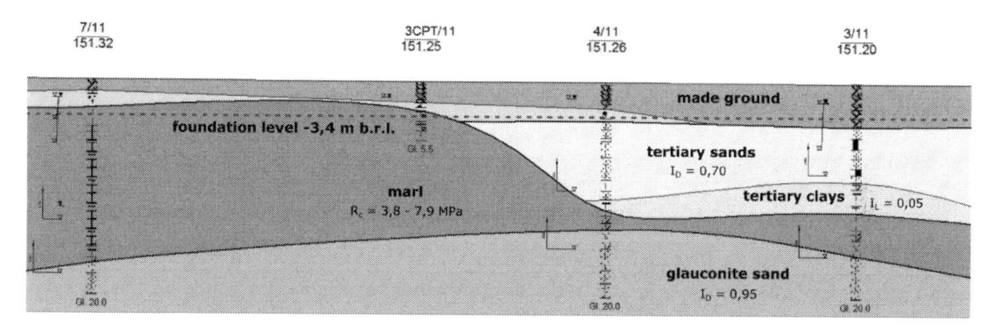

Figure 2. Simplified representation of ground conditions based on geotechnical cross-section along the fragment of the cooling tower's perimeter.

Table 1. Derived mean values of selected parameters based on geotechnical investigation report.

Stratigraphy	Description	Stratum	Soil state I_D/I_L [-], Unconfined compressive strength R_c [MPa]	Cohesion c' [kPa]	The angle of shear resistance ϕ' [°]	Constrained modulus M [MPa], (Deformation modulus E [MPa])	Shear stiffness modulus G_{max} [MPa]
	Fine sandy deposits	$IIIa_1$	0,70	0,0	31,8	90,0	77,6
		$IIIa_2$	0,55	0,0	30,9	75,0	-
	Clayey deposits (with coal)	$IIIe_1$	0,03	26,6	16,8	23,5	
		$IIIe_2$					
	Clayey deposits	$IIIf_1$	0,00	39,2	18,7	60,0	70,9
		$IIIf_2$	0,05	23,0	16,9	50,0	-
Tertiary		$IIIf_3$					
	Marly clays,	IVa_1	–0,03 / R_c = 1,5	34,0	34,4	108	257
	Weathered marl	IVa_2	0,05 / R_c = 0,2	26,7	36,9	60,0	150
Cretaceous	Soft carbonate rocks (marl, clayey marl)	IVb	R_c = 6,7	39,1	41,5	E = 390	239

to erosion and tectonic influences in the past. The north-eastern part of the cooling tower has been located directly above this zone (Fig. 2), where soils of different stiffness are present: soft rock (marl and weathered marl characterized by different level of jointing), clayey deposits, and alluvial sands with local interbeddings of organic soils.

As the SLS has been of primary concern, design values of geotechnical parameters equal to their derived mean values were used. For selected strata, these derived values of basic parameters are presented in Tab. 1. While values provided for tertiary deposits have been identified with relatively high level of certainty, some of the parameters for cretaceous marls and weathered marls exhibited a significant scatter of the results. Additionally, the soft rock behavior is also affected by jointing, which level at the area is highly variable and does not present any correlation with depth or spatial distribution. In the light of those factors, for the purpose of the analysis, the mean values were assumed as sufficiently representative. The glauconite sands and stiff clays / claystone located underneath the layer of marl has been included in the model, but it has been considered to be of lower significance for the analyzed problem.

4 SOIL-STRUCTURE NUMERICAL ANALYSIS

4.1 *General considerations*

In view of the size and the shape of the structure, as well as spatial distribution of geotechnical strata, FEM analysis with ZSoil (2012) software has been conducted for three-dimensional model in order to specify the optimal foundation type and provide guidance for structural design assumptions. To consider the influence of the superstructure stiffness, the analysis included the reinforced concrete (RC) shell of varying thickness.

The analysis of the problem has been considered as a deformation problem. As the ring foundation has been designed above the stabilization level of deeper groundwater level, the impact of groundwater on the stress-state has been accounted for in the unit weight of the specific soil materials, as a buoyant unit weight. Consolidation of the subsoil has not been considered, and only staged construction has taken into account time-dependency of displacement increment. Therefore, obtained results are considered as total settlements. As the construction process had taken several months, this approach has been considered as an acceptable simplification.

The dimensions of the model, excluding the height of the cooling tower itself, were assumed as 200 m × 200 m × 36 m, which allowed to limit the number of elements in use. The horizontal dimensions allowed to set boundary conditions at the distance of over 40 m each way from the ring foundation. The vertical dimension, even though the area has been investigated up to 50 m of depth, has been limited to 36 m, where a relatively horizontal layer of very stiff clay / claystone has been encountered in most boreholes.

The model consisted of 67819 Enhanced Assumed Strain (EAS) elements for continuum, 2368shell elements of varying thickness parameter for the concrete shell of the cooling tower, and 216 beam elements for 36 columns supporting the shell. The precast beams between heads of the columns, which were used as the starters for the climbing formwork during the construction, were not modeled explicitly, but were accounted for in the thickness parameters of the lowest shell elements.

The ring foundation has been represented by volumetric elements, with material defined as linear-elastic and with constant Young modulus value of E = 30 GPa. Between the foundation and

a)

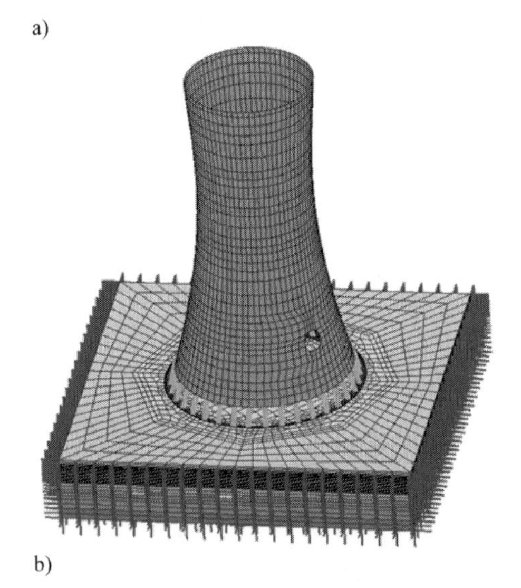

b)

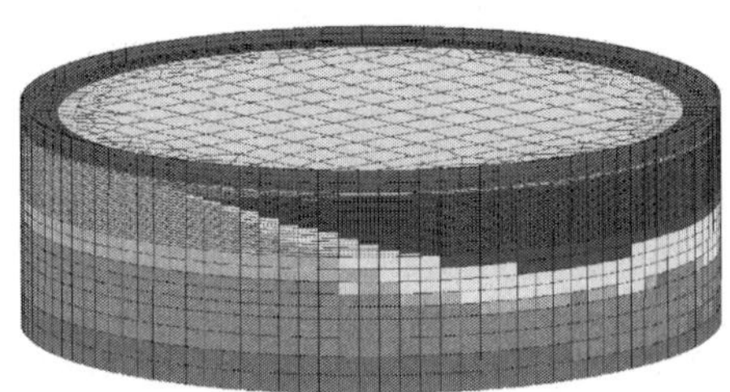

Figure 3. Numerical model of: a) the cooling tower; b) stratification of the subsoil directly under the foundation.

the soil, standard contact interface elements were used.

Boundary conditions for deformation were set to allow vertical movement at external vertical surfaces of the model, and were blocked in all directions at the bottom of the model.

4.2 *Geotechnical ground model in FEM analysis*

With the exception of structural elements, material properties were assigned to the elements of the model with the use of simple kriging procedure available in the FE code, based on 39 available soil profiles, most of which were located directly along the perimeter of the ring foundation. The kriging procedure took into account the sedimentation sequence and utilized the exponential variogram function; layers with thickness lower than 0,25 m, which corresponds to the lowest vertical dimension of volumetric elements, were ignored in this procedure. However, available geotechnical data had been reviewed and softer strata of relatively low thickness has been considered implicitly.

Although the original geotechnical investigation report defined a number of different strata, only the most significant ones were taken into account in the numerical model, as some were of low importance for the problem at hand; some strata have been generalized or merged to limit the scope of materials defined in the FE model. Finally, thirteen distinct materials in total were defined for the soil, marl rock, and the weathered rock. To consider significant aspects of soil-structure interaction, a number of these strata were modeled using Hardening Soil with small strain stiffness (HSs) constitutive model to account for the stiffness nonlinearity of the subsoil. Parameters for the analysis were chosen directly or by means of correlations, based on in-situ and laboratory soil test results, including triaxial, BET, CSWS/SASW, CPTU, and DMT. Parameters were derived according to recommendations provided in the software manual (Obrzud & Truty 2012) and based on local experiences (Godlewski & Szczepański 2015).

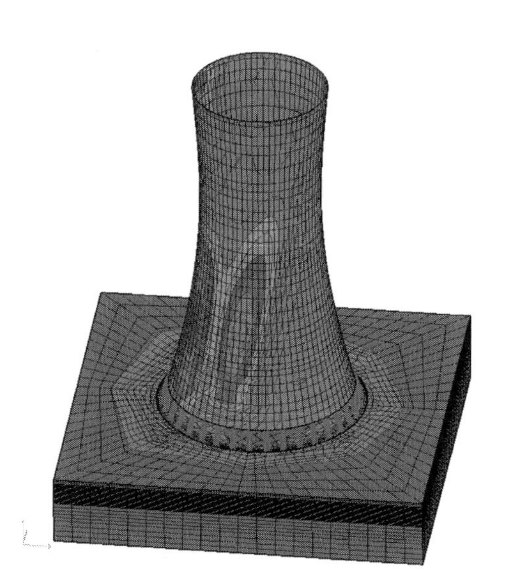

Figure 4. Displacement map for the entire model with visible deformation of the shell due to differential settlement.

4.3 *Calculations*

During calculations, either Full Newton-Raphson or Broyden–Fletcher–Goldfarb–Shanno (BFGS) algorithms were used as non-linear solvers; their choice for each stage depended on the balance between time-optimization and numerical stability of the performed calculations.

For the purpose of this paper, the problem has been recalculated to obtain detailed displacement history at specific nodes, without modification of original design assumptions. The analysis was conducted on a workstation with a 64-bit Windows 10 OS, Intel Xeon E5-2630 v4 CPU, 128 GB of RAM, and a solid state drive, in order to decrease calculation time as much as possible. The total calculation time for all the construction stages took approximately 4h. Such execution time can be considered as acceptable for design purposes, thus supporting the validity of FEM application in geotechnical design, despite introduced simplifications.

4.4 *Results*

As a result, predicted deformations and their distribution were evaluated (Fig. 4). The analysis has proven the applicability of shallow foundation as the most optimal solution. Based on the results, local soil replacement was conducted in areas where organic interbeddings were present at shallow depth below the foundation level. Additionally, a construction of a sand layer of 0,5 m thickness directly under the foundation was recommended as a part of a drainage system and to provide better stress redistribution under the foundation over soft rock.

5 INPUT FOR STRUCTURAL DESIGN

Historically, ring foundations of cooling towers were treated as continuous strip foundations that could be designed as infinite beams on elastic subgrade, characterized by a linear modulus of subgrade reaction. The stiffness of the shell was not taken into account in that approach (Ledwoń & Golczyk 1967).

Nowadays, the use of advanced numerical methods, in both structural and geotechnical design, allows for more detailed analysis of soil-structure interaction. However, both approaches, one putting more emphasis on a structure and the other on its interaction with the soil, still have their limitations. On one hand, geotechnical analysis with the use of finite element method and advanced constitutive models with non-linear stiffness, as well as spatially variable soil distribution, provides better representation of real interaction of a structure with the subsoil. On the other hand, it has limitations concerning different loading combinations as it is more demanding computationally, and the design of structural elements is a domain of structural engineers using dedicated software. Usually, the soil behavior is then represented in a simplified way by the modulus of subgrade reaction under the foundation of a 3D structural model.

The most basic definition of a modulus of a subgrade reaction k_z [kPa/m] is presented by Eq. (1). Its value is not a unique characteristic of a soil and depends on the shape, dimensions and stiffness of loaded area. Due to non-linear soil behavior, it is also influenced by the stress range caused by the load from the structure.

$$k_z = \frac{q}{s} = \frac{\sigma_{q;z}}{u_y} \tag{1}$$

where: q = load per unit area [kN/m^2]; s = settlement on elastic subsoil [m]; $\sigma_{q;z}$ = vertical stress below the foundation [kPa]; u_y = vertical displacement of the foundation [m].

The use of simplified methods for estimation of the moduli of subgrade reaction, which are available in literature for design of RC structures, is justified only in the case of simple and fairly homogeneous geotechnical conditions. Due to variable stratification at the site, the derivation of k_z values based on FEM displacement analysis was necessary (Fig. 5). Despite its simplicity, to some extent, it allowed to take into account the complex ground conditions, spatial variability of the strata, as well as the superstructure's stiffness and non-linear loading increment.

Finally, two most unfavorable sets of moduli of subgrade reaction were derived for structural analysis. Firstly, the lowest possible values derived from FEM analysis, as the more unfavorable condition for the design of the ring foundation. Secondly, based on simplified approach (PN-81/B-03020) resulting in higher values of moduli for rock formation, the larger difference over the entire foundation was assumed. It corresponded to the most unfavorable ground conditions for the design of cooling tower's shell if the rock below the part of the foundation level was intact, exhibiting higher stiffness. However, the limitation of simplified approach do not consider the load redistribution resulting from the stiffness of the superstructure and the deformation compatibility between the foundation and the soil.

6 SETTLEMENT ESTIMATION AND MONITORING RESULTS

One of the difficulties with the SLS analysis is the choice of limiting values for specific limit states. While no specific values are defined by the stakeholders or general provisions provided, the choice of limiting value is not always straightforward. However, contrary to parameters guiding the occurrence of ULS, SLS performance can be measured directly. Especially, since every calculation model is just a simplified representation of the reality and the uncertainty exists in regard to its reliability of prediction.

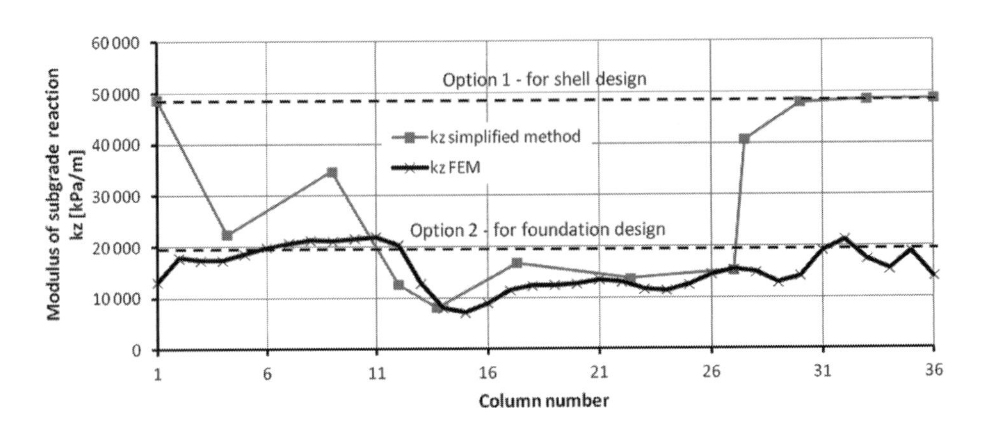

Figure 5. Calculated moduli of subgrade reaction according to Winkler's theory.

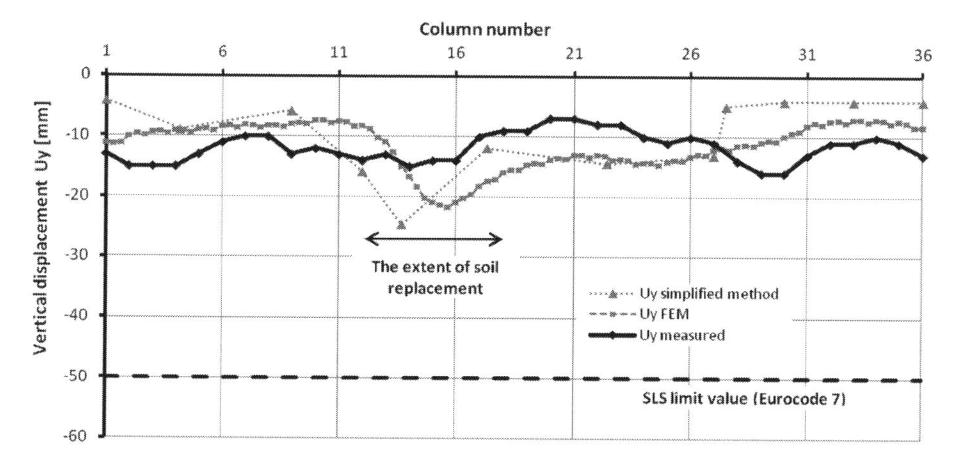

Figure 6. Vertical displacement of the foundation.

In order to monitor the behavior of the structure in the context of the ground conditions, as well as to validate the assumptions undertaken in the design, leveling pins were installed at each and every one of 36 columns of the cooling tower. The base measurement was taken when the permanent load (ring foundation, columns, and first ring of the shell) stood at 31% of the total self-weight of the structure. Afterwards, with monthly intervals, settlement measurements for all columns have been taken. The last available measurement was taken after reaching maximum height of the cooling tower, thus reaching 100% of the self-weight of the structure. The measured values are presented at Fig. 6, in comparison to values predicted for the self-weight of the structure and the limiting value of 50 mm presented in Polish NNA to Eurocode 7 (PN-EN 1997–1: 2008/Ap2: 2010), which was assumed as a general limiting criterion in the design.

Considering that the analysis had been performed at the design stage, prior to the construction and obtaining the measurements, the results are considered as satisfactory. It should be noticed that under the foundation between columns no. 36 to 1, the water inflow and outflow systems have been located. Additional earthworks, may have been responsible for increased settlements in that area of the foundation. Furthermore, between column no. 12 and 18, due to presence of interbeddings of soils with some organic content and higher deformability, local soil replacement has been conducted at the site.

By analyzing the differences (Fig. 6) between the prediction and the observed behavior, it can be concluded that displacements at the area of marl presence has been slightly underestimated, while it has been overestimated for the area dominated by tertiary soils. It follows the general design tendency for over-conservatism when dealing with standard and well investigated soils, which are considered as weaker; in comparison, rock is often considered as slightly more uncertain subsoil for shallow foundation than initially assumed.

At Fig. 7, the example of increase in vertical displacement of column no. 29, which exhibited the largest settlement, is presented. Regular and detailed measurements give important insight into the behavior of the structure, as well as provide an engineer with validation of the design assumptions and valuable experience as a reference for future designs.

7 DISCUSSION AND CONCLUSIONS

As most research papers containing case studies of foundation settlements are often back-analyzed, the geotechnical and structural designers are faced with the challenge of conducting the analysis under the conditions of uncertainty, with the measurement results obtained much later for validation purposes only. Furthermore, this uncertainty, time schedule of the project, and cost constrains are often significant limiting factors in the decision process of a design. However, to ensure sufficient reliability of prediction, the scope of the analysis, the level of ground investigation, as well as the monitoring, should depend on the level of complexity of a soil-structure interaction problem, which is defined in Eurocode 7 through geotechnical categories; it should also include the risk profile of the project and its expected consequences of failure.

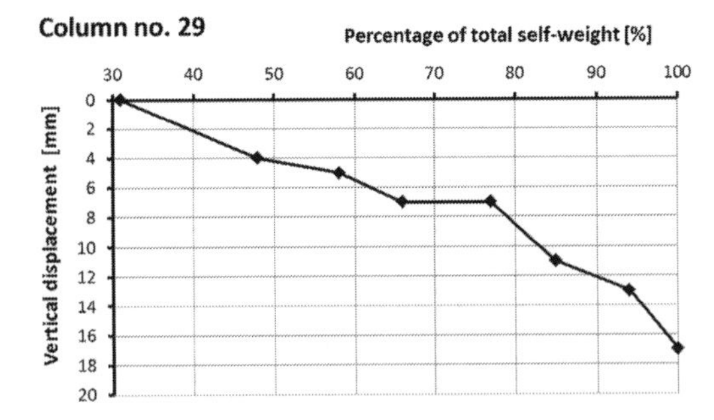

Figure 7. Measured vertical displacement of the column no. 29.

The presented example of the cooling tower, proves that the choice of optimal, economic and safe foundation is possible even in the case of difficult ground conditions. By ensuring appropriate level of analysis and sufficient geotechnical investigation, with additional testing during the design and execution phase, it can result in real cost and time savings for the investment, while maintaining sufficient margin of safety in compliance with design codes. Despite encountered conditions, due to risk-driven ground investigation and the use of advanced numerical methods, shallow foundation was proven as an adequate solution. For these complex soil-structure interaction conditions, non-standard approach to the estimation of moduli of subgrade reaction for structural design purposes had been implemented. It has proven to be correct and allowed to optimize the design while ensuring appropriate reliability of the structure. However, it has been possible primarily as a result of close cooperation between geotechnical and structural engineers at the design phase.

Interestingly, from structural point of view, the complexity of geotechnical conditions resulted only in approx. 10% increase in the amount of reinforcement necessary for the shell, compared to the cooling tower of power unit no. 6 with similar operating parameters but founded entirely over the soft rock stratum.

Although the use of advanced numerical methods should be encouraged among designers, it shall also be noted that it may give a false sense of security if used carelessly and without sound engineering judgment to evaluate plausibility of the obtained results. Most designers tend to use those methods as a confirmation of their initial assumptions and expectations in reference to their previous experience.

In the case presented herein, the settlement measurements have proven the validity of the chosen approach. Measured values of displacements were close to those predicted by FEM analysis, and the accuracy of prediction is considered as satisfactory for design purposes. Additionally, monitoring carried out during the construction provided not only validation, but also invaluable data about real behavior of such structure for future reference. It is of significant importance, especially, in the case of advanced numerical methods like FEM, to verify calculation results by comparing them with monitoring data.

Finally, some aspects of the FEM implementation in design practice still requires addressing. From practical point of view, the balance between the accuracy of prediction and the available time constrain is of the most importance. Sometimes, the most appropriate calculation model is not necessarily the best one available. Furthermore, from research perspective, the analysis using non-linear constitutive models for concrete may be a subject of future study of soil-structure interaction problem, as the time-dependent stiffness increase of continuously concreted spatial structure might have influenced the load distribution on the foundation.

REFERENCES

Baecher, G.B. & Christian, J.T. 2003. *Reliability and Statistics in Geotechnical Engineering*. John Wiley & Sons.

EN 199-1: 2008 *Eurocode 7. Geotechnical design—Part 1. General rules*. CEN.

Godlewski, T. & Szczepański, T. 2015. *Geotechnical testing methods for stiffness parameters estimation*. ITB guideline, Warsaw, Poland: ITB.

Harte, R. & Wittek, U. 2009. Recent developments of cooling tower design. *Proc. of Int. Assoc. for Shell and Spatial Structures Symp*. 198–210, Valencia, Spain.

Ledwoń. J. & Golczyk, M. 1967. *Cooling Towers*. Warsaw, Poland: Arkady.

Obrzud, R. & Truty, A. 2012. *The Hardenind Soil Model—a practical guidebook. Z_Soil.PC 100701 report*. Lausanne, Switzerland: Zace Services Ltd.

PN-81/B-03020: *Building soils. Foundation bases. Static calculation and design*. Warsaw, Poland: PKN.

PN-EN 1997-1: 2008/Ap2: 2010 *Eurocode 7. Geotechnical design—Part 1. General Rules. National normative annex*. Warsaw, Poland: PKN.

Simpson, B. 2011. Reliability in geotechnical design—some fundamentals, *Proc. Of 3rd Int. Symp. on Geotechnical Safety and Risk*, Munich.

Vardanega, P.J. & Bolton, M.D. 2015. Design of Geostructural Systems, *ASCE-ASME Journal of Risk and Uncertainty in Engineering Systems*, ASCE.

VGB PowerTech. 2010. Structural design of cooling towers. VGB Guideline VGB-R 610e.

ZSoil.PC. 2012. [Computer software]. Zace Services Ltd, Lausanne, Switzerland.

Numerical Methods in Geotechnical Engineering IX – Cardoso et al. (Eds)
© 2018 Taylor & Francis Group, London, *ISBN 978-1-138-33203-4*

Consideration of numerical methods in next generation Eurocode 7 (EN 1997)—current state of the amendment

A.S. Lees
Senior Application Technology Manager, Tensar International and Director, Geofem Limited, Nicosia, Cyprus

H. Walter
HTLuVA Salzburg, Salzburg, Austria

ABSTRACT: According to Mandate 515 of the European Commission "Mandate for amending existing Eurocodes and extending the scope of structural Eurocodes" and the response of CEN/TC250 "Towards a second generation of EN Eurocodes" Eurocode 7 should be extended, among other issues, to cover numerical methods in more detail. Among the tasks listed are: "Establish clear rules for using advanced numerical methods in day-to-day practice for inclusion in EN 1997-1. Such rules have been regularly sought during stakeholder feedback. This work would make use of the background research being undertaken by SC7/EG4. (…)"

Currently the first part of EN 1997, EN 1997-1, is being redrafted. It will contain general aspects of geotechnical design, including sections about "Basis of geotechnical design", "Geotechnical Analysis", "Ultimate limit states" etc. The third and, expected to be, final draft will be available in May 2018.

This paper gives an overview over the goals of the updated version of EN 1997 and the steps and organisation of its development.

The section "Geotechnical analysis" – with focus on the subsection "Numerical methods" – will be presented in more detail. For the verification of ultimate limit states a dual factoring approach will be recommended, requiring two analyses with different combinations of partial safety factors. Among other recommendations, the need for validation of the models will be highlighted with the recommended level of validation depending on the geotechnical category. The updated EN 1997-1 will be open to reliability-based analyses.

1 INTRODUCTION

The complete Eurocode suite of European Standards is about to be redrafted. The basis for this is Mandate M/515 of the European Commission (EC) "Mandate for amending existing Eurocodes and extending the scope of structural Eurocodes" and the response "Towards a second generation of EN Eurocodes" by CEN, the European Committee for Standardization. The work programme developed "focuses on ensuring the standards remain fully up to date through embracing new methods, new materials, and new regulatory and market requirements. Furthermore, it focuses on further harmonization and a major effort to improve the ease of use of the suite of standards for practical users." (CEN, 2015)

The full TC (Technical Committee) 250 work programme is comprised of 79 distinct tasks in four overlapping phases. Six tasks refer to EN 1997 (all parts) and are to be accomplished by so-called project teams (PTs) supported and controlled by TC 250 / Subcommittee SC 7, organised into Work Groups (WGs) and Task Groups (TGs).

The work programme should be finished in 2020 if funding by EC and EFTA is provided in time.

Among the tasks listed by CEN TC250 were the establishment of "clear rules for using advanced numerical methods in day-to-day practice." This reflected the fact that the implementation of many aspects of the current EN 1997 using numerical methods is unclear, in spite of the increasing use of such tools in everyday design.

There are some advantages to performing geotechnical design using numerical methods which include simultaneous checking for limit states in multiple forms, verifying serviceability and ultimate limits states (SLS and ULS) with one analysis model (using incremental methods) and more accurate simulation of ground behaviour and soil-structure interaction

In preparation for the work programme, TC250/ SC7 formed a number of *Evolution Groups* comprised of experts from across Europe to recommend to the sub-committee improvements to EN 1997 in key technical areas. These areas included characteristic values, water pressures and pile foundations, for example. *Evolution Group 4* (EG4) provided recommendations in the area of numerical methods.

The new draft of EN 1997 is comprised of three parts, "General Rules", "Ground Investigation" and "Geotechnical Constructions".

The new part 1, EN 1997–1, covers the basis of design of geotechnical constructions, including rules for analysis, both for serviceability limit states (SLS) and ultimate limit states (ULS). One of the subsections sets rules for numerical analyses which is the focus of this contribution.

All the information given here about the contents of the draft of EN 1997-1 and other parts of the revised Eurocode is preliminary and subject to change. It is based on the contents of the draft versions available in November 2017 (CEN 2017a, b).

2 GENERAL ASPECTS

Several categories of Eurocode-users have been identified and the category "Practitioners—competent engineers" has been selected as the primary audience of the Eurocode suite.

Main requirements for the amendment according to the Mandate are:

- Rules for assessment and strengthening of existing structures
- Rules for robustness
- Reduction of the number of Nationally Determined Parameters
- Improvement of ease of use
- Incorporation of recent developments.

The whole revised Eurocode suite is based on EN 1990 "Basis of structural and geotechnical design". EN 1990 gives the principles and requirements for safety, serviceability and durability of structures that are common to all Eurocode parts and are to be applied when using them. Since the content of EN 1990shall not be repeated in other Eurocode parts, EN 1990 will have to be used much more frequently by engineers than now.

3 DRAFT OF EN 1997

3.1 *Time line*

A final draft of the revised EN 1997–1 "General Rules" should be available in May 2018. Further stages before publication of the standard are a public "Enquiry", possibly followed by corrections and finally a "Formal Vote" according to CEN rules (CEN 2016).

A first draft of the revised EN 1997–2 "Ground Investigation" and of EN 1997–3 "Geotechnical Constructions" should be available in May 2018 and a final document in May 2020.

3.2 *EN 1997-1 – tasks*

Project Team PT1 have been in charge mainly of reorganising the content in three parts and harmonisation of the design. Their findings concern mainly

- factoring combinations for ULS design
- groundwater effects
- review of National Annexes
- review of design provisions in execution standards.

Project Team PT 2 has been responsible for

- draft of EN 1997–1
- reduction in number of National Choices
- enhanced ease of use
- reliability discrimination (geotechnical complexity, validation methods, etc.)
- groundwater pressures
- numerical models
- alignment with EN 1990.

These project teams worked concurrently to produce a first draft in April 2017 and a second draft in October 2017.

3.3 *EN 1997-3 – tasks*

The drafting of EN 1997-3 will be accomplished by two project teams: Project Team PT 4 is responsible for spread and pile foundations, slopes and ground improvement, whereas Project Team 5 deals with retaining structures, anchors and reinforced soil. Among their tasks is adding widely accepted calculation models to Eurocode 7 so that users do not have to resort to non-normative (typically national, not international) documents (CEN 2015).

3.4 *Organisation of EN 1997-1*

The amended EN 1997-1 contains the following main clauses:

- Basis of design
- Materials
- Groundwater
- Geotechnical analysis
- ULS
- SLS

- Execution
- Testing
- Reporting

The subheadings of "basis of design" are mostly identical to the main headings of EN 1990 and specify the application of the rules in EN 1990 to geotechnical structures. The headings "geotechnical analysis", "ULS" and "SLS", again complement the corresponding headings in EN 1990 and are more in common with the other Eurocodes. Consequently, EN 1997should be easier to use by those accustomed to the other Eurocodes. Clause "geotechnical analysis" contains the sub-clause about numerical methods.

The most significant changes to "basis of design" and "geotechnical analysis" compared to the current version of EN 1997 are summarised in the following sections of this paper.

4 BASIS OF DESIGN

Consequences classes—introduced in Annex B of the current EN 1990 – have been moved to the main text of the latest draft of EN 1990 and govern the required level of reliability for structural and geotechnical design. Five classes have been defined (CC0 to CC4) but classes CC0 and CC4 are beyond the scope of the Eurocodes.

The concept of geotechnical categories has been revised. they are now explicitly based on consequence classes on the one hand, while on the other hand, geotechnical complexity classes with three levels have been introduced which are independent of the consequences of failure. They are established based on criteria like soil properties, construction technique and soil-structure interaction. The classification is based on the most adverse design situation and may be modified during the design process if justified, e.g. by new information from ground investigation. The assignment to a geotechnical category is based on a table combining consequence classes and geotechnical complexity.

The required amount of ground investigation, validation of analysis models, design checking and execution supervision, and monitoring, etc. is based on the geotechnical category. (Geotechnical category 3 is not necessarily beyond the scope of EN 1997 anymore.)

Depending on the consequence class, partial safety factors may be amplified or reduced by a factor, either on actions or on ground strength properties.

For the verification of limit states four methods are foreseen:

- Verification by the partial factor method
- Verification by prescriptive measures
- Verification assisted by testing
- Verification by the Observational Method

5 GEOTECHNICAL ANALYSIS

This clause is new in the amended EN 1997-1 and covers material contained in section "2.4 Geotechnical design by calculation", subsection "2.4.1 General" of the current EN 1997-1 as well as the material about numerical methods. Among the general requirements for calculation models are:

- documentation of assumptions, including idealisations and limitations
- demonstration of existence of sufficient ductility for calculation models that assume ductile mechanisms
- review and validation according to the geotechnical category.

Suitable measures for validation are listed in Table 1.

Four types of calculation models are distinguished: empirical models, limit equilibrium methods, limit analysis methods and numerical methods.

Table 1. Suitable measures for validation (CEN 2017a).

Geotechnical category (GC)	Measures for validation depending on GC
GC 1	- Literature reference that the calculation model has been used for similar conditions - Local experience shows that the calculation model is suitable for the local conditions
-	- When using calculation models contained in EN 1997-3, confirmation that the design falls within the limits of application stated in EN 1997-3.
GC 2	- Documentation showing that the assumptions for the calculation model used are relevant for the specific site and structure
GC 3	- Calibration of the calculation model for the specific site - Sensitivity analyses for all relevant parameters.

6 NUMERICAL METHODS

Most of the material in this sub-clause of EN 1997-1 implements the recommendations of EG4. In consultations with users of EN 1997 during the re-drafting process, it has often been suggested that more guidance on the use of numerical methods should be provided and more prescription on the selection of constitutive models, for instance. This is not possible because EN 1997 is not a guidance document but rather a set of rules for safer design. Given the almost limitless scope of numerical methods and unforeseeable scientific developments in this field during the lifetime of the next generation of Eurocodes, it has been possible to provide only a rather generic set of rules. Heavy reliance is still placed on the competency of designers using such analysis tools.

6.1 *General*

The importance of user competency is reaffirmed in the general section on numerical methods, along with validation of analysis outputs and the use of parametric studies to help determine the reliability of outputs.

With conventional analysis methods, e.g. limit equilibrium or limit analysis, once the calculation method has been selected, taking due cognisance of its assumptions, limit state verifications are affected by a small number of parameters. Simple application of partial factors is usually sufficiently reliable for routine design by deterministic methods. However, due to the complexity of numerical methods, there are many influences on the prediction of limit states and simple reliance on partial factors could be dangerous. It is therefore a requirement to consider the sensitivity of limit state verifications to a number of aspects, including:

- discretisation of geometry, including discontinuities;
- initial stress states;
- preceding construction stages;
- boundary conditions;
- drainage conditions (including permeability);
- ground constitutive behaviour (including stiffness, dilatancy, anisotropy, yield criteria and flow rules);
- strength and stiffness of structural elements.

6.2 *SLS*

A short section on SLS hints at the improved predictions of deformations that can be obtained with advanced constitutive models with non-linear stiffness, for example. It also highlights the difference between SLS verification and best-estimate predictions of deformations (using estimates of mean values of material properties). A common error by some users of numerical methods is to compare the output from SLS verification analyses with site monitoring data, for example, expecting reasonable agreement even though cautious values of ground and structure parameters, geometry, sequencing, etc and full permanent and variable loading may have been adopted in the model. Such comparisons should be made with best-estimate predictions.

6.3 *ULS*

Partial factors are applied to actions, material parameters and resistances in the Eurocodes. Depending on the approach to verifying ULS, values greater than one are applied to one or more of these. There are pros and cons to each approach as summarised in the following three sub-sections. More detail and illustrative examples are provided by Lees (2013, 2017).

6.3.1 *Material factoring approach (MFA)*

Factors are applied to input parameters (material strength and some actions) such that design values of output are obtained directly. MFA is suited to verifying ULSs involving ground failure but its effect on structural forces is rather unpredictable, so it is less suited to verifying ULS of structural elements alone.

6.3.2 *Action effect factoring approach (EFA)*

Action effects are the output from calculations, such as pile axial load or wall bending moment. Except for a small factor on variable actions, input parameters remain at their characteristic values so EFA has the advantage of allowing the numerical analysis to be performed with more representative input parameters. Outputs are factored and then compared with resistances to verify ULS. It is suited to the verification of structural ULS but may provide insufficient reliability in cases where lower than expected ground strength has a significant effect on structural forces, e.g. forces in stabilisation measures to a marginally stable slope. In such cases, MFA provides more reliable design values of structural forces.

6.3.3 *Resistance factoring approach (RFA)*

Verification of ULS by numerical methods can involve RFA in two forms. Firstly, resistances can be determined independently by, for example, full-scale testing or independent calculation, and outputs of structural force compared with resistance to verify ULS (e.g. design value of soil nail force from numerical analysis obtained by EFA compared with design value of resistance measured by pull-out test). Secondly, resistance can be calculated by numerical methods by simulating a particular failure form, e.g. increasing the vertical displacement of a spread foundation until failure occurs to obtain bearing resistance as output. Its design value can then be compared with the design value of applied load to the foundation to verify the ULS of bearing failure.

The former is suited to ULS verification in structures such as axially-loaded piles, ground anchors, soil nails and rock bolts where resistance is often measured by testing and the use of MFA would be less reliable due to the uncertainty in soil-structure interface parameters. The latter is suited to cases where verification of the ULS of particular, relatively simple failure forms is desired but not suited to more general verification of multiple failure forms.

6.3.4 *Dual factoring approach*

Since no single factoring approach is sufficiently robust to verify ULS by numerical methods across a range of geotechnical structure types, a dual factoring approach to the determination of design values of structural forces is recommended in the new draft of EN 1997–1, with ground failure verified by MFA, as described in Figure 1. In many cases, design values of structural forces obtained by EFA will be the most onerous and these will be used to verify structural ULS.

However, in some cases, design values obtained by MFA will be more onerous (e.g. in marginally stable slopes) and these should be sued to verify structural ULS, hence the dual (EFA and MFA) approach. The same applies for design values of axial force in piles, soil nails, ground anchors and rock bolts, the most onerous of which would be

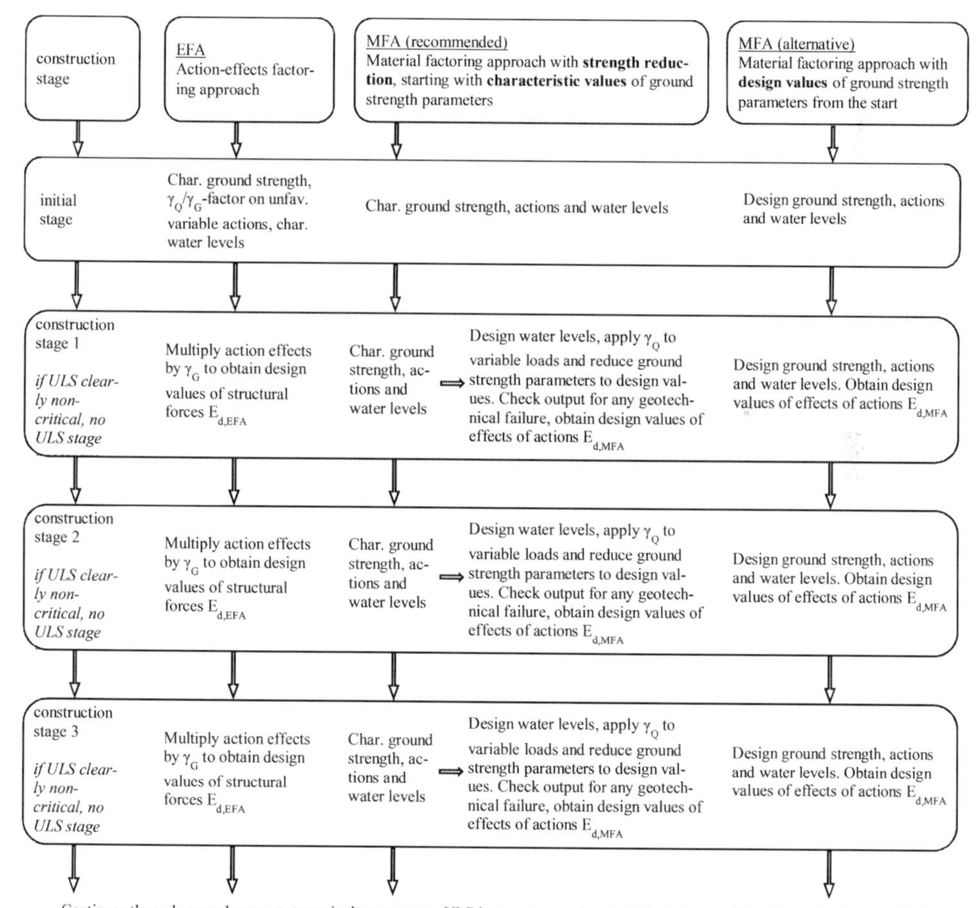

Figure 1. Procedures for ULS analyses with numerical methods.

compared with design values of axial resistance for these structures.

6.3.5 *Applying partial factors to input parameters*

The first stage of many simulations using numerical methods in geotechnical engineering involves establishing *in situ* stresses. Were these established with factored ground strength could lead to unrealistic stresses that influence outputs. Similarly, analyses with multiple construction stages using factored ground strength and actions may have hard-to-predict consequences on subsequent constructions stages with higher or lower degrees of conservatism than intended.

Consequently, rather than apply factors at the start of an analysis, it is recommended in the new draft of EN 1997-1 to run analyses with characteristic values throughout the construction sequence and to reduce ground strength, apply factors to actions and adopt design water regimes in separate adjunct stages at critical phases in the construction sequence.

The strength reduction may be performed simply by substituting the material model parameters for those with a lower, factored strength or by means of a stepwise strength reduction procedure, if available (Potts and Zdravković, 2012; Tschuchnigg et al, 2015).

Once the design value of ground strength is reached, ULS in the ground is verified by obtaining equilibrium in the calculation and interpreting from deformation output that a failure mechanism was not obtained. Design values of structural forces from MFA are also obtained. Strength reduction may be continued beyond this point to determine the most critical geotechnical failure.

It is permitted to combine ground strength reduction with structural strength or resistance reduction to help identify potentially critical collapse mechanisms of combined ground and structure failures.

6.4 *Numerical methods in EN 1990*

The current draft of EN 1990 contains a subclause about non-linear analysis in the clause "Structural analysis and design assisted by testing" which addresses mainly numerical methods. The recommendations there deal with the necessity of validation of the models against benchmarks, of basic tests on materials, of physical reference tests and mesh sensitivity tests. Further it is stated that key parameters should be entered into the limit state function as characteristic values and other parameters should be entered as mean values. A sensitivity study should be carried out—in case of non-linearity between the resistance and the variables influencing the resistance—to determine the most sensitive input parameter and how to apply the partial factors given in the Eurocodes.

Although some of these recommendations are rather vague, they are in line with the requirements and recommendations in EN 1997.

6.5 *Partial factors for ULS*

In the draft of EN 1990 (CEN 2017b) the partial factors on actions are grouped into four so-called "Design Cases".

In Design Case 1 the partial factors on actions are larger than 1.0, e.g. 1.5 on variable actions. Design Case 2 is a combined verification of strength and static equilibrium and requires two calculations. In geotechnics, it is used for verification of safety against uplift. Design Case 3 has a partial factor of 1.0 on permanent loads and is used in connection with the material factoring approach. Design Case 4 has partial factors on action effects instead of on the actions directly.

In Design Cases 1, 2 and 4 the partial factors should be modified by a consequence factor for consequences classes 1 and 3.

In the draft of EN 1997-1 (CEN 2017a), which covers geotechnical resistances, three sets of partial material factors are foreseen: factors of 1.0, factors larger than 1.0 or factors larger than 1.0 multiplied by a consequence factor. For conventional analyses, the combination of Design Case and set of partial material factor will be specified in EN 199-3 for each of the types of geotechnical construction.

The partial factors depend also on the type of design situation: for transient situations, the partial material factors may be reduced by an additional factor under certain circumstances; however, the recommended value of this factor is 1.0. For accidental and seismic design situations, there is a separate table of partial material factors.

If additional factors to the partial material factors are applied, care must be taken that the resulting combined factor is not less than 1.0.

The dual factoring approach for numerical methods involves Design Case 3, i.e. factors on variable actions combined with factors on material parameters (partial factor larger than 1.0 and multiplied by a consequence factor) and Design Case 4 (EFA). The least favourable of the resulting effects of actions are design actions for the structural elements, including anchors, piles, soil nails etc.

6.6 *Reliability based methods*

Whereas the draft of EN 1990 has a strong focus on the semi-probabilistic concept in the partial safety format and has strong restrictions on the use of reliability-based methods for verification of limit states, EN 1997-1 will be more open in this respect.

A more general treatment of geometrical entities in geotechnics (e.g. joint spacings and directions) than originally foreseen has been permitted in the main text of EN 1990. (The informative Annex C of EN 1990 contains a section about reliability-based methods.)

7 SUMMARY

The latest draft of part 1 of Eurocode 7, EN 1997-1 "Geotechnical design—General Rules", contains a new section on numerical methods. For ultimate limit state design two combinations of partial safety factors should be analysed to help ensure sufficient reliability against ultimate limit states occurring in the ground and structural elements for different types of geotechnical constructions.

Other requirements and recommendations in EN 1997-1 and EN 1990 "Basis of structural and geotechnical design" which are relevant for numerical analyses were also covered in this paper.

ACKNOWLEDGEMENTS

This contribution is based on work by the members of Evolution Group 4 and Project Teams 1 and 2 and the corresponding Task Groups as well as numerous comments and suggestions by the engineering community. All this input is gratefully acknowledged.

REFERENCES

CEN 2015 Tender documents https://www.nen.nl/Normontwikkeling/Eurocodes-2020.htm.

CEN 2016: https://boss.cen.eu/developingdeliverables/EN/Pages/default.aspx.

CEN 2017a: prEN 1997-1, Eurocode 7: Geotechnical design—Part 1: General rules, draft October 2017.

CEN 2017b: prEN 1990, Eurocode: Basis of structural and geotechnical design—Main element—Complementary element, draft October 2017.

Lees, A. (2013) Using numerical analysis with geotechnical design codes, in Modern Geotechnical Design Codes of Practice (eds. Arnold, Fenton, Hicks, Schweckendiek and Simpson). IOS Press, Amsterdam.

Lees, A. (2017). The use of geotechnical numerical methods with Eurocode 7, *Proceedings of the ICE Engineering Analysis and Computational Mechanics*. In press.

Potts, D.M. and Zdravkovic, L. (2012) Accounting for partial material factors in numerical analysis, *Géotechnique* **62**(12): 1053–1065.

Tschuchnigg, F., Schweiger, H.F., Sloan, S.W., Lyamin, A.V. and Raissakis, I. (2015) Comparison of finite-element limit analysis and strength reduction techniques, *Géotechnique* **65**(4): 249–257.

Shallow and deep foundations

Numerical Methods in Geotechnical Engineering IX – Cardoso et al. (Eds)
© 2018 Taylor & Francis Group, London, *ISBN 978-1-138-33203-4*

Analysis of piled foundations partially embedded in rock

J.R. Garcia
Federal University of Uberlândia, Minas Gerais, Brazil

P.J.R. Albuquerque & R.A.A. Melo
University of Campinas, São Paulo, Brazil

ABSTRACT: The behavior of piled foundations is one of the most commonly reviewed topics in foundation engineering, which is demonstrated by the number of publications available to the geotechnical community. However, when it comes to piles partially embedded into rocks, fewer studies exist. This paper analyses the behavior of three piled foundations comprising 9, 16 and 25 seven-meter long piles, spaced 3, 4 and 5 times the pile diameter (30 cm), respectively. The height of the surface foundation element crowning the piles varies by 0.5 m, 1 m and 2 m. When piles are either supported or partially embedded in material with rocky geomechanical characteristics, it is critical to carry out analyses by means of numerical tools so that the stress-strain behavior can be appropriately understood. In this article, for each situation analyzed in terms of piled groups and variation of heights of the block, the piles under analysis were supported both on a layer of rock and also partially embedded in 1 m, 2 m and 3 m of rock. The results of the numerical analyses showed that the block height influenced load distribution on top of the piles. On the other hand, the partial embedding into rock was shown to be influential on load distribution and transfer to the massif. Apparently, the results indicated an optimal pile length, beyond which no contribution was observed to the strength of the foundation element, but rather an effect that started to overload some piles and relieve others.

1 INTRODUCTION

1.1 *Overview of rock foundations*

Various engineering designs are developed on a daily basis to be employed beneath countless types of loads on top of the superstructure. Such designs must meet technical criteria and conditions of appropriate load capacity and settling so as to ensure a satisfactory safety factor. In most cases, settlement is the limiting factor in the search for an optimized geotechnical project intended to minimize settlement to the maximum extent. However, such a condition is physically impossible since, in order to develop the resistances of the foundation element, this element needs to displace in the ground. In this sense, settlement can be significantly reduced when geotechnical projects are dimensioned for groups of piles in deep foundations, with piles supported by layers of rock, provided that the loads of the structure are properly transferred (Figure 1). According to Mussara (2014), there is little information or lack of definition on embedding of piles into rock. This same author points out that some works have revolutionized the engineering of foundations embedded in rocks, such as those of Hobbs (1974), Rosenberg; Journeaux (1977), Horvath; Kenney (1979), Williams (1980), Pells; Rowe; Turner (1980), Rowe; Armitage (1987) and Carter; Kulhawy (1988). Despite the technological advances in underground excavation and in research equipment, some points remain challenging for designers and builders who choose to utilize piles embedded in rocks. Gannon et al. (1999) describe some of these issues, as follows:

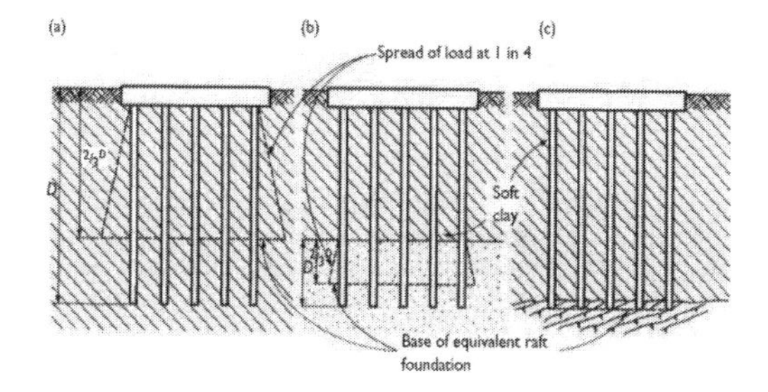

Figure 1. Load transfer to soil from pile groups: (a) Group of piles predominantly supported by shaft friction (b) Group of piles driven through soft clay to combine shaft friction and end bearing in stratum of dense granular soil (c) Group of piles supported in end bearing on hard rock stratum (Tomlinson and Woodward 2008).

- The control and study of geological formations do not get the required attention yet;
- Research is costlier when more sophisticated field and laboratory trials are used;
- The behavior of rock-embedded piles is not yet fully understood;
- The properties of the soil surrounding the piles can be substantially modified by the pile execution method.

Several researchers have recently studied the process of load transfer in piles built into rock by means of analytical and numerical models. (Carrubba, 1997; Zhang; Ernst; Einstein, 2000; Ng et al., 2001; Singh et al., 2017 and Chen et al., 2017). According to Carrubba (1997), the use of piles drilled and supported in rock is one of the best solutions when layers of soil overlap the rock at shallow depths. In these cases, a considerable load capacity is obtained due to skin friction in the rock, even with small displacements of the pile. However, Singh et al. (2017) point out that oftentimes the level at which the resistant rock stratum is situated is too deep. In these cases, an analysis is required, since the use of piles embedded into rock might not be economical.

In case of existence of a rock stratum, the key parameters are the strength of such rock (RQD) and the load distribution model of the sections of soil and sections of rock. Consequently, the structural strength of the pile has a greater influence on the behavior of this foundation (Garcia et al. 2016).

Tomlinson and Woodward (1995) present the operating assumptions in three proposals: (a) Group of piles predominantly supported by shaft friction; (b) Group of piles driven through soft clay, combining shaft friction and end bearing, supported in stratum of dense granular soil; (c) Group of piles supported at the bottom of the end bearing in hard rock stratum (Figure 1). When embedded into rock, the foundation elements carry the loads down directly to the area anchored in the rock (Garcia; Albuquerque; Melo, 2013; and Mussara, 2014).

2 FOUNDATIONS EMBEDDED IN ROCK

2.1 *Piles embedded in rock*

According to Rocha (1977), uniaxial strength and cohesion are the most appropriate characteristics to set the boundary between soils and rocks. According to Rocha, the angle of friction is an ineffective parameter to tell soils and rocks apart because higher soil values and lower rock values for the angle of friction share the same range.

The lower strength limit established by a specific classification does not eliminate the difficulty to set a clear boundary between very soft and hard rocks as well as very hard and cohesive soils because the materials in this transition domain can behave as soils or rocks depending basically on the manner stress is applied (Hencher 1993).

To predict the lateral strength of a pile embedded in rock, the analyses of the constitutive rupture model must incorporate coupling between shaft friction and regular displacement models (Pease; Kulhawy, 1984; Seidel; Haberfield, 1995). The constructive model should also enable a description of the behavior of lateral strength from initial loads through complete mobilization and ensuing rupture. The models require numerical accuracy of certain parameters that are not typically evaluated in engineering. It is common to use parameters such as rock cohesion (c), friction angle (ϕ) and uniaxial compression strength (qu) of rocks, as obtained from Equation 1 and Table 1 (Horvath; Kenney; Trow, 1980; Williams; Pells, 1981; Amir, 1986; Rowe; Armitage, 1987; Kulhawy; Phoon, 1993).

$$q_{ult} = q_u = 2 \cdot c \cdot \tan\left(45^\circ + \frac{\phi}{2}\right) \quad (1)$$

The stress-strain parameters (E_i) for rocks were obtained from the correlations presented by Hoek; Carranza; Corkum (2002), and from uniaxial compression strength (q_u).

According to Tomlinson and Woodward (1995), in some cases, because piles are supported on layers of hard soil or rock, the settling of this foundation element is severely reduced. Therefore, the pile magnitude may not be sufficient to develop skin friction along the pile. When the rock layer underlies soft clays and loose sands, the negative friction effect may occur due to weight, depending on the thicknesses of these layers.

2.2 *Piled foundations*

Piled foundations can be split into piled groups and piled rafts, depending on how the block-soil

Table 1. Reduction of strength parameters for a rock mass (Kulhawy and Goodman 1987).

	Rock property		
RQD (%)	Uniaxial stress resistance	Cohesion (c)	Friction angle (ϕ)
0–70	0.33	0.10	30°
70–100	0.33–0.80	$0.10 \cdot q_u$	30°–60°

contact is taken into consideration when calculating capacity and interaction of the foundation.

According to Sales (2000), the term 'piled raft' was set to specify the foundation system involving the association of a surface foundation element (raft or foot) with one pile or a group of piles, both being responsible for the performance of the foundation in terms of load capacity and settlement.

The parameters that influence the behavior of the piled raft, referring to load capacity and the susceptibility to settling, are related not only to its geometry (raft and piles), but also to the massif into which the piles are driven.

Tomlinson and Woodward (1995) present the operating assumptions in three proposals: (a) Group of piles predominantly supported by shaft friction (b) Group of piles driven through soft clay to combine shaft friction and end bearing in stratum of dense granular soil (c) Group of piles supported in end bearing on hard rock stratum. These authors suggest a calculation method named 'equivalent raft' to estimate settlement, as shown in Equation 2:

$$w = \frac{\mu_1 \cdot \mu_0 \cdot q \cdot B}{E_s} \qquad (2)$$

Where: μ_1 and μ_0 are coefficients tabulated or obtained by abacuses; B is the raft width; E_s is the Young's modulus and q is the load applied.

The Poisson's ratio is assumed to be equal to 0.5. The μ_1 and μ_0 factors, which are related to the equivalent piled raft geometry, the thickness of the compressible soil layer and the piled raft length / width ratio are presented by Christian and Carrier (1978). However, in most natural soil and rock formations, the strain modulus increases with depth so that calculations for conditions based on a constant modulus result in overestimated settlement. Thus, further analysis is required in cases where deep piled foundations are partially embedded in rock or soil of high resistance to compression.

3 NUMERICAL MODELING

In this paper, the numerical modeling was performed on a quarter of the problem due to the symmetry along the axis of the piles and raft, resulting in a rectangular 20 m × 20 m section block. The depth was a function of the length embedded in rock (Figure 2).

These dimensions were attributed after testing performed to make sure that the boundary conditions assigned at the ends of the model could be considered no displaceable or else had very low displacements and, therefore, could not affect the

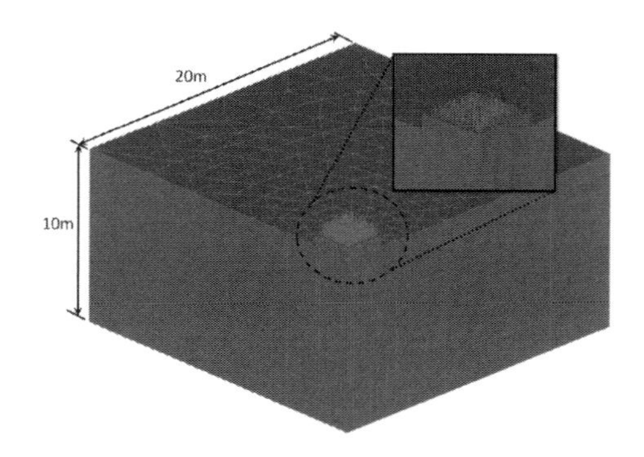

Figure 2. Model of finite elements for piled foundations.

Table 2. Strength and strain parameters of soil and rock.

Material	E [kPa]	γ [kN/m³]	c' [kPa]	ϕ' [(]	ν [-]
Soil	16	16	20	25	0.3
Rock	100	22	50	50	0.2

E: Young's modulus; γ:unit weight; c': cohesion; ϕ':friction angle; and ν: Poisson's ratio.

Table 3. Strength and strain parameters of concrete.

Material	E [kPa]	γ [kN/m³]	R_c [MPa]	R_t [MPa]	ν [-]
Concrete	25,000	25	25	2.5	0.2

R_c: compression strength; R_t: tensile strength;

results of the analyses. The elastoplastic model used varied according to the stresses applied, following a non-linear stress vs. strain behavior. The mesh of finite elements included quadratic interpolation triangular shaped elements, which were extruded every meter in depth.

The properties attributed to the different layers of soil and rock followed the Mohr-Coulomb criterion. For materials with fragile behavior, the parabolic model was used, as in the case of concrete. The CESAR v.5 software program (manufactured by Itech-soft) was used for numerical analyses. The parameters used in analyses for soil and rock are shown in Table 2. The parameters for concrete are shown in Table 3. These were adopted for a qualitative evaluation, with no intent of analyzing a specific case.

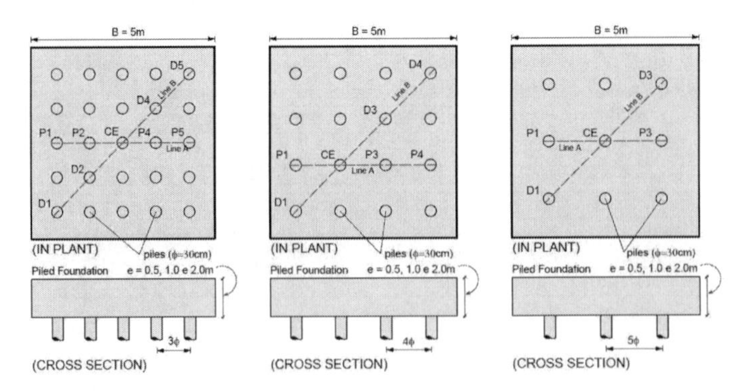

Figure 3. Piled foundations under study.

4 METHODOLOGY

The analyses were carried out based on the behavior of three piled foundations composed of 9, 16 and 25 seven-meter long piles, spaced 3, 4 and 5 times the diameter (30 cm), respectively. The height of the foundation element crowning the piles varied by 0.5 m, 1 m and 2 m. For the situation under analysis in terms of piled groups and variation in block heights, the piles were analyzed while supported on a layer of rock and partially embedded 1 m, 2 m and 3 m into the rock. The piled foundations were subjected to an evenly spread load of 100 kPa to simulate a water reservoir built on a surface, as shown in Figure 2. The variation in raft thickness made it possible to assess the influence of the stiffness of the raft and of the piles when supported directly on the ground and in a situation in which the piles under the raft were partially embedded in rock. The dimensions of the foundations under analysis, pile layout, spacing, raft thickness and pile diameter are shown in Figure 3. Also, load and set-up analyses were performed according to lines A and B.

5 RESULTS AND ANALYSIS

The results obtained in terms of stress-displacement of the behavior of piled foundations were compiled and analyzed according to the relative embedding in the rock layer, number of piles and thickness of the crowning block. These parameters were related to relative stiffness (k), which is considered the ratio between the load and the corresponding displacement.

The embedding into rock (L_r/L), which is the ratio between the length into rock and the total length of the pile, was compared to the relative stiffness (k) that increases with the increase in the ratio of embedding into rock and for the ratio between thickness (e) and width (B) of the block equal to 0.5 (Figure 4). This, in turn, affects the behavior of the piles beneath the block as a function of their arrangement. The stiffness of the center piles showed a significant increase up to the first meters of embedding ($L_r/L \cong 0.14$) of the order of 42 kN/mm. When the embedded ratio in rock increased to 0.29, the stiffness value increased by approximately 20% (Figure 4).

On the other hand, there was a difference in this increase as to the pile arrangement. The corner and border piles continued to increase their relative stiffness because of the increased ratio of rock embedding. However, the stiffness of the center piles remained unchanged for the piled groups of 9 and 16 piles, whereas there was a reduction in stiffness for the group of 25 piles. For the case where the piled group was supported by a rock layer ($L_r/L = 0$), relative stiffness varied from 3 to 20 kN/mm. In case of occurrence and increase in the embedding ratio, the relative stiffness of the group of 9 piles was significantly affected, followed by the groups of 16 and 25 piles.

On the other hand, there was a difference in this increase as to the pile arrangement. The corner and border piles continued to increase their relative stiffness because of the increased ratio of rock embedding. However, the stiffness of the center piles remained unchanged for the piled groups of 9 and 16 piles, whereas there was a reduction in stiffness for the group of 25 piles. For the case where the piled group was supported by a rock layer ($L_r/L = 0$), relative stiffness varied from 3 to 20 kN/mm. In case of occurrence and increase in the embedding ratio, the relative stiffness of the group of 9 piles was significantly affected, followed by the groups of 16 and 25 piles.

With the increase in the number of piles from 9 to 16, the net area of contact of the soil block was reduced; however, the behavior of relative stiffness was similar to the group with 9 piles (Figure 5). On the other hand, the corner piles were the ones that displayed higher values of stiffness and load absorption, followed by the border and center piles, respectively. By increasing the number of piles from 16 to 25, the net contact area was reduced again, accord-

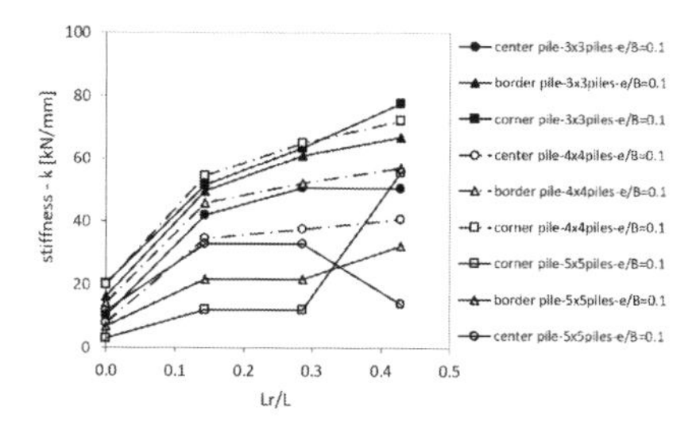

Figure 4. Variation in stiffness as a function of the embedding ratio (e/B = 0.1).

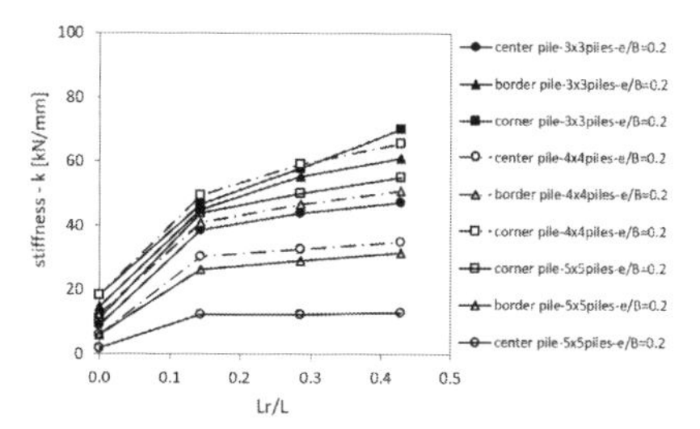

Figure 5. Variation in stiffness as a function of the embedding ratio (e/B = 0.2).

ing to the result shown in Figure 6. This figure shows that stiffness remained below 40 kN/mm for all piles for a rock embedding ratio smaller than 0.3.

The stiffness of piles beneath the block was significantly influenced by the partial embedding of the lengths of piles into the rock. On the other hand, the embedding efficiency was higher for L_r/L values equal to 0.14, or at most, 0.29.

Figure 5 displays an analysis of the behavior of piled groups for a greater thickness of the block, that is, ratio e/B = 0.2. The groups were appropriately affected by the increased embedding ratio. No localized increases or reduction were seen. This same figure shows that for the group of 25 piles, the efficiency of the embedding ratio remained practically unchanged for values above 0.14. Likewise, the groups of 16 and 9 piles were also weakly affected for values of L_r/L above 0.14 (Figure 5).

The behavior of the piled groups for a greater thickness of the block, that is, ratio e/B = 0.4, is shown in Figure 6. The groups were appropriately affected by the increased embedding ratio. No localized increase or reduction was seen. The same figure shows that, for the group of 25 piles, the efficiency of the embedding ratio remained practically unchanged for values above 0.14. The groups of 16 and 9 piles were also weakly influenced for values of L_r/L above 0.14 (Figure 5).

The piled groups were analyzed separately in terms of variation of block thickness and the ratio of rock embedding. For the group of 9 piles, an increase in ratio e/B above 0.2 weakly influences the values of relative stiffness (Figure 7).

For the group of 16 piles, the increase above 0.2 in the e/B ratio produces little reduction in values of relative stiffness for the center and border piles (Figure 8). The center piles were slightly influenced by the increase of e/B values, particularly from 0.2 to 0.4. On the other hand, the relative embedding into rock reduced the discrepancy between stiffness of piles, due to their respective arrangement in the block (Figure 8), same as with the group of 9 piles (Figure 7).

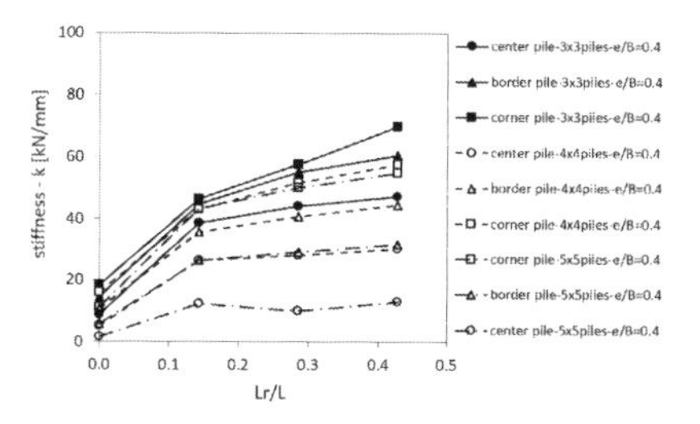

Figure 6. Variation in stiffness as a function of the embedding ratio (e/B = 0.4).

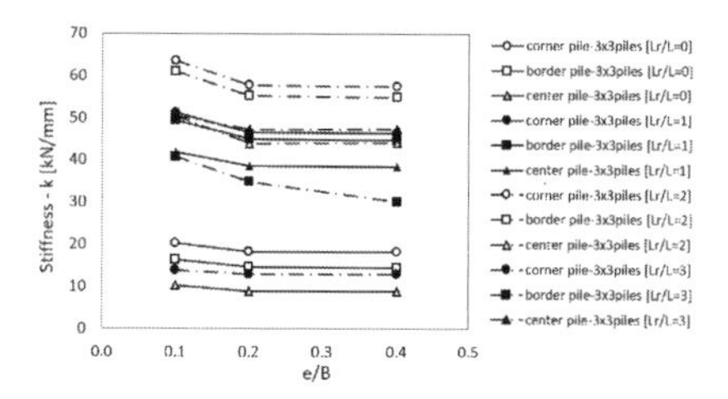

Figure 7. Variation in stiffness as a function of the embedding ratio for the group of 9 piles.

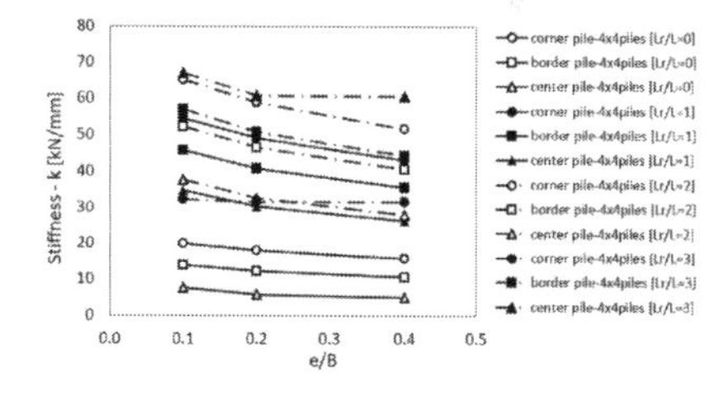

Figure 8. Variation in stiffness as a function of the embedding ratio for the group of 16 piles.

For the group of 25 piles (Figure 9), the increase above 0.2 in the e/B ratio did not affect the reduction of relative stiffness values of the center and corner piles. The border piles with L_r/L equal to 3 were influenced by the increase in e/B values, particularly in the 0.2 to 0.4 range. This same figure shows that, for values of e/B between 0.1 and 0.2, the piles were significantly influenced by the ratio of rock embedding, and stiffness values between corner and center piles were inverted (Figure 9).

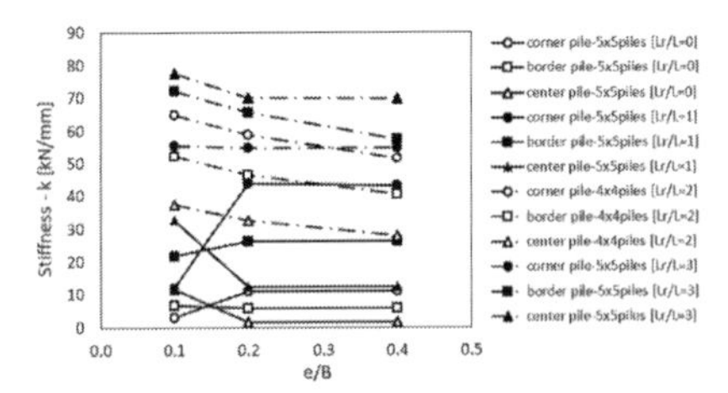

Figure 9. Variation in stiffness as a function of the embedding ratio for the group of 25 piles.

6 CONCLUSIONS

The following conclusions can be drawn from the analyses:

Piled foundations partially embedded in rock tend to even out vertical displacements, thus they can minimize problems with differential settlement in structures;

The stiffness of piles beneath one block is significantly affected by the partial embedding of the pile lengths into rock. On the other hand, the embedding efficiency is higher for Lr/L values equal to 0.14, or at most, 0.29.

The stiffness of the center piles shows a significant increase up to the first few meters of embedding ($L_r/L \cong 0.14$) of the order of 42 kN/mm for this work.

The relative stiffness of piles is affected by the thickness of the block (e) for values between 0.1 and 0.2 of the block width (B). Above these values, the influence of block thickness is either null or poor.

The reduction in the net contact area of the block, i.e., for groups with larger amounts of piles, the partial embedding of piles into rocks reduces stiffness slightly. However, groups of 25 piles, for e/B values between 0.1 and 0.2, are significantly influenced by the ratio of embedding into rock and stiffness values between corner and center piles are inverted.

ACKNOWLEDGMENTS

The authors wish to thank the São Paulo Research Foundation (Fundação de Amparo à Pesquisa no Estado de São Paulo—FAPESP) for the support granted to acquire the LCPC-Cesar software program made by Itech Software.

REFERENCES

Carrubba, P. 1997. "Skin Friction on Large-Diameter Piles Socketed into Rock." *Canadian Geotechnical Journal* 34 (2): 230–40. doi:10.1139/t96–104.

Carter, J. P., and F. H. Kulhawy. 1988. "Analysis and Design of Drilled Shaft Foundation Socketed into Rock." New York. doi:EL-5918.

Chen, Jin-Jian, Fan-Yun Zeng, Jian-Hua Wang, and Lianyang Zhang. 2017. "Analysis of Laterally Loaded Rock-Socketed Shafts Considering the Nonlinear Behavior of Both the Soil/Rock Mass and the Shaft." *Journal of Geotechnical and Geoenvironmental Engineering* 143 (3): 6016025. doi:10.1061/(ASCE)GT.1943–5606.0001610.

Gannon, J., G. Masterton, W. Wallace, and D. Wood. 1999. *Piled Foundations in Weak Rock*. London: Ciria.

Garcia, Jean Rodrigo, Paulo José Rocha de Albuquerque, Osvaldo de Freitas Neto, and Rodrigo Álvares de Araújo Melo. 2016. "Análise Numérica de Radier Estaqueado Composto Por Estacas Parcialmente Embutidas Em Rocha." In *XVIII Congresso Brasileiro de Mecânica Dos Solos E Engenharia Geotécnica, O Futuro Sustentável Do Brasil Passa Por Minas*. Belo Horizonte: ABMS. doi:10.20906/CPS/CB-08–0020.

Garcia, Jean Rodrigo, Paulo José Rocha de Albuquerque, and Rodrigo Álvares de Araújo Melo. 2013. "Experimental and Numerical Analysis of Foundation Pilings Partially Embedded in Rock." *Rem: Revista Escola de Minas* 66 (4): 439–46. doi:10.1590/S0370–44672013000400006.

Hobbs, N. 1974. "Factors Affecting the Prediction of Settlement of Structures on Rock with Particular Reference to the Chalk and Trias." In *Settlement of Structures—Conference British Geotechnical Society at Cambridge*, 579–654. London.

Horvath, R. G., and T. C. Kenney. 1979. "Shaft Resistance of Rock-Socketed Drilled Piers." In *Symposium on Deep Foundations*, 182–214. New York: ASCE.

Mussara, Marcello Duarte. 2014. "Análise de Comportamento de Estaca Barrete Embutida Em Rocha." Escola Politécnica da Universidade de São Paulo.

Ng, Charles W. W., Terence L. Y. Yau, Jonathan H. M. Li, and Wilson H. Tang. 2001. "Side Resistance of Large Diameter Bored Piles Socketed into Decomposed Rocks." *Journal of Geotechnical and Geoenvironmental Engineering* 127 (8): 642–57. doi:10.1061/(ASCE)1090–0241(2001)127:8(642).

Pells, P. J., R. K. Rowe, and R. M. Turner. 1980. "An Experimental Investigation into Sideshearfor Socketed Piles in Sandstone." In *Proceedings of the International Conference on Structural Foundations on Rock*, 291–302. Sidney: ARRB Group Limited.

Rosenberg, P., and N. L. Journeaux. 1977. "Friction and End Bearing Tests on Bedrock for High Capacity Socket Design: Reply." *Canadian Geotechnical Journal* 14 (2): 272–272. doi:10.1139/t77–029.

Rowe, R.K., and H.H. Armitage. 1987. "Design Method for Drilled Piers in Soft Rock." *Canadian Geotechnical Journal* 24 (1): 126–42. doi:10.1139/t87–011.

Singh, A. P., T. Bhandari, R. Ayothiraman, and K. Seshagiri Rao. 2017a. "Numerical Analysis of Rock-Socketed Piles under Combined Vertical-Lateral Loading." *Procedia Engineering* 191. Elsevier B.V.: 776–84. doi:10.1016/j.proeng.2017.05.244.

———. 2017b. "Numerical Analysis of Rock-Socketed Piles under Combined Vertical-Lateral Loading." *Procedia Engineering* 191. Elsevier B.V.: 776–84. doi:10.1016/j.proeng.2017.05.244.

Tomlinson, M., and J. Woodward. 2008. *Pile Design and Construction Practice*. 5thed. London and New York: Taylor & Francis Group.

Williams, A. F. 1980. "The Design and Performance of Piles Socketed into Weak Rock." Monash University.

Zhang, Lianyang, Helmut Ernst, and Herbert H Einstein. 2000. "Nonlinear Analysis of Laterally Loaded Rock-Socketed Shafts." *Journal of Geotechnical and Geoenvironmental Engineering* 126 (11): 955–68. doi:10.1061/(ASCE)1090–0241(2000)126:11(955).

Numerical Methods in Geotechnical Engineering IX – Cardoso et al. (Eds)
© 2018 Taylor & Francis Group, London, *ISBN 978-1-138-33203-4*

The bearing capacity of shallow foundations on slopes

S. Van Baars
University of Luxembourg, Luxembourg, The Grand Duchy of Luxembourg

ABSTRACT: In 1920 Prandtl published an analytical solution for the bearing capacity of a soil under a strip load. Over the years, extensions have been made for a surrounding surcharge, the soil weight, the shape of the footing, the inclination of the load, and also for a slope. In order to check the current extensions of a loaded strip footing next to a slope, many finite element calculations have been made, showing that these extensions are often inaccurate. Therefore new factors have been proposed, which are for both the soil-weight and the surcharge slope bearing capacity, based on the numerical calculations, and for the cohesion slope bearing capacity, also on an analytical solution.

1 INTRODUCTION

In 1920, Ludwig Prandtl published an analytical solution for the bearing capacity of a soil under a strip load, p, causing kinematic failure of the weightless infinite half-space underneath. The strength of the half-space is given by the angle of internal friction, ϕ, and the cohesion, c. The original drawing of the failure mechanism proposed by Prandtl can be seen in Figure 1.

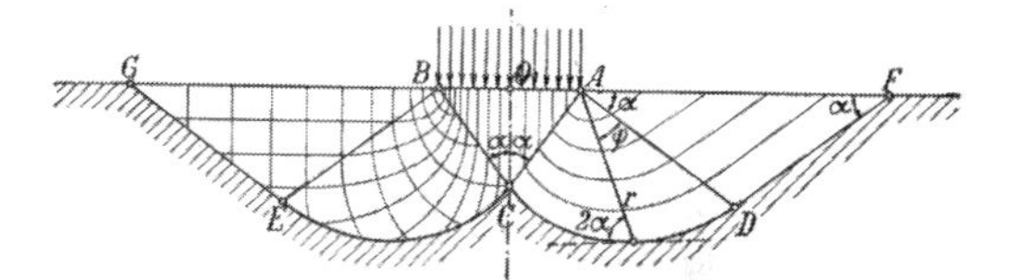

Figure 1. The Prandtl-wedge failure mechanism (Original drawing by Prandtl).

The lines in the sliding soil part on the left indicate the directions of the maximum and minimum principal stresses, while the lines in the sliding soil part on the right, indicate the sliding lines with a direction of $\alpha = 45° - ½\phi$ in comparison to the maximum principal stress. Prandtl subdivided the sliding soil part into three zones:

1. Zone 1: A triangular zone below the strip load. Since there is no friction on the ground surface, the directions of the principal stresses are horizontal and vertical; the largest principal stress is in the vertical direction.
2. Zone 2: A wedge with the shape of a logarithmic spiral, in which the principal stresses rotate through $½\pi$ radians, or 90 degrees, from Zone 1 to Zone 3. The pitch of the sliding surface of the logarithmic spiral equals the angle of internal friction; ϕ, creating a smooth transition between Zone 1 and Zone 3.
3. Zone 3: A triangular zone adjacent to the strip load. Since there is no friction on the surface of the ground, the directions of principal stress are horizontal and vertical; the largest principal stress is in the horizontal direction.

The solution of Prandtl was extended by Hans J. Reissner (1924) with a surrounding surcharge, q, and was based on the same failure mechanism. Albert S. Keverling Buisman (1940) and Karl Terzaghi (1943) extended the Prandtl-Reissner formula for the soil weight, γ It was Terzaghi who first wrote the bearing capacity with three separate bearing capacity factors for the cohesion, surcharge and soil weight. George G. Meyerhof (1953) was the first to propose equations for inclined loads, based on his own laboratory experiments. Meyerhof was also the first in 1963 to write the formula for the (vertical) bearing capacity p_v with bearing capacity factors (N), inclination factors (i) and shape factors (s), for the three independent bearing components; cohesion (c), surcharge (q) and soil weight (γ), in a way it was adopted by Jørgen A. Brinch Hansen (1970) and it is still used nowadays:

$$p_v = i_c s_c c N_c + i_q s_q q N_q + i_\gamma s_\gamma \tfrac{1}{2} \gamma B N_\gamma. \tag{1}$$

Prandtl (1920) solved the cohesion bearing capacity factor:

$$N_c = \left(K_p \cdot e^{\pi \tan\phi} - 1\right)\cot\phi \text{ with: } K_p = \frac{1+\sin\phi}{1-\sin\phi} \tag{2}$$

Reissner (1924) solved the surcharge bearing capacity factor with the equilibrium of moments of Zone 2:

$$N_q = K_p \cdot \left(\frac{r_3}{r_1}\right)^2 = K_p \cdot e^{\pi \tan \phi} \text{ with: } r_3 = r_1 \cdot e^{\frac{1}{2}\pi \tan \phi} \qquad (3)$$

Keverling Buisman (1940), Terzaghi (1943), Caquot and Kérisel (1953, 1966), Meyerhof (1951; 1953; 1963; 1965), Brinch Hansen (1970), Vesic (1973, 1975), and Chen (1975) subsequently proposed different equations for the soil-weight bearing capacity factor N_γ. Therefore the following equations for the soil-weight bearing capacity factor can be found in the literature:

$$\begin{aligned} N_\gamma &= \left(K_p \cdot e^{\pi \tan \phi} - 1\right)\tan(1.4\phi) \quad \text{(Meyerhof)}, \\ N_\gamma &= 1.5\left(K_p \cdot e^{\pi \tan \phi} - 1\right)\tan \phi \quad \text{(Brinch Hansen)}, \\ N_\gamma &= 2\left(K_p \cdot e^{\pi \tan \phi} + 1\right)\tan \phi \quad \text{(Vesic)}, \\ N_\gamma &= 2\left(K_p \cdot e^{\pi \tan \phi} - 1\right)\tan \phi \quad \text{(Chen)}. \end{aligned} \qquad (4)$$

The problem with all these solutions is that they are all based on associated soil ($\psi = \phi$). Loukidis et al (2008) noticed that non-dilatant (non-associated) soil is 15% - 30% weaker than associated soil, and has a rougher failure pattern. Van Baars (2015, 2016a, 2016b) confirmed these results with his numerical calculations and showed that, for non-dilatant soil, the following lower factors describe better the bearing capacity:

$$N_q = \cos^2 \phi \cdot K_p \cdot e^{\pi \tan \phi} \qquad (5)$$

$$N_c = \left(N_q - 1\right)\cot \phi \text{ with: } N_q = \cos^2 \phi \cdot K_p \cdot e^{\pi \tan \phi} \qquad (6)$$

$$N_\gamma = 4 \tan \phi \cdot \left(e^{\pi \tan \phi} - 1\right) \qquad (7)$$

The difference between the analytical solution and the numerical results has been explained by Knudsen and Mortensen (2016): The higher the friction angle, the wider the logarithmic spiral of the Prandtl wedge and the more the stresses reduce in this wedge during failure. So, the analytical formulas are only kinematically admissible for an associated flow behaviour ($\psi = \phi$), which is completely unrealistic for natural soils. This means for higher friction angles as well that, a calculation of the bearing capacity of a footing based on the analytical solutions (Equations 2–3), is also unrealistic.

Therefore, in this study, the bearing capacity factors and the slope factors will be calculated with the software Plaxis 2D for a bi-linear constitutive Mohr-Coulomb (c, ϕ) soil model without hardening, softening, or volume change during failure (so the dilatancy angle $\psi = 0$).

2 MEYERHOF & VESIC

Shallow foundations also exist in or near slopes, for example the foundation of a house or a bridge (Figure 2). Meyerhof was in 1957 the first to publish about the bearing capacity of foundations on a slope. He wrote: "*Foundations are sometimes built on sloping sites or near the top edge of a slope.... When a foundation located on the face of a slope is loaded to failure, the zones of plastic flow in the soil on the side of the slope are smaller than those of a similar foundation on level ground and the ultimate bearing capacity is correspondingly reduced*".

Meyerhof published a failure mechanism of a footing in a slope (Figure 3 above) and introduced figures with reduced bearing capacity factors (Figure 3 below). The problems with these figures are:

- it is unclear if the figures are based on non-associated flow behaviour, or not,
- the figures are not explained and cannot be verified,
- the figures are only for purely cohesive or purely frictional soil,

Figure 2. Footing of a house and bridge near a slope.

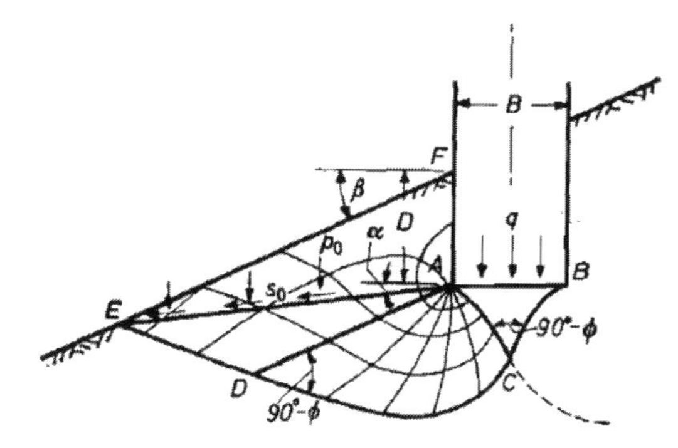

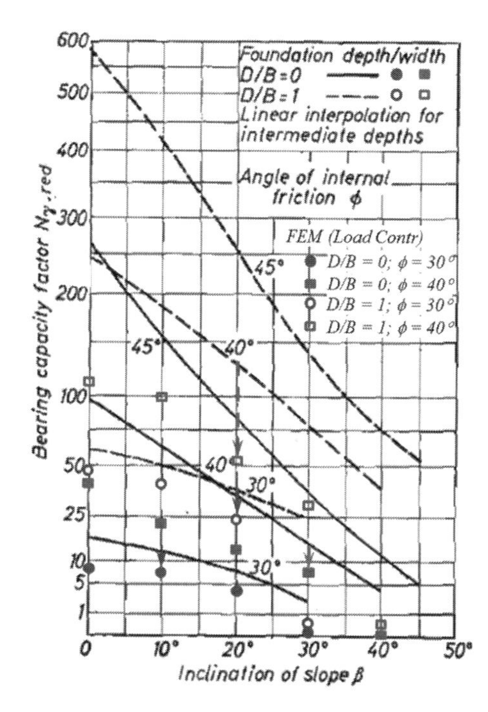

Figure 3. Above: failure mechanism of a footing in a slope. Below: reduced bearing capacity factors (according to Meyerhof, 1957).

- the important angle α (see line EA in the figure on the left) is never solved,
- the slope bearing capacity does not go to "0" for $\beta = \phi$, and
- the reduced bearing capacity factors of Meyerhof are too high according to the results of Finite Element Model (FEM load controlled) calculations, made in this article (see the added points in Figure 3 below).

Brinch Hansen (1970) also worked on the influence of a slope. Vesic (1975) combined the work of Meyerhof and Brinch Hansen and proposed the following bearing capacity equation:

$$p = \lambda_c c N_c + \lambda_q q N_q + \lambda_\gamma \tfrac{1}{2} \gamma B N_\gamma, \tag{8}$$

with the following slope factors:

$$\lambda_c = \frac{N_q \lambda_q - 1}{N_q - 1} \quad (\phi > 0),$$
$$\lambda_c = 1 - \frac{2\beta}{\pi + 2} \quad (\phi = 0), \tag{9}$$
$$\lambda_q = \lambda_\gamma = (1 - \tan\beta)^2.$$

The angles in these equations are in radians.

It is remarkable, if not to say impossible, that these slope factors do not depend on the friction angle ϕ, and that the surcharge slope factor λ_q and the soil-weight slope factor λ_γ are identical.

Another mistake is that the cohesion slope factor N_c is solved based on the assumption that Equation 6 about the relation between the cohesion bearing capacity N_c and the surcharge bearing capacity N_q, is also valid for inclined loading, and also for loading near a slope ($\lambda_c N_c = (\lambda_q N_q - 1)\cot\phi$). This assumption was published first by De Beer and Ladanyi (1961). Vesic (1975) calls this "*the theorem of correspondence*", and Bolton (1975) calls this "*the usual trick*". The relation between N_c and N_q in Equation 6 is coincidently valid for vertical ultimate loads without a slope ($N_c = (N_q - 1)\cot\phi$), but the assumption that this is also the case for inclined loading and loading near a slope, is not correct, according to the results of the numerical calculations, and also according to the analytical solution given later in this paper.

This indicates that not only the inclination factors, but also the slope factors proposed by Vesic, are incorrect and should not be used.

3 MODERN RESEARCH & GERMAN NORMS

Over the years quite some people have published about the bearing capacity of footings on a slope, but that was mostly limited to, or purely cohesive slopes (Azzouz and Baligh, 1983; Graham et al., 1988, Georgiadis, 2010, Shiau et al (2011) or purely non-cohesive slopes (Grahams et al., 1988), in a geotechnical centrifuge (Shields et al. 1990), or dedicated to even more complex cases like seismicity (Kumar and Rao, 2003; Yamamoto, 2010), reinforced soil (Alamshahi and Hataf, 2009; Choudhary et al., 2010) or 3D load cases near slopes (Michalowski, 1989; De Butan and Garnier, 1998), while the more simple non-seismic, non-reinforced, 2D situation is still not fully understood.

Chakraborty and Kumar (2013) were one of the first to make a more general study, but unfortunately only used, as most researchers, the lower bound finite element limit analysis with a non-linear optimization. They also did not present slope correction factors. The same applies to Leshchinsky (2015), who used an upper-bound limit state plasticity failure discretisation scheme.

The currently most used slope correction factors, are the following slope correction factors mentioned in the German design norm (in fact the German Annex to Eurocode 7 "Geotechnical Engineering"):

$$\begin{aligned} &\lambda_c = \frac{N_q e^{-\alpha} - 1}{N_q - 1} \quad (\phi > 0), \\ &\lambda_c = 1 - 0.4 \tan\beta \quad (\phi = 0), \\ &\lambda_q = (1 - \tan\beta)^{1.9}, \\ &\lambda_\gamma = (1 - 0.5\tan\beta)^6, \end{aligned} \tag{10}$$

in which: $\alpha = 0.0349 \cdot \beta \cdot \tan\phi$.

The angles in these equations are in degrees and to avoid slope failure: $\beta \leq \phi$.

There is no reference or any background information in the German design norm about these factors, which is a major problem. It is also remarkable, for these slope correction factors in the German norm, if not to say impossible, that the surcharge slope factor and also the soil-weight slope factor do not depend on the friction angle.

Because of these problems, the bearing capacity near slopes has been studied with the well-established and validated Finite Element Model Plaxis. First load controlled calculations have been made, and second, comparisons have been made between these Finite Element calculations and the results of the German design norm, the results of Bishop slip circle calculations (with the program "GEO5" from "Fine Civil Engineering Software") and, for the cohesion slope factor λ_c, also the results of the analytical solution proposed in this article.

4 SLOPE FACTORS

4.1 *Cohesion slope factor λ_c*

For two different friction angles ϕ = 0°, 30° and four different slope angles β = 0°, 10°, 20°, 30°, the failure mechanism for a cohesive (c = 10 kPa), weightless (γ' = 0 kN/m^3) soil has been calculated with numerical calculations (FEM) and compared with the Prandtl failure mechanism, see Figure 4.

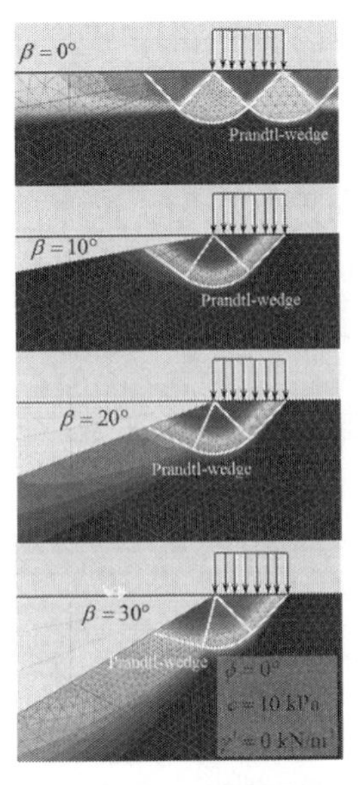

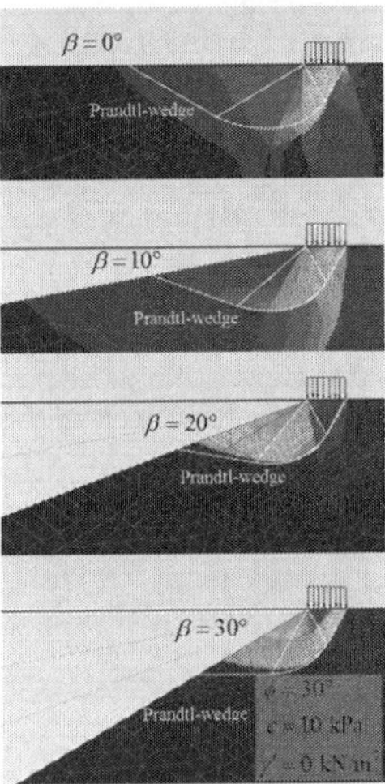

Figure 4. Failure mechanism: Prandtl-wedge versus FEM (Incremental displacement plots).

This figure shows that a Prandtl-wedge with a reduced Zone 2 (the logarithmic spiral wedge) describes in general the failure mechanism.

Because of this, it is also possible to derive an analytical solution for the cohesion slope factor, in the same way as the derivation of the cohesion inclination factor i_c (see Van Baars, 2014):

$$\lambda_c = \cos\beta \cdot \left(e^{-2\beta\tan\phi} - \frac{2\beta}{2+\pi} \cdot e^{-\pi\tan\phi} \right) \quad \beta \leq \phi \tag{11}$$

The results of this analytical solution, the German design norm and the Bishop's slip circle

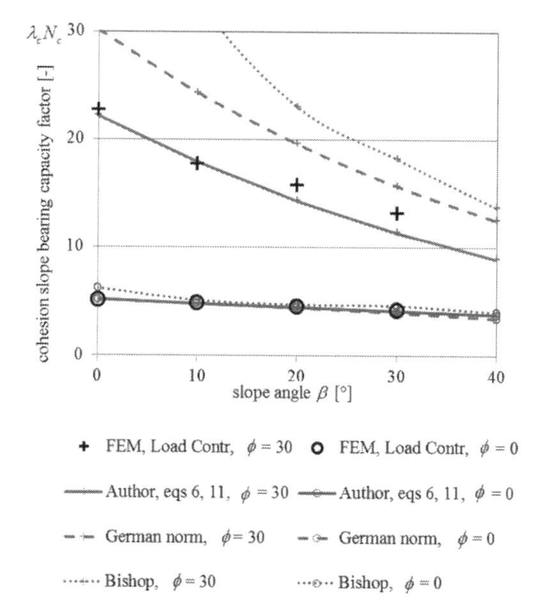

Figure 5. Cohesion slope factor (Analytical solution, German norm and Bishop versus FEM).

method have been plotted in Figure 5, together with the results from the Finite Element calculations. Figure 5 shows that the Bishop calculations are only correct for a zero friction angle. The analytical solution functions very well. The German norm would function just as good, if not the Prandtl solution for large dilatancy (Equation 2), but the solution for zero dilatancy (Equation 6), would have be used. The reason for this is because:

$$\frac{N_q e^{-0.0349 \cdot \beta \tan\phi} - 1}{N_q - 1} \approx \cos\beta \cdot \left(e^{-2\beta\tan\phi} - \frac{2\beta}{2+\pi} \cdot e^{-\pi\tan\phi} \right) \quad (12)$$

4.2 *Soil-weight slope factor λ_γ*

A mistake which can be found in the publication of Meyerhof (1957), but also in many recent publications, is the assumption that the failure mechanism in a purely frictional soil (N_γ), is a Prandtl-wedge with a reduced logarithmic spiral-wedge, which is according to the numerical calculations not the case. Also plots of the incremental displacements of the FEM calculations, indicating this failure mechanism, show that this approach is not correct for purely frictional soil (Figure 6).

Because of this, it is not possible to derive the soil-weight slope factor λ_γ, in a similar way as was done for the cohesion slope factor λ_c. Although,

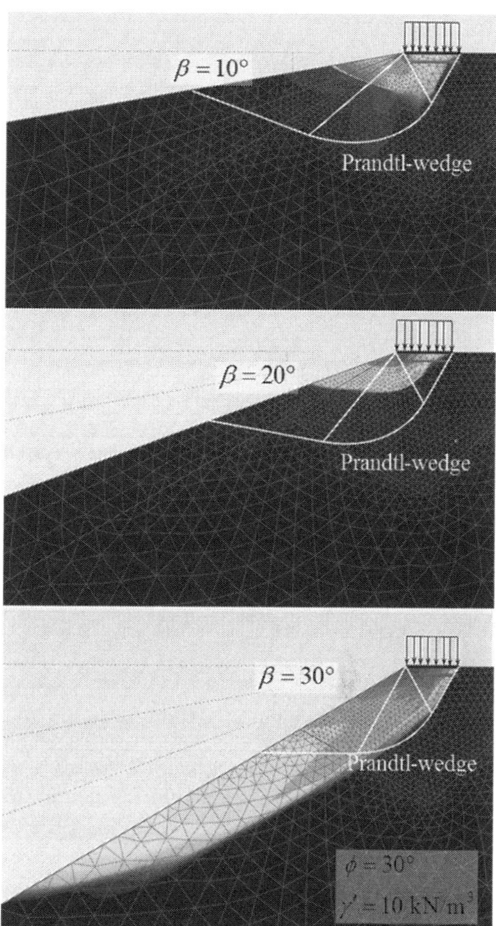

Figure 6. Failure mechanism: Prandtl-wedge versus FEM (Incremental displacement plots).

a simple approximation can easily be made, for example:

$$\lambda_\gamma = 1 - \left(\frac{\beta}{\phi}\right)^{2/3} \quad \beta \leq \phi \quad (13)$$

The results of this equation, the German design norm and the Bishop's slip circle method have been plotted in Figure 7, together with the results from the Finite Element calculations. This figure shows that the analytical approximation functions reasonably well. The German design norm does not fit. The Bishop calculations do not fit at all. An important reason for this is the fact that the slip circle in the Bishop calculations has been forced not to cross the foundation plate,

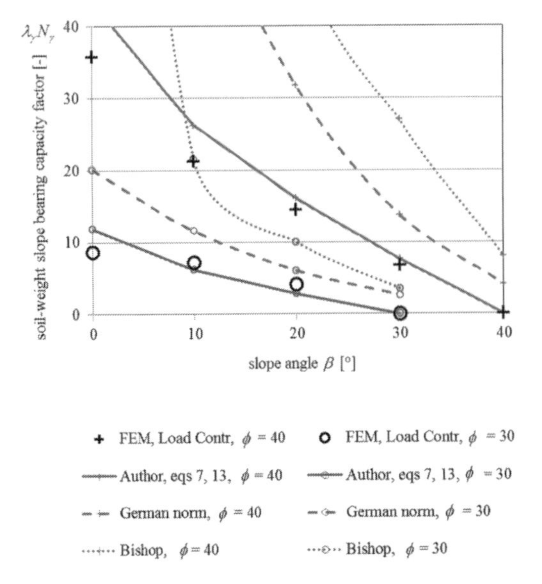

Figure 7. Soil-weight slope factor (Approximation, German norm and Bishop versus FEM).

while the FEM calculations show that the soil slides somewhere below the plate (see especially Figure 6 for $\beta = 20°$), which causes the plate to tumble over. This tumbling failure mechanism however, is not part of the Bishop calculation method.

4.3 *Surcharge slope factor* λ_q

In order to see the influence of having a shallow solid footing, additional finite element calculations have been made for a relative depth of $D/B = 1$. This relative depth creates an additional bearing capacity mostly due to the surcharge of $q = \gamma' D$, but also due to a larger slip surface, of which the influence is difficult to quantify.

Plots of the incremental displacements of the FEM calculations, indicating the failure mechanism, show that the failure mechanism of a shallow foundation in a frictional soil with self-weight, is an extended Prandtl-wedge with a reduced logarithmic spiral-wedge, see Figure 8.

Since Figure 8 shows that in general the failure mechanism looks like an extended Prandtl-wedge with a reduced logarithmic spiral-wedge, it seems possible to derive an analytical solution for the surcharge slope factor, which is purely based on the reduced logarithmic spiral-wedge. This would give the following surcharge slope factor:

$$\lambda_q = e^{-2\beta \tan\phi} \quad \beta \le \phi, \tag{14}$$

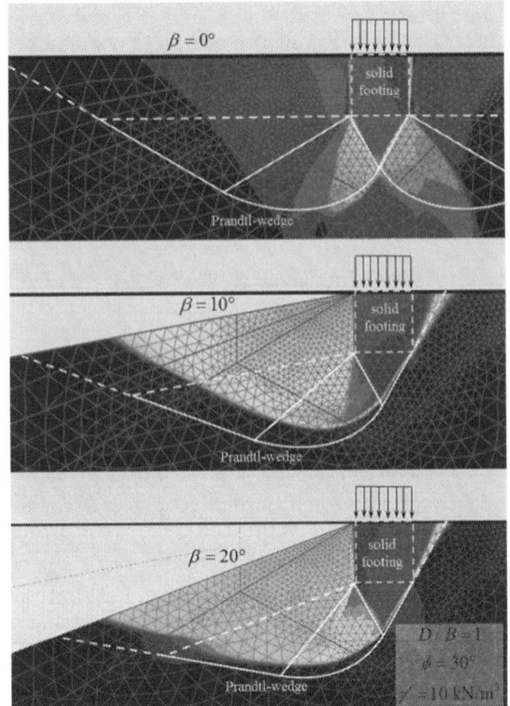

Figure 8. Failure mechanism: Prandtl-wedge versus FEM (Incremental displacement plots).

which is the same equation for the surcharge slope factor as the one proposed by Ip (2005). The problems with this solution are:

- it neglects the extension (dashed lines) due to the depth, and
- the surcharge slope factor is not "0" for $\beta = \phi$, which is the same problem as for the slope factors of Meyerhof in Figure 3.

It is therefore better to make, in this case, a simple approximation for the surcharge slope factor λ_q, for example:

$$\lambda_q = 1 - \left(\frac{\beta}{\phi}\right)^{3/2} \quad \beta \le \phi. \tag{15}$$

The results of this equation, the German design norm and the Bishop's slip circle method have been plotted in Figure 9, together with the results from the Finite Element calculations.

This figure shows that the Bishop calculations fit better for steeper slopes this time, but still not for gentile slopes. The German design norm and especially the analytical approximation are functioning reasonably well.

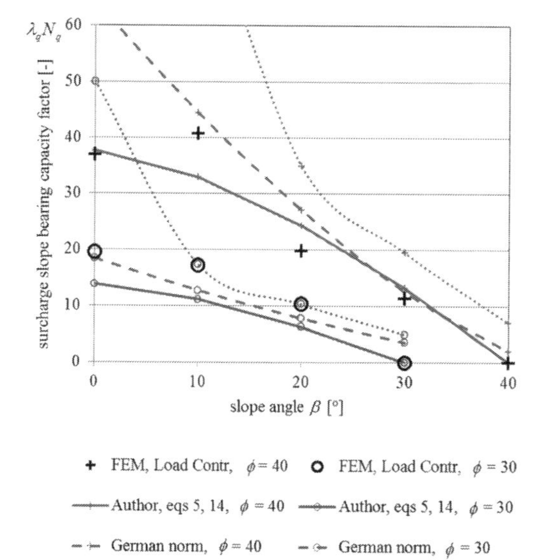

Figure 9. Surcharge slope factor (Approximation, German norm and Bishop versus FEM).

5 CONCLUSIONS

A large number of finite element calculations of strip footings next to a slope have been made in order to check the failure mechanisms, and to check the bearing capacities, first for the currently used equations for the slope bearing capacity factors, proposed by the German Annex of the Geotechnical Eurocode, and second for the bearing capacity calculated with the Bishop slip circle stability calculation method. These calculations proof that both the German slope factors and the Bishop calculations are often inaccurate.

Therefore new factors have been proposed, which are for both the soil-weight and the surcharge slope bearing capacity, based on the numerical calculations, and for the cohesion slope bearing capacity, also on an analytical solution.

REFERENCES

Alamshahi, S. and Hataf, N. (2009) Bearing capacity of strip footings on sand slopes reinforced with geogrid and grid-anchor, *Geotextiles and Geomembranes*, Volume 27, Issue 3, June 2009, pp. 217–226.

Azzouz, A.S. and Baligh, M.M. (1983) Loaded areas on cohesive slopes, *Journal of Geotechnical Engineering*, vol. 109, pp. 724–729.

Bolton, M.D. (1979) *Guide to Soil Mechanics*, Macmillian Press, London, p. 330.

Brinch Hansen, J.A. (1970) Revised and extended formula for bearing capacity, *Bulletin No 28*, Danish Geotechnical Institute Copenhagen, pp. 5–11.

Caquot, A., and Kerisel, J. (1953) Sur le terme de surface dans le calcul des fondations en milieu pulverulent. *Proc. Third International Conference on Soil Mechanics and Foundation Engineering*, Zurich, Switzerland, 16–27 August 1953, Vol. 1, pp. 336–337.

Caquot, A., and Kerisel, J. (1966) *Traité de Méchanique des sols*, Gauthier-Villars, Paris, pp. 349, 353.

Chen, W.F. (1975) Limit analysis and soil plasticity, Elsevier.

Chakraborty, D. And Kumar, J (2013) Bearing capacity of foundations on slopes. *Geomechanics and Geoengineering, An International Journal*, 8(4), pp 274–285.

Choudhary, A.K., Jha, J.N., Gill, K.S. (2010) Laboratory investigation of bearing capacity behaviour of strip footing on reinforced flyash slope, *Geotextiles and Geomembranes*, Volume 28, Issue 4, August 2010, pp. 393–402.

De Beer, E.E. and Ladany, B. (1961) Etude expérimentale de la capacité portante du sable sous des fondations circulaire établies en surface, *Proceedings 5th Intern. Conf. Soil Mech. Found. Eng.*, Paris, Vol 1, pp. 577–581.

De Buhan, P. and Garnier, D. (1998), Three Dimensional Bearing Capacity Analysis of a Foundation near a Slope, *Soils and Foundations*, 38(3), pp. 153–163.

Georgiadis, K. (2010), Undrained Bearing Capacity of Strip Footings on Slopes, *J. Geotechnical and Geoenvironmental Engineering*, 136(5), pp. 677–685.

Graham, J., Andrews, M. and Shields, D.H. (1988), Stress Characteristics for Shallow Footings in Cohesionless Slope, *Canadian Geotechnical Journal*, 25(2), pp. 238–249.

Ip, K.W. (2005) Bearing capacity for foundation near slope, *Master-thesis*, Concordia University, Montreal, Quebec, Canada. pp. 61–64.

Keverling Buisman, A.S. (1940). *Grondmechanica*, Waltman, Delft, the Netherlands, p. 243.

Knudsen, B.S. and Mortensen, N. (2016) Bearing capacity comparison of results from FEM and DS/EN 1997–1 DK NA 2013, *Northern Geotechnical Meeting 2016*, Reykjavik, pp 577–586.

Kumar, J. and Rao, V.B.K.M. (2003) Seismic bearing capacity of foundations on slopes, *Geotechnique*, 53(3), pp. 347–361.

Lambe. T.W. and Whitman, R.V. (1969), *Soil Mechanics*, John Wiley and Sons Inc., New York.

Leshchinsky, B. (2015). Bearing capacity of footings placed adjacent to c′-ϕ′ slopes. *Journal of Geotechnical and Geoenvironmental Engineering*, 141(6).

Loukidis, D., Chakraborty, T., and Salgado, R., (2008) Bearing capacity of strip footings on purely frictional soil under eccentric and inclined loads, *Can. Geotech. J.* 45, pp 768–787.

Meyerhof, G.G. (1951) The ultimate bearing capacity of foundations, *Géotechnique*, 2, pp. 301–332.

Meyerhof, G.G. (1953) The bearing capacity of foundations under eccentric and inclined loads, in *Proc. III intl. Conf. on Soil Mechanics Found.* Eng., Zürich, Switzerland, 1, pp. 440–445.

Meyerhof, G.G. (1957) The ultimate bearing capacity of foundations on slopes, *Proc. of the 4th Int. Conf. on*

Soil Mechanics and Foundation Engineering, London, August 1957, pp. 384–386.
Meyerhof, G.G. (1963) Some recent research on the bearing capacity of foundations, *Canadian Geotech. J.*, 1(1), pp. 16–26.
Meyerhof, G.G. (1965) Shallow foundations, *Journal of the Soil Mechanics and Foundations Division ASCE*, Vol. 91, No. 2, March/April 1965, pp. 21–32.
Michalowski, R.L. (1989). Three-dimensional analysis of locally loaded slopes. *Geotechnique*, *39*(1), 27–38.
Prandtl, L. (1920) "Über die Härte plastischer Körper." Nachr. Ges. Wiss. Goettingen, *Math.-Phys. Kl.*, pp. 74–85.
Shiau, J., Merifield, R., Lyamin, A., and Sloan, S. (2011) Undrained stability of footings on slopes. *International Journal of Geomechanics*, 11(5), 381–390.
Shields, D., Chandler, N., Garnier, J. (1990) Bearing Capacity of Foundations in Slopes, *Journal of Geotechnical Engineering*, Vol 116, No 3, pp. 528–537.
Reissner, H. (1924) "Zum Erddruckproblem." *Proc., 1st Int. Congress for Applied Mechanics*, C.B. Biezeno and J.M. Burgers, eds., Delft, the Netherlands, pp. 295–311.
Terzaghi, K. (1943) *Theoretical soil mechanics*, John Wiley & Sons, Inc, New York.
Van Baars, S. (2014) The inclination and shape factors for the bearing capacity of footings, *Soils and Foundations*, Vol.54, No.5, October 2014.
Van Baars, S. (2015) The Bearing Capacity of Footings on Cohesionless Soils, *The Electronic Journal of Geotechnical Engineering*, ISSN 1089–3032, Vol 20.
Van Baars, S. (2016a) Failure mechanisms and corresponding shape factors of shallow foundations, *4th Int. Conf. on New Development in Soil Mech. and Geotechn. Eng.*, Nicosia, pp. 551–558.
Van Baars, S. (2016b) The influence of superposition and eccentric loading on the bearing capacity of shallow foundations, *Journal of Computations and Materials in Civil Engineering*, Volume 1, No. 3, ISSN 2371–2325, pp. 121–131.
Vesic, A.S. (1973) Analysis of ultimate loads of shallow foundations. *J. Soil Mech. Found. Div.*, 99(1), pp. 45–76.
Vesic, A.S. (1975) Bearing capacity of shallow foundations, H.F. Winterkorn, H.Y. Fang (Eds.), Foundation Engineering Handbook, Van Nostrand Reinhold, New York (1975), pp. 121–147.
Yamamoto, K. (2010), Seismic Bearing Capacity of Shallow Foundations near Slopes using the Upper-Bound Method, *Int. J. Geotechnical Engineering.*, 4(2), 255–267.

Numerical Methods in Geotechnical Engineering IX – Cardoso et al. (Eds)
© 2018 Taylor & Francis Group, London, ISBN 978-1-138-33203-4

The failure mechanism of pile foundations in non-cohesive soils

S. Van Baars
University of Luxembourg, Luxembourg, The Grand Duchy of Luxembourg

ABSTRACT: In 1920 Prandtl published an analytical solution for the bearing capacity of a centric loaded strip footing on a weightless in-finite half-space, based on a so-called Prandtl-wedge failure mechanism. Meyerhof and Koppejan extended the logarithmic spiral part of the Prandtl-wedge and presented this as the failure mechanism for the tip of a foundation pile in non-cohesive soils. The numerical calculations made in this article show however that the failure zone (plastic zone) below a pile tip, is far wider and deeper than the Prandtl-wedge and that there is failure both in and out of the standard *x-y* plane, but most of the failure is due to an out-of-plane, circumferential or cleaving failure mechanism. Therefore, this failure mechanism is different from the Prandtl-wedge failure mechanism. Around the pile tip, there are circular thin zones with no out-of-plane failure. In these thin zones, the tangential (out-of-plane) stresses are relatively high due to large shear strains, formed during previous shearing or sliding of the soil.

1 INTRODUCTION

In 1920, Ludwig Prandtl published an analytical solution for the bearing capacity of a soil under a strip load, p, causing kinematic failure of the weightless infinite half-space underneath. The strength of the half-space is given by the angle of internal friction, ϕ, and the cohesion, c. The original drawing of the failure mechanism proposed by Prandtl can be seen in Figure 1.

The lines in the sliding soil part on the left indicate the directions of the maximum and minimum principal stresses, while the lines in the sliding soil part on the right, indicate the sliding lines with a direction of $\alpha = 45° - ½\phi$ in comparison to the maximum principal stress.

Prandtl subdivided the sliding soil part into three zones:

1. Zone 1: A triangular zone below the strip load (*ABC*). Since there is no friction on the ground surface, the directions of the principal stresses are horizontal and vertical; the largest principal stress is in the vertical direction.

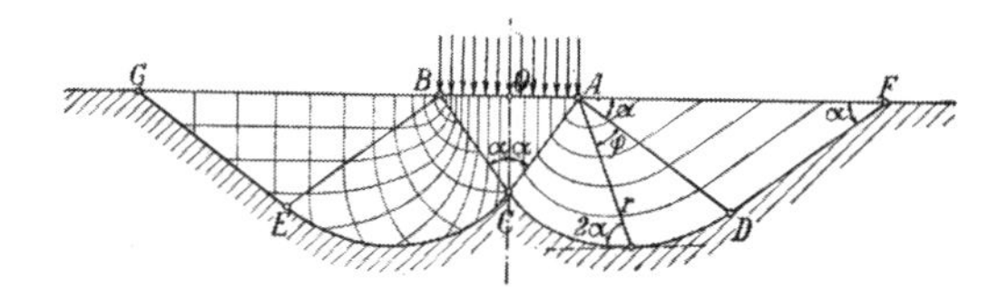

Figure 1. The Prandtl-wedge failure mechanism (Original drawing by Prandtl).

2. Zone 2: A wedge with the shape of a logarithmic spiral (*ACD*), in which the principal stresses rotate through ½π radians, or 90 degrees, from Zone 1 to Zone 3. The pitch of the sliding surface of the logarithmic spiral equals the angle of internal friction; ϕ, creating a smooth transition between Zone 1 and Zone 3.
3. Zone 3: A triangular zone adjacent to the strip load (*FAD*). Since there is no friction on the surface of the ground, the directions of principal stress are horizontal and vertical; the largest principal stress is in the horizontal direction.

The failure mechanism proposed by Prandtl has been validated by laboratory tests, for example those performed by Jumikis (1956), Selig and McKee (1961), and Muhs and Weiß (1972).

The solution of Prandtl was extended by Hans J. Reissner (1924) with a surrounding surcharge, q, and was based on the same failure mechanism. Albert S. Keverling Buisman (1940) and Karl Terzaghi (1943) extended the Prandtl-Reissner formula for the soil weight, γIt was Terzaghi who first wrote the bearing capacity with three separate bearing capacity factors for the cohesion, surcharge and soil weight. George G. Meyerhof (1953) was the first to propose equations for inclined loads, based on his own laboratory experiments. Meyerhof was also the first in 1963 to write the formula for the (vertical) bearing capacity p_v with bearing capacity factors (N), inclination factors (i) and shape factors (s), for the three independent bearing components; cohesion (c), surcharge (q) and soil weight (γ), in a way it was adopted by Jørgen A. Brinch Hansen (1970) and it is still used nowadays:

$$p_v = i_c s_c c N_c + i_q s_q q N_q + i_\gamma s_\gamma \tfrac{1}{2} \gamma B N_\gamma. \quad (1)$$

The use of shape factors would suggest that only a certain small correction is needed to go from the Prandtl-wedge plane strain failure mechanism to an axisymmetric failure mechanism. Since this could be incorrect, and especially for deep foundations, the main purpose of this study is to check the failure mechanism for axial-symmetric pile foundations.

2 AXISYMMETRIC FAILURE VERSUS PLANE STRAIN FAILURE

The success of the analytical solution for shallow foundations inspired researchers to try something similar for pile foundations. The first who applied the Prandtl-wedge for pile foundations were Keverling Buisman (1935, 1940) and Meyerhof (1951), but they forgot about the shape factors, because these were first published by Meyerhof in 1963.

Meyerhof proposed a failure mechanism near the pile tip, which is a sort of Prandtl-wedge type failure, but with, instead of a triangular shaped zone 3, an extended logarithmic spiral shaped zone 2, which continues all the way up to the pile shaft, see the right side of Figure 2.

Other researchers, for example Van Mierlo & Koppejan (1952) (Figure 3) and Caquot and Kerisel (1966) have adopted this extended logarithmic spiral shape failure mechanism around the pile tip.

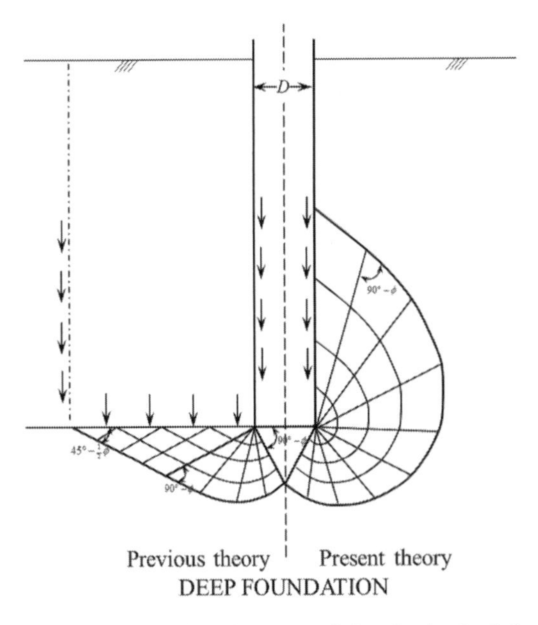

Figure 2. Prandtl-wedge approach for circular loaded areas.

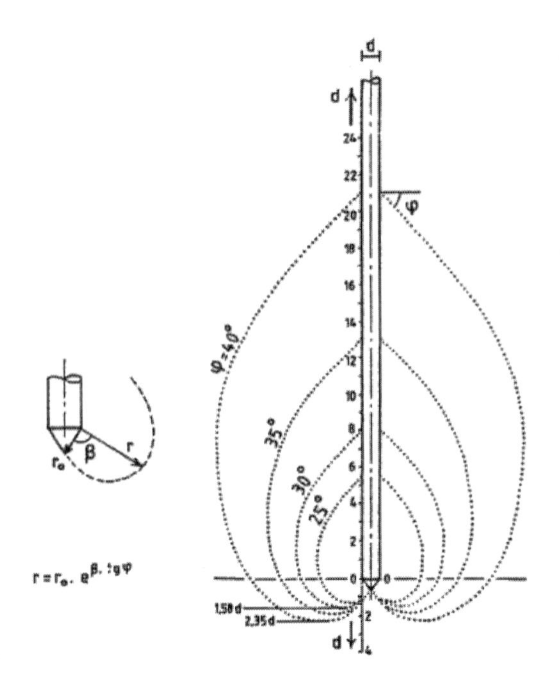

Figure 3. Logarithmic spiral shape failure mechanism around a pile tip (Van Mierlo & Koppejan, 1952).

There is a risk in assuming a Prandtl-wedge shaped failure mechanism for circular shaped loaded areas (Van Baars, 2014, 2015, 2016, 2017). The reason for this is that for axisymmetric failure, the ground surface area of zone 3 of the Prandtl wedge (i.e. the triangular shaped zone not below, but adjacent to the load), becomes too big, creating too much support due to the surcharge, so before this Prandtl-wedge failure mechanism can occur, already another mechanism occurs.

For a strip load (plane strain failure) the lowest principal stress, in zone 3 of the wedge, is the vertical stress, which is zero without a surcharge, and the largest principal stress is the horizontal stress (perpendicular to the load).

For a circular load (axisymmetric failure) however, the lowest principal stress, in zone 3 of the wedge, is not the vertical stress, but the tangential or circumferential stress (out of the x-y plane), which can be even zero.

Due to this cleaving failure mechanism, the bearing capacity stress for a circular load will be far less than for a strip load. Therefore an additional purpose of this study is to check the out-of-plane (tangential) stresses for axi-symmetric pile foundations.

3 FAILURE MECHANISM OF PILE FOUNDATIONS

For the determination of the bearing capacity of the tip of a foundation pile nowadays, still the general concept of Meyerhof is used, see for example Vesic (1967), or Fang (1990).

In many countries, the design of the pile bearing capacity is based on the bearing capacity (q_c-value) measured with the cone of the Cone Penetration Test (CPT). The thin CPT cone, as a model test, has a smaller diameter D and is unfortunately far more sensitive for the discontinuities of the subsoil as real piles. Therefore, every CPT-based method needs a rule for "smoothening" the measured discontinuities over a certain distance, called the "influence zone". The distance over which this "smoothening" rule must be applied is, in case of the Koppejan's method (Van Mierlo & Koppejan, 1952), based on this logarithmic spiral shape failure mechanism.

Because of this logarithmic spiral, the failure zone is, in the Koppejan's method, assumed to reach from 0,7 D to 4 D below the pile tip, until 8 D above the pile tip. This failure mechanism is however incorrect. Although the soil, below the level of the pile tip, can rotate away from the pile, a slip failure along this logarithmic spiral, above the level of the pile tip, is impossible, since this soil cannot rotate towards the pile and will not finally disappear in the pile.

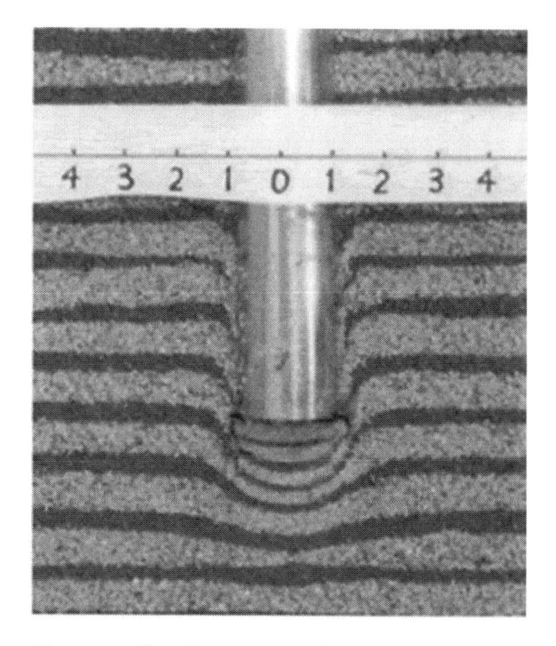

Figure 4. Global failure below pile tip in crushable sand. (Picture from Kyushu University, Geotechnical Engineering Research Group).

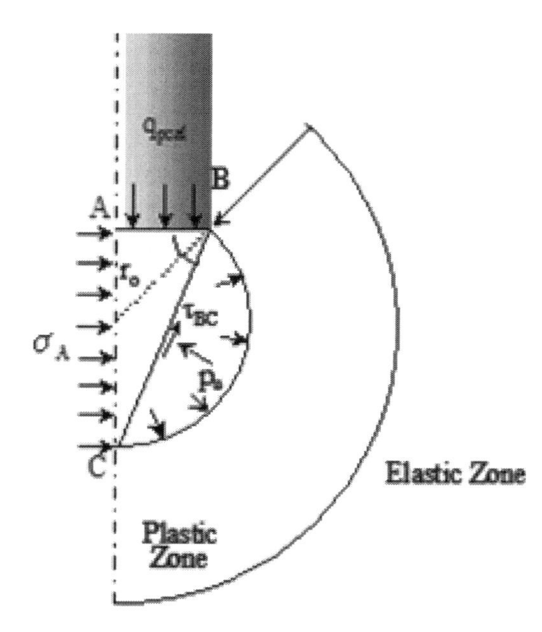

Figure 5. Global failure zone below pile tip. (Picture from Kyushu University, Geotechnical Engineering Research Group).

Laboratory model tests, see Figure 4, show another failure mechanism: a global, balloon shaped, failure zone, mostly below the pile tip. See also Figure 5. This shows that it is not correct to assume a Prandtl-wedge type failure mechanism near the tip of a foundation pile, or to derive the smoothing zone from the shape of a logarithmic spiral.

4 NUMERICAL CHECK

In order to check the axisymmetric failure mechanism for pile foundations, finite element calculations have been made with Plaxis 2D. A Mohr-Coulomb soil model has been used for a fine, wide mesh with 15-node elements. The soil has a Young's modulus of E = 50 MPa and a Poisson's ratio of $\nu = 0.3$. The effective soil weight γ' = 10 kN/m^3, the angle of internal friction $\phi = 35°$ and the horizontal earth pressure coefficient K_0 = 0.5. The groundwater level is at the ground surface. The pile has a high Young's modulus of E = 30 GPa, a Poisson's ratio of $\nu = 0.10$, and roughness coefficient of $R = 0.70$. The pile has a diameter of 0.40 m and is 10 m long.

A problem is that Plaxis 2D does not distinguish between out-of-plane, and in-plane, plots of the relative normal and shear stress ratios. This makes it difficult to study the failure mechanism. A trick has been used to solve this; the Mohr-Coulomb soil model has been programmed again as a User Defined Soil Model and compiled as a Dynamic Link Library (this is a special tool in Plaxis), but in this case also the relative normal stress ratios and relative shear stress ratios were defined as State Parameters, independently for the out-of-plane and in-plane situations. These can easily be plotted by Plaxis.

The pile loading was modeled by a large displacement (20% of the pile diameter). Figure 6 shows, for this circular pile, the relative shear stresses in the *x-y* plane (in-plane), in which the τ_{xy} is the shear stress, or the radius of the Mohr circle, in this *x-y* plane, and τ_{max} is the maximum allowable radius of a Mohr circle touching the Coulomb envelope.

Figure 7 shows the same relative shear stress, but now not in the *x-y* plane, but out of this plane (out-of-plane). This means that the lowest principle stress is in the tangential, circumferential or *z*-direction, and the highest principle stress is in the *1*-direction, which lays somewhere in the *x-y* plane. The used shear stress τ_{1z} is therefore the radius of the Mohr circle in this *1-z* plane.

The figures 6 and 7 show that there is failure both in-pane and out-of-plane, but most of the failure is due to out-of-plane failure (the failure zone, in which $\tau/\tau_{max} = 1$, is much bigger). The only zone where this is the opposite, is the small zone just above and around the pile tip; here there is mostly in-plane failure. The shape of the zone with, both in- or out-of-plane, failure is round as a ball. So, for pile foundations, the failure mechanism is different from the Prandtl-wedge failure mechanism.

This balloon shaped failure zone is mostly below, and not above, the level of the pile tip; from 5 to 6 D below the pile tip, until 2 to 3 D above the pile tip.

Also interesting in figure 7 are the circular thin zones around the pile tip, where there is no

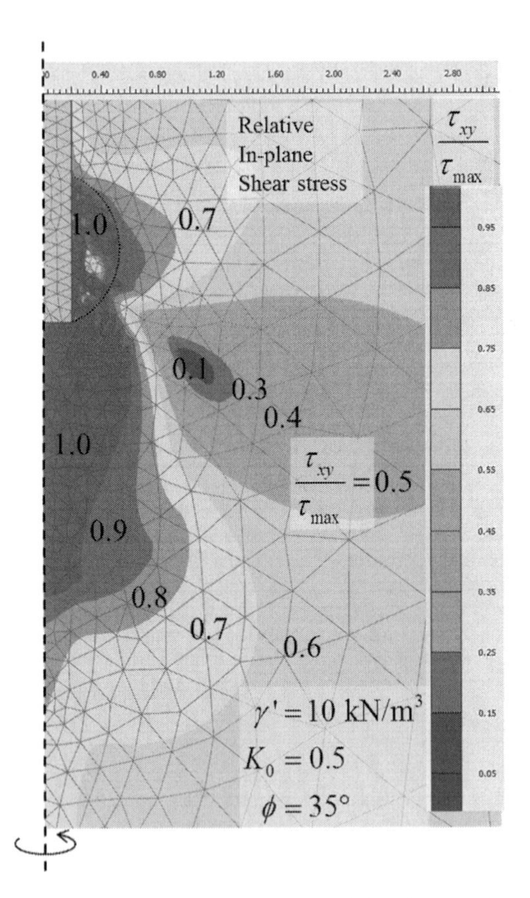

Figure 6. Relative shear stress, around the pile tip (only in-plane).

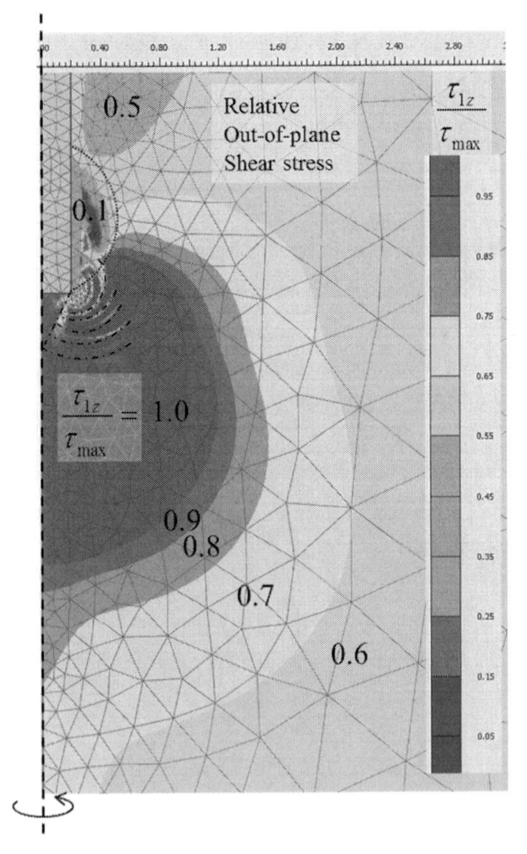

Figure 7. Relative shear stress, around the pile tip (only out-of-plane).

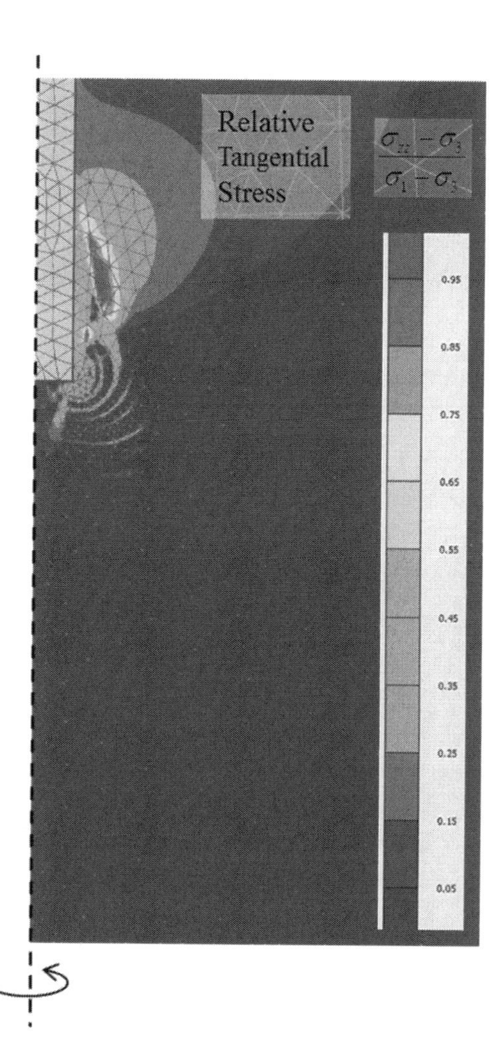

Figure 8. Relative tangential stress around the pile tip.

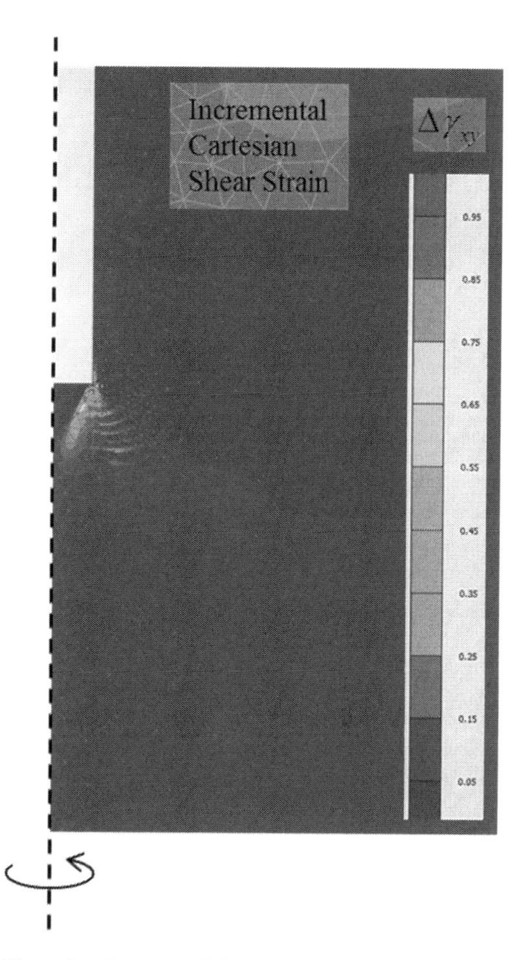

Figure 9. Incremental shear strain around the pile tip.

out-of-plane failure (but still in-plane failure!). The explanation for this follows from figure 8.

Figure 8 shows the relative tangential stress around the pile tip $(\sigma_{zz} - \sigma_3)/(\sigma_1 - \sigma_3)$. At the thin circular zones without out-of-plane failure, the tangential (out-of-plane) stresses are relatively high. Due to this high tangential stress, there is no failure out-of-plane failure anymore, since therefore a low tangential stress was needed.

Figure 9 shows the reason for the relatively high tangential (out-of-plane) stresses. The zones of the relatively high tangential (out-of-plane) stresses, are exactly the zones of relatively large (in plane) shear strains. These shear zones were created, because the inner circles rotate faster than the outside circles of the soil below the pile tip. So it seems that, this in-plane shearing or sliding allows the soil also to shear partially in tangential direction (the lowest stress direction), which increases the stresses in tangential direction.

5 CONCLUSIONS

The numerical calculations made in this article show that the failure zone (plastic zone) below a pile tip, is far wider and deeper than a Prandtl-wedge. These calculations also show that there is, for this axisymmetric pile tip, failure both in and out of the standard x-y plane, but the out-of-plane (tangential) failure zone is clearly larger than the in-plane failure zone. Therefore, this failure mechanism is different from the Prandtl-wedge failure mechanism.

Also interesting are the circular and diagonal thin zones around the pile tip, where there is no out-of-plane failure (but still in-plane failure!). The reason for the existence of these zones is that, in these thin zones without out-of-plane failure, the tangential (out-of-plane) stresses are relatively high. The relatively high tangential (out-of-plane) stresses in these zones are caused by the relatively large shear strains.

In these thin zones, the soil is shearing or sliding, because the inner circles rotate faster than the outside circles of soil below the pile tip.

This shearing or sliding allows the soil also to shear partially in tangential direction (because it is the direction with the lowest stress), which increases the stresses in tangential direction.

REFERENCES

Brinch Hansen, J.A. (1970). Revised and extended formula for bearing capacity, *Bulletin 28, Danish Geotechnical Institute*, Copenhagen, 5–11.

Caquot, A., and Kerisel, J. (1966). *Traité de Méchanique des sols*, Gauthier-Villars, Paris, 391.

Fang, Hsai-Yang (1990). *Foundation Engineering Handbook*, Kluwer, Norwell-USA / Dordrecht, the Netherlands.

Jumikis, A.R. (1956). Rupture surfaces in sand under oblique loads, *Journal of Soil Mechanics and Foundation Design*, ASCE, 82, 1, 1–26.

Keverling Buisman, A.S. (1935). De Weerstand van Paalpunten in Zand. *De Ingenieur*, 50 (Bt. 25–28), 31–35. [In Dutch].

Keverling Buisman, A.S. (1940). *Grondmechanica*, Waltman, Delft, the Netherlands, 227, 243.

Meyerhof, G.G. (1951). The ultimate bearing capacity of foundations, *Géotechnique*, 2, 301–332.

Meyerhof, G.G. (1953). The bearing capacity of foundations under eccentric and inclined loads, *Proc. III intl. Conference on Soil Mechanics and Foundation Engineering*, Zürich, Switzerland, 1, 440–445.

Meyerhof, G.G. (1963). Some recent research on the bearing capacity of foundations, *Canadian Geotechnical Journal*, 1(1), 16–26.

Muhs, H. and Weiß, K. (1972). Versuche über die Standsicheheit flach gegründeter Einzelfunda-mente in nichtbindigem Boden, *Mitteilungen der Deutschen Forschungsgesellschaft für Bodenmechanik (Degebo) an der Technischen Universität Berlin*, Heft 28, 122.

Prandtl, L. (1920). Über die Härte plastischer Körper. Nachrichten von der Gesellschaft der Wissenschaften zu Göttingen, Mathematisch-Physikalische Klasse, 74–85.

Reissner, H. (1924). Zum Erddruckproblem. *Proc., 1st Int. Congress for Applied Mechanics*, C.B. Biezeno and J.M. Burgers, eds., Delft (the Netherlands), 295–311.

Selig, E.T. and McKee, K.E. (1961). Static behavior of small footings, *Journal of Soil Mechanics and Foundation Design*, ASCE, 87, 29–47.

Terzaghi, K. (1943). *Theoretical soil mechanics*, J. Wiley, New York, USA.

Van Baars, S. (2014) The inclination and shape factors for the bearing capacity of footings, *Soils and Foundations*, 54, 5, 985–992.

Van Baars, S. (2015). The Bearing Capacity of Footings on Cohesionless Soils, *The Electronic Journal of Geotechnical Engineering*, Vol 20, No 26.

Van Baars, S. (2016). Failure mechanisms and corresponding shape factors of shallow foundations, *Proc. 4th Int. Conf. on New Development in Soil Mech. and Geotechn. Eng.*, Nicosia, Cyprus, 551–558.

Van Baars, S. (2017) The axisymmetric failure mechanism of circular shallow foundations and pile foundations in non-cohesive soils, *Journal of Computations and Materials in Civil Engineering*, Vol 2, No 1, 1–15.

Van Mierlo, J.C. and Koppejan, A.W. (1952) *Lengte en draag-vermogen van heipalen*, Bouw, No. 3.

Vesic, A.S. (1967) *A study of bearing capacity of deep foundations*, Final Report Project B-189, Georgia Institute of Technology, Atlanta, 231–236.

Vesic, A.S. (1975). Bearing capacity of shallow foundations, H.F. Winterkorn, H.Y. Fang (Eds.), *Foundation Engineering Handbook*, Van Nostrand Reinhold, New York, USA, 121–147.

Numerical Methods in Geotechnical Engineering IX – Cardoso et al. (Eds)
© 2018 Taylor & Francis Group, London, ISBN 978-1-138-33203-4

Long-term settlements induced by a large mat foundation

A. Sanzeni, F. Colleselli & V. Cortellini
DICATAM, University of Brescia, Brescia, Italy

ABSTRACT: A facility was constructed near Trento (Italy) in 2011–2012 to house the medical equipment for proton-beam tumour therapy. The facility's most important building had a 46 × 30 m concrete mat and weighed approximately 370 MN. The site is characterized by the presence of post-glacial soils (Upper Pleistocene): a superficial 14 m thick layer of gravel and sand, a 4 m layer of soft clayey silt and a deep deposit of sand. A monitoring system with topographical measurements and deep extensometers was installed, and a number of 2D finite element analyses were conducted during construction to evaluate the soil response under the true load history over a long period. The comparison between the experimental observations over a 4-year period and the result of the simulations demonstrated the relevance of the consolidation process of the deep clayey soil layer to simulate the foundation response over time.

1 INTRODUCTION

The "Proton Therapy Centre" (PTC) is a facility of the Trento Hospital in the region of Trentino-Alto Adige in Northern Italy. The Centre was constructed between July 2011 and January 2013 and it consists of two adjacent buildings, namely the hospital main building and the "bunker", where the most important facilities and equipment, such as the rooms for the therapy, the cyclotron and the beam line, are located (Figure 1).

The site elevation is approximately between 189 m and 194 m above sea level. The hospital's main building is a two-storey edifice, with a 78 m × 42 m concrete mat foundation at an elevation of 187.8 m above sea level. The Proton Beam building, also called the "bunker", is a two-storey structure, founded on a 46 m × 30 m concrete mat with three levels: 1) the core structure (where the tumour therapy equipment is located), 16.5 m × 27.5 m wide, founded at an elevation of 187.4 m a.s.l.; 2) the remaining part of the building, founded at an elevation of 191.3 m a.s.l.; 3) some areas of limited extension, located at an intermediate level. The total weight of the main hospital building is approximately 274 MN (27,400 t); the total weight of the bunker is approximately 369 MN (36,900 t).

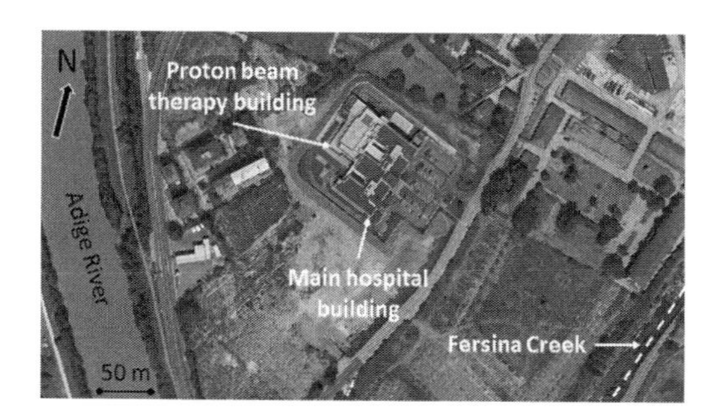

Figure 1. Aerial photograph of the area and identification of the two adjacent buildings of the Proton Beam Therapy Centre.

The site, previously part of a military area, is located roughly 2 km South-West of the city of Trento and only a few hundred meters from the banks of the Adige River, one of the longest (410 km) and widest (12 000 km^2 basin) in Italy. The local geology features recently deposited soils from the Upper Pleistocene, mainly alluvial deposits of sand and gravel with thin layers of fine-grained soils.

Because of the relevance of the Centre, the site has been extensively investigated and the construction process has been meticulously monitored: the load history of both buildings has been recorded and settlements were observed in detail both at the ground level and deep into the foundation soil. Aside from the design process, during the construction of the Centre a study was undertaken to analyze the response of the soil and of the buildings over time. A number of numerical simulations we executed during the final stages of construction to simulate the settlement evolution under the true load history and the results were compared with observations derived from the monitoring system.

2 SOIL CONDITIONS AND FOUNDATION DESIGN

2.1 *Soil conditions*

The site is located in the Adige River Valley, a few hundred meters from the riverbanks and near the current course of the Fersina Creek (Figure 1). From a geological point of view, the local deposits of alluvial soil originated during the Upper Pleistocene and consist mainly of sands (initially deposited by the Adige River), and gravels (subsequently deposited by the Fersina Creek); these two units are separated by a layer of predominantly silty-clayey soil.

The construction site has been extensively investigated through geotechnical campaigns carried out in 2007, 2010 and 2012, which consisted of 30–50 m deep borings, SPT tests, geophysical down-hole tests and laboratory tests on disturbed and undisturbed samples. Figure 2 shows the plan view of the Centre and the location of the boreholes; Figure 3 reports the results of SPT tests and shows a schematic soil profile.

From the ground surface (approximately 189 m a.s.l.) the following layers were encountered:

- Medium to loose medium-fine sand, N_{SPT} = 10–20, the thickness of this layer varied depending on the level of the ground surface (generally 4–5 m thick);
- Medium to dense gravel, N_{SPT} = 20–70, approximately 7 m thick;
- Medium-fine sand and clayey silt, N_{SPT} = 5–20, Pen = 50–150 kPa, Tor = 25–50 kPa, with a thickness in the range of 4–5 m (the thickness of the upper sand layer was approximately 1 m);
- Medium to loose medium-fine sand; N_{SPT} = 15–30 with a thickness up to 15–20 m; occasionally around 43–44 m b.g.l. a 1–2 m thick layer of silty sand was found.

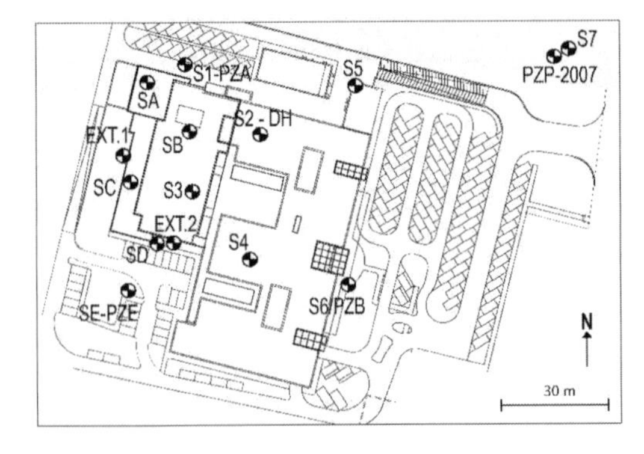

Figure 2. Plan view of PTC and location of boreholes.

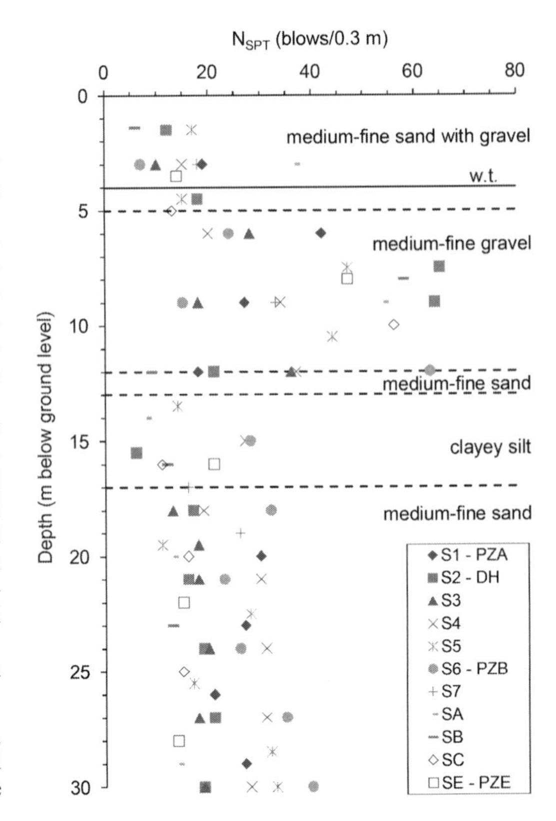

Figure 3. SPT test results and schematic soil stratigraphy.

The pore pressure regime is primarily governed by the level of the nearby Adige River and the water table is set to an elevation of 185 m a.s.l.

The results of laboratory tests performed on disturbed and undisturbed samples indicated that the soil layer encountered between 13 m and 17 m b.g.l. consists mainly of clayey silt (2–10% sand, 60–80% silt, 20–35% clay) with medium-low plasticity (liquid limit w_L = 40–55%, plastic limit w_P = 30–35%, plasticity index I_P = 10–20%), and thus can be classified as CL-ML-MH (according to the USCS, ASTM D2487–11). The in situ void ratio ranges between 0.915 and 1.020, the unit weight between 18.0 and 19.0 kN/m^3, and the water content between 36 and 39%. The results of a number of oedometer tests indicated the following characteristics: Overconsolidation ratio OCR = 1.5–1.8; Primary Compression Index C_C = 0.20–0.40; constrained/oedometric modulus E_{oed} = 6–10 MPa; ratio Ca/Cc = 0.030–0.032 (Mesri & Castro, 1987); average saturated hydraulic conductivity k = 1.0 × 10E-09 m/s.

2.2 *Foundation structural features and requirements (bunker)*

The installation and operation of the advanced equipment of the proton-beam therapy has imposed important limitations on the performance of the building under working conditions. In particular, a maximum differential settlement of 3 mm was accepted over the length of the beam line (approximately 27 m, along the longitudinal section). Therefore a stiff structure was designed. In particular, the "bunker" is a two-storey building, founded on a 46 m × 30 m concrete mat, whose inner core (where most of the therapy equipment is located) is 16.5 m × 27.5 m wide and founded at 187.4 m a.s.l.. The construction sequence of the basement adopted the following stages: 1) excavation to an elevation of 187.4 m a.s.l.; 2) formation of a 1.2 m thick concrete mat; 3) casting of monolithic blocks of lean concrete and of gantry walls; 4) formation of a second 1.3 m thick concrete mat to reach an elevation of 192.5 m a.s.l.; 5) casting of internal and external walls. The total weight of the bunker is approximately 36.9 MN (36,900 t). The following Figure 4 illustrates an isometric section of the bunker basement and a schematic section of the foundation. It is worth noting that the adjacent hospital building and a 15 m wide, 3.4 m thick embankment (to raise the elevation of the ground surface near the PTC) were simultaneously constructed respectively on the East and West sides of the bunker.

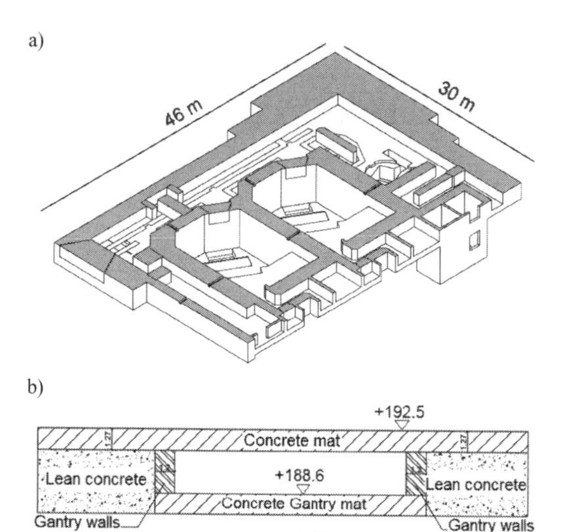

Figure 4. a) Isometric section of the bunker foundation system; b) schematic cross section of the bunker inner core foundation.

2.3 *Geotechnical design*

During the design process, the main concern with the construction of the PTC was the prediction of vertical settlements and the judgment over their acceptability. A first estimate of the vertical settlements under working conditions was performed using the classic analytical methods by Schmertmann and Schmertmann et al. (1970, 1978) and by Burland & Burbridge (1985).

In both cases the calculation neglected the presence of the adjacent building and embankment; also, the contribution of all the layers of soil under the foundation was considered in "drained conditions", therefore neglecting the time needed for the consolidation of the clayey silt layer (13–17 m b.g.l.). The method by Schmertmann predicted an average settlement s = 0.200 m over a three-year period, while the method by Burland and Burbridge predicted s = 0.123 m. Later, the method by Burland and Burbridge was discarded, due to its limited applicability in stratified soils (for which it is difficult to average a representative value of the SPT test) and the Schmertmann method was applied for estimating the settlement of predominantly coarse soil layers, while the consolidation settlement of the clayey silt layer was calculated with the one-dimensional method (Terzaghi et al., 1996). This analytical approach lead to estimate a 88 mm immediate and a 105 mm consolidation settlement in approximately one year. In this case, using the theory of 1-D consolidation, a period of approximately 100–150 days was estimated for completing 90% of the consolidation.

3 MONITORING AND LOAD HISTORY

A monitoring system was set up before construction which included deep extensometers and leveling measurements. Moreover, the loading history of both buildings was thoroughly recorded during the construction. Thus the history of settlement can be related to the history of construction and load application.

3.1 *Surface and deep monitoring system*

From almost the beginning of the construction, the PTC buildings were subjected to the measurement of vertical settlements. A net of twelve measuring points was installed around the buildings and precise levelling measurements began around October 2011 (approximately 3 months after the start of construction). To fully appreciate the response of the soil below the buildings, two deep extensometer chains were also installed near the bunker. In

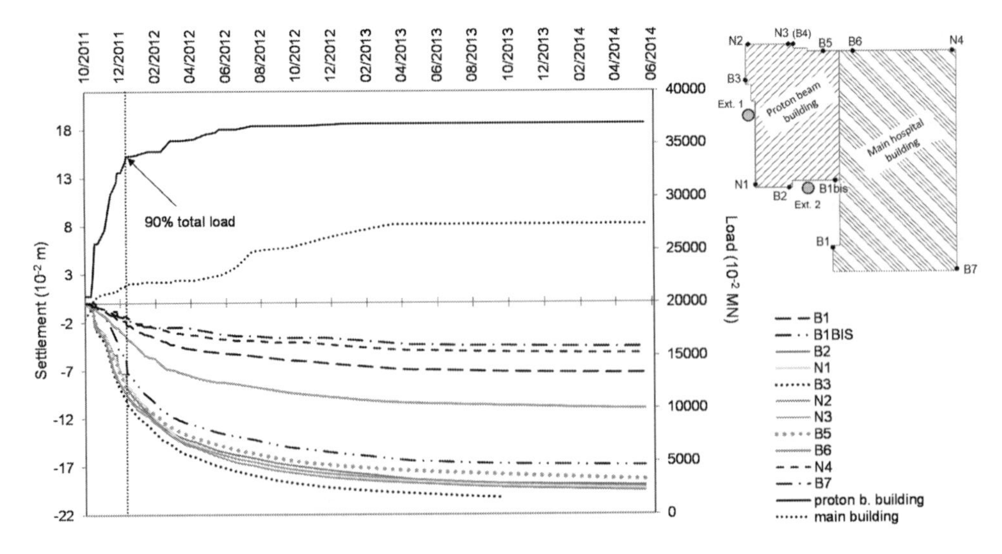

Figure 5. Schematic plan view of the PTC buildings and location of selected topographic points and extensometer chains. Load history and settlement history.

particular, extensometers were installed into the clayey silt layer between 13 and 17 m b.g.l. (plan view in Figure 5).

3.2 *Load history*

Along with the monitoring system, the progress of construction has also been observed. This has been carried out very carefully for both buildings and has allowed to record the loads that have been progressively applied to the soil during the casting of concrete and construction of all structural elements, and even during the phases of application of the finishes and installation of the medical equipment and furnishings.

The progress of construction in terms of load history and the settlements' evolution are presented in Figure 5. It is worth noting that the recording started approximately three months after the beginning of construction (when permanent topographic points could be established), during which the early stages of construction concerned the excavation of some 2–3 m of soil (to reach an elevation of 187.4 m) and the cast of the foundations. The progress of construction of the bunker was initially notably fast, since 50% of the total loads was applied in 2 months, and almost 90% was applied within approximately 4 months of activity (in December 2011); the remaining 10% of loads was applied in a much longer period of one year (the construction ended between December 2012 and January 2013). Meanwhile, almost 78% of the total load of the adjacent hospital building was applied within the first four months of construction, whilst fifteen more months were required to end the construction. In any case, the evolution of settlements with time appeared to be much longer than initially predicted.

4 MODELING AND INTERPRETATION OF SOIL RESPONSE

Based on the early results of the monitoring system and on the comparison with the expected response, a number of numerical simulations we executed during the final stages of the construction to analyze and reproduce the soil response under the true load history and to understand and predict the evolution of settlements in the future. The analyses were conducted in plain strain conditions with the finite element Plaxis 2D commercial software (Plaxis BV, Delft, The Netherlands).

4.1 *Mesh, construction sequence, constitutive laws and soil parameters*

The construction of the PTC and the soil response were simulated with a plain-strain, 15-node triangular element model with variable mesh fineness, ideally representing the transverse cross-section through the center of the PTC buildings (East-West direction). The model was 150 m wide and 50 m deep; the geotechnical units and pore pressure regime were defined according to the results of the geotechnical investigation campaigns (Figure 3). The construction of the "bunker" was simulated by modelling the simplified foundation arrangement presented in Figure 4: the main structural elements were modeled as non-porous soil layers with assigned elastic properties and unit weight of the concrete. The remaining structures of the bunker were modeled by applying uniformly distributed loads onto the foundation elements to progressively reach the maximum value of the applied loads according to the recorded load history. The construction of the main hospital building and of the embankment, on the West and East sides of the bunker, were also simulated with the gradual application of uniformly distributed loads. The maximum values of the applied contact pressure are 65 kPa, 245 kPa and 84 kPa for the embankment, the bunker and the main sanitary building respectively (Figure 6). The computation replicated the construction sequence and included the following main stages: the application of initial stress conditions with reference to the original ground level (189 m a.s.l); the excavation of soil to reach the foundation level (187.4 m), the activation of the bunker foundation elements and of 90% and 78% of the total load due to the bunker and to the main hospital building; the gradual application of remaining loads and the complete dissipation of excess pore pressure; finally a 20-year stage of creep was performed. Two sets of analyses were performed: a first-approach simulation with a simple constitutive model for every soil layer, namely the linear-elastic, perfectly-plastic model with Mohr-Coulomb failure criterion (named Mohr-Coulomb, MC, model in the software library); and a set of advanced analyses with two soil models. In the advanced simulations the mechanical behavior of predominantly coarse soils was described with an elastic-plastic, rate independent model with isotropic hardening and stress-dependent stiffness (the Hardening Soil, HS, model), whereas the cohesive soil between 13 and 17 m was modeled with a constitutive

Table 1. Relevant geotechnical parameters for the simplified numerical analysis with MC model.

Soil Layer	soil type	ϕ' °	E′ MPa	e -	k m/d
1	medium-fine sand	29	30	0.50	1.0
2	gravel	36	50	0.50	10.0
3	fine sand	29	35	0.50	1.0
4	clayey silt	32	6	1.00	$8.6\ 10^{-4}$
5	medium sand	29	35	0.50	1.0
6	medium sand	30	45	0.50	1.0
7	medium sand	33	55	0.50	1.0
8	sandy silt	32	20	1.00	0.01

γ_{sat} = 19–20 kN/m³

Table 2. Relevant parameters for the advanced numerical analyses.

Soil L.	soil model	E'_{ref} MPa	λ^* -	κ^* -	μ^* -	OCR -
1	HS[1]	40	-	-	-	-
2	HS	65	-	-	-	-
3	HS	45	-	-	-	-
4	SSC	-	0.06304	0.01739	0.00162	1.5
5	HS	50	-	-	-	-
6	HS	60	-	-	-	-
7	HS	75	-	-	-	-
8	HS	27	-	-	-	-

[1]other common HS parameters: m = 0.5, $E'_{ur} = 3E'_{ref}$

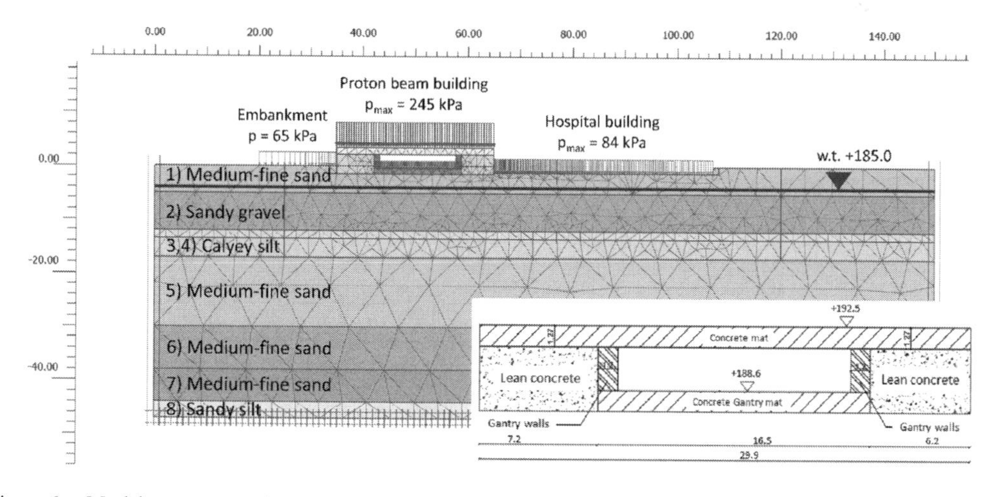

Figure 6. Model geometry and mesh, soil layers and applied loads.

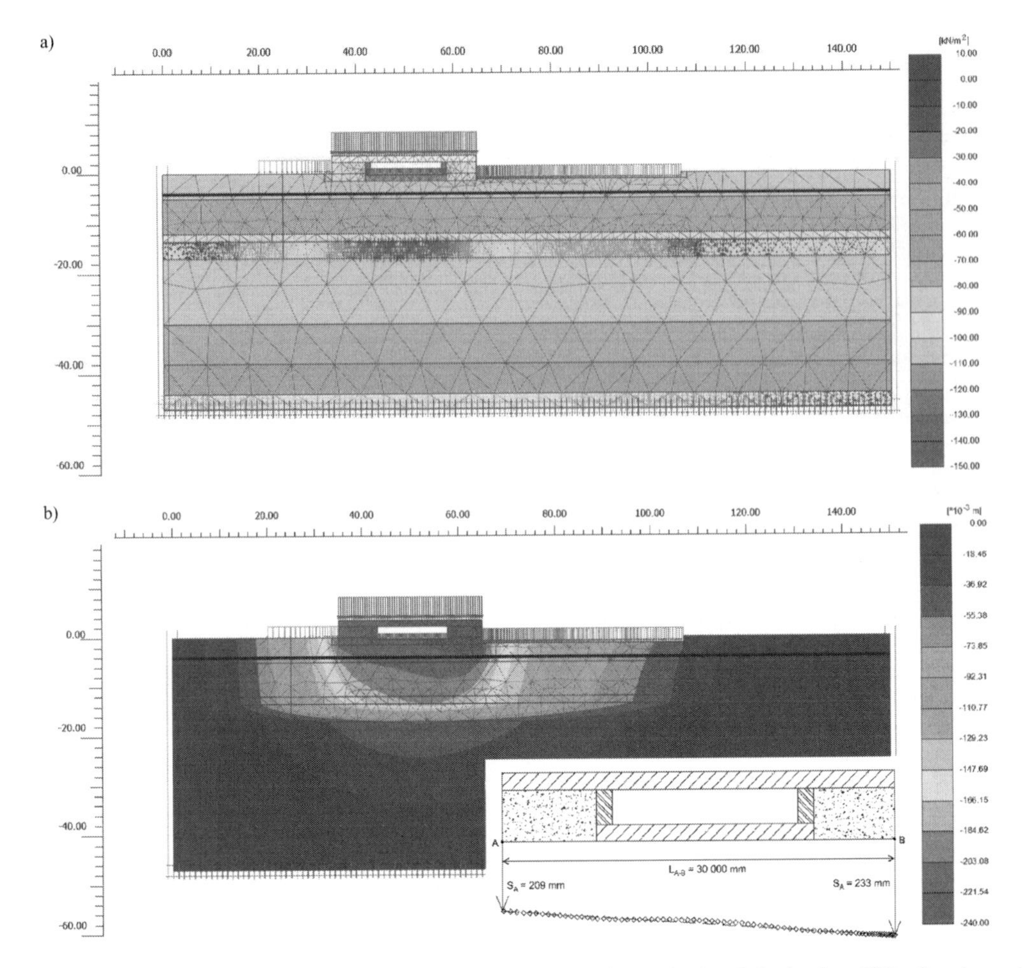

Figure 7. a) Excess pore pressure distribution after application of 90% loads of the bunker, 78% of main sanitary building and embankment; b) vertical settlement distribution at the end of the construction sequence (advanced numerical simulation).

law capable of computing high compressibility (logarithmic compression behavior) and time-dependent compression (the Soft Soil Creep, SSC, model). The most relevant parameters adopted for the numerical simulations were derived from the results of laboratory and in situ tests and are presented in Table 1 and Table 2.

4.2 *Results of numerical analyses*

The results of the numerical simulations are illustrated in Figure 7 and Figure 8. Figure 7 (a) presents the distribution of excess pore pressures at the beginning of the consolidation stage, obtained with the advanced numerical simulation after the application of 90% load of the bunker, 78% of the adjacent building and 100% of the embankment.

Table 3. Comparison between observed and computed vertical settlement of the PTC bunker.

Time (days)		Vertical settlement (mm) (Point A Fig. 9)		
	Analytical m.	MC	HS-SSC	Measurements
0	74	138	91	92
200	181	213	190	169
400	193	214	200	187
800	216	214	207	194

The corresponding computed excess pore pressure varies in the range of 30 kPa to 150 kPa. Figure 7 (b) shows the distribution of computed

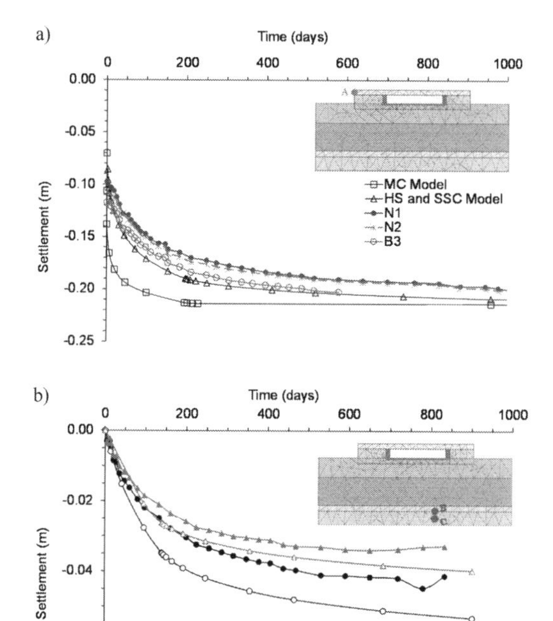

Figure 8. Comparison between observed and computed consolidation settlements at the ground level (a) and in the clayey silt layer (b).

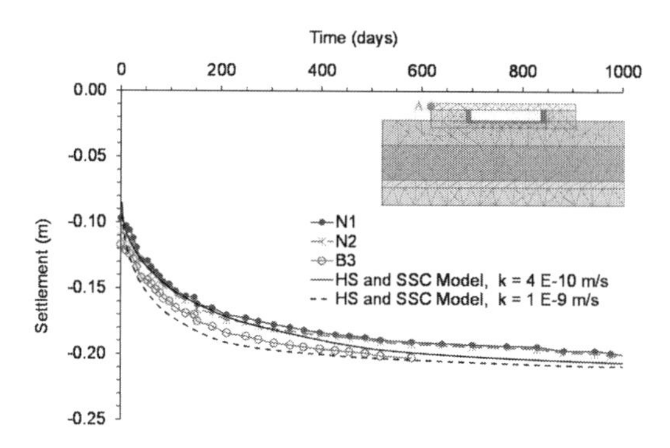

Figure 9. Comparison between observed and compute consolidation settlement, effect of the permeability of the clayey silt.

vertical settlements at the end of the consolidation. The results indicated that most of the settlements occurred in the first 20 m of soil and that the East side of the bunker settled more (233 mm) than the West side (209 mm). The foundation showed an almost rigid response with a computed angular distortion of 1/1300 over the length of the transverse cross section.

5 COMPARISON AND CONCLUSIONS

5.1 *Comparison with the observed behavior*

The comparison between numerical simulations and measurements obtained by the monitoring system is presented through Table 3, Figure 8 and Figure 9.

The values reported in Table 3 indicate that the analytical methods underestimated the observed immediate settlement and overestimated the effect of time on the long-term response, furthermore the time of consolidation (100–150 days) was not properly estimated. Regarding the numerical simulations, the first-approach analysis (adopting the MC model for all soil layers) has overestimated the immediate settlement and could not reproduce well the response of the soil over time. The analysis performed with the HS and SSC advanced soil models was capable of correctly estimating the amount of immediate settlement and the evolution of consolidation settlement over time. This behavior is also illustrated in Figure 8 (a) which shows the comparison between the observed and the computed vertical settlement of point A (on the East side of the bunker) as a function of the time from the beginning of the consolidation of the clayey silt layer (January 2012) up to November 2014.

On the other hand, all numerical simulations appeared to slightly overestimate the total settlement and its distribution at the foundation level of both buildings. This feature is most notable when comparing computed settlements with measurements obtained from the extensometers located in the clayey silt layer, as presented in Figure 8 (b). Finally, the result of a creep analysis over a 20-year period indicated a negligible contribution (less than 10 mm) to the overall settlement of the PTC buildings.

To further improve the fitting of the numerical model, another advanced numerical analysis was performed adopting a slightly different value of permeability for the clayey silt layer (13–17 m b.gl.). This value (4.0E-10 m/s) was estimated from the result of particle size distribution and plasticity tests of the clayey silt, using the semi-empirical method developed by Sanzeni et al. (2012) to estimate the specific surface of fine-grained soils and to improve the capability of the Kozeni-Carman formula (Carman, 1939). The result of this analysis is represented in Figure 9 and indicates that the long-term response of the soil can be explained with the extension of the consolidation process, rather than the occurrence of secondary compression phenomena.

5.2 *Conclusions*

During the final stages of the construction of the PTC in Trento (Italy) a study has been conducted to understand the response of the buildings in terms of vertical settlements under the true load history.

The classic analytical methods were capable of estimating the average settlement but could not predict the duration of the consolidation process.

The 2D numerical analyses allowed to overcome some limitations of the classic analytical methods and confirmed the satisfactory performance of the bunker in terms of settlement distribution. However the simulation conducted with a simple soil model could not reproduce the amount of immediate settlement and the duration of the consolidation process. The numerical analyses performed with advanced soil models were able to capture the observed soil response over time. Finally, the behavior of the soil below the PTC foundations over a 4 year period appeared to be primarily affected by the duration of the consolidation process, with limited or negligible influence by phenomena of secondary compression.

REFERENCES

ASTM D2487-11, Standard Practice for Classification of Soils for Engineering Purposes (Unified Soil Classification System), ASTM International, West Conshohocken, PA, 2011, www.astm.org

Burland, J.B. & Burbridge, M.C. 1985. Settlement of foundations on sand and gravel. *Proc. Inst. Civ. Eng.*, Part 1, 78: 1325–1381.

Carman, P. 1939. Permeability of saturated sands, soils and clays. *The Journal of Agricultural Science*, 29(2): 262–273.

Mesri, G. & Castro, A. 1987. Ca/Cc concept and K0 during secondary compression. *Journal of Geotechnical Engineering*, ASCE, 113(3): 230–247.

Sanzeni, A., Colleselli, F., Grazioli, D. 2012. Specific surface and hydraulic conductivity of fine-grained soils. *Journal of Geotechnical and Geoenvironmental Engineering*, ASCE, 139(10): 1828–1832.

Schmertmann, J.H. 1970. Static cone to compute static settlement over sand. *Journal of the Soil Mechanics and Foundations Division*, ASCE, 96, SM3, Proc. Paper 7302: 1011–1043.

Schmertmann, J.H., Hartman, J.P., Brown, P.R. 1978. Improved strain influence factor diagrams. *Journal of the Geotechnical Engineering Division*, ASCE, 104, GT8: 1131–1135.

Terzaghi, K., Peck, R.B., Mesri, G. (1996). *Soil mechanics in engineering practice*. 3rd edition. John Wiley & Sons, NY: 106–108.

Numerical Methods in Geotechnical Engineering IX – Cardoso et al. (Eds)
© 2018 Taylor & Francis Group, London, ISBN 978-1-138-33203-4

Design simulation of deformations during short-term loading of soft clay

A. Berglin
Sigma Civil AB, Stockholm, Sweden

A.B. Lundberg & S. Addensten
ELU Konsult AB, Stockholm, Sweden

ABSTRACT: Geomaterials typically exhibit non-linear elasticity during deformations over significant ranges of shear strain. The design of geomechanical structures is normally carried out both for the ultimate limit state, in which the load-deformation behaviour may not be of significant interest, and for the serviceability limit state, where realistic modelling of the load-deformation behaviour of the soil is essential. A suitable estimation of soil deformations depends on the loading geometry and the soil properties, and consequently the idealization process is very significant in practical design. This idealization concerns both the material behaviour and the model geometry, and a suitable compromise between model refinement and simplicity should be the aim of the design process. Here the idealization procedure is demonstrated for a practical case study for a medium-risk project, in which the formwork of a concrete bridge was placed on soft clay covered with gravel rather than the standard solution using a piled platform for temporary use. The elastic properties of the clay were assessed from triaxial tests and empirical correlations, and the time-dependent deformation response during loading was simulated by a small-strain finite element model in 2D plane strain geometry. The calculated deformations are compared to field measurements, and the suitable idealization of field conditions and model soil types for similar cases of routine design for medium-risk projects is discussed.

1 INTRODUCTION

Numerical simulations are increasingly being applied to practical problem and frequently play an important part of design in geotechnics. This tendency has followed the practice in structural design, where finite element models of structural systems are ubiquitous in practical design. There are however some stark differences between structural and geotechnical engineering which need to be considered in practical design and are related to the project risk.

Structural systems are in most cases well defined. The load-deformation properties of structural members are well known, and the bearing capacity of structural components are possible to assess both with precise calculation and representative testing.

In geotechnical design, an important and highly significant part of the design process is the model definition and idealization: assessment of the soil strata, and the determination of the soil properties of subsequently analysis of the load-deformation response.

The model definition in geotechnical design is dependent on the specific site instigation and preferred method of analysis. These are in turn governed by the risk involved in the specific design, something which is highly dependent on the environment and organization of the project design.

Most published material of geotechnical design involves highly detailed design of high-risk projects in which very detailed and refined models are available, e.g. Burland (1989), Gonzales et al (2012), Simpson et al (1979). In contrast, much of the daily practical design involved low- to medium-risk projects, in which time for analysis and budget for site investigation is quite limited, and a suitable compromise must be found between model detail and the risk involved in the project.

A design simulation of the short-term deformations of a shallow foundation on clay during casting of the concrete superstructures of a light-rail bridge is therefore elaborated. The aim is to elucidate some of the compromises and challenges in design in a medium-risk project, including site investigation, laboratory testing and model definition for numerical modelling, as well as the general risk assessment.

2 DEFORMATION ANALYSIS IN SLS DESIGN

2.1 *Design method*

Deformation assessment of geomaterials in SLS design can be carried out by several approaches:

- Empirical correlations derived from previous experience for similar field conditions.
- Analytical formulations of deformations, e.g. Atkinson (2000).
- Numerical simulations encompassing the stress-strain response with some suitable formulation of the geomaterial, e.g. Benz, (2007).

The use of empirical correlations is often limited to standardized geometries and uniform soil conditions, and analytical methods require some limitations of material uniformity and boundary conditions. On the other hand, the availability of fast personal computer and user-friendly computer programs means that numerical simulations are increasingly considered as an essential tool in practical design, especially for medium- and high-risk projects, e.g. Gonzales et al (2012).

2.2 *Numerical simulation of soil deformation*

Deformation analysis of geomaterials in SLS design needs to incorporate the stress-strain behaviour of the material. Typically, the numerical formation is carried out with a finite element program, and a constitutive model of the soil model is implemented in the program, e.g. (Jardine et al 1986). The accuracy of the numerical simulations is consequently dependent on the suitability of the constitutive model, (Simpson 1992).

The deformation behaviour of soil is highly non-linear, with an initial high-stiffness that is decreasing with the shear strain level, (Atkinson 2000, Benz 2007, Burland 1989). The small strain stiffness can be measured by different means, e.g. (Jardine et al 1984, Long et al 2017, Viggiani & Atkinsson 1995). A realistic assessment of soil deformation through numerical simulations should preferably include some formulation of the small-strain stiffness, (Burland 1989, Clayton 2011).

Different constitutive models of small-strain deformation of soils have been proposed, e.g. Benz et al (2009), Jardine et al (1986), Simpson (1992). These comprise some formulation of the small-strain stiffness and the stiffness reduction through shear strain. A small-strain model available for commercial practice is the Hardening Soil small model, described in Benz (2007), in which the small-strain properties are formulated through the initial shear modulus $G_{initial}$ and the shear strain with 70% of the initial shear modulus $\gamma_{0.7}$, which are fitted to an empirical reduction model by Santos & Correia (2001).

Figure 1. Outline of the viaduct with bridge supports.

3 CASE STUDY OF SLS DESIGN FOR A MEDIUM-RISK PROJECT

Here a medium-risk project suitable for numerical analysis of deformations in SLS design is described, and suitable detail of soil testing and numerical modelling relative to the risks and costs involved is discussed.

The specific project is a light-rail viaduct located outside of the center of Stockholm, consisting of a concrete slab bridge with several segments that are supported on concrete columns founded on end-bearing piles. The viaduct forms part of the light-rail ring system connecting the different subway lines around Stockholm. Figure 1 shows the layout of the site and the outline of the multi-span bridge with the bridge support columns. The driven piles are installed by driving and bearing capacity is verified through dynamic pile testing, (Andersson et al 2016). The structural bearing capacity of the pile is calculated with the methodology described in Lundberg et al (2016).

The concrete slab is casted on elevated forms, which need to be supported by a temporary structure before the concrete has cured. The construction method was not specified in the contract and, the contractor consequently could consequently decide how to arrange the temporary structures while accepting the risk of the design. The contractor faced the alternatives of temporary piling, which is expensive but ensures that the casting forms are not deforming during casting, or placing the casting forms on a layer of compacted gravel. The latter alterative would include some deformations which would occur during the during of the concrete, and the consideration of this option required an assessment of the possible deformations of the casting structure. After preliminary

Figure 2. The temporary support of the casting forms.

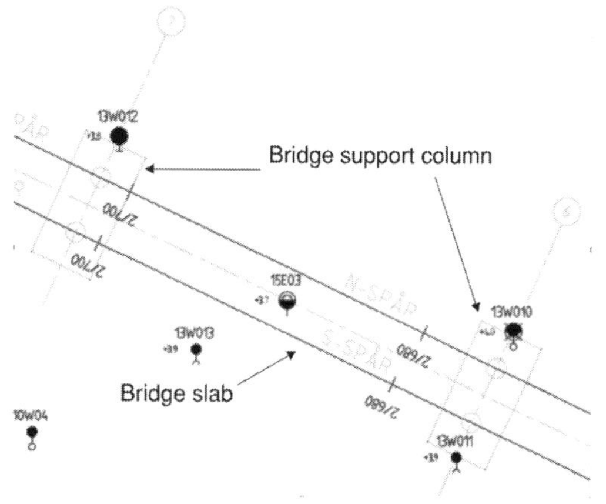

Figure 3. Location of site investigation boreholes.

design, including numerical simulations of the possible deformation, the contractor settled on the alternative with casting forms placed on a layer of compacted gravel layer on top of the clay, shown in Figure 2.

4 SOIL PROPERTIES

4.1 *Geological history and clay properties*

The soil strata consist of soft Holocene clay, covered with some gravel fill, and located on top of glacial moraine and bedrock. The Holocene post-glacial clay was deposited after the melting of the glacial ice, starting at around 10000 years BCE. The sedimentation of the clay particles occurred in a marine-brackish environment in the Ancylus sea in the Eastern Scandinavian area, resulting in high-plasticity clay, which is typical for the region, Lundberg & Li (2015). The area gradually started to rise from the water during isostatic lift, and was still covered by water around 1000–2000 years BCE.

4.2 *Site investigation*

The site investigation covered the whole stretch of the viaduct. Here the part related to the bridge-span where the simulation was carried out is examined. The site investigation of this specific span of the bridge comprised boreholes for soil sampling, total sounding to assess the soil strata and the location of the rock face, and field vane tests to estimate the undrained shear strength of the clay layer. The locations of the boreholes are shown in Figure 3 relative to the outline of the bridge. The sampling location where a triaxial test was carried out is in the center of the bridge span.

Table 1. Index properties of the clay layer.

Depth	Properties
1.5 m	Bulk density $\rho = 1.67$ t /m^3 Water content $w_n = 69\%$ Liquid limit LL = 67% Sensitivity $S_t = 17$
2.5 m	Bulk density $\rho = 1.58$ t /m^3 Water content $w_n = 80\%$ Liquid limit LL = 63% Sensitivity $S_t = 22$
3.5 m	Bulk density $\rho = 1.54$ t /m^3 Water content $w_n = 93\%$ Liquid limit LL = 72% Sensitivity $S_t = 26$
5.5 m	Bulk density $\rho = 1.64$ t /m^3 Water content $w_n = 65\%$ Liquid limit LL = 61% Sensitivity $S_t = 10$

4.3 *Laboratory testing*

Laboratory testing consisted of index tests, constant rate of strain (CRS) tests and 3 triaxial tests. The index properties of the clay are shown in Table 1. The CRS tests assessed the compressibility of the soil and were used to assess the cell pressure for the triaxial tests, and showed that the OCR of the clay layers varied between 1.2–1.4, i.e. the clay is normally consolidated to lightly overconsolidated, typical characteristics in the area, (Lundberg 2017). The triaxial are discussed further later in text in the context of soil parameter choice in the constitutive model for the numerical simulation.

5 LOAD CONDITIONS DURING CONCRETE CASTING

The load for the SLS design consisted of the concrete before curing. The temporary structure and the forms resulted in a negligible load. Casting was carried out by a concrete pump standing beside the temporary structure.

5.1 *Preparatory work*

The area was covered with gravel before work on the superstructure commenced to support the construction machines during pile driving and casting of the bridge support columns. The gravel was subsequently carefully compacted with vibratory rollers, and the temporary support was placed on wooden footings on the gravel as shown in Figure 2. The formwork for the concrete was then built on top of the temporary structure before casting of the concrete.

5.2 *Load during casting*

The load was transferred to the clay from the concrete casting forms to the through the temporary support structure shown in Figure 2. The temporary structure was founded on 400 × 400 mm^2 wooden footings placed on the compacted gravel on top of the clay. The footing loads ranged from 25 kN at the edge of the bridge to 50 kN at the middle of the bridge. The loads were varying along the bridge due to changes in the cross-section.

5.3 *Field measurements*

During the casting process, field measurements of the vertical displacement of the bridge forms were carried out at two locations (11 and 12) along the bridge, according to Figure 4. The accuracy of the measurement method (precision levelling) was +/- 0.5–1.0 mm. The measurements were carried out immediately after casting, and 21 days after casting.

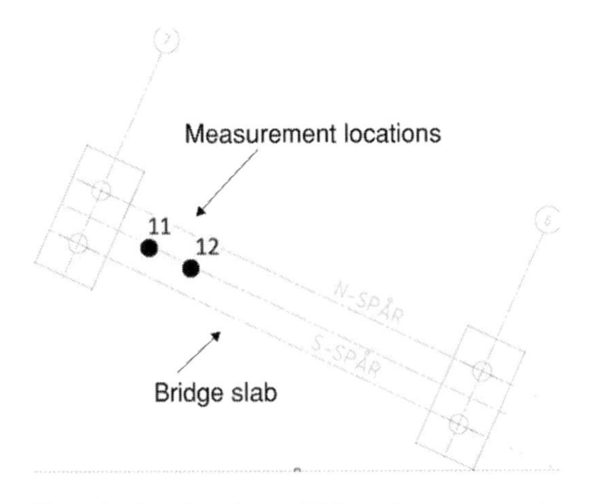

Figure 4. Location of vertical deformation measurements.

6 NUMERICAL MODELLING

Numerical modelling of the soil and superstructure comprised selection of a suitable constitutive model, parameter choice for the soil model, geometry and load idealization, and construction and execution of the numerical simulation.

6.1 *Soil models and parameter choice*

The selection of a suitable soil model and soil parameter choice was essential for an accurate simulation of the soil deformations during loading. The numerical simulations were carried out with the geotechnical software Plaxis 2D, (Brinkgreve & Vermeer, 1998). The 2D simulation was carried out with 15-noded triangular elements in plane strain.

A small-strain model was needed because of the non-linear deformation behaviour of the soil, (Atkinson 2000, Burland 1989). The small-strain stiffness model Hardening Soil-small (HS-small), which includes an implementation of the stiffness reduction model of Santos & Correia (2001) was used for the simulation. The model is described in detail in Benz (2007) and Benz et al (2009). Reference simulations were also carried out with the Mohr-Coulomb material model with linear elasticity.

The performance of advanced soil models is dependent on the parameter choice, frequently from quite limited test data, (Benz 2009). Since the simulations were carried out for a medium-risk project, small-strain testing was not executed, and the model parameter choice were based on the three triaxial tests performed on clay samples from the site.

These triaxial tests were anisotropically consolidated undrained in compression (CAUC). Anisotropic consolidation was carried out to 80% of the overconsolidation stress (at $\sigma_3'/\sigma_1'=0.78$) and the load was subsequently reduced to the effective vertical stress at the soil sample level. Shearing could then proceed with a strain rate of 0.7%/h to a total axial strain of around 12%.

An initial estimation of the parameters in the model was carried out manually, and subsequently an automatic curve-fit was automatically executed in the soil testing module of the program Plaxis (Brinkgreve & Vermeer 1998). The soil parameters were then manually inspected and assessed relative to the other laboratory testing and site investigation to assure the validity of the automatic parameter choice in the curve-fit. A triaxial test and the curve-fit with automatic model parameters are shown in Figure 5–7, where the axes show the ($s' = (\sigma_1' + \sigma_3')/2$), the shear stress ($t = (\sigma_1' - \sigma_3')/2$) and the maximum principal axial strain (ε_1). Figure 6 shows the triaxial test and the curve fit in a $\varepsilon_1 - t$ diagram, Figure 7 in a $s' - t$ diagram, and Figure 8 in a logarithmic $\varepsilon_1 - t$ diagram for and estimation of the small-strain behaviour.

Figure 5 shows that the curve-fit provides a good fit to the stress-strain curve, with a suitable fit to the strain-softening occurring after the peak shear strength of the clay. The stress-path curve in Figure 6 however shows more divergence between the test and the curve-fit. Since the model is mainly related to deformation and not ultimate bearing capacity, this may not influence the model simulation much. Figure 7 shows the limitations of the triaxial test in the logarithmic scale, in which instant deformation occurs after anisotropic consolidation, thereby not capturing the small-strain behaviour in the test. The curve-fit is shown from 0.01% strain to adjust for this feature.

Figure 8 shows all three triaxial tests that were carried out for the project. The test denoted 15E01 was sampled at 2.5 m depth, test 15E03 at 2.5 m depth and test 15E09 at 3.5 m depth. Test 15E03 was chosen for the curve fit since it showed a lower peak strength compared to the other triaxial tests. The small-strain stiffness response of the triaxial tests were reasonably similar, and the test 15E03 was chosen as a conservative value since this tests showed a higher flexibility compared to test 15E01 and 15E09, as shown in Figure 8.

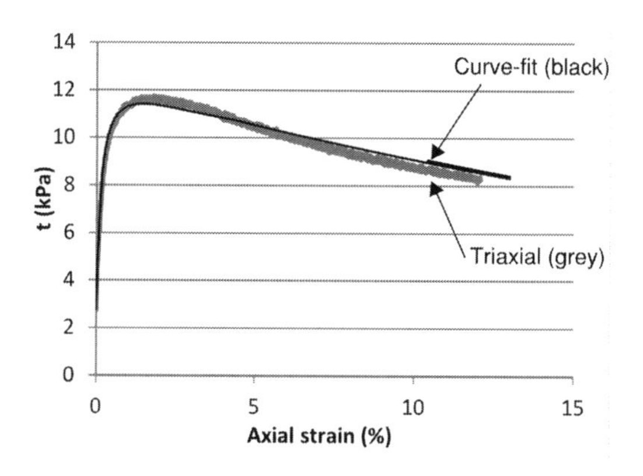

Figure 5. Shear stress- axial strain curve for the triaxial test and automatic curve-fit.

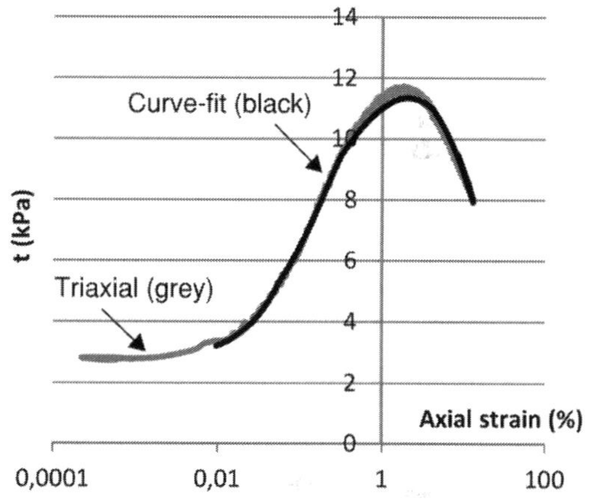

Figure 7. Shear-stress and logarithmic strain for the triaxial test and the curve-fit.

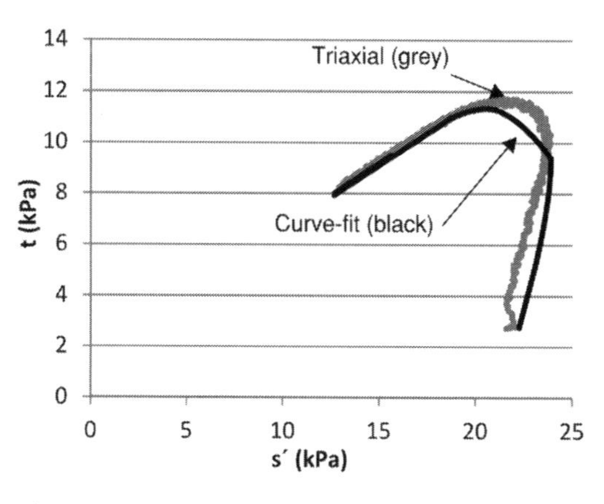

Figure 6. Stress path curve for the triaxial test and the curve-fit.

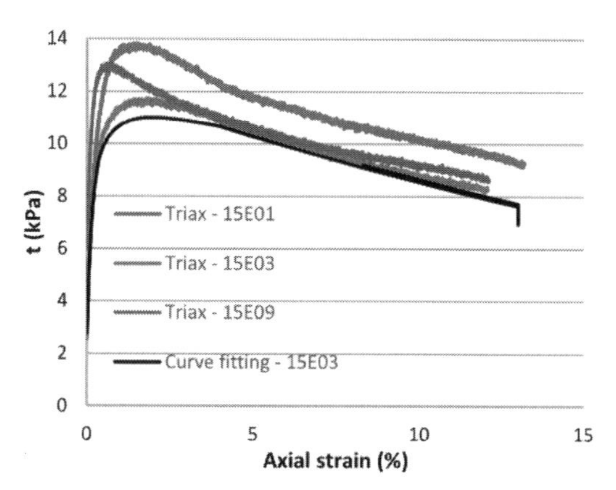

Figure 8. All three triaxial tests and the curve-fit.

6.2 *Model parameters*

The soil strata was idealized to a layer of compacted gravel and gravel fill covering the clay, which was located on top of the bedrock. During construction, the gravel layer was gradually built up around the bridge span on top of the gravel fill and compacted before the concrete forms were constructed.

The layer of compacted gravel and the gravel fill was modelled by the Mohr-Coulomb soil model, which was assumed to result in a suitable idealization of the material. Standardized parameters, shown in Table 3, were obtained by recommendation from the Swedish Transportation Authorities manual for geomechanics (Trafikverket 2013). The soil model parameters inserted into the numerical simulation are shown in Table 2. The hydraulic conductivity was assumed to be isotropic.

To capture the stiffness behaviour during loading, the clay was modelled by the hardening soil model with small strain. Soil parameters are shown in Table 4. Note the relatively large value of shear strain at 70% of $G/G_{0,Ref}$, resulting in a high stiffness for most of the deformation spectrum. A reference model of the clay layer with the Mohr-Coulomb model was executed with the soil parameters shown in Table 5.

Table 3. Soil parameters for the compacted gravel.

Soil parameter	Value
Saturated weight	$\gamma_{sat} = 16\ kN/m^3$
Friction angle	$\phi = 45°$
Poisson's ratio	$v = 0.23$
Elastic modulus	$E = 50000\ kPa$
Dilation angle	$\psi = 15°$
Earth pressure coefficient	$K_0 = 0.3$
Hydraulic conductivity	$k_x = 0.0596 \cdot 10^{-3}\ m/day$

Table 4. Soil parameters for the HS-small model.

Soil parameter	Value
Saturated weight	$\gamma_{sat} = 20\ kN/m^3$
Initial void ratio	$E_{init} = 2.02$
Friction angle	$\phi = 30°$
Ref. cohesion	$c'_{ref} = 1.5\ kPa$
Poisson's ratio	$v = 0.33$
Reloading Poisson's ratio	$v_{ur} = 0.2$
Ref. Elastic modulus	$E_{50,\ Ref} = 10200\ kPa$
Ref. Oedometric modulus	$E_{oed,\ Ref} = 5400 Pa$
Reloading Elastic modulus	$E_{ur} = 29000\ kPa$
Stiffness coefficient	$M = 0.91$
Dilation angle	$\psi = -2.9°$
Earth pressure coefficient	$K_0 = 0.78$
Small strain shear modulus	$G_{0,\ Ref} = 24000\ kPa$
Shear strain for G/G = 0.7	$\gamma_{0.7} = 0.043$
Asymptotic yield ratio	$R_f = 0.9$
Hydraulic conductivity	$k_x = 0.0596 \cdot 10^{-3}\ m/day$

6.3 *Geometry, 2D-model*

The 2D-model was carried out for the plane strain loading for the measurement locations shown in Figure 5. Since the loading was not completely symmetric, the full 2D-section was modelled and symmetry around the centerline was not included in the model. Before casting, part of the cast was excavated and refilled with gravel. The load was

Table 5. Soil parameters for the MC-model.

Soil parameter	Value
Saturated weight	$\gamma_{sat} = 20\ kN/m^3$
Friction angle	$\phi = 30°$
Poisson's ratio	$v = 0.33$
Elastic modulus	$E = 10500\ kPa$
Dilation angle	$\psi = 0°$
Earth pressure coefficient	$K_0 = 0.78$
Hydraulic conductivity	$k_x = 0.0596 \cdot 10^{-3}\ m/day$

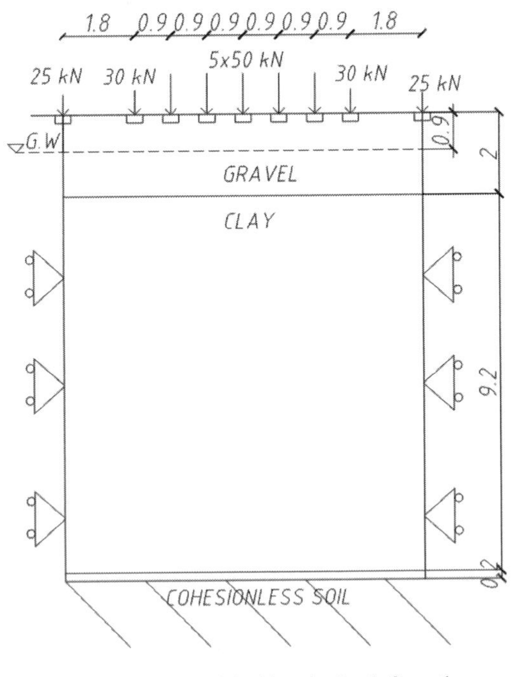

Figure 9. The 2D-model with point loads from the concrete casting forms.

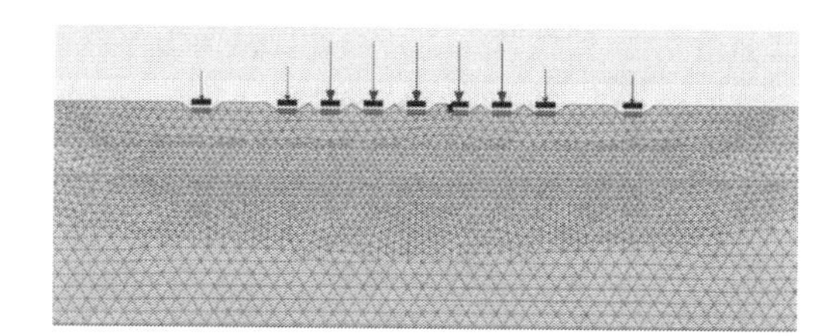

Figure 10. The mesh of the top 5 m of the model with.

applied as point loads on top of the gravel. Part of the 2D-model is shown in Figure 9.

6.4 *Mesh generation*

The mesh was generated automatically by the program Plaxis 2D, (Brinkgreve & Vermeer 1998). 15-noded triangular elements were used for the mesh. A high mesh density was specified in the fill and the compacted gravel, as well as 2 m into the clay layer, after which a constant mesh size was used to the rock face. The mesh for the top 5 m is shown in Figure 10.

6.5 *Numerical simulation for prediction in design*

A 2D-model was prepared for design, thereby constituting a Class A prediction, (Lambe 1973). The 2D plane strain model was used to predict settlements at specific cross-sections along the length of the bridge in order to help adjust the casting forms along the bridge span.

7 CALCULATED AND MEASURED VERTICAL DEFORMATIONS

7.1 *Measured deformations*

The deformations were measured with precision levelling on top of the casting forms according to Figure 4. The vertical displacements were recorded immediately after casting and after 21 days, after the concrete had cured. These are shown in Table 6.

7.2 *Calculated deformations*

The deformations were calculated through the numerical model in 2D-geometry. The load stages consisted of a K_0-stage to initialize the stress state in the soil. The loads were subsequently applied on top of the gravel in a plastic load step without complete pore-fluid interaction, followed by consolidation of the clay layer for 21 days. The calculation was carried out for location 11 and 12 for the 2D-model. The simulation was carried out with both HS-small model and the reference Mohr-Coulomb model.

Table 6. Measured deformations.

Location	After casting	21 days	Total deformation
11	4 mm	9 mm	13 (+/– 1 mm)
12	5 mm	9 mm	14 (+/– 1 mm)

Table 7. Normalized deformations, location 11.

Model	Normalized deformations $\delta_{model}/\delta_{field}$ (–)
2D—MC	1.43
2D—HSsmall	1.25

Table 8. Normalized deformations, location 12.

Model	Normalized deformations $\delta_{model}/\delta_{field}$ (–)
2D—MC	1.09
2D—HSsmall	0.98

The normalized simulated vertical deformations ($\delta_{model}/\delta_{field}$) are shown in Table 7 for load case 1 and Table 8 for load case 2.

8 DISCUSSION

The normalized deformations in table 7 and 8 show consistent overprediction of the deformations, except for the predicted deformations of the HS-small 2D-model for location 12. The deformations for load case 1 show and overprediction of 20–40%, with around 20% for HS-small model and larger overestimation of the Mohr-Coulomb model. The deformations for load case 2 show a good prediction with HS-small in 2D plane strain. However, the accuracy of the field measurements was also around 5–8% of the total displacement, which should be considered in the subsequent interpretation of the simulations.

The disparity between the normalized deformations at location 11 and 12 is possibly the result of time-dependent distribution of the surcharge load from the casting form. The soil at measurement location 11 is subjected to higher load due to the larger bridge area of the cross-section by the bridge column. The load during casting is initially hydrostatical, but during the curing process, the stiffness of the concrete increases rapidly. After 1 day the Elastic modulus of the concrete is around 9 GPa, and after 3 days around 18 GPa, according to Park et al (2010). The assumption that the load is hydrostatically distribution during the casting process is therefore a large simplification. This idealization also influences the results of the simulation at measurement location 12. The investigation the influence of the gradual hardening of the concrete superstructure, a model of the concrete hardening process is required, but this was outside the scope of the current project.

9 CONCLUSIONS

Design of a short-term temporary structure through numerical model and soil testing has been carried out. The simulation was conducted for 2D in a Class A prediction. The model simulation shows general overprediction of the expected deformations, with good fit for the HSsmall model, not exceeded the measured deformations of more than 20%. Since the design was carried out for a medium-risk project, no elaborate small-strain laboratory testing was carried out, and the stress-strain deformation behaviour was conducted with the basis of CAUC-triaxial testing. The overprediction likely results from some sample disturbance, (Lundberg 2017). But allowing for this, a suitable prediction could be carried out for practical design which was on the safe side, i.e. the estimated deformations were and overprediction of reality. The simulation shows that quite advanced numerical modelling can be executed for routine design for medium-risk projects, and suggests some suitable ways to validate and estimate the robustness of the model.

ACKNOWLEDGEMENTS

The project management at Veidekke AB are kindly acknowledge for their cooperation in the project. The authors also thank the Swedish Geotechnical Institute for very fruitful discussions about laboratory testing of clays.

REFERENCES

Andersson, A. Axelsson, G. & Lundberg, A.B. 2016 Reliability analysis of pile groups based on dynamic pile testing, *Nordic geotechnical conference 2016*, 503–512

Atkinson, J.H. 2000. Non-linear soil stiffness in routine design. *Géotechnique*, 50(5), 487–508.

Benz, T. 2007. Small-strain stiffness of soils and its numerical consequences (Vol. 5). Univ. Stuttgart, Inst. f. Geotechnik.

Benz, T., Vermeer, P.A., & Schwab, R. 2009. A small-strain overlay model. *International journal for numerical and analytical methods in geomechanics,* 33(1), 25–44.

Brinkgreve, R.B.J., & Vermeer, P.A. 1998. Plaxis finite element code for soil and rock analysis. Plaxis BV, The Netherlands.

Burland, J.B. 1989. Ninth Laurits Bjerrum Memorial Lecture:" Small is beautiful"—the stiffness of soils at small strains. *Canadian geotechnical journal,* 26(4), 499–516.

Clayton, C.R.I. 2011. Stiffness at small strain: research and practice. *Géotechnique*, 61(1), 5–37.

Dos Santos, J.A., & Correia, A.G. 2001. Reference threshold shear strain of soil. Its application to obtain an unique strain-dependent shear modulus curve for soil. In *Proceedings of the international conference on soil mechanics and geotechnical engineering*, 267–270

Gonzalez, N.A., Rouainia, M., Arroyo, M., & Gens, A. 2012. Analysis of tunnel excavation in London Clay incorporating soil structure. *Géotechnique*, 62(12), 1095.

Graham, J., & Shields, D.H. 1985. Influence of geology and geological processes on the geotechnical properties of a plastic clay. *Engineering geology*, 22(2), 109–126.

Jardine, R.J., Potts, D.M., Fourie, A.B., & Burland, J.B. 1986.

Jardine, R.J., Symes, M.J., & Burland, J.B. 1984. The measurement of soil stiffness in the triaxial apparatus. *Géotechnique*, 34(3), 323–340.

Lambe, T.W. 1973. Predictions in soil engineering. *Géotechnique*, 23(2), 151–202.

Long, M., Wood, T., & L'Heureux, J.S. 2017. Relationships Between Shear Wave Velocity and Geotechnical Parameters for Norwegian and Swedish Sensitive Clays. *In Landslides in Sensitive Clays.* 67–76. Springer International Publishing.

Lundberg, A.B. 2017. Sample Disturbance in Deep Clay Samples. In *Landslides in Sensitive Clays*. 133–143. Springer International Publishing.

Lundberg, A.B., & Li, Y. 2015. Probabilistic characterization of a soft Scandinavian clay supporting a light quay structure. *Geotechnical safety and risk V.* IOS Press, Amsterdam, 170–175.

Lundberg, A.B., Resare, F., & Axelsson, G. (2016). Numerical Modelling of Inclined Piles in Settling Soil. 13th *Baltic sea geotechnical conference,* 137–142

Park, H.G., Hwang, H.J., Kim, J.Y., Hong, G.H., Im, J.H., & Kim, Y.N. 2010. Creep and effective stiffness of early age concrete slabs. *Proceeding of the Fracture Mechanics of Concrete and Concrete Structures–Assessment, Durability, Monitoring and Retrofitting of Concrete Structure,* 751–754.

Schanz, T., Vermeer, P.A., & Bonnier, P.G. 1999. The hardening soil model: formulation and verification. Beyond 2000 in computational geotechnics, 281–296. Wiley. London.

Simpson, B. 1992. Retaining structures: displacement and design. *Géotechnique,* 42(4), 541–576.

Simpson, B., O'Riordan, N.J., & Croft, D.D. 1979. A computer model for the analysis of ground movements in London Clay. *Géotechnique*, 29(2), 149–175.
Studies of the influence of non-linear stress–strain characteristics in soil–structure interaction. *Geotechnique,* 36(3), 377–396.

Travikverket. 2013. Trafikverkets tekniska råd för geokonstruktioner. TDOK 2013:0668. Borlänge. (In Swedish)

Viggiani, G., & Atkinson, J.H. 1995. Stiffness of fine-grained soil at very small strains. *Géotechnique*, 45(2), 249–265.

Wood, T. 2015. Re-appraisal of the dilatometer for in-situ assessment of geotechnical properties of Swedish glaciomarine clays. In *DMT 2015 3rd international conference on the flat dilatometer*, Rome, Italy, June.

Wood, T. 2016. On the small strain stiffness of some Scandinavian clays and impact on deep excavation. Chalmers University of Technology.

Numerical Methods in Geotechnical Engineering IX – Cardoso et al. (Eds)
© 2018 Taylor & Francis Group, London, ISBN 978-1-138-33203-4

An elastoplastic 1D Winkler model for suction caisson foundations under combined loading

S.K. Suryasentana, B.W. Byrne & H.J. Burd
University of Oxford, UK

A. Shonberg
Ørsted Wind Power, London, UK

ABSTRACT: Most existing Winkler models use non-linear elastic soil reactions to capture the non-linear behaviour of foundations. These models cannot easily capture phenomena such as permanent displacement, hysteresis and the influence of combined loading on the failure states. To resolve these shortcomings, an elastoplastic Winkler model for suction caisson foundations under combined loading is presented. The proposed model combines Winkler-type linear elastic soil reactions with local plastic yield surfaces to model the non-linear soil response using standard plasticity theory, albeit in a simplified One-Dimensional (1D) framework. The results demonstrate that the model reproduces the appropriate foundation behaviour, comparing closely to Three-Dimensional Finite Element (3DFE) analyses but with the advantage of rapid computation time.

1 INTRODUCTION

1.1 *General background*

Winkler models are widely used to design deep foundations such as piles. However, in recent work (Gerolymos and Gazetas, 2006; Varun et al., 2009; Suryasentana et al., 2017), Winkler models have also been developed for shallow foundations such as caisson foundations. While these design methods may not be as accurate as more rigorous approaches such as the three-dimensional finite element (3DFE) method, Winkler models have the advantages of being relatively fast and easy to use.

Winkler models simplify the three-dimensional (3D) foundation-soil interaction problem into a more tractable one-dimensional (1D) problem, with the foundation replaced by a beam and the soil continuum by Winkler 'springs' (also termed as soil reactions in this paper). To simulate the non-linear response of soil, the Winkler models adopt non-linear elastic soil reactions. Examples of such models include the *p-y* and *t-z* methods (API, 2010; DNV, 2014) used to design laterally and axially loaded piles respectively.

Nevertheless, there are shortcomings with these existing non-linear Winkler models. For example, the non-linear elastic soil reactions used in the *p-y* or *t-z* methods for piles cannot reproduce observed cyclic loading phenomena such as permanent displacement or hysteresis. Moreover, they cannot account for combined loading effects on the failure state.

1.2 *Proposed model*

Recently, Suryasentana et al. (2017) developed a 1D Winkler model, calibrated against 3DFE analyses, to accurately predict suction caisson behaviour in linear elastic soil for six degrees of freedom (dof) loading. However, this model can only be applied to loading conditions where the soil response can be approximated as linear elastic.

This paper extends the 1D Winkler model developed in Suryasentana et al. (2017) to allow predictions of non-linear caisson behaviour in undrained clay under combined planar vertical V, horizontal H and moment M loading. The extension involves coupling linear elastic soil reactions with local plastic yield surfaces, which are calibrated against rigorous 3DFE failure state analyses.

The governing mechanics of the proposed 1D model is based on the same elastoplasticity framework used in 3DFE analyses. This allows straightforward reproduction of the 3DFE predictions, but with higher efficiency due to dimensionality reduction. Consequently, the proposed 1D model allows fast and accurate solutions of caisson behaviour in elastoplastic soil for design assessment under fatigue, serviceability and ultimate limit states (FLS, SLS, ULS). This enables an efficient design process, with the 1D model used to quickly shortlist potential designs from a large candidate space, before further refinement is conducted with 3DFE analyses.

2 METHODS

2.1 *1D model*

The 1D model adopted in this paper is similar to that detailed in Suryasentana et al. (2017) and it is briefly described as follows. The 1D model is a simplified representation of the original 3D caisson-soil interaction problem, where the caisson structure and soil continuum are replaced by a 1D rigid body and Winkler-type soil reactions respectively.

Figure 1 shows a schematic diagram of the original 3D caisson-soil problem and the 1D model representation. There are two types of soil reactions in this model: distributed soil reactions that act along the caisson skirt (referred to as the 'skirt soil reactions' and indicated as h^{skirt}, m^{skirt} and v^{skirt} in Figure 1) and concentrated soil reactions that act on the caisson base, including the soil plug (referred to as the 'base soil reactions' and indicated as h^{base}, m^{base}, v^{base} in Figure 1). h^{skirt}, v^{skirt} and m^{skirt} represent the distributed horizontal force, vertical force and rotational moment along the skirt length, while h^{base}, v^{base} and m^{base} represent the concentrated horizontal force, vertical force and rotational moment at the base.

There are, however, a few notable differences between the 1D model adopted in this paper and that described in Suryasentana et al. (2017). First, as this paper is only concerned with planar *VHM* loading, there are only 3 components (v, h, m) for each soil reaction, which correspond to the vertical w, horizontal u and rotational θ degrees of freedom (dof). Second, unlike the linear elastic soil assumed in Suryasentana et al. (2017), the current paper assumes a linear elastic-perfectly plastic soil. This gives an ultimate limit to the soil response, which the previous 1D model was not able to capture. To simulate this behaviour, the 1D model in this paper couples the linear elastic soil reactions with local plastic yield surfaces. These local yield surfaces are a direct analogy of the elemental yield surfaces in the soil reactions space (consisting of v, h, m components). Just as the canonical yield surfaces determine the set of allowable elemental stress states, the local yield surfaces determine the set of allowable soil reaction states.

The mechanics of the coupled soil reactions-yield surfaces model can be explained by standard plasticity theory. For soil reaction states lying inside the local yield surface, the soil response is linear elastic with the incremental response given by:

$$\delta \boldsymbol{p} = \boldsymbol{k}_e \delta \boldsymbol{u} \tag{1}$$

where $\boldsymbol{p}$ = soil reactions $\{v, h, m\}$, $\boldsymbol{k}_e$ = elastic stiffness matrix and $\boldsymbol{u}$ = local displacements $\{w, u, \theta\}$. $\boldsymbol{k}_e$ can be obtained from Suryasentana et al. (2017) as the caisson dimensions ($L/D = 1$) and elastic soil properties ($\nu = 0.49$) adopted in this paper are identical. However, for simplicity and faster numerical convergence, the coupling terms between h and m in $\boldsymbol{k}_e$ are ignored (the exclusion of these coupling terms will mainly impact the accuracy of the elastic horizontal and rotational predictions). Thus, $\boldsymbol{k}_e$ for the skirt and base soil reactions are as follows:

$$\boldsymbol{k}_e^{skirt} = G\begin{bmatrix} 4.28 & 0 & 0 \\ 0 & 6.51 & 0 \\ 0 & 0 & 1.17D^2 - 0.12zD \end{bmatrix} \tag{2}$$

$$\boldsymbol{k}_e^{base} = GD\begin{bmatrix} 2.4 & 0 & 0 \\ 0 & 1.17 & 0 \\ 0 & 0 & 0.42D^2 \end{bmatrix} \tag{3}$$

where G = shear modulus of soil, D = caisson diameter, z = depth below ground level (see Figure 1).

When the soil reaction states reach the local yield surface, the soil response becomes elastoplastic, with incremental behaviour given by:

$$\delta \boldsymbol{p} = \boldsymbol{k}_{ep} \delta \boldsymbol{u} \tag{4}$$

where $\mathbf{k}_{ep}$ = elastoplastic stiffness matrix. By convention, the local yield surface $f(\boldsymbol{p})$ is defined as

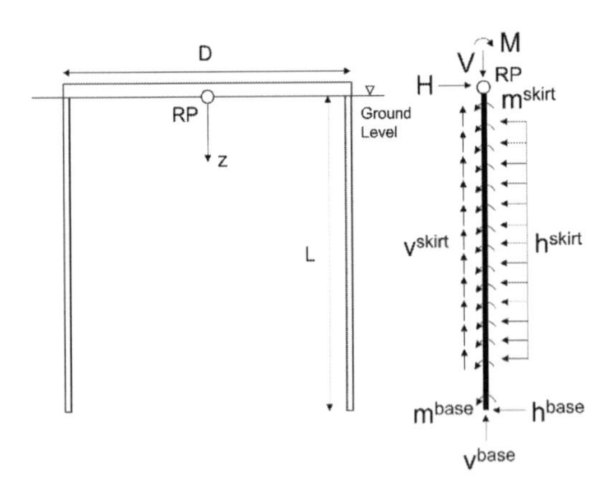

Figure 1. Schematic diagram of an embedded suction caisson foundation (left) and its corresponding simplified 1D representation (right), where RP is the loading reference point. v, h and m are the vertical, horizontal and rotational soil reactions.

follows: $f < 0$ for states inside the yield surface, $f = 0$ for states on the yield surface, and $f > 0$ for inadmissible states outside the yield surface.

When elastoplastic yielding occurs, permanent plastic displacements accumulate with the total displacement increment $\delta \boldsymbol{u}$ composed of elastic and plastic parts:

$$\delta \boldsymbol{u} = \delta \boldsymbol{u}_e + \delta \boldsymbol{u}_p \tag{5}$$

The elastic displacement increment $\delta \boldsymbol{u}_e$ is determined by the soil reaction increment through:

$$\delta \boldsymbol{u}_e = \boldsymbol{K}_e^{-1} \delta \boldsymbol{p} \tag{6}$$

The plastic displacement increment $\delta \boldsymbol{u}_p$ is determined by the flow rule:

$$\delta \boldsymbol{u}_p = \lambda \frac{\partial g}{\partial \boldsymbol{p}} \tag{7}$$

where $g(\boldsymbol{p})$ is a plastic potential function and λ is a non-negative, scalar plastic multiplier. When yielding occurs, the incremental soil reaction $\delta \boldsymbol{p}$ must remain on the local yield surface. This is enforced by the consistency condition:

$$\left(\frac{\partial f}{\partial \boldsymbol{p}} \right)^{\mathrm{T}} \delta \boldsymbol{p} = 0 \tag{8}$$

Following the conventional approach for linear elastic-perfectly plastic models, $\mathbf{k}_{ep}$ is obtained from:

$$\boldsymbol{k}_{ep} = \boldsymbol{k}_e - \frac{\boldsymbol{k}_e \left(\frac{\partial g}{\partial \boldsymbol{p}} \right) \left(\frac{\partial f}{\partial \boldsymbol{p}} \right)^{\mathrm{T}} \boldsymbol{k}_e}{\left(\frac{\partial f}{\partial \boldsymbol{p}} \right)^{\mathrm{T}} \boldsymbol{k}_e \left(\frac{\partial g}{\partial \boldsymbol{p}} \right)} \tag{9}$$

For this paper, an associated flow rule is assumed i.e. $g(\boldsymbol{p}) = f(\boldsymbol{p})$. The local yield surface $f(\boldsymbol{p})$ is calibrated using the limiting soil reactions extracted from the 3DFE analyses, which is described in Section 2.3.

The 1D model was implemented numerically using the Galerkin finite element methodology, where two-noded 1D soil elements (each with a linear shape function and two Gauss points) representing the skirt soil reactions are tied to two-noded 1D caisson rigid bar elements. The base soil reaction is represented by a lumped model tied to the bottom node of the deepest caisson element. The explicit Runge-Kutta (4, 5) algorithm (Dormand and Prince, 1980) was used for the integration process during elastoplastic behaviour and the full Newton-Raphson procedure was used to obtain the system solution.

2.2 *3DFE model*

The 3DFE analyses were carried out using the finite element program ABAQUS v6.13 (Dassault Systèmes 2010). The 3DFE model consists of a suction caisson foundation (of unit diameter D and unit skirt length $L = D$) embedded in homogeneous soil, which is similar to that used in Suryasentana et al. (2017).

The mesh domain is set as $8D$ for the diameter and $6D$ for the depth, which was verified to be large enough to avoid boundary effects for load capacity predictions. Mesh convergence analyses were also carried out to determine the required mesh fineness. Due to symmetry of the problem, only half of the caisson and soil domain was modelled. A typical mesh of the 3DFE model is shown in Figure 2.

The soil was defined as weightless, homogeneous and linear elastic-perfectly plastic. The soil is assumed to obey the von Mises yield criterion with an associated flow rule. The Young's modulus E of the soil is set as $1000\sqrt{3}s_u$ (where s_u is the undrained shear strength) and the Poisson's ratio is set as 0.49.

Fully-integrated, linear, brick elements C3D8H were used to model the soil elements. The caisson was modelled as being entirely rigid using rigid body constraints. The caisson reference point was set at RP, as shown in Figure 1. Contact breaking between the caisson and soil was prevented using tie constraints at the caisson-soil interface. Displacements were fixed in all directions at the bottom of the mesh domain and in the radial directions at the periphery.

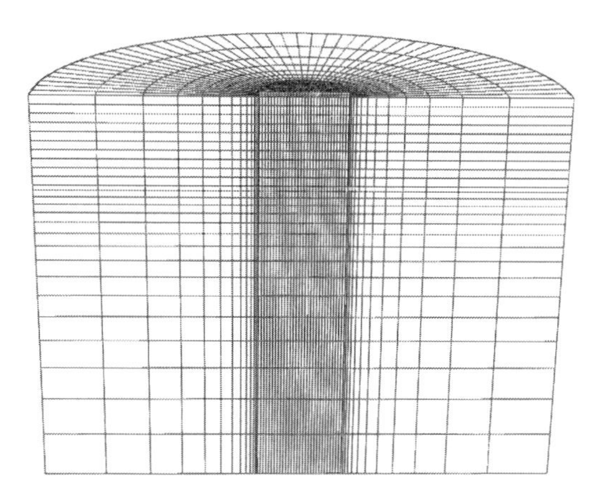

Figure 2. Mesh used for the 3DFE analyses. The diameter and depth of the mesh domain is set as $8D$ and $6D$ respectively.

2.3 *Calibration of local yield surfaces*

To calibrate the local yield surfaces in the proposed 1D model, a series of 3DFE analyses were carried out to obtain the limiting soil reactions. These analyses involved the determination of the global *VHM* failure envelope of the caisson-soil interaction problem (Bransby and Yun, 2009; Gourvenec and Barnett, 2011; Vulpe, 2015). This was done using mixed load and displacement control, where load control was used in the V load space while displacement control was used in the *HM* load space. In total, four vertical loads ($V/V_0 = 0, 0.25, 0.5, 0.75$ where V is the vertical load applied at RP and V_0 is the uniaxial vertical load capacity) were applied before displacement probes were applied in the *HM* load space. This determines the *HM* failure envelopes at fixed levels of V. It is hypothesized that, just as there exists a failure envelope that limits the global load space, there also exists a limiting envelope in the soil reactions space (termed as 'local yield surface' in this paper), which can be identified using the limiting soil reactions extracted from the 3DFE analyses.

The limiting soil reactions were extracted from the 3DFE results at the end of each displacement probe, corresponding to a global failure state of the caisson-soil interaction problem. For simplicity, the limiting skirt soil reactions are assumed to be constant along the skirt and are computed as the average of the soil reactions along the skirt.

To represent the local yield surface, an ellipsoid function $f(\boldsymbol{p})$ is adopted:

$$f = \left(\frac{v}{v_0}\right)^2 + \left(\frac{h}{h_0}\right)^2 + \left(\frac{m}{m_0}\right)^2 + \alpha\left(\frac{hm}{h_0 m_0}\right) - 1 \qquad (10)$$

where v_0, h_0 and m_0 are the limiting uniaxial vertical, horizontal and moment soil reactions (i.e. the uniaxial capacities in the soil reactions space) and α is a parameter that governs the rotation of the ellipsoid in the *hm* space. This ellipsoid function was adopted as it has favourable theoretical properties such as global convexity.

The unknown parameters v_0, h_0, m_0 and α were identified by running least squares regression against the limiting soil reactions extracted from the 3DFE results. The best-fit parameters for the skirt and base local yield surfaces are shown in Table 1.

The local yield surface contours generated by Equation 17 and the best-fit parameters in Table 1 are compared against the 3DFE limiting soil reactions in Figure 3. Although the global vertical load V is fixed while the *HM* failure envelope is probed, the distribution of the vertical load between the skirt and base soil reactions is not constant. Thus,

Table 1. Best-fit parameters for the skirt and base local yield surfaces.

Pa—rameter	Skirt	Base
v_0/s_u	A^{skirt}	$9.1A^{base}$
h_0/s_u	$2.07A^{skirt}$	$1.34A^{base}$
m_0/s_u	$0.19A^{skirt}D$	$0.72A^{base}D$
α	−1.23	-0.47

where D is the caisson diameter, A^{skirt} (skirt surface area per metre length basis) = πD and $A^{base} = \pi D^2/4$

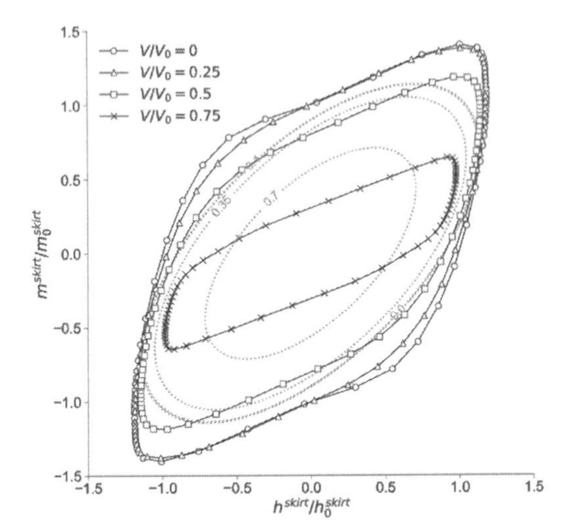

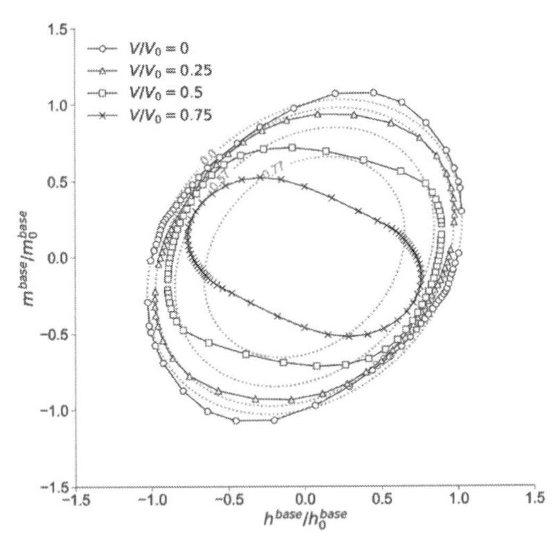

Figure 3. Comparison of the 3DFE limiting skirt and base soil reactions (as depicted by the markers in the figure) against the local yield surface contours (as depicted by the grey dotted lines) predicted by Equation 17 and Table 1. The average v/v_0 values (for each dataset corresponding to a fixed V) are shown in the contour labels and they are used in Equation 17 to produce the contours.

each of the limiting soil reactions is associated with a different v/v_0 value. To simplify the process, the average of these v/v_0 values (for each dataset corresponding to a fixed V) are used in Equation 17 to predict the hm contours for each V/V_0; their values are shown in the contour labels in Figure 3. For $V/V_0 = \{0, 0.25, 0.5, 0.75\}$, $v/v_0 = \{0, 0.1, 0.35, 0.7\}$ for the skirt soil reactions and $\{0, 0.32, 0.57, 0.77\}$ for the base soil reactions.

It was observed that the predicted local yield surface contours are good approximations to the limiting base soil reactions at low vertical loads. However, at higher vertical loads ($V/V_0 \$ 0.5$), there is less agreement as the ellipsoid function cannot capture the change in yield surface geometry. The fit is less than ideal for the limiting skirt soil reactions as they do not conform closely to an ellipsoidal shape.

However, although these simplified local yield surfaces do not match well on a local level, they produce reasonably accurate global predictions (as will be shown in Section 3.2).

2.4 *Evaluation of models*

To compare the predictions between the 1D model and the 3DFE model, three types of evaluations were implemented. First, the uniaxial load capacities (which are the load capacities under the application of V, H and M individually) were evaluated to assess the accuracy of the models for simple loading cases. Next, the influence of combined loading on failure states was assessed by using the 1D and 3DFE models to find the failure envelopes of the caisson in the VHM load space. Finally, a single cyclic load test was simulated using both models to assess the capability of capturing permanent displacement and hysteresis.

3 RESULTS

3.1 *Uniaxial load capacities*

Table 2 shows the uniaxial vertical V_0, horizonal H_0 and moment load capacities M_0 predicted by the 1D and 3DFE models. It is evident that the 1D model predictions agree very well with the 3DFE model predictions, with the largest difference being only 1.51% for the horizontal load capacity H_0.

Figure 4 compares the global load-displacement predictions under uniaxial loading, where w_{RP}, u_{RP} and θ_{RP} are the vertical, horizontal and rotational displacements of the loading reference point RP. As observed, the 1D and 3DFE model predictions tend to reach the load capacity at different displacements.

Under pure vertical loading, the 1D model reaches load capacity at a smaller displacement

Table 2. Comparison of the uniaxial global load capacities predicted by the 1D and 3DFE models.

Capacity	1D	3DFE	Diff (%)
$V_0/A^{base}s_u$	13.12	13.12	0.00
$H_0/A^{base}s_u$	6.01	5.93	1.51
$M_0/A^{base}Ds_u$	3.7	3.7	-0.07

where D is the caisson diameter and $A^{base} = \pi D^2/4$

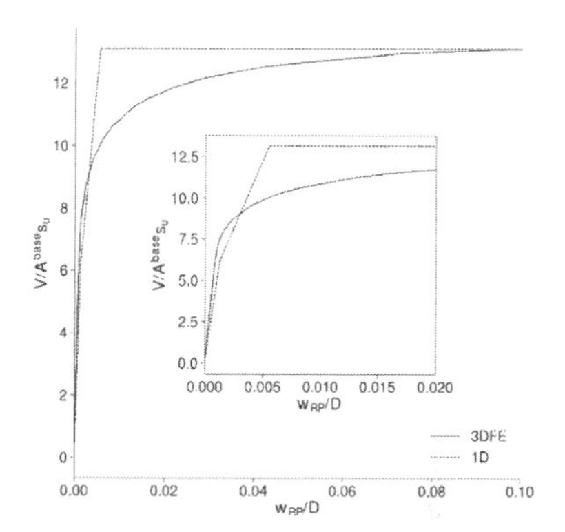

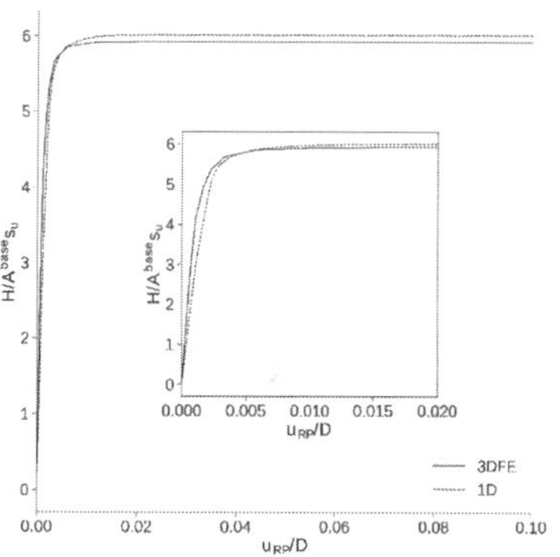

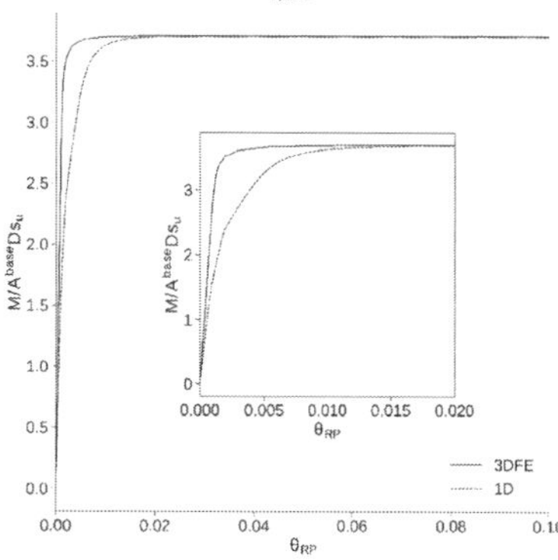

Figure 4. Comparison of global load-displacement predictions. The close-up insets focus on the results at small displacements.

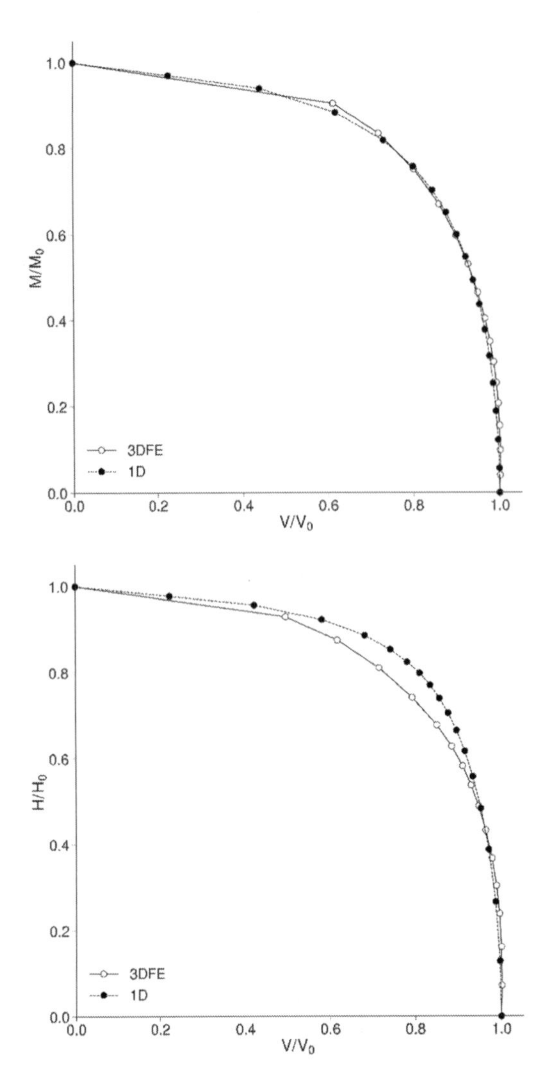

Figure 5. Comparison of global failure envelope predictions in the *VH* and *VM* load space.

than the 3DFE model. Moreover, it can be seen from the close-up inset that the 1D model load-displacement prediction is bilinear. The first linear response is the elastic response while the second linear response occurs when the base soil reaction has reached its local yield surface but the skirt soil reaction remains elastic.

Under pure horizontal loading, the 1D model predicts u_{RP}/D of 0.1 and θ_{RP} of 0.129, which compares well with the 3DFE model predictions of u_{RP}/D of 0.1 and θ_{RP} of 0.139. Similarly, under pure moment loading, the 1D model predicts θ_{RP} of 0.1 and u_{RP}/D of 0.05, which compares well with the 3DFE model predictions of θ_{RP} of 0.1 and u_{RP}/D of 0.0556.

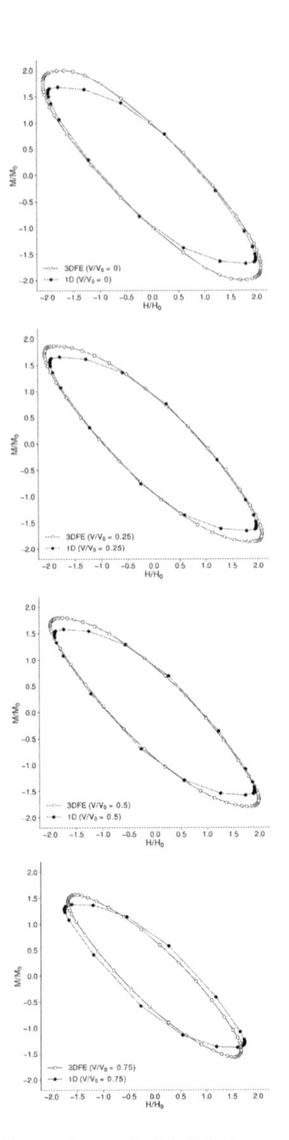

Figure 6. Comparison of global failure envelope predictions in the *VHM* load space.

Under pure horizontal or moment loading, the 1D model load-displacement predictions are not bilinear as both horizontal *h* and rotational *m* soil reactions occur during these loadings. The influence of combined *h* and *m* loading forces the soil reaction path to track on the local yield surface during elastoplastic yielding, until the global load capacity is reached.

3.2 *Failure envelopes*

Figure 5 compares the predictions of the *VH* and *VM* failure envelopes of the caisson in normalised forms, where the loads are normalised by their respective uniaxial capacities. The 1D model predictions of the *VH* and *VM* failure envelopes match the 3DFE results very well, albeit with a slight overprediction for the *VH* failure envelope for some load cases.

Next, Figure 6 compares the predictions of the *VHM* failure envelopes in normalised forms. Despite the poor match of the local yield surfaces at the local level (see Figure 3), the 1D model predictions of the global *HM* envelope under fixed *V* loads match the 3DFE predictions reasonably well.

Nevertheless, given the mismatch (especially that of the skirt local yield surface) in Figure 3, it is encouraging to see that the global failure envelope predictions are not too sensitive to the accuracy of these local yield surfaces. Furthermore, most loading scenarios are in the quadrants where *H* and *M* have the same sign. Thus, the mismatch in the quadrants where *H* and *M* have different signs are of less practical concern.

3.3 *Cyclic loading*

To assess whether the 1D model can simulate hysteretic behavior, a single cycle of positive and negative vertical displacements (w_{RP}/D = ±0.05) was prescribed onto the caisson. Figure 7 shows the comparison of the global load-displacement behavior under this cyclic loading. It is clear that the 1D model is able to simulate hysteresis, although the 1D model predictions is a rather crude piecewise linear approximation of the 3DFE model predictions.

By comparing the displacements at zero *V* load, it is evident that the permanent displacement

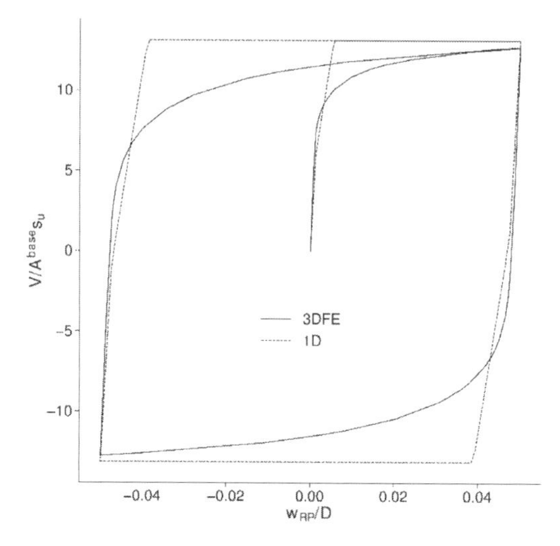

Figure 7. Comparison of global load-displacement behavior under cyclic vertical loading.

predictions of the 1D and 3DFE models are in good agreement. The 1D model comes with a built-in capability for simulating effects such as permanent displacement and hysteresis. This is not surprising as both the 1D and 3DFE models are based on fundamentally the same elastoplasticity concepts, but with different measures of 'stress' and 'strain'.

4 DISCUSSION

The 1D model ignores much of the detail of the original 3D continuum-based problem, with the aim of appropriate simplification to provide a fast proxy to the original problem. Despite the simplifying abstractions, the loss in accuracy is minimal, relative to the large gains in computational efficiency. For example, the 3DFE model took about 28 hours in total to run the analyses presented in Section 3. By contrast, the 1D model took about 0.8 hours in total, yielding a time saving of 97%.

This computational efficiency is very important for design optimization involving multiple foundations, such as that for an offshore wind farm. Whilst 3DFE is perhaps practical for design projects involving only a few foundations, it is clearly impractical when there are hundreds of foundations. A tool such as the trained 1D model offers the 3DFE accuracy but with much higher efficiency, and therefore allows more of the design space to be explored.

Furthermore, the proposed 1D model offers advantages over existing macro-element models for shallow foundations (e.g. Cassidy 2004, Salciarini et al. 2011). Given the localised nature of the soil reactions and the yield surfaces, the 1D model may be simply adapted to non-homogeneous or multi-layered grounds with arbitrary yield strength profiles. This contrasts with macro-element models, which can only be adapted to ground profiles similar to that in the original calibration. In other words, the 1D model is a more generalised model by comparison with the macro-element model.

The focus of this paper is a presentation of the mapping process from a 3DFE elastoplastic continuum model to a 1D elastoplastic Winkler model, and a demonstration of the accuracy of the approach. As such, generalised formulations of the yield surfaces, although established for caissons of L/D ≤2, are not presented. They will be described in future publications.

There are, of course, some observed limitations with the 1D model. For example, the load predictions under purely vertical loading is bilinear. This could be resolved by adding multiple or nested local yield surfaces but this increase in accuracy comes at the expense of increased computational effort. Also, there is room for improvement for the global failure envelopes predicted by the 1D model and this can be achieved by adopting a more expressive function with more parameters to represent the local yield surface. However, while there are ready solutions to these limitations, it is advisable to consider whether the additional complexity balances the aim of providing a rapid but approximate solution to the caisson-soil interaction problem for preliminary designs, which can then be refined using more advanced 3DFE analyses.

5 CONCLUSION

The main concern with Winkler models that use non-linear elastic soil reactions to approximate the soil continuum response is that they do not easily reproduce observed phenomena such as permanent displacement, hysteresis and influence of combined loading on failure states. This paper resolves this shortcoming by proposing a 1D Winkler model that couple linear elastic soil reactions with local plastic yield surfaces that limits the allowable soil reaction states. The results indicate that the proposed 1D model compares favourably with the 3DFE model predictions in terms of accuracy across a range of loading states. The principal advantage, however, is efficiency, as it takes only 3% of the computational time required by the 3DFE model. Furthermore, unlike macro-element models which can only be used for ground profiles that are similar to the original calibration, the 1D model can be used for non-homogeneous or multi-layered grounds with arbitrary yield strength profiles, making it a more general, and arguably, useful model. Thus, the proposed 1D model offers an efficient method to predict realistic, non-linear behavior of caissons in elastoplastic soil.

REFERENCES

API. 2010. RP 2 A-WSD—Recommended Practice for Planning, Designing and Constructing Fixed Offshore Platforms. Washington: American Petroleum Institute

Bransby, M.F. & Yun, G.J. 2009. The undrained capacity of skirted strip foundations under combined loading. *Géotechnique*, 59(2), pp.115–125.

Cassidy, M.J., Martin, C.M. & Houlsby, G.T. 2004. Development and application of force resultant models describing jack-up foundation behaviour. *Marine Structures*, 17(3–4), pp.165–193.

DNV. 2014. OS-J101 - Design of Offshore Wind Turbine Structures. Oslo: Det Norske Veritas.

Dormand, J.R. & P.J. Prince. 1980. A family of embedded Runge-Kutta formulae. *Journal of Computational and Applied Mathematics*, Vol. 6, pp.19–26.

Gerolymos, N. and Gazetas, G., 2006. Development of Winkler model for static and dynamic response of caisson foundations with soil and interface nonlinearities. Soil Dynamics and Earthquake Engineering, 26(5), pp.363–376.

Gourvenec, S. & Barnett, S. 2011. Undrained failure envelope for skirted foundations under general loading. *Géotechnique*, 61(3), pp.263–270.

Salciarini, D., Bienen, B. & Tamagnini, C. 2011. A hypoplastic macroelement for shallow foundations subject to six-dimensional loading paths. *Proceedings of the 2nd international symposium on computational geomechanics (COMGEO-II), USA.*

Suryasentana, S.K., Byrne, B.W. Burd, H.J., & Shonberg, A. 2017. Simplified Model for the Stiffness of Suction Caisson Foundations Under 6 DOF Loading. *Proceedings of SUT OSIG 8th International Conference, London, UK.*

Varun, Assimaki, D. and Gazetas, G., 2009. A simplified model for lateral response of large diameter caisson foundations—Linear elastic formulation. Soil Dynamics and Earthquake Engineering, 29(2), pp.268–291.

Vulpe, C., 2015. Design method for the undrained capacity of skirted circular foundations under combined loading: effect of deformable soil plug. Géotechnique, 65(8), pp.669–683.

Numerical Methods in Geotechnical Engineering IX – Cardoso et al. (Eds)
© 2018 Taylor & Francis Group, London, ISBN 978-1-138-33203-4

Analysis of friction piles in consolidating soils

N. O'Riordan, A. Canavate-Grimal, S. Kumar & F. Ciruela-Ochoa
Arup Geotechnics, London, UK

ABSTRACT: Increasing urbanisation, finite land resources and the associated demand for fresh water can combine to produce large regional subsidence troughs. UNESCO Working Group on Land Subsidence has been raising awareness of the issues for many years, however the problem of robust design of foundations under these conditions remains. This article focuses on the particular case of the Mexico City Basin. The presence of deep deposits of highly compressible clays in the Mexico City basin, coupled with regional subsidence that can exceed 350 mm/year due to deep water extraction, presents a unique challenge to foundation design. For structures founded on deep foundations a key concern is the emergence as a result of the ongoing consolidation that varies with depth. The consideration of aspects like down-drag and formation of the gap under the pile cap are also found to be critical for deep foundation design in this area of high seismicity.

This paper presents details of the Finite Element (FE) analysis carried out to assess the short-term and long-term behaviour of the piled foundation solution. The FE analysis was carried out using Hardening Soil model with Small Strain Stiffness (HSSmall) available in commercial FE software. The parameters of the HSSmall model were obtained by carrying out detailed calibrations using the extensive ground investigation data available for the New Mexico City International Airport (Nuevo Aeropuerto Internacional de la Ciudad de México or NAICM) site. The analysis considers modelling of the site history (changes to the piezometric profile in particular) to achieve the current state of stress and future long term conditions based on the continuance of groundwater extraction at current rates. Analytical procedures are described so that the assessment of pile behaviour for a wide range of consolidating soils and water extraction rates can be made.

1 INTRODUCTION

Regional or land subsidence produces a ground deformation which might aggravate the flood risk and induce forces on utilities, roads and structures foundation affecting their performance. The Mexico City Basin is a paradigm on the management of those risks because of the high regional subsidence rates, the area extension and the large population.

The soft ground conditions coupled with the continuation of regional subsidence due to groundwater extraction from the deep aquifer system present a considerable challenge for the foundation design in the Mexico City lacustrine zone (Gobierno del Distrito Federal 2016). Figure 1 presents the three common foundation types (Rodríguez Rebolledo et al 2015) in the lacustrine zone of Mexico City (i) box-type shallow foundation only fit for low-rise buildings, (ii) raft foundations enhanced by friction piles and (iii) foundations supported on long end bearing piles.

A fundamental property of the Mexico City soil is the comparatively high stiffness in unload-reload behaviour (Ovando Shelley 2011; Díaz Rodríguez

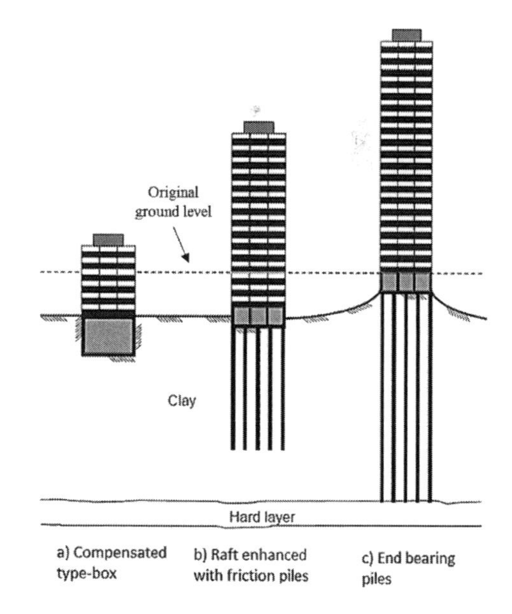

Figure 1. Behaviour of common types of foundations in Mexico City subjected to regional subsidence (Rodríguez Rebolledo et al 2015).

2003). The stiffness of the soil when effective stresses are kept below the yield stress is at least ten times the stiffness observed when the yield stress is exceeded. The box type shallow foundations are compensated foundations in which the weight of the building is intended to equalise the weight of soil and water excavated. This means that the building can then be founded on the soft and compressible upper-level soils while accommodating the same rate of regional subsidence affecting the surrounding landscape.

Raft foundations enhanced with friction piles are used to support heavier structures than the box shallow foundation type and reduce differential settlement. The friction piles reduce the soil deformation and transmit stresses to the deeper strata of lower compressibility (Zeevaert 1983). To ensure that the foundation settles at the same rate as the regional subsidence, the friction piles are designed to operate at their ultimate load capacity to minimize the development of negative skin friction. If negative friction develops on the upper part of the pile, there is the risk of a gap formation between the underside of the raft and the soil surface. The gap formation poses a risk especially for seismic events as the ground bearing is missed and the loads are carried only by the piles. The time should be accounted in the design because the consolidating soils gain strength due to the increase of effective stress. The presence of hard layers underlying the compressible soil should be also considered when designing the length of the piles. If there is not enough clearance between the toe of the pile and the hard layer, the pile might hit the top of the hard stratum affecting the foundation performance.

Foundation on long end bearing piles is the conventional approach for heavy structures. On a subsiding soil, the end bearing piles will be subjected to negative skin friction and the gap formation between the underside of the raft and the bearing ground is very likely to happen. As the piled foundation settles less than the surrounding soil, an apparent protrusion of the structure is bound to happen. In order to achieve more efficient designs, the downdrag force resulting from the negative skin friction caused by the regional subsidence should not be taken into account in the pile capacity determination. In line with Fellenius (2017) and the Mexico City Technical Norms (Gobierno del Distrito Federal 2016), when a pile reaches its ultimate capacity, the pile moves downwards relative to the soil and therefore negative skin friction cannot build up. The effect of the downdrag force on the pile is an increase of the stresses within the pile and the settlements due to the increasing compression of the bearing unit. An accurate prediction of the foundation settlement evolution during the service life is required for the architectural and utility design. Therefore, a thorough study of the ground conditions, paying particular attention to the pore water pressure evolution with time, is required to understand the complexity of foundation design in subsiding soft soils. Section 2 provides detail of the former Lake Texcoco which used to be an extensive undeveloped area adjacent to the east of Mexico City. The New Mexico City International Airport (*Nuevo Aeropuerto Internacional de la Ciudad de México* or NAICM) is currently being constructed on that site.

The design of a piled foundation in the lacustrine zone of Mexico City requires the assessment of the long term behaviour. Despite there are approximate methods to assess the effect of a subsiding soil on the piled foundation, advanced numerical methods are required to explore the total soil-structure interaction, especially for the seismic load case. The constitutive model should be able to capture the progressive soil hardening occurring when the pore water pressure reduces. Those models are computationally intense for that reason it is advisable to tackle the analyses increasing progressively the model detail. This paper presents in Section 3 two simplified axisymmetric analyses of short friction and long bearing piles under a large raft. This approach provides insight of the long-term foundation behaviour useful for the early stages of the design.

2 MEXICO BASIN BACKGROUND

2.1 *Stratigraphy*

The Basin of Mexico occupies an area of 10,000 km^2. The Basin is a predominately flat lacustrine plain with a typical elevation of about 2,250 m MSL. The stratigraphy in the lacustrine zone comprises typically an upper made ground and a desiccated crust, the Upper Clayey Formation (*Formación Arcillosa Superior* or FAS) which can be 30 m thick, a 0 to 5 m thickness of dense sands/volcanic glass known as the Hard Layer (*Capa Dura* or CD) followed by the Lower Clayey Formation (*Formación Arcillosa Inferior* or FAI) which can reach 50 m deep. Underlying the FAI a series of alluvial sand and gravels, cemented with clay and calcium carbonate called Deep Deposits (*Depósitos Profundos* or DP) up to 110 m deep. The FAS is an extremely soft and weak soil characterised with a typical lower bound undrained shear strength profile of $S_u = 7+1.05z$ (kPa), where z is the depth below the top of FAS unit. Further details on FAS properties can be found on Díaz Rodríguez (2003) and O'Riordan et al. (2017).

Below the DP, alluvial fans and debris flows (lahars) interbedded with volcanic pumice and ash known as the Tarango Formation develop. The fresh water for the Mexico City area population is abstracted from this formation and it is named as the aquifer unit in this paper.

2.2 *Hydrogeology*

From a hydrogeological perspective, the system comprising the FAS, CD, FAI and DP can be regarded as an aquitard from which brine was pumped at certain locations (Rudolph et al 1989). The ground comprising the Tarango Formation is an aquifer which supplies fresh water to the Mexico City population since 1930. The water has typically been pumped from the lower aquifers below a depth of 120 m, above which depth well casings are grouted in place, to minimize salt intrusion (CONAGUA 2015). One effect of pumping groundwater has been the lowering of the piezometric profile to less than hydrostatic within some soil units. The resulting consolidation of the soils in the Formacion Tarango and above has led to regional subsidence as high as 350 mm/yr (Comité Técnico Proyecto Texcoco 1969).

Current deep well records show a steady depressurization of about 7 kPa/yr in the soils below 120 m and on the NAICM site the CD layer is losing pressure at a rate of 3 kPa/yr.

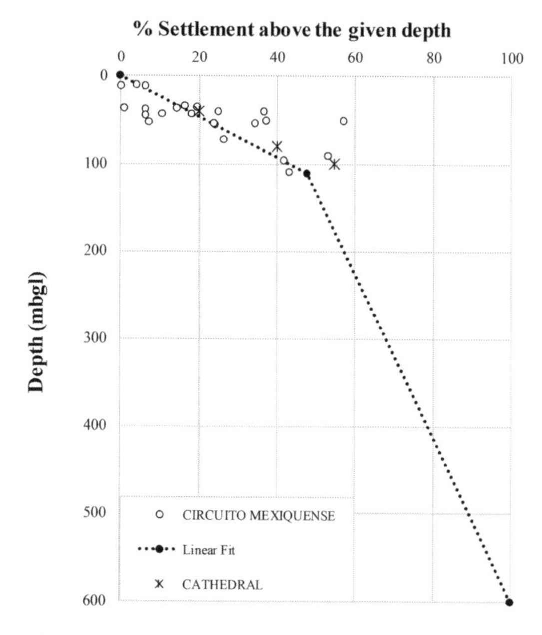

Figure 2. Depth regional subsidence participation percentage. (source: the Authors).

Santoyo & Ovando (2008) reported that in 1992 all the regional subsidence was produced by the compression of the upper 80 m of the Mexico City stratigraphy. The same authors reported that between 2003 and 2007 the contribution to regional subsidence below 80 m was between 43 and 72%. Those observations tie up well with the deep settlement marker data nearby the Circuito Mexiquense (internal communication) showing that about 50% of the total settlement is due to the compression of the soil below 100 m (see Figure 2).

3 NUMERICAL MODELLING

3.1 *Analyses set up*

Simplified analyses of a raft enhanced with short friction piles and a raft on long end bearing piles are presented. The modelling of the complete pile-raft system is a computationally intense task and a single pile analysis subjected to the ground subsiding conditions might suffice for preliminary design purposes.

The proposed analysis simulates the behaviour of a pile located within a large raft away from the edges. An axisymmetric model reduced to its tributary area is shown in Figure 3. A similar approach is proposed by Rodríguez Rebolledo et al (2015).

Reduced strength interface is used between raft and the bearing soil is used to allow the gap formation of gap. The soil strength is also reduced in the pile-soil interface.

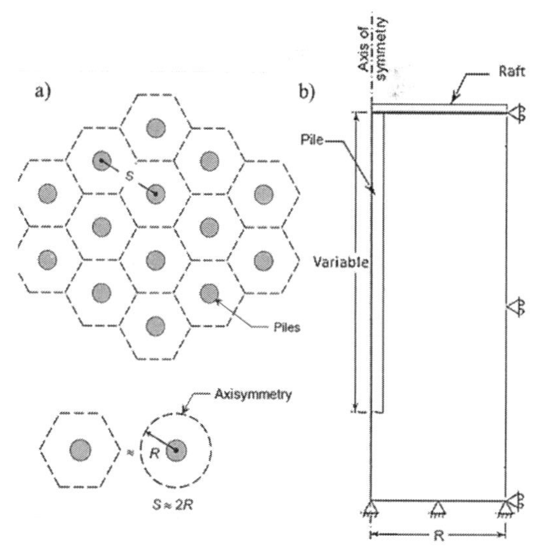

Figure 3. Model of a piled raft. a) Infinitely large pile raft supported by piles. b) Numerical model (Rodríguez Rebolledo et al 2015).

The long-term behaviour of a piled foundation within a subsiding soil is controlled by the evolution of the pore water pressure. The reduction of the pore water pressure in the different calculations staged is further discussed in the companion paper by O'Riordan et al. (2018). The raft enhanced with short piles is located at the former Texoco Lake and it was subjected to an industrial brine abstraction during a large period. The regional subsidence arising from groundwater abstraction from deep aquifers was then back analyzed from a point in time, 1930, when records show that significant pumping started. The brine pumping between 1945 and 1995 by Sosa Texcoco, in which wells were installed across the site at approximately 200 m centres to depths between 30 and 60 m and pumped at 1 l/s (Santoyo Villa et al. 2005), was also simulated. The progression in water pressure profiles on Texcoco Lake down through the soil column is given on Figure 4. This process enables a check on current soil properties, 85 years after the start of pumping, future consolidation of the soils, and associated changes in soil stiffness and strength to be estimated. Figure 4 also presents the estimated pore water pressure profile by the end of the structure service life (75 years).

The Plaxis Hardening Soil Small Strain model, abbreviated as HSSmall, is used to model the soil. The model parameters are shown in Table 1 for the different soil units at OCR = 1 at year 1930. The piles are assumed to behave elastically. The creep or secondary settlements is not considered explicitly as it is embedded within the continuous consolidation process.

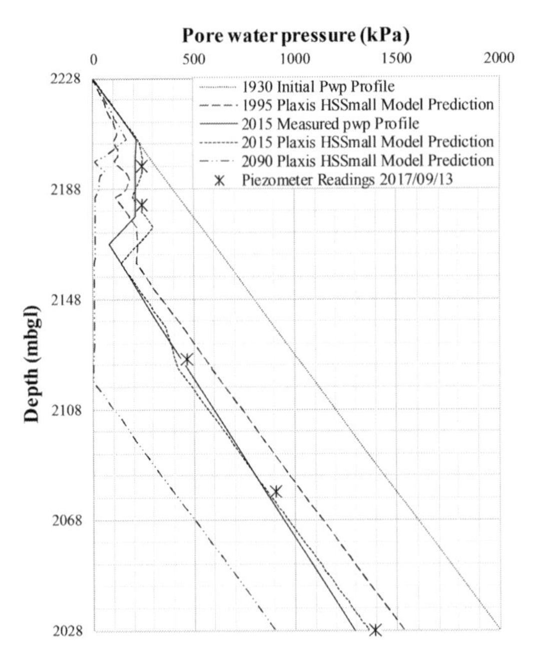

Figure 4. Water pressure profiles evolution for Raft enhanced with friction piles (Texcoco Lake).

3.2 *Raft enhanced with friction piles*

The analyses of a nearly compensated raft foundation at 5 m depth enhanced using driven 12 m and 18 m long square piles embedded within the FAS spaced 4 m are presented. The interaction between the piles and the soil happens in the upper unit FAS. However, the soil up to the aquifer unit has been considered in order to obtain a realistic estimation of the subsidence settlements. The current conditions have been reproduced using the parameters shown in Table 1 and modelling the consolidation process since 1930.

The length of the piles should be designed to ensure that the piles do not hit the hard layer (CD) underlying the FAS during the service life.

The geometry, the upper boundaries and meshing are shown in Figure 5 together with the long-term settlements.

The piles enhancing a raft foundation reduce the bearing pressure on the soil limiting the settlements and the soil degradation. However, if the pile is not yielding or working at its ultimate capacity, there is a risk of a gap formation between the underside of the raft and the ground surface. The consequence of gap formation was not explicitly considered by Rodríguez Rebolledo et al. (2015) for foundations of type b) in Figure 1. Measurements of performance of such a foundation under seismic loading by Mendoza et al. (2000) show how during two events in 1997, a little over 12 months after construction, the measured loads in piles reduced and were progressively transferred to the raft-soil interface. If the soil is strong enough at the raft-soil interface, this produces a 'fail-safe' mechanism. It follows that if a gap forms over time, piles will attract all the seismic load from the superstructure and plunging failure could occur until the underside of the raft re-makes contact with the soil: the mechanism is no longer 'fail-safe'. Figure 5 shows that the risk of large gap formation using 21 m long piles is very high whereas that risk is much reduced if shorter, 18 m long piles are used.

Figure 6 shows the long term load on the 21 m and 18 m long piles. It is readily seen that negative skin friction is developed on both piles, and the neutral plane rises close to the head of the shorter pile. As the piles are not yielding, the foundation will settle at a lower rate than the surrounding soil giving rise to an apparent protrusion. The determination of the differential settlement between the foundation and the neighbouring ground is out of the scope of this paper.

Table 1. HSSmall model properties.

HSS mall model properties										
Soil Unit	Layer Thick.	Unit Weight γ_t	Coeff of permeability /k_v	E_{oed}^{ref}	E_{ur}^{ref}	m	c^{ref}	Friction Angle	γ_{70}	G_0^{ref}
		(kN/m³)	(m/s)	(kN/m²)	(kN/m²)		(kN/m²)	(deg)	(kN/m²)	
1 Crust	1.5	14	1.00E-07	10,000	10,000	-	10	30	-	-
2 FAS	28.5	12	4.00E-09	300	6,000	0.8	0.1	41	1.00E-03	1.25E+04
3 CD	1.5	17	5.90E-05	1,250	6,250	0.8	0.1	38	1.00E-04	2.60E+04
4 FAI	11.5	13.1	2.00E-09	375	3,000	1	0.1	36	1.00E-03	1.25E+04
5 SES	11	14	1.00E-06	1,000	5,000	1	0.1	35	1.00E-03	2.05E+04
6 FAP	13	13.5	4.00E-09	800	5,000	1	0.1	35	1.00E-03	2.00E+04
7 SEI	23	16	1.00E-06	1,850	10,000	1	0.1	35	1.00E-03	4.00E+04
8 FAP 2	15	13.5	4.00E-09	900	5,450	1	0.1	35	1.00E-03	2.25E+04
9 SEI 2	15	16	1.00E-06	2,000	10,000	1	0.1	35	1.00E-03	4.10E+04
All soil units:	$k_v = k_h$ $E_{50}^{ref} = E_{oed}^{ref}$		$\upsilon = 0.2$ $p^{ref} = 100$		OCR (1930) = 1 Dilatancy = 0					

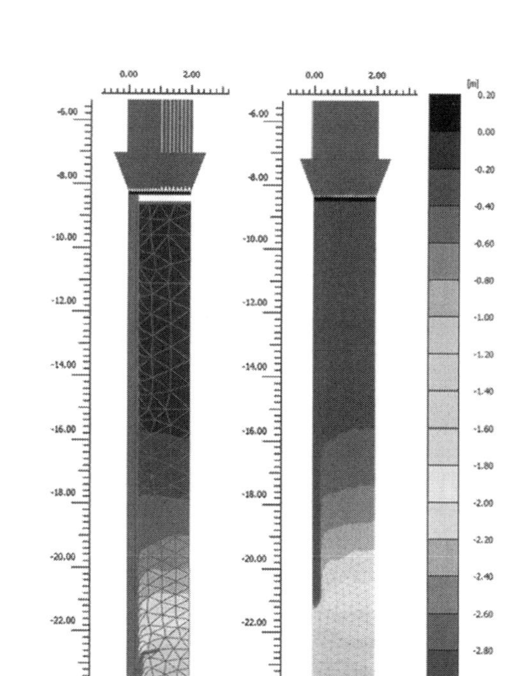

Figure 5. (a) 21 m long pile and (b) 18 m long pile foundation long-term settlements.

3.3 *Raft on long end bearing piles*

The simplified analysis of a raft sitting at 35 m depth on 1.8 m diameter bearing piles founded at 80 m depth spaced 14 m is presented.

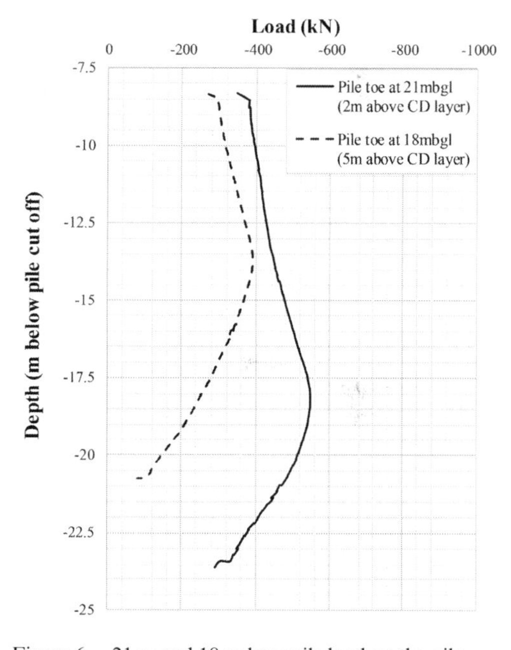

Figure 6. 21 m and 18 m long pile load on the pile.

The current and future pore water pressure profile used in this analysis is presented on Figure 7. These correspond to an area located within downtown Mexico City. In this case, instead of generating the current stress conditions from the pore water evolution, an OCR = 1.8 on the upper FAS has been considered whereas in the lower units OCR has been set to 1.

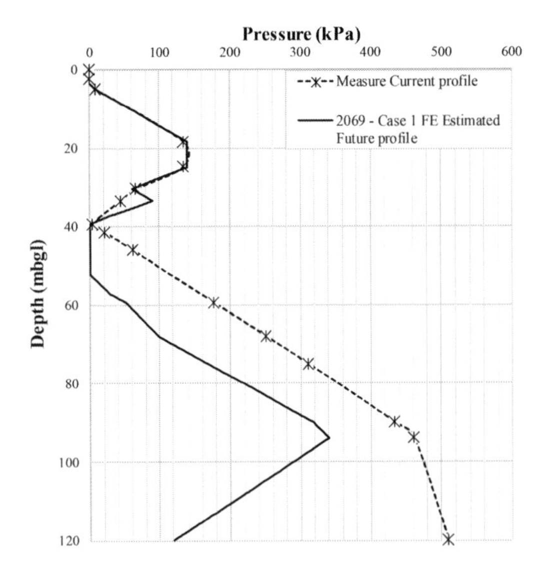

Figure 7. Pore water pressure profile considered for the end bearing analysis (downtown Mexico City).

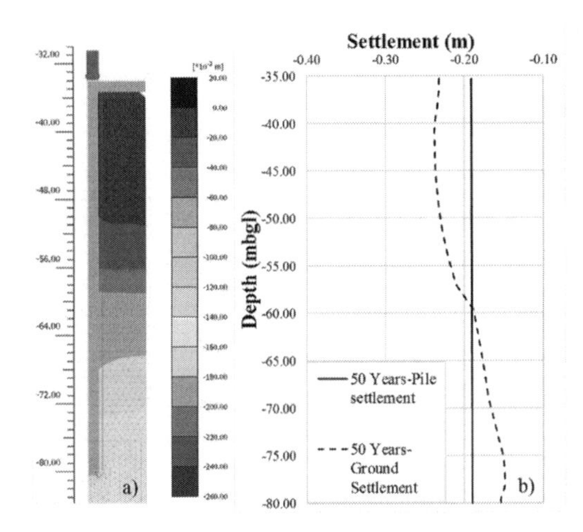

Figure 8. (a) 80 m deep end bearing pile long-term settlement and (b) comparison soil-pile long-term settlement.

Figure 8 present the long term settlements (50 year service life considered). The neutral plane is found to be at about 60 m depth. The piles restrain the movement of the soil under the foundation causing an increasing apparent protrusion as it settles less than the neighbouring soil above 60 m depth. As expected in an end bearing foundation, a gap will form between the soil and the underside of the raft is expected before the end of the building service life (approximately 60 mm gap). No

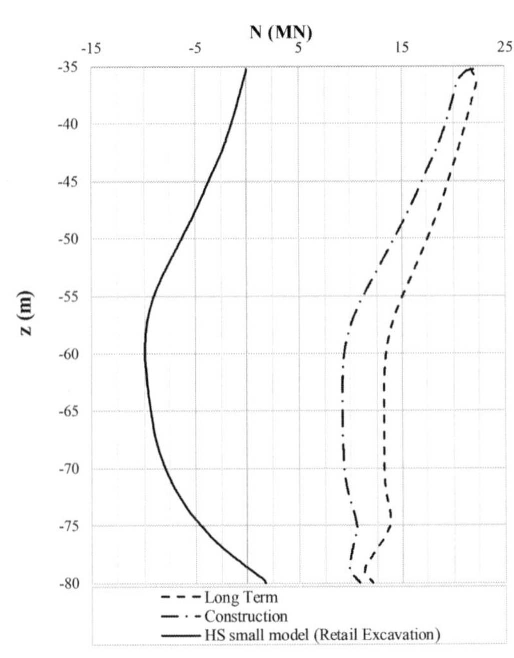

Figure 9. Depth regional subsidence participation percentage. (source: the Authors).

bearing contribution should be considered in the raft design.

The forces acting on the pile are given in Figure 9 after the excavation, after the building construction and in the long term. The large excavation intended (35 m to formation level) will induce tension forces up to 10MN which must be accounted for in the pile structural design. The piles are subjected to negative skin friction on the long term. However, the development of the downdrag force is limited due to the close spacing between piles.

4 CONCLUSIONS

The presented analyses of a raft enhanced with short friction piles and a raft on long end bearing piles have demonstrated that a simplified axisymmetric model could provide an affordable computational cost meaningful preliminary results for the foundation design in consolidating soils, namely:

- Short and long-term foundation settlements;
- Likelihood of gap formation between the ground and the underside of the raft;
- Likelihood of apparent protrusion; and
- Evolution of axial forces on the pile.

The success of the analysis on consolidating grounds depends on the sound knowledge of the past and future pore water pressure conditions. This paper outlines a fundamental approach when back analysing the regional subsidence alongside with the groundwater past conditions to achieve reliable predictions.

ACKNOWLEDGEMENTS

We thank Grupo Aeroportuario de la Ciudad de México (GACM), FR+EE and Foster & Partners for permission to publish some of the results of ground investigations for NAICM.

REFERENCES

Comité Técnico Proyecto Texcoco. 1969. Nabor Carrillo. El Hundimiento de La Ciudad de México y Proyecto Texcoco/ The Subsidence of Mexico City and Texcoco Project. Mexico City: Secretaria de Hacienda y Crédito Público (SHCP).

CONAGUA. 2015 Private Communication.

Díaz Rodríguez, J A 2003. Characterization and Engineering Properties of Mexico City Lacustrine Soils. In Characterization and Engineering Properties of Natural Soils Pp. 725–755.

Fellenius, BH 2017. Basics of Foundation Design (The Red Book).

Gobierno del Distrito Federal 2016. Normas Técnicas Complementarias Para Di-seño y Construcción de Cimentaciones, vol.96–BIS. Gaceta Oficial Del Distrito Federal, 17 de Junio 2016.

Mendoza, MJ, Romo, MP, Orozco, M, and Domínguez, L 2000. Static and Seismic Behavior of a Friction Pile-Box Foundation in Mexico City Clay. Soil and Foundations 40(4): 143–154.

O'Riordan, N, Cañavate-Grimal, A, Francisco Ciruela, F, and Kumar, S 2017. The Stiffness and Strength of Saltwater Lake Texcoco Clays, Mexico City. In. 19th International Conference on Soil Mechanics and Geotechnical Engineering (ICSMGE 2017), Seoul

O'Riordan, N, Kumar, S, Cañavate-Grimal, A, and Francisco Ciruela, F 2018. Estimation and Calibration of Input Parameters for Lake Texcoco Clays, Mexico City. In. 9th European Conference on Numerical Methods in Geotechnical Engineering (NUMGE 2018), Porto

Ovando Shelley, E 2011. Some Geotechnical Properties to Characterize Mexico City Clay. In. Pan-Am CGS Geotechnical Conference, Toronto.

Rodríguez Rebolledo, JF, Auvinet, GY, and Martínez Carvajal, HE 2015. Settlement Analysis of Friction Piles in Con-solidating Soft Soils. Dyna 82(192): 211–220.

Rudolph, DL, I Herrera, and R Yates 1989. Groundwater Flow and Solute Transport in the Industrial Well Fields of the Texcoco Saline Aquifer System near Mexico City. Geofísica Internacional 28(2): 363–408.

Santoyo Villa, E, and Ovando Shelley, E 2008. Mexico City's Cathedral and Sagrario Church. Geometrical Correction and Soil Hardening 1989–2002. Dirección General de Sitios y Monumentos del Patrimonio Cultural (CONACULTA).

Santoyo Villa, E, Ovando Shelley, E, Mooser, F, and León, E 2005. Síntesis Geotécnica de La Cuenca Del Valle de México. Mexico DF: TGC Geotecnia.

Zeevaert, L 1983. Foundation Engineering for Difficult Subsoil Conditions. New York: Van Nostrand Reinhold.

Numerical Methods in Geotechnical Engineering IX – Cardoso et al. (Eds)
© 2018 Taylor & Francis Group, London, ISBN 978-1-138-33203-4

Soil-foundation contact models in finite element analysis of tunnelling-induced building damage

W.N. Yiu, H.J. Burd & C.M. Martin
University of Oxford, UK

ABSTRACT: Finite element analysis provides a potentially useful option for assessing the likely extent of damage to existing buildings, caused by nearby tunnelling activities in an urban environment. The paper describes a 3D finite element model for masonry buildings founded on shallow strip footings, affected by single or twin tunnels in a typical London geological formation. The building is represented in the model as an isolated single facade. The paper focuses particularly on modelling the soil-foundation contact behaviour. The soil-foundation interface is represented by a Coulomb friction model, with a contact formulation allowing gapping. Results indicate that the normal traction redistribution along the length of the footing, the relative horizontal movement between the soil and the footing, and the tensile strain induced in the building facade are sensitive to the assumed value of the friction coefficient. The computed horizontal strains in the footing are typically smaller than those in the adjacent soil.

1 INTRODUCTION

Tunnel construction typically induces local ground movements. In an urban environment, these ground movements may cause damage to nearby buildings, via interactions between the building foundations and the ground.

Typical building damage assessments in the UK are initially conducted using a relatively simple analytical approach in which the building is modelled as an elastic beam that deforms according to the greenfield tunnel-induced ground movements estimated at foundation level (Burland & Wroth 1975). Bending and diagonal strains are computed using beam theory and, notably, horizontal greenfield strains in the soil are combined additively with those developed in the building by beam action to estimate the maximum tensile strain induced in the building (Boscardin & Cording 1989). If this approach indicates a risk of significant damage, more refined calculations are required to carry out additional, more detailed assessments.

Finite element methods provide a numerical means of predicting the response of buildings to nearby tunnelling activities. Robust modelling procedures, incorporating all of the key features of the problem, are required to obtain results that are sufficiently reliable for practical building damage risk assessment purposes. Building foundation details, however, are often excluded in the development of finite element models for damage assessment purposes. One reason for this could be uncertainties in the form, materials and condition of the foundation, particularly for historical buildings.

Previous 3D finite element studies (Yiu et al. 2017b) demonstrate that the geometry and properties of a strip footing have a significant effect on the predicted building damage for a simplified masonry building; this suggests that appropriate models of foundation behaviour should be incorporated in finite element modelling of tunnel-induced damage. A key difficulty, however, is that the foundation details are often difficult and expensive to obtain by site survey. Although it may be possible to deduce the foundation geometry and material from architectural history documents, engineering judgement is inevitably needed to determine the appropriate foundation details, and to assess the likely condition of the foundation, for finite element modelling purposes.

The current paper explores the role of soil-foundation interaction modelling in the context of 3D finite element analysis of tunnel-induced building damage. The results of the study provide an indication of the extent to which variations in the assumed contact model affect the computed strains in the building. Studies of this sort provide a means of understanding the likely influence of uncertainties in the soil-foundation contact behaviour in finite element models being developed for practical damage assessment purposes.

The models presented in the paper are applied to masonry buildings founded on shallow continuous strip footings, subjected to local ground movements caused by shallow tunnelling. The modelling procedures for the 3D finite element analyses are described in Yiu et al. (2017a), but these calculations are extended in the current paper to study the influence of soil-foundation contact behaviour on the overall performance of the building. A simplified building modelling approach—referred to as '3D isolated facade analysis' – is adopted in which each building facade is assumed to be independent of the rest of the building, and only the front facade is modelled. This approach is consistent with models typically employed in simplified assessments of building damage (e.g. Burland & Wroth (1975)).

By conducting a sensitivity study on the interface friction model, the paper aims (i) to develop an improved understanding of the role of the foundation in transmitting displacements and strains from the soil to the building, and (ii) to evaluate the importance of incorporating a realistic representation of the soil-foundation interaction behaviour in finite element models for building damage assessment studies.

2 EXAMPLE TUNNEL-SOIL-BUILDING SCENARIO

The configuration adopted in this study consists of a masonry building founded on a typical London geological profile, as shown in Figure 1.

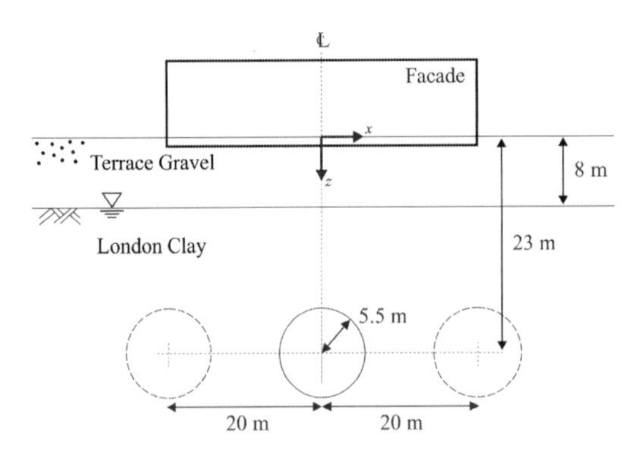

Figure 1. Basic model specification and tunnel configuration (solid line for single tunnel; dashed line for twin tunnels).

The dimensions of the building and the layout of the openings in the front facade are based on previous work (Burd et al. 2000). Two building heights are considered: a two-storey variant (H = 8 m), as shown in Figure 2, and a similar three-storey variant (H = 12 m). The thickness of the masonry wall is assumed to be 0.215 m, which corresponds to the thickness of two standard UK bricks. The building is assumed to be founded on a continuous strip footing of width 1 m and thickness 0.5 m with an embedded depth of 1 m below the ground surface (Fig. 3).

The assumed soil conditions consist of a layer of Terrace Gravel overlying London Clay. This is a simplified version of typical central London conditions. The groundwater table is assumed to be located at the surface of the clay, and undrained conditions are assumed for the clay.

The tunnel geometry and dimensions are based on the London Crossrail platform tunnels, with a diameter of 11 m and depth of 23 m (Skinner et al. 2014). The tunnels are assumed to be orthogonal to the building with zero eccentricity and two configurations are considered: a 'single tunnel' case where one tunnel passes under the mid-span of the building, and a 'twin tunnels' case where parallel tunnels pass underneath the two ends of the building (Fig. 1). A volume loss of 15% is assumed for each tunnel, corresponding to the moderately conservative level (Dulake 2011) adopted in the Crossrail project.

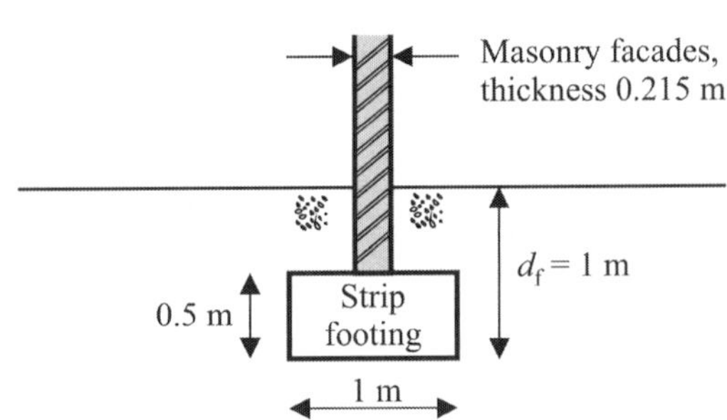

Figure 3. Geometrical details of facade and foundation.

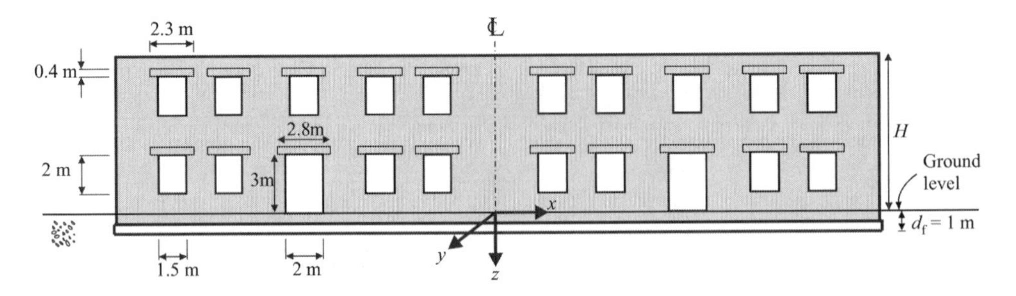

Figure 2. Details of facade openings and lintels, also showing the coordinate axes.

3 MODELLING PROCEDURES

Rather than modelling the whole building, a simplified approach 3D isolated facade analysis (as described above) is adopted for the current parametric study. In addition to the merits of simpler mesh generation and shorter computation time, the 3D isolated facade approach has been demonstrated in previous work (Yiu et al. 2017a) to be more conservative for building damage assessment than modelling a complete building. Although the facade is modelled as a planar structure, 3D modelling is required to capture the interaction between the foundation and the ground. The strip footing is explicitly represented in the 3D analysis to model the transmission of displacement and strain from the soil to the facade, and to allow definition of a contact property model (see Section 3.2).

The mesh generation and numerical calculations were performed using the finite element analysis program Abaqus v2016 (Dassault Systèmes Simulia Corp., Providence, RI, USA). Brief descriptions of the finite element modelling procedures are provided below, but emphasis is placed on the soil-foundation-facade interaction. Further details of the general modelling procedures adopted in the study, and the values adopted for the parameters of the various material models, are given in Yiu et al. (2017a).

3.1 *Finite element meshes*

The meshes for the single and twin tunnel cases are shown in Figure 4. The depth of the soil mesh

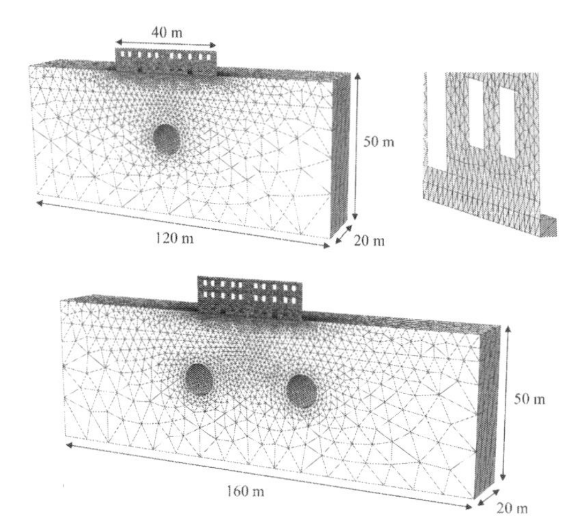

Figure 4. Example meshes employed for the isolated facade analyses. Top left: building height $H = 8$ m, single tunnel; top right: detail showing facade and foundation; bottom: building height $H = 12$ m, twin tunnels (elements internal to the tunnels have been removed for clarity).

follows Yiu et al. (2017a). Symmetry about the mid-plane of the facade is exploited. The feasibility of exploiting the symmetry about the facade centreline is realised for the current calculations; this symmetry is not utilised, however, since the current meshes are part of a series of analyses of eccentric tunnels in previous work (Yiu et al. 2017a). The nodal degrees of freedom on the base boundary are fixed in all directions and all vertical boundaries have roller supports.

Six-noded shell elements (Abaqus element type STRI65) with five Gauss points through the thickness are used to model the facade and lintels. The foundation and the soil are modelled using ten-noded tetrahedral elements (Abaqus element type C3D10).

3.2 *Soil-foundation interface*

The choice of soil-footing contact model is of particular interest in tunnelling-induced building damage assessment calculations, since the contact model has a key role in governing the relative movement between the foundation and the ground. For footings founded in granular soil, gapping and sliding beneath the footing have been observed in centrifuge model tests Farrell et al. 2014).

In the current modelling the interface between the gravel and the masonry footing is assumed to be frictional. The surface-to-surface finite-sliding formulation between two deformable bodies available in Abaqus is adopted. By pairing potential contact surfaces and defining contact conditions, appropriate contact elements are generated automatically by Abaqus. The computed normal and shear tractions can be extracted as nodal contact variables associated with the integration points.

Zero tensile strength and zero penetration in the normal direction (also known as the 'hard' pressure-overclosure relationship in Abaqus) are adopted to allow for potential gapping. The classical isotropic Coulomb friction model, available as a built-in model in Abaqus, is employed to represent the shear behaviour with an assumed friction coefficient μ. Sliding occurs when the shear traction exceeds the shear strength of the interface, which is given by the product of the friction coefficient and the local contact pressure.

To study the sensitivity of the computed building damage to the soil-footing interaction model, a range of interface behaviours from completely smooth to fully rough, represented by four friction coefficient values $\mu = [0, 0.3, 0.6, \infty]$, are adopted. The interface model with various friction coefficients is defined using Abaqus Keywords *SUR-

FACE INTERACTION and *FRICTION, and the contact model is applied on the interface using *CONTACT PAIR. The value $\mu=\infty$ is defined using *FRICTION, ROUGH without a numerical value.

3.3 *Foundation-facade interface*

In a real building of the type being modelled, each masonry wall is supported directly on a strip footing (Fig. 3). However, since the facade is modelled with shell elements and the footing uses continuum elements, an unrealistic concentrated load would be induced in the elements representing the footing at the point of contact if the facade shell elements were connected to a line of nodes along the top of the footing continuum elements. To avoid unrealistic stress concentrations, in the current analysis the facade is connected to the footing by adopting the embedded element technique, available as a built-in feature in Abaqus. In this approach, the facade mesh is extended to the base of the footing and the additional facade shell elements are attached to the host footing continuum elements using the Abaqus keyword *EMBEDDED ELEMENT. This ties the translational degrees of freedom of the embedded elements to those of the host elements. A diagram illustrating the finite element representation around the foundation is shown in Figure 5, to be compared with the geometric details in Figure 3. The meshing of the facade and the foundation is shown in Figure 4 (top right).

3.4 *Material constitutive models*

The material constitutive models employed in the analysis are summarised in Table 1. These constitutive models, and the chosen parameter values, are the same as those employed in Yiu et al. (2017a). The model adopted for the Terrace Gravel is an extended Mohr-Coulomb model proposed by Doherty and Muir Wood (2013); it was originally devised specifically to model foundation problems and is therefore regarded as suitable for the current work. The undrained behaviour of the London Clay is represented by a total stress multiple yield surface model developed by Houlsby (1999). This model incorporates the small strain non-linear effects that are known to be important in the analysis of ground movements around tunnels. The constitutive models for the Terrace Gravel and London Clay have both been implemented as Abaqus UMATs. Stress update calculations for the Doherty and Muir Wood (2013) model are conducted using the automatic sub-stepping procedures described in Sloan et al. (2001). Stress update calculations for the Houlsby (1999) model are conducted via a closed form expression. The assumed values of the lateral earth pressure coefficient K_0 in the Terrace Gravel and London Clay are 0.425 and 1.0 respectively.

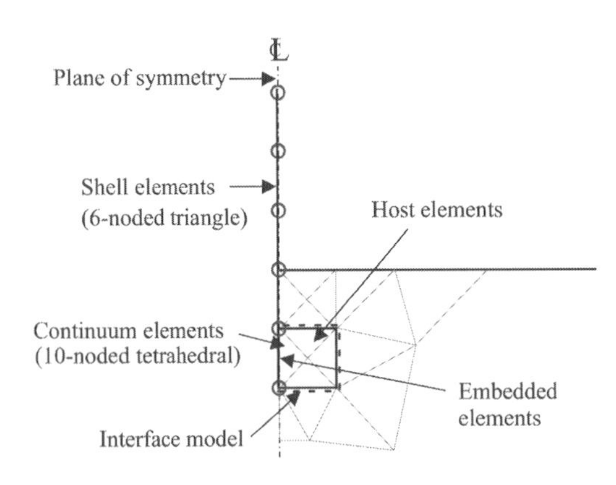

Figure 5. Finite element representation of the facade, the foundation and the soil-foundation-facade interfaces.

Table 1. Material constitutive models employed in the analysis.

Material	Constitutive model	Original reference
Masonry (facade & footing)	concrete damaged plasticity	Lee and Fenves (1998)
Lintel	linear elastic	–
Terrace Gravel	extended Mohr-Coulomb	Doherty and Muir Wood (2013)
London Clay	multiple yield surface kinematic hardening	Houlsby (1999)

3.5 *Calculation sequence*

Geostatic stresses are first initialised with the prescribed values of K_0. Then, the building stiffness and weight are added to the model. Finally, construction of the tunnel(s) is modelled in a single step (rather than as a staged excavation procedure).

The tunnel excavation process is modelled using the applied displacement boundary condition method (Amorosi et al. 2014). This method can achieve a predefined volume loss directly, and the resulting greenfield settlement profile is typically found to be close to the expected Gaussian curve.

4 RESULTS

In the results presented below, all displacement and strain data are the net values induced by the tunnelling process. All the nodal data at the interface are extracted along the centreline of the base of the footing ($y = 0$, $z = 1$ m). Note that the plots are not perfectly symmetric about $x = 0$ owing to the asymmetric meshes. The sign convention for strain is tension positive.

4.1 *Tractions at the soil-footing interface*

The normal tractions (compression positive) on the footing base for the facade of height $H = 8$ m before and after tunnelling for smooth ($\mu = 0$) and rough ($\mu = \infty$) interfaces are plotted in Figures 6 and 7 for the single and twin tunnel cases respectively. Construction of the tunnel causes the normal tractions at the base of the footing to redistribute, towards the edges of the facade in the single tunnel case, and towards the centre of the facade in the twin tunnel case. As the friction coefficient increases, the extent of normal traction redistribution becomes greater. Figure 8 shows the shear tractions on the footing base corresponding to the same cases that are plotted in Figures 6 and 7. As expected, for a smooth interface ($\mu = 0$), the shear tractions at the soil-footing interface are zero. For a rough interface ($\mu = \infty$) and also for intermediate values of μ (not shown in the figure), the shear tractions typically act towards the tunnel location.

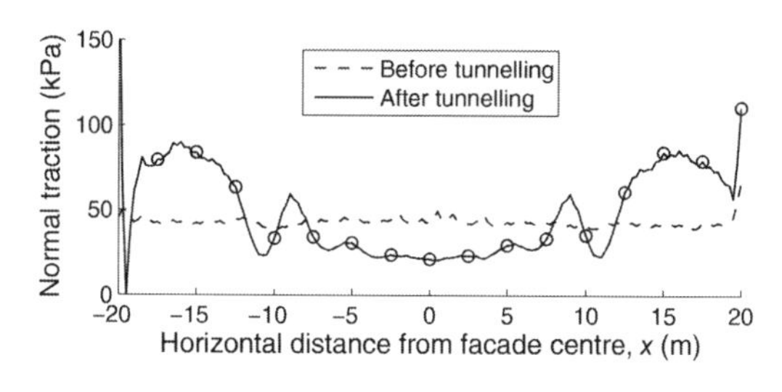

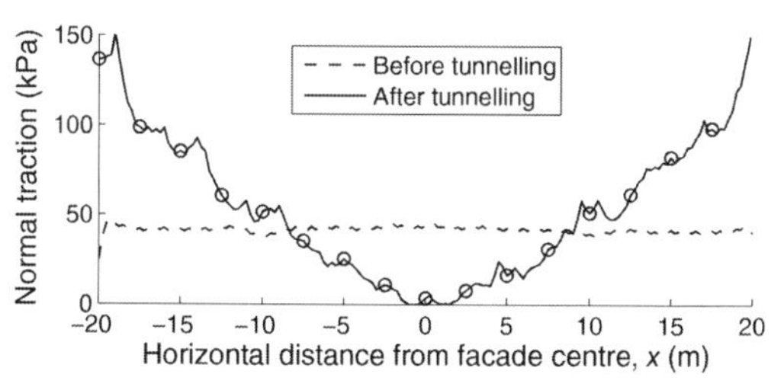

Figure 6. Normal traction on foundation base for facade of height $H = 8$ m, before and after tunnelling for single tunnel. Top: $\mu = 0$, bottom: $\mu = \infty$.

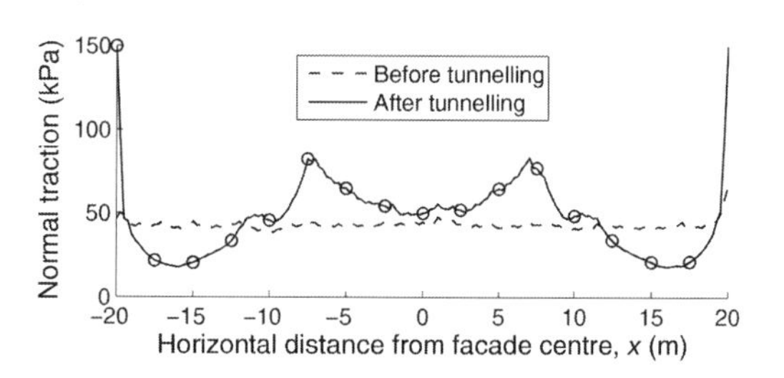

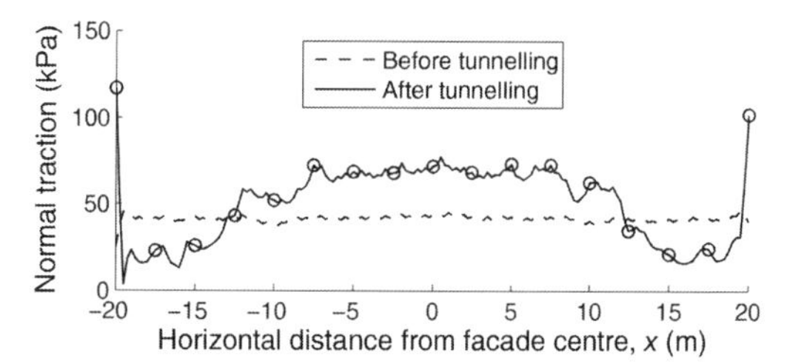

Figure 7. Normal traction on foundation base for facade of height $H = 8$ m, before and after tunnelling for twin tunnels. Top: $\mu = 0$, bottom: $\mu = \infty$.

4.2 *Displacement profiles*

Figure 9 shows the vertical displacement profiles of the base of the footing, and of the adjacent soil, for the facade with $H = 8$ m. In these cases the masonry facade deforms in a relatively flexible manner and gapping does not occur beneath the footing. It is noted that as an expected consequence of the weight and stiffness of the building, the ground displacement differs from the greenfield case (in which the building is absent from the model). For the single tunnel case, as the friction coefficient increases, the displacement profile flattens, causing the differential settlement to reduce. This effect is a combined consequence of the normal traction redistribution and the beneficial inward-acting shear tractions that develop at the base of the footing as a result of the tunnel-induced ground movements (Section 4.1). This effect is less noticeable in the twin tunnel case. The results for the facade with $H = 12$ m exhibit similar behaviour.

The horizontal displacements of the footing and the adjacent ground at the soil-footing interface are plotted in Figures 10 (single tunnel) and 11 (twin tunnels). As the problem is symmetric about $x = 0$, only the profiles corresponding to the right half of the span are shown. For both tunnel configurations, the friction coefficient has a significant effect on the horizontal displacement profile. For greenfield condition, when the building is absent, the soil moves towards the tunnel centre(s) in both cases. For a smooth interface ($\mu = 0$), for which the shear tractions at the footing-soil interface are

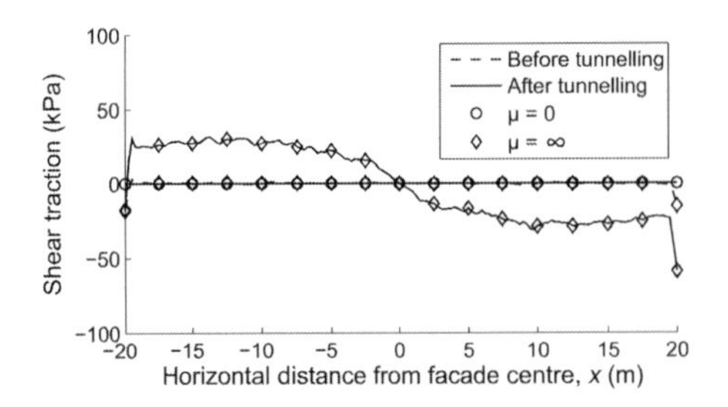

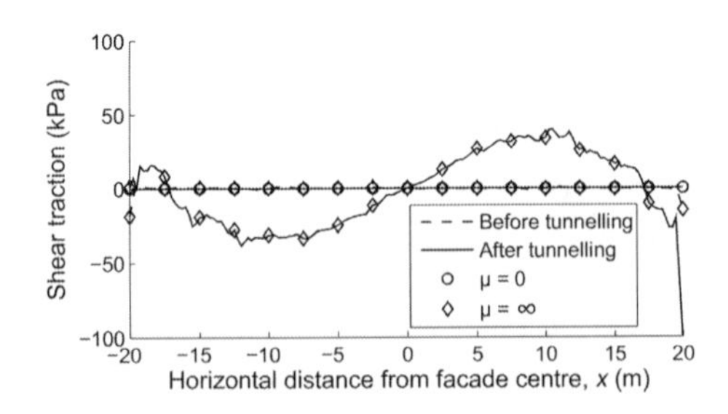

Figure 8. Shear traction on foundation base for facade of height $H = 8$ m, before and after tunnelling, for $\mu = [0, \infty]$. Left: single tunnel, right: twin tunnels.

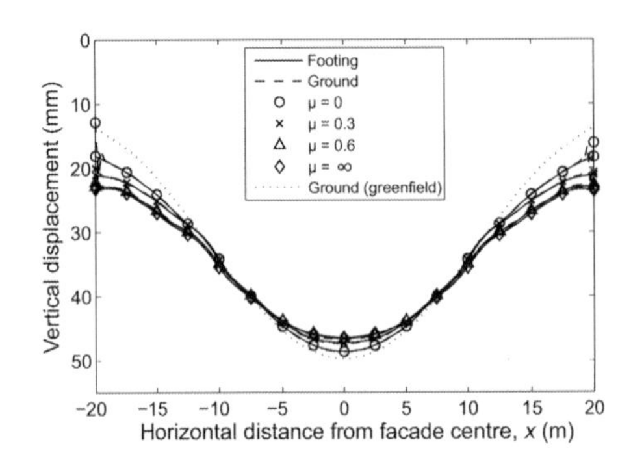

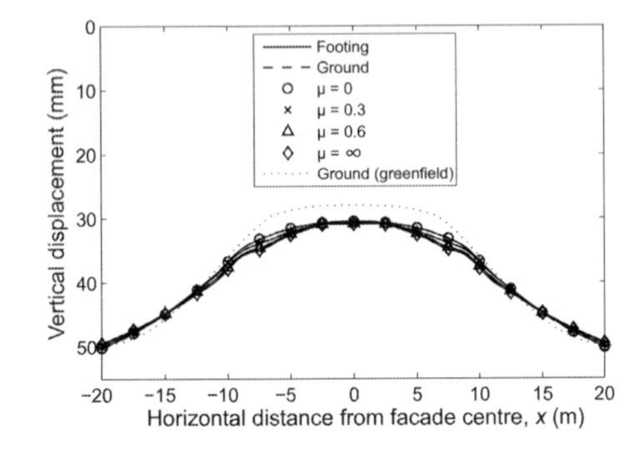

Figure 9. Vertical displacements at foundation base induced by tunnelling for facade of height $H = 8$ m with different friction coefficients, compared with greenfield vertical displacements. Left: single tunnel, right: twin tunnels.

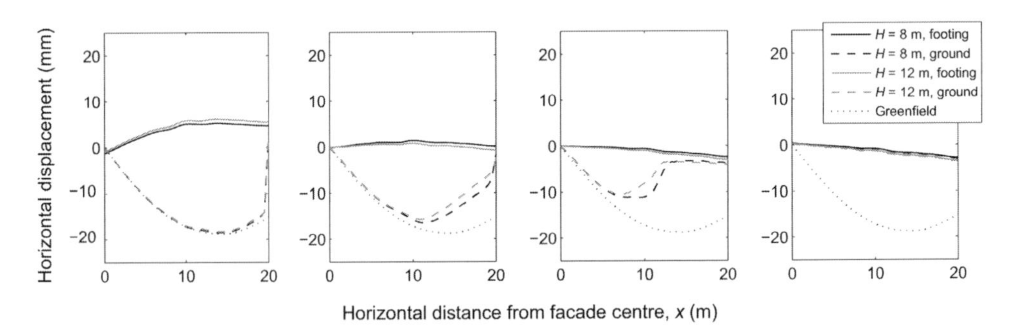

Figure 10. Horizontal displacements at foundation base level induced by tunnelling for single tunnel. From left to right: $\mu = [0, 0.3, 0.6, \infty]$.

zero (Fig. 8), the soil has a horizontal displacement profile that is close to the greenfield case. The horizontal displacements of the base of the footing are smaller in magnitude and opposite in sign to those in the adjacent soil due to the beam action of the building. Significant sliding therefore occurs at the soil-footing interface in both tunnel cases.

For larger values of the friction coefficient, as expected, the relative horizontal movement between the footing and the ground reduces. At $\mu = 06$, for example, in the single tunnel case, sliding displacements at the soil-footing interface near the ends of the facade ($x \approx 15$–20 m) become relatively small. Similar behaviour is observed at the mid-span ($x = 0$) of the facade for the twin tunnel case. Frictional effects appear to be more significant in the hogging regions than in regions where the building deforms in sagging. Moreover, as the friction coefficient increases, the footing gradually moves in the same direction as the soil. In these cases, the base of the footing is compressed under sagging and extended under hogging—which is different from the conventionally assumed beam behaviour. For a fully rough interface ($\mu = \infty$), the two surfaces essentially stick together and the horizontal displacements are substantially smaller in magnitude than in the greenfield case.

4.3 *Horizontal strains at the soil-footing interface*

The horizontal strains, shown in Figures 12 (single tunnel) and 6 (twin tunnels), are extracted directly from Abaqus at the nodes of the relevant elements. As the building is more rigid than the nearby soil, the strains in the footing are relatively small compared with those in the ground, consistent with the horizontal displacement profiles (Figs 12 and 13). For a smooth interface ($\mu = 0$), the strains at the base of the sagging section of the building are tensile, and in the hogging section they are compressive. As a consequence of the coupling between the soil and the foundation (when $\mu > 0$), the resulting horizontal movements are significantly modified. As the friction coefficient increases, the horizontal strains in the footing gradually become more compressive in the sagging section of the building and more tensile in the hogging section.

4.4 *Tensile strains in the facade*

For tunnelling-induced damage assessment, building damage is typically evaluated in terms of the tensile strain (ϵ^t) developed in the facade (e.g. Burland and Wroth (1975) and Boscardin and Cording (1989)). A convenient approach for extracting the tunnel-induced strains in the facade is to attach a layer of 'ghost' membrane elements with very small stiffness to the surface of the building facade prior to the simulation step of the tunnel construction process. The computed strains within the ghost elements then provide a direct indication of the tunnel-induced strains.

For practical assessment purposes, the use of a single metric to indicate the likely degree of facade damage provides a convenient approach. The use of the computed maximum tensile strain based on the results of finite element analysis presents a difficulty, however, since relatively large strains

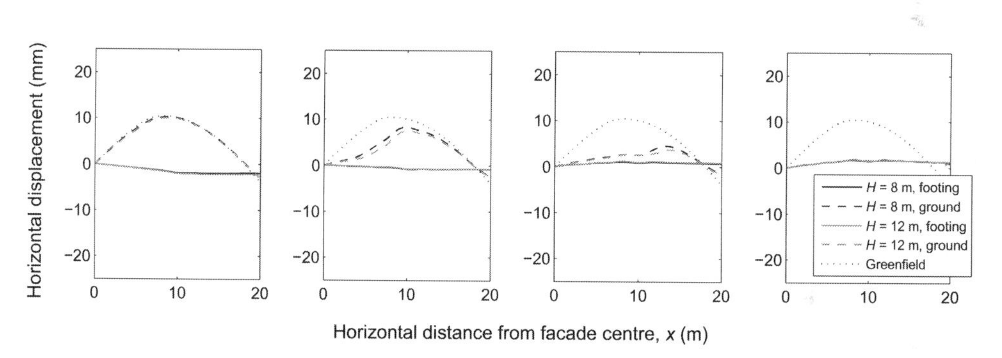

Figure 11. Horizontal displacements at foundation base level induced by tunnelling for twin tunnels. From left to right: $\mu = [0, 0.3, 0.6, \infty]$.

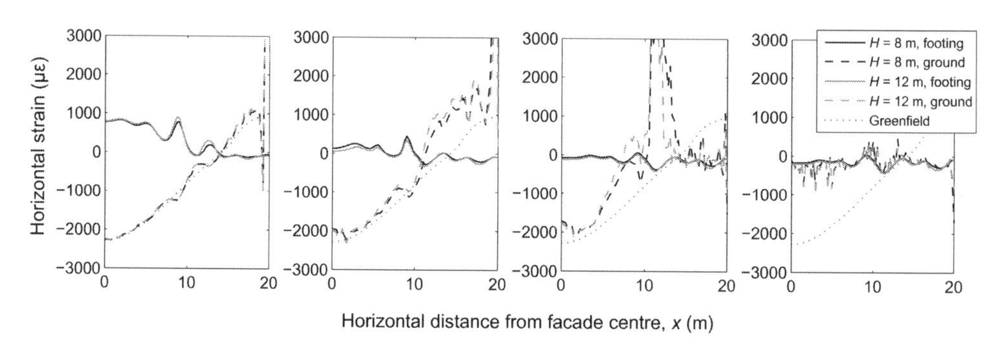

Figure 12. Horizontal strain at foundation base level induced by tunnelling for single tunnel. From left to right: $\mu = [0, 0.3, 0.6, \infty]$.

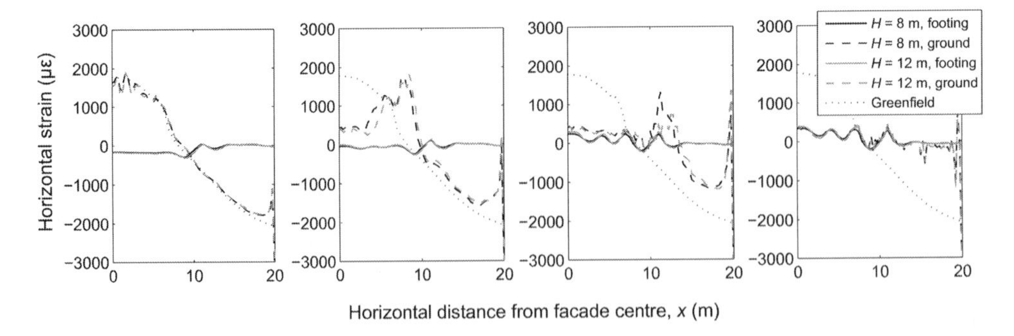

Figure 13. Horizontal strain at foundation base level induced by tunnelling for twin tunnels. From left to right: $\mu = [0, 0.3, 0.6, \infty]$.

develop close to the corners of any openings in the facade. The magnitude of the strains becomes increasingly large as the distance to the corner is reduced; as a consequence, the magnitude of the computed maximum tensile strain developed in the building depends on the density of the elements around the openings. To obtain an assessment approach that is relatively insensitive to mesh density, a procedure has been developed (Yiu et al. 2017a) in which the characteristic strain at the 99th percentile (ϵ_{99}^t) is adopted as the damage metric. This is defined as the value of the major principal strain that is not exceeded in 99 percent of the area of the masonry in the facade.

Figure 14 shows contour plots of the major (i.e. most tensile) strains induced in the 8 m high facade due to tunnelling, for the twin tunnel case, obtained from the ghost elements. Strain concentrations around the openings and the lintels are clearly visible, and it is apparent that the value of the soil-footing friction coefficient has a considerable influence on the facade strain field. When $\mu = 0$, high strain concentrates at the top of the facade and, regardless of the shear bands, there is an approximately uniform gradient of strain variation over the facade height. This roughly corresponds to beam action with the neutral axis located at the bottom of the beam. However, as the friction coefficient increases, the facade becomes increasingly coupled to the ground. In this case, relatively high strains develop in the bottom part of the facade and the neutral axis appears to shift to a lower position.

The significance of the value of friction coefficient employed in the analysis is illustrated in terms of the characteristic strains ϵ_{99}^t in Figure 15. Results obtained for the building with $H = 8$ m are compared with the tallervariant with $H = 12$ m. The facade with $H = 8$ m experiences values of characteristic strain that increase gradually with increasing friction coefficient, for both

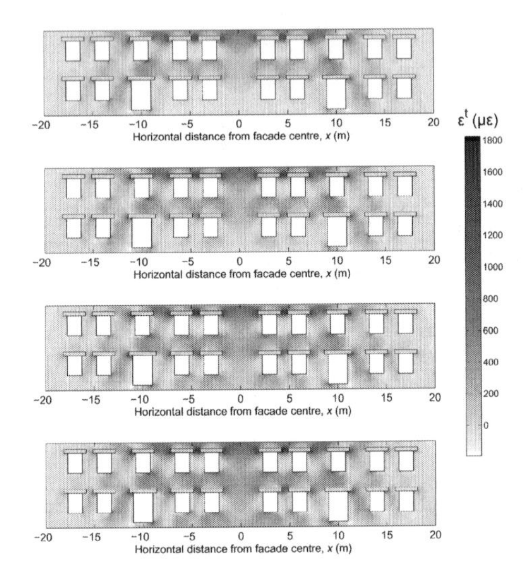

Figure 14. Major principal strains ϵ^t in facades of height $H = 8$ m induced by twin tunnels with different friction coefficients. From top to bottom: $\mu = [0, 0.3, 0.6, \infty]$.

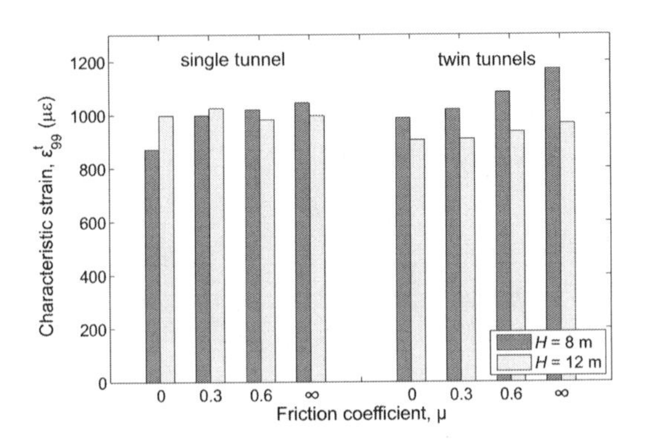

Figure 15. Characteristic strains ϵ_{99}^t induced in facades of height $H = 8$, 12 m with different friction coefficients for single tunnel and twin tunnels.

tunnel configurations. Conversely, the characteristic strain for the the taller building (H = 12 m) is relatively insensitive to variations in the friction coefficient.

It is noted that the characteristic strains generally increase as the friction coefficient increases, although the vertical differential displacements and the horizontal strains tend to decrease. This might indicate that higher tensile strains have developed around the facade openings (but not at the corners where the mesh-dependent values are discounted using characteristic strain).

5 CONCLUSIONS

Since the details of an existing foundation are often difficult to assess in a practical situation, it is important to understand the effect of various modelling assumptions to be used in numerical damage assessment calculations. The results of the 3D finite element calculations presented in this paper indicate that, for the scenarios considered, the building deformation caused by nearby tunnelling is influenced by the soil-foundation contact model in the following ways:

i. The differential settlements decrease with increasing friction coefficient in the single tunnel case, but are relatively unaffected in the twin tunnel case.
ii. Horizontal sliding along the base of the footing becomes increasingly constrained as the friction coefficient increases. The horizontal strains in the base of the footing are significantly affected, in both magnitude and direction, by the friction coefficient. The computed strains in the footing are much less than those in the soil corresponding to the greenfield case.
iii. The characteristic tensile strains in the facade are relatively unaffected by the value of the friction coefficient. Adopting a fully rough interface provides a conservative and convenient estimate of tensile strains for the purpose of tunnelling-induced building damage assessment.

REFERENCES

Amorosi, A., D. Boldini, G. de Felice, M. Malena, & M. Sebastianelli (2014). Tunnelling-induced deformation and damage on historical masonry structures. *Géotechnique 64*(2), 118–130.

Boscardin, M.D. & E.J. Cording (1989). Building response to excavation-induced settlement. *Journal of Geotechnical Engineering 115*(1), 1–21.

Burd, H.J., G.T. Houlsby, C.E. Augarde, & G. Liu (2000). Modelling tunnelling-induced settlement of masonry buildings. *Proc. Instn Civ. Engrs Geotech. Engng 143*(1), 17–29.

Burland, J.B. & C.P. Wroth (1975). Settlement of buildings and associated damage. Technical report, Building Research Establishment, Department of the Environment.

Doherty, J.P. & D. Muir Wood (2013). An extended Mohr-Coulomb (EMC) model for predicting the settlement of shallow foundations on sand. *Ge´otechnique 63*(8), 661–673. Dulake, C. (2011). Crossrail design and construction / Cross-rail Entwurf, Bemessung und Ausfu¨hrung. *Geomechanics and Tunnelling 4*(5), 592–604.

Farrell, R.P., R.J. Mair, A. Sciotti, & A. Pigorini (2014). Building response to tunnelling. *Soils and Foundations 54*(3), 269–279.

Houlsby, G.T. (1999). A model for the variable stiffness of undrained clay. *Proc. Int. Symp. on Pre-Failure Deformation Characteristics of Soils, Turin 1*, 443–450.

Lee, J. & G. Fenves (1998). Plastic-damage model for cyclic loading of concrete structures. *Journal of Engineering Mechanics 124*(8), 892–900.

Skinner, H., V. Potts, A. St. John, & D. McGirr (2014). Stabilisation of soil pillar adjacent to Eastern Ticket Hall at Bond Street Crossrail Station. In M. Black, C. Dodge, and U. Lawrence (Eds.), *Crossrail project: infrastructure design and construction*, pp. 419–431. ICE Publishing.

Sloan, S.W., A.J. Abbo, & D. Sheng (2001). Refined explicit integration of elastoplastic models with automatic error control. *Engineering Computations 18*(1/2), 121–154.

Wongsaroj, J., K. Soga, & R.J. Mair (2007). Modelling of longterm ground response to tunnelling under St James's Park, London. *Géotechnique 57*(1), 75–90.

Yiu, W.N., H.J. Burd, & C.M. Martin (2017a). Finite element modelling for the assessment of tunnel-induced damage to a masonry building. *Ge´otechnique 67*(9), 780–794.

Yiu, W.N., H.J. Burd, & C.M. Martin (2017b). Soil-building interaction in finite element analysis of tunnelling-induced building damage. In *Proc. IV International Conference on Computational Methods in Tunneling and Subsurface Engineering*, pp. 381–388. ECCOMAS.

Numerical Methods in Geotechnical Engineering IX – Cardoso et al. (Eds)
© 2018 Taylor & Francis Group, London, *ISBN 978-1-138-33203-4*

Finite element modelling of extent of failure zone in *c*-ϕ soil at the cutting edge of open caisson

J.T. Chavda & G.R. Dodagoudar
Indian Institute of Technology Madras, Chennai, India

ABSTRACT: Open caissons are sunk into the ground by removal of soil within the caisson shaft. During sinking of caisson, the stresses in the soil at the cutting edge increase and result in the bearing failure of soil. The extent of soil failure in the excavation side of the open caisson is termed as influence zone. In this paper, the finite element analysis is carried out to study the effect of geometric configuration (radius ratio and tapered angle of the cutting edge), strength parameters (c' and ϕ'), unit weight of soil and surcharge on the extent of the influence zone. The caisson considered in the study is having a radius ratio of 0.8, steinning thickness of 1 m and cutting edge with a tapered angle of 45°. The reaction offered by the soil at the cutting edge is also evaluated. The identification of the extent of failure zone at the cutting edge of the caisson helps in devising proper excavation strategy in the field for the controlled sinking of the caisson.

1 INTRODUCTION

The open caisson is a reinforced concrete structure, a form of top-down construction in which the concrete steinning along with well curb is sunk into the ground by removal of the soil within the caisson shaft. The open caissons are like pile foundations which provide solid and massive foundations for transferring the heavy loads of the structures to a soil stratum deep below. In India, open caissons are known as Well Foundations. The famous historical monument, Taj Mahal, Agra is supported on well foundation. The Mehtab Burj in the Mehtab garden near Taj Mahal, St. Andrew's Church in Chennai (Kurian 2005), Larsen and Toubro headquarter in Chennai and many other structures are supported on well foundations. The open caisson is classified as a deep foundation to support bridge piers, bridge abutments, heavy structures, coastal protection walls, jetties, ports, onshore structures, launch and reception pits for the tunnel boring machines and isolated structures like chimneys, transmission line towers, wind turbines, etc. The caissons can be sunk through various types of soils like sandy soils, non-stiff clays, boulder soils and to a limited extent in the rocky strata. The sinking of caisson is an important aspect in successful completion of a project. A proper method of excavation of soil at the curb level of the caisson is required for the controlled vertical sinking of the caisson.

The practicing engineers have faced many problems in the execution of sinking of open caisson. They are tilting, shifting, sudden collapse and freezing of caisson i.e. no further sinking. Many remedial measures have been taken during the execution of sinking of caissons (Salamov et al. 1965, Ter-Galustov et al. 1966, Belous 1968, Opershtein 1969, Bolyachevskii and Chumakov 1975, Baitsur and Zakharov 1975, Maslik et al. 1978, Saha 2007, Panda 2015). In order to avoid such problems, the detailed site investigation, engineering judgment and informed decisions are necessary. Chandler et al. (1984) presented the construction technology of caisson foundation for the successful completion of Jamuna river crossing project in Bangladesh. Using the case study example, they narrated the occurrence of liquefaction due to the removal of water within the caisson. The removal of water has caused an increase in excess pore water pressure leading to the liquefaction at the base of Caisson No. 11. Nonveiller (1987) describes various potential rupture surfaces, construction technique and the design of 60 m deep open caisson of outer diameter of 30 m for the pumped-storage hydropower plant. The author has reported the application of lubrication system using bentonite slurry to avoid the side friction thereby allowing the sinking of caisson effectively. The author has also reported two case studies of unsuccessful open caissons. The failure is attributed to lack of detailed soil investigation and planning.

Katti and Dewaikar (1977) studied the contact pressure distribution at the base and sides of the modelled circular well. The model was instrumented with pressure cells which were embedded in clayey soil subjected to vertical and radial loading. Alampalli and Peddibotla (1997) conducted experiments to study the settlement and deflection behaviour of open-ended caissons in sandy soil. Kumar and Rao (2010) presented the experimental results of earth pressure mobilization in caissons embedded in soft clay. Based on their studies, a contiguous wall made of embedded caissons in soft clay is proposed to serve as a coastal protection wall.

IS: 9527, Part 1 (1981) and IS: 3955 (1965) recommend the inclination of the inner face of the cutting edge with respect to the vertical between 30° and 45° depending on the type of strata. In the case of stiff clays, an angle of 30° and in the case of sands or soft clays, an angle of 45° is recommended. Tomlinson (2001) recommends various tapered angles depending on the soil types. The flatter cutting angles are recommended for sand compared to clay. Some studies have been carried out to evaluate the bearing capacity factor N_γ using limit equilibrium theory for tapered cutting edges (Solov'ev 2008, Yan et al. 2011).

From the above review, it is noted that the construction of open caissons has faced with many difficulties. Also, a few studies were carried out on the settlement and deflection behaviour of the caisson, the failure mechanism beneath the cutting edge and the bearing capacity of caisson with tapered cutting edge. The bearing failure at the cutting edge is important to permit the controlled sinking of the open caisson. Recently, Royston et al. (2016) conducted the experiments to identify the influence zone at the cutting edges in the sand under plane strain condition. Using Particle Image Velocimetry technique, soil displacement contours are plotted for varying cutting edge angles of 30°, 45°, 60°, 75° and 90°. Generally, the circular open caissons are considered as axisymmetric problems.

The influence zone in the soil at the cutting edge of the open caisson is identified in the study. The influence zone depends on the radius ratio of open caisson, angle of cutting edge, steinning thickness, soil strength parameters (c' and ϕ'), unit weight of soil and surcharge at the top level of the cutting edge. The radius ratio of the open caisson is the ratio of the inner radius of caisson to its outer radius. The finite element (FE) analysis is carried out to investigate the influence zone in the c-ϕ soil using different values for the above mentioned parameters. The reaction offered by the soil at the cutting edge of the open caisson is also evaluated for different tapered angles of the cutting edge.

2 FE SIMULATION OF FAILURE MECHANISM AT CUTTING EDGE OF THE OPEN CAISSON

2.1 *Failure model, interface, surcharge and influence zone*

The FE analysis is carried out using PLAXIS 2D program with axisymmetric formulation. The cutting edge is modelled as a soil polygon with Linear Elastic model assigned with a very high stiffness to capture the near rigid behaviour. The soil is assumed as elastic-perfectly plastic with Mohr-Coulomb failure criterion. The Mohr-Coulomb material model is used in the analyses. The material parameters assigned to the soil and cutting edge are given in Table 1. The interface between the outer face of the cutting edge and the soil (Fig. 1b) is modelled appropriately. The interface has a reduction factor (R_{int}) of $R_{int} = 0.5$ and the same is used to simulate the real field conditions. A surcharge (q) of 50 kPa is applied radially outward from the outer face of the caisson to simulate the overburden soil. The vertical incremental displacement is applied on the nodes representing the top width of the cutting edge. When the incremental displacements are assigned, the cutting edge applies the pressure on the soil and results in the deformation of the soil elements. The caisson cutting edge is considered as almost rigid and hence it initiates failure only within the soil elements. This fact establishes the radial and vertical extent of failure in the soil. The incremental displacement contours are plotted as output to investigate the failure zone in the soil. The failure zone depends on the angle of cutting edge, radius ratio of the caisson, friction angle and cohesion of soil, unit weight of soil and surcharge due to overburden at the level of the cutting edge acting radially in outward direction from the steinning of the caisson (surcharge = γh, see Fig. 1a). The extent of the failure zone around the cutting edge (i.e. a contour of the incremental displacement having approximately zero value), beneath the cutting edge and within the caisson is evaluated, i.e. the distance from the inner face of the caisson to the failure plane. The distance X is measured from the point where the inner face of the steinning and

Table 1. Input parameters: Soil and cutting edge.

Model	γ(kN/m^3)	c' (kPa)	ϕ'	ψ	E	ν
MC[1]	20	10	20°	0°	200 MPa	0.35
LE[2]	20	-	-	-	300 GPa	0.15

[1] MC—Mohr-Coulomb model for soil;
[2] LE—Linear Elastic model for cutting edge, ψ - dilation angle, E, ν - Young's modulus and Poisson's ratio of the soil.

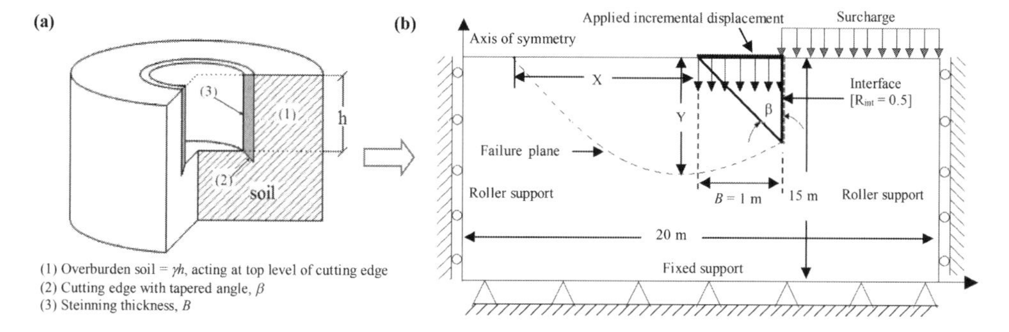

Figure 1. Open caisson with cutting edge: (a) 3D problem (b) 2D representation—axisymmetric and schematic view of model for open caisson with dimensions and boundary conditions.

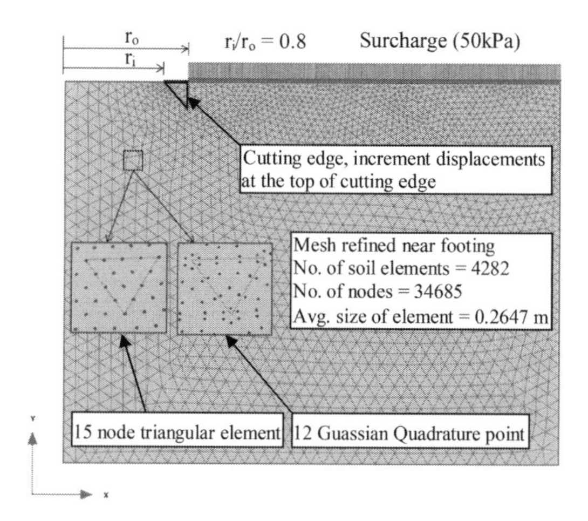

Figure 2. Finite element mesh and mesh details.

cutting edge meet and towards the axis of symmetry. The maximum depth of influence from the top level of the cutting edge is measured as Y in the vertical direction. The distances, X and Y (Fig. 1b), are normalized by dividing them with the steinning thickness (B). The extent of these ratios, X/B and Y/B, are termed as the normalized influence zone, where X is a radial zone of influence, Y is a vertical zone of influence and B is $r_o - r_i$. In all the cases studied, the steinning thickness (B) is taken as 1 m. The variations in the normalized influence zone with varying radius ratio, tapered angle of the cutting edge, friction angle and cohesion of the soil, unit weight of the soil and surcharge are investigated.

2.2 *FE mesh and boundary conditions*

The FE mesh and element details are depicted in Figure 2. The boundary conditions are introduced as roller supports at the sides to allow the movement only in the vertical directions and the lower boundary is fixed in both the directions (see Fig. 1b). The FE mesh dimensions are adopted as 20 m in the radial direction and 15 m in the vertical direction. The formation of plastic regions is checked for the radius ratio of 0.9. These FE mesh dimensions are finalized based on the formation of plastic regions well within the FE domain.

3 RESULTS AND DISCUSSION

The FE simulations have been performed to identify the zone of bearing failure beneath the cutting edge of the open caisson. The incremental vertical displacements are applied to simulate the extent of fail-ure zone. The failure pattern in the soil is observed to be similar in all the cases studied. The soil tends to have plastic failure starting from the outer face of the cutting edge towards the central axis of the caisson. The extent of failure within the soil changes with the variations in the strength parameters of the soil, unit weight of the soil and surcharge. It also depends on the tapered angle (β) of the cutting edge, radius ratio and steinning thickness. The extent of the normalized influenced zones (failure zones), X/B and Y/B, are plotted for varying parameters which influence the failure zone. For example, Figure 3 depicts the incremental displacement contours and displacement vectors plot for the soil having cohesion of 10 kPa. The extent of the failure zone is clearly identified from the figure. Similarly, the influence zone with varying friction angle of the soil, unit weight of the soil and surcharge is also studied. The best-fit curves are plotted to investigate the trend in the zone of influence with varying parameters. The reaction at the cutting edge offered by the soil is also evaluated.

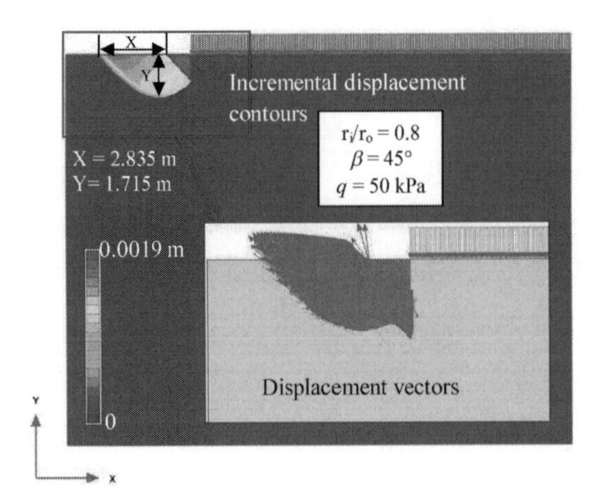

Figure 3. Radial and vertical extent of influence zone for the soil with cohesion = 10 kPa.

3.1 *Influence of radii ratio*

The radius ratios can range from 0 to 1. The open caisson with radius ratio of 0 represents solid cylinder (pile) and radius ratio of 1 represents circular caisson with infinitely small steinning thickness. The dimensions (i.e. external diameter and steinning thickness) of the open caisson are designed based on the intensity of loading coming from the super structure, the soil type, site conditions, ground water, velocity and discharge of water in river, scouring depth, etc. In the present study, the effect of varying radius ratio of the open caisson on the development of failure zone is studied. The radius ratio is varied from 0.5 to 0.9 with an increment of 0.1. The other parameters are considered as follows: angle of cutting edge = 45°, surcharge = 50 kPa, friction angle = 20°, cohesion = 10 kPa and unit weight of soil = 20 kN/m^3. Figure 4a depicts the effect of varying radius ratios of the open caisson on the radial and vertical extent of the failure zone in the soil at the cutting edge. It is observed from the figure that the extent of failure zone in vertical direction increases with the increase in the radius ratio. At radius ratios of 0.5, 0.6 and 0.7, the extent of failure zone in radial direction reached the central axis of the caisson. Hence, the extent of failure zone in radial direction is plotted for radius ratios of 0.8 and 0.9 only. It is noted that at the radius ratio of 0.7 and below, the entire soil mass within the caisson has failed (i.e. the influence zone is from inner face of the cutting edge to the central axis of the caisson) and heaved with progress in sinking of caisson.

3.2 *Influence of tapered angle of the cutting edge*

The effect of varying tapered angle (β) of the cutting edge on the development of failure zone is studied. The radius ratio is kept constant as 0.8 and the tapered angles are varied as 30°, 45°, 60°, 75° and 90°. The tapered angle of 30° represents the sharp cutting edge, whereas the tapered angle of 90° represents the cutting edge with flat bottom. The other parameters are as follows: surcharge = 50 kPa, friction angle = 20°, cohesion = 10 kPa and unit weight of soil = 20 kN/m^3. Figure 4b depicts the effect of varying tapered angles of the cutting edge on the radial and vertical extent of the failure zone in the soil at the cutting edge. It is observed from the figure that the extent of failure zone in radial and vertical directions reduces with the increase in the tapered angles of the cutting edge.

3.3 *Influence of friction angle of soil*

The effect of varying friction angle of the soil on the development of failure zone is studied. The friction angle is varied from 5° to 30° with an increment of 5°. The other parameters are as follows: radius ratio = 0.8, angle of cutting edge = 45°, surcharge = 50 kPa, cohesion = 10 kPa and unit weight of soil = 20 kN/m^3. Figure 5 depicts the effect of varying friction angle of the soil on the radial and vertical extent of the failure zone in the soil at the cutting edge. It is observed from the figure that the extent of failure zone in radial and vertical directions increases with the increase in the friction angle of the soil. It is noted that the zone of failure is larger for the soil having higher friction angle.

3.4 *Influence of cohesion of soil*

The effect of varying cohesion of the soil on the development of failure zone is studied. The cohesion of the soil is varied from 0 to 30 kPa with an increment of 10 kPa. The other parameters are: radius ratio = 0.8, angle of cutting edge = 45°, surcharge = 50 kPa, friction angle = 20° and unit weight of soil = 20 kN/m^3. Figure 6 depicts the effect of varying cohesion on the radial and vertical extent of the failure zone in the soil at the cutting edge. It is observed from the figure that the extent of failure zone in radial and vertical directions increases with the increase in the cohesion of the soil. The zone of failure in the soil at the cutting edge is larger for the soil having higher cohesion.

3.5 *Influence of unit weight of soil*

The effect of varying unit weight of the soil on the development of failure zone is studied. The unit weight of the soil is varied from 14 to 22 kN/m^3 with an increment of 2 kN/m^3. The other parameters are: radius ratio = 0.8, angle of cutting edge = 45°,

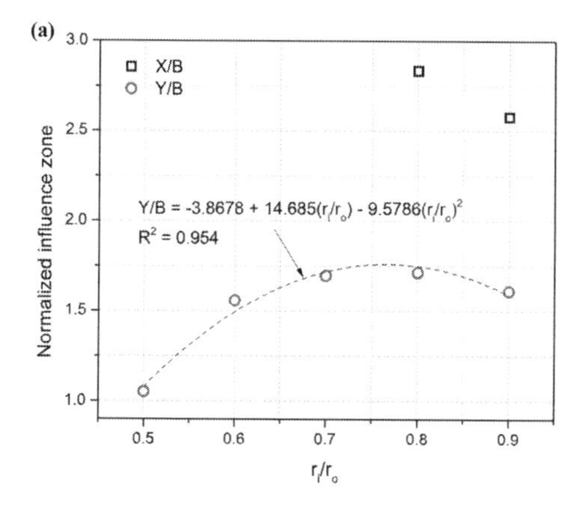

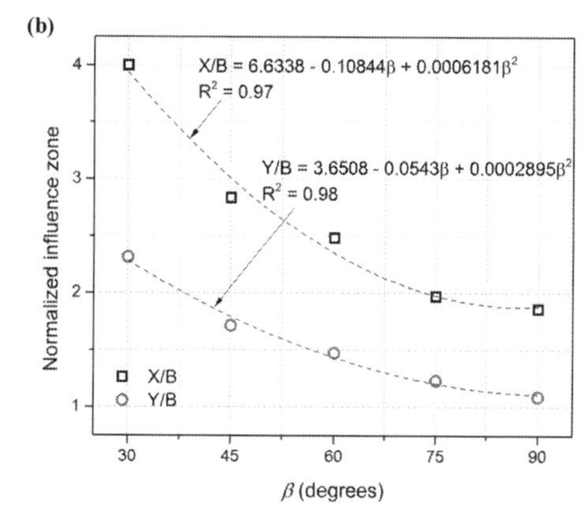

Figure 4. Radial and vertical extent of influence zone with varying: (a) radii ratio (b) tapered angles of cutting edge.

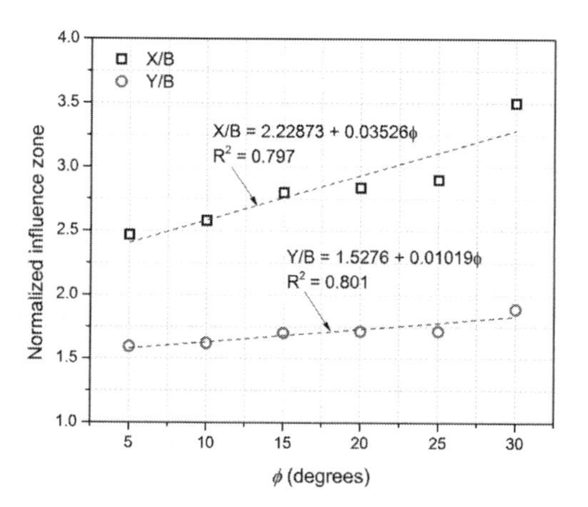

Figure 5. Radial and vertical extent of influence zone with varying friction angle of soil.

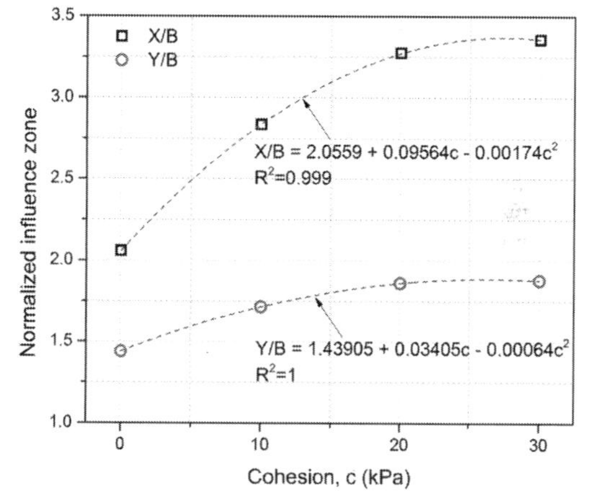

Figure 6. Radial and vertical extent of influence zone with varying cohesion of soil.

surcharge = 50 kPa, friction angle = 20° and cohesion = 10 kPa. Figure 7 depicts the effect of varying unit weight of the soil on the radial and vertical extent of the failure zone in the soil at the cutting edge. It is observed from the figure that the extent of failure zone in radial and vertical directions reduces with the increase in the unit weight of the soil. The zone of failure in the soil at the cutting edge is larger for the soil having lower unit weight.

3.6 *Influence of surcharge at the level of cutting edge of the open caisson*

As the caisson sinks, the overburden pressure at the top level of the cutting edge increases due to self-weight of the soil. The higher surcharge will restrict the failure in soil within the caisson i.e. on the excavation side. The FE analysis is carried out to investigate the effect of surcharge on the radial and vertical extent of the failure zone in the soil. The surcharge is varied from 0 to 150 kPa with an increment of 50 kPa. The other parameters are: radius ratio = 0.8, angle of cutting edge = 45°, friction angle = 20°, cohesion = 10 kPa and unit weight of soil = 20 kN/m³. Figure 8 depicts the effect of surcharge on the radial and vertical extent of failure zone in the soil at the cutting edge. It is seen from the figure that the extent of failure zone in radial and vertical directions increases with the increase in the surcharge. As the caisson sinks to a greater depth, the zone of influence extends to a larger area.

The effect of surcharge on the vertical and radial extent of failure zone in the soil is also studied with the cutting edge having a tapered angle of 75°. For this case, the interface between the cutting edge and the soil is considered as completely rough

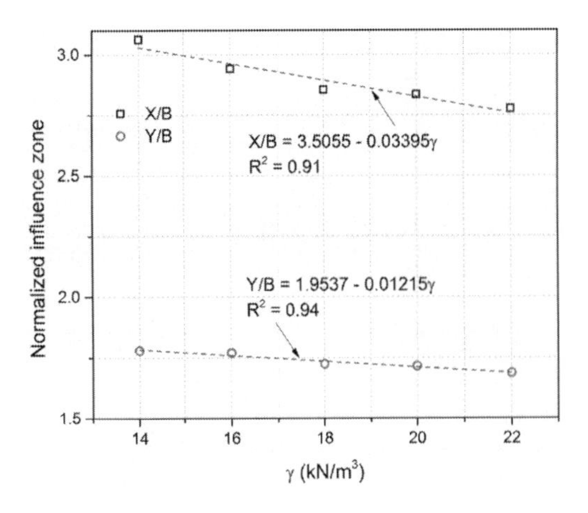

Figure 7. Radial and vertical extent of influence zone with varying unit weight of soil.

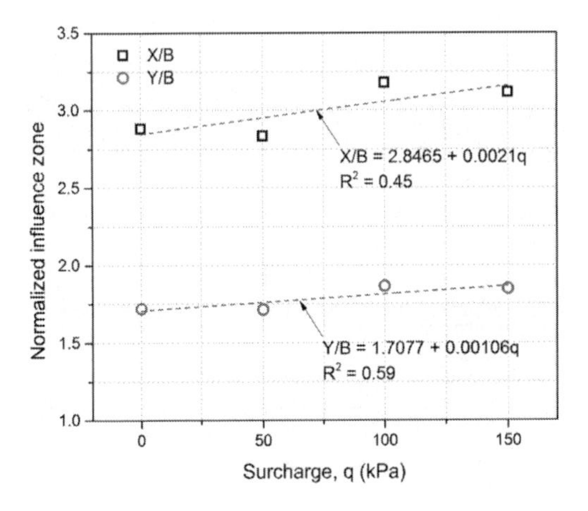

Figure 8. Radial and vertical extent of influence zone with varying surcharge at the level of cutting edge.

(i.e. rigid interface). Figures 9a, b depict the incremental displacement contour plots for the caisson with $\beta = 75°$ without a surcharge (i.e. top level of the cutting edge is at the ground level) and with a surcharge of 50 kPa (for example, caisson has sunk to 2.5 m in the soil having $\gamma = 20$ kN/m³) at the top level of the cutting edge respectively. For the caisson with cutting edge angle = 75°, the failure zone in the soil is found to extent towards the excavation and overburden sides when no surcharge is considered (Fig. 9a). With the consideration of surcharge of 50 kPa, the failure zone in the soil is restricted towards the excavation side only (Fig. 9b). It is observed from Figures 8 and 9 that the extent of radial and vertical zones of influence in the soil increases with the increase in the surcharge.

3.7 *Reaction at the cutting edge with varying β*

In the FE simulations, the incremental displacements are applied over the top width of the cutting edge, as shown in Figure 1b and Figure 10, until the formation of the fully developed failure surfaces beneath the cutting edge. Figure 11 depicts the different configurations of the cutting edge with varying tapered angles with constant top width (r_o - $r_i = B = 1$ m). It is noted that the same configuration has been used for determining the effect of varying tapered angles of the cutting edge on the radial and vertical extent of the failure zone in the soil at the cutting edge. When the state of plastic flow occurs, the pressure beneath the cutting edge remains constant whereas the displacement keeps on increasing until the ultimate failure is reached. This fact can be identified from the load-displacement plot (Fig. 12), wherein the displacement is continued to increase after reaching the resultant ultimate loads (failure loads) acting in the radial and vertical directions. From this observation, it can be noted that the failure criterion is dependent on the

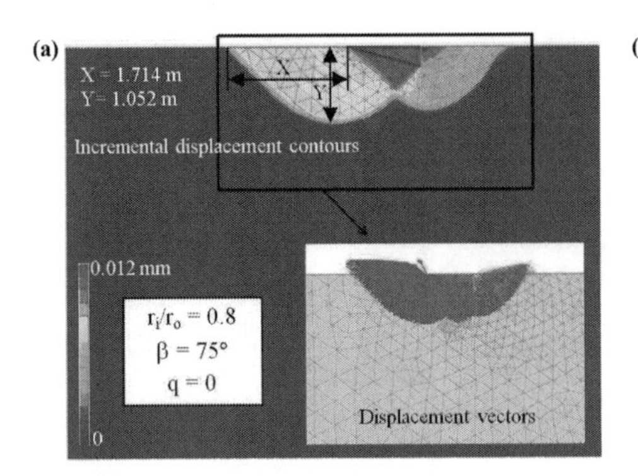

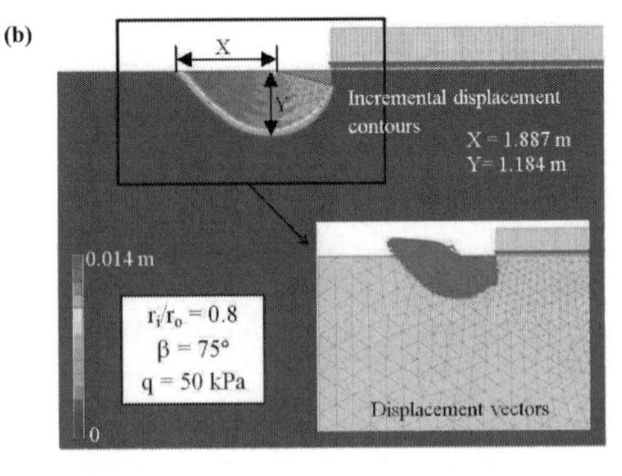

Figure 9. Radial and vertical extent of influence zone with varying surcharge: (a) $q = 0$ (b) $q = 50$ kPa.

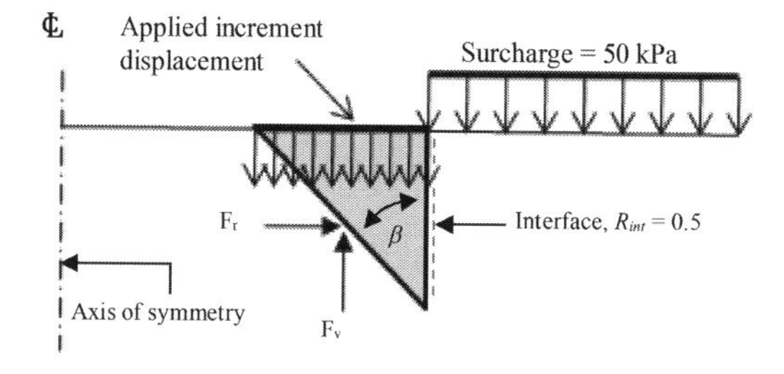

Figure 10. Schematic view of cutting edge with reaction components.

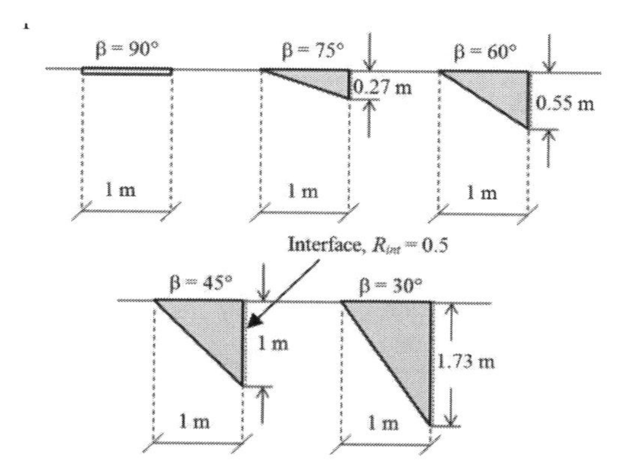

Figure 11. Configurations of cutting edge with different tapered angles.

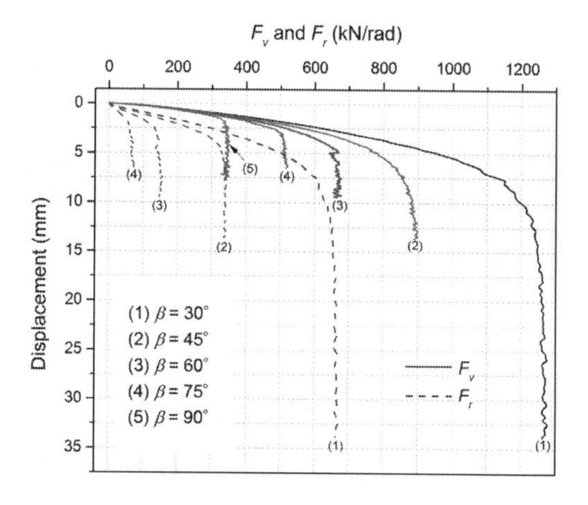

Figure 12. Load vs. displacement plots for cutting edge with varying β (surcharge = 50 kPa).

steady plastic flow of the soil beneath the cutting edge. Corresponding to the different tapered angles of the cutting edge, the ultimate loads (F_r and F_v) are determined. Figure 12 depicts the load vs. displacement plots for the cutting edge having tapered angles i.e. $\beta = 30°, 45°, 60°, 75°$ and $90°$. The failure load obtained from the axisymmetric modelling is expressed in kN/rad. The other parameters considered in the FE simulations are: radius ratio = 0.8, cohesion = 0, friction angle = 25°, unit weight of soil = 20 kN/m^3 and surcharge = 50 kPa. The interface at the outer contact surface of the cutting edge and the soil is modelled using a reduction factor of R_{int} = 0.5. The interface between the soil and the tapered length of the cutting edge is rough. From the figure it is noted that the cutting edge having tapered angle of 30° has provided a higher ultimate load as compared to the cutting edge with flat bottom. From Figure 11, it is seen that with increase in the cutting edge angles (30° to 90°, i.e. steeper to

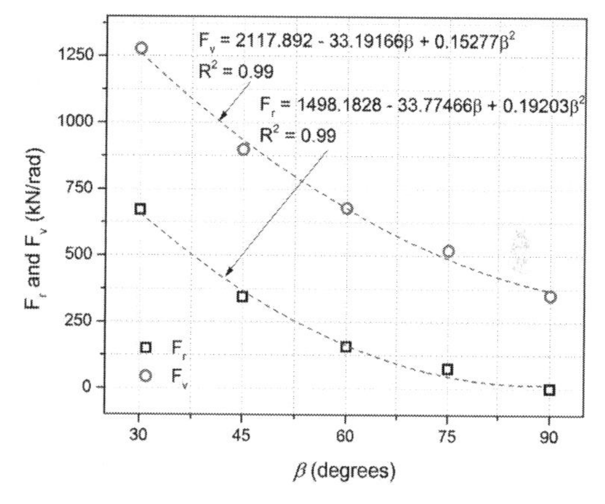

Figure 13. Radial and vertical reactions at the cutting edge of the caisson.

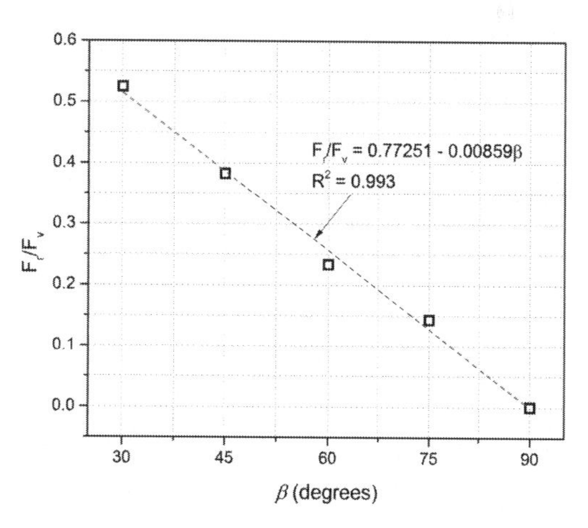

Figure 14. Ratio of radial and vertical reactions at the cutting edge of the caisson.

flat bottom), the contact area with the soil reduces for the constant steinning thickness. This fact is attributed to the reduction in the ultimate vertical and radial loads with the increase in the tapered angles of the cutting edge. Figure 13 depicts the decreasing trend in the resultant ultimate radial and vertical loads with the increase in the tapered angles of the cutting edge. The variation of the ratio of radial to vertical failure loads (F_r/F_v) versus the increase in the tapered angles of the cutting edge is shown in Figure 14. It is seen from the figure that the reaction offered by the soil at the cutting edge is dependent on the tapered angle of the cutting edge. The ultimate radial resistance (load) is zero when the cutting edge is modelled as flat bottom (i.e. $\beta = 90°$). The correlation depicted in Figure 14 is linear and the F_r/F_v ratio reduces to zero as the cutting edge takes the $\beta = 90°$. In addition, it is worth noting that for the cutting angle of 30°, F_r is 0.53 F_v which results in large hoop tension in the cutting edge of the caisson. This observation should be taken to advantage while designing the cutting blade of the open caisson.

4 CONCLUSIONS

The FE simulations have been carried out to identify the zone of influence in the soil at the cutting edge of the open caisson. The extent of the failure zone in the radial and vertical directions is evaluated for the varying radius ratio of the open caisson, different tapered angles of the cutting edge, different friction angles and cohesion of the soil, different unit weights of the soil and surcharge at the top level of the cutting edge. The effect of surcharge on the failure zone in the soil at the cutting edge with a tapered angle of 75° is also examined. The ultimate radial (F_r) and vertical (F_v) reactions are evaluated for the cutting edge with varying tapered angles of 30° to 90° with an increment of 15°. The results of the present study have significance in the field application of the sinking process of the open caisson by providing a clear picture of the extent of the failure zone around the cutting edge. This information help plan the proper excavation strategies so that the controlled sinking can be achieved in the field. Based on the results of the present study, the following conclusions are arrived at:

- The radius ratio of the open caisson, tapered angle of the cutting edge, strength parameters (c' and ϕ') and unit weight of the soil affect the influence zone in the soil at the cutting edge. With increase in the radius ratio and strength parameters (c' and ϕ') of the soil, the extent of the failure zone in the vertical and radial directions increases. With increase in the tapered angle of the cutting edge and unit weight of the soil, the extent of the failure zone in the vertical and radial directions reduces.
- With the progress in sinking of the caisson, the overburden pressure at the cutting edge increases due to the increased depth of the soil at the outer side of the caisson. The higher surcharge restricts the failure in the soil to take place within the caisson i.e. on the excavation side. As the sinking progresses, with increase in surcharge, the extent of the radial and vertical zones of influence in the soil at the cutting edge increases. The cutting edge with a flatter angle of 75° with no surcharge has failure in the soil towards the excavation and overburden sides. For the case with a surcharge of 50 kPa, the failure in the soil occurs only within the caisson i.e. on the excavation side.
- With the application of the incremental displacements over the top width of the cutting edge, the formation of the fully developed failure surfaces beneath the cutting edge takes place and reactions are developed in the radial and vertical directions at the cutting edge of the caisson. The resultant ultimate radial and vertical reactions are determined for the cutting edge with different tapered angles (30° to 90°). The ultimate radial and vertical failure loads reduce with the increase in the angles of the cutting edge. This is attributed to the reduction in the contact area between the soil and the cutting edge with the increase in the tapered angles (30° to 90°, i.e. steeper to flat bottom). For the case of cutting edge with flat bottom, the resultant ultimate radial load reduces to zero. The relationship between the ratio of F_r/F_v and the tapered angle of the cutting edge is linear. However, the ratio F_r/F_v reduces to zero as the tapered angle of the cutting edge approaches 90°. The large hoop tensile stresses are developed in the cutting edge of the caisson for the steeper tapered angles. This information has to be used in the design of cutting edge of the open caisson.
- The interface between the soil and the tapered length of the cutting edge is considered as fully rough in the FE simulations. Hence, the results show that the cutting edge with steeper tapered angle has higher ultimate reaction compared to the cutting edge with flatter bottom. In order to confirm these findings, the experimental investigations are needed and the work in this direction is currently underway.

REFERENCES

Alampalli, S. & Peddibotla, V. 1997. Laboratory investigation on Caissons-deformations and vertical load distributions. *Soils and Foundations*, 37(2): 61–69.

Baitsur, A.I. & Zakharov, I.B. 1975. Experimental investigations of anchoring of open caissons by means of radial piles (for purposes of discussion). *Soil Mechanics and Foundation Engineering*, 12(5): 296–299.

Belous, N.P. 1968. Deformation of soils and settlement of reference points at the zone where a caisson is sunk. *Soil Mechanics and Foundation Engineering*, 5(5): 359–361.

Bolyachevskii, B.I. & Chumakov, I.S. 1975. Construction of precast open caissons with forced regulation of their sinking. *Soil Mechanics and Foundation Engineering*, 12(6): 362–365.

Chandler, J.A., Hinch, L.W., Fair, R.I., Hughes, D.A., Peraino, P. & Rowe, P.W. 1984. Jamuna River 230 kV crossing, Bangladesh. Part 2: Construction. *IEE Proceedings C—Generation, Transmission and Distribution*, 131(7): 319–332.

IS: 9527, Part 1, 1981. Code of practice for design and construction of Port and Harbour structures, Part 1 Concrete Monoliths. Bureau of Indian Standards, New Delhi.

IS: 3955, 1965. Code of practice for design and construction of Well Foundations. Bureau of Indian Standards, New Delhi.

Katti, R.K. & Dewaikar, D.M. 1977. Studies on circular well foundations in Clay media. Proceeding of *5th Southeast Asian Conference on Soil Engineering*, Bangkok, Thailand, 2–4 July 1977, pp 41–54.

Kumar, N.D. & Rao, S.N. 2010. Earth pressures on caissons in marine clay under lateral loads—a laboratory study. *Applied Ocean Research*, 32(1): 58–70.

Kurian, N.P. 2005. *Design of Foundation Systems—Principles and Practices*. Narosa Publishing House, New Delhi, India.

Maslik, V.P., Kiselev, V.I & Chizhov, I.I. 1978. New method of sinking open caissons in a double thixotropic jacket. *Soil Mechanics and Foundation Engineering*, 15(2): 98–99.

Nonveiller, E. 1987. Open caissons for deep foundations. *Journal of Geotechnical Engineering*, ASCE, 113(5): 424–439.

Opershtein, V.L. 1969. Waterproofing of the walls of open caissons. *Soil Mechanics and Foundation Engineering*, 6(2): 115–118.

Panda, J. 2015. Correction of Tilt in Well No. 5 of Railway Bridge across River Luna in Odisha: A Case Study. *Indian Highways*, 43(2): 19–27.

Royston, R., Phillips, B.M., Sheil, B.B. & Byrne, B.W. 2016. Bearing capacity beneath tapered edges of open dug caissons in sand. *Proceedings of Civil Engineering Research in Ireland*, Galway, Ireland.

Saha, G.P. 2007. The sinking of well foundations in difficult situations. *Journal of Indian Road Congress*, 68(2): 123–131.

Salamov, K.P., Bernar, D.N. & Korsunskii, E.V. 1965. Construction of the underground rooms of continuous steel casting units by the open caisson method (from the seminar on foundation engineering at the exhibition of achievements of the national economy). *Soil Mechanics and Foundation Engineering*, 2(3): 167–171.

Solov'ev, N.B. 2008. Use of limiting-equilibrium theory to determine the bearing capacity of soil beneath the edges of caissons. *Soil Mechanics and Foundation Engineering*, 45(2): 39–45.

Ter-Galustov, S.A., Ponomarenko, A.I., Opershtein, V.L. & Ivanov, V.D. 1966. Experience in sinking an open caisson in a thixotropic lining. *Soil Mechanics and Foundation Engineering*, 3(2): 128–131.

Tomlinson, M.J. 2001. *Foundation Design and Construction*. 7th ed., Prentice Hall, Harlow, England.

Yan, F.Y., Guo, Y.C. & Liu, S.Q. 2011. The Bearing Capacity Analyses of Soil beneath the Edge of Circular Cassion. *Advanced Materials Research*, 250: 1794–17

Numerical Methods in Geotechnical Engineering IX – Cardoso et al. (Eds)
© 2018 Taylor & Francis Group, London, ISBN 978-1-138-33203-4

Numerical 3D analysis of masonry arch bridge cracks using jointed rock model

Balint Penzes, Hoe-Chian Yeow & Peter Harris
COWI UK, London, UK

Christopher Heap
Network Rail, London, UK

ABSTRACT: Historical masonry arch bridges are an elementary part of the European infrastructure landscape. However, due to increased loads and various changes in the use of these structures; their behavior under these changing conditions needs to be understood in order to ensure safety of the end users. Monitoring of the performance of such structures allows practitioners the opportunity to understand the soil-structure interaction in a more accurate way in order to simulate better their actual performance.

This paper describes a 3D finite element model, which analyses the interaction of the masonry structure with its backfill, embankment and foundation. Jointed Rock model was used to model the anisotropic nature of the masonry, which allows the determination of more accurate stress and strain concentrations where cracks and discontinuities in the masonry structure occur. The modelling approach is described in the paper and the computed behavior is compared with a masonry arch bridge where measurements have been undertaken. The findings of this type of study can be used to confirm safety of any operational changes, plan future maintenance requirements and monitoring and repair regime of masonry arch bridges.

1 INTRODUCTION

The list of masonry arch bridges, viaducts and various pathways is remarkably long. Just in the United Kingdom there are more than 60 000 highway masonry bridges in use. Most of these were built between the 17th and 19th centuries (Cox & Hallsal 1996). Since then the traffic loads increased dramatically, hence most of these historical structures now operate far beyond their original estimated design loads. Therefore, the assessment, continuous monitoring and proper maintenance of the existing masonry infrastructure is an enormous and crucial task. (McKibbins et al. 2006 and Sowden 1990).

In general a masonry arch bridge is a complex structure in which the main load bearing elements under the road pavement are the masonry arch and abutment together with the wing walls and pilasters. The loads applied to the bridge are transmitted through the road pavement and fill material to the arch. It is then distributed along the arch to the abutment and the foundations (Robinson 2000 and Smith 2008).

As the overall structural behavior, load-carrying capacity depends on the interaction between the embankment, backfill and the masonry structure itself it is important to analyze the coupled soil-structure interaction (Callaway et al. 2012 and Gilbert et al. 2013).

In the current study a detailed 3D finite element analysis is described. The work was prompted by the need to understand a recently occurred partial wing wall collapse of the Grove Nook Lane Bridge on 1st of August 2016 (RAIB 2017). However, here the main focus is not on the investigation of the collapse, but rather on the complex soil-structure analysis, which has also taken into account the anisotropic nature of the masonry with Jointed Rock (JR) model and the comparison with the monitored cracks and discontinuities of the structure.

2 CASE STUDY

The road bridge was originally built around 1840 as a single span brick arch bridge over two rail tracks. The cutting was later widened to the north in 1868 and an additional arch was constructed to add another two rail tracks which was opened to traffic in 1893. The collapsed South-East wall was a part of the original arch.

Figure 1 shows the extent of the collapsed South-East wall indicated by the dashed line and

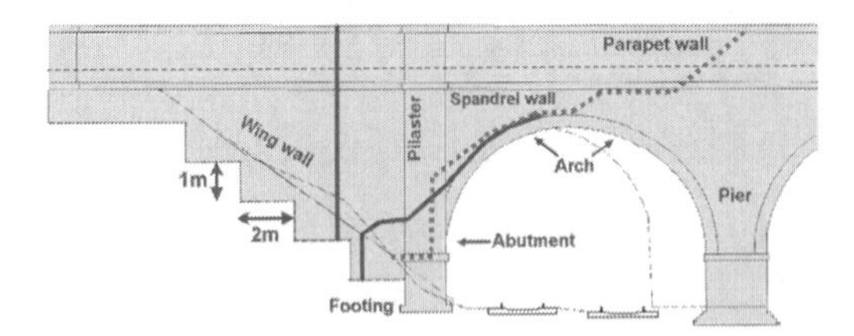

Figure 1. Elevation of the collapsed South-East wing wall with indicated cracks (solid) and collapse line (wall structure to the left of the dashed line).

the southern edge of the structure. All structural elements of the bridge; the arches, piers, abutments, spandrel walls, footings and parapets were constructed from brickwork. It was also confirmed that, unlike the South-West wall, the collapsed wing wall had a uniform thickness without any footing at the base. The wall height reduces a meter every two meters up the slope of the South cutting (Figure 1).

Figure 1 also shows a long vertical crack along the parapet wall to the bottom of the wing wall and an inclined crack developing from the bottom of the wing wall, through the pilaster and the spandrel wall before ending at the crown of the arch. The cracks have been continuously monitored since 2006 without much development. However, in 2015 an extended deep settlement of approximately 100 mm occurred on the road surface directly above the vertical crack at the toe of the parapet wall, which was repaired in the same year. An average inclination (12° to the vertical) of the parapet wall was also recorded, which occurred since 2014.

The abutment of the modelled area was constructed in a cut where the original soil mass behind the masonry structure was kept untouched and the slope of the cutting was constructed around this abutment. During the construction the original soil mass was exposed and stood vertically without any support (which means approximately 8.4 m free standing soil mass).

3 METHODOLOGY

The assessment of the collapse of the wing wall is undertaken using 3D numerical modelling with and without the observed cracks.

The 3D model without the cracks is used to confirm the envisaged current state of the structure and determine potential vulnerable areas of the wall. This model allows the determination of highly stressed locations and potential crack patterns based on the stress strain distribution on the masonry structure.

The second 3D model with the inclusion of cracks is then used to assess the situation before the collapse and assess how it changes the behavior of the wing wall.

The method of assessment is only restricted to quasi-static loading on the bridge structure. Any dynamic source of loading on the wall has not been considered.

3.1 *Applied constitutive soil models*

The masonry structure is an anisotropic composite of mortar and bricks with significant resistance against compression, but with much smaller tension capacity. Furthermore, along the mortar layers different shear planes can develop with respect to vertical and horizontal directions. Therefore, the Jointed Rock model implemented in Plaxis (Brinkgreve et al. 2015), an anisotropic elastic-perfectly-plastic model, was used to simulate the structural parts of the masonry bridge.

In the JR model a maximum of three joint directions can be defined with their weak planes. Plasticity can only occur along these planes when the stress state reaches its Mohr-Coulomb failure criteria for the specific plane.

The JR model has been used for modelling of various anisotropic materials, as fractured rocks (Le Cor et al. 2014 and Stelzer 2015) and weathered rock slopes (Vijayalakshmi & Neelima 2011). It was also used for masonry structures before in geotechnical numerical models (Amorosi et al. 2015 and 2016).

The various soil types were modelled with elastic-perfectly plastic Mohr-Coulomb (MC) soil model.

4 FINITE ELEMENT MODEL

In any numerical analysis of soil-structure interaction problems, modelling assumptions and simplifications are inevitable. Detailed modelling of all features of a structure is preferred but under most circumstances less critical details are sacrificed in order to achieve numerical stability and efficiency. The following assumption and simplifications have been applied in the numerical analysis.

The masonry structure was assumed as a continuous volume element with Jointed Rock model to simulate the behavior of the brick and mortar interaction. Through the modelling process the observed cracks have been incorporated as a thin zone of masonry elements with Jointed Rock model and reduced shear strength along the plane of the crack.

The slope of the cutting was modelled as clay with reduced cohesion (as the stress state has been changed by unloading and it was exposed to weathering) compared to the mudstone in the abutment.

4.1 *Geometry, mesh and boundary conditions*

A simplified quarter model was used in order to limit the size of the numerical model. The bridge was analyzed along the longitudinal symmetry axis of the bridge and the vertical symmetry axis between the two arches. With this model it was possible to optimize the calculation time and also enable the inclusion of a lot of modelling details.

Figure 2 shows the geometry of the developed Plaxis 3D model. The thickness of volume element of the wing wall and the parapet wall are 530 mm and 360 mm, respectively. The pilaster is 1.55 m wide and 600 mm thick at the base. The arch was modelled as a 450 mm thick volume element.

At the initial stage of the work, different mesh density was tested in order to ensure accurate results and numerically stable model (Lees 2016). A coarser mesh was generated at the boundaries, while a finer mesh was used for the structural volume elements (coarseness factor of 0.5) to ensure proper soil-structure interaction assessment. Meanwhile, the cracks were refined with 0.1 coarseness factor as significant plasticity was expected. All in all, around 400 000 finite elements have been generated with 1.05 m average element size associated with 1.0 coarseness factor.

The boundary conditions were defined as normally fixed deformation constrains perpendicular to each lateral planes and fully fixed at the bottom of the soil layers.

4.2 *Ground condition and applied parameters*

Figure 3. shows the analyzed abutment behind the wing wall. Which comprised 1 m of Fill material, overlying 1.6 m grey mudstone (Mudstone 1) over 0.3 m thick Limestone layer at a depth 2.6 m below the road surface. Based on the site inspection below the bridge structure the mudstone at depth was assumed slightly stiffer (Mudstone 2). Furthermore, it was observed during the site inspection work that at the arch structure stiffer fill material was used above the crown and behind the pilaster.

The strength parameters of the abutment was defined to allow construction of the masonry structure with the natural soil mass behind the wall kept unsupported, as it was noted in the construction history. However, apart from the strength parameter of the fill material and the clay of the slope in front of the wing wall, other parameters are not considered critical to the behavior of the wall.

The assumed soil and backfill properties are shown in Table 1. All soil types were modelled as drained Mohr-Coulomb soil material and the soil is assumed to be dry.

The assumed properties for the masonry (Table 2.) are based on laboratory test results of the mortar and the brick of the bridge.

The applied joint directions are aligned to the shear planes defined by the horizontal and vertical mortar layers. The laboratory tests of the mortar and brick were used to derive the shear strength of the masonry structure. The derived shear strength of the masonry structure in horizontal plane is 150 kPa (c'_1) while in the vertical plane it is 500 kPa (c'_2). For the cracks, the shear strength is reduced to 12 kPa along the direction and surface of the cracked structure.

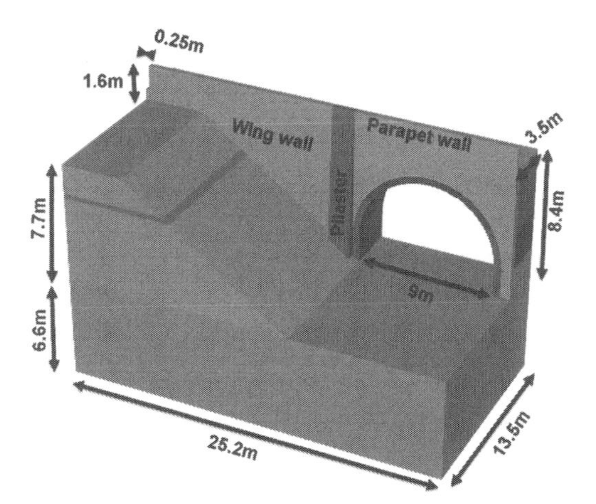

Figure 2. Geometry of the simplified quarter model of the bridge maybe change to mesh plot.

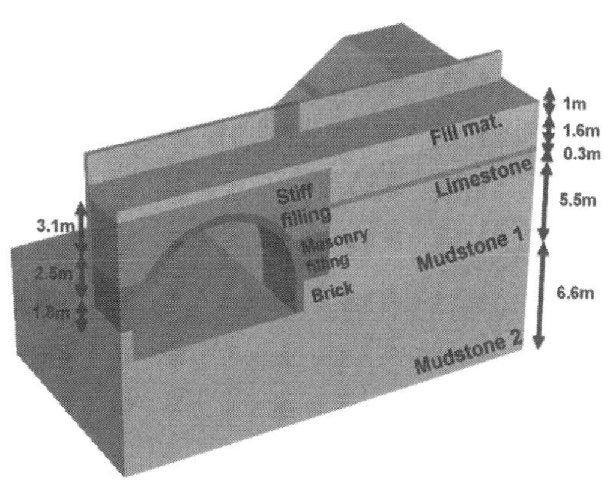

Figure 3. Ground conditions of the abutment and backfill of the bridge.

Table 1. Soil and backfill properties.

	Eff. friction	Cohesion	Young's modulus
Soil type	φ' [deg]	c' [kPa]	E' [MPa]
Fill	32	1	50
Mudstone 1	35	150	100
Limestone$*_1$	40	200	150
Mudstone 2	35	150	170
Clay (slope)$*_2$	-	25	100
Stiff fill	36	5	90
Masonry fill	32	100	120

$*_1$ 25 kN/m³ unit weight was assumed (the rest 22 kN/m³)
$*_2$ The slope was modelled as clay with reduced strength as the stress state has been changed by unloading and it was exposed to weathering

Table 2. Jointed Rock masonry properties.

	Young's modulus*		Strength	Unit weight
Material	E_1	E_2 [MPa]	f_k [kPa]	γ[kN/m³]
Masonry	4075	2037.5	8150	22

*$\nu = 0.2$ was used for E_1 and E_2 as well.

4.3 *Sequence of modelling*

After the calculation of the initial stresses, the modelling procedure was divided into three main parts. Namely, the construction of the abutment and the bridge, the backfilling and finally a loading phase with water pressure applied on the full height of the abutment, locally behind the wing wall to assess the deformations. The water pressure was applied due to the known presence of a cast iron water main (RAIB 2017).

5 RESULTS AND DISCUSSION

After the construction phases the main load was the full water pressure, assuming water infiltration, behind the wing wall until the pilaster within the abutment.

The two model cases, with and without implemented cracks, were compared based on this most onerous loading phase.

5.1 *Bridge model without cracks*

Figure 4 shows the total displacements of the wall. The maximum displacements are on the top of the parapet wall (38 mm) and at the maximum deformations of the wall, pavement settlements

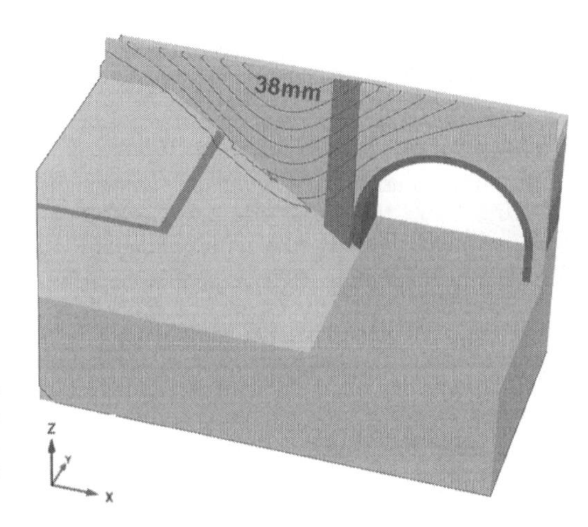

Figure 4. Total displacements with no cracks on the wing wall.

Figure 5. Deviatoric strains at the back of the wind wall.

occurred, as the fill material followed the horizontal deformations of the wall towards the slope (Penzes et al. 2016).

The deformed shape and the local settlements behind the wall were in agreement with the previously monitored deformations (Penzes et al. 2016).

The results show a typical active wedge of the fill material in the top 1 m of the retained soil while below that the competent ground of mudstone and limestone did not impose any earth pressure while the wall deformed. This deformation characteristic was in good agreement with the observed post failure ground profile, where the upper fill stratum has a more gentle surface profile (representing an active wedge) with the much steeper or near vertical profile below (RAIB 2017).

The computed shear strains of the wing wall were also carefully assessed in order to confirm they matched with the observed cracks on site. Figure 5 shows the computed deviatoric shear strains superimposed by the observed cracks in the wall (dashed lines).

It shows that the concentration of higher strain is in good agreement with the locations of the observed cracks (Penzes et al. 2016). This is especially so in the case of the transverse cracks at the pilaster and the upper part of the arch. In addition, there is also a fairly extended, higher concentration of strains along a vertical zone. The vertical crack on the wing wall (vertical dashed line) is in the centre of this zone.

Additional concentration of shear strains can be seen at the upper steps of the wing wall close to the parapet wall. In fact, Figure 4 and Figure 5 show that the computed deformation of the wing wall is likely to similar to the mode of the collapse, i.e. the displacements involve the entire lateral extent of the wing wall to the South (Figure 1).

Based on these results it is concluded that the cracks observed in 2006 are consistent with the behavior computed in the FE model.

However, it is also important to note that, the analysis shown overall stability of the structure, which also agrees with the fact that the wall was stable for more than 120 years since its construction.

5.2 *Bridge model with cracks*

As the analysis showed consensus between the observed shear strains and the existing cracks, in the following model these cracks were included as 100 mm thick volume elements with reduced shear strength. The thickness was chosen in order to be able to model them in a numerically stable way.

For the FE model with cracked wall Figure 6 shows the contours of displacements of the wing wall. The maximum computed displacement of 47 mm occurred on the top of the wall at the direct vicinity of the vertical crack, which

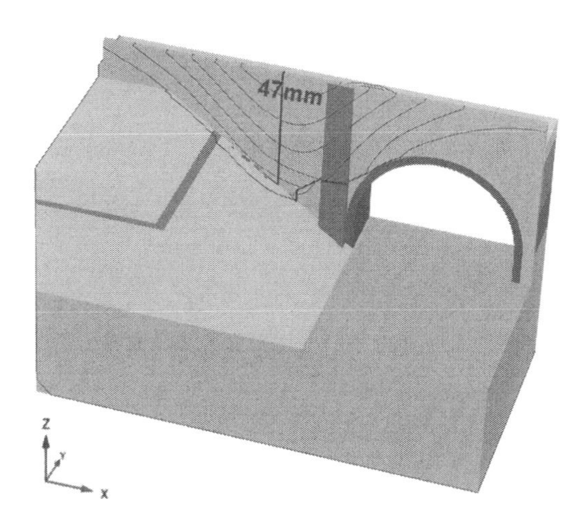

Figure 6. Total displacements with cracks on the wing wall.

resulted similar settlements in the fill material as observed on site. It is also clear that no deformations occurred along the vertical crack, which is in good agreement with the fact, that no substantial crack opening was observed (Penzes et al. 2016).

A major difference of the deformed shape between the cracked and uncracked models is the significant movement along the horizontal crack of the pilaster. From there, along the inclined crack the deformations can extend further, which represents again fairly well the boundary of the collapse above the arch (Figure 1).

In addition, it is important to note, that even with the increased (20% larger maximum displacements) and more extended deformations no stability loss was computed.

In an attempt to reach the observed failure, further investigation was undertaken, by studying the sensitivity of the structure, reducing the shear strength of the cracks and the shear strength of the slope. It was found that, although some additional displacement was computed, none of the above led to the instability of the wall, which entails to an assumption that additional load was necessary for the collapse (Penzes et al. 2016).

6 CONCLUSION

The 3D numerical analysis in which the soil-structure interaction, the anisotropic nature of the masonry structure were included, represented well the monitored deformations on the bridge.

Based on the wing wall analysis with Jointed Rock model, local deviatoric shear strain increments were identified at the area of the existing cracks. This confirms the good consistency between the simulation and the case study.

With the implementation of the cracks, as thin volume elements, additional displacements and closer deformation pattern was calculated compare to the observed collapse zone. However, no instability was computed, which leads to the assumption that additional loads were necessary for the collapse.

ACKNOWLEDGEMENTS

The authors highly acknowledge Network Rail and COWI UK for the opportunity and support to present this case study and the results of the numerical analysis.

REFERENCES

Amorosi, A., Boldini, B., de Felice G. & Malena, M. 2016. An integrated approach for geotechnical and structural analysis of the Nynphaeum of Genazzano. Taylor & Francis Group, *Structural Analysis of Historical Structures, Van Balen & Verstrynge (Eds)*. London, UK, ISBN: 9781138029514.

Amorosi, A., Boldini, D., Felice, G., Lasciarrea, W.G., Malena, M. 2015. Geotechnical and structural analysis of the Ninfeo di Genazzano. *Rivista Italiana di Geotecnica*. N–49. 29–44.

Brinkgreve, R.B.J., Kumarswamy, S. & Swolfs, W.M. 2015. *Plaxis 3D An. Edition. - Manual*. Plaxis. Delft, Netherlands.

Callaway, P., Gilbert, M. & Smith, C.C. 2012. Influence of backfill on the capacity of masonry arch bridges. *Proceedings of the Institution of Civil Engineers*: Bridge Engineering, 165 (3). 147–157. ISSN 1478–4637

Cox, D. & Halsall, R. 1996. *Brickwork Arch Bridges*. Brick Development Association. Berkshire, UK

Gilbert, M., Smith, C.C., Hawksbee, S.J., Swift, G.M. & Melbourne, C. 2013. Modelling soil-structure interaction in masonry arch bridges. *Proceedings of the 7th International Arch Bridges Conference*, Split, Croatia.

Le Cor, T., Merrien-Soukatchoff, V., Rangeard, D., Rescourio & C., Simon, J. 2014. Comparison of different modelling of an excavation realized in fractured and weathered rocks. Taylor & Francis Group. *Numerical Methods in Geotechnical Engineering—Hicks, Brinkgreve & Rohe (Eds)*: 851856. London, UK, ISBN: 9781138001466.

Lees A. 2016. *Geotechnical Finite Element Analysis*. ICE Publishing, London, UK, ISBN: 9780727760876.

McKibbins, L., Melbourne, C., Sawar, N. & Gaillard, S.C. 2006. *Masonry arch bridges: conditions appraisal and remedial treatment*. CIRIA. London, UK.

Penzes, B., Yeow, H.-C. & Harris, P. 2016. *Numerical Analysis of the Grove Nook Lane Bridge Collapse*. Technical Report. COWI UK Ltd. London, UK.

RAIB 2017. *Partial collapse of a bridge onto open railway lines at Barrow upon Soar, Leicestershire 1 August 2016*. Rail Accident Investigation Branch, Dep. for Transport. Derby, UK.

Robinson, J. 2000. *Analysis and Assessment of Masonry Arch Bridges*. PhD Dissertation. University of Edinburgh, Edinburgh, UK.

Smith, C. 2008. Geotechnical Aspects Of Masonry Arch Bridges. *Advances in Transportation Geotechnics*, N-2222, 459–466. Nottingham, UK.

Sowden, A. 1990. The maintenance of brick and stone masonry structures. Witwell Ltd. Southport, UK

Stelzer, M. 2015. *Numerical Studies on the PLAXIS Jointed Rock Model*. Mather Thesis. Graz University of Technology. Graz, Germany.

Vijayalakshmi,.R. & Neelima, S.D. 2011. Numerical Modeling of rock slopes in Siwalik Hills Near Manali Region: Case Study. *Proceedings of Indian Geotechnical Conference* N-006, 803–806. Kochi, India.

Numerical Methods in Geotechnical Engineering IX – Cardoso et al. (Eds)
© 2018 Taylor & Francis Group, London, ISBN 978-1-138-33203-4

Finite element analysis of soft boundary effects on the behaviour of shallow foundations

C.X. Azúa-González
School of Engineering, Cardiff University, Cardiff, UK
Department of Civil Engineering and Geosciences, Delft University of Technology, Delft, The Netherlands

C. Pozo & A. Askarinejad
Department of Civil Engineering and Geosciences, Delft University of Technology, Delft, The Netherlands

ABSTRACT: The response of a shallow foundation has been investigated by numerical simulations in a series of 1*g* small scale tests using the Finite Element Method. In this numerical study, special attention has been given to the influence of soft boundaries as a measure to counteract boundary effects, since limitations of space were present in the containers used for the experiments. These experiments were carried out on rigid shallow foundations on sand using strongboxes, which are routinely used in thesmall beam centrifuge at Delft University of Technology. Two soil constitutive laws were used: 1) the well-known linear elastic perfectly plastic model, and 2) a hypoplastic model. Soft boundaries have been modelled as a continuum, while soil-soft boundary interaction has been addressed by zero thickness interface elements. Model parameters have been back-figured from free field experiments, since boundary effects were considered negligible in these kinds of experiments. Finally, comparisons between numerical and experimental data showed hypoplasticity performed better than the elasto-plastic model to reproduce some aspects of mechanical boundary effects.

1 INTRODUCTION

The response of a shallow foundation is generally evaluated under so-called free field conditions, i.e. other structures or barriers, if present, are far enough to not interfere with the system response. Under these circumstances, classical plasticity solutions are preferred to assess ultimate capacity, whereas physical and numerical modelling techniques are less widely used. However, under constrained conditions the ultimate capacity predicted by analytical expressions as well as the load-displacement (LD) response may be affected. These changes in the foundation response may be properly analyzed using advanced numerical and physical modelling techniques. Among other circumstances, constrained conditions may appear e.g. when small scale tests are performed to evaluate the response of a shallow foundation and space limitations are present in the facilities, as it is the case of the small beam centrifuge at Delft University of Technology (Allersma 1994) or potentially in other geo-centrifuges around the world with similar characteristics. Under such limitations, boundary effects should be evaluated with special care, and mitigation strategies should be sought to avoid erroneous conclusions.

Various methodologies have been proposed in the literature to evaluate and potentially minimize hard mechanical boundary effects. These methodologies may entitled within three broad categories: empirical, numerical, and experimental methods. The reader may refer to Pozo (2016) for a review of these categories. An illustrative example of experimental methods comes from Pozo et al. (2016), where the influence of hard mechanical boundaries on the response of a shallow foundation was studied through 1*g* small-scale tests. In this experimental study, the influence of boundary effects on LD curves and strain field within the soil was evaluated. The evolution of volumetric and shear strain fields was monitored through Particle Image Velocimetry (PIV) analyses during penetration of the foundation using the MATLAB module GeoPIV-RG (Stanier et al. 2016). These series of experiments were performed in two strongboxes of different size, each one representing a free field and constrained condition, respectively. Through these experiments, Pozo et al. (2016) proved that the implementation of Soft Boundaries (SB) adjacent to strongbox walls was a suitable strategy to reduce mechanical boundary effects under constrained conditions.

In this paper, exemplary tests from Pozo et al.(2016) and Pozo (2016) have been investigated numerically using the Finite Element Method (FEM), aiming to reproduce the influence of SB on the behaviour of a small-scale shallow foundation. Predictions are restricted to 1*g* conditions, since sound experimental evidence has been collected under this situation; however, future research may include similar analyses under *Ng* conditions. The Finite Element (FE) software package PLAXIS (Brinkgreve et al. 2016) was used for this task. At first, the response under free field conditions is sought to be reproduced, in the sake of back-figuring model parameters. Later, comparisons are carried out between shear band predictions against experimental evidence collected through PIV analyses. Within this paper, stress and strain variables are presented within Soil Mechanics sign convention.

2 PHYSICAL MODELLING

2.1 *Soil characteristics*

Merwede River sand was used in this series of tests. It is a uniform silica sand which consists of sub-angular to sub-rounded particles with a mean particle size d_{50} of 0.92 mm and a specific gravity G_s of 2.65 (Pozo 2016). The sand has a mean uniformity coefficient of 1.3; whereas the minimum and maximum void ratio have been reported to be $e_{min} = 0.52$ and $e_{max} = 0.72$. Later, soil packing state will be referred by its relative density $RD = 100 \cdot (e_{max} - e_0)/(e_{max} - e_{min})[\%]$, where e_0 is the initial void ratio.

2.2 *Experimental setup*

The experimental database produced by Pozo (2016) corresponds to tests developed in dry sand. Samples were prepared through the travelling pluviation method (Lo Presti et al. 1992) aiming to reproduce specimens with uniform relative densities within 55 to 93%. Two strongboxes were used, which were made up of aluminium, PVC and transparent acrylic glass. Hereafter, these strongboxes will be referred as FB (free field condition) and CB (constrained condition). es, the large longitudinal dimension of the strongbox FB ensured negligible activity within soil, adjacent to container walls upon penetration, i.e. a free field condition was ensured. A steel strip footing with dimensions 30 mm × 50 mm (height × width) was used. This footing covered the full strongbox breath, aiming to ensure plain strain conditions perpendicular to the front acrylic glass. A sketch

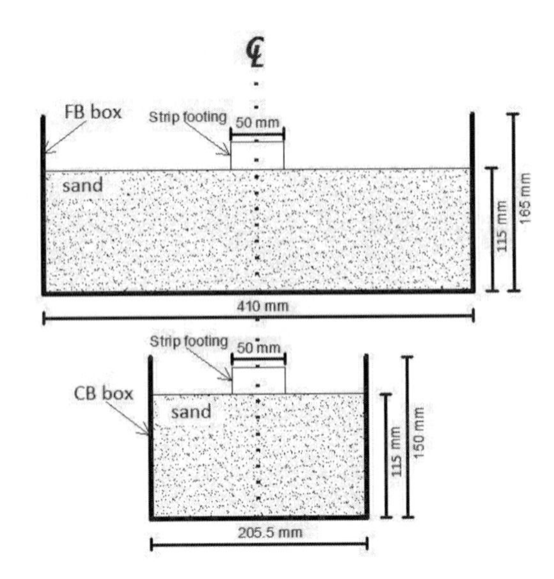

Figure 1. Sketch of free field (SB-F) and constrained (SB-S) strongboxes used for plain strain penetration tests.

of the experimental setup is shown in figure 1. diameter.

2.3 *Boundary conditions (BC)*

Two types of lateral BC were implemented in the experimental setup:

1. BC1: A hard boundary of aluminium or PVC was used. For the sake of simplicity, both materials were assumed to provide a large rigidity relative to the stiffness of the soil.
2. BC2: A layer of rubber material (shore hardness of 8–13*A*) was added between the sand and the hard lateral boundary to simulate a SB, where rubber padding thickness was set to 9 *mm*.

Independently of the lateral BC type, a sandpaper of a grit size of P150 (100 μm of average particle size) has been located between the soil and the soft or hard boundary. This procedure was aimed to enable a uniform roughness between the soil and lateral container boundaries. The FB-container was used only with a lateral boundary of the BC1 type, since a free field condition was ensured due to the large longitudinal dimension of the strongbox ($\text{length}_{\text{strongbox}} / \text{width}_{\text{footing}} = 8.2$). These type of tests will be referred as case 1. In addition, experiments performed within the CB-container combined with BC1 and BC2 lateral conditions will be referred as case 2 and case 3, respectively. The first combination aimed to mimic a constrained condition; whereas the latter sought to explore SB effects under constrained conditions. A summary of strongbox type and BC used for the three cases is shown in table 1.

2.4 *Load-displacement data and image acquisition*

Load-displacement data has been collected during footing penetration under displacement-controlled conditions (constant rate of 0.08 mm/s). In addition, this force-displacement data has been complemented through an image acquisition process (rate of 2 images/second) for further PIV analyses (Pozo 2016). The image acquisition process was carried out using a digital-single lens reflex camera (Canon EOS 750D) aided with continuous illumination through a lamp (~ 800 lumens of luminous flux) positioned 500 mm away from the strongbox; while the camera was located 300 mm away. An illustration of the data acquisition process is shown in figure 2.

2.5 *LD curves from experimental results*

Hard mechanical boundary effects were reflected in three features of the LD data in terms of pre-peak response, i.e. an increased bearing capacity and initial stiffness, and occurrence of peak state at a shorter penetration depth. While in terms of post-peak response, a sharper reduction of load and a lower residual bearing capacity was observed under constrained conditions. The three pre-peak aspects could be satisfactorily modified by the implementation of rubber paddings (case 3), i.e. the stiffness and bearing capacity was decreased, and peak state occurred at a larger penetration depth when compared to case 2. Nevertheless, post-peak response could not be modified significantly, as observed in figure 3.

Table 1. Strongbox type and BC for different cases.

Case	Strongbox type	lateral BC
1	FB	BC1
2	CB	BC1
3	CB	BC2

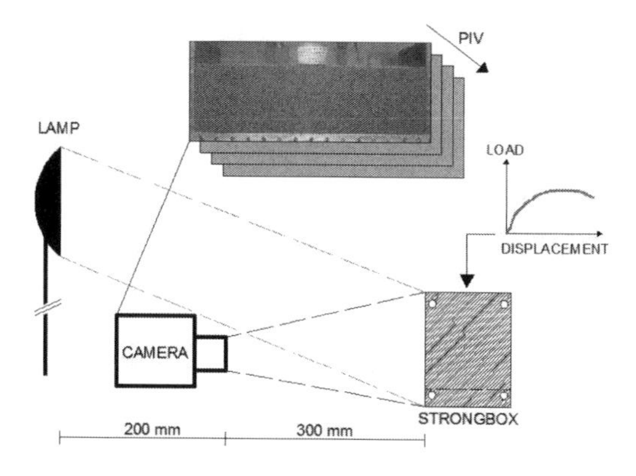

Figure 2. Sketch of parallel acquisition of illumination-aided images and load data during displacement-controlled tests.

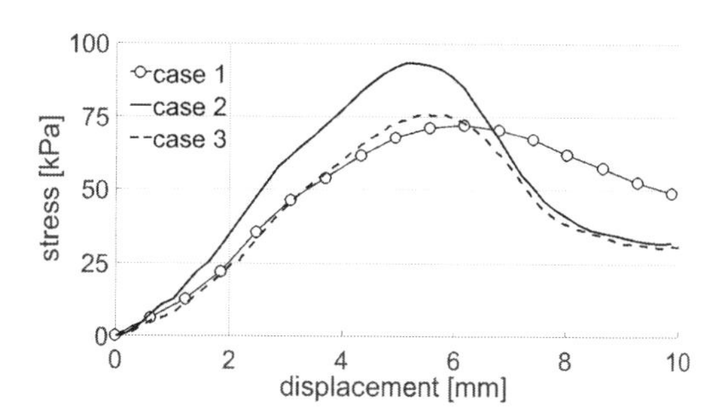

Figure 3. LD curves from experimental results.

3 NUMERICAL MODELLING

Plain strain FE models were used in all cases. Model domain was restricted laterally to half of the corresponding strongbox size due to the ideally symmetric configuration of the experiments, while full soil specimen depth (H_{soil} = 115 mm) was considered in the vertical axis. Soil and soft boundaries were modelled by means of 15-noded triangular elements. Regardless of the boundary type (BC1 or BC2), soil elements and virtual hard or soft boundaries were connected by means of 10-noded interface elements to mimic the influence of sandpaper on soil-SB interaction. Domain boundary conditions for case 1 and case 2 were set as normally fixed and fully fixed in the lateral and bottom boundaries, respectively. However, the bottom boundary was set as normally fixed in case 3, to allow horizontal movement at the base of the SB to be consistent with experimental observations through image analyses. For the sake of simplicity, the steel footing has been considered as fully rigid since a large bending stiffness compared to soil stiffness is expected under 1*g* conditions. Therefore, footing penetration has been modelled using uniform prescribed

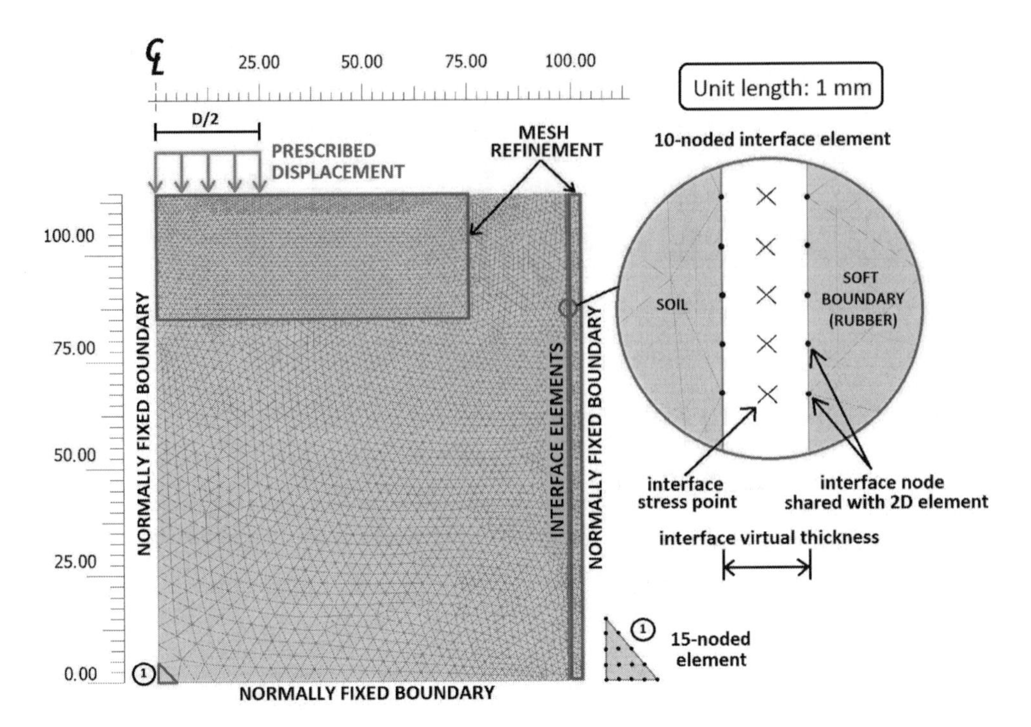

Figure 4. Elucidation of a typical (very fine) mesh for simulations corresponding to case 3 tests: 10207 elements, 82704 nodes.

displacements at the top of the specimen level as shown in figure 4.

3.1 *Constitutive laws*

3.1.1 *Soil*

Two constitutive models were chosen, namely 1) a linear elastic perfectly plastic (LEPP) model within a Mohr-Coulomb failure criterion and non-associative plasticity (Vermeer 1982), which has been adopted with a tension cut-off at the planes $\sigma_1' = \sigma_2' = \sigma_3' = 0$ in the principal stress space since a small cohesion $c' = 0.1\,kPa$ has been imposed to prevent premature soil collapse at the top level; and 2) the hypoplasticity model from von Wolffersdorff (1996).

Unlike most elasto-plastic constitutive laws, the hypoplastic model treats the void ratio e, and the Cauchy stress tensor σ' as state variables. This enables the model to capture the mechanical response at different packing and stress states within a unique parameter set. As other hypoplastic relations[1], this constitutive law is rate-independent and incrementally non-linear in the current strain rate $\dot{\varepsilon}$; and can be expressed by a single tensorial formulation as depicted in equation 1:

$$\dot{\sigma}'(\sigma', e, \dot{\varepsilon}) = \mathcal{L}(\sigma', e) : \dot{\varepsilon} + \boldsymbol{N}(\sigma', e)\,\dot{\varepsilon} \qquad (1)$$

where $\dot{\sigma}'$ is the (objective) Jaumman stress rate tensor, $\mathrm{L} = \mathrm{f_b f_e}(\mathrm{F^2 I} + \mathrm{a}^2 \hat{\sigma}' \hat{\sigma}')/(\hat{\sigma}' : \hat{\sigma}')$ is a fourth order tensor with $\mathfrak{I}$ being the fourth order unit tensor, and $\boldsymbol{N} = f_d f_b f_e F a(\hat{\sigma}' + \hat{\sigma}_d')/(\hat{\sigma}' : \hat{\sigma}')$ is a second order tensor. Both tensors are a function of the normalized stress tensor $\hat{\sigma}' = \sigma' / \mathrm{tr}(ma'')$, while N is also dependent on the deviatoric part of the normalized stress tensor $\hat{\sigma}_d' = \hat{\sigma}' - (1/3)\boldsymbol{I}$ with $\boldsymbol{I}$ being the second order unit tensor. Critical states are incorporated in the formulation by the scalar-valued function $F = f(\hat{\sigma}_d')$ and the parameter $a = \sqrt{3/8}\left(3/sin(\phi_{cv}') - 1\right)$. The scalar factors f_e, f_d cope with the influence of density; whereas f_b reflects the effect of stress level. Due to space limitations, the reader may refer to Herle and Gudehus (1999) for further details on parameter calibration.

3.1.2 *Soft boundary (SB)*

Rubber paddings have been modelled through isotropic linear elasticity, defined by two parameters,

[1] The term hypoplasticity may reference to three different theories, namely Karlsruhe-hypoplasticity, Dafalias-hypoplasticity and CLoE-hypoplasticity. The first meaning is adopted here.

namely the Young's modulus $E_{sb} = 0.271$ MPa as measured from unconfined compression tests (Pozo 2016), and Poisson's ratio ν_{sb} for rubber, which was adopted as 0.49 to address its widespread volumetric incompressibility.

3.2 *Soil—boundary interaction*

Interaction between soil and (rigid or soft) domain boundaries has been addressed using zero-thickness interface elements, which may represent a zone of soil with degraded stiffness and strength properties. This interface elements connect couples of nodes, enabling gapping or slipping to occur. The elastic relative displacements which may occur in these interfaces are governed by the interface elastic constants K_N and K_S; where $K_N = E_{oed,i} / t_v$ and $K_S = G_i / t_v$ are the interface elastic constants with t_v being the interface virtual thickness. Interface oedometric stiffness $E_{oed,i}$ and shear modulus G_i are computed from elastic properties of the adjacent soil including a degradation factor R and a Poisson's ratio $\nu_i = 0.45$, i.e. $G_i = R^2 G_{soil}$ and $E_{oed,i} = 2G_i(1-\nu_i)/(1-2\nu_i)$. In the sake of simplicity, the virtual thickness t_v has been set to 10% of the average element size.

In these interface elements, non-associative plasticity governs the flow rule within a yield surface of the Mohr-Coulomb type with a nil dilatancy angle ψ_i'. Strength reduction is achieved in a similar manner as carried out for the interface stiffness. In this case, the relations $tan(\phi_i') = tan(\phi')/R$ and $c_i' = c'/R$ have been adopted. It is worth to mention that when soil behaviour has been set to be hypoplastic, the interface yield locus has been removed to avoid convergence problems (by deactivating interface elements). Furthermore, when this advanced constitutive law has been used, a reference interface shear modulus has been estimated as $G_i^{ref} = R^2 E_{50}^{ref} / [2(1+\nu')]$, with $\nu' = 0.3$ and $E_{50}^{ref} = 600 RD[KPa]$ for a reference pressure of 100 kPa (Brinkgreve et al. 2010). In addition, a rate of stress dependency $m = 0.05$ (typical value for sands) has been adopted. The reduction factor R has been adopted as 0.75 independently of the soil constitutive law, in the sake ofsimulating a rough soil—boundary interface (as intended with the implementation of sandpaper in the experimental setup).

3.3 *Finite element mesh*

Three kinds of mesh density have been applied as available by default in PLAXIS, i.e. medium, fine and very fine mesh. The difference lies on the average element size l_{avg}, which can be computed as $l_{avg} = 0.06 r_e \sqrt{x_{max}^2 + y_{max}^2}$, where r_e takes the default values of 1.00, 0.67 and 0.50 for medium, fine and very fine mesh configurations, respectively; and x_{max} and y_{max} are the maximum horizontal and vertical lengths of the model domain, respectively. In addition, strategic mesh refinement zones have been adopted in the vecinity of the footing corner to improve the quality of displacement and strain fields, and within soft boundaries as shown in figure 4.

3.4 *Stress field initialization*

Soil initial stress field under 1g conditions has been generated using the so-called k_0 – procedure. In this procedure, only soil clusters are present and gravity loads are applied. These loads derive from the (dry) soil unit weight γ_{soil}. Required values of γ_{soil} have been estimated as a function of the initial void ratio, i.e. $\gamma_{soil} = \gamma_{water} G_s / (1 + e_0)$. Effective vertical stresses σ_{yy}' are calculated to comply with equilibrium in terms of body forces. Nevertheless, the effective lateral stress is back-figured from σ_{yy}' based on the coefficient of earth pressure at rest k_0, i.e. $\sigma_{xx}' = k_0 \sigma_{yy}'$. Required k_o – values have been estimated empirically as suggested by Jacky (1944), i.e. $k_0 \approx 1 - sin\phi'$, where ϕ' is the friction angle. Values of this parameter have been obtained using the empirical formula for sands $\phi' \approx 28 + 12.5\boldsymbol{RD}/100$ [°] (only for stress initialization), derived by Brinkgreve et al. (2010) using regression analysis on experimental data from Jefferies and Been (2006). This procedure enabled to obtain a realistic initial stress field, regardless of the chosen constitutive law. Soft boundaries, if present, were simulated using γ_{rubber} = 11.0 KN/m^3 as wished-in-place, by switching soil material models into soft boundary material. This initialization process caused inevitably artificial soil displacements, which were reset before further calculations were carried out.

3.5 *Displacement-controlled tests*

Footing displacement-controlled tests were modelled through sequential calculation phases involving small prescribed displacement increments, i.e. 1 mm per calculation phase. Within each calculation phase, the maximum prescribed displacement increment has been set to 1% per step to ensure that at least 100 steps are computed within each 1 mm of footing displacement. This procedure aided to reduce the required number of iterations per step. Soil collapse in the FEM simulations has been considered to be reached if the automatic load-increment routine is unable to proceed with increasing prescribed displacements (after trial-and-error a reasonable upper bound of maximum number of iterations per step was found to be ~ 60). Finally, Load-displacement curves have been elaborated by post-processing the vertical reaction force (per width) at model boundaries (induced by prescribed displacement increments only).

4 NUMERICAL RESULTS AND DISCUSSION

4.1 *Back-analysis of model parameters*

Tests corresponding to case 1 were chosen for inverse analysis, primarily because of the nearly negligible interaction of the strongbox boundaries with the adjacent soil. Model parameters were adjusted wisely by trial-and-error until a reasonably good agreement was achieved between experimental and numerical data.

In general, each parameter tend to dominate different features of the mechanical response. For instance, when the linear elastic perfectly plastic model was used, the Young's modulus controlled the slope of LD curves. Therefore, this parameter was changed upon an initial good agreement in terms of the secant slope of LD data, while the Poisson's ratio was set to 0.3 (typical value for soils). As an initial guess of strength parameters, the approximation $\phi' - \psi \approx 30^{\circ}$ (Brinkgreve et al. 2010) was adopted, while keeping a reasonable value of ϕ'. In general, final tuning between numerical and experimental data required refinement of all the parameters, since they interact between each other to some extent.

Similarly, when hypoplasticity was used, some parameters were fixed according to correlations from Herle and Gudehus (1999) and recommendations by Bauer (1996). Correlations for characteristic void ratios in terms of minimum and maximum void ratios were adopted, namely $e_{i0} = 1.2e_{max}$; $e_{c0} = e_{max}$; $e_{d0} = e_{min}$. The exponent β was set to 1.0, which may be valid for a wide range of sands. In addition, typical ranges were used to narrow possibilities during parameter adjustment, i.e. $0.10 < \alpha < 0.30$ and $0.18 < n < 0.40$.

The non-linearity of stiffness in LD curves could not be captured when the LEPP model was used, as observed in Figure 5. This can be explained by the fixed Young's modulus of the constitutive law (no stress dependency of stiffness is taken into account), and its incapacity to take into account the void ratio dependency of stiffness in the soil. Nevertheless, the average trend of LD response could be captured within a proper choice of Young's Modulus. Therefore, parameter selection was focused to force fitting in terms of peak state only. This drawback was observed to be less severe under looser samples, where non-linearity is less pronounced during pre-peak response. In addition, it is clarified that only pre-peak response was studied with the LEPP model since no softening behaviour is included in the formulation.

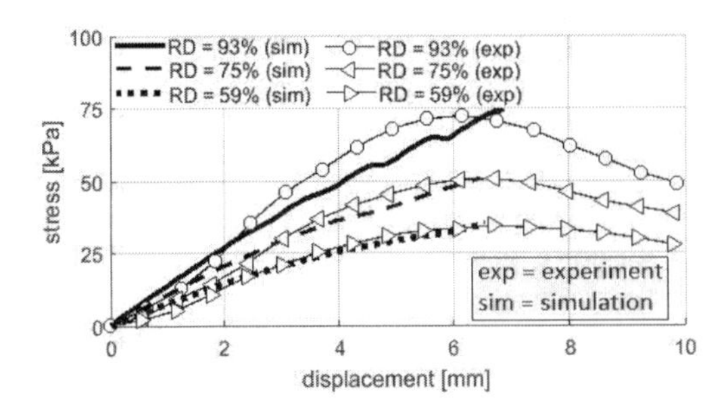

Figure 5. LD curves for case 1: LEPP model (very fine mesh).

On the other hand, the pre- and post-peak evolution of stiffness was intended to be captured when a hypoplastic model was used. Remarkably, strain softening-like behaviour was poorly captured by the hypoplastic model; unveiled by sudden load drops as shown in figure 6. In this case, it was difficult to match both pre- and post-peak behaviour for all relative densities within a unique parameter set. The chosen parameter set performed better for a dense packing state (RD = 92–93%) at the expense of a poor post-peak response for a looser state (RD = 59%). Parameters were chosen on purpose to perform better at a dense state because further numerical investigations are focused in dense samples only, where boundary effects may be more pronounced (Pozo et al. 2016). Tables 2

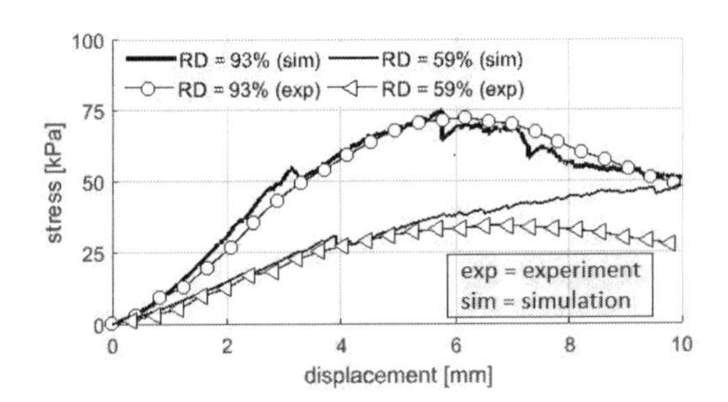

Figure 6. LD curves for case 1: hypoplasticity (very fine mesh).

Table 2. Soil constitutive model parameters.

LEPP model

RD [%]	E [kPa]	ν'	c'[kPa]	ϕ'[°]	ψ'[°]
93	800	0.3	0.1	48.00	25.00
75	650	0.3	0.1	45.00	21.50
59	470	0.3	0.1	42.70	18.60

Hypoplasticity

ϕ'_{cv} [°]	h_s [MPa]	n	α	β	e_{i0}	e_{c0}	e_{d0}
37.0	920	0.18	0.15	1.0	0.86	0.72	0.52

Table 3. Soil - (hard or soft) boundary interface parameters.

LEPP model					
RD [%]	G_i [kPa]	$E_{oed,i}$ [kPa]	c_i' [kPa]	ϕ_i' [°]	ψ_i' [°]
93	173	1904	0.075	39.80	0.0
75	141	1547	0.075	36.90	0.0
59	102	1119	0.075	34.70	0.0
Hypoplasticity					
RD [%]	G_i^{ref} [kPa]	$E_{oed,i}^{ref}$ [kPa]	p_{ref} [kPa]	M	
93	12072	132793	100	0.5	
59	7659	84245	100	0.5	

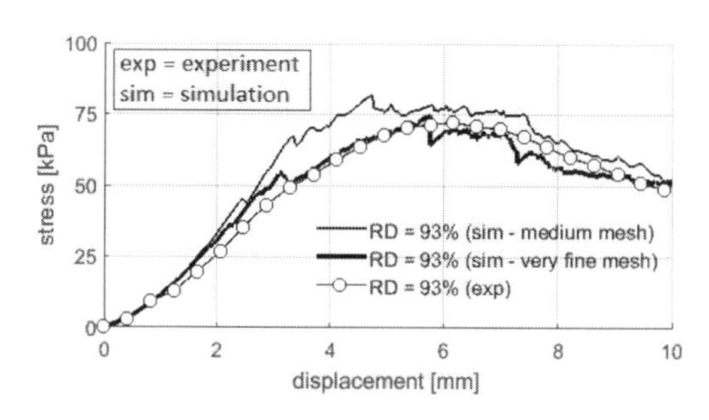

Figure 7. Mesh sensitivity for case 1 using hypoplasticity.

and 3 summarize back-figured model parameters and interface parameters, respectively.

The influence of mesh density was negligible when the linear elastic perfectly plastic model was used; nevertheless, mesh density was observed to have major effect with hypoplasticity, as observed in figure 7. Therefore, back-figured model parameters were chosen to ensure fitting when a very fine mesh configuration was used.

4.2 *Validation against Direct Shear (DS) tests*

Hypoplastic model parameters have been validated against experimental data on DS tests. These tests were carried out on (dry) medium dense sand samples (RD = 75%) under 16 kPa, 60 kPa and 103 kPa of normal load. A plotfor comparison between numerical simulations and experimental data is shown in figure 8. Numerical simulations were observed to approximate better the experimental results for tests under low normal loads; whereas for medium loads, the hypoplastic model showed underprediction of shear stress and an overprediction of dilatancy. Differences between predictions and experimental data has been quantified

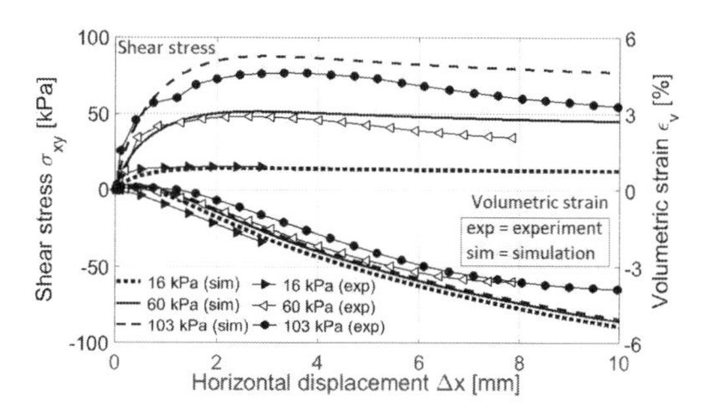

Figure 8. Shear stress and volumetric strain response on DS tests.

by the Root Mean Square Error (RMSE), which has ben normalized by the residual shear stress and volumetric strain after 10 mm of horizontal displacement. The normalized RMSEs in terms of shear stresses have been found to be 15.9%, 20.1% and 25.2% for samples under 16 kPa, 60 kPa and 103 kPa of normal load, respectively; whereas in terms of volumetric strains, RMSEs were determined to be 10.4%, 8.5% and 15.2% for samples under 16 kPa, 60 kPa and 103 kPa of normal load, respectively. These element test validations showed that the back-figured model parameters possessed better predictive capabilities in terms of volumetric strains than for shear stresses.

4.3 *Comparisons with PIV for cases 2 and 3*

Numerical predictions for a dense sand (RD = 93%) in terms of shear strains are presented in figure 9 (a) & (b) and (c) & (d) for cases 1 and 2, respectively. Figures 9 (a) and (b) show that shear band development for case 1 could be captured reasonably well by both constitutive laws. In both cases, shear band (horizontal) extension was in agreement with PIV results; however, shear band depth and inclination with respect to the horizontal plane near shear band tail were underestimated, whereas shear band inclination beneath the footing was overpredicted.

As discussed in the literature, shear band inclination θ with respect to the major principal stress direction may show dependency on the friction angle (Mohr-Coulomb theory: $\theta = \pi/4 - \phi'/2$) and on the dilatancy angle ((Roscoe (1970): $\theta = \pi/4 - \psi'/2$; Arthur et al. (1977): $\theta \approx \pi/4 - \phi'/4 - \psi'/4$). For both constitutive laws, the major principal stress direction showed the tendency to be rotated from a nearly vertical direction within the soil beneath the footing into a nearly horizontal direction within the soil near shear band tail. Therefore, this tendency of major principal stress direction rotation combined with

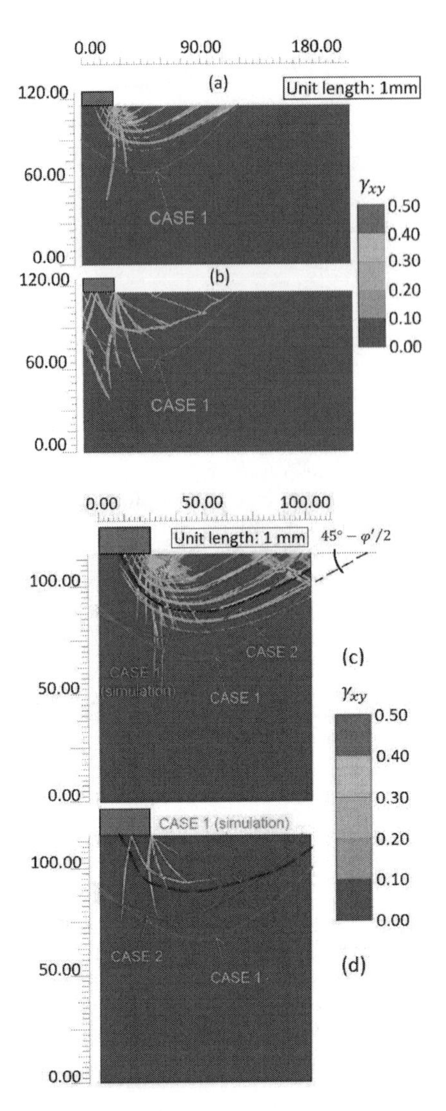

Figure 9. Shear strain field for cases 2 and 3 (RD = 93%): a) & c) LEPP model and b) & d) hypoplasticity. Shear band from experiments and case 1 simulations are shown with dashed lines.

too high values of friction angles (ϕ' and ϕ'_{cv}) and the overprediction of dilatancy (too high values of ψ' in LEPP parameters and overprediction of dilatancy as discussed in section 4.2) may have been the cause of an inadequate prediction of shear band inclination (underprediction of θ) within both constitutive laws. In addition, this poor predictive capability of both constitutive models in terms of shear band inclination may be a major contribution to the underprediction of shear band depth for both constitutive laws, since it affects shear band geometry.

In case 2, shear band depth was observed to be reduced through image analyses (see figure 9). Remarkably, only the hypoplastic model showed qualitatively this behaviour when compared to the prediction for case 1. On the other hand, a deeper shear band branch was predicted to be developed by the LEPP model. This resulted in reflected shear bands adjacent to the lateral boundary within an inclination of $45^{\circ} - \phi'/2$ ($\phi' = 48^{\circ}$), which may have erroneously suggested the occurrence of passive thrust.

4.4 *Effect of mechanical boundaries on LD curves*

4.4.1 *FEM simulation of case 2*

In this case, the LEPP model was not able to capture neither an increase in stiffness response and bearing capacity properly, nor a reduction in Peak State Penetration Depth (PSPD). Based on the experimental results, the peak secant stiffness was expected to be increased in 51.1%; the bearing capacity in 31.2%; and the PSPD to be reduced in 13.4%. However, the LEPP model predicted only 2.6% of increase in peak secant stiffness; an increase of 1.1% in bearing capacity; and a reduction of 1.4 in PSPD, i.e. hard mechanical boundary effects on the LD response were erroneously predicted to be negligible under a constrained condition. On the other hand, the hypoplastic model performed better in predicting an increased bearing capacity within 5.0% of error with respect to the experimental observation; however, it overpredicted the secant stiffness response by 35.8% and underpredicted the PSPD by 22.7%. In addition, no post-peak response could be captured since soil (numerical) collapse was obtained after the peak state occurred. Remarkably, the initial stiffness response until 2 mm of penetration was in good agreement with the experimental result when hypoplasticity was used; whereas a good agreement was observed only until 1 mm of penetration for the LEPP model. The poor predictive performance of the LEPP model in terms of key aspects of hard mechanical boundary effects could be attributed to the lack of stiffness stress dependency in its formulation. On the contrary, the better performance of hypoplasticity in capturing the influence of hard mechanical boundaries on the bearing capacity may be explained after the picnotropy and barotropy dependence of stiffness. Nonetheless, it should be noticed that poor capturing of strain-softening of the hypoplastic model was reflected again in sudden sharp load drops during strain localization. Finally, the premature (numerical) soil failure predicted at a penetration depth of 41 mm might be explained after this drawback of the model. Predictions of case 2 for a dense sand (RD = 93%) are shown in figure 10.

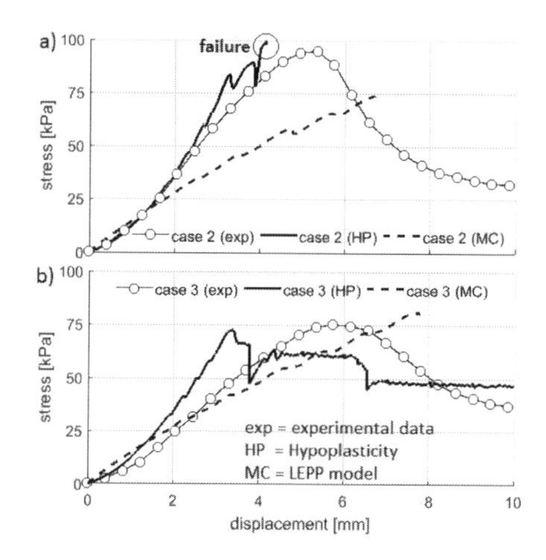

Figure 10. LD curves (RD = 93%): a) case 2, b) case 3.

4.4.2 *FEM simulation of case 3*

When case 3 was simulated using a LEPP model, the peak secant stiffness remained nearly unchanged within a reduction of 3.3% (percentage taken with respect to its numerical prediction of case 1). This supports the previous discussion on the insensitivity of the model to properly capture changes on LD stiffness as a boundary effect, due to the lack of pressure and void ratio dependency of stiffness in its formulation. Moreover, it erroneously predicted an increase of bearing capacity with respect to its numerical prediction for case 2 within 8.0%; contradictory to the expected reduction of bearing capacity when SB are implemented under constrained conditions (based on the experimental observations in section 2.5). This analyses revealed a poor predictive capacity of the LEPP model in terms of capturing hard and soft boundary effects.

On the other hand when case 3 was simulated using hypoplasticity, a reasonably good agreement in terms of bearing capacity was obtained. In this situation, the bearing capacity was predicted to be reduced with respect to its numerical prediction for case 2; consistent with experimental observations. This predicted bearing capacity was only 3.3% below the one measured experimentally in case 3. Nonetheless, the hypoplastic model overpredicted the initial stiffness, which resulted in an underprediction of the PSPD of 43.6% with respect to the experimental result. In this situation, a reduction of peak secant stiffness was captured as expected from experimental evidence; however, yet this reduction was quantified in 10.3% only, against a reduction of 28.9% as measured experimentally. Regarding post-peak response, the hypoplastic model overpredicted the residual load at 10 mm of footing penetration by a factor 25.2% with respect to the experimental result. This revealed hypoplasticity was able to predict most of the aspects of hard and soft boundary effects qualitatively. Moreover, it proved to have a good predictive capability of the influence of mechanical boundary effects in terms of bearing capacity. Nevertheless, a poor capacity to mimic strain-softening like behaviour was witnessed in case 3 as well, observed as sudden load drops which were persistent during post-peak response. In addition, the poor capacity of the model to capture properly the initial stiffness may be attributed to calibration-related problems. For instance, the parameter β which influences the size of the response envelop (therefore it affects the bulk and shear stiffness) has been adopted as 1.0 (typical for sands) during back-analysis because no further lab experiments were available at the moment numerical predictions were made. However, this dimensionless parameter controls the response under proportional stress paths such as oedometric compression (Herle and Gudehus 1999). Therefore, potential inaccuracies in the calibration of the β parameter may have influenced on the response of case 2 and 3, since these constrained conditions may be fairly considered to approach an oedometric condition. This suggests potential issues related to the uniqueness of the solution; however, such a discussion may exceed the scope of the current study.

5 CONCLUSIONS

A numerical investigation of the behaviour of a shallow foundation has been carried out using FEM, providing special attention to boundary effects. A LEPP and a hypoplastic model were used. Through this investigation, drawbacks of both constitutive laws have been revealed. The first constitutive law was unable to capture typical mechanical boundary effects, even after it proved to work reasonably well for free field conditions. Only the second law was able to reproduce some typical boundary effects, i.e. under constrained conditions, an increased ultimate capacity and larger stiffness was predicted, consistent with experimental evidence; however, premature soil failure prohibited to capture post-peak response (potentially attributed to the poor capacity of hypoplasticity to capture strain-softening like behaviour). When SB were modelled, a drop in ultimate capacity was observed, which was in agreement with experiments; nevertheless, the initial stiffness was overpredicted. As a result, the capacity of hypoplasticity to capture most of the aspects of hard and soft mechanical boundary effects has been shown.

ACKNOWLEDGEMENTS

Financial support provided to the first two authors by SENESCYT (Ecuador) is gratefully acknowledged. The authors would like to thank the assistance provided by Wim Verwaal and Arno Mulder during lab testing.

REFERENCES

Allersma, H. (1994). The University of Delft Geotechnical Centrifuge. In *Proceedings of the International Conference Centrifuge 94*, pp. 47–52.

Arthur, J.F.R., T. Dunstan, Q.A.J. Assadi, & A. Assadi (1977). Plastic deformation and failure in granular materials. *Géotechnique 27*, 53–74.

Bauer, E. (1996). Calibration of a comprehensive hypoplastic model for granular materials. Soils and Foundations. *Japanese Geotechnical Society 36*(1), 13–26.

Brinkgreve, R.B.J., E. Engin, & H.K. Engin (2010). Validation of empirical formulas to derive model parameters for sands. In *Numerical Methods in Geotechnical Engineering*, pp. 137–142.

Brinkgreve, R.B.J., S. Kumarswamy, & W. Swolfs (2016). *PLAXIS Manual*. Delft, Netherlands.

Herle, I. & G. Gudehus (1999). Determination of parameters of a hypoplastic constitutive model from properties of grain assemblies. *Mech. Cohes.-Frict. Mater. 4*, 461–486.

Jacky, J. (1944). The coefficient of earth pressure at rest. *Journal of Hungarian Architects and Engineers*, 355–358.

Jefferies, M. & K. Been (2006). *Soil Liquefaction: A critical state approach*. Abingdon, UK: Taylor & Francis.

Lo Presti, D.C., P. Sergio, & C. Virginio (1992). Maximum dry density of cohesionless soils by pluviation and by ASTM D 4353–83. A comparative study. *Geotechnical Testing Journal 15*, 180–189.

Pozo, C. (2016). Soft Boundary Effects (SBE) on the behaviour of a shallow foundation. Master's thesis, Delft University of Technology, Delft, Netherlands.

Pozo, C., Z. Gng, & A. Askarinejad (2016). Evaluation of Soft Boundary Effects (SBE) on the behaviour of a shallow foundation. In *3rd European Conference on Physical Modelling in Geotechnics*, Nantes, France, pp. 385–390.

Roscoe, K.H. (1970). The influence of strain in soil mechanics. *Géotechnique 20*, 129–170.

Stanier, S.A., J. Blaber, W.A. Take, & D. White (2016). Improved image-based deformation measurement for geotechnical applications. *Canadian Geotechnical Journal 53*(5), 727–739.

Vermeer, P.A. (1982). A five-constant model unifying well-established concepts. In *International Workshop on Constitutive relations for Soils, Grenoble*, pp. 175–198.

von Wolffersdorff, P.A. (1996). A hypoplastic relation for granular materials with a predefined limit state surface. *Mech. co- hesive - frictional mater. 1*, 251–271.

Numerical Methods in Geotechnical Engineering IX – Cardoso et al. (Eds)
© 2018 Taylor & Francis Group, London, ISBN 978-1-138-33203-4

Development of load-transfer curves for axially-loaded piles using fibre-optic strain data, finite element analysis and optimisation

L. Pelecanos
University of Bath, UK
Formerly: University of Cambridge, UK

K. Soga
University of California, Berkeley, USA
Formerly: University of Cambridge, UK

ABSTRACT: The load-transfer analysis method is a practical approach for analysing the deformation behaviour of foundation piles where the soil is modeled with a series of springs following a nonlinear response according to load-transfer (t-z and q-z) curves. Distributed Fibre Optic (FO) sensing following the Brillouin Optical Time-Domain Reflectometry/Analysis (BOTDR/A) methodology provides detailed information about the axial pile strains along the entire length of a pile. Direct integration and differentiation provide the corresponding values for vertical displacements and shaft friction which can therefore be used to develop relevant load-transfer curves. This paper presents an approach for developing such load-transfer curves for axially-loaded pile using distributed FO strain data.

1 INTRODUCTION

Pile load tests are performed to get the load-displacement relationship of a single foundation pile and its interaction with the ground. Useful information can be obtained from an instrumented pile (Bond et al., 1991; Jardine et al., 2013; Bica et al., 2014), such as load-transfer curves for axially-loaded piles. These curves represent the developed shear stress at the pile shaft (t-z) or normal stress at the end bearing (q-z) due to the vertical (axial) displacement of the pile relative to the far-field soil. Such information can be obtained from instrumented pile tests and provides an insight into the deformation and ultimate capacity of piles for future design (Frank & Zhao, 1982; Fahey & Carter, 1993; Honjo et al., 1993; Abchir et al., 2016).

This paper proposes a practical method for developing relevant load-transfer curves for axially-loaded piles using spatially-continuous FO strain data from a pile load test using the BOTDR/A technique. A hyperbolic relation is used to define the observed load-transfer curve which considers both the degradation of stiffness and hardening/softening. The proposed method is applied to a set of FO data from a recent pile test in London. The derived curves are subsequently used as input in forward-analyses of the pile load tests and a good comparison is obtained, thus verifying the accuracy of the proposed method.

2 FINITE ELEMENT MODEL

2.1 *Load-transfer analysis of pile-soil interaction*

The pile-soil interaction problem is shown in Figure 1: (a) vertically-loaded pile test is conducted in the field for various values of the applied load P, (b) distributed FO monitoring strain data are obtained for the distributed strain profiles for each load step, (c) monitoring data for the pile top load-displacement curve, (d) finite element (FE) beam-spring analysis model, (e) load-transfer curves.

The aim of this study is to use available monitoring data from pile tests in order to construct appropriate nonlinear load-transfer (for both pile shaft and base: t-z and q-z) curves for simple FE

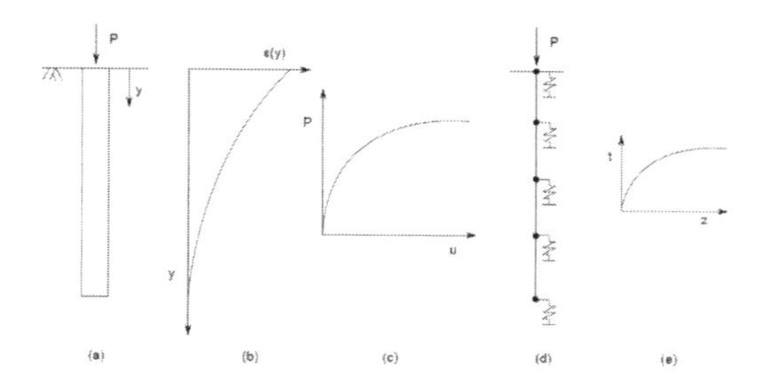

Figure 1. (a) axially-loaded pile, (b) axial strain profile, (c) top load-displacement, (d) FE model, (e) t-z curve.

beam-spring analysis models that can reproduce the pile behavior.

The field monitoring is important (Pelecanos et al., 2017b) and relevant data can be obtained using the Brillouin Optical Time-Domain Relfectometry/Analysis (BOTDR/A) technique which is able to provide spatially-continuous (i.e. distributed) strain data along the entire length of an optical fibre. The relevant background of the physical principle (i.e. theoretical physics, photonics and optics) and the associated experimental approaches followed for calibration can be found elsewhere (Horiguchi et al., 1995; Mohamad, 2007; Iten, 2011; Soga, 2014; Soga et al. 2015). A detailed description of the theory of distributed FO strain sensing and its applications in civil and geotechnical infrastructure may be found in Kechavarzi et al. (2016, 2019), while examples of application to piles are obtained from Klar et al. (2006), Ouyang et al. (2015, 2018) and Pelecanos et al. (2016, 2017a, 2018). More examples on monitoring soil-structure interaction may be found by Acikgoz et al. (2016, 2017), Cheung et al. (2010), Di Murro et al. (2016, 2019), Schwamb et al. (2014), Seo et al. (2017a,b) and Soga et al. (2017).

2.2 *Inverse analysis using FO data*

The proposed method assumes that the following monitoring information can be obtained from a pile load test:

i. distributed FO axial strain profiles with depth, y, for each load step (reading), j: $\varepsilon_a(y, j)$.
ii. top (y = 0) load-displacement curve, for each load step, j: P(u(y = 0), j).

Subsequently, the following information can be calculated for each load step, j:

a. Axial force, F_a(y,j):

$$F_a(y,j) = EA \cdot \varepsilon_a(y,j) \tag{1}$$

b. Shaft friction, SF(y,j):

$$SF(y,j) = \frac{1}{2\pi r} \cdot \frac{d}{dy}\left[EA \cdot \varepsilon_a(y,j)\right] \tag{2}$$

c. Vertical displacement, z(y,j):

$$z(y,j) = z(y = y_0, j) + \int_0^y \varepsilon_a(y,j)\,dy \tag{3}$$

where, E, A and r are the Young's modulus, cross-sectional area and radius of the pile. Young's Modulus, E, can be evaluated from the strain and load measured at the top of the pile. It is assumed here that E is constant with depth (i.e. constructed uniformly). In reality, this may not be the case and careful examination of the measured strain profile and the associated soil profile can lead to assessment of the quality of the test pile (Soga, 2014; Pelecanos et al., 2018).

At each point, i, along the depth of the pile, y = y_i, the evolution of shaft friction, SF(y = y_i, j), for all loadsteps, j, can be plotted against the externally applied load, P(j), or the experienced vertical displacement at that point, z(y = y_i, j). Alternatively, the corresponding values of the axial load transfer, t(y = y_i, j), can be plotted against the experienced vertical displacement at each point, u(y = y_i,j), along the depth of the pile, thus illustrating the field observed behaviour of load-transfer.

Subsequently, a nonlinear load-transfer (t-z or q-z) model can be calibrated against the field-observed t-z or q-z plots (i.e. the nonlinear model equation fitted in the field data using an appropriate optimisation scheme), thus deriving load-transfer curves for pile analysis from a particular field pile load test. The suitability of the derived load-transfer curves can subsequently be verified by comparing the results of a numerical pile FE analysis which employs the derived load-transfer curves against the observed field data, such as axial force, shaft friction of vertical displacement profiles.

2.3 *Nonlinear load-transfer model*

The nonlinear load-transfer (t-z or q-z) model used in this study is an extended version of the standard Hyperbolic model for loading-unloading (Pyke, 1979) that in addition may consider stiffness degradation and hardening/softening. The Degradation & Hardening Hyperbolic Model (DHHM) (Pelecanos et al., 2018; Pelecanos & Soga, 2017a; 2017b; 2018) has 4 distinct parameters which govern different aspects of the curve and its mathematical description is given by Equation 4.

$$t = \frac{k_m z}{\sqrt[d]{1 + \left(\frac{k_m}{t_m} z\right)^{hd}}} \tag{4}$$

where k_m is the maximum stiffness for displacement, z = 0 (units: [force/length^3]), t_m is the "maximum" value of shear stress, t (maximum only in the case of no hardening/softening, i.e. h = 0) (units: [force/length^2]), d is the degradation parameter (units: [-]), that governs the degradation of subgrade modulus, k, with displacement, z, and h is the hardening parameter (units: [-]), that mostly governs the model behaviour at large displacements, z.

This load-transfer relation is used to model the behaviour of nonlinear soil-springs in a FE beam-spring analysis, as shown in Figure 1 (c). These nonlinear springs define the global soil stiffness matrix, K_s, which along with the global pile stiffness matrix, K_p, and the externally applied load, P, contribute to global equilibrium, as described by Equation 5, in order to obtain global pile displacements, u.

$$([K_P]+[K_S])\cdot\{u\}=\{P\} \tag{5}$$

2.4 *Optimization algorithm*

The derivation of load-transfer curves involved calibration of the nonlinear model (Equation 4) through model-fitting against the experimental data by employing an optimisation scheme. The following optimisation scheme was adopted here: The optimal set of numerical analysis parameters $\{k_m, t_m, d, h\}$ is obtained by minimising the Objective Function (OF). This function is to be minimised and adopts the method of least squares of the difference between the t values of the model and the field data for each value of z, subject to the constraint of the model parameters being positive.

The optimisation problem was solved here by using the Levenberg-Marquardt (Levenberg, 1944; Marquardt, 1963) optimisation algorithm. The result of the optimisation problem gives the set of the model parameters that best-fit the field-observed data curve and which is essentially the calibration of the model.

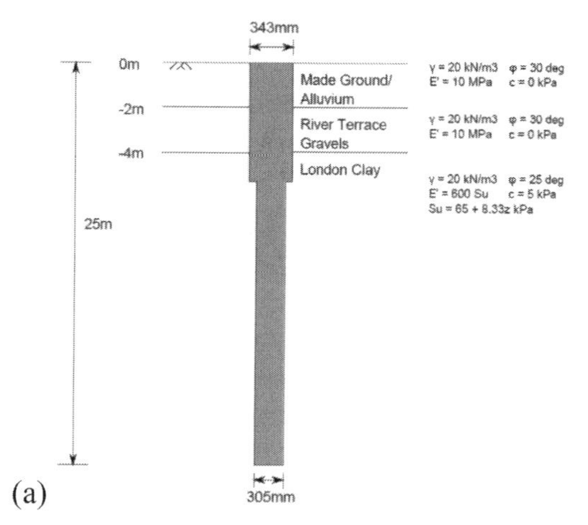

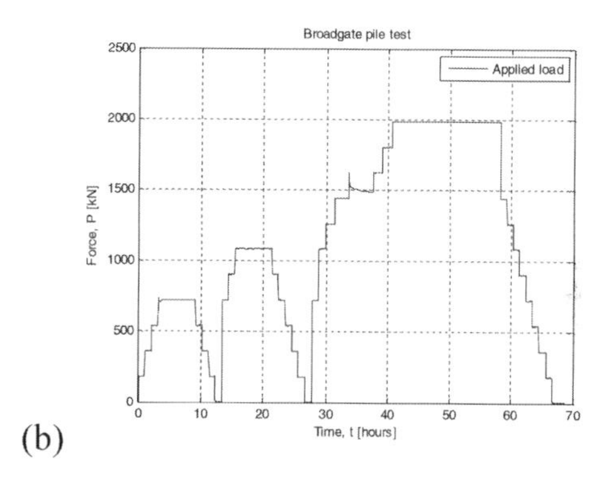

Figure 2. Pile load test: (a) pile geometry, (b) applied load.

3 APPLICATION IN LONDON

3.1 *Pile test & site description*

This procedure is subsequently applied to a recent case study of a maintained pile load test in London which was monitored with distributed FO sensing cables and had available data for both the top load-displacement curve and axial strain profiles with pile depth.

A vertical top-loaded test of a bored pile at a building site in London is considered (Gue et al., 2012). The monitored pile is 25 m long with a varying diameter of 343 mm for the top 6 m and 305 mm for the bottom rest 19 m. Some essential information about the pile test is shown in Figure 2. Local soil stratigraphy comprises of Made Ground, River Terrace Gravels and London Clay where the pile base is founded. The test consists of successive stepped loading-unloading cycles of 720, 1080 and 1985 kN total force applied at the top of the pile.

3.2 *Finite element model*

The back-analysis approach presented before is followed to derive the load-transfer curves (t-z and q-z) for this pile based on the available data from the pile load test. Here, the differentiation of axial strains for the derivation of axial shaft force profiles was done by dividing the pile into 2 different sections (see Table 1). Consequently, the nonlinear DHHM model relation was fitted in the back-calculated load-transfer curves providing the corresponding load-transfer model parameters. The values obtained from the optimisation procedure are listed in Table 1.

Figure 3 shows the back-calculated load-displacement curves. It is shown that the top soil layer exhibits a softer and of lower ultimate strength response than the second layer because the top layer is Made Ground and due to its small depth it has smaller confining pressure. Moreover, the stiffness and strength parameters of the second layer increase with the depth. Finally, the pile base

Table 1. Model parameters for the load-transfer curves.

	Depth	k_m	k_m/y	t_m	t_m/y	d	h
Layer	m	MN/m^3	$MN/m^3/m$	MN/m^2	$MN/m^2/m$	–	-
1	0–6	9	0	0.009	0	2	0.8
2	6–25	10	0	0.06	0.01	1.2	0.95
Base	25	12	-	0.5	-	2	1

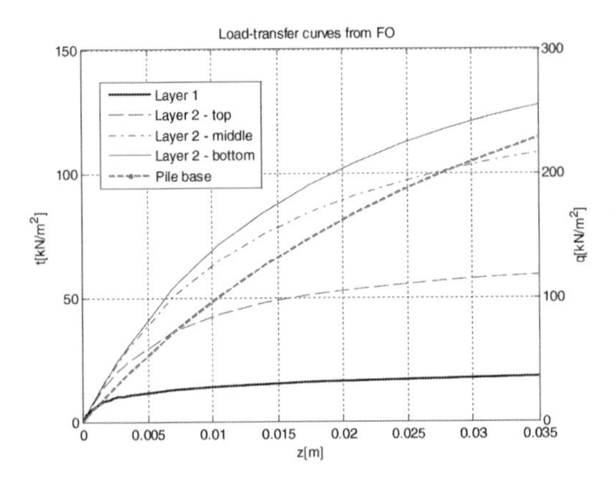

Figure 3. Back-calculated load-transfer curves for the soil.

behaves in a nonlinear way and exhibits a reasonably stiff response.

3.3 *Verification*

To verify the results of the curve derivation approach, a set of numerical FE analyses is run, using as input the load-transfer model parameters obtained from the derivation process (Table 1). Figure 4 shows the results of the "forward" FE analysis and their comparison with the true observed field test results. Top load-displacement and profiles of axial force and vertical displacement are presented. The vertical displacements are calculated directly from the monitoring distributed FO strain data by direct spatial integration (i.e. with respect to depth y), as described by Equation 3 (using the trapezium rule). It may be observed that a very good agreement is obtained between the predictions from the FE model and the field measurements, thus verifying the validity of the adopted load-transfer curves and therefore the applicability of the proposed curve derivation procedure. It is then shown that the monitored profiles of axial force (which are proportional to the measured axial strain) show a visible waviness which is a known characteristic of distributed FO

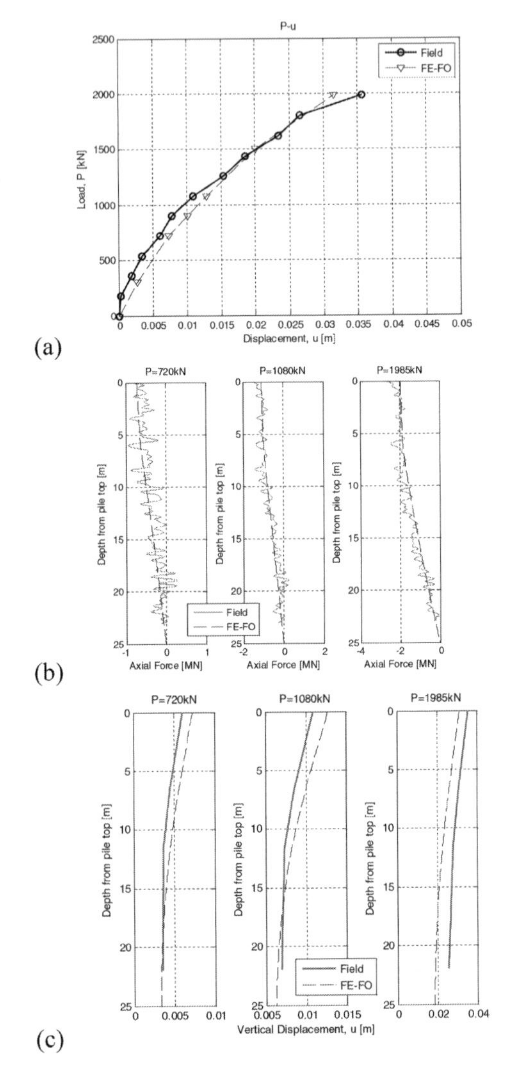

Figure 4. Verification: (a) top load-displacement, (b) axial pile force and (c) vertical pile displacement.

field measurements and this is within the error expected from this type of instruments, i.e. negligible. Additionally, as may be noted, this wavy profile does not have a major impact on the predicted displacements, as integration of strains leads to a smoother profile of displacements.

The latter verification exercise presented demonstrated that the back-analysis approach proposed is able to reproduce the observed field behaviour reasonably well. Therefore, this means that the proposed approach can be used to define the actual behaviour of similar axially-loaded piles. Finally, it is also suggested that this approach can be established as a standard method to post-process distributed fibre optic data and provide the relevant load-transfer curves for piles as part of a routine design. Such an approach can be followed for every instrumented pile load test in the future and hopefully the engineering community will be able to build a database of load-transfer curves that can be used in general pile design.

4 CONCLUSIONS

This paper presents a practical method to derive numerical load-transfer (t-z) curves for axially-loaded piles using distributed FO data. In particular, it concentrates on defining load-transfer curves for axially-loaded foundation piles using Brillouin Optical Time-Domain Reflectometry/Analysis (BOTDR/A). It shows a methodology to be applied and this is followed by an application to a recent pile load test in London, in which the pile was instrumented with distributed fibre optic sensors. It is shown that spatially-continuous strain data from BOTDR/A can offer important information to be used to derive relevant numerical models. It is envisioned that future structural health monitoring can assist in developing such computational models for foundation piles.

ACKNOWLEDGEMENTS

This work was carried out at the Centre for Smart Infrastructure and Construction (CSIC) of the University of Cambridge, which was funded by EPSRC and Innovate UK. Their contribution is gratefully acknowledged.

REFERENCES

Abchir, Z., Burlon, S., Frank, R., Habert, J. and Legrand, S., 2015. t–z curves for piles from pressuremeter test results. Géotechnique, 66(2), pp.137–148.

Acikgoz, MS, Pelecanos, L, Giardina, G, Aitken, J & Soga, K 2017, 'Distributed sensing of a masonry vault during nearby piling', Structural Control and Health Monitoring, vol 24, no. 3, p. e1872.

Acikgoz, MS, Pelecanos, L, Giardina, G & Soga, K 2016, 'Field monitoring of piling effects on a nearby masonry vault using distributed sensing.', International Conference of Smart Infrastructure and Construction, ICE Publishing, Cambridge.

Bica, A.V., Prezzi, M., Seo, H., Salgado, R. and Kim, D., 2014. Instrumentation and axial load testing of displacement piles. Proceedings of the ICE—Geotechnical Engineering, 167(3), pp.238–52.

Bond, A.J., Jardine, R.J. & Dalton, J.C.P., 1991. The design and performance of the Imperial College Instrumented Pile. Geotechnical Testing Journal, 14(4), pp.413–24.

Cheung, L, Soga, K, Bennett, PJ, Kobayashi, Y, Amatya, B & Wright, P 2010, 'Optical fibre strain measurement for tunnel lining monitoring', Proceedings of the ICE—Geotechnical Engineering, vol 163, no. 3, pp. 119–130.

Di Murro, V, Pelecanos, L, Soga, K, Kechavarzi, C, Morton, RF & Scibile, L 2016, 'Distributed fibre optic long-term monitoring of concrete-lined tunnel section TT10 at CERN.', International Conference of Smart Infrastructure and Construction, ICE Publishing, Cambridge.

Di Murro, V, Pelecanos, L, Soga, K, Kechavarzi, C, Morton, RF, Scibile, L, 2019, Long-term deformation monitoring of CERN concrete-lined tunnels using distributed fibre-optic sensing., Geotechnical Engineering Journal of the SEAGS & AGSSEA. 50 (1) (Accepted).

Fahey, M. & Carter, J.P., 1993. A finite element study of the pressuremeter test in sand using a non-linear elastic plastic model. Canadian Geotechnical Journal, 30(2), pp.348–62.

Frank, R. & Zhao, S.R., 1982. Estimation par les parametres pressiometriques de l'enforcement sous charge axiale de pieux fores dans des sols fins. Bull. Liaison Lab. Ponts Chaussees, 119, pp.17–24.

Gue, C.Y., Ouyang, Y., Elshafie, M.Z.E.B. & Soga, K., 2012. Fibre optic monitoring of a pile load test at Broadgate, London. Cambridge: University of Cambridge Engineering Department.

Honjo, Y., Limanhadi, B. & Wen-Tsung, L., 1993. Prediction of single pile settlement based on inverse analysis. Soils and Foundations, 33(2), pp.126–44.

Horiguchi, T, Shimizu, K, Kurashima, T, Tateda, M & Koyamada, Y 1995, 'Development of a distributed sensing technique using Brillouin scattering', Journal of Light-wave Technology, vol 13, no. 7, pp. 1296–1302.

Iten, M 2011, 'Novel applications of distributed fiber-optic sensing in geotechnical engineering', PhD Thesis, ETH, Zurich.

Jardine, R.J., Zhu, B.T. & Foray, Y.Z.X., 2013. Interpretation of stress measurements made around closed-ended displacement piles in sand. Geotechnique, 63(8), pp.613–27.

Kechavarzi, C, Soga, K, de Battista, N, Pelecanos, L, Elshafie, MZEB & Mair, RJ 2016, Distributed Fibre Optic Strain Sensing for Monitoring Civil Infrastructure, Thomas Telford, London.

Kechavarzi, C., Pelecanos, L., de Battista, N., Soga, K. 2019. Distributed fiber optic sensing for monitoring reinforced concrete piles. Geotechnical Engineering Journal of the SEAGS & AGSSEA. 50 (1) (Accepted)

Klar, A, Bennett, PJ, Soga, K, Mair, RJ, Tester, P, Fernie, R, St John, HD & Torp-Peterson, G 2006, 'Distributed strain measurement for pile foundations', Proceedings of the ICE—Geotechnical Engineering, vol 159, no. 3, pp. 135–144.

Levenberg, K., 1944. A method for the solution of certain nonlinear problems in least squares. Quarterly Journal of Applied Mathematics, 2, pp.164–68.

Marquadt, D.W., 1963. An algorithm for least-squares estimation of nonlinear parameters. Journal of the Society for Industrial and Applied Mathematics, 11(2), pp.431–41.

Mohamad, H 2007, 'Distributed optical fibre strain sensing of geotechnical structures. PhD Thesis', University of Cambridge, Cambridge, UK.

Ouyang, Y, Broadbent, K, Bell, A, Pelecanos, L & Soga, K 2015, 'The use of fibre optic instrumentation to monitor the O-Cell load test on a single working pile in London', Proceedings of the XVI European Conference on Soil Mechanics and Geotechnical Engineering, Edinburgh.

Ouyang, Y., Pelecanos, L., Soga, K. 2018, 'The field investigation of tension cracks on a slim cement grouted column'. DFI-EFFC International Conference on Deep Foundations and Ground Improvement: Urbanization and Infrastructure Development-Future Challenges, Rome, Italy. (Accepted)

Pelecanos, L, Soga, K, Hardy, S, Blair, A & Carter, K 2016, 'Distributed fibre optic monitoring of tension piles under a basement excavation at the V&A museum in London.', International Conference of Smart Infrastructure and Construction, ICE Publishing, Cambridge.

Pelecanos, L, Soga, K, Chunge, MPM, Ouyang, Y, Kwan, V, Kechavarzi, C & Nicholson, D 2017a, 'Distributed fibre-optic monitoring of an Osterberg-cell pile test in London.', Geotechnique Letters, vol 7, no. 2, pp. 1–9.

Pelecanos, L., Skarlatos, D. and Pantazis, G., 2017b. Dam performance and safety in tropical climates—recent developments on field monitoring and computational analysis. 8th International Conference on Structural Engineering and Construction Management (8th ICSECM), Kandy, Sri Lanka.

Pelecanos, L., Soga, K., Elshafie, M., Nicholas, d. B., Kechavarzi, C., Gue, C.Y., Ouyang, Y. and Seo, H., 2018. Distributed Fibre Optic Sensing of Axially Loaded Bored Piles. Journal of Geotechnical and Geoenvironmental Engineering. 144 (3) 04017122.

Pelecanos, L., & Soga, K. 2017a. Innovative Structural Health Monitoring Of Foundation Piles Using Distributed Fibre-Optic Sensing. 8th International Conference on Structural Engineering and Construction Management 2017, Kandy, Sri Lanka.

Pelecanos, L., & Soga, K. 2017b. The use of distributed fibre-optic strain data to develop finite element models for foundation piles. 6th International Forum on Opto-electronic Sensor-based Monitoring in Geo-engineering, Nanjing, China.

Pelecanos, L & Soga, K 2018, 'Using distributed strain data to evaluate load-transfer curves for axially loaded piles', Journal of Geotechnical & Geoenvironmental Engineering, ASCE (Under Review).

Pyke, R., 1979. Nonlinear soil models for irregular cyclic loadings. J. Geotech. Engng. ASCE, 105(6), pp.715–26.

Schwamb, T, Soga, K, Mair, RJ, Elshafie, MZEB, R., S, Boquet, C & Greenwood, J 2014, 'Fibre optic monitoring of a deep circular excavation', Proceedings of the ICE—Geotechnical Engineering, vol 167, no. 2, pp. 144–154.

Seo, H, Pelecanos, L, Kwon, Y-S & Lee, I-M 2017a, 'Net load-displacement estimation in soil-nailing pullout tests' Proceedings of the Institution of Civil Engineers—Geotechnical engineering, 170 (6) 534–547.

Seo, H & Pelecanos, L 2017b, 'Load Transfer In Soil Anchors—Finite Element Analysis Of Pull-Out Tests'. 8th International Conference on Structural Engineering and Construction Management 2017, Kandy, Sri Lanka.

Soga, K 2014, 'XII Croce Lecture: Understanding the real performance of geotechnical structures using an innovative fibre optic distributed strain measurement technology', Rivista Italiana di Geotechnica, vol 4, pp. 7–48.

Soga, K, Kechavarzi, C, Pelecanos, L, de Battista, N, Williamson, M, Gue, CY, Di Murro, V & Elshafie, M 2017, 'Distributed fibre optic strain sensing for monitoring underground structures—Tunnels Case Studies ', in S Pamukcu, L Cheng (eds.), Underground Sensing, 1st edn, Elsevier.

Soga, K, Kwan, V, Pelecanos, L, Rui, Y, Schwamb, T, Seo, H & Wilcock, M 2015, 'The role of distributed sensing in understanding the engineering performance of geotechnical structures', Proceedings of the XVI European Conference on Soil Mechanics and Geotechnical Engineering, Edinburgh.

Numerical Methods in Geotechnical Engineering IX – Cardoso et al. (Eds)
© 2018 Taylor & Francis Group, London, ISBN 978-1-138-33203-4

Numerical analysis of concrete piles driving in saturated dense and loose sand deposits

M. Aghayarzadeh, H. Khabbaz & B. Fatahi
University of Technology Sydney (UTS), Sydney, NSW, Australia

ABSTRACT: Many approaches and techniques are used to evaluate pile axial capacity ranging from static methods to dynamic methods, which are based on either the results of pile driving or numerical simulations, which require reliable constitutive models representing the real soil behaviour and the interaction between the pile and soil. In this paper, using PLAXIS software and different constitutive soil models including Mohr-Coulomb, Hardening Soil and Hypoplastic with Intergranular Strain models, the behaviour of concrete piles driven into saturated dense and loose sand deposits under a hammer blow is evaluated. The main objective of this study is to assess the influence of different factors including frequency of loading and Hypoplastic soil model parameters on the recorded velocity and pile head displacement. In addition, the concept of one-dimensional wave propagation induced by pile driving is discussed. It is indicated that using the Intergranular Strain concept, defined in Hypoplastic soil model, small strain behaviour of soil around the pile during driving can directly be captured. The results of this study reveals that considering the Hypoplastic model, incorporating the Intergranular Strain concept, can accumulate much less strains than the corresponding predictions excluding the Intergranular Strain, and hence predict the pile performance during driving more realistically.

1 INTRODUCTION

Despite the fact that pile design methods have considerably advanced within the past few years, the most fundamental aspect of pile design related to the axial bearing capacity, still heavily relies on empirical correlations. Many studies have been reported during the last decades to improve the prediction of pile axial and lateral capacity, as well as adopting dynamic methods during pile driving and testing for pile evaluation (Goble et al. 1967, Goble & Rausche 1970, Rausche et al. 1972, Fakharian & Hosseinzadeh 2011, Fatahi et al. 2014). In fact, pile head displacements, force and velocity traces, which are captured during pile driving and dynamic load testing, are used as the benchmarks of pile performance. Hence, the bearing capacity of a pile can be calculated by analysing these data, applying direct and indirect methods used in the pile driving analyser (PDA) device and CAPWAP software, respectively.

Some of the previous research studies tried to use the finite element method to analyse pile driving. Mabsout & Tassoulas (1994) and Mabsout et al. (1995) assumed a bored pile for the pile driving analysis with embedment depths of 6, 12 and 18 m, which were surrounded by saturated normally-consolidated clay. They employed a non-linear bounding surface-plasticity model with the aim of conducting a detailed analysis of pile driving, using a finite element technique by taking into account the non-linear behaviour of undrained clayey soil and tracing the penetration of the pile into the soil. Fakharian et al. (2014) simulated the signal matching process during dynamic load testing, using finite element and finite difference software packages in which Mohr-Coulomb soil model was assigned to the soil layers. In the latter study, the pile load testing procedure was simulated in more realistic conditions, so it was concluded that incorporating the radiation damping effect could results in more reasonable predictions. It should be noted that application of advanced soil models and consideration of nonlinear soil-pile interaction have been considered widely in other fields such as geotechnical earthquake engineering (e.g. Hokmabadi & Fatahi 2016, Fatahi et al. 2018).

In this study, using PLAXIS 2D version 2017, concrete pile performance, dynamically driven into saturated loose and dense sand has been scrutinised. In the analysis, the recorded force and velocity at the gauge location and the pile head displacement under a single hammer blow at the end of driving have been assessed, using different soil models ranging from Mohr-Coulomb to advanced soil models. The results, predicted through various soil models have been evaluated and compared. It should be noted that, the small strain behaviour of soil around the pile, which is observed during pile driving, has been assessed using an advanced soil model as a part of this study.

2 NUMERICAL MODEL CHARACTERISTICS

In this study, the axisymmetric finite element model was used in numerical simulation. Concrete pile as a volume pile element with a diameter of 0.4 m and a total length of 10 m was modelled numerically. A linear elastic model with an elastic modulus of 300 GPa, a Poisson's ratio of 0.2 and a unit weight of 25 kN/m^3 was assigned to the pile cluster. Whereas, elastic perfectly plastic Mohr-Coulomb (MC), Hardening Soil (HS) and Hypoplastic (HP) with Intergranular Strain (IGS) soil models were assigned to the soil cluster. In addition, viscous boundaries were used in the numerical model to simulate the geometric damping and the far-field boundaries. The hammer impact was simulated as a harmonic signal with an amplitude of 5 MPa, a phase of zero degree and a frequency of 50 Hz similar to what was reported in PLAXIS (2017).

In numerical simulations, loose and dense Baskarp sands were used as the soil deposit. As explained by Elmi Anaraki (2008), Baskarp sand is a uniform sand with a total unit weight of 20 kN/m^3, an initial void ratio of 0.83 and 0.65 representing the loose and dense conditions, respectively. The soil properties assigned for Hardening Soil and Hypoplastic soil models were selected based on Dung (2009) and Elmi Anaraki (2008) studies, while the equivalent Mohr-Coulomb soil model properties were obtained from Aghayarzadeh et al. (2018), who correlated the results of a drained triaxial test using the soil test facility defined in PLAXIS software. The soil properties corresponding to each soil model and interface parameters used in numerical simulation are summarised in Tables 1 to 4.

It should be noted that in order to simulate the interaction between the pile and soil, appropriate interface elements were considered. In all soil models, the interface strength and deformation parameters were assumed to be correlated to the surrounding soil parameters without consideration of any reduction factor for the sake of simplicity. In other words, for Mohr-Coulomb and Hardening Soil models the interface strength reduction factor (R_{int}) was assumed to be equal to one and for the Hypoplastic model, according to PLAXIS (2017), the interface parameters defined in Table 4 were considered. According to ASTM D 4945 (2010), the strain gauge and the accelerometer during the dynamic load testing should be mounted at least 1.5D (D is diameter of pile) below the pile head. In this study, force and velocity traces were recorded at 2D distance below the pile head. An illustration of the finite element model used in analysis is shown in Figure 1a. As shown in Figure 1b, half

Table 1. Baskarp sand properties for hypoplastic soil model with intergranular strain (after Dung 2009).

Parameters	Hypoplastic model with intergranular strain
ϕ_c (degree)	30
h_s (MPa)	4000
N	0.42
e_{d0}	0.548
e_{c0}	0.929
e_{i0}	1.08
α	0.12
β	0.96
m_T	2
m_R	5
R_{max}	0.0001
β_r	1
χ	2

Table 2. Baskarp sand properties for hardening soil model (after Dung 2009).

Parameters	Dense	Loose
E_{50}^{ref} (MPa)	40.5	31
E_{oed}^{ref} (MPa)	50	33
E_{ur}^{ref} (MPa)	121.5	93
ϕ (degrees)	37	31.3
ψ (degrees)	9	2
m	0.5	0.5
ν_{ur}	0.2	0.2
p^{ref} (kPa)	100	100

Table 3. Baskarp sand properties for Mohr-Coulomb soil model including both tangent and secant soil modulus (after Aghayarzadeh et al. 2018).

Parameters	Dense	Loose
E_i (MPa)	60	45
E_{50} (MPa)	33	24.75
υ	0.35	0.25
ϕ (degree)	37	31.3
ψ (degree)	9	2

Table 4. Interface parameters for hypoplastic soil model used in numerical modelling (after Aghayarzadeh et al. 2018).

Parameters	Dense	Loose
E_{oed}^{ref} (MPa)	50	33
c'_{ref} (kPa)	0.1	0.1
ϕ' (degree)	37	31.3
ψ (degree)	9	2
UD-Power	0	0
UD- P^{ref} (kPa)	100	100

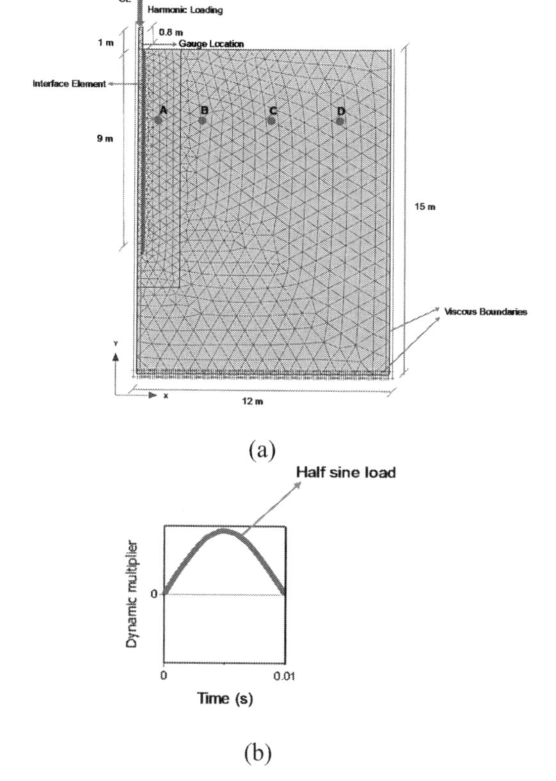

Figure 1. (a) Finite element model of the pile and the adjacent ground with the corresponding generated mesh and points A, B, C and D and their corresponding radiuses $r_A = 1$ m, $r_B = 3$ m, $r_c = 6$ m and $r_D = 9$ m (b) applied harmonic load.

sine load with a dynamic time interval of 0.01 s (i.e. 50 Hz as mentioned earlier) was applied on the pile head to simulate the hammer load.

The first version of the Hypoplastic constitutive law was proposed by Kolymbas (1985), describing the stress-strain behaviour of granular materials in a rate form. The Hypoplastic model can successfully predict the soil behaviour in the medium to large strain ranges. However, in the small strain range and upon cyclic loading it cannot predict the high quasi-elastic soil stiffness accurately. To overcome this problem, an extension of the Hypoplastic equation by considering an additional state variable, termed "Intergranular Strain (IGS)", was proposed by Niemunis & Herle (1997) to determine the direction of the previous loading. In fact, the Intergranular Strain concept enables to model small-strain-stiffness effects in Hypoplasticity and therefore adopted in this study.

3 PILE DRIVING SIMULATION

As explained by Masouleh & Fakharian (2008), one of the important advantages of pile driving and pile load testing simulation in finite element and finite difference software is that the radiation or geometric damping is automatically considered in numerical modelling. In fact, the travelling compressive or tensile wave along the pile shaft causes a relative displacement between pile and soil, which results in generation of shear wave in the adjacent soil that can propagate radially. For evaluating the radiation damping effect, in this study shear stress variations with time at a depth of 4 m and at different distances from the pile axis (i.e. 1, 3, 6 and 9 m) in both dense sand and loose sand were recorded (Figure 2). Figure 2 represents a rapid reduction of shear stress wave amplitude with distance from the pile skin, such that near the vertical boundaries, it is practically zero for both dense and loose sand. This finding not only confirms the soil inertia or radiation damping effect in finite element modelling, but also proves that the viscous boundary has been regarded far enough to prevent the wave reflection in the model.

During the pile load testing and pile driving, the pile head displacement is one of the most important factors that should be taken into account. In this paper the pile head displacement of concrete pile driven into the saturated dense and loose sand using three constitutive soil models are obtained and compared to each other, as illustrated in Figure 3.

Referring to Figure 3, it is evident that driving a pile into dense sand induces less displacement compared to loose sand. All employed constitutive soil models including Mohr-Coulomb, Hardening Soil and Hypoplastic with Intergranular Strain (IGS) delivered reasonably a similar trend. It is worth mentioning that in the study conducted by Aghayarzadeh et al. (2018) related to simulation of the static load testing, E_{50} (the secant modulus) was used in Mohr-Coulomb model and it showed

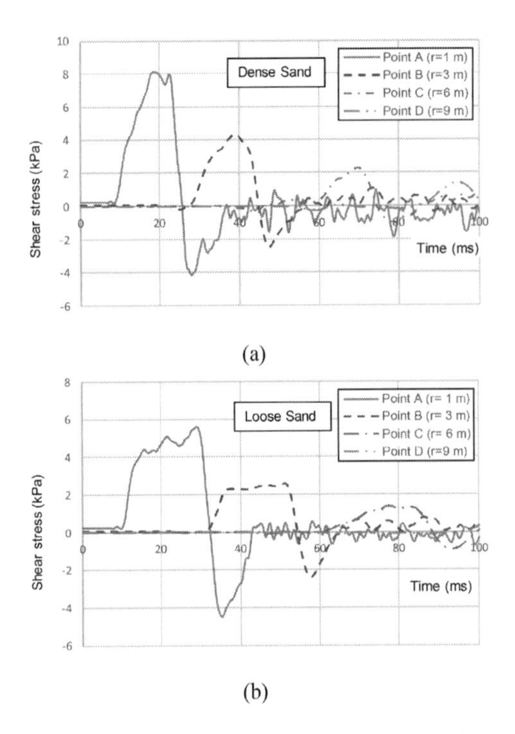

Figure 2. Variation of shear stress in soil at different distances from pile shaft (a) dense sand (b) loose sand (Hardening Soil model).

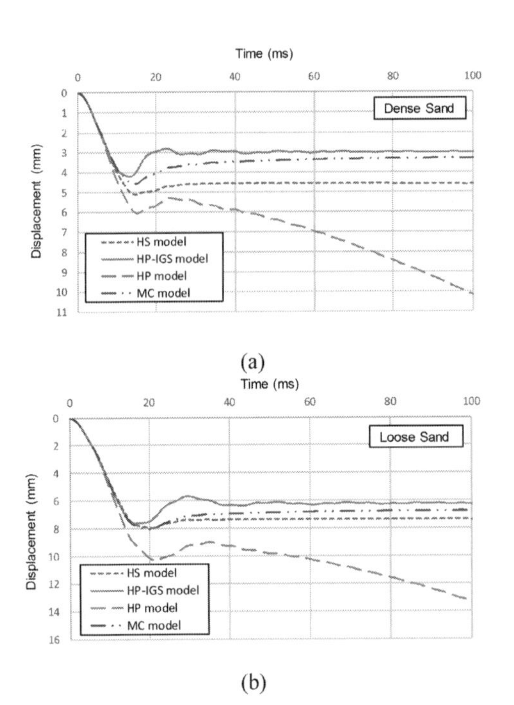

Figure 3. Pile head displacement (a) dense sand (b) loose sand.

a reasonable correlation with other soil models, hence in this study Mohr-Coulomb model was used embracing this elastic modulus. However, it can be seen that using Hypoplastic soil model without activating the Intergranular Strain generates an increase in the observed displacement of pile head with time. Since the stress wave induced by the hammer impact dissipates, then it is not expected that the displacement to increase significantly. It is evident that HP model with IGS activation yields much less strain compared to the case when the IGS is not applied. This is mainly attributed to the fact that IGS concept simulates the small strain behaviour which is dominant during the pile driving.

Generally, when a sudden axial force is applied on a pile, a stress wave is generated, which travels away from the point of force application. As long as stress waves at the gauge location, travel in only one direction and no reflections arrive at this point, the force in the pile is proportional to the velocity of particle motion; known as proportionality theory. The proportionality between these two curves (i.e. force and velocity versus time curves) is destroyed as soon as waves, caused by the soil resistance forces at the pile boundaries, reach the top of pile. The concept of one dimensional wave propagation generated by a hammer impact, comprising of induced compressive wave and reflected wave as a result of the soil resistance, is schematically shown in Figure 4. The proportionality relationship can be written as:

$$F(t) = Zv(t) \quad (1)$$

$V(t)$ = the velocity of particle motion (downward positive) at the point under consideration (m/s)

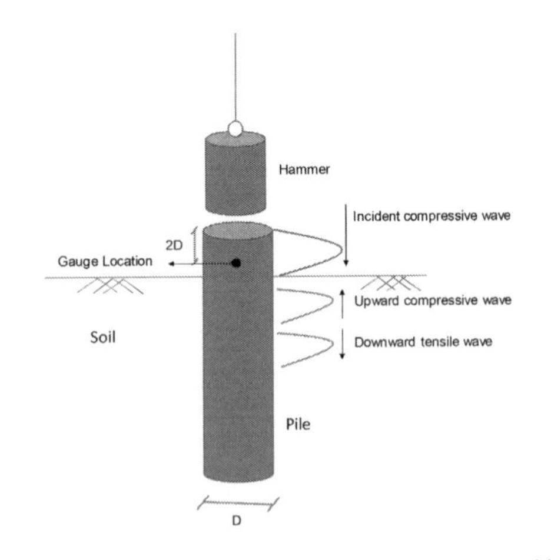

Figure 4. One dimensional wave propagation caused by hammer impact.

F (t) = the force (compression positive) at the point of interest where velocity is determined (kN)

$Z = EA/c$ = impedance or the proportionality coefficient between force and velocity (kN.s/m)

c = the velocity of propagation of the stress wave (m/s), which in concrete piles is regarded as approximately 3600 m/s

A = the pile cross-sectional area of pile (m^2)

E = the modulus of elasticity of pile (kPa)

The recorded velocity at the gauge location ($v(t)$) multiplied by the impedance (Z) of the pile is shown in Figure 5. As shown in Figure 5, velocity traces predicted by different soil models have a reasonable agreement. However, Hypoplastic model without considering IGS could not simulate the dissipation of velocity with time. As explained earlier, due to the small strain behaviour observed during driving by incorporating the Intergranular Strain (IGS), a more reasonable response and predictions are achieved.

Force and velocity × impedance traces, recorded at the gauge location, are compared in Figure 6. Referring to Equation (1), it is expected to observe a reasonable correlation between force and velocity × impedance traces particularly before the first peak of both curves. However, as it is illustrated in Figure 6, discapancies are observed. This is mainly attributed to the long stress wave period assumed for the hammer impact which causes overlapping between the incident and reflected waves within the observation time. For evaluating

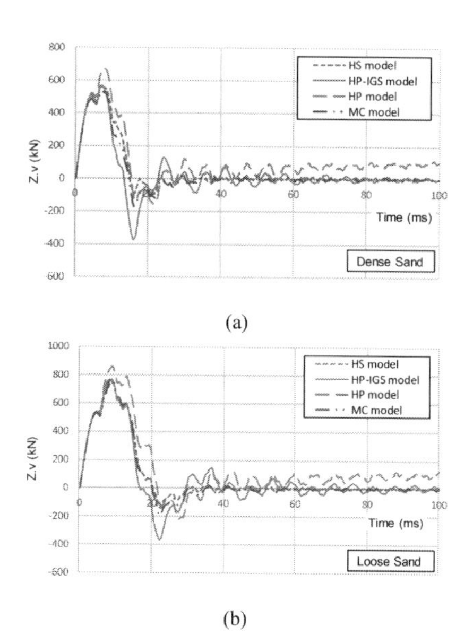

Figure 5. Impedance × Velocity variation with time, recorded at the gauge location for (a) dense sand (b) loose sand.

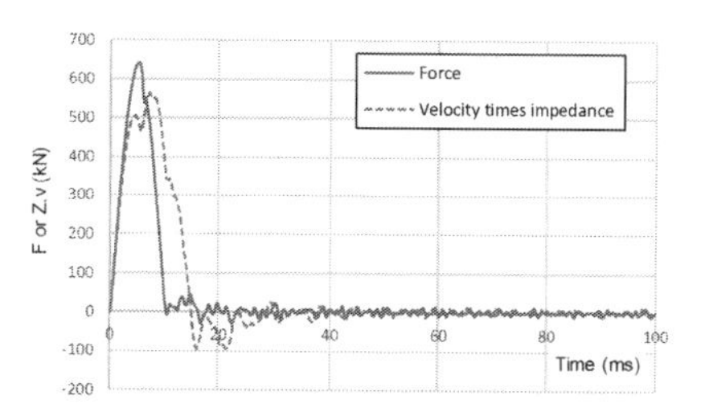

Figure 6. Velocity × impedance and force traces, recorded at the gauge location using hardening soil model for dense sand by applying harmonic loading with a frequency of 50 Hz.

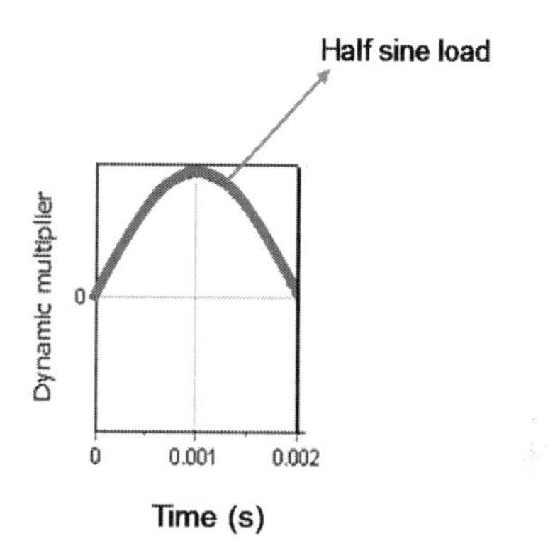

Figure 7. Applied harmonic load with a frequency of 250 Hz.

this wave overlapping effect, the frequency of loading increased from 50 Hz to 250 Hz. As shown in Figure 7, the half sine period or impact time decreased to 0.002 s. The corresponding force and velocity × impedance traces were recorded and shown in Figure 8.

As shown in Figure 8, by increasing the impact load frequency to 250 Hz, both force and velocity traces show reasonable correlations before the first peak (corresponding to t_1) since decreasing the impact time causes a decrease in the length of the induced compressive wave. As explained by Lowery et al. (1969), when the compressive stress wave reaches the toe of the pile, it will be reflected back up the pile in some manner depending on the soil resistance. If the toe of the pile is experiencing slight or no resistance from the soil, it will be reflected back up the pile as a tensile stress wave and these two waves may overlap at the certain points. On the other hand, if the soil is hard or very firm at the pile toe, the initial compressive stress wave traveling down the pile will be reflected back up the pile also as a compressive stress wave and again these two stress waves may overlap. In the case that toe is experiencing slight resistance

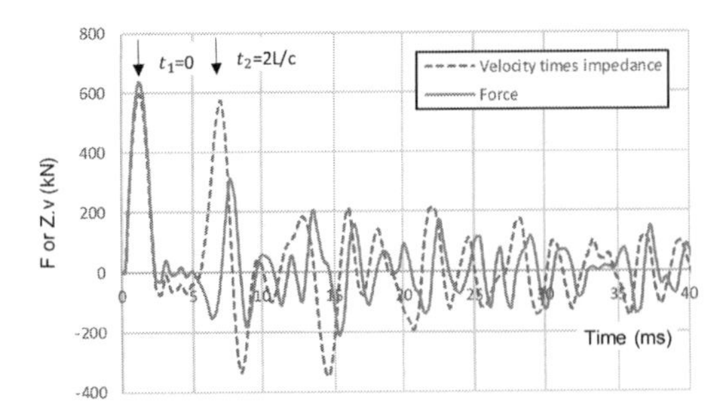

Figure 8. Velocity × impedance and force traces, recorded at the gauge location using hardening soil model for dense sand by applying a harmonic load with a frequency of 250 Hz.

and pile length is short compared to the length of the stress wave, the reflected tensile stress wave overlaps with the initial compressive stress wave coming down the pile. Since the net stress at any point is the algebraic sum of the two, they tend to cancel each other at the toe and hence little or no reflected tensile stress wave will occur. In addition, the length of the stress wave induced by ram impact can be calculated by the following equation:

$$L_s = ct_s \tag{2}$$

where L_s = the length of stress wave (m) t_s = the time of ram contact (impact time) (s) c = the velocity of stress wave (m/s).

By applying a load with a frequency of 50 Hz, the contact time of hammer and pile is regarded as 0.01 s, as shown in Figure 1b, generating a compressive stress wave with 36 m length that is considerably longer than pile length. However, when the frequency of loading increases to 250 Hz, the contact time of the hammer and pile decreases to 0.002 s while the length of corresponding stress wave would be 7.2 m, which is shorter than the pile length.

Referring to Figure 8, at time 2 L/c (the time corresponding to travelling down the wave and reflecting from pile toe to the gauge location) the force trace is negative and velocity × impedance is positive, which resembles the reflection of tensile stesss from the pile toe. In reality, the pile tip is embedded in a sandy deposit having a high densification potential during the driving process. On the basis of an analytical study conducted by Yang (2006), the influence zone of an axially loaded pile in clean sand and compacted silty sand can extend to 3.5D–5.5D and 1.5D–3D (D is the pile diameter) below the pile toe, respectively. However, in this study the the properties of soil below the toe were not changed and were basically assumed as the initial in-situ conditions. Thus, the pile toe showed a rather low resistance against penetration causing the reflection of tensile stresses.

4 FURTHER ASSESSMENT OF HYPOPLASTIC MODEL FOR PILE DRIVING

4.1 *Parametric study*

In this section, the influence of different paramters, defined in the Hypoplastic model with the Intergranular Strain concept on the obtanied pile head displacement during the pile driving is evaluated. As reported by Aghayarzadeh et al. (2018), among the Hypoplastic model parameters, the granular hardness (h_s) showed the least impact, whereas the critical friction angle (φ_c) indicated the most significant influence on the load-displacement curve, obtained during the simulation of static load testing. Herein, the effect of these two parameters are evaluated on the recorded pile head displacement during pile driving. To achieve this aim, the critical friction angle increased by 30%, while other Hypoplastic factors remained unchanged. As shown in Figure 9a, similar to the observed results during static load test, increasing the critical friction angle shows a considerable effect on the recorded displacement. Referring to Figure 9b, increasing the granular stiffness (h_s) by up to 30% does not show any notable influence on the predicted displacement.

The sensisitivity of the simulated response of the recorded pile head displacement during driving to the varaition of Intergranular Strain paramters are shown in Figure 10. The Intergranular Strain includes five model parameters:

M_R = paramter controlling the initial (very small strain) shear modulus upon 180 degree strain path reversal and in the initial loading

M_T = parameter controlling the initial shear modulus upon 90 degree strain path reversal

R = the size of the elastic range

β_r and χ = parameters controling the rate of degradation of the stiffness with strain

Reffering to Mašín (2015) in IGS extension, the stifness evolution is as follows:

$$E = m_R E_0 \qquad \varepsilon < R \tag{3}$$

$$E = E_0 + E_0(m_R - 1)(1 - \rho^{\chi}) \qquad \varepsilon > R \tag{4}$$

Where ρ is the magnitude of the Intergranular Strain rate and E_0 is the initial tangent modulus.

However as explained by Dung (2009), three of IGS parameters including m_T, m_R and R can be usually taken as constants. Therefore, in this study, only the variation of two parameters, namely β_r and χ, controlling the rate of degradation of the stiffness with strain, are considered and investigated.

In fact both parameters β_r and χ control the shape of stifness degradation curves (Mašín 2015, Niemunis & Herle 1997). Referring to Figure 10, it can be inferred that by increasing χ and decreasing

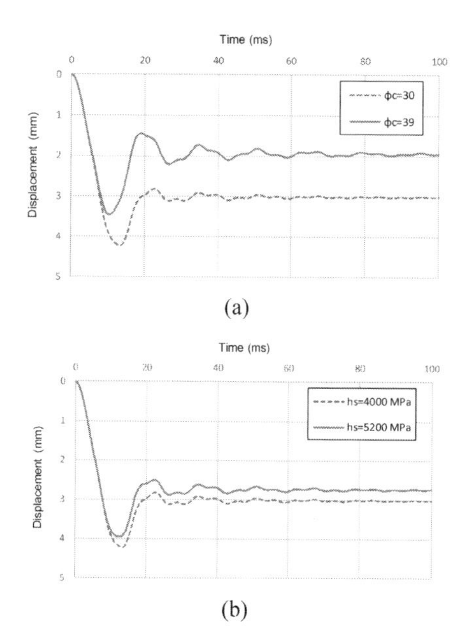

Figure 9. Influence of hypoplastic model parameters on the measured pile head displacement during the pile driving in dense sand (a) critical friction angle (b) granular hardness.

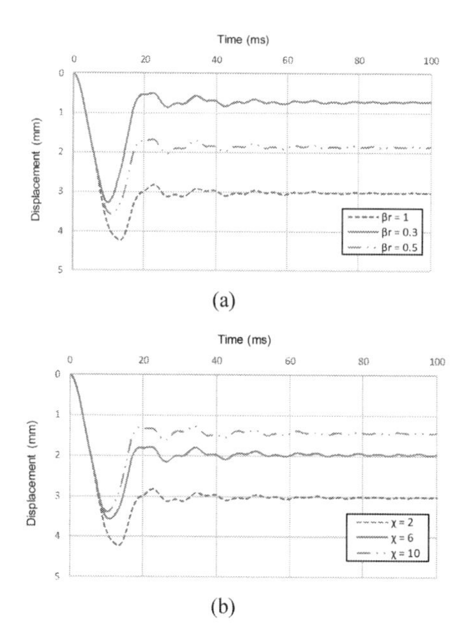

Figure 10. Influence of intergranular strain parameters defined in hypoplastic model on the measured pile head displacement during the pile driving in dense sand (a) β_r and (b) χ.

β_r the elastic range of stiffness degradation curves increases, and, hence pile head displacement shows more elastic deformation.

4.2 *Intergranular strain tensors*

In this section in order to assess the behaviour of piles during the driving and the extent to which Intergranular Strain (IGS) can impact the predictions, the pile was simulated in both dense and loose sand using Hypoplastic model with IGS. In fact the behaviour of pile during the driving consists of loading and unloading stages (Ng 2011, Ng et al. 2013). Herein the normalized Intergranular Strain tensors are depicted in Figures 11 and 12 for dense sand and loose sand at the end of the driving (unloading conditions), respectively. According to Mašın (2010) the normalized length of the Intergranular Strain tensor varies between 0 and 1, corresponding to the soil being inside the elastic range and swept-out of the small-strain memory, respectively. In other words, the normalized length of the Intergranular Strain tensor demonstrates how the small-strain stiffness is activated in different parts of the modelled geometry. In the case of the normalized length of the Intergranular Strain being equal to 1, the soil behaviour is governed by the basic Hypoplastic model. Figures 11 and 12 show that at the end of driving in the vicinity of the pile, the predicted values for normalized length of the Intergranular Strain tensor are very low and

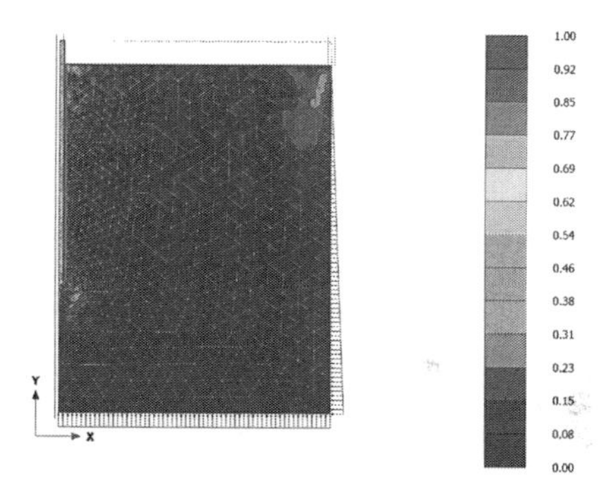

Figure 11. Concrete pile driving—normalised intergranular strain tensor for dense sand.

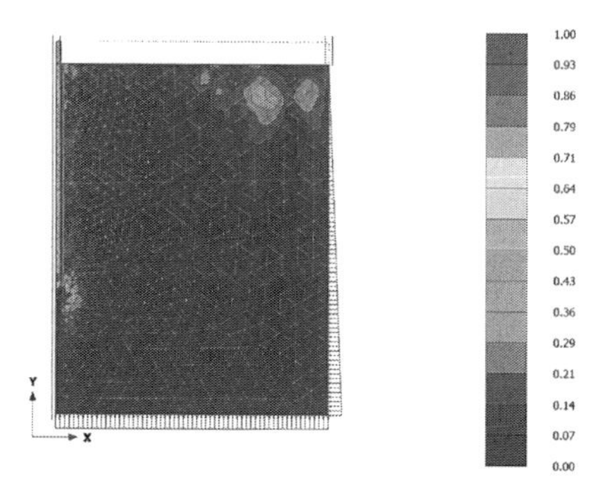

Figure 12. Concrete pile driving—normalised intergranular strain tensor for loose sand.

at these locations the soil elements behave elastically and show small strain behaviour. In addition, it is observed that dense sand shows more elastic behaviour compared to loose sand. This clearly indicates the presence of the small strain behaviour of soil around the pile during the propagation and dissipation of the induced wave in the pile.

5 CONCLUSIONS

In this study, it has been observed that the three target constitutive soil models, including Mohr-Coulomb, hardening soil and Hypoplastic with Intergranular Strain soil models, can yield similar trends in terms of the pile head displacement and the recorded velocity at the gauge location during pile driving process. However, introducing the Intergranular Strain concept provides more reasonable predictions and can accumulate significantly less strains.

In addition, the influence of frequency on pile driving was evaluated in this study. The results showed that by increasing the hammer impact time, the length of the induced stress wave increases and therefore induced and reflected waves may overlap considerably. This finding correlates well with the fact that in short piles (i.e. the length of the pile is considerably smaller than the stress wave length), the downward and upward travelling waves can overlap before the peak force or velocity at the gauge location is reached. Moreover, parametric study on Intergranular Strain was conducted on two key parameters of Hypoplastic model, including the critical friction angle and the granular hardness. It was found that the variation of critical friction angle impacted the predicted pile performance during driving significantly, whereas the effect of the granular hardness variation was negligible. Furthermore, parameters affecting the rate of degradation of the stiffness with strain defined in intergranular strain (IGS) concept could change the elastic range of soil. Finally, by evaluating the normalised Intergranualr tensors it was shown that by introducing the IGS in the model the soil in the vicinity of pile would indicate the small strain behaviour during the unloading stage as expected.

REFERENCES

Aghayarzadeh, M., Khabbaz, H. & Fatahi, B. 2018. Evaluation of concrete bored piles behaviour in saturated loose and dense sand during the static load testing. *5th GeoChina International Conference.* HangZhou, China (accepted).

ASTM 2010. Standard test method for high-strain dynamic testing of piles. *D 4945–08.* Philadelphia: ASTM.

Dung, P.H. 2009. *Modelling of installation effect of driven piles by hypoplasticity.* Master of Science thesis, Delft University of Technology.

Elmi Anaraki, K. 2008. *Hypoplasticity investigated.* Master of Science thesis, Delft University of Technology.

Fakharian, K. & Hosseinzadeh, A.I. 2011. Pile driving experiences in Persian Gulf calcareous sands. *Frontiers in Offshore Geotechnics II.* London: Taylor & Francis Group.

Fakharian, K., Masouleh, S.F. & Mohammadlou, A.S. 2014. Comparison of end-of-drive and restrike signal matching analysis for a real case using continuum numerical modelling. *Soils and Foundations,* 54, 155–167.

Fatahi, B., Basack, S., Ryan, P., Zhou, W.-H. & Khabbaz, H. 2014. Performance of laterally loaded piles considering soil and interface parameters. *Geomechanics and Engineering,* 11, 501–516.

Fatahi, B., Van Nguyen, Q., Xu, R. & Sun, W.-j. 2018. Three-Dimensional response of neighboring buildings sitting on pile foundations to seismic pounding. *International Journal of Geomechanics,* 18, 04018007.

Goble, G. & Rausche, F. 1970. Pile load test by impact driving. *Highway Research Board Annual Meeting, Washington, DC.*

Goble, G.G., Scanlan, R. & Tomko, J.J. 1967. Dynamic studies on the bearing capacity of piles. *Ohio Highway Engineering Conference Proceedings.*

Hokmabadi, A.S. & Fatahi, B. 2016. Influence of foundation type on seismic performance of buildings considering soil–structure interaction. *International Journal of Structural Stability and Dynamics,* 16, 1550043.

Kolymbas, D. 1985. A generalized hypoelastic constitutive law. *Proc. XI Int. Conf. Soil Mechanics and Foundation Engineering.*

Lowery, L.L., Hirsch, T., Edwards, T.C., Coyle, H.M. & Samson, C. 1969. *Pile driving analysis: State of the art,* Texas Transportation Institute, Texas A & M University.

Mabsout, M.E., Reese, L.C. & Tassoulas, J.L. 1995. Study of pile driving by finite-element method. *Journal of geotechnical engineering,* 121, 535–543.

Mabsout, M.E. & Tassoulas, J.L. 1994. A finite element model for the simulation of pile driving. *International Journal for numerical methods in Engineering,* 37, 257–278.

Mašın, D. 2010. PLAXIS implementation of Hypoplasticity.

Mašín, D. 2015. Hypoplasticity for practical applications. *PhD course on hypoplasticity.* Zhejiang University.

Masouleh, S.F. & Fakharian, K. 2008. Application of a continuum numerical model for pile driving analysis and comparison with a real case. *Computers and Geotechnics,* 35, 406–418.

Ng, K.W. 2011. *Pile setup, dynamic construction control, and load and resistance factor design of vertically-loaded steel H-Piles,* Iowa State University.

Ng, K.W. & Sritharan, S. 2013. Improving dynamic soil parameters and advancing the pile signal matching technique. *Computers and Geotechnics,* 54, 166–174.

Niemunis, A. & Herle, I. 1997. Hypoplastic model for cohesionless soils with elastic strain range. *Mechanics of Cohesive-frictional Materials,* 2, 279–299.

PLAXIS, B. 2017. Plaxis 2D Version. 2017. *Reference Manual. Delft, The Netherlands.*

Rausch, F., Moses, F. & Goble, G.G. 1972. Soil resistance predictions from pile dynamics. *Journal of the soil mechanics and foundations division,* 98, 917–937.

Yang, J. 2006. Influence zone for end bearing of piles in sand. *Journal of geotechnical and geoenvironmental engineering,* 132, 1229–1237.

Slopes and cuts

Numerical Methods in Geotechnical Engineering IX – Cardoso et al. (Eds)
© 2018 Taylor & Francis Group, London, ISBN 978-1-138-33203-4

MPM modelling of static liquefaction in reduced-scale slope

P. Ghasemi
Department of Civil Engineering, University of Salerno, Italy

M. Martinelli
Deltares, Delft, The Netherlands

S. Cuomo & M. Calvello
Department of Civil Engineering, University of Salerno, Italy

ABSTRACT: The paper deals with the capability of Material Point Method (MPM) in simulating retrogressive failure in a case of very loose coarse-grained material. A well documented reduced-size slope experimental test taken from literature is used. The case study clearly shows the onset of static liquefaction, a retrogressive failure and rapid flow-like motion with a flat final ground surface. A hypoplastic constitutive model is used to simulate the liquefiable behaviour of the material. The results of the numerical analyses are in appropriate agreement with the experimental evidence, and also allow some general considerations about the capabilities of using MPM combined with advanced constitutive models to simulate a complex physical process.

1 INTRODUCTION

Instability mechanisms of slopes are mostly induced by rainfall and may include the formation of successive shear bands, progressive or retrogressive slides as cause/consequence of static liquefaction. Large deformations may also occur in the soil.

The analysis of slope behaviour has been conducted, in the literature, through a variety of approaches capable to adequately consider the slope geometric configuration, the soil behaviour and specific triggering factors and mechanisms. Cascini et al. (2010) compared the results of standard Limit Equilibrium Method (LEM) based on well-known slice methods (Janbu, 1954; Morgenstern and Price, 1965) to more sophisticated Finite Element Method (FEM) analyses and showed that the drained failure of shallow soil covers subjected to rainfall can be satisfactorily simulated by both approaches. Whereas, the accurate simulation of the pore water pressure evolution and the related increase in strain rate necessarily requires an advanced constitutive model, such as the Generalized Plasticity model used by Cascini et al. (2013). The aforementioned contributions neglect soil deformations (LEM) or generally consider them as "small" (FEM); that is reasonable if only the failure condition is under investigation.

When the complete evolution of the slope geometry and related large deformations must be analysed, other approaches have to be used, such as, the extended versions of FEM (e.g. Finite Element Method with Lagrangian Integration Points, FEMLIP, Cuomo et al., 2012; Prime et al., 2014) or other so-called meshless methods such as Smooth Particle Hydrodynamics (SPH, e.g. Pastor et al., 2009; Soga et al., 2015).

Material Point Method (MPM) is one of the computational methods able to simulate the entire failure and post-failure stages of slope instability caused by rainfall infiltration and/or groundwater table rising. MPM has been recently employed to simulate slope failure case studies with different triggering factors, soil materials, and failure modes. For example, Bandara and Soga (2015) used MPM to simulate progressive failure of a river levee caused by water infiltration. Similarly, Wang et al. (2016) illustrated capability of MPM to reproduce retrogressive or progressive slope failures of cohesive soil induced by an excavation. Yerro et al. (2016) carried out the simulation of internal progressive failure in deep-seated landslides in rock and examined the stability conditions and the post-failure behaviour of a compound landslide whose geometry was inspired by the Vajont landslide. In addition to slope failure, landslide propagation was modelled by Ceccato and Simonini (2016) and Calvello et al. (2018).

This paper proposes the modelling of flume tests for a well-documented reduced-size slope. Previous modelling of the failure stage of that slope was proposed by Cuomo et al. (2014) using LEM and FEM analyses including a simple non-associative elastic perfectly-plastic constitutive model. Those simple analyses were enough to explain the drained failure of the slope and triggering stage of the observed slide. Here, MPM and an advanced soil constitutive model are used to properly reproduce large displacements and liquefaction. The whole landslide evolution process is reproduced including propagation and deposition.

2 EXPERIMENT OF ECKERSLEY (1990)

Geometry, initial conditions and boundary conditions are clearly recognised as key factors in slope (in)stability (Cascini et al., 2013). Thus, extensive literature has dealt with tests of laboratory reduced-size slope models (so-called flume tests) or artificial real-sized slopes (centrifuge tests).

Eckersley (1990) provided one of the very first contributions about a comprehensive observation (in a reduced-scale slope) of slope deformations, retrogressive failure, build-up of pore water pressure after failure, and later propagation of failed material up to rest along a very gentle profile. Similar tests were later extended to more complex groundwater conditions, such as downward rainfall infiltration from ground surface and/or a downwards/upwards water spring from the bedrock (Lourenco et al., 2006; among others).

Particularly, Eckersley (1990) performed a series of tests in order to investigate static liquefaction and the influence of initial density in a 1 m high coal slope constructed in a glass-sided tank. The slope was brought to failure by seepage of water from a lateral boundary towards the toe of the slope. Water entered the slope from a constant head tank located behind, with a wire cage filled with coarse gravel to facilitate inflow procedure. Inclination of the ground surface was 36 degrees and the floor was extended to 1.5 m beyond the slope toe.

Direct sliding of the slope along the coal/floor interface was inhibited by #80 waterproof sandpaper glued to the floor. According to a preliminary test the angle of resistance of 30–36° was reported for this interface. Different tests with different values of the material density were conducted. The results of experiment #7 are herein considered. During the test, the slope experienced a rapid retrogressive flowslide. Failure development will be discussed in detail later.

The material used was coking coal extracted in mines of Bowen Basin area (northern Australia). The coal particles ranged from fine sand and silt to gravel with a specific gravity of around 1.34. It is interesting to note that the D_{10} value (0.06–0.3 mm) is similar to that of other materials involved in past flowslides (Eckersley, 1985). Given the pore-size distribution it is reasonable to assume that matric suction is negligible.

The observed failure started by shallow sliding adjacent to the glass in the wettest area forming saturated coal slurry. Figure 1 shows what was observed from the experiment as the failure proceeded.

Stage 1 comprised two fairly distinct shallow slides, over a period of 4 s, with each slide extending rapidly uphill by surface slipping of the overstepped dry coal face. In the next stage a slab of coal about 0.2 m thick comprising the whole face failed, moving the previous debris ahead (Fig. 1, 2nd stage).

In stage 3, a final deep compound slide initiated, pushing the previously failed coal horizontally, with the whole mass eventually decelerating and coming to rest.

The mentioned stages are the subject of the simulations conducted in the presenting paper. The main goal is to reproduce the overall process with particular attention to the three described failure stages as well as to the final shape of the slope.

3 MATERIALS AND METHODS

3.1 *Material Point Method*

MPM can be considered as an extension of the widely-used Updated Lagrangian Finite Element Method (UL-FEM). Similar to the UL-FEM, MPM makes use of a Lagrangian finite element mesh in which nodes are attached to a deforming body and consequently they move through

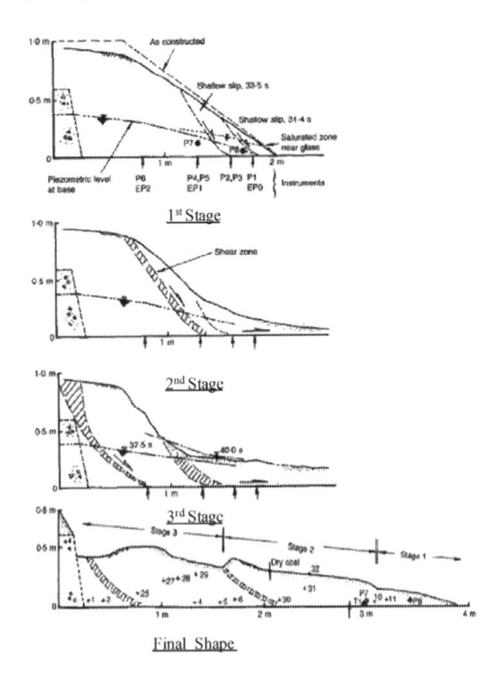

Figure 1. Experiment results: failure and post-failure stages (Eckersley, 1990).

space with the material. All computational data, such as state parameters and stresses, is stored in the material points at the end of each time step and the Lagrangian mesh is reset. Therefore, this computational method does not suffer from mesh distortion as the material points move through the underlying finite element mesh. At the beginning of the next step, the stored data in the material points are mapped back to the mesh based on the new locations of the material points..

In this paper, the single-point two-phase formulation proposed by Al-Kafaji (2013) for modelling saturated soils was adopted. It entails that the solid and fluid phases are represented by the same material point, with each phase associated to a fraction of the material point volume. The linear momentum conservations are solved for the liquid phase and for the mixture.

In order to simulate the propagation stage of a slope special attention should be paid to the contact between the basal surface and the soil particles. In this study, a frictional contact algorithm was adopted using a Mohr-Coulomb criterion (Bandara and Soga, 2015). It was realized that the vertical component of acceleration and velocity of soil particles, adjacent to the contact, could cause the separation between soil particles and the basal surfaces (Fig. 2a). Therefore, to avoid rebounding of the particles, after the correction of the velocity and acceleration by contact formulation, the vertical acceleration are set to zero for the soil along the contact. In this case the gap between two entities is significantly reduced (Fig. 2b).

Another numerical mitigation conducted in this study was related to the pore water pressure smoothing. In order to reduce the water pressure oscillation imposed by grid-crossing, at the end of each time step, the liquid pressure of the particles located within each element is averaged and re-assigned to the particles inside the element. In this case, special attention should be paid to the "mixed" elements (i.e. including both saturated and dry particles). Herein, the averaging and re-assigning of the pressure should be conducted only in terms of saturated particles. Otherwise, dry particles might carry liquid pressure, which could cause numerical difficulties due to the lack of liquid bulk modulus.

3.2 *Input data*

This study considered the onset of failure as the starting point of the model, with the initial condition of the phreatic line at the level for which the first failure was initiated in the experiment. A previous study (Cuomo et al., 2014), conducted using LEM and FEM, was already able to explain how the raising water table has led to the triggering of the slide.

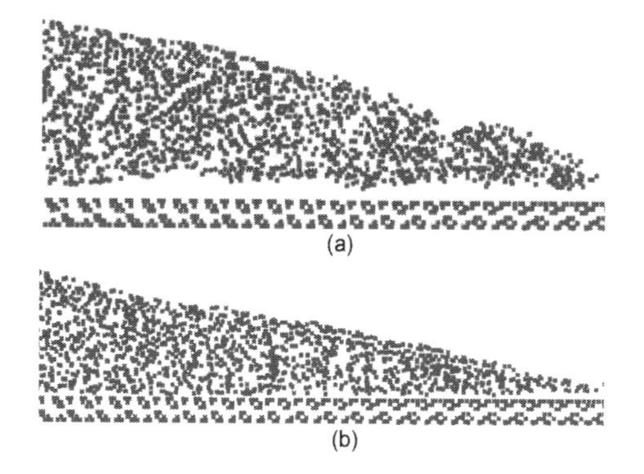

Figure 2. The correction of acceleration and velocity in contact. a) without correction b) with correction.

Figure 3 shows the model scheme as well as the initial condition considered in the computation. Two soil layers are considered, with the dry layer lying over the saturated layer. The dry layer was simulated by a one-phase material, whereas the saturated one was simulated by a two-phase material including solid skeleton and water. The plywood floor was simulated by a linear elastic material. Above the floor a frictional contact algorithm was adopted using a Mohr-Coulomb criterion (Bandara and Soga, 2015), with a friction coefficient equal to 0.5. This value corresponds to the one reported for the experiment (Eckersley, 1990). The wire cage at the rear of the slope was simulated by a linear elastic material.

A static water pressure was applied as boundary condition for the saturated layer to simulate the water pressure imposed from the reservoir. Due to the size of material used in the experiment, as mentioned before, the presence of suction was considered improbable.

The initial value for the effective stress and the pore water pressure were computed from a preliminary quasi-static simulation in which the horizontal and vertical velocities for the liquid were set to zero over the slope surface. These fixities were removed at further stages. It should be mentioned that the initial static simulation was performed using a non-associative Mohr-Coulomb model.

Then, the constitutive model was switched to hypoplasticity. This mitigated the numerical difficulties that the initial low-stress level would have caused to the hypoplastic model.

The computational mesh was composed of 4-node unstructured tetrahedral elements, and 4 material points were initialized in each element. The typical length of the element was 0.05 cm; the mesh included 13,062 elements.

In order to evaluate the soil response at certain locations of interest, 3 zones were tracked for each simulation. Each zone is represented by a small

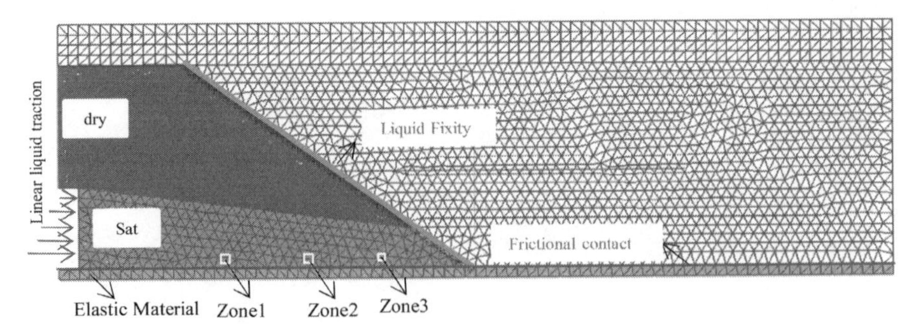

Figure 3. Geometry and spatial discretization of the computational domain.

square (Fig. 3), defined as fixed volume in the computational mesh, where some soil state parameters (e.g. pore water pressure, void ratio and effective stress) were averaged and reported as output. These zones, denoted as Zone 1, Zone 2 and Zone 3, were selected as follows.

Zone 1 is initially located close to the base beneath the crest, Zone 3 is situated at the toe of the slope, Zone 2 in the saturated part of the shear band expected in the second stage of failure (Fig. 3).

The basic idea of hypoplasticity was originally developed by Kolymbas (1985), with various aspects investigated in the past few years; (Buer and Wu, 1993, 1995), and refined to incorporate the Matsuoka-Nakai critical state stress condition (Volffersdorff, 1996), and intergranular strain concept (Niemunis and Herle, 1997).

According to the Elkadi (2013), the intergranular strain concept is essential for the modelling of cyclic loading in which elastic deformations occur. It especially occurs in dynamic numerical models, as the wave propagation through the soil will be reflected at each material point. It means that the process of loading and unloading cycles always happens during simulation. Therefore, the use of the hypoplastic model with intergranular strain will help to reduce the effect of ratcheting.

The simulation of the slope has been performed both not using and using the intergranular concepts. In the former case, the entire saturated part of the slope experienced a dramatic reduction of stress in the initial time steps, inconsistently with experimental evidence. Such circumstance is caused by ratcheting. Therefore, adopting the intergranular strain was preferred.

The employed model required 8 main material parameters, and 5 secondary parameters. The definition of each parameter is provided in the above mentioned papers. A comprehensive parametric analysis has been conducted in terms of initial void ratio, and it has been observed that higher values of void ratio lead to general collapse involving a deep failure surface. Whereas the slope tends to experience shallow sliding in the case of

Table 1. Constitutive parameters used for modelling.

α	β	$e0_d$	$e0_i$	m_r	m_t	R_{max}	βr
0.2	1.00	0.21	1.21	2.60	2.00	0.0007	0.1
χ	$e0_p$	hs	n	e_c			
1	0.98	93	0.07	0.93			

lower values of initial void ratio. Finally, the value of $e0_p = 0.98$ which corresponds to the void ratio equal to 0.58 for the soil located in Zone 1, allowed reproducing the retrogressive failure connected to static liquefaction. The different observed mode of failure in terms of different initial void ratio of the soil is consistence with what is reported in Eckersley (1990). The assumed values for the other parameters are given in Tab. 1.

This set of parameters was obtained through the application of an inverse analysis procedure, based on error function definitions, and SPQSO (Species-based quantum particle swarm algorithms) explained in Hosseinnezhad et al., (2014) and Cuomo et al., (2018), which is convenient for estimating a large set of parameters. The data set used for calibration is the one reported by Eckersley (1985).

4 RESULTS AND DISCUSSION

The mean effective stresses computed in the slope are shown in Fig. 4. They refer to the three time lapses during the post-failure stage of the experiment, respectively equal to 0.6 s, 1.0 s and 1.3 s. It is worth noting that mean effective stresses become very low initially in Zone 3 and close to Zone 2, and then in Zone 1 for the final time lapse. This spatial-temporal evolution of effective stress reduction is related to the onset and development of static liquefaction. First, liquefaction starts at the toe of slope, where stress levels are the lowest ones. Then, stresses reduces in the middle part of the slope, and

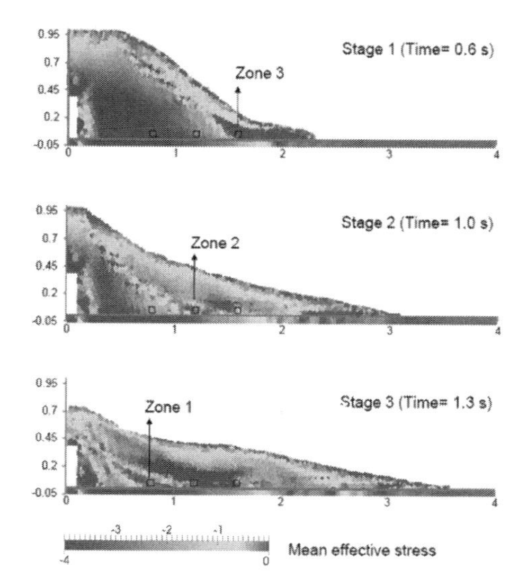

Figure 4. Mean effective stresses in the three considered post-failure stages.

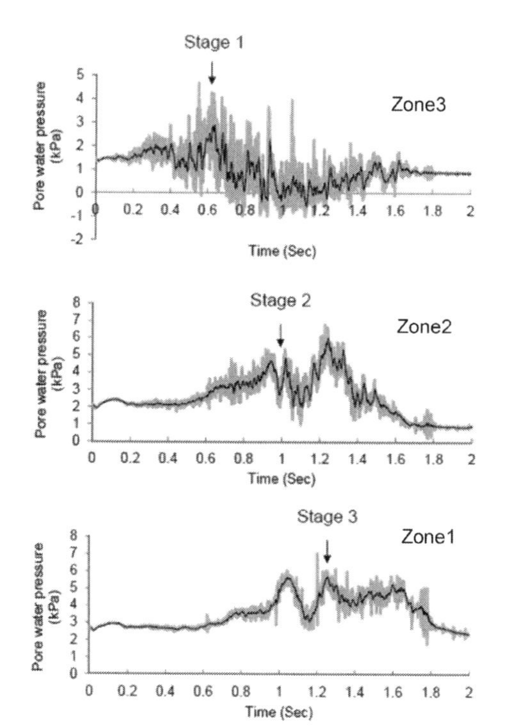

Figure 5. Evolution with time of the pore water pressures at three different fixed locations situated at the base of the slope (see Fig. 3).

finally at the rear. Globally, a retrogressive flowslide is simulated, as observed in the experiment. The low value of mean effective stress in these zones implies the occurrence of static liquefaction.

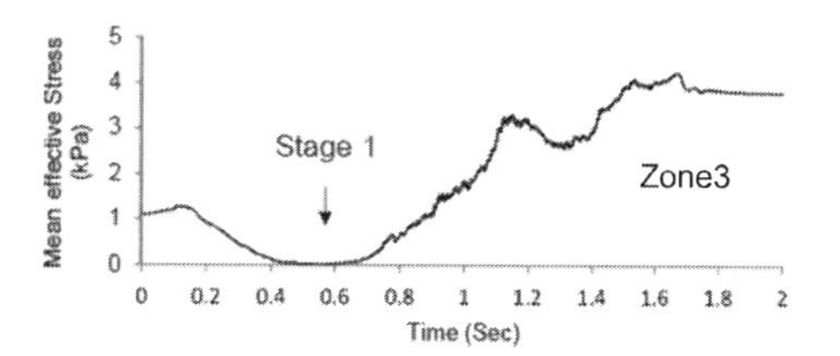

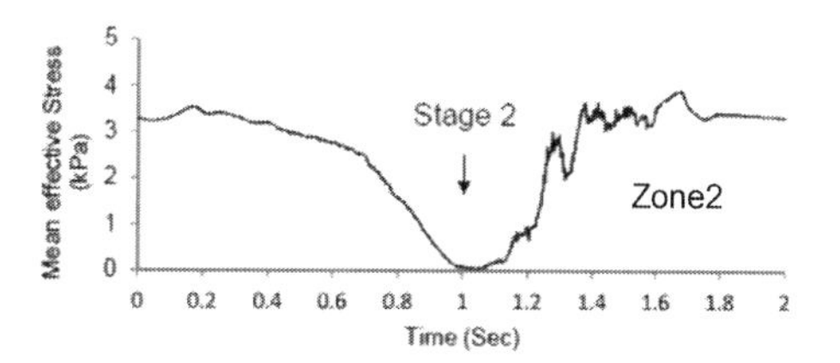

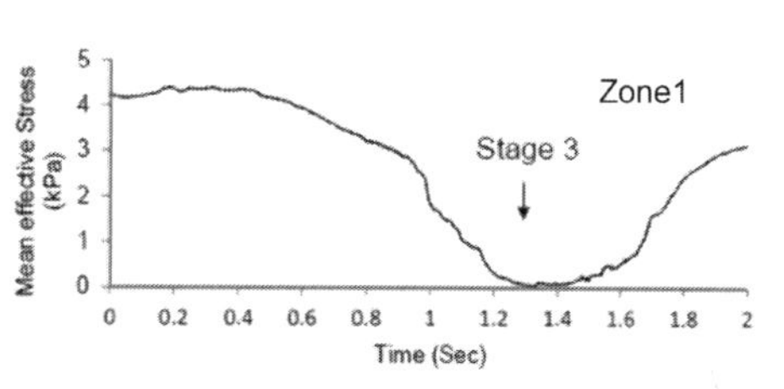

Figure 6. Evolution with time of mean effective stress at the three different fixed locations situated at the base of the slope (see Fig. 3).

In fact, the contractive behaviour of the material upon shearing leads to an increase of excess pore water pressures and a subsequent decrease of the effective stresses.

Figure 5 shows the changes of pore water pressure over time recorded in the three fixed zones specified in Figure 3. According to the results, at the onset of each failure stage the pore water pressure reached its maximum value, which in turn produced a minimum value of mean effective stress. Then, as the excess pore water pressure was vanishing as an effect of a consolidation process, the soil gained back its effective stress, and consequently its strength.

The time changes of the mean effective stress at the fixed target zones proved that the static liquefaction occurred upon shearing, and then a solid-like behaviour was recovered by the soil once the slope reached again a stable geometric configuration (Fig. 6).

5 CONCLUSIONS

In this paper a classical slope failure experiment was numerically simulated by means of a Material Point Method (MPM) approach and adopting a hypoplastic constitutive model. Significant care was adopted in looking at the behavior of the contact between dry and saturated material.

The failure was initiated by water seepage from a tank at the rear of the slope. Since the material was in loose conditions, it showed contractive behavior upon shearing; subsequently, static liquefaction occurred. The retrogressive failure caused by liquefaction in the coarse-grained material was addressed and successfully simulated.

The simulation results were in good agreement with the experimental ones. Therefore, it has been shown that the combination of MPM and hypoplastic model was able to well address this complex slope evolution process.

ACKNOWLEDGMENTS

MPM slope modelling was carried out using the Anura3D code (http://www.mpm-dredge.eu/). The research was developed within the framework of the Anura3D MPM Research Community.

REFERENCES

Al-Kafaji, I.K.A. (2013). Formulation of a dynamic material point method (MPM) for geomechanical problems. *PhD Thesis (Stuttgart University, Germany).*

Bandara, S., Soga, K. (2015).Coupling of soil deformation and pore fluid flow using material point method. *Computers and Geotechnics, 63*, 199–214.

Calvello, M., Ghasemi, P., Cuomo, S., Martinelli, M. (2018). Optimizing the MPM model of a reduced scale granular flow by inverse analysis. This conference.

Cascini, L., Cuomo, S., Pastor, M., Sacco, C. (2013). Modelling the post-failure stage of rainfall-induced landslides of the flow type. *Canadian Geotechnical Journal*, 50(9), 924–934.

Cascini, L., Cuomo, S., Pastor, M., Sorbino, G. (2010). Modeling of rainfall-induced shallow landslides of the flow-type. *Journal of Geotechnical and Geoenvironmental Engineering*, 136(1), 85–98.

Ceccato, F., Simonini, P. (2016).Study of landslide runout and impact on protection structures with the Material Point Method. In *INTERPRAEVENT 2016-Conference Proceedings*.

Cuomo, S., Ghasemi, P., Calvello, M., Hosseinezhad, V. (2018). Hypoplasticity model and inverse analysis for simulation of triaxial tests. This conference.

Cuomo S., Pastor M., Sacco C., Cascini L. (2014). Analisi delle fasi di rottura e post-rottura di frane tipo flusso. XXV CNG—AGI—Roma—ISBN 978–88–97517–05–4, pp. 479–485 (in Italian).

Cuomo, S., Prime, N., Iannone, A., Dufour, F., Cascini, L., Darve, F. (2013). Large deformation FEMLIP drained analysis of a vertical cut. *Acta Geotechnica*, 8(2), 125–136.

Eckersley, D. (1990). Instrumented laboratory flowslides. *Geotechnique, 40*(3), 489–502.

Eckersley, J.D. (1985). Flowslides in stockpiled coal. *Engineering Geology, 22*(1), 13–22.

Elkadi, A., & Msc, P.N. (2013). *Mpm validation with centrifuge tests: pilot case pile installation.* Technical Report 1206750-G05-HYE-GG01-Jvm, Deltares, Delft, The Netherlands.

Hosseinnezhad V, Rafiee M, Ahmadian M, Ameli MT. (2014) Species-based quantum particle swarm optimization for economic load dispatch. *International Journal of Electrical Power & Energy Systems*. 2014 Dec 31;63:311–22.

Janbu, N. (1954). Application of Composite Slip Surface for Stability Analysis. European Conference on Stability Analysis, Stockholm, Sweden.

Kolymbas, D., Herle, I., & Von Wolffersdorff, P.A. (1995). Hypoplastic constitutive equation with internal variables. *International Journal for Numerical and Analytical Methods in Geomechanics, 19*(6), 415–436.

Lourenco, S., Sassa, K. and Fukuoka, H. (2006). Failure process and hydrologic response of a two layer physical model: Implications for rainfall-induced landslides. *Geomorphology*, 731–2, 115–130.

Morgenstern, N.R. and Price, V.E. (1965). The analysis of the stability of general slip surfaces. *Geotechnique*, 151, 79–93.

Niemunis, A., & Herle, I. (1997). Hypoplastic model for cohesionless soils with elastic strain range. Mechanics of Cohesive–frictional Materials, 2(4), 279–299.

Pastor, M., Haddad, B., Sorbino, G., Cuomo, S. and Drempetic, V. (2009). A depth-integrated, coupled SPH model for flow-like landslides and related phenomena. *International Journal for Numerical and Analytical Methods in Geomechanics*, 33: 143–172.

Prime, N., Dufour, F., & Darve, F. (2014). Solid–fluid transition modelling in geomaterials and application to a mudflow interacting with an obstacle. *International Journal for Numerical and Analytical Methods in Geomechanics*, 38(13), 1341–1361.

Soga, K., Alonso, E., Yerro, A., Kumar, K., & Bandara, S. (2015). Trends in large-deformation analysis of landslide mass movements with particular emphasis on the material point method. *Géotechnique, 66*(3), 248–273.

von Wolffersdorff, P.A. (1996). A hypoplastic relation for granular materials with a predefined limit state surface. Mechanics of Cohesive-frictional Materials, 1(3), 251–271.

Wang, B., Vardon, P.J., Hicks, M.A., Chen, Z. (2016). Development of an implicit material point method for geotechnical applications. *Computers and Geotechnics, 71*, 159–167.

Wu, W., & Bauer, E. (1993). A hypoplastic model for barotropy and pyknotropy of granular soils. *Modern approaches to plasticity, 383*.

Wu, W., Bauer, E., & Kolymbas, D. (1996). Hypoplastic constitutive model with critical state for granular materials. *Mechanics of materials, 23*(1), 45–69.

Yerro, A., Pinyol, N.M., Alonso, E.E. (2016). Internal progressive failure in deep-seated landslides. *Rock Mechanics and Rock Engineering*, 49(6), 2317–2332.

Numerical Methods in Geotechnical Engineering IX – Cardoso et al. (Eds)
© 2018 Taylor & Francis Group, London, *ISBN 978-1-138-33203-4*

Thermomechanical modelling of rock avalanches with debris, ice and snow entrainment

P. Bartelt, M. Christen, Y. Bühler & O. Buser
WSL Institute for Snow and Avalanche Research, SLF, Davos Dorf, Switzerland

ABSTRACT: Rock avalanche exhibit two extreme flow forms. In cold, high mountain terrain they flow as mixed powder-type avalanches. The air-blast associated with the powder cloud can cause extensive, wide-ranging damage. This was the case in the recent Nepal avalanches in Langtang and Everest. In warmer climates, rock avalanches often transform into highly fluid, debris flows of rocks and mud. This was the case of the recent rock avalanche on Piz Cengalo, Switzerland. In both cases the inundation area is enlarged due to flow transformations, depending on the composition of the avalanche (rock, ice, snow) as well as the temperature of the flow material. Both dry fluidized and fluid lubricated flow regimes are possible depending on the initial and boundary conditions. The thermalmechanical properties of the material entrained by the avalanche often determines which flow regime dominates. Modelling avalanche speed and range therefore requires a thermomechanical model capable of accounting for the behaviour of different avalanche components with varying thermal properties (specific heat, melting temperature, latent heats). We present a thermomechanical model of avalanche flow capable of modelling both extreme cases, as well as mixed flow forms. The model is applied to simulate the dry mixed avalanche that destroyed the village of Langtang Nepal in 2015 as well as the Piz Cengalo rock/ice avalanche of 2017.

1 INTRODUCTION

In mountainous regions of the world, avalanches containing a mixture of rock and ice are a dangerous natural phenomena (Huggel et al. 2010, Faillettaz et al. 2015). One recent example is the co-seismic rock/ice avalanche that destroyed the village of Langtang, Nepal (Fujita et al. 2017, Nagai et al. 2017). Intense ground shaking during the Nepalese earthquake of 2015 destabilized hanging ice glaciers to start an immense avalanche (Kargel et al. 2016). The avalanche entrained considerable amounts of rocky debris and snow before reaching the valley bottom. Secondary debris flows were observed (Kargel et al. 2016). The village of Langtang, however, was destroyed by the air-blast that accompanied the avalanche of ice, snow and debris. Another recent example is the collapse of 3.1 mio m^3 of rock on the east face of Piz Cengalo, Switzerland in 2017 (Fig. 1). The unstable rock mass fell on a glacier, mobilizing an estimated 0.6 mio m^3 of ice. Although this avalanche was accompanied by a huge dust cloud, similar to Langtang, the airborne suspension did little damage. Several secondary debris flows released out of the water saturated deposits of the rock/ice avalanche. The debris flows inundated the village of Bondo. These recent examples indicate the wide variety rock/ice avalanche flow regimes are possible, varying from dry, disperse avalanches to fluid saturated debris flows (Huggel et al. 2005, Carey et al. 2015).

Figure 1. The Piz Cengalo rock/ice avalanche, Bondo, Switzerland 2017. Estimated release volume 3.1 mio m^3. Estimated ice content 0.6 mio m^3. Much of the ice melted by frictional heating (0.2 mio m^3) or became suspended in the powder cloud. Although visually impressive, the powder cloud did little damage. Secondary debris flows inundated the village of Bondo causing much damage.

2 AVALANCHE MASS

A model of rock/ice avalanches must account for a complicated mixture of four materials. In general, flowing avalanche mass M consists of rock M_r (density ρ_r) ice M_i (density ρ_i), fluid M_w (density ρ_i), and air M_a (density ρ_a) (Fig. 2). The mixture is defined by both source materials as well as the material entrained by the avalanche, including the air which is needed to form the powder suspension cloud. An additional complication arises because the ice can melt, reducing

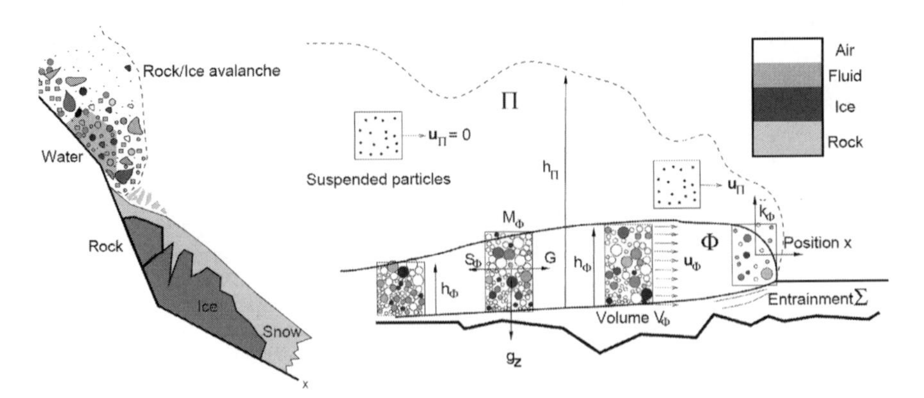

Figure 2. A volume of a rock/ice avalanche consists of rock, ice, fluid (water) and air. Rock, ice and snow can be entrained by the avalanche. Coordinate system. a) the components of a flow column. b) the avalanche core consists of gravity driven mass M_Φ and suspended mass M_Π.

the solid content of the mixture, while increasing the amount of fluid in the pore space of the granular solid material.

The size of the solid rock and ice mass can range between fine dust (particle diameters less than 0.1 mm) to large boulders (particle diameters greater than 1 m). The dust is created by the continual frictional grinding between larger particles, or during the entrainment process which involves the breakup of the running surface. The dust can be suspended in air, implying the force of gravity is offset by the Stokes-type drag forces. The movement of the larger particles is not controlled by air-drag, rather the frictional interaction with the ground. Solid mass in a rock/ice avalanche must be separated into two size categories. All mass that cannot be suspended (because the Stokes drag forces are small) belongs to the core of the avalanche. This mass is denoted M_Φ (Fig. 2b). Mass that can be suspended M_Π comprises the avalanche suspension cloud (Fig. 2b). The difference between the gravity driven mass M_Φ and the suspended mass M_Π leads to a natural segregation or layering in the avalanche (Fig. 3).

The fluid mass M_w is fixed in the interstitial space between particles and consequently bonded to the surface of the solid. It exists only in the avalanche core. The solid and fluid masses are therefore moving with the same mean velocity U_Φ (Fig. 2). Flow energy in the avalanche core is primarily dissipated by particle-particle and particle-ground interactions that are strongly mediated by the interstitial fluid. The fluid phase controls the magnitude of the flow friction. The total frictional force acting against the downward pull of gravity G is denoted S_Φ in the avalanche core (Fig. 2). We apply a Mohr-Coulomb type frictional rheology with cohesion (Bartelt et al. 2014, Bartelt et al. 2015).

Other sources of energy loss in the avalanche core include the entrainment of ice or debris and

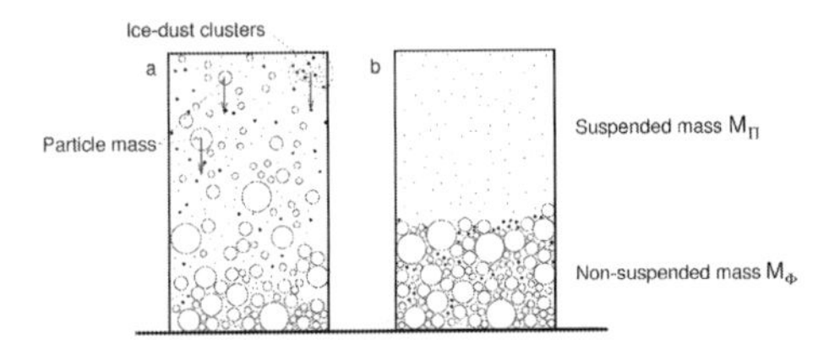

Figure 3. The avalanche consists of non-suspended mass M_Φ and suspended mass M_Π Stokes-type drag forces on the dust particles keep them suspended while the larger particles fall directly to the ground. The avalanche core and suspension cloud naturally segregate to from two independent flow regions.

the formation of the powder cloud, which extracts flow energy from the core. The suspended mass, containing both dust and air, is accelerated by the core and moves with the mean velocity U_Π. The inertia of the cloud is dissipated by air drag as the cloud moves into stationary air. Powder cloud drag is denoted S_Π. Other sources of dissipation are air-entrainment and turbulence. We track the mass of air M_Λ entrained by the avalanche. The velocity of the core and cloud can be approximately equal $U_\Phi \approx U_\Pi$ (during the formation phase) or unequal $U_\Phi \neq U_\Pi$ (during the runout phase). The core and cloud can therefore inundate different areas.

In a two-dimensional coordinate system fixed parallel to the mountain slope (directions x and y; velocity $U_\Phi = (u_\Phi, v_\Phi)$ and $U_\Pi = (u_\Pi, v_\Pi)$ this description of avalanche flow leads to the following balance equations of mass and momentum for the gravity driven avalanche core Φ,

$$\frac{\partial(M_\Phi)}{\partial t} + \frac{\partial(M_\Phi u_\Phi)}{\partial x} + \frac{\partial(M_\Phi v_\Phi)}{\partial y} = \\ = \dot{M}_{\Sigma\to\Phi} - \dot{M}_{\Phi\to\Pi} \tag{1}$$

$$\frac{\partial(M_i)}{\partial t}+\frac{\partial(M_i u_\Phi)}{\partial x}+\frac{\partial(M_i v_\Phi)}{\partial y}= \\ =\dot{M}_{\Sigma\to i}-\dot{T}_{i\to w}-\dot{M}_{i\to\Pi} \tag{2}$$

$$\frac{\partial(M_w)}{\partial t}+\frac{\partial(M_w u_\Phi)}{\partial x}+\frac{\partial(M_w v_\Phi)}{\partial y}= \\ =\dot{M}_{\Sigma\to w}+\dot{T}_{i\to w}-\dot{M}_{w\to\Pi} \tag{3}$$

$$\frac{\partial(M_\Phi u_\Phi)}{\partial t}+u_\Phi\frac{\partial(M_\Phi u_\Phi)}{\partial x}+v_\Phi\frac{\partial(M_\Phi u_\Phi)}{\partial y}+ \\ +M_\Phi g_z\frac{\partial h_\Phi}{\partial x}=G_x-S_{\Phi x}-\dot{M}_{\Phi\to\Pi}u_\Phi \tag{4}$$

$$\frac{\partial(M_\Phi v_\Phi)}{\partial t}+u_\Phi\frac{\partial(M_\Phi v_\Phi)}{\partial x}+v_\Phi\frac{\partial(M_\Phi v_\Phi)}{\partial y}+ \\ +M_\Phi g_z\frac{\partial h_\Phi}{\partial y}=G_y-S_{\Phi y}-\dot{M}_{\Phi\to\Pi}v_\Phi \tag{5}$$

and for the suspended powder cloud Π:

$$\frac{\partial(M_\Pi)}{\partial t}+\frac{\partial(M_\Pi u_\Pi)}{\partial x}+\frac{\partial(M_\Pi v_\Pi)}{\partial y}=\dot{M}_{\Phi\to\Pi} \tag{6}$$

$$\frac{\partial(M_\Lambda)}{\partial t}+\frac{\partial(M_\Lambda u_\Pi)}{\partial x}+\frac{\partial(M_\Lambda v_\Pi)}{\partial y}=\dot{M}_{\Lambda\to\Pi} \tag{7}$$

$$\frac{\partial(M_\Pi u_\Pi)}{\partial t}+u_\Pi\frac{\partial(M_\Pi u_\Pi)}{\partial x}+v_\Pi\frac{\partial(M_\Pi u_\Pi)}{\partial y}+ \\ +M_\Pi g_z\frac{\partial h_\Pi}{\partial x}=\dot{M}_{\Phi\to\Pi}u_\Phi-S_{\Pi x}-\dot{M}_{\Lambda\to\Pi}u_\Pi \tag{8}$$

$$\frac{\partial(M_\Pi v_\Pi)}{\partial t}+u_\Pi\frac{\partial(M_\Pi v_\Pi)}{\partial x}+v_\Pi\frac{\partial(M_\Pi v_\Pi)}{\partial y}+ \\ +M_\Pi g_z\frac{\partial h_\Pi}{\partial y}=\dot{M}_{\Phi\to\Pi}v_\Phi-S_{\Pi y}-\dot{M}_{\Lambda\to\Phi}v_\Pi \tag{9}$$

The definitions of the terms on the right-hand side of these equations are presented in Table 1.

Because the powder cloud contains only mass that can be suspended, the right-hand sides of the powder cloud momentum equations does not contain a gravity force. The initial velocity of the cloud is determined entirely by the initial momentum of the suspended mass (including both dust and air) ejected from the core, $\dot{M}_{\Phi\to\Pi}u_\Phi$ and $\dot{M}_{\Phi\to\Pi}v_\Phi$. Momentum is supplied to the cloud in both coordinate directions which allows the simulation of the lateral spreading of the cloud, an important problem in hazard mitigation studies (Dreier et al. 2016). Moreover, the density difference between dust particles and air is not the driving force of the powder cloud. We make the assumption that there is no relative velocity between the air and the dust. Therefore, in this model dust particles cannot settle. The dust therefore acts as a tracer to delineate the movement of the cloud, which is composed primarily of air.

3 THERMAL HEAT AND RANDOM KINETIC ENERGY

Because we must consider melting of ice it is necessary to track the temperature (thermal energy) of the avalanche core (Bartelt et al. 2006, Vera Valero et al. 2015). When the mean avalanche temperature exceeds the melting temperature of ice, a phase change is induced. If an avalanche flow volume contains no ice, the rock temperature can exceed the melting temperature. To model ice melting we assign two internal energies to the avalanche core: the heat energy E_Φ (J m^{-2})

Table 1. Definition of source and sink terms for Eqs 1–9.

Term	Definition
$\dot{M}_{\Sigma\to i}$	Ice mass entrained by avalanche core per unit time and area (kg m^{-2} s^{-1})
$\dot{M}_{\Sigma\to i}$	Water mass entrained by avalanche core per unit time and area (kg m^{-2} s^{-1})
$\dot{M}_{\Sigma\to r}$	Rock mass entrained by avalanche core per unit time and area (kg m^{-2} s^{-1})
$\dot{M}_{\Sigma\to\Phi}$	Total mass entrained by avalanche core per unit time and area (kg m^{-2} s^{-1})
$\dot{M}_{\Lambda\to\Phi}$	Air mass entrained by avalanche cloud per unit time and area (kg m^{-2} s^{-1})
$\dot{T}_{i\to w}$	Mass of ice melted by frictional dissipation per unit time and area (kg m^{-2} s^{-1})
$\dot{Q}_{i\to w}$	Internal energy required for melting ice per unit time and area (J m^{-2} s^{-1})
$\dot{Q}_{\Sigma\to\Phi}$	Heat energy produced during entrainment (J m^{-2} s^{-1})
$\dot{P}_{\Sigma\to\Phi}$	Fluctuation energy produced during entrainment (J m^{-2} s^{-1})
G_x	Gravitational force per unit area in x-direction (N m^{-2})
G_y	Gravitational force per unit area in y-direction (N m^{-2})
$S_{\Phi x}$	Resistance force (shearing) in the core per unit area in the $x-$direction (N m^{-2})
$S_{\Phi y}$	Resistance force (shearing) in the core per unit area in the x-direction (N m^{-2})
$S_{\Pi x}$	Resistance force (shearing) in the cloud per unit area in the x-direction (N m^{-2})
$S_{\Pi y}$	Resistance force (shearing) in the cloud per unit area in the x-direction (N m^{-2})

and the random kinetic energy R_Φ (J m^{-2}). The random kinetic energy is associated with random movements of the solid particles that is, all kinetic energy that is not represented by the slope parallel velocities $U_\Phi = (u_\Phi, v_\Phi)$. The heat energy E_Φ can be compared to microscopic fluctuation energy; the random energy R_Φ is the macroscopic fluctuation energy (granular temperature). The source of both internal energies is the frictional shear work $[S_\Phi U_\Phi]$ (Buser and Bartelt 2009). We introduce a partitioning coefficient α to divide the shear work into heat and random kinetic energy. When $\alpha = 0$, all shear work is dissipated immediately to heat. When $0 < \alpha \leq 1$, the shear work generates random kinetic energy R_Φ; this energy is dissipated to thermal heat at the rate βR_Φ. The parameter β accounts for the decay of random energy and is a function of the hardness properties of the rock and ice (Buser and Bartelt 2009). The random kinetic energy is thus a transitory, intermediate, mechanical energy that exists in the avalanche flow.

$$\frac{\partial(R_\Phi)}{\partial t} + \frac{\partial(R_\Phi u_\Phi)}{\partial x} + \frac{\partial(R_\Phi v_\Phi)}{\partial y} = \alpha[S_\Phi U_\Phi] - \beta R_\Phi - \frac{\dot{M}_{\Phi\to\Pi}}{M_\Phi} R_\Phi + \dot{P}_{\Sigma\to\Phi} \tag{10}$$

$$\frac{\partial(E_\Phi)}{\partial t} + \frac{\partial(E_\Phi u_\Phi)}{\partial x} + \frac{\partial(E_\Phi v_\Phi)}{\partial y} = (1-\alpha)[S_\Phi U_\Phi] + \beta R_\Phi - \frac{\dot{M}_{\Phi\to\Pi}}{M_\Phi} E_\Phi + \dot{Q}_{\Sigma\to\Phi} - \dot{Q}_{i\to w} \tag{11}$$

If solid mass is transferred from the avalanche core to the suspension cloud, then the random kinetic energy and thermal heat energy associated with the transferred mass $\frac{\dot{M}_{\Phi\to\Pi}}{M_\Phi}$ is likewise transferred to the suspension cloud. Thus, the formation of the avalanche cloud extracts different energies from the avalanche core. For completeness we also state the random kinetic energy and thermal heat energy of the powder cloud. We do not consider changes of heat energy E_Φ in the powder cloud caused by interactions with the environment (sensible heat fluxes, radiative cooling, air entrainment, ice dust sublimation, etc).

4 ENTRAINMENT

We treat entrainment as a fully plastic collision between the avalanche and entrained material. The avalanche core, moving at the speed U_Φ, collides with material initially at rest. At the end of this collision the entrained material is moving with the speed of the avalanche. The entrainment rate (defined by the amount of material originally at rest and now moving with the speed U_Φ) is denoted $\dot{M}_{\Sigma\to\Phi}$. This material can consist of pure rock, ice and fluid or any mixture of materials. Different bed materials are modelled using different constitutive relationships for $\dot{M}_{\Sigma\to\Phi}$.

The amount of energy the avalanche core looses per unit time during the entrainment process is

$$\dot{L}_{\Sigma\to\Phi} = \frac{1}{2}\dot{M}_{\Sigma\to\Phi} U_\Phi^2. \tag{12}$$

The energy losses can take two forms: the rise of internal heat energy and the production of random kinetic energy. Moreover, we produce both microscopic fluctuations (heat) $\dot{Q}_{\Sigma\to\Phi}$ and macroscopic granular fluctuations $\dot{P}_{\Sigma\to\Phi}$,

$$\dot{Q}_{\Sigma\to\Phi} = \frac{1}{2}(1-\epsilon)\dot{M}_{\Sigma\to\Phi} U_\Phi^2 + c_\Sigma \dot{M}_{\Sigma\to\Phi} T_\Sigma \tag{13}$$

and

$$\dot{P}_{\Sigma\to\Phi} = \frac{1}{2}(\epsilon)\dot{M}_{\Sigma\to\Phi} U_\Phi^2. \tag{14}$$

The model accounts for the increase in internal energy of the avalanche when it entrains a substrate with specific heat c_Σ and temperature T_Σ (Vera Valero et al. 2016). The The parameter $\epsilon = 0$ represents an entrainment process where there is no production of random kinetic energy; that is, all the work done to accelerate the entrained mass is dissipated entirely and immediately to heat. The value $\epsilon = 1$ characterizes an entrainment process where the energy lost by the avalanche is dissipated entirely to macroscopic fluctuations, typically the entrainment of hard, loose, granular materials (rock beds). This type of entrainment is usually associated with particle splashing and chaotic movements at the avalanche front. Both $\dot{P}_{\Sigma\to\Phi}$ and $\dot{Q}_{\Sigma\to\Phi}$ appear in the random kinetic energy and thermal balance equations (see above).

5 PHASE CHANGES: LUBRICATED FLOW REGIME

Mobility of rock/ice avalanches is strongly related to the presence of pore fluid in the granular solid. The amount of fluid can wet the particle surfaces, leaving some air in the pore space (which we define as an under-saturated flow) or the fluid can fill the pore space completely (no air, saturated flow, homogeneous suspensions, typically mud flows) or there can be more fluid than possible pore space (over-saturated flow, typically granular debris flows). Over-saturated flows can only exist when the configuration of solid material in the avalanche is not homogeneous. Here we consider only rock avalanches that are under-saturated, reaching the limit of fully-saturated homogeneous suspensions. The fluid mass with the avalanche has three possible sources: 1) the initially released mass, 2) the entrainment mass and 3) the creation of water by phase changes (melting of ice, including snow). The first two sources are defined by the initial and boundary conditions of a particular problem. Modelling phase changes requires tracking the internal energy (temperature) of the solid phase. Meltwater production $\dot{T}_{i\rightarrow w}$ is considered as a constraint on the flow temperature of the avalanche: the mean flow temperature T_{Φ} can never exceed the melting temperature of ice $T_m = 273.15$ K. The energy for the phase change is given by the latent heat L

$$\dot{Q}_{i\rightarrow w} = L\dot{T}_{i\rightarrow w} \tag{15}$$

under the thermal constraint such that within a time increment Δt

$$\int_0^{\Delta t} \dot{Q}_{i\rightarrow w} dt = M_{\Phi} c_{\Phi} \left(T_{\Phi} - T_m\right) \tag{16}$$

For $T_{\Phi} > T_m$, otherwise $\dot{T}_{i\rightarrow w} = 0$. Obviously, when the flow temperature of the avalanche does not exceed the melting temperature, no latent heat is produced $\dot{Q}_{i\rightarrow w} = 0$. Because we adopt a depth-average approach, the temperature of the core will not exceed the melting temperature T_m until all ice is melted in a core volume.

6 SLOPE PERPENDICULAR ACCELERATIONS: FLUIDIZED FLOW REGIME

The normal force N (reaction) on the running surface of the avalanche is given by Newtons law (Buser and Bartelt 2015, Bartelt and Buser 2018):

$$N = M_{\Phi}\left(g_z + \ddot{k}_{\Phi}\right) \tag{17}$$

where g_z is the gravitational acceleration in the slope perpendicular direction and $\ddot{k}_{\Phi}$ is the impulsive acceleration of the center-of-mass of an avalanche column. The location k_{Φ} is the distance between the running surface and center-of-mass (Fig. 4). The normal reaction on the running surface increases when the center-of-mass is accelerated upwards $\left(\ddot{k}_{\Phi} > 0\right)$ and can decrease to zero when the column is in free fall $\left(\ddot{k}_{\Phi} = -g_z\right)$. When $\left(\ddot{k}_{\Phi} = 0\right)$ the basal normal force is simply the weight of the flowing mass $N = M_{\Phi} g_z$.

The slope perpendicular acceleration of the center-of-mass is a strongly time-dependent function, depending on the interaction of the avalanche core with the basal boundary. For example, ground roughness creates sudden changes in slope perpendicular acceleration (jerks). The boundary jerks are related to changes in random kinetic energy, and therefore directly to the basal shear work, but also the production of random energy during entrainment. The effect of the acceleration is to expand the volume of the flowing avalanche (decrease the flow density). We term this expansion, fluidization, as space between the solid particles opens, producing a fluid type motion.

To model the expansion of the core we must predict time-dependent changes in slope perpendicular accelerations, or jerks $J(t)$. This is an external forcing function that depends on both the roughness of the basal surface and material properties of the flow material. One method to define this function is to make it directly proportional to the shear work rate (and therefore the production of random kinetic energy). According to Buser and Bartelt 2015, full-scale experiments with snow avalanches reveal that $J(t) \alpha S_{\Phi} U_{\Phi}$.

When the volume expands, air is drawn into the space between particles. The availability of air defines the drag acting against the volume expansion. If air is freely available (e.g. the avalanche front) then the drag is small; if air is not available (e.g. the avalanche interior) then the expansion drag is large. This implies that there are frictional processes that damp the jerking motion within the avalanche core. This concept leads to a system of three partial differential equations for the location of the center-of-mass k_{Φ}, velocity $\dot{k}_{\Phi}$ and acceleration $\ddot{k}_{\Phi}$ (Buser and Bartelt 2015, Bartelt et al. 2016),

$$\frac{\partial\left(\ddot{k}_{\Phi}\right)}{\partial t} + \frac{\partial\left(\ddot{k}_{\Phi} u_{\Phi}\right)}{\partial x} + \frac{\partial\left(\ddot{k}_{\Phi} v_{\Phi}\right)}{\partial y} = J(t) \tag{18}$$

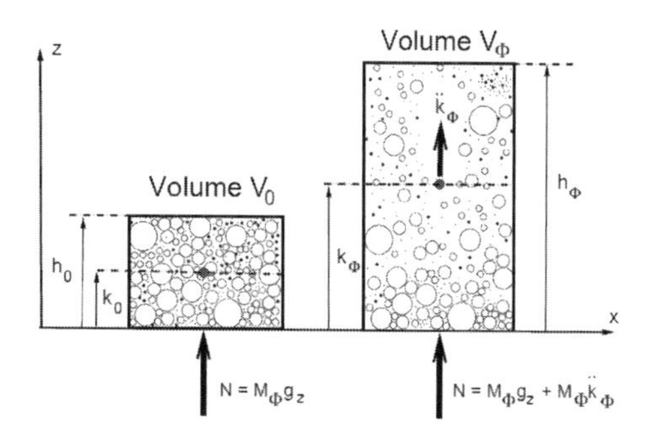

Figure 4. Slope perpendicular accelerations play an important role to fluidize the avalanche core Φ. When the center-of-mass k_{Φ} is stationary, the reaction at the base of the avalanche is equal to the weight. When the avalanche volume is expanding, the reaction increases in proportion to the slope perpendicular acceleration $\ddot{k}_{\Phi}$.

$$\frac{\partial(\dot{k}_\Phi)}{\partial t}+\frac{\partial(\dot{k}_\Phi u_\Phi)}{\partial x}+\frac{\partial(\dot{k}_\Phi v_\Phi)}{\partial y}=\ddot{k}_\Phi \quad (19)$$

$$\frac{\partial(k_\Phi)}{\partial t}+\frac{\partial(k_\Phi u_\Phi)}{\partial x}+\frac{\partial(k_\Phi v_\Phi)}{\partial y}=\dot{k}_\Phi \quad (20)$$

For now, we assume homogeneous distribution of mass in the column: $k_\Phi = 1/2\ h_\Phi$.

7 EXAMPLES

The depth-averaged equations were introduced into the RAMMS::AVALANCHE software system (Christen et al. 2010) and applied to simulate two recent rock/ice avalanche events.

7.1 *Langtang, Nepal 2015*

The air blast of a co-seismic rock/ice/snow avalanche killed more than 350 people and destroyed the village of Langtang, Nepal (Kargel et al. 2016, Fujita et al. 2017, Nagai et al. 2017). Intense ground shaking caused by the 7.8 magnitude earthquake of 25 April 2015 destabalized hanging glaciers on the mountain slopes above Langtang. The avalanche started at an elevation of between 6000 m - 6500 m and travelled 5000 m to reach the village (elevation 3500 m, drop height 3000 m). The avalanche core destroyed the outskirts of the village, before reaching the counter-slope. The powder suspension cloud, however, hit the village center, destroying most of the weakly constructed buildings (estimated air-blast pressure 5 kPa).

Simulations of the Langtang event were performed using a 20 m digital elevation model obtained from satellite images. We specified eight possible source areas, but found only one source area from which the mass reached the valley bottom, Fig. 5. Simulation mass was constrained by remote sensing measurements of deposition volumes. According to (Fujita et al. 2017) total deposition volume reached 14.4 mio m^3 (mean density 2200 kg/m^3). We calculated 14.7 mio mio m^3; however, we include mass outside the measurement zone. The calculations indicated that most mass was deposited at 5000 m, as observed by (Fujita et al. 2017). Deposition heights (> 10 m) and the overall deposition pattern matched the measurements, see (Fujita et al. 2017). The calculated avalanche reached a peak velocity of 90 m/s at 6000 m. When the avalanche core passed the village of Langtang, immediately before hitting the counter-slope, it was travelling at 70 m/s, in good agreement with the velocities (63 m/s), estimated by run-up calculations by (Fujita et al. 2017). The powder cloud inundated a larger area than the core, covering the village of Langtang completely. Air-blast impact pressures exceeded 5 kPa, but decreased substantially to 1 kPa as the cloud passed over the village. We specified a release temperature of -10C. Dissipated heating caused the avalanche to warm to melting temperatures near 0 C. The deposits contained some 150'000 m^3 of meltwater.

7.2 *Piz Cengalo, Switzerland 2017*

Approximately 3.1 mio m^3 of rock released on the the east face of Piz Cengalo on the 23rd of August 2017. The rock mass fell some 500 m (from 3000 m) directly on the glacier Vadrec del Cengalo and entrained some 0.3 mio m^3 to 0.6 mio m^3 of ice. The collision between the avalanching rock mass and glacier created a large suspension cloud of over 200 m height. The rock/ice mixture came to rest 3000 m from the release point on the 20 degree sloped runout zone in Val Bondasca. Observed deposition heights reached 30 m. Almost immediately aftern the event, muddy debris flows released from the rock/ice avalanche deposits, causing considerable damage to the village of Bondo. The event was spontaneously video recorded by climbers in the area. High resolution laser scans were performed from helicopters after the event to delimit the extent of the rock/ice avalanche deposits and to assess future danger.

Simulations were performed on a 10 m resolution digital elevation model. The video was used to pinpoint the release zone (Fig. 6). The first simulations were used to calculate the increase in temperature from frictional dissipation and ice entrainment. A17°C increase in temperature, relative to the initial temperature of the release mass was found. The next simulations were performed including meltwater (ice temperature constrained to 0°C). The initial temperature of the release mass is unknown; however, a slightly sub-zero temperature (-2°C), typical for August, is assumed. This result indicates that meltwater was clearly generated by frictional heating. We assumed the temperature of the entrained ice and avalanching rock to be the same. Below the glacier the avalanche entrained additional scree material (approx 1 mio m^3), which we assumed to have temperature above zero. The entrained material did not include any groundwater, which clearly leads to an underestimation of the amount of water in the avalanche deposits.

The calculated rock avalanche reached a peak velocity of 75 m/s. The entire event lasted 100 s (in agreement with seismic measurements). The avalanche entrained the observed 0.6 mio m^3 of ice and immediately began producing meltwater. The total calculated meltwater at the end of the event, reached 0.15 mio m^3. The meltwater accumulated in the frontal lobe of the avalanche deposits, at

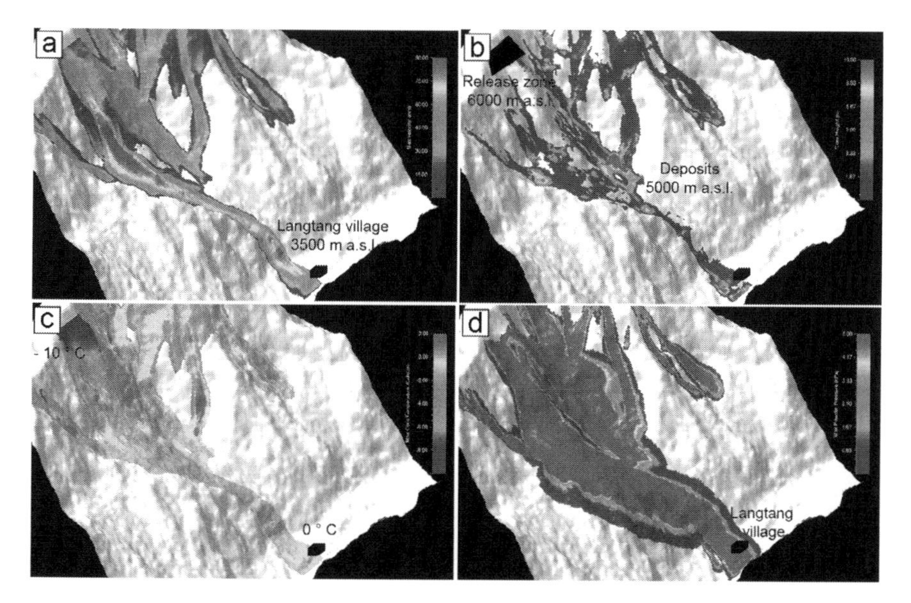

Figure 5. Simulation results of the Langtang ice avalanche, Nepal 2015. a) calculated maximum flow velocity. b) calculated avalanche deposits. c) calculated avalanche temperture. d) calculated air-blast pressure. Many source areas were released by seismic shaking, but only one source area reached the village. The avalanche obtained a peak velocity of 90 m/s.We assumed an initial temperature of T_{Φ} = -10°C. Frictional heating increased the temperature to T_{Φ} = -0°C; meltwater was produced. The avalanche core missed the village; however, the air-blast from the suspension layer was large. Calculated peak pressures exceeded 5 kPa. The location and distribution of deposits is in good agreement with the measurements presented in Fujita et al., 2017.

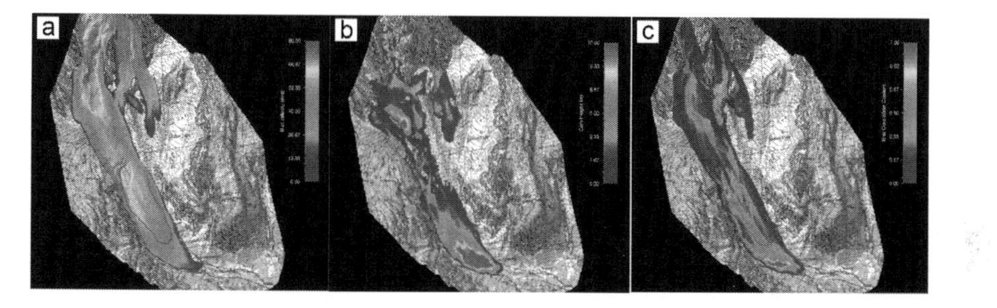

Figure 6. Simulation results of the Cengalo rock/ice avalanche, Switzerland 2017. a) maximum velocity. b) rock/ice deposits. c) meltwater. Debris flows initiated at the frontal lobe of the deposits and travelled to the drainage river.

the location from which the debris flow initiated. The model equations assume that the meltwater is homogenously distributed over depositions, which is clearly not the case. Approximately 10% - 20% water content (by volume) is needed to saturate the deposits (to create the debris flows) Ground water, or water droplets from the powder cloud, could easily make up for the missing water.

8 CONCLUSION

To simulate the physical complexity of rock/ice avalanches requires thermomechanical models that include the following processes:

1. Fluidization of the avalanche core Φ and formation of the dust suspension cloud Π,
2. Thermomechanical entrainment of ice and snow debris,
3. Melting of ice and snow in the avalanche core Φ,
4. Flow lubrication of the core Φ by meltwater,
5. Independent movement of the suspension cloud Π.

We adopted the only tractable numerical approach: depth-averaging. Slope perpendicular changes in accelerations that cause the fluidization of the avalanche are calculated by tracking the movement (location, velocity and acceleration) of the center-of-mass in the slope-perpendicular direction. The mechanical energies associated with the slope-perpendicular movements are constrained by energy balances that are given by the partitioning of internal heat and random mechanical energies. The system of differential equations is therefore always energy conserving, facilitating the prediction of avalanche flow temperature. The production of meltwater can therefore be calculated, if the initial (release) conditions and boundary (entrainment) conditions are known.

Clearly a full three-dimensional approach could be adopted. This would overcome some of the limitations of depth-averaging, specifically the assumptions of velocity and mass distribution over the height of the flow in both the avalanche core Φ and suspension cloud Π. The distribution of meltwater, for example, appears to be particularly important for the initiation of debris flows. At present, however, it is unlikely that the computational requirements for three-dimensional approaches are feasible for practical applications. The simulations could be performed in less than 10 minutes on standard personal computers.

REFERENCES

Bartelt, P.&O. Buser (2018). Avalanche dynamics by newton. reply to comments on avalanche flow models based on the concept of random kinetic energy. *Journal of Glaciology, 16.*

Bartelt, P., O. Buser, Y. Buehler, L. Dreier, & M. Christen (2014). Numerical simulation of snow avalanches: Modelling dilatative processes with cohesion in rapid granular shear flows. In Hicks, MA and Brinkgreve, RBJ and Rohe, A (Ed.), *NUMERICAL METHODS IN GEOTECHNICAL ENGINEERING, VOL 1*, pp. 327–332. 8th European Conference on Numerical Methods in Geotechnical Engineering (NUMGE), Delft Univ Technol, Delft, NETHERLANDS, JUN 18–20, 2014.

Bartelt, P., O. Buser, & K. Platzer (2006). Fluctuationdissipation relations for granular snow avalanches. *Journal of Glaciology 52*(179), 631–643.

Bartelt, P., O. Buser, C. Vera Valero, & Y. B¨uhler (2016). Configurational energy and the formation of mixed flowing/ powder snow and ice avalanches. *Annals of Glaciology 57*(71), 179–187.

Bartelt, P., C. Vera Valero, T. Feistl, M. Christen, Y. Buehler, & O. Buser (2015). Modelling cohesion in snow avalanche flow. *Journal of Glaciology 61*(229), 837–850.

Buser, O. & P. Bartelt (2009). Production and decay of random kinetic energy in granular snow avalanches. *Journal of Glaciology 55*(189), 3–12.

Buser, O. & P. Bartelt (2015). An energy-based method to calculate streamwise density variations in snow avalanches. *Journal of Glaciology 61*(227), 563–575.

Carey, J.M., G.T. Hancox, & M.J. McSaveney (2015, Oct). The january 2013 wanganui river debris flood resulting from a large rock avalanche from mt evans, westland, new zealand. *Landslides 12*(5), 961–972.

Christen, M., J. Kowalski, & P. Bartelt (2010). Ramms: Numerical simulation of dense snow avalanches in three-dimensional terrain. *Cold Regions Science and Technology 63*(1), 1–14.

Dreier, L. abd B¨uhler, Y., F. Dufour, G.C., & P. Bartelt (2016). Comparison of simulated powder snow velocities, volumes and flow widths with photogrammetric measurements. *Annals of Glaciology 57*(71), 371–381.

Faillettaz, J., M. Funk, & C. Vincent (2015). Avalanching glacier instabilities: Review on processes and early warning perspectives. *Reviews of Geophysics 53*(2), 203–224. 2014RG000466.

Fujita, K., H. Inoue, T. Izumi, Y. Satoru, A. Sadakane, S. Sunako, K. Nishimura, W. Immerzee, J. Shea, S.T.B.D. Kayastha, R., H. Yagi, & A. Sakai (2017). Anomalous winter-snow-amplified earthquake-induced disaster of the 2015 langtang avalanche in nepal. *Nat. Hazards Earth Syst. Sci. 17*, 749–764.

Huggel, C., N. Salzmann, S. Allen, J. Caplan-Auerbach, L. Fischer, W. Haeberli, C. Larsen, D. Schneider, & R. Wessels (2010). Recent and future warm extreme events and high-mountain slope stability. *Philosophical Transactions of the Royal Society of London A: Mathematical, Physical and Engineering Sciences 368*(1919), 2435–2459.

Huggel, C., S. Zgraggen-Oswald, W. Haeberli, A. Kääb, A. Polkvoj, I. Galushkin, & S.G. Evans (2005). The 2002 rock/ice avalanche at kolka/karmadon, Russian caucasus: assessment of extraordinary avalanche formation and mobility, and application of quickbird satellite imagery. *Natural Hazards and Earth System Sciences 5*(2), 173–187.

Kargel, J.S., G.J. Leonard, D.H. Shugar, U.K. Haritashya, A. Bevington, E.J. Fielding, K. Fujita, M. Geertsema, E.S. Miles, J. Steiner, E. Anderson, S. Bajracharya, G.W. Bawden, D.F. Breashears, A. Byers, B. Collins, M.R. Dhital, A. Donnellan, T.L. Evans, M.L. Geai, M.T. Glasscoe, D. Green, D.R. Gurung, R. Heijenk, A. Hilborn, K. Hudnut, C. Huyck, W.W. Immerzeel, J. Liming, R. Jibson, A. Kääb, N.R. Khanal, D. Kirschbaum, P.D.A. Kraaijenbrink, D. Lamsal, L. Shiyin, L. Mingyang, D. McKinney, N.K. Nahirnick, N. Zhuotong, S. Ojha, J. Olsenholler, T.H. Painter, M. Pleasants, K.C. Pratima, Q.I. Yuan, B.H. Raup, D. Regmi, D.R. Rounce, A. Sakai, S. Donghui, J.M. Shea, A.B. Shrestha, A. Shukla, D. Stumm, M. van der Kooij, K. Voss, W. Xin, B. Weihs, D. Wolfe, W. Lizong, Y. Xiaojun, M.R. Yoder, & N. Young (2016). Geomorphic and geologic controls of geohazards induced by nepals 2015 gorkha earthquake. *Science 351*(6269).

Nagai, H., M. Watanabe, N. Tomii, T. Takeo, & S. Suzuki (2017). Multiple remote-sensing assessment of the catastrophic collapse in langtang valley induced by the 2015 gorkha earthquake. *Nat. Hazards Earth Syst. Sci. 17*, 1907–1921.

Vera Valero, C., K.W. Jones, Y. Bühler,&P. Bartelt (2015). Release temperature, snow-cover entrainment and the thermal flow regime of snow avalanches. *Journal of Glaciology 61*(225), 173184.

Vera Valero, C., N. Wever, Y. B¨uhler, L. Stoffel, S. Margreth, & P. Bartelt (2016). Modelling wet snow avalanche runout to assess road safety at a high-altitude mine in the central andes. *Natural Hazards and Earth System Sciences 16*(11), 2303–2323.

Numerical Methods in Geotechnical Engineering IX – Cardoso et al. (Eds)
© 2018 Taylor & Francis Group, London, ISBN 978-1-138-33203-4

Effect of failure criterion on slope stability analysis

F. Tschuchnigg & H.F. Schweiger
Computational Geotechnics Group, Institute for Soil Mechanics and Foundation Engineering, Graz University of Technology, Graz, Austria

M. Sallinger
SKAVA consulting ZT-GmbH, Innsbruck, Austria

ABSTRACT: Stability analyses in geotechnical engineering are generally based on a Mohr-Coulomb failure criterion. However, this model neglects the intermediate principal stress and leads in some circumstances to conservative results related to the factor of safety. This paper shows on the one hand the effect of the Matsuoka-Nakai failure criterion on the factor of safety of slopes and it compares on the other hand results of slope stability analysis obtained from different calculation methods, namely limit equilibrium analysis, finite element limit analysis and displacement finite element Strength Reduction Technique (SRFEA). The latter approach could suffer from numerical instabilities when using non-associated plasticity and finite element limit analysis has the disadvantage that it is limited to associated plasticity. Therefore, a modification of the so-called Davis approach is presented, where the reduced strength parameters are related to the actual degree of non-associativity. The presented studies show first, that this enhanced procedure yield FoS in close agreement with SRFEA using a non-associated flow rule and secondly, that the modified Davis approach is suitable to prevent numerical instabilities. However, important for practitioners, the obtained factors of safety with the improved Davis procedure are still slightly conservative.

1 INTRODUCTION

For slope stability analyses it is common to use shear strength parameters of the soil to define the factor of safety. The limit equilibrium methods (LEM) by Janbu (1954), Bishop (1955) and Morgenstern & Price (1965) are based on the method of slices and have a wide tradition in slope stability analysis, but assumptions e.g. regarding the shape of the failure mechanism or the forces acting between the slices are required.

Therefore, displacement based finite element analyses (SRFEA) became increasingly popular over the last decades. But, as shown in Tschuchnigg et al. (2015a), large differences between friction angle φ' and dilatancy angle ψ' may lead to numerical problems without any clear definition of the factor of safety. Stability may also be assessed using finite element limit analysis (FELA), which are based upon the limit theorems of plasticity, that predict failure by optimising the applied loads with consideration of the stress equilibrium equations, the stress-strain relationship and the kinematic compatibility of the problem (e.g. Sloan 2013).

It has been shown previously that FELA, LEM and SRFEA yield similar results when employing a Mohr-Coulomb failure criterion and assuming associated plasticity ($\psi' = \varphi'$) (Tschuchnigg et al. 2015a). As FELA is restricted to associated plasticity, Davis (1968) proposed reduced strength parameters to simulate a non-associated material behaviour. This approach is suitable if the FoS is expressed in terms of maximising loads for a given strength (Tschuchnigg et al. 2015b). If the definition of safety is based on the shear strength parameters of the soil, the mentioned approach could lead to very conservative results.

The paper shows the influence of the failure criterion according to Matsuoka-Nakai on the factor of safety of slope stability problems and investigates enhanced versions of the Davis approach (Davis 1968) to overcome the aforementioned problems. Finally, some numerical details when performing SRFEA are addressed.

2 FAILURE CRITERIA—MOHR-COULOMB VS MATSUOKA-NAKAI

The most frequently applied failure criterion for soils is the well-known Mohr-Coulomb criterion (MC), which can be formulated as followed:

$$(\sigma_1' - \sigma_3') = (\sigma_1' + \sigma_3')\sin\varphi' + 2c'\cos\varphi' \quad (1)$$

However, Eq. (1) neglects the intermediate effective principal stress σ_2'.

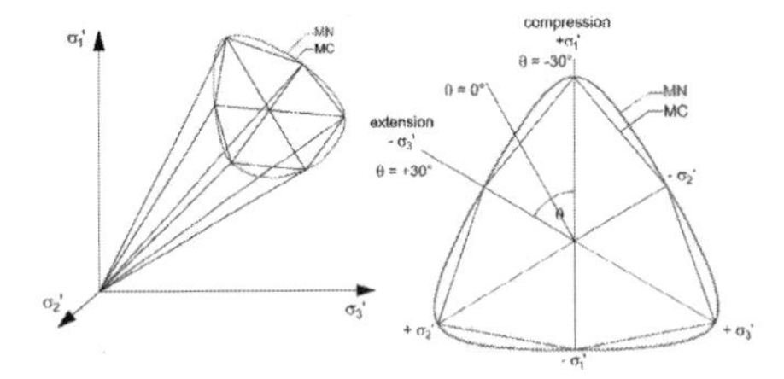

Figure 1. Matsuoka-Nakai and Mohr-Coulomb failure criterion in 3D principal effective stress space (left) and π-plane (right).

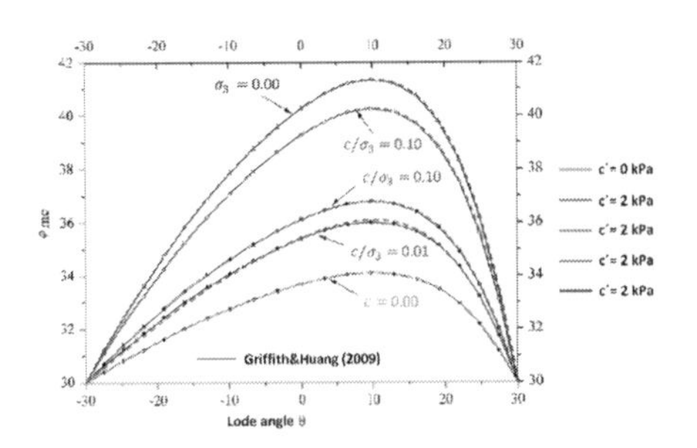

Figure 2. Matsuoka-Nakai failure criterion: Comparison with Griffith & Huang (2009).

The Matsuoka-Nakai failure criterion (MN) on the other hand considers the intermediate principal stress σ'_2 (Matsuoka & Nakai, 1974). The yield condition according to MN circumscribes in the deviatoric plane (π-plane) the hexagon of the Mohr-Coulomb criterion. Thus, the MN and MC predict the same maximum deviatoric stress q′ for stress paths in triaxial compression and triaxial extension (see Figure 1).

The friction angle based on the Mohr-Coulomb failure criterion is assumed constant rather than dependent on the lode angle θ (Figure 1). On the contrary, the Matsuoka-Nakai failure criterion varies the friction angle depending on θ. Moreover, the cohesion affects the shape of the MN yield surface, thus the equivalent friction angle φ'_{MC}.

The used version of the MN failure criterion was verified with an example presented by Griffith & Huang (2009), where the equivalent friction angle φ'_{MC} is worked out for different ratios of effective cohesion c′ to minor effective principal stress σ'_3. From Figure 2 follows that the computed values agree perfectly with the curves published by Griffith & Huang (2009).

3 USED METHODS TO COMPUTE THE SLOPE INCLINATION AT FAILURE

3.1 *Limit equilibrium consideration*

The acting forces on an arbitrary slice of a cohesionless infinite slope are considered in the following to compute a statically admissible state. The forces acting on such a slice are the gravity load of the soil G, the reaction force Q and the load due to the surrounding soil E (see Figure 3). Since an infinite slope is considered, the line of action as well as the magnitude are equal for both earth pressures (E_l and E_r). Consequently, the slope inclination at failure can be calculated only by means of G and Q. Where the destabilizing force $G_{//}$ is defined as:

$$G_{//} = G \sin \beta_s \tag{2}$$

The resistance force R depends on the normal force and the effective friction angle along the slip surface and the limit state is reached when R is equal to $G_{//}$ and so the slope inclination at failure computes to Eq. (4).

$$R = N \tan \varphi' = G \cos \beta_s \tan \varphi' \tag{3}$$

$$R = G_{//} \rightarrow G \sin \beta_s = G \cos \beta_s \tan \varphi' \rightarrow \beta_s = \varphi' \tag{4}$$

As a result, the slope inclination at failure is equal to the friction angle φ' in plane strain condition. More details related to the infinite slope model can be found in Fellin (2014). Additionally to the assumption of an infinite slope, limit equilibrium analysis using the Morgenstern-Price method (Morgenstern & Price, 1965) were carried out.

3.2 *Slope inclination at failure—Davis (1968)*

Based on the Mohr-Coulomb failure criterion and the assumption of a simple shear mechanism in the slip surface Davis (1968) proposed Eq. (5), which allows the determination of the slope inclination at failure including a dilatancy angle ψ' greater than zero and smaller than the friction angle. Obviously for $\psi' = 0$ Eq. (5) reduces to $\tan \beta_s = \sin \varphi'$ which gives much smaller slope inclinations at failure than Eq. (4).

$$\tan \beta_s = \frac{\sin \varphi' \cos \psi'}{1 - \sin \varphi' \sin \psi'} \tag{5}$$

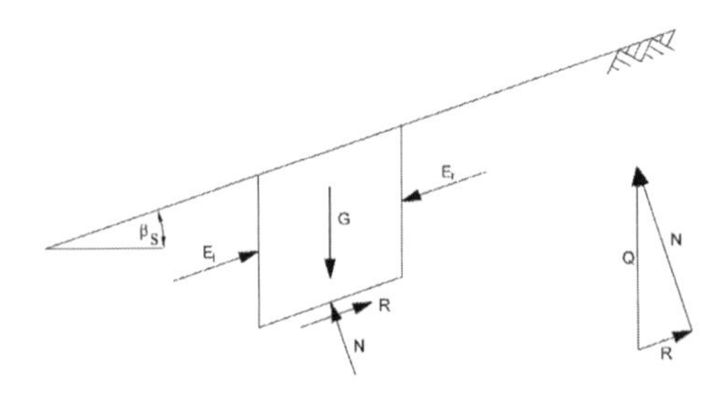

Figure 3. Arbitrary slice in an infinite slope (acc. to Fellin, 2014).

3.3 *Slope inclination at failure—Schranz & Fellin (2016)*

This approach uses a MN failure criterion, thus includes the "additional" strength because of the intermediate effective principal stress σ'_2. Based on numerical investigations of simple shear tests under constant normal stress and plane strain conditions Schranz & Fellin (2016) proposed Eq. (6), which is strictly speaking an extension of Eq. (5).

$$\tan\beta_s = \frac{\sin(1.085\varphi')\cos\psi'}{1-\sin\varphi'\sin\psi'} \tag{6}$$

3.4 *Strength reduction method (SRFEA)*

When performing a displacement based finite element analysis in combination with a strength reduction technique (see e.g. Tschuchnigg et al. 2015b) the FoS is obtained by simultaneously reducing tan φ' and c' until no equilibrium can be satisfied anymore (a Mohr-Coulomb failure criterion is generally assumed), and the computed FoS is defined as:

$$\text{FoS} = \frac{\tan\phi'_{\text{avail}}}{\tan\phi'_{\text{failure}}} = \frac{c'_{\text{avail}}}{c'_{\text{failure}}} \tag{7}$$

3.5 *Finite element limit analysis (FELA)*

The concept of limit analysis is based on the theorems of plasticity developed by Drucker et al. (1951, 1952), namely the lower and upper bound theorem. Both limit theorems assume a perfectly plastic material with an associated flow rule, and ignore the effect of geometry changes. Other than limit equilibrium methods, the approach considers the stress equilibrium equations, the stress-strain relationship and the kinematical compatibility throughout the whole soil body. Hence, the approach calculates true collapse loads, or in most cases, upper and lower bounds of the collapse loads.

The formulation used in this paper stems from the methods originally developed by Sloan (1988), Sloan (1989) and further improved by Lyamin and Sloan (2002a), Lyamin and Sloan (2002b), Krabbenhøft et al. (2005) and Krabbenhøft et al. (2007).

3.6 *Non-associated plasticity*

During an SRFEA performed with associated plasticity, both friction angle φ' and dilatancy angle ψ' are reduced at the same time. While highlighting the non-associated flow rule, it should be noted that as long as the reduced friction angle $\varphi'_{\text{red.}}$ is greater than the dilatancy angle ψ', the latter one is usually kept constant (at least in commercially available codes such as Plaxis 2D). At the point where $\varphi'_{\text{red.}} = \psi'$, both get reduced simultaneously (Brinkgreve et al. 2016).

The calculations presented in the following are performed with a user-defined version of the strength reduction procedure, where $\tan\varphi'$, $\tan\psi'$ and c' are reduced simultaneously. As a consequence the degree of non-associativity Λ (φ' - ψ') is also affected by the reduced value of ψ'. A shortcoming of SRFEA in combination with non-associated plasticity is that it may lead to numerical instabilities without any clear defined failure mechanism. This is particularly the case for materials with high friction angles in the range of $\varphi' > 40°$ (Tschuchnigg et al. 2015b).

FELA is restricted to associated plasticity, since stress and velocity characteristics are equal only for the case of an associated flow rule ($\psi' = \varphi'$). Therefore, the modified parameters c* and φ* have to be used as input parameters (in combination with associated plasticity).

The approach according to Davis (1968) is given in Eq. (8), Eq. (9) and Eq. (10) and is denoted as Davis A in the following. This original approach by Davis leads to very conservative results if the factor of safety is expressed by the strength parameters of the soil, therefore modified procedures (so called Davis B and Davis C) were developed to achieve a better agreement with non-associated SRFEA (Tschuchnigg et al. 2015b, Tschuchnigg et al. 2017).

$$c^* = \beta \cdot c' \tag{8}$$

$$\tan\varphi^* = \beta \cdot \tan\varphi' \tag{9}$$

$$\beta_{initial} = \beta_{failure} = \frac{\cos\psi' \cdot \cos\varphi'}{1-\sin\psi' \cdot \sin\varphi'} \tag{10}$$

In Davis B (Eq. 11) the strength reduction factor (β) is calculated based on the effective strength parameters applied in the current iteration, thus β changes during a SRFEA, because Λ changes ($\beta_{\text{initial}} \neq \beta_{\text{failure}}$).

$$\beta_{failure} = \frac{\cos\left(\tan^{-1}\left(\frac{\tan\varphi'}{FoS}\right)\right)\cdot\cos\left(\tan^{-1}\left(\frac{\tan\psi'}{FoS}\right)\right)}{1-\sin\left(\tan^{-1}\left(\frac{\tan\varphi'}{FoS}\right)\right)\cdot\sin\left(\tan^{-1}\left(\frac{\tan\psi'}{FoS}\right)\right)} \tag{11}$$

Davis C is similar to procedure B, except that ψ' is not affected by the strength reduction and is kept constant (Eq. (12)).

$$\beta_{failure} = \frac{\cos\left(\arctan\left(\frac{\tan\psi'}{FoS}\right)\right)\cdot\cos\psi'}{1-\sin\left(\arctan\left(\frac{\tan\psi'}{FoS}\right)\right)\cdot\sin\psi'} \tag{12}$$

4 EXAMPLE I—SLOPE AT FAILURE

In the following, the slope inclination at failure for homogeneous slopes with cohesionless material is investigated. The SRFEA are computed with the finite element code Plaxis 2D (Brinkgreve et al., 2016) and FELA are performed with Optum G2 (Krabbenhoft et al, 2016). The computed results are compared with the approaches mentioned in chapter 3.

The dimensions of the model are indicated in Figure 4, where L = 10/tan β_s. The slope has a height of 10 m but the slope inclination varies between 20° to 50°. The FE meshes used consist of 15-noded triangles including a local mesh refinement in the zone of the failure mechanism. A soil under drained conditions is considered with a unit weight of γ_{unsat} = 17.0 kN/m³ and a Poisson´s ratio of ν = 0.3. The dilatancy angle ψ' is varied (in SRFEA and FELA) with $\phi'/3$, $\phi'/4$, ϕ'-30° and 0°.

Figure 5 to 7 illustrate the slope inclination at failure (β_s) related to the effective friction angle. Figure 5 shows the results for ψ' = $\varphi'/3$, Figure 6 for ψ' = φ'-30° and Figure 7 for ψ' = 0. Results of different approaches are compared, namely FoS computed with Plaxis 2D (using failure criteria according to MC and MN), FELA (using an associated flow rule and the original Davis approach) and additionally results obtained with the classical derivation (Eq. (4)) and the approaches according to Davis (1968) and Schranz & Fellin (2016) (Eq. (5) and Eq. (6)). It has to be mentioned that the user defined Matsuoka-Nakai soil model suffered from numerical instabilities when higher slope inclinations and consequently larger amounts of non-associativity where considered.

However, the obtained results lead to the following conclusions. Comparing Figure 5, Figure 6 and Figure 7 it gets clear that the flow rule and the dilatancy angle play an important rule. The lower the dilatancy angle the higher the required friction angle. Nevertheless, the results also show that φ' (for a certain β_s) reduces significantly when using the MN failure criterion. When performing FEA with the MC failure criterion it seems that the corresponding limit friction angle at low slope inclinations agrees quite well with β_s obtained with Eq. (5). But if the slope gets steeper, the results show a better agreement with Eq. (6), which is based on numerical studies using the MN failure surface (Schranz & Fellin, 2016). But it has to be mentioned, that the considerations behind the equations Eq. (5) and (6) are based the assumption of a simple shear mechanism in the slip surface (see chapter 3).

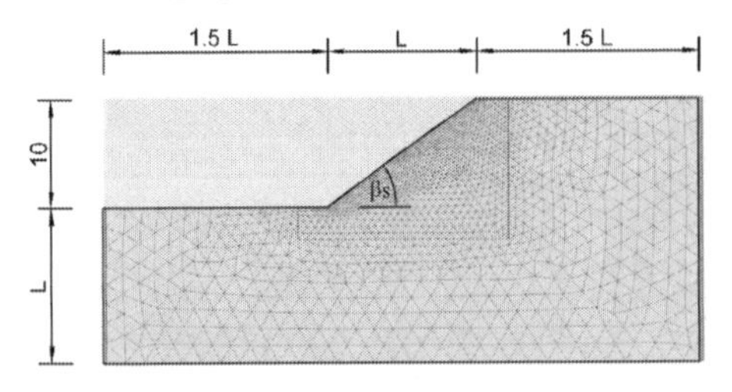

Figure 4. Finite element model and slope geometry.

The comparison between MC-Plaxis and MN-Plaxis show that the divergence increases slightly the more the dilatancy angle deviates from the friction angle (amount of non-associativity Λ). Observing a dilatancy angle of ψ' = $\phi'/3$ it can be seen that the friction angle obtained with MN-Plaxis is about 10% to 11% lower than with MC-Plaxis. However, with ψ'= 0° the difference is around 13%.

Table 1 illustrates the difference of required friction angles between the MC and the MN criterion, depending on the dilatancy angle and the slope inclination.

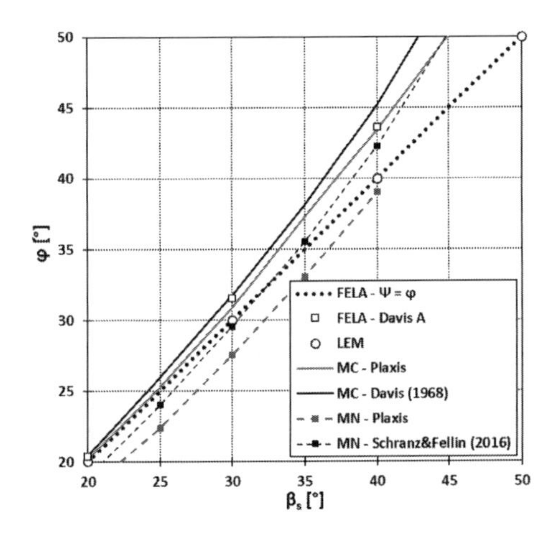

Figure 5. Comparison of limit states with ψ'= $\phi'/3$.

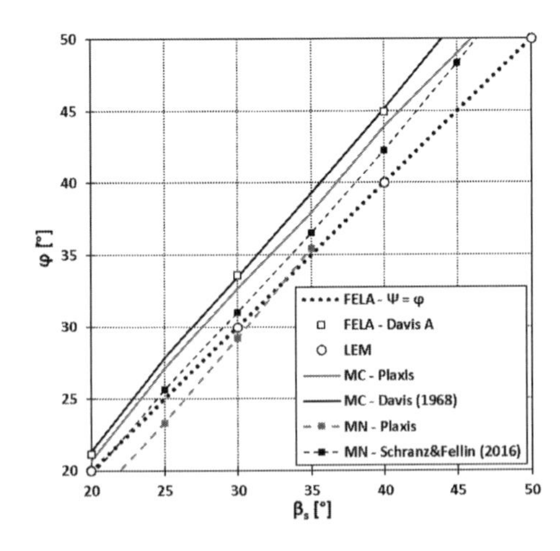

Figure 6. Comparison of limit states with ψ'= ϕ'-30°.

Finally, it seems that in some circumstances MN allows even steeper slopes than FELA or SRFEA assuming an associated flow rule (or considering Eq. (4)). Provided that the slope angle does not exceed ~35–40°. The same tendencies were also reported by Schranz & Fellin (2016).

5 EXAMPLE 2

The model size of the investigated slope (Figure 8) are related to the slope height $H = c'/(\gamma' \cdot 0.05)$ and the used parameters are listed in Table 2. Hence, a slope inclination of 26.6° is modelled and the FE model is discretised with 1004 15-noded triangles.

The computed factor of safety with the MC failure criterion is 1.40 and, when applying the MN the FoS = 1.62.

5.1 *Investigations related to the failure mechanism*

Preliminary studies using SRFEA showed that stress paths of stress points in the region of the failure mechanism, obtained during the strength reduction are different for the different failure criteria. A number of SRFEA using the Mohr-Coulomb failure criterion showed that the lode angle stays fairly constant during the calculation, whereas when applying the Matsuoka-Nakai criterion the lode angle tends to change significantly. In some analyses using the MN model the intermediate effective principal stress σ'_2 was about 45% larger compared with a MC model. Additionally also the variation of mean effective stress p′during the strength reduction is different for the MC and the MN failure criterion (Sallinger, 2017).

In the following, the plastic points as well as the failure mechanism obtained after the ϕ/c-reduction are investigated (for both MN and MC failure criteria). Therefore, several stability levels (values of FoS) are considered, which are controlled by a varying friction angle. The used effective cohesion and dilatancy angle are kept constant (see parameters in Table 2).

Figure 9 to 11 show that the plastic points with MC are almost similar in all calculations, which is not the case if the MN is used. With the MN model the zone of plasticity reduces with increasing factor of safety. Nevertheless, in all cases analysed the slip plane is well defined and intersects the foot of the slope. The distance Δl (see Figure 8) was worked out for different SRFEA and the outcome was that the onset of the failure mechanism does not diverge excessively for the different failure criteria used.

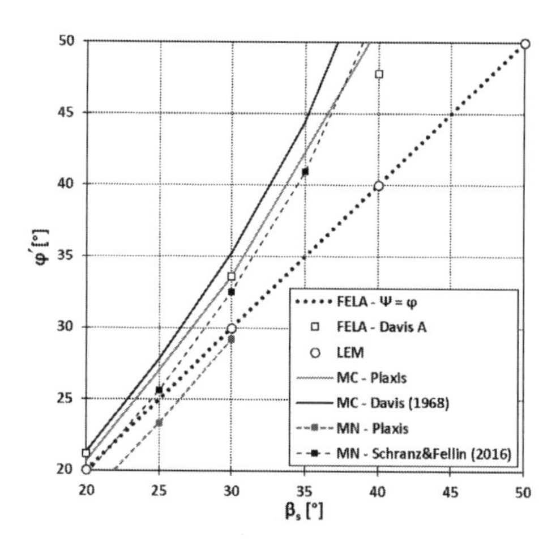

Figure 7. Comparison of limit states with $\psi' = 0°$.

Table 1. Difference of required friction angles at slope failure—MC vs MN (Plaxis analysis).

$\Delta\phi' = 100 \cdot (\phi'_{MC} - \phi'_{MN}) / \phi'_{MC}$ [%]

β [°]	$\psi' = \phi'/3$ [°]	$\psi' = \phi'/4$ [°]	$\psi' = \phi' - 30°$ [°]	$\psi' = 0°$ [°]
20	10.9	11.0	13.4	13.4
25	11.5	11.0	13.9	13.8
30	10.9	10.4	10.7	13.1
35	11.3	8.8	6.4	–
40	10.2	7.9	–	–

5.2 *Influence of soil strength on FoS*

The results plotted in Figure 12 illustrate the effect of c′ and φ' on the computed FoS. Since the considered slope has a moderate inclination of 1:2 the difference between LEM und SRFEA using a MC

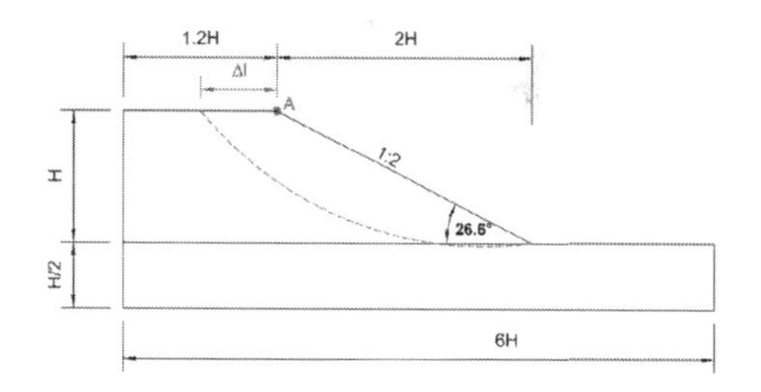

Figure 8. Slope geometry of Example 2.

Table 2. Input parameters Example 2.

Parameter	Value	Unit
γ'	16.0	[kN/m³]
ϕ'	21.1	[°]
c′	8.0	[kN/m²]
ψ'	0.0	[°]
ν'	0.2	[–]
H	10.0	[m]

failure criterion is almost negligible. However, it should be noted that for steep slopes, in combination with a large amount of non-associativity (Λ), LEM always predicts FoS which are on the unconservative side (see Tschuchnigg et al. 2017).

However, from Figure 12 also follows that the MN failure criterion predict significantly higher factors of safety. Figure 14 displays the influence of the failure criterion, where the positive ΔFoS value represents the difference between MN and MC ($\Delta FoS = (FoS_{MN}-FoS_{MC})/FoS_{MC}$). From the performed studies follows that the MN criterion predict roughly 15% higher values of FoS independent of c′.

Figure 13 on the other hand compares SRFEA results using non-associated plasticity with FELA applying the original (Davis A) and the enhanced versions of the Davis approach (Davis B and Davis C). Due to the soil properties (low friction angle) the flow rule does not have a big influence on the computed FoS, thus FELA (using associated plasticity) and SRFEA with a non-associated flow rule yield similar factors of safety. However, this fact

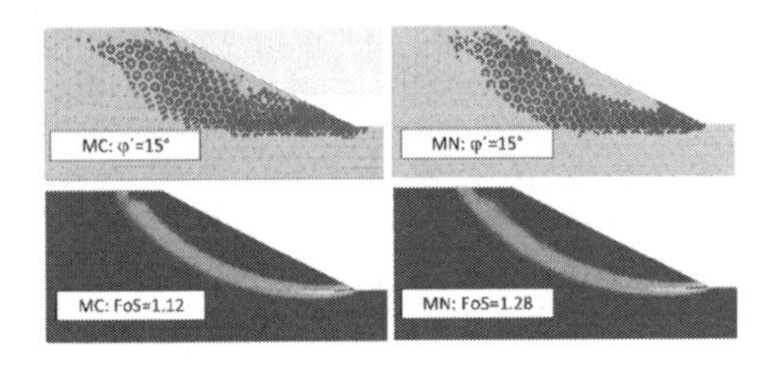

Figure 9. Plastic points and incremental deviatoric strains with MC (left) and MN (right) - $\phi' = 15°$.

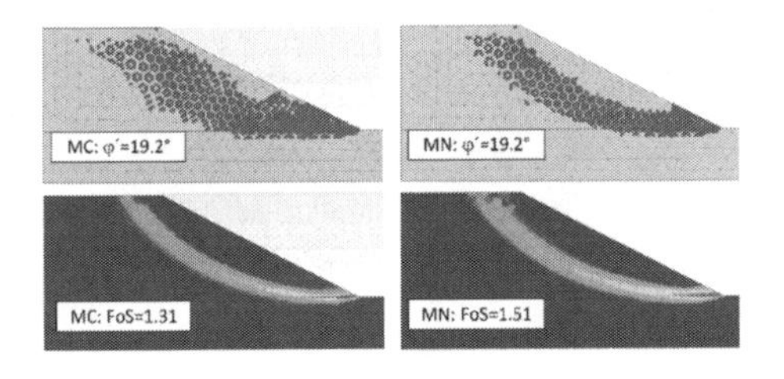

Figure 10. Plastic points and incremental deviatoric strains with MC (left) and MN (right) - $\phi' = 19.2°$.

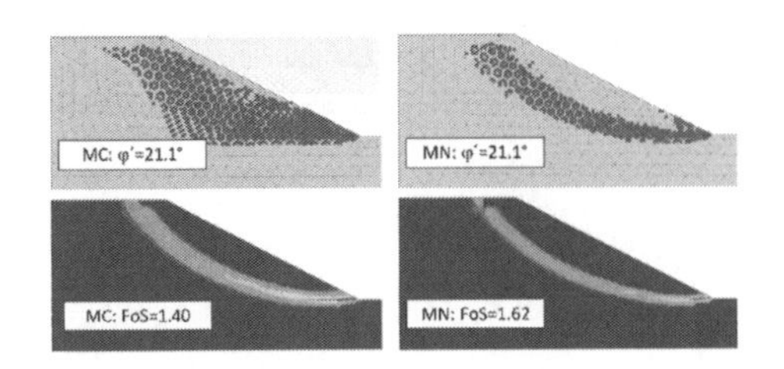

Figure 11. Plastic points and incremental deviatoric strains with MC (left) and MN (right) - $\phi' = 21.1°$.

is unknown until a SRFEA using non-associated plasticity is performed. Therefore, especially in combination with FELA, the aforementioned Davis procedure (Davis A) is generally applied in practice.

The obtained results clearly show that the predicted FoS using Davis A are very conservative, whereas the FoS computed with Davis B are much less conservative. Figure 14 highlights the difference between SRFEA and FELA for the original and the enhanced Davis procedure ($\Delta FoS = (FoS_{DAVIS}-FoS_{MC})/FoS_{MC}$).

This example demonstrated the big advantage of the enhanced Davis procedures presented in chapter 3, namely the results are still conservative but are much closer to FoS calculated with SRFEA using non-associated plasticity.

6 NUMERICAL STUDIES ON SRFEA

6.1 *Effect of iterative settings*

In the following, the influence of the tolerated error ε_{tol} and the Msf value on the performance of SRFEA (with MC failure criterion) is investigated. The slope geometry is similar to example 2 (Figure 8) and the strength parameters used are $\varphi' = 19.2°$, $c' = 8.0$ kPa and $\psi' = 0$.

Figure 15 plots the FoS over conducted calculation steps. The lower the tolerated error the stronger the erratic behaviour of the FoS. But the analyses also show the tendency that the computed FoS decreases with decreasing ε_{tol} (Figure 15). Further studies with more advanced constitutive models showed a similar trend. Based on a number of SRFEA it can be concluded that the tolerated error has an influence on the FoS and the erratic nature of the solution (Sallinger 2017), but for the considered boundary value problems the standard setting of $\varepsilon_{tol} = 1\%$ can be classified as sufficient.

The Msf value specifies the increment of the strength reduction of the first calculation step. The

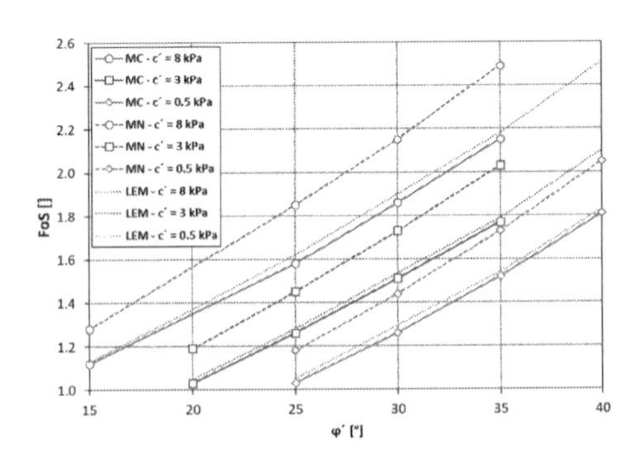

Figure 12. Effect of ϕ' and c′ - SRFEA (MN & MC) and LEM.

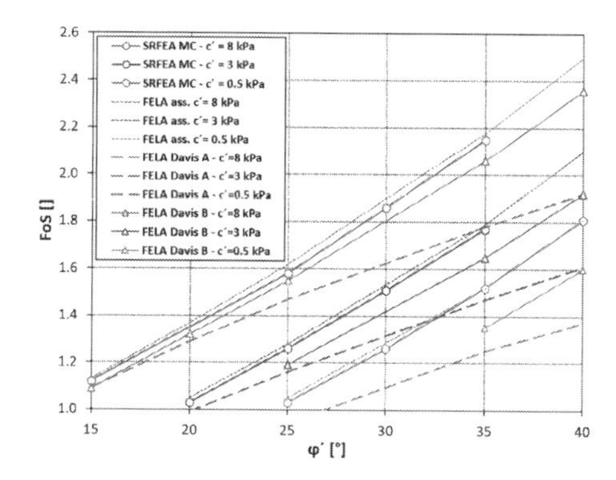

Figure 13. SRFEA vs FELA.

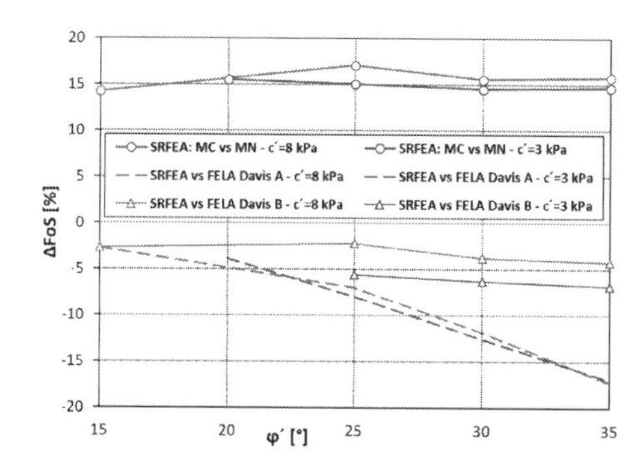

Figure 14. ΔFoS: SRFEA (MC vs MN) and FELA.

standard setting in the iterative procedure, when using Plaxis 2D (Brinkgreve et al. 2016) is 0.1.

As shown in Figure 16 the factor of safety is not affected by the magnitude of Msf. Since a lower Msf leads to a smaller increment of the strength reduction at the start of the calculation phase the typical peak at the beginning of a SRFEA is less pronounced (see Figure 16). However, studies with different constitutive models confirmed that the Msf value does not have an influence on both the obtained FoS and the oscillations of the solution (Sallinger 2017).

7 CONCLUSION AND OUTLOOK

Based on infinite slope considerations prior works investigated the effect of the failure criterion on slope stability analyses (Schranz & Fellin, 2016). There it has been shown, that the Matsuoka-Nakai failure criterion leads to higher factors of safety.

In this paper various cohesionless slopes were analysed by means of the displacement finite element method, the finite element limit analysis and the conventional limit equilibrium method in order to determine the strength parameters at ultimate limit state. It was found that the flow rule and consequently the dilatancy angle has a substantial influence on the results when performing SRFEA.

It could also be confirmed that the failure criterion has a significant influence on the resulting factors of safety. In SRFEA presented the difference in FoS between the Mohr-Coulomb and the Matsuoka-Nakai failure criterion is up to ~15%. A detailed investigation of the computed failure mechanism showed, that the incremental deviatoric strains obtained with the different failure criteria are almost identical.

The results presented in this paper revealed that limit equilibrium analysis (LEM), finite element limit analysis (FELA) and the strength reduction finite element analysis (SRFEA) assuming associated plasticity are in good agreement. Since the failure mechanism may varies during a φ-c reduction, when dealing with non-associated plasticity (oscillations) and, because FELA is restricted to the normality rule, the Davis (1968) approach was applied in combination with finite element limit analyses. The original concept (Davis A approach) yield very conservative results when the factor of safety is expressed in terms of strength parameters. Therefore, enhanced procedures Davis B and Davis C have been presented, where the reduction factor β is related to the actual degree of non-associativity. The results obtained with the enhanced Davis

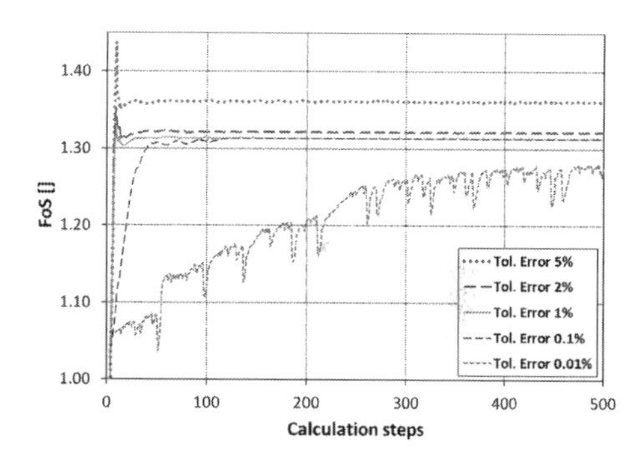

Figure 15. Effect of ε_{tol} on FoS.

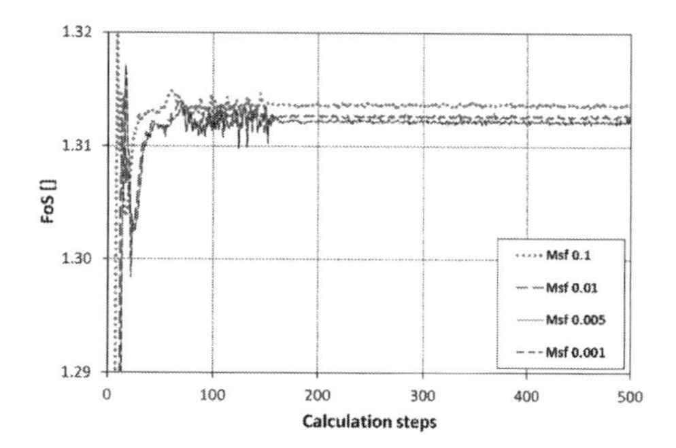

Figure 16. Effect of Msf on FoS.

procedures can be considered as more realistic and significantly less conservative than FoS obtained with original Davis approach. Another outcome of this study is that the numerical instabilities of non-associated SRFEA can be avoided by adopting the enhanced Davis procedure.

Finally, the influence of some iterative settings on the performance of strength reduction finite element analysis is investigated.

ACKNOWLEDGEMENT

We wish to thank Prof. Dr.-Ing. Thomas Benz for providing the user defined soil model with the Matsuoka-Nakai failure criterion.

REFERENCES

Bishop, A.W. (1955). The use of slip circles in the stability analysis of earth slopes. *Geotechnique* **5**: 7–17.

Brinkgreve, R.B.J., Swolfs, W.M. & Engin, E. (2016). *Plaxis 2D 2016 – user manual.* Delft, the Netherlands: Plaxis bv.

Davis, E.H. (1968). Theories of plasticity and the failure of soil masses. In *Soil mechanics: selected topics.* (ed. I.K. Lee), pp. 341–354. New York, NY, USA: Elsevier.

Drucker, D.C., Greenberg, W. & Prager, W. (1951): The safety factor of an elastic plastic body in plane strain, *Trans. ASME J. Appl. Mech. 7***3**, pp. 371–378.

Drucker, D.C., Prager, W. & Greenberg, H.J. (1952): Extended limit design theorems for continuous media. Q. *J. Appl. Math.* **9**, No. 4, pp. 381–389.

Fellin, W. (2014): The rediscovery of infinite slope model. *Geomechanics and Tunnelling;* **7**, No. 4, pp. 299–305.

Griffiths, D. & J. Huang, J. (2009): Observations on the extended Matsuoka-Nakai failure criterion. *Int. J. Numer. Anal. Meth. Geomech.*, No. **33**, pp. 1889–1905.

Janbu, N. (1954). Application of composite slip surface for stability analysis. In: *Proceedings of the European conference on stability of earth slopes,* Stockholm, vol. 3; p. 43–49.

Krabbenhøft, K., Lyamin, A.V., Hjiaj, M., Sloan, S.W. (2005). A new discontinuous upper bound limit analysis formulation. *Int J Numer Method Eng*; **63**, No. 7, pp. 1069–88.

Krabbenhøft, K., Lyamin, A.V., Sloan, S,W. (2007). Formulation and solution of some plasticity problems as conic programs. *Int J Solids Struct*; **44**, No. 5, pp.1533–49.

Krabbenhøft, K., Lymain, A.V. and Krabbenhøft, J. (2016). *Optum G2 2016 - User Manual.* Newcastle, Australia: Optum Computational Engineering.

Lyamin, A.V., Sloan, S.W. (2002a). Lower bound limit analysis using nonlinear programming. *Int J Numer Method Eng.*; **55**, No. 5, pp. 573–611.

Lyamin, A.V., Sloan, S.W. (2002b). Upper bound limit analysis using linear finite elements and nonlinear programming. *Int J Numer Anal Method Geomech*; **26**, No. 2, pp. 181–216.

Matsuoka, N. & Nakai, T. (1974). Stress-deformation and strength characteristics of soil under three different principal stresses. In *Proc., JSCE*, No. 232, pp. 59–70.

Morgenstern NR, Price VE. (1965): The analysis of the stability of general slip surfaces. *Géotechnique*, **15**, No1, pp. 79–93.

Sallinger M. (2017): Slope stability analysis by means of strength reduction technique. *Master Thesis*, Inst. of Soil Mechanics and Foundation Eng., Graz University of Technology.

Schranz, F. & Fellin, W. (2016). Stability of infinite slopes investigated with elastoplasticity and hypoplasticity. *Geotechnik* **39**, No. 3, pp. 184–194.

Sloan, S.W. (1988). Lower bound limit analysis using finite elements and linear programming. *Int J Numer Anal Method Geomech* **12**: 61–77.

Sloan, S.W. (1989). Upper bound limit analysis using finite elements and linear programming. *Int J Numer Anal Method Geomech* **13**: 263–282.

Sloan, S.W. (2013). Geotechnical stability analysis. *Geotechnique* **63**: 531–572.

Tschuchnigg, F., Schweiger, H.F., Sloan, S.W., Lyamin, A.V. & Raissakis, I. (2015a). Comparison of finite-element limit analysis and strength reduction techniques. *Géotechniques* **65**, No. 4, pp. 249–257.

Tschuchnigg F, Schweiger HF, Sloan SW. (2015b): Slope stability analysis by means of finite element limit analysis and finite element strength reduction techniques. Part I: Numerical studies considering non-associated plasticity. *Computers and Geotechnics*; **70**:169–77.

Tschuchnigg, F., Oberhollenzer, S., Schweiger, H.F. (2017). Finite element analyses of slope stability problems using non-associated plasticity. In *Proc. 15th IACMAG*, Wuhan.

Numerical Methods in Geotechnical Engineering IX – Cardoso et al. (Eds)

© 2018 Taylor & Francis Group, London,
ISBN 978-1-138-33203-4

2D and 3D rock slope stability analysis in an open-pit mine

J.G. Soto & C. Romanel
Department of Civil and Environment Engineering, PUC-Rio, Rio de Janeiro, Brazil

ABSTRACT: The main objective of this article is to investigate the stability of high rock slopes in a mining project in Peru. The analysis was performed by finite elements, considering bi and three-dimensional models, involving six different types of rock. Three-dimensional geometric effects in pseudo-static safety factors were also investigated, since the mine is situated in a region of high seismic activity, as well as the influence of the distance of the waste material from the edge of the excavation. The effects of waste dump on safety factor values were quite low, but the shape of the potential failure surface varied according to the distance from the excavation. It was also observed that the pseudo-static safety factors in 3D analyses are dependent on the orientation considered for the horizontal pseudo-static forces.

1 INTRODUCTION

Currently a major mining company in Peru is interested in establishing alternative workflows to explore the maximum mineral amount in a minimum time schedule, making as high as possible the performance of the investment. To achieve these goals, but maintaining the safety of operations, engineering analysis must be carried out to estimate the inclination of the open pit slopes and the maximum depth of excavation. The major purpose is to obtain the steepest angle of inclination to reduce the removal of waste material, thereby lowering the mining extraction costs. Of course, a gradual slope increasing may create risks of slope instability, compromising the safety of mining operations, with potential risk for loss of life as well as economic, environmental and social disasters with interruption or eventual closure of the mining facilities. The final excavation design is subject to factors such as the mechanical properties of the rock mass and characteristics of the regional geological structures.

In this paper, the slope stability of an open pit mine is investigated through numerical analysis using 2D and 3D finite element models. According to Wyllie & Mah (2004) and Cheng & Lau (2008) 3D effects influence the stability of slopes, especially in open pit mines where it is common the existence of curved slopes with various degrees of concavity. Numerical analyses of stability problems with attention to 3D effects were also published by Narendranathan (2013) and Kelesoglu (2016).

2 PROJECT DESCRIPTION

The mining project is situated in the Southwest region of Peru, 400 meters above the sea level and the depth of the open pit mine varies from 800 to 900 m. The area investigated in this research embodies the main mineral deposit, with 2.8 km long and 2.1 km wide.

2.1 *Rock properties*

The mineral deposit is formed by six types of rock (Figs. 1 to 7), each one geomechanically classified according to results from field and laboratory tests. The estimated values for the intact material are listed in Table 1, where γ is the unit weight, σ_{ci} is the uniaxial compressive strength and m_i represents a material constant of the generalized Hoek-Brown failure criterion (Hoek et al., 2002).

Table 2 presents the characteristics of the rock mass, obtained through the Rocdata software (Rocscience, 2017) with the following input parameter for each rock type: γ, σ_{ci}, m_i, GSI (Geological Strength Index) and D, a disturbance factor dependent on the degree of perturbation to which rock masses has been subjected by blast damage or stress relaxation. In the present investigation D = 0.85.

In Table 2 E_{rm} means the deformation modulus of the rock mass, ν is the Poisson's ratio, c and ϕ the parameters of the Mohr-Coulomb failure criterion (cohesion and friction angle, respectively). Values of GSI were obtained from the Rock Mass Rating

Table 1. Geotechnical parameters of intact rocks.

Intact rock	γ (kN/m^3)	m_i	σ_{ci} (MPa)
Altered zone	27.5	22	30.7
Limonite	27.5	9	184
Andesite	27.5	25	183
Vulcanic	27.3	20	203
Dolomite	29.7	15	55
Limestone	27.5	11	83

Table 2. Geotechnical parameters of the rock mass.

Rock mass	GSI	E_{rm} (MPa)	ν	c (MPa)	ϕ (°)
Altered zone	55	9239	0.31	4.80	36.7
Limonite	50	5832	0.30	1.13	39.9
Andesite	55	8454	0.31	2.90	43.4
Vulcanic	52	7771	0.30	1.95	44.7
Dolomite	55	3176	0.30	2.36	24.8
Limestone	50	4911	0.30	3.51	17.8

Figure 1. Plan of lithographs in the project site.

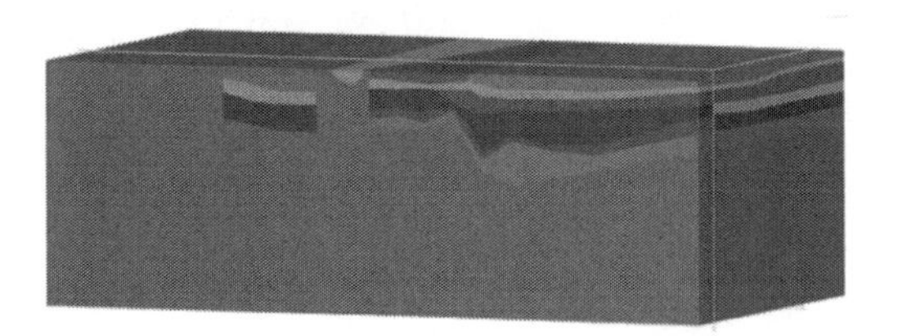

Figure 2. Cross section A-A'.

(RMR) system (Bieniawski, 1989) considering for RMR > 23 the following correlation:

$$GSI = RMR - 5 \quad (1)$$

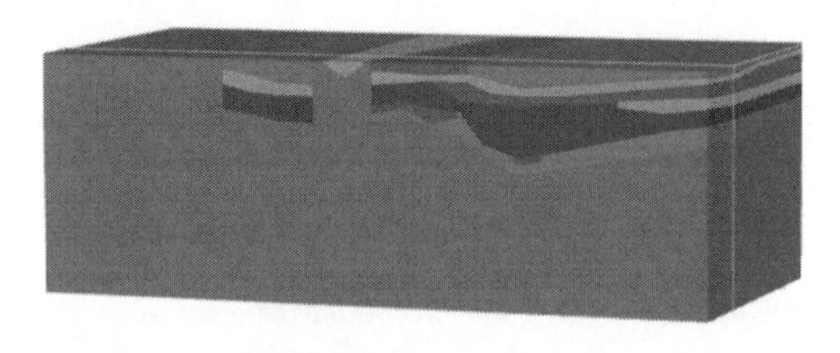

Figure 3. Cross section B-B'.

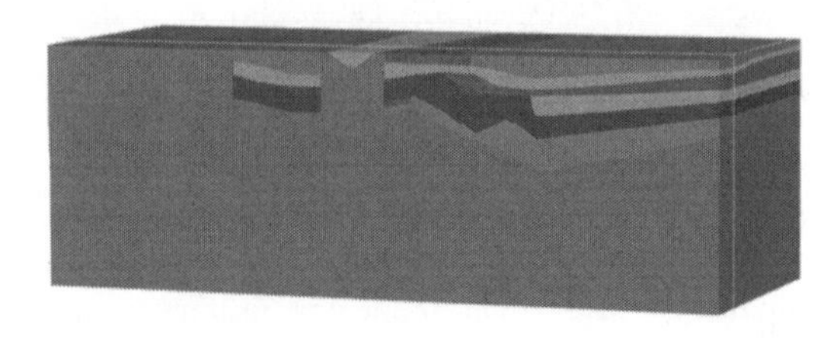

Figure 4. Cross section C-C'.

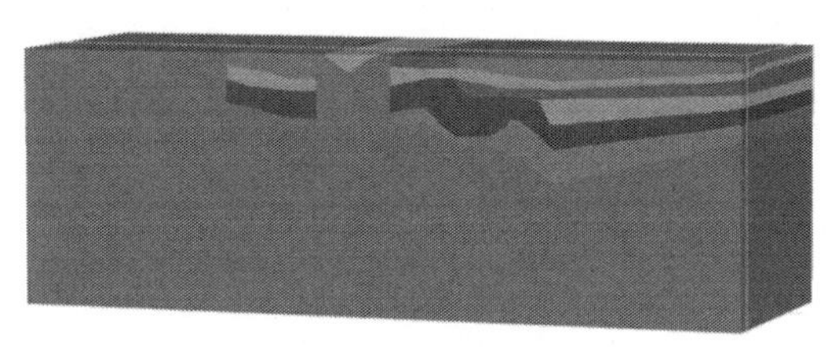

Figure 5. Cross section D-D'.

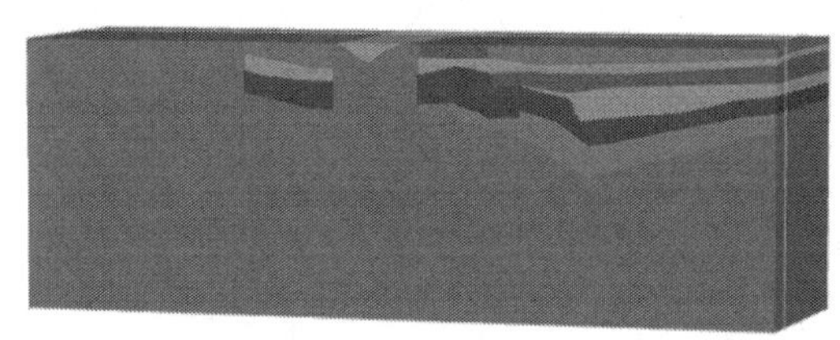

Figure 6. Cross section E-E'.

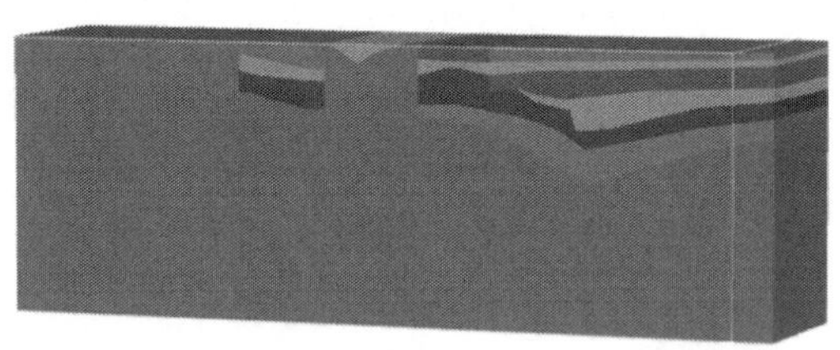

Figure 7. Cross section F-F'.

The 208 m high waste dump was considered as an elastic material with unit weight $\gamma = 18$ kN/m^3, E = 8.000 MPa and $\nu = 0.22$.

3 NUMERICAL MODELS

3.1 *Mesh geometry and boundary conditions*

3.1.1 *2D analysis*

In the 2D finite element analysis, thc plane strain problem has been represented by a mesh size following Sjöberg (1999) recommendations: vertical dimension between 2 and 3 times the average depth of excavation H =864 m (3.5H in Fig. 8) and horizontal dimension between 3 and 4 times the maximum width of excavation L = 1990 m (4.5 L in Fig. 8). The boundary condition at the base of the mesh was prescribed in terms of zero displacements and for the lateral boundaries null horizontal displacements were assumed. The numerical analysis was performed with quadratic triangular finite elements (6 nodes) using the software Plaxis 2D 2017. The selected cross section for the 2D stability analysis is shown in Figure 9

3.1.2 *3D analysis*

The boundary conditions applied to the faces of the 3D model were as follows: bottom face with displacements prevented in the 3 orthogonal directions x, y, z; side faces parallel to the xz plane with displacement component $u_y = 0$ but free in directions x and z; side faces parallel to the yz plane with displacement component $u_x = 0$ but free in directions x and z (Fig. 10.). The initial stresses before excavation were generated by the gravitational force using the software Plaxis 3D 2016 with quadratic tetrahedral elements (10 nodes).

At the beginning of the modeling process, several tests considering different discretization densities for the 3D meshes were carried out, in order to guarantee the feasibility of the numerical results without compromising the computational processing time. From such preliminary tests, it could be concluded that 3D meshes with 60 to 110 thousand nodes were the most suitable for the problem under investigation.

In numerical simulations, 8 excavation phases were considered, reaching a final depth of 864 m, according to the excavation sequence indicated in Figure 11. The overall slope of the 3D model varied between 42° and 45°, as shown in Figure 12.

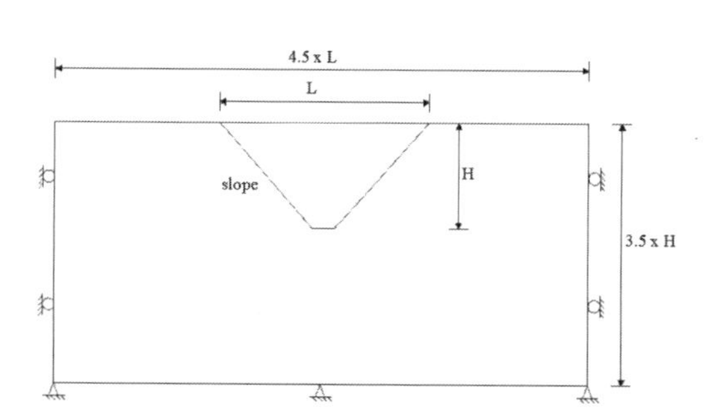

Figure 8. Geometry of the 2D finite element model.

Figure 10. Geometry of 3D finite element model (lithology also shown).

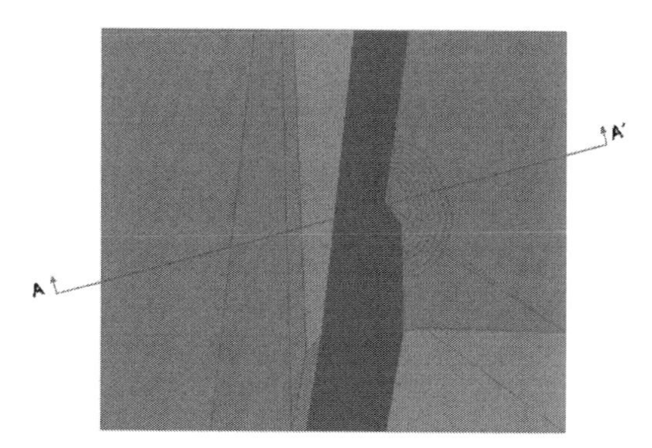

Figure 9. Selected cross section for 2D stability analysis.

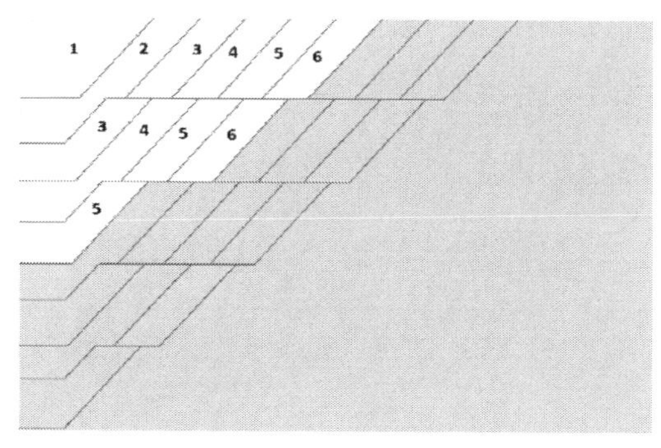

Figure 11. Sequence of excavation phases with enlargement of the open pit as the excavation deepens.

3.1.3 *Pseudo-3D analysis*

A pseudo-3D model has been built considering the same dimensions as the 2D finite element mesh, but with a thickness of 2000 m, and variation of rock types along the three dimensions (Fig. 13). The selected cross section for 2D analyses lies at the middle of the sector used in the pseudo-3D analysis. The boundary conditions are similar to the ones set for the true 3D analysis.

4 SLOPE STABILITY ANALYSIS

Figure 14 refers to the evolution of the safety factors computed in the 2D, pseudo-3D and 3D models after each step of excavation, computed through the shear strength reduction technique (Griffiths & Lane, 1999). For the last excavation phase the positions of the potential failure surfaces are shown in Figures 15 to 17.

The observed tendency of the safety factors indicates that 2D analyses are conservative, not incorporating the 3D effects from the geometry of the open pit neither the influence of the spatial variation of rock types. The values for the 3D analysis resulted between 14% and 40% higher than those computed with the bidimensional model.

On the other hand the discrepancies beween the safety factors calculated with the pseudo and the true 3D models are not significant, with 3D values oscillating within –4% to 8% with respect to the pseudo-3D results.

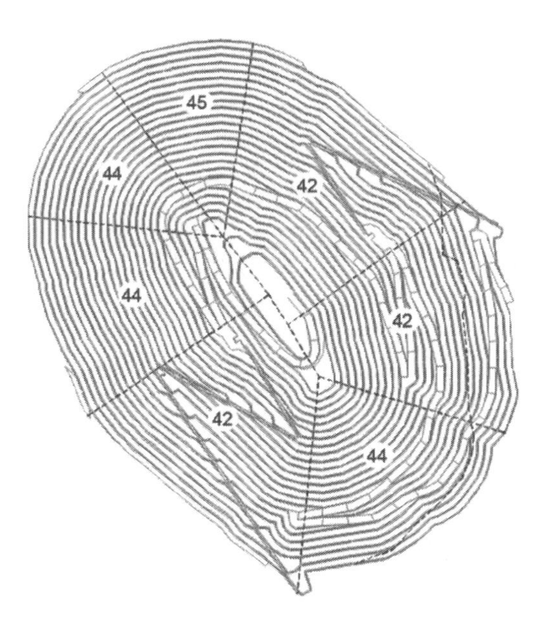

Figure 12. Global angles of the slopes around the bottom of the open pit mine.

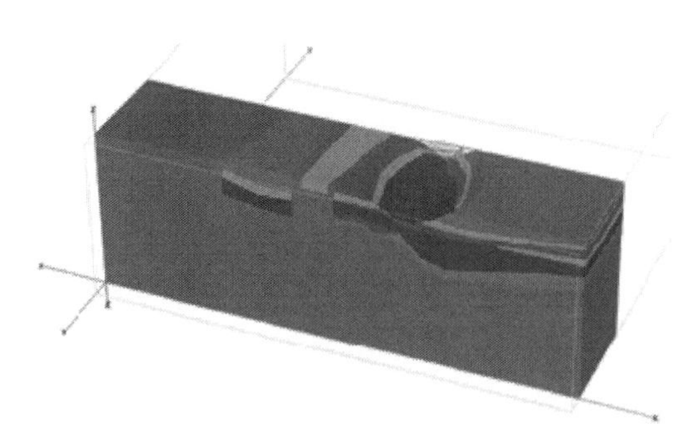

Figure 13. Geometry of the pseudo-3D finite element model.

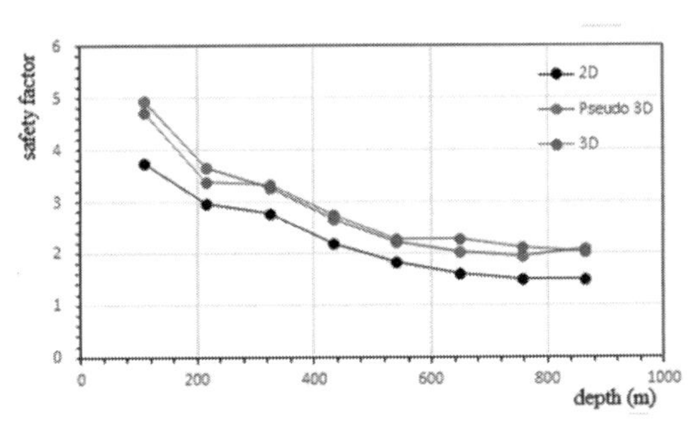

Figure 14. Evolution of the safety factor in 2D, pseudo-3D and 3D analyses with the excavation deepening.

Figure 15. Potential failure surface after the last excavation phase in 2D analysis.

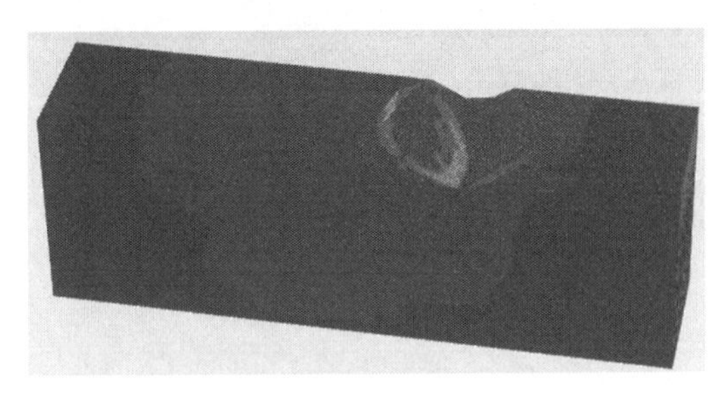

Figure 16. Potential failure surface after the last excavation phase in pseudo-3D analysis.

Figure 17. Potential failure surface after the last excavation phase in 3D analysis (lithology also shown).

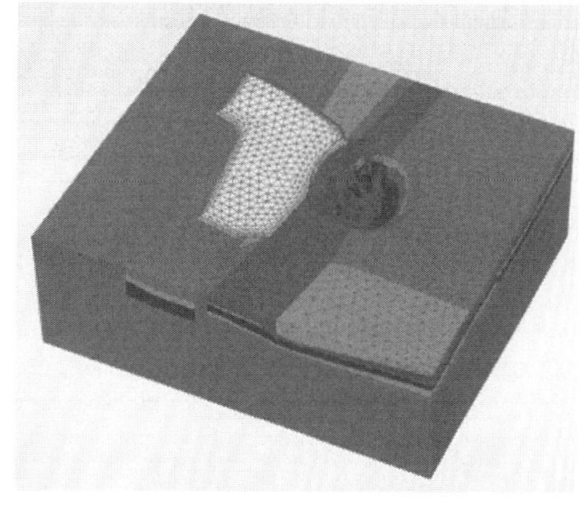

Figure 18. Potential failure surface considering a waste dump 325 m from the edge of the excavation.

Table 3. Safety factors from the last excavation phase with and without the waste dump.

Excavation Phase	Depth (m)	No waste dump	Waste dump A	Waste dump B
1	108	4.73	5.73	5.73
2	216	3.39	3.71	3.71
3	324	3.34	3.29	3.29
4	432	2.74	2.75	2.75
5	540	2.28	2.30	2.30
6	648	2.28	2.10	2.10
7	756	2.09	2.01	2.01
8	864	2.03	2.09	2.09

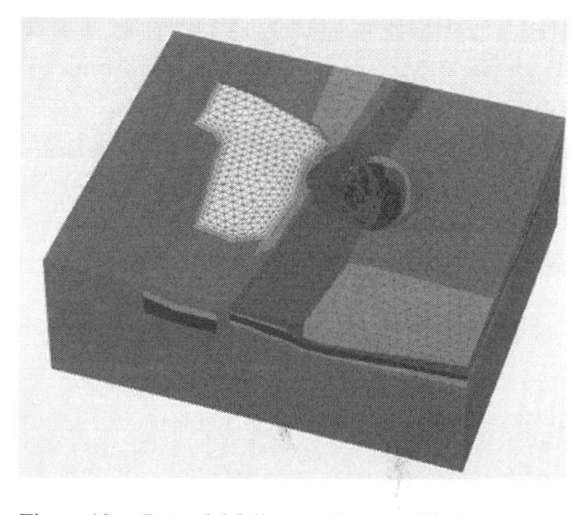

Figure 19. Potential failure surface considering a waste dump 650 m from the edge of the excavation.

4.1 *Influence of the waste dump distance*

It is expected that the presence of a waste dump near the excavation should have an influence on the slope stability. The safety factors calculated with the 3D finite element model were compared, considering a 208 m high waste dump situated 325 m (waste dump A) and 650 m (waste dump B) away from the edge of the excavation. The values determined in both cases are listed in Table 3, while Figures 18 and 19 illustrate the position of the potential failure surfaces.

For a better analysis of the numerical results, the images in Figures 18 and 19 were redrawn in order to make comparisons clearer. In the last excavation phase, although the values of the safety factors were calculated practically the same in all cases, the potential failure surface corresponding to the waste dump at 325 m from the excavation edge encompassed a larger region than that determined considering the absence of the dump pile (Fig. 20). For the waste dump distant 650 m from the excavation boundary, the potential failure surface showed a tendency to return to the previous failure surface determined without the waste dump.

4.2 *Pseudo-static safety factor*

Since the project is situated in a region of high seismic activity, pseudo-static analysis was also performed considering a horizontal seismic coefficient $k_H = 0.2$ equivalent to 50% of the maximum acceleration of the probable maximum earthquake.

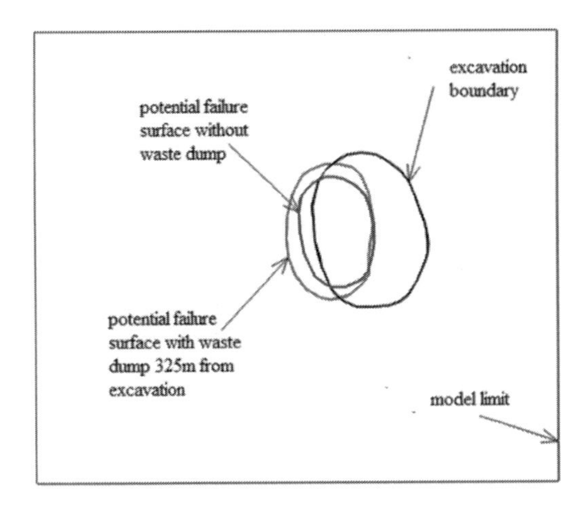

Figure 20. Position of potential failure surfaces considering absence and existence of waste dump 325 m from the excavation.

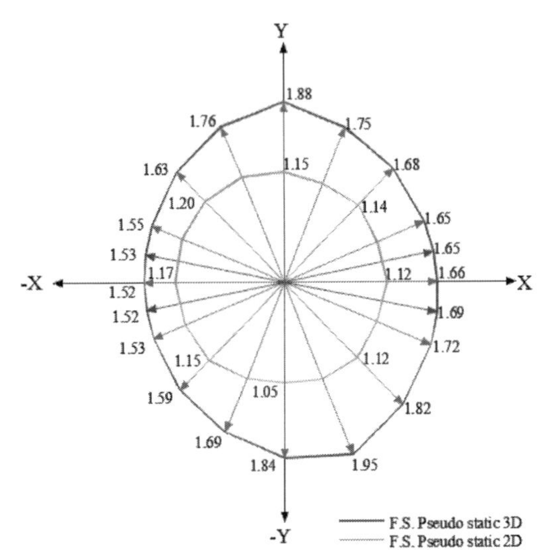

Figure 21. Variation of pseudo-static safety factor according to the direction of the horizontal pseudo-static force.

For the 2D and 3D analyses the direction of the pseudo-static forces was modified in order to verify the influence of its orientation on the pseudo-static safety factors.

Figure 21 shows the variation of the safety factor with the direction of the horizontal pseudo-static force, where the length of each vector represents the value of the safety factor in that direction. It can be observed that the lower 3D values occur within a range of directions next to the x-axis.

Table 4 shows the position of the potential failure surface, depending on the direction of the pseudo- static force. The blue line represents the edge of the excavation and the red line indicates the potential failure surface. The pseudo-static forces in the directions labelled Y, XY, X, X-Y give rise to failure surfaces on the right side of the excavation while the pseudo-static forces marked as -Y, -X-Y, -X and -XY directions create failure surfaces on the opposite side.

Safety factors for the same cross section considered in the previous 2D analysis (Fig. 9) are shown in Table 5 for the 2D and 3D models. Due to the

Table 4. Safety factor, position and shape of the failure surface according to the direction of the pseudo-static force.

Direction of pseudo-static force	Safety factor	Position and shape of failure surface and excavation
Y	1.88	
X,Y	1.68	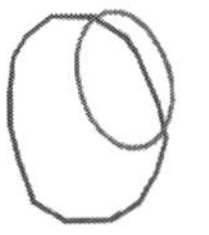
X	1.66	
X, -Y	1.82	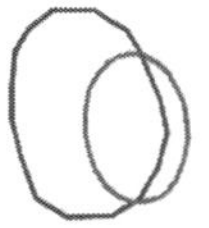
-Y	1.84	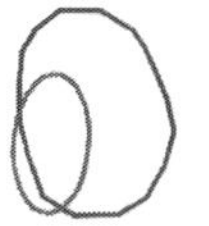
-X,-Y	1.59	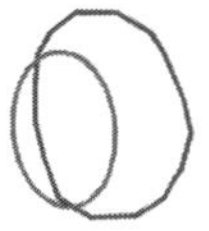
-X	1.52	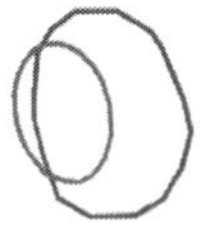
-X,Y	1.63	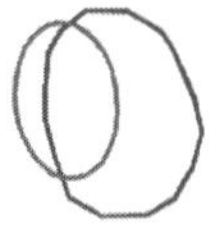

Table 5. Static and pseudo-static safety factors after the excavation phases.

Description	Phase	Depth (m)	2D	3D
Static analysis	1	108	3.75	4.73
	2	216	2.97	3.39
	3	324	2.79	3.34
	4	432	2.19	2.74
	5	540	1.82	2.28
	6	648	1.61	2.28
	7	756	1.50	2.09
	8	864	1.48	2.03
Pseudo static analysis	1	108	2.89	2.59
	2	216	2.82	2.57
	3	324	2.32	2.55
	4	432	1.56	2.01
	5	540	1.29	1.71
	6	648	1.16	1.70
	7	756	1.08	1.57
	8	864	1.12	1.52

low values computed in the 2D analysis after the last three excavation phases, a complete dynamic analysis would be usually recommended (based on 2D results) to better investigate the equilibrium conditions of the rock slope under seismic load.

4.3 *Influence of excavation geometry*

Real slopes are not infinitely long and rectilinear; they are curved both in elevation and in plan. The effects of curvature of a slope can only be determined by means of three-dimensional analyses.

Hoek & Bray (1981) have observed that lateral constraints, due to the material on either side of the potential failure surface, increase as the slope gradually becomes more concave. When the radius of curvature R of the slope is smaller than its height H, those authors suggested increasing in 10° the inclination angle used in conventional 2D analyses. For radius of curvature greater than twice the slope height, the inclination angle used by 2D analysis should be kept.

The influence of the 3D geometry on values of the safety factor was evaluated comparing results computed with different degrees of inclination (different ratios between slope height H and radius of curvature R) between 47° and 57°, in the 2D and in the 3D model (Fig. 22).

The analysis considered the open pit mine in a homogenous rock mass with the following properties: unit weight 27.3 kN/m^3, cohesion $c = 1.9$ MPa, friction angle $\phi = 44.7$ °. Table 6 lists the calculated safety factors, showing that for $H/R \geq 1$ the value 1.83 computed with 47° of inclination in the 2D model is a good approximation of the safety factor 1.95 obtained in the 3D model with 57°, an inclination 10° higher, thus confirming the suggestion previously given by Hoek and Bray (1981).

Figure 22. Three dimensional model of an open pit mine in homogeneous rock mass.

Table 6. Safety factors for different slope inclinations in a homogeneous rock mass.

Degree	2D	3D	H/R
47	1.83	2.39	1.00
48	1.79	2.40	1.04
49	1.78	2.36	1.07
50	1.72	2.24	1.10
51	1.70	2.19	1.14
52	1.65	2.13	1.18
53	1.58	2.06	1.22
54	1.57	2.04	1.26
55	1.53	2.01	1.30
56	1.48	1.99	1.34
57	1.44	1.95	1.38

5 CONCLUSION

Safety factors calculated with 3D models are generally higher than the corresponding values obtained in 2D models, which partially explains the predominance of 2D analyses in the practice of geotechnical engineering, besides the greater complexity of constructing 3D models, especially in cases involving variable lithology, and the required time to compute them.

However, especially for high slopes in open pit mines, a 3D analysis is recommended because results from 2D models may be very conservative and, in consequence, may generate anti-economic designs that will increase the amount of waste to be removed.

Alternatively, it should be also taken into account the suggestion given by Hoek and Bray (1981) to increase in 10º the inclination angle used in conventional 2D analyses for curved slopes whose radius of curvature R is smaller than its height H, as confirmed by the safety factors calculated in this research (Table 4) for an homogeneous rock mass.

The position of the potential failure surface determined in 3D analysis with the presence of a waste dump nearby the excavation suggests that the best location for the waste pile is outside the region encompassed by the potential failure surface determined on the hypothesis of absence of the waste dump.

The pseudo-static safety factors are also influenced by the orientation given to the pseudo-static forces. In 2D analysis these horizontal forces are parallel to x-axis, leading to results generally lower than those computed with a 3D model (Fig. 21).

REFERENCES

Bieniawski, Z.T. 1989. *Engineering rock mass classification.* New York: John Wiley & Sons, 251p.

Cheng, Y.M. & Lau, C.K. 2014. *Slope instability analysis and stabilization: new methods and insight*. CRC Press. 438p.

Griffiths, D.V. & Lane, P.A. 1999. Slope stability analysis by finite elements. *Géotechnique* 39(3): 487–403.

Hoek, E. & Bray, J.D. 1981. *Rock slope engineering.* CRC Press. 368p.

Hoek, E.; Carranza-Torres, C. & Corkum, B. 2003. Hoek-Brown failure criterion – 2002 edition. *NARMS-TAC Conference*. Toronto (1):267–273.

Kelesoglu, M.K. 2016. The evaluation of three-dimensional effects on slope stability by the strength reduction method. *KSCE Journal of Civil Engineering* 20(1): 229–242.

Narendranathan, S. & Thomas, R.D.H. 2013. The effect of slope curvature in rock mass shear strength derivations for stability modeling of foliated rock masses. 47th US Rock Mechanics/Geomechanics Symposium. San Francisco, California. 8p.

Rocscience. 2017. Rocdata – *Rock, soil and discontinuity strength analysis*. v.5

Sjöberg, J. 1999. *Analysis of large scale rock slopes.* Doctoral thesis. Lulea University of Technology. 682p.

Wyllie, D.C. & Mah, C. 2004. *Rock slope engineering – civil and mining,* CRC Press. 456p.

Numerical Methods in Geotechnical Engineering IX – Cardoso et al. (Eds)
© 2018 Taylor & Francis Group, London, ISBN 978-1-138-33203-4

Increased efficiency of finite element slope stability analysis by critical failure path detection

R. Farnsworth
Colorado School of Mines, CO, USA

A. Arora
Indian Institute of Technology (BHU), Varanasi, India

D.V. Griffiths
Professor of Civil Engineering, Colorado School of Mines, CO, USA

ABSTRACT: A new and efficient method of factor of safety determination has been developed, wherein after a small number of iterations, and for a given strength reduction factor which is typically higher than the true factor of safety, the software will look for a critical failure path through the slope. If found, the software then determines that the slope has failed and no additional iterations are needed. By doing this, most iterations in the non-convergent range are eliminated. The paper describes an algorithm that searches for a continuous path of Gauss points across the slope that have failed and are deforming plastically. The avoidance of unnecessary iterations, leads to significant computational savings.

1 INTRODUCTION

The slope stability problem is one of the oldest in geotechnical engineering (e.g. Abramson et al. 2001). The stability of a slope is determined by calculating its Factor of Safety, which is the ratio of the shear strength of the slope to the shear stresses developed by gravity. The finite element method (e.g. Griffiths & Lane 1999) has gained great popularity for computing the Factor of Safety. It has been shown to be accurate and robust, and requires no *a priori* judgement on the part of the user as to the shape or location of the failure mechanism, which comes out naturally from the analysis.

Two dimensional plane strain models constructed from 8-node quadrilateral elements with reduced integrations are used in this study. To calculate the factor of safety, the Mohr-Coulomb strength parameters, c' and ϕ', are modified by a Strength Reduction Factor (SRF):

$$c'_f = c' / SRF \tag{1}$$

$$\tan(\phi'_f) = \tan(\phi') / SRF \tag{2}$$

It is clear from the equation that the SRF value is inversely proportional to the strength parameters, which means that the strength of soil decreases when SRF increases and vice versa. The aim is to determine the smallest SRF value for which the slope fails, at which stage the SRF *is* the Factor of Safety for the slope. The slope is said to have failed when the soil is weakened enough to be unable to support the loads imposed by gravity.

A curiosity of slope stability analysis by finite elements is that the criterion for failure has historically been based on convergence within some limited number of iterations. This threshold is not usually determined by a failure mechanism or crest displacements; it is dictated by the modeler. Raising the iteration ceiling in an FE slope stability analysis is tempting for the user since it can lead to better-defined failure mechanisms, but this can cost a significant amount of computational time, especially with SRF values that are higher than the true factor of safety. A specialized adaptive algorithm called FGA was developed to reduce the number of SRF trials in the non-convergent range (Farnsworth 2015), but this optimization did not attempt to reduce iterations once a non-convergent SRF was encountered. To allow for efficient modeling with a large iteration ceiling, the paper therefore describes a method for automatic critical failure path detection.

The critical path (cp) is the first continuous path through a slope under a particular loading configuration to experience material failure. In the case of finite elements with an elastic-perfectly plastic constitutive law, material failure occurs when the stress at a calculated point reaches the strength and the point behaves plastically instead of elastically. As the SRF value is increased, calculated points in the model (called Gauss points because of their use in numerical integration) begin to experience material failure. As these Gauss points fail, they displace more, placing additional stress on adjacent points. Throughout the iteration process for a given SRF value, failed points continue to shed more and more stress on surrounding Gauss points until either the stress is balanced by surrounding non-failed Gauss points or a critical path forms and the slope experiences global failure. Global failure occurs once all of the Gauss points on a critical path have failed, enabling a mechanism to form when all strength along that path has been exhausted.

Current finite element modeling techniques use an iteration ceiling to distinguish between convergent and non-convergent trials of strength reduction factor. While this method of detecting non-convergence usually produces good results, it requires in many cases a significant number of iterations on a slope that has already experienced global failure. As a means of improving the performance of finite element slope modeling, it is proposed that critical path detection could provide a means of detecting a failed slope without needing to reach the iteration ceiling.

2 MODELING A CRITICAL PATH

An algorithm was developed for efficient detection of a critical path through a slope using finite element analysis. A critical path in this algorithm is considered to be any continuous path of adjacent Gauss points, from the uphill ground surface to the toe or downhill ground surface, which have failed by reaching their plastic limit. An important property of a critical path is that it is highly unlikely to enter and exit at the same boundary. For example, a critical path cannot both enter and exit the top of a slope. Rather, the path would start at one end of the slope and end at the other. The challenge is then to develop an algorithm that can efficiently detect a continuous path of failed Gauss points connecting two distinct boundaries.

To implement this algorithm, two important arrays were added to the finite element source code which is closely based on Program 6.4 (Smith, et al. 2014). One is Bnd, which is an array containing the indices of all Gauss points on the upper boundary of the model. The points along the top of the slope are assigned a boundary number of 1 and the points along the bottom of the slope are assigned a boundary number of 3 (See Figure 1). The Bnd array is important as a screening tool. There are significantly fewer boundary points than total

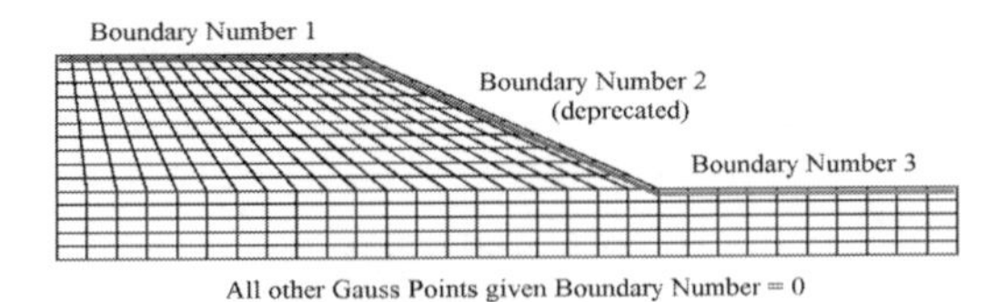

Figure 1. Boundary Number Assignments.

Gauss points in the model. If none of the Gauss points along the boundary have failed, then there is no possible way that a critical path could have formed. Furthermore, since a critical path passes through the boundaries on both the top and the bottom of the slope, the algorithm can skip searching for a critical path if failed points have not been detected in both boundaries. In short, time-consuming searches can be avoided if no failed points are detected among the relatively few boundary points.

The second array is the FA array, which stores information about each Gauss point. Among the information stored is the point's failure state (failed or not failed), boundary number, and whether or not the point has been visited in the FLook subroutine. Immediately before running FCheck each time FCheck is called, the visited state of each point is set to FALSE. The observant reader might recognize that boundary points are identified twice: once in Bnd and once in FA. The purpose of identifying the boundary points twice is for more time-efficient lookup. The Bnd array is optimized for critical path screening, while the FA array is optimized for the FLook subroutine.

The FCheck subroutine calls FLook on every failed boundary Gauss point. The FLook subroutine is a recursive algorithm that searches for a critical path beginning at the top of the slope. Knowing that the shape of the path is likely to be somewhat circular, and knowing that the top of the slope (as required by the program) is required to be on the left, the FLook subroutine was optimized for counterclockwise searching. Once a failed boundary is detected, FLook searches for adjacent failed Gauss points in the following order: down, right, up, then left. Once an adjacent failed point is found, then FLook is called on that point, which continues recursively until either a critical path is identified or until all adjacent failed points have been tested and no path found. FLook marks each tested point as visited in the FA array so that FLook is never called twice on the same point within a given FCheck trial, since multiple searches on a point are redundant.

The functions that are called by FLook to locate adjacent Gauss points in each direction can calculate the index of the adjacent point without having to search for it, thanks to the structured nature of the finite element mesh. As a guiding principle for this algorithm, searches are slow and index references are fast. Everything that does not absolutely require searching is optimized for index referencing.

The critical path search algorithm (cp algorithm) was incorporated into two versions of the FE slope program, one using the FGA algorithm for SRF advancement, and the other using the Bisection Method (Griffiths 2015).

3 TESTING FOR STABILITY

The program was tested for its agreement with classical FE slope programs. There was no concern that the program would overestimate the factor of safety because the iteration ceiling still exists. The test was to determine whether or not the software would underestimate the factor of safety by finding a critical path that would later stabilize. To eliminate this issue, boundary number 2 was removed from the analysis (See Figure 1). In the earliest of tests, it was found that the model tended to form a small failure path joining the top horizontal boundary and the slope boundary, which could not be considered a "true" critical failure path. When only the top and bottom slope boundaries were considered, the factor of safety calculated using cp version was within 0.01 of classical FE techniques for 90% of the 100 slopes tested. The remaining slopes were within about 0.1 of the classical FE techniques.

The shape of the failure mechanism is also an important indicator of stability. A cp algorithm could not be trusted if it produced the correct factor of safety, yet failed to produce the correct critical path. A homogeneous 2:1 soil slope was analyzed for failure path using the Simplified Bishop's Method (e.g. Duncan and Wright 2005) leading to a factor of safety of about 1.6 and a critical failure path passing through the toe similar to that shown in Figure 2. The same slope was analyzed using FE analysis with cp detection. Figure 3 shows the resulting failure path. The failure path passes through the toe of the slope. It is also interesting to note that an incomplete critical path was forming tangent to the base of the slope, although the toe failure controls the factor of safety in this case.

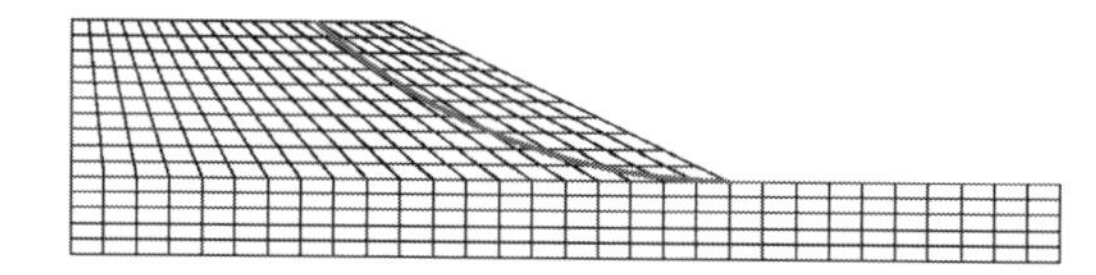

Figure 2. Theoretical circular failure path through a homogeneous 2:1 slope. Minimum FS of 1.59 calculated using the Bishop's Method. Additional slope properties given in Figure 6.19 of Smith, Griffiths, and Margetts, 2014.

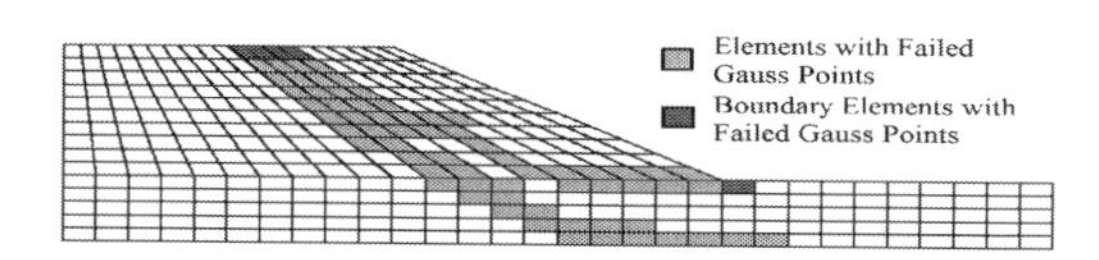

Figure 3. Strength failures identified by FE critical path detection. There are two distinct failure paths, one passing through the base, which is incomplete, and one passing through the toe, which is a complete critical path.

4 TESTING FOR EFFICIENCY

Both cp versions of the FE slope program were tested for efficiency against the original FE slope program and the FGA algorithm on 100 randomly generated slopes. The randomly generated slopes were either homogeneous, layered, or with randomly distributed soil properties. They were either drained or undrained, and they were either "steep" or "shallow" (slope angle > or < 45° respectively).

The results, shown in aggregate in Table 1, were somewhat mixed, dependent on the type of slope modeled. The cp-bisection version of the FE program was the overall fastest with an average 16% savings in computation time from the original FGA version (10,344 seconds total simulation time for cp-bisection, 12,267 second for original-fga). The cp-bisection version was especially efficient at finding the factor of safety of steep slopes, with a 39% time improvement. It also provided significant

Table 1. Aggregate performance results from 100 randomly generated slopes.

			Modeling time		
			Original fga	cp fga	cp bis
Type of slope		Trials	sec	sec	sec
All		100	12,267	10,898	10,344
Soil Distr.	Homog.	18	1245	972	1153
	Layered	41	5342	5010	5397
	Random	41	5679	4917	3794
Drainage	Drained	53	5833	4960	5125
	Un-Drained	47	6434	5938	5218
Slope	Shallow	59	7115	6327	7199
	Steep	41	5151	4571	3145
Steep and homog.		7	918	709	250
Steep and random		14	3097	2671	1320

Distr. = distribution. Homog. = homogeneous.

savings on slopes with randomly distributed soil properties at 33% faster. The best performance for the cp-bisection algorithm was on steep, randomly distributed slopes (57% improvement) and steep, homogeneous slopes (73% improvement). The cp-bisection version actually performed slightly slower than the original FGA version on layered and shallow slopes. On these types of slopes, the cp-FGA version was the fastest.

5 CONCLUSION

Critical path detection is a promising means of improving finite element slope stability software performance. With an efficient algorithm design, critical path detection can eliminate unnecessary iterations on already-failed slopes, saving an average of 15% calculation time. On specific types of slopes, the improvements are much more dramatic. Steep slopes at an angle greater than 45° show an average of 39% improvement in calculation time, and steep slopes that are also homogeneous are improved by 73%. The calculated factor of safety does not significantly change between critical path detection and former versions of the software, showing a high degree of reliability in results.

Future optimizations could be made to this algorithm. First, because critical path detection removes much of the disadvantage of SRF trials in the non-convergent range, the SRF-stepping algorithm could be re-optimized. For example, a bisection algorithm could be coupled to an exponential SRF stepping scheme to bound the slope factor of safety as quickly as possible. A second optimization could involve the introduction of a serviceability failure limit. Critical path detection is effective at finding the strength failure of a slope, but there are also conditions where ultimate limit state conditions may not have been met but serviceability considerations have already implied an unacceptable design.

REFERENCES

Abramson, L.W., Lee, T.S., Sharma, S., and Boycee, G.M. 2001. *Slope Stability and Stabilization Methods*. NY: John Wiley & Sons.

Duncan, J.M. and Wright, S.G. 2005. *Soil Strength and Slope Stability*. NY: John Wiley and Sons.

Farnsworth, R. 2015. *Algorithms for Increased Efficiency of Finite Element Slope Stability Analysis*. Ann Arbor, Michigan: ProQuest.

Griffiths, D.V. 2015 Slope64: A program that performs slope stability analysis by finite elements
inside.mines.edu/~vgriffit/slope64

Griffiths, D.V., and Lane, P.A. 1999. Slope Stability Analysis by Finite Elements. *Geotechnique* 49: 387–403.

Smith, I.M., Griffiths, D.V. and Margetts, L. 2014. *Programming the Finite Element Method, Fifth Edition*. West Sussex: John Wiley & Sons.

Numerical Methods in Geotechnical Engineering IX – Cardoso et al. (Eds)
© 2018 Taylor & Francis Group, London, *ISBN 978-1-138-33203-4*

Stabilisation of excavated slopes with piles in soils with distinctly different strain softening behaviour

S. Kontoe & D.M. Potts
Imperial College London, UK

F. Summersgill
Tony Gee and Partners LLP, formerly Imperial College London, UK

Y. Lee
Hyundai Engineering and Construction Co., Gyeonggi-do, Republic of Korea

ABSTRACT: The majority of existing design procedures for slope stabilization with piles treat the pile only as an additional force or moment acting on the critical slip surface of the un-stabilised slope, effectively ignoring any interaction of the pile with the evolution of the failure mechanism. This paper presents a numerical investigation that challenges this assumption, demonstrating the importance of the soil-pile interaction. Two dimensional plane-strain hydro-mechanically coupled finite element analyses were performed to simulate the excavation of a slope, considering materials with both a strain softening and non-softening response. The impact of pile position and time of pile construction on the stability of a cutting were parametrically examined, comparing and contrasting the findings for the different material types. The results suggest that an oversimplification during design regarding the soil/pile interaction could either entirely miss the critical failure mechanism (unconservative) or provide a conservative stabilisation solution.

1 INTRODUCTION

The stabilization of slopes with a row of discrete vertical piles is a widely used method which has been successfully employed to remediate failure of existing slopes and to stabilize potentially unstable slopes created by widening transport corridors (Carder 2005; Ellis et al. 2010).

The majority of the existing design procedures for horizontally loaded vertical stabilisation piles employ the displacements and/or critical slip surface of the unstabilised slope as the basis for the design of the stabilisation scheme. The p-y method uses the expected soil displacements to calculate pile reaction (e.g. Baguelin et al. 1977, Georgiadis & Georgiadis 2010). In a limit equilibrium or limit analysis design procedure, the pile is treated only as an additional force or moment located where the critical slip surface and pile coincide (Hassiotis et al. 1997).

These methods as well as hybrid methods, which combine limit equilibrium with finite element (FE) analysis (e.g. Kourkoulis et al. 2012), assume that the presence of a pile will not affect the failure mechanism.

Recently (Summersgill et al. 2017a) challenged this assumption demonstrating the importance of soil-pile interaction on the developed failure mechanism for a cutting in a strain-softening soil. This paper extends the work of Summersgill et al. (2017a) considering both strain-softening and non-softening soils. The influence of pile location and time of pile construction on the stability of a cutting are investigated, comparing and contrasting the findings for non-softening and strain softening soils.

2 DESCRIPTION OF THE NUMERICAL MODEL

2.1 *Analysis arrangement and assumed ground conditions*

The impact of pile position and time of pile construction on the stability of a cutting was examined for a generic slope geometry with dimensions known to be unstable in London Clay without any stabilisation measures (Potts et al. 1997, Ellis & O'Brien 2007). The slope is 10 m in height with a 1 in 3 vertical to horizontal slope angle. The adopted model represents one half of a symmetric excavation and the width of the cutting (16 m) is chosen to correspond to a typical 2 to 3 lane motorway.

All the analyses were undertaken with the Imperial College Finite Element program ICFEP (Potts & Zdravković 1999) in plane strain conditions, adopting eight-noded isoparametric elements with reduced integration. An accelerated modified Newton-Raphson scheme with a substepping stress point algorithm was employed to solve the nonlinear finite element equations (Potts & Zdravković 1999). It should be noted that a coupled consolidation formulation is employed, which allows the modelling of the generation of excess pore fluid pressures during excavation and their subsequent equilibration with time. The mesh configuration is shown in Figure 1, together with the associated mechanical and hydraulic boundary conditions. Before the excavation of the slope, initial stresses are specified in the soil using a bulk unit weight of γ= 18.8 kN/m^3 and a uniform coefficient of lateral earth pressure, K_0 = 2.0. The pore water pressures are hydro-static with 10 kPa suction specified at the ground surface, following the average height expected for the phreatic surface in the UK (Vaughan & Walbancke 1973). The bottom and side boundaries are impermeable. The permeability, k of the soil is modelled as isotropic and is linked to the mean effective stress, p', using the non-linear relationship in Equation 1 (Vaughan 1994).

$$k = k_0 e^{-bp'} \tag{1}$$

The slope was excavated in horizontal layers over 0.25 years. This unloads the soil surrounding the excavation and the low permeability of the soil creates negative pore water pressures. After excavation, 10 kPa suction is applied at the free boundary as indicated in Figure 1. The point of failure is defined as the last increment of the analysis that will converge with a time step of 0.01years (Potts et al. 1997).

In addition to the previously described coupled consolidation analyses, where appropriate, Factor of Safety (FoS) analyses were also performed either immediately after excavation or at a set time post excavation. Potts & Zdravković (2012) showed that the most consistent approach in computing the Factor of Safety in FE analysis is to start the analysis with the characteristic strength and at relevant stages of the analysis to gradually increase the safety factor until failure in the soil is fully mobilised. In each increment of the analysis a larger factor of safety is adopted until failure is reached. The factor of safety is applied by reducing the strength properties of the soil by a factor, F, as shown in Equations *2* & 3:

$$c' = \frac{c'_{in}}{F_s} \tag{2}$$

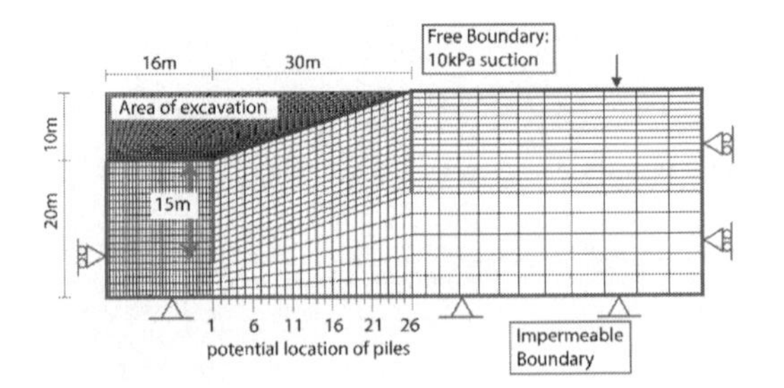

Figure 1. FE mesh with mechanical and hydraulic boundary conditions.

$$\varphi' = tan^{-1}\left(\frac{\tan\varphi'_{in}}{F_s}\right) \tag{3}$$

where F_s is the current Factor of Safety and c'_{in} and φ'_{in} are the values of cohesion and angle of shearing resistance respectively at the beginning of the FoS analysis. In the present study the strength reduction started either once the excavation was completed or at a set time post excavation. The FoS was incrementally increased from an initial value of 1.0 until the factored strength resulted in an unstable slope. The time and associated pore pressure changes are stopped on the increment prior to the first factor of safety increment and the remaining FoS analysis is drained. In each increment of the analysis, it is the current strength that is factored and so when a strain softening model is employed, the reduced strength is factored. The strain softening behaviour is therefore still captured by the FoS analyses.

2.2 *Strain-softening soil*

London Clay was chosen as a representative strain-softening material, which was simulated with a variant of the Mohr-Coulomb model (Potts et al. 1990). This is an elasto-plastic model in which softening behaviour is facilitated through a variation of the angle of shearing resistance ϕ', and the cohesion intercept c' with the deviatoric plastic strain invariant. The material properties are summarised in Table 1 and are based on the work of Kovacevic (1994) and Potts et al. (1997) who simulated a realistic failure time and mechanism that agreed with field data for cutting slopes of the same dimensions in London Clay. The adopted permeability values are from subsequent work of Hight et al. (2007).

It is well established that the conventional FE solution of strain softening problems, in which strain localisation occurs within a zone of limited thickness, can lead to numerical instability and significant mesh dependency of the solution (Galavi

& Schweiger 2010, Summersgill et al. 2017b). Recently Summersgill et al. (2017c) showed that the use of nonlocal regularisation in slope stability problems reduces significantly the mesh dependency and leads to consistent failure mechanisms irrespective of the adopted element size. The nonlocal plastic strain, which regulates the reduction in strength, is given by Equation 4 as proposed by Eringen (1981) and Bazant et al. (1984).

$$\varepsilon^{p*}(x_n) = \frac{1}{V_\omega} \iiint \left(\omega(x_n') \varepsilon^p (x_n') \right) dx_1' dx_2' dx_3' \quad (4)$$

where ε^p is the accumulated plastic deviatoric strain tensor, * denotes the nonlocal parameter, x_n is the global coordinate at which the calculation of the nonlocal plastic strain, ε^{p*} is required, whereas x_n' refers to all the surrounding locations, i.e. the location of reference strain, with n = 1,2,3. Therefore, $\varepsilon^p(x_n')$ equals the reference (i.e. local) strain at the reference location. The weighting function, $\omega(x_n')$ is defined for all the reference locations, but it is centred at the location x_n. The adopted weighting function uses the G&S modifications (Galavi & Schweiger 2010) (see Equation 5) which limits the central concentration of strains.

$$\omega(x_n') = \frac{\sqrt{(x_n' - x_n)^T (x_n' - x_n)}}{DL^2} \exp\left[-\frac{(x_n' - x_n)^T (x_n' - x_n)}{DL^2} \right] \quad (5)$$

The integral of the weighting function in the three dimensions x_1, x_2 and x_3 is referred to as the reference volume, V_ω, as shown in Equation 6. This is used to normalise the calculation of the nonlocal strain and for the G&S distribution is equal to approximately one. The defined length, DL, influences the distribution of the weighting function, with a higher DL resulting in a wider slip surface with a lower maximum nonlocal strain.

$$V_\omega = \iiint \omega(x_n') dx_1' dx_2' dx_3' \quad (6)$$

In all the analyses presented herein a DL = 1 m was used, which in conjunction with the peak and residual strength deviatoric plastic strain limits of 5% and 20% respectively (see Table 1), leads to a representative softening rate for London Clay, as demonstrated by Summersgill et al. (2017c). An additional nonlocal parameter, the radius of influence, was also used to restrict the area of the reference space for the nonlocal calculations and therefore increase numerical efficiency. The radius of influence was set at 3 multiples of DL (i.e. 3 m) based on the recommendations of Summersgill et al. (2017c), who found this ratio to be a suitable compromise between accuracy and time saving.

2.3 *Non-softening soil*

A small parametric study was conducted for a slope without piles in order to choose appropriate strength parameters for the non-softening analyses that would allow qualitative comparison with the results of strain-softening analyses for the same slope geometry without piles (Lee 2015). It was shown that adopting the non-softening strength values listed in Table 1 in a cutting without piles, it leads to failure 46 years after excavation. This time of failure is comparable to the time of failure of 40 years computed by Summersgill et al. 2017c for a cutting of the same geometry in strain-softening soil (see corresponding strain-softening properties in Table 1). Apart from strength, identical soil properties were employed for the softening and non-softening analyses.

2.4 *Pile simulation*

The mesh has been designed to allow the placement of vertical piles in 26 different locations between the toe and crest of the slope (Figure 1). The length of the pile can be varied at 1 m intervals up to 15 meters. In these analyses the pile is wished in place immediately after excavation of the slope, with the exception of the pile construction time investigation.

The pile is modelled using a single column of beam elements placed between the solid quadrilateral elements. These elements are of zero thickness and model the bending behaviour of the pile using the specified stiffness, density, cross sectional area, A and second moment of inertia, I. The simulated pile diameter is 0.9 m with a spacing

Table 1. Soil properties.

Property	Assumed value
Bulk Unit weight, γ	18.8 kN/m^3
Poisson's ratio, μ	0.2
Young's modulus, E	25(p' +100) (min 4000 kPa)
Angle of dilation, ψ	0
Coefficient of permeability, k	$k_0 = 5 \times 10^{-10}$ m/s, b = 0.003 m^2/kN
Strain-softening soil	
Peak strength (bulk)	$c_p' = 7$ kPa, $\varphi_p' = 20°$
Residual strength	$c_r' = 2$ kPa, $\varphi_r' = 13°$
Plastic deviatoric strain, E_d	peak 5%, residual 20%
Non-softening soil	
Strength parameters	$c' = 2$ kPa, $\varphi' = 19°$

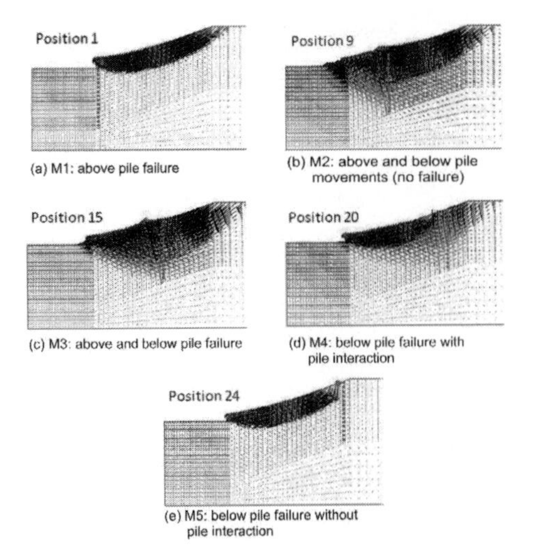

Figure 2. Accumulated plastic deviatoric strain contours for the last increment of the non-softening analyses.

of 2.7 m or three diameters. The calculated A and I were divided by the pile spacing to account for the total quantity of soil that would be supported by a discrete pile in a row. A Young's modulus of 14GPa and a density of 2400 kg/m^3 were specified. A linear elastic constitutive soil model is employed and the maximum bending moment is monitored to identify potential plastic hinge formation.

3 RESULTS

For a slope without any stabilisation piles, failure occurred 46 and 40 years after excavation was complete for the non-softening and strain-softening analyses respectively. For the non-softening analysis, the contours of accumulated plastic deviatoric strain showed the development of a clear, relatively shallow slip surface initiating at the toe of the slope and extending towards its crest. On the other hand for the strain-softening analysis the strain contours showed two potential slip surfaces initiating below the toe of the slope and extending into and towards the crest of the slope. The shallower slip surface became critical and was comparable with the one developed in the non-softening analysis. Inserting stabilisation piles that interact with the potential slip surfaces changed the failure mechanism and time to slope failure for both types of soils considered.

3.1 *Influence of pile location*

A 15 m long pile was placed in each of the 26 locations between the toe and crest of the slope in Figure 1. These locations are spaced 1.2 m apart. The position of the pile was found to have a significant influence on the pile and slope failure mechanism. For each set of analyses (i.e. non-softening, strain-softening) five groups of failure mechanisms were identified by the pattern of slope and pile movements which are summarized in Tables 2 and 3. The vectors of incremental displacement for the final increment of each analysis give an example of each mechanism in Figures 2 and 3 for the non-softening and the strain softening analyses respectively. The sizes of the arrows are relative to the largest incremental displacement for each analysis, but not proportional between the analyses due to the large difference in the size of displacements depending on the mechanism. It should be noted that each plot refers to the final converged increment of the analysis before failure, with the exemption of mechanism 2 (pile positions 7–12) for the non-softening analyses. In the latter case no slope failure was identified 465 years after excavation and therefore the results refer to the last increment of the analysis for the considered time period (i.e. 465 years).

For both soil types there is very limited or no interaction between the pile and the critical slip surface for mechanisms 1 and 5 (see corresponding pile locations in Tables 2 & 3). As a consequence,

Table 2. Failure mechanisms depending on the pile location for the non-softening analyses.

Mechanism	Pile Position	Type of failure
1	1–6	above pile failure
2	7–12	above and below pile movements (no failure)
3	13–18	above and below pile failure
4	19–23	below pile failure with pile interaction
5	24–26	below pile failure without pile interaction

Table 3. Failure mechanisms depending on the pile location for the strain-softening analyses.

Mechanism	Pile Position	Type of failure
1	1	above pile failure
2	2–6	extended toe and above pile failure
3	7–14	above and below pile failure
4	15–20	below pile failure with some pile interaction
5	21–26	below pile failure without pile interaction

mechanisms 1 and 5 have low values of factor safety and time to failure as shown in Figure 4.

In Mechanism 4, for both sets of analyses, the pile did interact to some extent with the failure mechanism and was subjected to horizontal displacements, as it can be seen by the horizontal arrows at the location of the pile and smaller movements of the soil upslope (Figs. 2d & 3d). For the non-softening soil this resulted in an increase in the time to failure from 46 years (no piles) to between 80 and 174 years after excavation (see Figure 4a). The improvement is more modest for the strain softening soil, increasing the time to failure from 40 years (no piles) to between 57 and 98 years (see Figure 4b). However, in the latter case this is still likely to be an inadequate improvement in stability for the required lifetime of transport slopes.

Mechanism 2 (pile locations 2–6, close to the toe) provides the most significant improvement in the time to failure for the case of the strain softening soil, as the pile is an intergral part of the failure mechanism. The pile interaction and hence the improvement in terms of both FoS and time to failure are more limited for Mechanism 3 for the strain-softening soil.

On the contrary, Mechanism 2 for the non-softening soil pile locations 7–12, which correspond to mid-slope locations, provide the optimum stabilisation, as failure was not depicted during the considered time period of the analyses (465 years).

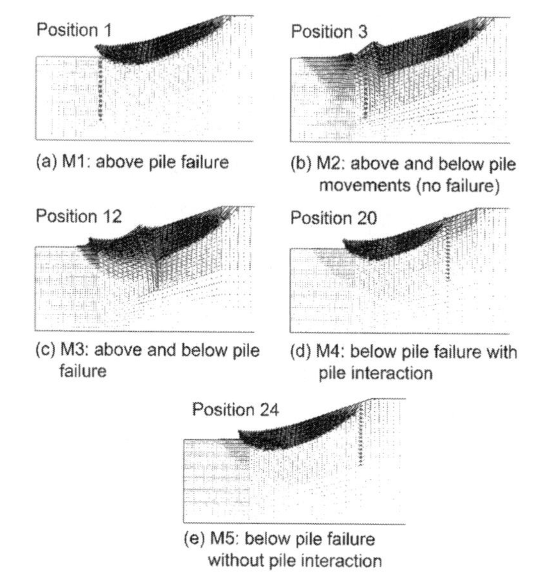

Figure 3. Accumulated plastic deviatoric strain contours for the last increment of the strain softening analyses.

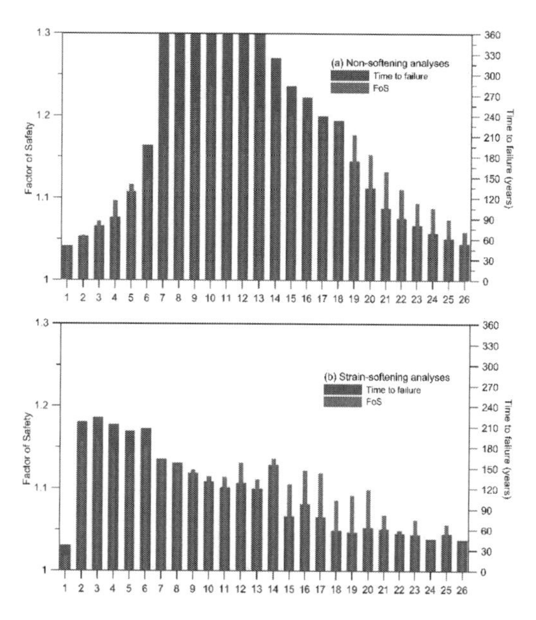

Figure 4. Factor of Safety and time to failure for all pile locations, (a) non-softening soil (b) strain softening soil.

Mechanism 3 for the non-softening soil corresponds to a time to failure in excess of 225 years.

Overall, for the examined cases, it can be concluded that the presence of the pile does not prevent soil movement, which occurs as a reaction to slope excavation and high initial lateral soil stresses. The pile is most effective when it interacts with the development of the slip surface and affects the failure mechanism. This raises an obvious question about the suitability of limit equilibrium based design approaches which assume that the failure mechanism is not modified by the presence of the piles. In addition, if a strain softening material is modelled with constant strength properties, chosen as intermediate values between peak and residual, this will not capture the pile and soil interaction appropriately. For the strain softening soil the optimum pile location is close to the toe, as it delays the development of progressive failure. For the non-softening soil the optimum pile position in the middle of the slope in agreement with previous studies.

3.2 *Influence of pile construction time*

In all the analyses presented so far the pile is constructed immediately after slope excavation. This models the situation of employing piles as a preventative measure to extend the stability period of the slope if the design dimensions make the slope vulnerable to a subsequent failure. If the pile stabilisation system is used to remediate an existing slope that has been identified as unstable then, the pile will be placed possibly decades after excavation. The time of pile construction can play a significant role in the effectiveness of the stabilisation and is examined parametrically herein by constructing the 15 m long pile at location 9 at different periods of 10, 20, 30 and 40 years after slope excavation.

For the non-softening soil the time of pile construction does not seem to affect the time to slope failure, as in all cases failure was not depicted during the considered time period of 465 years (Figure 5a). Furthermore the evolution of the mid-slope horizontal displacement is also fairly insensitive to the time of pile construction.

On the other hand for the strain softening soil, the time of pile construction affects significantly the time to failure, with values varying from 52 years to 142 years, Figure 5b. The greatest improvement in stability occurs when the pile is placed immediately after slope excavation. When the pile is placed 10 years after the slope excavation, there is still a large improvement in stability, while a pile placed 20 years after slope excavation almost doubles the lifetime of the slope prior to failure. A pile placed at 30 or 40 years however, only increases slope stability by just over 10 years.

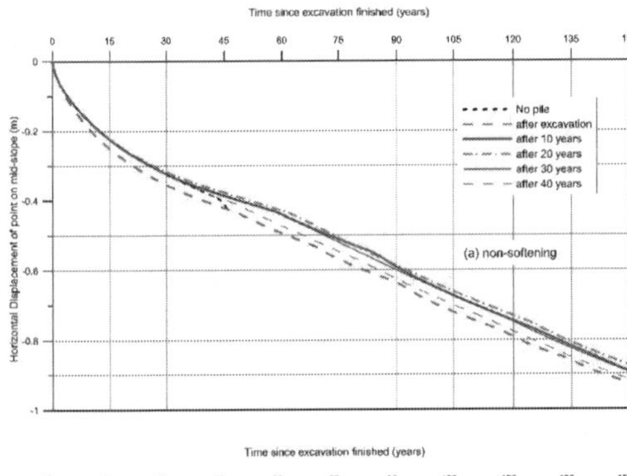

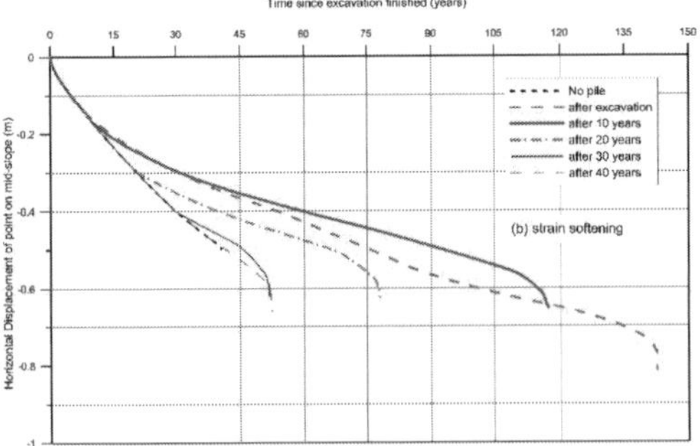

Figure 5. Horizontal displacement of mid-slope over time for pile position 9 after varied periods of consolidation, (a) non-softening soil (b) strain softening soil.

4 CONCLUSIONS

A parametric study was undertaken to systematically examine the impact of pile position and time of pile construction on the stability of a cutting in both non-softening and strain softening materials. London Clay was chosen as a representative strain softening material. To avoid the mesh dependency seen in FE analyses involving strain localisation (e.g. Summersgill et al. 2017c), a nonlocal strain softening model was employed.

The parametric study showed that the pile position has a large influence on the stabilising effect of the pile. The pile should be designed to interact with all potentially critical slip surfaces. Based on the computed FoS values and time to failure, it was shown that for slopes in strain softening materials the pile should be placed between the midslope and the toe of the slope, although not exactly at the toe of the slope. For slopes in non-softening soils the optimum pile locations are at the mid-slope area in agreement with previous studies.

The time of pile construction was found to be very influential on the contribution of the pile to slope stabilization for the strain-softening cases, but to have a minimal impact on the stability of the non-softening cuttings. In the former case, it was found that if the slip surface had developed beyond the pile location prior to pile construction, the pile did not significantly alter the existing strain distributions and the original failure mechanism persisted. This occurs as a result of the strain softening behaviour of the soil encouraging the continued development of a slip surface once it has begun.

The analyses results demonstrate that the interaction of a vertical stabilisation pile and slope is complex, particularly in strain softening materials where the developed failure mechanism evolves significantly with time. The results clearly show that construction of a pile does not provide a single stabilizing action at the intersection with the critical slip surface of the unstabilised slope. Moreover, the pile is most effective in extending the stability of the slope when the failure mechanism is significantly altered by the presence of the pile. This was true both for the strain-softening and the non-softening cases. These findings challenge some of the current design procedures for these piles, which treat the pile only as an additional force or moment and simplify soil/pile interaction. An oversimplification during design could miss the critical failure mechanism (unconservative) or provide a conservative stabilisation solution.

ACKNOWLEDGEMENTS

Part of this work relates to the PhD research of the second author at Imperial College London supported through an Industrial Cooperative Award in Science and Technology (CASE) jointly funded by the Engineering and Physical Sciences Research Council (EPSRC) and Geotechnical Consulting Group LLP. This support is gratefully acknowledged.

REFERENCES

Baguelin, R., Frank, R. & Said, Y.H. (1977) Theoretical study of lateral reaction mechanism of piles. Géotechnique. 27 (3), 405–434.

Bazant, Z.P. Belytschko, T.B. & Chang, T.P. (1984) Continuum model for strain softening. J. Engng. Mech. 110(12): 1666–1692.

Carder, D. (2005) Design guidance on the use of a row of spaced piles to stabilise clay highway slopes. Transport Research Laboratory (TRL), Report TRL632, ISBN 1-84608-631-0.

Ellis, E.A. & O'Brien, A.S. (2007) Effect of height on delayed collapse of cuttings in stiff clay. Proceedings of the Institution of Civil Engineers - Geotechnical Engineering. 160 (2), 73–84.

Ellis, E.A., Durrani, I.K. & Reddish, D.J. (2010) Numerical modelling of discrete pile rows for slope stability and generic guidance for design. Géotechnique. Vol. 60 (3), 185–195.

Eringen, A.C. (1981) On nonlocal plasticity. Int. J. Eng. Sci., 19: 1461–1474.

Galavi, V. & Schweiger, H.F. (2010) Nonlocal multilaminate model for strain softening analysis. Int. J. Geomech. 10: 30–44.

Georgiadis, K. & Georgiadis, M. (2010) Undrained lateral pile response in sloping ground. Journal of Geotechnical and Geoenvironmental Engineering. 136 (11), 1489–1500.

Hassiotis, S., Chameau, J.L. & Gunaratne, M. (1997) Design Method for Stabilization of Slopes with Piles. Journal of Geotechnical and Geoenvironmental Engineering. 123 (4), 314–323.

Higgins, W., Vasquez, C., Basu, D. & Griffiths, D. (2013) Elastic Solutions for Laterally Loaded Piles. Journal of Geotechnical and Geoenvironmental Engineering. 139 (7), 1096–1103.

Hight, D.W., Gasparre, A., Nishimura, S., Minh, N.A., Jardine, R.J. & Coop, M.R. (2007) Characteristics of the London Clay from the Terminal 5 site at Heathrow Airport. Géotechnique. 57 (1), 3–18.

Kourkoulis, R., Gelagoti, F., Anastasopoulos, I. & Gazetas, G. (2011) Slope Stabilizing Piles and Pile-Groups: Parametric Study and Design Insights. Journal of Geotechnical and Geoenvironmental Engineering. 137 (7), 663–677.

Kourkoulis, R., Gelagoti, F., Anastasopoulos, I. & Gazetas, G. (2012) Hybrid Method for Analysis and Design of Slope Stabilizing Piles. Journal of Geotechnical and Geoenvironmental Engineering. 138 (1), 1–14.

Kovacevic, N. (1994) Numerical analysis of rockfill dams, cuts slopes and road embankments. Ph.D. thesis. PhD Thesis, Imperial College London.

Lee, Y.S. (2015) Numerical modelling of non-softening material cut slopes with piles for stabilisation, MSc thesis, Imperial College London.

Potts D.M. Dounias G.T. & Vaughan P.R. (1990) Finite element analysis of progressive failure of Carsington embankment. Geotechnique, Vol. 40 (1), pp. 79–101.

Potts, D.M. & Zdravković, L. (2012) Accounting for partial material factors in numerical analysis. Géotechnique. 62 (12), 1053–1065.

Potts, D.M., Kovacevic, N. & Vaughan, P.R. (1997) Delayed collapse of cut slopes in stiff clay. Géotechnique. 47 (5), 953–982.

Potts, D.M. & Zdravković, L. (1999) Finite Element analysis in geotechnical engineering: Theory Thomas Telford, London.

Summersgill FC, Kontoe S, Potts DM, (2017a) Stabilisation of excavated slopes in strain-softening materials with piles, Geotechnique, available online ahead of print, doi: 10.1680/jgeot.17.P.096

Summersgill FC, Kontoe S, Potts DM, (2017b) Critical Assessment of nonlocal strain-softening methods in biaxial compression. Int. J. Geomech. (ASCE) Vol.17 (7), doi: 10.1061/(ASCE)GM.1943-5622.0000852.

Summersgill FC, Kontoe S, Potts DM, (2017c) On the use of nonlocal regularisation in slope stability problems, Computers and Geotechnics, Vol: 82, Pages: 187–200, ISSN: 0266-352X.

Summersgill, F.C. (2015) Numerical modelling of stiff clay cut slopes with nonlocal strain regularisation. PhD Thesis Imperial College London.

Vaughan, P.R. & Walbancke, H.J. (1973) Pore pressure changes and the delayed failure of cutting slopes in over-consolidated clay. Géotechnique. Vol. 23 (4), 531–539.

Vaughan, P.R. (1994) Assumption, prediction and reality in geotechnical engineering. Géotechnique. Vol.44 (4), 573–609.

Numerical Methods in Geotechnical Engineering IX – Cardoso et al. (Eds)
© 2018 Taylor & Francis Group, London, ISBN 978-1-138-33203-4

Groundwater flow modelling for the design of a dry dock

R. Ramos & L. Caldeira
Laboratório Nacional de Engenharia Civil, Lisbon, Portugal

E. Maranha das Neves
Instituto Superior Técnico, Universidade de Lisboa, Lisbon, Portugal

ABSTRACT: This work focuses on 3D groundwater modelling as a predictive tool to assist in the design and optimization of the dewatering system to be installed during the construction and operation phases of a dry dock. In order to simulate the 3D groundwater flow conditions, finite-difference numerical models in steady-state were generated. The simulations allowed for the evaluation of the hydrogeological conditions and the quantification of water inflows to the excavation and drainage system planned under the designed bottom concrete slab. Vertical upward hydraulic gradients and total hydraulic gradients were analysed as well, aiming to prevent potential limit states related to hydraulic heave and internal erosion.

1 INTRODUCTION

One of the greatest challenges in ground construction design is the groundwater flow, which may produce instabilities and equilibrium loss during the excavation and operational stages that could lead to an increase on both construction time and costs, as well as structural damages (Cashman & Preene 2017). Therefore, it is essential to implement solutions that ensure safety conditions at the construction site and structures implemented.

Although there are several approaches for the evaluation of groundwater flow into a construction site such as analytical, empirical and numerical methods (Hassani et al. 2018), numerical models have the advantage of allowing the identification of complex flow patterns and areas with significant flows. Nevertheless, they still require an appropriate conceptual model based on geological and hydrogeological input data (Zhou & Li 2011).

In the last decades, groundwater flow modelling has been widely used, with great success, as a powerful tool for predicting water flows into mines, tunnels and excavation pits (Montenegro & Odenwald 2009; Font-Capó *et al.* 2011; Pujades *et al.* 2014; Surinaidu *et al.* 2014; Liang *et al.* 2017).

This paper addresses the use of numerical methods for 3D groundwater flow modelling and their application to the construction and operational aspects of the design of a large dry dock for building and repairing ships to be used on offshore oil exploitation activities.

For the construction in dry conditions, the groundwater flowing into the excavation must be appropriately pumped and the hydraulic failure modes must be prevented, namely hydraulic heave and internal erosion.

Regarding the operational phase, it is important to properly design a bottom drainage system and corresponding pumping system, in order to economically design the dock structure. Therefore, information about flow rates is essential for a drainage solution design.

2 STUDY AREA

2.1 *Location and geological conditions*

The shipyard is located in a small bay on the Paraguaçu River, near Salvador, Brazil (Fig. 1).

According to the geological and geotechnical investigations, the study area is characterised by a lithological sequence composed of fine to medium-grained, reddish, fluvial-aeolian Jurassic sandstones extended down to large depth (>100 m), overlain by alluvium (soil) deposited during the Holocene transgression. At the top of the sandstone, a residual soil layer of about 5 m thickness occurs. The alluvial deposits are composed of silty sands, sandy clays and silty clays, exhibiting low density/consistency. Their thickness ranges from 1 to 14 m at the construction site. However, it can reach 38 m Southeast of the dry dock, due to the presence of a palleovalley carved in the sandstone.

Natural groundwater flow is controlled by the water level of the Paraguaçu River and its tributary, situated at North and West of the study area, respectively. Nearby the study area, the mean tide ranges from circa 1.40 m to 2.70 m, in neap

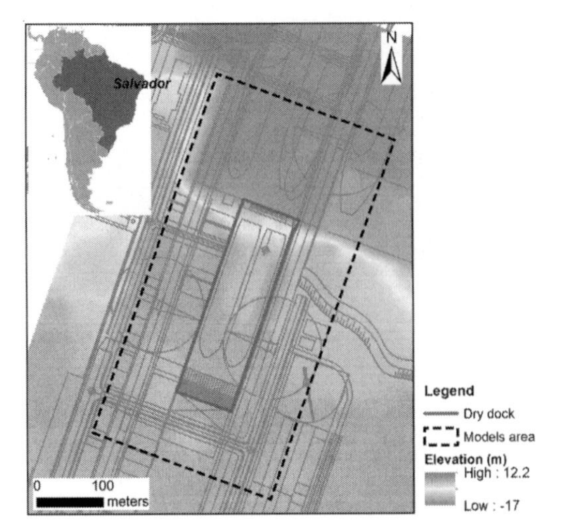

Figure 1. Study area.

and spring tides, respectively (BMA 2009). Water levels approximately at elevation of +3 m were observed in the study area during the geotechnical investigations.

2.2 *Description of the construction site*

The structural solution for the dry dock (300 m long, 87.5 m wide and 15.5 m deep) is composed of diaphragm walls with anchors, a gate on the entrance side and a ground cut slope on the opposite side (to allow future extension of the dock) adjacent to the deepest diaphragm wall. Downstream of the dock gate, additional lateral walls were designed for future quays construction.

The dock bottom, at elevation –10 m, comprises a 0.60 m thick concrete slab, under which a granular filter and a drainage system will be placed.

On the surrounding area, a 2 m high embankment creates a platform at the elevation 4.5 m.

To perform the excavation down to the elevation –11 m, an embankment cofferdam with a central impervious curtain was also planned.

The described design solutions are presented in Figure 2.

3 NUMERICAL MODELLING

In numerical modelling in stationary conditions, the main aspects to be considered include the geometry of the dry dock structure, the stratigraphy of the ground and the ground geotechnical properties, such as hydraulic conductivity and boundary conditions.

In order to simulate the 3D groundwater flow conditions for the final excavation stage and operational phase of the dry dock, two predictive numerical models were generated using Groundwater Modelling System (GMS) version 7.1, which contains the groundwater model MODFLOW.

MODFLOW is a 3D, cell-centred, finite difference, saturated flow model, developed by the United States Geological Survey (McDonald & Harbaugh 1988).

3.1 *Conceptual model*

The existing geological and geotechnical conditions lead to the assumption of an existing unconfined and porous aquifer system, with an infinite lateral extension and variable thickness.

The hydraulic conductivity in saturated conditions of the two geological units was evaluated through specific geotechnical ground investigation

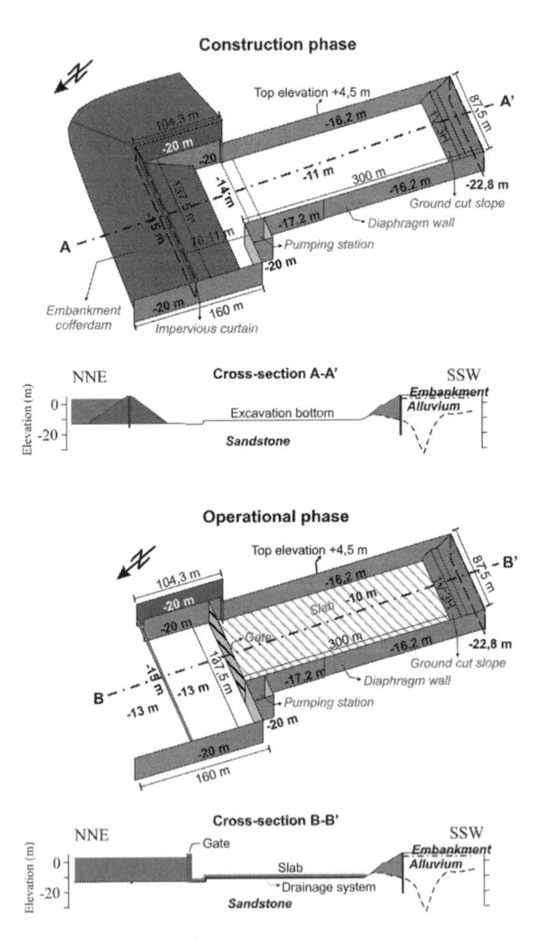

Figure 2. Aspects of the design solutions for the construction and operational phases of the dry dock.

Table 1. Hydraulic parameters used in the numerical modelling.

Unit	Horizontal hydraulic conductivity (m/s)	Vertical hydraulic conductivity (m/s)
Alluvium	5×10^{-6}	1×10^{-6}
Sandstone + residual soil	1×10^{-6}	1×10^{-6}

works. Analysis of existing data allows estimating the same hydraulic conductivity value (10^{-6} m/s) for both units.

As the deposition of the alluvial sediments induces anisotropy in the soil, different hydraulic conductivities were considered for the horizontal and vertical directions in this formation. A slightly higher value was assumed in the horizontal direction. For the sandstone, an isotropic hydraulic behaviour was considered.

The simulation of groundwater flow conditions in the construction site was carried out with the parameters displayed in Table 1.

3.2 *Model discretization and boundary conditions*

At an initial stage, a geological model was developed based on the available data, comprising the geometry and thickness of the recognised geological units. Subsequently, numerical models with a volume of 290 m (x) per 560 m (y) per 104.5 m (z) were generated taking into account the geological model.

These numerical models are composed of prismatic cells with a cross section of 5 m per 5 m and variable height. A total of 25,984 cells were distributed over four layers which in turn were required for the establishment of the geometry of the aquifer and structural designed solutions (Fig. 3).

Layer 1 constitutes the upper layer of the models and encompasses the alluvial deposits at the surrounding area of the dry dock and at the ground cut slope. In the construction phase, it also includes the sandstone rock material at the excavation bottom and the embankment cofferdam at the downstream area. In the operational phase, it also comprises the bottom slab and the top of the sandstones at the downstream area and extends down to the foundation level of the cofferdam curtain. Layer 2 develops until the foundation level of the diaphragm walls (with the exception of the cofferdam curtain) and, in the remaining area, it comprises the total thickness of the residual soil. Layer 3 extends from the bottom of layer 2 down to the elevation –50 m. Finally, layer 4 exhibits a uniform thickness of 50 m and occurs between –50 m and –100 m ground elevation.

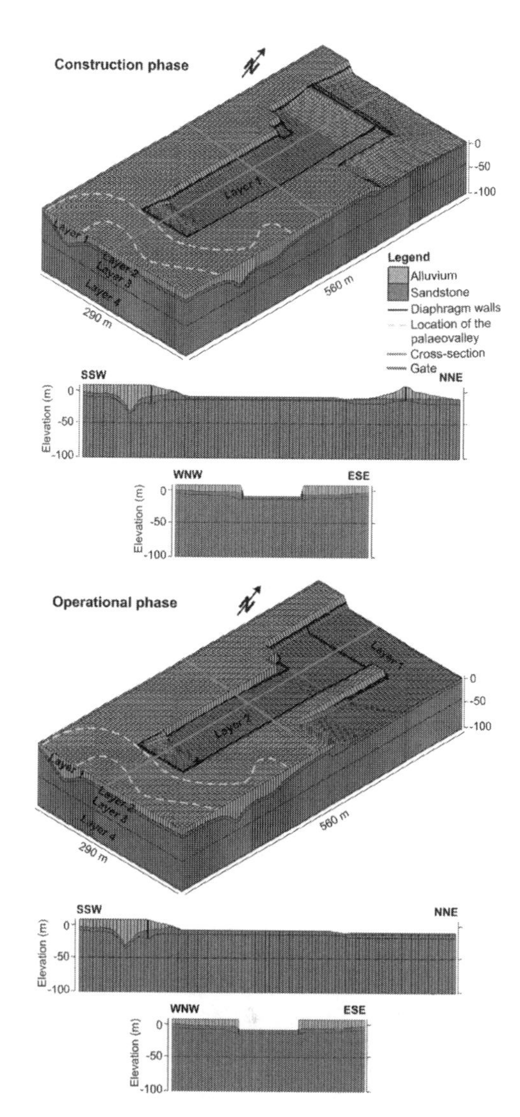

Figure 3. 3D grid created for each phase, with the location of the palleovalley, and their representative profiles.

In what concerns boundary conditions rendering the final excavation stage, constant heads of –14.1 m were considered at the pumping station and the dock gate areas and of –11.1 m in the remaining cells intersected by the excavation bottom. Downstream of the embankment cofferdam as well as at the perimeter limits of the top layer, a constant head of +3 m was input

The model for the operational phase is characterized by cells with no flow (inactive) in the area matching with the concrete slab and constant heads of –13.8 m at the pumping station and the dock gate areas and of –10.8 m in the remaining cells underneath the bottom slab. In addition, constant heads of +3 m were admitted at the perimeter limits of the top layer and downstream of the dry dock.

In order to assess the groundwater flow conditions of the two design situations, the models were simulated in steady-state.

4 RESULTS

4.1 *Prediction model for the final excavation stage*

During the construction works of the dry dock, groundwater is expected to be permanently pumped, to guarantee water level lowering within the excavation perimeter. Therefore, for the final excavation stage, flow conditions similar to those obtained in the numerical simulation, as presented in Figures 4–5, are anticipated.

According to the results, groundwater will flow into the construction site through the excavation bottom, skirting the base of the concrete diaphragm walls. It is also expected that groundwater seepage down to at least the elevation –100 m will be affected (Fig. 5).

As a result of the drawdown induced by pumping, the total heads outside the construction site will drop towards the impervious structures, mainly in the eastern and western sides where the water level will reach the elevation –5 m. On the south area, the pumping effect will be lower due to

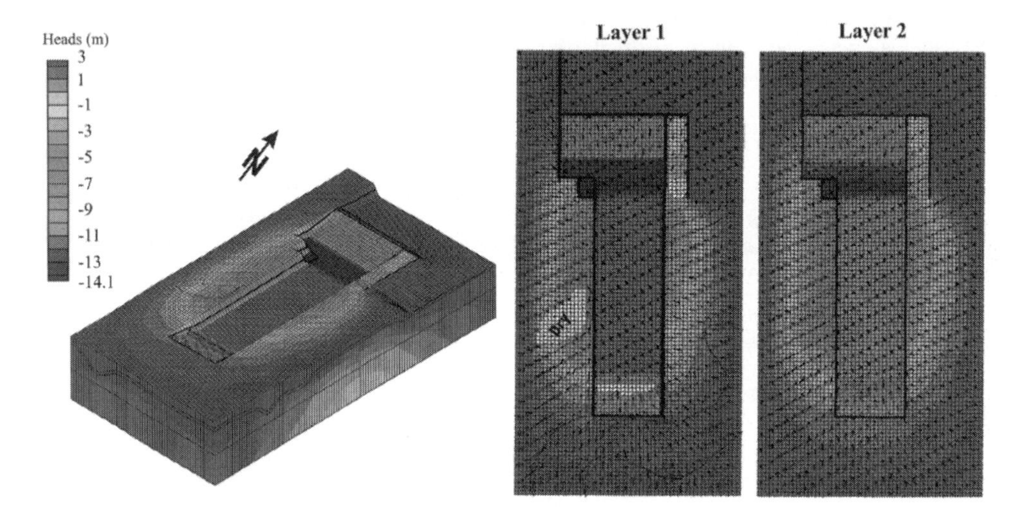

Figure 4. Spatial distribution (3D and 2D) of the predicted heads and flow directions for the final excavation stage.

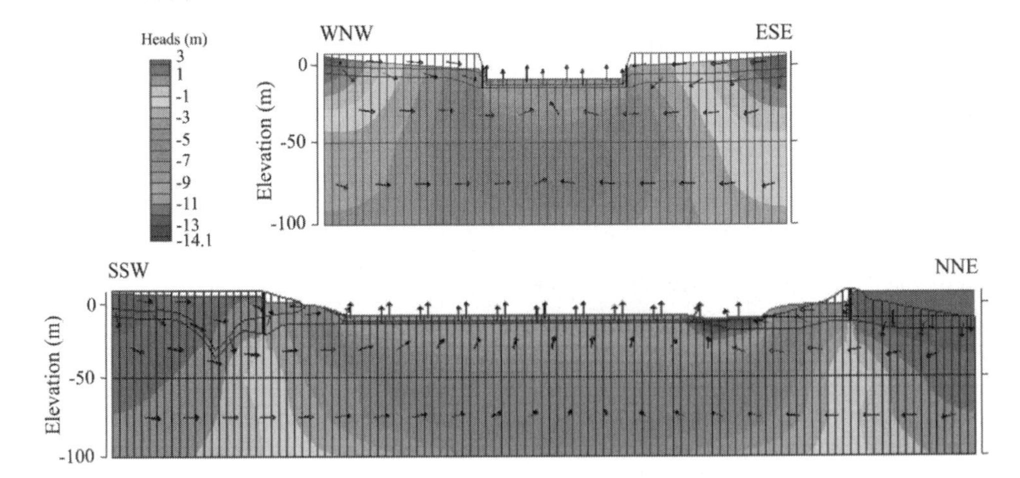

Figure 5. Cross-sections of the predicted heads and flow directions of the final excavation stage. Location displayed in Figure 3.

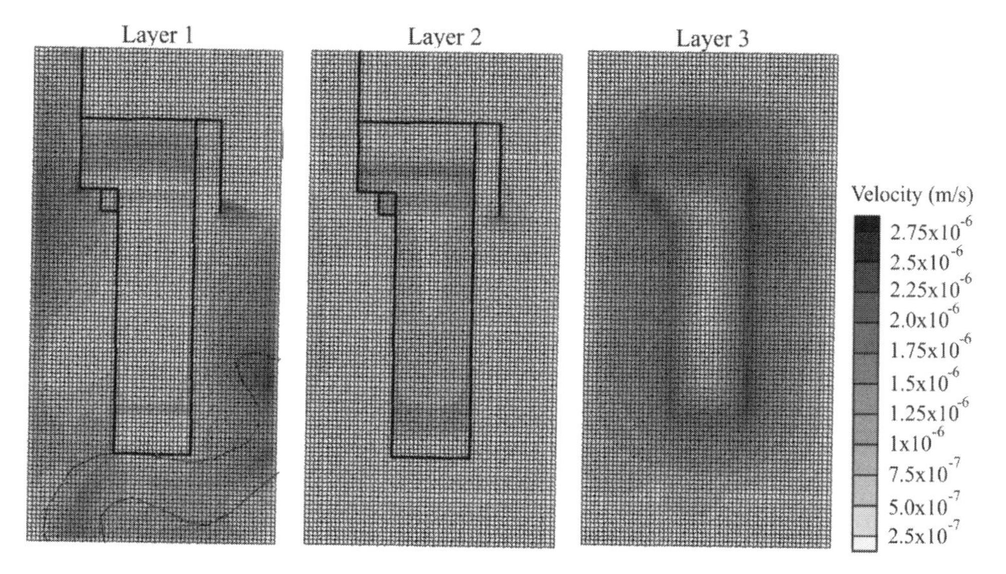

Figure 6. Flow velocities in layers 1, 2 and 3 of the representative model of the final excavation stage.

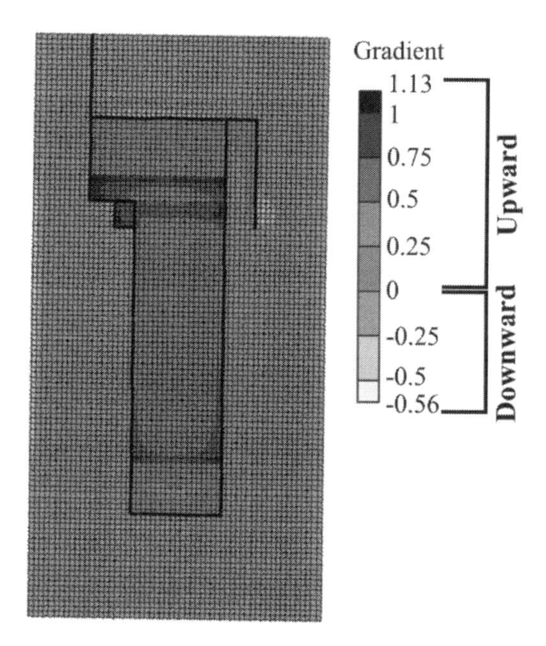

Figure 7. Expected gradients generated by the vertical flow during the final excavation phase, in layer 2.

the presence of the palaeovalley (high thickness of alluvium) combined with a greater distance to the induced pumping. In this area, maximum drawdown of about 2 m is expected in the vicinity of the southern diaphragm wall.

Within the excavation area, total heads of about –3 m are expected at the ground cut slope and inner cofferdam shell, whereas at the zones of the pumping station and at the bottom upstream of the cofferdam, total heads of –14.1 m are likely to occur. The remaining area should be characterized by total heads of –10.8 m.

Results also suggest that the maximum velocities of the flow (Fig. 6) in the soil (layer 1) will be attained at the palleovalley and northwest of the construction site close to the pumping station area and adjacent walls. In the sandstone, the highest velocities will be achieved at the base of the eastern and western diaphragm walls, underneath the cofferdam curtain, at the toe of the ground cut slope and at the toe of the inner cofferdam shell (layer 3).

Based on the 3D numerical simulation, a flow rate of 6×10^{-3} m^3/s should be pumped during the excavation works of the dry dock, aiming to ensure the construction under dry conditions. This value is considered low taking into account the dimensions of the work site and the water levels in the surrounding area and it is related with the hydraulic characteristics of the geological formations. Nevertheless, to design the pumping system during construction, it is important to assess the spatial distribution of flow rates. Thus, for the final excavation stage, it is verified that about 35% of the total flow rate corresponds to the water that needs to be pumped in the deepest zone, located at the pumping station area and at the toe of the inner cofferdam shell.

In what concerns total hydraulic gradients, the results suggest the generation of the highest values at the base of the cofferdam curtain and of the western and eastern diaphragm walls, ranging between 1.5 and 2.15 (Layer 3). Within the excavation area, maximum gradients will vary between 0.5 and 1 nearby the cut slope toe, the inner cofferdam shell toe, the pumping station area and the deepest bottom area at elevation –14 m (Layer 2).

It should be noticed that, in most of the sites (cells), all three directions (X, Y and Z) are relevant to characterize the groundwater flow. Which means that the results described before would not be easily identified with the same detail using 2D numerical models, particularly nearby the edges and corners of the impervious structures, the cut slope, and the cofferdam and at different bottom elevations at the excavated area.

The analysis of the spatial distribution of the vertical hydraulic gradient allowed for the identification of the pumping station, the deepest bottom zone, the cut slope toe as the areas of maximum vertical gradients generated by the upward flow (Fig. 7). Magnitudes ranging between 0.75 and 1.13 are anticipated for these gradients (in layer 2).

Nevertheless, layer 2 does not represent the excavation bottom, which means that the maximum values of the vertical gradient will not occur close to the water exit. Nearby the excavation bottom (layer 1), maximum vertical gradients of about 0.76 are expected to occur at the deepest bottom zone.

4.2 *Prediction model for the operational phase*

Once the construction is completed, groundwater will concentrate beneath the concrete slab, generating flows that will be collected by the drainage system after passing through the filter.

The results of the numerical model representing this phase are shown in Figures 8–9.

According to the simulation, seepage conditions very similar to those obtained in the previous model will occur during the operational phase if a drainage system under the concrete slab is installed. The exception is verified in the dock gate zone and adjacent downstream area due to the demolition of the embankment cofferdam, the excavation down to the elevation –13 m and the subsequent expected submersion. As a consequence, total heads slightly higher (maximum 1 m) than those obtained in the previous phase are expected for the southwestern half of the study area. On the northeast half of the model, total head differences will be larger, reaching a maximum value of 16.1 m outside the dry dock, in the proximity of the dock gate.

In what concerns to flow velocities, this phase will also be characterized by a slight increase (~2 x 10^{-6} m/s) underneath the dock gate (Layer 3), due to an increase of the hydraulic gradient produced by the water level rise downstream of the dry dock.

During the operational phase, a flow rate similar to the previous phase (of about 6×10^{-3} m^3/s) should be drained in order to ensure the dissipation of the water pressures near the base of the concrete slab. About 27% of this value corresponds to the flow rate that must be drained at the dock gate and pumping station zones.

In regards to total hydraulic gradients, the results suggest the occurrence of highest values underneath the dock gate, the western and eastern diaphragm walls and the ground cut slope, ranging from 1.25 to 3 (Layer 3). However, it will be achieved a maximum vertical upward hydraulic gradient lower than the previous model. The

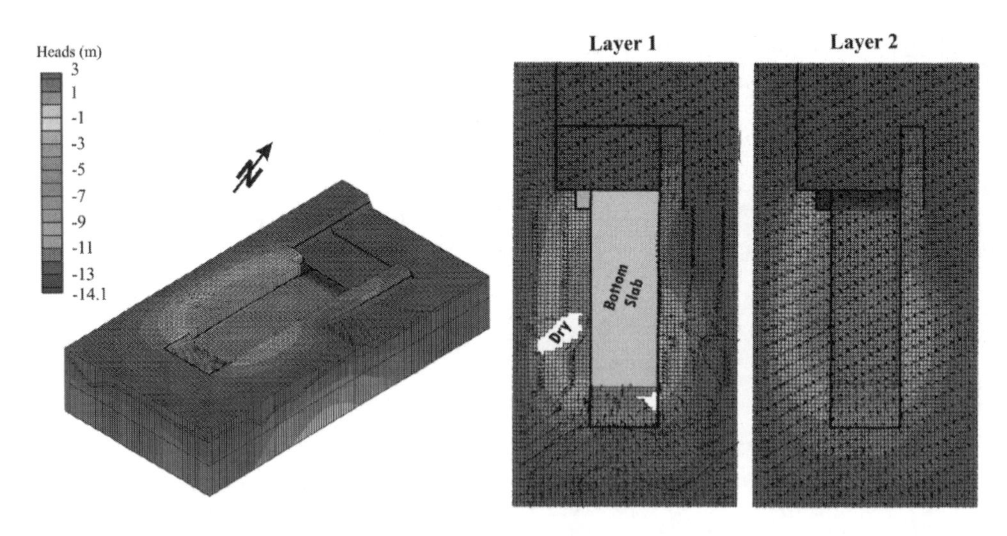

Figure 8. Spatial distribution (3D and 2D) of the predicted heads and flow directions for the operational stage.

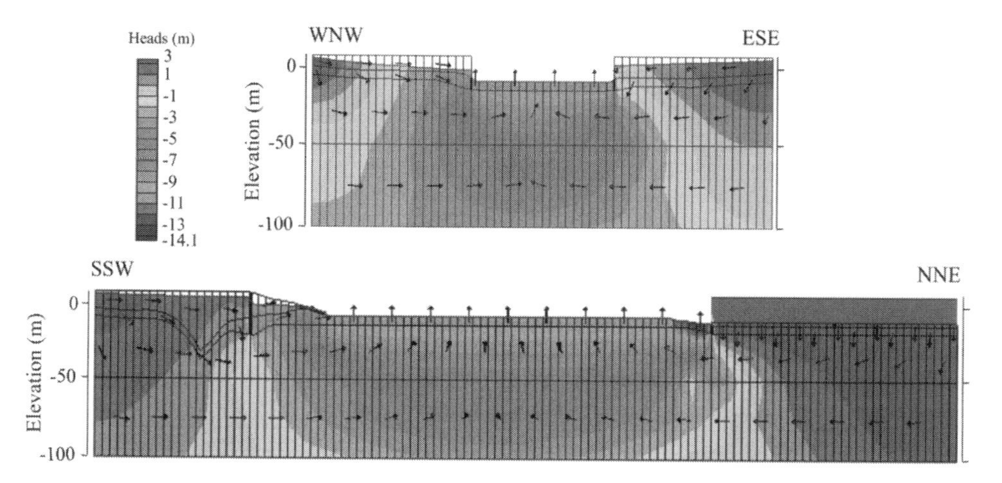

Figure 9. Cross-sections of the predicted heads and flow directions of the operational stage. Location of the profiles displayed in Figure 3.

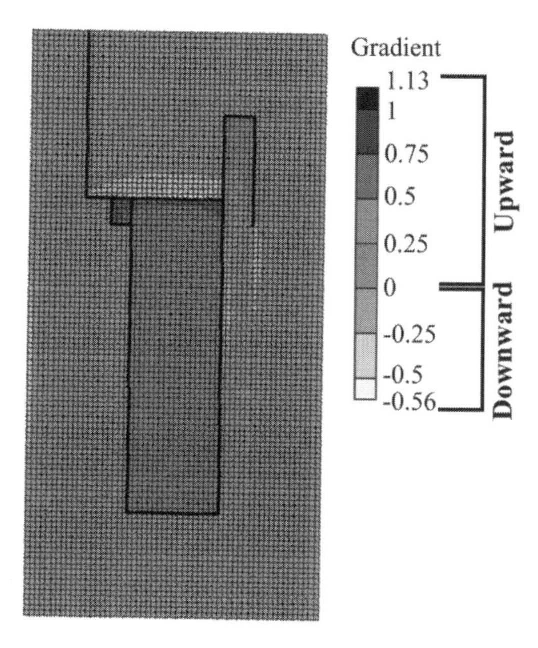

Figure 10. Expected gradients generated by the vertical flow during the operational phase, in layer 2.

highest values will occur in the vicinity of the dock gate and at the pumping station area and will vary between 0.5 and 0.67 (Fig. 10).

5 CONSIDERATIONS ABOUT THE DESIGN AND HYDRAULIC SAFETY

The results presented beforehand were used to design pumping, drainage capacity, filters and drains for the construction and operational phases. Nevertheless, they also allow the analysis of the relevant ultimate limit states (ULS) associated with the water flow conditions generated during the construction and operation of the dry dock. For this analysis, quantification and direction of the water gradients, already evaluated, are used for the calculation of seepage forces (their value and direction) which are needed to evaluate hydraulic safety. In this case, failure due to heave and internal erosion are the relevant ULS to be considered. However, their analysis is beyond the scope of this study.

6 CONCLUSIONS

The developed hydrogeological numerical modelling allowed for the determination of the three-dimensional distribution of the total heads and flow velocities, as well as the hydraulic gradients for the final excavation and operational phases of a dry dock. Accordingly, the most problematic zones were identified.

Flow rates were also estimated for the design and optimization of the bottom reinforced concrete slab, pumping system and filter-drainage system.

In what concerns the excavation stage, a drawdown of about 8 m will be created by pumping, in the eastern and western areas outside the construction site. In the southern area, a reduction of the pumping effect will be observed due to the presence of a palleovalley. Outside the construction site, maximum velocities will be attained at the palleovalley and in the vicinity of the pumping station area, while inside the work site they will be achieved underneath the cofferdam curtain, at the base of the eastern and western walls and at the toe of both slopes.

Regarding the operational phase, groundwater flow conditions very similar to those obtained for the construction stage are expected if a drainage system beneath the concrete slab is considered. The exception will occur in the dock gate zone and adjacent downstream areas as a result of the removal of the embankment cofferdam and the subsequent submersion. This phase will also be characterized by a slight increase in velocity underneath the dock gate due to the increase of hydraulic gradient.

During both phases, flow rates of about 6×10^{-3} m^3/s should be pumped/drained aiming to avoid instabilities and equilibrium loss in the construction site. In addition, the maximum hydraulic gradients generated by the upward flow are expected to be higher during the construction phase and to occur at the pumping station area, at the deepest bottom zone and at the toe of both slopes (cut slope and inner shell cofferdam), ranging from 0.75 to1.13.

As the simulated groundwater flow is inherently 3D (in the direction X, Y and Z), the results presented in this paper would be difficult to observe with the same detail using 2D numerical modelling, especially where edges and corners of the impervious structures and special geotechnical structures prevail.

Finally, the expected gradients will allow the safety analysis of the structure during its operation as well as during construction.

REFERENCES

BMA 2009. *Estudo de impacto ambiental e relatório do impacto ambiental do Estaleiro do Paraguaçu (Bahia)*. Relatório de impacto ambiental – RIMA.

Cashman, P.M. & Preene, M. 2017. *Groundwater Lowering in Construction: A Practical Guide to Dewatering*. Second Edition, CRC Press.

Font-Capó, J., Vázquez-Suñé, E., Carrera, J., Martí, D., Carbonell, R., Pérez-Estaun, A. 2011. Groundwater inflow prediction in urban tunneling with a tunnel boring machine (TBM). *Engineering Geology* 121: 46–54.

Hassani, A.N., Farhadian, H. Katibeh, H. 2018. A comparative study on evaluation of steady-state groundwater inflow into a circular shallow tunnel. *Tunnelling and Underground Space Technology* 73: 15–25.

Liang, Z., Ren, T., Ningbo, W. 2017. Groundwater impact of open cut coal mine and an assessment methodology: A case study in NSW. International Journal of Mining Science and Technology 27: 861–866.

McDonald, M.G & Harbaugh, A.W. 1988. *A modular three-dimensional finite-difference ground-water flow model: Techniques of Water-Resources Investigations of the United States Geological Survey*. Book 6, Chapter A1, 586 p.

Montenegro, H. & Odenwald, B. 2009. Analysis of Spatial Groundwater Flow for the Design of the Excavation Pit for a Ship Lock in Minden Germany. 2nd International FEFLOW User Conference. 14–18 September, Potsdam, Germany.

Pujades, E., Vàzquez-Suñé, E., Carrera, J., Jurado, A. 2014. Dewatering of a deep excavation undertaken in a layered soil. *Engineering Geology* 178: 15–27.

Surinaidu, L., Gurunadha Rao, V.V.S., Srinivasa Rao, N., Srinu, S. 2014. Hydrogeological and groundwater modeling studies to estimate the groundwater inflows into the coal Mines at different mine development stages using MODFLOW, Andhra Pradesh, India. *Water Resources and Industry* 7–8:49–65.

Zhou, Y. & Li, W. 2011. A review of regional groundwater flow modeling. *Geoscience Frontiers* 2(2): 205–214.

Numerical Methods in Geotechnical Engineering IX – Cardoso et al. (Eds)
© 2018 Taylor & Francis Group, London, ISBN 978-1-138-33203-4

Influence of periodical rainfall on shallow slope failures based on finite element analysis

A. Chinkulkijniwat, S. Horpibulsuk & S. Yubonchit
Center of Excellence in Civil Engineering, School of Civil Engineering, Institute of Engineering, Suranaree University of Technology, Nakhon Ratchasima, Thailand

ABSTRACT: Although the rainfall intensity-duration thresholds for initiation of slope failure (ID thresholds) based on the historical slope failure data is commonly used to assess slope failure, critical influence factors triggering shallow slope failures are often disregarded. Sets of parametric study were therefore performed through finite element model to investigate the effect of saturated permeability of soil and antecedent rainfall on instability of shallow slope. The hydrological mode during the failure depends on the magnitude of rainfall intensity comparing with the infiltration capacity at soil saturation state. The rate of reduction in safety factor increases with an increasing in rainfall intensity, only in a range of lower than the infiltration capacity at soil saturated state. As such the saturated permeability of the soil plays an important role in the shallow slope failure. It was found also to govern a range of rainfall intensity and duration at failure state on ID thresholds. The antecedent rainfall was found to play role on instability of the shallow slope. It controls the initial stability of slope, which results in the different linear relationship of ID thresholds.

1 INTRODUCTION

A simple tool for mitigating the landslides disaster is relied on the critical rainfall concept, namely ID thresholds, which is a relationship between intensity and duration of rainfall for the initiation of slope failure (Aleotti 2004; Caine 1980; Calcaterra et al. 2000; Cannon & Gartner 2005; Chien et al. 2005; Corominas 2000; Crosta & Frattini 2001; Guzzetti et al. 2007). The advantages of this concept are their simplicity and rapid assessment.

Although the ID thresholds is simple and easy to use to assess the failure of slope, understanding of the critical mechanism triggering the failure of slope is often neglected. In addition, understanding the rainfall-induced slope failure problem needs coupled flow simulation and mechanical deformation modeling, especially in unsaturated ground water flow environment. Hence, analysis of these relevant problems requires a powerful tool to conduct a series of numerical experiments those simulate the problem in a coupling of hydrological-mechanical manner. Recently, many attempts have been developed to employ algorithms that simulate the coupling of ground water flow and mechanical deformation. Rutqvist et al. (2002) and Rutqvist (2011) linked two computer codes, TOUGH2 (Pruess et al. 1999) and FLAC3D (Itasca 2009), to conduct 3D hydrologic-mechanical analyses, including multiphase (gas and liquid) and multicomponent (air and water) processes. (Chinkulkijniwat et al. 2014) carried out the analysis of 3-D compressed air tunneling advancement using the method proposed by Rutqvist et al (2002). Galavi et al. (2009) implemented a fully coupled hydro-mechanical analysis based on Biot's consolidation theory for saturated and unsaturated soils in PLAXIS code. Hamdhan & Schweiger (2013) conducted a slope stability analysis subjected to rainfall infiltration and rapid drawdown condition in PLAXIS code. However, few attempts have conducted the analysis of rainfall-induced slope failure on a shallow slope in a fully coupling hydrological-mechanical manner. Moreover, the existed attempts performed the analysis on the slope subjected to a specific period of rainfall. In the other word, analysis on the slope subjected to a certain rainfall continuously until the arrival of slope failure had never been conducted. Therefore, the failure conditions of shallow slope under various conditions of the influence factors have not been investigated to date.

2 THEORETICAL BACKGROUNDS

Finite element PLAXIS code with a fully coupled flow-deformation analysis (Brinkgreve et al. 2010) was used in this study. The application of the code to rainfall infiltration related problems was verified by Hamdhan & Schweiger (2013). Mohr-Coulomb failure criterion is assigned to this study. The shear strength of soil related to unsaturated conditions is obtained by combining Bishop's effective stress concept (Bishop & Blight 1963) and Mohr-Coulomp failure criterion.

$$\tau = c' + (\sigma_n - u_a)\tan\phi' + \chi(u_a - u_w)\tan\phi' \qquad (1)$$

where τ is shear strength of unsaturated soil, σ_n is total normal stress, u_a is pore air pressure, u_w is pore-water pressure, (σ_n-u_a) is net normal stress, u_a-u_w is metric suction, c' is effective cohesion, ϕ' is internal soil friction angle and χ is scalar multiplier which is assumed as effective degree of saturation(θ_e) in this study.

As for the hydrological process, Richard's equilibrium used to simulate transient flow through unsaturated soil, which can be expressed as:

$$\frac{\partial\theta}{\partial t} = k_x\frac{\partial^2 h}{\partial x^2} + k_y\frac{\partial^2 h}{\partial y^2} \qquad (2)$$

where θ is volumetric water content, h is the total head, k_x and k_y are the unsaturated coefficient of permeability in the x—and y—directions, and t is time.

The permeability in unsaturated soil is highly dependent on soil-water characteristics (SWC). The SWC is a relationship between moisture content and pressure head which can be explained by van Genuchten model (van Genuchten 1980) and the permeability function is explained by van Genuchten-Mualem model (Mualem 1976). Equations 3 and 4 are respective van Genuchten and van Genuchten-Mualem models.

$$\theta_e = \frac{\theta - \theta_{res}}{\theta_{sat} - \theta_{res}} = \left[1 + \left(\alpha\left|h_p\right|\right)^n\right]^{-m} \qquad (3)$$

$$k_r(\theta_e) = \theta_e^{0.5}[1-(1-\theta_e^{1/m})^m]^2 \qquad (4)$$

where θ_e is effective degree of saturation, θ is degree of saturation, θ_{res} is residual saturation at very high value of suction, θ_{sat} is the saturation of saturated soil, h_p is matric suction head, k_r is the relative permeability coefficient. α [kPa^{-1}], m and n are fitting parameters, and according to Mualem hypothesis Mualem (1976), m is assigned as $(n-1)/n$.

These two group of material parameters including shear strength parameters (i.e. c', ϕ') and three hydraulic related parameters (i.e. α, n, k_{sat}) are the required parameters to perform an analysis of rainfall-induced slope failures in PLAXIS. In this study, relevant parameters were obtained from previous research works and they will be discussed in the following section.

3 MATERIALS AND METHODS

The shallow slope failures is common found in many parts of the world. The geological setting in each hazardous area is different depended upon climate conditions, rate of weathering, slope geometry etc. As such, this study gathered the soil properties reported from the relevant literatures in slope failures, including Jotisankasa & Mairaing 2010, Jotisankasa & Vathananukij 2008, Bordoni et al. 2015, Dahal et al. 2008, Oh & Lu 2015, Vieira et al. 2010, and Godt & McKenna 2008. Generally, it is found that the parameter α is ranged from 0.016 to 0.360 kPa^{-1}, the desaturation parameter n is ranged from 1.290 to 2.780, θ_{sat} is ranged from 0.286 to 0.480 and θ_{res} is ranged from 0 to 0.250. Besides, the saturated permeability of soil is ranged from 1.0×10^{-6} to 2.1×10^{-4} m/s. As for the shear strength parameters, the cohesion and friction angle of the soils range from 0.0 to 17.60 kN/m^2 and from 32° to 38.6°, respectively.

Previous literatures (Li et al. 2013, Rahardjo et al. 2007, Rahimi et al. 2010) reveal that the saturated permeability plays a major role on the stability of slope. Hence, the saturated permeability is being focused in this study. While, the other parameters, including c', ϕ', α, and n were kept constant at 6.74 kN/m^2, 33.6°, 0.162 kPa^{-1} and 1.564 respectively. The magnitude of these parameters were deducted from the average value of the parameters reported in the references mentioned above. In this study, variation of the saturated permeability is represented by type of soil, i.e. the soils A, B and C stand for low k_{sat} ($= 10^{-6}$ m/s), medium k_{sat} ($= 10^{-5}$ m/s) and high k_{sat} ($= 10^{-4}$ m/s) drain ability, respectively.

Twelve cases of simulation run are conducted to evaluate the effect of periodical rainfall on time to failures of shallow slope. Rainfall event in this series were separated into two events subjected to the same intensity. In the first event, rainfall duration is prescribed as 24 hr. While the second event is assigned until the arrival of slope failure. In addition, the time between rainfall events (t_b) as prescribed as 2 days. The rainfall intensity between first and second rainfall events was assumed as zero, which means that

Table 1. Summary of numerical experiments.

Rainfall intensity related to soil type i (mm/hr)			t_b (day)	Total
A	B	C	–	12
0.36	1	1	∞, 2	
3.6	10	10		

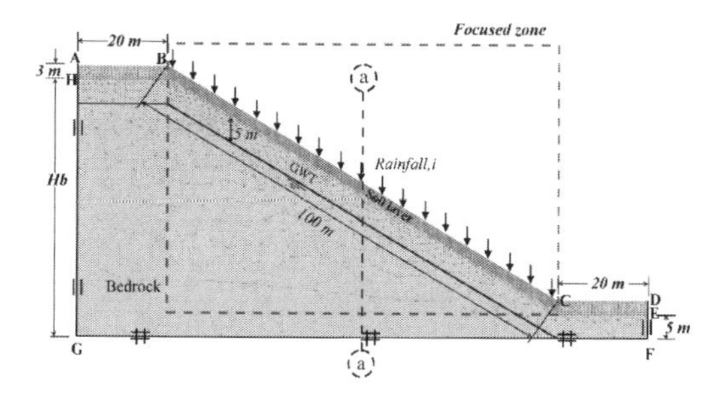

Figure 1.

the evaporation is disregarded. In this study, the effect of time between rainfall events is evaluated under three rainfall intensities and two types of soil, while slope angle was also kept constant as summarized in table 1.

4 SET UP OF EXPERIMENTS

Slope geometry, boundary conditions and fixity used in this study are shown in figure 1. The slope model is divided into two layers, which represent the general shallow slope in stratigraphy. Length of the slope is 100 meters and bedrock layer is covered by soil layer with a shallow thickness of 3 meters, which gives in ratio of soil depth to slope length of 3/100. Standard fixities were prescribed to allow only vertical movement along the boundary sides, while lateral and vertical movements were fixed at bottom boundary. 15-node triangles finite element mesh is used in the problem. The finer elements were generated at the soil layer, and the finest mesh was generated at interested zone.

A prescribed flux, which relates to the desired intensity of rainfall, was assigned along the slope surface BC. Based on the rainfall intensity specified in Table 2, a range of pore pressure was between –0.05 m and 0.05 m. By this maximum pore water pressure of 0.05 m, the pounding water due to the excess of rainfall intensity over the drainage capacity at soil saturation state could be developed up to 5 cm. over the slope surface. While the minimum pore water pressure of –0.05 m was used to represent a depth where the negative flux due to evaporation is no longer occur. The boundaries AB and CD were assigned as no flux boundary, while the boundaries AHG, DEF and GF were prescribed as impervious boundary.

The initial conditions of the model is according to the depth of the groundwater table of 8 m below the soil surface, hence the initial pore water pressure is ranging from –50 to –80 kPa. Besides,

Table 2. Material parameters required in PLAXIS.

Parameter	Symbol	Soil layer		
General parameters				
Material model	–	Morh Coulomb		
Drainage type	–	Undrained A		
Dry unit weight	γ	17.36 kN/m^3		
Total unit weight	γ_{sat}	17.36 kN/m^3		
Strength parameters				
Cohesion	c'	6.74 kPa		
Friction angle	φ'	33.62°		
Flow parameters				
Material model	–	Van Genuchten		
Soil type	–	*A*	*B*	*C*
Permeability (m/s)	k_{sat}	1×10^{-6}	1×10^{-5}	1×10^{-4}
n	n	1.564		
α	α	0.162 kPa^{-1}		

parameters required for PLAXIS is summarized in table 2.

5 NUMERICAL RESULTS AND DISCUSSIONS

5.1 *General mechanism of rainfall-induced shallow slope failures*

Two cases of simulation results were used to explain the shallow slope failure mechanism. Rainfall intensity of 10 and 36 mm/hr added were assigned to soil type B ($k_{sat} = 1 \times 10^{-5}$ m/s) corresponding to the slope angle of 30°. For all rainfall intensities, the slope stability was calculated under continuous rainfall until the slope failure occurred, which is indicated by FS of 1.0.

Figure 2 shows the variation of FS with time of rainfall. As rainfall intensity of 10 mm/hr, the reduction of FS is shown as a small rate during the first period of rainfall. But the FS suddenly decreases to critical value after 48 hours. In case of rainfall intensity equals to saturated permeability (saturated permeability of soil type B is 1×10^{-5} m/s = 36 mm/hr), the FS decreases since beginning of rainfall period to the critical.

Figure 3 shows pore water pressure profile (section a-a in Figure 2) related to rainfall intensity of 10 and 36 mm/hr. In case of rainfall intensity of 10 mm/hr, it illustrates that the pore water pressure can be divided into two main stages. For the first stage, due to the rainfall infiltration process, the pore water pressure increases from initial value to that of about –3 kPa. The stability of slope as shown in Figure 2 is still stable (higher than 1.0) during this process, because the matric suction involved with negative pore water pressure remains in soil slope.

After end of first stage, rising of water table results in positive pore water pressure in second stage, when infiltration of rainfall reach to soil-bedrock interface located at depth of 3m below soil surface. The positive pore water pressure causes the reduction in soil shear strength especially in zone of soil-bedrock interface, and then slope failures can be occurred.

Only having one stage of pore water pressure profile is shown in the case of rainfall intensity of equal to saturated permeability. It can be seen that pore water pressure increases from initial value to zero since rainfall started, which means that the matric suction is completely dissipated during rainfall infiltration process. It is therefore resulting in drop of FS and then slope failures as shown in Figure 2.

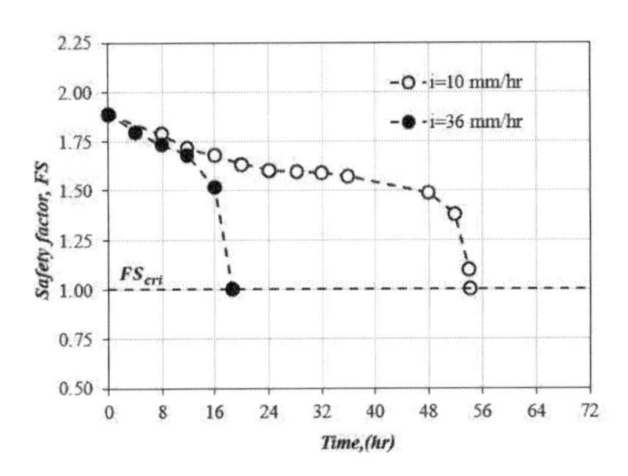

Figure 2. Relationship between safety factor and time under rainfall intensities.

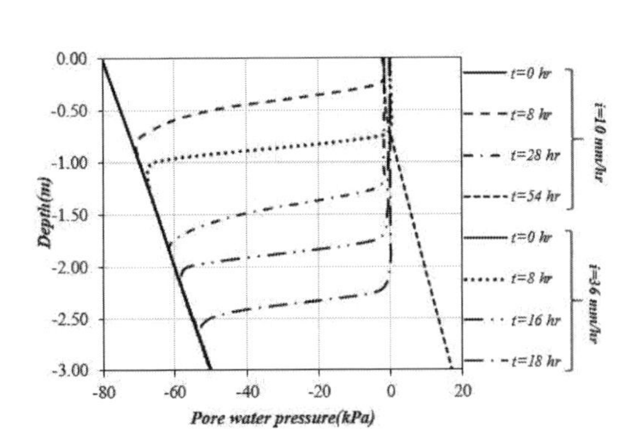

Figure 3. Pore water pressure profile under different rainfall intensity of soil type B.

5.2 *Rainfall thresholds for initiation of shallow slope failure (ID thresholds)*

Rainfall thresholds for initiation of shallow slope failure (ID thresholds) used for predicting shallow slope failure has been proposed by many previous researchers, for example Aleotti 2004, Caine 1980, Calcaterra et al. 2000, Cannon & Gartner 2005, Chien et al. 2005, Corominas 2000, Crosta & Frattini 2001, Guzzetti et al. 2007. The concept of ID thresholds developed based on historical slope failure data. Generally, the ID thresholds is established by plotting the relationship between rainfall intensity I_f versus duration D_f at failure states of slope on double logarithm relationship, which can be explained as:

$$I_f = a + cD_f^m \tag{5}$$

where: a, c, and m are model parameters, which represent the curvature, interception and gradient of ID threshold, respectively.

The model parameters for ID thresholds from many researchers have obtained by statistical approach, which is unable to interpret the effect of possible factors triggering rainfall-induced slope failures. This study attempts to establish the relationship and explain the effect of focused factors on ID thresholds based on finite element analysis. All cases of simulation were evaluated stability subjected to continuous rainfall until the slope failed. The D_f was collected, when slope failure is taken place.

Figure 4 shows ID thresholds for all simulation runs. It shows that the relationship between I_f and D_f stands on the same negative linear relationship on double logarithm scale for all of soil types.

The effect of periodical rainfall on ID thresholds is also presented in Figure 4. The time

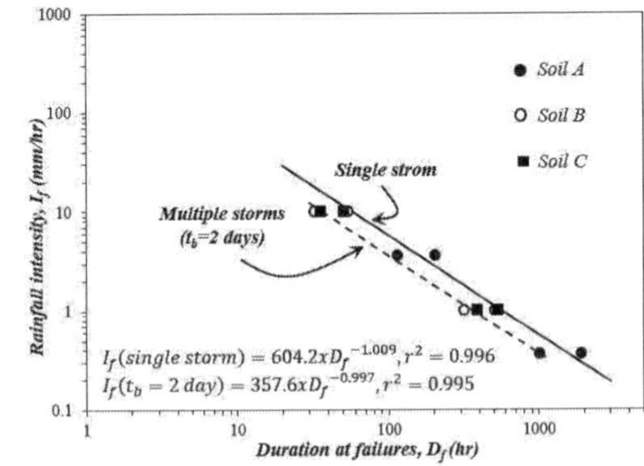

Figure 4. ID thresholds with varying in time between rainfall events.

between rainfall events was prescribed as 2 days. At the same rainfall intensity, the D_f decreases with short period between two rainfall events. In other word, the faster failure in shallow slope can be occurred, when the time between rainfall periods is shorter. And the effect of time between rainfall events results in the fitting parameter of ID thresholds, the parameter *c* increases with increasing in time between rainfall events, while the parameter *m* seems to be constant.

6 CONCLUSIONS

Some numerical investigations were carried out varying magnitude of rainfall intensity until the slope failure takes place. The conclusions in this study are drawn as:

1. Under a certain slope geometry, shallow slope failure can be triggered under either the rainfall infiltration or the rising of water table modes depending on the soil properties and rainfall intensity. The soil saturated permeability is one of the primary factored control the range of rainfall intensity. The lower rainfall intensity compared to the saturated permeability, the higher stability of the slope during infiltration stage, this is because of the remained matric suction, and hence the slope failure is possibly found during the rising of water table. For rainfall intensity equals to or greater than saturated permeability, the matric suction will completely disappear during infiltration stage, and hence the slope failure is possible found during the infiltration state.
2. The periodical rainfall plays an important role on failure state of shallow slope. The time to slope failure (D_f) subjected to assigned rainfall intensity (I_f) decreases with a short period between rainfall events. Furthermore, it also results in the fitting parameter representing interception of rainfall intensity-duration for initiation of shallow slope failures (ID thresholds).

REFERENCES

Aleotti P. 2004. A warning system for rainfall-induced shallow failures. *Engineering Geology* 73(3–4):247–265.

Bishop AW. & Blight GE. 1963. Some aspects of effective stress in saturated and partly saturated soils. *Géotechnique* 13(2):177–197.

Bordoni M., Meisina C., Valentino R, Bittelli M. & Chersich S. 2015. Site-specific to local-scale shallow landslides triggering zones assessment using TRIGRS. *Natural Hazards and Earth System Science* 15(5):1025–1050.

Brand E.W. 1984. State-of-the-art report of landslides in Southeast Asian. In*: 4th International Symposium on Landslides*, Toronto, Canada 1:377–384.

Caine N. 1980. The rainfall intensity-duration control of shallow landslides and debris flows. *Geografiska Annaler Series A-physical Geography* 62:23–27.

Calcaterra D, Parise M, Palma B. & Pelella L. 2000. The influence of meteoric events in triggering shallow landslides in pyroclastic deposits of Campania. *Proceedings 8th International Symposium on Landslides*, (Bromhead E, Dixon N, Ibsen ML, eds) Cardiff: AA Balkema 1:209–214.

Cannon S.H. & Gartner J.E. 2005. Wildfire-related debris flow from a hazards perspective. In: *Debris flow Hazards and Related Phenomena* (Jakob M, Hungr O, eds). Springer Berlin Heidelberg:363–385.

Chien Y.C., Tien C.C., Fan C.Y., Wen C.Y. & Chun C.T. 2005. Rainfall duration and debrisflow initiated studies for real-time monitoring. *Environmental Geology* 47:715–724.

Corominas J. 2000. Landslides and climate. Keynote lecture—In: *Proceedings 8th International Symposium on Landslides*, (Bromhead E, Dixon N, Ibsen ML, eds) Cardiff: AA Balkema 4:1–33.

Crosta G.B. & Frattini P. 2001. Rainfall thresholds for triggering soil slips and debris flow. In: *Proceedings 2nd EGS Plinius Conference on Mediterranean Storms* (Mugnai A, Guzzetti F, Roth G, eds) Siena:463–487.

Dahal R.K., Hasegawa S., Nonomura A., Yamanaka M., Masuda T. & Nishino K. 2008. Failure characteristics of rainfall-induced shallow landslides in granitic terrains of Shikoku Island of Japan. *Environmental Geology* 56(7):1295–1310.

Dai F., Lee C. & Wang S. 1999. Analysis of rainstorm-induced slide-debris flows on natural terrain of Lantau Island, Hong Kong. *Engineering Geology* 51(4):279–290.

Godt J.W. & McKenna J.P. 2008. Numerical modeling of rainfall thresholds for shallow landsliding in the Seattle, Washington, area. *Review in Engineering Geology* 20:121–136.

Green W.H. & Ampt C.A. 1911. Studies on soil physics: flow of air and water through soils. *The Journal of Agricultural Science* 4:1–24.

Guzzetti F., Peruccacci S., Rossi M. & Stark C.P. 2007. Rainfall thresholds for the initiation of landslides in central and southern Europe. *Meteorology and Atmospheric Physics* 98:239–267.

Hamdhan I.N. & Schweiger H.F. 2013. Finite Element Method–Based Analysis of an Unsaturated Soil Slope Subjected to Rainfall Infiltration. *International Journal of Geomechanics* 13(5):653–658.

Jotisankasa A. & Mairaing W. 2010. Suction-Monitored Direct Shear Testing of Residual Soils from Landslide-Prone Areas. *Journal of Geotechnical and Geoenvironmental Engineering* 136(3):533–537.

Jotisankasa A. & Vathananukij H. 2008. Investigation of soil moisture characteristics of landslide-prone slopes in Thailand. *Proceeding of International Conference on Management of Landslide Hazard in the Asia-Pacific Region, Sendai Japan:1–12.*

Keefer D.K., Wilson R.C., Mark R.K., Brabb E.E., Brown W.M.-I., Ellen S.D., Harp E.L., Wieczorek G.F., Alger C.S. & Zatkin R.S. 1987. Real-time

landslide warning during heavy rainfall. *Science* 238:921–925.
Mualem Y. 1976. A new model predicting the hydraulic conductivity of unsaturated porous media. *Water Resources Research* 12:513–522.
Oh S. & Lu N. 2015. Slope stability analysis under unsaturated conditions: Case studies of rainfall-induced failure of cut slopes. *Engineering Geology* 184:96–103.
Rahardjo H., Li X.W., Toll D.G. & Leong E.C. 2001. The effect of antecedent rainfall on slope stability. *Geotechnical and Geological Engineering* 19:371–399.
Rahardjo H., Nio A.S., Leong E.C. & Song N.Y. 2010. Effects of Groundwater Table Position and Soil Properties on Stability of Slope during Rainfall. *Journal of Geotechnical and Geoenvironmental Engineering* 136(11):1555–1564.
Rahardjo H., Ong T.H., Rezaur R.B. & Leong E.C. 2007. Factors controlling instability of homogeneous soil slopes under rainfall. *Journal of Geotechnical and Geoenvironmental Engineering* 133(12):1532–1543.
Rahimi A., Rahardjo H. & Leong E.-C. 2010. Effect of hydraulic properties of soil on rainfall-induced slope failure. *Engineering Geology* 114(3–4):135–143.
Rahimi A., Rahardjo H. & Leong E.C. 2011. Effect of Antecedent Rainfall Patterns on Rainfall-Induced Slope Failure. *Journal of Geotechnical and Geoenvironmental Engineering* 137(5):483–491.
Sirangelo B. & Braca G. 2004. Identification of hazard conditions for mudflow occurrence by hydrological model. Application of FLaIR model to Sarno warning system. *Engineering Geology* 73:267–276.
Sirangelo B., Versace P. & De Luca D.L. 2007. Rainfall nowcasting by at site stochastic model PRAISE. *Hydrology and Earth System Science* 11:1341–1351.
van Genuchten M.T. 1980. A closed-form equation for predicting the hydraulic conductivity of unsaturated soil. *Soil Science Society of American Journal* 44:615–628.
Versace P., Sirangelo B. & Capparelli G. 2003. Forewarning model of landslides triggered by rainfall. *3rd International Conference on Debris-flow Hazards Mitigation: Mechanics, Prediction and Assessmen*t, Davos, Switzerland 2:1233–1244.
Vieira B.C., Fernandes N.F. & Filho O.A. 2010. Shallow landslide prediction in the Serra do Mar, São Paulo, Brazil. *Natural Hazards and Earth System Science* 10(9):1829–1837.
Wilson R.C., Mark R.K. & Barbato G. 1993. Operation of a real-time warning system for debris flows in the San Francisco Bay area, California. *International Conference Hydraulics Division, ASCE 2*:1908–1913.

Numerical Methods in Geotechnical Engineering IX – Cardoso et al. (Eds)
© 2018 Taylor & Francis Group, London, ISBN 978-1-138-33203-4

Slope instabilities triggered by creep induced strength degradation

A. Kalos & M. Kavvadas
School of Civil Engineering, National Technical University of Athens, Greece

ABSTRACT: This paper investigates delayed slope failures in stiff structured clays under dry conditions, via 2D finite element analyses. The soil is described by a new rate-dependent constitutive model, which includes structure and its degradation by creep strains including post-peak strain softening by creep shear strains. The analyses show that in a slow creeping slope, the stress states are initially below the failure envelope and thus the slope is stable. However, in highly sheared zones, creep strains accumulate causing gradual strength degradation. As the process continues, the degrading envelope can reach the stress state in some locations and, then, in situ shear stresses decrease to be compatible with the ever-degrading failure envelope. Eventually, this condition occurs in a continuous zone and the slope fails due to tertiary creep. Tertiary failure under dry conditions has the form of an earthflow, as the soil mass above the slip surface suffers significant shear strains.

1 INTRODUCTION

Landslides involve the downward movement of significant ground volumes under gravity loads. Topography, geology, geotechnical conditions, prior loading history, exogenous processes due to late severe climatic changes and creep compound destabilizing agents endangering marginally stable natural slopes and man-made cuts. Mountainous terrains and hilly topographies have long been known to suffer from creep-induced landslides (i.e., Swanston and Swanson (1976), Tavenas and Leroueil (1981), Vulliet and Hutter (1988), Potts et al. (1990), Kováčik (1991), Chigira and Kiho (1994), Irfan (1994), Ellis and O'Brien (2007), De Vita et al. (2013), Angeli et al. (2016), Taboada et al. (2017)).

Delayed failure naturally occurs under high shear stress levels (i.e., Bishop (1966), Bishop and Lovenbury (1969)) due to tertiary creep-induced strength damage. Slow creep-induced landslides tend to accelerate when shear strength degrades to residual reaching a point where the slope cannot sustain its own stress field without failing.

Constitutive models including creep rupture, usually address only the undrained response of geomaterials (e.g. Sekiguchi (1984), Adachi et al. (1987), Yin and Graham (1999), Yin et al. (2010), Yin et al. (2011), Ou et al. (2011), Sivasithamparam et al. (2015), Jiang et al. (2016)). These models cannot predict long-term creep failures, as undrained conditions are not relevant in these cases. Constitutive models capable to describe tertiary creep rupture under fully drained and/or partially drained conditions cannot systematically reproduce primary, secondary and tertiary creep stages for all sustained levels of loading (e.g., Dragon and Mroz (1979), Al-Shamrani and Sture (1998), Shao et al. (2003), Pellet et al. (2005)), especially in cases where significant stress redistributions take place due to viscous phenomena.

This paper studies the capabilities of a new Time-dependent constitutive Model for cohesive Structured soils (TMS; Kalos (2014)) to successfully give creep-induced failures in dry conditions. The model combines the advanced structural components of inviscid soils with volumetric and shear viscous effects; it assumes that volumetric viscous strains build on strength due to the ageing process, while viscous shear strains damage shear strength by degrading the Strength Cone (S-Cone). The novel creep-induced damage mechanism proves sufficient in triggering slope failure under dry conditions in a marginally stable slope, without excluding cases where the slope safety factor is significantly higher.

2 TMS MODEL FOR DRAINED CREEP RUPTURE

The TMS model is an incremental plasticity rate-dependent model for cohesive structured soils giving tertiary creep rupture in drained conditions. The extension of inviscid elastoplasticity to account for rate effects imposes the dependence of the time scale on the stress state.

Next follows a brief description of the TMS model, where "dot" over symbols to indicate time-rate, while differential symbols "*d*(..)" denote small (infinitesimal) increments of the corresponding quantities. The model is described via effective stresses and all "prime" symbols (indicating effective stresses) have been dropped for simplicity. Bold-face symbols indicate tensorial quantities, like the effective stress tensor ($\boldsymbol{\sigma}$) and the stress deviator (s) which is defined as: $\boldsymbol{s} = \boldsymbol{\sigma} - \sigma \mathbf{I}$, where $\sigma \equiv p = \frac{1}{3}\, \boldsymbol{\sigma}{:}\mathbf{I}$ is the mean effective stress and (I) is the unit isotropic tensor. The symbol ":" denotes the scalar inner product of two tensorial quantities.

The infinitesimal strain increments ($d\boldsymbol{\varepsilon}$) during a short-time interval (dt) consists of an elastic (reversible $d\boldsymbol{\varepsilon}^e$) and an inelastic part (irreversible $d\boldsymbol{\varepsilon}^i$), which translates to $d\boldsymbol{\varepsilon}=d\boldsymbol{\varepsilon}^e+d\boldsymbol{\varepsilon}^i$; the irreversible strain increment is compounded of a plastic ($d\boldsymbol{\varepsilon}^p$) and a viscous ($d\boldsymbol{\varepsilon}^v = \dot{\boldsymbol{\varepsilon}}^v\, dt$) component, expressed in the form: $d\boldsymbol{\varepsilon}^i = d\boldsymbol{\varepsilon}^p + \dot{\boldsymbol{\varepsilon}}^v\, dt$. The model works on an incrementally linear isotropic poro-elasticity, which associates the elastic strain increment ($d\boldsymbol{\varepsilon}^e$) with the respective causal effective stress increment ($d\boldsymbol{\sigma}$) via the tangent elastic stiffness tensor ($\mathbf{C}^e$), $d\boldsymbol{\sigma} = \mathbf{C}^e : d\boldsymbol{\varepsilon}^e$.

The TMS model uses three characteristic surfaces in the effective stress hyperspace consisting of the isotropic axis (σ) and the deviatoric hyperplane ($\boldsymbol{s}$):

i. the internal Plastic Yield Envelope (PYE). Stress paths inside it (reversible paths) cause elastic (reversible) strains only, while stress paths originating inside the PYE reaching the PYE, either move carrying the PYE with it (irreversible path) or retreat from it moving inside the PYE (reversible path);
ii. the external Structure Envelope (SE) bounding the PYE and all accessible structured states. Stress (irreversible) paths reaching the SE cause the PYE to align with the SE at the conjugate stress point and the stress state to either move carrying the SE and PYE with it (irreversible path) or retreat from the SE and the PYE moving inwards (reversible path); and
iii. the Intrinsic Envelope (IE) which corresponds to an "equivalent" intrinsic (structureless) material state with same mean effective stress (σ) and void ratio (e) as the current structured state.

The Structure Envelope (SE) is an ellipsoidal surface similar in shape to the yield surface of the well-known Modified Cam-Clay model (Roscoe and Burland (1968)), given in the following formula:

$$F(\boldsymbol{\sigma},a,d,c) \equiv \frac{1}{c^2}\boldsymbol{s}:\boldsymbol{s} + (\sigma - a + d)^2 - a^2 = 0 \qquad (1)$$

The SE is centered on the isotropic axis at $(a-d)\mathbf{I}$ from the stress origin to account for the tensile strength of the geomaterial due to structure. Parameters (a) and ($c\ a$) are the half-sizes of the SE ellipse along the isotropic and deviatoric axes respectively (c is the SE eccentricity parameter). The Plastic Yield Envelope is analogous in shape to the Structure Envelope scaled by a constant similarity ratio ξ and is enclosed within (or tangent to) the SE at all times and stress states.

The Intrinsic Strength Envelope (IE) is an ellipsoidal surface described by a function analogous to the SE, centered at a^* along the isotropic axis, with intrinsic eccentricity (c^*) and null tensile strength:

$$F^*(\boldsymbol{\sigma},a^*,c^*) \equiv \frac{1}{(c^*)^2}\boldsymbol{s}:\boldsymbol{s} + (\sigma - a^*)^2 - (a^*)^2 = 0 \quad (2)$$

The half-size a^* of the IE depends on mean stress (σ) and specific volume ($V = 1 + e$) in a way similar to the Modified Cam-Clay model. The graphical representation of the implemented characteristic surfaces (PYE, SE, IE) and associated strength cones (S-Cone, CS-Cone) are shown in Figure1.

The viscous strain component is compounded of a volumetric and a shear part, similar to the definitions of Kavazanjian and Mitchell (1977) and Hsieh et al. (1990):

$$\boldsymbol{\varepsilon}^v = \sum d\boldsymbol{\varepsilon}^v = \sum \dot{\boldsymbol{\varepsilon}}^v\, dt = \sum \left(\tfrac{1}{3}\dot{\varepsilon}^v \mathbf{I} + \tfrac{\sqrt{6}}{2}\dot{\varepsilon}^v_q\, \boldsymbol{s}/\sqrt{\boldsymbol{s}:\boldsymbol{s}} \right) dt \, (3)$$

The TMS model uses the classical semi-logarithmic creep law to describe the volumetric creep shear strain-rate:

$$\dot{\varepsilon}^v = \frac{\psi}{V\, t_0} \exp\left(-\frac{V}{\psi}\varepsilon^v \right) \qquad (4)$$

where the material constant $\psi = C_{ae}/ln10$ is a function of the classical secondary compression index Cae. Reference time t_0 denotes the start of creep, which gives the validity for expression (4) ($t \geq t_0$).

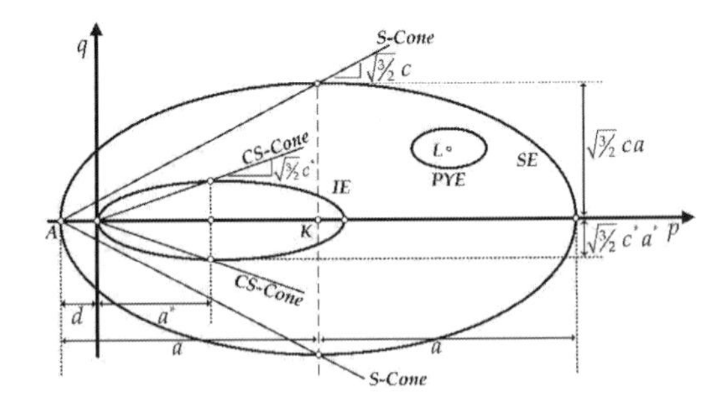

Figure 1. Characteristic surfaces of the TMS model (PYE, SE, IE) and associated strength cones (S-Cone, CS-Cone) in the p–q space.

The ageing process (Bjerrum (1967)) is included through the increase of the SE half-size a with viscous volumetric strains. A slightly modified Singh-Mitchell formula (Mitchell et al. (1968)) describes the viscous shear strain-rate:

$$\dot{\varepsilon}_q^v = \frac{2A\sinh(\bar{a}\,D^r)}{\left[1+\dfrac{(1-m)\,\varepsilon_q^v}{2A\sinh(\bar{a}\,D^r)\,t_0}\right]^{\frac{m}{1-m}}} \quad for \quad m \neq 1 \quad (5a)$$

$$\dot{\varepsilon}_q^v = \frac{2A\sinh(\bar{a}\,D^r)}{\exp\left[\dfrac{1}{2A\sinh(\bar{a}\,D^r)\,t_0}\,\varepsilon_q^v\right]} \quad for \quad m = 1 \quad (5b)$$

where $D = q/q_F$ is the shear stress level defined as the ratio of shear stress "q" over the shear strength "q_F", A, m and $\bar{a}$ are the material constants determined experimentally (Singh and Mitchell (1968)). The formulae above give a viscous shear strain-rate: (i) tending to zero at large times, at faster rates as (m) increases and (D) decreases; and (ii) increasing with increasing shear stress level (D) at a specific time, becoming maximum at failure $D = 1$. The Singh-Mitchell formulae address primarily the stationary creep phase and secondarily the primary. Tertiary creep is not predicted through the Singh-Mitchell creep law and requires advanced constitutive numerical tools associated with the creep-induced damage of the Strength Cone. The material constant "r" is a weighting factor giving higher shear viscous strains at stress levels close to failure (where $D = 1$) for $r > 1$. For $r = 1$ the formulae above reduce to the well-known Singh-Mitchell creep model expressions.

As the material state may change without changes of stress and strain due to creep, stresses and strains do not complete the full spectrum of state variables. The hardening variables compound state variables controlling the shape and position of the characteristic surfaces. The evolution of hardening variables is described through hardening rules working on irreversible straining, both viscous and plastic. The TMS uses: (i) an isotropic hardening rule controlling the half-size a of the SE; and (ii) three kinematic hardening rules controlling the tensional transpose of the SE to tensile strengths of the structured geomaterial, the position of the PYE inside the SE and the eccentricity of the SE.

The half-size a of the SE degrades towards the intrinsic half-size a^* of the IE with plastic straining. The isotropic hardening law works on the structure ratio $B = a/a^* \geq 1$ defined as the ratio of the SE half-size over the IE half-size, with B degrading exponentially from an original B_o to the residual B_{res} with plastic strains; the residual value B_{res} may reach unity depending on the sustained microstructural residual bonding ($B_o \geq B_{res} \geq 1$). The TMS model reduces to the well-known Modified Cam-Clay model when the structure is eliminated $B_o = B_{res} = 1$ and the similarity ratio $\xi = 1$ setting the PYE to coincide with the SE. The IE half-size a^* may shrink or expand with plastic strains in a way similar to the Modified Cam-Clay, by thus affecting the half-size of the SE a. The ageing process is included in the time-dependent formulation through the increase of the IE half-size a^* with volumetric viscous strains.

The TMS model uses three kinematic hardening rules to further control the evolution of the S-Cone "c", the tensional transpose "d" of the SE to the tensile regime to account for the effect of structure and the secondary anisotropy through the location of the PYE within the SE. The framework ensures uniqueness of the Critical State at all times, reached at large shear strains where the material structure has been fully degraded to the intrinsic, while additional deformation occurs at a constant volume and constant mean effective stress (only shear strains continue to build).

The SE eccentricity parameter c degradation from its initial short-term value c_{in} towards its unique critical state value c^* is controlled by a kinematic hardening law. The hardening rule uses an exponentially decaying formula degrading the S-Cone to the CS-Cone with irreversible strains (both plastic and viscous). This transition of the S-Cone due to viscous strains causes stress states lying above the CS-Cone to sustain failure as the SE elongates (potentially also causing hardening or softening) until the material cannot withstand its own stress field. Stress states below the CS-Cone may sustain only primary and secondary creep without failing. This creep-induced damage mechanism can successfully emulate tertiary creep rupture in drained conditions and acts supplementary to the classical structure degradation mechanism.

The degradation of the tensile strength, relayed through the tensional transpose "d" of the SE along the isotropic axis from its initial value d_{in} to zero, is controlled by another exponentially decaying hardening rule working on irreversible shear straining. The final kinematic hardening law controls the movement of the PYE towards its conjugate point on the SE. The TMS model uses the Kavvadas and Amorosi (2000) hardening rule to control this secondary anisotropy which ensures that the PYE and the SE align upon a unique tangent point when the stress state reaches the SE. At that point forth, additional loading carries the SE and the PYE aligned towards their new position due to sustained isotropic and/or kinematic hardening.

Incremental visco-plasticity similarly to classical incremental plasticity, uses a flow rule to give the magnitude and direction of the plastic strain increment $d\boldsymbol{\varepsilon}^p = d\Lambda\, \boldsymbol{P}$, where $d\Lambda$ is the plastic multiplier and $\boldsymbol{P}$ is the plastic potential assumed equal to the gradient $\boldsymbol{Q}$ of the PYE, as the TMS works on an associated flow rule. The complete definitions of all constitutive equations can be found in Kavvadas and Kalos (2017).

The next section studies the capabilities of the model to give slope failure by means of the creep-induced damage mechanism under drained conditions.

3 DRY CREEP-INDUCED SLOPE FAILURE

Slopes failing after considerable times without apparent reason associated with annual evapotranspiration cycles and/or man-made acts degrading the factor of safety (i.e., through excavation of the toe) are mainly attributed to creep. Such delayed slope failures in structured cohesive soils are usually manifested in distinct failure surfaces. However, as the presence of water combined with the low permeability of such soils generates substantial interparticle excess porewater pressures this study focuses on the capabilities of the TMS creep-induced damage mechanism to trigger slope instability in "dry" conditions. Dry conditions aim to deconvolve the generation of excess pore-pressures due to irreversible straining, which gives further inelastic straining on dissipation, from the effect of the viscous damage mechanism during the pre-failure stage towards the full activation of the failure mechanism.

The TMS model was implemented in the commercial Finite Element code SIMULIA ABAQUS via an external User MATerial subroutine (UMAT). Two-dimensional (2D) numerical analyses of a theoretical slope were performed using 8-noded continuum plane strain elements (CPE8). The investigated slope gives a factor of safety equal to 1.10 for Mohr-Coulomb parameters c = 15 kPa (cohesion) and ϕ = 25° (effective friction angle). The slope is 30 m high inclined at 33.5° from the horizontal. All lateral boundaries have been placed 90 m away from both the crest and the toe to ensure minimal boundary effects, while the bedrock (bottom constrains) has been placed 30 m below the toe level to allow for possible activation of deeper failure surfaces. The slope is formed via excavation of an initially levelled soil deposit. After all stresses have reached an equilibrium (at time t_o = 1.1 days), creep is introduced until the slope fails. The creep-induced damage mechanism giving slope failure in drained conditions degrades the SE eccentricity parameter "c" from c_{in} = 0.9 (corresponding to a friction angle of ϕ = 28°, slightly higher than ϕ = 25° to ensure that possible low over-consolidated states are enclosed within the SE) to c^* = 0.46 (for a friction angle of ϕ = 15°).

Figure 2 shows the total displacement magnitude U (in m) contour plots at failure, after time $t_F/t_0 \cong 204$. Displacement profiles are also shown at four locations: (a) at the CREST; at points (b) CR10 and (c) CR20 located at a horizontal distance of 10 m and 20 m respectively from the CREST along the ridgeline; and (d) at point CL10 located on the slope at a horizontal distance of 10 m downhill. Numerical analysis gives maximum displacement is 0.675 m at the CREST, while locations extending radially from it experience lower displacement levels with increasing distance. The slip surface is shown in the dark grey line in Figure 2 through points A, B, C and D. Accompanying set of points E-F-G-H serves the purpose of ensuring the validity of the proposed TMS framework as the stress states need to lie above the CS-Cone for the model to give tertiary creep rupture. The constitutive parameters used in the numerical analyses are summarized in Table 1.

The shear plastic (denoted SDV5) and viscous (denoted SDV7) irreversible strains are depicted via contour plots in Figures 3 and 4 at failure. As the plastic shear strains are minimal, it is the viscous shear part that leads to failure by means of the proposed creep-induced damage mechanism. The accumulation of the shear viscous strain measure damages the SE eccentricity parameter c causing the S-Cone to degrade, by thus increasing the stress level D. This transition of the S-Cone towards the CS-Cone causes the viscous part to increase further setting the stress state closer to the CS-Cone, which signifies failure.

The total displacement magnitude time histories are shown in Figure 5. The majority of surficial time histories (at signified points CREST, CL10,

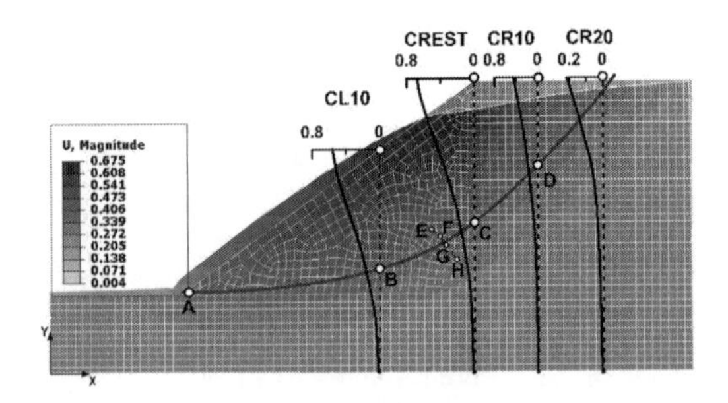

Figure 2. Contour plots of total displacement magnitude U (in m) at failure (normalized failure time $t_F/t_0 \cong 204$) and total displacement profiles with depth at four locations: (a) at the CREST; at points (b) CR10 and (c) CR20 located 10 m and 20 m respectively away from the crest along the ridgeline; and at point CL10 located at a horizontal distance of 10 m downslope.

CR10) show significant increase with time, while point CR20 lying at the close proximity of the slip surface gives a more gradual increase characteristic of stationary creep. Points A-B-D on the slip surface show a rather ductile response with depth, with C experiencing higher displacement levels as it lies closer to the CREST. Figure 5 shows no evidence of a curvature transition point. This suggest that the slope experiences ductile failure resembling somewhat of an earthflow as the entire region above the slip surface deforms, while flowing on the slip surface. The initially slow creeping slope transforms at around $t/t_0 \cong 22$ to a creep-induced landslide.

Table 1. TMS constitutive model parameters.

N_{iso}^*	$B_0 = B_{res}$	c_{in}	c^*	OCR	ν	λ
3.6	1	0.909	0.463	3	0.25	0.10
κ	$\eta_v^p = \eta_q^p$	δ	γ	d_{in} (kPa)	r	ξ
0.05	150	5	1	10	3.6	1%
t_0 (days)	$\theta_q^p = \vartheta_q^p$	$a_1^v = a_2^v$	ψ	A	m	$\bar{a}$
1.1	0	200	$3\cdot10^{-5}$	$4.8\cdot10^{-3}$	0.85	3.6

Figure 3. Plastic shear strain contour plot (denoted SDV5) at failure (normalized failure time $t_F/t_0 \cong 204$).

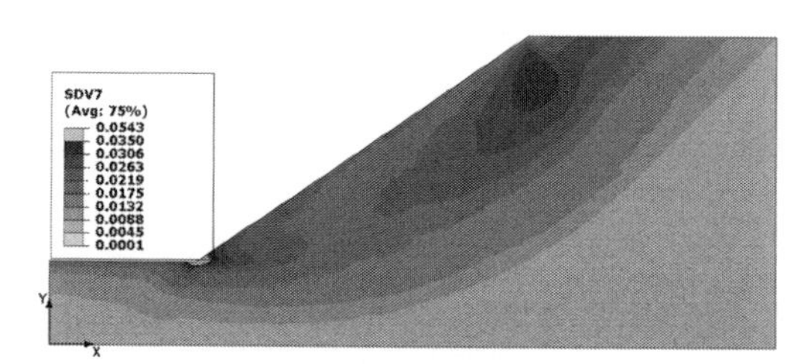

Figure 4. Viscous shear strain contour plot (denoted SDV7) at failure (normalized failure time tF/t0≅204).

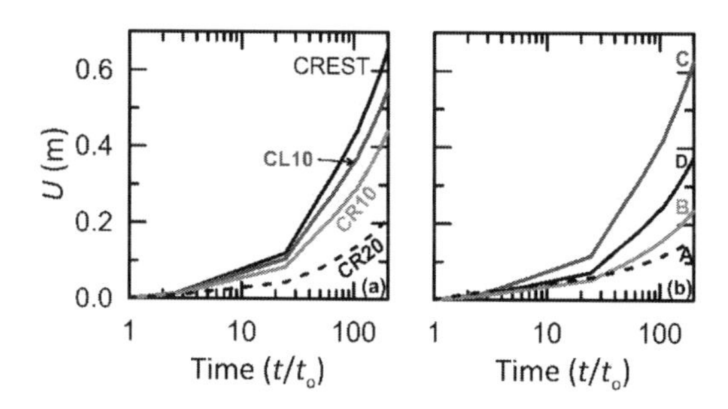

Figure 5. Displacement magnitude time histories: (a) at the surficial points CREST, CR10, CR20 and CL10 and (b) at points A, B, C and D on the slip surface (shown in Figure 2).

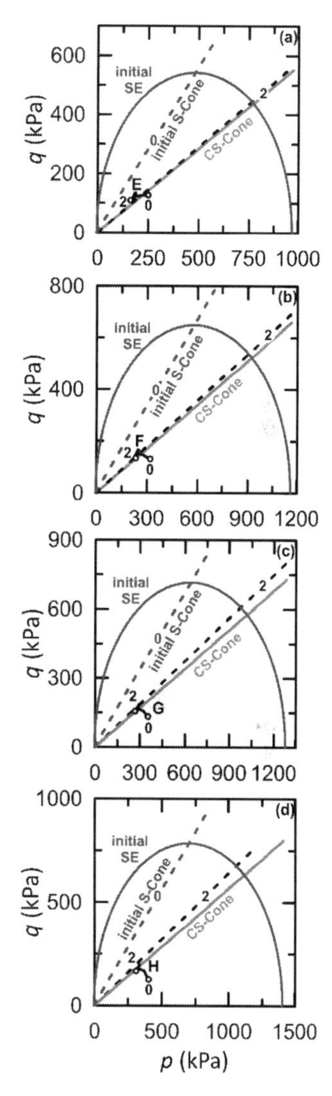

Figure 6. Stress paths at points (a) E, (b) F, (c) G and (d) H depicted in black solid lines (identified in Figure 2). The stress paths start at 0 and fail at 2. The initial S-Cone, the failure S-Cone (denoted 2) and the CS-Cone are shown in dotted dark-grey, dotted black and solid light-grey lines respectively. The Figure also shows the initial SE in the solid dark-grey line.

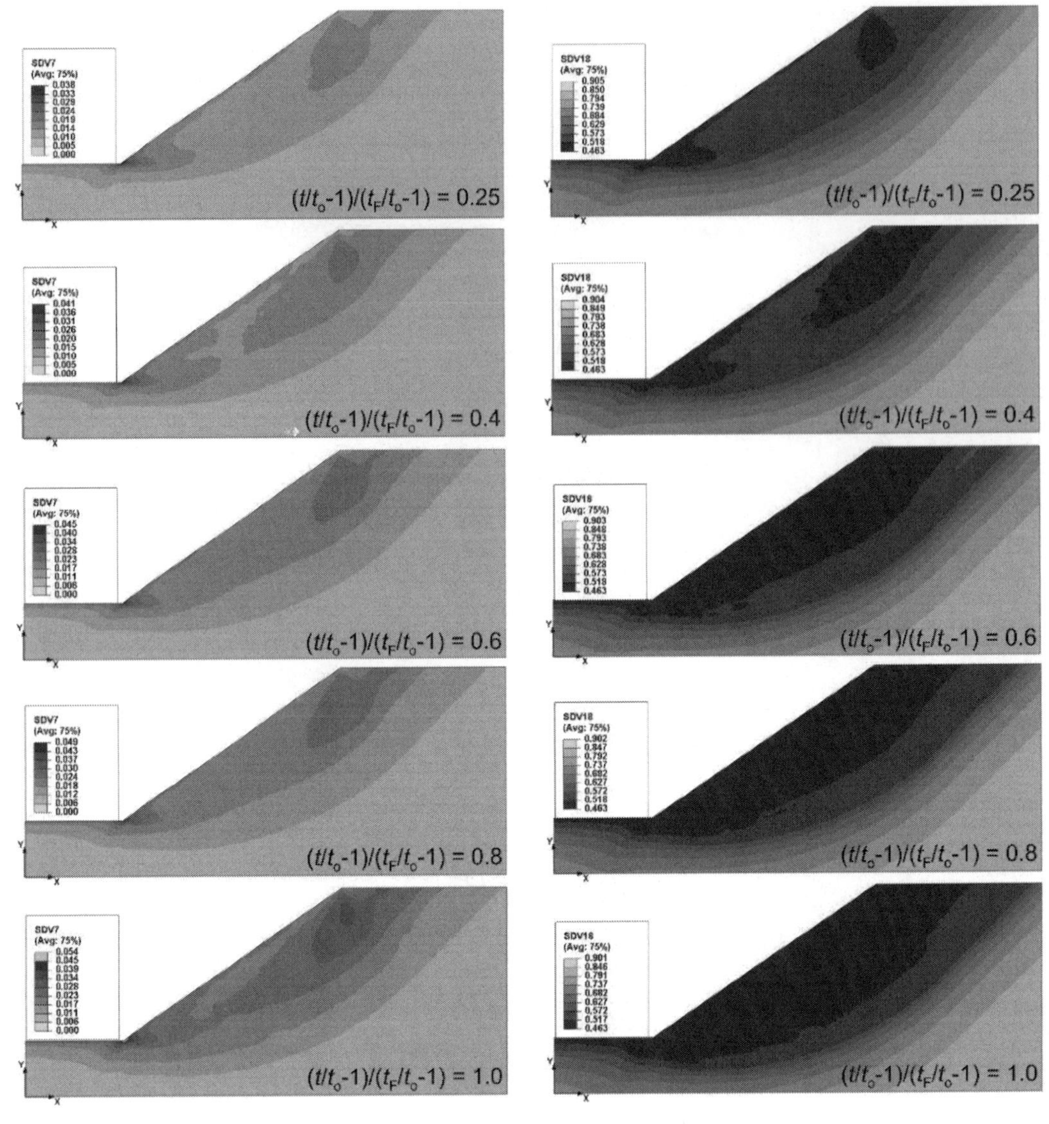

Figure 7. Evolution of the deviatoric viscous strain measure (denoted SDV7) with normalized time $[(t - t_0)/(t_F - t_0)]$.

Figure 8. Evolution of the SE eccentricity parameter "c" (denoted SDV18) with normalized time $[(t - t_0)/(t_F - t_0)]$.

The stress paths at a line perpendicular to the slip surface of Figure 2 (denoted as E-F-G-H) are shown in Figure 6. The stress paths originate at "0" (at time $t = t_0$) and fail at "2" (at time $t = t_F$). The initial S-Cone is shown in the dotted dark-grey line crossing the initial SE at its peaks. The S-Cone at the time of failure (denoted "2") is shown in the dotted black line while the CS-Cone is depicted in the solid light-grey line. All stresses (at the time of failure "2") lie higher than the CS-Cone proving that failure occurs by means of the novel creep-induced damage mechanism. As the S-Cone degrades towards the CS-Cone the stress state moves closer towards the current S-Cone ultimately placing stress on it. As further time passes the stationary stress state cannot sustain the stress drop necessary to reach equilibrium, which signifies failure.

Figure 7 to 9 show the evolution of the deviatoric plastic strains (Figure 7) damaging the SE eccentricity parameter c (Figure 8) which inflates the total displacement magnitude (Figure9). The evolution by means of contour plots at different

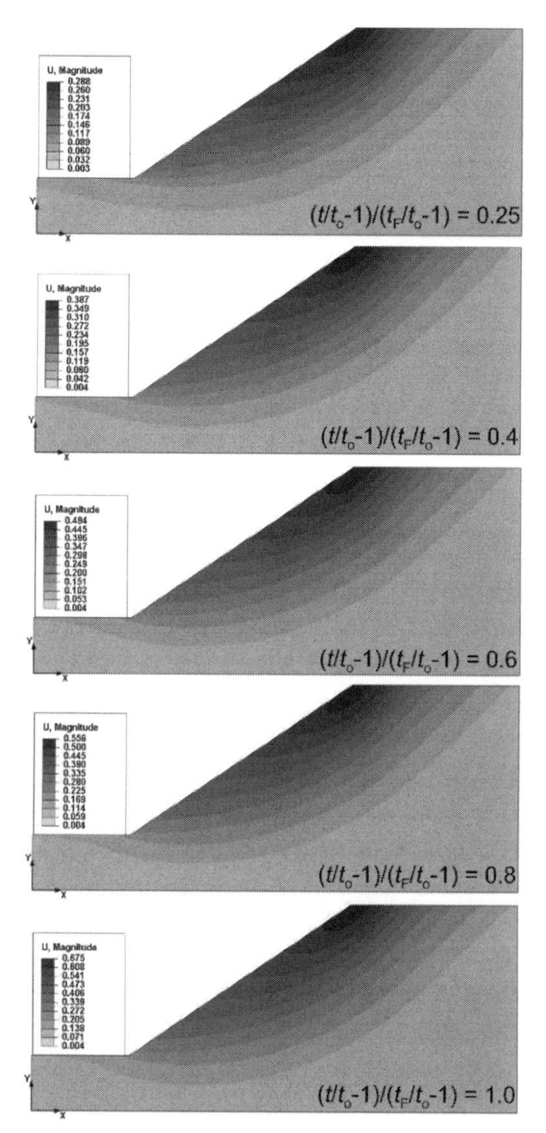

Figure 9. Evolution of the total displacement magnitude U (in m) with normalized time $[(t - t_0)/(t_F - t_0)]$.

normalized times. The selected time scale $(t - t_0)/(t_F - t_0)$ gives unity at failure where $t = t_F$ and 0 at the time origin $(t = t_0)$. Figure 7 shows that even from an early normalized time as $(t - t_0)/(t_F - t_0) = 0.25$ the inflated viscous shear strain around the toe and at the crest cause the SE eccentricity parameter to degrade close to its critical value c^* (Fig. 8), giving a maximum magnitude of 0.288 m at the crest. As time elapses viscous strains accumulate causing further damage to the SE eccentricity c. At normalized time $(t - t_0)/(t_F - t_0) = 0.4$ the branch originating from the crest propagating downwards appears to extend towards the toe, until it unites (at time $(t - t_0)/(t_F - t_0) = 0.6$) with the secondary segment which originates at the toe and propagates backwards and uphill. After that time the degraded zone continues to extend downwards reaching its final settlement. This progressive failure mechanism causes the displacements to inflate as it gives creep-induced slope failures by means of an earthflow.

4 CONCLUSIONS

Landslides can undeniably damage public and private infrastructure, lifelines, endanger economic growth and sustainability and even severely harm marginally stable ecosystems. Delayed slope instabilities failing after considerable times without apparent reasoning are attributed to creep. This paper investigates such slope failures and their underlying mechanisms. The study uses the novel TMS model giving drained creep rupture to describe the time-dependent response of structured cohesive soils.

The TMS is an incremental plasticity time-dependent constitutive model incorporating principles of critical state soil plasticity and represents a major advancement over the existing set of advanced constitutive time-dependent frameworks, as it can systematically emulate the primary, secondary and tertiary creep phases under increasing stress levels both in drained and undrained conditions. This feature stems from the incorporated creep-induced damage mechanism degrading the S-Cone towards critical with accumulating irreversible viscous shear strains. This mechanism is supplementary to the classical structure degradation damage giving stress rebound.

The creep-induced damage mechanism proved sufficient in giving slope instability under drained conditions. The progressive failure mechanism propagates from the crest downwards to unite with a secondary segment originating from the toe. Failure is manifested by means of a flowing soil mass moving along a slip surface, which is indicative of an earthflow rather than a classical distinct failure surface where the soil mass moves as an un-deformed rigid block downwards due to gravitational body forces.

ACKNOWLEDGEMENTS

This research was funded by the State Scholarships Foundation (IKY) under the "IKY Fellowships of Excellence for Postgraduate Studies in Greece – Siemens Program" in the framework of the Hellenic Republic – Siemens Settlement Agreement.

REFERENCES

Adachi, T., Oka, F. & Mimura, M. 1987. Mathematical structure of an overstress elasto-viscoplastic model for clay. Soils and Foundations, 27, 31–42.

Al-Shamrani, M. A. & Sture, S. 1998. A time-dependent bounding surface model for anisotropic cohesive soils. Soils and Foundations, 38, 61–76.

Angeli, M., Bromhead, E., Gasparetto, P., Marabini, F. & Pontoni, F. Stop-start landslides and the creep phenomena. Landslides and Engineered Slopes. Experience, Theory and Practice: Proceedings of the 12th International Symposium on Landslides (Napoli, Italy, 12–19 June 2016), 2016. CRC Press, 325.

Bishop, A. W. 1966. The Strength of Soils as Engineering Materials. Géotechnique, 16, 91–130.

Bishop, A. W. & Lovenbury, H. T. Creep characteristics of two undisturbed clays. Proceedings of 7th International Conference of Soil Mechanics and Foundation Engineering, 1969. 29–37.

Bjerrum, L. 1967. Engineering geology of Norwegian normally-consolidated marine clays as related to settlements of buildings. Geotechnique, 17, 83–118.

Chigira, M. & Kiho, K. 1994. Deep-seated rockslide-avalanches preceded by mass rock creep of sedimentary rocks in the Akaishi Mountains, central Ja-pan. Engineering Geology, 38, 221–230.

De Vita, P., Carratù, M., La Barbera, G. & Santoro, S. 2013. Kinematics and geological constraints of the slow-moving Pisciotta rock slide (southern Ita-ly). Geomorphology, 201, 415–429.

Dragon, A. & Mroz, Z. A model for plastic creep of rock-like materials accounting for the kinetics of fracture. International Journal of Rock Mechanics and Mining Sciences & Geomechanics Abstracts, 1979. Elsevier, 253–259.

Ellis, E. & O'brien, A. 2007. Effect of height on delayed collapse of cuttings in stiff clay. Proceedings of the Institution of Civil Engineers-Geotechnical Engineering, 160, 73–84.

Hsieh, H., Kavazanjian Jr, E. & Borja, R. 1990. Double-yield-surface Cam-clay plasticity model. I: theory. Journal of Geotechnical Engineering, 116, 1381–1401.

Irfan, T. Y. 1994. Mechanism of creep in a volcanic saprolite. Quarterly Journal of Engineering Geology and Hydrogeology, 27, 211–230.

Jiang, J., Ling, H. I., Kaliakin, V. N., Zeng, X. & Hung, C. 2016. Evaluation of an anisotropic elastoplastic–viscoplastic bounding surface model for clays. Acta Geotechnica, 1–14.

Kalos, A. 2014. Investigation of the nonlinear time-dependent soil behavior. PhD, School of Civil Engineering, National Technical University of Athens

Kavazanjian, E. & Mitchell, J. K. A general stress-strain-time formulation for soils. Proceedings of 9th International Conference on Soil Mechanics and Foundation Engineering. Tokyo: Japanese Society of Soil Mechanics and Foundation Engineering, 1977. 113–120.

Kavvadas, M. & Amorosi, A. 2000. A constitutive model for structured soils. Geotechnique, 50, 263–273.

Kavvadas, M. & Kalos, A. 2017. A Time-dependent plasticity Model for Structured soils (TMS) simulating drained tertiary creep. Computers & Geotechnics Journal, (under review).

Kováčik, M. 1991. Slope deformations in the flysch strata of the West Carpathians. Landslides—Glissements de Terrain, 1, 139–144.

Mitchell, J. K., Campanella, R. G. & Singh, A. 1968. Soil creep as a rate process.

Ou, C., Liu, C. & Chin, C. 2011. Anisotropic visco-plastic modeling of rate-dependent behavior of clay. International Journal for Numerical and Analyti-cal Methods in Geomechanics, 35, 1189–1206.

Pellet, F., Hajdu, A., Deleruyelle, F. & Besnus, F. 2005. A viscoplastic model including anisotropic damage for the time dependent behaviour of rock. In-ternational journal for numerical and analytical methods in geomechanics, 29, 941–970.

Potts, D., Dounias, G. & Vaughan, P. 1990. Finite element analysis of progressive failure of Carsington embankment. Géotechnique, 40, 79–101.

Roscoe, K. H. & Burland, J. B. 1968. On the generalized stress-strain behaviour of wet clay. Engineering Plasticity, Cambridge University Press, Cam-bridge, pp. 535–609.

Sekiguchi, H. 1984. Theory of undrained creep rupture of normally consolidated clay based on elasto-visco-plasticity. Soils and Foundations, 24, 129–147.

Shao, J.-F., Zhu, Q. & Su, K. 2003. Modeling of creep in rock materials in terms of material degradation. Computers and Geotechnics, 30, 549–555.

Singh, A. & Mitchell, J. K. 1968. General stress-strain-time function for soils. Journal of Soil Mechanics & Foundations Division, 21–46.

Sivasithamparam, N., Karstunen, M. & Bonnier, P. 2015. Modelling creep behaviour of anisotropic soft soils. Computers and Geotechnics, 69, 46–57.

Swanston, D. N. & Swanson, F. J. 1976. Timber harvesting, mass erosion, and steepland forest geomorphology in the Pacific Northwest. Geomorphology and engineering, 4, 199–221.

Taboada, A., Ginouvez, H., Renouf, M. & Azemard, P. Landsliding generated by thermomechanical interactions between rock columns and wedging blocks: Study case from the Larzac Plateau (Southern France). EPJ Web of Conferences, 2017. EDP Sciences, 14012.

Tavenas, F. & Leroueil, S. 1981. Creep and failure of slopes in clays. Canadian Geotechnical Journal, 18, 106–120.

Vulliet, L. & Hutter, K. 1988. Viscous-type sliding laws for landslides. Canadian Geotechnical Journal, 25, 467–477.

Yin, J.-H. & Graham, J. 1999. Elastic viscoplastic modelling of the time-dependent stress-strain behaviour of soils. Canadian Geotechnical Journal, 36, 736–745.

Yin, Z.-Y., Chang, C. S., Karstunen, M. & Hicher, P.-Y. 2010. An anisotropic elastic–viscoplastic model for soft clays. International Journal of Solids and Structures, 47, 665–677.

Yin, Z.-Y., Karstunen, M., Chang, C. S., Koskinen, M. & Lojander, M. 2011. Modeling time-dependent behavior of soft sensitive clay. Journal of Geotechnical and Geoenvironmental Engineering, 137, 1103–1113.

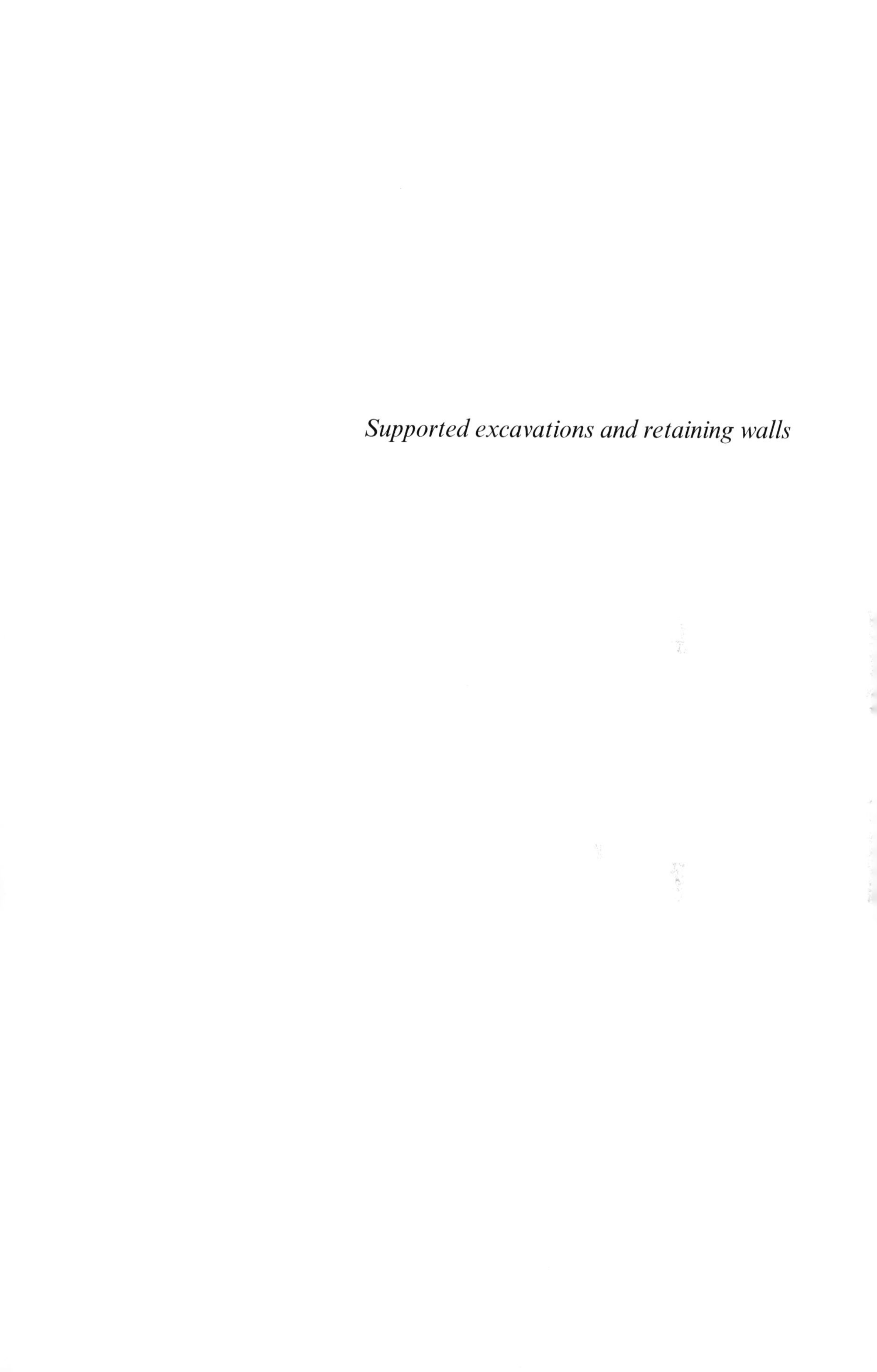

Supported excavations and retaining walls

Numerical Methods in Geotechnical Engineering IX – Cardoso et al. (Eds)
© 2018 Taylor & Francis Group, London, *ISBN 978-1-138-33203-4*

Numerical analyses of a pile wall at the toe of a natural slope

J. Castro, J. Cañizal, A. Da Costa, M. Miranda & C. Sagaseta
University of Cantabria, Santander, Spain

J. Casanueva
CMC Ingenieros, Santander, Spain

ABSTRACT: The construction of an industrial unit required a 10-meter-deep excavation at the toe of a steep natural slope with signs of previous instabilities. The adopted solution was a soldier pile wall with several levels of ground anchors. On the other hand, the upper natural slope was reinforced using horizontal drains, driven steel piles (rail segments) and ground anchors. Bolted steel meshes were also used to avoid shallow instabilities. Limit equilibrium and finite element analyses were performed for the design of those remediation measures. Finite element analyses provided useful information about bending moments in the piles and anchor forces. Detailed instrumentation, namely inclinometers, anchor load cells and topographical surveying, was used to monitor the performance of the wall and the slope. The field measurements were compared with the finite element predictions along all the construction period, showing a general good agreement.

1 INTRODUCTION

The construction of an industrial unit required a 10-meter-deep excavation at the toe of a steep natural slope with signs of previous instabilities. This required the design and construction of retaining structures, as well as an intensive monitoring of the slope and the surrounding structures.

The industrial unit, 280 m by 90 m, with the long sides in NW-SE direction, is located between the toe of the slope and a river (Fig. 1). On the northeast part, a building with basement was constructed. It was in that particular area, where the deeper excavations were needed. To this end, a soldier pile wall with variable height and characteristics was built. Different wall profile types were identified and labelled (Figs 1 and 2). In Zones 3 and 3', maximum excavation depth was 8.5 m, pile diameter was 1 m and center-to-center pile spacing was 1.2 and 1.1 m, respectively.

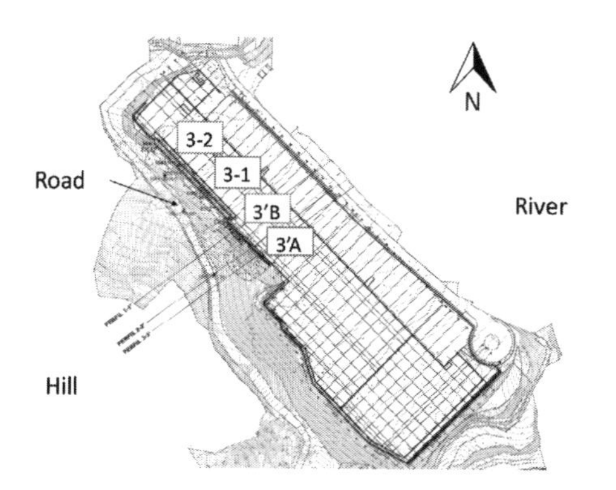

Figure 1. Top view of the industrial unit and the different zones of the pile wall under consideration.

The signs of slope instability existing before and during the construction process are described in Section 2. The geotechnical characterization of the area is presented in Section 3. Limit state analyses and the corrective measures adopted are shown in Section 4. Finite element calculations are exposed in Section 5. Finally, monitoring results are depicted in Section 6 and some conclusions are presented.

2 PREVIOUS SIGNS OF SLOPE INSTABILITY

Levelling works were performed during the development of the industrial estate and prior to the construction of the unit. As part of these works, a riprap wall was constructed against the existent slope on the northeast of the plot (Pile wall zones 3-1 and 3-2) (Fig. 2). Two major landslides occurred close to this wall in December 2010 (Fig. 3a) and in March

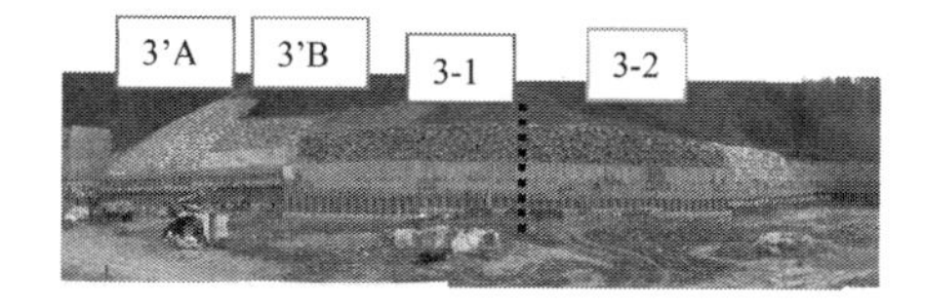

Figure 2. General view of the pile wall. Critical zone in April 2014, before reaching the maximum level of excavation.

2013 (Fig. 3b). The first landslide was initially fixed building a riprap lining. Later, taking into account the geotechnical study, an anchored concrete gravity wall was constructed at the lower part (Zone 3'B). In 2013, a slide occurred after a heavy rainy period. The adopted remediation measures were: i) to clean up the debris, ii) to set back the slope about 2–3 meters, iii) and to place a riprap lining above the anchored wall to refill the area. (Fig. 2).

Monitoring and control systems were set in place due to the instability signs. In September 2011, topographic tracking targets were placed on the top part of the riprap wall in Zones 3-1 and 3-2. Movements of 2 cm were registered in these targets at the beginning of the construction (April 2013). This, in addition to the presence of instability signs, motivated the installation of inclinometer pipes at the backside of the riprap wall.

Bored pile construction finished in December 2013 and the excavation started immediately afterwards. Significant movements were recorded during the excavation in front of the soldier pile wall during winter 2014. This led to reconsider the initial proposed solution.

3 GEOTECHNICAL CHARACTERIZATION

Geotechnical studies shown unfavorable geotechnical environment and topography for the stability of the natural slope at the back of the pile wall, particularly in the 3'A zone.

Figure 3. Landslides prior to the construction of the industrial unit.

The site is within the influence of the Laredo-La Peña fault, which is also NW-SE oriented. According to National geological map, in the eastern side of the fault, where the site is located, the ground consists on a clay and silt matrix including large rock boulders, possibly due to Keuper diapiric phenomena in this weakest area. The presence of these blocks may explain the difficulties in defining the location of the bedrock, as well as, the discontinuities found in the limestone when placing the ground anchors and during pile construction for the retaining wall. Old landslides were detected and confirmed by the exploration boreholes performed. Just outside the site, there is a river running NW-SE. It is oriented in the same direction as the fault and the building area (Fig. 1). On both sides, two tributary streams flow to the main river.

Site investigation and geotechnical parameters presented here are based on: i) the results of 3 boreholes performed in the slope, and ii) on the information gathered during the anchor and pile construction. The first two boreholes (BH1 and BH2) were part of the general site investigation for the industrial park project. They are in the areas corresponding to 3-1 and 3-2 wall profiles. The third borehole (BH3) was performed later in May 2014. It was located in the most critical area labelled as 3'A (Fig. 2).

Sandy silt, silty clay with gravel, and heavily altered sandstone gravel are materials from previous slides found in the slope. At different depths, fractured limestone blocks, or the limestone bedrock can be reached. Particularly, borehole BH3, found the bedrock at 10 m depth. However, close to this area, the bedrock was not found when installing the ground anchors. This high scatter, made it hard to define the bedrock's depth along the plot profile.

The other two boreholes (BH1 and BH2) allowed sampling and common characterization tests (SPT, particle-size distribution, Atterberg limits…). Besides, five pressuremeter tests, 2 C-U triaxial tests and 1 C-D reversal direct shear test were carried out. In Table 1, some soil parameters (including strength parameters) derived from the laboratory tests are presented. The residual strength measured in the reversal direct shear test was 18°.

4 LIMIT STATE ANALYSES

4.1 *Calculation assumptions and parameters*

Slope stability analyses were performed assuming circular failure surfaces and using the method of slices. Morgenstern-Price's method and the commercial numerical code Slope/W (Geoslope 2012) with the assumption of semi-sinusoidal contact stress distribution between slices were employed. Strength representative parameters of the soil were

Table 1. Soil parameters from BH1 and BH2.

Depth		c	ϕ	LL	PI	w
(m)	Test	(kPa)	(°)	(%)	(%)	(%)
4.0–4.6	CU Triax.	40	23.0	32	12	15
9.0–9.6	CU Triax.	10	25.8	29	11	23
7.1–7.4	Dir. shear	25	24.5	52	24	25

estimated taking into account triaxial test results combined with those of direct shear test (Table 1). Residual parameters (friction angle of 18°), were not considered because the analyzed failure surface involves the whole mass of soil. Therefore, residual parameters could be just representative for a potential ancient slip surface, but not for the whole mass of soil.

On the other hand, it is common for soils to be more altered at shallow depths and therefore, to have lower resistance. This usually affects the apparent cohesion. To define the altered layer thickness and its resistance, back-analyses of recent landslides were carried out. They are presented in Section 4.2. A summary of the geotechnical units and their parameters employed in the calculations are shown in Table 2.

4.2 *Back-analyses of recent slides*

The aim of these back-analyses was to validate and improve the understanding of the geotechnical parameters and hypotheses employed in the numerical calculations. Two previous slides were analyzed: one of them occurred in March 2013 near the 3 A wall profile (Fig. 3b), and the other is one of the seasonal slides on the slopes placed at the edges of the unpaved road that runs along the slope (Fig. 1).

The back-analysis of the March 2013slide confirmed the critical slope stability situation because it predicted a safety factor equal to 1 (Fig. 4) with the parameters from Table 2. High ground water table (GWT) level was considered in the analysis, corresponding to the period when the slide happened (March and after heavy rains). The depth of the failure surface was around 3–4 meters, coinciding with a main flow surface subparallel to the slope. As already mentioned, it was considered that the material forming the slope was more altered in the top. Hence, two materials were considered (Fig. 4): One, named silt (shallow), was used for the first 5 m depth with a lower apparent cohesion (5 kPa). The other, named silt (deep), with apparent cohesion of 20 kPa (Table 2). Both geotechnical units have the same internal friction angle (26°). At the lower part of the slope during the levelling works performed to construct the industrial park, the surface material was removed. Therefore, at this lower part all the profile has the properties of the

Table 2. Material parameters for the limit state analyses.

		γ/γ_{sat}	c	ϕ
Name	Model*	(kN/m³)	(kPa)	(°)
Bedrock	Undrained	27	2000	–
Silt (shallow)	M-C	20	5	26
Silt (deep)	M-C	20	20	26
Riprap	M-C	19	5	47
Wall (concrete)	Undrained	25	1000	–

* Material model considered for the numerical code (Slope/W).

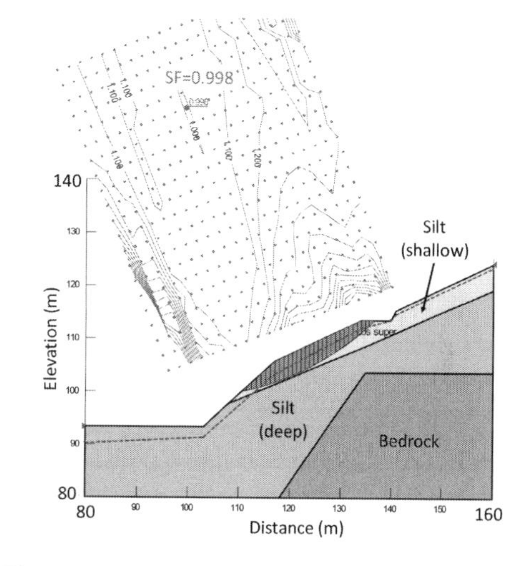

Figure 4. Back-analysis of March 2013 landslide.

deep material. This helps to explain that the landslide did not affect the toe of the slope (Fig. 4), as it was seen during the remediation works.

Seasonal instabilities appeared in the slopes at the edges of the unpaved road that runs across the slope. They were reactivated during the rainy seasons. The gradient of the slope was between 2H:3V and 1H:2V. Back-analyses provided a safety factor equal to 1.19, which confirmed the marginal stability. Nonetheless, this was just an approximated calculation due to all the existing uncertainties: detailed and exact slope profile, exact position of the GWT, small size of the instabilities.

4.3 *Slope stability analysis during the excavation*

Monitoring results during basement excavation, in April 2014, showed worrying slope movements. This critical slope stability situation was confirmed by the limit equilibrium analysis performed (safety factor = 1.12). For the analysis, the critical cross-section of the wall was deduced as the type 3'A, where the natural slope is steeper (Figs 1 and

2). The failure surface was not expected to go under the wall or through it because of the high resistance of the concrete (1-m-diameter piles and center-to-center spacing of 1.1 and 1.2 m) and the slope geometry and location of the bedrock. Hence, the length of the retaining wall and the ground anchors are not relevant in this analysis. Therefore, the wall was modelled considering only concrete resistance, with no anchors (Fig. 5).

4.4 *Corrective measures*

To improve slope stability two corrective measures were taken in place:

- Lowering down the GWT with sub-horizontal drains of around 25 m length (Fig. 6), above the retaining pile wall and using the unpaved road that runs across the slope.
- Driving steel rail segments (piles) braced on top with an anchored concrete beam. This solution was adopted because the unpaved road provided easy access for its construction. Besides, a quick protection measure for the unpaved road was also wanted as cracks started to appear on it.

The steel rails were placed in two staggered rows at 0.6 m distance and 1 m spacing between rails. The rails were 5.8 m length and with a weight of 54 kg/m. Predicted length of each rail (pile) was 10 meters but refusal was achieved at variable depths due to the irregular bedrock depth, varying from 7 to 30 m. The rails were placed on the outer side of the unpaved road. It was not possible to construct the anchors on that side; therefore, they were placed on the inner edge of the unpaved road (Fig. 6). 7 ground anchors of 400 kN were constructed with 3.5 m distance between them. They were placed at an angle of 30 °. Their free length was 8 m and a bonded length of 16 m. The anchor heads were braced with a concrete beam and this beam was attached to the rail wall with tendons (Gewi ϕ50) pre-stressed at 180 kN.

Rails were simulated in the Slope/W code as a 1-m-thickness equivalent soil with a shear strength of 300 kPa, which corresponds to a 100 kN/m reaction force at the top of the rails (introduced by the anchor load) and a passive resistance of 200 kN/m at the bottom of the rails (caused by the embedment in the bedrock).

Table 3 summarizes the safety factors (SF) obtained for the initial situation (April 2014, high GWT and worrying slope movements), lowering down the GWT, driving steel rails and combining the two corrective measures, which allows to have an acceptable SF of 1.44. In those analyses, shallow slip circles that affect the riprap lining were not considered. For those shallow slip circles, the SF was lower than 1.25. Therefore, a mesh of steel cables fixed by long bolts was proposed. That gives a SF of 1.45 for both deep and shallow slip circles (Fig. 5).

The mesh of steel cables also avoids eventual block falls from the riprap lining.

5 FINITE ELEMENT SIMULATIONS

The poor stability conditions of the slope during the construction (basement excavation and pile wall anchoring) caused important wall deflections, which in turn, increased the anchor loads well beyond the pre-stressed (initial) loads, particularly in the first row of anchors.

Finite element simulations were performed using the commercial code Plaxis 2D 2012 (Brinkgreve et al. 2012) to study mainly the pile wall deflections, bending moments and the anchor loads. An example of the numerical model and the finite element mesh is shown in Figure 7. The analysis may be considered a type B prediction (Lambe 1973) because it was done during construction; available data of anchor loads and wall deflections were used to calibrate the model and the numerical simulations were used to propose corrective measures and predict future anchor loads, wall deflections and bending moments.

The parameters used in the finite element simulations are summarized in Table 4. The Hardening Soil (HS) model (Schanz et al. 1999) was used to model the behavior of the soil (silt), both at shallow

Table 3. Summary of the factors of safety for the two major corrective measures and their combination.

Analysis	Safety factor
April 2014 (high GWT)	1.12
Lowering GWT (horizontal drains)	1.33
With driven rails	1.21
Horizontal drains and driven rails	1.44

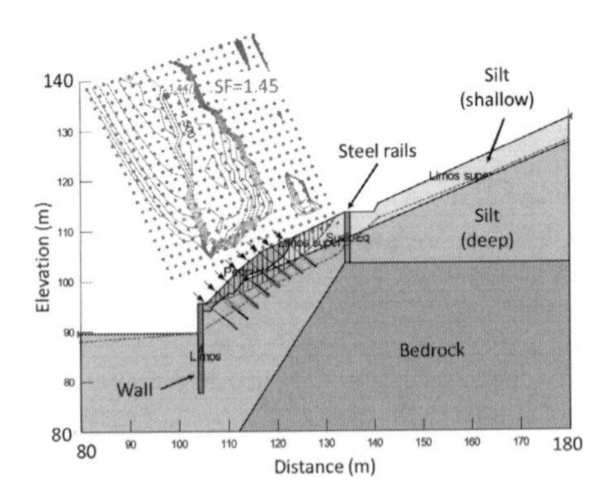

Figure 5. Stability analysis of the slope with corrective measures (horizontal drains, steel rails and bolted mesh of steel cables). Shallow slip circle.

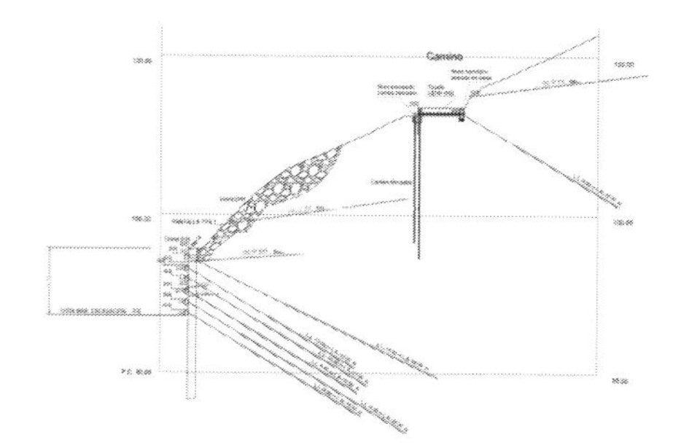

Figure 6. Sketch of corrective measures (long bolts and mesh of steel cables not shown).

and deep levels, to better match the displacements and anchor loads during the excavation process. The values of the cohesion of the silt units were slightly increased with respect to the limit equilibrium analyses (Table 2) to avoid local numerical problems caused by the poor stability conditions of the natural slope.

Finite element simulations confirmed the poor stability conditions of the natural slope and the increase in the anchor loads and the bending moments in the pile wall. The initial design of the pile wall planned four row of anchors (called An1, An2, An3a and An3b). After the finite element simulations, it was decided to add 2 additional row of anchors (called In1 and TopR). In1 was an intermediate row of anchors at the level of the current excavation when problems were detected (April 2014). This intermediate row was not enough to decrease the loads of the existing anchors to acceptable values, and therefore, it was decided to install another additional row at the top of the retaining wall (TopR). The installation of the top anchors required temporary protection of the pile wall with geomembranes and geotextiles and backfilling the excavation up to the needed level (Fig. 8). In the end, it was a satisfactory solution that helped to reduce the anchor loads to acceptable levels. Lowering the GWT was also important to reduce anchor loads.

The finite element analysis was used to predict the anchor loads during the different construction phases (excavation and installation of the different row of anchors). The agreement between numerically simulated values and measured values was good (Fig. 9).

6 FIELD MONITORING

The retaining pile wall in Zones 3 and 3' was instrumented to control the behavior of the system. The following devices were installed:

- Load cells in ground anchors (one per row of anchors) in Zones 3 and 3'.
- Inclinometer pipes on the back of the pile wall (4 inclinometer pipes in Zone 3' and 6 in Zone 3).
- Control targets for topographic surveying of ground displacements on the riprap lining and on the small anchored wall that had been constructed during the development of the industrial estate in Zone 3 (Fig. 2).

During basement excavation, the displacements measured at the control targets and at the inclinometer pipes were noticeable until May 2014, with maximum horizontal displacement rates of around 0.5 mm/day at inclinometer pipes I-610

Table 4. Soil parameters for the finite element analyses.

	Bedrock	Silt (shallow)	Silt (deep)	Rockfill
Model	Elastic	HS	HS	M-C
γ/γ_{sat} (kN/m³)	27	20	20	19
p'_{ref} (kPa)	–	100	100	–
m	–	0.5	0.5	–
E^{ref}_{oed}(MPa)	40000	20	20	30
E^{ref}_{50}(MPa)	–	20	20	–
E^{ref}_{ur}(MPa)	–	60	60	–
ν_{ur}	0.3	0.2	0.2	0.3
c (kPa)	–	20	30	5
ϕ (°)	–	26	26	47
ψ (°)	–	0	0	5
OCR	–	1	1	–

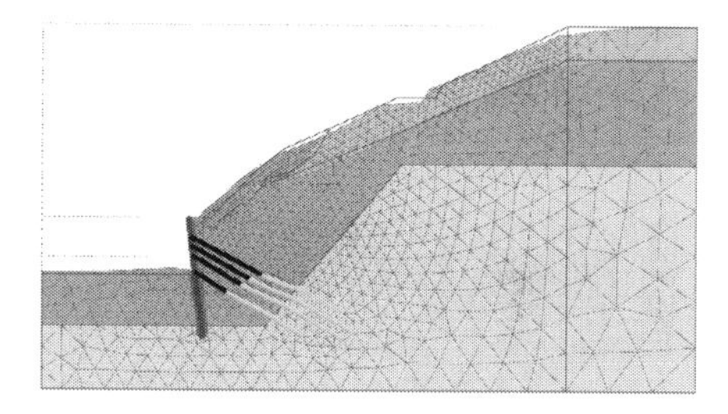

Figure 7. Finite element model and mesh. Predicted displacements (amplified by a scale factor of 20) in April 2014 (before adopting corrective measures).

Figure 8. Execution of the top row of ground anchors with temporary pile wall protection and backfilling of the excavation.

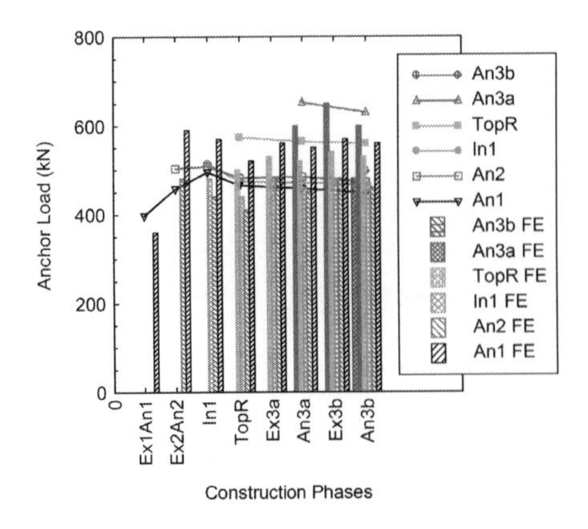

Figure 9. Anchor loads during construction process. Comparison between finite element simulations and measured values.

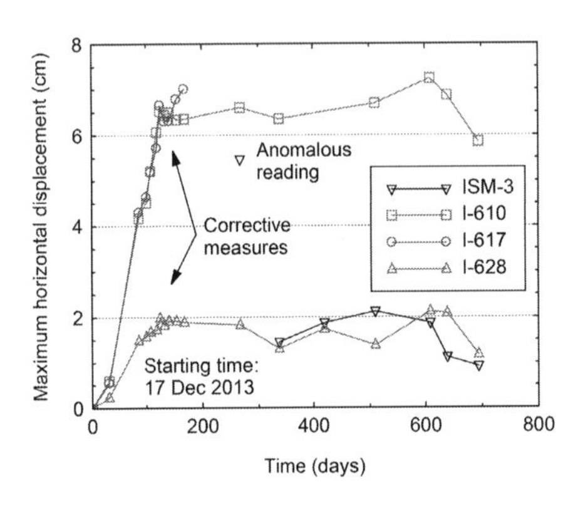

Figure 10. Evolution of maximum horizontal displacements measured at the inclinometer pipes.

and I-617 (Fig. 10). The profile of horizontal displacements measured at the inclinometer pipes showed maximum horizontal displacement at the top and null displacements at the depth that corresponds to the bottom of the pile wall. That indicated a proper embedment of the pile wall and the problems caused by the stability of the slope at the back of the pile wall.

In the summer of 2014, most of the corrective measures were implemented (long drain pipes, additional rows of anchors in the pile wall and driving of steel rails in the natural slope, Fig. 6). From that date onwards, the measured displacements were nearly constant (Fig. 10), but for some slight oscillations, which could be caused by the accuracy of the measurements, seasonal variations and additional works, such as the installation of the mesh of steel cables fixed by long bolts in the summer of 2015 or the completion of the building basement and the corresponding concrete floor slabs.

7 CONCLUSIONS

The construction of an industrial unit required a 10-meter-deep excavation at the toe of a steep natural slope with signs of previous instabilities. The designed solution was a soldier pile wall with several rows of ground anchors. During excavation, important horizontal displacements were measured and corrective measures were adopted to increase the stability of the natural slope, such as long drain pipes to lower the GWT, additional rows of ground anchors at the retaining pile wall and driving rail tracks in the natural slope. The shallow stability of the slope was additionally increased using bolted steel meshes.

Limit equilibrium and finite element analyses were performed for the design of those remediation measures. Finite element analyses provided useful information about pile wall deflections, bending moments and anchor forces. Detailed instrumentation, namely inclinometers, anchor load cells and topographical surveying, was used to monitor the performance of the pile wall and the slope. The field measurements were compared with the finite element predictions, showing a general good agreement.

ACKNOWLEDGEMENTS

The authors acknowledge the cooperation and help provided by all the professionals and institutions involved in the construction works presented in this paper (site manager, construction company, project manager…). The authors wish to recognize in particular the company that owns the industrial unit and its project manager.

REFERENCES

Brinkgreve, R.B.J., Engin, E., Swolfs, W.M. 2012. *Plaxis 2D 2012 Manual*. Delft: Plaxis bv.

GeoSlope. 2012. *Stability Modeling with SLOPE/W. An Engineering Methodology*. July 2012 Edition. Calgary: GeoSlope Int. Ltd.

Lambe, W.T. 1973. Predictions in soil engineering. *Géotechnique* 23: 149–202.

Schanz, T., Vermeer, P.A., Bonnier, P.G. 1999. The hardening-soil model: Formulation and verification. In R.B.J. Brinkgreve (ed), *Beyond 2000 in Computational Geotechnics*: 281–290. Rotterdam: Balkema.

Numerical Methods in Geotechnical Engineering IX – Cardoso et al. (Eds)
© 2018 Taylor & Francis Group, London, ISBN 978-1-138-33203-4

Inverse analysis of horizontal coefficient of subgrade reaction modulus for embedded retaining structures

Ping He & Weidong Wang
Department of Geotechnical Engineering, Tongji University, Shanghai, China

Zhonghua Xu & Jing Li
Shanghai Underground Space Engineering Design and Research Institute, Shanghai, China
Shanghai Engineering Research Center of Safety Control for Facilities Adjacent to Deep Excavations, China

Zili Li
Civil and Environment Engineering, University College Cork, Republic of Ireland

ABSTRACT: Embedded retaining structures are widely adopted for supporting deep excavations in urban areas. Their stability primarily relies on the lateral passive resistance of the ground. Usually, the ground resistance is considered as a set of elastic soil springs (i.e. Winkler's springs approach) following routine engineering design code. Such simplified subgrade reaction method is capable of simulating general soil-structure interaction and predicts retaining structure behavior largely in line with field observations. However, in practice, it's difficult to determine the horizontal coefficient of subgrade reaction modulus, i.e. parameter m (the stiffness of soil springs $k = mz$) due to lack of experiments and soil uncertainties. In this study, an inverse analysis was conducted to determine the stiffness of soil springs m against the inclinometer data of lateral wall displacements for a deep excavation case in Shanghai clay. The inverse calculation of m involved an iterative computation process aiming to minimize the discrepancy between the computed wall displacements and the field measurements. In particular, sensitivity analyses were conducted to evaluate the effect of horizontal soil springs in different soil layers on the lateral wall displacements. Results show the m values of the soil layers play an important role on the lateral displacement, while the optimization of m may greatly improve computational accuracy and efficiency. The findings derived from the computed results and the comparison with the field measurements will provide some guidance for the design of embedded retaining wall in Shanghai clay.

1 INTRODUCTION

Embedded retaining walls have been extensively used for deep excavations in urban areas to control the deformations in the surrounding soil. For a deep excavation project, one of the main challenges is the evaluation of the retaining wall internal forces and displacements, as it requires to account for soil-structure interaction, staged soil excavation sequence and many other construction factors. As an advanced modelling approach, Finite Element Method (FEM) is able to simulate complex ground characteristics and construction details, but in turn may cost significant computational resources. Moreover, it is difficult to determine the input geotechnical parameters due to the complexity of soil. In engineering practice, Subgrade Reaction Method (SRM), which simplifies complex soil-interaction as a set of springs, is more widely adopted than time-consuming FEM analyses, particularly at the early-stage design of a retaining wall. In the SRM, the key parameter is the spring stiffness for soil-structure interaction; that is the horizontal coefficient of subgrade reaction modulus (i.e. parameter m). The m is influenced by numerous factors, such as the soil stratigraphy, the excavation support systems, ground improvement and so forth. In practice, the parameter m is either derived from time-consuming and expensive tests (e.g. horizontal loaded monotonic pile experiment) or determined by empirical relationships (Monaco & Marchetti 2004). In a more recent guideline, the Shanghai Technical Code for Excavation Engineering (2010) only provides a considerably large range of the m value for different soil types, which may end up with an arbitrary selection by different civil engineers. This suggest that the parameter m still demands dedicated investigation for engineering design in Shanghai clay.

As an alternative, inverse analysis of model parameters (e.g. the parameter m) is widely used in excavation based on displacement observations. In the inverse analysis, the model parameters were calculated iteratively as to match the field measurements. Wang et al. (1998) proposed a back-analysis method for m value in deep excavation by using a nonlinear optimization method. Xu et al. (2014) proposed a back analysis of the proportional coefficient of horizontal resistance in SRM for deep excavations. Kobayashi et al. (2008) identified the subgrade reaction coefficient of a foundation soil in an open pier by inverse analysis. Finno and Calvello (2005) used inverse analysis to optimize the parameters of Harding Soil model for the finite element model of a 12.2 m deep excavation through Chicago glacial clays. Rechea et al. (2008) compared the results from two different inverse analysis minimization algorithms, i.e. gradient method and genetic algorithm, against field measurements collected in a practical excavation project.

This paper proposed an inverse analysis methodology for parameter m in the SRM. Based on the inverse analysis method, a deep excavation case in Shanghai clay was investigated to determine the m value of different soil layers. In the inverse analysis, the inclinometer data of later wall displacement was used to calibrate the calculation model.

2 INVERSE ANALYSIS METHODOLODY

2.1 *Subgrade reaction method*

The schematic calculation model of Subgrade Reaction Method was shown in Figure 1. It assumes that the retaining structure is in a 2D plane strain condition. The retaining wall was modelled as vertically placed elastic beam with unit width, and the struts were modelled as spring supports. The water pressure and effective earth pressure was applied on the wall as the external loading. There are two approaches to calculate the loading. One is considering the water and earth pressure separately based on the effective stress method, whereas the other calculates together based on total stress method. In this study, the effective stress method was adopted according to the Shanghai Technical Code for Excavation Engineering (2010). The water pressure distribution was considered as hydrostatic increasing along the depth below the water table and remained constant under the water table inside the excavation (shown in Fig.1). The distribution of earth pressure was calculated by Rankine active earth pressure above the excavation surface and remained rectangle distribution underneath the excavation surface (shown in Fig.1). The soil underneath the excavation surface was modelled as a set of springs. Based on Winkler's theory for a beam on an elastic foundation, the flexural differential equation of the beam can be expressed as:

$$EI\frac{dy^4}{dz^4} = -ky + p(x) \tag{1}$$

where EI = bending stiffness of the wall; y = lateral displacement of the wall; z = depth; $p(x)$ = the distribution of lateral loading, including water pressure, earth pressure and surcharge at the ground surface; k = subgrade reaction modulus; ky = the earth resistance force. There are different methods based on the different distribution assumption of subgrade reaction modulus changing with depth. For simplicity, the subgrade reaction modulus k is assumed to increase linearly with the depth (i.e. the m method), which can be expressed as:

$$k = mz \tag{2}$$

where m = horizontal coefficient of subgrade reaction

Modulus(m = 0 for the beam above the excavation surface). Substituting Eq.(2) into Eq.(1):

$$EI\frac{dy^4}{dz^4} = -mzy + p(x) \tag{3}$$

For a multi-propped retaining wall, the wall is usually divided into several sections, and the

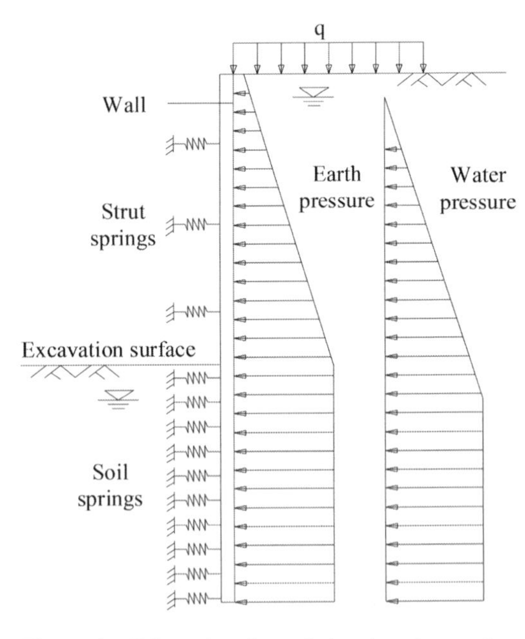

Figure 1. Schematic of vertical subgrade reaction method.

flexural differential equation is established for each section. The details of the calculation procedure can be found in Xiao et al. (2004).

2.2 *Inverse analysis procedure*

To conduct the inverse analysis of parameter *m* in the SRM, Abaqus and UCODE (Poeter & Hill 1998) was combined together to obtain the optimized values based on the field observation of lateral wall displacements. Figure 2 shows a detailed flowchart of the parameter *m* optimization algorithm. The SRM was programmed by Abaqus subroutine and an initial *m* value was attempted to execute the Abaqus calculation model. In the next step, the input *m* values was calculated by UCODE through an iterative process as to match the computed lateral wall displacement against the observed data. UCODE, which was first designed for inverse modelling of ground-water models, is a universal inverse code that can be used in conjunction with other application models. It performs inverse modeling by calculating the parameter values that minimize a weighted least-squares objective function. The weighted least-squares objective function $S(b)$ can be expressed as:

$$s(b)=\left[y-y'(b)\right]^{T}\omega\left[y-y'(b)\right] \tag{4}$$

where b = vector of the estimated parameters, herein is the m value; y = vector of the wall deflection observations; $y'(b)$ = vector of the wall deflection computed results of model; ω = weight matrix of every observation ($\omega_{ii} = 1/\sigma_i^2$, σ_i is the measurement error). The aim is to optimize the estimated parameters to minimize the objective function.

It is necessary to conduct sensitivity analysis to facilitate the iteration computation thereafter. The sensitivity matrix, X, is computed using a forward difference method based on the changes in the computed solution due to slight perturbations of the estimated parameter values. It examines the influence of the estimated parameters on the calculated results, which can be expressed as:

$$X=\frac{\partial y'(b)}{\partial b}=\frac{y'(b+\Delta b)-y'(b)}{(b+\Delta b)-b} \tag{5}$$

Parameter values that minimize the objective function are calculated using normal equations. In *UCODE*, the nonlinear regression analysis is accomplished with the modified Gauss–Newton method. The normal equations and the iterative process can be expressed as:

$$\left(C^{T}X_r^{T}\omega X_r C+Im_r\right)C^{-1}d_r=C^{T}X_r^{T}\omega\left(y-y'(b_r)\right) \tag{6}$$

$$b_{r+1}=\rho_r d_r+b_r \tag{7}$$

where C = diagonal scaling matrix with elements c_{jj} equal to $((X^T\omega X)jj)^{-1/2}$; X_r = sensitivity matrix; I = identity matrix; ω = weight matrix; m_r = Marquardt parameter (Marquardt 1963) used to improve regression performance; d_r = vector used to update the parameter estimates b; r = iteration number; ρ_r = damping parameter, computed as the change in consecutive estimates of a parameter normalized by its initial value, but it is restricted to values less than 0.5.

After performing the modified Gauss–Newton optimization, the optimized m value will be obtained if it satisfies the convergence criteria. The final solution of input parameters are then used to run the finite element model and evaluate the wall displacements.

UCODE can simultaneously calibrate multiple input parameters, but it's unlikely to estimate every parameters due to high computational cost. In UCODE, the composite scaled sensitivity, css_j, which define the relative significance of the input parameters, is expressed as:

$$css_j=\left[\sum_{i=1}^{ND}\left(\left(\frac{\partial y_i'}{\partial b_j}\right)b_j\omega_{ii}^{1/2}\right)^2/ND\right]^{1/2} \tag{8}$$

where ND is the number of observations, y_i' is the ith simulated value; b_j is the jth estimated

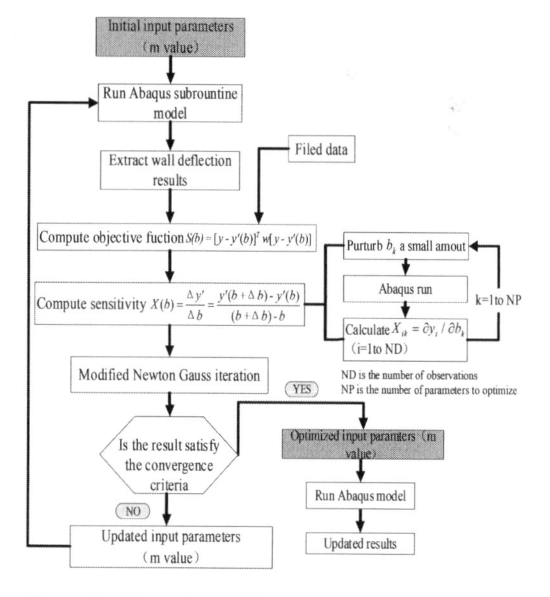

Figure 2. Flow chart of inverse analysis using Abaqus and UCODE.

parameter; the other terms in the equation are the same as aforementioned. The composite scaled sensitivity illustrates the total amount of information provided by the observations.

3 EXCAVATION PROJECT

3.1 *Project description*

The Changfeng hotel excavation project was select-ed to test the inverse analysis methodology. Changfeng hotel excavation was located at the cross of west Yan'an road and Panyu road in Shanghai, China. The main structures consisted of a 52-storey building and a 5-storey podium building, with three level basement for underground garage and equip-ment. The depth of excavation in the main building and podium part were 13.5 m and 12.0 m, respective-ly. The excavation area was approximately 7000 m^2. The site layout and inclinometer instrumentation of the excavation were shown in Figure 3.

Bottom up excavation method was adopted for the main building and south podium area, while top down method was used for the north of the podium area. The diaphragm walls of 800 mm in thickness were constructed to sustain lateral earth pressure due to excavation, and later the walls were used as a part of permanent basement structures. Three level con-crete struts were adopted for the supporting system, and slab were propped for the top down area. The schematic of retaining system was shown in Figure 4.

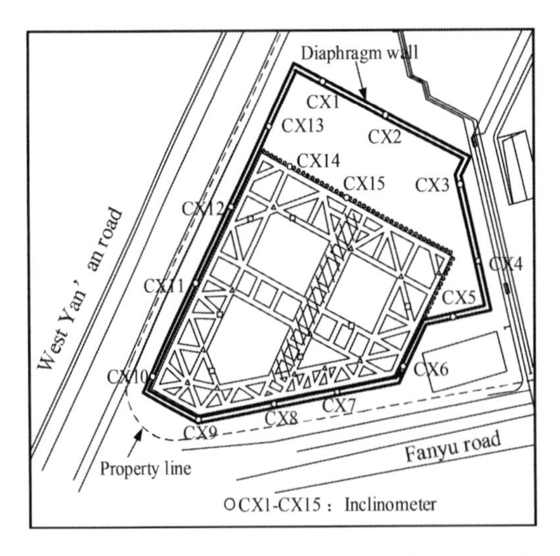

Figure 3. Layout and instrumentation of the excavation.

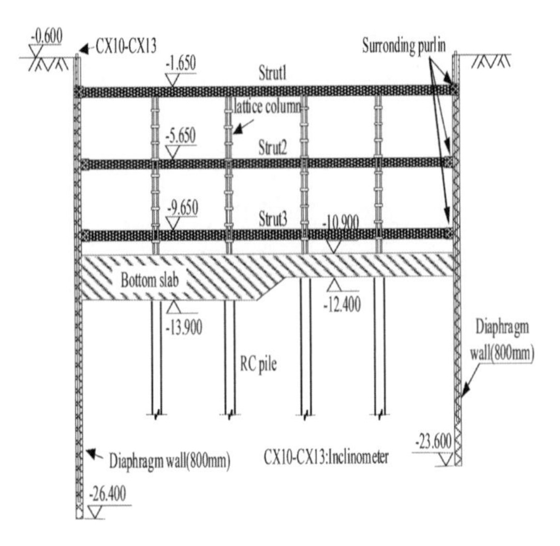

Figure 4. Schematic of retaining system.

Table 1. Excavation sequences.

Stage	Model sequence
1	Excavated to –2.35 m
2	Cast 1st strut and excavated to –6.35 m
3	Cast 2nd strut and excavated to –10.35 m
4	Cast 3st strut and excavated to the base

Table 2. Parameters of the soil layer.

Soil layer	Unit weight (kN/m3)	c (kPa)	ϕ (°)	m (MN/m^4)
1 Fill	18.0	0	22	1.5
2 Clay	18.0	20	17	3
3 Silty clay	17.5	13	15	2
4 Mundy clay	16.6	14	10.5	1.5
5-1-1 Clay	17.3	11.5	15.5	3
5-1-2 Silty clay	17.9	8.5	23.5	3

There were 15 inclinometers (designated CX1-CX15) installed to measure the lateral wall displacement during the excavation. Figure 3 shows that CX5, CX6, CX9, CX10 were located at corners of the ex-vacation, which might not be suitable for comparative analysis due to the corner effect. CX8, CX15 were installed near the trestle construction area and might get affected by heavy vehicle loading. Therefore, CX7, CX11, CX12 were selected for the following analysis due to the integrity monitoring data.

3.2 *Abaqus calculation model*

The excavation was modelled by SRM programmed with Abaqus subroutine. The retaining wall was modelled using the beam element B21 with interval of 0.5 m between nodes. This is consistent with the monitoring data points in the inclinometers. The struts supports were modelled with Spring2 element. The horizontal soil springs were modelled with element SPINGA. The side of the soil springs were fixed in *x* and *y* direction as the boundary conditions. The excavation sequences are shown in Table 1.

During the process of excavation, soil spring stiffness changed with the excavation sequences. This was achieved by the Abaqus user subroutine UFIELD. As aforementioned, the stiffness increased linearly with the depth. The change in external loading during the excavation sequences was achieved by the Abaqus user subroutine DLOAD. The 20 kPa surcharge was also taken into consideration when calculating the earth pressure. The water table outside of the excavation was 0.5 m below the ground surface and 1 m below the excavation base inside the excavation. UCODE requires users to write control documents according to a certain format to achieve inverse analysis, including the main documents such as *filename.uni*, *filename.pre* and *filename.ext* etc. The compilation of control documents can be referred to Calvello (2002) and Li (2016).

The soil layer and parameters used for the analysis are listed in Table 2, where the *m* value is the initial assumption values according to the Shanghai Technical Code for Excavation Engineering (2010).

3.3 *The sensitivity of the m value*

Prior to inverse analysis *m* value, a sensitivity analysis was conducted to examine the impacts of different layer of the *m* value on the calculation results. The composite scaled sensitivity was calculated based on three selected inclinometer data (CX7, CX11, CX12).

Figure 5 shows the sensitivity of the *m* values of the six soil layers (m1, m2, m3, m4, m5-1-1, m5-1-2) to the objective function. In this figure, the first layer and second layer, i.e. m1 and m2, has relatively insignificant influence on the wall displacement results. This is may be attributed to the small thickness of the first and second layer within only 2 m. Instead, the m3, m4, m5-1-1, m5-1-2 was selected to inverse analysis in the following section.

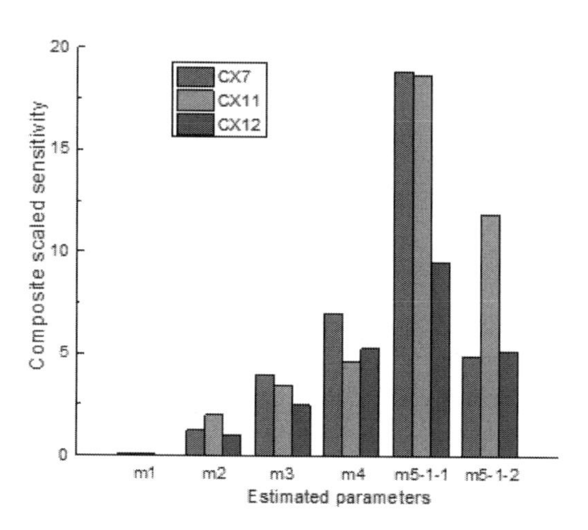

Figure 5. Sensitivity results of *m* for different soil layers.

3.4 *Inverse analysis results*

Figure 6 compares the final stage deflection of inclinometers CX7, CX11 and CX12 with the inverse analysis results. The monitored deflection shows a typical bulging shape as expected in multi-strutted deep excavations, and the maximum displacements often occurred near the excavation surface (Ng et al. 2012). The computed inverse results match well with the field data, confirming the capability of the inverse analysis technique to determine *m* value.

The values of parameter *m* derived from inverse analysis are shown in table 3. There is indeed a difference between the results of *m* values for the same soil layer, but the magnitude is negligible. Such discrepancy is probably due to the influence by the excavation sequence of soil. It is also noted that the ratio of the range to the average value of each group is less than 40%, suggesting that the average value of the back analysis results can represent the

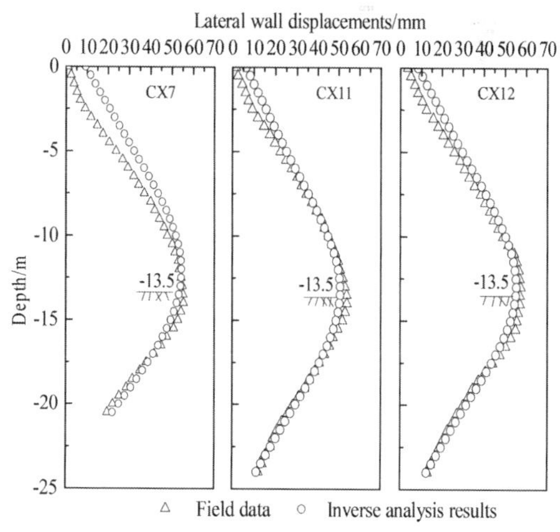

Figure 6. Comparison between the field measurements of wall displacements and the inverse results at final excavation stage.

Table 3. Inverse analysis results of m.

Item	m_3 (MN/m^4)	m_4 (MN/m^4)	$m_{5\text{-}1\text{-}1}$ (MN/m^4)	$m_{5\text{-}1\text{-}2}$ (MN/m^4)
CX7	1.36	1.23	0.96	1.61
CX11	1.78	1.73	1.42	1.56
CX12	1.69	1.80	1.39	1.33
Range r	0.42	0.57	0.46	0.28
Average a	1.61	1.59	1.26	1.5
r/a	26.1%	36.8%	36.6%	18.7%
Code	1~2	1~2	2~4	2~4

resistance level of each soil layer. Compared with the empirical value range given by the code (2010), the m value obtained by back analysis for the fifth layer is obviously smaller.

4 CONCLUSIONS

An inverse analysis method was proposed to calibrate parameter m in Subgrade Reaction Method (SRM) by finite element modelling. The guidelines of the inverse analysis are briefly summarized as follows: (a) Selecting the reasonable lateral wall displacements field data to initiate the inverse analysis; (b) Programming SRM by Abaqus subroutine; (c) Compiling the necessary files in UCODE to combine the Abaqus and UCODE; (d) Calculating the m value by UCODE through an iterative process as to match the computed lateral wall displacement with the field data. In this study, an excavation case in Shanghai clay was adopted to determine the m value for different soil layers. Based on the computational results and the comparison against field measurements, the following conclusions can be drawn:

1. The calculation model results derived from the optimized m value show good agreement with the measured lateral wall displacement, demonstrating the capability of the inverse method to determine the m value. This suggests that the proposed method in this paper may be widely adoptable in the analysis of other similar retaining wall projects.
2. Prior to inverse analysis, sensitivity analysis of the parameters should be first conducted to examine their impacts on the computed results. This sensitivity analysis not only improves the iteration efficiency, but also ensures the accuracy of back analysis results.
3. There still remains a slight difference between m values for different inclinometer locations. This may be induced by the construction process at different locations. For simplicity, the average of the m value was considered as suitable parameters of the soil layers.
4. After the analysis of wall displacements, the optimized m value can be used for subsequent prediction of the lateral wall displacements. As a future study, similar inverse analysis of m values can be conducted on many more excavations in a specific region (e.g. Shanghai clay). The accumulated knowledge will establish a comprehensive database of m values for the engineering design of the embedded retaining wall in that region.

REFERENCES

Calvello, M. 2002. Inverse analysis of supported excavations through Chicago glacial clays. PhD thesis, Evanston: Northwestern University.

DGTJ08-61-2010 J11577-2010, Technical code for excavation engineering.

Finno, R.J. & Calvello, M. 2005. Supported Excavations: Observational Method and Inverse Modeling. Journal of Geotechnical & Geoenvironmental Engineering, 131(7):826–836.

Kobayashi, N., Shibata, T., Kikuchi, Y. & Murakami, A. 2008. Estimation of horizontal subgrade reaction coefficient by inverse analysis. Computers & Geotechnics, 35(4):616–626.

Li, J. 2014. Back analysis of the "m" values in vertical elastic subgrade beam method for deep excavations in Shanghai. Master thesis, Shanghai: Tongji University.

Marquardt, D.W. 1963. An Algorithm for Least-Squares Estimation of Nonlinear Parameters. Journal of the Society for Industrial & Applied Mathematics, 11(2):431–441.

Monaco, P. & Marchetti, S. 2004. Evaluation of the coefficient of sub grade reaction for design of multi-propped diaphragm walls from DMT moduli. International Conference on Site Characterization ISC.

Ng, C.W.W., Hong, Y., Liu G.B., Liu T. 2012. Ground deformations and soil–structure interaction of a multi-propped excavation in Shanghai soft clays. Géotechnique, 62(10):907–921.

Poeter, E.P. & Hill, M.C. 1998. "Documentation of UCODE, a computer code for universal inverse modeling." U.S. Geological Survey Water-Resources investigations Rep. No. 98-4080, USGS.

Rechea, C., Levasseur, S. & Finno, R. 2008. Inverse analysis techniques for parameter identification in simulation of excavation support systems. Computers & Geotechnics, 35(3):331–345.

Wang, X.D., Huang, L.P., Ruan, Y.P. & Xu, J.L.1998. Back-Analysising m value of subgrade reaction in excavation. Journal of Nanjing Architectural and Civil Engineering Institute, (2):48–54.

Xiao, H.B., Tang, J.C., Li, Q.S. & Luo, Q.Z. 2004. Analysis of multi-braced earth retaining structures. Structures & Buildings, 157(5):355–356.

Xu, Z.H., Li, J. & Wang, W.D. 2014. Back analysis of propor tional coefficient of horizontal resistance in vertical elastic subgrade beam method for deep excavations. Rock and soil mechanics, 35(s2):398–404.

Numerical Methods in Geotechnical Engineering IX – Cardoso et al. (Eds)
© 2018 Taylor & Francis Group, London, ISBN 978-1-138-33203-4

Performance based design for propped excavation support systems in Qom, Iran

E. Ghorbani, M. Khodaparast & A. Moezy
Department of Civil Engineering, Faculty of Engineering, University of Qom, Qom, Iran

ABSTRACT: Using an efficient method to restrict the magnitude of the neighboring ground deformation in stabilizing deep urban excavations is highly recommended. The Stress Based Design of a retaining wall does not consider an allowable deformation range for the neighboring structures; therefore, Deformation or Performance Based Design (PBD) presents a more appropriate criterion in designing deep excavations. In the current study it is endeavored to approach a PBD method considering Top-Down construction in the geotechnical conditions of Qom in Iran. Geodetic surveys are applied by use of Total Station TS02 to monitor the field performances. According to the geodetic analysis, numerical model is calibrated. By dividing the geotechnical conditions into four different geotechnical types and taking the structural elements as variables in the calibrated model, exponential graphs are presented which enables to design excavation support systems in the studied geotechnical conditions on the basis of performance.

1 INTRODUCTION

Ground deformation is the most important factor in evaluating the performance of supporting system in excavations in urban areas. According to Boone et al. (1999), 45 buildings in the vicinity of deep excavations out of 50 have suffered negligible to serious damages.

The sequences in propped excavation support systems like Top-down method are more flexible rather than anchored excavation support systems. Flexibility in execution empowers the contractor to lighten or strengthen any part of the supporting system at any time in all the stages of stabilizing process in order to increase safety or economy of the project. This is consistent with the concept of PBD approach. Previous studies show the efficiency of Top-Down method in underground constructions (Long, 2001; Li et al., 2014; Paek and Ock, 1996; Wang et al., 2010; Ghavidel Aghdam, 2011; Hong et al., 2010; Huang et al., 2014; Lee et al., 1999; Wang et al., 2006).

Numerical models are usually calibrated with field monitoring results using trial and error methods (Finno and Calvello, 2005). After the calibration, it is possible to evaluate different excavation support systems and geotechnical conditions by means of numerical studies.

Determining the appropriate monitoring tool for ground deformation measurement depends on many factors including stabilizing method, excavation geometry, atmospheric conditions, numerical model, study budget, ease of utilization and etc. Different monitoring tools have been implemented to monitor ground deformation in deep excavations in previous studies (Finno et al., 2002; Finno and Blackburn, 2005; Finno and Tu, 2006; Finno, 2007). It is necessary to identify the limitations, characteristics, advantages and disadvantages in choosing monitoring tool. Total survey station is a simple and applicable tool which can be used with optical prisms to monitor displacements and is easy and quick to implement. Lower costs besides, the ease of utilization in this method without having difficulty in installing tools or digging pits make it admirable. The estimated accuracy for this monitoring tool is ranged between 0.01 to 10 millimeters which depends on method of utilization, type of the tool and the prism, atmospheric inconsistency, environmental conditions, user experience and etc. (Finno and Hassash, 2006; Song, 1999; Yang et al., 2004; Yuan et al., 2012; Luo et al., 2016; Sabzi and Fakher, 2013; Dunnicliff, 1998).

Numerical study is an accurate method in estimating the lateral deformation of excavation wall (δ_h) but it is not an appropriate method in predicting the vertical settlement of adjacent ground (δ_v) (Finno, 2007). Researchers have proposed empirical correlations between these two parameters (Peck, 1969; Bowels, 1988; Ou et al., 1993; Clough and O'Rourke, 1990; Hsieh and Ou, 1997; Long, 2001; Kung et al., 2007).

Different methods are proposed to correlate the vulnerability of the adjacent structure with the ground settlement. The most famous relations are those presented by Boscardin and Cording (1989), Burland (1997), Day (1998) and Boone (2001).

Some neighboring buildings tolerate great deformations but some others are insufficient even to the smallest deformations. Therefore, designing an excavation based on performance can present an applicable tool. By calculating the lateral movement using numerical models and estimating the ground settlement using empirical correlations, it is possible to design an excavation based on the performance of neighboring structures. Most of the previous studies are applicable in soft to very soft clays and less is discussed in dense sand or stiff clay conditions (Clough and O'Rourke, 1990; Long, 2001; Kung et al., 2007; Wang et al., 2010; Marr and Hawkes, 2010; Bryson and Zapata, 2010; Bryson and Zapata, 2012).

In this article it is endeavored to create a framework in order to reach a PBD approach considering top-down construction in dense sand or stiff clay geotechnical condition. Accordingly, to tackle with the excavation hazards in the city of Qom, field monitoring, numerical model calibration and extensive numerical studies were performed in a way to reach graphs similar to those are referred in past studies which correlate the supporting system stiffness to the maximum allowable lateral movement and consequently maximum allowable ground settlement. By estimating the maximum allowable ground settlement adjacent to the excavation, the required system stiffness is predictable.

2 STUDY AREA

Qom is located in the central Iran and in a distance of 125 kilometers south of Tehran. According to the studies performed by Sahel Consultant Co. for the subway project, Line A in 2011, the groundwater table is generally in a depth of 30 to 40 meters below the ground. 825 meters borehole were excavated by Sahel Consultant Co. in Line A and a number of 297 standard penetration tests (SPT), 9 down hole tests, 69 pressure meter tests (PMT), 19 plate load tests (PLT), 8 direct shear tests and 29 Lefranc permeability tests were implemented. The performed field and laboratorial investigations in the study area show that the soil layers are divided into four different types. The strength and stiffness parameters for these four layers are presented in Table1.

In this article, two excavation projects in two different regions are studied. As it is illustrated in Fig.1, the direction of the subway project in Line A passes the boundary of the two excavation projects in the stations A5 (Police Sq.) and A6 (Amaryaser).

The two projects are in a distance of 3.5 kilometers from each other. It is noticeable that the subway tunnel excavation did not reach the study area during excavation, construction and monitoring of the projects.

In the Police Sq. project an area of 600 m^2 is excavated to a depth of 10 meters below the ground and stabilized using top-down construction. Site investigations were implemented by digging three 21 meters mechanical boreholes (Paydar Gostar Soil Mechanics Co., 2013). According to the site investigations implemented by Paydar Gostar Soil Mechanics Co. (2013) and Sahel Consultant Co. (2011), it can be claimed that the overall condition to the depth of 21 meters in this project is comprised of the three layers of Qc-1, Qf-2 and Qf-1.

In the Amaryaser project an area of 800 m^2 is excavated to a depth of 24 meters below the ground and stabilized using top-down construction. Site investigations were implemented by digging two 40 meters mechanical boreholes and two 30 meters manual pits (Pey Bonyan Ista Consulting Co., 2014). According to the soil classifications presented by Pey Bonyan Ista Consulting Co. (2014) and investigations implemented by Sahel Consultant Co. (2011) it is inferred that the geotechnical condition to the depth of 40 meters in this project can be attributed to the layers of Qc-2, Qc-1 and Qf-2 alternately. Further information on geotechnical condition and field investigation can be reached in Ghorbani (2017).

Table 1. Strength and stiffness parameters for the four soil layers (Sahel Consultant Co., 2011).

Soil properties		Qc-1	Qc-2	Qf-1	Qf-2
Cohesion (kg/cm^2)	c'	0.05 ± 0.05	0.15 ± 0.02	0.1 ± 0.02	0.2 ± 0.05
Friction angle (degree)	φ'	36 ± 3	32 ± 3	30 ± 3	27 ± 4
Elastic modulus (kg/cm^2)	Loading	300 ± 50	220 ± 50	190 ± 50	180 ± 50
	Unloading	900 ± 50	700 ± 50	570 ± 50	600 ± 50
Poisson's ratio (u)		0.3	0.3	0.32	0.3
Dry unit weight (g/cm^3)		1.7 ± 0.1	1.6 ± 0.1	1.55 ± 0.1	1.6 ± 0.1
Total unit weight (g/cm^3)		2 ± 0.1	1.93 ± 0.1	1.85 ± 0.1	1.95 ± 0.1

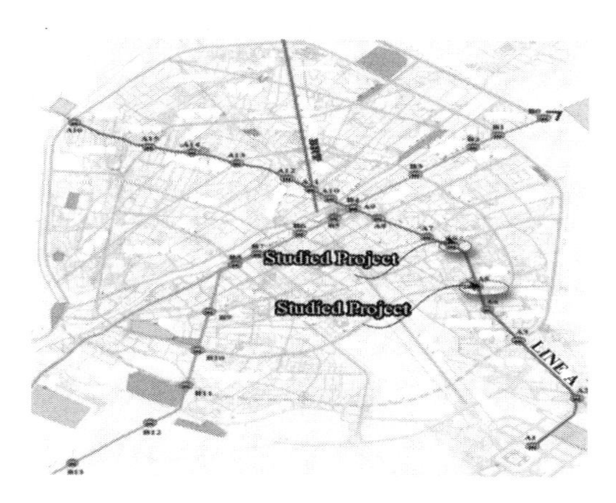

Figure 1. Subway lines and the studied projects on the map of Qom.

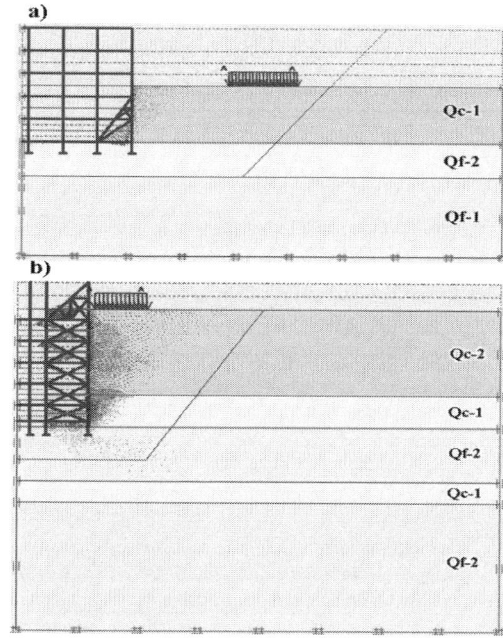

Figure 2. Modeled geometry in a) Police Sq. and b) Amaryaser projects.

3 DATA AND METHOD

Semi top-down construction was implemented in the studied projects. In semi top-down method concrete slab of substructure floors are cast from bottom to up after the construction of the other structural elements. If it were necessary the substructure floors in some regions could be carried out before the excavation to restrict the amount of deformation (Paek and Ock, 1996).

Numerical studies are implemented in this study to achieve a framework which correlates maximum ground deformation with stiffness of excavation supporting system in the geotechnical condition of Qom. By taking the advantage of symmetrical geometry and symmetrical construction sequence, only half of the excavation is modeled in 2D plain-strain conditions by means of 2D PLAXIS VER. 8.5. Numerical models were performed using hardening soil (HS) model which is an advanced elastoplastic multiyield surface model. In this model failure is defined by the Mohr-Coulumb failure criteria (Schanz and Vermeer, 1998; Schanz et al., 1999; PLAXIS Manual, 2014).

In this model default software values are used for unloading-reloading modulus (E_{ur}) and oedometeric tangent modulus (E_{oed}). The exponent m is assumed 0.5 for Qc-1 layer, 0.6 for Qc-2 layer, 0.8 for Qf-1 layer and 1 for Qf-2 layer. The value of this parameter is suggested to be ranged from 0.5 to 0.75 for geotechnical conditions like Qc-1 and Qc-2 and from 0.8 to 1 for conditions like Qf-1 and Qf-2 (Schanz and Vermeer, 1998; Janbu, 1963; Von Soos, 1990). The modeled geometry for Police Sq. and Amaryaser projects are shown in Fig.2a and b respectively.

In this article calibration is performed in Police Sq. geometrical conditions; so validation of the calibrated parameters can be evaluated in Amaryaser geometry. After the model is calibrated, numerical studies can be implemented by assuming the structural parameters as variables. In this study, 60 different finite element models are performed for each geotechnical condition. Consequently, maximum value of lateral movement can be correlated with excavation supporting system stiffness for Qc-1, Qc-2, Qf-1 and Qf-2 geotechnical conditions. Parametric studies are applied within these four soil types. As a result of 240 different numerical models implemented in these conditions, exponential equations are achieved which correlate the maximum lateral movement with supporting system stiffness.

4 RESULTS AND DISCUSSION

As the reference points are assumed to be fixed in place, their relative displacement from each other should always be zero which makes it possible to evaluate the accuracy of total survey station. Relative displacement of reference points are measured 115 and 147 times in Police Sq. and Amaryaser Projects respectively. According to Finno and Blackburn (2005), accuracy of total station is estimated to be equal with 2σ which σ is the calculated standard deviation obtained from measured data. This boundary covers 94% and 93%

Table 2. Relative importance of parameters obtained from sensitivity analysis.

Number	Parameter	Relative importance
1	$\varphi 1$	42.2%
2	c2	17.1%
3	E2	16.7%
4	R	10.6%
5	c1	8.5%
6	E1	2.6%
7	$\varphi 2$	2.3%

of data in projects of Police Sq. and Amaryaser respectively. The results show that the accuracy of total station is 3 mm for Police Sq. and 3.8 mm for Amaryaser projects (Ghorbani, 2017).

Statistical studies show that the obtained monitoring data using Total Station TS02 have been collected with a good quality. However use of a more accurate tool with a higher accuracy in high risk excavations is recommended. Sensitivity analysis is conducted on geotechnical parameters in order to calibrate the numerical model. Studied geotechnical parameters in HS model are secant Young's modulus at the 50% stress level, friction angle, cohesion, exponential parameter and the bottom boundary of the numerical model. Default values are assigned for all the other HS model parameters. A decreasing multiplier of R is applied to column stiffness to account for soil disturbance induced by column installation which is considered as a variable in sensitivity analysis. In order to simplify the numerical models in sensitivity analysis, geotechnical layers of Qf-2 and Qf-1 were merged and named as Qf-1. Accordingly, Qc-1 layer describes geotechnical conditions to the bottom of excavation depth and Qf-1 layer describes geotechnical conditions from bottom of excavation to bottom boundary of the model. Geotechnical parameters indexed 1 are those which belong to Qc-1 layer and the ones indexed 2 are those which belong to Qf-1 layer. Sensitivity analysis was performed by conducting 73 different models. According to the obtained results inconsistent parameters were discounted from calculations. Relative importance of studied parameters in sensitivity analysis is shown in Table2. It is evolved that the parameters of ϕ_1 and ϕ_2 with a relative importance of 42% and 2% are the most and the least effective parameters in this numerical model.

According to the performed sensitivity analysis and by discounting the parameters of c_1 and c_2 from calculations, Eq.1 was inferred from the analysis (Fig.3). By using Eq.1 it is possible to calculate the maximum lateral movement of excavation wall by varying the value of parameters of E_1, E_2, ϕ_1, ϕ_2 and R without processing any new numerical model. The inferred equation performed an expressive role in calibrating the numerical model with results obtained from field monitoring in Police Sq. project.

$$\delta_{h-max} = 0.155E_1 + 42.67\varphi_1 + 6.689E_2 + 0.13\varphi_2 + 2.68R - 89.4 \quad (1)$$

After the calibration of geotechnical parameters in Police Sq. geometry, obtained results were validated in Amaryaser conditions. Validation results for one of the sections in Amaryaser project is shown in Fig.4. In Fig.4 numerical results are compared with monitoring results from 18th of May until 22nd of September in 2016.

Small discrepancies observed in comparisons are attributed to monitoring tool inaccuracy and simplifications applied in numerical models. One of the insufficiencies of field monitoring using total survey station is inaccessibility of points in depth lower than bottom of excavation.

Fig.4 shows that the numerical models have predicted deformation trend and maximum lateral movement of excavation wall in different stages with an appropriate precision. Accordingly validation results confirm that the calibrated parameters for the geotechnical layers of Qc-1, Qc-2, Qf-1 and Qf-2 are estimated accurately. Site investigations implemented in different regions of Qom approve that about 75% of the city is limited to the discussed geotechnical layers. Therefore it is possible to assign the calibrated parameters to a large number of excavation projects executed in Qom. Values of calibrated parameters are shown in Table3. Additional details can be found in Ghorbani (2017).

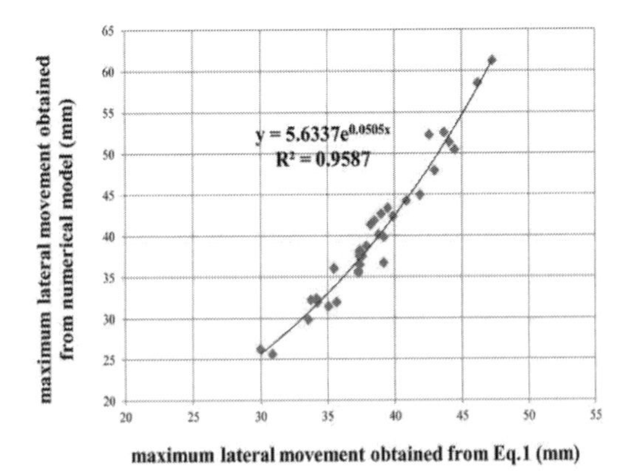

Figure 3. Estimation of maximum lateral movement of excavation wall using Eq.1.

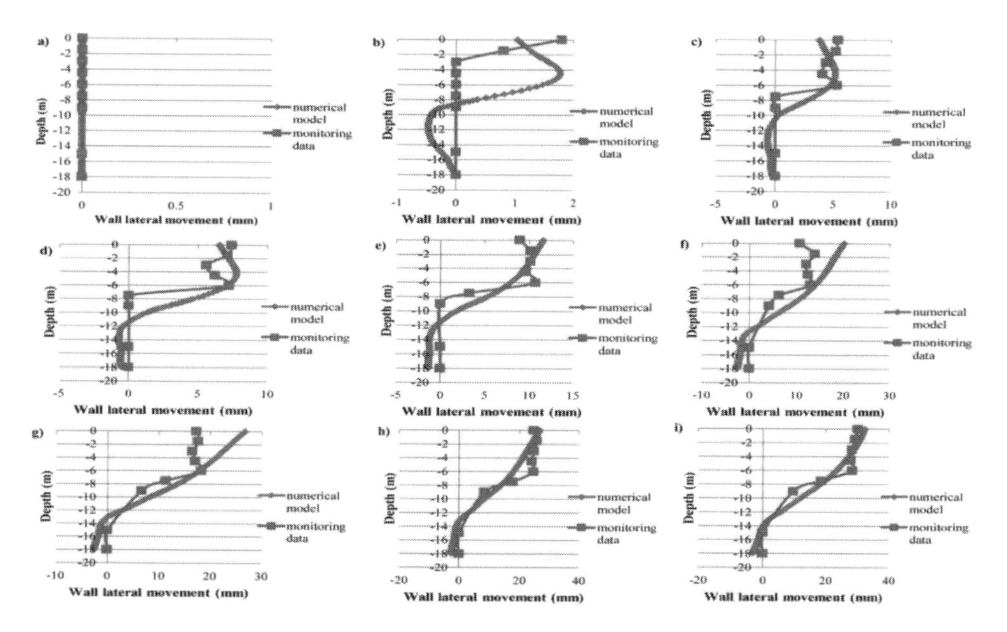

Figure 4. Validation results in Amaryaser project in a) 5/18/2016 b) 6/2/2016 c) 6/16/2016 d) 7/3/2016 e) 7/18/2016 f) 7/28/2016 g) 8/21/2016 h) 9/5/2016 i) 9/22/2016.

Table 3. Calibrated parameters for the four geotechnical layers.

Soil properties		Qc-1	Qc-2	Qf-1	Qf-2
Cohesion (kg/cm^2)	c'	0	0.08	0.05	0.1
Friction angle (degree)	φ'	32	28	25	25
Elastic modulus	Loading	300	280	110	105
(kg/cm^2)	Unloading	900	840	330	315
Dry unit weight (g/cm^3)		1.85	1.65	1.6	1.6
Total unit weight (g/cm^3)		1.95	1.75	1.75	1.75

In this study in order to design supporting elements based on deformation control, structural parameters were assumed as variables in the four geotechnical layers. Moment inertia (I) and bending stiffness (E) of retaining wall (wall rigidity), horizontal and vertical support spacing (S_H and S_V), in-plane span length (S_P), adjacent overburdens (Q), pile embedment below the excavation depth (T), excavation width (w) and excavation height (H) are taken as structural variables. In addition a 10 meters long overburden is assumed in a distance of 14 meters from the edge of excavation. Overburden amount is taken 0, 10, 20 and 30 kPa as four extra different numerical models. All the numerical models are processed in two different depths of 6 and 10 meters which have resulted in 60 different numerical models for each geotechnical layer. Accordingly, a PBD framework can be obtained for the four geotechnical layers using 240 numerical models.

The obtained values of maximum lateral movement are normalized with respect to the height of the excavation. Comparisons show that the Qc-2 layer has the least maximum lateral movement among the four geotechnical layers. Regarding the negligible cohesion values reported for the Qc-1 layer, highest maximum lateral movements are resulted for this layer in numerical studies.

The relative importance of studied variables for each geotechnical layer is shown in Table4. The results are consistent with former studies (Clough and O'Rourke, 1990; Bryson and Zapata, 2012). According to numerical studies, Qc-2 layer has the least pliability from structural variables among the four geotechnical layers which is due to its inherent stability; there is no discernable difference in general behavior of different supporting systems implemented in these kinds of geotechnical layers (Clough and O'Rourke, 1990). Increase in supporting system stiffness may sometimes reach a point which will not decrease ground deformations any more. In such an especial condition if the lateral movement exceed the allowable ground deformation, inevitably excavation depth should be limited or general geotechnical conditions should be improved.

In order to correlate maximum lateral movement with supporting system stiffness, the

Table 4. Relative importance of structural variables in the studied geotechnical layers.

Structural variable	Relative importance of structural variables in different geotechnical conditions			
	Qc-1	Qc-2	Qf-1	Qf-2
I	2.74%	2.55%	0.58%	0.06%
E	1.23%	0.51%	0.47%	0.16%
S_H	9.6%	5.1%	12.81%	11.09%
S_V	49.38%	35.71%	37.25%	25.34%
w	23.32%	20.41%	19.79%	36.43%
T	6.86%	10.2%	13.97%	11.09%
S_P	6.86%	25.51%	15.13%	15.84%

relative stiffness parameter (R) which was introduced by Bryson and Zapata (2012) is reformed to adjust the current studies. By making slight changes in definition of R, the parameters of pile embedment length (T), excavation width (w) and in-plane span length (S_P) are taken into account (Eq.2). This parameter is dimensionless and is governed by the stiffness of soil and supporting structure. The parameters of E_s, E, S_H, S_V, H, I, S_P, w and T in Eq.2 are elastic modulus of soil, bending stiffness of wall, horizontal support spacing, vertical support spacing, excavation height, moment inertia of wall, in-plane span length, excavation width and pile embedment length respectively.

$$R = E_s / E \times S_H S_V H / I \times S_P / wT \tag{2}$$

Presented graphs in Fig.5 and 6 correlate the normalized maximum lateral movement with relative stiffness of supporting system in geotechnical conditions of Qc and Qf respectively. Slight dispersion is remarkable in Qf-1 and Qf-2 layers (Fig.6). Although the general trend in data obtained from numerical studies in these layers is comprehensible. According to data obtained from numerical studies, Eq.3, 4, 5 and 6 are presented which correlate the normalized maximum lateral movement (δh/H%) with relative stiffness of supporting system (R) in geotechnical layers of Qc-1, Qc-2, Qf-1 and Qf-2 respectively.

$$\delta h / H\% = 0.0034 e^{0.0029R} \tag{3}$$

$$\delta h / H\% = 0.0005 e^{0.0031R} \tag{4}$$

$$\delta h / H\% = 0.0011 e^{0.0115R} \tag{5}$$

$$\delta h / H\% = 0.0007 e^{0.0174R} \tag{6}$$

Existed dispersion in Qf-1 and Qf-2 layers emphasizes the necessity of further studies on effective structural parameters. Monitoring real case studies in similar geotechnical conditions and comparing results with numerical models can be helpful in developing current correlations.

In order to evaluate the precision of the presented equations, eight hypothetical models were processed in Amaryaser geometry conditions. Obtained results from hypothetical models are compared with those in graphs of Fig.5 and 6. Validations show that the maximum lateral movement can be predicted with cautions as a first order estimation in engineering practices in similar geotechnical conditions.

After estimating maximum lateral movement using Fig.5 and 6, maximum vertical settlement of the adjacent ground is predictable by means of empirical correlations. Hsieh and Ou (1997) defined Eq.7 and Eq.8 which estimate vertical settlement in a distance adjacent to the excavated site. In these relations δ_v is adjacent ground settlement, d is distance from edge of the excavation wall, H_e is excavation depth and δ_{vm} is the maximum value of ground settlement. The ratio of maximum vertical settlement (δ_{vm}) to maximum lateral movement (δ_{hm}) is about 0.75 in stiff clay or medium dense sand condition (Clough and O'Rourke, 1990; Hsieh and Ou, 1997).

$$\delta_v = \left(-0.636\sqrt{d/H_e} + 1\right)\delta_{vm} \quad for\, d/H_e \le 2 \tag{7}$$

$$\delta_v = \left(-0.171\sqrt{d/H_e} + 0.342\right)\delta_{vm} \quad for\, 2 < d/H_e \le 4 \tag{8}$$

As a result, the proposed procedure predicts maximum lateral and vertical ground movements by using geotechnical and supporting system parameters and therefore allows to design supporting system elements on the basis of allowable ground deformations. Limiting conditions in this method is defined by the maximum allowable ground settlements and so this framework is recommended for urban areas regarding the acceptable damage level of adjacent buildings.

Design procedure in the proposed PBD approach in dense sand or stiff clay geotechnical conditions can be defined by the following 6 steps.

1. Determining type of the geotechnical condition (it should be one of the studied geotechnical conditions studied herein)
2. Determining the allowable damage level of adjacent buildings and the allowable adjacent ground settlements by evaluating vulnerability of neighboring structures
3. Calculating maximum allowable lateral movement of excavation wall using empirical correlations obtained from former studies (it is 1.33 times bigger than maximum vertical settlement in stiff clay and dense sand geotechnical conditions)

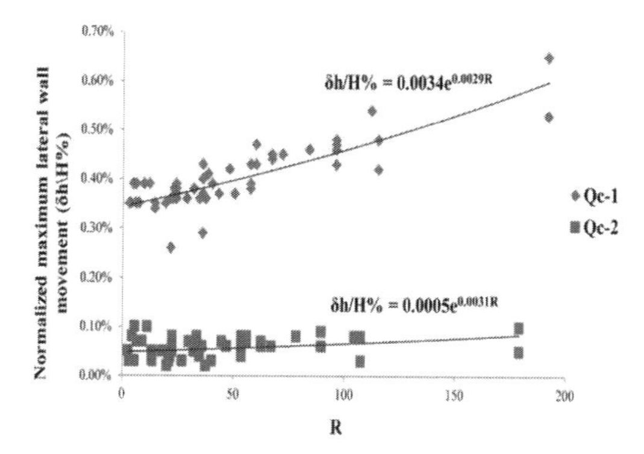

Figure 5. Normalized maximum lateral movement in Qc geotechnical layers.

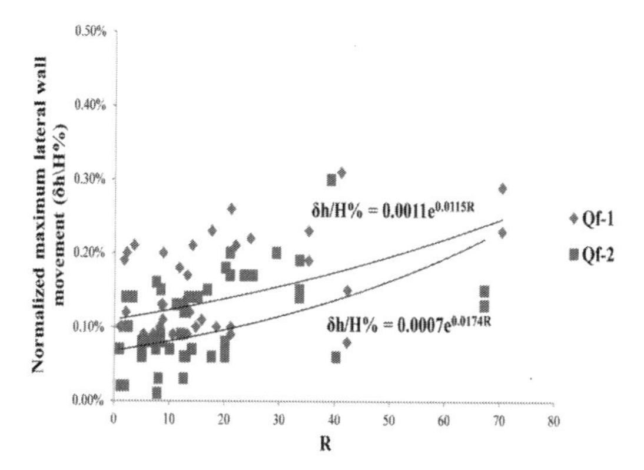

Figure 6. Normalized maximum lateral movement in Qf geotechnical layers.

4. Calculating the required system stiffness according to the geotechnical condition and excavation depth using Eq. 3, 4, 5 and 6 and Figs. 5 and 6
5. Estimating structural parameters regarding the required system stiffness obtained from step 4 in a way which satisfies Eq.2
6. Checking structural parameters regarding the limit equilibrium correlations to confirm its sufficiency against structural failure

It is noticeable that the presented method is only applicable in the studied geotechnical conditions which should be confirmed by field investigations. Resulted PBD is applicable in stabilizing subway stations of Line A in Qom subway project and any other excavation project which is executed in one of the studied geotechnical conditions.

5 SUMMARY AND CONCLUSIONS

Excavation support systems are usually designed on the basis of limit equilibrium method in the regulations. The output in this concept is sufficient in resisting structural failure but may lead to excessive ground deformations which will result in different damage levels in adjacent buildings. In order to overcome this deficiency a PBD approach in dense sand and stiff clay of Qom in Iran is presented. The following conclusions are made on the basis of the work presented herein.

- The simple field studies presented herein helps the contractor to reconsider the construction phases in order to optimize the stabilization process.
- Geotechnical parameters of the study area were calibrated and validated in this research.
- Resulting graphs evolved from this study are capable to determine the required stiffness of excavation support systems based on the maximum allowable deformation of adjacent structures in the studied geotechnical conditions.
- Results obtained from this study are applicable in stabilizing subway stations of Line A in Qom subway project and any other excavation project which is executed in one of the studied geotechnical conditions.

ACKNOWLEDGEMENTS

We thank Sahel Consultant Engineers Co. for their generosity in sharing geotechnical data. We would also like to show our gratitude to Mr. Sakhai, laboratory administrator of University of Qom for his assistance in our field monitoring.

REFERENCES

Boone, S.J. 2001. Ground movement-related building damage. *Journal of Geotechnical Engineering*, Volume 122, No. 11, pp. 886–896.

Boone, S.J., Westland, J. & Nusink, R. 1999. Comparative evaluation of building responses to an adjacent braced excavation. *Canadian Geotechnical Journal*, Volume 36, No. 2, pp. 210–223.

Boscardin, M.D. & Cording, E.G. 1989. Building response to excavation induced settlement. *Journal of Geotechnical Engineering*, Volume 115, No. 1, pp. 1–21.

Bowles, J.E. 1988. *Foundation analysis and design*. 4th ed. McGraw-Hill Book Company, New York.

Bryson, L.S. & Zapata, D.G. 2010. Direct approach for designing an excavation support system to limit ground movement. *2010 Earth Retention Conference 3*, ASCE Library, Ondokuz Mayis University, pp.154–161. DOI: 10.1061/41128(384)12.

Bryson, L.S. & Zapata, D.G. 2012. Method for estimating system stiffness for excavation support walls. *Journal of Geotechnical and Geoenvironmental Engineering*, Volume 138, No. 9, pp. 1104–1115.

Burland, J.B. 1997. Assessment of risk of damage to building due to tunneling and excavation. *Earthquake Geotechnical Engineering*, Ishihara Edition, Balkema, Rotterdam, pp. 1189–1201.

Clough, G.W. & O'Rourke, T.D. 1990. Construction-induced movements of in situ walls. *In Proceeding, Design and Performance of Earth Retaining Structures*, ASCE Special Conference, Ithaca, New York, pp. 439–470.

Day, W. 1998. Discussion on Ground movement-related building damage. *Journal of Geotechnical and Geoenvironmental Engineering*, Volume 124, No. 5, pp. 426–465, URL: http://worldcat.org/oclc/3519342

Dunnicliff, J. 1998. *Geotechnical instrumentation for monitoring field performance*. John Wiley and Sons, A Wiley Interscience Publications.

Finno, R.J. 2007. Use of monitoring data to update performance predictions of supported excavations. *Seventh International Symposium on Field Measurements in Geomechanics*, pp. 1–30. DOI: 10.1061/40940(307)3.

Finno, R.J. & Blackburn, J.T. 2005. Automated monitoring of supported excavation. *Geotechnical Application for Transportation Infrastructure*, pp. 1–12. DOI: 10.1061/40821(181)1.

Finno, R.J., Bryson, S. & Calvello, M. 2002. Performance of a stiff support system in soft clay. *Journal of Geotechnical and Geoenvironmental Engineering*, Volume 128, No. 8, pp. 660–671.

Finno, R.J. & Calvello, M. 2005. Supported excavation: Observational method and inverse modeling. *Journal of Geotechnical and Geoenvironmental Engineering*, Volume 131, No. 7, pp. 826–836.

Finno, R.J. & Hassash, Y.M.A. 2006. Integrating tools to predict monitor and control deformation due to excavations. *Geo Congress*, DOI: 10.1061/40803(187)75.

Finno, R.J. & Tu, X. 2006. Selected topics in numerical simulation of supported excavation. *International Conference of Construction Processes in Geotechnical Engineering for Urban Environment*, Th. Triantafyllidis, ed., Bochum, Germany, Taylor & Francis, London, 3–20.

Ghavidel Aghdam, O. 2011. Assessment of underground construction using Top-Down method. *First Asian and 9th Iranian Tunneling Symposium*, Iran, Tehran (In Persian).

Ghorbani, E. 2017. Designing an excavation considering Top-Down method on the basis of deformation control using finite element approach based on monitoring data. University of Qom, Master's Thesis (In Persian).

Hong, W.K., Kim, J.M., Lee, H.C., Park, S.C., Lee, S.G. & Kim, S.I. 2010. Modularized Top-Down construction technique using suspended pour forms (Modularized RC System Downward, MRSD). *The Structural Design of Tall and Special Buildings*, Wiley Online Library, pp. 802–822. DOI: 10.1002/tal.521.

Huang, Z.H., Zhao, X.S., Chen, J.J. & Wang, J.H. 2014. Numerical analysis and field monitoring on deformation of the semi-Top-Down excavation in Shanghai. *New Frontiers in Geotechnical Engineering*, pp. 198–207.

Hsieh, P.G. & Ou, C.Y. 1997. Shape of ground surface settlement profiles caused by excavation. *Canadian Geotechnical Journal*, Volume 35, pp. 1004–1017.

Janbu, N. 1963. Soil compressibility as determined by oedometer and triaxial tests. *Proceeding Europe Conference on Soil Mechanics and Foundation Engineering*, Wiesbaden, 19–25.

Kung, G.T.C., Juang, C.H., Hsiao, E.C.L. & Hashash, Y.M.A. 2007. Simplified model for wall deflection and ground surface settlement caused by braced excavation in clays. *Journal of Geotechnical and Geoenvironmental Engineering*, Volume 133, No. 6, pp.731–747.

Lee, H.S., Lee, J.Y. & Lee, J.S. 1999. Noneshored formwork system for Top-Down construction. *Journal of Construction Engineering and Management*, Volume 125, No. 6, pp. 392–399.

Leica Geosystems. 2008. *Leica FlexLine TS02/TS06/TS09 user manual*. Leica Geosystems AG, Heinrich-Wild-Strasse, CH-9435 Heerbrugg, Switzerland, www.leica-geosystems.com.

Li, M.G., Chen, J.J., Xu, A.J., Xia, X.H. & Wang, J.H. 2014. Case study of innovative Top-Down construction method with channel-type excavation. *Journal of Construction Engineering and Management*, Volume 140. DOI: 10.1061/(ASCE)CO.1943–7862.0000828.

Long, M. 2001. Database for retaining wall and ground movements due to deep excavation. *Journal of Geotechnical and Geoenvironmental Engineering*, Volume 127, pp. 203–224.

Luo, Y., Chen, J., Xi, W., Zhao, P., Qiao, X., Deng, X. & Liu, Q. 2016. Analysis of tunnel displacement accuracy with total station. *Measurement*, Volume 83, pp. 29–37.

Marr, W.A. & Hawkes, M. 2010. Displacement based design for deep excavations. *2010 Earth Retention Conference 3*, ASCE Library, Hacettepe University, pp 82–100.

Ou, C.Y., Hsieh, P.G. & Chiou, D.C. 1993. Characteristics of ground surface settlement during excavation. *Canadian Geotechnical Journal*, Volume 30, No. 5, pp. 758–767.

Paek, J.H. & Ock, J.H. 1996. Innovative building construction technique: Modified Up-Down method. *Journal of Construction Engineering and Management*, Volume 122, No. 2, pp. 141–146.

Paydar Gostar Soil Mechanics Co. 2013. Geotechnical investigation report, 10529, 24 and 25 (In Persian).

Peck, R.B. 1969. Deep excavation and Tunneling in soft ground. *In Proceeding of the 7th International Conference on Soil Mechanics and Foundation Engineering*, Mexico City, pp. 225–290.

Pey Bonyan Ista Consulting Co. 2014. Geotechnical investigation report, 8826 and 8827.1 (In Persian).

PLAXIS Manual. 2014. *material models*. PLAXIS publications, www.PLAXIS.nl

Sabzi, Z. & Fakher, A. 2013. A field investigation into the performance of inclined struts connected to adjacent buildings during excavation. *Modares Civil Engineering Journal*, Volume 13, Issue 4, pp. 27–43, (In Persian).

Sahel Consultant Engineers Co. 2011. Engineering service for Qom subway project—Line A, Geology studies in determining tunnel direction, SCE 2000 UNGR TUN EG RP-B0 (In Persian).

Schanz, T. & Vermeer, P.A. 1998. On the stiffness of sands. *Geotechnique, Pre-failure Deformation Behaviour of Geomaterials, Institution of Civil Engineers*, Great Britain, pp. 383–387.

Schanz, T., Vermeer, P.A. & Bonnier, P.G. 1999. The hardening soil model: Formulation and verification. *Beyond 2000 in Computational Geotechnics- 10 Years of PLAXIS*, Balkema, Rotterdam, ISBN 90 5809 040 X
Song, Y. 1999. Inspection on the precision of 3D deformation observation by the free stationing method. *Geotech. Investing. Surv.*, Volume 1, pp. 61–63.
Von Soos, P. 1990. *Properties of soil and rock*. In Grundbautaschenbuch part 4. Ernst and Sohn, Berlin (In German).
Wang, J.H., Xu, Z.H., Di, G.E. & Wang, W.D. 2006. Performance of a deep excavation constructed using the united method: Bottom-Up method in the main building part and Top-Down method in the annex building part. *Underground Construction and Ground Movement*, pp. 385–392.
Wang, J.H., Xu, Z.H. & Wang, W.D. 2010. Wall and ground movement due to deep excavations in Shanghai soft soils. *Journal of Geotechnical and Geoenvironmental Engineering*, Volume 136, No. 7, pp. 985–994.
Yang, S.L., Liu, W.N., Wang, M.S., Huang, F. & Cui, N.Z. 2004. Study on the auto-total station for monitoring analyzing and forecasting tunnel country rock deformation. *Journal of China Railway Society*, Volume 3, pp. 93–97.
Yuan, H., Liu, C.L., Lu, J., Deng, C. & Gong, S. 2012. The principle and accuracy analysis of non-contact monitoring for tunnel based on free station of total station. *Geotech. Investig. Surv.*, Volume 8, pp. 63–68.

Numerical Methods in Geotechnical Engineering IX – Cardoso et al. (Eds)
© 2018 Taylor & Francis Group, London, *ISBN 978-1-138-33203-4*

Three-dimensional effects of nail arrangement on soil-nailed convex corners

M. Sabermahani, M.R. Nabizadeh Shahrbabak & M. Mohammadi Bagheri
School of Civil Engineering, Iran University of Science and Technology, Tehran, Iran

ABSTRACT: Prediction of soil-nailed wall deformation has been of great interest to many researchers. Excavation overall stability and performance are highly affected by such deformation especially during construction as well as the exposure to external loading. Utilizing numerical models is one the common approaches to investigate the behavior of the soil-nailed walls particularly at convex corners. The focus of this study is to assess the deformation in existence of different soil nail arrangements, namely parallel and perpendicular. To this end, three-dimensional finite element models constructed with the above-mentioned arrangements are generated. The results show that the parallel arrangement enhances the control of horizontal displacements at the convex corners due to its better anchorage of the soil-nailed wall to the back soil body, while the perpendicular arrangement improves the behavior of the vertical displacements caused by its fully reinforced zone generated with overlapping nails.

1 INTRODUCTION

In today constructions, which perform in municipal areas, limitation in excavation geometry usually is inevitable due to reasons likewise, land sacristy, increasingly need of basement parking floors, and existed buildings density. Limited geometries can cause convex or concave corners, which need more consideration in their stability design and calculation process. An official-commercial complex is an example of limited geometry excavation located in Tehran, Iran. As shown in Figure 1, due to neighbor properties geometry this excavation has two convex corners that should be stabilized.

In this project, soil-nailed wall system was used as a permanent stabilization system and excavation construction finished 2 years ago. After 2 years, convex corners horizontal and vertical deformations caused tension cracks in shotcrete wall facing as shown Figures 2–3, so failure risk level reached an alarming state. After all, the mentioned case is an example of daily challenges that geotechnical engineers are dealing with corners in excavation design, calculation, and construction process, yet there is a technical-practical gap in comprehensive studies of convex corners deformation behavior.

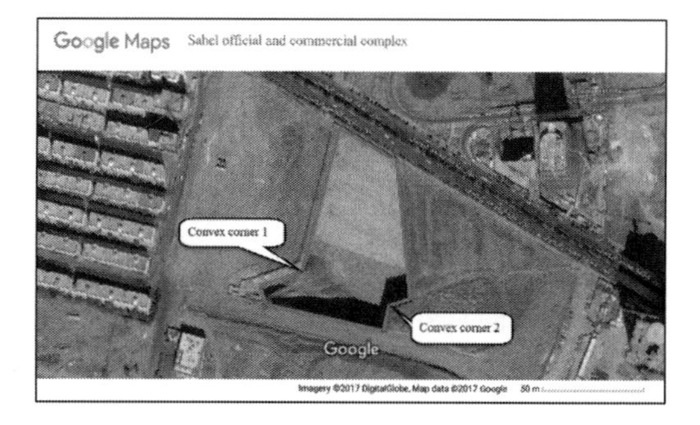

Figure 1. Excavation plan of official-commercial complex.

Figure 2. Tension cracks of shotcrete wall facing in convex corner No.1 (image by authors).

2 LITERATURE REVIEW

One of the most important parameters affecting the soil mass overall stability and serviceability conditions is the deformation caused by excavation.

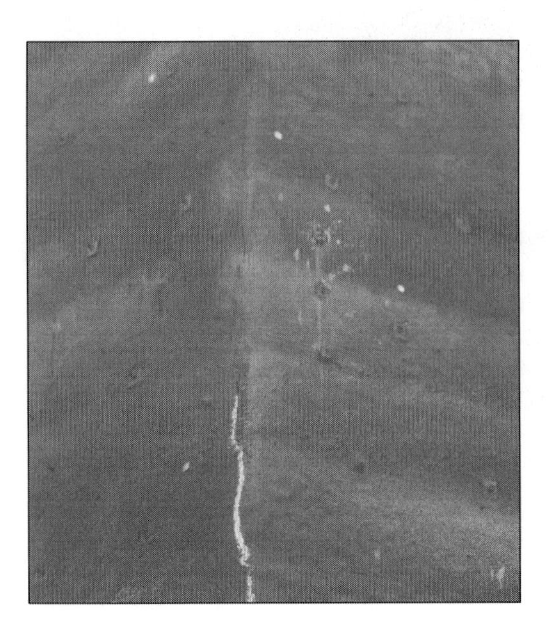

Figure 3. Tension cracks of wall shotcrete facing in convex corner No.2 (image by authors).

Various soil reinforcement methods have been used to control and reduce wall deformations; however, one of the most useful methods is soil nailed wall system. Soil nailing is an in situ reinforcing method with steel bar elements in the shape of ordinary or hollow rods. To make the system properly-operative, the tensile force in reinforcing elements should be mobilized by lateral displacement of the soil mass which is introduced as passive behavior (Lazarte et al., 2003). According to FHWA guideline of soil nailed walls (Lazarte et al., 2003), horizontal and vertical deformation of a soil nailed wall depends on several factors including wall height, H, wall geometry, soil type, horizontal and vertical nail spacing, maximum height of excavation at any level, factor of safety, nails length to the wall height ratio, nails inclination degree and surcharge loads. On the other hand, soil nail arrangement has a significant effect on wall deformations in excavation, including convex corners.

Various methods including limit state equilibrium method, finite element method, and finite difference method are available for numerical analysis and software modeling of nailed wall systems. Mittal (2006) evaluated the effective parameters on nailed walls overall safety factor using limit state equilibrium and analytical friction circle method (Mittal, 2006). Discussion on factors affecting soil-nailed walls behavior and deformation evaluations often has been implied by assuming plane strain and two-dimensional conditions (Lima et al., 2003, Lima et al., 2004, Singh and Babu, 2010, Babu et al., 2002, Fan and Luo, 2008). Generally in Three-dimensional modeling only a limited sector of nailed wall is modeled containing a column of soil nails to represent soil-nail arrangement in the third dimension (Kim et al., 2013, Yang and Drumm, 2000, Zhang et al., 2011, Smith and Su, 1997). NG and Lee (2002), evaluated soil nailed system stiffness effect on the tunnel facing deformations using three-dimensional modeling (Ng and Lee, 2002).

However, a few researches have been conducted to investigate corners deformation behavior. In a fundamental study conducted by Ou & Shiau (1998), effect of excavation model boundaries has been evaluated. Furthermore, based on their field observations and three-dimensional analysis of the case histories presented in their research, it was found that short wall deformations are smaller than long wall deformations also wall deformation decreases with decreasing distance from the corner (Ou and Shiau, 1998). Pan et al. (2008) analyzed spatial effected deformation of corners in composite soil nailing walls. The results indicate that maximum deformations occur on the side near convex corners, and deformation of the concave corner is less than the convex corner (Pan et al., 2008).

The recent study conducted by Zhao et al. (2015) on the excavation concave and convex corners influence on adjacent buildings shows areas which affected by soil settlements near the concave and convex locations have much different pattern from areas were near middle length of the excavation wall. They figured out that settlement pattern in the corner location of excavation is a three-dimensional surface and cracks in neighboring buildings induced by significant differential settlement and torsional deformation can appear when buildings are located in the main affected-areas of the corner effect of excavation (Zhao et al., 2015).

Most studies reported before, can be categorized into two main groups: fundamental and case studies. Fundamental studies are trying to determine three-dimensional behavior of soil-nailed walls; also, case studies are aiming to describe three-dimensional behavior of walls. However, the main aim of this paper is to evaluate the effects of nail arrangement on three-dimensional deformation behavior of soil-nailed walls in convex corners. For this purpose, two different nail arrangement has been applied to three-dimensional convex corner models with variable length of the corner.

3 RESEARCH METHOD

3.1 *Models geometry and soil nails arrangement*

To perform three-dimensional models, 10 meters-high walls with variable corner length were

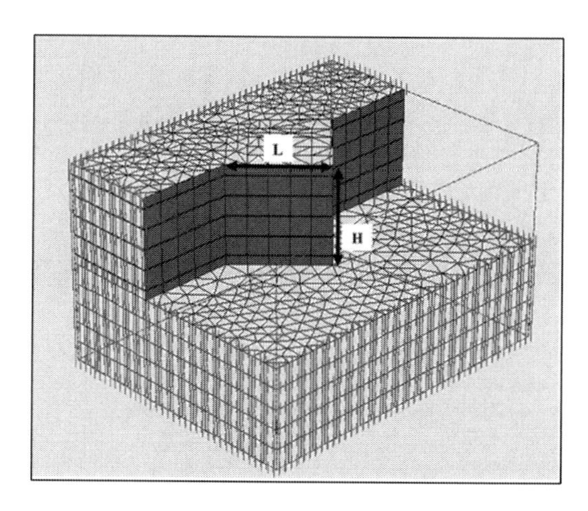

Figure 4. Proposed geometry for the three-dimensional modeled wall (H and L are wall height and corner length of corner respectively).

assumed. In this study, an ordinary corner was modeled in the excavation with no adjacent walls or limiting boundaries, although in practical aspects corners are located nearby other corners, local surcharges and adjacent structures, which effects corner deformation. Therefore, to avoid effective factors in net corner deformations a unique corner is modeled as shown schematically in Figure 4.

In this study corner angle is equal to 45 degrees and the corner length varies from $0.2 \times H$ to $2.0 \times H$ (2 meters to 20 meters) to investigate the effect of nail arrangement in varying convex corner length on wall deformations. Two nail arrangements are introduced in this study, perpendicular arrangement (PDD) and parallel arrangement (PRR). In PDD arrangement soil nails are located perpendicular to the corner wall, and in PRR arrangement soil nails are located parallel to the corner bisector (45-degree inclination with corner wall). These two arrangements are the most common nail arrangements in the practical fields. PDD and PRR arrangements are shown in Figures 5 and 6, respectively. In current research, each PDD and PRR nails arrangement is applied on various corner lengths (10 models, 2 to 20 meters length), so 20 three-dimensional models are created and analyzed by finite element method.

In order to evaluate PDD and PRR nail arrangement effects on soil nailed wall deformations in plain strain analysis without any corners (to simulate two-dimensional analysis), an extended length of the corner should be considered. By extending the corner length, two models are produced as shown in Figures 7 and 8.

In this study, the convex corner length varies from 2 to 20 meters and the length variation

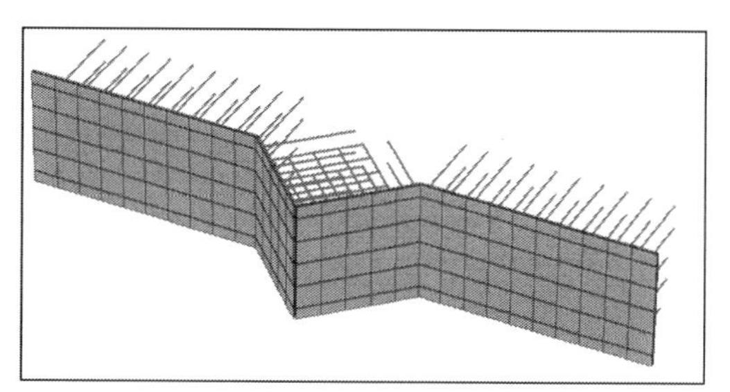

Figure 5. PDD arrangement of soil nails in the convex corner (In order to avoid soil nail elements intersection, left soil nails are located 10 cm below the right ones).

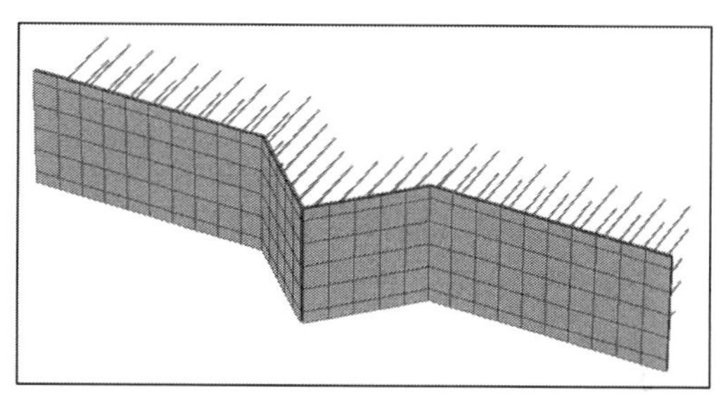

Figure 6. PRR arrangement of soil nails in the convex corner.

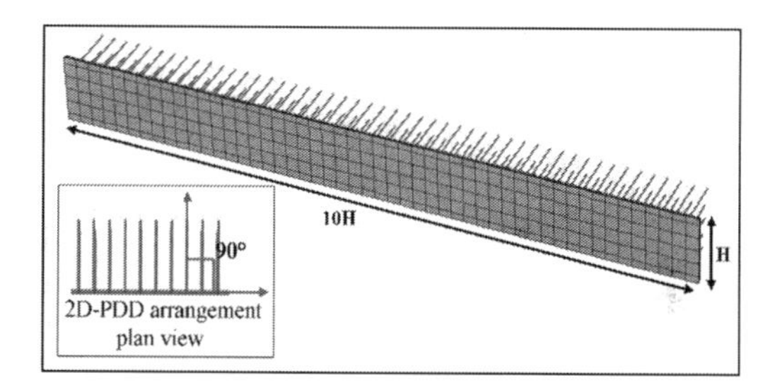

Figure 7. Geometry of PDD plain strain model (2D-PDD).

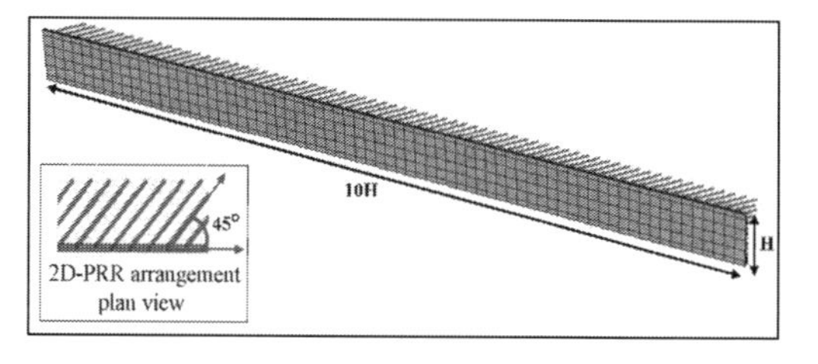

Figure 8. Geometry of PRR plain strain model (2D-PRR).

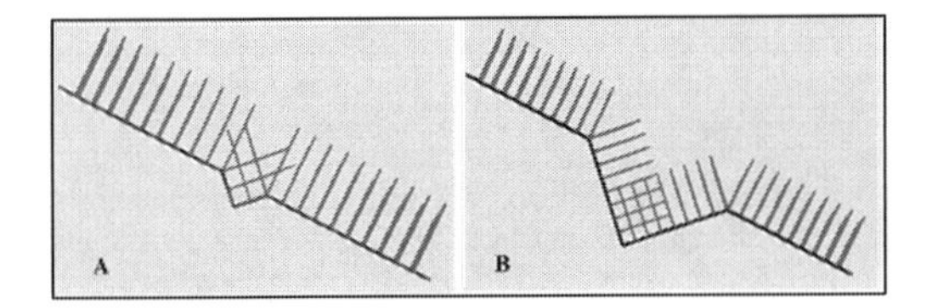

Figure 9. Plan view of PDD walls (corner lengths are equal to (a) $0.4 \times H$, and (b) $1.6 \times H$).

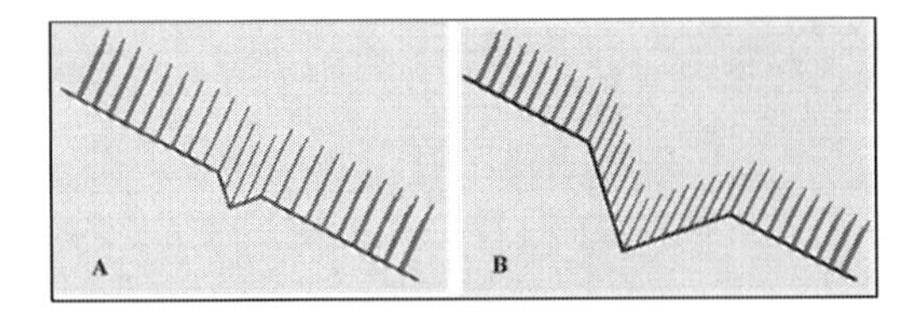

Figure 10. Plan view of PRR walls (corner lengths are equal to (a) $0.4 \times H$, and (b) $1.6 \times H$).

causes the wall different behaviors. Selected walls with various corner length are shown in Figures 9 and 10.

3.2 *Soil, nails and shotcrete facing properties*

In the modeling procedure, material properties including soil mechanical properties (presented in Table 1), nails and, shotcrete facing properties (presented in Table 2) should be defined. These mechanical properties are assumed based on authors experience of general condition of Tehran soil and commonly used types of nail and shotcrete facing elements. Soil material is modeled by hardening soil small strain model in the software.

3.3 *Three-dimensional model analysis*

After creating the geometry and defining the materials mechanical properties, Three-dimensional soil nailed wall models are analyzed. To study deformation behavior of the corner, models are analyzed by finite element method and deformations are calculated in each elements.

Three axes are considered to evaluate horizontal deformation of the corner, as shown in Figure 9. These three axes are located in corner head (HC axis), middle of corner length (MC axis), and at the end of corner length (EC axis). On the other hand, to evaluate vertical deformation of the corner, two points are assumed on the corner length. The points are located on the top of the wall in corner head point (HCP point) and end of corner length point (ECP point). These points are shown in Figure 11.

Table 1. Mechanical properties of desired soil.

	Mechanical properties				
Parameter	$\gamma\left(\frac{kN}{m^3}\right)$	$E(MPa)$	$\varphi(°)$	$C\left(\frac{kN}{m^2}\right)$	υ
Value	19	30	37	5	0.35

Table 2. Mechanical properties of desired soil-nails element and shotcrete facing.

Material	Mechanical Properties	
Soil-nails elements	Bar diameter (m)	0.032
	Borehole diameter (m)	0.1
	$E_s(GPa)$	200
	$E_g(GPa)$	20
	$\gamma\left(\frac{kN}{m^3}\right)$	32.72
	υ	0.2
Shotcrete facing	Facing thickness (m)	0.1
	$E_s(GPa)$	200
	$E_{shotcrete}(GPa)$	20
	$\gamma\left(\frac{kN}{m^3}\right)$	21.35
	υ	0.2
	$EI(kN.m^2)^{*}$	1,832
	$EA(kN)^{**}$	2,198,000

* Flexural strength
**Axial stiffness

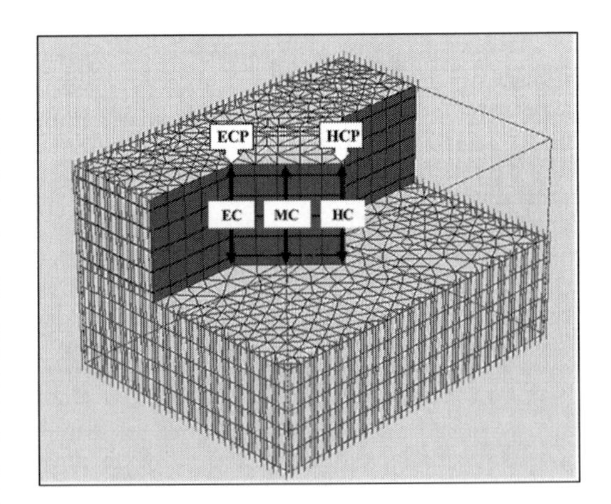

Figure 11. Location of the horizontal displacement evaluation axes (HC, MC, and EC) and the vertical displacement evaluation points (HCP, and ECP).

In this section creation and definition of three-dimensional models is described, and their calculated deformation by finite element method will be used for analysis. In the next section result of models analysis are presented and discussed.

4 RESULTS

4.1 *Horizontal deformations*

Horizontal deformation is measured on three axes located on the corner wall in order to evaluate corner effects on soil nailed wall horizontal deformation. Figure 12, shows maximum horizontal deformation of HC (head of the corner) axis for various corner length in both PDD and PRR arrangements.

As shown in Figure 12, due to the better anchorage of soil nailed wall to the backside soil body mass in PRR arrangement, maximum horizontal deformation of the corner head in PDD arrangement is greater than occurred deformation in PRR arrangement, while deformations decrease with an increase of corner length in both PDD and PRR arrangement. In both PDD and PRR arrangements, by increasing corner length, the major region of deformations moves to the middle of the wall, because corner walls tend to act as plain strain walls. These movements can cause high amount of deformations and is the main reason for deformation decrease in shotcrete, because of its tensile strength that pulls back soil body while it is under tension from a region located in the middle of corner length.

Figure 13, shows maximum horizontal deformation of MC (middle of the corner length) axis in various corner length for PDD and PRR arrangement.

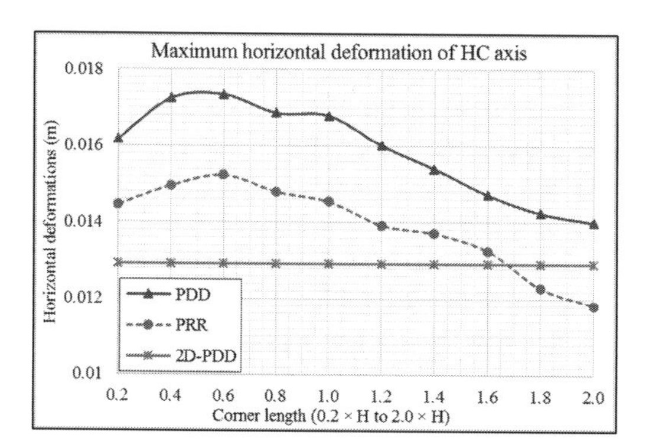

Figure 12. Maximum horizontal deformation of HC axis for PDD and PRR arrangement versus various corner lengths.

As results shown in Figure 13, when the corner's length is less than its height, in PRR arrangement models maximum horizontal deformation occurring in middle of the corner length is lower than deformations in PDD arrangement models. This can be explained by mentioned reason for better anchorage of soil nailed wall to the main soil body, However when the corner length is equal or greater than its height, PRR arrangement deformations start to increase and become greater than PDD arrangement deformations. It is obvious that PDD arrangement deformations tend to 2D-PDD deformations because a long-length wall behavior inclines to a plain strain wall.

2D-PRR arrangement maximum horizontal deformation is about 0.06 m and is proportionally a high amount because soil nails are located with a 45-degree inclination to the wall and their complete bearing capacity will be mobilized. As results indicate, PRR arrangement maximum horizontal

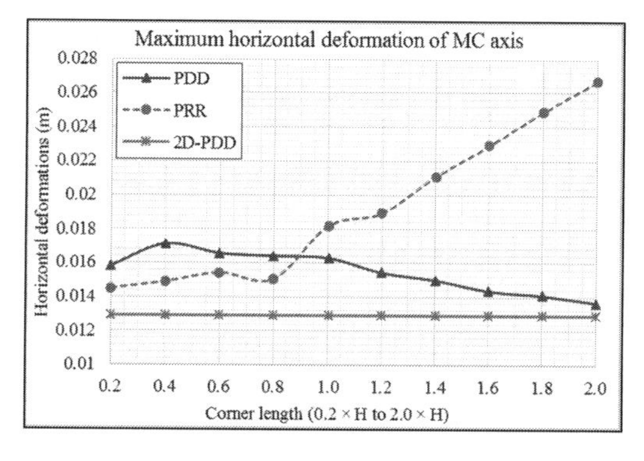

Figure 13. Maximum horizontal deformation of MC axis for PDD and PRR arrangement versus various corner lengths.

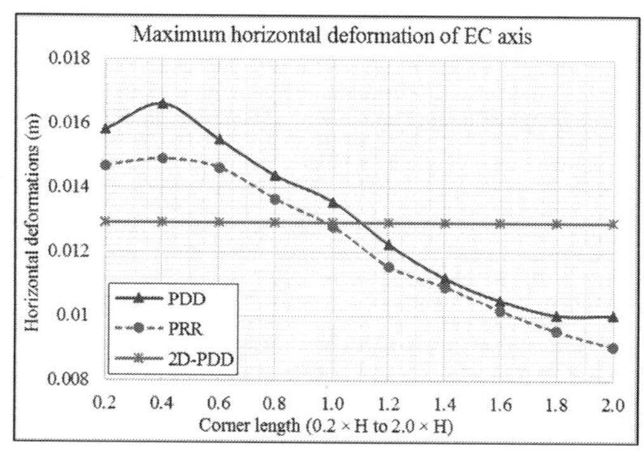

Figure 14. Maximum horizontal deformation of EC axis for PDD and PRR arrangement versus various corner lengths.

deformation inclines to the 2D-PRR arrangement deformation, so as a significant point; the tendency of long walls behavior to plain strain walls plays a key role in the identification of the convex corner behavior.

Figure 14 shows maximum horizontal deformation of EC (end of the corner length) axis in various corner length for PDD and PRR arrangement. The importance of this axis is because it connects soil-nailed corner to a more stabilized region.

Figure 14 shows a significant point about corner horizontal deformations. Not only Horizontal deformations in PRR arrangement are lower than PDD arrangement horizontal deformations (Because of better soil nailed wall anchorage to the main soil body in PRR arrangement), but also horizontal deformations in both arrangements decrease with corners length increase. When corner wall length is small and its head is located near the adjacent walls, main deformations concentration causes a large amount of deformations in EC axis, and when the corner wall length starts to increase, main deformations concentration moves to the corner head and consequently causes deformation decrease in EC axis.

4.2 *Vertical deformations*

Vertical deformations are measured in two points of corner geometry in order to evaluate corner effects on soil nailed wall vertical deformation. Figure 15, shows vertical deformations of HCP (head of the corner point) point in various corner length for both PDD and PRR soil nail arrangements.

Figure 15 indicates the corner head point vertical deformations (settlement). It is obvious that vertical deformation of the corner head in both PDD and PRR arrangements increases as the corner length increase. One of the main reasons for this increase is growth in the distance of the corner head point from the main body of back-soil. Despite the decrease in horizontal deformations of the corner (caused by tension strength of shotcrete facing) that mentioned in the previous section, vertical deformations increase because of corner's less stability. Another significant point is PRR arrangement vertical deformations (settlements) are greater than in PDD one. This can be due to a tendency of PRR arrangement wall to move toward corner sides (this mentioned result would be explained afterward).

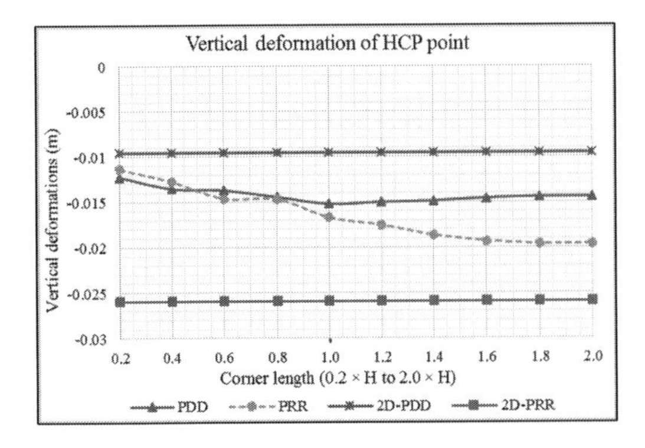

Figure 15. Vertical deformation of HCP point for PDD and PRR arrangement versus various corner lengths.

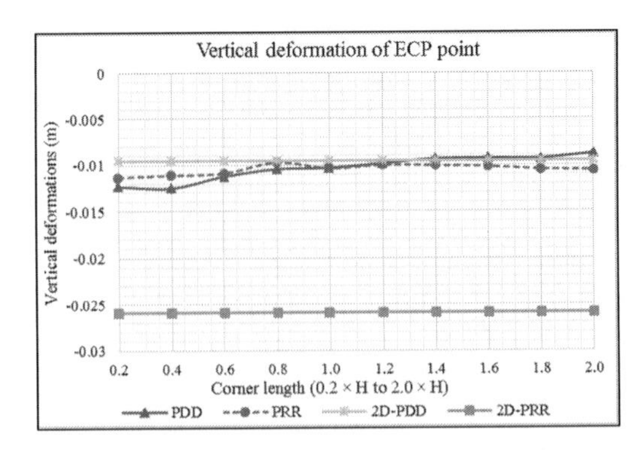

Figure 16. Vertical deformation of ECP point for PDD and PRR arrangement versus various corner lengths.

Figure 16, shows vertical deformation of ECP (end of the corner point) point in various corner length for PDD and PRR arrangement.

Figure 16 indicates the corner endpoint vertical deformations (settlement). PDD and PRR arrangements vertical deformation in the end of corner length decreases with the corner length increase and reach to the 2D-PDD values. One of the main reasons for this decrease is growth in the distance of the corner head point from the main body of back-soil like the previous section. In small corner lengths, the corner has an undesirable effect on vertical deformations and can cause greater amounts, but in advance with an increase in corner length, end of the corner length turns to a heavily reinforced zone and results in a lower amount of deformation.

4.3 *Supplementary results*

In previous sections, results are indicated by charts. It is essential to investigate more about the differences between PDD and PRR nail arrangement deformations of the convex corners. Therefore horizontal displacement contours of PDD and PRR nail arrangements are shown in Figure 17. In addition, figure 18 provides vertical displacement contours PDD and PRR nail arrangements.

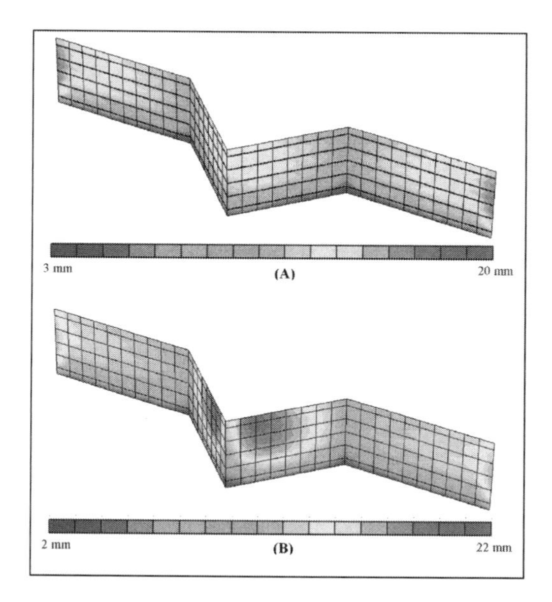

Figure 17. Horizontal displacement contours in a convex corner with length equal to 1.6 × H (a) PDD, and (b) PRR arrangement.

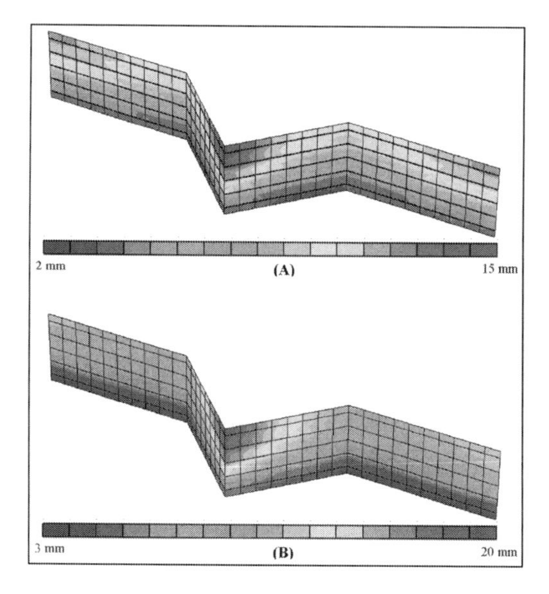

Figure 18. Vertical displacement contours in a convex corner with length equal to 1.6 × H (a) PDD, and (b) PRR arrangement.

Figures 17 and 18, show a significant result in two different arrangements deformation pattern.

In PDD soil nails arrangement, major displacement occur near the head of the corner since a fully reinforced zone consisting of overlapped soil nails, behaves like a rigid body in front of the corner and

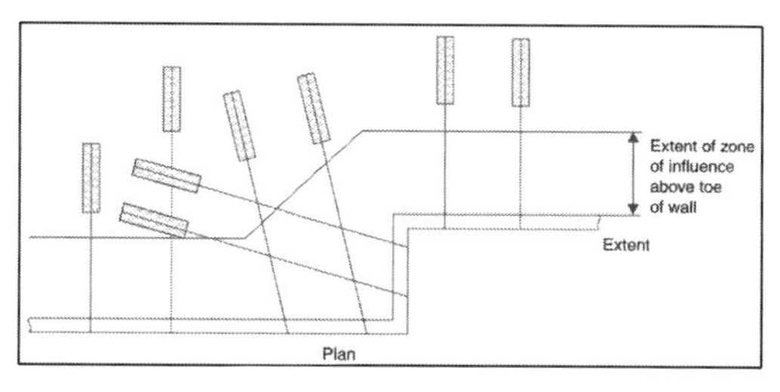

Figure 19. Burland (2012) suggested plan for proper arrangement of ground anchors (Burland, 2012).

is not restrained to the backside soil body. In this arrangement, there is an unreinforced zone in the head of the corner. On the other hand, in PRR soil nails arrangement, major deformations occur near the corner length since the side walls are not restrained enough to act as a rigid zone, so they tend to move toward sides of the corner. In this arrangement head of the corner is properly braced to the backside soil body. The mentioned point can be the most significant result of this study. These results show a good agreement with Burland (2012) suggestions about the introduction of extent influence zone in anchored walls while they have corners (Shown in Figure 19.) (Burland, 2012).

5 CONCLUSION

This study is an initial evaluation of different soil nails arrangement effects on the convex corners deformations. Thus, two main arrangements likewise common arrangements in the practical field is generated in form of 20 finite element models. PDD and PRR arrangements are with 0 degrees and 45 degrees inclination of the corner walls, respectively. The major conclusions of this study are presented below:

- PRR arrangement has better performance for controlling convex corner horizontal deformations due to its better anchorage to the back soil body.
- Formation of a fully reinforced zone with overlapping nails in PDD arrangement, leads to a poor horizontal deformations control in convex corners, since the fully reinforced zone, acts as a rigid block on the back-soil body and tends to slide towards the convex corners head point.
- In the lengthy corner wall models, soil nails with PRR arrangement are unable to restrain sidewalls, tending to move toward the corner sides. These models are similar to 2D-PRR arrangement models with large deformations. To control PRR arrangement horizontal deformations in the lengthy corner wall models, using tieback elements is efficient.

- In short length convex corner walls (less than the corner height), PRR soil nail arrangement and tieback elements restraining two sidewalls, could be an appropriate system for excavation stabilization. On the other hand, using PRR soil nail arrangement in head of the corner and PDD soil nail arrangement along its length in lengthy convex corner walls (more than the corner height), could be an appropriate system for excavation stabilization.

REFERENCES

Babu, G.S., Murthy, B.S. & Srinivas, A. 2002. Analysis of construction factors influencing the behaviour of soil-nailed earth retaining walls. *Proceedings of the ICE-Ground Improvement,* 6, 137–143.

Burland, J.B. 2012. ICE Manual of Geotechnical Engineering, Volume 2-Geotechnical Design, Construction and Verification, ICE publishing.

Fan, C.-C. & Luo, J.-H. 2008. Numerical study on the optimum layout of soil–nailed slopes. *Computers and Geotechnics,* 35, 585–599.

Kim, Y., Lee, S., Jeong, S. & Kim, J. 2013. The effect of pressure-grouted soil nails on the stability of weathered soil slopes. *Computers and Geotechnics,* 49, 253–263.

Lazarte, C., Elias, V., Espinoza, R. & Sabatini, P. 2003. Geotechnical engineering circular no. 7: Soil nail walls. *Federal Highway Administration, Washington, DC.*

Lima, A., Gerscovich, D. & Sayão, A. 2004. Considerations on the soil nailing technique for stabilizing excavated slopes. *stress,* 10, 0.69.

Lima, A.P., Sayão, A. & Gerscovich, D.M.S. Considerations on the Numerical Modeling of Nailed Soil Excavations. Proc. of the 4rd Int. Workshop on Applications of Computational Mechanics in Geotechnical Engineering, 2003. 143–151.

Mittal, S. 2006. Soil Nailing Application in erosion control–an experimental study. *Geotechnical & Geological Engineering,* 24, 675–688.

Ng, C. & Lee, G. 2002. A three-dimensional parametric study of the use of soil nails for stabilising tunnel faces. *Computers and Geotechnics,* 29, 673–697.

Ou, C.-Y. & Shiau, B.-Y. 1998. Analysis of the corner effect on excavation behaviors. *Canadian Geotechnical Journal,* 35, 532–540.

Pan, H., Zhou, C.-F. & Cao, H. 2008. Analysis of spatial effectand deformation of corner of composite soil nailing walls. *ROCK AND SOIL MECHANICS-WUHAN-,* 29, 333.

Singh, V.P. & Babu, G.S. 2010. 2D numerical simulations of soil nail walls. *Geotechnical and Geological Engineering,* 28, 299–309.

Smith, I. & Su, N. 1997. Three-dimensional FE analysis of a nailed soil wall curved in plan. *International Journal for Numerical and Analytical Methods in Geomechanics,* 21, 583–597.

Yang, M.Z. & Drumm, E.C. 2000. Numerical analysis of the load transfer and deformation in a soil nailed slope. *Geotechnical special publication,* 102–116.

Zhang, M., Wang, X.-H., Yang, G.-C. & Wang, Y. 2011. Numerical investigation of the convex effect on the behavior of crossing excavations. *Journal of Zhejiang University SCIENCE A,* 12, 747–757.

Zhao, W., Chen, C., Li, S. & Pang, Y. 2015. Researches on the Influence on Neighboring Buildings by Concave and Convex Location Effect of Excavations in Soft Soil Area. *Journal of Intelligent & Robotic Systems,* 79, 351.

Numerical Methods in Geotechnical Engineering IX – Cardoso et al. (Eds)
© 2018 Taylor & Francis Group, London, ISBN 978-1-138-33203-4

Design of anchored sheet pile walls in cohesionless soil

S. Krabbenhoft
Aalborg University, Civil Engineering Department, Aalborg, Denmark

R. Christensen
Ramboll, Consulting Engineers, Aarhus, Denmark

ABSTRACT: In Denmark design of sheet pile wall is often carried out using Brinch Hansen's earth pressuretheory. This design is often verified by a calculation using Finite Element Methods, typically based on Plaxis 2D. The Mohr Coulomb failure criterion is often used in the design with Plaxis. If the fill consists of cohesionless material a constant value is used for the friction angle when using the Mohr Coulomb failure criterion. However research has shown that the friction angle is stress dependent and the finite element software OptumG2, which in some ways is similar to Plaxis, besides the traditional constitutive models, also features a constitutive model called the GSK Model taking into account the stress dependency of the friction angle. This paper accounts for the results obtained by the three approaches in the design of a sheet pile wall. The study, which has been performed under the assumptions of both an associative and nonassociative flow rule, shows reasonable accordance between most of the Plaxis and OptumG2 results, and also that considerable savings are possible when applying the GSK model.

1 INTRODUCTION

There are a large number of methods—both empirical, semi empirical and analytical—available for the design of sheet pile walls. An account and comparison of some of the methods still used in sheet pile design has been given by amongst others Brinch Hansen (1953) and Day & Potts (1989).

In Denmark by far the most common method is known as "The Danish Method" (not to be confused with the method called the Danish Rules, which dates back to 1923), which is a limit equilibrium method based on Brinch Hansen's earth pressure theory (1953) and although it is suitable for hand calculations, design according to this method today is generally carried out by applying the software Spooks (1996).

As outlined by Brinkgreve & Swolfs (2008) one of the advantages of the finite element method is the possibility of applying more advanced soil models than the traditional Mohr Coulomb model and also both deformations and the construction process can be realistically simulated. This study comprises the study of an anchored sheet pile wall in cohesionless soil using the following three methods:

1. Brinch Hansens earth pressure theory, (Spooks)
2. Finite element design using Plaxis (1998),
3. Finite element design using OptumG2 (2014)

The Brinch Hansen and Plaxis calculations are based on the Mohr-Coulomb model and the OptumG2 calculations are performed using both the Mohr-Coulomb model and a new soil model called the GSK model, which can be seen as a Mohr-Coulomb model taking into account the stress dependency of the friction angle.

Over the past 2–3 decades commercial software packages based on numerical methods such as the finite element method have become within reach for consulting companies and the purpose of this study is to compare the results obtained by a well-established method known for leading to a very economic design with the results obtained by the finite element method using two different, but in some respect similar softwares. The effect of non-associativity is studied and a comparison between the Mohr Coulomb failure criterion and a nonlinear failure criterion—the GSK model—has been made. Only sheet pile walls with one anchor row are considered.

2 THE DESIGN METHODS

2.1 *The Brinch Hansen method (Danish Method)*

The design is based on the Mohr Coulomb failure criterion both soil and sheet pile being perfectly plastic. The method, which is a limit equilibrium method, assumes both soil and sheet piles to be perfectly plastic, and as such it should—in contrast to some of the other methods currently in use—be well suited for ultimate limit state (ULS) design as prescribed by Eurocode 7, (2004).

In the failed state the wall, which is considered rigid, is assumed to rotate about a point (or more points) and one or more yield hinges in the wall may develop. Between the yield hinges the wall is rigid. During failure, yield lines in the soil develop, and as the soil is considered non-dilatant the yield lines must—for kinematic reasons—have the shape of either a straight line or a circle. The soil volumes between the wall and the yield lines are assumed to be in an elastic state; i.e. rigid.

The location of the point of rotation of the soil volume is unknown but must—because the soil is rigid—for kinematic reasons lie on a straight line perpendicular to the point of rotation of the wall. This point very often coincides with the anchoring point and therefore will be known. The stresses and hence forces in the rupture line can be found by means of Kötters equation (Kötter 1903) and iteratively the location of the rotational point of the wall and subsequently the forces and point of action of the forces on the wall can be found.

According to the theory any failure mode ensuring compatibility between the movements of the structure and the surrounding soil can be chosen, provided that the equilibrium equations are fulfilled, and in this project the three failure modes shown in Figure 1 are being dealt with.

The method is described in detail by Brinch Hansen (1953).

The method, has been subject to a lot of criticism, but is known for leading to very economic structures, many of which are still in use today and it is still the preferred method amongst consulting engineers in Denmark.

The shortcomings of The Danish Method have been reported by amongst others Mortensen & Steenfelt (2001) and Hansen (2001).

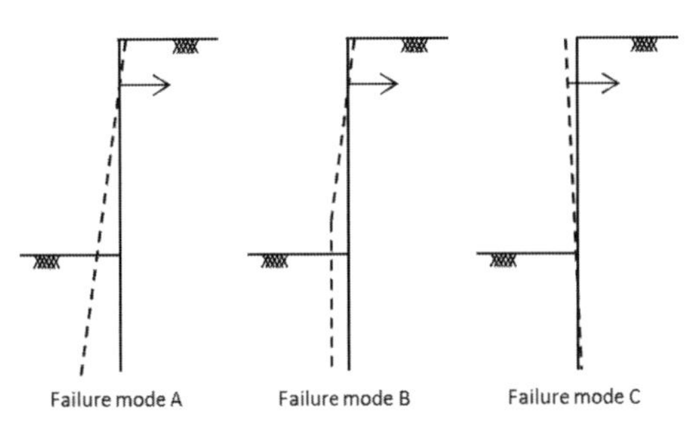

Figure 1. Failure modes A, B and C—Danish Methods. Failure mode A and C is composed of one rigid wall section with rigid and non-rigid anchor. Failure mode B is composed of two rigid wall sections connected by yield hinges in failure with rigid anchor.

An example of a line rupture of the A type is shown in Figure 2. This kind of rupture is assumed in cases where the point of rotation is located in the interval 0.52h to 1.26h where h is the height of the wall. In Figure 2 the rotational point is located at a distance 0.8h above the bottom of the wall. The wall is considered completely rough.

2.2 *The finite element method (Plaxis and OptumG2)*

The theory of the finite element method is given in numerous textbooks on the subject, and for a more detailed study the reader is kindly referred to these.

In finite element analysis using the displacement method, the soil domain which is regarded as a continuum, is divided into discrete elements having common nodes. In the most common formulation—the displacement method—for this assembly of elements a global equation based on the stress—strain relationships for the soil, the compatibility conditions and the equilibrium equations a relationship between the nodal forces and displacements can be established.

One of the key points of the finite element method, regardless of the formulation, is the possibility of ascribing different soil characteristics to the individual elements and as a result even complex soil conditions and more advanced soil models can be dealt with.

Another major advantage of the finite element method is its "built in" ability to automatically find the most critical failure mechanism and hence the "correct" bearing capacity of the structure being dealt with.

The displacement method for elasto-plastic analysis has been implemented in Plaxis, and in OptumG2 elasto-plastic calculations are carried

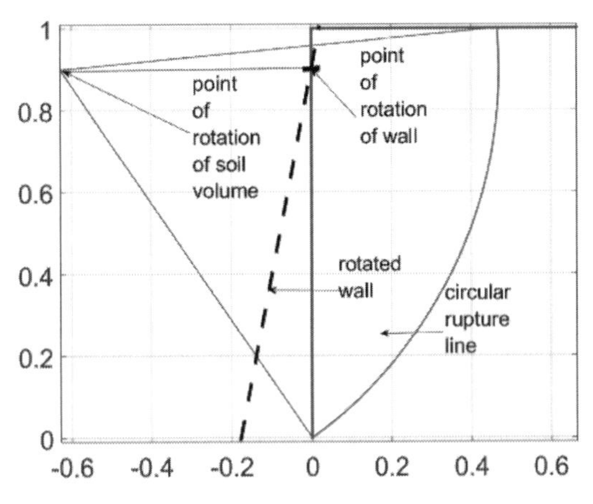

Figure 2. Line rupture type a according to Brinch Hansen.

out using variational principles, (minimizing the potential energy of a structural system).

Both methods always lead—although it cannot be proved mathematically—to an unsafe solution and therefore always mesh refinements and convergence studies must be performed. This drawback can be remedied by the application of finite element based limit analysis in which the two plastic bound theorems (lower and upper bounds, Chen 1975) are very useful in that, by performing both upper and lower bound analysis the true solution can be bracketed from above and below.

Limit analysis has been implemented in OptumG2, and using the programs adaptive remeshing facility the true solution can be found in most cases within an accuracy of about 1% and limit analysis of a given structure is performed much faster than the equivalent elasto-plastic analysis. When the same parameters are used the two methods must end up providing the same results.

Limit analysis—in contrast to elasto-plastic analysis—gives no information of the deformations of the structure and only the mode of collapse together with the collapse load is found.

These trends render limit analysis suitable for ULS calculations while elasto-plastic analysis can be used for both ULS and SLS analysis.

In the present project Plaxis' feature of $\varphi - c'$ reduction has been used. In this type of analysis one finds a safety factor (SF) defined by the equation SF = $\tan(\varphi_{input})/\tan(\varphi_{required})$ where φ_{input} is the input value of the friction angle and $\varphi_{required}$ is the limit value of the friction angle rendering the structure unstable.

The limit analysis with OptumG2 is performed applying the programs "strength reduction" feature, which is similar to Plaxis' $\varphi - c'$ reduction. In OptumG2 it is possible not only to reduce the strength of the soil but also the strength of the structural elements, i.e. both the sheet pile wall and the anchor simultaneously or the sheet pile wall and the anchor separately. In this project the required bending moment was minimized. The elasto-plastic analysis is performed by "staged construction" in which the construction of the wall is simulated and in this case the soil in front of the wall was simulated being excavated.

3 THE MOHR-COULOMB MODEL

The best way of finding the friction angle of the soil is considered to be the triaxial test, but for both practical and economic reasons, this method is very often excluded. Instead knowledge of the soil strength is provided by in situ tests such as SPT, CPT etc. and on the basis of such tests the friction angle and/or the relative density of a cohesionless soil can be estimated. In this project the starting point for the evaluation of the soil strength is the latter because it is considered in relation to strength the most essential and unambiguous quantity.

There exists a number of empirical equations converting the relative density to a friction angle and in this project is was chosen to use an equation from the Danish Code of Practice (1985) giving the value of the nonassociated, triaxial friction angle (NAF = non associated flow rule):

$$\varphi_{tr,NAF} = 30 - \frac{3}{U} + (14 - \frac{4}{U})I_D \tag{1}$$

where I_D is the relative density and U is the coefficient of uniformity. For a medium dense, uniform sand I_D is taken = 0.60 and U = 2.1 and this yields

$$\varphi_{tr,NAF} = 30 - \frac{3}{2.1} + (14 - \frac{4}{2.1})0.6 = 35.8$$

Using a guideline by Schmertmann (1976) would have resulted in a value for uniform fine sand of 36.5° and for uniform medium sand of 38.5° and this clearly demonstrates the uncertainty which exists in the determination of the friction angle.

The angle of dilation ψ is found by the well known rule of thumb: $\psi = \varphi - 30$ and this yields $\psi = 5.8°$.

This triaxial angle may be modified to plane strain conditions as the actual structure is considered to be a 2 D structure. The difference between axisymmetry and plane strain conditions has been allowed for in Eurocode 7 by the wording:"many geotechnical parameters are not true constants but depend on stress level and mode of deformation".

The characteristic, plane, non associated angle of friction is found by an equation suggested by Stakemann (1976) and Bonding (1977).

$$\varphi_{pl,NAF} = (1 + 0.163 I_D)\varphi_{tr,NAF} = (1 + 0.163 \cdot 0.6)35.8 = 39.3$$

Limit state analysis is based on an associated flow rule and consequently an associated friction angle must be used. If this should be done and how it should be done is a point of controversy in the geotechnical community, but it is a matter of fact that for example the equations for the bearing capacity factors are based on associated plasticity and therefore it seems obvious, that an associated friction angle must be used.

Several studies such as Michalowski (1997), Frydman & Byrd (1997), Yin et al. (2001) and Loukidis & Salgado (2009), have shown, that the angle of dilation has a considerable impact on the bearing capacity in that the bearing capacity increases with an increase in the angle of dilation, and also it should be kept in mind that all the usual values of the earth pressure coefficients for zone ruptures, for example the ones given in the annex C of the Eurocode 7 have been found on the basis of associated plasticity. It has been suggested by Davis (1968) and Hansen (1979) to take the nonassociativity into account by modifying the nonassociated friction angle by means of the below equation:

$$\varphi_{tr,AF} = \text{atan}\left(\frac{\sin\varphi_{tr,NAF}\cos\psi}{1-\sin\varphi_{tr,NAF}\sin\psi}\right) \quad (2)$$

(AF = associated flow rule).

The equation, which is sometimes referred to as the "Davis equation or Davis approach" may in some cases be too much on the safe side, (Hansen 1979 and 2001) but the theory has shown remarkably good agreement between model tests and the Brinch Hansen theory for the calculation of earth pressures for rupture figures containing line ruptures, (Hansen & Steenfelt 1976). On these grounds the calculation of earth pressures in the present project in the case of limit analysis and Brinch Hansen theory (the Danish method) are based on a friction angle modified according to the Davis equation; i.e.:

$$\varphi_{tr,AF} = \text{atan}\left(\frac{\sin\varphi_{tr,NAF}\cos\psi}{1-\sin\varphi_{tr,NAF}\sin\psi}\right) = \text{atan}\left(\frac{\sin 35.8\cos 5.8}{1-\sin 35.8\sin 5.8}\right) = 31.8$$

The characteristic, plane associated angle of friction is found by Stakemann's equation (1976)

$$\varphi_{pl,AF} = (1+0.163 I_D)\varphi_{tr,AF} = (1+0.163\cdot 0.6)31.8 = 34.9$$

The design value of the plane, associated friction angle using a partial factor = 1.25 (Eurocode 7):

$$\varphi_{pl,AF,d} = \text{atan}\left(\frac{\tan(34.9)}{1.25}\right) = 29.2$$

Both Plaxis and OptumG2 are able of performing elasto-plastic calculations based on nonassociative conditions and in these cases the design value of the friction angle is found as:

$$\varphi_{pl,NAF,d} = \text{atan}\left(\frac{\tan(39.3)}{1.25}\right) = 33.2$$

Table 1 shows the values of the soil parameters employed in the MC model.

The soil unit weights $\gamma/\gamma_{saturated}$ were taken equal to 18/20 kN/m^3 in all calculations and in all analysis the wall friction angle was taken = 2/3·friction angle as indicated in Eurocode 7.

Table 1. Design values of soil parameters.

Software	Type of calculation	Friction angle φ	Dilation angle ψ
Spooks	Danish Method	29.2	0
Plaxis	Elasto-plastic	33.2	5.8
	Elasto-plastic	29.2	29.2
OptumG2	Elasto-plastic	33.2	5.8
	Elasto-plastic	29.2	29.2
	Limit analysis	29.2	29.2

4 THE GSK MODEL

The Mohr-Coulomb soil model assumes, that the angle of friction can be regarded as a constant only being dependent on the soil characteristics, (primarily the relative density) but it has been reported by amongst others Ponce and Bell (1971), Bolton (1986), Okamura et al. (2002), Cerato & Lutenegger (2007), Krabbenhoft et al. (2012) and Jahanandish et. al. (2012), that the friction angle is also affected by the stress level at failure, and that the greater the stress level, the smaller the friction angle.

The GSK model, which can be applied to a cohesive—frictional material, takes into account the fact, that the friction angle is stress dependent; that is the failure envelope becomes curved instead of being linear as with the Mohr Coulomb model. The yield function is similar to the linear Mohr Coulomb and is expressed by the following equation: $\tau = \sigma\cdot\tan\varphi_s$ where φ_s is the secant angle in $\sigma - \tau$ space. Using principal stresses the yield function for the GSK model for cohesionless soil is formulated in the below equations where compressive stresses are taken positive:

$$\begin{aligned} \sigma_1 &= a_2\sigma_3 + k\left(1-\exp\left(-\frac{a_1-a_2}{k}\sigma_3\right)\right) \\ a_1 &= \frac{1+\sin\varphi_1}{1-\sin\varphi_1} \Leftrightarrow \sin\varphi_1 = \frac{a_1-1}{a_1+1} \\ a_2 &= \frac{1+\sin\varphi_2}{1-\sin\varphi_2} \Leftrightarrow \sin\varphi_2 = \frac{a_2-1}{a_2+1} \\ k &= \frac{2c\cos\varphi_2}{1-\sin\varphi_2} \Leftrightarrow c = \frac{k(1-\sin\varphi_2)}{2\cos\varphi_2} \end{aligned} \quad (3)$$

The GSK model can be described by the three parameters φ_1, φ_2 and c, and these have been found on the basis of values from a database of peak friction angles established by Andersen & Schjetne at The Norwegian Geotechnical Institute in Oslo (2013).

Table 2. Triaxial peak friction angles—NGI tests.

	I_D				
σ_3 [kPa]	0.20	0.40	0.60	0.80	1.00
11.25	35.5	38.0	41.0	44.0	47.5
26.25	34.8	37.2	40.1	43.0	46.5
56.25	34.4	36.5	39.0	42.0	45.0
112.5	34.0	36.0	38.0	40.9	43.6
262.5	33.2	34.5	36.8	39.1	42.2
2062.5	32.4	33.3	34.7	36.8	39.7

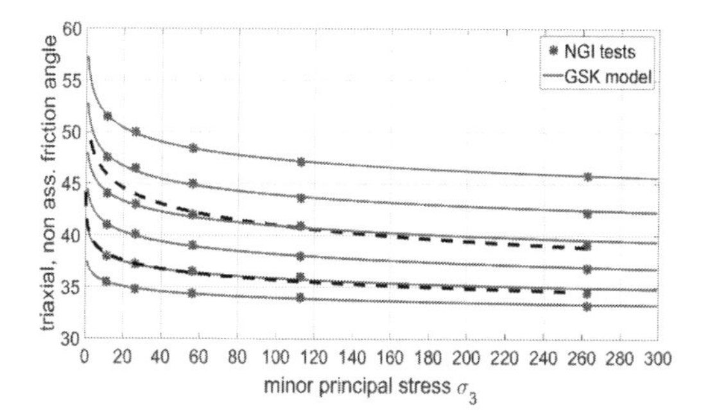

Figure 3. Minor principal stress versus friction angle—NGI tests, GSK model and Bolton equation.

The database contains results from more than 500 triaxial compression tests on 54 different sand from 38 different sites. The mean particle size $D_{50} = 0.23$ mm and the coefficient of uniformity $C_u = 1.95$. The quartz content was 85% on the average, and for most of the sands the grain shape has been reported and varied from rounded to angular.

The results of the NGI tests giving the peak, triaxial angle are shown in Table 2.

The results of the NGI tests together with the GSK model values found by regression analysis are shown in Figure 3; the lowermost curve represents $I_D = 0.20$ values and the uppermost I_D values = 1.20. The results have been compared with the Bolton equation (Bolton 1986) for relative densities 0.40, and 0.80, (the broken lines), and as can be seen, the Bolton results conform with the NGI and GSK values, though the Bolton friction angles at very small values of the minor principal stress are somewhat greater than the NGI values. The GSK yield function is shown in $\sigma - \tau$ space in Figure 4. The equation of the asymptote: $\tau = c + \sigma \cdot \tan\varphi_2$.

The values of the three parameters φ_1, φ_2 and c can be expressed by the following equations :

$$\begin{aligned} \varphi_1 &= A_1 I_D{}^2 + B_1 I_D + C_1 \\ \varphi_2 &= A_2 I_D{}^2 + B_2 I_D + C_2 \\ c &= A_3 I_D{}^2 + B_3 I_D + C_3 \end{aligned} \quad (4)$$

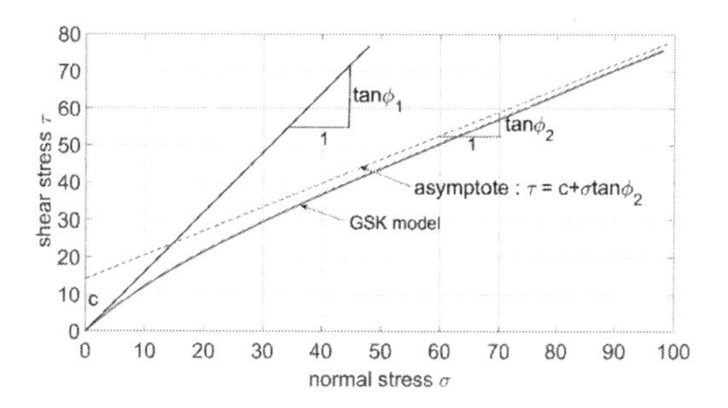

Figure 4. GSK failure function in $\sigma - \tau$ space.

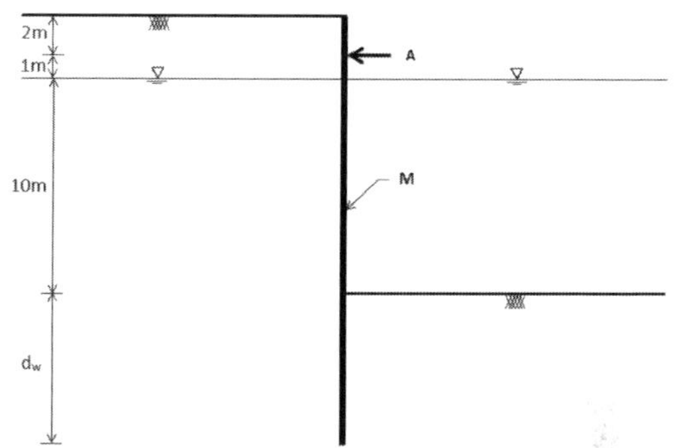

Figure 5. Layout of structure.

For triaxial conditions in the non associated case the coefficients attain the following values:

$$\begin{aligned} &A_1 = 2.60,\ B_1 = 12.4,\ C_1 = 33.3 \\ &A_2 = 6.00,\ B_2 = 3.50,\ C_2 = 31.4 \\ &A_3 = -8.10,\ B_3 = 24.4,\ C_3 = 0.0 \end{aligned} \quad (5)$$

To convert these to the associated triaxial case one can use the Davis equation or the nonassociated values of φ_1 and φ_2 can be multiplied by a factor = 0.886 and the value of c is multiplied by 0.77.

To convert to the plane strain case the φ values are multiplied by $(1 + 0.163I_D)$ and c by $(1+0.45I_D)$. It must be stressed, that the GSK model has, as far as the authors know, not as yet been applied to "real life structures" but it has—in combination with application of the Davis equation—provided results which are in very good accordance with 1 g model tests, (Krabbenhoft et al. 2012)

5 THE SHEET PILE WALL

The layout of the structure evaluated in the study is shown in Figure 5. The Plaxis analysis were performed using 2000 "15-noded elements" with local refined mesh in the failure zone and the OptumG2 analysis were performed using 4000 elements and 3 adaptivity steps.

The sheet pile wall was chosen as an Arcelor Mittal (2013) steel sheet pile AZ 13–700, steel grade S 355 GP with the following characteristics:

Cross sectional area A:	123 cm^2/m
Elastic section modulus W_{el}:	1305 cm^3/m
Plastic section modulus W_{pl}:	1540 cm^3/m
Moment of inertia I:	20540 cm^4/m
Design yield moment $M_{d,pl}$:	497 kNm/m
Normal stiffness EA:	2835000 kN/m
Bending stiffness EI:	43134 kNm^2/m

The design yield moment of the sheet pile is determined without considering possible buckling.

A partial factor = 1.1 has been applied to the characteristic steel strength.

The anchor consists of a steel threadbar, diameter 47 mm, length 15 m. The characteristic yield force = 1648 kN. The anchors are spaced 2.80 m apart giving a design force = 436 kN/m. A partial factor = 1.35 was applied.

6 RESULTS OF THE ANALYSIS

The results of the analysis are shown in Table 3,4,5 and 6. In the tables the anchor force, maximum bending moment, driving depth and factor of safety are designated A, M, d_w and SF. In all calculations it has been assumed, that vertical equilibrium is not a problem, and it has been verified, that the resulting vertical force on the wall is acting

Table 3. Spooks results, $\varphi = 29.2$, $\psi = 0.0$.

Mode of failure	A [kN/m]	M [kNm/m]	d_w [m]	SF
A	359	515	3.20	NA
B	231	519	3.40	NA
C	181	715	4.30	NA

Table 4. Plaxis results.

Type of analysis, φ/ψ	A [kN/m]	M [kNm/m]	d_w [m]	SF
EP 29.2/29.2	317	386	3.20	1.019
EP 33.2/5.8	278	389	3.20	1.06

Table 5. OptumG2 results, MC soil model.

Type of analysis, φ/ψ	A [kN/m]	M [kNm/m]	d_w [m]	SF
EP 33.2/5.8	278	398	3.20	1.05
EP 29.2/29.2	316	470	3.20	1.00
LA 29.2/29.2	330	468	3.20	NA

Table 6. OptumG2 results, GSK soil model.

Type of analysis, φ_1, φ_2, c	A [kN/m]	M [kNm/m]	d_w [m]	SF
EP, non-associated 39.4/33.1/11.9	221	254	2.50	1.00
EP associated 34.4/29.0/9.2	282	320	2.50	1.04
LA, associated 34.4/29.0/9.2	298	306	2.50	NA

downwards being of the order of 100 kN/m, which certainly can be resisted by soil.

The LA values in Table 5 and 6 are taken as the average of upper and lower bounds. Numerical lower bound is 1–3% less than numerical upperbound.

The results in Table 3 show, as regards the maximum bending moment and driving depth, very little difference between the results for failure mode A and B. The difference in anchor force in the two cases is however remarkable, as the anchor force in mode A is about 55% greater than the one in mode B. Failure mode C, which is almost similar to the free earth support method, gives—because the anchor is not fixed—smaller anchor force, but greater values of bending moment and driving depth.

The values of the nonassociated analysis (φ/ψ = 33.2/5.8) produced by Plaxis and OptumG2 shown in Table 4 and 5 are very similar, and they are smaller than the LA values in Table 5, indicating that application of the Davis approach for correcting the nonassociated friction angle may be somewhat on the safe side. The EP values in Table 5 in the associated case ((φ/ψ = 29.2/29.2) are in good accordance with the LA values in Table 5 and this is, what must be expected.

The values of the LA case in Table 5 are in reasonable accordance with the values of mode A in

Table 3, the discrepancies in both anchor force and moment being less than 10%, whereas the Plaxis results from Table 4 in the associated case—especially the maximum moment—are considerably smaller than the equivalent Optum values in Table 5.

Table 6 shows the results from both the associated and nonassociated GSK analysis. As with the MC values in Table 5 the nonassociated values for both anchor and moment are the smaller and the LA values and the associated EP values are close to each other.

A comparison with the values in Table 5 shows, that the GSK model requires smaller driving depth, and the maximum moments are between 30 and 36%, and the anchor forces between 10 and 20% smaller for the GSK model. This implies that substantial savings are possible, when applying the GSK model. The reduction in especially the bending moment is due to the fact, that this is caused by the active soil pressure at the back of the wall, and as the stress level here is relatively low, there is a considerable increase in the friction angle resulting in smaller values of the soil pressure coefficient.

It must be noted, that all the calculated values in this study are based on the magnitude of the peak friction and dilation angle and possible postpeak reduction in these is not considered, and this is—maybe surprisingly—in accordance with current design practice.

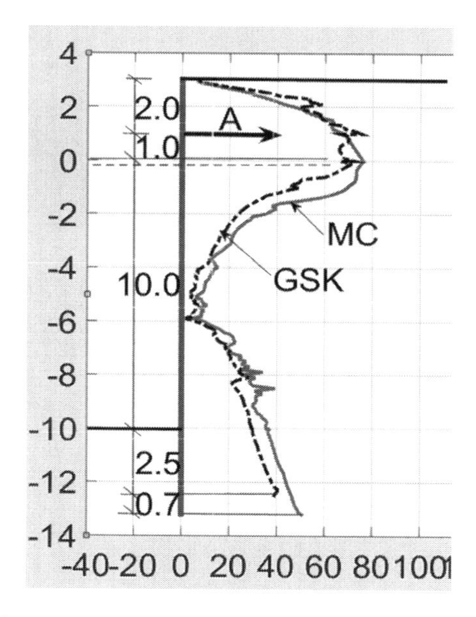

Figure 6. Active normal stresses on wall for MC and GSK in LA analysis OptumG2. Stresses in kPa and heights in metres.

7 CONCLUSION

A study of three methods for the design of sheet pile walls has been conducted. For the purpose of comparing the three methods, these have been used to find the necessary pile length, the maximum bending moment and the anchor force for a specific sheet pile wall. The basis of each method has been shortly explained, and it has been found, that in the nonassociated case the two finite element programs are in good agreement with each other. In the associated case deviations of approximately 20% between the maximum bending moments produced by the two finite element programs have been found, the Plaxis values being the smaller. The Brinch Hansen method produces results which in most cases agree reasonably well with the Optum values in the associated case. The GSK model, which takes into account the stress dependency of the friction angle, produces results, which for a given relative density of the soil, are smaller than the Mohr Coulomb model, this mainly being due to the low stress level on the active side of the wall.

REFERENCES

Andersen, K.H & Schjetne, K. 2013. Database of friction angles of sand and consolidation characteristics of sand, silt and clay. *Journal of Geotechnical and Geoenvironmen tal Engineering.*

ArcelorMittal, 2013. *Steel foundation solutions for projects. Steel sheet piling. General catalogue 2013.*

Bolton, M.D. 1986. The strength and dilatancy of sands. *Geotechnique, 36(1), 65–78.*

Bonding, N. 1977. Triaxial state of failure in sand. *DGI Bulletin 26, 1977*

Brinch Hansen, J. 1953. *Earth Pressure Calculation*, The Danish Technical Press, Copenhagen.

Brinkgreve, R.B.J & Swolfs, W.M 4th Conference on Advances and Applications of GID, Ibiza 2008

CEN. (2004). *Eurocode 7: "Geotechnical design—Part 1: General rules. Standard EN 1997 – 1"*. European Committee for Standardization (CEN), Brussels, Belgium.

Chen, W.F. !975. *Limit analysis and soil plasticity.* Elsevier Science Publishers, BV, Amsterdam, The Netherlands.

Code of practice for foundation engineering. Copenhagen 1985. *DGI-Bulletin No. 36.*

Davis, E.H. 1968. Theories of plasticity and failure of soil masses. *In Soil Mechanics*, selected topics. Edited by I.K. Lee, Butterworths, London, 341–380.

Day, R.A. & Potts, D.M. 1989. A comparison of design methods for propped sheet pile walls. *SCI Publication 077.*

Frydman, S. & Burd, H.J. (1997). "Numerical studies of bearing-capacity factor N_γ". *Journal of geotechnical and geoenvironmental engineering.*, Vol. 123, No. 1, January 1997.

Hansen, B. & Steenfelt, J. 1976. J. Brinch Hansen's earth pressure theory tested by experiments in a pin model.

Proc. Fifth Budapest Conf. Soil Mech. Found. Eng., Budapest, 447–458.
Hansen, B. 1979. Design parameters in geotechnical engineering, British geotechnical Society, London,1979.
Hansen, B. 2001. *Advanced theoretical soil mechanics.* Danish geotechnical society.
Jahanandish, M., Veiskarami, A. & Ghahramani, A. (2012). "Effect of Foundation Size and Roughness on Bearing Capacity Factor, N_γ, by Stress Level-Based ZEL Method". *Arab J Sci Eng* (2012) 37:1817–1831
Kötter, F. 1903. Die Bestimmung des Druckes an Gekrümmten Gleitflachen, *Sitzungungsber. Kgl. Presuss. Akad. Der Wiss.,Berlin.*
Krabbenhoft, S., Clausen, J. & Damkilde, L. 2012. The bearing capacity of circular footings in sand. Comparison between model tests and numerical simulations based on a nonlinear Mohr failure envelope. Hindawi Publishing Corporation. *Advances in civil engineering. Volume 2012.*
Loukidis, D. & Salgado, R. (2009). "Bearing capacity of strip and circular footings in sand using finite Elements". *Computers and Geotechnics* 36 (2009) 871–879.
Michalowski, R.L. (1997). "An estimate of the influence of soilweight on bearing capacity using limit analysis". *Soils Found.*, 37(4), 57–64.
Mortensen, N. & Steenfelt, J. 2001. Danish plastic design of sheet pile walls revisited in the light of FEM. XVth International conference on soil mechanics and geotechnical engineering, Istanbul, August 2001.
OPTUM CE. 2014. *Optum G2, analysis and theory,* Optum Computational Engineering, Newcastle, New South Wales, Australia.
PLAXIS 1998. *Finite Element Code for Soil and Rock Analyses. User's Manual, A.A. Balkema/Rotterdam/ Brookfield.*
Ponce, V.M. and Bell, J.M. (1971). "Shear strength of sand at extremely low pressures". *J. Soil Mech. Found.* Div., ASCE, Vol. 47, No. SM4, 625–637
Schmertmann, J.H. 1976. Guidelines for cone penetration Test; Performance and design, Washington D.C.: Department of Transportation. Federal Higway Administration, 1976.
SPOOKS W.1.11. 1996. *User's Manual,* The Danish Geotechnical institute.
Stakemann, O. 1976. Brudbetingelse for G12-sand og plane modelforsøg (Failure conditions for G12-sand and plane strain model tests; in Danish). *Internal Memo I.M.* 1976–1, Danish Geotechnical Institute.
Yin, J.H., Wang, Y.J. & Selvadurai, A.P.S. (2001). "Influence of nonassociativity on the bearing capacity of a strip footing". *Journal of geotechnical and geoenvironmental engineering.*, Vol. 127, No. 11, November 2001.

Numerical Methods in Geotechnical Engineering IX – Cardoso et al. (Eds)
© 2018 Taylor & Francis Group, London, ISBN 978-1-138-33203-4

Back-analysis of Crossrail deep excavations using 3D FE modelling—development of BRICK parameters for London Clay

Y. Chen & G. Biscontin
University of Cambridge, UK

A.K. Pillai & D.P. Nicholson
Ove Arup & Partners, UK

ABSTRACT: New underground stations up to 40 m deep for the Crossrail project in the London have been constructed. Excavations were supported by diaphragm walls or secant pile wall, using either the top-down or bottom-up construction sequence. To investigate the effects of these deep excavations on the existing adjacent buildings and tunnels, 3D FE models have been developed. The non-linear elastic plastic soil constitutive model—BRICK was used to model the behavior of London Clay and Lambeth Group. A back-analysis of the field inclinometer data from Tottenham Court Road Station—Western Ticket Hall was carried out using the 3D FEM. Revised BRICK parameters were derived from the back-analysis and applied in three other Crossrail excavations. The back analysis of four excavations has provided an intensive review on the BRICK model parameters for London Clay. The reliability of the BRICK model in London Clay to assess most probable ground movements has been demonstrated.

1 INTRODUCTION

The demand for space in cities has been expanding with the urban population growth. This also affects underground development, particularly where new tunnels continue to be constructed close to existing tunnels and buildings' foundations. This increased underground congestion leads to additional difficulties for the construction of new underground structures in urban areas. In addition, increasing asset protection requirements for existing tunnels and piles have made excavation work even more complex. In order to address these issues for future excavation design, an accurate three-dimension finite element model (3D FEM) would be valuable for designers. The implementation of 3D FEM in design for excavations in London has been previously reported: Crossrail Liverpool Street Station—Moorgate Shaft (Zdravkovic, Potts and John, 2005; Farooq *et al.*, 2015); King's Place development (Yeow and Feltham, 2008; Ellison, 2012); and One New Change basement (Fuentes, *et al.*, 2010).

The soil constitutive model BRICK for London Clay was originally published by Simpson (1992) and the parameters were later reviewed by Pillai (1996). A more recent reassessment of BRICK parameters for London Clay was reported by Pillai and Fuentes (Pillai *et al.*, 2011). These are referred to as moderately conservative parameters.

The Crossrail project included the construction of eight new underground stations in the central London area. This paper presents four excavation case histories collected from these newly constructed stations. They are shown on the Crossrail route map in Figure 1.

During the original design stage in 2010, the ground movements and asset protection assessments for these new stations were conducted in 3D FEM (by LS-Dyna®) using the BRICK soil model with the parameters updated by SCOUT (2007). This subsequent back-analysis of these Crossrail excavations has adopted the same analytical 3D FE models, but slightly modified for the as-built conditions. The aim is to back-analyse the case history performance of these excavations and develop most probable BRICK parameters for London Clay. These parameters give a most probable estimate of wall deflections for use in the observational method. (Hardy *et al.*, 2017)

2 BRICK PARAMETERS

2.1 *Classic parameters*

The initial set of BRICK parameters for London Clay was proposed by Simpson (1992). The BRICK parameters have been updated for London Clay through projects in London Clay by Pillai (1996), and SCOUT (2007). These updated

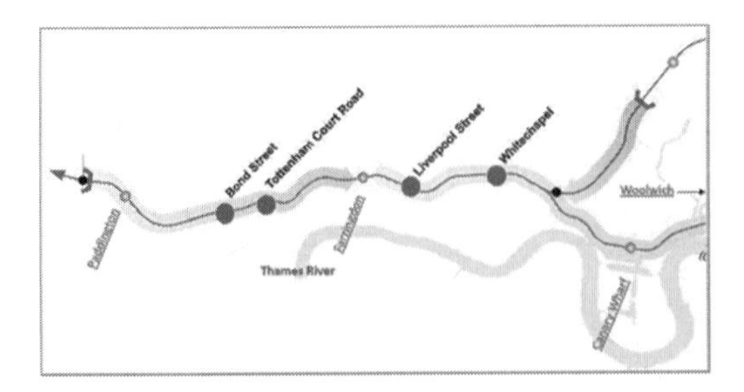

Figure 1. Crossrail route map at the central London area—locations of four excavation case histories (Crossrail, 2017).

sets of parameters were considered as 'Characteristic' and 'Most Probable' by Pillai *et al.*(2011) and shown in Table 1. The definition of the latest BRICK parameters is given in Ellison *et al.*(2012).

The 'Most Probable' BRICK parameters were implemented in the initial Crossrail 3D FEM asset protection assessments in 2010. The predicted ground movements associated with the new tunnel boring machine (TBM) work and the deep excavations have been used for the existing structural damage assessments. They met the asset protection criteria specified by Crossrail or the relevant third parties.

2.2 *Calibration using Crossrail laboratory test data*

In preparation for this back-analysis, a calibration study has been undertaken using the laboratory triaxial testing data from boreholes along the 20-mile long Crossrail alignment in the central London.

London Clay is commonly divided into five units (A to E) (King, 1981), and the triaxial test results have been assessed using these London Clay units. Variations in the test results also depend on the quality of the samples and the test procedures. The BRICK parameters for London Clay Units A3 & A2 have been calibrated and presented in Table 2.

Details of calibration laboratory testing data will be given in Chen (2018). Lower small strain stiffness values have been observed for London Clay Unit C/B from the triaxial laboratory testing. However, the limited number of the sample tested and the large range of values made calibration difficult. Therefore, the laboratory test calibrated BRICK parameters for the London Clay Unit A3 have been applied to the Unit C/B. Similarly, there are very few quality test results for Lambeth Group, so the laboratory test calibrated BRICK parameters for the London Clay Unit A2 have been adopted for the Lambeth Group.

Table 1. Latest updated BRICK parameters.

String length $L_{(b)}$ Characteristic[1]	Most Probable[2]	G_t/G_{max}
0.0000304	0.000030	0.92
0.0000608	0.000075	0.75
0.000101	0.00015	0.53
0.000121	0.00040	0.29
0.00082	0.00075	0.13
0.00171	0.0015	0.075
0.00352	0.0025	0.044
0.00969	0.0075	0.017
0.0222	0.02	0.0035
0.0646	0.06	0

Note: 1. Characteristic BRICK parameters by Pillai (1996), $\lambda^* = 0.1$, $\kappa^* = 0.02$, $\iota = 0.0019$, $\nu = 0.2$, $M_u = 1.3$, $\beta^G = \beta^\varphi = 4.0$. 2. Most Probable BRICK parameters by SCOUT (2007), $\lambda^* = 0.1$, $\kappa^* = 0.02$, $\iota = 0.00175$, $\nu = 0.2$, $M_u = 1.3$, $\beta^G = \beta^\varphi = 4.0$.

Table 2. Calibrated BRICK parameters from triaxial tests.

String length $L_{(b)}$ LC—A3	LC—A2	G_t/G_{max}
0.0000304	0.000030	0.92
0.0000608	0.000075	0.75
0.000101	0.00015	0.53
0.000121	0.00040	0.29
0.00082	0.00075	0.13
0.00171	0.0015	0.075
0.00352	0.0025	0.044
0.00969	0.0075	0.017
0.0222	0.02	0.0035
0.0646	0.06	0

Note: for LC-A3: $\lambda^* = 0.1$, $\kappa^* = 0.02$, $\iota = 0.0019$, $\nu = 0.2$, $M_u = 1.3$, $\beta^G = 4.0$, and $\beta^\varphi = 2.0$. For LC-A2: $\lambda^* = 0.1$, $\kappa^* = 0.01$, $\iota = 0.0015$, $\nu = 0.2$, $M_u = 1.3$, $\beta^G = \beta^\varphi = 5.0$. (Chen, 2018)

3 REVISED BRICK PARAMETERS

3.1 *Calibration using inclinometer data—through TCR-WTH FEM back-analysis*

These "calibrated" BRICK parameters based on triaxial testing have been used to calculate wall deflections at the Tottenham Court Road Station—Western Ticket Hall (TCR-WTH) deep shaft excavation. However, differences of up to a factor of two remain between the measured and the calculated deflections.

After a careful review of the inclinometer data and the as-built construction details including excavation levels, wall, prop and waling stiffness, a number of back-analysis iterations have concluded that a stiffer set of BRICK parameters was needed to replicate the inclinometer data. Another set of revised London Clay BRICK parameters was developed based on the inclinometer data for TCR—WTH back-analysis and presented in Table 3. These are referred to as "revised" BRICK parameters.

3.2 *TCR-WTH deep box excavation*

The TCR-WTH deep shaft is almost in a rectangular shape, measuring about 41 m × 31 m in plan. The maximum excavation depth is approximate to 29.5 m. The original design adopted a 1.0 m thick diaphragm wall with temporary props for a bottom-up construction sequence. The site layout and section are shown in Figure 2.

To minimize the excavation duration, the observational method was adopted to modify the design during construction, allowing the lowest level of temporary props to be omitted. This reduced costs and saved about four weeks on the programme. (Yeow, 2014) The as-built construction sequence is summarised in Table 4.

The 3D FEM model for the TCR-WTH was updated with the as-built construction sequence and the most probable structural properties for the temporary props and the waler beams. All soil elements and the piles were modelled using solid elements. The retaining walls, capping beams, slabs, walers and other walls were modelled as shell elements. The temporary props were modelled using beam elements. The updated 3D FEM for TCR-WTH back-analysis is shown in Figure 3.

A ShapeArrayAccel (SAA) was installed in the TCR-WTH diaphragm wall panels to measure the wall deflection. SAA-3 measured the maximum

Table 3. Revised BRICK parameters based on inclinometer data.

String length $L_{(b)}$ LC-A3 & LC-A2	G_t/G_{max}
0.000030	0.92
0.000075	0.75
0.00015	0.60
0.00025	0.50
0.0005	0.35
0.0010	0.25
0.0015	0.15
0.0020	0.05
0.0033	0.01
0.06	0

Note: $\lambda^* = 0.1$, $\kappa^* = 0.01$, $\iota = 0.0015$, $\nu = 0.2$, $M_u = 1.3$, $\beta^G = \beta^\phi = 5.0$. (Chen, 2018)

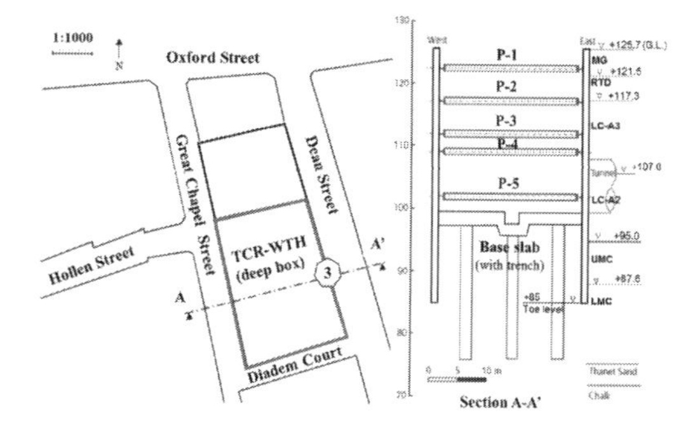

Figure 2. TCR-WTH layout plan and section (Bologna, 2017, Yeow, 2014).

wall deflection profile during the excavation and was adopted for the back-analysis. The location of SAA-3 is indicated in the layout plan in Figure 2.

A comparison of the wall deflection profiles at stage 2, stage 4 and stage 6 in Table 4 is presented in Figure 4. At each stage, the calculated deflection profile by the calibrated BRICK parameters (shown as the broken lines in Figure 4) and the calculated deflection profile by the revised BRICK parameters (shown as the solid lines in Figure 4) were compared with the inclinometer data. The predictions by the revised BRICK parameters have shown a significantly improved agreement with the inclinometer data.

In the TCR-WTH 3D FEM back-analysis, the Lambeth Group has been modelled by the revised BRICK parameters using a slightly higher over consolidation ratio (OCR) in order to stiffen the soil and control the wall deflection within the Lambeth Group. Details of this OCR adjustment will be given in Chen (2018).

In order to validate the TCR-WTH revised BRICK parameters for wall deflection calculation, three additional Crossrail deep excavations have been back-analysed using the revised BRICK parameters.

4 CASE HISTORIES

4.1 *Bond Street Station—Western Ticket Hall*

The Bond Street Western Ticket Hall (BS-WTH) is not far from the TCR-WTH site. This box measures 28 m × 56 m in plan and the maximum excavation depth is about 26 m from existing ground level. The BS-WTH box is surrounded on four sides by local roads with buildings beyond the roads. The layout of the site and the section are shown in Figure 5.

The box is supported by a 1.2 m thick diaphragm wall and constructed using a top-down excavation

Table 4. Summary of as-built construction sequence—TCR-WTH.

Stage	Description	Formation level (m ATD)
1	Excavate 1	+121.6
2	Install P-1 & excavate 2	+116.4
3	Install P-2 & excavate 3	+111.1
4	Install P-3 & excavate 4	+108.1
5	Install P-4 & excavate 5	+101.0
6	Excavate 6 & cast base slab	+96.8

After (Yeow, 2014; Chen *et al.*, 2015)

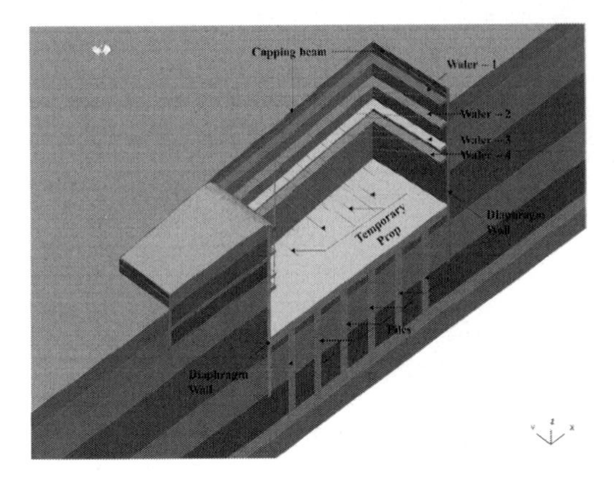

Figure 3. TCR-WTH 3D FEM—section along axis Y.

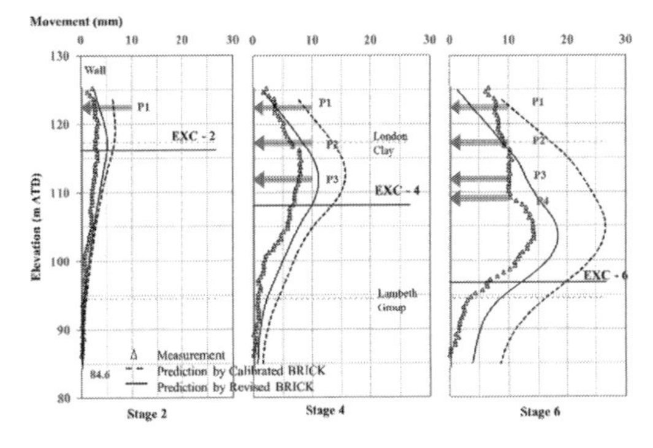

Figure 4. TCR-WTH SAA-3 wall deflection profiles: measurements compared with calculations.

sequence. Permanent reinforced concrete slabs were installed during excavation and provided the required lateral support to the retaining wall. The TBM went past during the box excavation. Some compensation grouting work was also conducted as the box excavation progressed. The as-built construction sequence is summarised in Table 5.

The 3D FEM of the BS-WTH was originally undertaken for the asset protection studies. The analysis included the existing tunnels and buildings, together with the new Crossrail TBM tunnels and SCL tunnels, see Figure 6. The back-analysis of BS-WTH has adopted the same 3D FE model with the revised BRICK parameters based on the TCR-WTH inclinometer data.

The ground movement case study by Bologna (2017) provides details of the construction of the Crossrail TBM tunnels. These run parallel to the BS-WTH box diaphragm walls in the longitudinal direction, see Figure 6. The tunnelling effects have been included in the inclinometer measurements but not modelled in the 3D FEM.

It is assumed that the tunneling effect has less impact on the walls along the axis perpendicular to the tunnel. Hence the inclinometer IN-15 installed in the middle of one short side next to Gilbert Street as indicated in the site layout plan in Figure 5, was selected for the back-analysis comparison. The calculated wall deflection profiles were compared with the measured deflection profiles from inclinometer IN-15 in Figure 7. In addition, the inclinometer data was only available from the

Table 5. Summary of as-built construction sequence—BS-WTH.

Stage	Description	Formation level (m ATD)
1	Excavate 1 & cast GF slab	+118.5
2	Excavate 2 & cast L-1 slab	+112.2
3	Excavate 3 & cast L-2 slab	+106.8
4*	Westbound TBM transit	As above (+106.8)
5*	Eastbound TBM transit	As above (+106.8)
6	Excavate 4 & cast L-4 slab	+101.0
7	Excavate 5 & cast L-5 slab	+95.0

Note: stage 4 & 5 excluded in the 3D FEM reanalysis (Bologna, 2017)

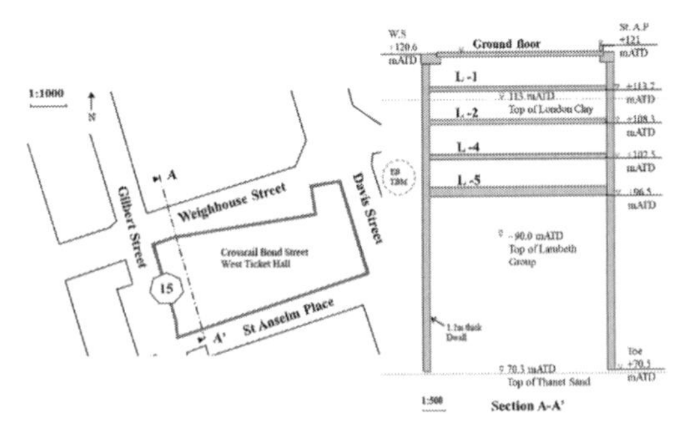

Figure 5. BS-WTH layout plan and section (Bologna, 2017).

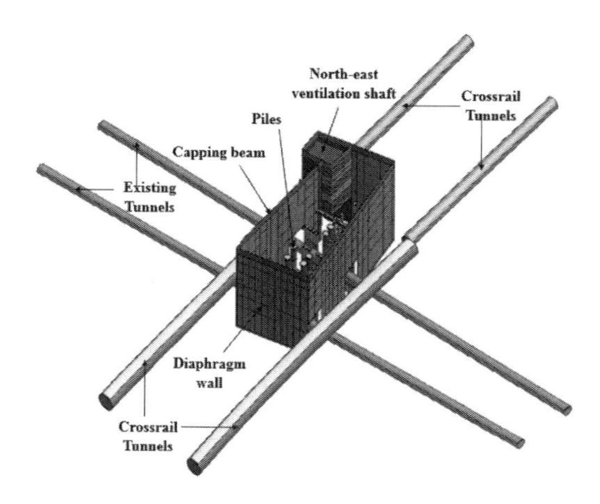

Figure 6. BS-WTH 3D FEM model.

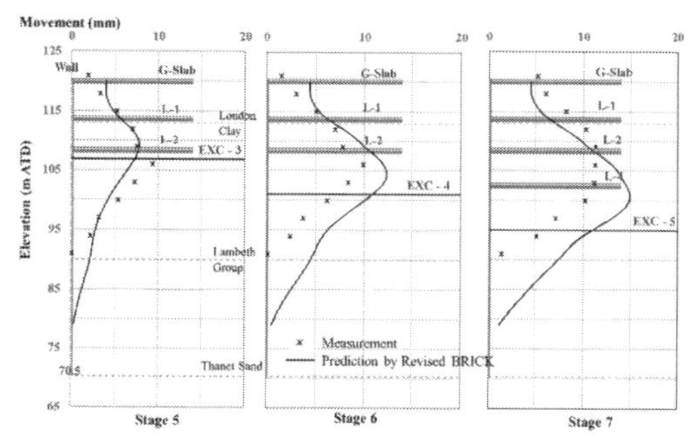

Figure 7. BS-WTH IN-15 wall deflection profiles: measurements compared with calculations using revised BRICK parameters.

beginning of the TBM tunnelling works (stage 4 in Table 5). Therefore, the wall deflection comparison was carried out for stage 5 to stage 7. Measured and calculated wall deflection profiles are compared in Figure 7 showing reasonable agreement.

4.2 *Liverpool Street Station—Moorgate Shaft*

The Moorgate Shaft is part of Liverpool Street Station (LIS-MS) structure. It is an octagonal shape, about 35 m × 35 m in plan with the maximum excavation depth of about 39 m from existing ground level. The site layout plan and the section are presented in Figure 8.

The shaft is formed by a 1.2 m thick diaphragm wall in a depth of about 52 m. Permanent propping to the shaft is provided by a series of ring beams forming the top-down excavation sequence. In the East-West direction, two cross-walls act as propping walls for the lower level during excavation. The initial design included two-levels of temporary props below the lowest ring beams (RBs) that would be removed after the installation of the base slab.

Due to the delayed site obstruction clearance work, it became necessary to speed up the shaft construction to enable the twin TBMs to pass through the MS on time. The observational method was adopted to modify the design. Several excavation stages were successfully combined. The lowest two-levels of temporary propping was omitted.

The as-built construction sequence is presented in Table 6. The 3D FE model for the LIS-MS was updated with the as-built conditions as shown in Figure 9.

As the diaphragm walls at East-West direction have been affected by clay inclusions at the end

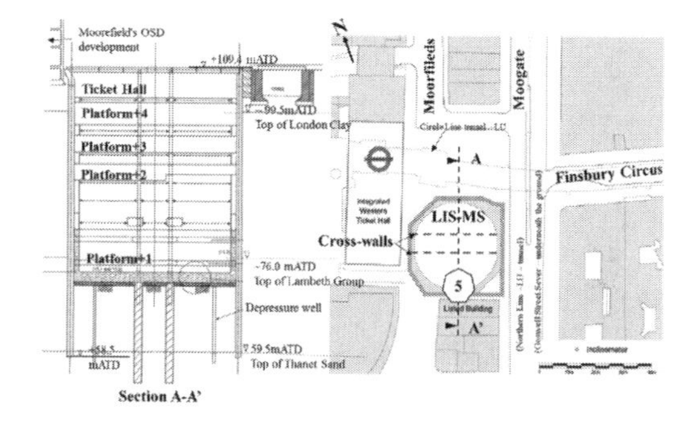

Figure 8. LIS-MS layout plan and section, after (Farooq et al., 2015).

of the two cross-walls (Chen *et al.*, 2015), the wall along the North-South direction was reviewed in the 3D FEM back-analysis. Measurements taken at inclinometer IN-5 on the south side of diaphragm wall are shown in Figure 10. These measured deflections were compared with the calculated wall deflection using the TCR-WTH revised BRICK parameters. Consideration was also given to an adjacent development which took place behind the south side diaphragm wall.

The comparisons at construction stage 2, stage 5 and stage 7 as listed in Table 6 are presented in Figure 10. Good agreement can be seen. There is some variation in the top 20 m of the wall deflection profiles, which is believed mainly attributed to the adjacent development.

At stage 5 and stage 7, the maximum predicted wall deflection value is slightly less than the measured value (within 1.5 mm). This could be associated with the selection of the Lambeth Group BRICK parameters. Noted that the final excavation stage occurred in the Lambeth Group clay.

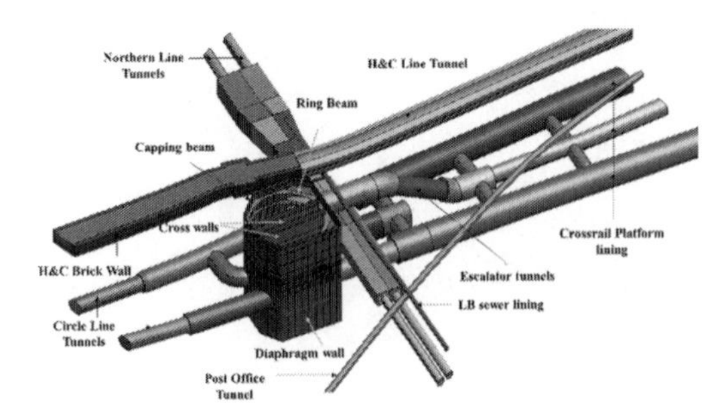

Figure 9. LIS-MS 3D FEM model.

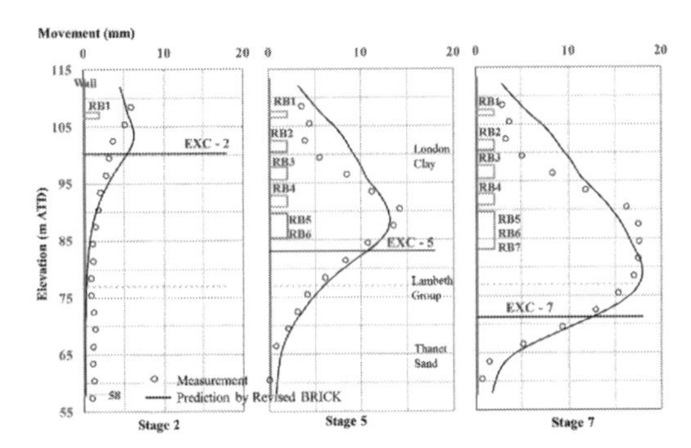

Figure 10. LIS-MS IN-5 wall deflection profiles: measurements compared with calculations using revised BRICK parameters.

4.3 *Whitechapel Station—Durward Street Shaft*

The Whitechapel Station, Durward Street Shaft (WS-DSS) was excavated in similar ground condition to the LIS-MS. The DS-DSS shaft is a hexagon shape, measuring about 59 m × 32 m in plan with a maximum excavation depth of about 29.5 m from ground surface level. The perimeter diaphragm walls were 1.2 m thick and extended to a level of +70.1 mATD. The site layout plan and the section are shown in Figure 11.

A bottom-up sequence was adopted for the shaft construction as summarised in Table 7.

In order to speed up the excavation programme to cater for the TBM arrivals, the observational method was adopted from Stage 4 in Table 7 to modify the later excavation stages. The maximum wall deflection profiles at the WS-DSS measured from Inclinometer IN-3 was used to compare with the calculated wall deflection from the initial back-analysis. The back-analysis results showed that the early removal of the lowest level of props was acceptable without exceeding the design wall deflection criteria. (Mills, 2016)

Table 6. Summary of as-built construction sequence of LIS-MS.

Stage	Description	Formation level (m ATD)
1	Capping beam & excavate -1	+106.1
2	Cast RB-1 & excavate -2	+100.3
3	Cast RB-2 & excavate -3	+94.4
4	Cast RB-3/4 & excavate -4	+87.6
5	Cast RB5/6 & excavate -5	+83.5
6	Cast RB7 & excavate -6	+79.1
7	Excavate – 7	+71.3
8	Cast base slab	–

(Farooq *et al.*, 2015; Chen *et al.*, 2015)

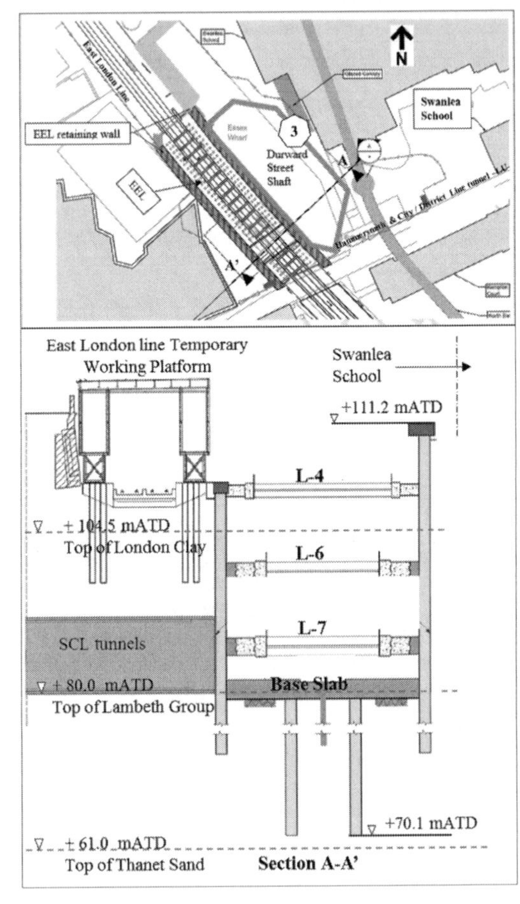

Figure 11. WS-DSS layout plan and section, after (Mills, 2016).

In this subsequent back-analysis, the revised BRICK parameters from the TCR-WTH have been applied in the same 3D FE model for the WS-

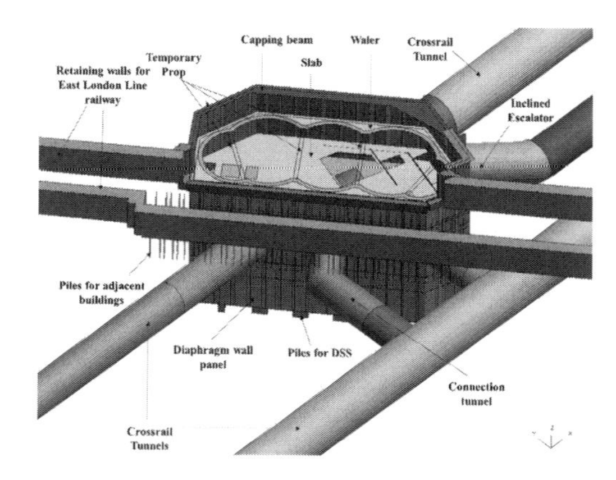

Figure 12. WS-DSS 3D FEM model.

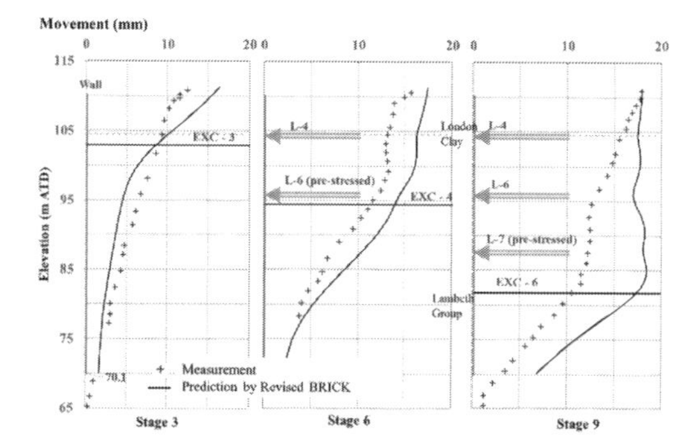

Figure 13. WS-DSS IN-3 wall deflection profiles: measurements compared with calculations using revised BRICK parameters.

DSS as shown in Figure 12. The calculated wall deflection profiles are compared with the inclinometer IN-3 data at stage 3, stage 6 and stage 9, as shown in Figure 13.

During the early cantilever excavation stages (stage 1 to stage 3 in Table 7), the maximum of up to 9.2 m excavation below ground level occurred, but the relativly large deflection at the deeper level was observed from IN-3 monitoring data. For example, about 5 mm movement was recorded at a depth of 10 m below the excavation level, and around 2.5 mm displacements in the Lambeth Group which is about 30 m below the excavation level. An inclinometer data review on IN-3 may explain these large displacements.

During the subsequent excavation stages when the excavation was propped, the calculated wall deflection profiles are similar to the measured profiles. A significant variance between measurements and predictions is observed in comparison at stage 9, in which the monitoring data has been

Table 7. Summary of as-built construction sequence of WS-DSS.

Stage	Description	Formation level (m ATD)
1	Excavate -1	+109.0
2	Excavate -2	+103.9
3	Excavate -3	+102.9
4	Install prop L-4 at +104.3	As above
5	Excavate -4	+94.4
6	Pre-stressed prop L-6 at +95.8	As above
7	Excavate -5	+86.6
8	Pre-stressed prop L-7 at +87.5	As above
9	Excavate -6	+81.7
10	Cast base slab	As above

(Mills, 2016)

restrained between the bottom in the Lambeth Group to the propping level L–6. This could be associated with the selection of the Lambeth Group BRICK parameters. Noted that the final excavation is immediately above the Lambeth Group clay.

5 SUMMARY & DISCUSSION

5.1 *Summary of Crossrail excavation case histories*

Four Crossrail excavation case histories have been collected and back-analysed. The measured maximum wall deflections ($\delta_{H,Max}$) are summarised in Table 8 together with the corresponding excavation depths (H_e). This data is plotted onto historical data in Figure 14 and shows the Crossrail case histories have a ratio of $\delta_{H,Max}$ to H_e of about 0.05%. This is smaller than previous experience.

5.2 *Discussion*

The soil stiffness was investigated using laboratory small strain triaxial testing data from the Crossrail ground investigations. Initially, the BRICK parameters for different London Clay Units were calibrated against these laboratory test. However, when the calibrated BRICK parameters were used in the 3D FEM, the calculated wall deflections were significantly larger than the inclinometer data. This implied that the laboratory tested small strain stiffness was not large enough.

The revised BRICK parameters were derived from the back-analysis of the TCR-WTH incli-

Table 8. Summary of maximum wall deflection over excavation depth.

Case	$\delta_{H,Max}$ *(mm)*	H_e *(m)*	$\delta_{H,Max}/H_e$ *(%)*	Remark
TCR-WTH (SAA-3)	14.24	29.5	0.05%	B/L = 31/41 (1.32)
[1]BS-WTH (IN-15)	11.29	26.0	0.04%	B/L=28/56 (2.00)
LIS-MS (IN-5)	17.58	38.8	0.05%	B/L = 35/35 (1.00)
[2] WS-DSS (IN-3)	16.20	29.5	0.06%	B/L = 32/59 (1.84)

Note: 1). BS-WTH measurement from IN-15 is on the short side of the excavation box, it may not represent the maximum wall deflection for this excavation case; 2). WS-DSS measurement from IN-3, the maximum movement is taken from below +105 mATD to mimic the cantilever excavation effect.

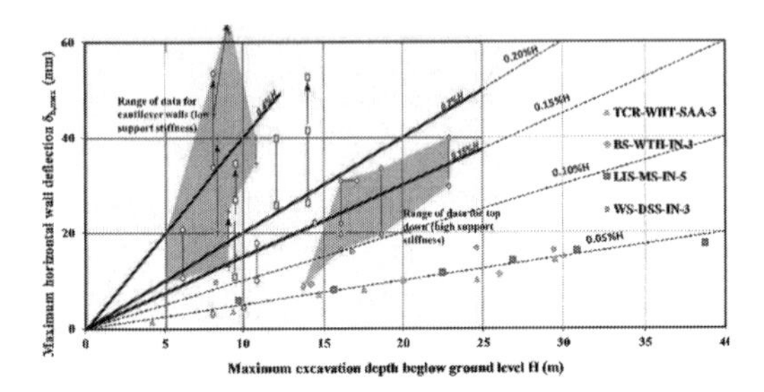

Figure 14. Observed maximum lateral wall deflections from excavations in London Clay, after Ciria C760 (Gaba et al., 2017).

nometer data. Higher small strain stiffnesses were needed to match the inclinometer results compared with the triaxial laboratory test data. The in-situ self-boring pressure (SBP) testing data (Crossrail, 2009) was also reviewed and found to be stiffer than the triaxial test data. This indicates that the small strain triaxial laboratory testing did not accurately reflect the in-situ soil behavior.

The Lambeth Group was initially modeled using the TCR-WTH revised BRICK parameters. This lead to the back-analysed wall deflections being slightly overestimated at the LIS-MS and the WH-DSS case histories. In these two cases, the Lambeth Group was encountered at the bottom of the excavation. The further studies are needed to confirm the BRICK parameters for the Lambeth Group.

The Crossrail case histories show consistent ratio of $\delta_{H,Max}$ to H_e of about 0.05%. This is a smaller ratio compared with the historical data reported in Ciria C760 (St John *et al.*, 1992; Gaba *et al.*, 2017). It seems the stiff retaining wall and propping systems used on Crossrail achieved less-wall deflection in both the bottom-up and top-down methods compared with previous excavations in London Clay.

6 CONCLUSIONS

Back-analysis of four Crossrail deep excavation case histories have been carried out using the 3D FEM. The inclinometer wall deflection profiles have been used for the comparisons. Initially, the BRICK soil constitutive model parameters were calibrated using triaxial laboratory testing data. This led to overestimated wall deflections. Subsequently, revised BRICK parameters were developed, based on Crossrail TCR-WTH inclinometer data to fit the wall deflections.

All these back analyses have been carried out assuming undrained soil properties during the excavation work.

Particular care is required to interpretate the reliable inclinometer data.

The revised BRICK parameters for London Clay have been tested at three other Crossrail deep excavations, that had either bottom-up or top-down construction sequences. Good agreement has been achieved.

1. The revised BRICK parameters are considered to be 'most probable' or best estimate parameter values at this stage and can be used with the observational method in London Clay to predict wall deflections;
2. The assumption of using the revised London Clay BRICK parameters for the Lambeth Group is valid, although it may slightly overestimate the wall deflection when the excavation is close to or in the Lambeth Group;
3. The four Crossrail case histories recorded wall deflections which were consistently smaller than previous excavations in London Clay. Hence less impace on the surrouding structures. The reason for this needs further study.

NOTATION

H_e	excavation depth
$\delta_{H,Max}$	maximum wall lateral displacement
G_t	shear modulus
G_{max}	the maximum shear modulus
G/G_{max}	the proportion of material in BRICK
λ^*	the slope of the isotropic normal compression line in lnv - lnp' space
κ^*	the slope of the swelling line in lnv - lnp' space
ι	parameter controlling elastic stiffness
ν	Poisson's ratio
M_u	constant in the Drucker-Prager modification (μ)
β^G	the overconsolidation parameter for stiffness in BRICK model
$\beta^{\varnothing}$	the overconsolidation parameters for strength (failure angle) in BRICK model
$L_{(b)}$	the array of string length in BRICK model
OCR	over-consolidation ratio
SBP	self-boring pressure
B	the width of the excavation
L	length of the excavation
TBM	tunnel boring machine
SCL	sprayed concrete lining
RB	ring beam

REFERENCES

Bologna, P. (2017) *Benchmarking empirical methods of prediction of ground movement for deep excavation.* MSc thesis, University of Cambridge.

Chen, Y. (2018) *Application of new observational method on deep excavation retaining wall design in London Clay*', PhD thesis in preparation, University of Cambridge.

Chen, Y. *et al.* (2015) Application of the observational method on Crossrail projects, in *Crossrail Lessons Learnt Conference, London. UK.* London.

Crossrail, (2009) *Geotechnical Sectional Interpretative Report; Report 1&2, 3, 4*; internal references: 1D0101-G0G00–00549, 550 and 551.

Crossrail, (2017) rout map visited in November 2017, web link: *http://www.crossrail.co.uk/route/maps/regional-map*

Ellison, K.C. (2012) *Constitutive Modelling of a Heavily Overconsolidated Clay.* PhD thesis, University of Cambridge.

Farooq, I. *et al.* (2015) An innovative Verification Process speeds construction of Crossrail's Moorgate shaft, *Geotechnics ICE.*

Fuentes, R., Pillai, A. and Devriendt, M. (2010) Short term three dimensional back-analysis of the One New Change basement in London, pp. 1–6. Available at: http://discovery.ucl.ac.uk/1318407/

Gaba, A.R. et al. (2003) Embedded retaining walls: guidance for economic design, Proceedings of the ICE - *Geotechnical Engineering.* doi: 10.1680/geng.2003.156.1.13.

Hardy, S. *et al.* (2017) New observational method framework and application, *Proceedings of the 19th International Conference on Soil Mechanics and Geotechnical Engineering, Seoul, Korea.*

King, C. (1981) The stratigraphy of the London Clay and associated deposits. Tertiary Research Special Paper, No. 6. Rotterdam, Backhuys.

Mills, C. (2016) Project in focus—Crossrail Whitechapel Station, *Proceedings of Basement and Underground Structures*, October, London, UK.

Pantelidou, H. and Simpson, B. (2007) Geotechnical variation of London Clay across central London, *Géotechnique*, 57(1), pp. 101–112. doi: 10.1680/geot.2007.57.1.101.

Pillai K.A. (1996) *Review of BRICK model of soil behavior.* MSc dissertation, Imperical College, London.

Pillai, K.A. *et al.* (2011) Back-analysis of a basement with a raft foundation in overconsolidated stiff clay, *Proceedings of 15th European Conference on Soil Mechanics and Geotechnical Engineering*, Athens, Greece. Available at: http://discovery.ucl.ac.uk/1323408/

SCOUT. (2007) D18 Report on Observational Method under the Framework of Eurocodes. European Commission Sixth Framework Programme. Project no: 516290. Re—port ref no: SB.DTTAP-NOT-6.045. Arup Geotechnics.

Simpson, B. (1992) Retaining structures: displacement and design, *Geotechnics*, 4, pp. 541–576.

St. John, H.D., Potts, D.M., Jardine, R.J., and Higgins, K.G. (1992) Prediction and performance of ground response due to the construction of a deep basement at 60 Victoria Embankment. *Proc., Instn, of Civ. Engrs.*, Part1, Vol.92, pp. 449–465.

Yeow, H. (2014) Application of observational method at Crossrail Tottenham Court Road station, *Geotechnical Engineering*, 167(GE2), pp. 182–193.

Yeow, H. and Feltham, I. (2008) Case histories back analyses for the application of the Observational Method under Eurocodes for the SCOUT project, in *6th international conference on case histories in Geotechnical Engineering*, Arlington, VA, pp. 1–13.

Zdravkovic, L., Potts, D.M. and John, H.D.S. (2005) Modelling of a 3D excavation in finite element analysis, *Géotechnique*,55(7), pp. 497–513. doi: 10.1680/geot.2005.55.7.497.

Numerical Methods in Geotechnical Engineering IX – Cardoso et al. (Eds)
© 2018 Taylor & Francis Group, London, ISBN 978-1-138-33203-4

Numerical analysis of an unsymmetrical railcar unloading pit and connection trench

G. Pisco, C. Fartaria & R. Tomásio
JETsj, Geotecnia, Lisboa, Portugal

J. Costa & J. Azevedo
TECNASOL, Fundações e Geotecnia, Lisboa, Portugal

ABSTRACT: A numerical analysis was performed simulating the deep excavation and dewatering effects on retaining walls of an unsymmetrical railcar unloading pit and trench. With a depth of 22 meters and an internal diameter of 45 meters the circular pit has a 12 meters width connecting with the trench, which is non collinear with the pit center. A 3D Finite Element Model using PLAXIS software was conducted in order to access both general and local effects in the design of the pit structural elements, due to singularities introduced by the trench opening on the west side of the pit wall and the trench excavation. Both geotechnical and hydrogeological characteristics of site were considered as well as the main construction stages covering the excavation sequence and the groundwater flow analysis.

1 INTRODUCTION

Deep excavations comprise often a very complex soil-structure interaction. A diversity of aspects must be accessed in order to achieve a safe design, such as geometry, construction sequences, water flow and stress/deformation states. In common practice, 2D finite elements models are widely used to analyze geotechnical structures as a plane strain or an axisymmetric approach are often suitable.

However, and usually due to geometric singularities, some structures require a three-dimensional analysis in order to access more accurately global and local soil-structure interaction (Hou et al. 2009). In this paper is presented and discussed the numerical analysis of a circular pit deep excavation connected to a non collinear trench, using both 2D and 3D finite element models.

2 SITE CONDITIONS AND STRUCTURE DESCRIPTION

2.1 *Geological and hydrogeological conditions*

An extensive site investigation campaign was carried out in order to access the ground stratigraphic profiling and evaluate the relevant material geomechanical and hydrogeological properties. The findings showed the existence of a modern clayed-silty fill under an alluvium unit composed by an upper layer of clayey silt/silty clay overlying an organic clay, with approximately 10 m of thickness. Underneath those superficial formations was observed a sequence of clayey materials with increasing stiffness with depth up to 40 m where was found the Ordovician bedrock, mainly composed by very stiff to hard dark grey to black claystone.

At the base of the alluvium materials, mostly composed by fine grain materials, the presence of a sandy layer was observed. With thickness ranging from 4 to 8 m, the interpretation of this layer behavior was fundamental to the structure's design due to its high permeability when compared with the remaining formations. The upper alluvium and lower clayey layers were characterized by low permeabilities ranging from 10^{-8} to 10^{-10} m/s, while the sandy layer permeability was evaluated as 10^{-5} m/s. Being the sandy layer confined between low permeable soil layers, the water in this layer is subjected to positive pressures which could comprise the stability of the base of the excavation by heave effect (Ou 2006).

Water level measurements were accessed continuously indicating a ground water table level ranging between 1 to 6 m depth being related to the tides.

2.2 *Retaining wall structures*

The structure is part of a complex for aluminum ore extraction and distribution, comprising a main railcar unloading pit and the conveyor system trench. The pit has a circular shape with a diameter of 44.5 with an approximate maximum depth of 22 m. A 12 m width trench is connected from the bottom of the pit to the surface level, over a length of 122 m.

Due to site geological, geotechnical and hydrogeological conditions, particularly the position of the phreatic level, a 1 m thick peripheral diaphragm wall was designed allowing the excavation of both pit and trench. An isometric representation of the retaining structure is illustrated in Figure 1.

An approximately circular pit was shaped with the execution of consecutive rectangular D-Wall panels. In order to ensure the D-Wall panels' behave as a continuous structure and also to increase the radial stiffness of retaining structure, five horizontal levels of reinforced concrete rings were designed (Fig. 2). The fifth ring, placed below the mat foundation level, provide positive support of the D-Wall panels and carry circumferential compression induced by the trench opening. As an additional security measure to prevent eventual lateral water flow through the retaining structure, 800 mm diameter jet-grouting columns at the backside of each panel joint, were executed.

In order to avoid lateral instability of the pit D-Wall panels on the trench side, during the trench excavation, four bracing slabs were executed in the backside of those to ensure the transfer of the outward thrust to the trench's side D-walls. The horizontal equilibrium of the trench retaining structure was assured by a set of concrete struts (Fig. 3).

In order to avoid hydraulic failure of the low permeable silty clay layers during the excavation works, 7 pressure relief vertical drillings were executed inside the pit and the trench. This drills shall intersect the sandy layers located beneath the silty clay ones, relieving the hydrostatic pressure of the confined aquifer. To minimize the water flow towards the excavation, the D-Wall panels were design with an average depth of 45 m, assuring a 3 m embedment in the low permeability Ordovician formation.

To ensure the proper foundation of the future structures and to prevent water inflow a mat foundation were design comprising both the pit and the trench assuring the vertical equilibrium of the long term water pressure that will settle in the bottom of the mat elements, assured with a set of self-drilling micropiles.

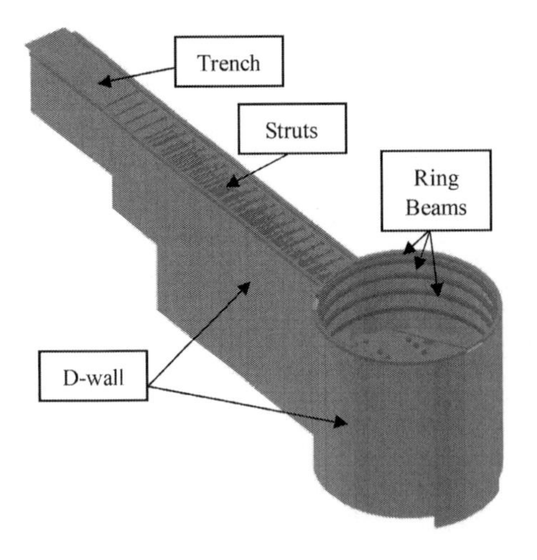

Figure 1. Pit and trench retaining wall isometric view.

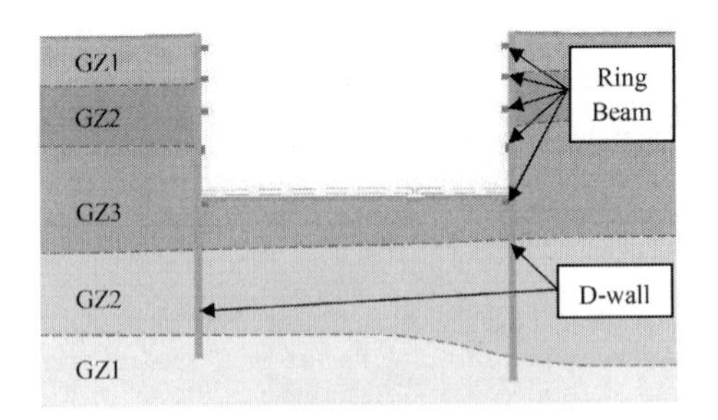

Figure 2. Pit retaining wall cross section.

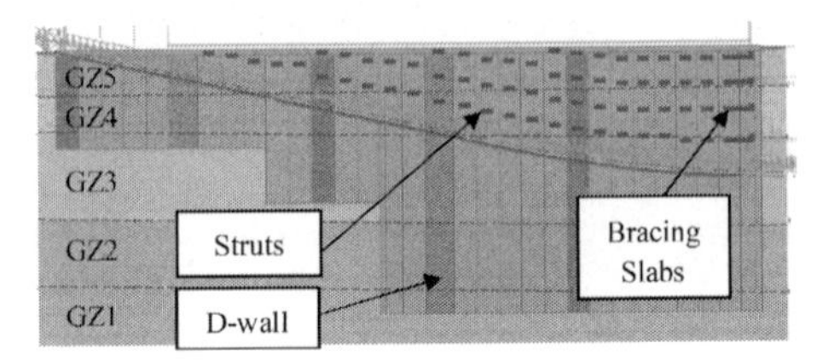

Figure 3. Trench retaining wall longitudinal cross section.

3 NUMERICAL MODELS

3.1 *General information*

A series of 2D and 3D analyses were carried out using the finite element model program PLAXIS. As a first approach 2D models were developed and later used to validate the 3D global model results and to design the sections that were not under the influence of the pit/trench connection.

The soil behavior was simulated with the Hardening-Soil model (Brinkgreve et al. 2017), an elastic-plastic model with a multi-surface yield criterion that simulates the increasing of soil stiffness with deformation. The following tables (Tab. 1 and Tab. 2) summarize the main parameters that regulated the behavior of the soil formations. Due to the type of structure and execution technology, it was considered a R_{int} parameter of 0.67 for all the soil layers.

Despite the low-permeability layers, in typical excavation works, the critical condition is the long term behaviour of the soils. Therefore, a drained analysis was performed for both excavation and permanent stages in a conservative design option.

Within each model a hydrostatic analysis was performed considering the water level one meter below the soil surface. The water level variation and hydrostatic pressures were studied in agreement with staged construction sequence and the relief vertical drilling. Additionally, flow analyses were performed on 2D models in order to access the water flow at the bottom of the excavation and estimate surface surrounding settlements.

3.2 *2D Axisymmetric pit retaining wall model*

As a first approach a 2D axisymmetric analysis of the pit retaining wall was performed using plate elements with D-wall equivalent flexural stiffness, and fixed-end-anchors with the equivalent flexural stiffness of the ring beams. An image of the 2D axisymmetric model mesh is illustrated in Figure 4.

Simulating a perfect circular wall, the opening and the trench effect were not included in this analysis. Another drawback of this model is the impossibility of evaluate the level of the radial axial forces on the wall, which have a main role in circular structures that mobilize the arch effect.

Apart from the mentioned limitations, this model represented an important tool in the validation and calibration of a global 3D model. The results of the 2D model were compared with the ones from the 3D global model, further from the opening, where was expected an identical behavior of the soil structure-interaction.

3.3 *2D Plane strain trench retaining wall models*

Considering the trench geometry, a set of 2D plane strain finite element models were performed using plate elements with D-wall equivalent flexural stiffness and fixed-end-anchors with the equivalent axial stiffness of concrete struts. The connection

Table 1. Soil layers and permeability properties.

Soil Layers	Depth m	k ms^{-1}
GZ5 -Alluvium	0–7.5	3.36×10^{-6}
GZ4 - Silty clay 1	7.5–15.5	1×10^{-4}
GZ3 - Silty clay 2	15.5–29.0	2.1×10^{-10}
GZ2 - Medium sand	29.0–41.0	1.6×10^{-5}
GZ1 - Claystone	>41.0	5.56×10^{-7}

Table 2. Soil geomechanical parameters.

Soil Layers	γ_{sat}	c'	φ'	E^{ref}_{50}	E^{ref}_{oed}	E^{ref}_{ur}	m
	kN/m³	kPa	°	MPa	MPa	MPa	–
GZ5	17	3	21	6	6	18	0.8
GZ4	22	8	34	30	30	90	0.8
GZ3	20	35	25	20	20	60	0.8
GZ2	21	0	38	100	100	300	0.5
GZ1	22	40	38	150	150	450	1.0

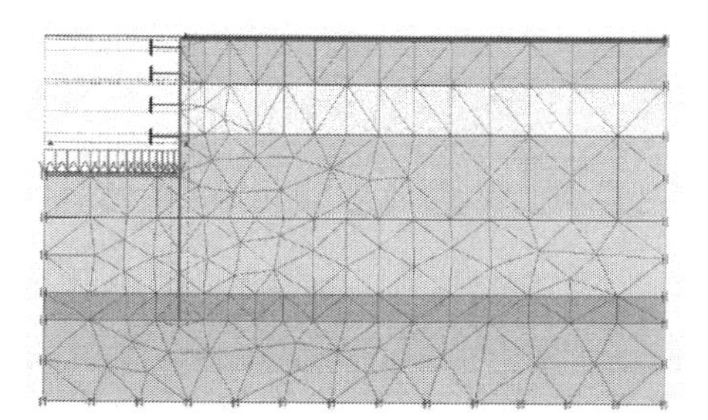

Figure 4. 2D Axisymmetric pit retaining wall model—mesh.

between struts and walls was not design to tension forces but only to shear the fixed-end-anchors, so tension axial stiffness was set to 0.

Since the trench has variable excavation heights, five different models were analyzed in order to simulate the highest excavation point for each lateral support scenario, ranging from four to zero levels of struts. Figure 5 illustrates the 2D plane strain model mesh representing a trench section with 3 levels of struts.

Due to the overall longitudinal dimension of the structure the 2D model results were used at the design of the majority of trench's D-wall panels, while the global 3D analysis was used to define in which panels the 2D results were not valid. As in the axisymmetric model, the trench 2D model was used also for the validation of the 3D global model.

3.4 *3D Global retaining wall model*

3.4.1 *Model definition*

In order to access both general and local effects in the structural design of the pit and trench, that were not possible to model with 2D models, a 3D global finite element model was developed. Figure 6 illustrates the 3D global model mesh.

With a geometry of 200 m × 150 m and 66.5 m depth, the 3D model mesh has 137692 finite elements and 204068 nodes. The mesh density was optimized using local coarseness features. To reduce the model size and to avoid numerical errors some simplifications were assumed, such as trench length reduction as well as assuming a constant subgrade level at the maximum excavation depth.

The D-Wall panels for both pit and trench retaining wall panels and bracing slabs were modeled trough plate elements with equivalent flexural stiffness. As the constructive technology of D-walls impose a non-monolithic connection between successive panels, a hinged connection was modeled between plate elements (Fig. 7). The capping beams, trench concrete struts and the pit ring beams were modeled with beam elements with equivalent stiffness. In Tables 3 and 4 the model plate and beam elements are summarized.

3.4.2 *Model stage construction*

The model stage construction was defined considering the excavation sequence prescribed with the following stages: 1) establishment of the initial stress field; 2) activation of the D-wall plate elements and its hinged connections; 3) activation of the capping beam elements; 4) 1st excavation stage including simulation of dewatering effects; 5) activation of 1st level ring beam, strut beam and bracing slab plate elements; 6) replication of previous two stages for each excavation level down to the 5th excavation level; 7) deactivation of plate elements simulating trench/pit opening.

3.4.3 *Model singularities*

The main aspects captured with the 3D model were the local and overall behavior introduced by the trench excavation on the west side of the pit as well as, in a later stage, the trench/pit opening. The resultant imbalance in the back of the pit retaining walls was account and balanced with bracing slabs elements. The relative orientation of the trench to the pit was another important issue, since they were not concentric. In the 3D model, the D-wall panels were defined by plate elements with width dimensions as executed and were consecutively disposed in order to simulate the real geometrical conditions of the pit retaining wall.

The 3D global model allowed the modeling of the T-panels that connect the trench and pit's wall. For constructive reasons, the web and the flange cage was placed disjointedly so both translational and rotational degrees of freedom were released in those connections. Figure 8 illustrates the model at the zone of interception pit/trench with the visible opening, pinned connections between panels and bracing slabs.

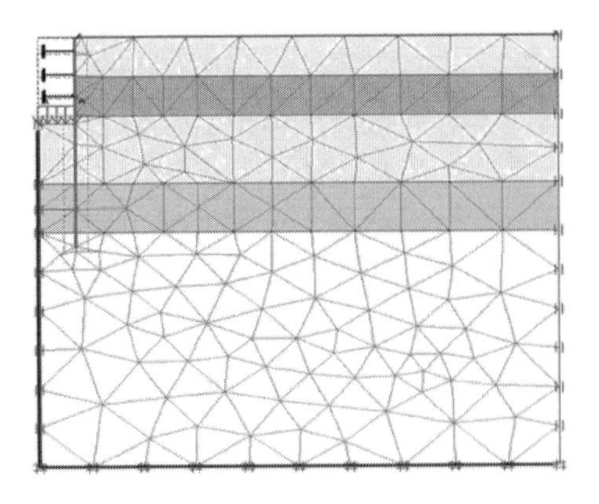

Figure 5. 2D Plane strain trench retaining wall model—mesh.

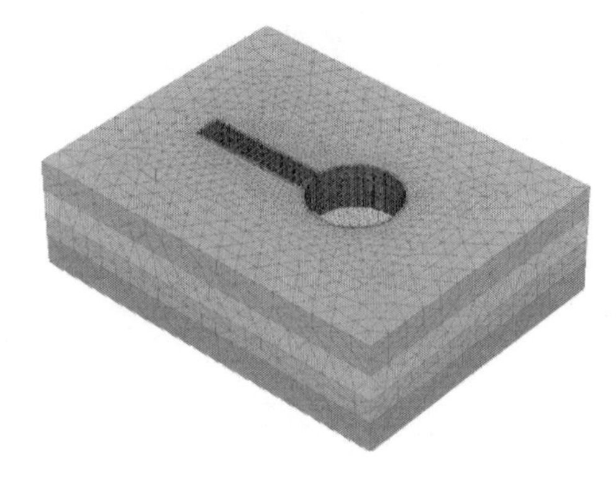

Figure 6. 3D Global model isometric view—mesh.

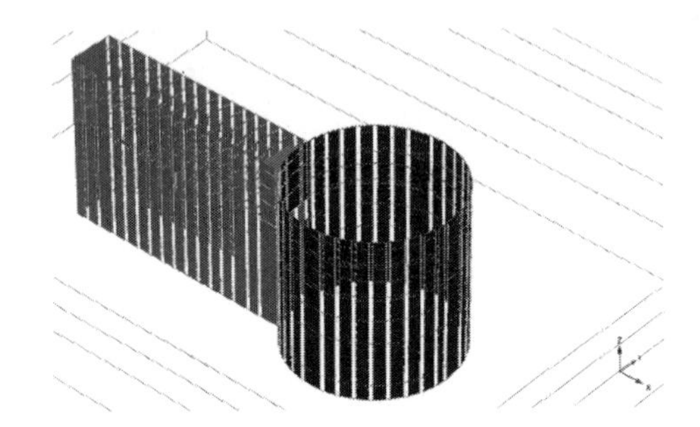

Figure 7. 3D Global model isometric view—plate elements.

4 NUMERICAL MODEL RESULTS

4.1 *2D vs. 3D Pit retaining wall model results*

The pit retaining wall stresses and deformations results of the 2D axisymmetric and of the 3D global models were compared, particularly at the opposite side of the trench where its influence was expected to be minor. The main results of the two models, for that section and concerning the final excavation stage, are illustrated in the figures 9, 10 and 12.

Regarding the retaining wall bending moments, the results obtained in both models (Fig. 9) show a reasonable approximation concerning diagrams and maximum values reached. Bending moments diagrams are very similar, the peak values for both positive and negative moments are well defined and located at approximately the same depths in both models, showing a value difference of about 20%.

The horizontal displacement diagram shows a very good agreement between the two models, being estimated maximum displacement values almost identical (Fig. 10).

To set an order of magnitude for the expected circumferential stresses obtained from both 2D and 3D models, a simple calculation was performed. When a circular ring is radially load, the hoop stresses are equal to the product between its radius and distributed load (Fig. 11).

Table 3. Plate elements properties.

Structural Element	d	γ	E	ν_{12}	G_{12}
	m	kN/m³	MPa	–	MPa
D-Wall	1.0	25	34	0.2	14.17
Bracing Slab	0.85	25	34	0.2	14.17

Table 4. Beam elements properties.

	A	γ	E	I_3	I_2
Structural Element	m²	kN/m³	MPa	m⁴	m⁴
Struts (200 × 85 mm)	1.7	25	34	0.102	0.567
Ring 4 (100 × 80 mm)	0.8	25	34	0.043	0.067
Rings 1,2,3 and 5 (50 × 120 mm)	0.6	25	34	0.072	0.013
Capping beam (1000 × 1000 mm)	1.0	25	34	0.083	0.083

For the calculation of the p load was considered the effect of earth pressures, ground water pressures and surcharges. In order to simplify the calculation, the stresses were calculated at the depth were higher hoop forces were expected and for an equivalent soil layer with average geomechanical parameters of all soil layers. With a radius of 22,25 m and an estimated load of 365 kPa it was predicted a maximum hoop stress value of 8121 kN/m.

Regarding the normal stress in the tangential direction, referred as hoop forces, although values obtained in 2D model were higher than the 3D model ones, their magnitude is similar and within the order of magnitude predicted by the simplified calculation (Fig. 12). The related diagrams are in agreement in both models and, as expected, the hoop forces increase with depth.

4.2 *2D vs. 3D Trench retaining model results*

The trench retaining wall stresses and deformations results between 2D plane strain and 3D global models were compared, particularly on a section near the trench/pit connection were one can predict a major difference due to the connection singularity. The main results of the two models, for that section and concerning the final excavation stage, are illustrated in the figures 13, 14 and 15.

Regarding the 2D model results, it was observed a typical behavior of a retaining wall with multiple passive horizontal supports (Fig. 13).

Results from the 3D model at a section with the same horizontal supports and excavation depth showed identical deformations and stresses (Figs. 14 and 15) despite the expected difference.

From the previous assessment, one can conclude that the ring beams and bracing slab elements allow an enough stable pit overall behavior and a lower impact in trench retaining walls.

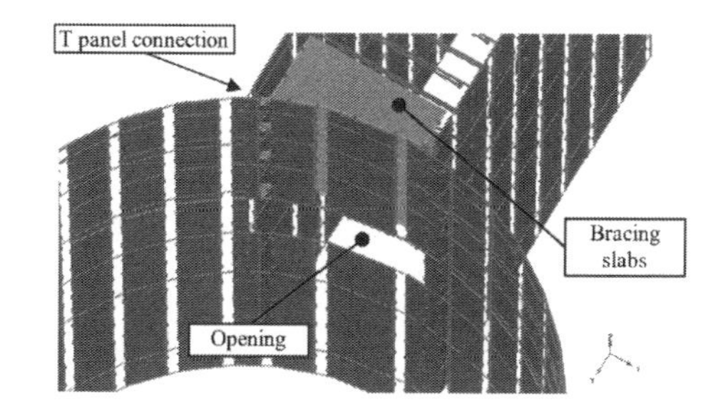

Figure 8. 3D global model view of the trench/pit connection.

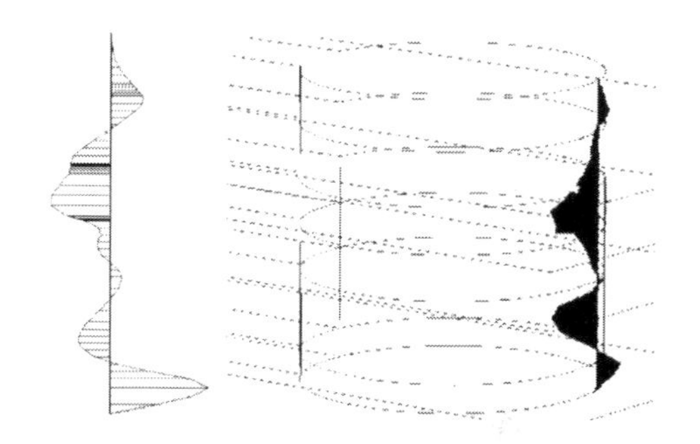

Figure 9. Pit retaining wall vertical bending moment diagram: 2D axisymmetric model [maximum value: 272 kNm/m] (left) and 3D global model [maximum value: 345 kNm/m] (right).

4.3 *3G Global retaining wall model results*

Although the 3D global model was developed aiming mainly to validate previous models and simplified calculations, as well as to determine the overall behavior of the retaining structure, it was also essential in the design of local singularities.

The trench/pit connection opening was modeled at a final stage and its effect was observed within the near elements (Fig. 16).

The design of the structural elements affected by the opening was validated with simplified calculations, particularly the nearest ring beams, where an increase in its axial forces was expected and had to be accessed (Fig. 17).

The 3D model was also fundamental to obtain bracing slabs stresses since they transfer the outward thrust to the trench retaining walls, equilibrating the imbalance caused by the trench excavation. The model results assessment allowed also to confirm the vital role of this elements in the overall stability of the retaining wall structures.

Regarding its design, due to relevant axial loads in both plan directions, a combined bending axial

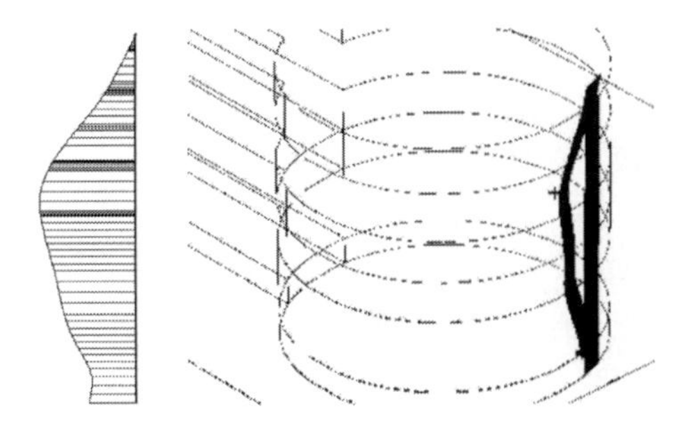

Figure 10. Pit retaining wall horizontal displacements: 2D axisymmetric model [maximum value: 5 mm] (left) and 3D global model [maximum value: 6 mm] (right).

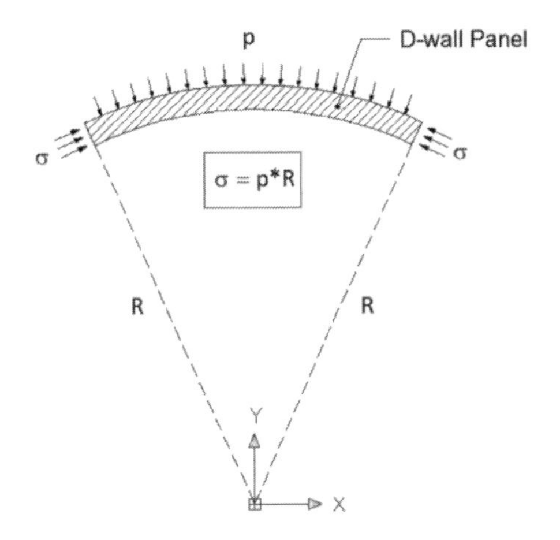

Figure 11. Simplified calculation of the hoop stresses in the D-wall panels.

forces interaction was used at the reinforcement design. Also, the absolute value of torsional bending has been totally added to flexural bending moments on both directions. Absolut values of vertical shear and horizontal shear has been totally added, considering a total shear acting out of plane, while bracing slab/d-wall interface connection was also design based on model results.

The effect caused by the opening was also observed on bracing slabs elements and conforming the 3D global model importance since the trench/pit connection opening lead to a significant increase of the axial load, estimated of about 40% (Fig. 18).

4.4 *Hydrostatic pressures*

Since the magnitude of water inflow at the bottom of the excavation isn't substantial due to the depth of the retaining wall the water pressure in both sides of the wall is unbalanced. To account for water horizontal pressures on the wall it was considered a hydrostatic analysis.

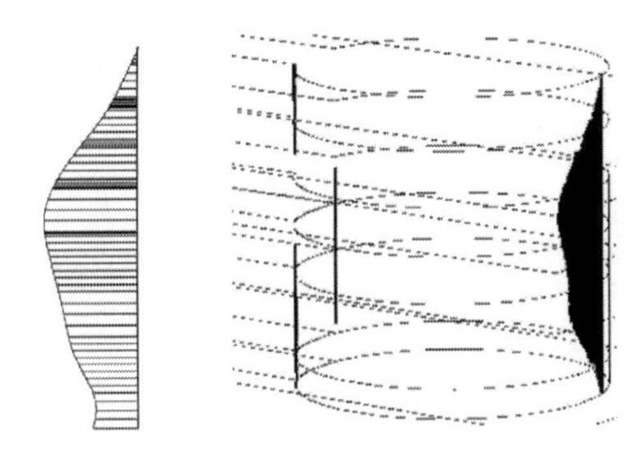

Figure 12. Pit retaining wall hoop forces diagram: 2D axisymmetric model [maximum value: 7864 kN/m] (left) and 3D global model [maximum value: 6270 kN/m] (right).

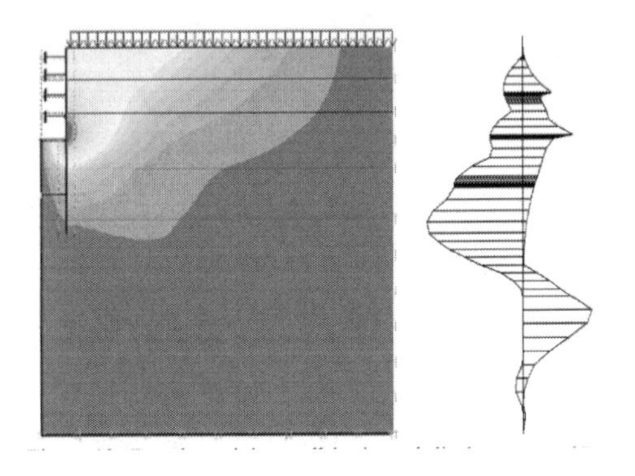

Figure 13. Trench retaining wall horizontal displacements: 2D plane strain model [maximum value: 41 mm] (left); and bending moment diagram [2381 kNm/m] (right).

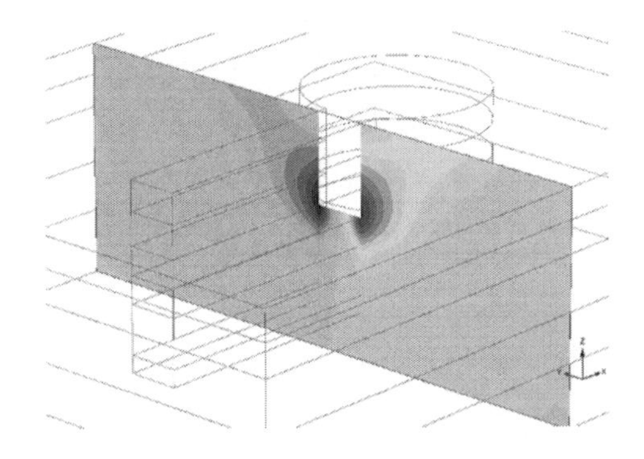

Figure 14. Trench retaining wall horizontal displacements: 3D global model [maximum value: 35 mm].

However, the permeable soil layer below the excavation is under positive hydrostatic pressure increasing the possibility of a hydraulic failure by heave of the bottom of the excavation. Thus, with the decreased of soil height above the sandy layer due to the excavation, pressure relief drills were prescribed to reduce the hydrostatic vertical pressure under the less permeable layers.

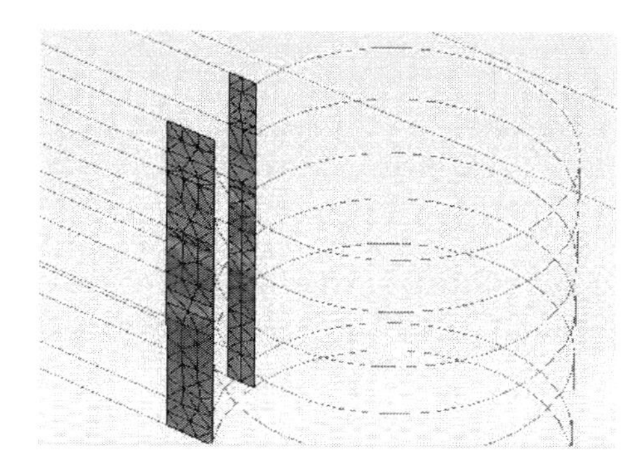

Figure 15. Trench retaining wall vertical bending moment diagram: 3D global model [maximum value: 2552 kNm/m].

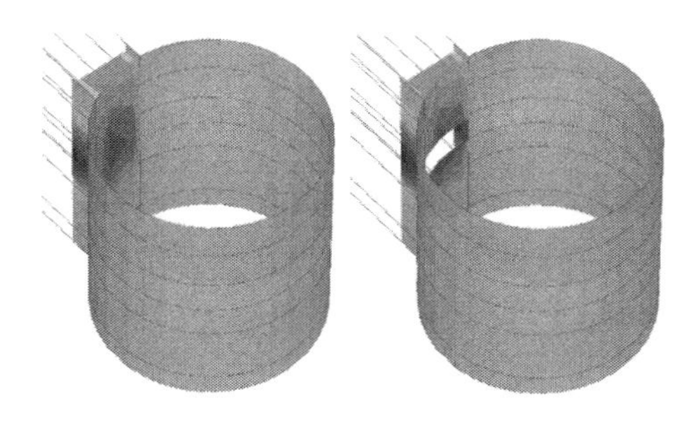

Figure 16. Retaining wall horizontal displacements: 3D global model before (left) and after the opening (right).

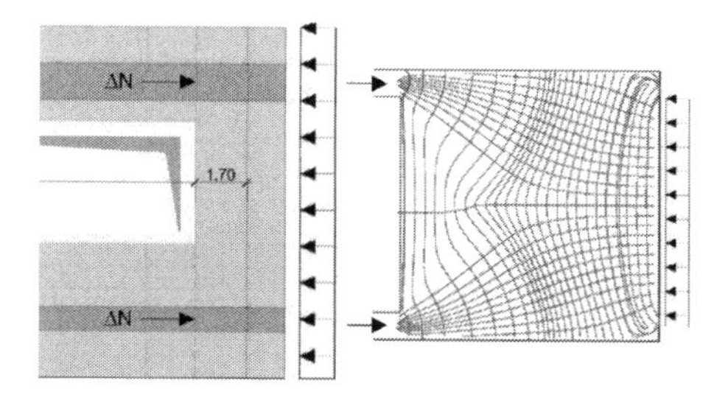

Figure 17. Simplified model for hoop forces increase due to opening (left). Expected stress path due to opening (right).

The constructive stages of the numerical analyses were defined taking into account this procedure and confirming its effectiveness on the prevention of heave failure. The execution of the pressure relief drills was set from the 2nd excavation stage until the final one. Up to the 2nd excavation stage, the water pressure inside the permeable layer was kept constant. On the excavation side, the water head was placed at temporary subgrade elevation, using a linear interpolation of pressures in a soil layer beneath the excavation bottom ranging between zero pressure (top of the layer) and hydrostatic pressure (layer bottom).

From the 2nd excavation phase until the final one, the water head of the permeable layer was lowered to the level of its center line simulating the dewatering effects of the pressure relief drills. Additionally, using the methodology described above, was set a water pressure interpolation layer beneath the excavation bottom level (Fig. 19).

4.5 *Flow analysis*

Although the majority of the soil layers have relatively low permeabilities, due to the dimension of the excavation and the presence of significantly more permeable sandy layer, which could lead to relevant inflow, 2D seepage analyses were conducted.

As a first measure to control the water inflow, the D-wall panels' embedment length is located in a very low permeable soil layer, beneath the sandy permeable layer. Nevertheless, the water hydrostatic pressure imbalance will lead to a flow towards the excavation base. The flow analysis performed within the 2D axisymmetric pit retaining wall model showed a lateral water income mostly through the permeable soil layer as expected and a flow bellow the D-wall embedment toward the excavation base (Fig.20). For the final excavation stage, it was obtained, at the base of the excavation, a flow value of approximately 95 L/day, an allowable value regarding the water pumping systems and the neighborhood conditions.

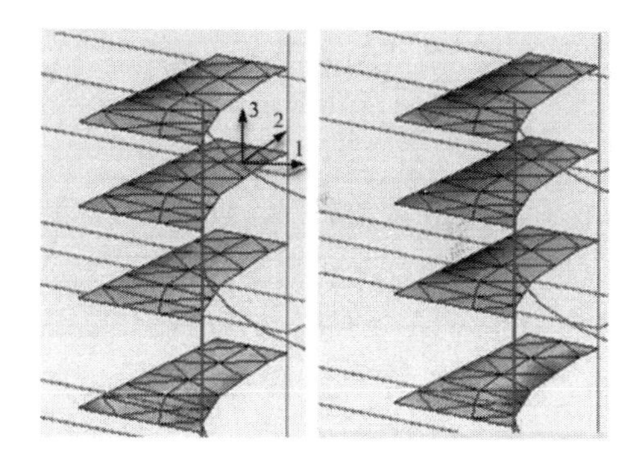

Figure 18. Bracing slabs axial forces N2 before [maximum value: 3665 kNm/m] (left) and after the opening [maximum value: 5232 kNm/m] (right).

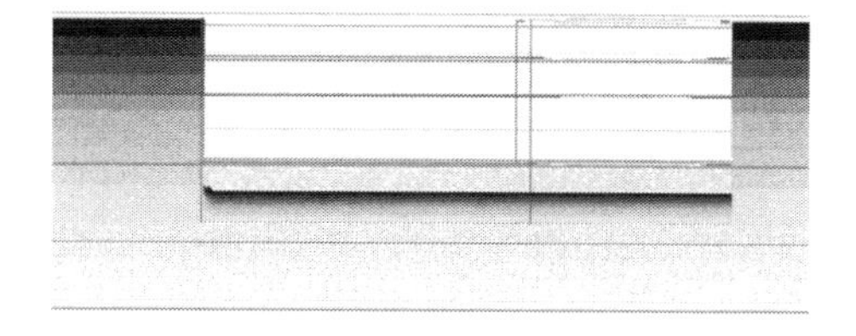

Figure 19. Active pore pressure with dewatering effects: 3D global model—Longitudinal trench and pit section at ultimate excavation stage.

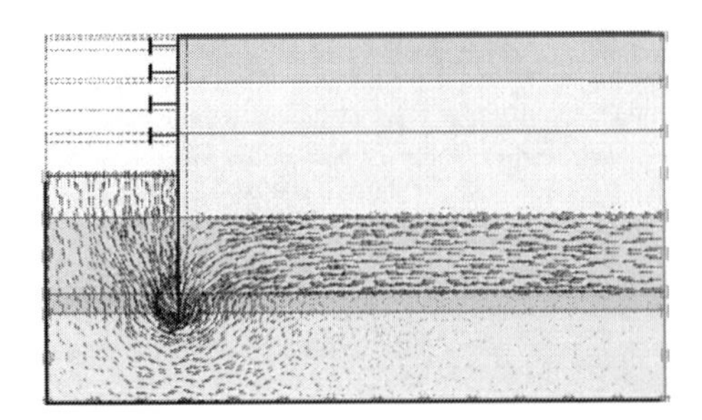

Figure 20. 2D Axisymmetric pit retaining wall model—Flow analysis.

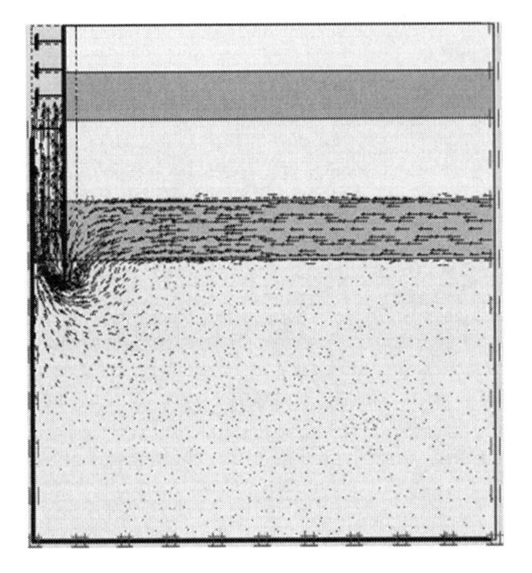

Figure 21. 2D Plane strain trench retaining wall model—Flow analysis.

In order to evaluate the water inflow at the excavation base along the trench, similar flow analyses were conducted within several 2D plane strain models comprising trench cross sections with excavation bottoms at different depths.

The behavior observed within the permeable layer was identical in comparison with the previous model. However, since the trench model has a smaller area for the water inflow, the flow analysis shows a higher concentration around the retaining wall bottom level (Fig. 21). Considering the flow analyses results of all 2D plane strain trench models, it was calculated the water inflow at the bottom of the excavation totalizing an estimated value of 45 L/day.

Apart from the estimation of the water inflow at the excavation base, through the performed analyses, the efficiency of the D-wall panels' embedment length was confirmed.

The flow analyses performed were limited to the 2D models since it was not expected to have more accurate results using 3D models as the excavation depth and area of water possible inflow was the same.

5 CONCLUSIONS

Finite element analyses have been carried out on both 2D and 3D models simulating a deep excavation and the necessary dewatering effects for the execution of an unsymmetrical railcar unloading pit and its connection trench. As a first approach, 2D analyses were performed for both pit and trench retaining wall modeling. Those analyses covered separately the pit, exploiting its cylindrical symmetry to establish an axisymmetric model, and the trench, due to its overall length modeled within a plane strain analysis.

Due to the noncollinearity between the trench alignment and the pit center, the overall stability and behavior of the structure, as a whole, needed to be accessed. Also the connection between those structures ought to be design which would not be possible to do within the 2D models. A 3D global model was developed, being validated and calibrated with 2D numerical analyses results.

The numerical analyses results between 2D and 3D models showed, in generally, a good agreement and allowing some conclusions on the validity of the 2D models within some parts of the structure. The 3D model was particularly useful on the analysis of some structural elements, allowing a better understanding of the overall behavior. Concerning the flow analysis, the 2D models revealed to be adequate.

Since 3D calculation can be time consuming, 2D prelaminar analyses can be performed, even when a 3D analyses is geometrically crucial, enabling some initial design calculations and consequentially increasing the level of development of the structure conception for a subsequent implementation in a 3D model. Rather than opposites, 2D and 3D calculations showed to be useful complementary tolls in the design of complex tridimensional structures.

The instrumentation readings validate the model results, showing a good agreement between the measurements and simulated values, both in terms of displacements and water inflow.

REFERENCES

Brinkgreve, R.B.J. & Kumarswamy, S. & Swolfs, W.M. 2017. *Plaxis—Material Models Manual 2017*, Delft: Plaxis bv

Hou,Y.M. & Wang, J.H. & Zhang, L.L. 2009. Finite-element modeling of a complex deep excavation in Shanghai. *Acta Geotechnica* 4:7–16. Shanghai, China

Ou, C.Y. 2006. *Deep Excavation—Theory and Practice.* London: Taylor & Francis Group

Numerical Methods in Geotechnical Engineering IX – Cardoso et al. (Eds)
© 2018 Taylor & Francis Group, London, ISBN 978-1-138-33203-4

Numerical analysis of a tied-back wall in saturated cohesive soils

K. Ninanya & J.C. Huertas
Essential Material Consulting, Company at Lima, Peru

H. Ninanya & C. Romanel
Department of Civil and Enviromental Engineering, Pontifical Catholic University of Rio de Janeiro, Rio de Janeiro, Brazil

ABSTRACT: The need for deeper urban excavations has imposed to geotechnical engineers the challenge of balancing high horizontal forces with occurrence of minimum displacements in soil as well as in the structures nearby. In many cases, a tied-back earth retaining wall is the recommended technical solution. The use of ground anchorage, as a direct extension of the rock anchoring technique, is intensely executed throughout the world. This work aims to evaluate a hydromechanical geotechnical problem solving the coupling between soil mass and water in the tied-back wall system. The safety factor of the tied-back wall and the displacement field generated by the construction of the retaining wall system are analyzed through the finite element method. The saturated cohesive soil is modeled by the Mohr-Coulomb constitutive law and effects due to variations of width and depth of the wall, anchors inclination and bulbs stiffness are also considered within a parametric study.

1 INTRODUCTION

The basic purposes of excavation support systems are to provide stability and minimize movements of the adjacent ground. Stability means that excavation support system can remain open without causing very large movements that endanger surrounding structures and does not represent a threat to construction operators.

Prediction of excavation-induced ground movements is an essential part in the design of deep excavation, where a previous estimation of parameters is necessary. Existing methods for estimation of the excavation-induced ground movements are mainly separated into three categories i.e. empirical methods (Peck 1969, Clough and O'Rourke 1990), numerical methods (Mana and Clough 1981, Hashash and Whittle 1996, Pakbaz et al. 2009) and the analytical solutions (Roscoe and Burland 1968, Ghaboussi and Sidarta 1997). Each of these methods can be applied in accordance with site conditions and complexity of the construction. However, the numerical analysis in particular can take account of non-linear soil behaviour and complex construction procedure. Particularly two-dimensional finite element method is used since late 1960 s (e.g. Clough and Duncan 1969)) by researchers and practitioners due to its relative simplicity, flexibility and availability of computational facilities to simulate different field conditions and excavation techniques.

In the present investigation, 2-D Finite element model is developed using PLAXIS to predict the deformation pattern of surrounding ground due to deep surface excavation and to represent the performance of a Tied-back diaphragm wall on the stress distribution and deformation characteristics of ground below adjacent structure at vulnerable locations. The analysis is carried out considering non-linear behavior of soil using Mohr-coulomb failure criteria. A permanent particular construction near the sea on saturated soft clays is idealized. For the design of the retaining system the active and passive earth pressures were determined through Rankine (1857) for defining the tied-back anchors and structural wall dimensions. The effect of anchor-soil-structure interaction system and the effects of the own excavation was numerically verified and supporting system is analyzed in terms of bending moment, shear force and admissible displacement of Tied-back wall at the level of excavation.

2 TIED-BACK WALL

Tied-back walls are used in both temporary and permanent structures. This structural system is a very versatile excavation retaining system for deep excavation in areas surrounded by major infrastructures, guaranteeing that limited lateral displacement are not exceeded. It can be used for water retainable structures and constructed to their full depth in a single phase or in a progressive construction of horizontal, vertical ribs, concrete fillings and anchors inserted and stressed as a new layer is added (Xanthakos 1991). Information on selecting particular wall type can be found in FHWA-SA-96-038 (Sabatini et al. 1997).

According to Sabatini et al. (1999), a range of potential failure conditions should be considered in design of anchored walls for limiting movements of the soil and the wall while providing a practical and economical basis for construction. All these potential failure and their mechanisms as shown in Figure 1.

Anchors can provide the necessary stabilizing forces which are transmitted back into the soil at a suitable distance behind the active soil zone loading the wall (see Figure 2(a)), establishing, in turn, the minimum distance behind the wall at which the anchor bond length is formed. The anchor bond length must extend into the ground in order to intersect any critical failure surfaces which could pass behind the anchors and below the base of the wall (see Figure 2(b)).

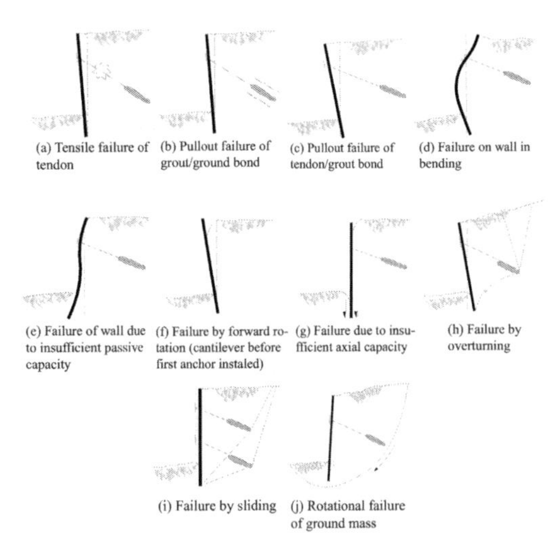

Figure 1. Potential failure conditions to be considered in design of anchored walls.

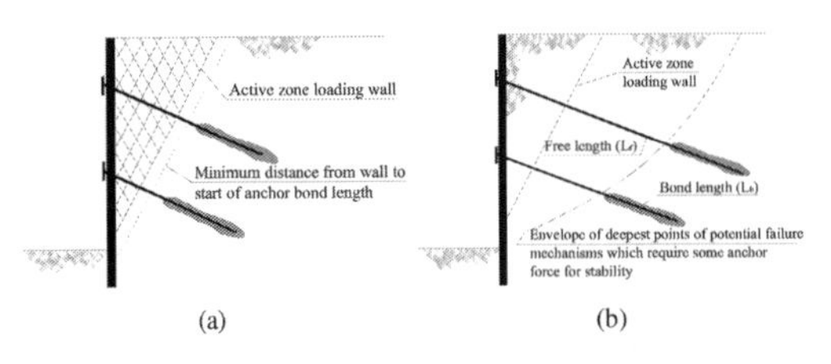

Figure 2. Contribution of ground anchors to wall stability (Modified from Sabatini et al. (1999)).

3 RANKINE'S THEORY

The methods commonly used nowadays for design practice are derived from classical earth theories developed by Coulomb (1773) and Rankine (1857). Therefore, an adequate design and construction of a tied-back wall involves a thorough knowledge of the lateral forces that are present between the structural system and the soil masses being retained. These lateral forces are caused by lateral earth pressure.

3.1 *Active and passive pressure*

Active earth pressure is associate with the movement of the wall away from the backfill and passive pressure is associated with the wall moving into the backfill. The value for the coefficient of passive and active lateral earth pressure for a cohesive soil can be determined through the Eq. 1.

$$K_{p,a} = tan^2(45 \pm \phi/2) \pm \frac{2c'}{\sigma'_v} tan(45 \pm \phi/2) \quad (1)$$

where φ' and c' are effective stress strength parameters. Likewise, the magnitude of passive and active lateral pressure forces are presented in Eq. 2.

$$E_{p,a} = \frac{1}{2} \lambda h^2 K_{p,a} \quad (2)$$

where λ is the unit weight of the soil and h is the depth of evaluation.

4 CASE STUDY OF A TIED BACK WALL

The tied-back wall project is located in Brazil, near an Atlantic seashore. The excavation depth and width are 20.5 *m* and 25 *m*, respectively. The wall length and width measure 35.5 *m* and 0.8 *m*, respectively, and is anchored to the adjacentsoil mass by 11 tied-back anchors, distributed along the first 20.5 *m* of the wall.

Diaphragm wall is a reinforced concrete structure constructed in situ by panel. The wall is usually designed to reach very great depth. Reinforced concrete guide walls of 1 *m* depth are mainly used to assist the trenching operation. The reinforcement cages are then lowered into the slurry filled trench, with each unit spliced to the other, to form a continuous cage to the required depth. Lastly, placing of concrete is done.

Additionally, a distributed loading of 25 kN/m^2 was considered on the ground surface, which represents an overburden, as can be detailed in Figure 3.

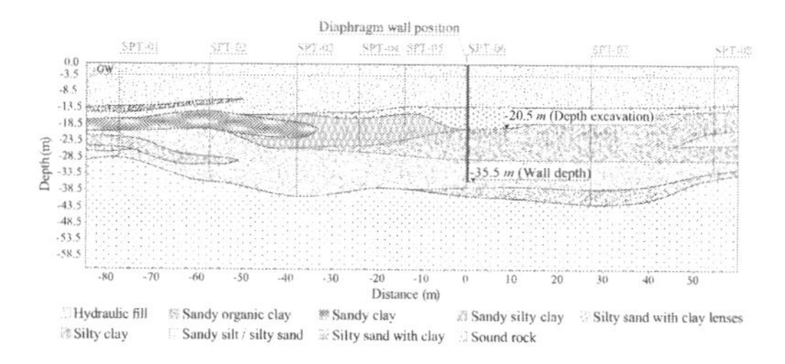

Figure 3. Geomechanical model.

4.1 *Investigation and soil condition*

Comprehensive geotechnical investigations (in situ and laboratory testing) were performed to get a realistic value of the geotechnical parameters, among physical properties, strength of the Mohr-Coulomb model, stiffness and hydraulic parameters. All this kind of parameters as well as the interface strength reduction coefficient R_{inter} PLAXIS can be shown in Table 5. From the soil characterization the area in study is represented for heterogeneous layers.

The mean sea level (MSL) is located 2.7 *m* below the ground surface. The seabed soils and surface materials that comprise the region of the wall are formed mainly by marine silt and clays, which consists of upper soft silty clays and bottom stiff silty clays layers and medium-dense-to-dense cemented clayey sand layer. the areas of thick deposition.

5 TIED-BACK WALL DESIGN

The U.S. Army Corps of Engineers and the Federal Highway Administration (FHWA) produced a series of reports about the design of earth retaining system. Works by Pfister et al. (1982), Strom & Ebeling, and Ray (2009) helped to clarify approaches to tied-back wall design, discussing assumptions of different models, design methodologies, required data and performance of tied-back walls and their components.

According to Ray (2009), methods in approximate increasing sophistication include: (a) Rigid model, (b) Winkler model, and (c) Finite Element method (FEM). In the FEM, soil can be nonlinear, anchor and wall can have interface elements, wall and structural elements modeled as elastic.

FEM has been applied for the present case, considering a set of assumptions and construction steps, as follows:

Step 1 - Define design parameters (establishing design factors of safety), all geometry, construction constraints and evaluate properties of in situ soil and rock.
Step 2 - Determine total lateral earth pressures distribution acting on back of wall for final wall height (that include water and surcharge)
Step 3 - Calculate horizontal ground anchor loads and wall bending moments. Afterwards, check/adjust the preliminary dimensions and evaluate required anchor inclination based on right-of-way limitations (appropriate anchoring strata)
Step 4 - Resolve each horizontal anchor load into a vertical force component and a force along the anchor, and evaluate horizontal spacing of anchors based on wall type. Then, calculate individual anchor loads
Step 5 - Check the preliminary number of tied-back tiers with respect to wall depth, and select type of ground anchor
Step 6 - Evaluate vertical and lateral capacity of wall below excavation subgrade
Step 7 - Evaluate overall stability of the anchored system
Step 8 - Estimate maximum lateral wall movements and ground surface settlements.

The required design factor of safety for anchored shoring system is about 1.5. It is worth mentioning that the design process began with a preliminary estimation for the prestressing load based on the lateral pressure distribution (Rankine's theory). Then, the design was checked by the finite element analysis verifying wall movements, anchor's position and pre-stress level.

5.1 *Active and passive lateral pressure*

The lateral pressure was estimated taking into account the Rankine method (Section 3).

Through the equation 1, the coefficient of passive and active lateral earth pressures were calculated to obtain the lateral pressure distribution along the wall for final wall height, as shown in Figure 4. It is notable that the active lateral pressures developed up to silty clay layer whilst passive lateral pressures begin from the level of the excavation up to the tip of the wall.

5.2 *Anchors' forces, length and type of bar*

From lateral pressure distribution, anchors' forces and length can be estimated. Different sets of inclinations and heights of wall elements (see Figure 5) were verified for each anchor. The wall element was defined in 2.6 *m* width.

The tied-back anchors consist of a free length, L_f, and bond length, L_b. The bond lengths, referred to the bulb, can also be determined by using the pull-out capacity of soil anchors (Eq. 3), provided in the Bustamante & Doix (1985) method, where P_{ult} is the pull-out capacity, that, in

Table 1. Soil design parameters.

Parameters	Hydraulic fill	Sandy silt	Sandy clay	Sandy silty clay	Silty clay	Silty sand with clay	Young residual soil	Altered rock	Sound rock
γ_{sat} (kN/m^3)	18	19	16.5	17	16.5	17.5	20	22	24
E (kN/m^2)	5E+04	1.6E+05	4.5E+04	7.5E+04	4.5E+04	4.5E+04	1.8E+05	1E+06	5E+06
c (kPa)	1	5	3	7	3	2	12	500	2000
φ $(°)$	30	35	25	27	25	29	35	50	50
k (m/day)	8.6E-01	8.6E-03	8.6E-05	8.6E-04	8.6E-05	8.6E-04	3.6E-04	8E-03	8E-08
R_{int}	0.65	0.65	1	0.65	1	0.65	0.65	0.65	1

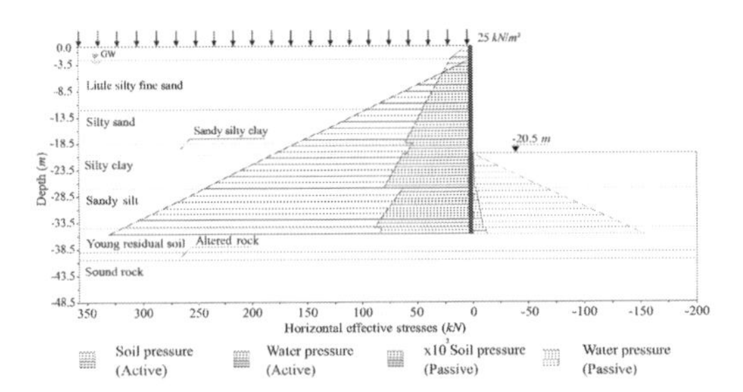

Figure 4. Active and passive lateral pressure.

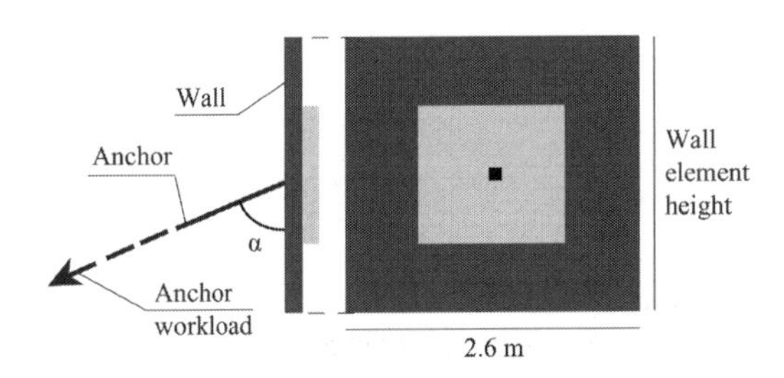

Figure 5. Wall element characteristics.

this case correspond to the force F in Table 2, D is the borehole diameter, and, q_s is the unit friction between ground and bonded length.

$$P_{ult} = \pi \cdot D \cdot L_b \cdot q_s \tag{3}$$

Table 2 shows the calculated bond length for each anchor. However, in order to have general value, it was adopted 9 m as the bond lengths. The inclination, vertical and horizontal separation and the number of tied-back were explored, in conjunction with the anchors' lengths. From this, a number of 11 tied-back were defined. The final set of inclination, length and depth of the tied-back anchors is also shown in Table 2.

Table 2. Forces in anchors.

Anchor	Depth (m)	α (°)	Wall element height (m)	F (kN)	$L_b^{(1)}$ (m)	L_f (m)
1	1.475	10	2.95	113.0	4.4	33.0
2	4.200	10	2.50	267.3	6.7	30.0
3	6.575	20	2.25	440.3	9.1	27.0
4	8.575	20	1.75	465.5	6.5	25.0
5	10.200	20	1.50	484.6	7.5	24.0
6	11.700	25	1.50	582.8	8.4	28.0
7	13.200	25	1.50	660.8	8.9	28.0
8	14.700	25	1.50	741.8	10.3	25.0
9	16.200	25	1.50	822.7	10.3	21.0
10	17.700	25	1.50	903.6	11.3	21.0
11	19.475	25	2.05	1347.1	11.3	18.0

(1) $L_b = 9\ m$ (adopted)

Table 3. Bar properties.

Anchor	Workload (kN)	Section area (mm^2)	Failure Force (kN)	Yield Force (kN)
1	200	642	453	378
2	340	1140	805	671
3,4	510	1781	1258	1048
5	600	2027	1432	1193
6,7	700	2288	1616	1347
8,9	860	2858	2019	1682
10,11	1000	3491	2466	1918

Failure stress = 72 kg/mm^2, Yield stress = 60 kg/mm^2.

Thereafter, bars were selected in function on the commercial availability, by taking into account superior workloads than calculated in Table 2. Table 3 shows the structural characteristics for the

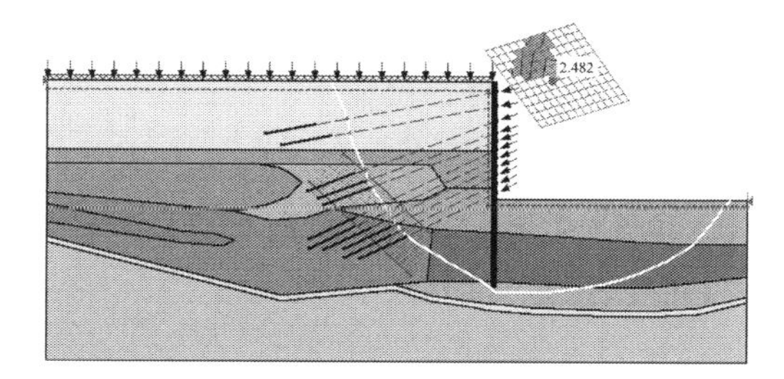

Figure 6. Global stability analysis (failure surface in white line).

different bar kinds, section area, yield and failure stress as well as the incorporation forces. It is worth noticing that these information were required to define the anchors.

5.3 *Stability analysis*

The global stability of tied-back wall system was obtained through limit equilibrium analysis, the most common procedure used in the geotechnical engineering to calculate the safety factor. This type of analysis takes into consideration all possible failure surface shown in Figure 1.

SLOPE/W (Krahn et al. 2004) software was employed to perform the factor of safety using Morgenstern & Price (1965) method, which considers both shear and normal interslice forces, satisfies both moment and force equilibrium conditions and assumes for a variety of user-selected interslice force functions. The result factor of safety of the tied back wall system was around 2.48, as shown in Figure 6.

6 NUMERICAL ANALYSIS

In order to predict ground movements, finite element analysis (FEA) using the PLAXIS program (Brinkgreve et al. 2014) was performed. The design is "checked" by FEA to verify assumptions concerning wall movement, anchor placements and pre-stress levels. It is worth noticing that due the fact that PLAXIS works with strength reduction it is not possible to determine the safety factor of the system.

6.1 *Modelling*

A finite element modeling was carried out to analyze the staged excavation and dewatering for one of the underground stations. The ground was discretized using fifty-nodded triangular plane strain elements. While the wall was modeled as a plate-element having flexural rigidity (Table 1) representing reinforced concrete. The model boundary was extended approximately three times (excavation side) and four times (behind wall) the width of excavation to minimize the effect of artificial boundaries on the ground movement. These model boundary distances were based on few trials considering the effect on deformation of excavation area.

The complete excavation was performed in 24 different stages in order to accurately reproduce the field procedure. Stage 1 involves the construction of the wall and activation of the corresponding static surface loads while in the stage 2 is excavated the first 2.2 m and stage 3 is installed and pre-stressed the first anchor. Basically, the following stages consist of excavating and installing anchors one by one until generate the last phase in which is incremented the distributed load to 25 kN/m^2. It is important to take into account in each excavation stage the definition or position of the groundwater level.

The mesh is formed by 4793 triangular elements with an average element size of 1.5 m, considering also an interface along the wall length in order to model the soil-structure behavior. Furthermore, standard boundary condition are adopted with $u_x = 0$ on vertical sides and $u_x = u_y = 0$ at the horizontal base.

In the present study, soils and rocks were modeled considering the Mohr-Coulomb elastoplastic constitutive model to represent the mechanical behavior, due to lack of relevant soil parameters.

6.2 *Interface strength, R_{int}*

This finite element study was undertaken using interface elements. Day & Potts (1998) investigated the effect of interface properties on the behavior of a vertical retaining wall and the deformation of the ground through a series of finite element analyses. Results showed that the active and passive pressures on wall depend on the maximum value of friction angle but not on the properties of interface element, whereas interface had some effect on the ground surface deformation. The interface element provides a useful means to model reduced interface friction and non-associated interface behavior in finite element analysis of retaining walls.

The interface element allows the development of relative displacements between soil and structure. In PLAXIS, the parameter that defines the interface is a reduction factor, R_{int}, which relates friction angles of both soil to soil with the structure. If the interface is rough, the R_{int} is equal to 1.0, otherwise, the R_{int} assumes values less than unit. The magnitude and nature of the interaction are modeled by choosing an adequate value for the reduction of resistance of the interface. More

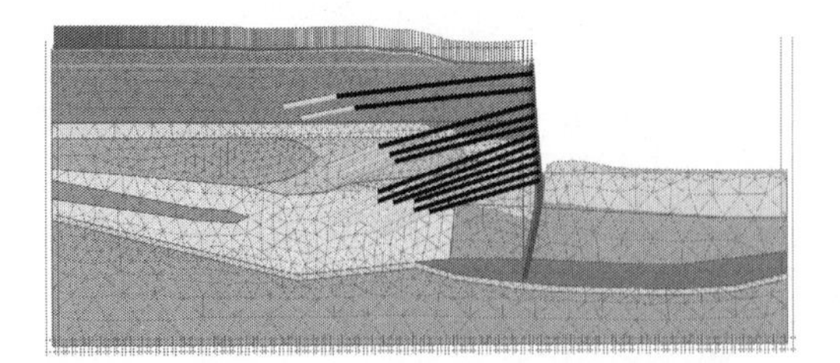

Figure 7. Deformed mesh in Plaxis (scaled up 20 times).

(2003) suggests R_{int} values between 0.5 to 1.0, depending on the type of both soil and material of the contact element (Table 1). This factor relates the strength of the interface (friction and adhesion) to strength of the soil (friction and cohesion).

6.3 *Analysis*

In simulating, the systematic excavation process, the initial vertical state of stress were first created by running on in situ stress condition with the assumption of wished-in-place wall. The lateral stress was created by considering stress condition aspresented in (Table 1). The excavation process was then modeled by adding and removing elements at corresponding steps. The soil strata was considered to behave in drained condition (plastic calculation type) to obtain the final deformation even in less permeable soils.

Model analysis was performed considering initial groundwater level at 2.7 m below the ground surface with ground water flow condition, initial geometrical configurations and initial state of stress. Figure 7 illustrates the corresponding deformed mesh for the final stage, with structural components and geometry of soil model and boundary conditions.

7 RESULTS AND DISCUSSION

According to the results obtained in this study, there is a tendency to increase the maximum value of horizontal displacements (Figure 8) with the advance of excavation stages. On the other hand, the vertical displacements at the ground surface only are considerable near to the tied back (up to −27 *m*), with maximum value of −0.16 *m* at −8 *m* from the tied back wall (Figure 9). The values after finishing excavation showed a significant change due mainly to magnitude of overburden (25 kN/m^2). The variation of the anchor force was not presented, however, a reduction of horizontal displacements are expected when the prestressing load is greater (Fernandes et al. 1993).

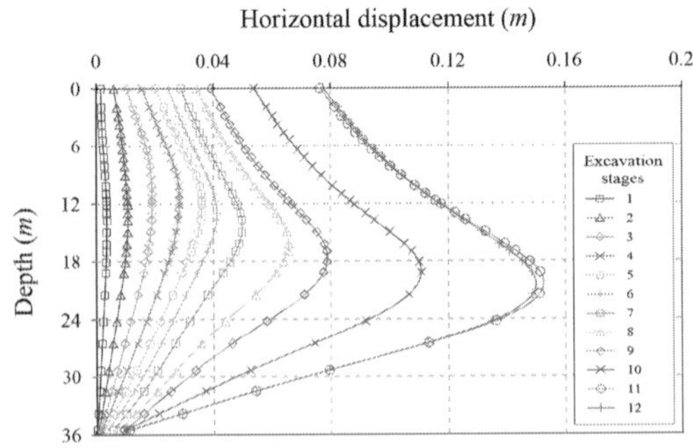

Figure 8. Horizontal displacements along wall.

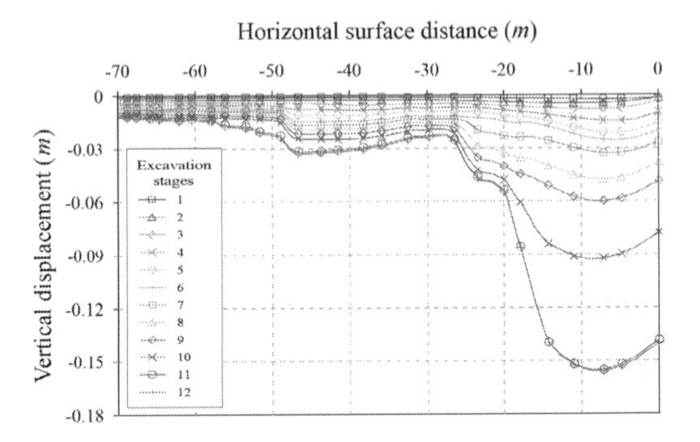

Figure 9. Vertical displacements at the ground surface.

Figure 10 shows the variation of horizontal displacements, shear forces and bending movements of wall along the depth with respect to the ground deformations. Maximum horizontal movement (0.15 *m*) of wall in located at the level of excavation. The tip of the diaphragm wall is founded on residual soil with high resistance parameters. Bending moment and shear force in the wall are dependent on the location of tied-back members. The maximum values of shear force (600 *kN/m*) and bending moment (2223 *kN/m/m*) are placed at the level of excavation and at the excavated side. This values were then verified by the structural design.

Figure 11 shows the development of shear strain in ground. It can see that the effects of excavation are more pronounced at the level of excavation and behind wall because of active and water pressures, supported by the forces of tied-back anchored in the wall.

Furthermore, the groundwater head distribution is also presented in Figure 12. Flux velocity vectors that occurs after complete excavation (steady state flow condition) are also examined. The wall was designed to allow the minimum quantity as possible of water flowing by the altered rock. However, the maximum gradient values appear close to the wall at the level of the excavation. Fig-

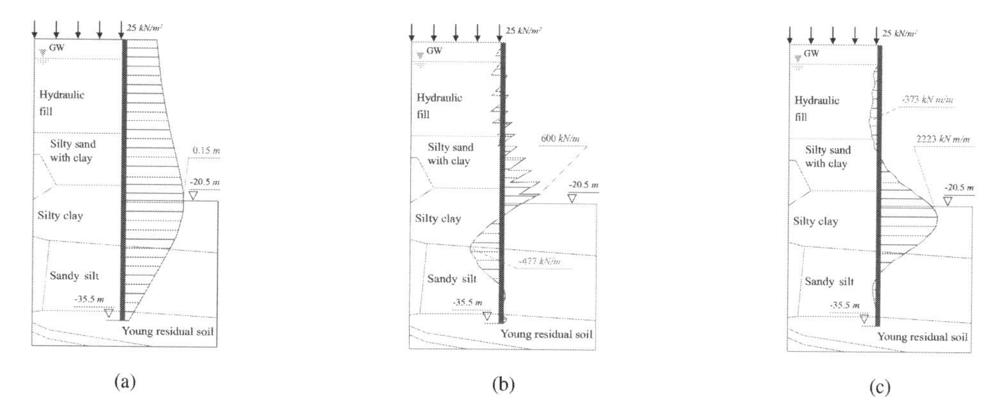

Figure 10. (a) Total horizontal displacements, (b) shear forces and (c) bending moments along the wall.

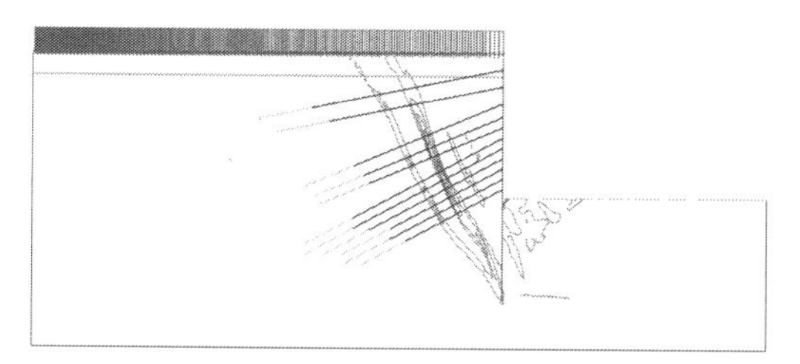

Figure 11. Distribution of shear strains in ground and after complete excavation.

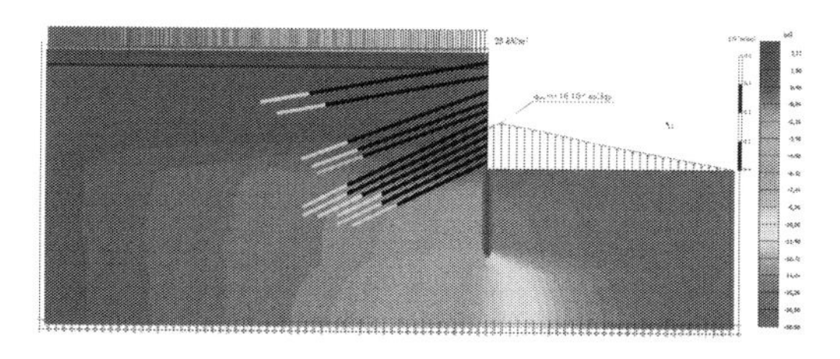

Figure 12. Flow velocity vectors at the level of the excavation.

ure 11 suggest an output flow of 0.0037 $m^3/day/m$, corresponding to a flow rate 0.16×10^{-3} m^3/day.

7.1 *Parametric study*

In order to verify the influence of the support system design on the displacement and force distributions in the D-wall, the thickness and depth increase of wall, and variations in anchor's inclination and bulb's stiffness were examined.

7.1.1 *Effect of wall thickness*

Two additional values of D-wall's thickness were considered (0.8, 0.9 and 1.0 *m*) without modification of tied-back incorporation loads.

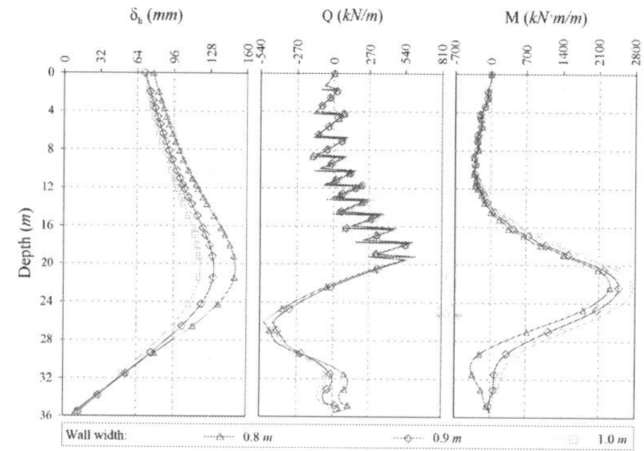

Figure 13. Effect of wall width: total horizontal displacement (δ_h), shear forces (Q) and bending moments (M) along the wall.

The results show that the influence of stiffness of D-wall was more significant than the stiffness of backfill soil. There is a decrease in horizontal displacements with the increase of wall's stiffness, though the soil stiffness remains same. This effectis more visible at the level of excavation where compressible clay soil is present (Figure 13(a)). Figures 13(b) and (c) present shear force and bending moment distribution for three wall's thickness, indicating that the maximumpositive bending happens for the thickest (1 *m*). The differences in shear forces are not much relevant in all cases.

7.1.2 *Effect of wall depth*

For the wall thickness 0.8 *m* and remaining without change the tied-back incorporation loads, two additional wall's depth (36.5 *m* ± 2.7%) were verified.

According to Yoo & Lee (2007), the wall bending stiffness influences not only the magnitudes of ground movements, but also the pattern of movements. In this case, maximum horizontal

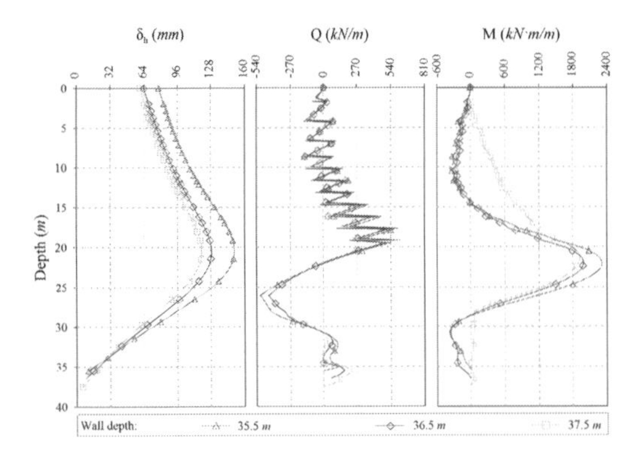

Figure 14. Effect of wall depth: total horizontal displacement (δ_h), shear forces (Q) and bending moments (M) along the wall.

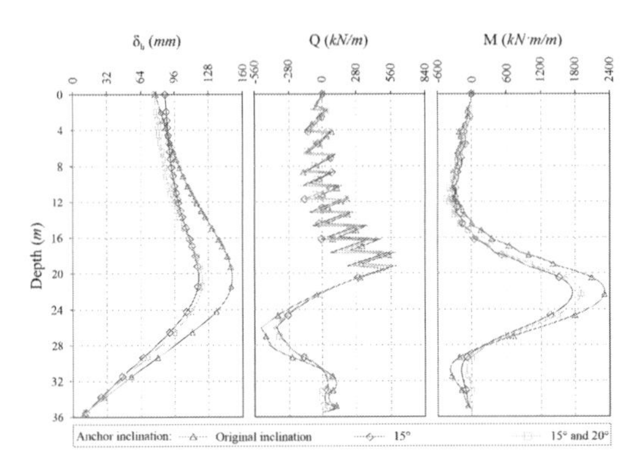

Figure 15. Effect of anchor inclination: total horizontal displacement (δ_h), shear forces (Q) and bending moments (M) along the wall.

displacements decrease as the increase of the wall depth. as shown in Figure 14 (a), while the variation of the shear forces showed in Figure 14 (b) is not so notable. The bending moment increases as the increase of the wall depth, due to the tendency of the wall to become more slender and embedding.

7.1.3 *Effect of anchor inclination*

Three sets of anchor inclinations were compared. The first set correspondents to the designed anchors, which has the firsts two with $\alpha = 15^{\circ}$, the next three with $\alpha = 20^{\circ}$, and the last six with $\alpha = 25^{\circ}$. The second setis representing by anchors with same $\alpha = 20^{\circ}$ inclinations. And, in the third set, the firsts six are with $\alpha = 15^{\circ}$ and the next five with $\alpha = 20^{\circ}$. From the results, the designed set produced higher horizontal displacements (Figure 15 (a)), shear forces (Figure 15 (b)) and bending moments (Figure 15 (c)) responses in comparison with the others two sets. Consequently,

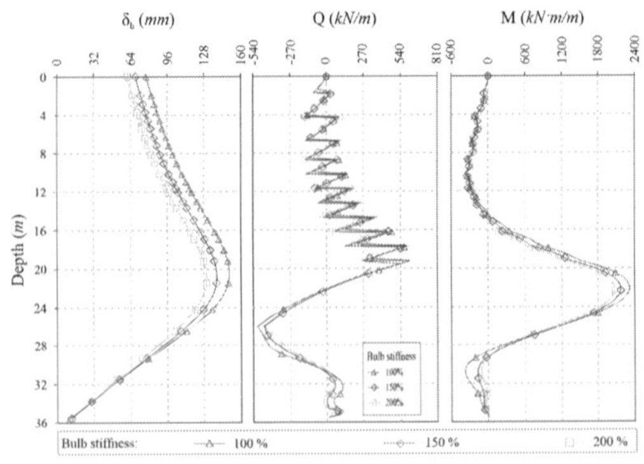

Figure 16. Effect of bulb stiffness: total horizontal (δ_h), shear forces (Q) and bending moments (M) along the wall.

the anchor inclination analysis revels an important designing factor in the tied-back support system, mainly when is necessary to reduce displacement and bending moments.

7.1.4 *Effect of bulb stiffness*

Finally, the influence of bulb stiffness was analyzed by considering three situations: bulb stiffness (a) 100%, (b) 150% and (c) 200%. From results (Figure 16), the maximum horizontal displacements reduced with the increase of bulb stiffness, but on the other hand, there are not significant variation of the shear forces and bending moment distribution in the wall. Therefore, it can be inferred that the effect of increase the bulb stiffness is not determinant in the tied-back D-wall design.

8 CONCLUSIONS

The present study provides an understanding of the effect of tied-back diaphragm wall technique to reduce the ground movements during deep excavations, from which some important inferences can be done.

From the displacements and force results, the tied-back D-wall can be a good technique to hold stable (FS = 2.482) a deep excavation and surface adjacent sites.

Predicted amount of horizontal (D-wall) and vertical displacements (ground surface) of the numerical model were approximately 0.15 *m*. These values are reasonably well because the magnitude of overburden on surface and compressibility characteristics of the soil profile. For the all studied cases, the tendency of the maximum horizontal displacements and force position is at the level of excavation.

According to parametric study, a rather reduction of displacements can be achieved by increasing wall thickness and depth, but, a reduction in shear forces and bending moments in wall did not have the same effect. A quite reduction in displacements and bending moments can be achieved with the variation of tied-back anchor inclinations.

Although numerical simulation in PLAXIS had used the Mohr-Coulomb model to representing the non-linear behavior of soils, a more realistic soil model that consider the small-strain stiffness and unloading process stress-dependency of soils should be required, making necessary more complete field and laboratory parameters.

Results of numerical simulation are not more than working hypothesis that have to be subjected to confirmation or modification during construction. Thus, numerical simulation can be incorporated as an integral part of the observational method to allow maximum economy and assurance of safety during construction.

REFERENCES

Brinkgreve, R., E. Engin, & W. Swolfs (2014). Plaxis 2014. *PLAXIS bv, The Netherlands.*

Bustamante, M. & B. Doix (1985). Une méthode pour le calculdes tirants et des micropieux injectés. *Bull Liaison Lab Ponts Chauss* (140).

Clough, G. & J. Duncan (1969). Finite element analyses of port allen and old river locks. Technical report, California Univ Berkeley Inst of Transportation and Traffic Engineering.

Clough, G.W. & T.D. O'Rourke (1990). Construction induced movements of insitu walls. In *Design and performance of earth retaining structures*, pp. 439–470. ASCE.

Coulomb, C.A. (1773). Essai sur une application des règles de maximis et minimis `a quelques probl`emes de statique relatifs à l'architecture. *Mem. Div. Sav. Acad.*, 7.

Day, R. & D. Potts (1998). The effect of interface properties on retaining wall behaviour. *International Journal for Numerical and Analytical Methods in Geomechanics 22*(12), 1021–1033.

Fernandes, M.M., A. Cardoso, J. Trigo, & J. Marques (1993). Bearing capacity failure of tied-back walls—a complex case of soil-wall interaction. *Computers and Geotechnics 15*(2), 87–103.

Ghaboussi, J.&D. Sidarta (1997). New method of material modelling using neural networks. In *6th International Symposium on Numerical Models in Geomechanics*, pp. 393–400.

Hashash, Y.M. & A.J. Whittle (1996). Ground movement prediction for deep excavations in soft clay. *Journal of Geotechnical Engineering 122*(6), 474–486.

Krahn, J. et al. (2004). Stability modeling with slope/w: An engineering methodology. *GEOSLOPE/W International Ltd. Calgary, Alberta, Canada.*

Mana, A.I. & G.W. Clough (1981). Prediction of movements for braced cuts in clay. *Journal of Geotechnical and Geoenvironmental Engineering 107* (ASCE 16312 Proceeding).

More, J.Z.P. (2003). An´alise num´erica do comportamento de cortinas atirantadas em solos. *Master⊠s Dissertation-Departamento de Engenharia Civil, Universidade Cat´olica do Rio de Janeiro, Rio de Janeiro.*

Morgenstern, N. & V.E. Price (1965). The analysis of the stability of general slip surfaces.

Pakbaz, M., R. Mehdizadeh, M. Vafaeian, & K. Bagherinia (2009). Numerical prediction of subway induced vibrations: case study in iran-ahwaz city. *J. Appl. Sci 14*(9), 1812–5654.

Peck, R.B. (1969). Deep excavations and tunneling in soft ground. *Proc. 7th Int. Con. SMFE, State of the Art*, 225–290.

Pfister, P., G. Evers, M. Guillaud, & R. Davidson (1982). Permanent ground anchors: Soletanche design criteria. *NASA STI/Recon Technical Report N 83.*

Rankine, W.M. (1857). On the stability of loose earth. *Philosophical transactions of the Royal Society of London 147*, 9–27.

Ray, R.P. (2009). Design practice for tieback excavations in the u.s. *ISSMGE International Seminar on Deep Excavation and Retaining Structures, Budapest ISBN: 978-963-06-6665-7*, 72–98.

Roscoe, K. & J. Burland (1968). On the generalized stress-strain behaviour of wet clay.

Sabatini, P., V. Elias, G. Schmertmann, & R. Bonaparte (1997). Geotechnical engineering circular number 2: Earth retaining systems. *Washington, DC: FHWA.*

Sabatini, P., D. Pass, & R.C. Bachus (1999). Geotechnical engineering circular no. 4: ground anchors and anchored systems. Technical report.

Strom, R.W. & R.M. Ebeling (2002). *Methods used in tieback wall design and construction to prevent local anchor failure, progressive anchorage failure, and ground mass stability failure.* US Army Corps of Engineers, Engineer Research and Development Center, Information Technology Laboratory.

Xanthakos, P.P. (1991). *Ground anchors and anchored structures.* John Wiley & Sons.

Yoo, C. & D. Lee (2007). Deep excavation-induced ground surface movement characteristics–a numerical investigation. *Computers and Geotechnics 35*(2), 231–252.

Numerical Methods in Geotechnical Engineering IX – Cardoso et al. (Eds)
© 2018 Taylor & Francis Group, London, ISBN 978-1-138-33203-4

A parametric study of efficiency of buttress walls in reducing the excavation-induced tunnel movement

K.H. Law
KH Geotechnical Sdn Bhd, Kuala Lumpur, Malaysia

ABSTRACT: Deep excavations give rise to movements in the surrounding ground with consequent potential for damage to surrounding structures and buried services. When the excavations are in the vicinity of existing tunnels, it is necessary to assess the influence of these excavations on existing tunnels such as displacements of the lining and additional loads on the lining. This paper presents a finite element parametric study of the application of buttressed diaphragm wall in reducing tunnel movements caused by a 22.5 m deep excavation in Kuala Lumpur. This study aims to examine the influence of the geometry of buttress walls (spacing, thickness, and length) on the displacement of the tunnel and diaphragm wall via a series of analyses. The results show that it is more efficient by increasing the buttress walls length as compared to reducing the buttress walls spacing. The thickness of buttress walls play insignificant role in the reduction in the wall and tunnel displacement. The relative effectiveness of the sequentially removing-off and not removing-off buttress walls on the movement of the tunnel and diaphragm wall is also studied. The study demonstrates that the buttress walls should not be sequentially removed during the excavation stages in minimising the tunnel movement.

Keywords: Deep Excavation, Tunnel Movement, Buttressed Diaphragm Wall, Three-Dimensional Analysis

1 INTRODUCTION

In the built-up urban environment, there is an increasing demand for construction of both commercial and residential buildings near the metro line. One of the critical concerns in constructing an excavation near existing tunnel is the impact of the associated ground movements on the tunnel. A case history of the excessive displacement and damage to existing tunnel due to adjacent excavation has been reported by Chang *et al.* (2001). Thus, it is necessary to assess the influence of these excavations on existing tunnels such as displacements of the tunnel and additional forces induced on the tunnel's structure.

For a deep excavation, the ground improvements and strengthening of lateral support systems are commonly adopted to control movements induced by deep excavation. According to the Malaysia Railway Protection Act (1998), the maximum movement of the tunnel caused by an adjacent construction should not exceed 15 mm. Given the stringent requirement on the tunnel movement, the above methods may not be the adequate protective measures to reduce the excavation-induced tunnel movements within the allowable limit of 15 mm. Therefore, an additional auxiliary measure such as installing the buttress walls, cross walls or ground improvement may be required to control movements induced by deep excavation to a tolerable limit.

Buttress wall is a concrete wall constructed perpendicular to the diaphragm wall to a certain length and depth before excavation (Lim *et al.* 2016). The successful excavation case histories with the buttress walls in reducing the excavation-induced movements have been reported (Ou *et al.* 2008; Hsieh *et al.* 2011; Hsieh *et al.* 2015).

This study aims to examine the influence of the geometry of buttress walls (spacing, thickness and length) on controlling the displacement of the tunnel and diaphragm wall for a deep excavation project in Kuala Lumpur, Malaysia via a series of 3D finite element analyses. The relative effectiveness of the sequentially removing-off and not removing-off buttress walls on the displacement of the tunnel and diaphragm wall is also studied.

2 PROJECT DESCRIPTION AND SITE GEOLOGY

Figure 1 shows the site location of the excavation-tunnel interaction problem investigated in this paper.

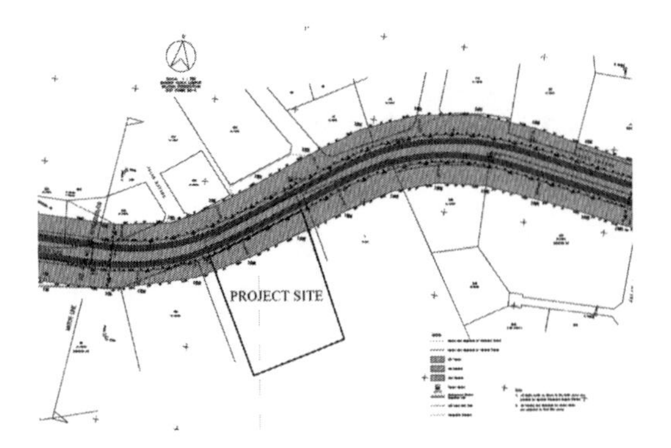

Figure 1. Site location plan.

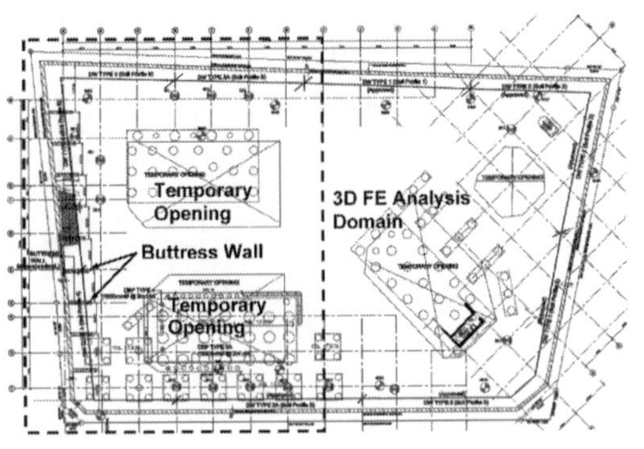

Figure 2. Diaphragm wall and buttress wall layout.

Figure 2 shows the diaphragm wall and buttress wall layout of the project site. The project site is situated in Kuala Lumpur city centre. It is a mixed commercial and residential development with a five-level underground basement. The geometry of the basement excavation is approximately 85 m × 135 m in plan with the total excavation depth of 22.5 m below existing ground level. The excavation is carried out using the top-down construction method in six excavation stages. The excavation is carried out near the existing Light Rail Transit (LRT) twin bored tunnels which run parallel to the north side of the excavation, and the distance from the supported diaphragm wall to the nearest inbound tunnel axis is 6.3 m. The buried depth of the existing twin tunnels is 15.5 m from the ground to the tunnel axis, see Figure 3.

Geological map of Kuala Lumpur, Malaysia (1993) indicates that the site is underlain by weathered residual soils of Kenny Hill Formation overlying the Limestone Formation. Based on the boreholes located at the alignment of northern boundary wall, the subsoil condition within this area could be divided into three distinctive soil layer based on the standard penetration test (SPT-N) result as shown in Figure 3. The SPT-N value is increasing with depth. A hard material with SPT-N more than 50 blows/300 mm is encountered at depth approximately 22.5 m below existing ground level. The residual soil mainly comprises of silty sand and sandy silt material. The measured groundwater table is located at about 5 m below ground surface.

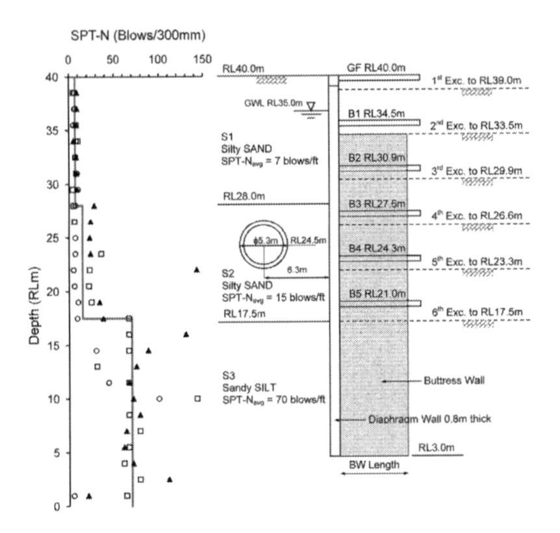

Figure 3. Cross section profile of excavation stages.

For simplicity, only half of the excavation area was modelled as shown in Figure 2. The excavation plane was assumed to be a regular rectangular to simplify the modelling procedure. Thus, the modelled excavation length and the excavation width were approximately 67.5 m and 85 m, respectively. The four vertical side boundaries of the mesh were restrained from movement in the horizontal direction but were free to move vertically, and the bottom boundary of the mesh was entirely restrained from movement in both the horizontal and vertical directions. The distance between the diaphragm wall and the outer boundary of the mesh was set at five times of the final excavation depth to minimise the boundary effect. The temporary slab openings at each floor level to facilitate the removable of the soil were also taken into consideration in 3D finite element analyses.

3 3D FINITE ELEMENT ANALYSES

For numerical simulations carried out in this study, a commercially available 3D finite element program PLAXIS 3D was used. Figure 4 shows the 3D finite element mesh used in the parametric study.

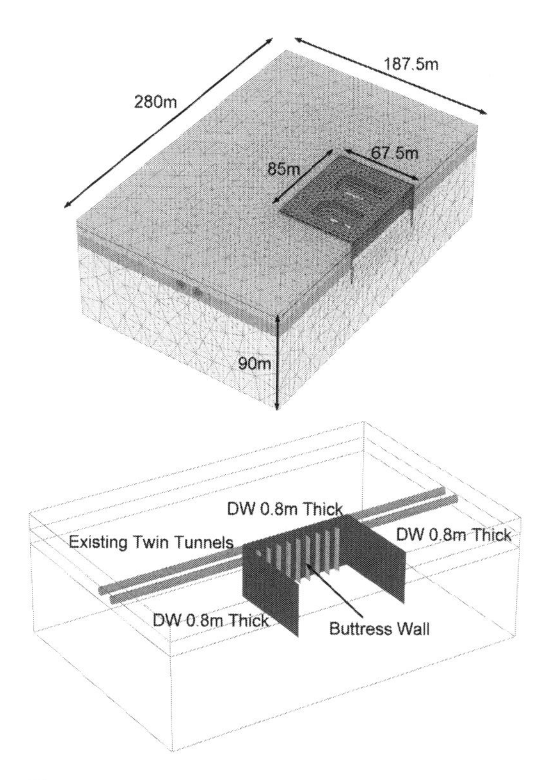

Figure 4. 3D finite element mesh of the model.

Ten-node tetrahedral elements were used to model the soil volume. The stress-dependent stiffness model, the Hardening Soil (HS) model (Schanz *et al.*, 1999) was adopted to simulate the soil behaviour under drained condition. Table 1 summarises the input parameters for the soils modelled. The soil stiffness parameters of HS model were based on empirical correlation as reported by Law *et al.* (2014).

Six-node plate elements were used to model the diaphragm wall, the buttress wall, the concrete slab, and the tunnel lining. Also, 12-node interface elements were applied to model the soil-wall interaction behaviour with an interface reduction factor, R_{inter} of 0.67 was adopted to model the interaction between soils and structural elements. The structural elements, such as diaphragm walls, buttress walls, concrete slabs, and tunnel linings are assumed to behave as linear elastic material. The analyses assumed that the diaphragm wall and buttress wall are "wished-in-place" and hence, do not consider local changes in stresses or soil properties associated with trench excavation and concreting process. Table 2 summarises the input parameters for the structural elements modelled.

In this study, a series of parametric analyses were carried out to explore the effects of buttress wall geometry (spacing, thickness and length) on existing tunnel and diaphragm wall movements. Also, the relative effectiveness of the sequentially removing-off and not removing-off buttress walls on the displacement of the tunnel and diaphragm wall is also compared. The sectional view of the excavation-tunnel problem investigated in this study is shown in Figure 3. The diaphragm wall and buttress wall thickness and depth were set to 0.8 m and 37 m respectively. In total, fifteen 3D analyses were conducted as summarised in Table 3.

Table 1. Effective shear strength and stiffness parameters.

Soil Profile	C' (kPa)	φ' (°)	ψ (°)	$E^{ref*}_{50} = E^{ref}_{50}$ (MPa)	E^{ref}_{ur} (MPa)
S1	1	29	0	14	42
S2	3	30	0	30	90
S3	15	35	0	105	315

Table 2. Structural elements input parameters.

Structures	Concrete Grade (MPa)	Wall Thickness (m)	$E_1 = E_2$ (kN/m²)
Diaphragm wall	35	0.80	28×10^6
Buttress wall	15	0.80	18×10^6
Tunnel Lining	60	0.18	36×10^6
Slab	30	0.275	26×10^6

4 RESULTS OF 3D ANALYSES

The inbound tunnel is much closer to the excavation, and it is more vulnerable to damage during the basement excavation. Thus, in the present analyses, only responses of the inbound tunnel are presented.

4.1 *Influence of the buttress wall spacing*

Figure 5 shows the wall deflection profiles at the final stage of excavation from 3D analyses for different buttress wall spacing at the centre of the wall running parallel to the existing LRT tunnels. As this figure indicates, the deflection of the buttressed diaphragm wall reduces considerably, compared with the condition without buttress walls. The result indicates that the wall deflection decreases as the buttress walls spacing reduces. The rate of reduction of wall deflection is somewhat consistent and correlated linearly to buttress walls spacing.

Table 3. Range of parameters considered in parametric study.

Parameter	Range
BW spacing (m)*	4, 6, 8, 10
BW length (m)**	3, 6, 9, 12
BW thickness (m)***	0.6, 0.8, 1.0, 1.2, 1.5
Progressive removing-off buttress wall	–
Reference analysis	No BW

*BW length and thickness are set to 6 m and 0.8 m respectively. **BW spacing and thickness are set to 8 m and 0.8 m respectively. ***BW length and spacing are set to 6 m and 8 m respectively.

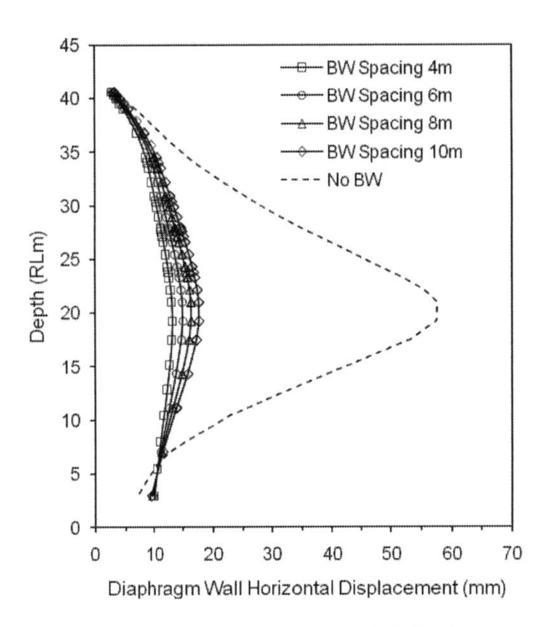

Figure 5. Effect of BW spacing on wall deflection.

Figure 6 shows the inbound tunnel displacement at the final stage of excavation for buttress walls spacing of 4 m, 6 m, 8 m and 10 m. As this figure indicates, the tunnel displacement increases linearly with buttress walls spacing. Similar to wall deflection, the tunnel displacement decreases linearly with increasing number of buttress walls. Without installing the buttress walls as an auxiliary measure, the tunnel displacement would have exceeded the allowable limit of 15 mm.

4.2 *Influence of the buttress wall length*

Figure 7 shows the wall deflection profiles at the final stage of excavation from 3D analyses for dif-

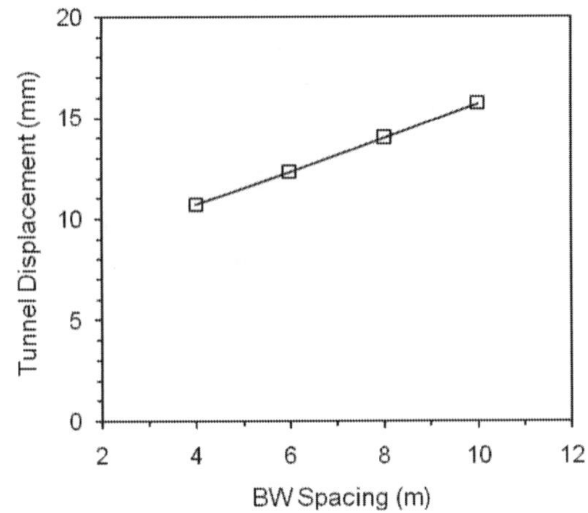

Figure 6. Effect of BW spacing on tunnel displacement.

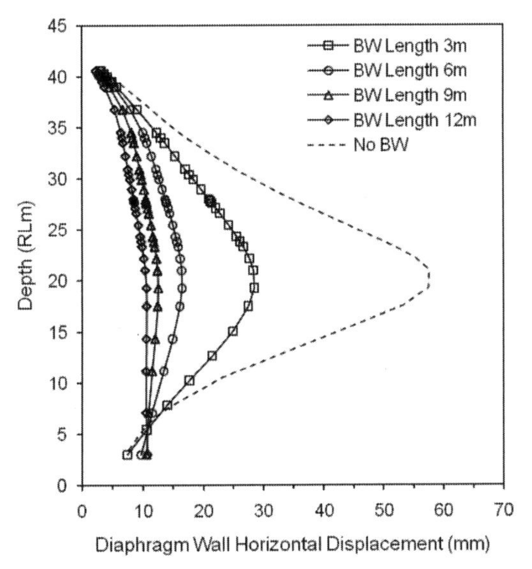

Figure 7. Effect of BW length on wall deflection.

ferent buttress wall length with the buttress walls spacing fixed at 8 m. It is evident that the wall deflection decreases with the increase of the buttress wall length. However, the rate of reduction in the wall deflection decreases as the buttress wall length increases, especially when buttress length reaches 9 m and above. As shown in Figure 7, a long buttress wall performs better than a short one. A substantial wall deflection (about 60 mm) is observed when the excavation is carried out without buttress walls. This figure shows that the deflection profile of buttressed diaphragm wall would gradually change from bulging mode to toe-translating mode when the buttress wall length is increased, and its behaviour is consistent with the finding from Ou *et al.* (2008).

Figure 8 shows the inbound tunnel displacement at the final stage of excavation for buttress walls length of 3 m, 6 m, 9 m and 12 m. Similar to wall deflection, the tunnel displacement decreases non-linearly with the increase of the buttress wall length. The efficiency in reducing the tunnel displacement seems to reduce when the buttress wall length reaches 9 m and above. Figure 8 suggests that there is no advantage to adopt buttress wall length of more than 9 m as the rate of reduction of tunnel displacement is insignificant from the economic and practical point of view.

According to Hsieh *et al.* (2016), the mechanism of buttress walls in reducing movement mainly comes from the frictional resistance developed between buttress walls and adjacent soils rather than from the combined bending stiffness of diaphragm wall and buttress walls. For 3 m length buttress walls, the relative displacement between the adjacent soil and buttress wall is small (both the adjacent soil and buttress walls move almost the same amount), causing low mobilisation of frictional resistance, hence, has a poor effect in reducing the wall and tunnel displacements.

To compare the relative effectiveness of buttress wall spacing and buttress wall length, the efficiency of the buttress walls as a tunnel displacement reduction (TDR) ratio is defined as

$$TDR = \frac{\delta_{hm} - \delta_{hmb}}{\delta_{hm}} x100\% \quad (1)$$

where δ_{hm} is the maximum tunnel displacement without buttress walls, and δ_{hmb} is the maximum tunnel displacement with buttress walls. An increase in the value of TDR implies a higher efficiency in reducing the tunnel displacement. Thus, the higher the value of TDR the more efficient of the buttressed diaphragm wall. Using this ratio, the relative effectiveness of the spacing and the length of the buttress walls are evaluated here.

Figure 9 shows the relationship between the TDR and the buttress walls surface area. The buttress wall surface area is defined as contact surface area between buttress walls and adjacent soil. Two scenarios have been considered here: (1) reducing the buttress walls spacing, (2) increasing the buttress walls length. As can be observed from Figure 9, with the same amount of the buttress walls surface area in contact with soil, increasing the buttress walls length yielded a higher value of TDR as compared to reducing the buttress walls spacing. In other words, to reduce diaphragm wall or tunnel displacement, it is more efficient by increasing the buttress walls length rather than reducing the buttress walls spacing.

4.3 *Influence of the buttress wall thickness*

Figure 10 shows the inbound tunnel displacement at the final stage of excavation for buttress wall thickness of 0.6 m, 0.8 m, 1.0 m, 1.2 m and 1.5 m for buttress wall length of 3 m and 6 m. For the cases considered here, the buttress walls spacing was fixed at 8 m. Results show that the thickness of buttress wall plays a relatively minor role in reducing the tunnel displacement. An increase of 100% of buttress wall thickness (from 0.6 m to 1.2 m) only yielded a reduction of 14% and 11.5% tunnel displacement for buttress wall length of 3 m and 6 m respectively.

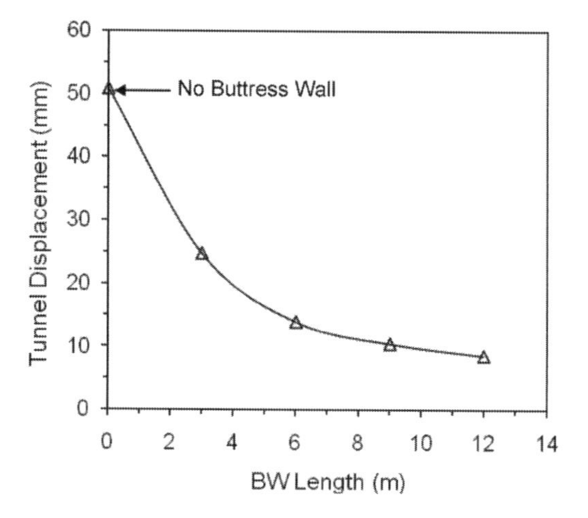

Figure 8. Effect of BW length on tunnel displacement.

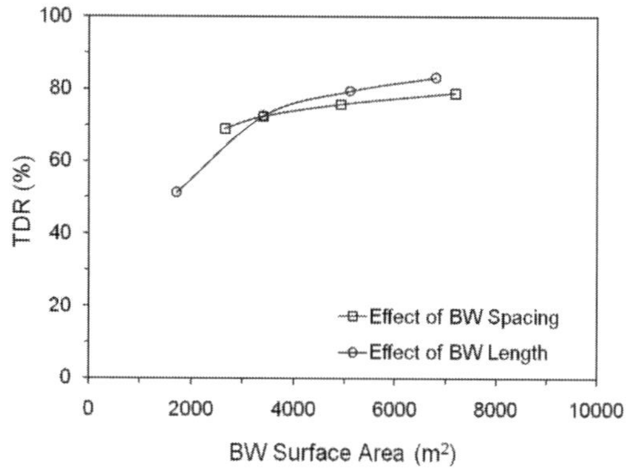

Figure 9. Relationship between TDR and BW surface area.

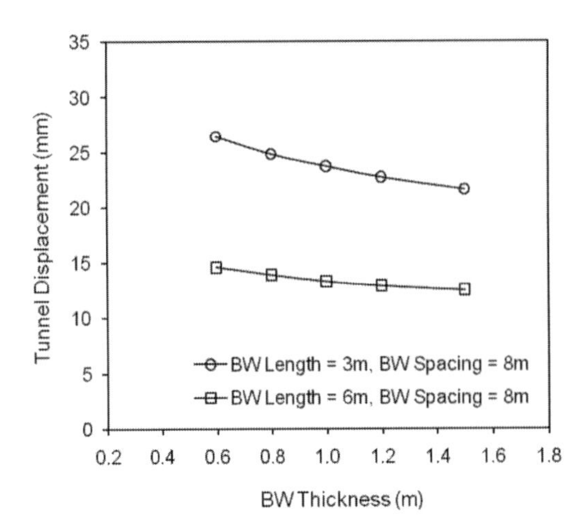

Figure 10. Effect of BW thickness on tunnel displacement.

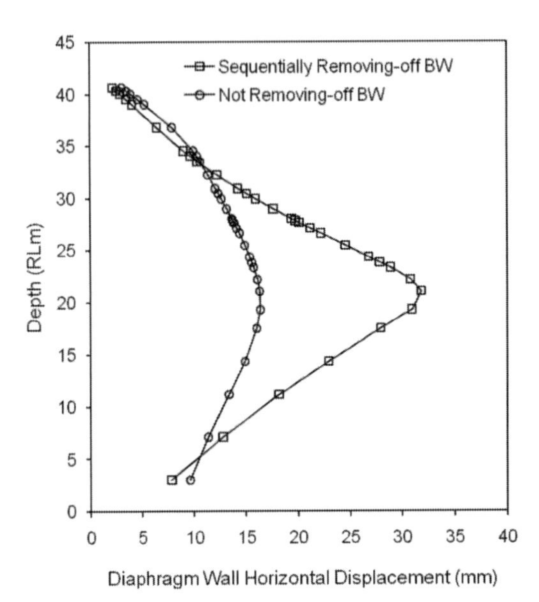

Figure 11. Comparison between not removing-off and sequentially removing-off buttress walls on wall displacement.

4.4 *Influence of the progressive removal of buttress wall*

Figures 11 and 12 compare the relative effectiveness of the sequentially removing-off and not removing-off buttress walls on the diaphragm wall and tunnel displacement respectively. For the cases considered here, the buttress walls spacing and length are fixed at 8 m and 6 m respectively. As can be seen from Figures 11 and 12, the buttressed diaphragm wall is more efficient in restraining the

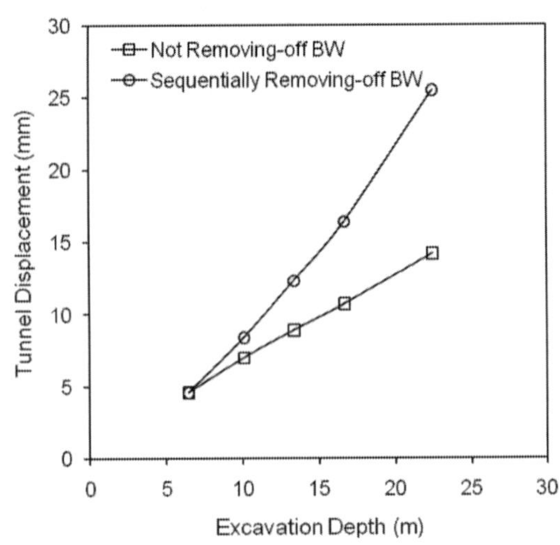

Figure 12. Comparison between not removing-off and sequentially removing-off buttress walls on tunnel displacement.

tunnel displacement if the internal buttress walls are not demolished during each excavation stage. When the buttress walls are sequentially demolished during excavation, the effect of reducing the wall and tunnel displacement mainly comes from frictional resistance developed between the side surface of buttress walls and adjacent soil. If the buttress walls are demolished after completion of basement slabs construction, the reduction of wall and tunnel displacement mainly comes from both the frictional resistance developed between the side surface of buttress walls and adjacent soil and combined bending stiffness of diaphragm wall and buttress walls, even though the later plays a minor role as compared to the former, see Section 4.4. Hence, to achieve optimum performance in minimising the wall and tunnel movement, buttress walls should not be sequentially removed-off during the excavation stages.

5 CONCLUSIONS

Based on the numerical study in this paper, the following conclusion can be drawn:

1. Based on the parametric study results, buttress wall of 6 m length, at 8 m spacing and 0.8 m thick has been adopted as an additional auxiliary measure for this particular project to control excavation-induced tunnel movement within a tolerable limit of 15 mm.
2. The wall and tunnel displacement decrease as the buttress wall spacing reduces.

3. The amount of reduction in wall and tunnel displacement increased with the increasing length of the buttress wall.
4. Under a condition of the same amount of area of buttress wall in contact with adjacent soil, increasing the buttress walls length has a better effect than reducing the buttress walls spacing in reducing the wall and tunnel displacement.
5. Buttress wall thickness plays a relatively minor role in reducing the wall and tunnel displacement. The reduction of wall and tunnel displacement mainly derives from the frictional resistance developed between the side surface of buttress walls and adjacent soil.
6. To achieve optimum performance in reducing wall and tunnel movement, buttress walls should not be demolished along with excavation of soil.

REFERENCES

Chang, C.T., Sun, C.W., Duann, S.W., and Hwang, R.N. 2001. Response of a Taipei Rapid Transit System (TRTS) tunnel to adjacent excavation. *Tunnelling and Underground Space Technology* 16(3): 151–158.

Hsieh, H.S., Wu, L.H., Lin, T.M., Cherng, J.C., and Hsu, W.T. 2011. Performance of T-shaped diaphragm wall in a large scale excavation. *Journal of GeoEngineering* 6(3): 135–144.

Hsieh, P.G., Ou, C.Y., Lin, Y.K., and Lu, F.C. 2015. Lessons learned in design of an excavation with the installation of buttress walls. *Journal of GeoEngineering* 10(2): 63–73.

Hsieh, P.G., Ou, C.Y., and Hsieh, W.H. 2016. Efficiency of excavations with buttress walls in reducing the deflection of the diaphragm wall. *Acta Geotechnica* 11(5): 1087–1102.

Law, K.H., Hashim, R., and Ismail, Z. 2014. 3D numerical analysis and performance of deep excavations in Kenny Hill formation. *Proc. 8th European Conference Numerical Methods in Geotechnical Engineering (NUMGE), Delft*: 759–764.

Lim, A., Hsieh, P.G., and Ou, C.Y. 2016. Evaluation of buttress wall shapes to limit movements induced by deep excavation. *Computers and Geotechnics* 78: 155–170.

Ou, C.Y., Teng, F.C., Seed, R.B., and Wang, I.W. 2008. Using buttress walls to reduce excavation-induced movements. *Geotechnical Engineering, ICE*. 161(4): 209–222.

Schanz, T., Vermeer, P.A., and Bonnier, P.G. 1999. The hardening soil model: formulation and verification. *Beyond 2000 in Computational Geotechnics*. Balkema, Rotterdam: 281–296.

Numerical Methods in Geotechnical Engineering IX – Cardoso et al. (Eds)
© 2018 Taylor & Francis Group, London, ISBN 978-1-138-33203-4

Soil parameter identification for excavations: A falsification approach

Wang Ze-Zhou, Goh Siang Huat & Koh Chan Ghee
Department of Civil and Environmental Engineering, National University of Singapore, Singapore
ETH Zurich, Future Cities Laboratory, Singapore-ETH Centre, CREATE, Singapore

Ian F.C. Smith
Applied Computing and Mechanics Laboratory (IMAC), School of Architecture, Civil and Environmental Engineering (ENAC), Swiss Federal Institute of Technology (EPFL), Lausanne, Switzerland
ETH Zurich, Future Cities Laboratory, Singapore-ETH Centre, CREATE, Singapore

ABSTRACT: Soil-parameter identification is a long-standing topic that has received considerable attention over the years. Traditional identification approaches usually follow the least-squares criteria. These approaches aim to find a single set of parameter values that produces the minimum absolute error between numerical predictions and measurements. However, these approaches have several drawbacks, particularly their inability to accommodate systematic modelling uncertainty. A recent data-interpretation methodology, Error-Domain Model Falsification (EDMF), is used to overcome these drawbacks. A key feature of EDMF is the explicit representation of systematic uncertainties originating from model simplifications. This paper presents the application and adaptation of this methodology in the area of soil parameter-identification. A synthetic excavation case is used as an illustration. Using EDMF, accurate soil properties can be identified with the inclusion of systematic uncertainties, which leads to improved predictions of the excavation performance. The residual minimization approach, however, returns inaccurate soil properties and biased predictions. The present study shows that EDMF offers a promising data-interpretation methodology that can yield multiple sets of admissible parameter values, which in turn help engineers to appreciate and understand the possible variations in soil properties and predictions that are compatible with field measurements.

1 INTRODUCTION

Many, if not most, excavations are monitored. Collected measurement data includes, but is not restricted to, retaining wall deflection, ground settlement and strut load. Coupled with numerical modelling techniques, such as the finite element method, an inverse analysis can be performed to identify soil parameter values that explain these measurements.

A complete inverse analysis needs measurement data, a model, a data interpretation methodology and an ad-hoc optimisation technique. There are several data interpretation methodologies, such as the residual minimisation approach, the maximum likelihood approach and the Bayesian inference approach. The studies of Finno & Calvello (2005), Hashash et al (2006), Levasseur et al (2008) and Reche et al (2008) are based on the residual minimisation approach. Ledesma et al (1996) and Gorral (2013) reported studies using the maximum likelihood approach, while Cividini et al (1983), Juang et al (2012) and Qi & Zhou (2017) adopted the Bayesian inference approach.

In numerical simulations, a perfect model is never achieved. There are always some discrepancies between models and the real cases. In excavations, besides soil parameter uncertainty, which is usually the most significant component, model simplifications and omissions (Einstein & Baecher, 1982) may also introduce additional discrepancies. Finno & Calvello (2005) reported that the three-dimensional effects in their study, though small, affected the accuracy of the soil parameter values identified. However, no attempt was made to address this issue. In the aforementioned studies, apart from Qi & Zhou (2017), soil parameter uncertainty was the only source of uncertainty considered. Hashash et al (2011) reported an inverse analysis using a 3D model for an approximately square-shaped excavation. For such problems, significant corner constraint effects may be encountered, and hence a 2D model would no longer yield reasonable results (Lee et al, 1998). With a 3-D model, however, parallelisation and multiple PCs were needed owing to the substantial computational time involved.

One way to achieve a balance between efficiency and accuracy is to employ an error term to represent such discrepancies due to model simplifications. A plane strain ratio (PSR), defined as the ratio of the maximum computed wall deflection from a 3D analysis normalised by that computed using a plane strain simulation (Ou et al, 1996) (Finno et al, 2007), can be used to adjust the plane strain predictions to account for three-dimensional effects. However, the use of PSR is constrained by its inability to correct wall deflections other than the maximum deflection point typically encountered at the centre of the excavation. In an inverse analysis, all measurement points at different depths along the wall deflection profile are usually considered, as the measured behaviour at one location alone may not provide sufficient information (Qi & Zhou, 2017). In the present study, instead of adopting a deterministic error term such as the plane strain ratio, an uncertainty distribution is employed to represent model simplifications. Details will be provided in later sections. In addition, measurement data at all measurement locations are considered.

A recent data interpretation methodology, Error-Domain Model Falsification (EDMF) (Goulet & Smith, 2013), is employed in this study. This methodology has been applied and shown to be effective in bridge engineering (Goulet et al, 2010), leak detection (Moser et al, 2015), wind simulation improvement (Vernay et al, 2014) and construction activities monitoring (Soman et al, 2017). Previous studies have demonstrated that this methodology is robust with respect to changes in correlations and has the ability to explicitly include modelling uncertainties.

The application and adaptation of this methodology in geotechnical excavations is presented in this paper. A synthetic excavation case is used as an illustration. The comparison between the residual minimisation approach and the EDMF approach is presented and discussed.

2 ERROR-DOMAIN MODEL FALSIFICATION

EDMF, which is intended to identify plausible physics-based models using information provided by field-response measurement data, was initially proposed by Goulet & Smith (2013). Plausible models are defined by n_θ parameter values and a model class, which is denoted by G_k.

For each measurement location $i \in \{1, \ldots, n_v\}$ on the structure of interest, in which n_v represents the number of measurement locations in a wall deflection profile, R_i denotes the real response at location i and $\hat{y}_i$ denotes the measured response at the same location. Predictions $g_{i,k}$ (Θ'_k) of the model class, which is typically based on a finite element analysis, are derived at location x_i by assigning Θ'_k, which corresponds to the true parameter value, to the model class. A perfect model class is never achieved. There are always some discrepancies between model predictions and the real problem that the model describes because of parameter uncertainty and model simplifications and omissions. Parameter uncertainty is addressed by parametrizing the important parameters. Model simplifications and omissions are represented mathematically as modelling uncertainties, denoted as $U_{i,gk}$ at location i. In addition, $U_{i,\hat{y}}$ represents measurement uncertainty at location i. Equation (1) can then be employed to estimate the true behaviour of the structure incorporating modelling uncertainty, measurement uncertainty, model predictions and measurements.

$$g_{i,k}(\Theta'_k)+U_{i,g_k}=R_i=\hat{y}_i+U_{i,\hat{y}} \quad \forall_i \in \{1,\ldots,n_y\} \tag{1}$$

Upon rearrangement, the following is obtained:

$$g_{i,k}(\Theta'_k)-\hat{y}_i=U_{i,ck} \tag{2}$$

Where $U_{i,ck}$ is a random variable representing the difference between $U_{i,y}$ and $U_{i,gk}$ at location i.

The left term of Eq. 2 represents the difference between model predictions and the measurement data at location i. This term is typically called the residual r_i. The probability density function (PDF) describing the error in the model class $f_{Ui,gk}(u_{i,gk})$ can be estimated using values reported in the literature, engineering judgement and/or local experience.

The implementation of EDMF starts with the definition of an initial model set, which contains n_Ω model instances $\Omega_k = \{\theta_{k,m}, m = 1, \ldots, n_\Omega\}$. Threshold bounds are then defined by computing the interval $\{u_{i,low}, u_{i,high}\}$ that represents a probability equal to $\varnothing_d^{1/n_v}$ for the combined PDFs $f_{Ui,c}(u_{i,c})$ at each measurement location i. This is seen in the following equation:

$$\varnothing_d^{1/n_v}=\int_{u_{i,low}}^{u_{i,high}} f_{u_{i,c}}\left(u_{i,c}\right)du_{i,c} \quad \forall_i \in \{1,\&,n_v\} \tag{3}$$

A value of 0.95 for the confidence level $\varnothing_d \in [0,1]$ is commonly employed. The confidence level $\varnothing_d$ is adjusted using the Sidák correction to take into account the fact that n_v measurement locations are simultaneously considered. The Sidák correction creates a hyper-rectangular acceptance region. Under this scheme, the value of the correlation between sensor locations is no longer needed. This is particularly important because it is often difficult to determine the correlation values between sensor locations. Goulet & Smith (2013) has shown that this scheme is conservative. Falsification is then performed according to the following equation:

$$\Omega''_k = \{\theta_k \in \Omega_k \,|\forall_i \in \{1,\ldots,n_v\}\, u_{i,low} \leq g_{i,k}(\theta_k)-\hat{y}_i \leq u_{i,high}\} \tag{4}$$

where the candidate model set (CMS), Ω''_k, is made up of all the initial model instances except those that have been falsified. An instance θ^*_k of a model class G_k is a candidate model if, for each sensor location $i \in \{1, \ldots, n_y\}$, the residual r_i value falls inside the threshold bounds derived from Equation (3). All model instances that belong to the CMS are assigned a constant probability as it is often difficult to justify a more sophisticated distribution in practical situations.

In EDMF, predictions are made with all model instances of a candidate model set. In large-scale, multi-staged excavations, predictions may be made for different aspects of the problem, either in time or in space. For example, at an early or intermediate point in the excavation process, it is often useful and desirable to predict the field responses of subsequent excavation stages. This is different from some earlier applications of EDMF, for example bridge engineering, wherein predictions are typically made using the same model class G_k. In contrast, excavation configurations are different at different stages, and hence each stage of excavation has its associated model class. Accordingly, predictions for a different excavation stage are made using a different model class G_l that corresponds to that particular stage. Predictions at the same measurement location i of a different excavation stage l, denoted as $Q_{i,l}$, are expressed by:

$$Q_{i,l} = g_{i,l}(\theta''_k) + U_{i,cl} \quad \forall_i \in \{1, \ldots, n_y\} \tag{5}$$

where $U_{i,cl}$ is a vector that represents the difference between $U_{i,y}$ and $U_{i,gl}$ of model class G_l at location i, and $\theta''_k = \{\theta^*_k \mid \theta_k \in \Omega''_k\}$ is the set of vectors containing parameter values in the CMS identified from model class G_k.

3 SYNTHETIC EXCAVATION CASE STUDY

3.1 *Material properties*

As shown in Figure 1 and 2, this analysis considers a 25 m deep excavation whose plan is 50 m in length and 40 m in width. The excavation is supported by concrete diaphragm walls embedded to a depth of 40 m in a single layer of soil, together with four layers of centre and corner struts. The walls and struts are modelled using elastic plate elements and node-to-node anchors respectively, while the soil is modelled as an elastic-perfectly plastic Mohr Coulomb material whose stiffness varies linearly with depth. The structural and soil material parameters are summarised in Table 1. The soil response is modelled as undrained, using the 'almost incompressible' effective stress approach, also known as Undrained Method A (Whittle and Davis, 2006; COI, 2005).

3.2 *Finite element model*

This excavation, braced internally, is excavated and supported at equal vertical intervals of 5 m. The formation level is 8.5 m below the fourth layer of strut. It is assumed that wall installation is 'wish-in-place'.

The excavation comprises 5 stages in total, with stage 1 corresponding to the initial cantilever phase and stage 5 corresponding to the excavation to formation level. This is illustrated in Figure 2. Each strutting phase and excavation phase are combined into a single phase. For example, stage 2 refers to the installation of the first layer of strut and the excavation to 7 m below ground level. This setting is reasonable because there is no pre-stress in the struts. The installation of the struts will not cause any significant stress change in the system.

Both 2D and 3D analyses are performed. In the 2D model, only the centre struts are modelled. The effective width assigned to the struts is 20 m. This is a conservative assumption. The omission of the corner struts in the 2D model is one source of model uncertainty, which will be explained later. Roller supports are assigned to the vertical boundaries in both the 3D and 2D models, with the soil domain extending up to 100 m behind the

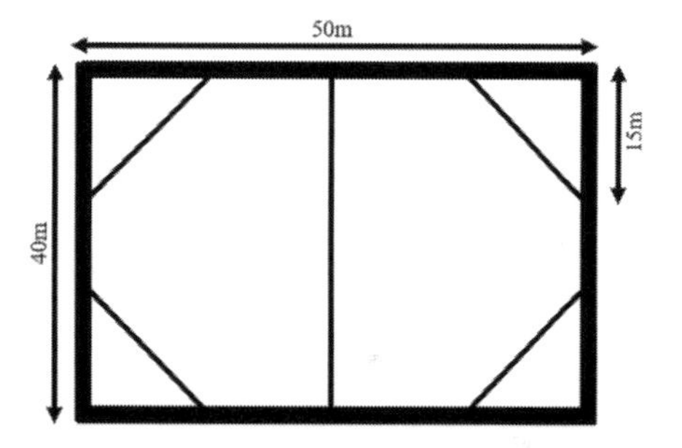

Figure 1. Excavation plan view and bracing configuration.

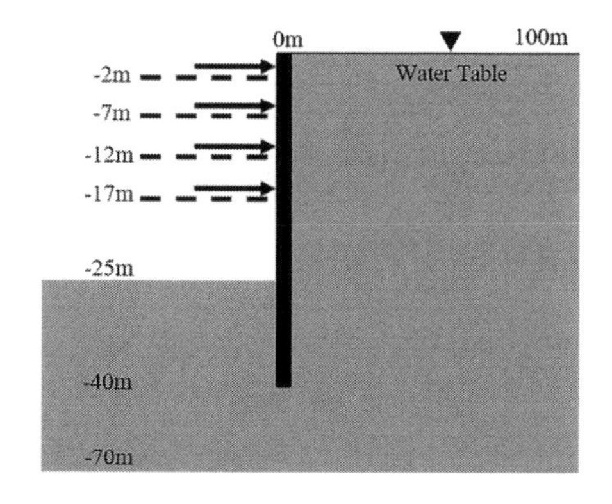

Figure 2. Excavation sequence and bracing configuration (not to scale).

Table 1. Parameter values used to generate synthetic excavation-induced wall deflection data.

Member	Property	Value
Plate	E_1 (MPa)	25E3
	E_2 (MPa)	25E3
	v_{12}	0.2
	Thickness (m)	1.0
Centre Strut	EA (MN)	20E3
Corner Strut	EA (MN)	10E3
Soil	E' (MPa)	35
	E'_{inc} (MPa/m)	0.5
	v_1	0.2
	c' (MPa)	0
	Φ' (°)	30
	Ψ (°)	0
	R_{inter}	0.9

retaining walls. Both 2D and 3D models are fully fixed at the base, which is 30 m below the toe of the retaining wall. This setting complies with what was suggested by Ou & Shiau (1998). No flow condition is assigned to the soil body. Dewatering in the excavated zone is achieved through setting the soil clusters in the excavated zone as dry. Mesh convergence studies have been performed to arrive at an optimal mesh size.

3.3 *Generation of synthetic measurement data*

The excavation geometry in this case study indicates that three-dimensional effects may be expected, the influence of which may be captured using a 3D model. Synthetic wall deflection profiles at the mid-section of this pit are generated by performing a three-dimensional finite element analysis using Plaxis 3D (Brinkgreve et al, 2013).

Measurement errors are usually expected in real field data. In order to capture such measurement errors in this synthetic example, the computed deflection data from the Plaxis 3D simulation output is perturbed with measurement errors, following the equation reported by Mikkelsen (2003). The synthetic wall deflection profiles are shown in Figure 3. These synthetic profiles are used as benchmark data against which the predicted profiles obtained using the identified parameters from 2-D analysis will be compared.

3.4 *Identification framework*

Owing to the high computational demand associated with running a 3D model, the parameter identification exercise is performed by carrying out 2D finite element analyses using the commercial software Plaxis 2D AE (Brinkgreve et al, 2014). This is consistent with industry practice, where 2D analy-

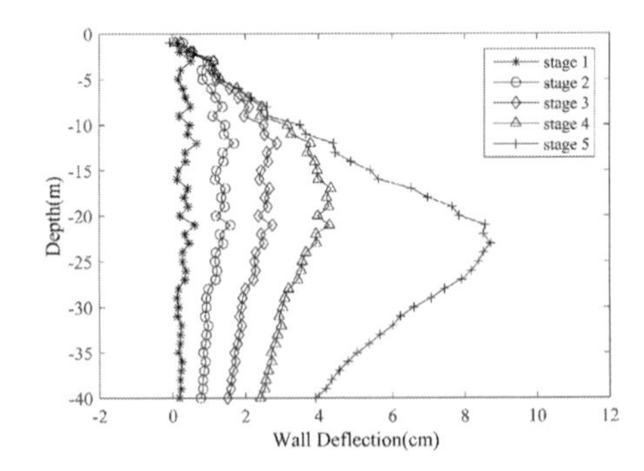

Figure 3. Synthetic wall deflection measurements.

ses are more commonly performed due to computational time and memory constraints.

A staged excavation analysis is conducted to simulate the construction process on site. In the real-world practice, measurement data is usually collected as the excavation work proceeds, which makes it possible to implement a progressive identification exercise. For example, upon collecting the field measurement data at the end of stage 1, a parameter identification exercise can be performed. As the excavation progresses, additional measurement data can be progressively incorporated into the inverse analysis to refine the identified parameter values.

For the synthetic excavation considered in this case study, the computed wall deflections from the 3D analyses serve as a substitute for the real-world field measurement data. Henceforth, the authors shall refer to round 1 as the parameter identification exercise performed using the synthetic wall deflection data of excavation stage 1, while round 2 refers to the identification exercise performed using the combined synthetic wall deflection data of excavation stages 1 and 2. In the same way, round 5 refers to the identification exercise performed using the accumulated synthetic wall deflection data from excavation stages 1 to 5.

In the present EDMF exercise, after the synthetic wall deflection data in excavation stage 1 is used for the Round 1 falsification, the resulting candidate model set will form the initial model set for the next round of falsification corresponding to excavation stage 2. With the inclusion of additional deflection data from subsequent excavation stages, falsification is performed only using the candidate model set obtained after the previous round of falsification. With more synthetic wall deflection data available for falsification, the size of the candidate model set is expected to be reduced. After each round of identification, predictions can be made with the identified parameter values for the subsequent excavation stages. For example, after round 1, which uses the synthetic wall deflection data of stage 1, predictions for stages 2 to 5 can be made.

Sensitivity analysis, following the concept of composite scaled sensitivity (Finno & Calvello, 2005), has been performed before the identification exercise. The results indicate that the stiffness parameters E' and E'_{inc} consistently exhibit high composite scaled sensitivities across all five excavation stages. In contrast, the sensitivity values associated with the soil friction angle Ø' show a spike only at excavation stage 5, which is due to a significant zone of soil yielding that develops at the final stage of excavation. Therefore, in the current study, the soil parameters E' and E'_{inc} are adopted as the unknown parameters to be identified, which will be carried out using both the residual minimisation and EDMF approaches. The ability of the two methods to identify the values of the E' and E'_{inc} parameters adopted in the synthetic example will be presented and compared in the following sections.

4 RESIDUAL MINIMISATION

The residual minimisation approach usually follows the weighted least-squares criterion, which aims to find a single set of parameter values that produces the minimum absolute error between numerical predictions and measurements. Finno & Calvello (2005) reported its application to an excavation problem. The objective function is expressed as:

$$S(\theta^*_k) = [\hat{y} - g_k(\theta^*_k)]^T \omega [\hat{y} - g_k(\theta^*_k)] = r^T \omega r \qquad (6)$$

where θ^*_k = parameter values; $\hat{y}$ = vectors containing measured responses; $g_k(\theta^*_k)$ = vector containing predictions given θ^*_k; ω = weight matrix; r = vector of residuals. In this study, the weight matrix is a diagonal matrix with values corresponding to the inverse of the inclinometer error variance, with the inclinometer error following the equation reported by Mikkelsen (2003). No correlation between sensor locations is assumed. The θ^*_k values that lead to the minimum $S(\theta^*_k)$ are the optimized final parameter values.

The search space for the two parameters to be identified, E' and E'_{inc}, is defined in Table 2. The search space is asymmetrical with respect to the true values.

Figure 4a shows the predicted wall deflections for Stages 1 to 5 obtained using the E' and E'_{inc} values identified from the minimisation exercise involving the stage 1 cantilever phase. It is observed that reasonable predictions of the 2D wall deflections of stage 1 to 4 can be obtained by using only the synthetic data of stage 1. However, wall deflection is underestimated at stage 5 by approximately 20% at the depth of maximum deflection. The minimization exercise is then repeated for subsequent excavation stages, using the synthetic wall deflection data generated from the 3D analyses. Figure 4b shows the predicted wall deflections obtained after including all synthetic wall deflection data. In general, the predicted wall deflections up to stage 4 are quite reasonable. However, the predicted wall deflections at stage 5 are always smaller than the synthetic response, even when the accumulated synthetic wall data for all stages are utilized in the identification process. Figure 5 shows the E' and E'_{inc} values identified after each round of identification. It is observed that the true parameter combination is never identified correctly.

One would intuitively expect that the predicted wall deflections at stage 5 should agree well with the synthetic 3D wall deflection data as all synthetic data is included in the identification process. However, Figure 4 shows otherwise. This can be explained by the concept of modelling uncertainty. A 2D plane strain model is adopted to identify responses resulting from a 3D problem. Given this excavation configuration, significant corner constraint effects can be expected. In addition, corner struts are not explicitly modelled in the 2D model. In this situation, both parameter uncertainty and modelling uncertainty are expected. However, the modelling uncertainty component is not explicitly included in the identification process. Therefore, parameter values need to be artificially modified to compensate for the omis-

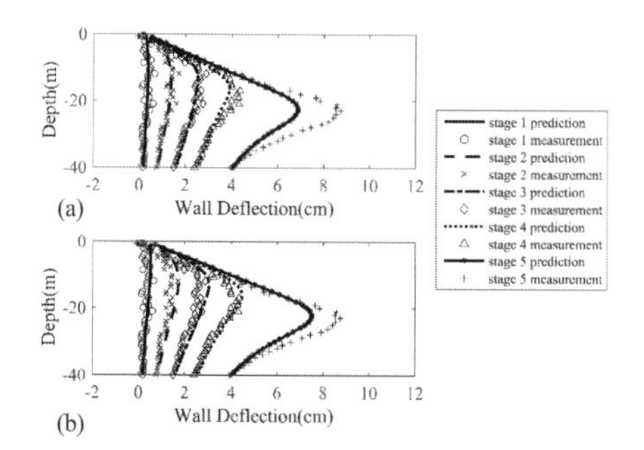

Figure 4. Wall deflection predictions after (a) 1st round of identification (b) 5th round of identification.

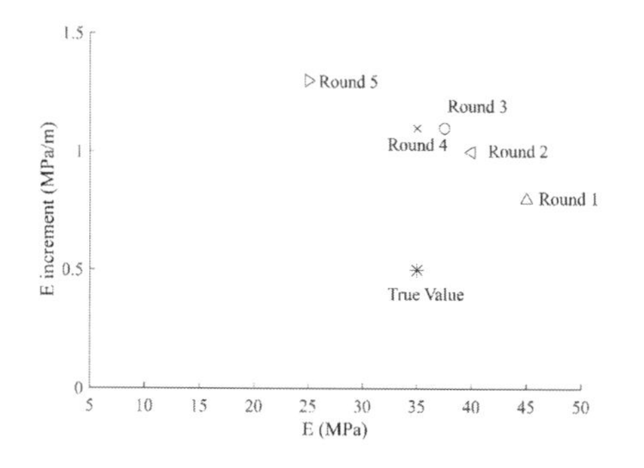

Figure 5 Parameter values identified after each round of falsification.

sion of modelling uncertainty. A similar explanation was also highlighted by Einstein & Baecher (1982). In the present synthetic example, parameter values are artificially adjusted to fit the 3D wall deflection data as shown in Figure 4. However, differences in the mechanism in 2D and 3D models cannot be fully compensated by the modification of parameter values, leading to the underestimation of wall deflection shown in Figure 4.

5 EDMF

5.1 *Quantification of modelling uncertainty*

As described earlier, EDMF supports the explicit inclusion of modelling uncertainties originating from model simplification. In the present study, the quantification of modelling uncertainty due to plane strain assumption is achieved by comparing the predicted wall deflection of a 2D and a 3D analysis. Two 3D simulations are conducted in order to estimate the modelling uncertainty. Table 3 summarised the parameter values used for these two 3D simulations, denoted as Case 1 and Case 2.

Cases 1 and 2 correspond respectively to the lower and upper bounds of the soil stiffness parameters shown previously in Table 2, and hence their results should approximately serve to define the bounds of the modelling uncertainty. Figure 6 present the 2D and 3D predicted wall deflections of Cases 1 and 2 respectively. Some observations are presented below.

Table 4 shows the comparison of PSR at 20 m depth and at stage 5 in both cases. It is shown that, using the same finite element model, PSR depends on soil parameter values. Similar trends are also noted at other measurement depths. Second, given the same soil parameter values, PSR varies across different stages of the excavation. Table 5 presents the PSR at 20 m depth with Case 2 parameter values at all 5 stages. This observation is in agreement with Ou et al (1996), who highlighted that PSR is a function of excavation depth. Third, PSR, given the same parameter values and excavation stage, is different at different measurement depth. Table 6 presents the PSR values computed at stage 5 with Case 2 parameter values at different measurement depths.

Based on these observations, one can tell that PSR is a function of excavation depth, measurement depth and parameter values. Theoretically, a deterministic PSR should be used to correct the 2D predictions to reflect the three-dimensional effects, but the aforementioned observations indicate that the determination of PSR depends on parameter values that are unknown beforehand. In addition, this synthetic example is configured in a way such that modelling simplification arises solely from the plane strain assumption. In reality, there are many sources of modelling uncertainty. This makes it almost impossible to arrive at a deterministic error model. Therefore, a uniform probability density function is employed to describe the variation of modelling uncertainty. Every measurement depth within a given stage of excavation is associated with a modelling uncertainty. In this example, there are 5 stages of excavation and 40 measurement points. Therefore, there are 200 modelling uncertainties. These 200 modelling uncertainties are interpolated from the two sets of simulations listed in Table 3.

Table 2. Definition of searching space.

Parameter	Range
E' (MPa)	5–50
E'_{inc} (MPa/m)	0–1.5

5.2 *Initial model set*

The search space is discretized to facilitate the identification process. E' is sampled every 2.5 MPa, and E'_{inc} is sampled every 0.1 MPa/m. Based on grid based sampling technique, there are in total 304 cases, which take approximately 10 hours to evaluate.

5.3 *Identification results*

Figure 7 shows the candidate models identified after each round of identification. The true parameter values are also indicated in the figure. It is observed that a progressive falsification is achieved with the inclusion of additional measurement data after each stage of excavation. The number of candidate models in the candidate model set reduces after each round of identification. It is also observed that the true parameter values are always contained within the candidate model set across all rounds of identification.

The orientation of the candidate model set in Figure 7 indicates that E' and E'_{inc} are negatively

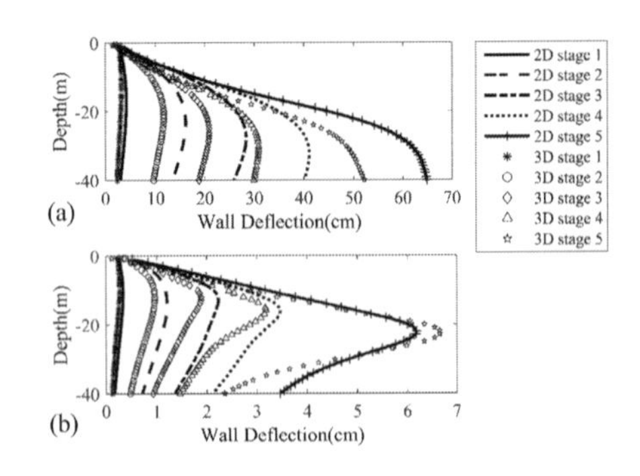

Figure 6. 2D and 3D computed wall deflections of (a) Case 1 (b) Case 2.

Table 3. Parameter values in 3D model for uncertainty quantification.

	E' (MPa)	E'_{inc} (MPa/m)
Case 1	5000	0
Case 2	50000	1500

Table 4. Comparison of PSR at 20 m depth and stage 5 between 2 cases.

	Depth (m)	2D (cm)	3D (cm)	PSR
Case 1	20	44.55	34.22	0.77
Case 2	20	5.41	5.93	1.10

Table 5. Comparison of PSR at 20 m depth in different stages.

	Depth (m)	2D (cm)	3D (cm)	PSR
Stage 1	20	0.26	0.19	0.74
Stage 2	20	1.00	0.75	0.75
Stage 3	20	1.80	1.41	0.78
Stage 4	20	2.86	2.55	0.90
Stage 5	20	5.41	5.93	1.10

Table 6. Comparison of PSR at different depth in stage 5.

	Depth (m)	2D (cm)	3D (cm)	PSR
Stage 5	1	0.53	0.14	0.26
Stage 5	10	2.77	2.95	0.94
Stage 5	20	5.41	5.93	1.10

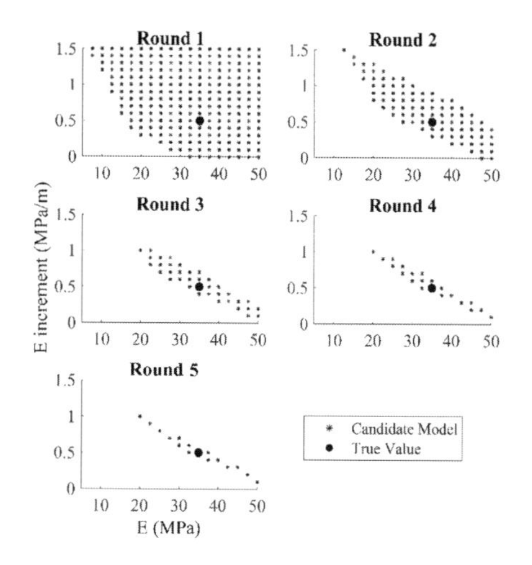

Figure 7. Identified candidate models after every round of identification.

correlated, which is consistent with what Bernal (2006) presented. Although identical or nearly identical results can be obtained with different combinations of E' and E'_{inc}, there is a lack of information to determine the exact dependence between these two parameter values. Although E' and E'_{inc} are numerically dependent, it is not necessary that they are physically dependent. Therefore, it may not be reasonable to relate these two parameters using the coefficient of correlation that is determined solely from numerical parametric studies. For this reason, EDMF, focuses on the identification of a set of feasible parameters. Unless additional information is obtained, all parameters in the candidate model set are considered as correct parameters.

Figure 8a and 8b show the predicted wall deflections of stages 1 and 5 respectively after the 1st and 5th round of falsification. It is observed that predictions are consistent because measurements are always within the bounds defined. As additional measurement data is included, more parameter values are falsified, leading to a narrower predicted range.

As predictions are made using all candidate models in the EDMF scheme, the results can be 'averaged'

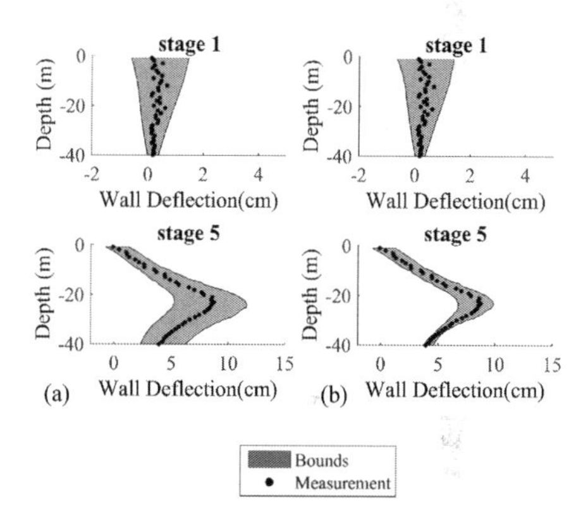

Figure 8. Bounds of predicted wall deflection of selected stage (a) after 1st round of falsification (b) 5th round of falsification.

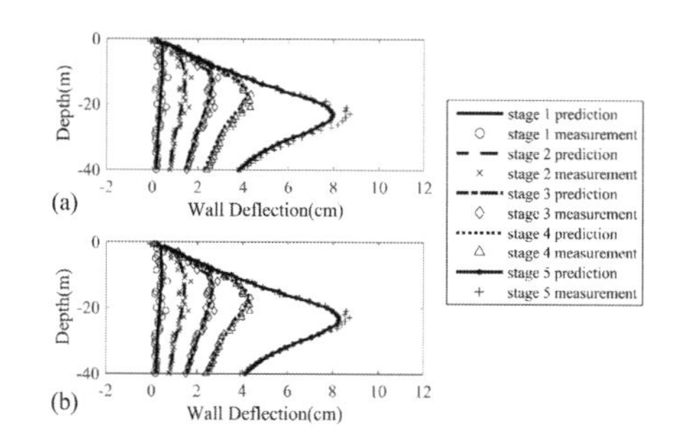

Figure 9. Mean EDMF predicted wall deflection to all 5 stages after (a) 1st round of identification (b) 5th round of identification.

to produce a mean prediction. Figure 9 presents the mean predictions for all 5 stages. It is seen that promising predictions are obtained at all 5 stages.

6 CONCLUSION

The importance of incorporating modelling uncertainty in the parameter identification process is clearly demonstrated. The EDMF accurately identifies the parameter values, which in turn leads to improved predictions. Great simplification is achieved as only two three-dimensional simulations are performed.

In conclusion, EDMF, which embraces the inclusion of modelling uncertainty, is a robust methodology for soil parameter identifications in excavations. In the real world practice, other sources of modelling uncertainties can be identified and included in the analysis.

REFERENCES

Brinkgreve, R.B.J., Engin, E. and Swolfs, W.M., 2013. PLAXIS 3D 2013 user manual. *Plaxis bv, Delft*.

Brinkgreve, R.B.J., Engin, E. and Swolfs, W.M., 2014. PLAXIS 2D AE manual. *AA Balkema*.

Cividini, A., Maier, G. and Nappi, A., 1983, October. Parameter estimation of a static geotechnical model using a Bayes' approach. In *International Journal of Rock Mechanics and Mining Sciences & Geomechanics Abstracts* (Vol. 20, No. 5, pp. 215–226). Pergamon.

COI, 2005. Report of the Committee of Inquiry into the incident at the MRT Circle Line worksite that led to collapse of Nicoll Highway on 20 April 2004. *Ministry of Manpower, Singapore*.

Jofré, G.A.C., 2013. *Methodology for updating numerical predictions of excavation performance* (Doctoral dissertation, Massachusetts Institute of Technology, Department of Civil and Environmental Engineering).

Einstein, H.H. and Baecher, G.B., 1982. Probabilistic and statistical methods in engineering geology I. Problem statement and introduction to solution. In *Ingenieurgeologie und Geomechanik als Grundlagen des Felsbaues/Engineering Geology and Geomechanics as Fundamentals of Rock Engineering* (pp. 47–61). Springer, Vienna.

Finno, R.J. and Calvello, M., 2005. Supported excavations: observational method and inverse modeling. *Journal of Geotechnical and Geoenvironmental Engineering*, *131*(7), pp.826–836.

Finno, R.J., Blackburn, J.T. and Roboski, J.F., 2007. Three-dimensional effects for supported excavations in clay. *Journal of Geotechnical and Geoenvironmental Engineering*, *133*(1), pp.30–36.

Goulet, J.A., Kripakaran, P. and Smith, I.F., 2010. Multimodel structural performance monitoring. *Journal of structural engineering*, *136*(10), pp.1309–1318.

Goulet, J.A., Coutu, S. and Smith, I.F., 2013. Model falsification diagnosis and sensor placement for leak detection in pressurized pipe networks. *Advanced Engineering Informatics*, *27*(2), pp.261–269.

Goulet, J.A. and Smith, I.F., 2013. Structural identification with systematic errors and unknown uncertainty dependencies. *Computers & structures*, *128*, pp.251–258.

Hashash, Y.M., Marulanda, C., Ghaboussi, J. and Jung, S., 2006. Novel approach to integration of numerical modeling and field observations for deep excavations. *Journal of Geotechnical and Geoenvironmental Engineering*, *132*(8), pp.1019–1031.

Hashash, Y., Song, H. and Osouli, A., 2011. Three-dimensional inverse analyses of a deep excavation in Chicago clays. *International Journal for Numerical and Analytical Methods in Geomechanics*, *35*(9), pp.1059–1075.

Hsein Juang, C., Luo, Z., Atamturktur, S. and Huang, H., 2012. Bayesian updating of soil parameters for braced excavations using field observations. *Journal of Geotechnical and Geoenvironmental Engineering*, *139*(3), pp.395–406.

Ledesma, A., Gens, A. and Alonso, E.E., 1996. Estimation of parameters in geotechnical backanalysis—I. Maximum likelihood approach. *Computers and Geotechnics*, *18*(1), pp.1–27.

Lee, F.H., Yong, K.Y., Quan, K.C. and Chee, K.T., 1998. Effect of corners in strutted excavations: Field monitoring and case histories. *Journal of Geotechnical and Geoenvironmental Engineering*, *124*(4), pp.339–349.

Levasseur, S., Malécot, Y., Boulon, M. and Flavigny, E., 2008. Soil parameter identification using a genetic algorithm. *International Journal for Numerical and Analytical Methods in Geomechanics*, *32*(2), pp.189–213.

Mikkelsen, P.E., 2003, September. Advances in inclinometer data analysis. In *Symposium on Field Measurements in Geomechanics, FMGM*.

Moser, G., Paal, S.G. and Smith, I.F., 2015. Performance comparison of reduced models for leak detection in water distribution networks. *Advanced Engineering Informatics*, *29*(3), pp.714–726.

Ou, C.Y., Chiou, D.C. and Wu, T.S., 1996. Three-dimensional finite element analysis of deep excavations. *Journal of Geotechnical Engineering*, *122*(5), pp.337–345.

Ou, C.Y. and Shiau, B.Y., 1998. Analysis of the corner effect on excavation behaviors. *Canadian geotechnical journal*, *35*(3), pp.532–540.

Qi, X.H. and Zhou, W.H., 2017. An efficient probabilistic back-analysis method for braced excavations using wall deflection data at multiple points. *Computers and Geotechnics*, *85*, pp.186–198.

Bernal, C.R., 2006. Inverse analysis of excavations in urban environments. ProQuest.

Rechea, C., Levasseur, S. and Finno, R., 2008. Inverse analysis techniques for parameter identification in simulation of excavation support systems. *Computers and Geotechnics*, *35*(3), pp.331–345.

Soman, R.K., Raphael, B. and Varghese, K., 2017. A System Identification Methodology to monitor construction activities using structural responses. *Automation in Construction*, *75*, pp.79–90.

Vernay, D.G., Raphael, B. and Smith, I.F., 2014. Augmenting simulations of airflow around buildings using field measurements. *Advanced Engineering Informatics*, *28*(4), pp.412–424.

Whittle, A.J. and Davies, R.V., 2006. Nicoll Highway Collapse: Evaluation of Geotechnical Factors Affecting Design of Excavation Support System. *Proceedings of the International Conference on Deep Excavations, Singapore, 16 pp*.

Numerical Methods in Geotechnical Engineering IX – Cardoso et al. (Eds)
© 2018 Taylor & Francis Group, London, ISBN 978-1-138-33203-4

Numerically derived P-Y curves for rigid walls under active conditions

I. El-Chiti, G. Saad, S.S. Najjar & S. Alzoer
Department of Civil and Environmental Engineering, American University of Beirut, Lebanon

ABSTRACT: This paper aims at investigating the relationship between lateral earth pressure and displacement (p-y relationship) for rigid basement walls that support sand backfill for active states of loading. PLAXIS 2D analyses are conducted to identify and characterize the components of the p-y relationship at different depths with particular emphasis on the effects of the interface friction angle between the wall and the soil and the sand properties. The soil hardening model that is based on the Duncan-Chang hyperbolic stress-strain relationship is utilized in the numerical analysis. The results of the study show that (1) active conditions are mobilized at different wall displacements along the height of the wall, (2) the displacement required for mobilizing active conditions at a given depth is a function of the depth of the point below the ground surface and the properties of the sand, and (3) the p-y relationship for a frictional wall could be determined from the p-y relationship of an equivalent frictionless wall and the reduction in vertical stress at the soil/wall interface due to friction.

1 INTRODUCTION AND BACKGROUND

For structures with underground basement walls, the soil-structure-interaction between the side soil and the walls affects the response of the system, particularly under cyclic loading conditions. In the context of performance-based design, there has been interest in quantifying the relationship between the lateral earth pressure and the wall displacement using the concept of p-y curves.

Briaud & Kim (1998) were the first to recommend p-y relationships for the analysis and design of tie-back walls. These p-y relationships were calibrated/back calculated using data collected from full scale tests on walls in sand. Briaud & Kim (1998) state that the lateral earth pressure that is exerted by the soil on the wall is bounded by the active and passive earth pressure conditions. Based on the data collected, they recommend that the active earth pressure could be assumed to be mobilized at wall movements of 1.3 mm (away from the retained soil). El Ganainy & El Naggar (2009) and Saad et al. (2016) adopted this p-y relationship as the backbone curve for the lateral pressure-lateral deflection relationship used for modeling the side soil in their analysis of the response of buildings with underground stories.

In reality, the relationship between lateral earth pressure and wall displacement is expected to be complex and is affected by the height of the wall, the relative density of the backfill material, the interface friction between the wall and the soils, the non-linearity of the soil response, and the type of wall movement (translation and/or rotation). Such complexities are not captured by traditional lateral earth pressure methods that are aimed exclusively at predicting the stresses behind the wall at displacements that are large enough to assume that full active conditions are mobilized at all depths behind the wall. Such traditional methods replace the soil by an earth pressure distribution that is evaluated as the product of the vertical effective stress and the active earth pressure coefficient K_a. Examples of these models include the infamous Rankine and Coulomb theories (Das 2015), in addition to upper and lower bound solutions (Chen 1975, Soubra & Macuh 2002, Lancellotta 2002, and Paik & Salgado 2003) that are based on limit analysis and soil arching principles.

More recently, several FEM-based methods (Loukidis & Salgado 2012 and Elchiti et al. 2017) investigated the relationship between lateral earth pressure and wall displacement while varying a small set of design parameters. The analysis conducted in Loukidis & Salgado (2012) was limited to gravity walls while the analysis presented in Elchiti et al. (2017) targeted rigid basement walls that were simplistically assumed to be fixed at their base (supported on rock).

There is a need for realistic and simplified models that could describe the p-y relationship for rigid retaining walls to be used as input in robust soil-structure-interaction problems in the context of performance based design. Such p-y relationships could be incorporated in commercial structural analysis software in the form of non-linear springs that describe the lateral pressure versus lateral displacement relationship. This is consistent with the p-y curves concept commonly used to model the reaction of the soil for laterally loaded piles.

The objective of this paper is to investigate the static soil-structure interaction between rigid walls and sand backfill using finite element analyses for active states of loading while enforcing a realistic modeling of the soil support at the base of the wall. The main goal is to identify and characterize the components of the p-y relationship at different depths with particular emphasis on the effects of the interface friction angle between the wall and the soil.

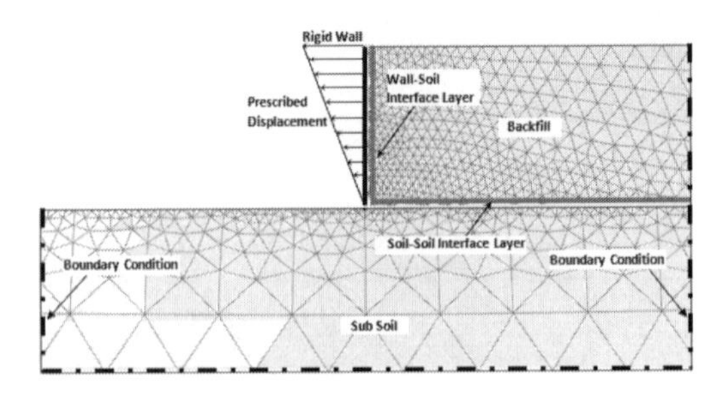

Figure 1. Finite element model adopted.

2 FINITE ELEMENT MODEL

2.1 *Mesh and boundary conditions*

The finite element model of the rigid retaining wall and the backfill material was built in Plaxis 2D and presented in Figure 1. The model consists of a rigid retaining wall subjected to a varying horizontal prescribed displacement, a semi-infinite half space backfill soil medium, a semi-infinite half space subsoil medium, and two interface soil layers acting on the boundaries of the backfill with the rigid wall and the subsoil layer. The rigid wall is modeled as a rectangular plate with an axial (EA) and flexural (EI) stiffness of 6.25×10^6 kN/m and 32.5 kN-m^2/m, respectively. The wall is moved incrementally away from the backfill soil in a rotating manner using a varying prescribed displacement. The prescribed displacement restrains the wall in the vertical direction, allows no bending in the wall to occur, and ensures a rigid rotation of the wall with a bottom movement of 20% that of the top.

The soil continuum, both subsoil and backfill, are meshed using 15-noded plain-strain triangular elements of varied sizes ranging from very fine triangular elements at locations of stress concentration to course triangular elements at boundary conditions. The length of the backfill soil together with the height and length of subsoil continuum are critically chosen to ensure a semi-infinite behavior of the soil continuum with no limitation imposed on the model by the boundary conditions. Thin interface layers are modeled at both backfill/wall and backfill/subsoil contact surfaces to allow for a reduced interface friction angle between both materials and a possible differential movement. The left and right boundary conditions of the soil continuum are restricted to move in the vertical direction while allowing free vertical movement of the soil. The bottom edge of the subsoil is restrained from movement both in the vertical and horizontal directions.

The analysis consisted of enforcing a lateral displacement field along the height of the wall and determining the associated evolution of the lateral earth pressures on the wall with lateral wall displacement. Elchiti et al. (2017) investigated the effect of several conditions of wall movement on the resulting p-y relationship. Several conditions of wall movement where adopted including pure wall rotation around a fixed lower boundary and a combination of translation and rotation, whereby the lower end of the wall was moved as a percentage (10% and 20%) of the upper wall movement. They observed that the resulting p-y relationships at different depths were found to be slightly sensitive to the assumed base wall movement condition. To cater for realistic inevitable displacements that would occur at the base of the wall in design problems involving rigid basement walls, the displacement field adopted in this study involved a combination of rotation around the base of the wall and a translation at the base that is taken as 20% of the top wall displacement.

2.2 *Constitutive soil model*

In this study, the Hardening soil model was adopted to model the constitutive response of the soil in the FE analysis. The Hardening model is an advanced non-linear model that assumes a hyperbolic relationship between stress and strain. The Hardening model adopts isotropic hardening and includes two yield surfaces to differentiate between shear and isotropic loading (Shanz et al. 1999). The final bounding surface is based on the Mohr-Coulomb failure envelope with a non-associated flow rule using the dilatancy angle as the non-normality parameter. With respect to its stiffness behavior, the model considers stress and strain level dependency of the moduli. A hyperbolic stress—strain relation is considered between the deviatoric stress and the major principal strain to account for strain dependency, while a power law is adopted to consider stress dependency, and the user specifies the moduli at a certain reference pressure. In this study, the parameters utilized in the Hardening soil model (Table 1) were adopted from Skeini (2015) and are consistent with the behavior of medium dense and dense Ottawa sand.

3 TEST RESULTS AND ANALYSIS

3.1 *P-Y response for cases of frictional wall*

For typical concrete basement walls, the interface friction coefficient between concrete and sand is expected to fall anywhere between 0.6 and 0.9 of the friction coefficient of the sand. In this paper, the interface friction coefficient for the cases involving wall friction is realistically assumed be equal to 0.8 of the friction coefficient of the sand.

The variations of the lateral earth pressure with wall displacement at depths ranging from 1.0 to 8.0 m, in a 10-m high wall are presented in Figures 2 (medium dense sand) and 3 (dense sand) for analyses involving a frictional soil-wall interface. Also plotted on the figures is the theoretical "active" earth pressure condition that corresponds to the Coulomb method for computing K_a for frictional walls. The curves on Figures 2 and 3 represent p-y curves in the sense that they represent relationships between the lateral earth pressure at a depth D versus the wall displacement at that particular depth D. At zero wall displacement, the lateral earth pressures correspond to the at-rest condition. As the wall is displaced laterally, the lateral earth pressures are reduced with displacement and the rate of reduction in lateral stress with displacement is observed to decrease systematically until fully active conditions are mobilized.

The results on Figures 2 and 3 lead to the following observations: (1) the initial reduction in the lateral pressure with displacement is steep and linear, with a slope that is independent of depth, (2) following the linear portion, the rate of reduction in lateral stress with displacement decreases gradually and is consistent with the expected reduction in the stiffness of the soil with displacement for the hyperbolic stress-strain relationship in the hardening model, (3) fully active stress conditions are attained at wall displacements that increase systematically with depth, and (4) the lateral stresses that are observed at larger wall deformations in the FEM results correlate well with the theoretical active lateral pressures that are predicted using the theoretical Coulomb Ka value.

More importantly, the p-y relationships presented in Figures 2 and 3 for frictional basement walls indicate that the adoption of a depth-independent fixed lateral displacement of 1.3 mm as a criterion for full mobilization of active earth pressure in published p-y curves (ex. Briaud & Kim 1998, El Ganainy & El Naggar 2009 and Saad et al. 2016) may not be completely representative of the true response. The 1.3 mm displacement criterion is presented graphically on Figures 2 and 3. While this criterion may be more-or-less applicable in

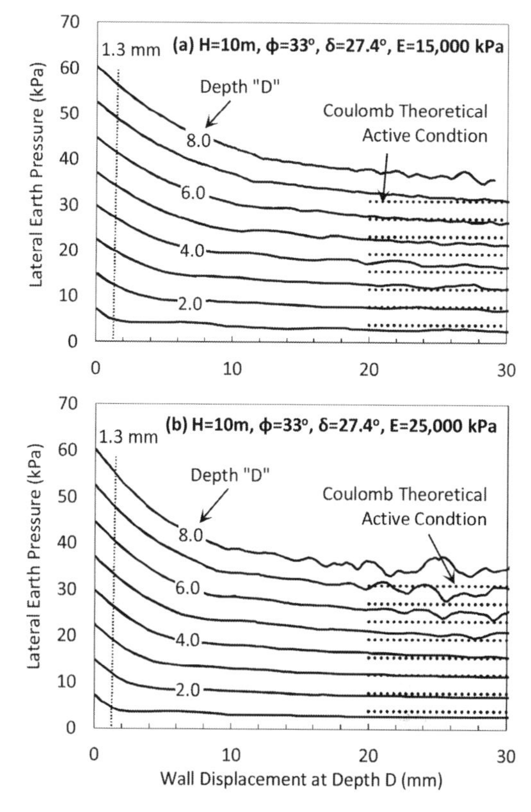

Figure 2. P-Y relationship for frictional wall cases with medium dense sands.

Table 1. Soil properties in constitutive model.

Parameter	Hardening Soil Model
Soil Unit Weight, $\gamma (kN/m^3)$	16.5, 18.0
Initial Void Ratio, e_{int}	0.615
Secant stiffness, E_{50}^{ref} (kN/m^2)	15000–50000
Tangent stiffness, E_{oed}^{ref} (kN/m^2)	15000–50000
Unload/reload stiffness, $E_{ur}^{ref} (kN/m^2)$	200000
Power for stress dependency of E, m	0.4
Cohesion, c'_{ref} (kN/m²)	1.0
Friction Angle, φ' (°)	33, 36
Dilation Angle, ψ(°)	4, 7
Poisson's Ratio, v'	0.2
Reference stress, p_{ref} (kN/m^2)	100
Failure ratio, R_f	0.95

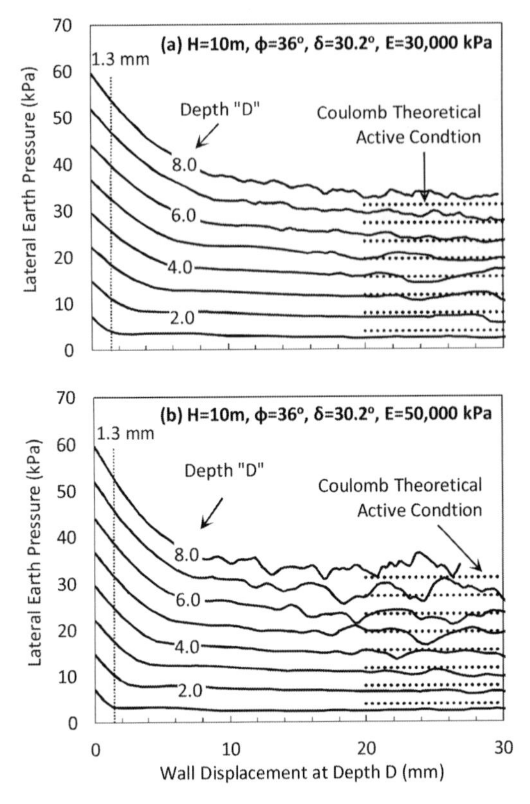

Figure 3. P-Y relationship for frictional wall cases with dense sands.

predicting the displacement required to approach active conditions at very shallow depths (D = 1 m), the p-y curves at larger depths indicate that displacements as high as 10 mm may be required for depths as high as 8 m for cases involving medium dense sands.

The effect of density on the p-y relationship for identical wall cases is clearly illustrated when comparing the results on Figure 2 (medium dense) to the results on Figure 3 (dense sand). The p-y curves for dense sands are observed to be initially steeper (faster reduction in lateral stress with displacement) than those of medium dense sands. In addition, the displacement required to approach active stresses in dense sands are smaller (1.5 to 8 mm) than medium dense sands (2 to 15 mm). It is interesting to note that at any given density, the p-y curves are shown to be also sensitive to the assumed modulus of elasticity as reflected in Figures 2a (E = 15,000 kPa) and 2b (E = 25,000 kPa) and Figures 3a (E = 30,000 kPa) and 3b (E = 50,000 kPa). As the modulus of elasticity increases, the steepness of the initial portion of the p-y curve increases and the displacement required to approach active conditions decreases.

3.2 *P-Y response for cases of frictionless wall*

Finite element analyses were also conducted for cases involving zero friction between the wall and the soil. The resulting p-y relationships at depths ranging from 1.0 to 8.0 m in a 10-m high wall are presented in Figures 4 (medium dense) and 5 (dense), respectively. Also plotted are the theoretical "active" earth pressure conditions that correspond to the Coulomb method for computing K_a for frictionless walls.

The p-y curves on Figures 4 and 5 exhibit a response that is clearly different than that observed for the cases involving soil/wall friction. The p-y curves in the frictionless cases are smoother, with an initial response that is less steep and a reduction in stiffness that is gradual starting from the at-rest condition all the way to the fully active condition. The displacements required to approach fully active conditions are clearly larger than the displacements observed in the frictional wall cases.

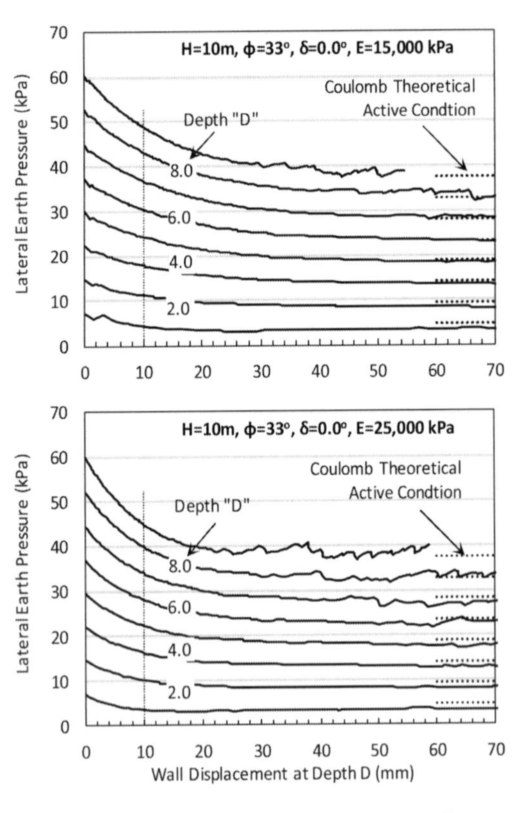

Figure 4. P-Y relationship for frictionless wall cases with medium dense sands.

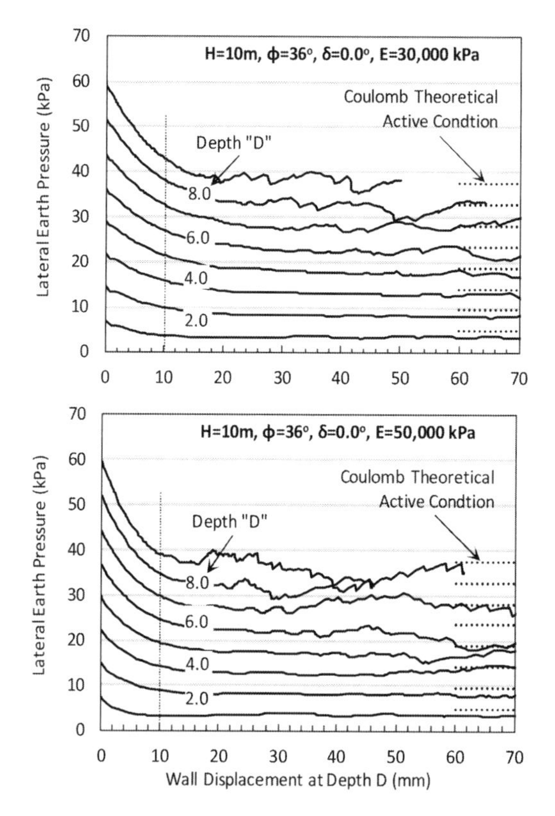

Figure 5. P-Y relationship for frictionless wall cases with dense sands.

The displacements required to approach active conditions are clearly greater than 10 mm in all the frictionless wall cases that are analyzed in this study, particularly for the medium dense cases where larger displacements are needed.

3.3 *Deriving the P-Y curves of frictional wall from P-Y curves of frictionless wall*

Results in Sections 3.1 and 3.2 indicate a significantly different p-y response for frictional and frictionless wall conditions, respectively. In this section, an effort is made to investigate the interrelationship and interdependency between the two p-y relationships. This is achieved by analyzing the development/mobilization of the frictional stresses at the wall-soil interface with wall displacement at any depth D and linking it to the changes that occur to the vertical stresses in the elements within the soil/wall interface as the wall is displaced horizontally. It is hypothesized that the difference between the two p-y relationships for frictionless and frictional walls could be explained by the changes in the vertical stresses due to friction along the soil/wall interface.

To test the above hypothesis, the p-y relationships for identical frictional and frictionless wall conditions are plotted together on Figures 6 and 7 for medium dense sands and dense sands, respectively. Also plotted on the figures are the variations observed in the vertical stresses and the frictional stresses at the interface between the soil and the wall at the same depths. For illustration, results are only presented for cases involving representative depths of 3.0 m and 6.0 m, respectively.

A comparison between the p-y curves for the frictional and frictionless wall cases confirms the observations made in Section 3.2 regarding the clear differences in the p-y responses. The p-y response of the frictional wall is initially steeper and shows relatively quicker mobilization of active conditions compared to the frictionless case. The p-y curves for the frictional wall cases also show a clear abrupt change in the p-y response at relatively small displacements as indicated in Figures 6 and

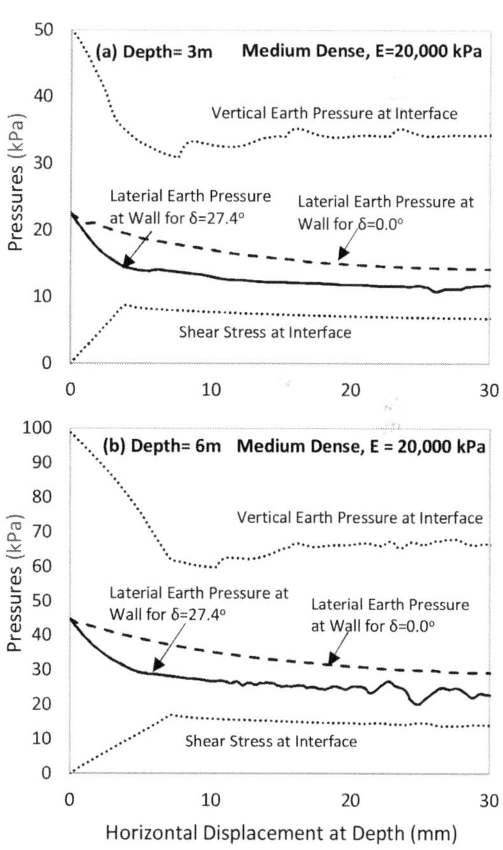

Figure 6. Correlation between lateral earth pressure of frictional and frictionless wall and the vertical and shear stress at the interface for the case of medium dense sand.

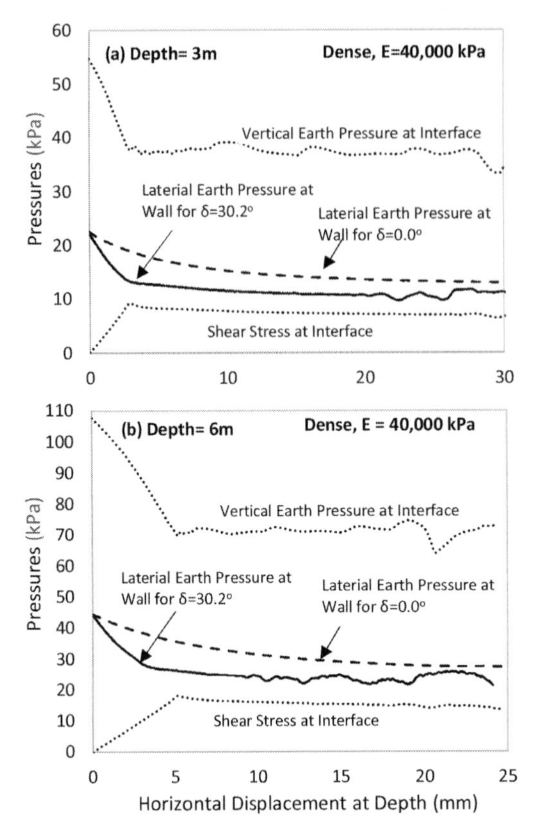

Figure 7. Correlation between lateral earth pressure of frictional and frictionless wall and the vertical and shear stress at the interface for the case of dense sand.

7 (~ 5 to 7 mm for medium dense sands and ~ 3 to 5 mm for dense sands).

On the other hand, the variations in the vertical stresses and the wall friction stresses with displacement show that the frictional stresses increase with displacement resulting in a corresponding decrease in the vertical stresses. At a given wall displacement that is unique for any given density and depth, the frictional stresses reach their maximum values and the vertical stresses reach their minimum values. Not surprisingly, the displacements at which the maximum friction stresses and the minimum vertical stresses are reached correspond clearly with the discontinuity that is observed in the slope of the p-y curve in the frictional wall cases. This indicates that the difference between the p-y relationships in the frictional and frictionless wall cases is directly related to the variation of the frictional stresses with displacement.

To determine the relationship between the lateral earth pressure for a frictional wall and that of the lateral earth pressure for a frictionless wall, an analysis of the free body diagram of the failure wedge behind a rigid wall at active conditions for granular soil was studied as indicated in Figure 8a. Since the failure wedge at the limit state is in static equilibrium, the forces acting on the wedge constitute a closed force triangle as shown in Figure 8b (Coulomb theory). From the law of sines, we have:

$$P'_a = W' \frac{\sin(\theta'_{cr} - \phi)}{\sin\left(90 - (\theta'_{cr} - \phi) + \delta\right)} \tag{1}$$

All the symbols in Equation 1 are clearly defined in Figure 8. The "prime" after each symbol is used to designate cases with a frictional interface. For the cases involving frictionless walls (δ = 0), Eq. (1) becomes:

$$P_a = W \frac{\sin(\theta - \phi)}{\sin\left(90 - (\theta_{cr} - \phi)\right)} \tag{2}$$

Using the geometry of the wedge and some trigonometric relationships, the ratio P'_a / P_a can be written as:

$$\frac{Pd}{Pa} = g.\left[\frac{1}{\cos(\delta)\left[1 + \dfrac{\tan(\delta)}{\tan(90 - (\theta'_{cr} - \phi))}\right]}\right] \tag{3}$$

where g is given by:

$$g = \left[\frac{\tan(\theta'_{cr})}{\tan(\theta_{cr})}\right] \cdot \left[\frac{\sin(\theta'_{cr} - \phi)}{\sin(\theta_{cr} - \phi)}\right] \cdot \left[\frac{\sin(90 - (\theta_{cr} - \phi))}{\sin(90 - (\theta'_{cr} - \phi))}\right] \tag{4}$$

For friction angles ranging from 33 to 36 degrees, the variable g in Eq. (4) can be approximated as 1.0. For granular soil having a soil friction angle of $\phi = 36°$ retained by a rough concrete wall having an interface friction angle $\delta = 30.2°$, the corresponding θ_{cr} and θ'_{cr} are equal to 63.0 and 58.75 degrees, respectively. Therefore, Eq. (3) can now be approximated as

$$\frac{Pd}{Pa} = \left[\frac{1}{\cos(\delta)\left[1 + \dfrac{\tan(\delta)}{\tan(90 - (\theta'_{cr} - \phi))}\right]}\right] \tag{5}$$

Using the force triangle presented in Figure 8b, both tan(δ) and $\tan\left[90 - (\theta'_{cr} - \phi)\right]$ are given as Pd_v / Pd_h and $(W' - Pd_v) / Pd_h$, respectively.

Replacing both parameters in Eq. (5), the above equation can be rewritten as:

$$P'_{ah} = P_a - P_a.\frac{Pa'_v}{W'} \tag{6}$$

By expressing the force terms in Eq. (6) such that $P'_{ah} = K'_{ah}.\left(\frac{\gamma h^2}{2}\right)$, $P_a = k_a.\left(\frac{\gamma h^2}{2}\right)$, $Pa'_v = \tau\left(\frac{h}{2}\right)$, and $W' = \left(\frac{\gamma h^2}{2}\right).\frac{1}{\tan(\theta'_{cr})}$ and substituting back in equation 6, one gets the following relationship:

$$K'_{ah}.\gamma h = k_a.\gamma h - k_a \tau_a \tan(\theta'_{cr}) \tag{7}$$

where

K'_{ah} = active coefficient of earth pressure $\delta \neq 0$.

k_a = active coefficient of earth pressure $\delta = 0$.

τ_a = interface shear at active condition at depth h to be given by FE model.

θ_{cr} = the angle which the failure plane makes with the horizontal axis for $\delta = 0$.

Equation 7 expresses the active lateral earth pressure behind a wall with $\delta \neq 0$ as the active lateral earth pressure behind a wall with $\delta = 0$ minus the term $k_a.\tau_a.tan(\theta'_{cr})$, whereby τ_a represents frictional stresses along the soil-wall interface.

In this paper, Eq. (7) will be used to predict the lateral earth pressure in the frictional wall as a function of the lateral earth pressure in the frictionless wall and the mobilized friction at the interface in the frictional wall. This relationship will be assumed to be applicable at all levels of wall displacement despite the fact that it was derived for active conditions. The variation of the frictional stresses with displacement will be obtained from the results of the finite element analyses and used as input in Equation 7 to predict the lateral stresses for the frictional wall case.

To test the applicability of Equation 7, two cases involving medium dense and dense sands are considered as indicated in Figures 9 and 10, respectively. For both cases, the p-y curves that resulted from the finite element analysis of the cases involving friction between the wall and the soil are plotted against p-y curves that were predicted from the p-y curves of the identical frictionless wall cases while incorporating the effects of friction at the interface as per the relationship derived in Equation 7.

A comparison between the predicted (Equation 7) and numerically derived (FE) p-y curves on Figures 9 and 10 indicates an excellent match between the curves at all levels of wall displacement and for almost all depths considered. The only exception is the p-y relationships predicted for depth of 7.0 m and 8.0 m where some discrepancy is observed in the predicted versus numerically derived curve, particularly at larger wall

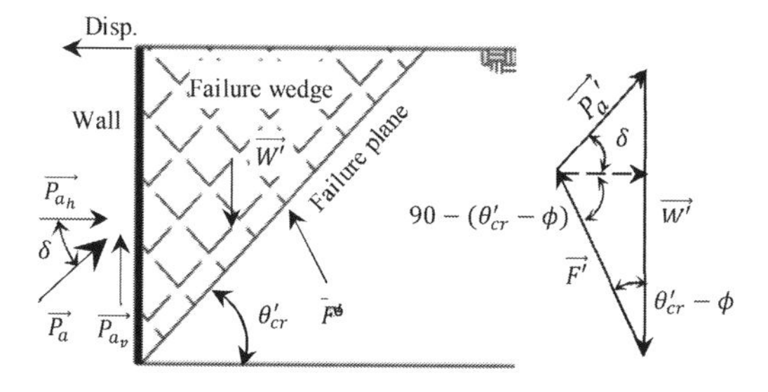

Figure 8. Coulomb's active pressure (a) failure wedge (b) force triangle.

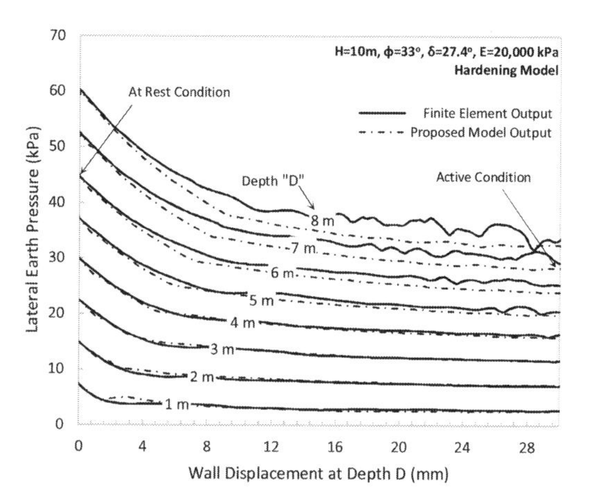

Figure 9. Comparison between predicted (Equation 7) and numerical p-y relationship for a 10-m wall in medium dense sands.

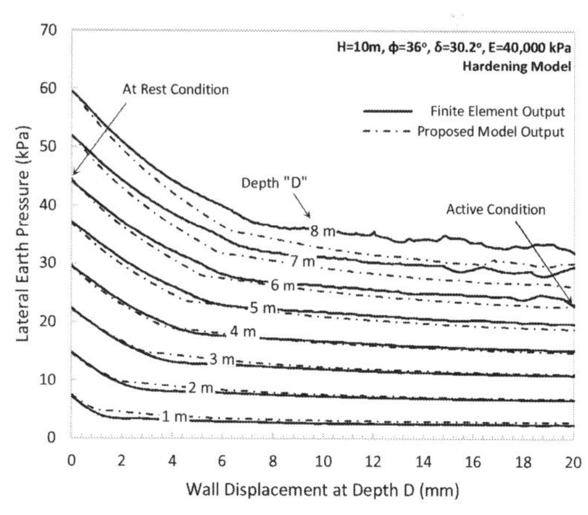

Figure 10. Comparison between predicted (Equation 7) and numerical p-y relationship for a 10-m wall in dense sands.

displacements. For all practical purposes, this discrepancy is considered to be relatively small and insignificant. It could be concluded with certainty that the relationship that is modeled mathematically in Equation 7 is capable of producing realistic and representative model predictions for the p-y relationships of frictional rigid walls supporting cohesionless backfill. The model requires as input the p-y curves of an equivalent frictionless wall and the variation of the frictional stresses with wall displacement. These input relationships were obtained in this paper from the numerical finite element analyses. Analytical models could be derived in future studies to represent the p-y relationships of frictionless walls and the friction-displacement relationship needed as input to Equation 7.

3.4 *Conclusions*

In this paper, finite element analyses were conducted to model the mobilization of active earth pressures behind rigid frictional and frictionless basement walls. The intent is to investigate the possibility of generating p-y relationships that could be used in modeling soil response analogous to those used in pile analysis and design. Based on the results, the following conclusions can be made:

1. The use of the Hardening soil model resulted in p-y relationships that are realistic and consistent.
2. A significantly different p-y response was observed for frictional and frictionless wall conditions. The p-y response of the frictional wall is initially steeper and shows relatively quicker mobilization of active conditions compared to the frictionless case. The p-y curves for the frictional wall cases show a clear abrupt change in the p-y response at relatively small displacements in the order of 5 to 7 mm for medium dense sands and 3 to 5 mm for dense sands.
3. The difference between the two p-y relationships for frictionless and frictional walls was traced back to the changes in the vertical stresses due to friction along the soil/wall interface. The displacements at which the maximum friction stresses and the minimum vertical stresses are reached correspond clearly with the discontinuity that is observed in the slope of the p-y curve in the frictional wall cases.
4. The Coulomb active wedge theory was used to predict the lateral earth pressure in the frictional wall at any wall displacement as a function of the lateral earth pressure in the frictionless wall and the mobilized friction at the interface in the frictional wall at the same displacement. The relationship was proven to be capable of producing realistic and representative model predictions for the p-y relationships of frictional rigid walls supporting cohesionless backfill. The model requires as input the p-y curves of an equivalent frictionless wall and the variation of the frictional stresses with wall displacement.

REFERENCES

Briaud, J. & Kim, N. 1998. Beam-column method for tieback walls. *J. of Geotechnical and Geoenvironmental Engineering* 124(1): 67–79.

Chen, W.F. 1975. Limit analysis and soil plasticity. *Elsevier Science Publishers.*

Das, B.M. 2011. Principles of foundation engineering. *Cengage Learning, Stamford,* CT.

Elchiti, I., Saad, G., Najjar, S.S. & Nasreddine, N. 2017. Investigation of active soil pressures on retaining walls using finite element analysis. *Proceedings of the GeoFrontiers Conference,* March. 12–15, Orlando, Florida, USA.

El Ganainy, H. & El Naggar, M. 2009. Seismic performance of three-dimensional frame structures with underground stories. *Soil Dynamics and Earthquake Engineering* 29(9): 1249–1261.

Lancellota, R. 2002. Analytical solution of passive earth pressure. *Geotechnique* 52(8): 617–619.

Loukidis, D. & Salgado, R. 2012. Active pressure on gravity walls supporting purely frictional soils. *Canadian Geotechnical Journal* 49: 78–97.

Piak, K. H. & Salgado, R. 2003. Estimation of active earth pressure against rigid retaining walls considering arching effects. *Geotechnique* 53(7): 643–653.

Saad, G., Najjar, S.S. & Saddik, F. 2016. Seismic performance of reinforced concrete shear wall buildings with underground stories. *Earthquakes and Structures* 10(4): 965–988.

Schanz, T., Vermeer, P.A. & Bonnier P. G. 1999. The hardening soil model: Formulation and verification. Beyond 2000 in computational geotechnics: 10 years of PLAXIS International. *Proceedings of the International Symposium beyond 2000 in Computational Geotechnics,* Amsterdam, the Netherlands, 18–20 March, Balkema, Rotterdam.

Soubra, H. & Macuh, B. 2002. Active and passive earth pressure coefficients by a kinematical approach. *Proceedings of the Institution of Civil Engineers - Geotechnical Engineering* 155(2): 119–131.

Numerical Methods in Geotechnical Engineering IX – Cardoso et al. (Eds)
© 2018 Taylor & Francis Group, London, ISBN 978-1-138-33203-4

Two dimensional upper and lower-bound numerical analysis of the basal stability of deep excavations in clay

T. Santana, M. Vicente da Silva, A.N. Antão & N.C. Guerra
UNIC, Departamento de Engenharia Civil, Universidade NOVA de Lisboa, Portugal

ABSTRACT: The basal stability of deep excavations in clay is often analysed using the classical approaches of Terzaghi (1943) and Bjerrum & Eide (1956). Terzaghi's approach seems to be appropriate to excavation with small depth to width ratios, h/B, whereas Bjerrum & Eide give better results for greater h/B ratios. Not many attempts have been made to unify the two situations. In the present paper, the problem is analysed using upper and lower-bound considering two two-dimensional cases: a basic case, where the rigid stratum was considered deep, not affecting the stability of the excavation, and a case considering a rigid stratum at a certain depth. The embedded length of the wall was assumed null. A proposal for the stability numbers at failure is presented, for a wide range of h/B ratios.

1 INTRODUCTION

The problem of the basal stability of deep excavations in clayey soils under undrained conditions has been addressed by several authors. For a case such as the one shown in Figure 1a, considering an excavation of depth h and width B performed under undrained conditions in a homogeneous clayey soil with unit weight γ and undrained shear strength c_u, basal instability can occur if the total stresses imposed by the supported soil, above the base of the excavation, cause the failure of the soil below, due to insufficient strength. The classical approaches to this problem are those from Terzaghi (1943) and from Bjerrum & Eide (1956). These methods are based on limit equilibrium principles and are developed from a basic case of no wall embedded length and considering a deep stratum, which means that (Figure 1a) d is large and does not affect the mechanism.

Collapse occurs, according to Terzaghi's method, when the stability number $\gamma h/c_u$ is:

$$\frac{\gamma h}{c_u} = 5.7 + \sqrt{2}\frac{h}{B} \tag{1}$$

using Terzaghi's method, and:

$$\frac{\gamma h}{c_u} = (2+\pi)\left(1 + 0.34\arctan\frac{h}{B}\right) \tag{2}$$

using Bjerrum and Eide's method, which is based on the depth correction for shallow foundations proposed by Skempton (1951).

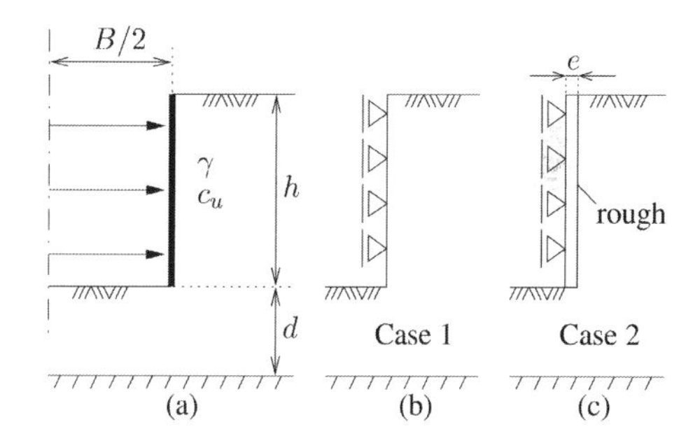

Figure 1. Geometry of the problem (a); cases analysed: Case 1 (b) and Case 2 (c).

In both methods the failure mechanism is affected by the rigid stratum when the depth d (Figure 1a) is:

$$d < \frac{\sqrt{2}}{2}B \tag{3}$$

and in this case equations (1) and (2) remain valid by replacing B by $\sqrt{2}d$:

$$\frac{\gamma h}{c_u} = 5.7 + \frac{h}{d} \tag{4}$$

in Terzaghi's method and:

$$\frac{\gamma h}{c_u} = (2+\pi)\left(1 + 0.34\arctan\frac{h}{\sqrt{2}d}\right) \tag{5}$$

in Bjerrum and Eide's method.

These methods have also been extended to the cases of wall embedded length, soil-to-wall adhesion, soil anisotropy, etc, but these cases are beyond the scope of the present paper.

Numerical limit analysis has been applied by Ukritchon et al. (2003) to this problem, using both upper (UB) and lower-bound (LB) methods. Santos Josefino et al. (2010) have also applied to the same problem the upper-bound method using a numerical implementation and results have significantly improved the previous upper-bound ones.

In the present paper, the calculations presented by Santos Josefino et al. (2010) are extended to other geometries and are complemented by lower-bound numerical calculations using the finite element program *mechpy*, developed within the authors' research team.

2 THE FINITE ELEMENT PROGRAM

The finite element limit analysis code is implemented in *mechpy*, a platform written in Python language for the development of non-conventional finite element formulations (Deusdado 2017). The limit analysis module of this software is an evolution of the *sublim3D* program, also developed within the authors' research team (Vicente da Silva and Antão 2008, Vicente da Silva 2009, Antão et al. 2012). As is well-known, the determination of plastic collapse loads can be done based on the Limit Analysis theorems. From a numerical viewpoint, this approach leads to a non-linear optimization problem. The optimization technique used in *mechpy* is the Alternating Direction Method of Multipliers (ADMM). Its iterative solution scheme is based on an operator splitting algorithm, which is suitable to efficiently solve large-scale variational problems with parallel processing. It has been applied successfully to several geotechnical problems, including earth pressures and deep excavations (Antão et al. 2008, Antão et al. 2011, Deusdado et al. 2015). This software is capable of producing either lower or upper bounds depending on the type of the finite element used. In this work, 6-noded triangles (with the usual 3 corner nodes and 3 midside nodes) are used to obtain strict upper bounds. In these elements, the velocity field approximation is assumed to be quadratic and the approximation of both the strain rate and stress fields to be linear. Whereas to obtain strict lower bounds, a more peculiar 7-noded triangular element is used (with 2 nodes associated to each edge of the element, plus an additional node located at the baricenter). In this element the strain rate and stress fields approximations are assumed to be linear. Conversely, the velocity field is incoherent and discontinuous. It is worth mentioning that in the current formulation the velocity, strain rate and stresses fields are independently approximated.

3 NUMERICAL ANALYSIS

The geometry of the problem analysed is presented in Figure 1. Two cases were considered: Case 1, considering a smooth interface and not modelling the wall itself, and Case 2, which considers the wall and a rough soil-to-wall interface.

A set of calculations was initially performed considering Case 1 and Case 2 assuming depth d to be large enough to have no influence on the mechanism and on the results. Different h/B ratios were assumed (between 0.2 and 3). For Case 2 several e/h ratios were considered (in the range 0.005 to 0.1).

Another set of calculations was performed, using Case 1, considering different d/h ratios (from 1/3 to 3).

4 RESULTS OBTAINED FROM CASE 1 FOR DEEP RIGID STRATUM

Results obtained from Case 1 for deep rigid stratum are included in Table 1 (column 2) and are represented in Figure 2. This figure also represents other available solutions, such as those from Terzaghi (1943) and from Bjerrum & Eide (1956) and the numerical limit analysis results from Ukritchon et al. (2003).

Both the table and the figure allow the following remarks: i) the values obtained for the stability numbers are between Terzaghi's and Bjerrum and Eide's solutions; ii) the results obtained are slightly closer to Terzaghi's solution for small values of h/B and to Bjerrum and Eide's for greater values of h/B; iii) upper and lower-bound solutions significantly improve the ones from Ukritchon et al. (2003); iv) upper and lower-bound results are quite close, with

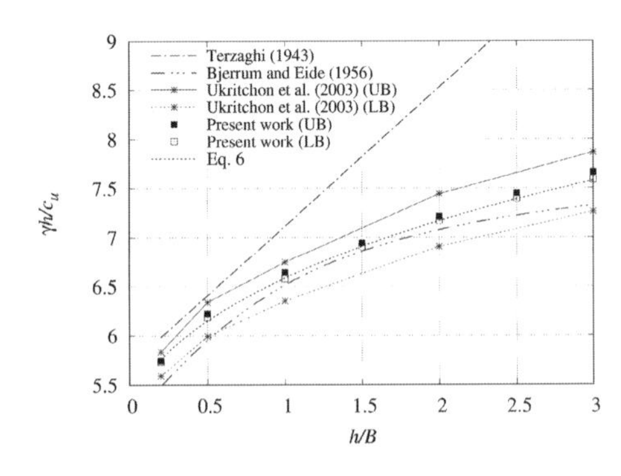

Figure 2. Results of the stability numbers obtained from Case 1. Comparison with other solutions.

Table 1. Results of the stability number $\gamma h/c_u$ obtained from the numerical upper and lower-bound calculations for Cases 1 and 2.

	Case 1	Case 2				
h/B		e/h				
		0*	0.005	0.01	0.02	0.1
Lower-bound (LB)						
0.2	5.733	5.739	5.736	5.739	5.744	5.744
0.5	6.186	6.196	6.199	6.213	6.240	6.372
1	6.579	6.580	6.597	6.621	6.671	7.005
1.5	6.908	6.911	6.936	6.968	7.023	7.462
2	7.168	7.173	7.201	7.240	7.304	7.815
2.5	7.393	7.399	7.430	7.474	7.545	8.118
3	7.589	7.596	7.630	7.678	7.756	8.382
Upper-bound (UB)						
0.2	5.745	5.740	5.748	5.750	5.756	5.842
0.5	6.222	6.232	6.236	6.248	6.272	6.402
1	6.642	6.645	6.666	6.688	6.735	7.076
1.5	6.944	6.949	6.974	7.004	7.066	7.506
2	7.214	7.222	7.250	7.286	7.360	7.874
2.5	7.449	7.458	7.489	7.530	7.613	8.190
3	7.659	7.667	7.698	7.747	7.839	8.470

*obtained by linear regression (Figure 5)

the maximum error margin being around 1%. This error was evaluated dividing the difference between the upper and the lower-bound stability numbers by the lower-bound ones.

Results from Case 1 should, therefore, be almost exact. The following proposed equation fit the results:

$$\frac{\gamma h}{c_u} = (2+\pi)\left\{1 + 0.815\arctan\left[0.359\left(\frac{h}{B}\right)^{0.55}\right]\right\} \quad (6)$$

This equation was obtained using the lower-bound values, for an estimate on a (slightly) conservative side. This equation is also shown in Figure 2 and, as it can be seen, it represents an accurate approximation to the obtained numerical results.

5 RESULTS OBTAINED FROM CASE 2 FOR DEEP RIGID STRATUM. INFLUENCE OF THE WALL THICKNESS

Results obtained from Case 2 for deep rigid stratum, for different values of the wall thickness ratio e/h, are shown in Table 1 (columns 4 to 7). The evolution of collapse values of the stability number with h/B for two e/h ratios are represented in Figure 3 (only two ratios were represented in

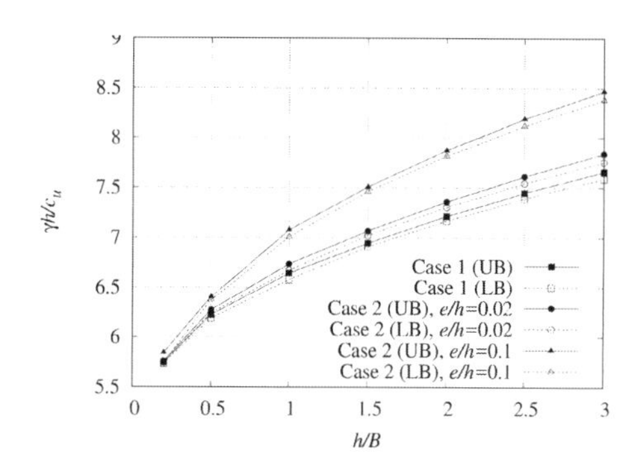

Figure 3. Results of the stability numbers obtained from Case 1 and from Case 2 as a function of h/B for two e/h ratios.

the figure to keep it legible). It can be seen from the table and the figure that a greater thickness of the wall leads to greater values of the stability number. This is expected, as agreater wall thickness means that a greater base of the wall is in contact with the soil below, which results in a more restricted problem and therefore in greater collapse loads.

Figure 4 shows the yield condition zones (for LB calculations) and plastic deformation zones (for UB calculations) for the different e/h ratios analysed for Case 2 and also for Case 1, for h/B = 2. From these figures mechanisms corresponding to the collapse situations can be inferred. It can be seen that greater wall thickness involve larger volumes of soil in the mechanisms and for the lesser thickness the mechanisms obtained are practically the same and quite close to the mechanism for Case 1. The mechanisms also show that the wall has no other effect besides the one at the base, as the soil behind it always moves as a rigid block.

Similar mechanisms between Case 1 and Case 2 for the lesser e/h ratio results in similar values of the stability number. This can be seen in Table 1, comparing values in columns 2 and 4. Also, the results from this table are represented in Figure 5 as a function of e/h, for different h/B ratios and a linear relation is found between e/h and $\gamma h/c_u$. For each one of the h/B ratio (and for both LB and UB results) a linear regression was performed and the obtained straight lines are represented in the figure. Results from Case 1 are also plotted and it can be seen that they match the y-intercept of the linear regression lines. This can also be seen in Table 1, comparing columns 2 and 3.

Case 1, therefore, represents a limit situation for Case 2 when e/h tends to zero. For this limit situation the most conservative results are obtained and in the next calculations only this case will be considered.

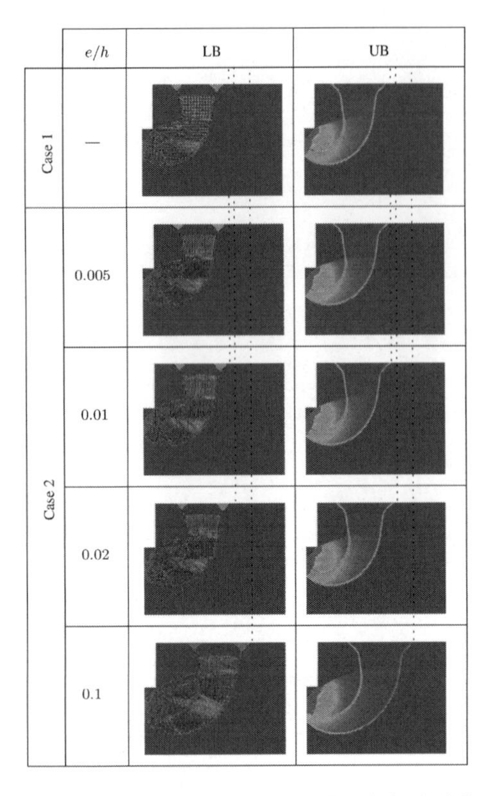

Figure 4. Yield condition zones (LB) and plastic deformation zones (UP) for $h/B = 2$ obtained for Case 1 and for Case 2, for different e/h ratios.

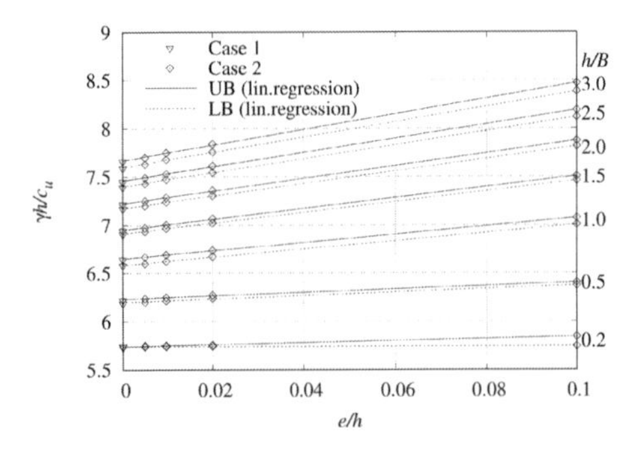

Figure 5. Results of the stability numbers obtained from Case 2 as a function of e/h for different h/B ratios. Lines are the result of linear regression and results from Case 1 are shown for $e/h = 0$.

6 INFLUENCE OF THE DEPTH OF THE RIGID STRATUM

Considering Case 1, a series of numerical LB and UB calculations were performed, considering different d/h ratios. Results are presented in Table 2 and in Figure 6. The last column in Table 2 shows the results for deepstratum (not affecting the results) and are the same shown in Table 1 for Case 1.

It can be seen, again, that results from LB calculations are quite close to the ones from UB calculations. The finite element meshes do not always have the same refinement; larger problems, involving greater soil volumes, were modelled using slightly larger finite elements. This could be the reason for some apparent very small inconsistencies on the results presented in table 2, usually on the third decimal place. It can also be seen that for a given d/h ratio values of the stability number are constant with h/B up to a certain h/B ratio. Although with significant differences in the values, the same type of results are obtained from Terzaghi's and Bjerrum and Eide's methods. This is caused by the fact that for larger excavations (lesser h/B ratios), mechanisms are controlled by the d/h ratio and are therefore independent of the excavation width. For narrower excavations their width becomes relevant. This is also true for Terzaghi's and Bjerrum and Eide's solutions, as also represented in Figure 6 and can be observed in the mechanisms inferred from the plastic deformation zones obtained from the UB calculations represented in Figure 7. In this Figure only the UB results are shown, but similar results from representing the yield condition zones are obtained from LB calculations.

Results obtained from the numerical calculations, as can be seen in Figure 6, are therefore represented by a horizontal line for a given d/h ratio, but this line merges smoothly in the curve

Table 2. Results of the stability numbers $\gamma h/c_u$ obtained from the numerical upper and lower-bound calculations for Case 1 for different d/h ratios.

h/B	d/h				
	1/3	1/2	1	3	deep
Lower-bound (LB)					
0.2	7.679	7.153	6.512	5.886	5.733
0.5	7.679	7.153	6.512	6.196	6.186
1	7.679	7.153	6.616	6.595	6.579
1.5	7.679	7.153	6.908	6.908	6.908
2	7.678	7.262	7.168	7.168	7.168
2.5	7.700	7.431	7.393	7.393	7.393
3	7.785	7.605	7.599	7.627	7.589
Upper-bound (UB)					
0.2	7.760	7.205	6.537	5.894	5.745
0.5	7.760	7.205	6.537	6.210	6.222
1	7.760	7.205	6.643	6.620	6.642
1.5	7.760	7.201	6.944	6.944	6.944
2	7.751	7.317	7.214	7.214	7.214
2.5	7.777	7.493	7.449	7.467	7.449
3	7.870	7.676	7.658	7.681	7.659

for deep stratum represented in Figure 2. This smooth transition represents a situation where both the excavation width and the depth of the rigid stratum are affecting the results. If this smooth transition is not considered, the problem can be described by equation 6 for deep stratum and by the different values of the stability number that depend on d/h. Values of the stability numbers corresponding to different d/h ratios are represented in Figure 8. For this figure, additional calculations were performed for d/h equal to 1.5, 2, 2.5 and 3 using a small value of the h/B ratio of 0.1, corresponding to a very large excavation, to ensure that the excavation width does not affect the results.

Results represented in Figure 8 can be obtained with the following proposed equation (obtained using the lower-bound results):

$$\frac{\gamma h}{c_u} = 1 + \pi + \exp\left[-0.409 + 1.275\left(\frac{d}{h}\right)^{-0.246}\right] \qquad (7)$$

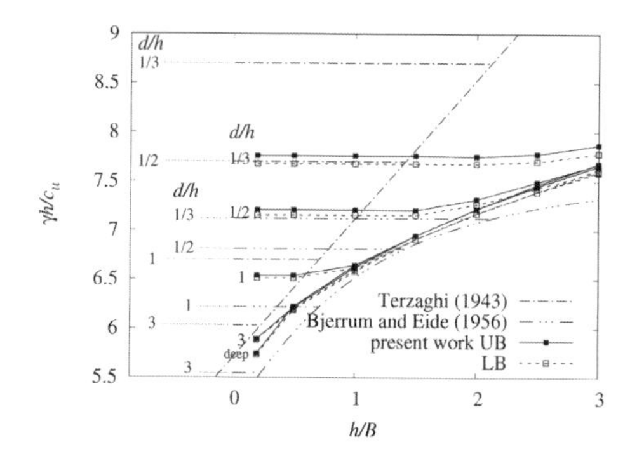

Figure 6. Influence of d/h ratio. Results of the stability numbers obtained in the calculations compared with other solutions.

7 PROPOSED EQUATION FOR BASAL INSTABILITY IN DEEP EXCAVATIONS IS CLAYEY SOILS

The quality of the numerical results obtained, that can be shown by the very small gap between the LB and the UB calculations, justifies a proposal for the basal instability of deep excavations in clayey soils, for the case with no wall embedded length and possible influence of rigid stratum below the excavation, based on equations 6 and 7.

Such proposal allows the determination for the collapse situation of the stability number, given the depth ratio, d/h, of the rigid stratum and the h/B ratio, using:

$$\left[\frac{\gamma h}{c_u}\right]\left(\frac{h}{B};\frac{d}{h}\right) = \max\left\{\left[\frac{\gamma h}{c_u}\right]_{\text{eq.}};\left[\frac{\gamma h}{c_u}\right]_{\text{eq.7}}\right\} \qquad (8)$$

Such equation results in the values represented in Figure 9, where the proposal is superposed with the results from the numerical calculations that allowed the proposal.

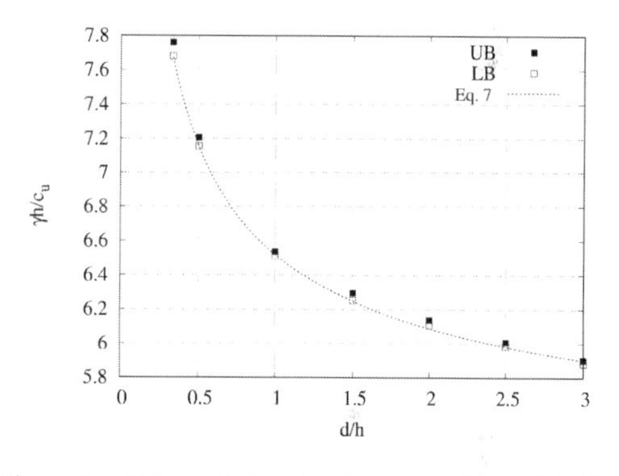

Figure 8. Values of $\gamma h/c_u$ for the cases where d/h influences the results (horizontal lines in Figure 6).

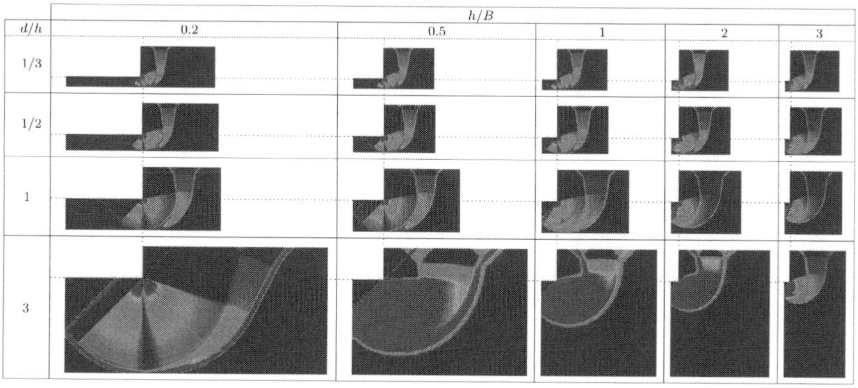

Figure 7. Plastic deformation zones obtained from the UB calculations for different d/h and h/B ratios.

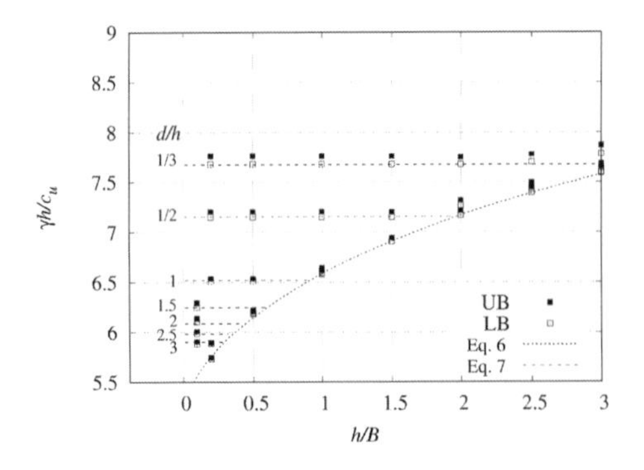

Figure 9. Proposal for the calculation of basal instability in clayey soils superposed with the numerical calculations that lead to such proposal.

8 CONCLUSIONS

The problem of the two-dimensional basal stability of deep excavations in clayey soils under undrained conditions was addressed with upper and lower-bound limit analysis using the finite element program *mechpy*, developed within the authors' researchteam.

The wall was assumed with no embedded length and a rigid stratum was considered at a variable depth. Two cases were initially analysed: one where the wall was not modelled (Case 1) and one other where the wall was modelled and different wall thicknesses were considered (Case 2). Case 1 was found to be a limit situation of Case 2 when the wall thickness tends to zero and to give more conservative results. The main study was therefore conducted using Case 1 only.

Results for deep rigid stratum (not affecting the stability numbers) and for different depths of the rigid stratum were obtained using UB and LB calculations. The two approaches—LB and UB—were found to give very close results, usually with a marginof less that 1%.

An equation for the calculation of the stability numbers for the case analyses was proposed, derived from the numerical results.

REFERENCES

Antão, A.N., N.M.C. Guerra, A.S. Cardoso, & M. Matos Fernandes (2008). Influence of tension cut-off on the stability of anchored concrete soldier-pile walls in clay. *Canadian Geotechnical Journal 45*(7), 1036–1044.

Antão, A.N., T. Santana, M. Vicente da Silva, & N.M.C. Guerra (2011). Passive earth-pressure coefficients by upper-bound numerical limit analysis. *Canadian Geotechnical Journal 48*(5), 767–780.

Antão, A.N., M. Vicente da Silva, N.M.C. Guerra, & R. Delgado (2012). An upper bound-based solution for the shape factors of bearing capacity of footings under drained conditions using a parallelized mixed f. e. formulation with quadratic velocity fields. *Computers and Geotechnics 41*, 23–35.

Bjerrum, L. & O. Eide (1956). Stability of strutted excavations in clay. *Géotechnique 6*(1), 32–47.

Deusdado, N. (2017). *Método do Lagrangeano Aumentado aplicado ao desenvolvimento de formulações paralelas de elementos finitos para análise limite.* Ph. D. thesis, FCT, Universidade Nova de Lisboa. In Portuguese; under evaluation.

Deusdado, N., A.N. Antão, M. Vicente da Silva, & N.M.C. Guerra (2015). Determinação de impulsos de terras através de implementação numérica dos teoremas est´atico e cinem´atico. In *CMNE 2015, Congresso de Métodos Numéricos em Engenharia*, Lisboa, 29 Junho - 2 Julho.

Santos Josefino, C., M.T. Santana, M. Vicente da Silva, A.N. Antão, & N.M.C. Guerra (2010). Twodimensional basal stability of deep exacavation in homogeneous clay deposit using upper bound numerical analysis. In *Geotechnical Challenges in Megacities*, Volume 2, Moscow, Russia, 7–10 June, pp. 614–621.

Skempton, A.W. (1951). The bearing capacity of clays. In *Proc. Building Research Congress*, Volume 1, pp. 180–189.

Terzaghi, K. (1943). *Theoretical Soil Mechanics* (2nd ed.). New York: John Wiley and Sons. 510 pages.

Ukritchon, B., A. Whittle, & S.W. Sloan (2003). Undrained stability of braced excavations in clay. *ASCE Journal of Geotechnical and Geoenvironmental Engineering 129*(8), 738–755.

Vicente da Silva, M. (2009). *Implementação Numérica Tridimensional do Teorema Cinemático da Análise Limite*. Ph. D. thesis, Universidade Nova de Lisboa. In Portuguese.

Vicente da Silva, M. & A.N. Antão (2008). Upper bound limit analysis with a parallel mixed finite element formulation. *International Journal of Solids and Structures 45*(22–23), 5788–5804.

Numerical Methods in Geotechnical Engineering IX – Cardoso et al. (Eds)
© 2018 Taylor & Francis Group, London, ISBN 978-1-138-33203-4

A method to consider the nonlinear behaviour of reinforced concrete in flexible earth-retaining walls: Preliminary results

J. Cândido Freitas
CONSTRUCT, Faculty of Engineering (FEUP), University of Porto, Porto, Portugal
Superior Institute of Engineering (ISEP), Polytechnic Institute of Porto, Porto, Portugal

M. Matos Fernandes & M.A.C. Ferraz
CONSTRUCT, Faculty of Engineering (FEUP), University of Porto, Porto, Portugal

J.C. Grazina
Faculty of Science and Technology (FCTUC), University of Coimbra, Coimbra, Portugal

ABSTRACT: This paper presents a numerical methodology to incorporate in finite element analyses of deep excavations the nonlinear behaviour of reinforced concrete walls. This method is based on the interaction between two well validated numerical models: one used for incremental analysis of geotechnical works (but assuming concrete as linear elastic) and the other used for nonlinear analysis of reinforced concrete structures taking into account the characteristics of the concrete and the steel reinforcement at each structural section. The analysis of an excavation in clay supported by a single propped diaphragm wall is presented as an example of application. It is shown how the consideration of the nonlinear behaviour of reinforced concrete influences the global performance of the retaining structure.

1 INTRODUCTION

Deep excavations in urban areas are usually supported by diaphragm or pile walls made of reinforced concrete (Liu et al., 2011; Ng et al., 2012; Finno et al., 2015).

The interaction between such structures and the supported soil is quite complex because their configuration evolves as the construction progresses and their deformations influence the magnitude and the distribution of the earth pressures and of the structural stresses.

The understanding of this problem has been considerably enhanced, from the late 1970s onward, by the use of finite element models (Clough and Tsui, 1974; Finno and Harahap, 1991; Matos Fernandes et al., 1993; Zdravkovic et al., 2005; Schafer and Triantafylidis, 2006; Hashash et al., 2011; Burlon et al., 2013; Dong et al., 2016).

In spite of the increasing progress and sophistication of these models along the last decades, a linear elastic response for the reinforced concrete wall is usually assumed. However, since concrete has a relatively small tensile strength, its mechanical behaviour becomes nonlinear for values of the mobilized stresses quite below the correspondent limit stresses (Bazant and Parameshwara, 1977).

The nonlinear response of reinforced concrete, induced by cracking, has been intensively studied in structural engineering and reliable numerical models are available to deal with such behaviour (Crisfield, 1997; Mohr and Bairán, 2010).

This paper presents a simple numerical methodology to consider the nonlinear behaviour of reinforced concrete in finite element analyses of deep excavations. In spite of its simplicity, the procedure consists of an interaction of two relatively advanced models: a finite element model for incremental geotechnical analysis and a *pure* structural model for nonlinear analysis of reinforced concrete structures that takes into consideration the characteristics of concrete and of steel reinforcement at each section.

In order to illustrate its potential, the presentation of the proposed methodology is complemented with an example of application, an excavation supported by a single propped diaphragm wall.

2 PROPOSED METHODOLOGY

2.1 *General*

The basic idea, suggested by Matos Fernandes (2010), was to use a geotechnical finite element model (so-called model GEO) and a model dealing with the nonlinear behaviour of reinforced concrete (so-called model RC) *just as they are*, and to develop a new computer code in order to make the two models to work and interact in parallel.

As shown by Figure 1, for each stage of the simulation of the excavation by Code GEO, the computed bending moment and axial stress diagrams of the retaining wall are applied to the sections modelled by Code RC, which accounts for the actual steel reinforcement at each section. Code RC then calculates the strains and cracking in the reinforced concrete wall and the corresponding adjusted stiffness at each section, which is transmitted to Code GEO, using an adjusted deformation modulus, to be used in the next construction stage.

2.2 *Model GEO*

Code GEO is a typical FEM code used to model 2D (plane stress, plane strain and axisymmetric) and 3D geotechnical works. It was adapted from the code FEMEP, developed at the Universities of Porto and Coimbra, Portugal (Almeida e Sousa et al., 2011).

The code incorporates several types of finite elements, allowing the simulation of the ground mass, structures and the soil-structure interfaces; associated to each element there is a specific activation and deactivation criterion, so various construction phases can be considered separately and sequentially.

Problems can be analysed in terms of total or effective stresses, including excess pore pressure generation and consolidation through a coupled formulation of equilibrium and flow equations.

The code permits the consideration of several constitutive models to characterize the nonlinear behaviour of the soil and the soil-structure interface, as well as geometric nonlinearity. To account for the nonlinear behaviour, an iterative algorithm based on the modified Newton-Raphson method is activated.

2.3 *Model RC*

Code RC is a nonlinear code for reinforced concrete analyses, based on the 1970's classic fibre method (Cohn and Ghosh, 1972; Chen and Shoraka, 1974). It corresponds to code FIBRAS, developed at the University of Porto using EVOLUTION framework (Ferraz, 2010).

In the fibre method, the moment-curvature relationship is obtained numerically by dividing the actual cross section into small elements called fibres (Figure 1) where uniaxial stress strain laws are used to describe the response of materials. Assuming that the section remains plane and normal to the longitudinal axis, the strain distribution over the section is linear, and consequently the strain on each fibre is related to the curvature of the section. Considering the nonlinear stress-strain relationships of concrete and reinforcing steel fibres into which each section is divided, the stresses are related to the applied bending moments and axial forces through the condition of equilibrium and are obtained using the tangent stiffness iterative step-by-step procedure.

An elastic-perfectly-plastic behaviour for compression and a model of tension stiffening for tension (Figure 2a) are considered for concrete, whereas a bilinear elasto-plastic behaviour (Figure 2b), both for tension and compression, is assumed for reinforcing steel.

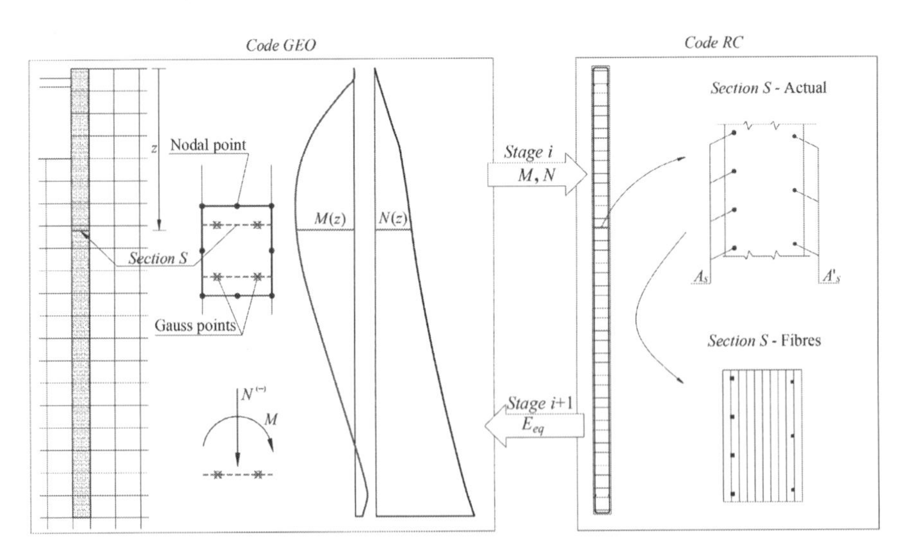

Figure 1. Interaction between the geotechnical model and the structural model.

The use of a tension stiffening model for tension in concrete is due to the fact that at a crack the full internal tensile force is carried by the reinforcement, whereas between cracks some amount of the tensile force is transferred through bond to the surrounding concrete. As a result, the reinforcement strains between cracks are smaller than the ones at the cracks. In brief, a cracked concrete member behaves as a member with a variable cross section, due to the highly reduced stiffness in the cracked zone. However, between the cracks the concrete in tension continues to contribute to the flexural stiffness, which reduces the curvature.

The use of the above mentioned tension stiffening model for the concrete under tension results in a relation between the curvature of the section and the bending moment as presented in Figure 3.

Three states (I, II and III) are denoted close to straight lines, whose inclination has the meaning of a bending stiffness. The so-called State I corresponds to the uncracked section; State II corresponds to a cracked section neglecting the tensile strength of concrete; State III corresponds to a situation where the tensile strength of both concrete and reinforcement steel is exhausted and the bending stiffness becomes residual (formation of a plastic hinge). The curve depicted in the figure reveals a smooth transition between states I and II as a result of the adopted tension stiffening model, which represents an approximation to the "real" behaviour.

Creep and shrinkage is not considered in this model.

2.4 *Interaction between Codes GEO and RC*

As shown in Figure 1, in Code GEO each horizontal pair of Gauss points of a finite element belonging to the retaining wall defines a section. Therefore, the n two-dimensional finite elements of the retaining wall in Code GEO give rise to $2n$ horizontal sections to be analysed by Code RC.

In a preliminary stage (Phase 0), Code RC computes the initial elastic bending stiffness of all sections, on the basis of the respective concrete and steel reinforcement, whereas Code GEO generates the at-rest state of stress. For each of the following stages, Code GEO conveys to Code RC the values of the incremental stresses in each section of the wall. Then, Code RC establishes the internal equilibrium of the sections taking into consideration the constitutive laws and the characteristics of concrete and steel, calculating the state of stress and strain for each individual fibre; this process leads to the calculation, for each section, of the new bending stiffness affected by cracking. The stiffness obtained by the adjustment of the deformation modulus, E_{eq}, is then introduced in Code GEO, to be used in the next construction stage.

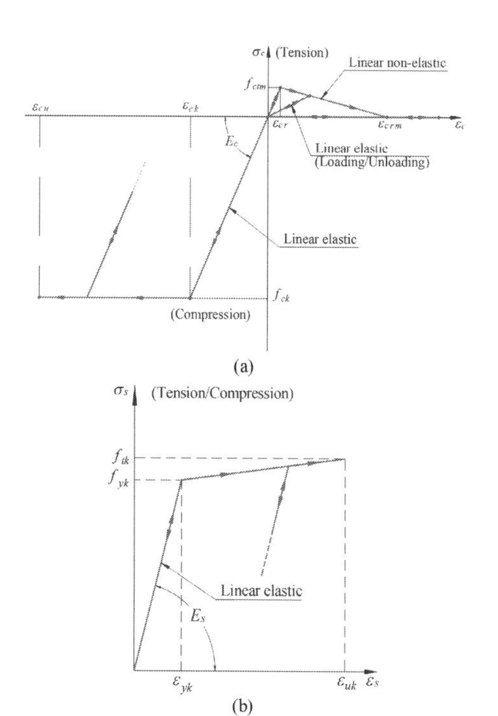

Figure 2. Constitutive law assumed in Code RC: (a) concrete; (b) steel.

3 EXAMPLE OF APPLICATION

3.1 *Geometry of the problem and basic input data*

Figure 4 represents a cross section of the numerical case study performed to illustrate the application of the proposed methodology: a symmetric excavation, 9 m deep and 40 m wide, supported by a reinforced concrete diaphragm wall, 0.8 m thick, propped at the top.

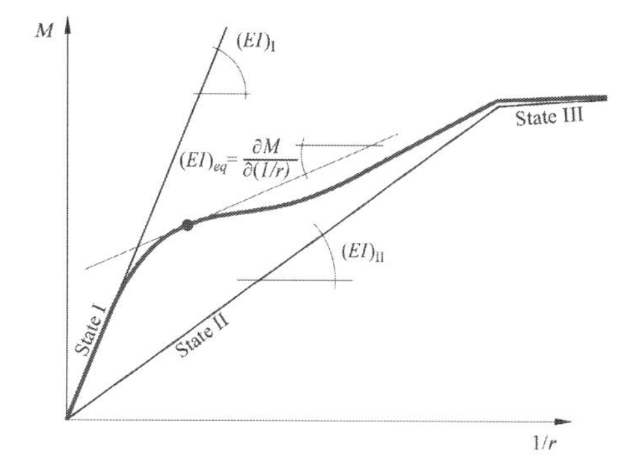

Figure 3. Relationship between curvature and bending moment at a reinforced concrete section evidencing the phenomenon of tension stiffening.

The ground is composed by a soft to medium clay deposit until a depth of 19 m, underlain by a stiff clay layer and, at 25 m depth, by bedrock. The wall tip penetrates 1 m in the stiff clay.

The analyses with Code GEO were performed under plane strain conditions and in total stresses, assuming undrained behaviour of both clays. The constitutive model adopted for the soils was a classic Tresca elastic-perfectly plastic law. Table 1 summarizes the geotechnical parameters of the analyses.

A similar constitutive law was adopted for the soil-wall interface, whose strength was taken equal to 2/3 of the undrained shear strength of the soil at the same depth, being mobilized for a tangential relative displacement of 1 mm.

Figure 5 shows the finite element mesh used in Code GEO, formed by 1092 eight-node isoparametric finite elements to represent the soil (1052 elements) and the wall (40 elements), 80 six-node joint elements to represent the soil-wall interface plus 4 six-node joint elements in the stiff clay and 1 two-node bar-element to simulate the prop level.

The left boundary was placed at the plane of symmetry of the excavation, whereas the right boundary was assumed 50 m behind the face of the excavation. The lower boundary was assumed at the top of bedrock.

It was admitted that the installation of the diaphragm wall does not alter the at-rest state of stress. The excavation was simulated by 18 stages of 0.5 m each; after the two first excavation stages, corresponding to 1.0 m depth, the level of props is installed (0.5 m from the top of the wall) with no pre-stress.

The level of props, assumed as linear elastic, corresponds to HEB320 steel sections, spaced 3.0 m on the longitudinal direction. Their effective axial stiffness was adopted equal to 50% of the theoretical stiffness, to account for eventual imperfections and gaps at the prop-wall interface.

With such input data and assumptions, a preliminary analysis with Code GEO has been performed assuming the diaphragm wall as linear elastic, with a bending stiffness corresponding to the actual concrete section (0.8 m thick). The maximum positive and negative bending moments obtained from this analysis were selected to calculate the steel

Table 1. Geotechnical parameters introduced in Code GEO.

Soil	γ (kN/m^3)	K_0	c_u (kPa)	E_u (MPa)	ν
Soft/ medium clay	17.8	0.6	5+3.43 ·z(m)	400·c_u	0.49
Stiff clay	19.6	0.7	200	400·c_u	0.49

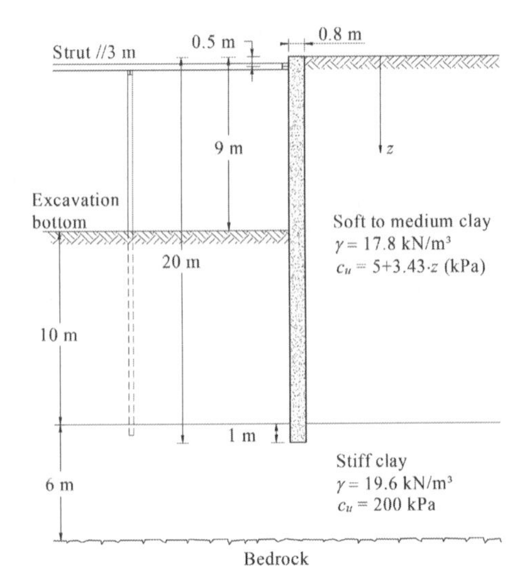

Figure 4. Example of application: cross section of the excavation and of the retaining structure.

reinforcement at both wall faces, considering a partial safety factor of 1.35 for permanent actions in the safety check for ultimate limit states, as recommended by Eurocode 7.

Table 2 summarizes the concrete and steel parameters adopted for steel reinforcement design and later introduced in Code RC. It is assumed that the steel reinforcements computed at the sections with maximum bending moments are kept constant along the full height of the wall.

As shown in Figure 6, the 40 finite elements that represent the wall in Code GEO lead to 80 horizontal sections to be analysed by Code RC. Each section is divided into 40 equal thickness concrete fibres, parallel to the neutral axis, plus a number of square shaped fibres, with an area equivalent to that of the steel rods, and placed at the same relative position in the actual section.

The geometrical and mechanical parameters (for concrete and steel) considered for the diaphragm wall correspond to an initial elastic bending stiffness, $(EI)_0$, of 1614 MNm^2/m.

3.2 *Discussion of results*

Two analyses have been performed, just differing with regard to the behaviour assumed for the diaphragm wall material:

i. analysis A, with the linear elastic wall, thus with constant bending stiffness;
ii. analysis B, considering the nonlinear behaviour of the reinforced concrete, with the bending stiffness of each section adjusted, stage by stage, to take into account the cracked part of the section.

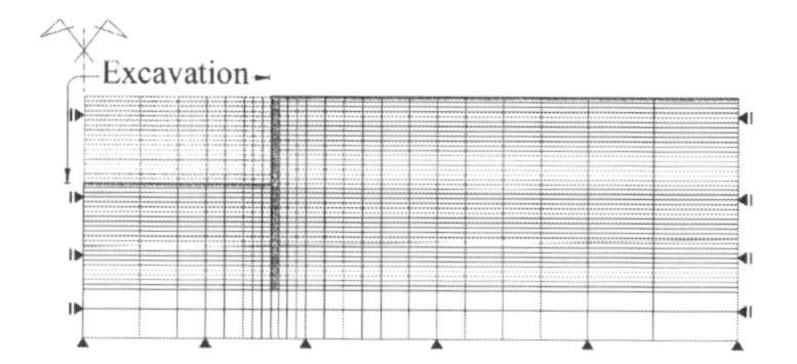

Figure 5. Finite element mesh for Code GEO.

Table 2. Structural parameters introduced in Code RC.

Concrete
Strength class: C30/37 Characteristic value of compressive strength: $f_{ck} = 30$ MPa Mean value of axial tensile strength: $f_{ctm} = 2.9$ MPa Modulus of elasticity: $E_c = 33$ GPa
Steel
Strength class: S400 Characteristic value of yield strength: $f_{yk} = 400$ MPa Modulus of elasticity: $E_s = 200$ GPa Steel reinforcement at the front face: $A_s = 80.4$ cm²/m (10φ32/m) Steel reinforcement at the back face: $A'_s = 25.1$ cm²/m (8φ20/m)

Naturally, in analysis A only the model GEO is used, whereas in analysis B the interaction between models GEO and RC is applied. Both analyses, at their start, assume the same elastic wall bending stiffness, $(EI)_0$.

Figure 7 includes the bending moment and axial force diagrams, as well as the horizontal wall displacement, at the completion of the excavation for both analyses. The position of section 29 at the depth of $z = 7.11$ m, where maximum positive bending moment occurs, is also indicated; this section will be latter object of a detailed analysis.

The analysis of the figure reveals that the consideration of nonlinearity induced by concrete cracking reduces the wall bending stiffness, which leads to lower wall bending moments (Figure 7a) and to higher wall deflections (Figure 7c).

On the other hand, it does not influence the distribution of axial forces (Figure 7b) because it does not affect the axial compressive stiffness. In this case, since there are no inclined and pre-stressed wall supports (such as anchors), axial stresses are small and just due to the self-weight of the wall and to the tangential stresses along both faces of the wall-soil contact.

The value of the strut load for analyses A and B are 293 kN/m and 267 kN/m, respectively, at completion of the excavation.

For the case of analysis B, Figure 8a and 8b illustrate the distribution of the wall bending moment

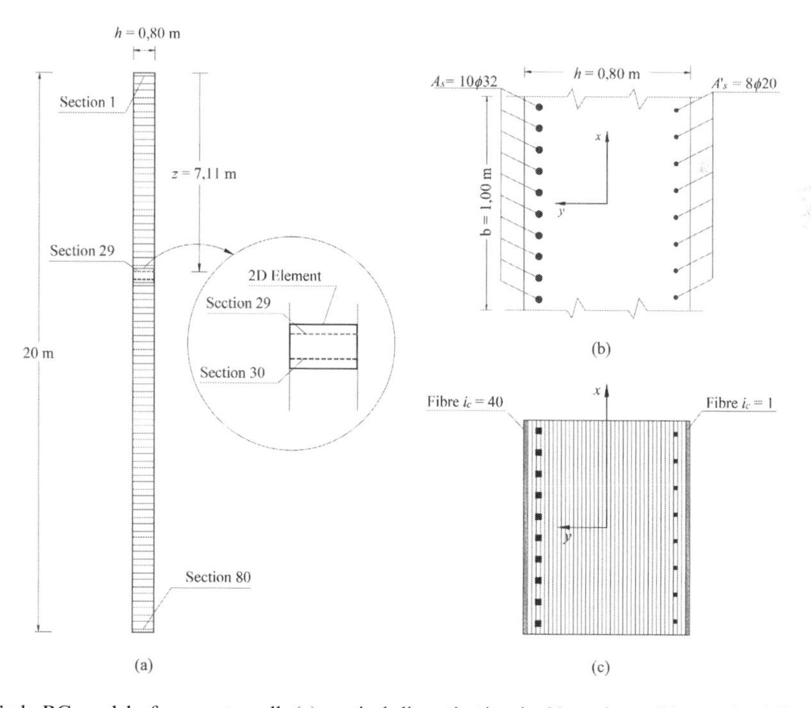

Figure 6. Code RC model of concrete wall: (a) vertical discretization in 80 sections; (b) actual reinforced concrete section; (c) discretization of the section in concrete and steel fibres.

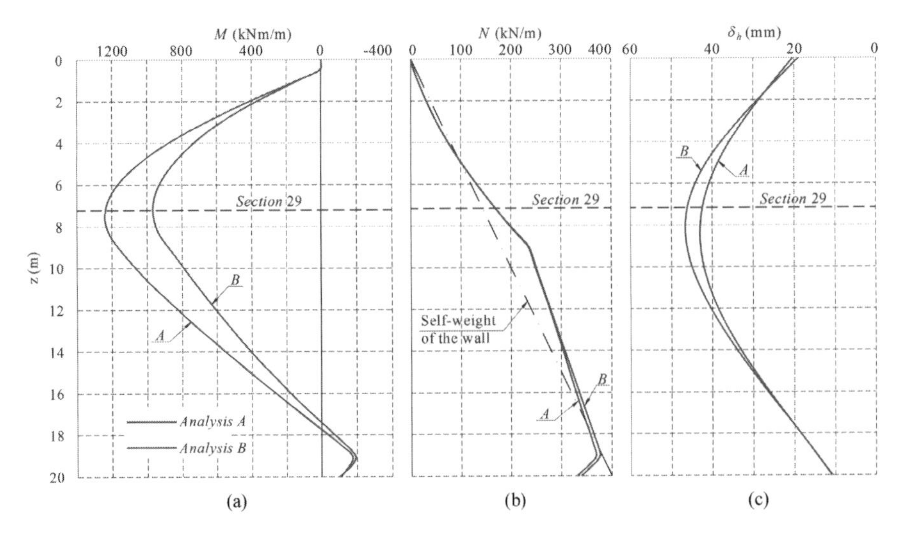

Figure 7. Bending moments (a), axial forces (b) and horizontal displacements (c) of the wall at completion of the excavation, for analyses A and B.

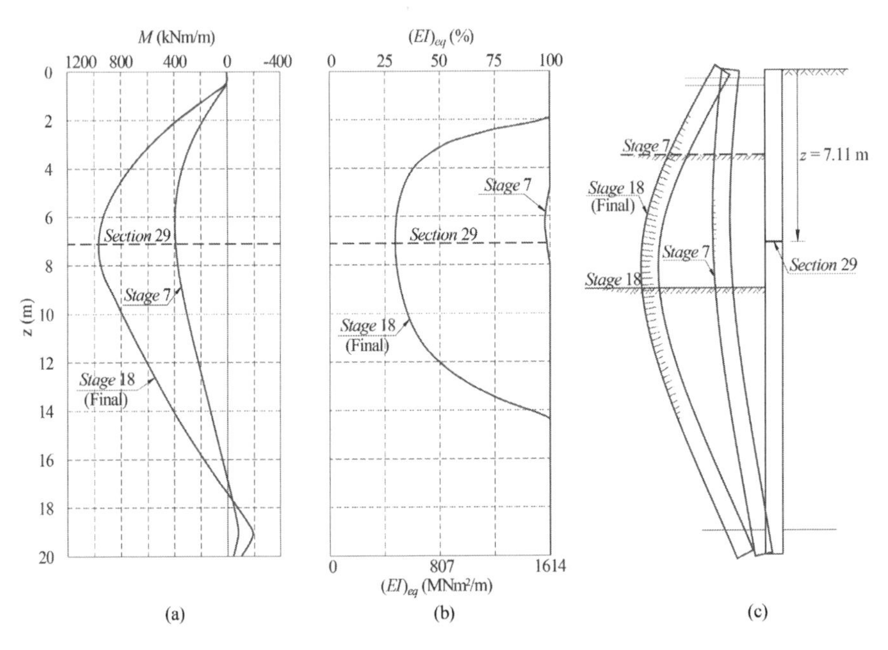

Figure 8. Wall bending moments (a), effective wall bending stiffness (b) and wall deflections (c) from analysis B (non-linear behaviour of concrete) for stages 7 and 18 (final).

and of the effective bending stiffness (expressed as a percentage of the initial elastic value) for stages 7 (excavation at 3.5 m depth, when first cracking occurs) and 18 (completion of the excavation).

As shown by the figure, at stage 18 the bending stiffness suffered a reduction of about 70% along 1/3 of the entire wall height. Moreover, approximately 60% of the wall sections suffered stiffness reduction by cracking at this final stage. The distribution of cracking together with the deflections of the wall may be observed in Figure 8c, for stages 7 and 18, as well.

For the same analysis B, Figure 9a presents the distribution of normal strain and stress in section 29, for stages 7 and 18 (negative values are adopted for compression). In the figure, fibre 1 is adjacent to the back wall face whereas fibre 40 is adjacent to the front face.

It can be observed that for stage 7, when cracking is still negligible, the distribution is almost linear, with the neutral axis (null values of stress and strain) at the centre of rigidity of the section. A quite distinct situation is obtained for stage 18, with more than 60% of the section under tensile stress, giving rise to cracks and to a pronounced reduction of the bending stiffness as previously referred, as well as a change on the neutral axis position.

The evolution of the normal stress at the two extreme fibres is illustrated in Figure 9b. Until stage 7 is reached, a linear evolution is observed, with equal values of compressive and tensile stress, the latter reaching the respective strength approximately at this stage. The start of cracking induces a radical change in that evolution with tensile stresses experiencing a progressive reduction, explained by the so-called tension stiffening phenomenon. Contrarily, the evolution of compressive stresses beyond stage 7 maintains the same initial linear trend, since the value of compression at the completion of excavation (stage 18) is just about 37% of f_{ck}, as can be observed in Figure 9b.

Finally, Figure 9c illustrates the location of stages 7 and 18 in the complete curvature *versus* bending moment diagram. It can be observed that stage 7 is still very close to State I, whereas stage 18 is about at midway from States I and II, which induces a drop in the bending stiffness of about 70%, as commented above. Nonetheless, since a common service situation is being analysed, even for this section of maximum bending moment, it can be seen that there is still a comfortable margin of safety with regard to the ultimate limit state.

Figure 9 contains only information about strain and stress in the concrete. With regard to steel reinforcement, the strain coincides with that of concrete in the same relative position in the section. The stress mobilized in the steel is always well below the respective yield strength.

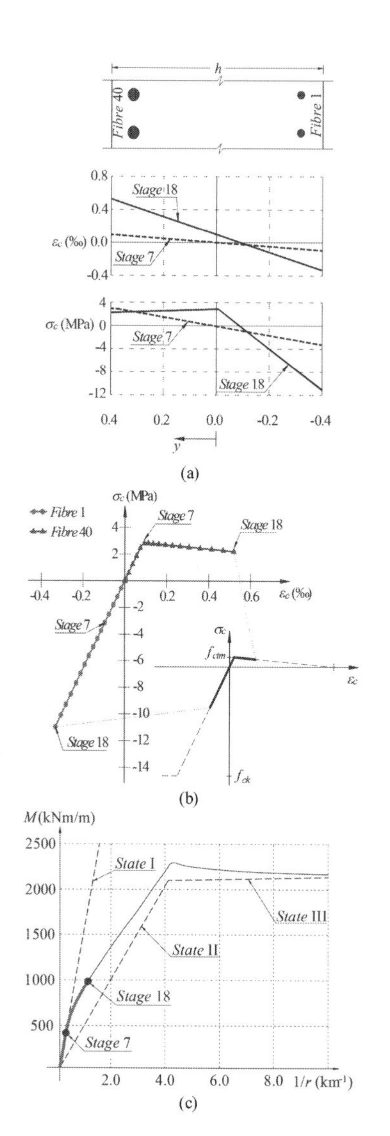

Figure 9. Results from analysis B for concrete section 29: (a) distribution of normal strain and stress, for stages 7 and 18; (b) normal stress versus normal strain for the extreme fibres, for all construction stages; (c) positions of stages 7 and 18 in the complete curvature versus bending moment diagram.

4 CONCLUSIONS

The paper presents a methodology to incorporate the nonlinear behaviour of reinforced concrete in finite element analyses of deep excavations supported by concrete walls. This methodology is based on the interaction between two well-validated numerical models: one used for incremental analysis of geotechnical works (but assuming concrete as linear elastic) and the other used for nonlinear analysis of reinforced concrete structures taking into account the characteristics of the concrete and the steel reinforcement at each structural section.

This methodology was applied to the study of an excavation supported by a reinforced concrete diaphragm wall propped at the top. The consideration of nonlinearity induced by concrete cracking reduces the wall bending stiffness, and this leads to lower wall bending moments and to higher wall deflections. It was illustrated how this methodology allows assessing the distribution of normal strain and stress in both the concrete and the steel, as well as cracking, in a given section of the retaining wall.

The results obtained are encouraging. The methodology seems to have great potential for the analysis of more complex problems, such as multi-propped walls or to the study of plastic hinge development, simulating the occurrence of ultimate limit states in certain sections of the retaining wall.

AKNOWLEDGEMENTS

This work was financially supported by: Project POCI-01-0145-FEDER-007457–CONSTRUCT—Institute of R&D In Structures and Construction funded by FEDER funds through COMPETE2020 – *Programa Operacional Competitividade e Internacionalização* (POCI) – and by national funds through FCT – *Fundação para a Ciência e a Tecnologia*; and Doctoral Scholarship SFRH/BD/72454/2010 through FCT.

The careful review of the paper by the Colleagues José Marques and Paulo Pinto is gratefully acknowledged.

REFERENCES

Almeida e Sousa, J., Negro, A., Matos Fernandes, M. & Cardoso, A.S. 2011. Three-dimensional nonlinear analyses of a metro tunnel in Sao Paulo porous clay, Brazil. *Journal of Geotechnical and Geoenvironmental Engineering, ASCE* 137(4):376–84.

Bazant, Z.P. & Parameshwara, D.B. 1977. Prediction of hysteresis of reinforced concrete members. *Journal of Structural Engineering, ASCE* 103(1):153–167.

Burlon, S., Mroueh, H. & Shahrour, I. 2013. Influence of diaphragm wall installation on the numerical analysis of deep excavation. *International Journal for Numerical and Analytical Methods in Geomechanics,* (37):1670–1684.

Chen, W.F. & Shoraka, M.T. 1974. Analysis and design of reinforced columns under biaxial loading. *Proceedings of Symposium on Design and Safety of Reinforced Concrete Compression Members, IABSE* (16):187–195.

Clough, G.W. & Tsui, Y. 1974. Performance of tied-back walls in clay. *Journal of Geotechnical Engineering, ASCE,* 100(12):1259–1273.

Cohn, M.Z. & Ghosh, S.K. 1972. The flexural ductility of reinforced concrete sections. *IABSE* (32):53–83.

Crisfield, M. 1997. Non-linear Finite Element Analysis of Solids and Structures. John Wiley & Sons, vol. 2, New Jersey, USA.

Dong, Y.P., Burd, H.J. & Houlsby, G.T. 2016. Finite-element analysis of a deep excavation case history. *Géotechnique* 66(1):1–15.

EN 1997–1. 2004. Eurocode 7: Geotechnical Design - Part 1: General Rules, CEN.

Ferraz, M.A.C. 2010. Models for the Evaluation of the Structural Behaviour of Bridges. *Ph.D thesis* (in Portuguese), Univ. Porto.

Finno, R.J. & Harahap, I.S. 1991. Finite element of HDR-4 excavation. *Journal of Geotechnical Engineering, ASCE* 117(10):1590–1609.

Finno, R.J., Arboleda-Monsalve, L. & Sarabia, F. 2015. Observed Performance of the One Museum Park West Excavation. *Journal of Geotechnical and Geoenvironmental Engineering, ASCE* 141(1):1–11.

Hashash, Y.M., Song, H. & Osouli, A. 2011. Three-dimensional inverse analyses of deep excavation in Chicago clays. *International Journal for Numerical and Analytical Methods in Geomechanics* (35):1059–1075.

Liu, G.B., Jiang R.J., Ng, C.W. & Hong, Y. 2011. Deformation characteristics of a 38 m deep excavation in soft clay. *Canadian Geotechnical Journal* 48(12):1817–1828.

Matos Fernandes, M. 2010. Deep urban excavations in Portugal: practice, design, research and perspectives. *Soils and Rocks* 33(3):115–142.

Matos Fernandes, M., Cardoso, A., Trigo, J. & Marques, J. 1993. Bearing capacity failure of tied-back walls - a complex case of soil-wall interaction. *Computers and Geotechnics* 15(2):87–103.

Mohr, S. & Bairán, J.M. 2010. A frame element model for the analysis of reinforced concrete structures under shear and bending. *Engineering Structures* 32(12):3936–3954.

Ng, C.W., Hong, Y., Liu, G.B. & Liu, T. 2012. Ground deformations and soil-structure interaction of a multi-propped excavation in Shanghai. *Géotechnique* 62(10):907–921.

Schafer, R. & Triantafylidis, T. 2006. The influence of the construction process on the deformation behaviour of diaphragm walls in soft clayey ground. *International Journal for Numerical and Analytical Methods in Geomechanics* (30):563–576.

Zdravkovic, L, Potts, D.M. & St. John, H.D. 2005. Modelling of a 3D excavation in finite element analysis. *Géotechnique* 55(7):497–513.

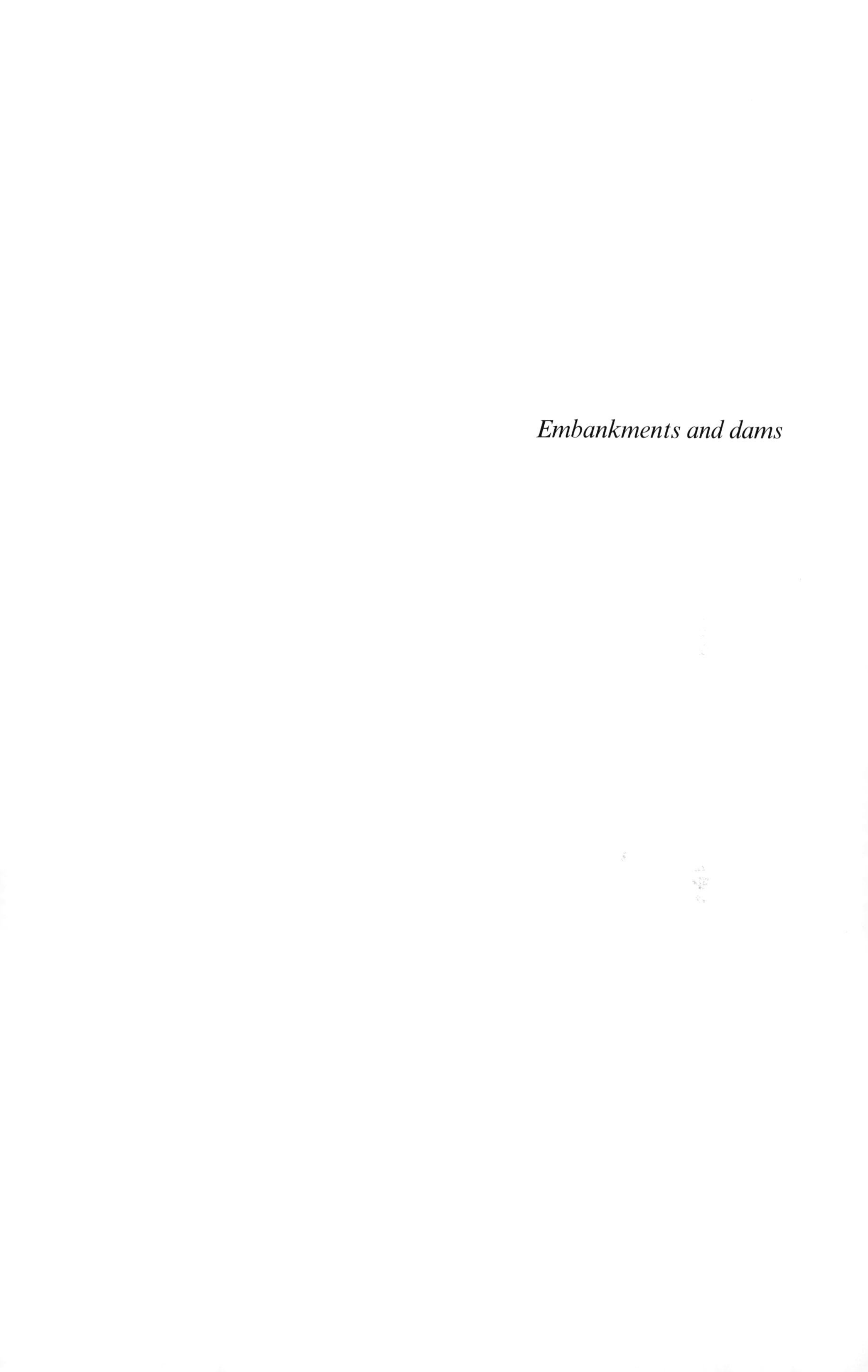

Embankments and dams

Numerical Methods in Geotechnical Engineering IX – Cardoso et al. (Eds)
© 2018 Taylor & Francis Group, London, *ISBN 978-1-138-33203-4*

A simplified finite element implementation of the Sellmeijer model for backward erosion piping

B.A. Robbins
U.S. Army Engineer Research and Development Center, Vicksburg, Mississippi, USA

D.V. Griffiths
Department of Civil and Environmental Engineering, Colorado School of Mines, Golden, Colorado, USA

ABSTRACT: Backward Erosion Piping (BEP) is a process by which erosion channels gradually progress upstream through the foundation of water retaining structures. The Sellmeijer piping model is perhaps the most commonly used model for assessing the hydraulics of BEP in which flow resistance of the pipe is determined from sediment transport and laminar flow equations. Typically, One-Dimensional (1D) rod finite elements are used to discretize the hydraulics in the erosion channels, and Two-Dimensional (2D) elements are used to discretize the groundwater flow. In this study, a simplified approach is developed in which quadrilateral elements are used throughout. Simulation results using both discretization schemes indicate the simplified approach closely approximates the traditional approach, with the quality of the approximation improving as the element size is reduced.

1 INTRODUCTION

Internal erosion refers to any process by which soil particles are gradually eroded from within or beneath water retaining structures. Accounting for 46 percent of historical dam failures (Foster et al. 2000), internal erosion is an issue of significant concern for geotechnical engineers. Internal erosion can be subdivided into four distinct mechanisms: (1) concentrated leak erosion, (2) backward erosion piping (BEP), (3) contact erosion, and (4) suffusion. BEP accounts for approximately one-third of all internal erosion failures (Richards & Reddy 2007), and is the focus of this study. For further details on the other erosion mechanisms, the interested reader is referred to Bonelli (2013) and ICOLD (2015).

The process of BEP is illustrated in Figure 1. BEP initiates at a downstream, unfiltered exit, often resulting in the formation of a sand boil where eroded material is being ejected. The erosion progresses upstream along the contact of a cohesive cover layer and erodible foundation sands. If erosion is allowed to progress to the upstream water source, the eroded pipe may quickly enlarge leading to failure of the embankment. The likelihood of progression has been demonstrated to be related to the local hydraulic gradients immediately upstream of the pipe (van Beek 2015). It is therefore necessary to model the influence of the pipe on the groundwater flow patterns in the foundation to determine the gradients in front of the pipe. Sellmeijer (1988) proposed a model in which a steady state hydraulic solution was obtained for a given pipe length. Finite element implementations of the Sellmeijer model (Sellmeijer 2006, Van Esch et al. 2013) have been developed that allow variable geometry and geology to be accounted for in a BEP analysis. However, these models use 1D rod elements for discretizing the pipe domain. This requires custom FEM software involving two element types, greatly limiting the accessibility of the models to geotechnical engineers. In this study, a simplified approach using solely 2D quadrilateral elements is developed and compared to the traditional approach using 1D elements. In the following sections, overviews of the governing equations, discretization approaches, and analysis results are presented.

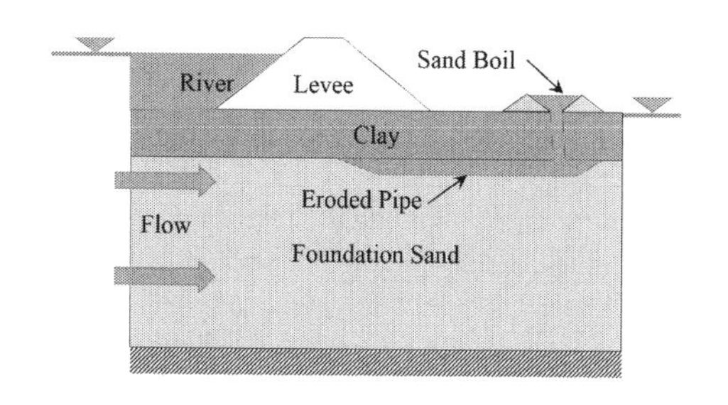

Figure 1. Illustration of backward erosion piping.

2 GOVERNING EQUATIONS

For steady, laminar flow, the flow through the soil is governed by the Laplace Equation, given in 2D by

$$\frac{\partial}{\partial x}\left(k_x \frac{\partial H}{\partial x}\right) + \frac{\partial}{\partial y}\left(k_y \frac{\partial H}{\partial y}\right) = 0 \tag{1}$$

where k_x and k_y denote the hydraulic conductivities in the x and y directions, respectively, and $H = y + p/(\rho g)$ denotes the total head (potential) in terms of the elevation head x and the pressure head given in terms of the pore pressure (p), the fluid density (ρ), and the acceleration of gravity (g).

In 2D, the eroded pipes are assumed to be similar to that of 1D flow passing through two parallel plates. For steady, laminar, 1D flow, the quantity of flow through the pipe (q_p) is readily determined through simplification of the Navier-Stokes equations as

$$q_p = -\frac{a^3 \rho g}{12\mu} \frac{dH}{dx} \tag{2}$$

(Sellmeijer 1988) where a is the depth of the eroded pipe and μ is the dynamic fluid viscosity. From continuity,

$$\frac{\partial q_p}{\partial x} + S = 0 \tag{3}$$

where S denotes a sink/source term. Substitution of Eq. 2 into Eq. 3 yields the differential equation governing the pipe flow:

$$\frac{a^3 \rho g}{12\mu} \frac{d^2 H}{dx^2} = S \tag{4}$$

In addition to satisfying Eq. 4, the flow in the pipe must be such that the grains in the bottom of the pipe are in equilibrium; otherwise, the pipe depth will change, and a steady solution is not obtained. The boundary shear stress exerted on the top and the bottom of the pipe channel is determined from static equilibrium to be

$$\tau = \frac{a}{2} \rho g \frac{dH}{dx}. \tag{5}$$

If the wall shear stress in the pipe exceeds a critical shear stress, τ_c, the grains are transported, resulting in deepening of the eroded pipe. The critical shear stress for cohesionless soils is known to be a function of the flow regime (Shields 1936); however, for consistency with the Sellmeijer model (Sellmeijer 1988, Van Esch et al. 2013), White's (1940) definition of critical shear stress will be used where

$$\tau_c = \eta \frac{\pi}{6} \gamma'_p d \tan(\theta) \tag{6}$$

with η, θ, γ'_p, and d denoting White's constant, the bedding angle of the sand, particle specific weight, and a representative grain diameter, respectively. Conceptually, η represents the proportion of boundary shear stress transferred to a single grain, and θ represents the rolling resistance of a grain of sand (from sediment transport theory, not to be confused with soil friction angle). In practice, however, both η and θ are used as model calibration constants. Recognizing that $\tau \le \tau_c$ must hold for a valid solution to be obtained, a final condition governing the pipe behavior becomes

$$a \frac{dH}{dx} \le \eta \frac{\pi \gamma'_p}{3\gamma_w} d \tan(\theta). \tag{7}$$

A valid solution must simultaneously satisfy Eqs. 1, 4, and 7.

3 FINITE ELEMENT DISCRETIZATION

In the following sections, two approaches are described for discretizing and solving the governing BEP equations. The first approach is the conventional approach suggested by Van Esch et al. (2013) in which the pipe is simulated by 1D, "rod" elements. The second approach is an approximation in which the pipe is represented by 2D quadrilateral elements.

3.1 *Discretization approach 1*

In the first approach, the pipe flow is simulated through 1D "rod" elements as illustrated in Figure 2a with the soil domain represented by 2D elements. Using the Galerkin weighted residual method, the finite element discretization of the coupled soil and pipe flow is (e.g. Smith and Griffiths 2004, Van Esch et al. 2013)

$$[K_e]\{H\} = \{Q\} \tag{8}$$

where $\{H\}$ and $\{Q\}$ represent the total head and net inflow/outflow at the FEM nodes and

$$[K_e] = \sum_{\Omega} \int_{\Omega_e} k_x \frac{\partial N_i}{\partial x}\frac{\partial N_j}{\partial x} + k_y \frac{\partial N_i}{\partial y}\frac{\partial N_j}{\partial y} d\Omega_e + \sum_{\Gamma} \int_{\Gamma_e} \frac{a^3 \gamma_w}{12\mu} \frac{\partial N_i}{\partial x}\frac{\partial N_j}{\partial x} d\Gamma_e \tag{9}$$

with the 2D soil domain being designated by Ω, the 1D pipe domain being represented by Γ, N_i designating the shape functions, and the subscript e designating element subdomains. The solution to Eq. 8 satisfies Eq. 1 and Eq. 4.

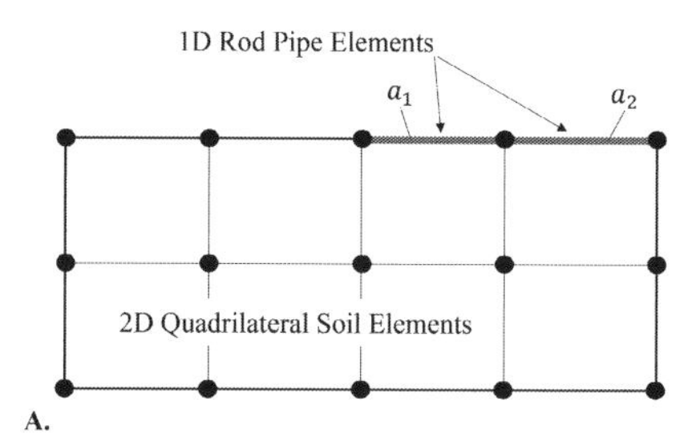

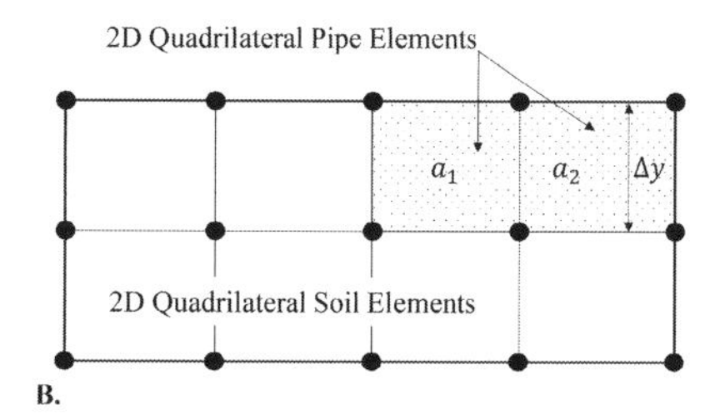

Figure 2. Finite element discretization of the piping problem using (A) discretization approach 1 and (B) discretization approach 2.

Given that a is also unknown in each pipe element, the solution must be iteratively solved, starting from an initial value of a in each pipe element. For each iteration, the inequality in Eq. 7 is checked element-by-element. If the equilibrium condition is violated, a is increased by a fraction of a grain diameter for the next iteration. Iterations continue until Eq. 7 is satisfied in all pipe elements. In this study, the pipe depth was initialized to $a = 5d$ and incremented by $d/2$.

3.2 *Discretization approach 2*

For the second approach, the pipe is represented by the first row of quadrilateral elements adjacent to the pipe path as illustrated in Figure 2b. That is, elements that were connected to 1D pipe elements in Approach 1 are now "pipe elements" in which a simplified approximation is made to try to obtain the same flow relations as before. Consider a single quadrilateral pipe element as shown in Figure 2b. Each quadrilateral pipe element has a scalar pipe depth, a, and element height, Δy, associated with it. The quantity of flow that should pass through the pipe channel in each element is given by Eq. 2.

Assuming that the flow in the pipe elements is horizontal due to the increased permeability, the computed amount of flow through the element, q_{FEM}, from Darcy's Law is approximately

$$q_{FEM} \approx -k_{pipe} \frac{\partial H}{\partial x} \Delta y \tag{10}$$

where k_{pipe} is the equivalent hydraulic conductivity of the pipe element. Considering the flow in the soil fraction of the pipe element to be negligible relative to the amount of flow passing through the pipe, the approximation $q_{FEM} \approx q_p$ can be made. Setting Eq. 2 equal to Eq. 10 yields

$$k_{pipe} = \frac{a^3 \rho g}{12 \mu \Delta y} \tag{11}$$

As before, the finite element discretization of Eq. 1 is given by Eq. 8 where $[K_e]$ is now given by

$$[K_e] = \sum_{\Omega^s} \int_{\Omega_e^s} k_x \frac{\partial N_i}{\partial x}\frac{\partial N_j}{\partial x} + k_y \frac{\partial N_i}{\partial y}\frac{\partial N_j}{\partial y} d\Omega_e^s + \sum_{\Omega^p} \int_{\Omega_e^p} k_{pipe} \left(\frac{\partial N_i}{\partial x}\frac{\partial N_j}{\partial x} + \frac{\partial N_i}{\partial y}\frac{\partial N_j}{\partial y} \right) d\Omega_e^p \tag{12}$$

with Ω^s and Ω^p denoting the portions of the 2D domain representing the soil and erosion pipe, respectively. The same iterative procedure is used in this approach to solve for a as before, but the iterations are now performed over the quadrilateral elements designated as pipe elements instead of the 1D, rod elements. Once Eq. 7 has been satisfied in all pipe elements, the iterations have converged to a valid solution for a given pipe length.

4 ANALYSIS RESULTS

A finite element program capable of performing calculations using both discretization schemes was developed using the Fortran libraries documented in (Smith & Griffiths 2004). In this section, the results obtained using the developed program for a simple test problem are presented.

The test problem of a simplified dam geometry presented in Van Esch et al. (2013) was used to evaluate the performance of the two discretization approaches. The geometry of the problem is presented in Figure 3 with a flownet illustrating the impact of the pipe. The pipe elements influence the flow net similarly to a "leaky boundary", causing the flow lines to exit the domain at an acute angle due to the unique pressure-discharge relationship imposed by Eq. 4. The foundation consists of a homogenous soil layer with no-flow boundaries (Neumann conditions) at the left, bottom, and right boundaries. An impervious structure resting on the foundation is simulated through a no-flow boundary as well. The horizontal ground surface both upstream and downstream of the structure are constant head (Dirichlet) boundary conditions. Foundation soil properties are provided in Table 1.

The analysis results presented in Van Esch et al. (2013) for a differential head of 1.85 m were repli-

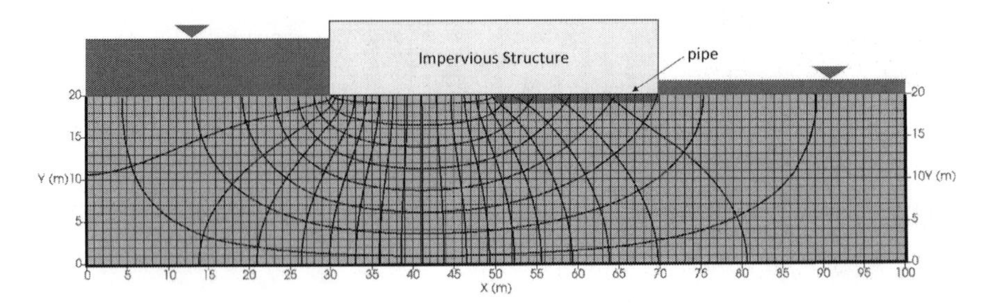

Figure 3. Simplified dam geometry (after Van Esch, Sellmeijer, & Stolle 2013).

Table 1. Soil properties for example problem.

Property	Assigned Value	Units
k_x,k_y	100.0	m/day
d	2.08×10^{-4}	m
γ_w	1.0×10^4	N/m^3
μ	1.0×10^{-3}	Ns/m^2
γ'_p	1.65×10^4	N/m^3
η	0.25	–
θ	37	degrees

cated using the Approach 1 discretization scheme. The total head at the downstream ground surface was set to 0 m. The 2D quadrilateral elements were squares with side length 0.5 m, while the 1D pipe elements were 0.5 m in length. As the focus of this study was solely on the hydraulic solution of piping (as opposed to assessing pipe progression), the pipe elements were manually activated from x = 61.5 m to x = 70.0 m to match the conditions simulated in Van Esch et al. (2013) for the given head difference. The computed pipe depth and total head profile in the vicinity of the pipe using Approach 1 are compared to the results from Van Esch et al. (2013) in Figure 4.

The pipe depths computed using Approach 1 were smaller than those obtained by Van Esch et al. (2013) despite the identical model formulation. The shallower pipe depths resulted in increased flow resistance in the erosion pipe leading to higher total head along the pipe profile (Figure 4b). While insufficient details are presented in Van Esch et al. (2013) to deduce the exact cause of the differences, the authors believe the differences may be due to differing element sizes, initial pipe height, and/or the size of pipe depth increments used. Rosenbrand (2016) demonstrated that the pipe depth computed using the program developed in Van Esch et al. (2013) can readily vary by the magnitude illustrated in Figure 4a due to these various factors. Identifying the precise cause of the differences between the two programs will be a topic of future study.

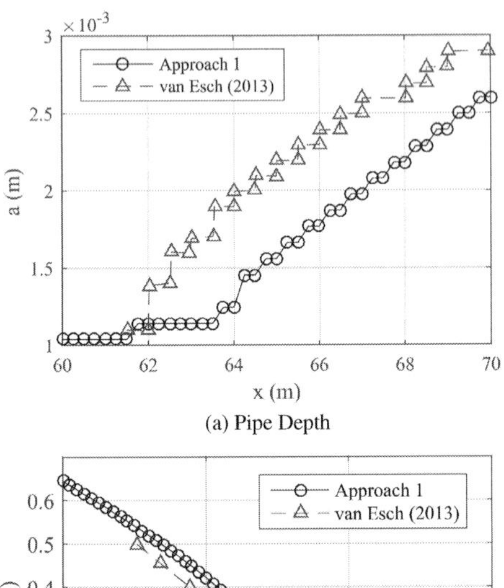

(a) Pipe Depth

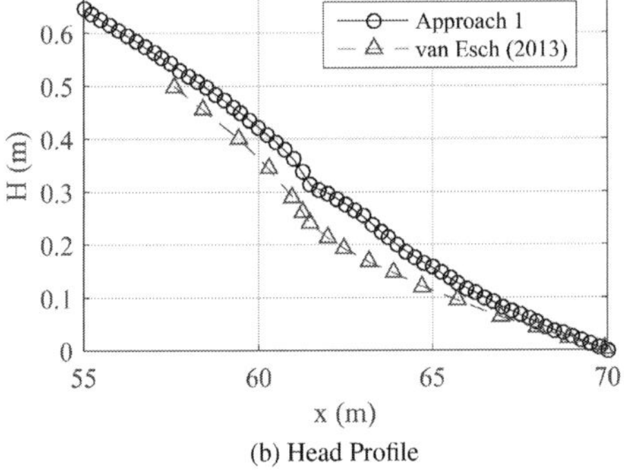

(b) Head Profile

Figure 4. Comparison of Approach 1 (a) pipe depth and (b) head profile to van Esch (2013) results.

Given that the Van Esch et al. (2013) model has been shown to be sensitive to element size, computations were performed using both Approach 1 and Approach 2 discretizations for element sizes of 0.25 m, 0.50 m, 1.0 m, and 2.0 m. In Approach 1 computations, 1D pipe elements were set to the same length as the quadrilateral element edges such that all 1D element nodes were connected to quadrilateral nodes. In order to precisely accommodate the coarser domain discretization corresponding to the larger elements, the eroded pipe was assumed to have progressed 10 m such that pipe elements were located from x = 60 m to the dam toe at x = 70 m. The upstream and downstream boundary conditions were fixed to the same total heads as before such that a differential head of 1.85 m was applied.

The results obtained using both Approach 1 and Approach 2 discretizations for the various element sizes are provided in Figure 5 and Figure 6, respectively. For each simulation, the total head along the base of the dam and the depth of the eroded pipe is plotted in the vicinity of the pipe. Additionally, the Approach 1 results obtained with 0.25 m elements are repeated in Figure 6 for purposes of comparison with Approach 2. The Approach 1 simulation with the smallest elements was selected for comparison, as it was considered to be the most accurate solution.

Figure 5a shows that the element discretization influenced the resolution of the head profile. However, it is somewhat surprising that the various element sizes resulted in nearly identical approximations of the pipe depth (Figure 5b). In contrast, Rosenbrand (2016) demonstrated that the pipe depth along the erosion channel appears to be sensitive to mesh size. Identifying the cause of this discrepancy will be a topic of further study.

Figure 6 shows that both the head profile and the pipe depth are sensitive to element size using Approach 2. In particular, the pipe depth and the head profile increasingly diverge from the Approach 1 solution as the element size is increased. This is due to the fact that the assumption of pipe flow constituting all of the flow in the elements designated as pipe elements becomes increasingly violated as elements become larger. With decreasing element size, the Approach 2 solution converges to the Approach 1 solution.

The fact that the Approach 2 solution converges to the Approach 1 solution as element size decreases is convenient for a number of reasons. As Approach 2 requires only quadrilateral elements, existing FEM models can readily be extended to simulate BEP without modification of the underlying FEM code. Additionally, Approach 2 naturally extends to three dimensions without further modification. Three-dimensional computations using Approach 1 require assigning a width dimension to all 1D pipe elements to properly compute flow quantities. Lastly, because Approach 2 simply requires the changing of hydraulic conductivity to represent the pipe flow, this approach can be used in conjunction with commercial software applications for which end users do not have the option to modify the FEM stiffness matrix assembly as required in Approach 1. This provides great ben-

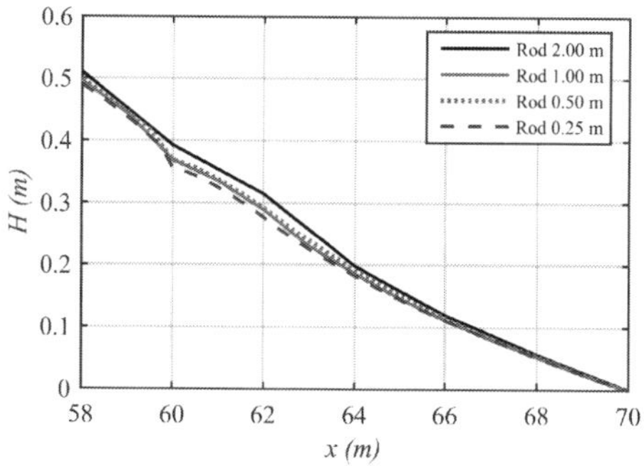

(a) Head Profile

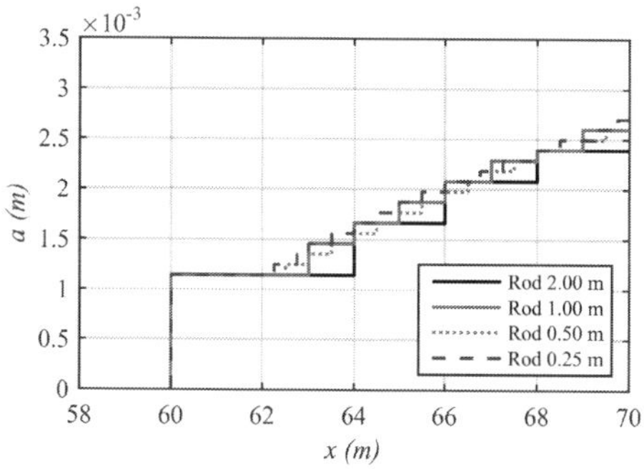

(b) Pipe Depth

Figure 5. Head profile (a) and pipe depth (b) results obtained for the test problem using Approach 1 with various element sizes.

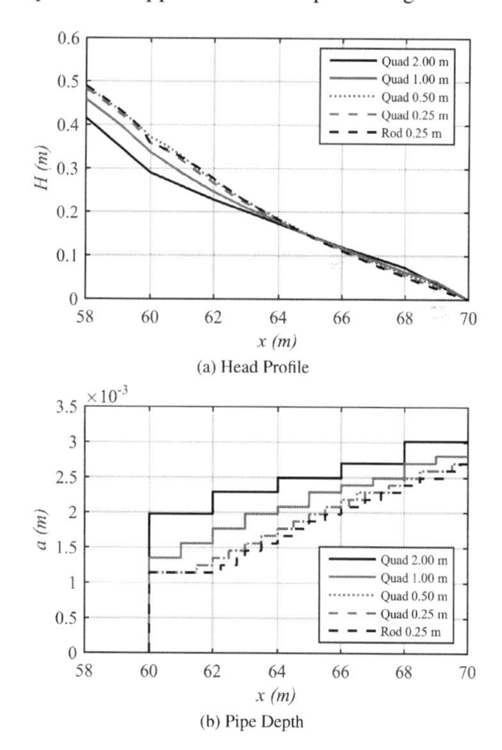

Figure 6. Head profile (a) and pipe depth (b) results obtained for the test problem using Approach 2 with various element sizes. Approach 1 result for 0.25 m elements plotted for comparison.

efit to geotechnical engineers comfortable with commercial software applications, as previous analyses can readily be modified to evaluate the potential for BEP.

5 CONCLUSIONS

A simplified finite element implementation of the Sellmeijer piping model was developed in which both the soil domain and the pipe domain were discretized using 2D quadrilateral elements. The simplified discretization approach was compared to the traditional discretization approach in which the pipe domain is represented by 1D rod elements. Simulation results using both discretization schemes indicate the simplified approach closely approximates the traditional approach, with the quality of the approximation improving as the element size is reduced. The results of this study indicate that BEP analysis can be conducted with any quadrilateral FEM program using the procedure described, greatly expanding the tools available for practicing engineers. Further, the simplified discretization approach is naturally extensible to three dimensions which will be a topic of further research.

Permission to publish was granted by the Director, Geotechnical and Structures Laboratory.

NOMENCLATURE

$[K_e]$ global conductivity matrix
Δy height of quadrilateral elements
η White's coefficient
μ dynamic fluid viscosity
ρ fluid (water) density
τ shear stress acting on erosion pipe walls
τ_c critical shear stress for incipient motion of foundation sand
θ bedding angle of sand being eroded
$\{H\}$ total head at finite element nodes
$\{q\}$ net flow at FEM nodes
a depth of erosion channel
d representative grain size
g gravitational acceleration
H total head
k_x hydraulic conductivity in the x-direction
k_y hydraulic conductivity in the y-direction
k_{pipe} equivalent hydraulic conductivity of 2D pipe elements
N_i shape functions used in FEM analysis
p pore pressure
q_p quantity of flow passing through the erosion channel
S sink/source term in continuity equation

REFERENCES

Bonelli, S. (2013). *Erosion in Geomechanics Applied to Dams and Levees*. Hoboken, NJ, USA: John Wiley & Sons, Inc.
Foster, M., R. Fell, & M. Spannagle (2000). The statistics of embankment dam failures and accidents. *Canadian Geotechnical Journal 37*(10), 1000–1024.
ICOLD (2015). Internal erosion of existing dams, levees and dikes, and their foundations. Volume 1: Internal erosion processes and engineering assessment. Bulletin 164, International Committee on Large Dams.
Richards, K.S. & K.R. Reddy (2007). Critical appraisal of piping phenomena in earth dams. *Bulletin of Engineering Geology and the Environment 66*(4), 381–402.
Rosenbrand, E. (2016). Validation piping module in DG Flow GUI. Report 1230003-002, Deltares, Delft, The Netherlands.
Sellmeijer, J. (1988). *On the Mechanism of Piping Under Impervious Structures*. Ph. D. thesis, Delft University of Technology.
Sellmeijer, J. (2006). Numerical computation of seepage erosion below dams (piping). *International Conference on Scour and Erosion*.
Shields, A. (1936). Application of similarity principles and turbulence research to bed-load movement. *Mitt. Preuss. Versuchsanst. Wasserbau Schiffbau 26*(5–24), 47.
Smith, I.M. & D.V. Griffiths (2004). *Programming the finite element method* (4th ed.). Hoboken, NJ, USA: JohnWiley & Sons, Inc.
van Beek, V.M. (2015). *Backward Erosion Piping: Initiation and Progression*. Ph. D. thesis, Technische Universiteit Delft.
Van Esch, J., J. Sellmeijer, & D. Stolle (2013). Modeling transient groundwater flow and piping under dikes and dams. In *3rd International Symposium on Computational Geomechanics (ComGeo III)*, pp. 9.
White, C.M. (1940, feb). The equilibrium of grains on the bed of a stream. *Proceedings of the Royal Society A: Mathematical, Physical and Engineering Sciences 174*(958), 322–338.

Numerical Methods in Geotechnical Engineering IX – Cardoso et al. (Eds)
© 2018 Taylor & Francis Group, London, ISBN 978-1-138-33203-4

Modelling a sand boil reactivation in the middle-lower portion of the Po river banks

M.F. García Martínez, G. Gottardi, M. Marchi & L. Tonni
Department DICAM, University of Bologna, Bologna, Italy

ABSTRACT: The Po River, which flows eastward through the Northern of Italy, is the largest and most important watercourse in the country. A system of major embankments safeguards over half of its length, existing since the XVI century. The stability of such structures has become a crucial issue following some significant flood events in the past. Among the most common collapse triggering mechanisms for Po river embankments, backward erosion piping turned out to be particularly menacing during the last most important high-water events (e.g. 1951, 1994, 2000 and 2014). In particular, the November 2014 high-water event led to the reactivation of a few relevant sand boils along the middle-lower stretch of this watercourse. This paper presents a 3D finite element model of the groundwater flow within a specific segment of the Po River, located in the province of Ferrara, which underwent piping reactivation during the 2014 event. The numerical model has been developed taking into account a detailed geotechnical characterization obtained from both site and laboratory investigations along with the available 2014 high-water event measurements. The effectiveness of the model is then validated with reference to a piping event recently occurred. Results are discussed with the aim of providing some insight into the most significant factors governing initiation of piping mechanism.

1 INTRODUCTION

Flowing eastward through the Northern Italy, the Po River is the longest watercourse in the country. Its basin has a large population and it embraces principal agricultural and industrial activities, which are a key source of national economic growth.

Major embankments safeguard over half of its length, existing since the XVI century. In order to provide flood protection as well as damage reduction, such man-made water-retaining structures have been progressively enlarged and extended upstream towards the main tributaries and their mechanical properties have been improved. In this way, the frequency of failure occurrence has progressively decreased and higher water levels can be sustained. On the other hand, such higher water levels create greater hydraulic gradients in foundation deposits, resulting in an increase of vulnerability toward piping phenomena.

Backward erosion piping, which consists in the formation of shallow pipes or channels at the interface between a sandy aquifer and an impermeable top layer (Van Beek et al. 2013), turns out to be one of the most threatening mechanisms of collapse for Po river embankments. The process starts with the formation of sand boils, which are often observed when there is a sufficient hydraulic head difference exiting across the structure (i.e. during high-water events) and an unfiltered exit is present downstream. Such open exit in the Po river embankment systems is very likely to be ascribed to the presence of a weak spot in the impermeable cover layer. That configuration is commonly referred in the literature to as "hole-type" exit (e.g. Van Beek 2014).

Sand volcanos will be observed when the flow is sufficient to make sand particles from the aquifer settle forming a ring around the boiling area (Van Beek et al. 2013). Ringing sand boils with sand sacks is a typical correction action usually taken to stop migration of sand particles.

In November 2014, a major high-water event, identified as the most significant in the last decades, led to the reactivation of a few relevant sand boils (see García Martínez et al. 2016) along the middle-lower stretch of the river. This paper focuses on the numerical 3D finite element modelling of the groundwater flow within a specific segment of the Po river, located in the province of Ferrara (Fig. 1), which has experienced piping phenomena during such event. Indeed, the detailed stratigraphic architecture provided by in situ tests seems to be compatible with the typical configuration required for the initiation of such mechanism.

The numerical model has been developed taking into account the geotechnical parameters obtained

Figure 1. a) View of the Po River Basin and location of sand volcano; b) Ringed sand volcano near Guarda Veneta, Province of Ferrara (Courtesy of Agenzia Interregionale per il fiume Po, AIPo).

from both in-situ and laboratory investigations, coupled with the available 2014 high-water event measurements. The effectiveness of the model is then verified with reference to a piping event recently occurred. The results are discussed with the aim of providing some insight into the most significant factors governing initiation of piping mechanism.

2 THE CASE STUDY

The analyses presented in this paper refer to a cross section located in the Province of Ferrara (Emilia Romagna Region) affected by the reactivation of a sand boil during the 2014 high-water event. Such sand boil (Fig. 1b) is located at about 35 m from the embankment toe.

2.1 *Geotechnical characterization of the embankment*

The mechanical characterization of the embankment foundation system has mainly relied on piezocone tests (CPTU) carried out along the cross-section, taking advantage of the amount of experience gained on the interpretation of in situ tests carried out in a similar environment, within the recent major project SISMAPO (Merli et al. 2015, Gottardi et al. 2015). Information obtained from boreholes and laboratory tests, part of such thorough project, have been also taken into account to aid in the interpretation.

Figure 2 shows the geometry of the selected cross section with the location of the three piezocone tests carried out in the February 2016 testing campaign. One of the tests is located in the floodplain area, a second one on the landside banquette and a third one a few meters from the toe embankment.

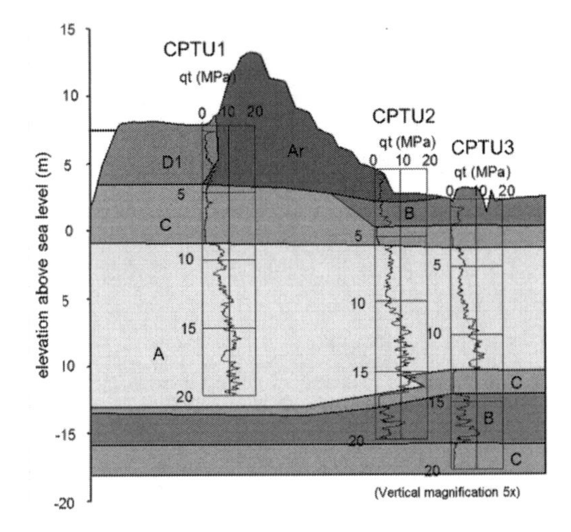

Figure 2. Stratigraphic cross-section and location of CPTU tests.

The soil stratigraphy has been determined using the classification framework developed by Robertson (2009), in which soil classes are defined in terms of the Soil Behaviour Type (SBT). Accordingly, results from the application of the method to the CPTU data have revealed the following stratigraphic soil units:

- Unit Ar (major embankment): alternation of sands, silty sands, sandy silts and clayey silts (SBT zones 3, 4 and 5);
- Unit D (floodplain deposits): alternation of clayey silts to sandy silts (SBT 4 and 5);
- Unit C: predominantly clays and silty clays (SBT 3 and 4);
- Unit B: alternation of silty mixtures (SBT 5, 4 and occasionally SBT 3);
- Unit A (aquifer): sands and silty sands (SBT 6 and 5).

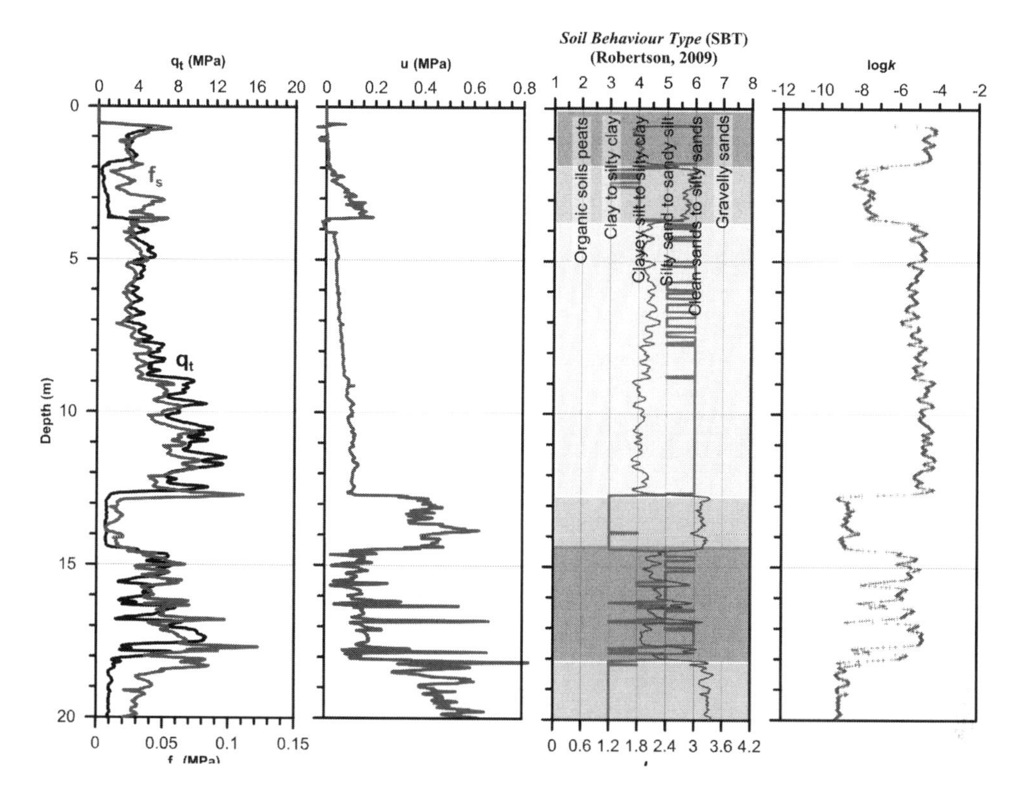

Figure 3. CPTU3 log profiles, soil classification results and profile of the logarithm of soil permeability *k (m/s)*.

Figure 3 shows the corrected cone resistance, q_t, the pore water pressure, u, and the SBT profiles for CPTU3, located at the toe of the embankment. It provides evidence of the typical configuration required for piping initiation, consisting of a top (impermeable) layer of about 3–4 m thick overlying a 9–12 m thick sandy aquifer. Erosion may then start when a weak point already exists in the confining layer or after cracks have been formed due to high water pressures (Van Beek 2015).

Relevant geotechnical parameters of the different soil units have been derived using some widely-used CPTU-based correlations. The coefficient of permeability, k, required for groundwater flow calculations, has been obtained using the relationship proposed by Robertson (2010):

$$\begin{array}{lll} 1.0 < I_c < 3.27 & k = 10^{(0.952-3.04 I_c)} & (m/s) \\ 3.27 < I_c < 4.0 & k = 10^{(-4.52-1.37 I_c)} & (m/s) \end{array} \quad (1)$$

Where I_c is the Soil Behaviour Type index developed by Robertson (2009). As it can be observed from the last plot (in terms of log$k$) in Figure 3, the hydraulic conductivity of the aquifer has turned out to be of the order of $10^{-5} \div 10^{-4}$ m/s.

2.2 *The numerical model of the groundwater flow*

The finite element code PLAXIS 3D has been used to simulate the groundwater flow causing the sand boil reactivation.

The cross-section geometry and soil layering configuration have been established based on topographic surveys and CPTU classification results. Furthermore, CPTU performed within the SISMAPO project in a cross-section located close to the selected one, have been also analysed in order to help in defining the stratigraphic model. Figure 4 shows the 3D model and the adopted finite element mesh, composed of 96514 tetrahedral 10-noded elements. The model length (y_{max}) has been set equal to 300 m.

As an initial attempt to simulate the event, the pre-existing exit configuration (Fig. 4) has been modelled as a cylindrical (vertical) pipe passing through the downstream top layers (at 35 m from the toe) and a relatively large region at the base of such pipe, both characterized by a high permeability value (k_{exit}). Due to the 3D nature of the phenomenon, the latter region may assume a rather orb-like shape (Fleshman & Rice 2014). Following a series of preliminary analyses, this region has been finally modelled as a 4-m diameter semi-spherical cluster. Such configuration is aimed at reliably simulating the presence of already eroded zones and better facilitating the flow.

The material model used in the analyses is the Mohr-Coulomb. Soil shear strength parameters have been deduced using some well-known empirical CPTU correlations previously validated on these soils (Gottardi et al. 2015). Table 1 shows the adopted values of shear-strength and permeability for the different soil units.

In order to perform the groundwater flow analysis, the variation of the river level during the November 2014 high-water period has been derived. The hydraulic boundary conditions (BC) for the (initial) steady state phase have been assigned considering the average river level over the days preceding the event in conjunction with available piezometric readings coming from a previous campaign (Severi & Biavati 2013). Such readings have revealed high piezometric heads in the aquifer during high-water periods, even at a considerable distance from the landside toe.

Figure 5 shows the hydraulic boundary conditions assumed for the transient analysis (from the 5th to the 25th November 2014), where the continuous blue line refers to the river level variations assigned to the left-hand side of the model (y_{min}). The dotted red line refers to the piezometric head at the right-hand side of the aquifer (y_{max}), as deduced from preliminary 2D numerical analyses based on a model long enough to allow neglecting piezometric head variations with time at y_{max}.

2.2.1 *Calibration of the numerical model*

Sensitivity analyses, aimed at identifying which input parameters turned out to be most relevant on the numerical response, have been carried out as a first step. They showed that the coefficient of permeability of the aquifer and of the outflow exit volumes, the diameter of the vertical pipe and the model width have a major influence on the results. At this stage of the study, attention has been focused in particular on two crucial aspects of the model, i.e. the diameter of the vertical pipe (Ø) and its coefficient of permeability (k_{exit}). The diameter of the semi-spherical region, the model width and the permeability of the aquifer have been set equal to 4 m, 40 m and $1.1 \cdot 10^{-5}$ m/s, respectively, the latter being derived from the application of Equation (1).

The model has been calibrated taking into account some existing criteria for sand boil reactivation, concerning both hydraulic gradients and flow velocity. Indeed, based on observations during the 1950 flood along Mississippi river, USACE (1956) observes that sand boils might occur at

Table 1. Angle of shearing resistance and coefficient of permeability for selected soil layers.

Unit	Φ' (°)	k (m/s)
A	35.9	$1.12 \cdot 10^{-5}$
B	35.0	$4.29 \cdot 10^{-6}$
C	30.4	$6.09 \cdot 10^{-9}$
D	32.4	$1.41 \cdot 10^{-7}$
Ar	33.1	$2.39 \cdot 10^{-7}$

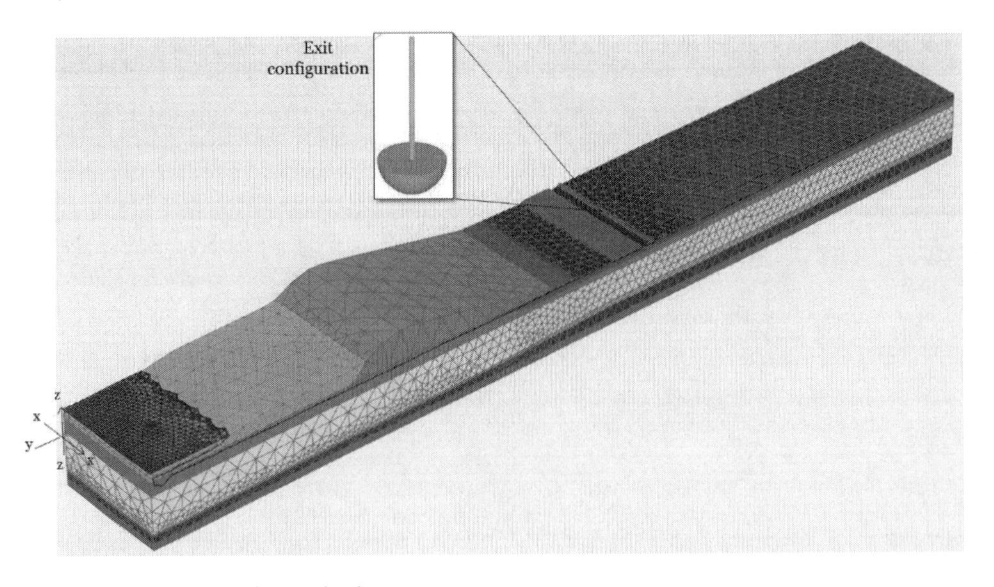

Figure 4. Discretized model of the embankment.

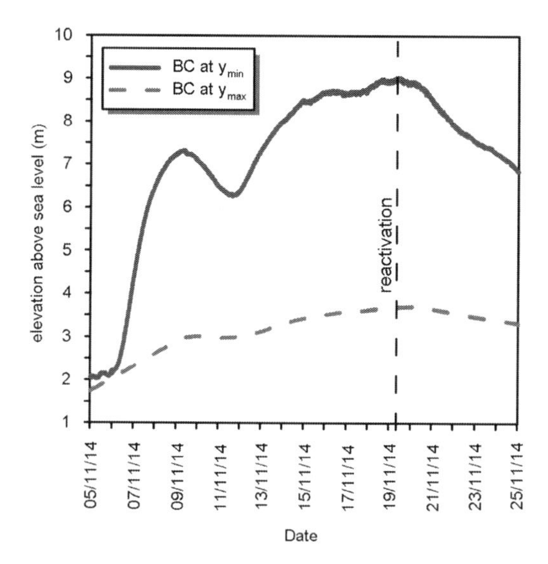

Figure 5. Evolution of hydraulic loads during the simulated high-water event.

upward hydraulic gradients higher than 0.5. On the other hand, the flow velocity in the vertical pipe should exceed the falling velocity of the grains in the sand-water mixture for sand boiling to take place (Van Beek 2015).

Such minimum flow velocity has been determined by the application of the Richardson & Zaki's (1954) semi-empirical equation:

$$v_z \geq w_t(1-c)^n \tag{2}$$

In which n is an empirically determined exponent dependent on the particle Reynolds number, ranging approximately between 2.4 and 4.65, c is the volumetric concentration and w_t is the settling velocity for small spherical particles. This latter variable can be determined by the Stokes law, which expresses w_t in terms of the square of the grain diameter. Assuming for this case of study $n = 4$, $c = 0.55$ and, according to a few available grain size distribution curves, $d = d_{50} = 0.18$ mm, it turned out that $v_z \geq 0.9$ mm/s.

Accordingly, on the basis of the adopted criteria for sand boil initiation, the model has been calibrated with respect to the reference values $i \geq 0.5$ e $v_z \geq 0.9$ mm/s.

Results of the model calculation, expressed in terms of average vertical gradient along the pipe, $\Delta h/L_{pipe}$, and representative vertical flow velocity, v_z, at the time when reactivation of sand boiling occurred, are shown in Figure 6. Different values of permeability of the pipe and the semi-spherical region, k_{exit}, have been investigated, assuming k_{exit} proportional to the permeability of the aquifer.

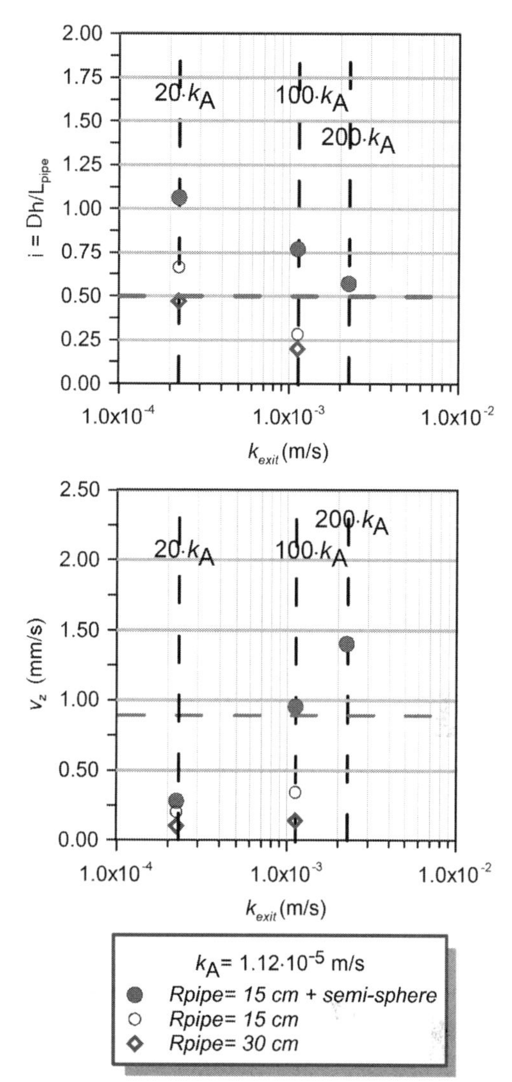

Figure 6. Predicted hydraulic gradients and flow velocity at the time when the sand boil reactivation occurred.

For useful comparison, results from analyses performed without considering a high-permeability semi-spherical cluster have been also included.

As expected, if k_{exit} is increases, the hydraulic gradient along the pipe decreases and flow velocity increases. Apparently, the latter parameter seems to increase more sharply if the semi-spherical region is included. Furthermore, for a constant value of k_{exit}, both gradient and velocity decrease for larger pipe diameters.

Criteria for sand boil reactivation are satisfied for Ø = 30 cm and k_{exit} in the range $1.1 \cdot 10^{-3}$ - $2.2 \cdot 10^{-3}$ m/s, which is about two order of magnitude higher than the permeability of Unit A,

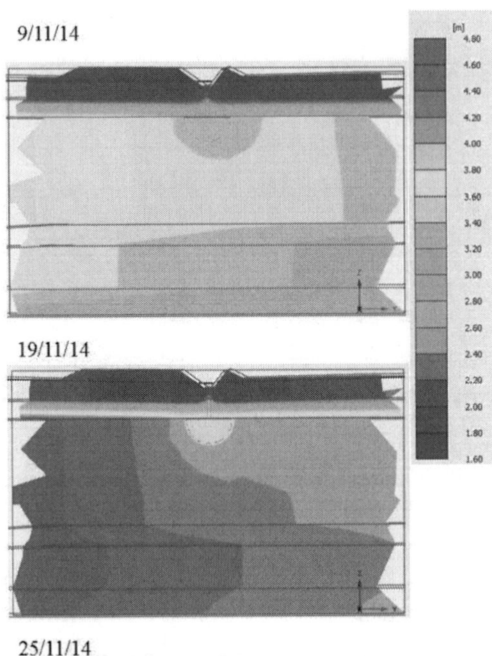

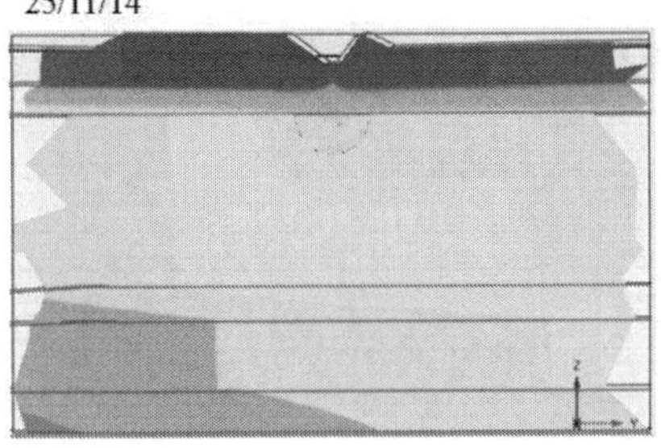

Figure 7. Groundwater heads in the y-z plane at different stages.

k_A. Based on field observations, greater diameters would be non-realistic. It is worth remarking here that in order to confirm the reliability of the calibration results, additional detailed investigations would be required.

Groundwater head results, displayed as shadings in the y-z plane passing through the pipe, are shown in Figure 7 for three different stages.

2.2.2 *Verification for the November 2016 high-water event*

In November 2016, high-water levels were recorded along the river. Although being particularly threatening in the middle-lower portion, the event did not reactivate the sand boil analysed in this paper. Figure 8 shows the river level evolution in the cross-section, from November 25 to December 3, 2016. As it can be observed, a unique peak of about 7.8 m a.s.l. can be individuated on November 29. If compared, river levels reached during the 2014 and 2016 events are rather different in terms of both trends and maximum levels.

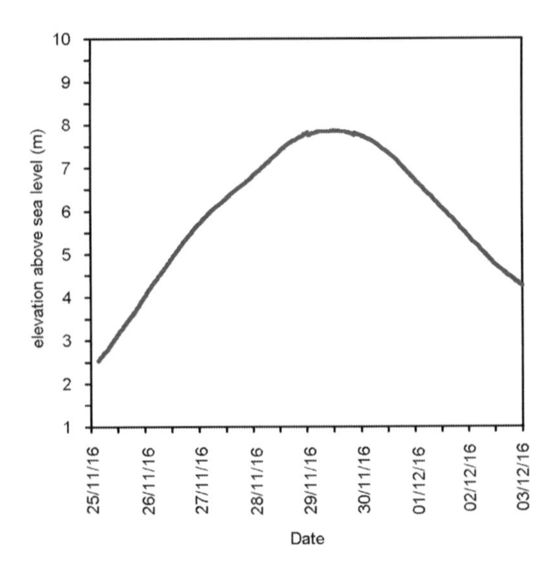

Figure 8. River level evolution during the November 2016 event.

The effectiveness of the model, as previously calibrated, has been assessed with respect to such recent event. The computed results, expressed in terms of average vertical gradient along the pipe and average vertical flow velocity at peak discharge, have turned out to be $\Delta h/L_{pipe} = 0.56$ and $v_z = 0.73$ mm/s, respectively. According to the criteria adopted in this study, they seem to confirmthe no-reactivation of the sand boil.

3 CONCLUSIONS

Backward erosion piping threatens Po river embankments stability in large portions of the middle-lower stretch of the watercourse. This paper tackles the issue of sand boil reactivation with reference to a specific river sector, located in the province of Ferrara (Northern Italy). A 3D numerical model of the groundwater flow causing the last reactivation, during the 2014 major high-water event, is presented.

Stratigraphic conditions of the cross-section passing through the sand boil have been mainly identified from CPTU data interpretation. Such configuration, consisting of a top impervious layer overlying a sandy aquifer, has turned out to be consistent with the typical one required for piping initiation.

Geotechnical parameters of the different soil units, including permeability, have been derived using some well-established CPTU-based correlations. Mean values of the relevant parameters have been adopted in this preliminary analysis, although a more advanced model, accounting for spatial variability of the subsoil across the section, might help in properly simulating the actual response of the embankment.

The pre-exiting downstream exit, required for the activation of the phenomenon, has been modelled as a vertical cylindrical pipe, passing through the impervious layers, ending in a 4 m-diameter semi-spherical region at the base. The coefficient of permeability of the outflow regions, k_{exit}, as well as the diameter of the pipe (Ø) have been derived from back-analysis, taking into account relevant criteria found in the literature for sand boil reactivation (e.g. USACE 1956).

The calibrated model has been then verified for the November 2016 high-water event, during which the sand boil did not reactivate. Indeed, the predicted hydraulic gradient along the pipe and the vertical flow velocity have turned out to be smaller than the threshold values assumed for sand boil reactivation, though close to them.

Although the proposed model seems to properly reproduce the response of the riverbank system, further information regarding the outflow zone geometry, grain size distribution of soils and groundwater monitoring data recorded during high-water periods would certainly help in reducing the uncertainties in the preliminary assumptions here made.

REFERENCES

Fleshman, M.S. & Rice, J.D. 2014. Laboratory Modeling of the Mechanisms of Piping Erosion Initiation. *J. Geotech. Geoenviron. Eng*. 140(6): 04014017.

García Martínez, M.F., Gragnano, C.G., Gottardi, G., Marchi, M., Tonni, L. and Rosso, A. 2016. Analysis of Underseepage Phenomena of River Po Embankments. *Procedia Engineering* 158: 338–343.

Gottardi, G., Marchi, M. and Tonni, L. (2015). Static stability of Po river banks on a wide area. Geotechnical Engineering for Infrastructure and Development. *Proc. XVI European Conf. on Soil Mechanics and Geotechnical Engineering* 4: 1675–1680.

Merli, C., Colombo, A., Riani, C., Rosso A., Martelli, L., Rosselli, S., Severi, P., Biavati, G., De Andrea, S., Fossati, D., Gottardi, G., Tonni, L., Marchi, M., García Martínez, M.F., Fioravante, V., Giretti, D., Madiai, C., Vannucchi, G., Gargini, E., Pergalani, F. and Compagnoni, M. 2015. Seismic stability analyses of the Po River Banks. *Engineering Geology for Society and Territory* 2: 877–880.

Richardson, J.F. & Zaki, W.N. 1954. Sedimentation and fluidisation: part 1. *Trans. Inst. Chem. Eng*. 32: 35–53.

Robertson, P.K. 2009. Interpretation of cone penetration tests – a unified approach. *Can. Geotech. J.* 46(11): 1337–1355.

Robertson, P.K. 2010. Estimating in-situ soil permeability from CPT & CPTu. In *Proc. of the 2nd Int. Symp. on Cone Penetration Testing (CPT'10)*, Huntington Beach, Calif.

Severi, P. & Biavati, G. 2013. Definizione del modello geologico e idrogeologico della zona arginale del fiume Po in destra idrografica da Boretto (RE) a Ro (FE). *Internal report*. Regione Emilia-Romagna, Servizio Geologico Sismico e dei Suoli.

USACE 1956. Investigation of underseepage and its control – Lower Mississippi river Levees. *Technical memorandum No. 3–424*, Volume 1, Waterways Experiment Station, Vicksburg, Mississippi.

Van Beek, V.M. 2015. Backward erosion piping: initiation and progression, PhD thesis, TU Delft, The Netherlands.

Van Beek, V.M., Bezuijen, A. & Sellmeijer, H. 2013. Backward erosion piping. In S. Bonelli (ed.),:*Erosion in geomechanics applied to dams and levees*: 193–269. London, UK and Hoboken, NJ, USA: Wiley.

Van Beek, V.M., Bezuijen, A., Sellmeijer, J.B. & Barends, F.B.J. 2014. Initiation of backward erosion piping in uniform sands. *Géotechnique* 64(12): 927–941.

Numerical Methods in Geotechnical Engineering IX – Cardoso et al. (Eds)
© 2018 Taylor & Francis Group, London, *ISBN 978-1-138-33203-4*

Finite element analysis of the monitored long-term settlement behaviour of Kouris earth dam in Cyprus

L. Pelecanos
University of Bath, UK

D. Skarlatos
Cyprus University of Technology, Cyprus

G. Pantazis
National Technical University of Athens, Greece

ABSTRACT: This paper presents a finite-element analysis of the long-term (>25 years) deformations of an earth dam. The aim is to examine the influence of (a) consolidation, (b) reservoir level changes and (c) seismic activity on the global deformations of embankment dams. The monitored long-term deformation behaviour of Kouris earth dam is analysed. Analysis of the field displacement data shows a consistent settlement trend with the time. Hydro-mechanical nonlinear finite-element analyses of the dam are conducted considering plasticity, small-strain stiffness and permeability. The stress history of the dam is modelled, including staged construction, reservoir impoundment, consolidation and seasonal reservoir level changes. It is shown that (a) seasonal reservoir level changes result in small fluctuations of dam settlements (i.e. vertical displacements), (b) horizontal dam crest displacements are insensitive to reservoir level changes and (c) the majority of the total settlements is due to soil consolidation.

1 INTRODUCTION

A large number of earth dams exist in the world and have different purposes, such as water supply, irrigation or generation of hydroelectric power. Their long-term operation and safety is important. The major threats for earth dam safety include earthquakes (Elgamal et al., 1990; Pelecanos et al., 2012a, b; 2013a, b; 2015; 2016; 2017a; 2018b), internal erosion (Bridle & Fell, 2013; Shire & O'Sullivan, 2013, 2016, Shire et al., 2013), reservoir level changes (Pytharouli & Stiros, 2005; Gikas & Sakellariou, 2008), hydraulic fracture etc. Climate variations are also important safety as large seasonal changes (e.g. hot summers and cold winters) result in significant reservoir level changes which further affect the water pressures within the dam and the entire stress regime. It is therefore essential to recognize the impact of climate changes on the response of earth dams and calculate its effects, especially if they become a threat for the dam's safety.

Many recent studies cover a wide range of dam behavior and operation, such as field monitoring (Kyrou et al., 2005; Dounias et al., 2012; Pelecanos et al., 2017b), laboratory experiments and computational simulations (Tedd et al., 1997; Alonso, 2005; Charles et al., 2008) for understanding dam performance and safety under various long-term conditions. Field monitoring is crucial for health monitoring of civil infrastructure (Acikgoz et al., 2018, 2017; Di Murro et al., 2016, 2019; Kechavarzi et al., 2016, 2019; Ouyang et al., 2015, 2018, Pelecanos et al., 2016; 2017c; 2018a; Seo et al., 2017a,b; Soga et al., 2015; 2017). Some researchers highlighted the importance of reservoir-induced dam deformations and tried to predict this interaction. However, this is not trivial and therefore more field data are required to better quantify the effects of reservoir level changes on dam deformations and safety.

This study presents a computational analysis of field monitored behavior of an instrumented earth dam, the Kouris dam, which is the largest dam in Cyprus. It aims to understand the effects of long-term environmental actions, such as reservoir level changes, on the behaviour of earth dams. The displacement data from over 25 years are available and nonlinear finite element analyses are performed to model the entire stress history of the dam, including construction, reservoir impoundment and reservoir level changes. The numerical analyses confirm the observed hypothesis that reservoir level changes affect the deformations of the dam crest.

2 KOURIS DAM

2.1 *Description of the dam*

Kouris dam is the largest and tallest dam in Cyprus. It is a zoned earth-rockfill operated by the Cyprus Water Development Department (WDD) and serves as the main water storage facility in the country. It was built during 1984–1988 and its embankment consists of a central clay core of low permeability, followed by thin layers of fine and coarse filters. The upstream shell consists of terrace gravels, which are adjacent to the upstream filters and then river gravels covered by rip rap on the upstream dam slope with a small cofferdam at the upstream dam toe. The downstream shell consists in its entirety by terrace gravels with talus deposits, which rest on a thin drainage gallery. Its crest is 570 m long and the embankment is 112 m high with the highest level of its reservoir at 102 m, and its total reservoir capacity is 115 million m3. Figure 1 shows a cross-sectional view with the various soil layers.

2.2 *Instrumentation*

Three independent instrumentation sets are installed on the dam to provide monitoring data about the deformations of the dam: (a) Embankment Crest Movement Indicators (ECMI), (b) a vertical geodetic network and (c) a three-dimensional (3D) geodetic network.

The first instrumentation network was installed after the construction of the dam in 1991 and consists of an observation pillar, (which serves as a fixed point) and six embankment crest movement indicators (ECMI) which were again, installed at the time of construction. The measurements of these points have been taken (irregularly, monthly or bi-monthly) since 1990 by the Cyprus WDD. The horizontal distance and the height difference of each ECMI from the pillar was determined using a Leica TC1101 total station, which provides an accuracy of ±2 mm ±2 ppm for the distance measurements and ±1" for the angle measurements.

Additionally, two modern geodetic networks were installed later, after the dam operated safely for more than 15 years, in 2006. The latter two networks were designed, deployed and managed by an academic team of the National Technical University of Athens and later of the Cyprus University of Technology. These consist of a vertical (1D) and

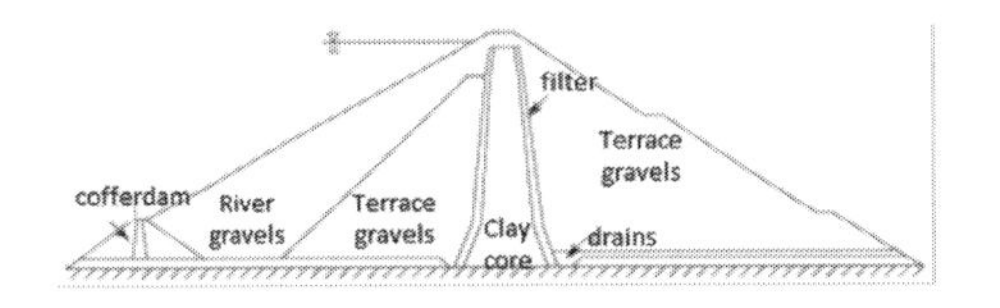

Figure 1. Geometry of the dam.

a three-dimensional (3D) control network. In theory, deformation monitoring is performed by the establishment, measurement and adjustment of such a network. The vertical control network was deployed in 2006 (Constantinou, 2013). It consists of 7 control points. Six of them (R1-R6) are bronze benchmarks, located along both sides of the wall on the dam crest road and have a distance of about 100 m between successive benchmarks. The seventh monitoring point is the pillar T2 which is located about 1 km away from the dam, at a distant location that is considered to be unaffected by the dam deformations and therefore can be considered as the fixed reference point of the network.

Three main periods of measurements (July and December 2006 (Temenos, 2007) and June 2012 (Pantelidou, 2013)) were performed to determine the elevation differences between the points by using either (a) the spirit leveling method, or (b) the accurate trigonometric heighting method. More details about these methods, their technical background and their validation are offered by Lambrou (2007) and Lambrou & Pantazis (2007, 2010). The latter two modern systems are considered extremely robust and much more accurate than the old initial system. The reason for installing these additional systems was to offer a more detailed deformation pattern of the dam structure. The installation location of the various instruments on the dam is shown in Figure 2.

2.3 *Monitoring data*

A wealth of monitoring data exists for Kouris dam, and therefore a number of important observations can be made. Obviously, not all of that data can be discussed here and therefore for brevity only the most relevant data are presented in Figure 3. Figure 3 (a) shows the time-histories of several points on the dam crest from the ECMIs, since the construction of the dam embankment. It is shown that there is a consistent downward displacement trend for all crest instruments. It may also be observed that there is an initial large value of crest settlements of which the increase rate tends to get smaller and almost reaching a plateau, i.e. no significant crest settlements after 25 years. Moreover, it is shown that the monitoring points close to the middle of the dam crest exhibit the largest values of dam settlement, which was expected, as this is the evidence from other similar field measurements (Kyrou et al., 2005; Dounias et al., 2012).

In addition, Figure 3 (b) shows a comparison of dam deformations (vertical crest displacements) as those where obtained from the three different and independent instrumentation sets over the period 2006–2012. This is the period that was monitored from all three instrumentation sets. It is may be observed that a very good comparison is obtained

Figure 2. Deformation instrumentation on the dam crest.

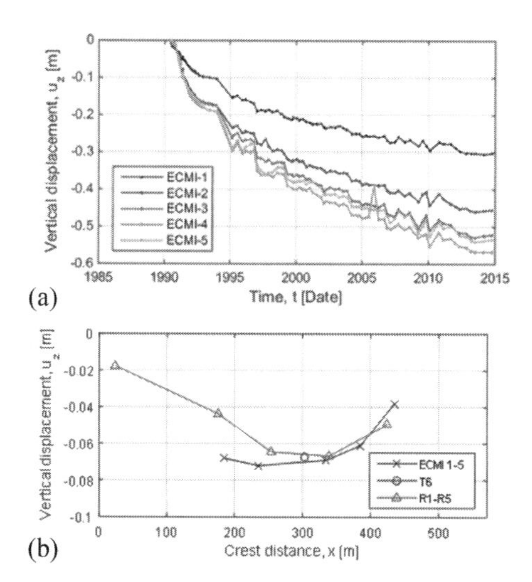

Figure 3. Monitoring data: (a) time-history of vertical settlements of the dam crest (b) vertical crest displacements along the axis of the dam between 2006–2012.

between the three different and independent monitoring data sets which confirms their accuracy and reliability. Some minor differences that can be observed are well within what might be considered instrument accuracy or "noise" and therefore the monitoring system is demonstrated to be performing well.

3 NUMERICAL ANALYSIS

3.1 *Finite element model*

To better understand the response of the dam to environmental changes, relevant finite element analysis was conducted. This was to explore the long-term dam behaviour and shed some light on the physical processes involved which governed the observed long-term settlements of the dam crest. More specifically, such an analysis would allow a comparison between the relative effects of soil consolidation and reservoir level fluctuations.

Two-dimensional (2D) plane-strain non-linear elasto-plastic transient coupled-consolidation finite element (FE) analyses were performed with the FE software ABAQUS/Implicit. The dam and the foundation underneath were discretized with 2540 finite 8-noded quadratic iso-parametric elements with 16 displacement and 4 pore water pressure degrees-of-freedom (CPE8RP). The FE mesh (Figure 4 (a)) includes the clay core, the core filters, the rockfill shells, the grout curtain and two layers of foundation material. The materials are modeled as elasto-plastic consolidating. The elasto-plastic constitutive model adopted is the Mohr-Coulomb (MC) coupled with a hyperbolic small-strain stiffness that dictates the non-linear stiffness degradation of shear modulus, G, with the shear strain, γ, (similar to Pelecanos et al. (2015)) given by Equation 1 and combined with a constant Poisson's ratio, ν.

$$\tau = \frac{G_{max} \cdot \gamma}{1 + \lambda \cdot \gamma} \tag{1}$$

Where, G is the soil shear modulus ($= E/2(1+\nu)$), E is the soil Young's modulus, G_{max} is the maximum value of G for zero shear strain, γ is the shear strain and λ is a model calibration parameter. The latter parameter dictates shear stiffness degradation with strain. However, due to lack of relevant experimental data (the design of the dam was performed in the 1970–80 s), the hyperbolic model was calibrated against appropriate empirical curves from the literature, using Vucetic & Dobry (1991) were used for the cohesive clay core and the foundation materials (including the grout curtain) and the Rollins et al. (1998) curves for the granular filters and the rockfill shells.

The analysis was performed over a number of steps in order to simulate the appropriate stress history of the dam. Firstly, initial stresses were generated with level ground (i.e. riverbed) and subsequently the embankment was constructed in successive layers within 12 months and finally the reservoir impoundment took place. Then, the operation of the dam was simulated including consolidation and the associated fluctuations of the reservoir which followed the monitored time-history of the level of the reservoir, as shown in Figure 4(b).

3.2 *Computed dam response*

The results of the FE analysis are shown in Figure 5. Figure 5 (a) shows the resultant displacements of the nodes at the case of full reservoir, and it can clearly be observed that most of the deformations are located close to the dam crest. Moreover, Figure 5 (b) shows the pore water pressure distribution within the dam at the time of full reservoir. This reveals the saturation of the upstream rockfill and the reduction of the hydraulic head within the centre due to the seepage through the low-permeability clay core and grout curtain. Figure 5 (c) presents the time-hisotry of vertical displacements at selected points in the dam (C: crest, E: mid-height of Embankment height, F1: shallow point in the foundation, F2: deep point in the foundation).

It is observed that there is generally a good match between the field-monitored (from ECMI-4) and numerically-predicted displacement time-histories. A general trend of downward movements can be observed which is attributed to the consolidation of the dam materials. This means that the reservoir fluctuations do not appear to have significant effect on the settlements. The vertical displacements within the embankment and the two foundation layers (Points E, F1 and F2) are presented as well. It may be observed that the consolidation settlement values of the foundation

Table 1. Material properties & model parameters.

	E	φ	c	k	λ
Material	MPa	deg	kPs	m/s	–
Filter	300	45	37	10^{-3}	2500
Rockfill	400	45	37	10^{-3}	2500
Foundation	700	45	35	10^{-8}	900
Clay core	300	45	40	10^{-9}	900

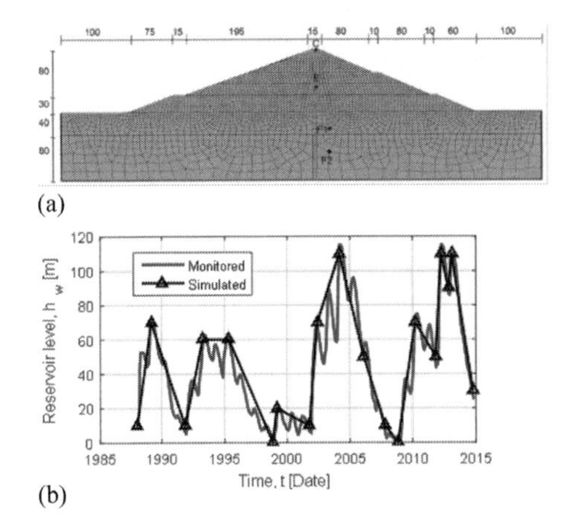

Figure 4. Numerical modelling: (a) FE mesh, (b) reservoir level changes.

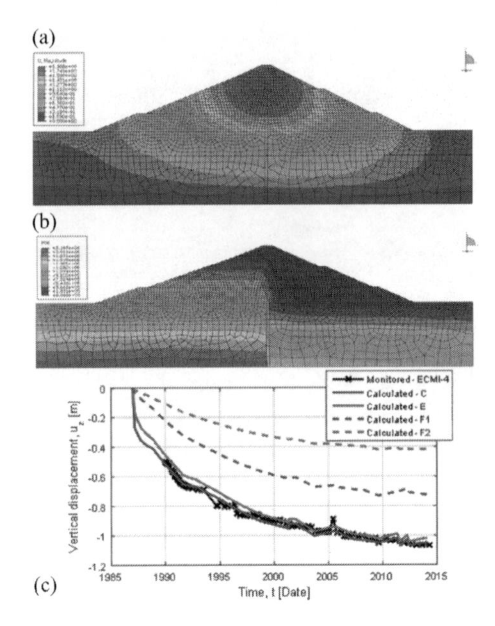

Figure 5. Analysis results: (a) resultant deformations, (b) pore water pressures and (c) time-histories of vertical crest displacements.

layers are significantly large and comparable to those of the dam crest, which may suggest that the consolidation settlements within both the dam embankment and the foundation dominate the total crest settlements. This is believed to be due to the dissipation of the excess pore water pressure built-up in the embankment and the low-permeability foundation because of the construction of the large dam.

It is also postulated that reservoir level fluctuations do not contribute very substantially and instead the dominant mechanism inducing long-term vertical crest displacements is consolidation of the soil. It is therefore concluded that there is no major influence from the reservoir fluctuations implying that the majority of displacements is due to long-term soil consolidation.

4 CONCLUSIONS

This paper explores the long-term settlement behaviour of earth dams due to climate changes leading to variations of the reservoir level. The study considers the long-term monitoring data from a well-instrumented dam along with relevant nonlinear finite element analyses.

The numerical model was analysed using the nonlinear finite element method, considering hydro-mechanical coupling and elasto-plastic soil behaviour with nonlinear elasticity prior to material yield.

The findings of this study may be summarised as follows:

a. Consolidation of the dam embankment and foundation materials following the construction of the dam appears to be the main reason for the long-term dam settlements.
b. There is an observed correlation between the dam displacements and reservoir level fluctuations which suggests seasonal variations of dam settlements.
c. The contribution of reservoir level fluctuations on the displacements of the dam was found to be small compared to that of long-term consolidation.

ACKNOWLEDGEMENTS

Assistance, data and information were provided by S. Patsali, and E. Kanonistis of the Cyprus Water Development Department. Students C. Temenos, G. Stavrou, C. Constantinou and A. Pantelidou of NTUA and CUT participated in the field measurements and data processing. This contribution is gratefully acknowledged.

REFERENCES

Acikgoz, MS, Pelecanos, L, Giardina, G, Aitken, J & Soga, K 2017, 'Distributed sensing of a masonry vault during nearby piling', Structural Control and Health Monitoring, vol 24, no. 3, p. e1872.

Acikgoz, MS, Pelecanos, L, Giardina, G & Soga, K 2016, 'Field monitoring of piling effects on a nearby masonry vault using distributed sensing.', International Conference of Smart Infrastructure and Construction, ICE Publishing, Cambridge.

Alonso, E.E., Olivella, S. and Pinyol, N.M., 2005. A review of Beliche Dam. Géotechnique, 55(4), pp. 267–285.

Bridle, R. and Fell, R. eds., 2013. Internal erosion of existing dams, levees and dykes, and their foundations. Bulletin 164, Volume 1: Internal Erosion Processes and Engineering Assessment.

Charles, J.A., 2008. The engineering behaviour of fill materials: the use, misuse and disuse of case histories. Géotechnique, 58(7), pp. 541–570.

Constantinou, C., 2013, Monitoring the defromation of Kouris dam in Cyprus by using GNSS measurements. Diploma thesis (in Greek), Cyprus University of Technology, Cyprus.

Di Murro, V, Pelecanos, L, Soga, K, Kechavarzi, C, Morton, RF & Scibile, L 2016, 'Distributed fibre optic long-term monitoring of concrete-lined tunnel section TT10 at CERN.', International Conference of Smart Infrastructure and Construction, ICE Publishing, Cambridge.

Di Murro, V, Pelecanos, L, Soga, K, Kechavarzi, C, Morton, RF, Scibile, L, 2019, Long-term deformation monitoring of CERN concrete-lined tunnels using distributed fibre-optic sensing., Geotechnical Engineering Journal of the SEAGS & AGSSEA. 50 (1) (Accepted).

Dounias, G.T., Anastasopoulos, K. and Kountouris, A., 2012. Long-term behaviour of embankment dams: seven Greek dams. Proceedings of the Institution of Civil Engineers-Geotechnical Engineering, 165(3), pp. 157–177.

Elgamal, A.W.M., Scott, R.F., Succarieh, M.F. and Yan, L., 1990. La Villita dam response during five earthquakes including permanent deformation. Journal of Geotechnical Engineering, 116(10), pp. 1443–1462.

Gikas, V. and Sakellariou, M., 2008. Settlement analysis of the Mornos earth dam (Greece): Evidence from numerical modeling and geodetic monitoring. Engineering Structures, 30(11), pp. 3074–3081.

Kechavarzi, C, Soga, K, de Battista, N, Pelecanos, L, Elshafie, MZEB & Mair, RJ 2016, Distributed Fibre Optic Strain Sensing for Monitoring Civil Infrastructure, Thomas Telford, London.

Kechavarzi, C., Pelecanos, L., de Battista, N., Soga, K. 2019. Distributed fiber optic sensing for monitoring reinforced concrete piles. Geotechnical Engineering Journal of the SEAGS & AGSSEA. 50 (1) (Accepted)

Kyrou, K., Penman, A. and Artemis, C., 2005. The first 30 years of Lefkara Dam. Proceedings of the Institution of Civil Engineers-Geotechnical Engineering, 158(2), pp. 113–122.

Lambrou, E., 2007, Accurate definition of altitude difference with geodetic total stations (in Greek), Technika Chronika, vol. 10, no. 1–2, pp. 37–46.

Lambrou, E. and Pantazis, G., 2007, A convenient method for accurate height differences determination, Sofia.

Lambrou, E. and Pantazis, G., 2010, Applied Geodesy, Athens: Ziti.

Lomb, N.R., 1976. Least-squares frequency analysis of unequally spaced data. Astrophysics and space science, 39(2), pp. 447–462.

Ouyang, Y, Broadbent, K, Bell, A, Pelecanos, L & Soga, K 2015, 'The use of fibre optic instrumentation to monitor the O-Cell load test on a single working pile in London', Proceedings of the XVI European Conference on Soil Mechanics and Geotechnical Engineering, Edinburgh.

Ouyang, Y., Pelecanos, L., Soga, K. 2018, 'The field investigation of tension cracks on a slim cement grouted column'. DFI-EFFC International Conference on Deep Foundations and Ground Improvement: Urbanization and Infrastructure Development-Future Challenges, Rome, Italy. (Accepted)

Pantelidou, A., 2013, Monitoring the vertical displacements of Kouris dam in Cyprus by usng modern accurate methods. Diploma thesis (in Greek), Cyprus University of Technology, Cyprus.

Pelecanos, L, Soga, K, Hardy, S, Blair, A & Carter, K 2016, 'Distributed fibre optic monitoring of tension piles under a basement excavation at the V&A museum in London.', International Conference of Smart Infrastructure and Construction, ICE Publishing, Cambridge.

Pelecanos, L, Kontoe, S & Zdravkovic, L 2012a, 'Numerical Analysis of the Seismic Response of La Villita Dam in Mexico'. 17th Biennial BDS (British Dam Society) Conference, Leeds, UK.

Pelecanos, L, Kontoe, S & Zdravkovic, L 2012b, 'Static and dynamic analysis of La Villita dam in Mexico'.

2nd International Conference on Performance-Based Design in Earthquake Geotechnical Engineering, Taormina, Italy,
Pelecanos, L., Kontoe, S. and Zdravković, L., 2013a. Numerical modelling of hydrodynamic pressures on dams. Computers and Geotechnics, 53, pp. 68–82.
Pelecanos, L., Kontoe, S., & Zdravkovic, L. 2013b. Assessment of boundary conditions for dam-reservoir interaction problems. 4th International Conference on Computational Methods in Structural Dynamics (COMPDYN), Kos Island, Greece.
Pelecanos, L., Kontoe, S. and Zdravković, L., 2015. A case study on the seismic performance of earth dams. Géotechnique, 65(11), pp. 923–935.
Pelecanos, L., Kontoe, S. and Zdravković, L., 2016. Dam–reservoir interaction effects on the elastic dynamic response of concrete and earth dams. Soil Dynamics and Earthquake Engineering, 82, pp. 138–141.
Pelecanos, L, Kontoe, S & Zdravkovic, L 2017a, 'Steady-state and transient dynamic visco-elastic response of concrete and earth dams due to dam-reservoir interaction' 6th International Conference on Computational Methods in Structural Dynamics and Earthquake Engineering, Rhodes, Greece.
Pelecanos, L., Skarlatos, D. and Pantazis, G., 2017b. Dam performance and safety in tropical climates – recent developments on field monitoring and computational analysis. 8th International Conference on Structural Engineering and Construction Management (8th ICSECM), Kandy, Sri Lanka.
Pelecanos, L, Soga, K, Chunge, MPM, Ouyang, Y, Kwan, V, Kechavarzi, C & Nicholson, D 2017c, 'Distributed fibre-optic monitoring of an Osterberg-cell pile test in London.', Geotechnique Letters, vol 7, no. 2, pp. 1–9.
Pelecanos, L., Soga, K., Elshafie, M., Nicholas, d. B., Kechavarzi, C., Gue, C.Y., Ouyang, Y. and Seo, H., 2018a. Distributed Fibre Optic Sensing of Axially Loaded Bored Piles. Journal of Geotechnical and Geoenvironmental Engineering. 144 (3) 04017122.
Pelecanos, L., Kontoe, S., Zdravković, L., 2018b. Nonlinear analysis of nonlinear dynamic earth dam-reservoir interaction during earthquakes. Journal of Earthquake Engineering. (In Press)
Pelecanos, L., & Soga, K. 2017a. Innovative Structural Health Monitoring Of Foundation Piles Using Distributed Fibre-Optic Sensing. 8th International Conference on Structural Engineering and Construction Management 2017, Kandy, Sri Lanka.
Pelecanos, L., & Soga, K. 2017b. The use of distributed fibre-optic strain data to develop finite element models for foundation piles. 6th International Forum on Opto-electronic Sensor-based Monitoring in Geo-engineering, Nanjing, China.
Pytharouli, S.I. and Stiros, S.C., 2005. Ladon dam (Greece) deformation and reservoir level fluctuations: evidence for a causative relationship from the spectral analysis of a geodetic monitoring record. Engineering Structures, 27(3), pp. 361–370.
Rollins, K.M., Evans, M.D., Diehl, N.B. and III, W.D.D., 1998. Shear modulus and damping relationships for gravels. Journal of Geotechnical and Geoenvironmental Engineering, 124(5), pp. 396–405.
Scargle, J.D., 1982. Studies in astronomical time series analysis. II-Statistical aspects of spectral analysis of unevenly spaced data. The Astrophysical Journal, 263, pp. 835–853.
Seo, H, Pelecanos, L, Kwon, Y-S & Lee, I-M 2017a, 'Net load-displacement estimation in soil-nailing pullout tests' Proceedings of the Institution of Civil Engineers - Geotechnical engineering, 170 (6) 534–547.
Seo, H & Pelecanos, L 2017b, 'Load Transfer In Soil Anchors – Finite Element Analysis Of Pull-Out Tests'. 8th International Conference on Structural Engineering and Construction Management 2017, Kandy, Sri Lanka.
Shire, T., Pelecanos, L., Bo, H. and Taylor, H., 2013. Current research in embankment dam engineering at Imperial College London. Dams and Reservoirs, 23(1), pp. 25–28.
Shire, T. and O'Sullivan, C., 2013. Micromechanical assessment of an internal stability criterion. Acta Geotechnica, 8(1), pp. 81–90.
Shire, T. and O'Sullivan, C., 2016. Constriction size distributions of granular filters: a numerical study. Géotechnique, 66(10), pp. 826–839.
Skarlatos, D., 1999. Orthophotograph production in urban areas. The Photogrammetric Record, 16(94), pp. 643–650.
Skarlatos, D. and Kiparissi, S., 2012. Comparison of laser scanning, photogrammetry and SFM-MVS pipeline applied in structures and artificial surfaces. ISPRS annals of the photogrammetry, remote sensing and spatial information sciences, 3, pp. 299–304.
Soga, K, Kechavarzi, C, Pelecanos, L, de Battista, N, Williamson, M, Gue, CY, Di Murro, V & Elshafie, M 2017, 'Distributed fibre optic strain sensing for monitoring underground structures - Tunnels Case Studies ', in S Pamukcu, L Cheng (eds.), Underground Sensing, 1st edn, Elsevier.
Soga, K, Kwan, V, Pelecanos, L, Rui, Y, Schwamb, T, Seo, H & Wilcock, M 2015, 'The role of distributed sensing in understanding the engineering performance of geotechnical structures', Proceedings of the XVI European Conference on Soil Mechanics and Geotechnical Engineering, Edinburgh.
Tedd, P., Charles, C.J.A., Holton, I.R. and Robertshaw, A.C., 1997. The effect of reservoir drawdown and long-term consolidation on the deformation of old embankment dams. Geotechnique, 47(1), pp. 33–48.
Temenos, C., 2007, Monitoring the deformations of Kouris dams in Cyprus. Diploma thesis (in Greek). Diploma Thesis, National Technical University of Athens, Greece.
Vucetic, M. and Dobry, R., 1991. Effect of soil plasticity on cyclic response. Journal of geotechnical engineering, 117(1), pp. 89–107.

Numerical Methods in Geotechnical Engineering IX – Cardoso et al. (Eds)
© 2018 Taylor & Francis Group, London, ISBN 978-1-138-33203-4

Hydromechanical analysis of gravity dam foundations

N. Monteiro Azevedo & M.L.B. Farinha
Laboratório Nacional de Engenharia Civil (LNEC), Lisboa, Portugal

G. Mendonça & I. Cismasiu
Faculdade de Ciências e Tecnologia (UNL-FCT), Caparica, Portugal

ABSTRACT: A fully coupled hydromechanical analysis of the water flow through the rock mass discontinuities of three hypothetical gravity dams of different height was carried out. It was assumed that the dams had similar rock mass foundations, and 2D discontinuous models were developed and analysed with the coupled model for dam foundation analysis Parmac2D-Fflow. This paper presents the results of studies done to analyse the influence that the model geometry and the aperture of foundation discontinuities have on the numerical discharges and hydraulic heads. Analysis was done assuming different behaviour scenarios: i) linear elastic and ii) non-linear behavior at the concrete/rock and rock/rock interfaces. Additionally, results obtained during cycles of filling and emptying the reservoir were used to establish the influence lines of the hydrostatic pressure on recorded discharges. The relevance of using fracture flow models of dam foundation rock masses that take into account the coupled hydromechanical behavior is highlighted.

1 INTRODUCTION

A gravity dam, like other gravity structures, is designed so that the forces acting on the dam are primarily resisted by the dam's self-weight (USACE 1983). The main loads in gravity dams are the self-weight and water pressures (hydraulic pressures and uplift). The typical cross section of a gravity dam is a triangle with a base-to-height ratio of 0.8, and thus the dam/foundation interface is very large. Uplift pressures at the base of the dam and hydraulic pressures play an important role, as they reduce the stabilizing effect of the structure weight. Such hydraulic pressures can occur in rock mass discontinuities or even in the dam body, if there is any deficiency in lift joints, or fissures in contact with the reservoir.

In gravity dams seepage through the rock mass foundation discontinuities takes place mainly in the upstream-downstream direction and there is a significant interdependence between the mechanical and hydraulic behaviour. Some of the first studies on coupled hydromechanical behaviour in fractured rocks were related to dam foundations (e.g., Londe and Sabarly 1966; Louis 1969; Louis and Maini 1970), and since then it has been the subject of extensive laboratory and in situ research.

There are two different modelling approaches to simulate hydromechanical coupling behaviour of fractured rock masses. The first is based on an equivalent continuum (e.g. Erban & Gell 1984) and the second on discrete fracture networks (e.g. Farinha et al 2017). In fracture flow models the discontinuities are explicitly represented, with their individual hydromechanical properties. The flow of water is usually assumed to occur only through the joints. The joint water pressure gives rise to changes in joint apertures and, consequently, changes in flow rates.

This paper presents the results of two-dimensional (2D) fully coupled hydromechanical analysis of the water flow through the rock mass discontinuities of three hypothetical gravity dams of different height. Analysis was done assuming two different behaviour scenarios: i) linear elastic and ii) non-linear behavior at the concrete/rock and rock/rock interfaces. Additionally, results obtained during cycles of filling and emptying the reservoir were used to establish the influence lines of the hydrostatic pressure on recorded discharges.

2 HYDROMECHANICAL DISCONTINUUM MODEL

2.1 *Fluid flow analysis with Parmac2D-Fflow*

The hydromechanical model used in the study presented here, Parmac2D-Fflow, is part of the program Parmac2D (Azevedo 2003), and allows the interaction between the hydraulic behavior and the mechanical behaviour to be studied in a fully-coupled way. The model uses interface elements between the various blocks of the mechanical domain, and seepage channels (SC) in the hydraulic domain, which coincide with the midplane of the interface elements. The interaction between the various blocks which represent both the rock mass foundation and the dam is always edge to edge. The apertures of the foundation discontinuities and water pressures are updated at every timestep, as described in Azevedo and Farinha (2015) and in Farinha et al. (2017). It is assumed that rock blocks are impervious and that flow takes place only through the set of interconnecting discontinuities. Water pressures are defined on the hydraulic nodes (HN), which are coincident with the mechanical nodes, and flow rates are calculated in the SC.

Flow is modelled by means of the parallel plate model (Louis 1969), and the flow rate per model unit width is thus expressed by the cubic law. The flow rate [(m³ s⁻¹)m⁻¹] in each seepage channel is given by:

$$Q_{SC} = \frac{1}{12\,\mu} a_{h,SC}^3 \,\rho_w\, g\, \frac{\Delta H_{SC}}{L} = k_{SC}\,\rho_w\, g\,\Delta H_{SC} \quad (1)$$

where $1/(12\ \mu)$ = theoretical value of a joint permeability factor (also called joint permeability constant), being μ the dynamic viscosity of the fluid; a = hydraulic aperture of the seepage channel; ρ_w = water density; g = acceleration of gravity; and ΔH = difference in piezometric head between both ends of the discontinuity; and L = length of the SC.

The hydraulic aperture to be used in Equation 1 is given by:

$$a_h = a_0 + \Delta u_n \quad (2)$$

where a_0 = hydraulic aperture at nominal zero normal stress and Δu_n = joint normal displacement taken as positive in opening. A maximum aperture, a_{max}, is assumed, and a minimum value, a_{min}, below which mechanical closure does not affect the contact permeability.

Each model block is internally divided into a mesh of linear shape triangular elements in order to take its deformability into account. Flow is dependent on the state of stress within the foundation.

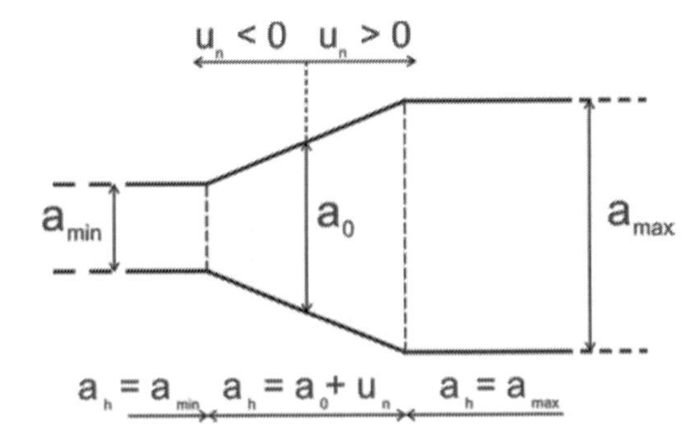

Figure 1. Hydraulic aperture.

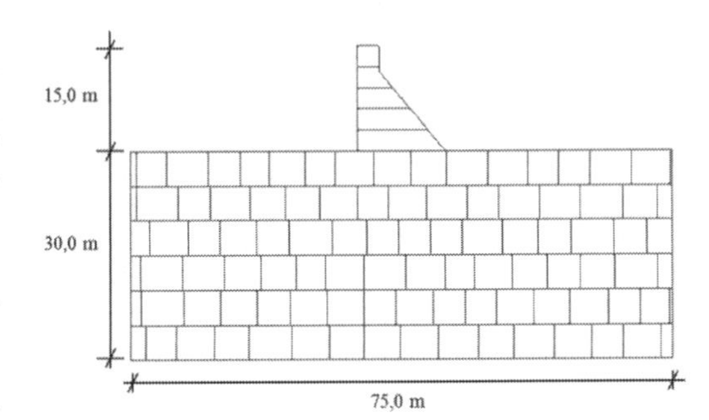

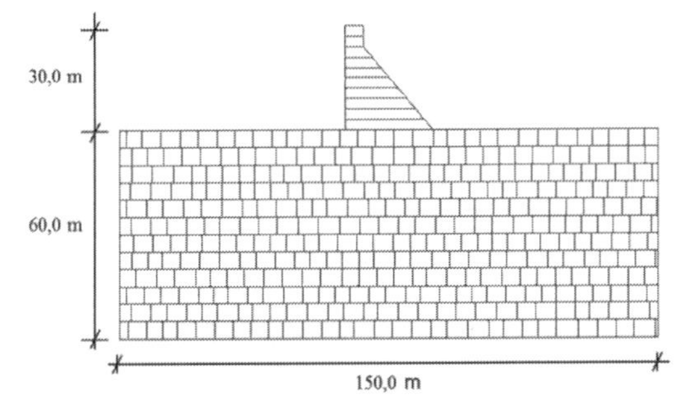

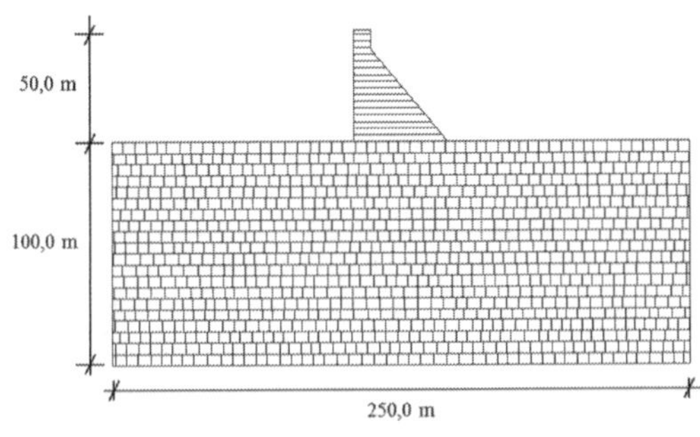

Figure 2. Geometry of the models of three hypothetical dams of different height.

2.2 *Geometry of the three different models*

The geometry of the different models that were developed is shown in Figure 2. Three different dam heights were considered: 15 m, 30 m and 50 m. In the foundation of each dam two sets of discontinuities were simulated: the first joint set is horizontal and continuous, with a spacing of 5.0 m, and the second set is formed by vertical cross-joints, with a spacing of 5.0 m normal to joint tracks and standard deviation from the

Table 1. Information about the mechanical models.

Dam height m	Number of deformable blocks	triangular elements	nodal points	interface elements
15	101	3873	3024	867
30	360	15433	11889	3559
50	1003	42793	32919	10096

Table 2. Information about the hydraulic models.

Dam height m	Number of hydraulic nodes	seepage channels
15	785	842
30	3157	3447
50	8886	9772

Table 3. Material mechanical and mass properties.

Material	E (GPa)	ν (-)	ρ (kg/m³)
Dam concrete	20.0	0.2	2400.0
Rock mass	12.0	0.2	2650.0

mean of 2.0 m. In concrete, a set of horizontal continuous discontinuities located 3.0 m apart was assumed to simulate dam lift joints.

Table 1 and Table 2 show information regarding both the mechanical and hydraulic models.

2.3 *Mechanical and hydraulic parameters*

Both dam concrete and rock mass blocks are assumed to follow elastic linear behaviour, with the properties shown in Table 3. In the nonlinear analysis, discontinuities are assigned a Mohr-Coulomb constitutive model, complemented with a tensile strength criterion (Table 4). A ratio of 0.5 is assumed between normal and shear stiffnesses. At the dam construction joints and at the concrete/rock mass interface cohesion and tensile strength were assigned 2.0 MPa. In rock joints, cohesion and tensile strength were assumed to be zero.

Table 5 shows the hydraulic properties of the seepage channels.

Properties are based on those used in the models developed for Pedrógão dam, which were validated against recorded data (Farinha 2010).

The following hydraulic apertures were considered in a base run: $a_0 = 0.1668$ mm, $a_{min} = 0.05$ mm and $a_{max} = 0.25$ mm. Afterwards, hydraulic apertures of $2a_0$ and of $a_0/2$ were also considered, in order to assess the influence of the initial aperture. In this case, $a_{min} = a_0/3$ and $a_{max} = 5\ a_0$.

The grout curtain was simulated by changing the permeability of the discontinuities in the grouted area, in such a way that it is 10 times less pervious than the surrounding rock mass.

2.4 *Boundary conditions*

Regarding the mechanical models, both horizontal and vertical displacements at the base of the models and horizontal displacements perpendicular to the lateral boundaries were prevented. In each model, a pressure equivalent to the hydrostatic pressure was applied on the bottom of the reservoir and at the upstream face of the dam.

Regarding hydraulic boundary conditions, the dam body and the lateral and bottom model boundaries are impermeable. It was assumed that there was no water downstream from the dam and that the reservoir was at the same level as the crest of the dam, therefore a constant pressure is assigned at the bottom of the reservoir corresponding to the water height. The drainage system was simulated assigning a hydraulic head along the drains equal to one third of the head upstream from the dam.

2.5 *Sequence of analysis*

Analysis was carried out in two loading stages, assessing firstly the mechanical effect of gravity loads with the reservoir empty, and secondly, applying the hydrostatic loading corresponding to the full reservoir to both the upstream face of the dam and reservoir bottom. In this second loading stage, mechanical pressure was first applied, followed by hydromechanical analysis.

3 RESULTS ANALYSIS

3.1 *Fluid flow analysis*

Results of the hydromechanical fluid flow analysis, with the reservoir at the same level as the crest of the dam are shown in Figures 3 and 4. Figure 3 shows the pseudo-equipotentials of piezometric head in the foundation of each dam. The term pseudo-equipotentials (Kafritsas 1987) is used due to the discrete nature of the flow, which takes place along the rock mass foundation discontinuities. For ease of analysis, Figure 4 shows the same information but in a different way. In this case, the percentages of hydraulic head contours within the dam foundation are presented. Percentage of hydraulic head is the ratio of the water head measured at a given

Table 4. Mechanical properties of interface elements.

Interface Element	Normal stiffness (GPa/m)	Shear stiffness (GPa/m)	Friction angle (°)	Coh. (MPa)	Tensile strength (MPa)
Concrete/concrete	12.0	6.0	35.0	2.0	2.0
Concrete /rock	12.0	6.0	35.0	2.0	2.0
Rock/rock	12.0	6.0	35.0	0.0	0.0

Table 5. Hydraulic properties.

Seepage channel	K_w (GPa)	K_{SC} ($MPa^{-1}\,s^{-1}$)
Concrete/rock	2.1	0.8300×10^8
Rock/rock	2.1	0.4150×10^8

level, expressed in metres of height of water, to the height of water in the reservoir above that level.

Both figures show that the loss of hydraulic head is concentrated below the heel of the dam, at the drainage and grout curtain's area. It should be noted that without drainage the hydraulic head decreases gradually below the base of the dam, as shown in Figure 5 for the lowest dam.

Figures also show that for the higher dams the loss of hydraulic head in depth has a tendency to be concentrated below the base of the dam.

3.2 *Water pressures along the dam/foundation joint*

The variation of water pressures along the dam/foundation joint for the three different dams is shown in Figures 6, 7 and 8. Each figure presents the results of non-linear analyses obtained with the hydraulic apertures of $a_0/2$, a_0, and $2a_0$, considering dams with no grout curtain or drainage system and dams with both grout curtain and drainage system.

The distribution of water pressures along the base of the dam is particularly relevant. At the design stage it is usually assumed that at the drainage line the head is equal to one third of the head difference between the heel and toe of the dam. Thus, a bi-linear distribution of the uplift pressure along the base of the dam is assumed. For an inoperative drainage system a linear distribution is usually assumed, i.e. a constant decrease in head between the heel and toe of the dam.

Figures show that the uplift pressures along the base of the dam are higher than those usually assumed at the design stage. Therefore, in gravity

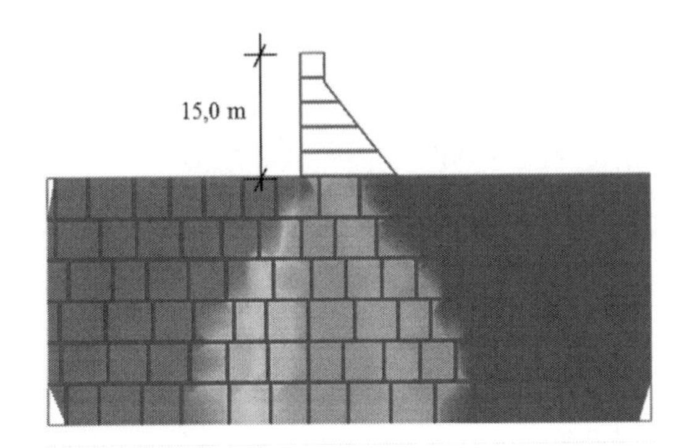

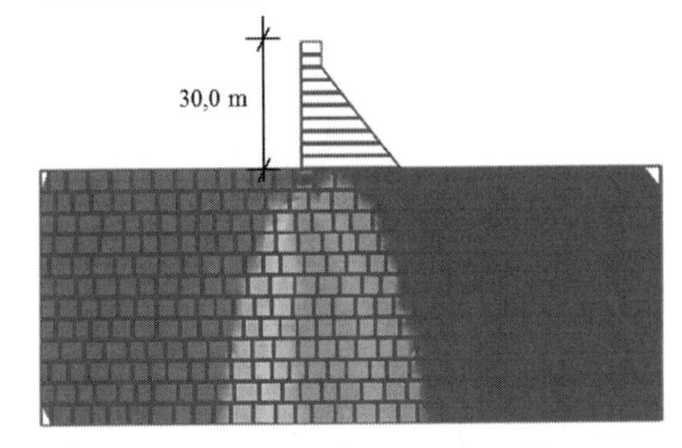

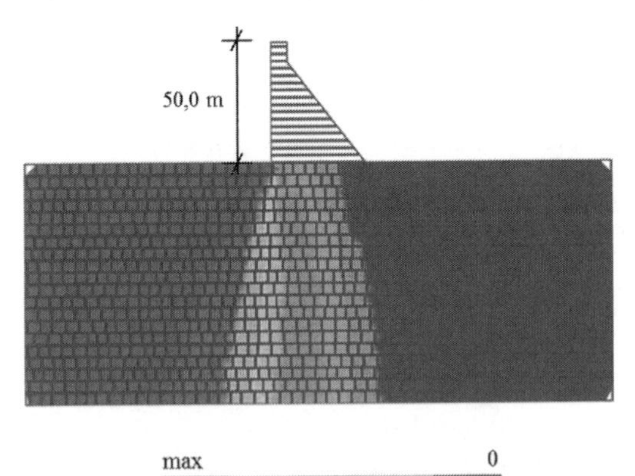

Figure 3. Pseudo-equipotentials contours of hydraulic head.

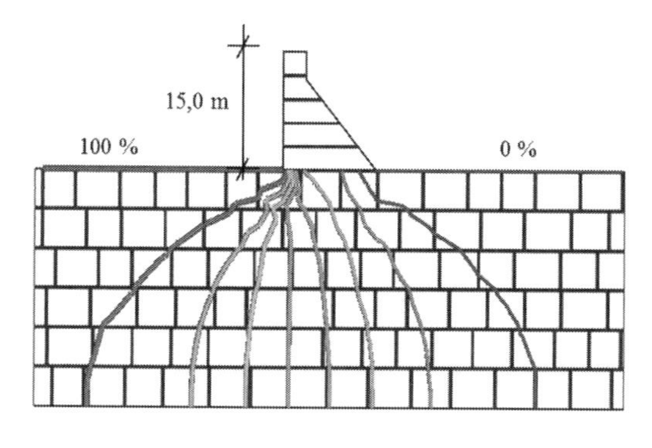

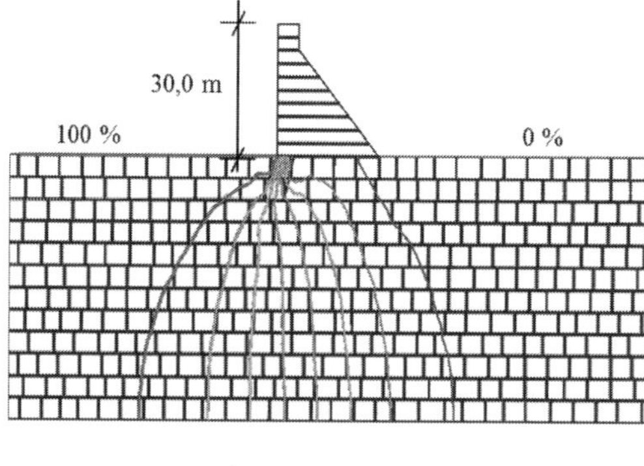

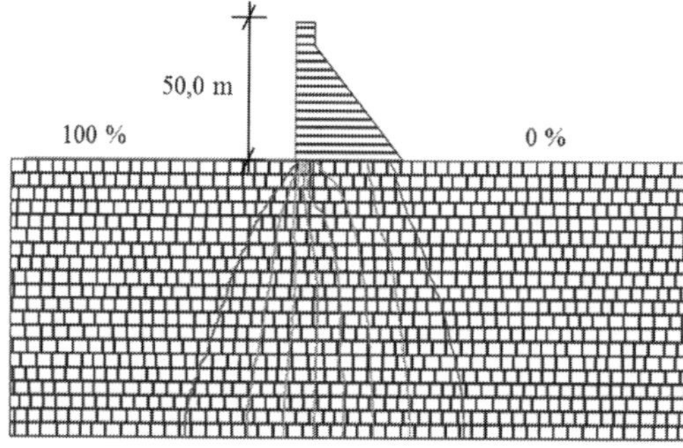

Figure 4. Percentage of hydraulic head for full reservoir.

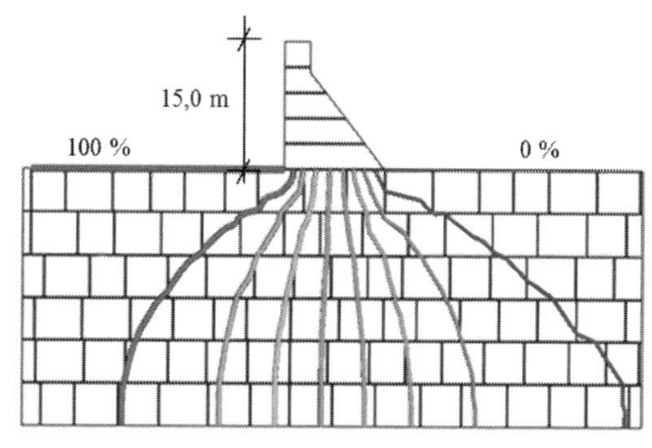

Figure 5. Percentage of hydraulic head for full reservoir assuming a dam foundation without grout curtain and without drainage system.

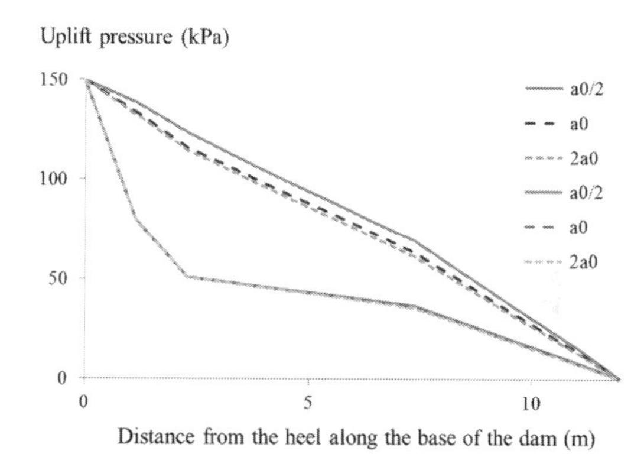

Figure 6. Water pressures along the base of the dam (dam 15 m high).

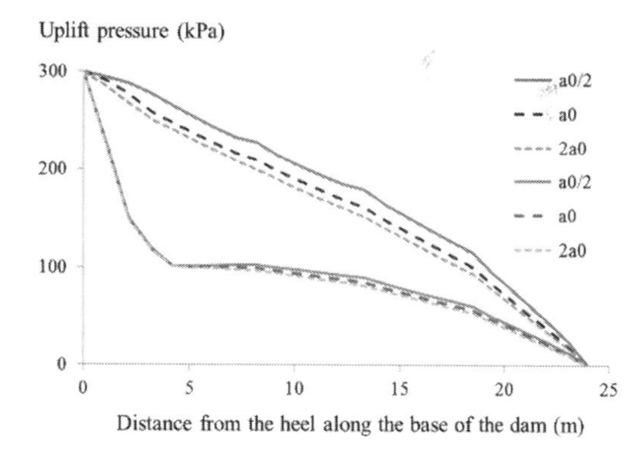

Figure 7. Water pressures along the base of the dam (dam 30 m high).

dam safety assessment non-linear hydromechanical analysis must be carried out.

The difference between the results of linear elastic and non-linear analysis increases with the dam height, however the difference is not relevant when a hydraulic aperture of $2a_0$ is considered.

3.3 *Quantity of water that flows through the model*

The quantity of water that flows through the models with full reservoir is presented in tables 6 and 7, assuming linear and non-linear behaviour, respectively. Results presented are those obtained in the analyses in which it is assumed that both grout curtains and drainage systems are installed within the dam foundation.

Results show that the quantity of water that flows through the models increases about ten times

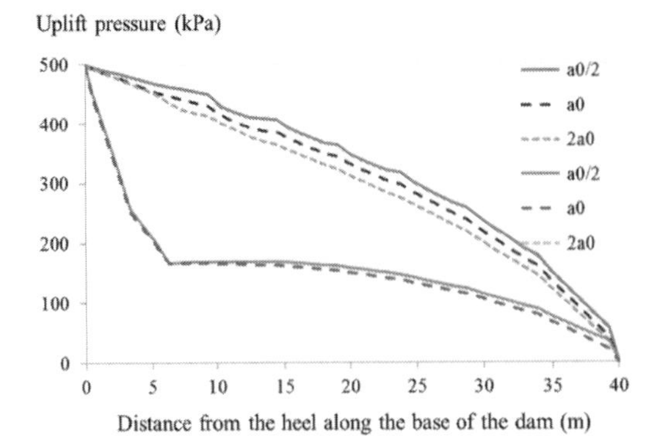

Figure 8. Water pressures along the base of the dam (dam 50 m high).

Table 6. Quantity of water that flows through the model: linear elastic analysis.

Dam height	Quantity of water (Lmin^{-1}) m^{-1}		
m	$a_0/2$	a_0	$2a_0$
15	0.053785	0.494029	4.214245
30	0.074790	0.833254	7.702266
50	0.076322	1.089814	11.37606

Table 7. Quantity of water that flows through the model: non-linear analysis.

Dam height	Quantity of water (Lmin^{-1}) m^{-1}		
m	$a_0/2$	a_0	$2a_0$
15	0.062366	0.531126	4.428751
30	0.085358	0.91286	8.411556
50	0.08851	1.300124	13.3804

when the aperture of the discontinuities doubles, and that it is greater when a non-linear analysis is carried out.

4 INFLUENCE LINE OF THE HYDROSTATIC PRESSURE ON RECORDED DISCHARGES

4.1 *Variations in discharges due to variations in reservoir level*

Due to the hydromechanical behaviour of rock masses, the permeability depends on the state of stress and strain within the foundation and, consequently, discharges vary with variations in the reservoir level and resulting changes in rock mass permeability.

Numerical simulations were carried out applying successive increments of water head at the reservoir bottom and of hydrostatic pressure in the upstream face of the dam so as to simulate the rising of water in the reservoir. For the three different dams, and for different hydraulic apertures, this load path was followed by an unload path, simulating the filling and subsequent emptying of the reservoir.

Figure 9 shows the drain discharges for the three dams of different height, assuming $a_0 = 0.1668$ mm.

4.2 *Adjustment of polynomial curves*

A series of different polynomial curves was adjusted to those shown in Figure 9, using the least square method, in order to define the function that better fits the numerical results (Figure 10).

From the eight polynomial functions that were adjusted to the calculated drain discharges it was found that a function of the type q = f (h^2, h) would roughly fit the numerical results for the lowest dam, 15 m high, and a function of the type q = f (h^4, h^2) would fit the numerical results for the highest dams, 30 and 50 m high.

4.3 *Functions that better fit the numerical results*

The functions that better fit the numerical results, for each dam of different height, are:

$$q = 6.78 \times 10^{-4}\, h^2 + 1{,}3131 \times 10^{-2}\, h \tag{3}$$

$$q = 5.41 \times 10^{-7}\, h^4 + 1{,}4769 \times 10^{-2}\, h \tag{4}$$

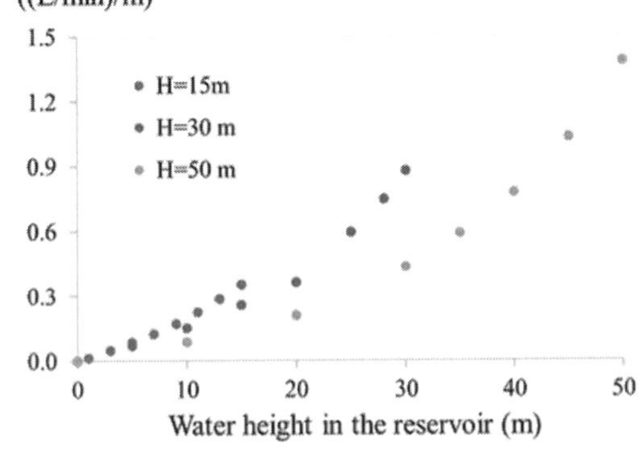

Figure 9. Calculated drain discharges at different reservoir levels for the three dams of different height.

$$q = 1.38 \times 10^{-7}\, h^4 + 1{,}0463 \times 10^{-2}\, h \qquad (5)$$

Equations 3, 4 and 5 fit the results obtained with the dams 15, 30 and 50 m high, respectively. Correlations coefficients of 0.999844, 0.999591 and 0.999662 were obtained.

The comparison between numerical results and those obtained with the adjusted polynomial curves is presented in Figure 11. Figure analysis clearly shows that the graphs of the adjusted curves follow adequately the curves which represent the variation in discharges due to variations in reservoir level.

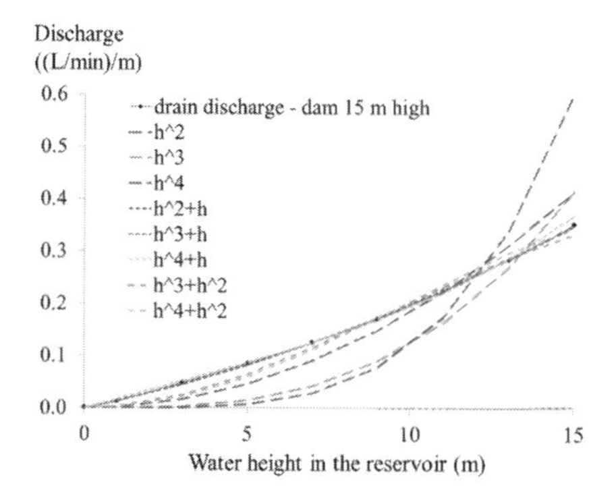

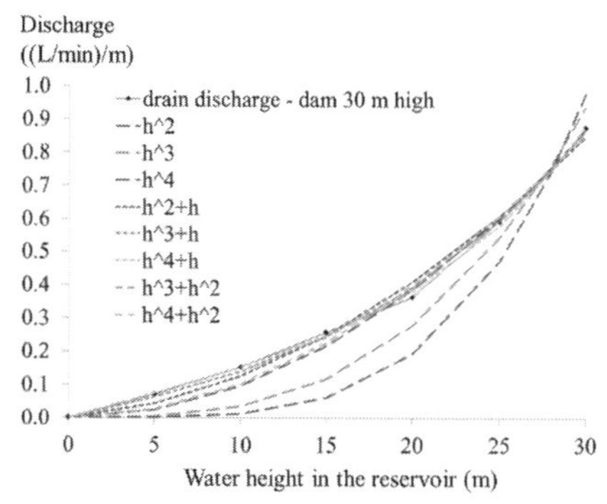

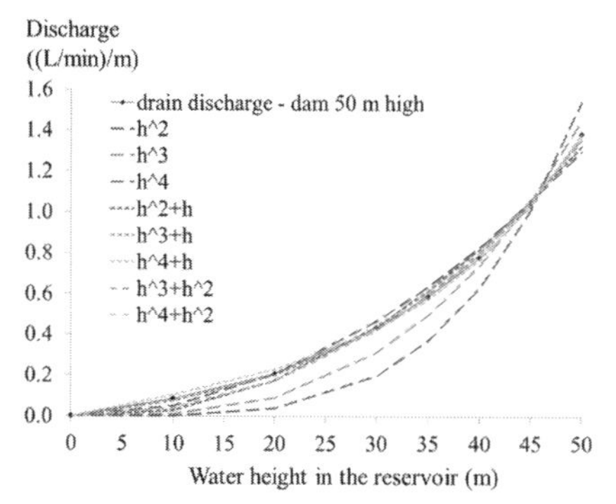

Figure 10. Adjustment of polynomial curves to the numerical results.

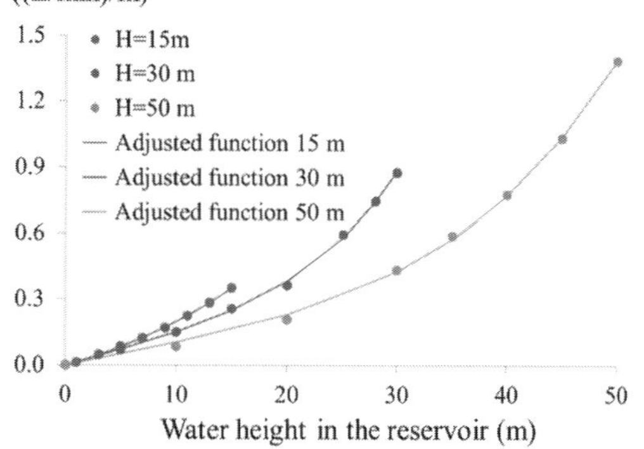

Figure 11. Functions that better fit the numerical results.

It must be highlighted that, in practice, the comparison of recorded discharges with those calculated using such functions may allow anomalous foundation behaviour to be identified.

5 CONCLUSIONS

This paper presents a study on seepage in gravity dam foundations using discontinuum models. Three hypothetical gravity dams of different height were considered, and analysis was done using a two dimensional model, Parmac2D-Fflow, which takes into account the coupled hydromechanical behaviour of rock masses. For each dam, the distribution of hydraulic head within the dam foundation, the variation of water pressures along the base of the dam and the quantity of water that flows through the models were analysed, assuming various apertures of foundation discontinuities and different behavior scenarios.

Results obtained clearly show the influence of the grout curtain and drainage system on the distribution of water pressures. The uplift water pressure along the dam base is consistent with what is usually prescribed in design codes, assuming a bi-linear uplift distribution to account for the relief drains. However, in non-linear analysis, water pressures downstream from the drainage line are slightly higher than those prescribed, which may compromise dam stability. The quantity of water that flows through the model depends mainly on the aperture of the foundation discontinuities and is higher for non-linear behaviour, when compared to the results obtained in a linear elastic analysis.

The study carried out allowed the shape of the influence line of the hydrostatic pressure on discharges to be obtained, for each dam height, and polynomial functions of which the graph adequately follows the shape of the above-mentioned lines were established. Further work is under way in order to establish a single polynomial function that takes into account not only the calculated discharges but also the dam height.

ACKNOWLEDGMENTS

The study presented here is part of a MSc thesis developed by the first author within the research project "DAMFA: Cutting-edge solutions for sustainable assessment of concrete dam foundations" which has been carried out jointly by LNEC and FCT/UNL with the main purpose of developing a numerical multiphysic integrated tool for the sustainable assessment of concrete dam foundations, taking into account the interaction between the mechanical, hydraulic and thermal behaviours.

REFERENCES

Azevedo N.M. 2003. A rigid particle discrete element model for the fracture analysis of plane and reinforced concrete. PhD Dissertation, Heriot-Watt University, Scotland.

Azevedo N.M. & Farinha, M.L.B. 2015. A hydromechanical model for the analysis of concrete gravity dam foundations. *Geotecnia - Journal of the Portuguese Society for Geotechnical Engineering* 133(March 2015):5–33 (in Portuguese).

Erban, P. & Gell, K. 1988. Consideration of the interaction between dam and bedrock in a coupled mechanic-hydraulic FE-program. *Rock Mech Rock Eng* 21(2):99–117.

Farinha M.L.B. 2010. Hydromechanical behavior of concrete dam foundations. *In situ* tests and numerical modelling. PhD Dissertation, Instituto Superior Técnico, Lisboa.

Farinha, M.L.B., Azevedo, N.M. & Candeias, M. 2017. Small displacement coupled analysis of concrete gravity dam foundations: Static and Dynamic Conditions. *Rock Mech Rock Eng* 50(2):439–464.

Kafritsas J.C. 1987. Coupled flow/deformation analysis of jointed rock with the distinct element method. PhD Dissertation, Massachusetts Institute of Technology, Cambridge.

Londe, P. & Sabarly, F. 1966. La distribution des perméabilités dans la fondation des barrages voûtes en fonction du champ de contraint. In *Proc. of the 1st Intern. Congr. on Rock Mechanics. 25 September - 1 October 1966,* Lisboa, Portugal, Vol.II, pp 517–522.

Louis C. 1969. A study of groundwater flow in jointed rock and its influence on the stability of rock masses. PhD Dissertation, University of Karlsruhe (in German), English translation, Imperial College Rock Mechanics Research Report nº10, London.

Louis, C. & Maini, Y.N. 1970. Determination of in situ hydraulic parameters in jointed rock. In *Proc. of the 2nd Intern. Congr. on Rock* Mechanics, 21–26 September 1970, Belgrade, USA, Vol.I, pp 235–245.

USACE 1983. Design of gravity dams on rock foundations: sliding stability assessment by limit equilibrium and selection of shear strength parameters. Technical Report GL-83-13, Washington DC.

Numerical Methods in Geotechnical Engineering IX – Cardoso et al. (Eds)
© 2018 Taylor & Francis Group, London, ISBN 978-1-138-33203-4

3D coupled hydromechanical analysis of dam foundations

M.L.B. Farinha, N. Monteiro Azevedo, N.S. Leitão, E. Castilho & R. Câmara
Laboratório Nacional de Engenharia Civil (LNEC), Lisboa, Portugal

ABSTRACT: The explicit formulation of a small displacement model for the coupled hydro-mechanical analysis of concrete dam foundations based on interface finite element technology is presented. The proposed 3D hydromechanical coupled model is based on a 2D coupled model that has been recently proposed. This 3D model, as in the 2D version, requires a thorough pre-processing stage to ensure that the interaction between the various blocks which represent the rock mass foundation and the dam is always face to face. The mechanical part of the model, though limited to small displacements, has the advantage of allowing an accurate representation of the stress distribution along the interfaces, such as rock mass joints. The hydraulic part and the mechanical part of the model are fully compatible. The verification and application examples presented in this paper show that the proposed 3D hydromechanical coupled model allows seepage flow through rock masses to be accurately simulated.

1 INTRODUCTION

It is well known that the majority of recorded failures of concrete dams have been due to problems in the foundation rock mass. In these rock masses the majority of the flow takes place through the foundation discontinuities, and the coupled hydromechanical behavior plays a relevant role. The assessment of the stability of concrete dams is usually based on limiting equilibrium procedures but nowadays more advanced analysis may be carried out with discontinuum models, which should take into account the hydromechanical behavior of concrete dam rock mass foundations.

The first finite element numerical models for coupled hydromechanical analysis of fracture media were presented in the early 1970s (e.g. Noorishad 1971). Several authors have subsequently presented either continuum or discontinuuum coupled hydromechanical models specifically developed for dam engineering (e.g. Wittke 1990, Erban & Gell 1988, Damjanac 1996, Callari et al. 2004). More recently, 3D hydromechanical coupled models have been mainly developed to simulate the complex problem of hydraulic fracturing, for oil and gas exploration, hot-dry rock geothermal energy investigations, and studies for nuclear waste disposal.

Numerical analysis presented in this paper is carried out using a hydromechanical model which is the extension to 3D of a previously developed 2D coupled model for dam foundation analysis (Azevedo & Farinha 2015), which has been fully validated in both static and dynamic conditions (Azevedo et al. 2017). The paper starts by briefly presenting the proposed formulation of the hydraulic part that has been implemented in the code Parmac3D-Fflow, followed by some verification examples. Finally, an application to an idealised concrete dam/foundation system is presented.

2 FORMULATION

2.1 *Mechanical part of the model*

The mechanical model is a model of discrete nature and uses an explicit solution algorithm based on the centred difference method (Azevedo 2003). The domain is divided into a group of blocks which interact with each other and this interaction between blocks must be face to face. The interfaces can slip and separate. Each model block is internally divided into a tetrahedral mesh in order to simulate the material deformability.

In 3D, the process of ensuring that the interaction between blocks is always face to face is much more complex than in 2D, where the interaction is edge to edge. Figure 1 shows the triangular interface element in small displacements used in the mechanical part of the 3D hydromechanical coupled model and Figure 2 shows the mesh of tetrahedra and the triangles at the interfaces between a system of six different blocks. Both figures show that the contact surfaces between adjacent blocks are fully compatible.

2.2 *Compatibility between the mechanical and the hydraulic part of the model*

The interfaces between the blocks are filled with water, which moves due to the fluid pressure field, from areas of higher energy to areas of lower energy. The flow through the interfaces is con-

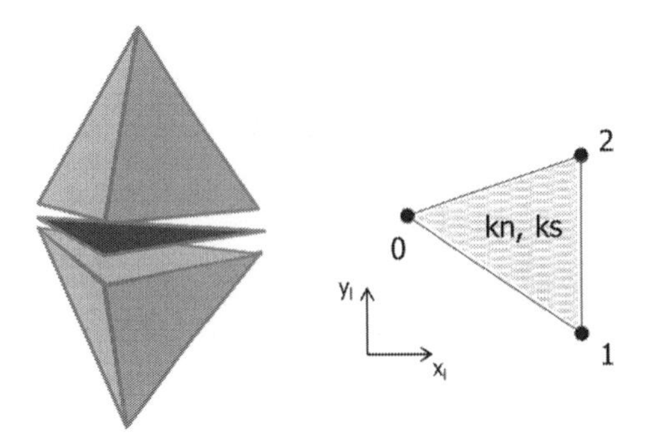

Figure 1. Triangle joint element in small displacements.

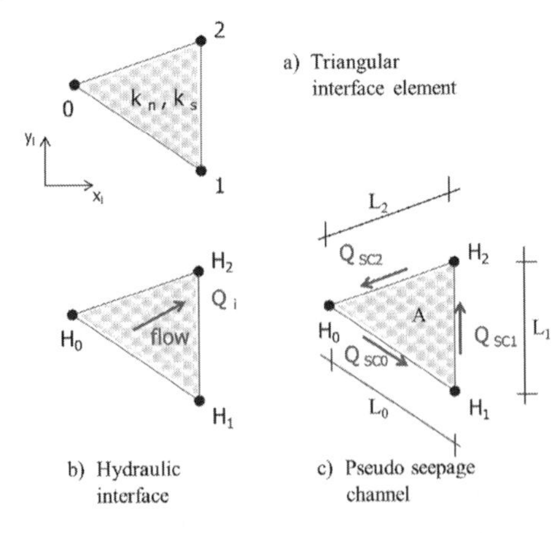

Figure 3. Mechanical and hydraulic part of the model. Formulation used in the hydraulic analysis.

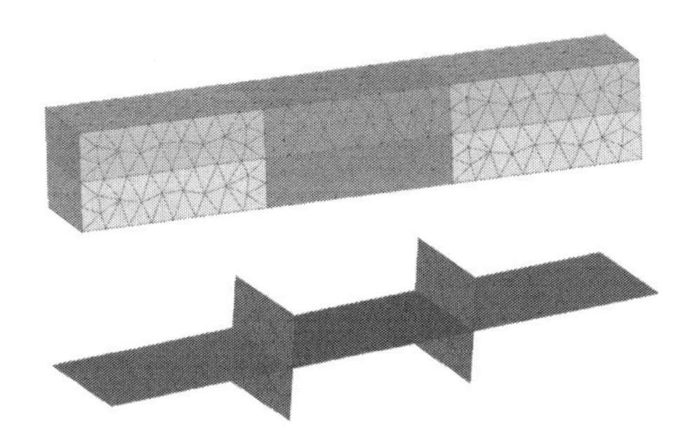

Figure 2. System of six different blocks: division into a mesh of tetrahedral elements and triangles at the interfaces between the blocks.

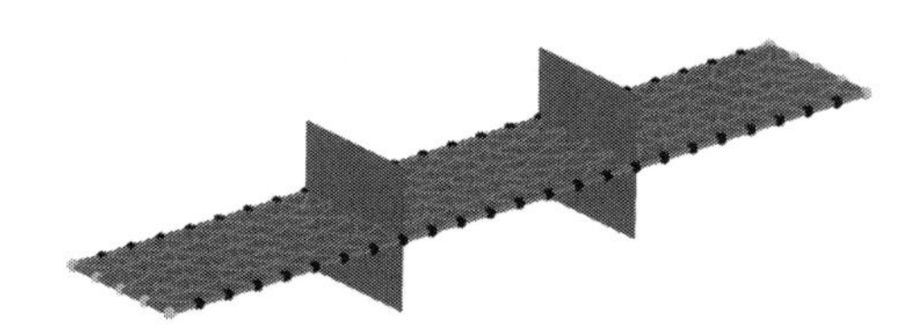

Figure 4. Horizontal discontinuity through which water flows: hydraulic super nodes and network of pseudo seepage-channels.

sidered laminar and it is assumed that the solid rock matrix is impervious. It should be noted that Louis (1969) and Barton et al. (1985) showed that the assumption of laminar flow in rock mass dam foundation discontinuities is valid. Flow is modelled by means of the parallel plate model, and the flow rate is expressed by the cubic law (Louis 1969).

The hydraulic model is superimposed on the mechanical model. Figure 3 shows both the mechanical part and the hydraulic part of the proposed interface model, which are fully compatible. Figure 3a) shows the triangular interface element used in the mechanical model for block interaction. Given the triangular interface element formulation, the hydraulic interfaces are created and each of the hydraulic super nodes (H) represents all the mechanical nodes that are found with the same coordinates.

A 2D hydraulic formulation directly based on FE technology (Yan & Zheng 2017), as shown in Figure 3b) could have been adopted, but it was decided to adopt an extension of the 2D model, which is based on simpler but numerically more robust unidirectional flow formulation. Figure 3c) shows the hydraulic super nodes and the unidirectional seepage channels, called pseudo seepage channels, located on the edges of the triangular hydraulic interfaces. For the pseudo seepage channels located at the edges of the associated triangular interface a pseudo width is calculated, in such a way that the total area of the pseudo seepage channels is equal to the area of the hydraulic interface:

$$(L_0 + L_1 + L_2)\, w = A \tag{1}$$

where L = length of each edge of the triangular interface; w = pseudo width, that has to be calculated for each triangular interface element; and A = area of the triangular interface element.

Figure 4 shows a horizontal discontinuity through which water flows, where the hydraulic super nodes and the network of pseudo seepage channels are represented.

The hydraulic aperture associated with each integration point (both ends of the pseudo seepage channels) is obtained as a function of the interface's normal displacement (mechanical aperture) and of three different parameters: a_0, a_{min} and a_{max}. As shown in Figure 5, a_0 is the hydraulic aperture for zero mechanical aperture, which represents the permeability of the medium when free of imposed stresses. For very high values of the interface aperture the value of a_{max} is used, which limits the permeability of the pseudo seepage channel up to a maximum value.

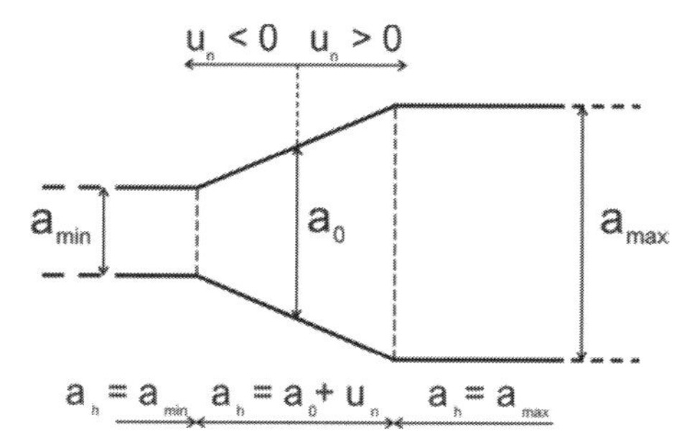

Figure 5. Hydraulic aperture.

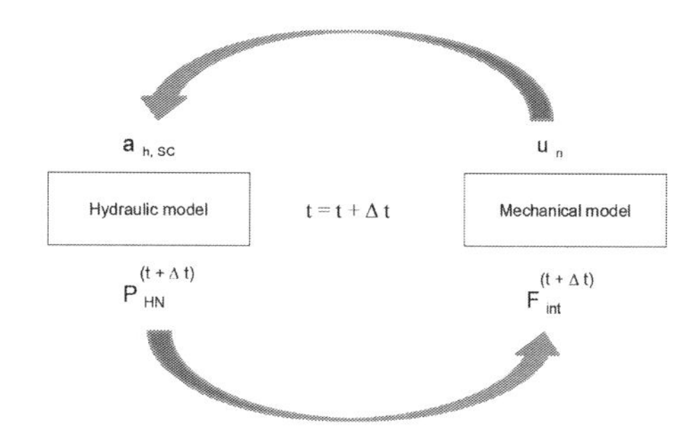

Figure 6. Hydromechanical calculation cycle.

For an interface element submitted to significant compression stresses, the value of a_{min} is used, which represents the permeability when the discontinuities are highly compressed.

2.3 *Coupled hydromechanical model*

The seepage-stress coupled model results from the coupling between the mechanical model and the hydraulic model. Figure 6 shows the calculation cycle of the hydromechanical model, which evolves over time through the interaction between both domains. At each timestep the hydraulic apertures are calculated taking into account the discontinuities' normal displacements calculated with the mechanical model. The water pressures calculated with the hydraulic model are then transferred to the mechanical model and are considered in the calculation of the internal forces at the discontinuities (effective stresses).

It is important to note that in the model presented here (Parmac3D-Fflow) there is a perfect superimposition between both the mechanical and hydraulic models (the nodal points of the mechanical model are at the same position as the nodal points of the hydraulic model), which makes it easier to define boundary conditions and optimises information transfer between the two domains.

3 VERIFICATION EXAMPLES

3.1 *Flow cases*

The ability of the Parmac3D-Fflow model to accurately predict the flow rates and water pressures was verified by carrying out two different hydraulic numerical analysis: i) flow through a rock mass isotropic medium, and ii) flow through two different permeability formations. For both cases the numerical results were compared with analytical solutions, which are presented in Yan & Zheng (2017). The performance of the coupled hydromechanical model was assessed considering a system of six impervious blocks and a saturated discontinuity.

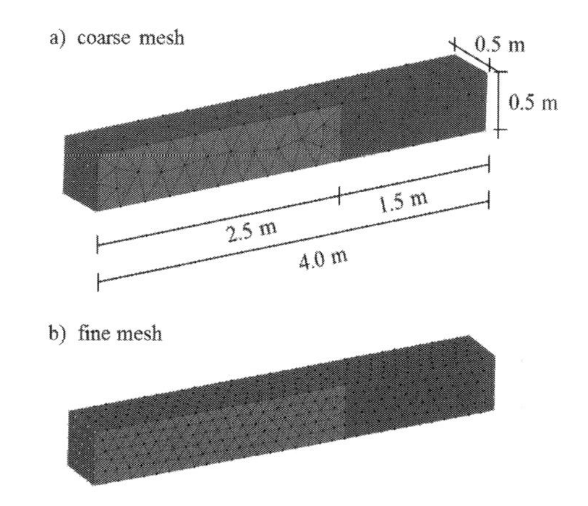

Figure 7. Meshes used in numerical analysis. Division into blocks and internal mesh.

3.2 *Flow through a rock mass isotropic medium*

The first studied case was the steady state seepage flow through a rock mass medium of which the boundaries are defined by a brick shape 4 m long, 0.5 m high and 0.5 m wide. A homogeneous and isotropic medium was assumed and firstly it was considered that the whole model had the same permeability; then two different permeability formations were considered, the first 2.5 m long and the second 1.5 m long. Figure 7 shows the two different meshes which were used: a coarse mesh with an average size of the tetrahedral edges of 0.25 m and a fine mesh with an average size of the tetrahedral edges of 0.125 m. In both cases, the domain is composed of two different blocks which are internally divided into tetrahedral meshes. Water flow takes place through the pseudo seepage channels located at the edges of the triangular interface elements, located between adjacent tetrahedral elements.

Figure 8 shows the hydraulic super nodes and the seepage channels considered in both meshes and Table 1 gives some information about both meshes.

The four boundaries at the front, rear, bottom and top are assumed to be impervious and the

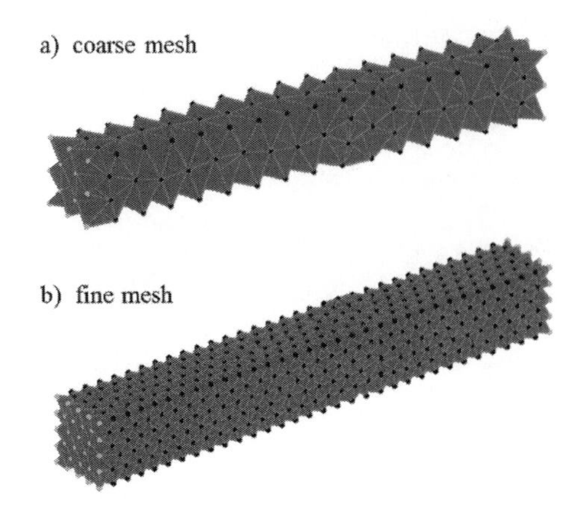

Figure 8. Hydraulic super nodes and network of pseudo seepage channels.

Table 1. Flow through a rock mass isotropic medium: characteristics of the 3D meshes.

	Average size of tetrahedral edges	
	0.25 m	0.125 m
Gridpoints	1832	11,196
Tetrahedra	458	2799
Triangular interfaces	738	4875
Hydraulic super nodes	177	841
Pseudo seepage channels	2214	14,625

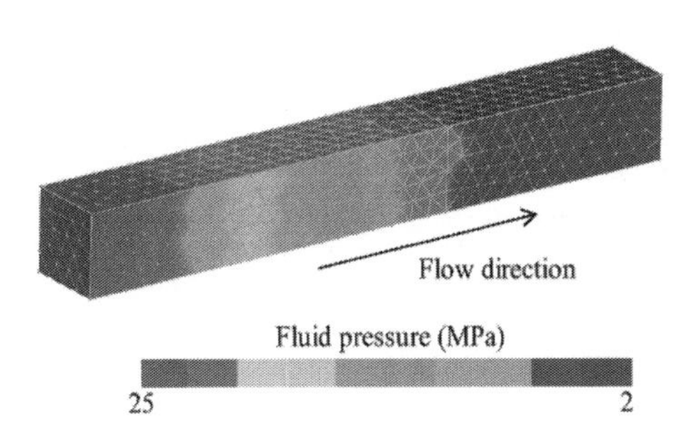

Figure 9. Field of water pressures (fine mesh).

water pressure is fixed at 25 MPa at the left boundary surface and at 2 MPa at the right boundary surface. Therefore, water flows under the pressure difference from the left hand side to the right hand side of the model.

The hydraulic aperture of the pseudo seepage channels is assumed to be $a_0 = 0.1668$ mm. In this case, a simple hydraulic analysis is carried out and thus the hydraulic aperture of the pseudo seepage channels remains constant throughout the calculation.

Figure 9 shows the field of water pressures obtained with the 3D hydraulic fine mesh model and Figure 10 shows the variation in water pressures along the central axis of the model in the direction of the flow. In this figure the results obtained with both the coarse and the fine meshes are presented along with the analytical solution. The bi-linear distribution of water pressures is due to the two different permeabilities. It can be seen that the numerical results are very close to the analytical solution, and that a better agreement is obtained with the fine mesh.

Table 2 shows the difference to the analytical solutions, in terms of flow rate and water pressure at the interface between the two domains with different permeability. Results show that the numerical water pressures are very close to the analytical solution and that the difference in flow rate is also small.

3.3 *System of six impervious blocks and a saturated discontinuity*

The performance of the coupled hydromechanical model was assessed with a group of six blocks separated by a horizontal discontinuity, through which water flows, and by two vertical discontinuities, which are impervious (Fig. 11). An elastic model is assumed in the horizontal discontinuity and no shear displacement is allowed at the vertical triangular interfaces. The Young's modulus, density and Poisson's ratio of each block are assumed to be 20 GPa, 2400 kg/m^3, and 0.2, respectively. The following parameters are assumed at the hori-

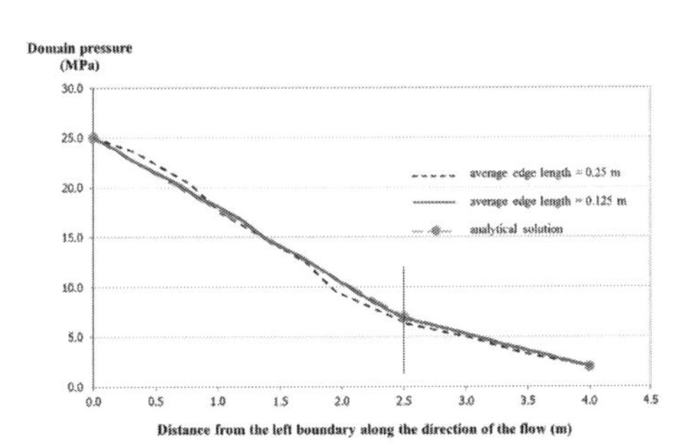

Figure 10. Variation in water pressures along the axis of the model in the direction of the flow.

Table 2. Flow rate and water pressure at the interface: difference to the analytical solution.

	Difference to the analytical solution (%)	
Model	Flow rate	Water pressure at the interface between the two domains with different permeability
Coarse mesh	–7.5%	–2.6%
Fine mesh	–1.6%	–0.2%

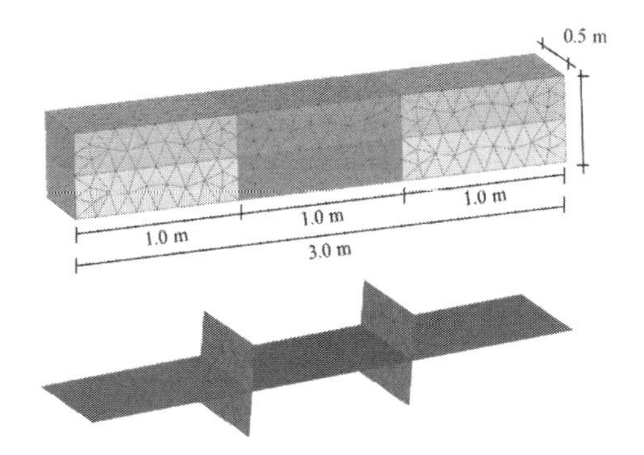

Figure 11. System of six blocks and a saturated horizontal discontinuity: internal mesh of tetrahedra and triangular elements at the interfaces.

Table 3. System of six impervious blocks and a saturated discontinuity: mesh characteristics.

Gridpoints	1987
Tetrahedra	2458
Triangular interfaces	330
Hydraulic super nodes	150
Pseudo seepage channels	726

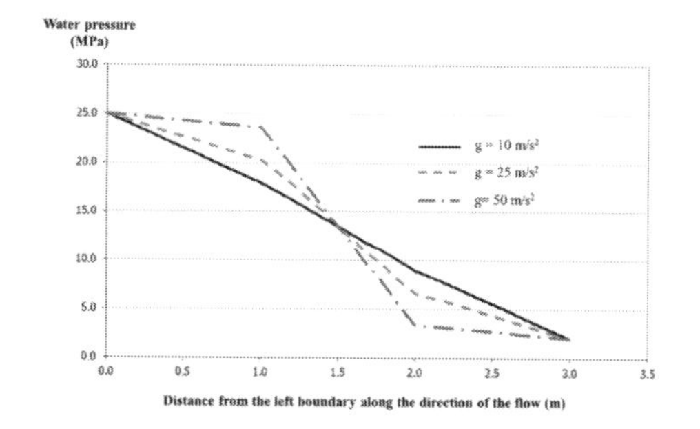

Figure 12. Variation in water pressure along the horizontal discontinuity.

zontal discontinuity: normal and shear stiffnesses of 200 GPa/m, joint aperture at nominal zero normal stress (a_0) of 8.34×10^{-8} m, and a_{min} equal to $a_0/3$. Mesh characteristics are shown in Table 3.

The model is fixed at the bottom. As in the examples presented in section 3.2, the water pressure is fixed at 25 MPa at the vertical left boundary surface and at 2 MPa at the vertical right boundary.

Firstly it was assumed that the three blocks at the top of the model had the same weight, and then the weight of the top middle block was increased, considering higher values of the acceleration of gravity (g).

Figure 12 shows the numerical water pressures calculated in the three different analyses along the central axis. It can be seen that when it is assumed that the weight of all the blocks is the same the water pressure decreases linearly. However, when the weight of the top block in the middle increases, the aperture of the discontinuity section below that block decreases, which has an influence on the water pressure. These results show that the coupled hydromechanical model is able to properly simulate a reduction in permeability due to changes in the mechanical aperture of part of the discontinuity.

It is worth mentioning that the initial hydraulic aperture of the triangular interface elements defines the permeability of the rock mass. Therefore, in practice, this parameter has to be calibrated so as to properly simulate not only flow through the foundation discontinuities but also the overall permeability of the rock mass.

4 APPLICATION TO AN IDEALISED DAM/FOUNDATION SYSTEM

4.1 *Numerical analysis*

The Parmac3D-Fflow proposed coupled model was used to simulate the hidromechanical behavior of an idealized dam/foundation system. It is assumed that the foundation is homogeneous and, in a simplified way, that the dam has no grout or drainage system.

Three different analyses were carried out: firstly, an uncoupled analysis in which it is assumed that the hydraulic aperture of the foundation discontinuities remains constant; secondly, an elastic coupled analysis; and thirdly, a coupled analysis that takes into account the non-linear behaviour of both the dam/foundation interface and of the dam foundation discontinuities.

4.2 *Model geometry*

The seepage-stress coupled model of an idealized dam/foundation system is shown in Figure 13. The foundation model has a base width of 170.0 m in the upstream-downstream direction and is 80.0 m deep. The model thickness is 20.0 m. The concrete dam is 40.0 m high.

The foundation model is divided into cube shaped blocks with an edge of 10.0 m, and the concrete dam is divided into brick shape blocks 10.0 m long, 10.0 m wide and 2.5 m high. Blocks are internally divided into tetrahedra.

It is assumed that the dam/foundation interface is at elevation 0.0 m and thus the top of the dam is at elevation 40.0 m, which is the water level assumed in the hydromechanical analysis presented here.

The hydraulic super nodes and the pseudo seepage channels at the dam foundation are also shown in Figure 13. The dam is assumed to be impervious.

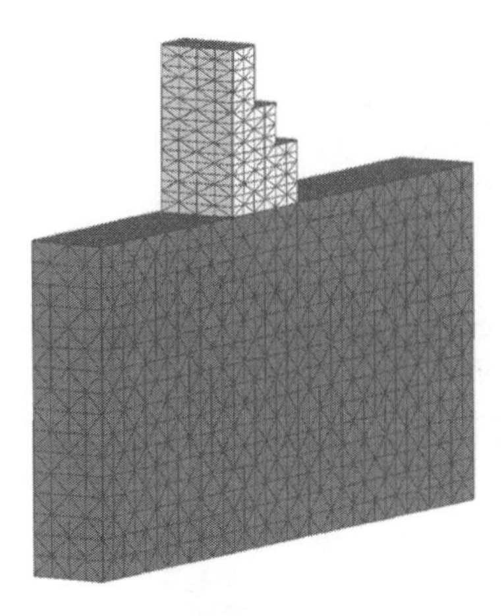

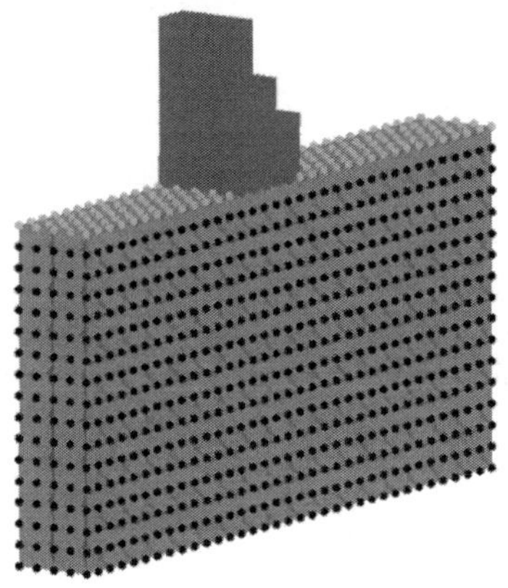

Figure 13. Idealised dam/foundation system: grid used for 3D hydromechanical analysis.

Figure 14 shows a detail of the hydraulic model, where the hydraulic super nodes and the pseudo seepage channels are clearly identified.

In the Parmac3D-Fflow model, the deformable blocks of the dam foundation are divided into 14,592 tetrahedral elements, with 58,368 nodal points and 27,440 plane interface elements. The superimposed hydraulic model has 2975 hydraulic super nodes, 24,672 hydraulic triangular interfaces and 74,016 pseudo seepage channels.

It is assumed that the concrete dam is impervious and therefore the permeability of the triangular interfaces located in this area of the model is zero (no pseudo seepage channels are adopted in the concrete area).

4.3 *Material properties*

Both dam concrete and foundation rock mass blocks are assumed to follow elastic behavior, with the properties given in Table 4. Discontinuities are assigned a Mohr-Coulomb constitutive model, with the properties shown in Table 5.

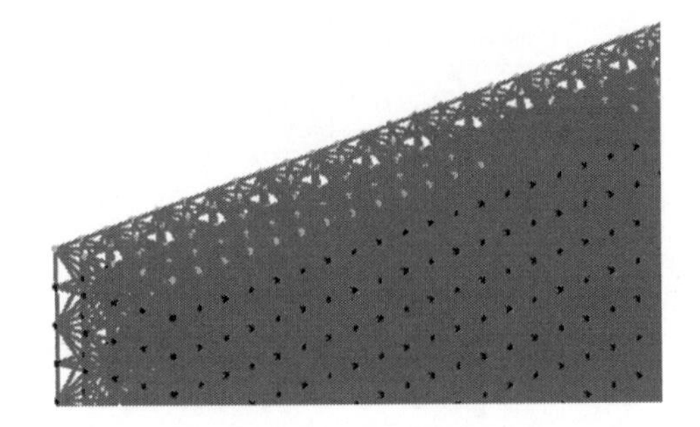

Figure 14. Idealised dam/foundation system: detail of the grid used for 3D hydromechanical analysis.

Table 4. Material mechanical and mass properties.

Material	E (GPa)	ν (–)	ρ (kg/m^2)
Dam concrete	30.0	0.2	2400.0
Rock mass	10.0	0.2	2650.0

The hydraulic apertures of the pseudo seepage channels are $a_0 = 0.1668$ mm, $a_{min} = 0.05$ mm and $a_{max} = 0.25$ mm.

4.4 *Sequence of analysis*

Analysis was carried out in two loading stages, assessing firstly the mechanical effect of gravity loads with the reservoir empty, and secondly, applying the hydrostatic loading corresponding to the full reservoir to both the upstream face of the dam and reservoir bottom. In this second loading stage, mechanical pressure was first applied, followed by hydromechanical analysis.

Figure 15 shows block deformation due to dam weight, hydrostatic loading and flow, obtained after the coupled nonlinear hydromechanical analysis.

4.5 *Boundary conditions*

Regarding the mechanical model, both horizontal and vertical displacements at the base of the model and horizontal displacements perpendicular to the lateral model boundaries were prevented.

Regarding hydraulic boundary conditions, zero permeability was assumed along the bottom and sides of the model. It was assumed that the reservoir was 40 m above the ground surface, and that there was no water downstream from the dam.

4.6 *Results of the numerical analyses*

4.6.1 *Pressure distribution along the base of the dam*

Figure 16 shows the distribution of water pressures along the dam/foundation interface calculated

Table 5. Triangular joint element elastic and strength properties.

Joint element	Normal stiffness (GPa/m)	Shear stiffness (GPa/m)	Friction angle (°)	Cohesion (MPa)	Tensile strength (MPa)
Concrete/concrete	30.0	12.0	50.0	2.0	2.0
Concrete /rock	30.0	12.0	50.0	2.0	2.0
Rock/rock	10.0	4.0	50.0	0.0	0.0

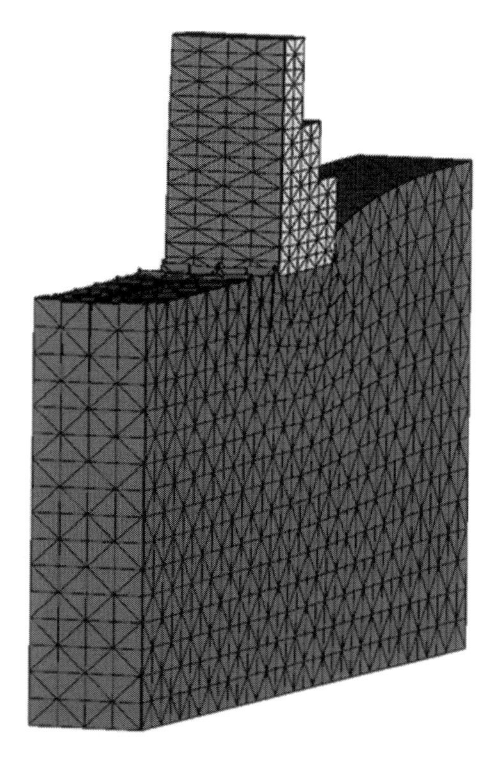

Figure 15. Block deformation due to dam weight, hydrostatic loading and flow (couple nonlinear).

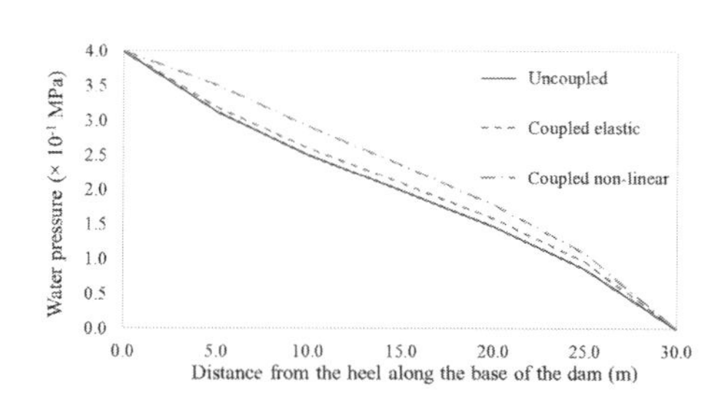

Figure 16. Water pressures along the base of the dam.

with the three different hypotheses. Water pressure varies from 0.4 MPa at the heel of the dam, given from the upstream water level, to zero at the toe of the dam, where the water is assumed to be at the ground level.

In the case of undrained foundations a linear distribution of the water pressures along the base of the dam is usually assumed, i.e. a constant decrease in head between the heel and toe of the dam. Numerical results presented in Figure 16 show that this may not be a conservative assumption, as, when compared to a linear distribution, higher water pressures are obtained.

The uncoupled analysis, in which the hydraulic aperture of the discontinuities remains constant, leads to the lowest water pressures. These are slightly higher when a coupled analysis is carried out, and further increase when the non-linear behaviour of the dam foundation discontinuities is taken into account. This difference in water pressures is due to the opening of dam foundation discontinuities below the heel of the dam and reduction in their aperture below the toe of the dam.

4.6.2 *Field of water pressures*

Figure 17 shows the pseudo-equipotentials of piezometric head in the dam foundation obtained in the different numerical analyses. Taking the uncoupled analysis as reference, figure shows that although the field of water head does not vary significantly when a coupled elastic analysis is carried out, there is, in fact, an increase in water head in depth upstream from the dam. This increase is considerably higher when a non-linear analysis is done. This is due to the opening of foundation discontinuities in the upstream area of the dam foundation.

4.6.3 *Quantity of water that flows through the model*

Even slight changes in the hydraulic aperture of the discontinuities can lead to significant changes in the quantity of water flowing through the dam foundation. Table 6 shows the quantity of water that flows through the foundation pseudo channels in each numerical analysis. Taking the uncoupled analysis as reference it can be concluded that the quantity of water that flows through the model decreases by about 2.14% when a coupled elastic analysis is carried out and increases by about 9.24% when the non-linear effects are taken into account.

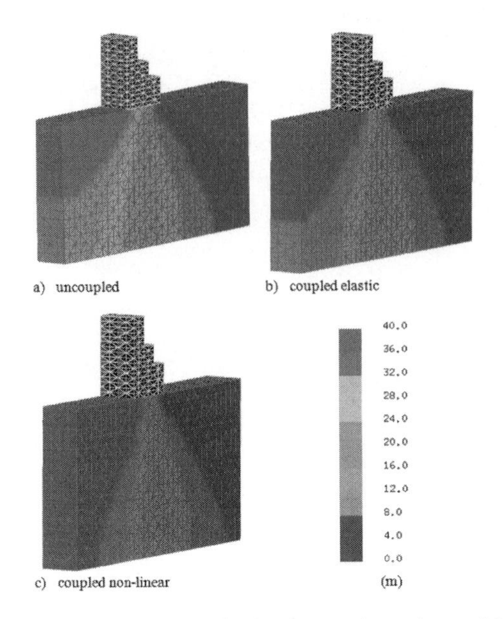

Figure 17. Piezometric head pseudo-equipotential contours.

Table 6. Quantity of water that flows through the model.

Numerical analysis	Quantity of water ($\times 10^{-4}$ m^3/s)
Uncoupled	1.3413
Coupled elastic	1.3124
Coupled non-linear	1.4653

5 CONCLUSIONS

The hydomechanical coupled model presented in this paper, Parmac3D-Fflow, is the 3D extension of the 2D coupled model that has been recently proposed. Due to its known numerical robustness simple unidirectional flows at the triangular interface edges were adopted. The coupled approach allows the interrelated effects of both mechanical and hydraulic mechanisms to be taken into account as the numerical computations proceed.

The analyses described above show the accuracy of the algorithm in uncoupled hydraulic analysis and indicate that the Parmac3D-Fflow can be used successfully to model the coupled hydromechanical behaviour of dam foundations.

Further work is under way not only to improve the validation of the 3D hydromechanical model presented here but also to perform analysis of arch dam foundations, taking into account the modelling of both the grout and drainage systems.

ACKNOWLEDGMENTS

The study presented here is part of the research project "DAMFA: Cutting-edge solutions for sustainable assessment of concrete dam foundations" which has been supported by LNEC with the main purpose of developing a numerical multiphysic integrated tool for the sustainable assessment of concrete dam foundations, taking into account the interaction between the mechanical, hydraulic and thermal behaviours.

REFERENCES

Azevedo N.M. 2003. A rigid particle discrete element model for the fracture analysis of plane and reinforced concrete. PhD Dissertation, Heriot-Watt University, Scotland.

Azevedo N.M. & Farinha, M.L.B. 2015. A hydromechanical model for the analysis of concrete gravity dam foundations. *Geotecnia - Journal of the Portuguese Society for Geotechnical Engineering* 133(March 2015):5–33 (in Portuguese).

Barton, N., Bandis, S., and Bakhtar, K. 1985. Strength, deformation and conductivity coupling of rock joints. *Int J Rock Mech Min Sci Geomech Abstr* 22(3): 121–140.

Callari C, Fois N, Cicivelli R (2004) The role of hydromechanical coupling in the behaviour of dam-foundation system. *Proc. VI World Congress on Computational Mechanics*, Pequim, China, pp 1-11.

Damjanac, B. 1996. A three-dimensional numerical model of water flow in a fractured rock mass. PhD Dissertation, University of Minnesota, Minneapolis, USA.

Erban P. & Gell K. (1988) Consideration of the interaction between dam and bedrock in a coupled mechanic-hydraulic FE-program. *Rock Mech Rock Eng* 21(2):99–117.

Farinha, M.L.B.; Azevedo, N.M. & Candeias, M. 2017. Small displacement coupled analysis of concrete gravity dam foundations: Static and Dynamic Conditions. *Rock Mech Rock Eng* 50(2):439–464.

Louis C. 1969. A study of groundwater flow in jointed rock and its influence on the stability of rock masses. Ph.D. Dissertation, University of Karlsruhe (in German), English translation, Imperial College Rock Mechanics Research Report n°10, London.

Noorishad J (1971) Finite element analysis of rock mass behavior under coupled action of body forces, fluid flow, and external loads. PhD Dissertation, University of California, Berkeley.

Yan, C. & Zheng, H. 2017. FDEM-flow3D: A3D hydromechanical coupled model considering the pore seepage of rock matrix for simulating three-dimensional hydraulic fracturing. *Computers and Geotechnics* 81:212–228.

Wittke, W. 1990. *Rock Mechanics. Theory and applications with case histories*. Springer Verlag, Berlin, Germany.

Numerical Methods in Geotechnical Engineering IX – Cardoso et al. (Eds)
© 2018 Taylor & Francis Group, London, ISBN 978-1-138-33203-4

An in-house VBA program to model the settlement and consolidation of thickened mine tailings

N. Raposo
Instituto Politécnico de Viseu, Viseu, Portugal

R. Bahia
Golder Associates Portugal, Porto, Portugal

A. Topa Gomes
Faculty of Engineering (FEUP), University of Porto, Porto, Portugal

ABSTRACT: The settlement and consolidation of mine tailings is a classical geotechnical problem, although mainly related to mine tailings engineering. Due to this specificity, commercial software have limitations on simulating the problem, therefore a VBA (Visual Basic for Applications) code was developed. The present paper presents this code, able to model mine tailings accretion, with the following main objectives: predict the void ratio at several depths, estimating its unit weight for properly evaluating the consolidation phenomena and calculate the pore pressure at any instant in order to compute the strength of the tailings. The code is based on the Finite Difference Method, with the particularity of using a unidimensional mesh with varying geometry. Apart from the detailed description of the software, the paper presents some applications to a tailings storage facility, comparing the *in situ* data with the results obtained by the developed code. Finally, limitations and future developments are discussed.

1 INTRODUCTION

Due to environmental and economic reasons, achieving higher geochemical and geotechnical stability is a mandatory issue. The deposition of mine tailings as thickened slurries or even at paste consistencies has become an alternative to subaqueous deposition, in the recent years. Tailings are usually formed by crushed hard-rock to sands or silts, with granulometries just above the clays size. Being originated on mill and metallurgic processes involving hydraulic processes and transport, they are frequently deposited as very loose slurries. As a consequence, after deposition, they settle and consolidate with very large strains, initially under saturated conditions but, after a certain time period, under unsaturated conditions.

There are several commercial software able to deal with the settling, consolidation and unsaturated consolidation but, hardly, in an integrated way. Moreover, they do not deal with the specific problem of predicting the tailings' void ratio and strength. In such context, it was decided do develop an in-house code in VBA (Visual Basic for Applications) due to the simplicity of this language and the possibility to integrate it with a normal spreadsheet, what is considered an advantage regarding practical industry environments.

The developed code has as main objectives to be able to estimate the void ratio and the unit weight along the entire thickness of the tailings. These two variables are fundamental when predicting the capacity of the deposit and play a decisive role regarding strength and liquefaction potential evaluation. The presented paper is based on the work of Raposo (2016).

For the presentation and validation of the code possibilities, it was applied to site data of a real tailings facility. The tailings have been pumped into the pond as loose slurries (subaqueous deposition) for more than two decades. After the tailings level reached the dam crest, the deposition continued by means of subaerial thickened tailings disposal, creating self-standing slopes and stack. Figure 1 presents a detail of freshly settled mine tailings, with very weak strengths, unable to support bare-foot walk, and highly deformable, which helps to understand the large strains of the material the in-house software is devoted to deal with.

2 DESCRIPTION OF THE CODE

The software developed has a very simple organization, illustrated in the flow chart of Figure 2. It is

Figure 1. Image of mine tailings after deposition (Raposo, 2016).

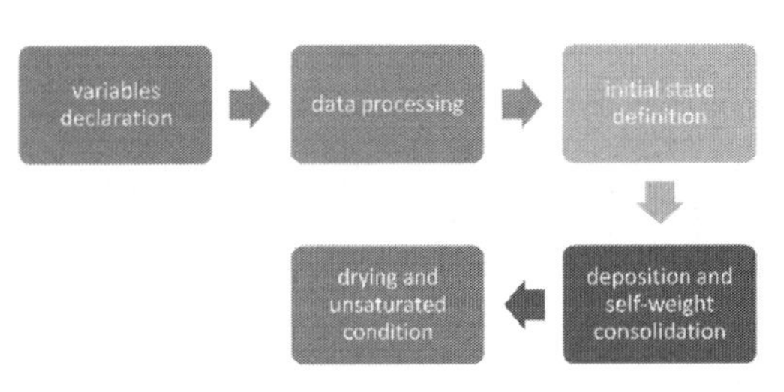

Figure 2. Flow chart of the in-house developed program with its 5 distinct modules.

constituted by 5 distinct modules: the first 3 modules support the main calculations performed in the modules "deposition and self-weight consolidation" and "drying and unsaturated condition.

In the module "variables declaration" all the variables are declared and the main vectors and arrays are defined. The second module, "data processing", reads from an auxiliary spreadsheet supporting the code the input information. This module also completes the dimensioning of all the arrays and starts the output preparation, depending on the geometry of the model.

The third module, "initial state definition", defines the control variables, namely: the time increment for the resolution of the consolidation equations (Δt), the initial height of the elements (equal for all the elements), and the "as deposited" solids content of each element (equal for all the elements of the same type of material). Presently, the code is prepared to consider the historic subaqueous deposition and the subsequent thickened tailings subaerial stacking. Additionally, the initial geometry of the model is defined by calculating, for each node, its initial position, piezometric height and total stress.

The module "deposition and self-weight consolidation ", whose working flow chart is presented in Figure 3, performs all the calculations that take into account the effects of self-weight consolidation and the additional settlement resulting from the ongoing deposition.

For this module two different time intervals are considered: the first, denominated

ΔT in the flow chart of Figure 3, defines the time increments for the deposition accretion and nodes increment (1 week is a reasonable time reference to produce a soft stack increment, as it produces a deposition in the order of some centimeters); the second, labeled Δt, corresponds to a much smaller time interval, (in most of the cases smaller than 1 hour) and is used in the numerical process of solving the consolidation problem. The variable

$T_{ma'x}$ represents the limit time for the calculations.

Inside the module "deposition and self-weight consolidation", a subroutine performs all the calculations and giving as output the hydraulic head

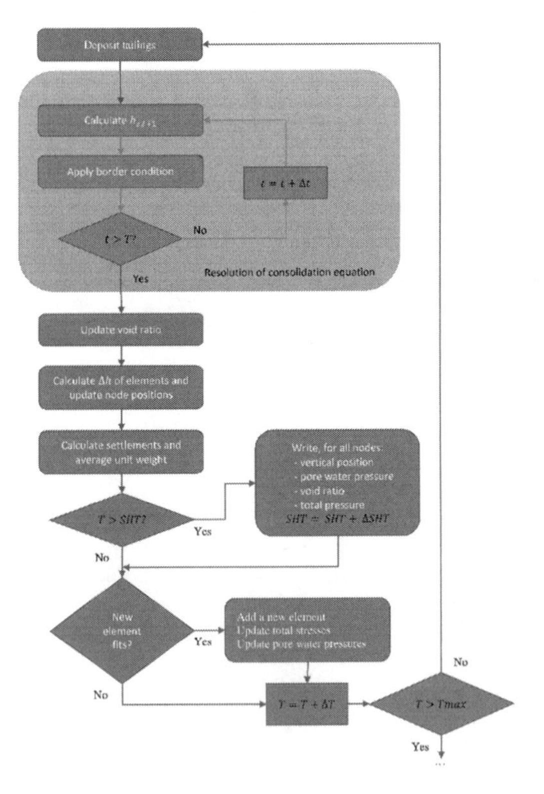

Figure 3. Flow chart of the module "Deposition and self-weight consolidation".

at any point. As the total stress is also known, it is possible to calculate the effective stress.

The calculation of the void ratio is completed using a subroutine where the compressibility law for each material is defined as a function of the effective stress.

The code is able to consider a linear or parabolic variation of the void ratio with the logarithm of the effective stress.

As it will be explained in the following section, the code adopts a resolution technique in which the position of the nodes changes with the consolidation, imposing a constant solids volume between two nodes. In such conditions the height of a given element is calculated by Equation 1:

$$\Delta_{Z_{i,T}} = \frac{\Delta_{Z_0}}{1+e_0}\left(1+e_{i,T}\right) \tag{1}$$

where $\Delta_{Z_{i,T}}$ represents the height of element i at instant T, Δ_{Z_0} stands for the initial height of the element, e_0 is the initial void ratio, and $e_{i,T}$ is the void ratio of the element at instant T. As previously referred, the initial height is set equal for all the elements, and the initial void ratio is a property of each material and deposition condition.

With the previous information, it is possible to calculate the average total unit weight of the deposit using Equation 2:

$$\gamma_{d,T}^{ave} = \frac{\Delta_{Z_{i,T}}}{1+e_0}\frac{nG\gamma_w}{Z_{top}-Z_1} \tag{2}$$

where n represents the number of elements, G is the specific gravity of solids, γ_w is the unit weight of water, and Z_{top} and Z_1 the coordinates of the top and base of the deposit, respectively. In case more than one material exist, Equation 3 should be used:

$$\gamma_{d,T}^{ave} = \frac{\Delta_{Z_{i,T}}}{z_{top}-z_1}\sum_j \frac{n_j\, G_j\, \gamma_w}{1+e_{0,j}} \tag{3}$$

Before the end of each cycle, the surface position is evaluated in order to check whether an additional element should be included. If this is the case, it is necessary to update the list of nodes, to add the weight of this additional element to the total weight of the deposit, and to update the piezometric height.

The last module, "drying and unsaturated condition", evaluates the drying effects on the tailings. During this process there is no deposition of new tailings, hence no new elements are added. In this scenario, the new loading condition is defined by the variation of the position of the water table, changing the hydraulic head. This process is controlled by an auxiliary subroutine defining the change in the hydraulic head with time.

3 RESOLUTION OF THE MAIN EQUATIONS

3.1 *Resolution of the consolidation differential equation*

The developed code uses the Finite Difference Method to solve the consolidation equation. It has the particularity of simulating the accretion process using a variable geometry mesh, gaining new elements as new tailings are deposited over the column of tailings consolidating under their self-weight. Moreover, the node coordinates are updated at each time step, as a function of the occurred settlements. It corresponds to the same Lagrangian coordinates as the ones used by Masala (1998), where the elements are deformable and movable, containing always the same particles.

The code adopts the large strain consolidation theory proposed by Gibson, Schiffman, & Cargill (1981) and is able to consider the variation on the permeability and compressibility depending on the stress state. At any instant t, the reduction rate of the total head, h, in a point at position z in the tailings deposit, may be calculated by Equation 4:

$$\frac{\partial h}{\partial t} = c_v \frac{\partial^2 h}{\partial z^2} \tag{4}$$

where c_v represents the coefficient of consolidation. The resolution of the differential equation adopts the scheme illustrated by Figure 4.

Assuming a small enough time interval, Δt, the first term of Equation (4) in a point at position z can be calculated using Equation (5):

$$\frac{\partial h}{\partial t} = \frac{h_{z,t+1}-h_{z,t}}{\Delta t} \tag{5}$$

Similarly, assuming the distance between 2 points as Δz, it results:

$$\frac{\partial^2 h}{\partial z^2} = \frac{h_{z+1,t}-2h_{z,t}+h_{z-1,t}}{\Delta z^2} \tag{6}$$

And, substituting Equations (5) and (6) in Equation (4) it results Equation (7):

$$\frac{h_{z,t+1}-h_{z,t}}{\Delta t} = c_v \frac{h_{z+1,t}-2h_{z,t}+h_{z-1,t}}{\Delta z^2} \tag{7}$$

Which may be rearranged to explicit the total head in the subsequent time interval, $t+1$, based on the total head at the point under analysis and points in the vicinity, at instant t, resulting in Equation (8):

$$h_{z,t+1} = h_{z,t} + \frac{c_v\ \Delta t}{\Delta z^2}(h_{z+1,t} - 2h_{z,t} + h_{z-1,t}) \quad (8)$$

Using the parameter β defined by Equation (9), Equation (8) can be simplified, resulting in Equation (10).

$$\beta = \frac{c_v\ \Delta t}{\Delta z^2} \quad (9)$$

$$h_{z,t+1} = h_{z,t} + \beta\,(h_{z+1,t} - 2h_{z,t} + h_{z-1,t}) \quad (10)$$

According to Crank (1975), in order to guarantee the convergence of the numerical process, β should be smaller than 0.5. In the present case a value of β of 0.4 is adopted.

When there are layers with different coefficients of consolidation, c_v, it is necessary to calculate different values for β, as the time increments, Δt, are uniform. Therefore, the material with higher coefficient of consolidation should be used to establish the time increment and, afterwards, the β for each material can be calculated.

It must be stated that the problem under analysis can be studied considering only 1 dimension, as the horizontal dimensions of the deposition areas of this kind of facilities are usually one to two orders of magnitude larger than the vertical dimension of the stacks.

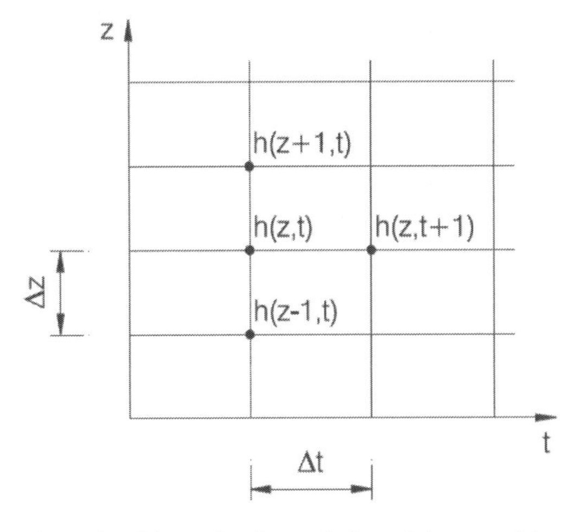

Figure 4. Scheme for the resolution of the consolidation equation using the Finite Difference Method.

3.2 *Boundary conditions*

For the resolution of the consolidation problem, it is necessary to know the boundary conditions, particularly at the top and base of the model. As, during the subaqueous deposition and subsequent subaerial accretion, there is always water ponding at the surface, or at least some run-off water bleeding from the tailings, for the upper boundary, it is assumed that the piezometric level is coincident with the free surface of the thickened mine tailings. Regarding the lower boundary condition, two hypotheses are possible:

- Impermeable border, imposing a total head in this border equal to the adjacent node;
- Permeable border, which corresponds to a constant piezometric level, inputted by the user.

3.3 *Relation between compressibility and effective vertical stress*

Typically, the compressibility of a normally consolidated fine soil is governed by the compression index, C_c, which relates the void ratio, e, with the effective vertical stress, σ'_v, by Equation 11.

$$e = e_1 - C_c\ \log(\sigma'_v) \quad (11)$$

Where, e_1 is the void ratio at the tailings formation, corresponding to a vertical effective stress of 1 kPa. Figure 5 presents the oedometer test results for a hydraulically deposited sample of thickened mine tailings with two line fits: the classical linear fit; and a parabolic fit.

From Figure 5 it results evident that the linear adjustment returns a good fit for stress levels above 100 kPa. Below this stress, the difference between the predicted and measured void ratio is significant. This difference is not relevant in most of the geotechnical problems, as stress levels much below 100 kPa are not common on most of the civil engineering applications. Though, this is not the case of the tailings deposits, as, at the moment of deposition, the stresses are very low. Therefore, the initial stretch is decisive to predict the tailings behavior from the very initial deposition. Due to the importance of this detail, and alternative quadratic equation, Equation 12, was adopted:

$$e = e_1 - e_a \log(\sigma'_v)^2 - e_b\ \log(\sigma'_v) \quad (12)$$

where e_a and e_b are two fitting parameters. Figure 5 shows that the test results are fitted thoroughly.

Finally, it must referred that several measurements of the formation void ratio were performed, *in situ* and in the lab, showing a high correlation

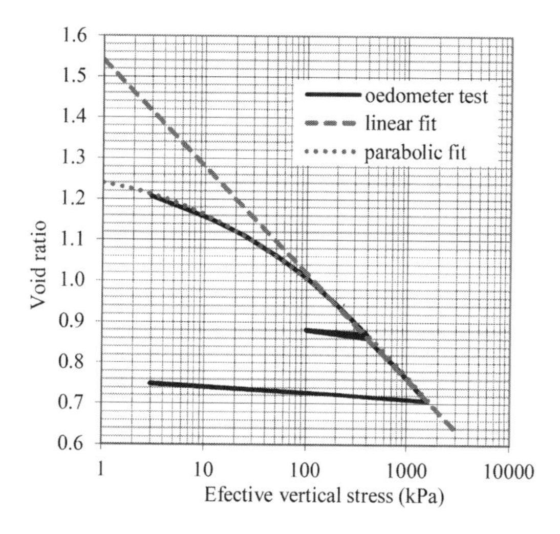

Figure 5. Example of an oedometer test in thickened mine tailings showing a linear and a parabolic fit.

with the grade of the thickening process. In the case of the chosen tailings facility the formation void ratios are 1.5 and 1.25 for the subaqueous slurry and the thickened tailings, respectively.

3.4 *Tailings accretion rate*

One of the main input parameters for the developed program is the accretion rate of the tailings. In the example adopted, both subaqueous and subaerial deposition occurred and the accretion rates varied significantly along the deposit lifetime. Accretion rates between 0.7 m/year and 8.0 m/year occurred and these values were used in the simulations.

The introduction of the accretion rate requires an iterative process due to the difference between deposition rate and *in situ* measured accretion rates. In reality, the *in situ* measured accretion rate includes the deposition rate plus settlements resulting from the load of additional tailings, which does not correspond directly to the deposition rate.

3.5 *Drying and unsaturated condition*

On several tailings facilities the climate may be considered dry, with annual evaporation much higher than the total annual precipitation and resulting net infiltration. Consequently, after the deposition is complete, there are long periods where the tailings loose water by evaporation, reaching an unsaturated condition.

The drying effect was considered on a simplified uncoupled way, as suction was determined independently of the consolidation effects, and vice-versa. The consolidation occurs during a period of days, while evaporation takes months, depending on the thickness of the layers. Therefore, it is reasonable to uncouple the phenomena and also very useful for simplicity purposes.

The stress state produced by the drying was determined without considering any deformations resulting from suction. Similarly, it were ignored the eventual changes in the tailings due to the secondary consolidation.

Under unsaturated conditions, the governing equations for the settling are not the ones resulting from the consolidation theory. The stress state produced by the evapotranspiration was calculated using an external application that used the diffusion theory, in the case, the explicit solution developed by Benson Ronald & Sill Benjamin (1991), using a parabolic approximation.

Another important aspect regarding unsaturated soils is the relation between water content and suction, typically given by the Soil-Water-Characteristic-Curve (SWCC). The SWCC of the tailings is also implemented in the external code. Further details of this software may be seen in Li et at. (2012).

4 VERIFICATION AND CALIBRATION OF THE DEVELOPED CODE

4.1 *Selection of the area used for calibration*

The area of the tailings facility with the higher thickness of tailings was selected to validate the routine. Figure 6 presents the vertical profile of this area, which also contains the thickness of the several layers, as well as the time gap corresponding to each deposition phase. In this area there were CPTU test results, with dissipation phases for measuring the equilibrium pore water pressures. There were also precise topographic measurements of the top surface of the tailings.

4.2 *Coefficient of consolidation of the tailings*

Raposo (2016) presented a thoroughly experimental characterization of the tailings' stacks for consolidation and stability analyses purposes. In that work, the coefficient of consolidation, c_v, is determined using the standard oedometer apparatus as well as the Rowe cell.

As referred in the previous section, in the area adopted for calibration, several CPTU tests with dissipation phases were performed, resulting in a large number of horizontal consolidation coefficients, c_h. Considering the results of the laboratory and the *in situ* tests, the following values were assumed for the vertical coefficient of consolidation for pulp (subaqueous slurry deposition) and

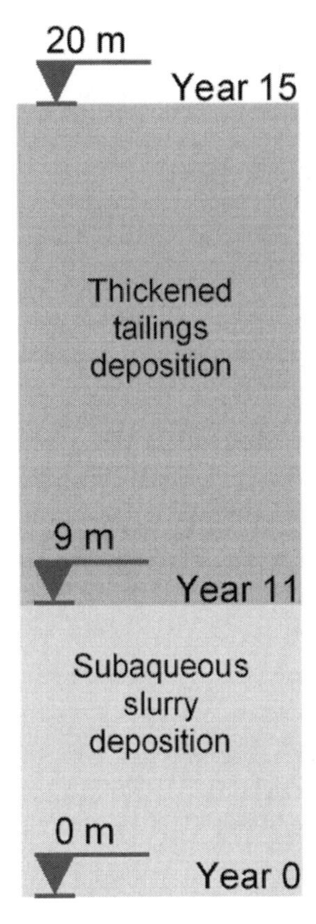

Figure 6. Vertical profile of the tailings stack area used for calibration purposes.

paste (thickened tailings deposition), respectively: $7\times10^{-6}\ m^2/year$ and $3\times10^{-6}\ m^2/year$.

4.3 *Model calibration*

As shown in Figure 6, Tailings deposition in the calibration area started as subaqueous slurry, lasted for 11 years and reached a total height of 9 m of pulp. From then on, the tailings were previously thickened, and thereafter deposited using a subaerial technique, totalizing 20 m of material. Table 1 summarizes the most relevant information regarding this calibration example.

Figure 7 presents the main interface of the software. The cells shaded in yellow require input from the user.

It must be referred that it is possible to consider different boundary conditions for the deposition phase and the drying phase, as they are sequential processes.

The local instrumentation in the calibration area measured a downwards gradient, corresponding to a limited infiltration in the local aquifer. This

Table 1. Deposition rates used for the calibration.

Time (years)	Deposition Rate (m/year)	Deposition type
0 to 10.7	0.7	Subaqueous (pulp)
10.7 to10.9	9.5	Subaqueous (pulp)
10.9 to 11.2	9.5	Subaerial (paste)
11.2 to 12.3	1.6	Subaerial (paste)
12.3 to 13.1	5.2	Subaerial (paste)
13.1 to 13.9	0.5	Subaerial (paste)
13.9 to 15.0	4.2	Subaerial (paste)
After 15.0	No deposition	Drying

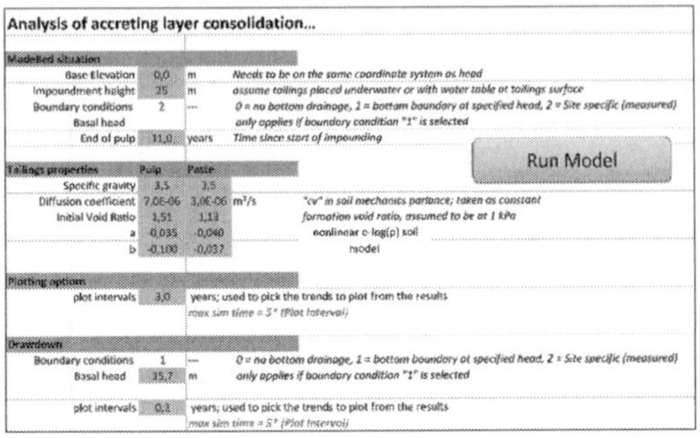

Figure 7. Aspect of the spreadsheet to input the data.

gradient is equivalent to 0.25 m/m and was used to simulate the consolidation process by imposing a hydraulic head at the base of the model corresponding to the piezometric level at the surface of the impoundment discounted by 25% of the thickness of the tailings.

Regarding the surface boundary condition, every time a new layer of tailings is deposited the piezometric level at surface is modified, by forcing it to be coincident with the top of the deposit.

4.4 *Analysis of the results - consolidation*

Regarding the output of the software, it is possible to follow the evolution of the more important parameters, namely, total stress, effective stress, water pressures, surface settlements, void ratio, state parameter, and dry unit weight.

As an example, Figure 8 presents the evolution of the surface level and the average dry unit weight of the tailings. It shows an initial difference between the predicted and measured surface position. Actually, due to numerical reasons, the process has to be initiated with a minimum number of elements. In the present case, this number is 10 elements, each one with a total height of 12.5 cm, therefore totalizing 1.25 m. Thus, this initial dif-

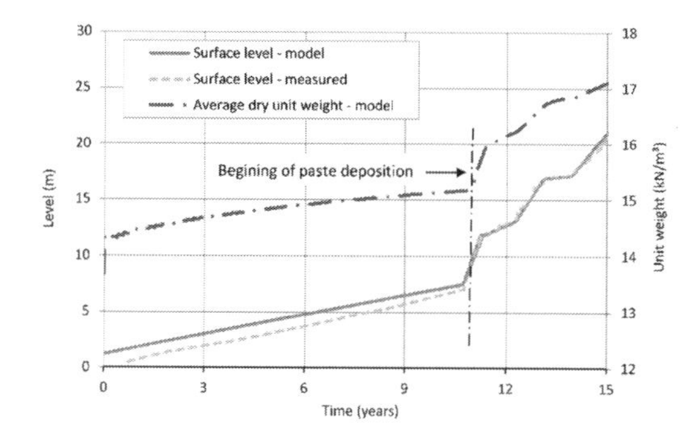

Figure 8. Surface level and average dry unit weight evolution of the tailings in the calibration area.

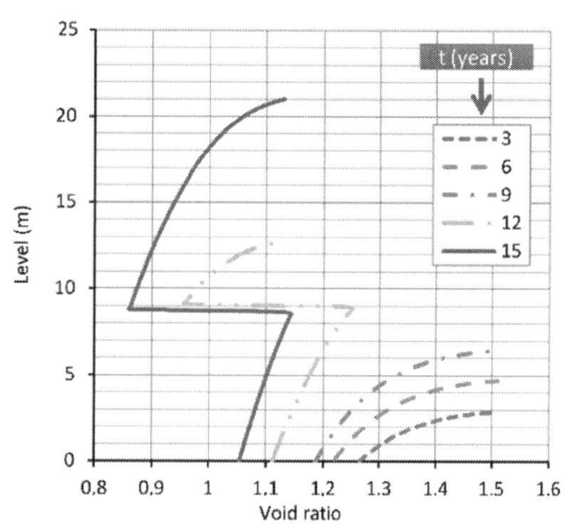

Figure 9. Times shots of tailings void ratio for the calibration area.

ference is artificial, and the program guarantees the convergence of the model and real measurements before the beginning of paste deposition. The initial abrupt variation of the unit weight of the tailings is also a result of the numerical process described and the large strains suffered by those initial elements.

Another interesting aspect to observe is the rapid increase of the unit weight after the beginning of paste deposition. Two main reasons contribute for this phenomenon: the thickened tailings deposited as paste have a smaller initial void ratio; the deposition rate increased significantly. It should also be emphasized the final agreement between the predicted and measured surface position.

Another possibility of the code is to determine the state of the tailings at any position and time. The following example shows five time shots, equally spaced, though the user might modify them. Figure 9 presents the evolution of the tailings stack void ratio during 15 years.

The first three curves have a very similar evolution, with a maximum void ratio at surface of 1.5 as this was assumed as the formation void ratio for the pulp. The last two curves show a sudden change of the void ratio at level 9 m, which corresponds to the transition between paste and pulp. For these different materials, even with the same vertical stress, the void ratio is significantly distinct.

Figure 10 presents the variation of the piezometric level for the same time intervals used in the previous figure. In the first years, due to the low deposition rate and reduced thickness of the tailings pile, there are no excess pore water pressures, and, therefore, the profiles are represented by a straight line. With further depositions, the situation changes drastically and it is clear for the final period (t = 15 years) the curvature of the profile, which indicates excess pore water pressures along the entire depth.

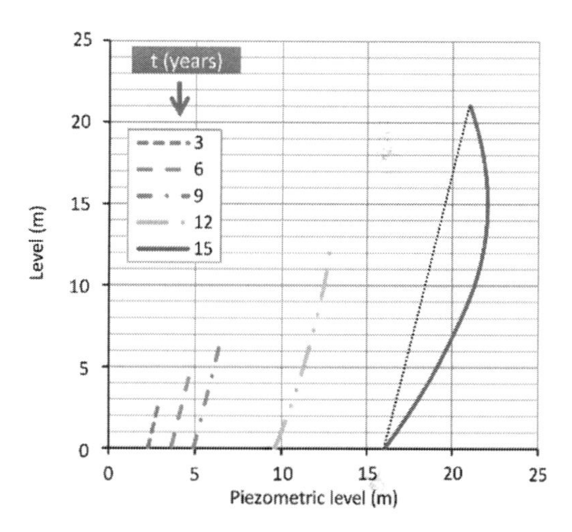

Figure 10. Vertical profile of the tailings piezometric heads for various time shots.

4.5 *Drying process and unsaturated condition*

As previously mentioned, the incorporation of the drying effects was performed on a simplified way, imposing on the surface the piezometric level corresponding to the suction obtained by an auxiliary numerical model.

For example, Figure 11 presents the evolution of the suction profile with depth for a drying period considering an evaporation rate of 6 mm/day. From Figure 11 it is possible to verify that, for example, for a drying period of 3 months (90 days), the suction at surface is approximately 180 kPa. With this information, a new simulation was performed, considering a drying period of 3 months, followed by a period of rain that restored the saturated condition in the tailings. Figure 12 presents the results of this new simulation, in terms of settlements and average dry unit weight of the pile.

For easiness of reading, the horizontal scale is not uniform, as the drying period is much shorter than the consolidation period. The consolidation associated with the drying period produces an additional settlement of 60 cm, which occurs mainly in the first four months, increasing the average dry unit weight of the pile from 17.1 to 17.6 kN/m^3.

Figure 13 exhibits a vertical profile of the void radio along the pile. Instant $t = 0$ coincides with the beginning of the drying process.

After an initial robust variation of the void ratio, it stabilizes approximately for $t = 0.4$ years. Afterwards, though still some variations occur, the differences are insignificant. As already mentioned, the code has also the possibility to represent the excess pore water pressures. In this case, as shown in Figure 14, it is possible to compare the situation with the steady state condition. Furthermore, if there are *in situ* measurements of the water pressures, for example with CPTU tests and piezometers, the numerical results may be compared with the *in situ* measurements.

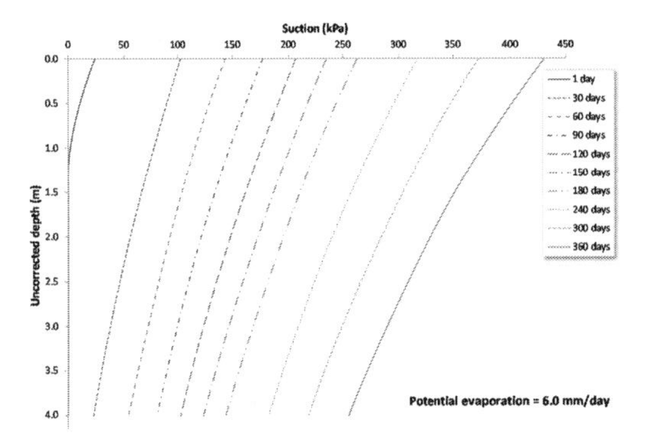

Figure 11. Results layout of the auxiliary software to evaluate drying – vertical suctions profile assuming a constant evaporation rate of 6 mm/day.

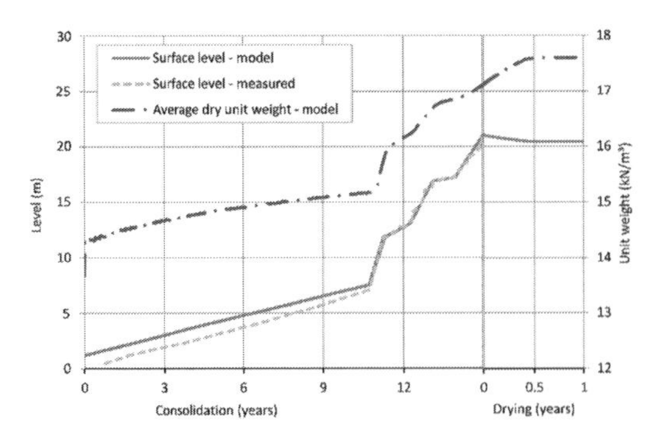

Figure 12. Surface evolution and average dry unit weight for the calibration area after the drying period.

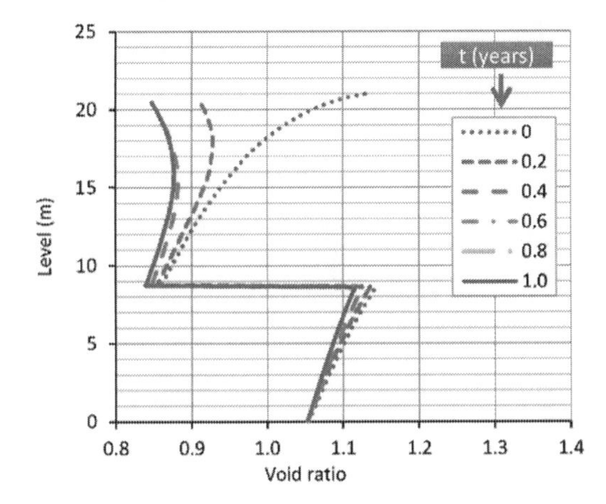

Figure 13. Vertical profile for the void ratio (e), after the drying period.

Regarding the results of Figure 14, the CPTU results presented coincide with instant $t = 0$ and they show a good agreement with the numerical results. It is important to refer that, during the drying process, negative excess pore water pressures develop, mainly close to the surface. After the drying period, the rain restored the saturation condition to the tailings.

4.6 *Comparison with the critical void ratio*

When the materials are prone to liquefaction, the evaluation of the void ratio, and its comparison with the critical void ratio, assumes paramount importance. The critical void ratio may be obtained through triaxial tests, for example. In such conditions, the program is able to plot, along a vertical profile, the comparison between the void ratio and the critical void ratio, as shown in Figure 15a. Figure 15 b shows the calculated state parameter plotted against values obtained with the CPTU tests.

The state parameter is the difference between the void ratio and the critical void ratio and it is possible to derive it from CPTU tests. The first aspect to mention is the coherence between the calculated state parameter and its value from *in situ* tests, in the upper layer, that corresponds to paste. At the bottom layer its visible a much higher variation in the *in situ* test results, which results from the typical segregation of the pulp in silty and sandy layers.

Another practical aspect relevant to mention regards the fact that the void ratio, almost along the entire profile, is above the critical state line.

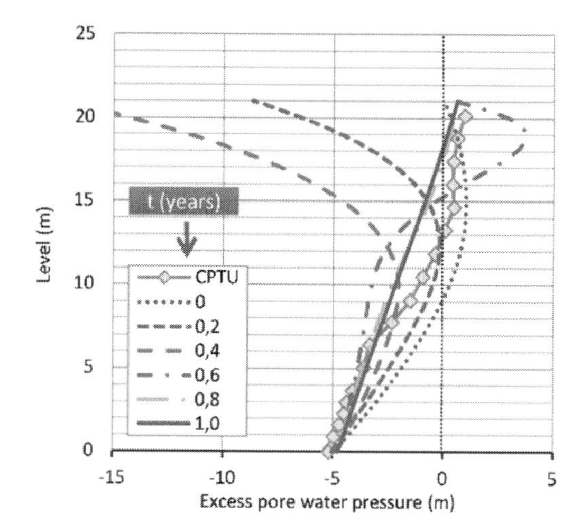

Figure 14. Vertical profile of the excess pore water pressures after the drying period.

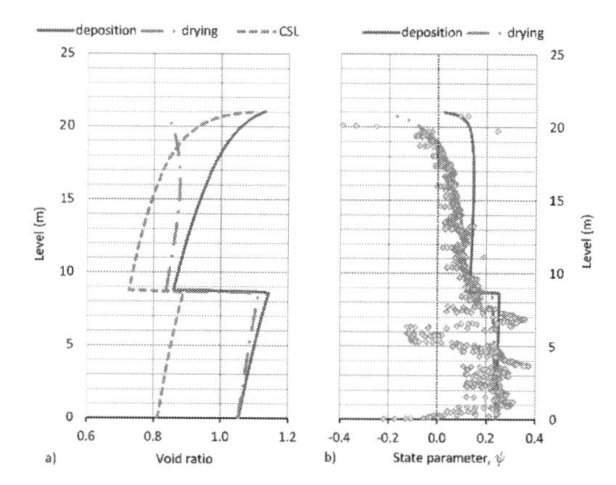

Figure 15. Vertical profile of the calibration area: a) void ratio *vs* critical void ratio; b) state parameter.

This indicates that the tailings will behave as a contractile material, thus, being prone to liquefaction.

5 CONCLUSIONS

The present paper presents an in-house numerical code, developed in VBA, to evaluate the settlement and consolidation of mine tailings, particularly paste and pulp. The paper describes the organization of the code and discusses numerical results, comparing them with the example of an area for which there were real *in situ* measurements. The results proved that the developed tool was adequate for control purposes, conducing to reliable results, when compared to *in situ* measurements, and allowing understanding the fundamental behavior of mine tailings.

ACKNOWLEDGEMENTS

This work was financially supported by: Project POCI-01-0145-FEDER-007457 – CONSTRUCT - Institute of R&D in Structures and Construction funded by FEDER funds through COMPETE2020 - Programa Operacional Competitividade e Internacionalização (POCI) – and by national funds through FCT - Fundação para a Ciência e a Tecnologia.

The authors would like to thank Mike Jefferies and Dawn Shuttle for the permission of using their code as a starting point for the developed code.

REFERENCES

Benson, R.E.J., & Sill, B.L. (1991). Evaporative Drying of Dredged Material. *Journal of Waterway, Port, Coastal, and Ocean Engineering*, 117(3): 216–234.

Crank, J. (1975). *The Mathematics of Diffusion* (Second Edition). Clarendon Press, Oxford.

Gibson, R.E., Schiffman, R.L., & Cargill, K.W. (1981). The theory of one-dimensional consolidation of saturated clays. II. Finite nonlinear consolidation of thick homogeneous layers. *Canadian Geotechnical Journal,* 18(2): 280–293.

Li, A.L., Been, K., Wislesky, I., Eldridge, T., & Williams, D. (2012). Tailings Initial Consolidation and Evaporative Drying after Deposition. *15th International Seminar on Paste and Thickened Tailings (Paste 2012)*, Sun City, South Africa.

Masala, S. (1998). *Numerical Simulation of Sedimentation and Consolidation of Fine Tailings.* (MSc), University of Alberta, Edmond, Alberta, Canada.

Raposo, N. (2016). *Thickened Mine Tailings Deposition. Experimental Characterization and Numerical Modeling (in Portuguese)* (PhD), University of Porto, Porto, Portugal.

Numerical Methods in Geotechnical Engineering IX – Cardoso et al. (Eds)
© 2018 Taylor & Francis Group, London, ISBN 978-1-138-33203-4

Deep foundations and ground improvement for the slope stability of the Disueri dam (Italy)

F. Castelli & M. Greco
Facoltà di Ingegneria e Architettura, Università di Enna KORE, Enna, Italy

ABSTRACT: The paper describes the study of the geotechnical challenges and safety measures adopted for the Disueri Dam's left side in order to restrict the subsidence phenomena detected during the geotechnical investigations carried out between 2013 and 2014. These investigations witnessed several karst cavities due to the dissolution of the gypsum composing the subsoil of dam's left side. As the dissolution process can't be stopped because the gypsum's volume can be isolated permanently, it was considered appropriate to restrict the dissolution's effect: the cavities will be filled and a structural mesh will be located at the base of the slope by bored piles connected at the head for the slope stabilization. To evaluate the effectiveness of the proposed solution, a 3D finite element simulation was carried out. The FEM numerical analysis permits to estimate the improvement of the slope stability in terms of displacements, both in static and seismic conditions.

1 INTRODUCTION

The Disueri dam, located in the town of Mazzarino, close to Caltanissetta city (Sicily), was built blocking the Gela river by a 48.00 m high masonry dam completed in 1948. Due to the dam's behaviour has never been fully satisfactory and the basin has almost completely buried, a new dam was built, immediately downstream of the old dam, creating a barrier of loose materials with a 55,60 m high core.

The volume is about 23 million m^3. The dissolution process involving the portion of the left bank subject to study, was predicted at the time of the construction of the new dam in the 70 s but it was thought could be inhibited through a surface protection, made with concrete slabs connected with joints, and with two deep screens of waterproofing injections, arranged transversely to the left bank, at the edges of the aforementioned surface protection.

The surveys carried out between 2013 and 2014 resulted in the finding of the inadequacy of the transversal screens, as well as in the dissolution of the gypsum accelerated, moreover, by the water circulation supplied by the Disueri basin. The effectiveness of the interventions aimed at the isolation of the gypsum is therefore not very adequate; it seems more reasonable to take measures to control the phenomenon as it evolves.

2 DESCRIPTION OF THE BASIN

As previously mentioned, the Disueri dam was built in two phases by the block of the Gela river. The new dam, immediately downstream of the old one between 1988 and 1997 built, consists of loose materials and a central waterproof sealing core.

The dam's banks are made of evaporitic limestone and the central core consists of clay.

The surface discharge device consists of two goblet spillways, shown in Figure 1, located on the left bank. The maximum height within which the water is kept inside the two goblet spillways is 161.00 m above sea level.

The dam rests on sulphurous rocks, which are affected by folds and fault lines, giving rise to an extremely complex structural order; different geological formations made up of deeply different rocks converge and partly overlap each other: tortonian clays, soft rocks of the chalky-sulfide series, including limestone and gypsum.

Figure 1. Goblet spillways.

Figure 2. Karst cavities.

Figure 3. Collapse and cracks of concrete slabs.

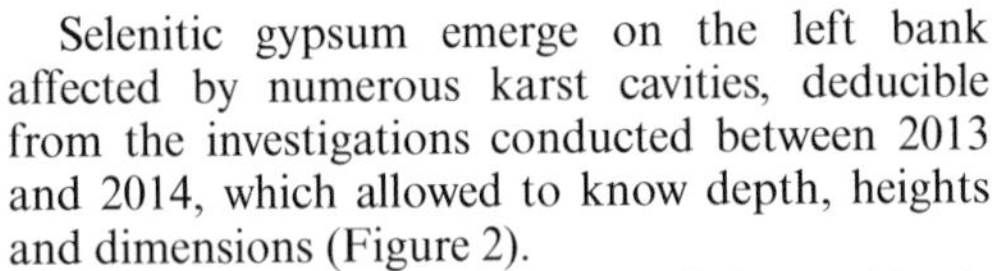

Selenitic gypsum emerge on the left bank affected by numerous karst cavities, deducible from the investigations conducted between 2013 and 2014, which allowed to know depth, heights and dimensions (Figure 2).

In particular, the lowest part of the cavities is 107.60 m above sea level, the highest is 177.66 m above sea level; the height varies between 0.30 m and 3.0 m. Some cavities are connected, the remaining ones are filled in whole or in part of soft material, silty-clayey material mixed with physical-chemical degradation products of the gypsum.

The existence on the left bank of gypsums and the consequent dissolution processes were already known in the design phase of the new dam: for this reason two deep screens of waterproofing injections, placed transversely to the banks and arranged at the edges of a surface protection of the gypsums were made. This protection was made with jointed concrete slabs and with a PVC membrane connected upstream to the slabs and downstream to an impermeable mat.

The cladding slabs were placed on a layer of varying thickness of assorted limestone material with a predominantly gravelly fraction.

The sealing of the coating, especially in the area above the plaster casts, is impaired for more than ten years.

3 INSTABILITY OF THE LEFT BANK

Since 30-06-2011 systematic measures of horizontal and vertical displacements were performed in a large area of the left bank, which includes both the area affected by the collapse and cracks of concrete slabs, shown in Figure 3, and the adjacent areas by means of measures topographic and inclinometer.

The left and right goblet spillways were also monitored and, in the summer of the year 2014, on the top of the goblet three points were materialized in order to detect the possible existence of rotations in the vertical plane.

Figure 4. Subsidence basin.

Figure 5. Landslide mechanism.

On the left bank the massive presence of outcropping selenitic gypsum experimental evidences showed affected by numerous large karst cavities.

Figure 6. The displacements converge towards the centre of the basin.

They are mainly present in the area between the lake and the berm at 158.50 m above sea level.

The evolution of these cavities is rapid due to the high dissolution rate of the gypsum (Valore, Ziccarelli, & Gambino 2011) under flowing water conditions (of the order of 2 g/l), which the formation of a large subsidence basin shown in Figure 4 and horizontal and vertical displacements caused. What abovementioned causes the formation of a real mechanism of breaking the bank (Valore & Muscolino 2013).

In particular, the section of the bank located downstream of the centre of the subsidence basin moves upwards, i.e. towards the area below the most depressed area of the basin.

In the upstream stretch subtended by the subsidence basin a landslide mechanism is evident (Figure 5) and along the barrier and the right edge of the landslide body there are numerous potholes.

The displacements of the bank converge towards the centre of the basin both from the mountain and from the valley (Figure 6). These displacements indicate the formation of a breaking mechanism.

4 PROPOSED INTERVENTION DESCRIPTION

Since the dissolution process can hardly be stopped due to the difficulty of implementing an isolation of gypsum's volume, as evidenced by previous interventions performed, it is considered appropriate to evaluate measures to buffer the effect of dissolution of gypsum.

The outputs of these evaluations lead to a design solution that primarily requires clogging of the cavities using cement mixtures.

For the stabilization of the slope, then, it is expected the construction of a structural mesh made with bored piles connected by beams at the head as schematically shown in Figure 7, arranged over the area in landslide.

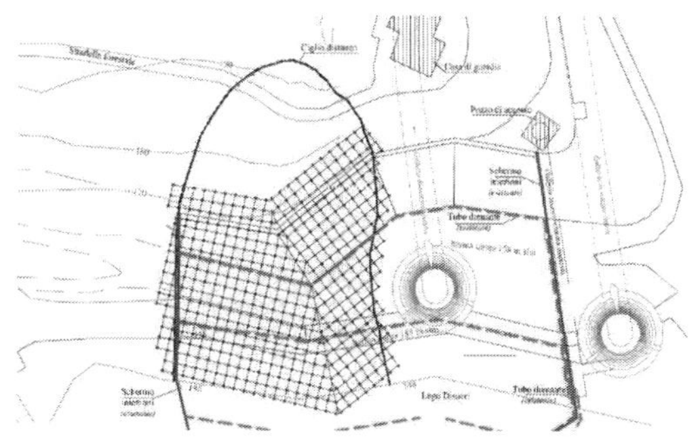

Figure 7. Structural mesh with bored piles connected by concrete beams.

Figure 8. Planimetric distribution of structural elements extend from altitudes 172 m above sea level up to a height of 148 m above sea level.

These structural elements extend from altitudes 172 m above sea level up to a height of 148 m above sea level according to the planimetric distribution of Figure 8, whose effectiveness can be verified by a suitable monitoring program (observational method).

It is also expected to remove the cladding slabs of intervention area and, where necessary, the subsequent restoration and/or replacement of drainage devices. Further, it appears essential to restore the lining of the bank and proceed superficial waterproofing of the same.

In the case of earth dams, among the various options available, the technique of covering the dam barrier with layers of bituminous conglomerate turns out to be the most suitable as it offers the following advantages: the roof covering has an excellent sealing; the multiple layers of bituminous conglomerate constitute a thick and solid protective barrier, with high resistance to damage; the flexible and plastic nature of the conglomerate gives the layer a remarkable ability to adapt to the collapse of the dam, so any failure phenomena are visible much earlier than the rigid waterproofing systems. Then the durability increases by several tens of years and the maintenance required is minimal.

The design choice to clog the cavities derives from the need to slow down the dissolution process; the larger cavities blockage hinders water filtration motion which causes such dissolution.

It is therefore provide concrete injections periodically. However, a monitoring program of the slope and of the exhaust devices is necessary because the problem can't be definitively resolved. Therefore the use of classical tools and techniques is foreseen, like pore pressures and deformations monitoring of the slope, preferably integrated by a remote monitoring through seismic accelerometers (MEMS-type sensors) placed on the beams connection.

Piles total number is equal to 336, their diameter is Φ 1000 mm and their lengths are equal to 35 m (from the lowest altitude of 148.00 m above sea level to the intermediate level of 158.00 m above sea level) and equal to 42 m (from 158.00 m above sea level up to the summit).

The piles are thus much dimensioned to ensure the overcoming of the sliding surface and the realization of an adequate gripping in the basic limestones. Meshes of size 5 × 5 m are foreseen and the connection of the piles is guaranteed on the head by a grid of 637 reinforced concrete beams having a 90 × 100 cm section of considerable rigidity designed to effectively distribute the stresses between the piles. Since the slope is characterized by a strong average inclination that exceeds 23° and it a conspicuous landslide of materials to below causes, it is foreseen the reprofiling, shown in Figure 9, for the global stability improvement, in order to give to the slope an inclination of about 18°–20° by the insertion of stairways (Figure 10) that allow the passage of service vehicles.

Appropriate measures are envisaged for interception upstream of runoff water with catchment devices, placed upstream of the area to be waterproofed.

In order to evaluate the effectiveness of the proposed solution, as well as the effects of the stabilization intervention which involves cavities clogging and piles insertion in the area affected by the subsidence, a three-dimensional numerical analysis with a finite element modelling (FEM) was performed by Plaxis 3D (Brinkgreve & Vermeer 2002) code (Figure 11).

The numerical modelling was performed considering two different conditions "before" and "after" stabilization intervention (Sabetta & Pugliese 1996), shown in Figures 12–13, corresponding respectively to the state of the art and after stabilization intervention.

The safety factor Fs (Baldovin, De Paola, & Morelli 2011) is evaluated for these conditions: it is smaller than the unit (Fs = 0.9408) in the first case shown in Figure 14. In the second case, shown in Figure 15, it increases more than twice, i.e. after the stabilization intervention with cavities filling and deep structural elements insertion (Fs = 2.324).

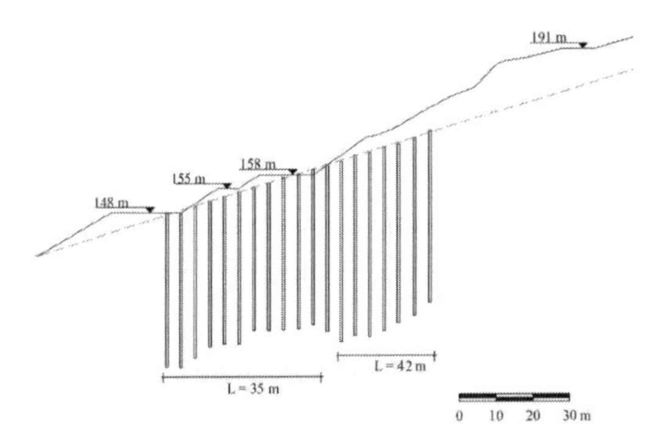

Figure 9. Slope profiling.

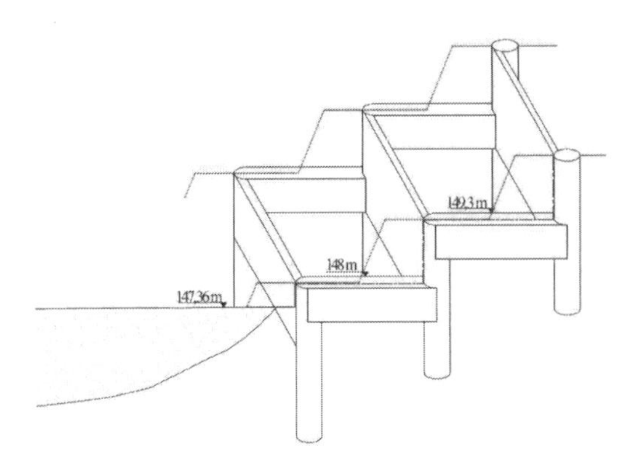

Figure 10. Stairways with piles and beams.

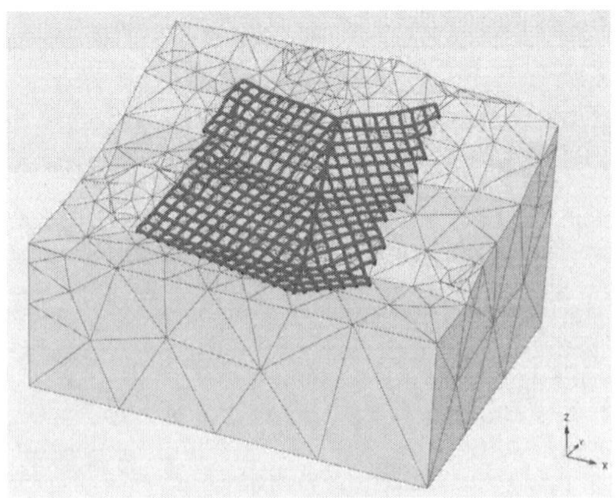

Figure 11. Finite element modelling.

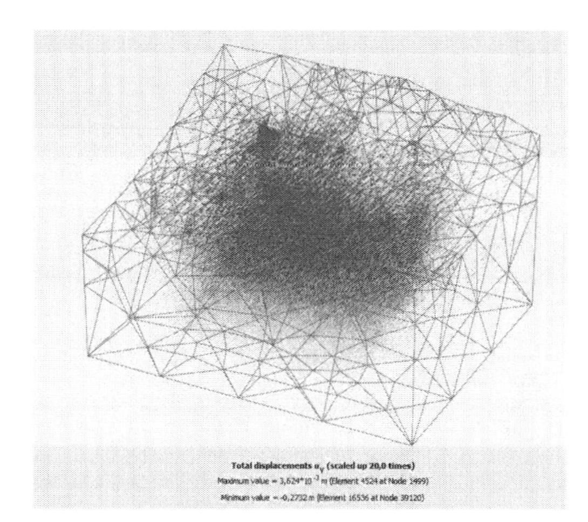

Figure 12. Numerical modelling before stabilization intervention.

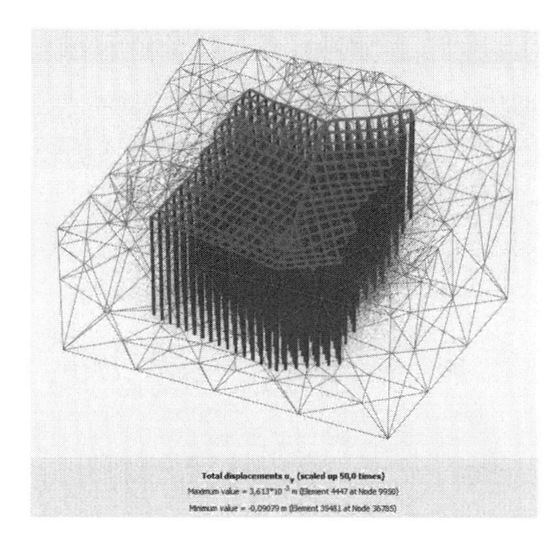

Figure 13. Numerical modelling after stabilization intervention.

Moreover, the significant reduction in displacements (about 1/4) in the landslide area is demonstrated, as shown in Figures 16a-16b.

It should be noted in this aim so that the stability conditions are guaranteed over time it is necessary that the foundation soils (gypsums) maintain the mechanical properties achieved as a result of the planned interventions. Useful indications in this direction will derive from the monitoring program (observational method).

For the purposes of the numerical modelling under seismic conditions (Castelli, Lentini, & Trifaró 2016, Castelli, Lentini, Trifaró, Greco, & Lombardo the "NTC2008" (D.M.14/01/2008)

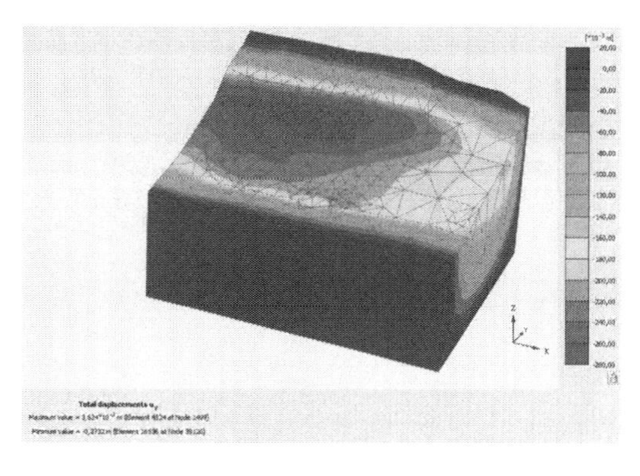

Figure 14. Actual safety conditions.

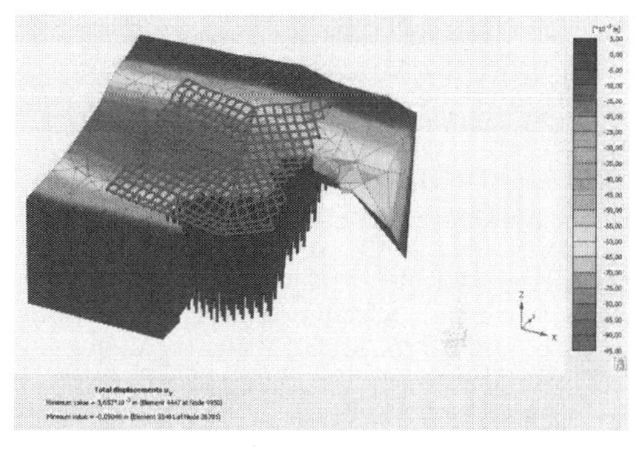

Figure 15. Safety conditions after stabilization intervention.

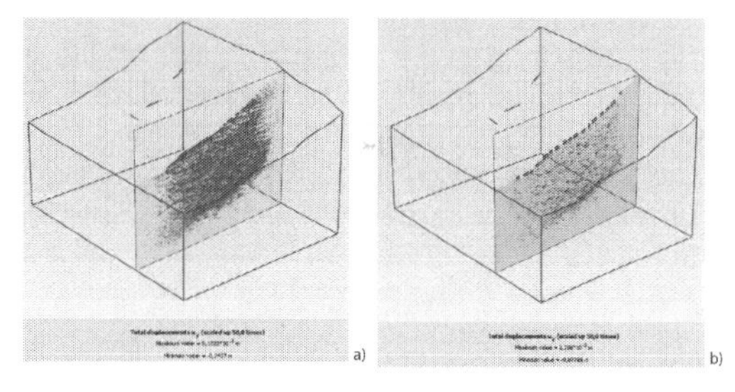

Figure 16. Displacements vector representation: a) Displacement to the actual conditions b) Displacement after stabilization intervention.

and "Decreto Ministeriale 26/06/2014" (D.M.26/06/2014) were adopted.

For the purposes of the seismic amplification the subsoil can be considered as a rigid soil, so the stratigraphic amplification coefficient is equal to 1. The topographic amplification coefficient concerning the topographic category slope (average inclination slopes 15°) is equal to 1.2.

In the numerical analysis the seismic action was evaluated through a pseudo-static approach, by evaluating an appropriate seismic coefficient k_h in the horizontal direction and a seismic coefficient k_v in the vertical direction.

For the stabilization purposes, the piles reinforcement must be extended to their total length and it must be effectively clamped into upper beams truss and into limestones located at the piles base.

By the proposed finite element study it was possible to know the stresses acting on each of the structural elements, highlighting that the most stressed piles are positioned at the lower altitudes.

Therefore, an accurate study was carried out on the values of the stress characteristics of the structural elements such as bored piles and connecting beams, as well as on the appropriate media operations for the calculation of the reinforcement of the same.

5 CONCLUSIONS

The Disueri dam, located in the town of Mazzarino, close to Caltanissetta city (Sicily) is affected by numerous karst cavities. Active dissolution and internal erosion by leakage water in the foundation of a dam may cause, in addition to permeability increase, a rapid degradation in the mechanical properties of the material bearing the dam body and associated structures.

The proposed analysis is based on a performance-based approach, which leads to appreciate how resistance forces increase following cavities filling and piles contribution to improving slope stability in terms of cumulative displacements also in seismic conditions.

The displacements of the reinforced slope are smaller significantly than non-stabilized slope case study.

The outputs show the effects of stabilization intervention connected to the structural mesh on the landslide slope and the consequent considerable reduction of the seismic-induced displacements; they also contribute to significantly increase the safety factor of the dam's left bank.

REFERENCES

Baldovin, E., A. De Paola, & G.L. Morelli (2011). Interventi di risanamento della Diga Giudea in localit´a Gello (PT) e verifiche dinamiche. *Rivista Italiana di Geotecnica n.2/2011*, 11–22.

Brinkgreve, R.B.J. & P.A. Vermeer (2002). *PLAXIS: Finite Element Code for Soil and Rock Analyses.* Balkema, Rotterdam / Brookfield.

Castelli, F., V. Lentini, & C.A. Trifaró (2016). 1D seismic analysis of earth dams: the example of the Lentini site. In Elsevier (Ed.), *Proceedings CNRIG2016 - VI Italian Conference of Geotechnical Engineering Researchers, Bologna, September 2016*, pp. 356–361.

Castelli, F., V. Lentini, C.A. Trifar´o, M. Greco, & C. Lombardo (2017). Pericolositá sismica e vulnerabilit´a di alcune dighe siciliane. In AGI (Ed.), *Proceedings XXVI Italian Geotechnical Conference "La Geotecnica nella Conservazione e Tutela del Patrimonio Costruito". Rome, June 2017*, pp. 1035–1042.

D.M.14/01/2008. *Norme Tecniche per le Costruzioni.*

D.M.26/06/2014. *Norme tecniche per la progettazione e la costruzione degli sbarramenti di ritenuta (dighe e traverse).*

Sabetta, F. & A. Pugliese (1996). Estimation of response spectra and simulation of non-stationary earthquake ground motion. In *BSSA 86.*

Valore, C., M. Ziccarelli, & C. Gambino (2011). The dissolution of gypsum of disueri reservoir left bank. In *XV ECSMGE. Athens, September 2011*, pp. 1205–1210.

Valore, C. and Ziccarelli, M. & S.R. Muscolino (2013). Spostamenti della sponda di un lago artificiale causati da carsismo da dissoluzione dei gessi. In *Incontro Annuale Ricercatori di Geotecnica, Perugia 2013.*

Numerical Methods in Geotechnical Engineering IX – Cardoso et al. (Eds)
© 2018 Taylor & Francis Group, London, *ISBN 978-1-138-33203-4*

Discrete element modelling of the failure mechanisms of Foz Tua arch dam foundation

M. Espada, J. Muralha & J.V. Lemos
National Laboratory for Civil Engineering (LNEC), Lisbon, Portugal

N. Plasencia, J.N. Figueiredo & D. Silva Matos
Energies of Portugal (EDP), Porto, Portugal

J.C. Marques
Faculty of Engineering of the University of Porto (FEUP), Porto, Portugal

ABSTRACT: This work presents an update study of the structural evaluation of the Foz Tua arch dam, recently built by EDP – Energias De Portugal in the north of Portugal, for foundation failure scenarios including the geological conditions encountered during the excavation works and geotechnical surveys. For this purpose, a three-dimensional discrete element model was developed using the 3DEC software to evaluate the potential failure mechanisms through the rock mass discontinuities, in which the main geological structures found in both banks were considered. The safety evaluation methodology based on the progressive reduction of the shear strength of rock discontinuities is presented, as well as the numerical results obtained with the discrete model. This study allowed the identification of potentially unstable rock blocks in the downstream slopes which, however, did not compromise the safety of the dam, verifying the compliance with safety standards.

1 INTRODUCTION

Safety assessment of large arch dams throughout their lifetime demands comprehensive studies for the characterisation of the rock mass foundation. The existence of natural rock discontinuities or local heterogeneities and the presence of joint water pressures strongly influence the behaviour of the dam rock mass foundation. The potential failure mechanisms are typically related with motions along weakness surfaces, i.e., along the rock discontinuities. For the identification of these failure mechanisms, the use of discontinuum models is required, in particular discrete element (DE) models where the rock mass singularities can be properly simulated (Lemos, 2011).

For design purposes, LNEC carried out in 2010 a safety assessment study for foundation failure scenarios of the Foz Tua arch dam (LNEC, 2010a). However, the geological and geotechnical conditions encountered during the excavation works, led to the re-evaluation of the safety conditions of the dam-foundation system (LNEC, 2015; Espada *et al.*, 2016). This work presents the main results of this safety re-evaluation assessment, aimed at analysing potential failure mechanisms through the rock mass discontinuities in the dam foundation, using a three-dimensional DE model developed in 3DEC software (Itasca, 2013).

The geological mappings undertaken during the excavation works, complemented with previous information from geological surveys and *in situ* and laboratory test results, allowed a detailed characterisation of the rock foundation, namely the definition of the different geotechnical zones, as well as the main geological structures that intersect it (faults, veins and joints). The actual ground topography surrounding the dam and the most relevant discontinuities identified in the geological mappings were included in the numerical model.

The safety evaluation methodology based on the progressive reduction of the shear strength parameters of discontinuities is presented, as well as the results of the 3DEC model.

2 THE FOZ TUA HYDROELECTRIC PROJECT

2.1 *General description*

The Foz Tua hydroelectric scheme is located on the Tua river, a right bank tributary of the Douro river, at the northern region of Portugal. The dam position is about 1100 m upstream of the river's confluence and the scheme has the following main elements:

- double arch concrete dam, 108 m high (between the theoretical lower point of the foundation and the crest, respectively at 64 m and 172 m) and 5 m thick at the crest, which is 275 m long; a spillway centered on the dam body, equipped with 4 gates 15.7 m wide each; a bottom discharge and an ecological flow device;
- underground hydraulic circuit, located on the right bank, with two independent hydraulic tunnels and two independent generators with a rated power of 262 MW;
- a shaft powerhouse located 500 m downstream of the dam.

Between the outlet and the Tua river mouth, a channel was excavated on the riverbed in order to allow adequate conditions for pumping operation.

The reservoir receives water from a hydrographic basin of 3809 km², has a capacity of 106.1 hm³, for a full reservoir level situated at elevation 170 m, and occupies an area of 420.9 ha.

Normally, in the exploration regime, the reservoir water level will oscillate between 170 m and 167 m, corresponding to a volume variation of 10 hm³. The extraordinary water level can reach the elevation of 162 m; at this situation, the usable reservoir volume is 28 hm³.

The dam is divided by 17 contraction joints (D1 to D8 and E1 to E9) forming 18 blocks with a length of 13.5 m to 17 m. The joints are defined by vertical planes radial to the crest arch. To provide the monolithism of the dam, the joints were injected with cement. Downstream there is a dissipation basin, separated from the dam by a vertical not injected joint. The Foz Tua hydroelectric scheme was mainly engineered by EDP technical staff, including the dam shape definition and the stability analysis for current load scenarios.

2.2 *Rock mass foundation*

The dam is located in a V-shaped valley section of the Tua river, inserted in a granitic rock mass outcrop surrounded by metamorphic rocks. The topographic, geological and geotechnical conditions are suitable for the selected type of dam.

Based on the geological surveys and on geotechnical tests presented on Tables 1 and 2 (EDP, 2010), good conditions for the dam foundation were anticipated. It was expected that the dam would be founded on the ZGB and ZGA zones, with the exception of faulted and tectonized zones where the geotechnical characteristics would correspond to those of the ZGC zone. It was also expected to find the worst characteristics at the upper right bank.

Table 1. Geological-geotechnical rock dam-foundation model.

Zone	W	F	RQD (%)	RMR
ZGC	W3 (W4-5)	F3, (F4-5)	0–75	30–50
ZGB	W2 (W3)	F2 (F3)	75–90	50–70
ZGA	W1-2	F1-2	>90	>70

Table 2. Rock mass dam-foundation geomechanical parameters.

ZG	UCS (MPa)	E_r (GPa)	ϕ (°)	c (MPa)	E_m (GPa)	E_d (GPa)
ZGC	95	36	41	0,22	-	7
ZGB	85	42	34	0,15	-	11
ZGA	133	44	38	0,09	32	19

UCS – uniaxial compression test; Er – deformability modulus of rock; ϕ – friction angle; c – cohesion; Em – deformability modulus of rock mass; Ed – dilatometric modulus.

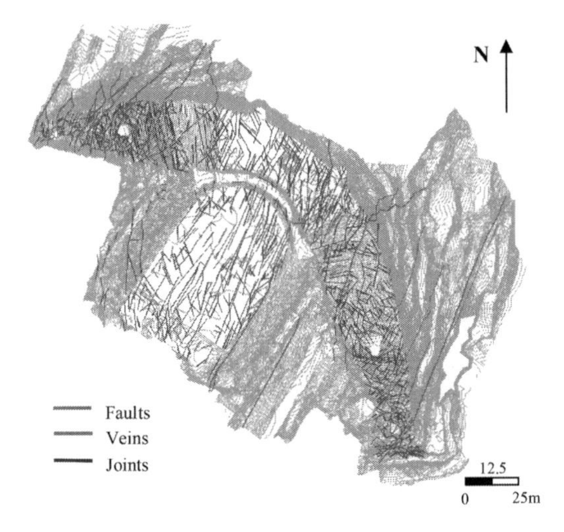

Figure 1. Geological mapping of the dam and dissipation basin (based in Espada *et al.* 2016).

The geological characteristics provided by the study phases were globally confirmed during the excavations. The geological mapping of the excavations for the dam and dissipation basin foundation is presented in Figure 1 with the representation of the geostructures. The NE-SW (N30° and N60°) subvertical structures are clearly identified. The geological faults N30° are the most frequent on both banks and on the bottom of the valley, but with a greater tendency for persistence and less

Table 3. Joints conditions accordingly to ISRM.

Joint set	F1		F2		F3		F4	
Bank	RB	LB	RB	LB	RB	LB	RB	LB
Strike	N75°E	N62°E	N32°E	N31°E	N24°E	N14°E	N12°W	N02°E
Dip	86°SE	89°SE	88°NW	83°SE	30°SE	34°NW	88°NE	87°NW
Persistence	1–3 m	3 m	3–10 m		3–10 m		3 m	
Weathering	W3		W3		W3		W3	
Opening	closed		closed or 0.5 mm		closed		0.5 mm	
Filling	without		without		without		without	
JRC	6–8	10–16	6–8		8–10	14–18	6–8	
Water	dry	dry to moist	oxides		dry		dry	dry to moist

"LB" – left bank; "RB" – right bank.

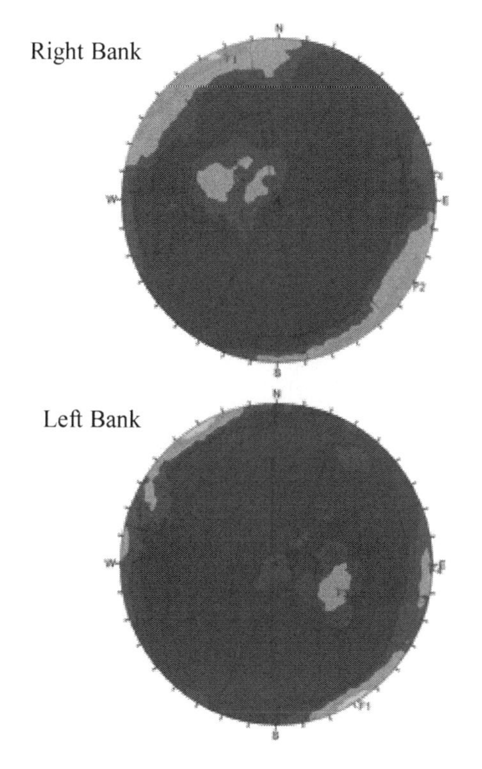

Figure 2. Joint statistical orientation analysis. Equal area contoured lower hemisphere (Schmidt concentration).

spacing on the right bank. The alignments are associated with areas of weathering, sometimes with clay, and fractured and parallel plans.

The statistical analysis of joints was performed according to ISRM criteria. The results are presented in Table 3 and Figure 2.

Next to the river, the subvertical extensive fractures are clear as well as the discontinuities dipping about 30° to the valley.

3 GENERATION OF THE NUMERICAL MODEL AND MATERIAL PROPERTIES

The three-dimensional DE model of the dam-foundation system was developed using the 3DEC software from Itasca.

The finite element mesh for the concrete dam was externally generated and then imported into 3DEC. It consists of 18 blocks separated by vertical contraction joints, composed by several 20-node solid brick elements. The concrete was assumed elastic, with a Young's modulus of 20 GPa and a Poisson's ratio of 0.20.

The generation of the rock mass foundation was carried out in several stages. Starting with the terrain elevation contours, and using the Rhinoceros software (Robert McNeel and Associates, 2014), the terrain surface surrounding the dam was generated, as shown in Figure 3a). Then, using again the Rhinoceros software, the mesh of polygons corresponding to the terrain surface generated in the previous step was obtained (Figure 3b). Finally, the 3DEC blocks were created extruding the terrain surface mesh polygons downwards, in the vertical direction, as shown in Figure 3c).

Next, the blocks below the dam were generated and the set of blocks correspondent to the mesh of polygons was extended upstream and downstream. Larger blocks were also created below and laterally, allowing to apply the boundary conditions at a certain distance from the area of interest. Figure 4 illustrates the finite element mesh of the dam and the generation process of the rock mass foundation.

After creating the geometry, the most important discontinuities identified in the geotechnical surveys performed during the excavation works were inserted in the 3DEC model. The selection of the rock discontinuities to be included in the model considered the development of possible failure mechanisms eventually generated by the geological structures.

Most of the faults and veins identified at the site in both banks were inserted in the model, but the representation of the joints is more complicated, because they are usually in larger numbers and it is not feasible to include them all due to the computational effort required. Hence, a global analysis was carried out in order to identify and select the more representative joints to be included in the model. The joint selection process was based on the detailed geological mappings collected during the excavation works. Figure 5 is a photograph of part of the left bank dam foundation excavation surface, where can be identified three different geological structures (fault, joint and vein) that were included in the model.

Finally, the mesh inside the rock blocks was automatically generated in 3DEC. The rock mass is represented by deformable polyhedral blocks that are internally divided into a finite element mesh of tetrahedral elements. As can be seen in Figure 6, the discontinuities were not indefinitely extended upstream, because a vertical joint was placed along the upstream edge of the dam concrete-rock interface. Besides the computational savings, this is a conservative simplification which accounts for the expected rock joint opening due to tensile stresses developed at the dam upstream heel.

It is also important to stress that the discontinuities were considered as continuous planes. Though it is usually adopted in safety assessment studies, this is a conservative hypothesis, since it is very difficult to account for the real persistence of discontinuities in DE models.

Figures 7 and 8 display the geological planes inserted in the rock mass foundation. The dam contraction joints, the concrete-rock interface, the vertical joint along the upstream edge and the auxiliary planes for the geometry construction are also depicted in the figures.

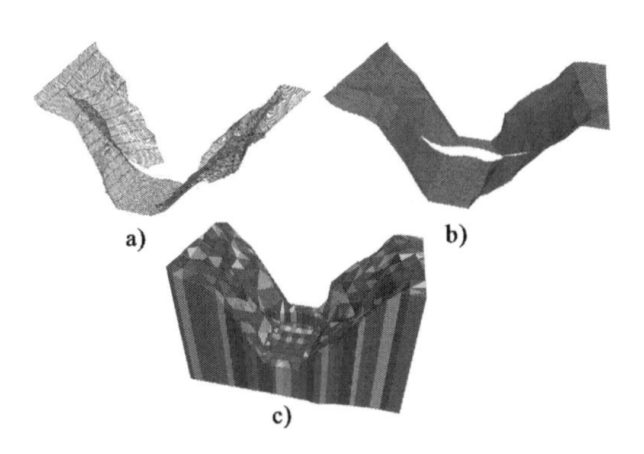

Figure 3. Downstream views of the terrain surface surrounding the dam with the ground contours a), the mesh of polygons b), and the blocks generated by the extrusion of the polygons c).

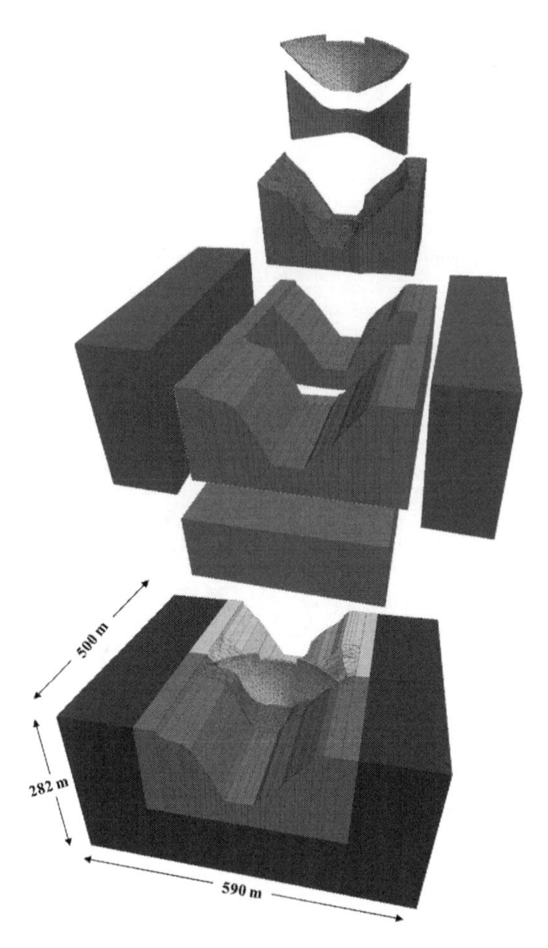

Figure 4. Block-components of the 3DEC dam-foundation model.

Figure 5. Left bank dam foundation excavation surface with the identification of three geological structures (Geoárea, 2014).

It was assumed that the rock mass is elastic and isotropic, with an elastic modulus of 24 GPa, a Poisson's ratio of 0.20 and a unit weight of 26.5 kN/m^3. These values resulted from the analysis of *in situ* and laboratory tests (LNEC, 2010b).

According to the expected and observed geological and geotechnical characteristics, the upper zones of the right bank display lower deformability modulus associated with higher weathering and fracturing intensity. Therefore, in those zones reduced elastic moduli of 12 GPa and 6 GPa were considered.

A purely frictional Coulomb behaviour, without cohesion or tensile strength, was assumed for the faults, veins and joints in the safety evaluation analysis. According to the shear test results carried out by LNEC (2010b), a friction angle of 38° was assigned to these geological structures. Joint normal and shear stiffnesses also resulted from the same experimental results: for faults and veins, k_n = 10 GPa/m and k_s = 2 GPa/m, and for rock joints, k_n = 20 GPa/m and k_s = 3 GPa/m. Similar shear strength parameters were considered for the vertical joint along the upstream edge. The foundation joint was also assigned a Mohr-Coulomb model, with a friction angle of 45°, a cohesion of 3 MPa and a tensile strength of 2 MPa. However, in safety evaluation analyses, cohesion and tensile strength of the foundation joint were set to 0. The joint normal stiffness was taken as 10 GPa/m, and the shear stiffness as 5 GPa/m. The vertical contraction joints between the dam cantilevers were assumed to have a purely frictional behaviour, with a friction angle of 40°, and no cohesion nor tensile strength.

It must be highlighted that the consideration of no cohesion and tensile strength in failure scenarios is imposed by Portuguese dam safety regulation (RSB, 2007), while the choice of the other shear strength parameters took into account, not just the results of *in situ* and laboratory tests, but also the experience acquired in similar studies.

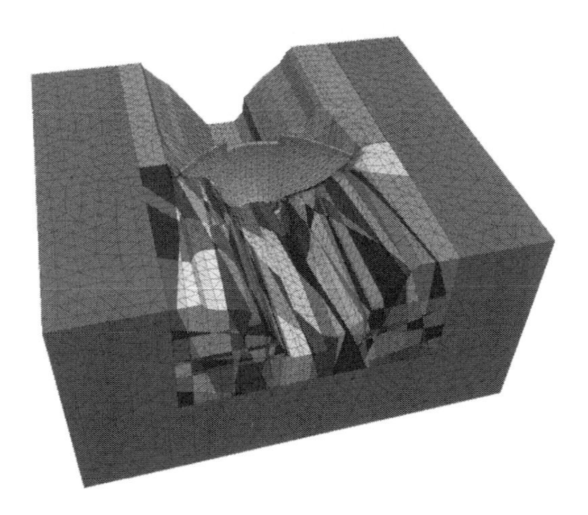

Figure 6. Finite element mesh of the numerical model.

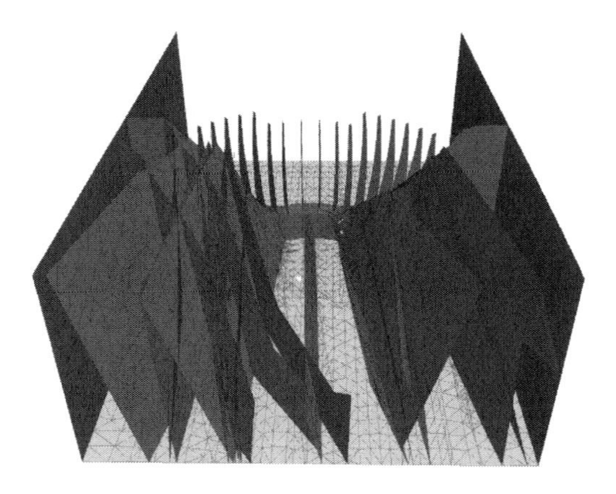

Figure 7. Downstream view with the faults and veins included in the 3DEC model.

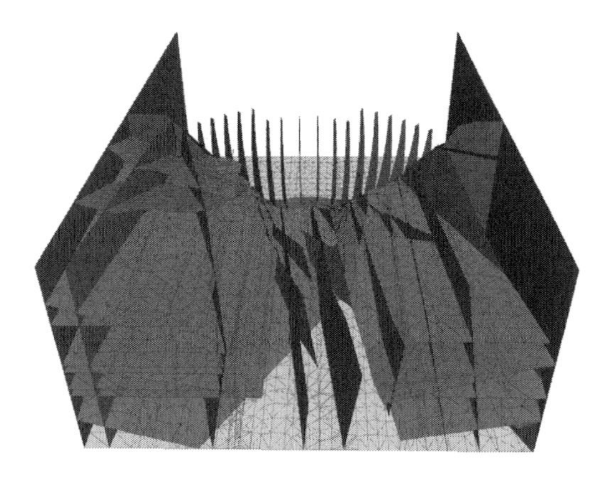

Figure 8. Downstream view with the joints included in the 3DEC model.

4 MODELLING PROCEDURE

The first step of the modelling procedure was to consider the *in situ* state of stress in the rock mass foundation, before dam construction. The initial normal vertical stress was assumed as the overburden weight and the initial normal horizontal stresses corresponded to 0.25 times the weight of the overlying ground. These data resulted from the analysis of geotechnical investigations conducted at the site.

Next, the simulation of dam construction with independent cantilevers was done. After equilibrium was reached, the closure of the vertical contraction joints was imposed, and then the hydrostatic pressure at the upstream face was applied. The normal storage level was set at elevation 170 m, i.e., 2 m below the dam crest.

In the next stage the uplift pressures along the concrete-rock interface and in the rock discontinuities were set. At the concrete-rock interface the uplift pressures were prescribed according to the usual design criterion, a bilinear diagram with 1/3 of the reservoir head at the drainage curtain location. For the rock discontinuities, simplified water pressure distributions were considered: the full reservoir head was considered upstream, whereas downstream a simplified pressure field defined in terms of a water table compatible with the valley slopes was imposed.

Figure 9 shows the displacements and the minimum and maximum principal stresses at the downstream face after applying all loads in the model, i.e., due to the joint effect of self-weight, hydrostatic pressure and of uplift pressures in the rock mass and along the concrete-rock interface. The highest displacements in the dam body were obtained between blocks E3-E7, in the left bank, reaching the maximum value of about 32.3 mm. Regarding the minimum principal stresses (compressive stresses), the maximum value of 6.7 MPa was obtained at the downstream face, near the foundation surface. The maximum tensile stress registered in the dam body was approximately

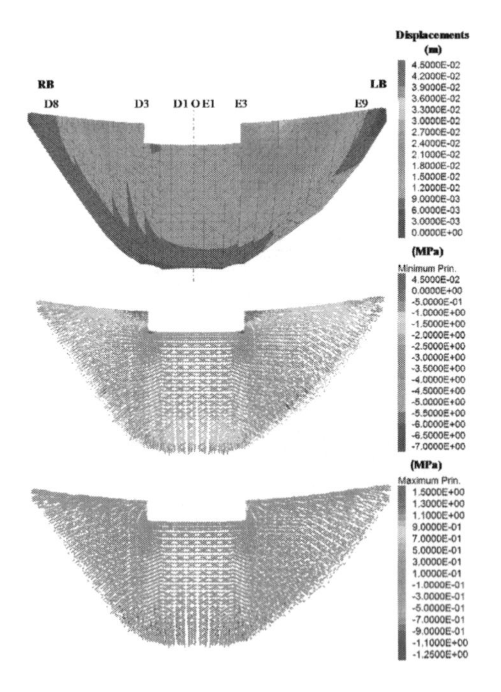

Figure 9. Displacements and minimum (compressive) and maximum (tensile) principal stresses at the downstream face, considering the joint effect of self-weight, hydrostatic pressure and uplift pressures.

1.1 MPa. These values are well below the concrete strength.

Finally, the safety evaluation procedure was conducted, which consists in the sequential reduction of the strength parameters of the rock discontinuities by increasingly larger factors until collapse takes place or displacement magnitudes reach unacceptable levels (Lemos, 2012). In this study, the reduction factor RF was only applied to the tangent of the friction angle of all faults, veins and joints, since a condition of no cohesion and no tensile strength is stipulated in the regulations.

5 FOUNDATION FAILURE SCENARIOS

The safety evaluation for failure scenarios involving displacements in rock discontinuities is addressed in this section. A complete analysis of the evaluation of the different foundation failure scenarios of the Foz Tua arch dam can be found in LNEC (2015) or in Espada *et al.* (2016).

The Portuguese dam code (RSB, 2007) requires that in safety assessment studies for failure scenarios, whether or not they include dynamic loads, the stresses in failure surfaces must satisfy the Mohr-Coulomb criterion, defined for null cohesion and tensile strength and residual values of friction angle, with minimum global safety factors between 1.2 and 1.5.

In this study, the strength reduction procedure consisted in progressively dividing the tangent of the friction angle of all discontinuities included in the 3DEC model by a reduction factor RF. The friction angle of the concrete-rock interface was not reduced during the procedure, but the effect of cohesion and tensile strength was neglected.

Figure 10 shows the evolution of displacements in the dam body at the top and at the base of two blocks in the right bank (D5-D6 and D8-D9) and in one block in the left bank (E5-E6), during the reduction of the tangent of the friction angle of all discontinuities. Figure 11 presents the displacement contours in the downstream rock foundation, before the reduction process, i.e., for RF = 1.0 ($\phi = 38°$), and for the reduction factors of 1.5 and 1.7. It can be seen in Figure 10 that the displacements progressively increase in the dam body, in both banks, up until RF = 1.7 ($\phi = 24.7°$). From this factor up to RF = 2.0 ($\phi = 21.3°$) a significant increase of the displacements is noted, mainly in the right bank blocks. This is due to a local failure mechanism, in the downstream rock slopes in the right bank, that starts to be recognisable for a reduction factor of 1.5 ($\phi = 27.5°$), as shown in Figure 11. It was verified that up until RF = 1.5 a redistribution of the stresses in the rock mass allows equilibrium to be reached, although some

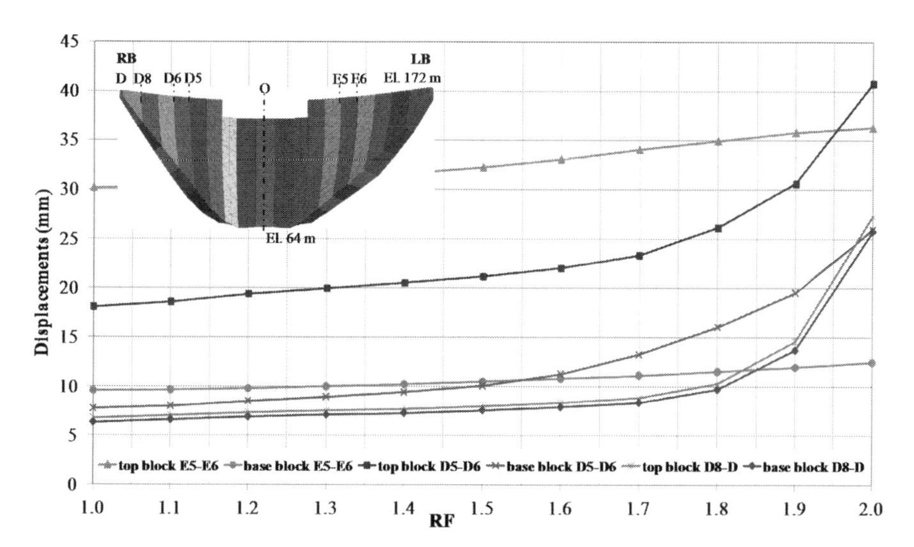

Figure 10. Evolution of displacements at the top and at the base in three dam blocks during the strength reduction process.

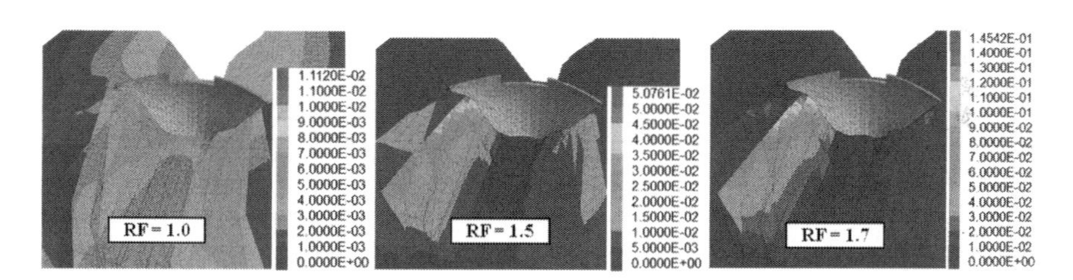

Figure 11. Contours of displacement (m) in the rock mass before the reduction process (RF = 1.0) and for RF = 1.5 and 1.7.

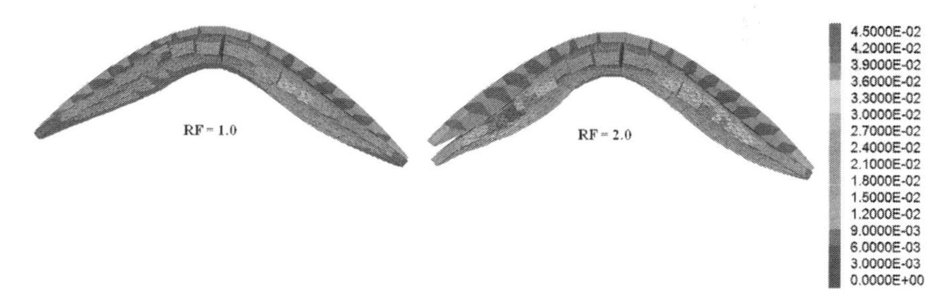

Figure 12. Contours of displacement (m) in the dam body for reduction factors of 1.0 and 2.0 (displacements amplified 500 times).

rock blocks show important displacements. However, from RF = 1.6 the equilibrium is no longer verified, and a significant increase of the maximum displacements in the rock mass occurs as the reduction factor increases.

The displacement contours in the dam body for the initial stage, before the reduction procedure, and for RF = 2.0 are displayed in Figure 12. As can be seen in this figure, the maximum displacements in the dam body in the left bank and in the right bank are, respectively, 37.6 and 41.4 mm, which represent an increase of about 28% in relation with the maximum value obtained at the initial stage (RF = 1.0).

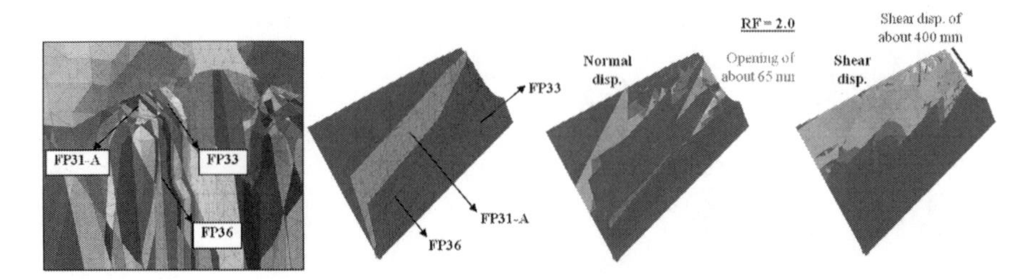

Figure 13. Normal and shear displacements in FP33, FP31-A and FP36 for RF = 2.0.

It was concluded that this failure mechanism involves rock blocks in the downstream slope of the right bank defined by the intersection of the sub-vertical faults FP33, FP31-A and FP36 (see Figure 13). Fault FP33 displays the higher normal displacements as the reduction factor increases, reaching about 65 mm for RF = 2.0. All faults show significant shear displacements during the analysis, reaching, for RF = 2.0, the maximum value of approximately 400 mm. Finally, it can be concluded that the shear along these faults controls this local failure mechanism in the rock mass, however it did not affect the global stability of the concrete dam.

6 CONCLUSIONS

The stability analysis for foundation failure scenarios of the Foz Tua arch dam was addressed using a 3DEC model of the dam-foundation system in this work. The use of DE models allows the representation of the different weakness surfaces, such as the complex rock jointing patterns or the concrete-rock interface, where potential failure modes can be developed.

The analysis showed that a safety assessment procedure, based on the progressive reduction of the shear strength parameters of rock discontinuities, provides a helpful insight into the effects of geological features on the structural behaviour of the dam.

Nevertheless, it is important to notice that these studies for the safety assessment in jointed rock masses using DE models, necessarily contain some simplifications and conservative assumptions. An obvious limitation lies in the representation of the real persistence of joints, which is still difficult to take into account in DE models. Additionally, the support system in the excavation slopes, which consisted of rock bolts and shotcrete, was not simulated in the 3DEC model.

This study allowed to conclude that a sliding failure mechanism involving three sub-vertical faults was obtained in the downstream rock slope of the right bank from RF = 1.7 ($\varphi = 24.7°$). This mechanism led to an increase of the displacements in the dam body, but it does not compromise the compliance with safety standards.

REFERENCES

EDP 2010. *Foz Tua hydroelectric power scheme – Design. Geological and geotechnical characterization of the site* (in Portuguese).

Espada, M., Muralha, J., Vieira de Lemos, J., Plasencia, N. & Paixão, J. 2016. *Safety assessment of the Foz Tua dam foundation failure scenarios*. Proc. of 15th National Geotechnical Congress, FEUP, Porto, Portugal (in Portuguese).

Geoárea 2014. *Foz Tua hydroelectric power scheme. Geological and Geotechnical Report of the dam and dissipation basin*. Geoárea (in Portuguese).

ITASCA 2013. *3DEC, Version 5.0, User's Guide*. Itasca Consulting Group, Minneapolis, USA.

Lemos, J.V. 2011. *Discontinuum models for dam foundation failure analysis*. Proc. of 12th ISRM Congress, Beijing, pp. 91–98.

Lemos, J.V. 2012. *Modelling the failure modes of dams' rock foundations*. Chapter 14 of MIR 2012 - Nuovi metodi di indagine, monitoraggio e modellazione degli amassi rocciosi (Ed. G. Barla), Politecnico di Torino, pp. 259–278.

LNEC 2010a. *Study of foundation failure scenarios for the Foz Tua dam*. LNEC, Report 211/2010: Lisbon (in Portuguese).

LNEC 2010b. *Foz Tua hydroelectric power scheme. Geological and geotechnical surveys – Geomechanical characterization*. LNEC, Report 196/2010: Lisbon (in Portuguese).

LNEC 2015. *Foz Tua hydroelectric power scheme. Update of the safety assessment study for foundation failure scenarios*. LNEC, Report 348/2015: Lisbon (in Portuguese).

Robert McNeel & Associates 2014. *Rhinoceros, Modeling tools for designers, Version 5.0*. Robert McNeel & Associates, Seattle, USA.

RSB 2007. *Regulamento de Segurança de Barragens e correspondentes portarias: I - Normas de projecto de barragens; II - Normas de observação e de inspecção de barragens*. Decreto-Lei 344/2007 de 15 de Outubro, Diário da República, Lisboa (in Portuguese).

Tunnels and caverns (and pipelines)

Numerical Methods in Geotechnical Engineering IX – Cardoso et al. (Eds)
© 2018 Taylor & Francis Group, London, ISBN 978-1-138-33203-4

Linear models for the evaluation of the response of beams and frames to tunnelling

A. Franza, S. Acikgoz & M.J. DeJong
Department of Engineering, University of Cambridge, Cambridge, UK

ABSTRACT: This paper investigates the Tunnel-Building Interaction (TSI); in particular, the contribution of shear structure deformations to TSI is addressed. An elastic solution is adopted based on beam elements connected to an elastic continuum. Bearing wall structures are modelled as Euler-Bernoulli and Timoshenko simple beams (located at the ground level and mid-height of the structure), whereas plane frames are adopted for the frames on a continuous strip footing. Tunnelling-induced displacements, deformations, and internal forces at the beam axes and foundation members are detailed. Results show that shear flexibility of Timoshenko beams and framed configurations alter the TSI mechanism and can lead to greater tunnelling-induced deformations. For frames, the stiffening effect of the higher storeys is minor. Finally, results also display the extent to which stiff elements experience larger tunnelling-induced internal forces with respect to flexible beams. The potential for structural failure within RC elements is estimated with a simplified procedure.

1 INTRODUCTION

The excavation of tunnels for transportation infrastructure and services results in ground movements. In urban areas, as the results of tunnel-structure interaction (TSI), these movements can deform and displace surface structures and, potentially, lead to a loss of their serviceability performance and structural failure.

Preliminary assessment of the effects of TSI is possible through the use of predicted greenfield ground movements and the modification factor approach for buildings on continuous shallow foundations (Franzius, Potts, & Burland 2006, Giardina, DeJong, & Mair 2015, Haji, Marshall, & Franza 2018, Mair, Taylor, & Burland 1996). Research has primarily dealt with low- and medium-rise masonry buildings on strip footings, whereas alternative framed superstructures and/or foundation configurations (e.g. separated footings) have received less attention. However, several studies (Goh & Mair 2014, Fargnoli, Gragnano, Boldini, & Amorosi 2015, Franza & DeJong 2017) indicated that tunnel-frame interaction differs from the case of bearing wall structures. Although risk assessments of RC buildings require estimating tunnelling-induced internal forces within frame elements (i.e. column/pier, beams), few studies performed structural analyses of frames in a systematic way (e.g. Cai et al. (2017)) and previous research focused on foundation displacements. The response of alternative structural configurations to tunnelling remains unclear and, thus, engineers need to carry out their 3D numerical modelling, which is time-consuming, during preliminary design stages (High Speed Two Limited 2014).

This paper aims to illustrate, through elastic models of tunnelling beneath surface bearing walls and framed buildings, the influence of shear deformability on tunnel-structure interaction (TSI). Results detail tunnelling-induced displacements and deformations; the distributions of internal forces within structural elements are also analysed.

2 BACKGROUND

2.1 *Greenfield tunnelling-induced movements*

In clays, tunnelling-induced vertical (u_z) and horizontal (u_x) ground movements at the surface can be estimated using the standard Gaussian curves displayed in Equation (1) and Equation (2), respectively (Mair, Taylor, & Bracegirdle 1993, Mair & Taylor 1997).

$$u_z = u_{z,max} \exp\left(-\frac{x^2}{2i^2}\right) \tag{1}$$

$$u_x = u_z \frac{x}{(1+0.175/0.325)\,z_t} \tag{2}$$

where x is the horizontal spatial coordinate, $u_{z,max} = 1.25R^2V_{l,t}/i$ is the maximum settlement,

$i = Kz_t$ is the horizontal distance of the settlement trough inflection point to the tunnel centreline, z_t is the tunnel axis depth, $K = 0.5$ is the width parameter, R is the tunnel radius, and $V_{l,t}$ is the tunnel volume loss parameter, which is the ground loss at the tunnel periphery per unit length of tunnel normalised by the tunnel area. Tunnelling-induced ground movements can also be modelled for varying soil conditions using either alternative empirical methods or analytical solutions (Marshall, Farrell, Klar, & Mair 2012, Loganathan & Poulos 1998).

2.2 *Tunnelling-induced structure deformations and the role of soil-structure interaction*

Semi-flexible structures react to tunnelling-induced ground displacements. To quantify the structure stiffness contribution on the reduction of structural deformations, Potts & Addenbrooke (1997) proposed the deflection ratio modification factors, $M^{DR,sag}$ an $M^{DR,hog}$, as follows

$$M^{DR,sag/hog} = \frac{DR_{sag/hog,bld}}{DR_{sag/hog,GF}} \tag{3}$$

where building deflection ratios are $DR_{sag,bld}$ and $DR_{hog,bld}$ (determined by the greenfield inflection point position, i), deflection ratios of the greenfield settlement trough are $DR_{sag,GF}$ and $DR_{hog,GF}$ (depending on the structure inflection point, i_{bld}).

The role of the foundation scheme is remarkable in the TSI. Structures with continuous horizontal foundation elements (e.g. continuous strip footings) have a significant axial stiffness and, consequently, tunnelling-induced structural horizontal strains ε_h axis are typically low (Dimmock & Mair 2008, Ritter, Giardina, DeJong, & Mair 2017). On the other hand, framed structures with columns/piers on separated footings can undergo significant differential horizontal displacements because of tunnelling, which are associated with a bending deformation of first-storey columns/piers rather than an overall axial deformation of the structure (Goh & Mair 2014). Additionally, Franza & DeJong (2017) illustrated that a complex combination of rotational and translational displacements (consisting of both vertical and horizontal movements) can occur for separated footings of framed structures.

TSI can also result in varying the shape of the structure settlement curve with respect to greenfield ground movements. One measure to quantify this is the difference between the structure inflection point, i_{bld}, and the inflection point of the greenfield settlement trough, i, as shown in Figure 1 (Farrell, Mair, Sciotti, & Pigorini 2014, Frischmann, Hellings, Gittoes, & Snowden 1994, Franza, Marshall,

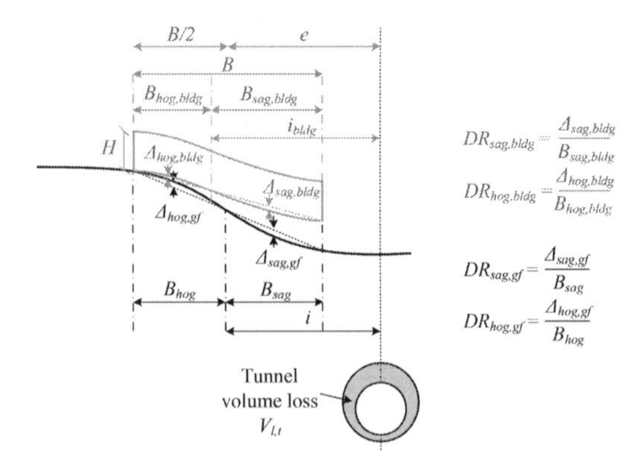

Figure 1. Nomenclature for the flexural TSI problem (Franza et al. 2017).

Haji, Abdelatif, Carbonari, & Morici 2017, Potts & Addenbrooke 1997). To quantify this, the modification factors $M^{B,sag}$ and $M^{B,hog}$ for the sagging and hogging region lengths are definedas follows

$$M^{B,sag/hog} = \frac{B_{sag/hog,bld}}{B_{sag/hog}} \tag{4}$$

where B_{sag} and B_{hog} are the lengths of the building in the sagging and hogging zones based on the greenfield settlement trough, whereas $B_{sag,bld}$ and $B_{hog,bld}$ are based on the building settlement profile. The numerical analyses from Potts & Addenbrooke (1997) indicate that this phenomenon depends on the structure shear flexibility.

To estimate the impact of the structure stiffness on M^{DR}, Mair (2013) proposed the following relative bending stiffness factors

$$\rho_{sag/hog} = \frac{EI}{E_s B^3_{sag/hog} L} = \frac{EI^*}{E_s B^3_{sag/hog}} \tag{5}$$

where EI is the bending stiffness of the superstructure (in kN.m^2), EI* is EI per running metre (in kN.m^2/m), E_s is a soil Young's modulus representative of the soil affected by the excavation, and L is the longitudinal length of the building in the tunnel axis direction. Equation (5) neglects that a structure can span across sagging and hogging regions and the effects of the structure weight.

Although Bilotta, Paolillo, Russo, & Aversa (2017) and Giardina, DeJong, & Mair (2015) displayed that the increase in weight can result in greater M^{DR}, this influence is secondary compared to the role of stiffness. Consequently, the use of elastic two-stage models is a viable approach to investigate TSI for ordinary structures and low-medium volume losses (Deck & Singh 2012).

In both research and practice, it is common to simplify the structure to equivalent linear elastic plates/beams to decrease computational costs during initial assessment. Generally equivalent beams/plate are identified by matching the bending stiffness *EI* (Franzius, Potts, & Burland 2006, Bilotta, Paolillo, Russo, & Aversa 2017). However, this approach results in equivalent beams/plates with a lower height-to-length ratio and, thus, to a decreased shear deformability with respect to the target building. Pickhaver, Burd, & Houlsby (2010) suggested a procedure to identify an equivalent Timoshenko beam for bearing wall structures with openings aiming to match the total structure stiffness (both bending and shear contributions). This might effectively capture the effect of openings in a simplified manner, but more thorough understanding of the influence of the shear flexibility on TSI is still needed.

3 PROPOSED ELASTIC MODELS

In this paper, an elastic two-stage analysis method is adopted that models the surface structure as either a simple beam or a plane frame resting on an elastic homogeneous half-space continuum subjected to tunnelling-induced ground movements (Klar, Vorster, Soga, & Mair 2005). In this study, a perfect bond condition is assumed between the soil and the structure (i.e. separation and relative soil-structure sliding are not allowed). As basic assumptions, the tunnel and structure presence does not influence, respectively, the response of the continuum and the tunnelling process. In addition, the presence of the tunnel is neglected when defining the response of the continuum to loading. Thanks to the superposition principle, the effects of structural loads and tunnelling are studied independently.

The structures considered in this work are simple beams (for bearing wall buildings) and frames on a continuous beam (for framed buildings on a continuous strip foundation), as shown in Figure 2. Rather than modelling the structure by means of an equivalent beam, the cross-sectional properties and Young's and shear moduli of the beam elements are chosen to represent the structural characteristics of the target building. To decouple the effects of bending and shear flexibilities, Euler-Bernoulli and Timoshenko beams are adopted for the simple beams and foundation elements whereas Euler-Bernoulli beams are used for the framed superstructures.

The influence of the offset between structure neutral axis and ground level is also investigated in this paper. Elastic models are solved for both simple beams and foundations elements located

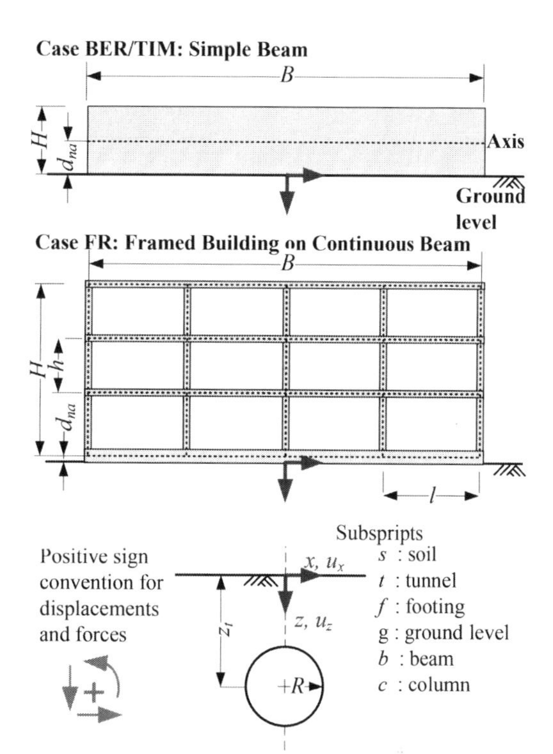

Figure 2. Studied configurations.

at the ground level (labelled 'GL' solutions) and at mid-height of the simple beam/foundation height (labelled 'NA' solutions). Note that for GL solutions, greenfield tunnelling-induced forces are applied at the simple beam/foundation axis, whereas for NA solutions these forces are applied at the bottom fibres of these beams. For framed structures, the offset between superstructure neutral axis and ground level is implicitly accounted for by modelling a plane frame.

The finite element method (FEM) is used to solve the TSI problem numerically. For the GL solutions, the equilibrium equation of the system, written in matrix form, is

$$\mathbf{K}\mathbf{u} = \mathbf{q}; \qquad \mathbf{K} = \mathbf{S} + \mathbf{K}_s; \qquad \mathbf{q} = \mathbf{p} + \mathbf{K}_s\mathbf{u}^{cat} \tag{6}$$

where $\mathbf{u}$ is the structure axis displacement vector ($\mathbf{u}^T = [\mathbf{u}_x \quad \mathbf{u}_z \quad \phi]$, in which $\mathbf{u}_x$ and $\mathbf{u}_z$ consist of the translational displacements along x and z, respectively, whereas φ contains rotations of the finite element nodes), $\mathbf{q}$ is the load vector of the structure ($\mathbf{q}^T = [\mathbf{q}_x\ \mathbf{q}_z\ \mathbf{m}]$, in which $\mathbf{q}_x$ and $\mathbf{q}_z$ are translational forces whereas $\mathbf{m}$ is the vector of applied bending moments), $\mathbf{u}^{cat}$ is the greenfield surface displacement due to tunnelling (a generic greenfield displacement field can be the input to the model), $\mathbf{p}$ is the external loading vector of the structure, $\mathbf{S}$ is the structure stiffness matrix, and $\mathbf{K}_s$ is the stiffness matrix of the soil equal to the inverse of the soil flexibility matrix defined with the elastic integrated forms of Mindlin's solutions given by Vaziri, Simpson, Pappin, & Simpson (1982).

For the NA solutions, the offset between the axis of the simple beam/foundation beam and ground level (d_{na}) is implemented by considering that the structure stiffness matrix $\mathbf{S}$ is given for forces and displacements with respect to its neutral axis, that the soil stiffness matrix $\mathbf{K}_s$ and tunnelling-induced forces $\mathbf{K}_s\mathbf{u}^{cat}$ are given with respect to the ground level (i.e. structure bottom fibre), and that there is a linear relationship between axis and ground displacements and forces. Primed symbols are used for the ground level reference system. For a given node,

$$u'_{i,z} = u_{i,z}; u'_{i,x} = u_{i,x} + \phi_i d_{na} \quad (7a)$$

$$q'_{i,z} = q_{i,z}; q'_{i,x} = q_{i,x}; m = q'_{i,x} d_{na} \quad (7b)$$

The equilibrium equation for NA solutions can be written as follows

$$\mathbf{K}^*\mathbf{u} = \mathbf{q}^*; \quad \mathbf{K}^* = \mathbf{S} + \mathbf{K}_s^*; \quad \mathbf{q}^* = \mathbf{p} + (\mathbf{K}_s\mathbf{u}^{cat})^* \quad (8)$$

where $\mathbf{K}_s$* and $(\mathbf{K}_s\mathbf{u}^{cat})$* are the soil stiffness matrix and the vector of tunnelling-induced forces, respectively, at the structure axis reference system; $\mathbf{K}_s$* and $(\mathbf{K}_s\mathbf{u}^{cat})$* were obtained from $\mathbf{K}_s$ and $(\mathbf{K}_s\mathbf{u}^{cat})$ using Equations (7a) and (7b) (Cheung & Nag 1968).

4 RESULTS

4.1 *Investigated scenarios*

For the elastic soil continuum, a Young's modulus $E_s = 100 MPa$ and a Poisson's ratio $\nu_s = 0.5$ were assumed. Tunnelling-induced surface ground movements in greenfield conditions were defined on the basis of Equations (1) and (2) for $V_{l,t}$ = 1%. To consider different ground movement characteristics, two tunnelling scenarios were investigated: 6 m diameter (D) tunnels with a depth to tunnel axis either z_t = 8 m or 20 m.

The structures were orthogonal to the tunnel and located centrally with respect to the tunnel centreline (corresponding to x = 0). In this work, the moment of inertia I of each element was taken relative to the geometric centroid of the full cross-section. A finite element size of 0.5 m was adopted.

The simple beams were representative of low-rise bearing-wall structures with the actual wall characteristics (rather than fictitious Young's moduli E and cross-sectional dimensions). The simple beams had a transverse length B = 20 m, a Young's modulus, E = 1 GPa, a Poisson's ratio ν = 0.3, and cross-sectional depth d_b = 9 m and width b_b = 5 m. Euler-Bernoulli and Timoshenko simple beams had matching bending stiffness (EI) and axial stiffness (EA); in this way, it is possible to isolate the effects of shear flexibility. Adopted simple beams are summarised in Table 1. The ratio between Young's and shear moduli was varied; E/G = 0.5, 2.6, and 12.5 were considered to cover the range of values typically adopted within the limiting tensile strain method. As detailed by Burland, Broms, & De Mello (1977), E/G = 2.6 is the value corresponding to the isotropic material with ν = 0.3, E/G = 12.5 was proposed for low shear stiffness structures (e.g. because of the presence of openings), and E/G = 0.5 considers the lack of horizontal stiffness.

RC frame buildings had a transverse length B = 20 m, a Young's moduli, E = 30 GPa, and a Poisson's ratio ν = 0.3. The foundation (strip footing) had cross-sectional depth d_f = 0.7 m and width b_f = 1.2 m. Columns and beams had squarecross-sections with a 0.5 m side length ($d_c = d_b = b_c = b_b$ = 0.5 m). The inter-storey heights were h = 3.2 m and the bays had a span of l = 4 m. Frames with 1, 3, 5, 10, and 15 storeys were modelled (e.g. 5-storey frame is labelled FRS5). The foundation was modelled with a Timoshenko isotropic beam (E/G = 2.6), whereas columns and beams with Euler-Bernoulli elements due to their high slenderness.

4.2 *Bearing wall buildings: Euler-Bernoulli and Timoshenko beams*

Tunnelling-induced vertical (u_z) and horizontal (u_x) displacements of the Euler-Bernoulli and Timoshenko beam axes predicted by the elastic GL (solid lines) and NA (dashed lines) solutions are displayed in Figures 3 and 4 for z_t = 8 and 20 m, respectively. Note that structural displacements are reported at the height of the neutral axis that is z = 0 and $z = -d_b/2$ for GL and NA solutions, respectively. For comparison, greenfield settlements at the ground level (z = 0) are also reported (dotted lines).

Structure settlement curves are shown in Figures 3(a) and 4(a), for which the beam is located,

Table 1. Implemented simple beams.

Label	Type	E/G (–)
BER	Euler-Bernoulli	2.6
TIM0.5	Timoshenko	0.5
TIM2.6	Timoshenko	2.6
TIM12.5	Timoshenko	12.5

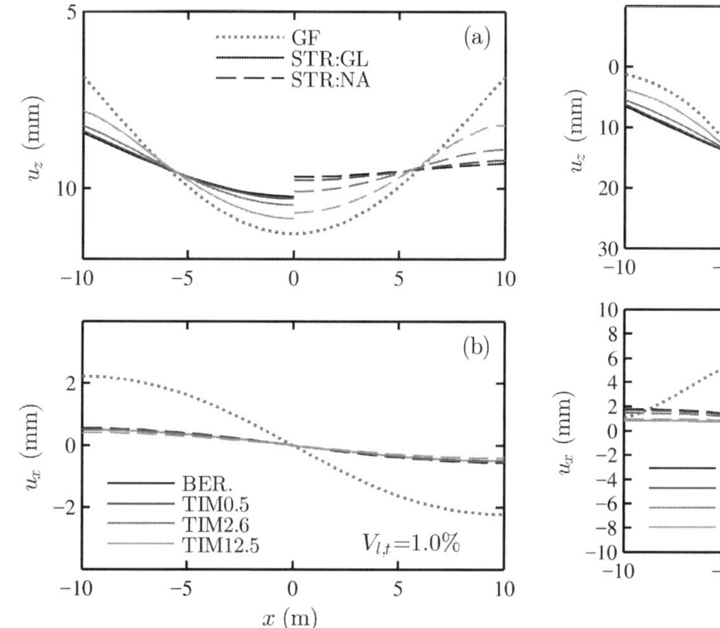

Figure 3. Greenfield and beam displacements for z_t = 20 m.

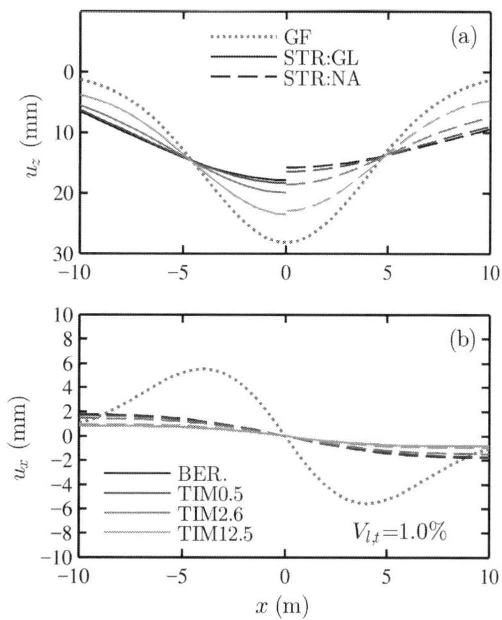

Figure 4. Greenfield and beam displacements for z_t = 8 m.

respectively, entirely in the greenfield sagging zone and across the sagging and hogging regions of the greenfield settlement trough. Firstly, beams with their axis at the ground level (GL, solid lines) are discussed; their flexural and axial response is almost decoupled (i.e. horizontal ground movements marginally affect axis settlements and structure deflection does not highly impact axial strains). Results confirm that BER beams are stiffer than TIM beams, which have additional shear flexibility. In addition, the shape of u_z curves is worth noting. In Figure 3(a), BER beams tend to deform entirely in sagging (similarly to the greenfield curve), whereas the BER beam in Figure 4(a) varies its shape of deflection (with respect to the greenfield settlement curve) to a fully sagging deformation mode. On the other hand, beams TIM2.6 and TIM12.5 with high shear flexibility have a deflection shape similar to the greenfield settlement trough when spanning across greenfield sagging and hogging zones (see Figure 4(a)), whereas their stiffness results in a hogging region at the structure edges and a narrow sagging region of the structure in Figure 4(a). i and i_{bld} values for GL solutions are reported in Table 2. In general, results illustrate that the increase in shear flexibility is associated with a decrease in i_{bld}. This mechanism impacts the structure deformations in terms of flexural modification factors, M^{DR} and M^{B}, as discussed later in the text. On the other hand, u_x displacement profiles of GL structures are overlapped in Figures 3(b) and 4(b) because the axial deformations of the beam only depend on the greenfield horizontal ground movements and the axial stiffness EA.

Secondly, beams NA accounting for the offset between the ground level and the beam axis are analysed for z_t = 8 m. In addition to the interaction mechanisms characterising the GL solutions, the effect of the offset ground level-beam axis resulted in the stiffening of structures NA with respect to beams GL, particularly for low shear flexibility structures BER and TIM0.5. This stiffening effect caused the decrease of the maximum tunnelling-induced structure settlements of the beams NA as well as the increment of hogging deformations at the edges of the beams. This is due to the fact that the offset d_{na} results in the coupling of rotational and horizontal degrees of freedom (Equations (7a) and (7b), e.g. horizontal forces at the ground level induce distributed bending movements along the structure axis) and a soil stiffness matrix $\mathbf{K}_s$* at the beam axis with additional terms corresponding to the rotations φ (Cheung & Nag 1968).

Bending deformations of structure BER and TIM0.5 mobilised the stiffness of the matrix $\mathbf{K}_s$* at the rotational degrees of freedom and resulted in high compressive reaction forces of the soil (as displayed in Figure 4(b) the structures NA undergo greater negative horizontal strains than GL beams). On the other hand, the effects of the axis offset on tunnelling-induced displacements are minor for structures TIM2.6 and TIM12.5 (with a significant shear deformability) because the deflection is dominated by shear which induces minor nodal rotations; for instance, u_x and u_z curves of TIM12.5 are close to values associated with BER GL beams at the ground level. Overall, for the NA solutions the influence of the offset d_{na} was only significant for low shear flexibility beams.

Modification factors M^{DR} and M^{B} associated with GL solutions are presented in Figure 5, which indicates that the greater the E/G ratio (i.e. the shear flexibility), the larger the deflection in both sagging and hogging M^{DR}. Furthermore, M^{B} tends to unity for high E/G in the case of structures spanning across the greenfield sagging and hogging regions (for z_t = 8 m), whereas for beams entirely located in the sagging zone (z_t = 20 m), $M^{B,sag} \leq 1$ and its value decreases with E/G. This is consistent with i_{bld}/i values reported in Table 2. These results give evidence of the need to account for the shear flexibility when assessing structure deformations due to TSI. In this way, the influence of the structure on the shape of the settlement trough could be more directly accounted for, rather than using a Timoshenko beam only to back-calculate strains Burland, Broms, & De Mello (1977), as is typically done in current damage assessment procedures.

Finally, internal forces (N = axial force, S = shear force, M = bending moment) induced by tunnelling at the simple beam axis (F_{tun}) are displayed in Figure 6 for GL solutions. Results illustrate that the magnitude of N is low compared to S, while M has a distribution that is qualitatively similar to the greenfield ground settlements. The BER beam, which has the lowest deflection ratios, has the highest inner forces. Also note that both S and M decrease in magnitude with E/G because of the increase in shear flexibility, whereas their profile shapes do not depend on the E/G ratio.

4.3 *Framed buildings*

Multi-storey concrete frame buildings were considered to investigate the effects of the framed configurations, the shear flexibility of the framed structures, and the stiffening effects of higher storeys in high-rise structures.

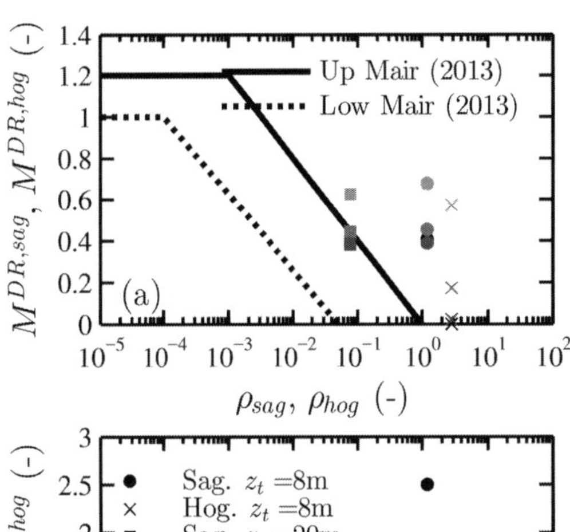

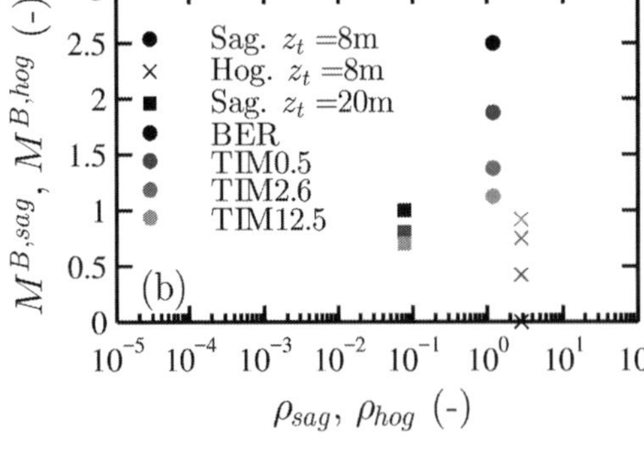

Figure 5. Modification factors for GL solutions of the simple beams.

Table 2. Greenfield and beam inflection point locations of GL solutions.

	z_t = 8 m		z_t = 20 m	
	i	i_{bld}	i	i_{bld}
	(m)	(m)	(m)	(m)
BER	4	10	10	10
TIM05	4	7.5	10	8
TIM26	4	5.5	10	7
TIM125	4	4.5	10	7

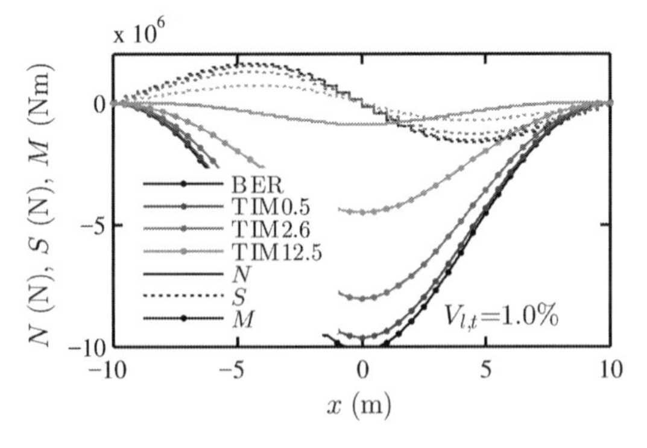

Figure 6. Inner forces of the simple beams for z_t = 8 m and GL solutions.

The tunnelling-induced displacements at the foundation level of the multi-storey frames are reported in Figures 7 and 8 for the shallow ($z_t = 8$) and mid-depth ($z_t = 20$) tunnels, respectively. Generally, foundation settlement profiles are similar to the Timoshenko beam results with high E/G ratio from the previous section; frames tend to have a decreased relative deflection at the base of ground level columns, caused by both a decrease of the central deflection of the structure and embedment at the ends, with respect to the greenfield settlement curve. This mechanism results in hogging deflections at the edges of the foundation beam for $z_t = 20$ m. Furthermore, settlements curves display that the stiffening contribution of upper storeys in high-rise frames is minor (compare settlement curves of FRS10 and FRS15). Although the decreasing effect of higher storeys in high-rise buildings agrees with the empirical formulas given by Haji, Marshall, & Tizani (2018), Figures 7 and 8 demonstrates that this effect remains true when the change in the shape of the foundation settlement profile with respect to the greenfield settlement trough is included in the analysis.

Finally, to address the risk of structural failure, tunnelling-induced internal forces (both forces and bending moments) within the finite elements can be analysed. For example, this section considers the internal forces in the foundation strip footing. Structural analyses of potential failure depends on pre-tunnelling service loads of the superstructure. In this section, service loads (self-weight and additional loads) were assumed equal to uniform distributed loads of $q = 50$ kN/m at the superstructure beams and $q = 21$ kN/m at the foundation beams. The pre-tunnelling internal force distributions (F_{ini}) and the tunnelling-induced internal forces (F_{tun}) are plotted in Figures 9(a) and 9(b), respectively, for $z_t = 8$ m. Interestingly, while F_{ini} increases with the number of storeys, tunnelling-induced internal force values at the foundation beams are relatively unaffected by the number of stories for the range of frames considered; this has implications on the potential for failure. To be acceptable, post-tunnelling forces ($F_{fin} = F_{ini} + F_{tun}$) should be limited to the capacity of the frame and foundation elements (i.e. F_{fin} should be within the envelopes of capacity). However, at preliminary stages, a simplified procedure (similar to the approach proposed by Cai, Verdel, & Deck (2017) based on the permissible internal forces, F_{per}, can be adopted for which F_{fin} should be less than the simple approximation of $F_{per} \approx 2 \times F_{ini}$, which neglects the interaction between bending moments and axial forces. In addition, for the bending moments the side of the tensile fibres should be considered. To perform this check, F_{fin} and F_{per} in terms of M and S are displayed in Figures 10(a) and 10(b), respectively. Results illustrate that the foundation of the low-rise frame FRS3 has greater

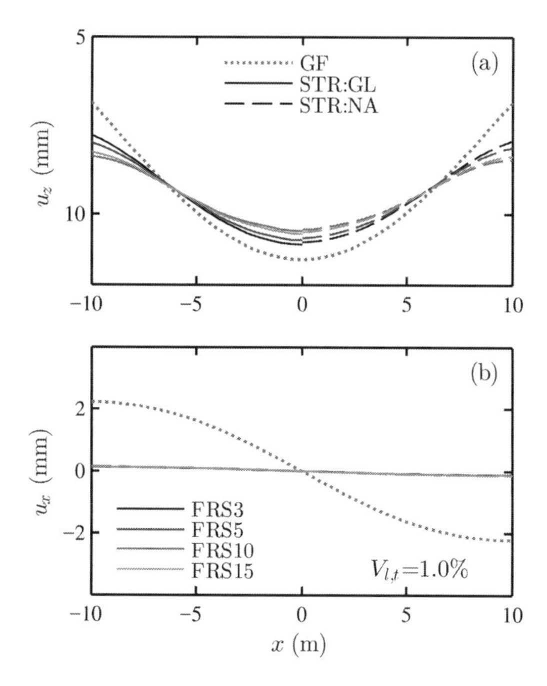

Figure 7. Greenfield and frame displacements for z_t = 20 m.

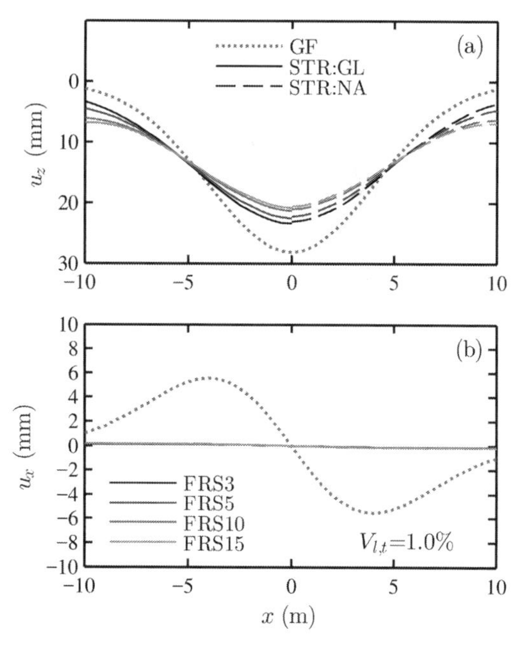

Figure 8. Greenfield and frame displacements for z_t = 8 m.

potential for failure than the high-rise building FRS15 (compare F_{fin} and F_{per} at the central and external spans) because F_{per} are greater due to the high level of vertical loads at the foundation level, and the incremental load induced by tunnelling is relatively smaller. Note that the compressive tunnelling-induced force could play a role in preventing the failure due to tunnelling. For specific structures, F_{per} values associated with the actual reinforcement of RC beams and foundations could be instead considered.

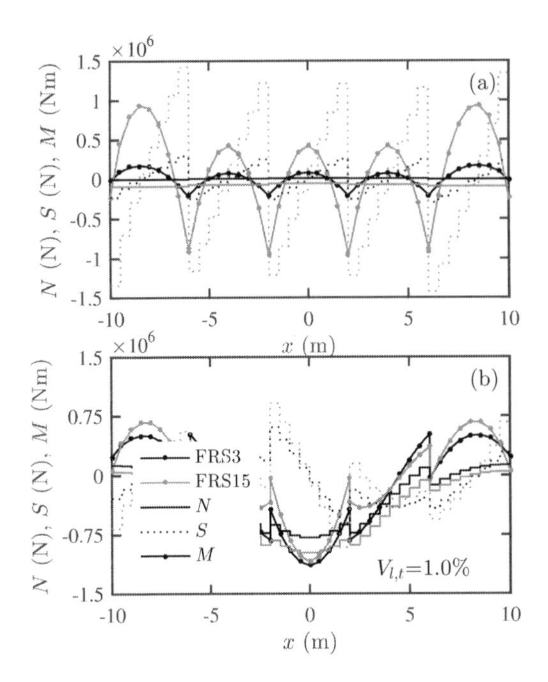

Figure 9. Inner forces of the foundation of the frame buildings due to (a) structure weight, F_{ini}, and (b) tunnelling for z_t = 8 m, F_{tun} (GL solutions).

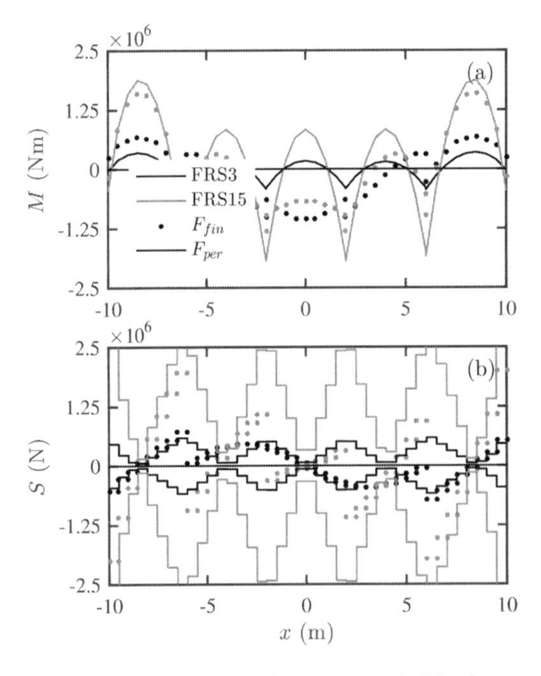

Figure 10. Comparison between permissible forces (lines) and post-tunnelling forces (markers) within the foundation beam: (a) bending moments, (b) shear forces (GL solutions).

5 CONCLUSIONS

Simple FEM numerical models have been presented in this paper to study tunnel-structure interaction. The proposed models allow consideration of structure characteristics and load conditions, and has low computational costs. As a result, the proposed method provides a tool that enables efficient large-scale parametric studies to improve understanding, as well as a tool for rapid initial assessment to evaluate tunnelling-induced deformations and internal forces of surface structures while accounting for soil-structure interaction.

The performed analyses dealt with the cases of tunnelling centrally beneath [i] bearing walls modelled by both Euler-Bernoulli and Timoshenko simple beams with varying shear flexibility and [ii] framed RC structures on continuous strip footings with increasing numbers of storeys. The following conclusions can be drawn.

- The shear flexibility, which results both from the local shear flexibility of structural elements and the global shear flexibility of framed configuration, altered the response of the structure both in terms of the magnitude and distribution of displacements and deformations (as discussed by Boldini, Losacco, Bertolin, & Amorosi (2016) for frames). For simple beams, shear flexibility significantly increased the total deflection induced by the excavation.
- For simple beams, the offset between the structure's neutral axis and the ground level resulted in coupling between flexural and axial deformations; this coupling is more significant for lower shear flexibility (e.g. Euler-Bernoulli beams).
- For framed buildings, outcomes confirmed that higher storeys in high-rise frames contribute marginally to the stiffening of the structure (Haji, Marshall, & Tizani 2018) and indicated that the potential for foundation structural failure was greater for low-rise structures than for high-rise frames.
- Stiff elements experience larger tunnelling-induced internal forces with respect to flexible beams (despite smaller deflections). This has

important implications in the assessment of the potential failure of RC structures. A simple procedure is adopted to compare post-tunnelling inner forces with permissible values based on the pre-tunnelling scenario (Cai, Verdel, & Deck 2017).

REFERENCES

Bilotta, E., A. Paolillo, G. Russo, & S. Aversa (2017). Displacements induced by tunnelling under a historical building. *Tunnelling and Underground Space Technology 61*, 221–232.

Boldini, D., N. Losacco, S. Bertolin, & A. Amorosi (2016). Modelling of Reinforced Concrete Framed Structures Interacting with a Shallow Tunnel. *Procedia Engineering 158*, 176–181.

Burland, J.B., B.B. Broms, & V.F.B. De Mello (1977). Behaviour of foundations and structures. In *Proceedings of the 9th International Conference on Soil Mechanics and Foundations Engineering*, Volume 2, Tokyo, pp. 495–546.

Cai, Y., T. Verdel, & O. Deck (2017, sep). Using plane frame structural models to assess building damage at a large scale in a mining subsidence area. *European Journal of Environmental and Civil Engineering*, 1–24.

Cheung, Y.K. & D.K. Nag (1968). Plates and beams on elastic foundations–linear and non-linear behaviour. *Géotechnique 18*(2), 250–260.

Deck, O. & A. Singh (2012). Analytical model for the prediction of building deflections induced by ground movements. *International Journal for Numerical and Analytical Methods in Geomechanics 36*(1), 62–84.

Dimmock, P.S. & R.J. Mair (2008). Effect of building stiffness on tunnelling-induced ground movement. *Tunnelling and Underground Space Technology 23*(4), 438–450.

Fargnoli, V., C.G. Gragnano, D. Boldini, & A. Amorosi (2015). 3D numerical modelling of soilstructure interaction during EPB tunnelling. *Géotechnique 65*(1), 23–37.

Farrell, R., R. Mair, A. Sciotti, & A. Pigorini (2014). Building response to tunnelling. *Soils and Foundations 54*(3), 269–279.

Franza, A. & M.J. DeJong (2017). A simple method to evaluate the response of structures with continuous or separated footings to tunnelling-induced movements. In I. Arias, J.M.

Blanco, S. Clain, P. Flores, P. Lourenc¸o, J.J. R´odenas, and M. Tur (Eds.), *Proceeding of the Congress on Numerical Methods in Engineering 2017*, Valencia, Spain, pp. 919–931.

Franza, A., A.M. Marshall, T. Haji, A.O. Abdelatif, S. Carbonari, & M. Morici (2017). A simplified elastic analysis of tunnel-piled structure interaction. *Tunnelling and Underground Space Technology 61*, 104–121.

Franzius, J.N., D.M. Potts, & J.B. Burland (2006). The response of surface structures to tunnel construction. *Proceedings of the ICE - Geotechnical Engineering 159*(1), 3–17.

Frischmann, W., J. Hellings, G. Gittoes, & C. Snowden (1994). Protection of the Mansion House against damage caused by ground movements due to the Docklands Light Railway Extension. *Proceedings of the Institution of Civil Engineers - Geotechnical Engineering 107*(2), 65–76.

Giardina, G., M.J. DeJong, & R.J. Mair (2015). Interaction between surface structures and tunnelling in sand: Centrifuge and computational modelling. *Tunnelling and Underground Space Technology 50*, 465–478.

Goh, K.H. & R.J. Mair (2014). Response of framed buildings to excavation-induced movements. *Soils and Foundations 54*(3), 250–268.

Haji, T.K., A.M. Marshall, & A. Franza (2018). Mixed empirical-numerical method for investigating tunnelling effects on structures. *Tunnelling and Underground Space Technology 73*, 92–104.

Haji, T.K., A.M. Marshall, & W. Tizani (2018). A cantilever approach to estimate bending stiffness of buildings affected by tunnelling. *Tunnelling and Underground Space Technology 71*, 47–61.

High Speed Two Limited (2014). HS2 Phase One information papers: property, compensation and funding - Ground settlement (C3).

Klar, A., T.E.B. Vorster, K. Soga, & R.J. Mair (2005). Soil-pipe interaction due to tunnelling: comparison between Winkler and elastic continuum solutions. *Géotechnique 55*(6), 461–466.

Loganathan, N. & H.G. Poulos (1998). Analytical prediction for tunneling-induced ground movements in clays. *Journal of Geotechnical and Geoenvironmental Engineering 124*(9),846–856.

Mair, R. (2013). Tunnelling and deep excavations: ground movements and their effects. In A. Anagnostopoulos, M. Pachakis, and C. Tsatsanifos (Eds.), *Proceedings of the 15th European Conference on Soil Mechanics and Geotechnical Engineering - Geotechnics of Hard Soils - Weak Rocks (Part 4)*, Amsterdam, the Netherlands, pp. 39–70. IOS Press.

Mair, R.J. & R.N. Taylor (1997). Theme lecture: Bored tunneling in the urban environment. In *14th International conference on soil mechanics and foundation engineering*, Hamburg, pp. 2353–2385. Balkema.

Mair, R.J., R.N. Taylor, & A. Bracegirdle (1993). Subsurface settlement profiles above tunnels in clay. *Géotechnique 43*(2), 315–320.

Mair, R.J., R.N. Taylor, & J.B. Burland (1996). Prediction of ground movements and assessment of risk of building damage due to bored tunnelling. In R.J. Mair and R.N. Taylor (Eds.), *Proceedings of the International Symposium on Geotechnical Aspects of Underground Construction in Soft Ground*, London, United Kingdom, pp. 713–718. Balkema, Rotterdam.

Marshall, A.M., R. Farrell, A. Klar, & R. Mair (2012). Tunnels in sands: the effect of size, depth and volume loss on greenfield displacements. *Géotechnique 62*(5), 385–399.

Pickhaver, J., H. Burd, & G. Houlsby (2010). An equivalent beam method to model masonry buildings in 3D finite element analysis. *Computers & Structures 88*(19), 1049–1063.

Potts, D.M. & T.I. Addenbrooke (1997). A structure's influence on tunnelling-induced ground movements. *Proceedings of the ICE - Geotechnical Engineering 125*(2), 109–125.

Ritter, S., G. Giardina, M.J. DeJong, & R.J. Mair (2017, aug). Influence of building characteristics on tunnelling-induced ground movements. *Géotechnique*, 1–12.

Vaziri, H., B. Simpson, J.W. Pappin, & L. Simpson (1982). Integrated forms of Mindlin's equations. *Géotechnique 32*(3), 275–278.

Numerical Methods in Geotechnical Engineering IX – Cardoso et al. (Eds)
© 2018 Taylor & Francis Group, London, ISBN 978-1-138-33203-4

Numerical modelling within a risk assessment process for excavations over a brick arch tunnel at Earls Court, London

B. Gilson, F. Mirada, C. Deplanche & M. Devriendt
Arup, London, UK

M. Scotter
Earls Court Partnership Limited (Joint Venture between TfL and Capco), London, UK

H. Jayawardena
London Underground Limited, Transport for London, UK

ABSTRACT: The development of Earl's Court Village, London, begun with the demolition of Earl's Court 1 (EC1) and Earl's Court 2 (EC2) exhibition centres, directly above London Underground Limited (LUL) running tunnels. Excavations as close as 150 mm above a brick arch running tunnel were required in order to remove deep transfer beams buried below the basement slab. Numerical modelling of the shallow cut-and-cover brick arch tunnel was undertaken to understand the stress change within and displacement of the tunnel lining. Acceptability criteria were agreed with the asset owner, LUL, and compared with the finite element model results. The excavation profile was adapted to provide sufficient confining stresses during each construction stage. Automated in-tunnel monitoring and a condition survey, including hammer tap testing, was used to validate the modelling parameters adopted. This case study sets out a holistic risk assessment approach incorporating numerical modelling and compares monitoring data to calculated movements in order to appraise brick arch tunnel modelling parameters and provide a precedent for future assessments of brick arch tunnels.

1 INTRODUCTION

Demolition of the Earl's Court Exhibition Centre forms the first phase of the Earl's Court Masterplan that is to comprise four new urban villages and High Street in west London, UK. The site contains a number of assets which are owned and operated by London Underground Limited (LUL), including four shallow cut-and-cover tunnels of the District Line and two deep level tunnels of the Piccadilly Line shown in Figure 1. The site also contains an overland railway, referred to as the West London Line, which is owned and operated by Network Rail (NR).

Ove Arup & Partners Limited (Arup) was commissioned by Earls Court Partnership Limited (ECPL), to provide outline engineering advice, including ground movement assessments, for the demolition of the exhibition centre buildings Earl's Court 1 (EC1) and Earl's Court 2 (EC2). Demolition to the lowest basement slab preceded the removal of portal beams that directly overlie the cut-and-cover district line tunnels that run underneath EC1. Collaborative working between LUL, Keltbray, ECPL and Arup led to the successful completion of the demolition works and portal removal.

A gap of approximately 150 mm existed between the underside of the portals and the crown of the brick arch lining, therefore, close to 100% of the tunnel over-burden was removed locally, at approximately 5 m and 35 m intervals at portal beam locations.

This paper focuses on finite element (FE) analysis that was used to demonstrate that the risk of impact of the portal removal works on the Down Ealing brick arch District Line tunnel was acceptable within the last train—first train access protocol agreed with LUL, based on the As Low As Reasonably Practicable (ALARP) principle.

2 LONDON UNDERGOUND ASSETS

LUL assets beneath the site include four shallow cut-and-cover tunnels of the District Line and two deep level tube tunnels approximately 17 m below ground level of the Piccadilly Line as presented in Figure 1.

District line Up Ealing and Down Putney covered way tunnels house double-track railways and

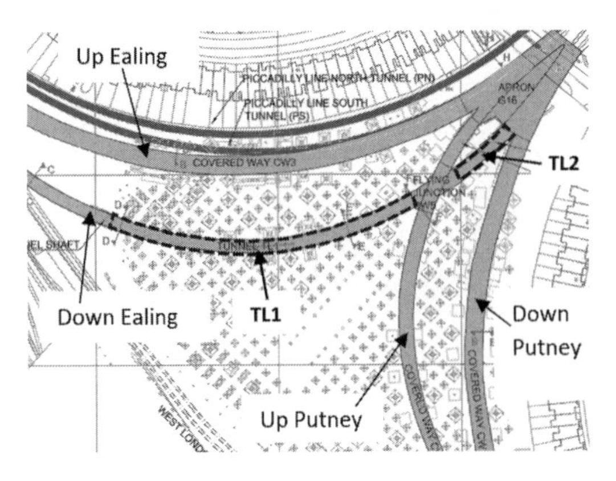

Figure 1. LUL assets beneath EC1.

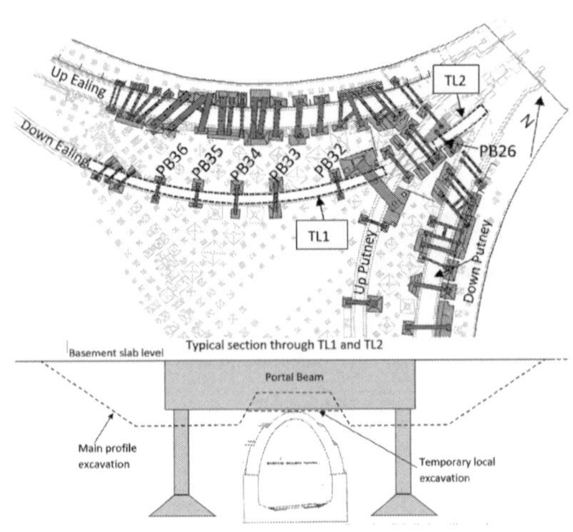

Figure 2. District Line Tunnels under portal frames at EC1 showing key portal beam numbers and a typical section at a portal location.

were constructed within open cuttings around 1869, see Figure 2. These were covered over as part of the construction of EC1. Down Ealing and Up Putney are two single-track railway tunnels constructed below the site in 1912, shortly prior to the construction of EC1 in the 1930s. The District Line tunnels merge at the northern end of EC1 and this area is covered by reinforced concrete slab, referred to as an apron. District Line tunnels are comprised of a number of individual component sections of varying material type and construction, most of which are reinforced concrete covered ways. Tunnel sections TL1 and TL2 (shown on Figure 1) are brick arches supported on concrete side walls. Between TL1 and TL2, Down Ealing tunnel passes under the Up Putney tunnel through a reinforced concrete flying junction. This paper focuses on the brick arch tunnel that houses the Down Ealing line (TL1 and TL2).

A review of LUL and Earl's court historical drawings, ground penetrating radar (GPR), photographic and intrusive surveys were undertaken to ascertain the geometry and assess the condition of the brick tunnel lining. Portal sections at each end of both TL1 and TL2 sections are shown to comprise six rings of bricks. The remaining lengths of the brick arch tunnel were understood to comprise up to six rings of brick at knee level reducing incrementally to three rings at the crown, see Figure 2. This was observed at several places through coring at the tunnel shoulder to comprise two rings of well laid bricks and 550 mm of less carefully laid bricks and mortar. The GPR surveys indicated possible areas of delamination in localised areas but the brickwork and mortar was reported to be good quality. A concrete base slab is shown to connect the sidewalls. Historical drawings indicated that the brick arch was constructed using a temporary support timber frame.

EC1, overlying LUL tunnels of the District and Piccadilly line, was constructed by 1937 and was founded on shallow footings. Where columns were located directly above existing tunnels, portal beams were constructed to transfer the load to either side of the tunnels onto pad footings founded below or at tunnel invert level (refer to Figure 2 where beams are shown in red with pad footings either side of the tunnels). A gap of c.150 mm left between the underside of each portal beam and the extrados of the tunnel crown was reported by Siddall (1937), although it was not clear the extent of any temporary works at the time. Photographs taken during the construction of EC1 showed trains running through the brick arch tunnel simultaneous to excavation above the tunnel portals, though the depths of excavations are unclear.

3 DEMOLITION WORKS

The first phase of demolition started in August 2015 with the removal of the EC1 and EC2 building superstructure. By August 2016 most of the superstructure was demolished except for the basement slab and the first-floor slab at the three peripheral nodes of EC1.

The second phase of demolition, the focus of this paper, comprised the removal of 61 portal beams, some of which were buried fully or partially beneath the basement slab. A single 4,300 tonne supercrane was used for the duration of phase 2 to safely lift the portal beams and process them away from the tunnels. This mitigated risk associated with dismantling portal beams weighting up to 1500 tons directly above live railway tunnels and operating numerous cranes in close proximity to the tunnels. This second phase of work involved a large excavation and engineered backfilling to construct the 2 m thick crane foundation, significant excavations directly above and around tunnels to expose the portal beams, and lifting the portal beams.

During the demolition of the superstructure in phase 1, unloading of the ground adjacent to the tunnels occurred predominantly below the tunnel invert level. Loading and unloading of the ground in the second phase took place above tunnel invert level and locally directly above the brick arch tunnel lining, resulting in significant change in stresses within the tunnel lining. Excavations above the brick arch tunnel were backfilled once the portals had been removed.

4 GROUND MOVEMENT RISK ASSESSMENT METHODOLOGY

An assessment of the risk of excavation and portal removal impacting the operation of the LUL tunnels at both ultimate limit state (e.g. tunnel collapse) and serviceability limit state (e.g. falling debris and clearances) was undertaken, informed by two-dimensional (2D) and three-dimensional (3D) FE analyses. Key criteria agreed with the project team comprised:

1. A minimum compressive stress in the tunnel lining in order to prevent bricks from falling out under very low confining stress, conservatively calculated assuming no beneficial effect of the mortar, or friction due to the weight of the brick acting on an inclined surface of 2.5 kPa.
2. A maximum compression in the lining equal to existing maximum calculated stresses of up to 800 kPa.
3. Tunnel clearances derived from LUL Standard S 1156, considered in addition to longitudinal deformation effects on clearances. This criterion was not found to govern the assessment.

2D FE analysis was used to evaluate the first two criteria to identify the risk of tunnel collapse or serviceability. These analyses are conservative, given that a 2D model considers a localised excavation to be infinitely wide parallel to the tunnel axis. Should the criteria not be met, the demolition methodology was reviewed to try to reduce the impact, such as reducing the extent of excavation. In all but one case, the demolition methodology could be refined by reducing the volume of excavation above the tunnels, in order to meet the criteria. In one case, for Portal Beam 26, consideration of the full 3D effects of the localized excavation was needed to justify that criterion 1) was met, and a 3D FE model was developed.

FE modelling included the inferred construction history of the tunnel lining to calculate the existing stresses within the lining. However, verification of the condition of the tunnel lining using hammer tap testing was undertaken in order to confirm that the internal lining was in compression and that the calculated minimum stresses compared with criterion 1) in particular could reasonably be used to assess the risk of bricks or other debris dropping out.

A previous core hole observed during hammer tap testing site visits showed that the brick arch lining varied locally from archive drawings. Additional coring was specified and undertaken and sensitivity studies undertaken to investigate the impact of the change in brick arch lining.

Following demonstration by calculation that the criteria were met, verification of the calculations by in-tunnel instrumentation and monitoring the tunnel lining was undertaken. A comparison of calculated and measured displacement of the tunnel lining was used to verify that the risk of impact to the operation of the tunnels had been mitigated as the calculations demonstrated.

An assessment of longitudinal deformation on this and the other LUL tunnels will be reported separately.

5 FE ANALYSIS

5.1 *Model parameters*

5.1.1 *Ground model*

Phase 1 of the site specific ground investigation was carried out in 2014 including four deep boreholes to a maximum depth of 70 m for the design of Earl's Court Village. Additional ground investigation including nine boreholes to a maximum depth of 40 m was carried out in 2016 along with ground water level monitoring for the design of the heavy lift crane foundation. The ground model presented in Table 1 was assumed for the assessment.

Soil parameters modelled are reported in Table 2. Standard penetration tests and undrained triaxial tests were used to derive the design values using empirical relationship between SPT N-values and vertical stiffness from Stroud (1989) and undrained shear strength and vertical stiffness from Hewitt (1988).

5.1.2 *Ground water*

The initial ground water level was assumed to be at initial ground level (i.e. +5 mOD). After construction of the brick arch Tunnels, the ground water level around the tunnel was assumed to be drained to tunnel invert level. The ground water assumption was confirmed by ground water level monitoring as part of 2016 ground investigation.

5.1.3 *Structural parameters*

TL1 and TL2, both brick arch sections of the Down Ealing tunnel, gently slope up to the northeast and are founded between -3 mOD and -1 mOD. The crown is located between +4 and +6 mOD. Modelled parameters are shown in Table 3.

A tension cut off was modelled in order to avoid any part of the lining experiencing tension, due to the lack of confidence in the brick lining withstanding tensile stresses.

BS EN 1996–1-1:2005suggests a short-term stiffness of 400 f_m to 1000 f_m where f_m is Masonry crushing strength. The adopted short term elastic modulus for masonry for the analysis is $E = 400f_k$, where f_k is the compressive strength of the masonry. This relationship is lower than the guidance recommended values to account for actual lining condition. Guidance values are based on experimental work carried out in laboratories where the quality of construction is much superior to what might be expected in real tunnel lining. Actual masonry lining thickness is often irregular and the quality of mortar joints is also uncertain both longitudinally and transversally. Separation between the rings at the crown is common, often due to relative shortening between an inner ring and its overlying slightly longer neighbour. Furthermore, over time the combined effects of water ingress and sulphur and other aggressive chemicals from locomotive smoke and groundwater may cause deterioration of masonry both by spalling by leaching out of mortar. The characteristic compressive strength $f_k = 4.8$ N/mm^2, as recommended in the London Underground Standard S1061 'Bridges and Structures Assessment Standard' (September 2016), was used to calculate the stiffness.

The strength of the masonry was set as the average shear strength between the shear strength values of 3 MPa for the bricks and 0.5 MPa for the mortar, based on in-house case study data.

The portal beam, columns and footings were not modelled in the 2D analyses, as the structure was considered to be beneficial in stiffening the ground and reducing the impact on the brick arch lining.

5.1.4 *Modelling software*

2D FE analysis was undertaken in Plaxis 2D incorporating the Mohr Coulomb constitutive model for both ground and the brick arch tunnel lining. 3D FE analysis was undertaken in LS Dyna incorporating the BRICK model (Simpson, 1992) for London Clay, with moderate-conservative parameters, and a Mohr Coulomb constitutive model for the other soil materials and brick arch lining.

Table 1. Stratigraphy assumed for the assessment.

Stratum	Top of stratum (mOD)	Thickness (m)
Made Ground	+7 or +8	1.0
RTD	+5.0	5.0
London Clay	+0.0	53.0
Lambeth Group	–53.0	23.0
Thanet Sands	–76.0	3.5
Upper Chalk	–79.5	–

Table 2. Ground parameters adopted.

	γ(kN/m^3)	E' (kPa)	E_u (kPa)	c_u (kPa)	c' (kPa)	φ' (°)
Made Ground	20	30,000	–	–	0	30
RTD	20	50,000	–	–	0	38
London Clay	20	300c_u	500c_u	90+7z	0	24

Notes: (1) A Mohr Coulomb constitutive model was used for ground materials in the 2D FE analysis. In the refined 3D FE analysis the BRICK constitutive model (Simpson, 1992) was used for London Clay with moderate-conservative parameters to account for the strain-stiffness dependency of the London Clay. (2) z is the depth below 0 mOD.

Table 3. Structural parameters adopted for the brick lining.

	γ(kN/m^3)	E' (kPa)	E_u (kPa)	c_u (kPa)	c' (kPa)	φ'
Brick masonry	20	1.92×10^6	–	–	1,700	0

5.2 *FE analysis results*

5.2.1 *Sensitivity studies*

A sensitivity study on key parameters was undertaken to understand the impact or variations on the results. Key parameters that were tested included the at rest earth coefficient (k_0), Young's Modulus of the Made Ground (E'), and the Young's Modulus (E') and strength (c') of the tunnel lining. The studies demonstrated that variation of the parameters within a plausible range did not affect the conclusions of the assessment.

5.2.2 *Plaxis 2D assessment—general case*

A significant reduction of the normal stresses in the lining was observed at the critical excavation stage. This corresponded to the reduction of the hoop forces due to the removal of overburden on top of the tunnel.

Two critical points around the brick arch lining were identified: the shoulder of the tunnel, where the minimum compression after excavation was calculated; and, the crown of the tunnel where the maximum compression occurred.

Initially, the stresses in the lining dropped towards zero indicating that tension had developed in the lining. The excavation profile was reviewed with the demolition contractor, Keltbray Group, and the depth and width was reduced in order to lessen the amount of overburden removed from above the tunnel.

Figure 3 presents the normal stresses in the lining for the critical section at the shoulder. The initial normal stresses in the tunnel show the outside of the lining in maximum compression. After the excavation to the main profile (~1 m above underside of the portal beam), and after the temporary local excavation (to fit the lifting beams), the whole section remained in compression, and the minimum compression was found significantly greater than the 2.5 kPa minimum compression criteria.

The final distribution of normal stresses in the lining has a similar shape compared to the initial distribution and is also significantly greater than the 2.5 kPa minimum compression criteria. Criteria 1) and 2) were therefore satisfied.

At the crown the calculated normal stresses were found to be below the calculated maximum stress at the initial condition, and to remain in significant compression. Furthermore, the bricks in the internal rings that increased relative to the existing condition were found to be in better condition following a review of the archive information and tunnel coring. Therefore, based on the conservative assumptions modelled (which were verified later with the hammer tap testing, see Section 5.3) the stress state at the crown was considered no worse than existing and criteria 1) and 2) were satisfied.

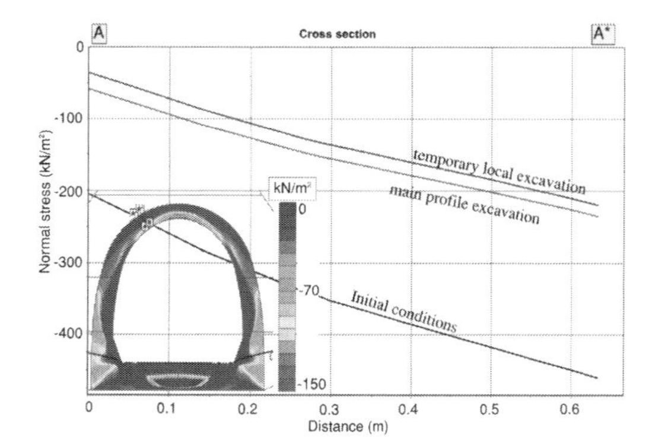

Figure 3. Normal stresses distribution at the shoulder (Section A-A*) at each stage of soil removal on top of the tunnel.

5.2.3 *Plaxis 2D assessment—Portal Beam 26*

Refinement of the demolition sequence and excavation extents around portal beam No. 26 could not be undertaken to demonstrate that the effect of overburden removal on the tunnel would be no worse than the general case. A 2D FE analysis of the excavation around portal beam No. 26 was undertaken and Figure 4 presents the normal stresses in the lining for the critical shoulder section. The initial conditions (ground level at +7 mOD) show the outside of the lining in maximum compression. After the excavation around the portal to +4 mOD most of the lining at the shoulder reached the tension cut off. In particular, the intrados of the tunnel lining between the shoulder and the crown was under no compression and therefore the analysis suggested that bricks forming the inner rings of the lining may fall if not well bonded by the mortar.

For this reason, a 2D analysis was insufficient to demonstrate that the stresses in the brick lining would remain within the proposed minimum and maximum tunnel lining compressive stress criteria.

5.2.4 *LS-DYNA 3D assessment—Portal Beam 26*

Subsequently, 3D FE analysis was undertaken to account for the 3D localised excavation and the existing lowered slab level on one side of the portal beam that could not be accounted for in 2D. An extract of the 3D LS Dyna model is shown in Figure 5.

Principal stresses within the tunnel lining at the main excavation stage to 1 m above the underside of the portal beam are shown in Figure 6. Figure 6(a) shows the minimum principal stress and the whole section of the lining remains in compression before starting the localised excavation to below the underside of the portal beam. Fig-

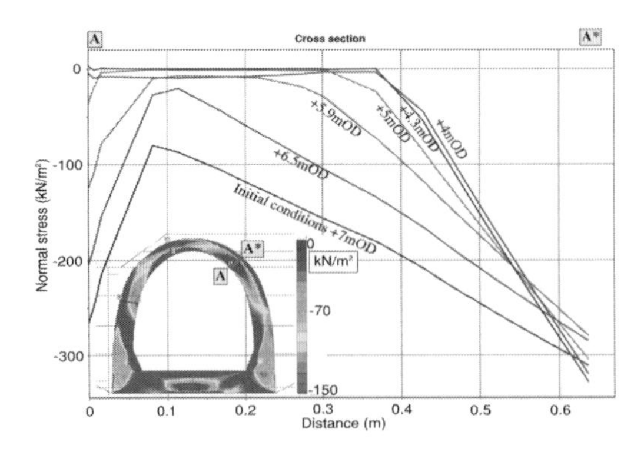

Figure 4. Normal stresses distribution at the shoulder (Section A-A*) at each stage of soil removal on top of the tunnel.

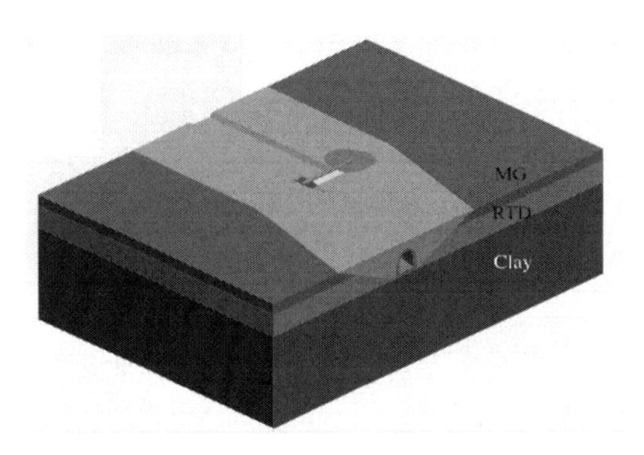

Figure 5. 3D FE model (LS-DYNA).

ure 6(b) shows the minimum principal stress in the lining after removing the portal beam. Although a reduction of the stresses in the shoulder is observed (as in the 2D FE), the whole section is still in compression, and the minimum compression is greater than the 2.5 kPa minimum compression criteria. The smaller reduction in principal stresses compared to the 2D FE analysis is due to modelling a localised excavation, rather than an infinitely long excavation as for the plane strain case, and due to modelling the longitudinal stiffness of the tunnel lining that distributes the stresses and reduces the peak changes. Based on the 3D analysis, criteria 1) and 2) are justified.

Figure 7 shows the calculated vertical displacements in the tunnel lining following excavation around the portal does not exceed 1 mm, compared with up to 5 mm calculated in the 2D case.

5.2.5 *Tunnel clearances*

Tunnel clearances were assessed by conservatively summing the reduction in clearance due to the end throw of carriages subject to longitudinal hogging or

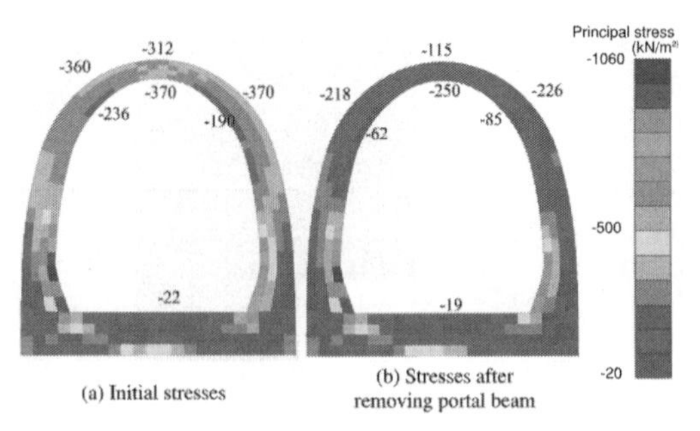

Figure 6. 3D assessment results: normal stresses in the lining.

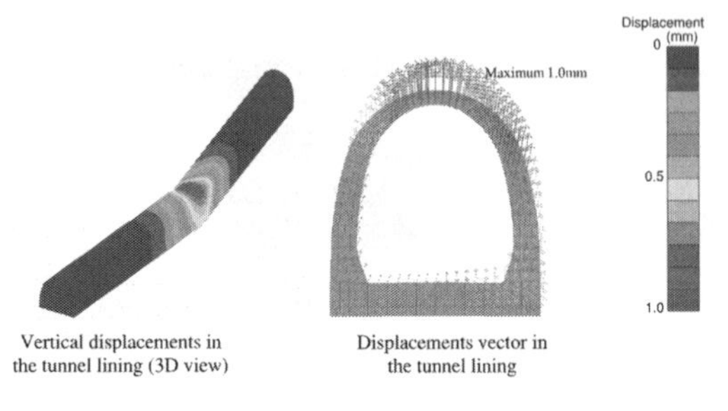

Figure 7. Vertical displacements in the tunnel lining after excavation around portals.

sagging and the worst case radial distortion of the tunnel lining. The maximum calculated radial distortion of 3.4 mm resulted in a reduced clearance of less than 7 mm. This was accepted by LUL based on the existing clearances within the tunnel being sufficient to accommodate this small change in clearance.

5.3 *Risk assessment and mitigations*

FE analyses were undertaken to inform an assessment of ultimate limit state and serviceability limit state risks to the operation of the tunnel.

The risk of tunnel collapse due to exceeding brick compressive stresses, the key ultimate limit state consideration, was mitigated by demonstrating that calculated compressive stresses in the tunnel lining. The risk of bricks or debris dropping out due to low confining stresses was mitigated by adjusting the demolition methodology as necessary to keep the minimum calculated lining stress above 2.5 kPa.

Sensitivity studies were undertaken to reduce the risk of the impact on the tunnel lining varying from the calculated impact due to non-representative modelled parameters.

A key modelling assumption was that the existing stresses in the tunnel lining are in compression and the brick lining is in a reasonable condition, without loose bricks, delamination or potential voiding. Loose bricks, delamination or voiding may have resulted in the compression stresses being significantly lower than calculated, resulting in the analysis conclusions of criterion 1) not being applicable to the tunnel lining.

To mitigate the risk of this occurring hammer tap testing was recommended to better understand the condition of the lining prior to excavation. Although some localised loss of mortar, some water ingress and evidence of previous remedial works such as local repointing were noted, overall the lining was observed to be in good condition.

The site observations verified the conclusions of the ground movement assessment except for the survey below Portal Beam No. 32 where hollow sounding bricks at the crown of the tunnel lining was observed. The crown was generally calculated to experience an increase in compression as excavation progresses that would close up joints, therefore, mitigation in the form of re-pointing brickwork was not undertaken.

In-tunnel monitoring was proposed to verify that the behaviour of the tunnel lining was consistent with the FE analysis. Automated monitoring was specified in order to provide near real time monitoring to inform the project team prior to the first trains running. It was anticipated that a maximum period of 2.5 hours could elapse between the tunnel lining displacing and the automated total station monitoring system reporting whether a trigger level had been breached. Given this risk that a delay of fault would result in monitoring data not being available before the first train entering service, it was agreed that the LUL Infrastructure Protection Inspector (IP Inspector) would accompany the driver of the first train and, at slow speed through the area of the beam (beam locations marked on tunnel wall), check for significant defects. It was not possible to undertake a full visual inspection of the tunnels before the first train passed through the tunnel.

Portals requiring a smaller excavation were lifted first in order to confirm the tunnel lining behaved as expected before moving on to higher risk portal removals.

6 MONITORING

6.1 *Strategy and methodology*

Monitoring of the displacement of the tunnel lining was used to confirm that the behaviour of the structure was consistent with the modelled behaviour. Monitoring from above ground could not be undertaken due to the depth of overburden remaining above the tunnel, unlike the covered way structures that intruded into the basement void.

Monitoring during demolition of the superstructure comprised manual in-tunnel monitoring of reflective retro targets on five point arrays at ten metre intervals on a two-monthly basis. Manual monitoring between lifting the portal beam and the first train entering service running was deemed not to be feasible, given the lifting exclusion zone within the tunnels and the narrow window for lifting within last train—first train access protocol agreed with LUL, providing an additional 30 minutes above the nominal engineering hours of 3 hrs and 5 minutes. Monitoring was also required during excavation around the portals within normal working hours when the tunnel was in use. Automated monitoring was therefore employed to provide near real time monitoring of five prism arrays at ten metre intervals, tying into the existing control network at either end of the brick arch tunnel.

The maximum period of 2.5 hours accepted between the tunnel lining displacing and plotting the measured displacement against project trigger levels, comprised a 1 hour interval between monitoring rounds, sighting each prism in the traverse and computing displacements from the network of in-tunnel total stations. It was acknowledged that a situation that prevented monitoring data from being available before the first train could occur, and measures were taken to reduce this scenario to ALARP, including the LUL IP Inspector cab ride.

A maximum total system error of 3.1 mm was calculated midway along the tunnel chainage and was deemed appropriate. Relative movements between monitoring points in a radial direction that provided the change in tunnel lining ovalisation were interpreted to within ±0.5 mm

Monitoring of the tracks was not undertaken. Instead, the displacement of the tunnel lining was used to trigger track monitoring using either a bance gauge or a track trolley depending on the length of track to be surveyed and the availability of the track trolley.

Additional benefits of automated monitoring included reducing the exposure of surveyors to night-time working within a rail environment and potential 'whole life' savings offered by an automated system given the proposed future construction works.

6.2 *Results and comparison with ground movement assessment*

Heave displacements between 3 mm and 4.5 mm were measured following excavations of portal beams 32, 33, 34, 35 and 36. The maximum heave occurred near portal 36 where spacing between portals is the smallest and the geometry is therefore more representative of the 2D infinitely long excavation modelled in the 2D FE assessment.

This compares to 7 mm heave calculated in the 2D FE analysis. The displacement at each monitoring point at the maximum displacement array is compared to the 2D FE analysis calculated displacement in Figure 9. At the same location, 1.7 mm elongation in the vertical direction was measured as compared to 3 mm calculated in the 2D FE analysis, shown in Figure 10.

At the location of portal beam 26, negligible heave and elongation were measured following the excavation around the portal. This compares to 1 mm heave and less than 1 mm elongation calculated in the 3D FE analysis for Portal beam 26. The smaller amount of heave that occurred at portal beam 26 can be attributed to two possible causes; the volume of excavation was smaller than modelled; and the tunnel lining was thicker and therefore stiffer at that location as exposed by structural coring through the brick arch lining.

Figure 8. Overview of hollow bricks in the tunnel crown with evidence of water ingress.

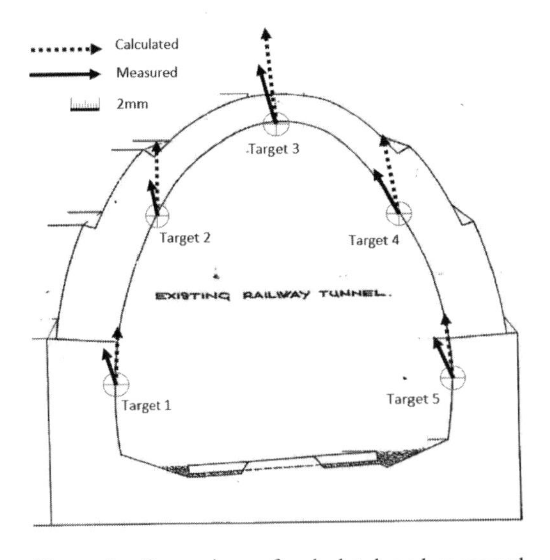

Figure 9. Comparison of calculated and measured maximum tunnel lining displacements.

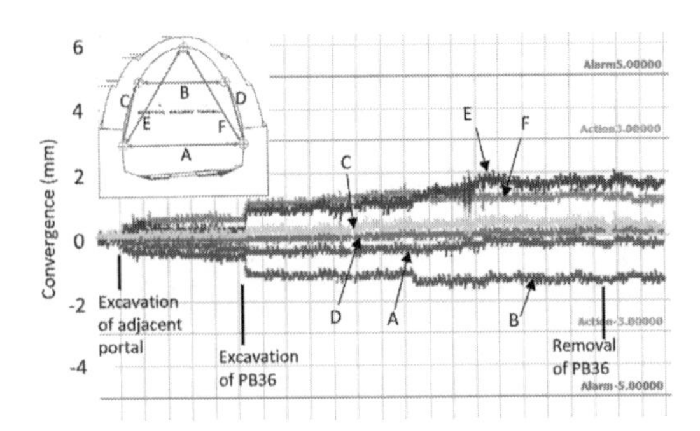

Figure 10. Measured tunnel lining elongation at Portal Beam 36.

7 RISK ASSESSMENT PROCESS SUMMARY

An assessment was carried out of the risk of portal removal works and associated excavations on the operation of LUL tunnels included FE analysis to compare the change in stresses within the tunnel lining with criteria relating to the minimum and maximum tolerable stresses. FE results were also used to demonstrate that tunnel clearances remained acceptable. Verification of the FE analysis results was undertaken before the site work by undertaking sensitivity studies, investigating the pre-existing condition of the tunnel lining using hammer tap testing, and coring in one location to reduce uncertainty regarding the lining thickness and construction type. Verification was undertaken during and following the site work by monitoring tunnel lining displacement hourly and using automated alerts to flag up trigger breaches to the project team. To mitigate the risk of excessive deformation of the lining occurring between lifting portals and the first train running but not being reported due to a fault with the monitoring system, the LUL IP Inspector rode in the cab of the first train each morning to undertake a visual inspection.

This holistic approach to risk assessment enabled the portals to be safely removed from directly above the tunnels within the 'last train—first train' hours, without any need for extended line closures.

8 CONCLUSIONS

An assessment of the risk of excavations required to remove portals impacting the operation of the brick arch tunnel included 2D and in one case 3D

FE analysis of the tunnel lining. Risk mitigation measures included refining the extent of excavation to keep calculated lining stresses within agreed criteria, verification of the existing stresses in the tunnelling using hammer tap testing, and automated in-tunnel monitoring. This holistic approach to risk assessment successfully demonstrated to the asset owner LUL that the risk of impacting the operation of the District Line was ALARP.

All portals were successfully lifted and the tunnel lining behaved as expected. Displacement of the lining was less than calculated as expected, given the conservatism that was built in to FE analysis, particularly the 2D idealization of the tunnel at the excavations.

Structural properties of the brick lining were based on previous case study data held by Arup and were verified to some extent by the calculation prior to and the monitoring during the works. The parameters were found to provide a reasonable basis for understanding the impact on the tunnel lining.

Automated in-tunnel monitoring verified the behaviour of the tunnel and no significant deterioration of the tunnel lining was observed.

This study at Earl's Court provides a useful case study for future assessments of the impact of ground movements on brick arch lining tunnels. FE modelling parameters assumed have been found to provide a reasonable understanding of the behaviour of the tunnel lining when compared with monitoring data and condition surveys.

ACKNOWLEDGMENTS

The authors wish to thank in particular: colleagues Peter Ingram, Simon Lindop, Tony Jones and Xiangbo Qui (Arup) for their input to this project, Chris Burns (ECPL) as our client responsible for the portal removal works, David Norris (Gardiner and Theobald), Jim O'Sullivan and Dave Rowe (Keltbray) and Chris Asbury (PC Monitoring).

REFERENCES

British Standard BS EN 1996-1-1:2005 – Design of masonry structures.

Hewitt. 1988. Settlement of structures on over consolidated clay. MEngSc thesis, University of Sydney.

LUL Standard S 1060. 2014. Bridges and structures inspection standard. Issue number A6.

LUL Standard S 1156. 2009. Gauging and clearances. Issue number A4.

Siddall, R. 1937. Earl's Court Exhibition Building, *The Engineer*.

Simpson B S. 1992. Retaining Structures: Displacement and design. 32nd—Rankine Lecture. *Geotechnique*. pp 541–576.

Stroud. 1989. The standard penetration Test—its application and interpretation. *Proceedings of the geotechnology conference on penetration testing in the UK, ICE.*

Numerical Methods in Geotechnical Engineering IX – Cardoso et al. (Eds)
© 2018 Taylor & Francis Group, London, ISBN 978-1-138-33203-4

Numerical modelling strategy to accurately assess lining stresses in mechanized tunneling

A. de Lillis, V. De Gori & S. Miliziano
Department of Structural and Geotechnical Engineering, Sapienza, University of Rome, Italy

ABSTRACT: The paper presents a 3-D numerical strategy developed to accurately simulate the soil-shield and the soil-grout-lining interaction processes associated with the gap closure in mechanized tunneling. The problem is particularly relevant for the structural design of pre-cast lining of very deep tunnels subjected to high stresses. The numerical model works in large strain mode and simulates the main features of the excavation and construction processes including face pressure, cutterhead overcut, shield conicity, annular tail void grouting and time-dependent hardening of the grout, installation of the lining inside the shield and jacks' forces. The proposed approach overcomes in a very simple and effective manner the numerical inaccuracies associated to the erroneous shape of the excavation profile numerically obtained due to non-symmetric pre-convergences.

1 INTRODUCTION

The structural design of linings for tunnels excavated by TBMs must consider the stress and strain modifications in the soil caused by the excavation process and by the shield-soil-lining interaction. Provided that the main excavation and construction features are properly simulated, 3D numerical modelling can yield realistic evaluations of the stresses induced in the pre-cast concrete segments of the tunnel lining. At the same time, three-dimensional modelling is increasingly manageable thanks to growing computational power.

Numerical modelling of mechanized tunneling is particularly complex because of a number of factors, both geometrical and mechanical, that need to be included in the analyses. In addition to the complexities of the soil behaviour, the main issues are the following: a) the face pressure applied by the front of the TBM and by the muck in the excavation chamber; b) the geometry of the TBM including the cutterhead overcut and the conicity of the shield, which influence on the deformation of the surrounding soil is often overlooked; c) the annular void behind the rear end of the machine; d) the tail void grouting and the hydro-mechanical properties of the grout; e) the structural system of the segmental lining. Proper modelling of mechanized tunneling is of utmost importance for the design of both shallow and deep tunnels; specifically for the design of the lining, it is particularly relevant in the case of deep tunnels, where the stresses induced by the soil-structure interaction can be very high.

Several three-dimensional numerical models for the simulation of mechanized tunneling have been proposed in the past years. Dias *et al.* (2000) developed a model where the TBM and the lining were simulated as rigid bodies, while the tail void grout was represented by a pressure boundary in the early stages and by elastic continuum elements in the hardened stage. Kasper & Meschke (2004) developed a 3D model for shield-driven tunnels accounting for the main features of the construction process. The same model was then used by the authors (2006) to investigate the influence of TBM operation parameters, specifically face pressure and grouting pressure, on the ground surface settlements and on the internal forces in the tunnel lining. In the cited model, face and grout pressure were both simulated as pressure boundary conditions varying linearly over the height. Lambrughi *et al.* (2012) developed a three-dimensional model for TBM-EPB excavation simulating the steering gap through very soft continuum elements and the grout pressure through continuum elements with an artificially modified isotropic stress state equal to the injection pressure.

The segmental nature of the lining has been investigated by several authors (for instance Arnau & Molins, 2012 and Do *et al.*, 2014), usually modelling the interaction among segments using spring elements. Kavvadas *et al.* (2017) developed a finite element 3D model and used it to simulate the segmental lining in three different ways: continuous shell without joints, shell with aligned joints and shell with staggered joints; the comparison of the internal forces in the lining showed a limited effect of the segmental nature of the lining.

Numerical analyses of tunneling problems are usually based on calculation grids initially designed to accurately reproduce the excavation profile. In cases of large ground deformation around the tunnel, though, when a step-by-step approach is adopted, the correct portion of ground to be removed can be misidentified by the numerical algorithm. This is due to the development of pre-convergences ahead of the tunnel face, induced by the stress release, which cause the grid nodes, initially set to belong to the excavation profile, to change their position.

This issue was approached by others, among whom Alsahly *et al.* (2013), who defined a complex algorithm able to automate the process of mesh generation near the tunnel face within the advancing process. The meshing technique proposed by the authors achieve a continuous adaptation of the finite element mesh to match the actual motion path of the shield machine in each excavation step.

The paper presents a simpler, computationally less expensive, approach developed to tackle this issue. The proposed strategy allows to define the starting calculation grid so that, following the pre-convergences, the actual profile of excavation is correctly reproduced without having to resort to re-meshing techniques after each excavation stage.

In the following, the influence of the proposed meshing technique on the numerical results is investigated through a comparison with a standard calculation grid. The results show that a correct identification of the excavation boundary has significant effects on the whole interaction process and on the state of stress arising in the lining.

2 NUMERICAL MODEL

The numerical model has been developed using the Finite Difference computer code FLAC3D (Itasca, 2012). The model accounts for the main features of mechanized tunneling such as face pressure on the excavation front, cutterhead overcut, shield conicity, annular tail void, installation of the lining inside the machine shield and the application of the hydraulic jacks' forces, tail void grouting and its hardening in time.

The model is fine-tuned for the simulation of deep TBM bored tunnels; as such, the effects of gravity can be neglected. TBMs destined to work under high cover depths are usually designed to minimize the blockage risk and the forces on the lining. To this aim, the cutterhead overcut and the shield conicity are willfully increased to magnify the steering gap and the stress release in the surrounding soil, which experiences significant deformations. It is apparent that, in such cases, the importance of an accurate identification of the excavation profile in the numerical analysis is highlighted.

The extent of the calculation domain has been optimized to minimize boundary effects through a series of sensitivity analyses aimed at guaranteeing sufficient accuracy while maintaining a manageable computational effort. The resulting grid size largely satisfies the minimum criteria proposed by previous studies such as Franzius & Potts (2005) or Zhao *et al.* (2012).

The tunnel diameter is about 4 m, the height of the model is 80 m, the width is 80 m (Fig. 1). The starting grid size in the longitudinal direction is equal to 58 m; after every excavation phase the length of the model is increased adding a soil slice to the end of the calculation grid to maintain a constant distance between the front boundary and the excavation face. The total excavation length is about 110 m. Every excavation step has a 1.2 m length, equal to the lining segment. Since the effects of gravity are neglected, only a quarter of the tunnel is modelled.

The TBM shield is modelled using continuum elements with very high stiffness; its geometrical properties are reported in Table 1.

The lining is modelled using continuum elements (Augarde & Burd, 2001), equivalent to a cylindrical shell, with a Young modulus of 37.2 GPa. Several studies report that neglecting the segmental nature of the lining does not have a significant influence on the internal forces and on the overall results (for instance Kavvadas *et al*, 2017). For the sake of simplicity, the hydraulic jacks' forces are modelled as a longitudinal uniform pressure on the last installed lining ring.

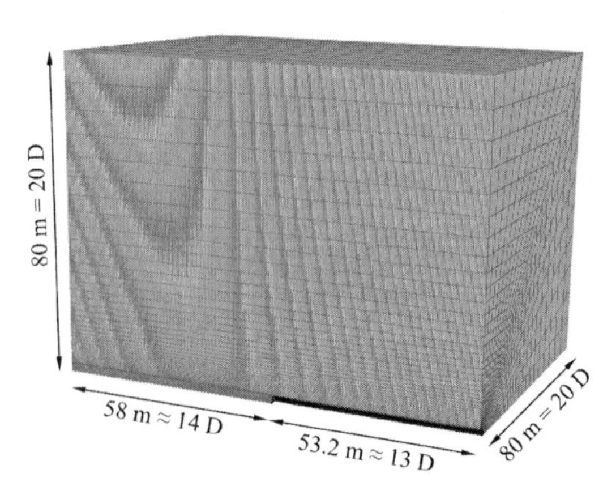

Figure 1. Numerical model.

To accurately simulate the soil-shield and the soil-grout-lining interaction processes associated with the gap closure, the algorithm works in large-strain mode, thus updating the position of the grid nodes at the end of each calculation step.

To the goal of this study, the very simple linear elastic perfectly plastic model with Mohr-Coulomb failure criterion, adopted for modelling the soil behavior, was adequate. The physical and mechanical parameters are reported in Table 2. The initial stress state is set to 2900 kPa in the vertical direction and 2050 kPa in the horizontal direction. The pore pressure is equal to 750 kPa.

The excavation process is simulated in detail accounting for the construction phases involved (Fig. 2). The sub-stages are the following: 1) the excavation advances one ring removing the corresponding soil slice directly ahead of the tunnel face; 2) the face pressure is applied to the new excavation front; 3) the machine shield moves forward. The last lining ring, activated during the previous excavation phase inside the shield, becomes exposed to the surrounding soil and a new lining ring is activated inside the shield; 4) the annular tail void is grouted at atmospheric pressure. The algorithm reads the position of the nodes both on the surrounding soil and on the shield and fills the empty space with low stiffness continuum elements (appropriate for fresh grout); 5) the stiffness of previously injected grout is increased following the hardening rule proposed by Kasper & Meschke (2006). To correlate the advancement of the excavation with time, a construction rate of 10 rings per day (12 m/d) is assumed.

The excavation is carried out in undrained condition, thus hypothesizing that the development of consolidation processes is negligible during the excavation. Once a stationary state is reached, roughly 8 diameters behind the excavation front, the excavation stops and the long-term behavior is simulated re-imposing the initial pore pressure throughout the calculation domain (the consolidation process was not simulated).

Table 1. TBM and lining geometry.

Cutterhead radius (m)	Shield front radius (m)	Tail radius (m)	Length (m)	Lining extrados radius (m)
2.06	2.045	2.015	6.4	1.95

Table 2. Main soil parameters.

E' (MPa)	ν' (–)	c' (kPa)	φ' (°)	ψ (°)	K_0 (–)
210	0.3	0	23	0	0.6

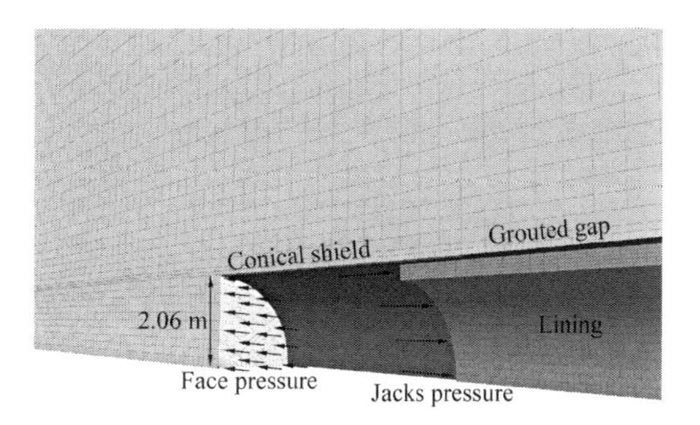

Figure 2. Simulation of the excavation process.

As 3D modelling, in addition to the strictly computational effort, is usually burdened by the significant amount of time needed to fine tune the models, the algorithm has been automated and made flexible to any change regarding the geotechnical context, the machine geometry and the specific construction procedures.

2.1 *Meshing technique*

In TBM bored tunnels the excavation is circular, as is the typical starting mesh in the proximity of the cutterhead. Figure 3a shows a circular grid designed to initially have nodes coincident with the cutterhead profile. Assuming an anisotropic geostatic stress state ($K_0 = 0.6$), the pre-convergences ahead of the tunnel face will be asymmetrical, higher at the tunnel crown and lower at the springline. When the tunnel front reaches a generic section, the typical excavation algorithm removes the portion of ground enclosed by the deformed mesh (gray area), which is: i) smaller than the actual excavation area; ii) elliptical in shape. Therefore, the remaining soil surrounding the TBM is at an asymmetrical distance from the shield, differently from what actually occurs during the excavation. The soil-lining interaction process is strongly influenced by the closing of this distance, which is related to the stress relaxation around the tunnel, as detailed in the following.

The proposed meshing technique is aimed at solving this issue with a very limited computational effort. The internal geometry of the defined mesh, relevant to the identification of the excavation profile, is reported in Figure 3b. It has a starting elliptical shape, designed to turn into a circle of the same radius of the cutterhead after the development of pre-convergences, when the generic section is to be excavated. To this aim, a preliminary analysis was conducted employing a standard

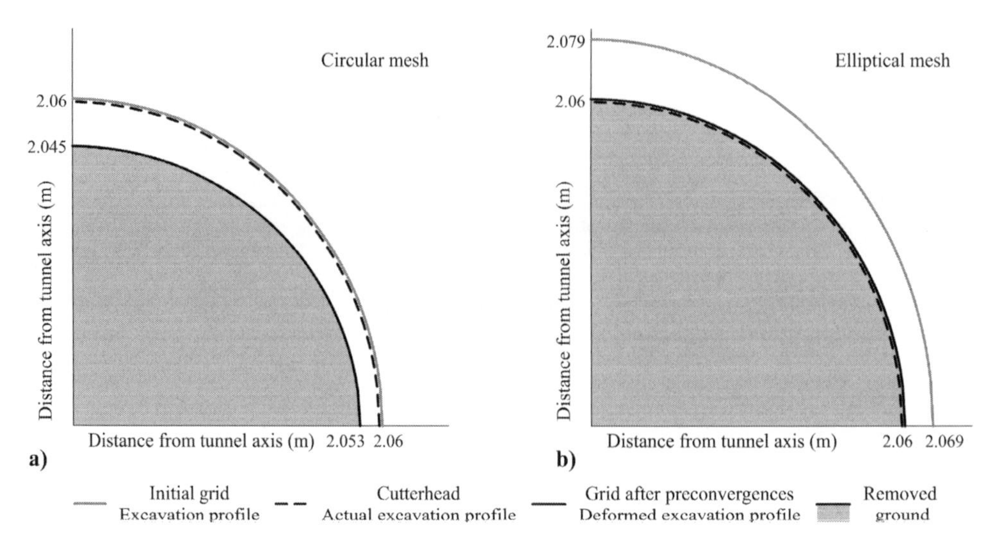

Figure 3. Excavation profile: a) circular mesh; b) elliptical mesh. Relative differences with the 2.06 m radius are magnified 20-fold.

circular mesh and measuring the pre-convergences developed along the excavation profile 1.2 m ahead of the tunnel face. The measured values can simply be added to the position of the nodes belonging to the excavation profile in the new starting grid, obtaining an elliptical shape, reflecting the fact that pre-convergences will be different based on the position along the profile of excavation: maximum at the tunnel crown and minimum at the springline. Since the pre-convergences of the newly designed grid are not the same of the circular one, to increase the accuracy of the procedure, it can be iterated (twice in this case).

A starting calculation grid improved following the described procedure allows to reproduce the actual excavation profile. Hence, the soil surrounding the shield will be equidistant from it at the moment of the excavation of a generic section. This allows to accurately simulate the subsequent interaction process, governed by the ground closure onto the grout-lining system.

3 NUMERICAL RESULTS

The described numerical model has been used to investigate the differences resulting from the use of a mesh designed as described in the previous paragraph, apt to accurately reproduce the actual excavation boundaries, compared to a standard one, typically circular in the case of mechanized tunneling. In this paragraph some numerical results will be illustrated showing the effects of the excavation in both undrained and drained conditions.

In Figure 4 the convergence longitudinal profiles of the tunnel crown and springline, obtained adopting the two different meshes are reported. Employing the circular mesh, the excavation boundary, which encloses the area of ground that will be removed, starts at a minor radial distance compared to what happens adopting the elliptical one. After the pre-convergences, the nodes of the circular mesh (Figure 4a) do not coincide with the cutterhead profile, $r = 2.06$ m, anymore and are closer to the TBM shield than those of the elliptical mesh (Fig. 4b). Between $d = 0$ m and $d = 6.4$ m the soil closes rapidly onto the machine shield in both cases, except for the crown of the elliptical mesh which does not perfectly adhere to the machine. Behind the tail of the TBM, $d > 6.4$ m, the convergence of the springline becomes greater than that of the crown, as opposed to the pre-convergence behavior. This is due to a drastic change in behavior from substantially elastic, ahead of the tunnel face, to largely plastic, behind the tail (see also Fig. 5).

Globally, the radial displacements are higher using the elliptical mesh. This is due to the significantly different displacements required to engage the shield in the two analyzed cases. As higher displacements are associated with higher stress release, this leads to the soil interacting with the shield, and then with the grout-lining system, at lower stress states. This, in turn, is bound to induce lower forces in the lining.

Figure 5 shows the magnitude of shear strains around the tunnel obtained using the elliptical mesh. The shear strains are concentrated at the side of the tunnel because of the combined action of

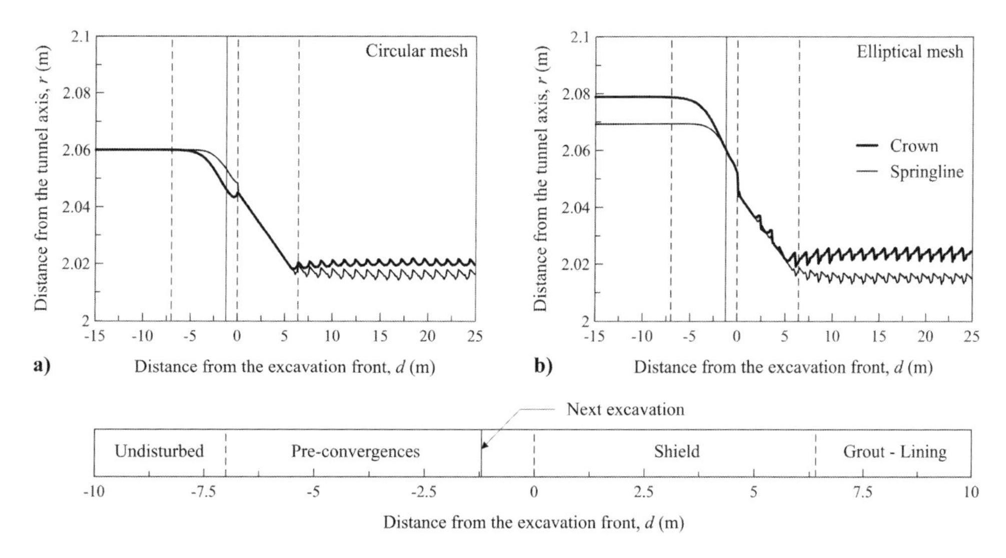

Figure 4. Convergence longitudinal profile: a) circular mesh; b) elliptical mesh.

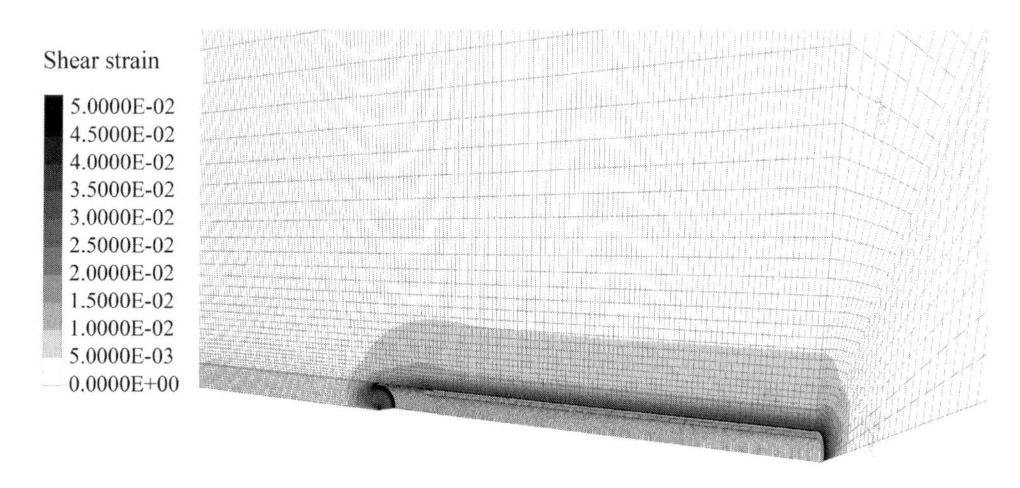

Figure 5. Shear strains contours.

stress relaxation and arching effect. Starting from an anisotropic geostatic stress state ($K_0 = 0.6$), at the side of the tunnel the major principal stress (vertical) increases because of the arching effect, while the minor (horizontal) decreases due to the excavation, resulting in an increase of the deviatoric stress; vice versa at the crown the stress release decreases the maximum principal stress and the arching effect increases the minimum one, thus reducing the deviatoric stress. The resulting stress path at the crown moves towards isotropic stress states before reaching plastic states, while at the springline it rapidly moves towards plastic states.

The distribution of the pore pressures after the excavation, obtained in undrained condition adopting the elliptical mesh, is represented in Figure 6. Near the tunnel the induced suction causes the pore pressure to reach negative values with a minimum of about -600 kPa at the tunnel crown. Along the vertical direction the excess pore pressure is always negative; along the horizontal transverse direction, a region of positive excess pore pressure develops, with a maximum of about 1200 kPa at about 1.5 diameters from the springline. This is due to the increase of the mean pressure associated with the formation of the arching effect. Adopting the circular mesh, the excess pore pressures induced by the excavation near the tunnel are in average about 10% lower of those obtained using the elliptical mesh. Because of the low front

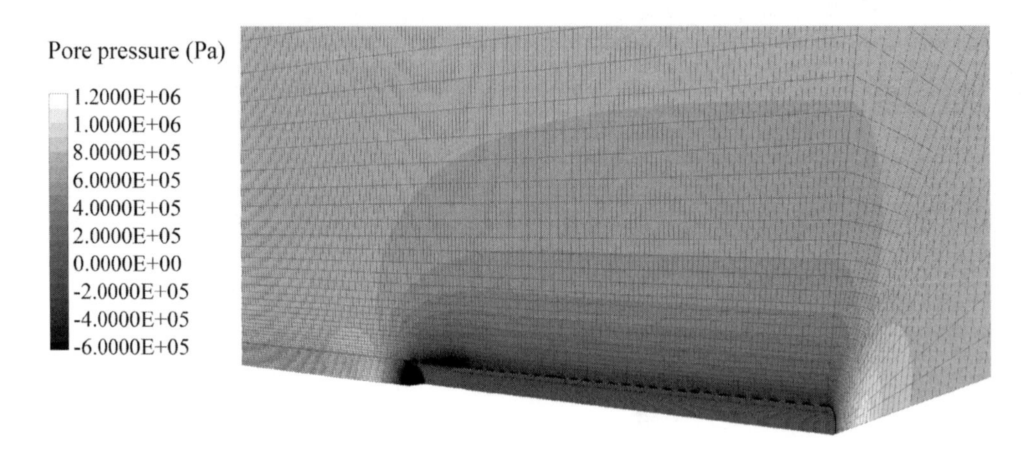

Figure 6. Pore pressures contours.

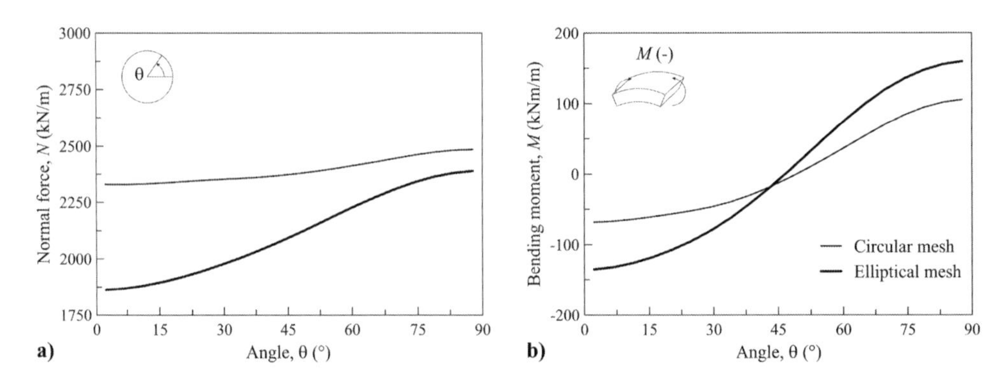

Figure 7. Internal forces in the lining in undrained conditions: a) normal force; b) bending moment.

pressure applied, consistent with a TBM operating in open mode, negative excess pore pressures develop at the excavation face.

The soil-lining interaction process in undrained conditions yields the lining stresses reported in Figure 7. The axial forces obtained adopting the circular mesh are 25% and 4% higher than those obtained adopting the elliptical mesh, at the springline and at the crown respectively. Bending moments instead are higher adopting the elliptical mesh: by 50% at the springline and by 34% at the crown. The absolute values are very little.

The internal forces can be associated with the convergence profiles seen in Figure 4b. The global difference between the starting and final position of the grid nodes is a measure of the stress relaxation, inversely related to the axial force. Moreover, the relative position difference between crown and springline, is related to the variability of the axial forces along θ and to the bending moments.

Globally, the effect of considering a starting circular mesh is a more homogeneous stress state of the lining with higher normal forces, less variable in the transverse section of the ring, and lower bending moments (lower load eccentricity).

The internal forces in the lining in long-term drained conditions are showed in Figure 8. The equalization of the pore pressure has a homogenizing effect on the stress state of the lining. The normal forces increase due to the dissipation of the significant negative excess pore pressures around the tunnel. This results in a slightly bigger increase in normal forces adopting the elliptical mesh. The difference between the results obtained with the two meshes decreases; the axial forces obtained with the circular mesh are greater by 13% at the springline while at the crown the results are essentially coincident. The bending moments experience a very small reduction and their ratio is practically the same it was in undrained conditions. Thus, the load eccentricity on the lining decreases slightly, as reported in Figure 9a.

The maximum compressive stresses in the lining, calculated assuming a homogeneous concrete

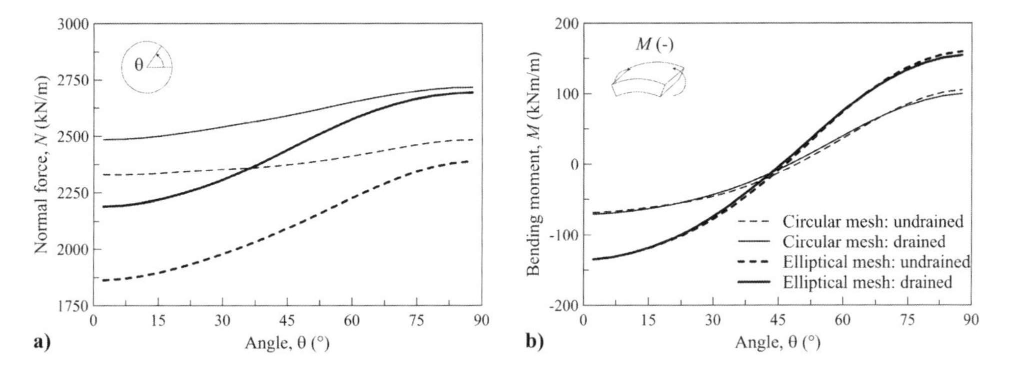

Figure 8. Internal forces in the lining in drained conditions: a) normal force; b) bending moment.

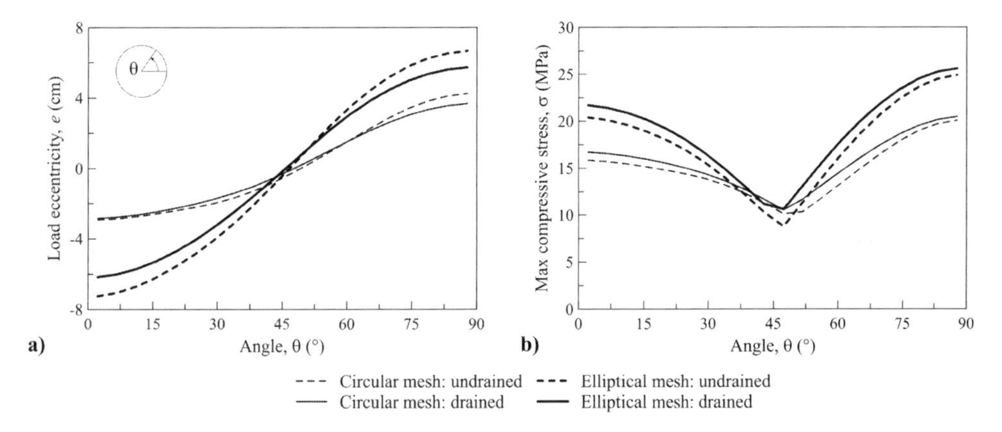

Figure 9. a) load eccentricity on the lining; b) maximum compressive stress.

section, are showed in Figure 9b. The differences between the results obtained using the circular mesh and the elliptical one are equal to 25% and 30%, at the crown and at the springline respectively. Said differences are nearly the same in both undrained and drained conditions.

4 CONCLUSIONS

The paper presents a very simple and effective numerical 3-D strategy to accurately simulate the soil-shield and the soil-grout-lining interaction processes associated with the gap closure in mechanized tunneling.

Working in large strain mode, the numerical model simulates the more important features of the excavation and construction processes including face pressure, cutterhead overcut, shield conicity, installation of the lining inside the shield and application of the jacks' thrusts, tail void grouting and evolution of the mechanical properties of the grout.

The proposed approach overcomes the numerical inaccuracies associated to the erroneous shape of the excavation profile numerically obtained due to non-symmetric pre-convergences, adopting a starting mesh, properly distorted, able to correctly reproduce the actual profile of excavation.

The comparison of the lining's compressive stresses obtained adopting the proposed mesh-design procedure and a standard circular mesh shows appreciable differences. In the investigated case, the use of the circular mesh leads to underestimate the compressive stresses by up to 30%, thus remarking the significant influence of the initial design of the calculation grid on the reliability of 3D models as either design or analyses tools.

Even though more sophisticated techniques may be adopted, the illustrated results show that even a simple procedure can effectively improve the numerical simulations.

REFERENCES

Alsahly, A., Stascheit, J. & Meschke, G. 2013. Three dimensional re-meshing for real time modeling of advancing process in mechanized tunneling. In J.P.M. Almeida, P. Dìez, C. Tiago, N. Pares (eds.), *Adaptive Modeling and Simulation, Lisbon 2013*.

Arnau, O. & Molins, C. 2012. Three dimensional structural response of segmental tunnel lining. *Engineering Structures* 44: 210–221.

Augarde, C.E. & Burd, H.J. 2001. Three-dimensional finite element analysis of lined tunnels. *International Journal of Numerical and Analytical Methods in Geomechanics* 25(3): 243–262.

Dias, D., Kastner, R. & Maghazi, M. 2000. Three dimensional simulation of slurry shield tunneling. In O. Kusakabe, K. Fujita, Y. Miyazaki (eds.), *Geotechnical Aspects of Underground Construction in Soft Ground, Tokyo 1999*. Rotterdam: Balkema.

Do, N.A., Dias, D., Oreste, P. & Djeran-Maigre, I. 2014. Three-dimensional numerical simulation for mechanized tunneling in soft ground: the influence of the joint pattern. *Acta Geotechnica* 9(4): 673–694.

Franzius, J.N. & Potts, D.M. 2005. Influence of mesh geometry on three-dimensional finite-element analysis of tunnel excavation. *International Journal of Geomechanics* 5(3): 256–266.

Itasca Consulting Group, Inc. 2012. FLAC3D Version 5.0, Fast Lagrangian Analyses of Continua in Three-Dimensions, User's manual. Minneapolis.

Kasper, T. & Meschke, G. 2004. A 3D finite element simulation model for TBM tunneling in soft ground. *International Journal for Numerical and Analytical Methods in Geomechanics* 28(14): 1441–1460.

Kasper, T. & Meschke, G. 2006. On the influence of face pressure, grouting pressure and TBM design in soft ground tunneling. *Tunnelling and Underground Space Technology* 21(2): 161–171.

Kavvadas, M., Litsas, D., Vazaios, I. & Fortsakis, P. 2017. Development of a 3D finite element model for shield EPD tunneling. *Tunnelling and Underground Space Technology* 65: 22–34.

Lambrughi, A., Rodrìguez, L.M. & Castellanza, R. 2012. Development and validation of a 3D numerical model for TBM-EPB mechanized excavations. *Computers and Geotechnics* 40:97–113.

Zhao, K., Janutolo, M. & Barla, G. 2012. A completely 3D model for the simulation of mechanized tunnel excavation. *Rock Mechanics and Rock Engineering* 45: 475–497.

Numerical Methods in Geotechnical Engineering IX – Cardoso et al. (Eds)
© 2018 Taylor & Francis Group, London, ISBN 978-1-138-33203-4

Tunnelling induced settlements—finite element predictions, soil model complexity and the empirical inverse Gaussian settlement curve

G. Marketos
Tony Gee and Partners LLP, Esher, UK

ABSTRACT: Underground tunnel construction leads to volume loss and stress release in the ground which translates to downwards movement at the ground surface. Many authors have reported that Finite Element analyses predict a much wider surface settlement curve than what is typically observed in the field (i.e. an inverse Gaussian curve). This is in contrast to Discrete Element approaches which have shown a much better match as they do not pre-specify the soil constitutive response. With this observation in mind, this paper starts from a simple 2D plane strain Finite Element that uses a linear elastoplastic soil model and then add levels of complexity to the soil material model (non-linearity, a small strain stiffness variation etc), one at a time, to observe the influence each has on the calculated settlement trough in the model. By comparing the calculated surface settlement trough each time with the inverse Gaussian curve (which has been shown to fit very well field data) it is possible to identify which components of more complex soil material models are most important in modelling tunnel construction problems, with a wider application to other soil-structure interaction problems.

1 INTRODUCTION—THE TUNNELLING SETTLEMENTS PROBLEM

1.1 *Type area*

Underground tunnel construction leads to volume loss and stress release in the ground which translates to downwards movement (settlement) at the ground surface. A number of authors (e.g. Peck, 1969, see also Mair et al. 1993) have observed that the settlement trough can be described very well by the inverse Gaussian curve of Eq. 1 (the red curve in Figure 2).

$$S = S_{max}\exp\left(-\frac{x^2}{2i^2}\right) = S_{max}\exp\left(-\frac{x^2}{2K^2H^2}\right) \quad (1)$$

Empirically it has been found that i = KH, where H is the depth of the tunnel axis and K is a non-dimensional trough width parameter, which usually varies between about 0.3 and 0.6 (see e.g. Rankin, 1988).

Many authors (e.g. Jurečič et al, 2012) have reported that Finite Element analyses often predict a much wider surface settlement curve than what is typically observed in the field (i.e. an inverse Gaussian curve). This is in contrast to Discrete Element approaches which, due to the fact that they model the soil as a collection of particles that interact through contact springs, do not pre-specify the soil constitutive response and so demonstrate that it is possible to reproduce the expected Gaussian settlement curve with a very idealised soil-like material (see e.g. Bym et al., 2013). The question therefore arises as to whether it is possible to reproduce realistic tunnelling settlement curves in relatively simple Finite Element models, and if so, what the necessary components of a soil constitutive model to do so are.

2 THE FINITE ELEMENT MODEL CONFIGURATION

The tunnel is modelled here under 2D plane strain conditions (i.e. zero extension or contraction in the out-of-plane direction) using the commercially available programme PLAXIS2D (2017 version). A pure volumetric compressive strain of 10% is applied to the circular tunnel area in a single step so as to model the creation of the tunnel, and the associated volume loss within a uniform continuum with a K_0 value of 1 (i.e. initially $\sigma_h = \sigma_v$ throughout). The model tunnel used here is 5 m in diameter and its centre lies at a depth of H = 25 m. The Finite Element model box, which is contained by rigid boundaries is 500 m wide and 250 m deep. Model boundaries are fixed in the normal direction, but on rollers in a direction parallel to them. Here no water is modelled to avoid complications—the water table is assumed to be well below the zone of influence of the tunnel. The case modelled can therefore be thought as dry.

In what follows the same model configuration will be used (shown in Figure 1), the only variation will be in the soil stress-strain model.

3 ISOTROPIC ELASTOPLASTIC MOHR COULOMB ANALYSIS (MODEL A)

The simplest elastoplastic soil model is an isotropic linear Mohr Coulomb model—this is routinely used in many engineering design calculations as often extra input parameters for more complex soil models are not available due to the practicality and cost of collecting them. This model needs a limited set of parameters, E (Youngs modulus), v (Poisson ratio), c' (cohesion) and φ' (friction angle). Initially there is no depth dependence of the elastic modulus.

The Poisson ratio was chosen as 0.25 so that the FE model is directly comparable to some plots shown in Bym et al. (2013). The E value was set as 50 MPa, the soil unit weight as 20 kN/m^3, and the friction angle (φ') was set to 25°, with c' = 0 initially. The results for this soil model are shown in Figure 2 below. The least-squares fit to the data using the formula of Equation (1) – an inverse Gaussian curve—has also been included in the plot.

The fit for Model A was not satisfactory—the Mohr Coulomb model gave a rather wide settlement trough, with a K value of 0.89, wider than what is typically observed in the field (K values of 0.3 to 0.6 – see Rankin, 1988). For a fuller discussion of this see model comparison section below.

4 NON-LINEAR HARDENING SOIL ANALYSIS (MODEL B)

A soil model which results in a non-linear stress-strain relationship up to a Mohr Coulomb—like failure limit as shown in Figure 3 below was then

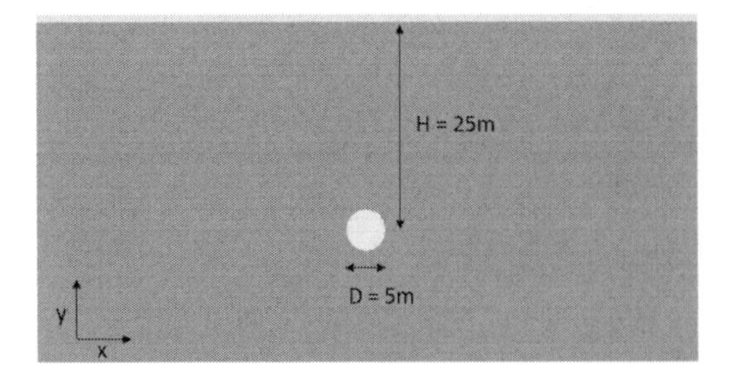

Figure 1. A sketch of the model tunnel used in PLAXIS.

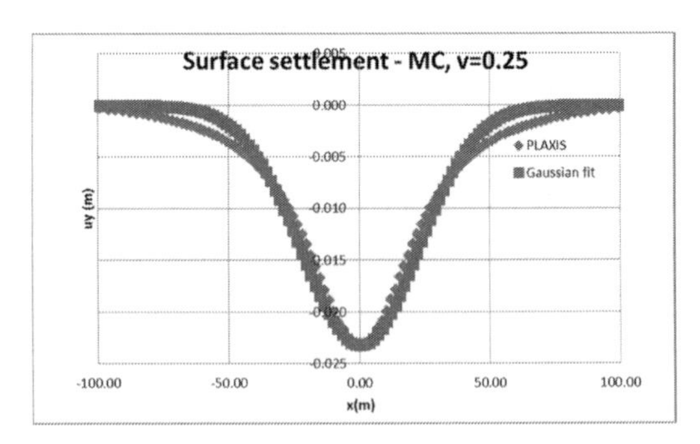

Figure 2. Model A: The settlement uy vs. horizontal distance from the tunnel centreline for the Mohr Coulomb soil—the Finite Element model prediction and the inverse Gaussian least-squares fit to the data.

used (the Hardening Soil model in PLAXIS). Note that this model mobilises plastic strain even before the Mohr-Coulomb failure limit gets reached, hence the non-linearity of the stress-strain curve and the higher stiffness when unloading—reloading (which is an elastic stiffness).

This model requires a few more input parameters, the meaning of some of which is shown in Figure 3. The E_{50} was set here to the Mohr-Coulomb value linear E value (50 MPa). E_{ur} = 3 E_{50} was used, so E_{ur} was chosen as 150 MPa. c' and φ' were set as the same as in the Mohr Coulomb analysis of Model A (0 and 25° respectively), and E_i was 1.82 E_{50} = 91 MPa. The stiffness was set to be independent with depth in model B (i.e. m, the parameter for the stress level dependency of stiffness was set to 0).

This still gave a relatively unsatisfactory fit to the data (see Fig. 4), and a settlement trough width parameter (K) of 0.75, still larger than the 0.3 to 0.6 that has been observed in the field. A soil model that incorporates small strain stiffness effects was tried next.

5 ANALYSIS WITH A SOIL MODEL THAT INCLUDES A STIFFNESS DEGRADATION CURVE AND IN WHICH STIFFNESS IS INDEPENDENT OF STRESS (MODEL C)

The next analysis was carried out with the HSsmall PLAXIS model. This has many of the features of the Hardening Soil model but additionally models the degradation of shear stiffness with strain. The extra input parameters needed, $\gamma_{0.7}$ (the shear strain level at which the secant shear modulus drops to 72.2% of its initial very small strain value) and G_0^{ref} (the initial or very small strain shear modulus) were set as 0.0001

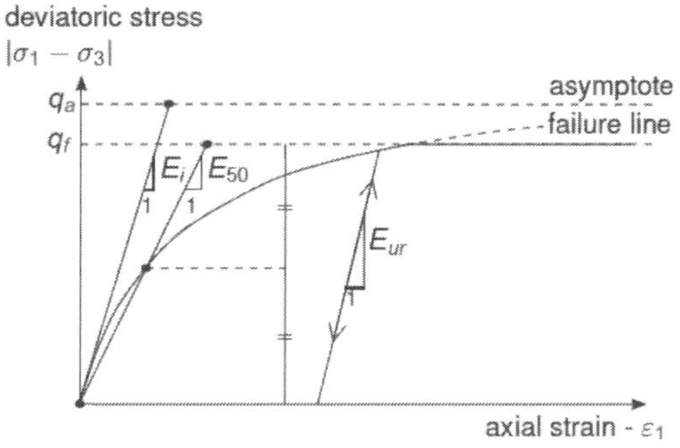

Figure 3. The hyperbolic stress-strain relation in primary loading for a standard drained triaxial test—A reproduction of Fig. 6.1 from the PLAXIS2D Material Models Manual.

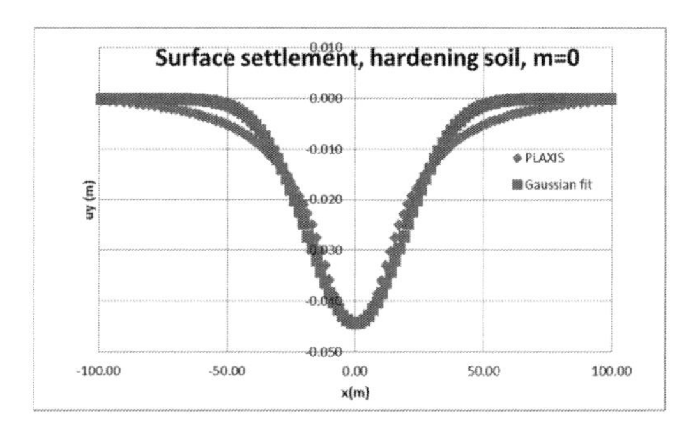

Figure 4. Model B: Settlement uy vs. horizontal distance from the tunnel centreline for the non-linear hardening soil model—Finite Element model prediction and the inverse Gaussian least-squares fit to the data.

and 312.5 MPa (= 5 G_{ur}, the unload-reload shear stiffness) respectively. Note that the G_{ur} value is what is used as the lower cut-off in the modulus degradation (reduction) curve shown below. So a factor of 5 was chosen to give an adequate span of moduli. The relevant plots for these model parameters as output by PLAXIS are shown in Figure 5 below.

The results for the HSsmall model were encouraging (see Figure 6), the settlement width parameter (K) has been brought down to 0.49, which is within the range of values observed in the field (0.3 to 0,6), and the quality of the Gaussian fit to the data has increased.

6 ANALYSIS WITH THE HSSMALL MODEL AND A STRESS DEPENDENCE OF STIFFNESS (MODEL D)

Adding a stiffness variation with depth (through a power exponent for stress-level dependency of stiffness of m = 1) gives the surface settlement curve shown in Figure 7 below. The quality of the fit is further increased, and the settlement trough has become narrower, as will be discussed in the comparison section that follows.

7 COMPARISON OF THE FINITE ELEMENTS SETTLEMENT TROUGHS

The results of all four models presented here are summarised in Table 1. The table includes best-fit values for i and K, parameters that appear in the empirically-derived settlement curve of Equation 1. Note that the range of values for K that are typically observed in the field lies between 0.3 and 0.6. The last column lists a metric of the quality of the fit, which has been calculated as:

$$\chi^2 = \frac{\Sigma(u_{FiniteElemet} - u_{fitted})^2}{u_{max}{}^2} \quad (2)$$

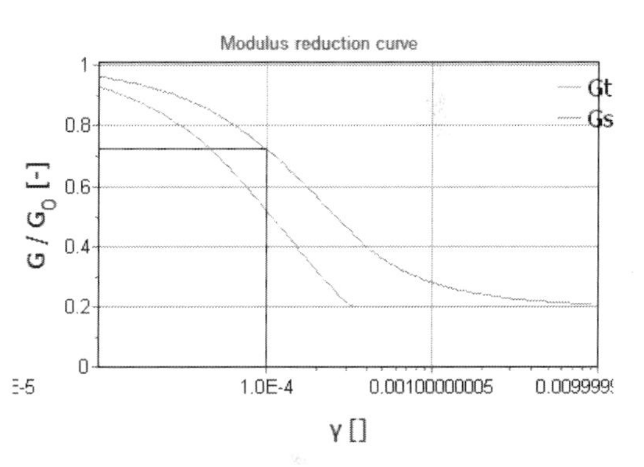

Figure 5. The stiffness degradation curve for the soil model input parameters used for model C.

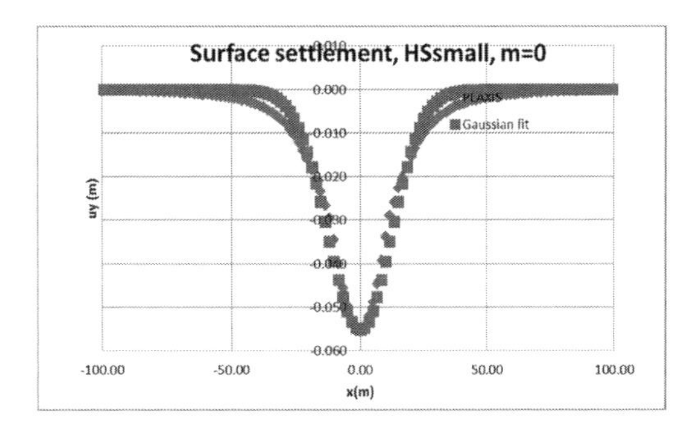

Figure 6. Model C: The settlement uy plotted against horizontal distance from the tunnel centreline for the HSsmall soil model with no stiffness variation—Finite Element model prediction and the inverse Gaussian least-squares fit to the data.

Table 1. A comparison of results for all models presented here.

Model ID	Soil model used	i (m) – settlement trough width	K—settlement trough parameter (= i/H)	χ^2 – Quality of fit (lower value better)
A	Mohr Coulomb	22.3	0.89	0.47
B	Hardening	18.8	0.75	0.59
C	HSsmall, m = 0	12.4	0.49	0.30
D	HSsmall, m = 1	8.2	0.33	0.13

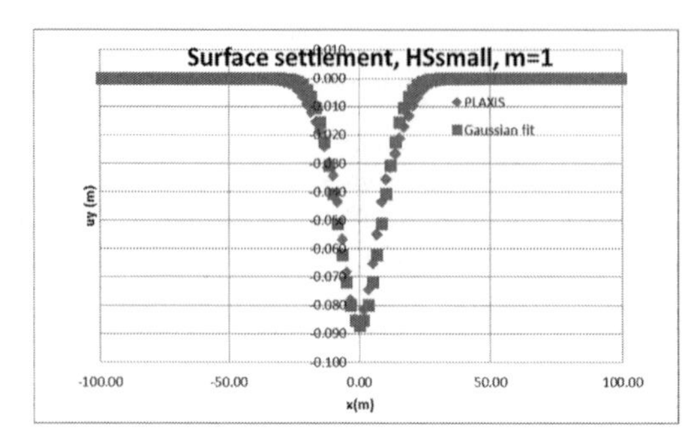

Figure 7. Model D: The settlement uy plotted against horizontal distance from the tunnel centreline for the HSsmall soil model with a stiffness variation with a power exponent (m) of 1 – PLAXIS prediction and the inverse Gaussian least-squares fit to the data.

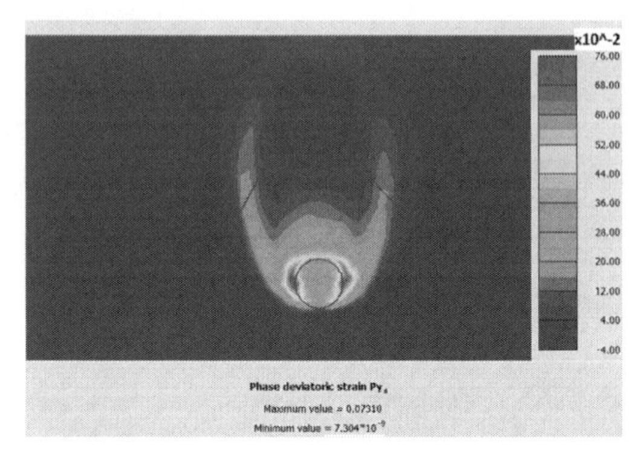

Figure 9. A plot of the deviatoric strains as obtained in the Finite Element simulation (model D).

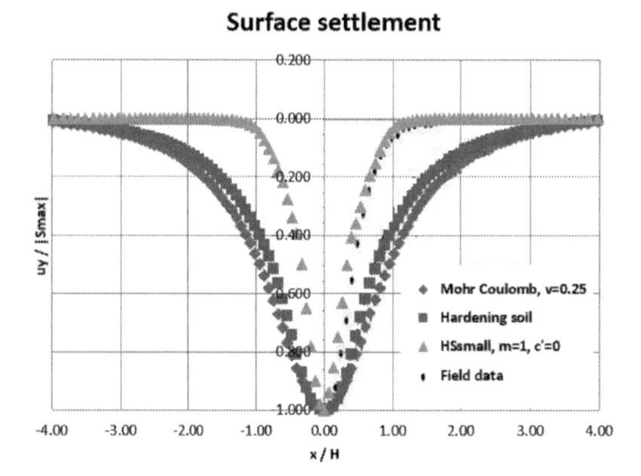

Figure 8. A non-dimensional plot of settlement versus horizontal distance which collects the data from simulations reported here and which also includes the field data shown in Figure 13 of Jurečič et al. (2013). Vertical settlement has been divided by the maximum settlement observed in each case (Smax) while horizontal distance x has been divided by the depth of the tunnel (H).

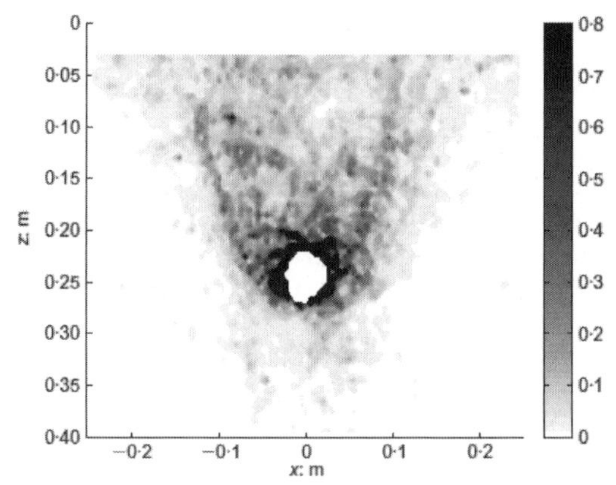

Figure 10. The deviatoric strain for a tunnel excavation simulation with the Discrete Element Method which models the soil as a collection of grains that interact with each other through springs at their mutual contacts. A reproduction of Figure 5 from Bym et al. (2013).

In order to be able to compare all FE model results presented here the curves of Figures 2, 4, 6 and 7 have been non-dimensionalised by dividing them by the maximum settlement observed in each model (Smax). The x axis has been further non-dimensionalised by the depth of the tunnel (H), which for the simulations here was 25 m. The curves have been plotted in Figure 8. The plot includes the data shown in Figure 13 of Jurečič et al. (2013) which were for a tunnel excavation in London clay, at a depth of 30.5 m.

As one can see it is only with the HSsmall model that the narrower settlement trough commonly observed in the field can be reproduced. This hints to the fact that the HSsmall model with a depth variation of stiffness sufficiently captures the relevant aspects of soil behaviour. These also include plastic strains before the attainment of the Mohr-Coulomb frictional limit, and the small stiffness variation shown in Figure 5. It also seems that inclusion of initial stiffness anisotropy is not critical—the HSsmall model (that does not use it) seems to be able to allow strains to evolve in a way that is very similar to results previously obtained by the author and his coworkers through a DEM simulation (see Figs. 9 and 10 respectively).

8 CONCLUSIONS

This work has used the relatively well constrained problem of ground surface displacements induced by tunnelling through soil to test a number of soil models of increasing complexity. Use of a linear elastoplastic Mohr Coulomb model resulted in a settlement trough that was wider than what observed in the field. Use of a model with a non-linear stress-strain response up to failure (ie a model that includes some plastic strains before reaching the Mohr-Coulomb failure) made the settlement trough narrower but still somewhat wider of what has been observed in the field. It was only with a model that includes a stiffness degradation curve (i.e. the variation from a high small-strain stiffness to a lower large-strain stiffness) that the empirically derived inverse Gaussian curve could be reproduced very well. This model also resulted in a settlement trough width within bounds as measured in the field, giving more confidence that the model adequately captures the aspects of soil behaviour that are relevant to this problem. Inclusion of a stiffness variation with stress level (i.e. with depth) made the fit of the Finite Element analysis by the Gaussian curve even better. Note that preliminary analysis has indicated that the settlement trough width is a function of the friction angle (ie the location of the ultimate Mohr Coulomb limit in the PLAXIS HSsmall soil model tried) – higher friction angles resulted in wider settlement troughs.

The work presented here has therefore essentially helped clarify which components of an advanced soil model might be necessary for modelling the tunnel settlement problems. It also indicates that the HSsmall PLAXIS model might capture well the variations of strains away from a soil region to which a disturbance is applied. Hence this observation might also be relevant to a wider array of soil-structure interaction problems in which accurate displacement predictions away from the structures might be needed.

REFERENCES

Bym, T., Marketos, G., Burland, J.B. & O' Sullivan, C. (2013). Use of a two-dimensional discrete-element line-sink model to gain insight into tunnelling-induced deformations. Géotechnique, vol. 63, pp. 791–795.

Jurečič, N., Zdravković, L. & Jovičić, V. (2013). Predicting ground movements in London Clay. Proceedings of the Institution of Civil Engineers—Geotechnical Engineering, vol. 166, no. 5, pp. 466–482.

Mair, R.J., Taylor, R.N. & Bracegirdle, A. (1993). Subsurface settlement profiles above tunnels in clays. Ge´otechnique, vol. 43, no. 2, pp. 315–320.

Peck, R. (1969). Deep excavations and tunnelling in soft ground. Proc. 7th Int. Conf. Soil Mech., Mexico, State of the Art, vol., pp. 225–290.

Rankin, W.J. (1988). Ground movements resulting from urban tunnelling: prediction and effects, Engineering Geology Special Publications, vol. 4, pp. 79–92. London, UK: Geological Society.

Numerical Methods in Geotechnical Engineering IX – Cardoso et al. (Eds)
© 2018 Taylor & Francis Group, London, ISBN 978-1-138-33203-4

Stress redistribution in the central pillar between twin tunnels

A.M.G. Pedro
ISISE, University of Coimbra, Coimbra, Portugal

J.C.D. Grazina
Civil Engineer, Coimbra, Portugal

J. Almeida e Sousa
University of Coimbra, Coimbra, Portugal

ABSTRACT: Given the construction constrains at ground surface in the major urban environments the use of underground to install infrastructure and transport networks has increased considerably in the last decades. However, this solution is also approaching its limit, with a substantial part of the underground already being occupied. In this scenario it has become more frequent to excavate new tunnels in close proximity to existing ones, leading to interaction problems that are necessary to take into account in the design stage. Evidences of such interaction have been reported in the literature and are strongly dependent on the distance between tunnels. However, most of these studies focus on the displacements measured at ground surface and scarce results are available in the literature about the stress redistribution and mobilised stresses that occur in the central pillar of soil. In this paper a numerical study of the construction of side-by-side twin tunnels excavated sequentially is performed with the purpose of further investigate this aspect. The results show that depending on the distance between tunnels the mobilised stress levels in the pillar rise with the excavation of the 2nd Tunnel, leading to an increase of the yielded area. A parametric study conducted also showed that the area affected by the excavation is dependent of the initial stress conditions considered in the analyses.

1 INTRODUCTION

The excavation of shallow tunnels inevitably induces deformations in the soil, which might cause damage on the buildings, depending on their conditions, location and on the magnitude of the movements. In major urban environments, given the density of the networks required and the existing constrains in the use of the subsoil, it is often required to excavate new tunnels in close proximity with existing ones. This scenario frequently leads to a considerable increase in the deformations of the soil due to the interaction that occurs between the tunnels. Evidences of such interaction have been reported in different case studies in the literature (e.g. Bartlett & Bubbers, 1970; Cording & Hansmire, 1975; Nyren, 1998; Withers, 2001; Cooper et al., 2002; Elwood & Martin, 2016) and according to Mair & Taylor (1997) it occurs because the soil where the 2nd Tunnel is going to be excavated is already disturbed, presenting smaller strength and stiffness due to the first excavation.

Generally, this interaction is evaluated based on the ratio between the volume loss associated with the excavation of the 2nd Tunnel, $V_{L,}2nd_T$ (obtained by the subtraction of the volume loss after the excavation of the 1st Tunnel to the final volume loss), and the value measured after the excavation of the 1st Tunnel, i.e. in greenfield conditions, $V_{L,greenfield}$. Based on this ratio ($V_{L,}2nd_T/V_{L,greenfield}$) interaction occurs if the values are higher than 1.0, while a ratio equal to 1.0 indicates that the excavation of the 2nd Tunnel was not influenced by the presence of the 1st Tunnel. Values smaller than 1.0 have not been reported in the literature, and although possible would imply that the excavation of the 2nd Tunnel occurred in more favourable conditions.

Naturally, one of the principal factors responsible for the interaction is the distance between tunnels, often designated as the pillar width, *L*. According with the numerical studies performed by Addenbrooke & Potts (2001) and the centrifuge tests conducted by Divall & Goodey (2015) an increase of up to 30% on the volume loss associated with the excavation of the 2nd Tunnel occurs for tunnels spaced of $L=1.5{\cdot}D$, being *D* the diameter of the tunnel. As expected, as the pillar width increases the interaction between tunnels decreases considerably, although it is still observed for *L*/*D* ratios higher than 4.0, according with those studies, though it is almost negligible for ratios higher than about 2.0 (Ghaboussi & Ranken, 1977; Kim et al., 1998; Pedro et al., 2017). Naturally, it is difficult to establish a unique minimum interaction distance as its value also depends on the soil characteristics, on initial stress conditions and on the geometry and construction methods employed in the excavation.

Despite the volume loss and the movements at ground surface being the most visible effects of the interaction between tunnels, it is particularly relevant to analyse the stress redistribution that occurs in the central pillar, as this can be considered the trigger of the interaction. However, this aspect has not been thoroughly investigated and few results are available in the literature (Ghaboussi & Ranken, 1977; Koungelis & Augarde, 2004; Gercek, 2005). From these, the study performed by Ghaboussi & Ranken (1977) is the most complete, with the authors showing that for *L*/*D* smaller than 2.0 the excavation of the 2nd Tunnel induces an increase in both vertical and horizontal stresses in the pillar, while the deviatoric stress remains almost unchanged. For higher ratios the results indicate a negligible interaction. However, this study was performed assuming a linear elastic model for the soil, which is an unrealistic assumption for its behaviour.

In this paper, a series of numerical analysis are carried out using the in-house finite element program, UCGeoCode, in its 2D version, in order to further investigate the stress redistribution and mobilised stress levels that occur in the central pillar due to the sequential excavation of side-by-side twin tunnels. The influence on the stresses of factors such as the distance of the pillar width and the initial stress conditions is also evaluated through an extensive parametric study.

2 NUMERICAL MODEL

2.1 *Geometry and parameters of the model*

The analyses were performed assuming plane strain conditions with the geometry considered for the numerical model schematically represented in Figure 1. Two circular tunnels with a diameter, *D*, of 6 m were located at a depth, *H*, of 19 m (cover of 16 m) and centred in the mesh. The distance between tunnels, *L*, in the analyses was varied between a minimum value of 1.5 m and a maximum value of 24 m, corresponding to extreme *L*/*D* ratios of 0.25 and 4.0, respectively. In total 6 scenarios with *L*/*D* ratios of 0.25, 0.50, 1.0, 2.0, 3.0 and 4.0 were analysed. In all models the lateral boundaries were placed at a distance measured from the axis of each tunnel of 72 m to both sides, in order to ensure no interference with the results. The bottom boundary was located 21 m below the invert of the tunnels. The boundary conditions were set so that at the top free movements could occur, while at the bottom the movements were restricted in both directions. On the lateral boundaries only vertical displacements were allowed to occur.

The lining of both tunnels was considered to be 0.3 m thick and was simulated through solid elements with a linear elastic behaviour, characterised by a Young's modulus of 20 GPa and a Poisson's ratio of 0.2.

For the soil a conventional Mohr-Coulomb failure criterion with the parameters displayed in Table 1 was adopted. For the stiffness a constant Poisson's ratio of 0.3 and a deformability modulus, *E*, that varied with the effective confining stress, σ_3', according to the Janbu (1967) law (Eq. 1) were considered.

$$E = A \cdot P_a \cdot \left(\frac{\sigma_3'}{P_a} \right)^n \tag{1}$$

The parameters *A*, P_a and *n* adopted in Equation 1 are presented in Table 1 and were kept constant throughout the analyses. However, the value of *E* was updated at the beginning of each stage based on the value of σ_3' determined in the previous stage.

Dry conditions and a geostatic initial stress profile, assuming reference values of 20 kN/m^3 for the unit weight, γ, and 0.6 for the at rest earth pressure coefficient, K_0, were considered for the reference analysis (Table 1). For the last study K_0 values of 0.45, 1.0, 1.4 and 1.8 were also adopted in order to evaluate the influence of this parameter in the results.

2.2 *Construction sequence*

The simulation of the excavation in full section of both tunnels was performed by employing the

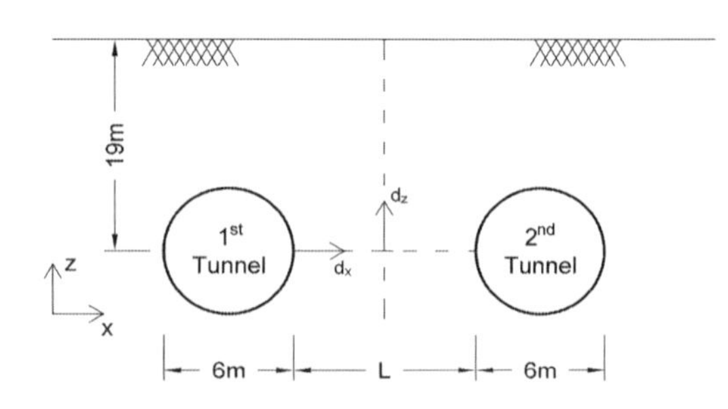

Figure 1. Geometry considered in the numerical model.

Table 1. Soil parameters considered.

γ (kN/m^3)	K_0	c' (kPa)	φ	ψ	A	P_a (kPa)	n	ν
20	0.6	5	30	5	400	100	0.5	0.3

stress relaxation method (Potts & Zdravković, 2001). In this method the 3D effects usually associated to the tunnel construction are simulated in two separate stages (Moller, 2006). In the first, the solid elements inside the tunnel are removed but only a percentage of the resulting stresses (α) is applied in the contour of the excavation. In the second stage the lining is installed, and the remainder of the unbalanced stresses is applied (1-.α) so that a final equilibrium state is achieved. In the present case a stress relief factor α of 0.6 was considered in all the excavation stages. A total of 4 stages were then modelled: 1) excavation of the left tunnel (1st Stage); 2) installation of its lining (2nd Stage); 3) excavation of the right tunnel (3rd Stage); and 4) installation of its lining (4th Stage).

3 STRESS REDISTRIBUTION IN THE CENTRAL PILLAR

In order to establish the pattern of the stress redistribution and mobilised stress levels that occur due to the excavation of the tunnels an initial reference analysis with a L/D of 1.0 was performed. Figure 2 plots the vertical (σ'_z) and horizontal (σ'_x) stresses normalised by the corresponding initial stresses $(\sigma'_{z,0};\sigma'_{x,0})$ predicted along the horizontal centreline of the pillar, as indicated in Figure 1, for the four stages modelled. The results show that the influence of the excavation of the 1st Tunnel extends beyond a d_x/L of 1.0, implying that the soil where the 2nd Tunnel is going to be constructed is affected by the first excavation. However, for the reference conditions and for a L/D of 1.0 this interference appears to be reduced, given that only an increase smaller than 10% is expected to occur in both vertical and horizontal stresses. As expected, in the area adjacent to the first excavation significant variations of stress occur. With the excavation of the 1st Tunnel it is possible to observe a reduction to about 0.8 of the vertical stress in the area immediately adjacent to the tunnel. However, this stress tends to increase sharply, reaching a maximum peak of about 1.5 just for a d_x/L of about 0.25 and then gradually decays until reaching steady stress conditions for a d_x/L of approximately 2.0. This behaviour appears to be unaffected by the application of the lining as the results from stage 1 (S1) and 2 (S2) are almost indistinguishable. As expected, the normalised horizontal stress shows a reduction to 0.4 immediately adjacent to the 1st Tunnel. Then, the horizontal stresses quickly increase with the distance reaching steady stress conditions also for d_x/L of 2.0. It is also possible to verify that the application of the lining causes a slightly increase of the horizontal stress along the entire horizontal centreline of the pillar.

With the excavation of the 2nd Tunnel a stress redistribution in the pillar occurs due to arching and an almost symmetrical behaviour of both vertical and horizontal stresses can be observed. By comparing the final results with those obtained after the excavation of the 1st Tunnel is possible to verify the interaction between tunnels, since the vertical stresses adjacent to the tunnels increased considerably, for up to 1.7, and only a decay to 1.4 is observed at half distance (d_x/L=0.5). A similar conclusion can be drawn based on the results of the horizontal stresses, although in this case the maximum differences are achieved at the middle of the pillar, with an increase to up 1.2 after the installation of the lining of the 2nd Tunnel.

A similar plot is presented in Figure 3 but for the normalised stresses observed at the vertical centreline of the pillar (considered at middle distance between the tunnels and positive upwards from the horizontal axis of the tunnels, as indicated in Figure 1). Also along this centreline it is possible to observe that the area of influence of the construction of the 1st Tunnel extends beyond the middle distance between tunnels with both vertical and horizontal stresses presenting variations. With the first excavation the vertical stresses show an increase at the depth of the axis of the tunnel of about 1.2. This increase is further amplified after the excavation of the 2nd Tunnel reaching about 1.5 and showing a significant interaction between tunnels. However, the area affected by the excavation at this vertical profile appears to be independent of the second excavation, extending from one diameter below the horizontal axis of the tunnels

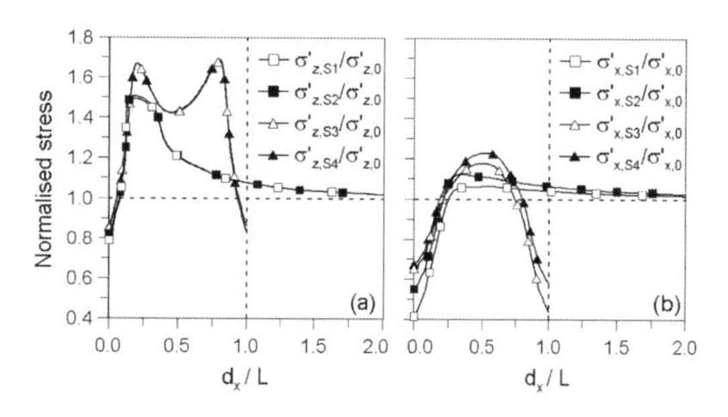

Figure 2. Normalised stresses along the horizontal centreline of the pillar: (a) vertical stress; (b) horizontal stress.

to about 1.5 above. The horizontal stresses show a more complex behaviour. After the excavation of the 1st Tunnel a slight increase is observed at the depth of the horizontal axis followed by a sudden decrease in both directions that reaches a minimum of 0.9 at about the depths of the crown ($d_z/D=0.5$) and of the invert ($d_z/D=-0.5$). After that the horizontal stresses tend to increase to steady stress conditions at higher depths and for higher values near the ground surface. This behaviour is amplified by the excavation of the 2nd Tunnel where higher values are achieved. The installation of the lining has a minimal impact probably because the distance to the tunnels is considerable in this vertical profile.

In order to observe more clearly the area affected by the construction of both tunnels the mobilised stress level, *MSL,* for each Gauss point of the mesh was determined through Equation 2.

$$MSL = \frac{J}{J_{max}} \tag{2}$$

In this equation J is the current mobilised deviatoric stress and J_{max} the maximum shear strength of the soil. The former depends on the current principal effective stresses $(\sigma_1', \sigma_2' \, and \, \sigma_3')$ and can be determined using Equation 3, while the latter is given by Equation 4.

$$J = \frac{1}{\sqrt{6}} \cdot \sqrt{(\sigma_1' - \sigma_2')^2 + (\sigma_2' - \sigma_3')^2 + (\sigma_3' - \sigma_1')^2} \tag{3}$$

$$J_{max} = \left(\frac{c'}{\tan \varphi'} + p' \right) \cdot \frac{\sin \varphi'}{\cos \theta + \dfrac{\sin \theta \cdot \sin \varphi'}{\sqrt{3}}} \tag{4}$$

where:

$$p' = \frac{\sigma_1' + \sigma_2' + \sigma_3'}{3} \tag{5}$$

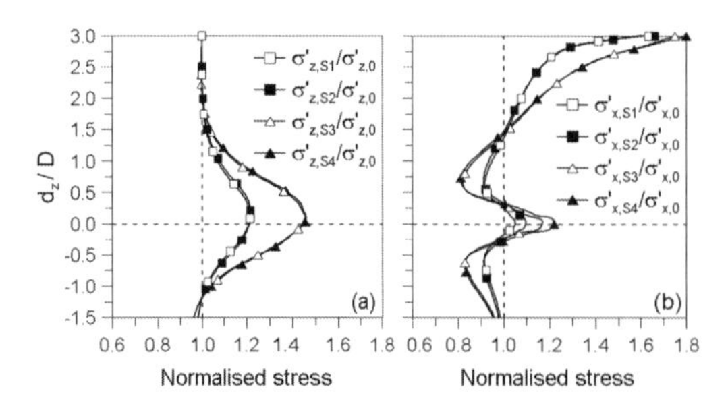

Figure 3. Normalised stresses along the vertical centreline of the pillar: (a) vertical stress; (b) horizontal stress.

$$\theta = \tan^{-1}\left[\frac{1}{\sqrt{3}} \cdot \left(2 \cdot \frac{(\sigma_2' - \sigma_3')}{(\sigma_1' - \sigma_3')} - 1 \right) \right] \tag{6}$$

with p' being the mean effective stress, θ the Lode's angle and c' and φ' the Mohr-Coulomb strength parameters (Table 1). The *MSL* can be considered as a direct measurement of the shear strength conditions of the soil. An extreme value of 1.0 implies that the soil strength has been fully mobilised and consequently the soil has yielded. In contrast, a value of 0 indicates that no shear is applied and the soil retains its full shear strength capacity.

The contours of *MSL* for the 2nd and 4th Stages of the reference analyses are depicted in Figure 4. In similarity with the results presented by Hoyaux & Ladanyi (1970) the excavation of the 1st Tunnel increases considerably the *MSL* on the lateral sides of tunnel (particularly at 45° in all directions from the tunnel centre), with some areas yielding due to the increase of the vertical stress and the decrease on the horizontal stress. In opposition, directly above and below the tunnel, a reduction of the *MSL* is observed due to the decrease of the vertical stress that becomes more similar to the horizontal stress. With the excavation of the 2nd Tunnel is possible to verify that the yielding area on the pillar increases considerably due to the interaction between tunnels. For the reference *L/D* of 1.0 only a small part of the pillar has not reached its ultimate capacity and failure was eminent if the lining was not applied. This result highlights the importance of the pillar width and of its stress conditions, since the increase of the *MSL* is directly related with higher movements and consequently higher risk of damaging buildings.

In Figure 5 is depicted the *MSL* obtained along the horizontal (a) and vertical (b) centrelines of the pillar for the four stages modelled. Superimposed in the figure is the initial *MSL*, which is uniform in the horizontal centreline (same depth) and increases in the vertical centreline due to the increase of the stresses with depth. The values of *MSL* on the horizontal centreline (Fig. 5a) show that with the first excavation (1st Stage) the area adjacent to the tunnel yields, while for greater distances the *MSL* tends to decay rapidly reaching the initial value for a distance of about $1.5 \cdot D$. It is interesting to note that after the installation of the support in the 2nd Stage the mobilised stress levels adjacent to the tunnel decrease considerably due to the increase of the horizontal stress (see Fig. 2).

The second excavation originates an almost mirrored distribution of the *MSL* along the pillar, with higher values near the tunnels and smaller at middle distance and immediately adjacent to the tunnels. However, the highest interaction does not occur at the horizontal centreline depth but at the depths of the crown and of the invert as can be seen in Figure 4 and also in Figure 5b). The analysis of the latter shows that near the ground surface the *MSL* did not changed significantly, remaining similar to the initial values in all stages. However, after the excavation of the 1st Tunnel, in a vertical extension of about 1.5·*D*, measured from the tunnel axis in both directions, it is possible to observe a substantial increase of the *MSL*, with the maximum values observed at the depth of the crown and invert. With the excavation of the 2nd Tunnel the *MSL* tends to increase considerably in that area and yielding is almost reached at the depth of the crown of the tunnel.

4 INFLUENCE OF THE PILLAR WIDTH

In order to further assess the relevance of the pillar width five additional analyses were performed with values of *L*/*D* of 0.25, 0.50, 2.0, 3.0 and 4.0. In these analyses all the other parameters were kept equal to those employed in the reference analysis.

The normalised vertical and horizontal stresses along the horizontal centreline of the pillar are plotted in Figure 6 for the final stage modelled. The analysis of the vertical stresses shows that interaction appears to occur for *L*/*D* smaller than 2.0 with an increase of about 1.2 observed at middle distance between tunnels. Only for higher values that increase becomes almost residual. In opposition, for values smaller than 2.0 significant increases in stresses are observed in the pillar. It is also interesting to note that the distribution of stresses changes with the decrease of *L*/*D* and instead of having an isolated stress peak near each tunnel a single but significant peak, reaching about 2.0, occurs at approximately mid-distance for a *L*/*D* of 0.50. That behaviour is caused by the small distance between tunnels (only 3 m) that does not allow for the formation of two individual peaks. Instead, the stresses originated by the second excavation are accumulated with those from the first excavation justifying the substantial increase observed. However, for the minimum *L*/*D* of 0.25 this behaviour is not observed and only a moderate peak near the 1st Tunnel is observed followed by a decay that culminates in a small reduction of stresses near the 2nd Tunnel. In this case, because the tunnels are spaced of only 1.5 m, the excavation of the 2nd Tunnel occurs precisely in the area where the stress concentration resulting from the 1st Tunnel was located (see Fig. 2a)), making its influence almost unnoticeable. The proximity between tunnels also results in the direct transmission of the stresses to the lining of the 1st Tunnel by arching, thus not overloading the soil in the pillar.

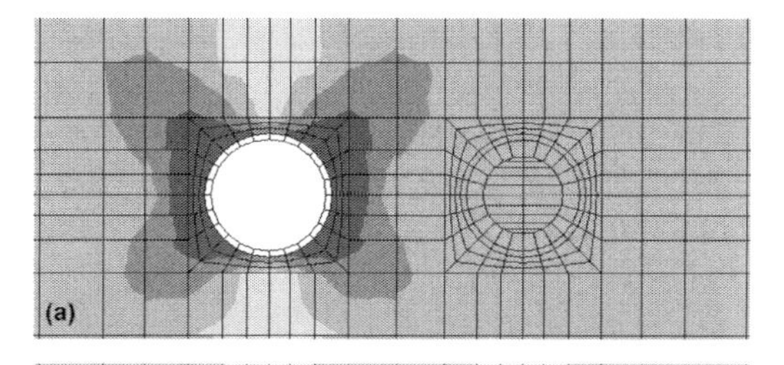

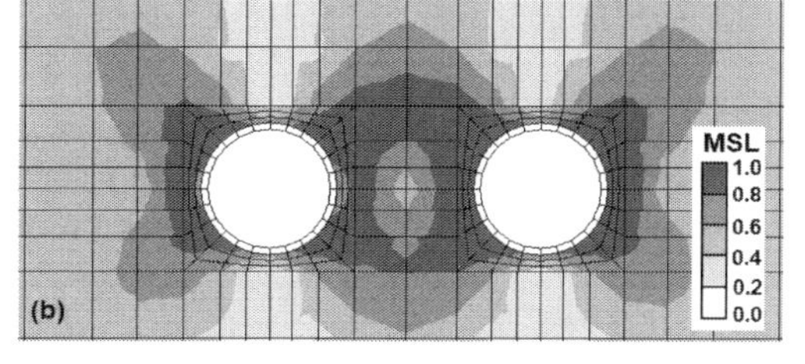

Figure 4. Contours of the *MSL* in the central pillar after the excavation: (a) 2nd Stage; (b) 4th Stage.

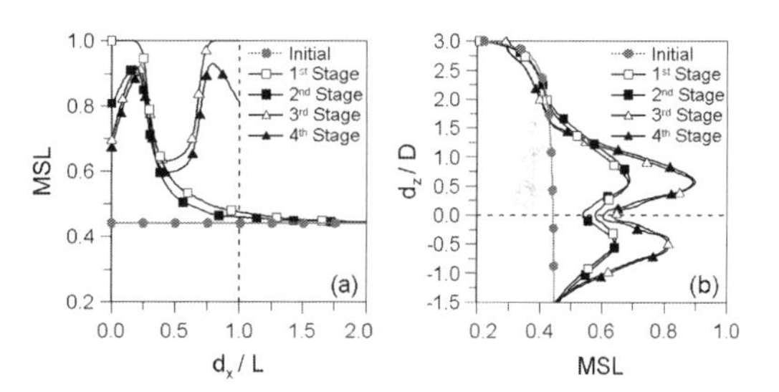

Figure 5. *MSL* along the centrelines of the pillar: (a) horizontal; (b) vertical.

Identical conclusions can be drawn by observing the changes in the horizontal stresses in Figure 6b. Also in this case the interaction between tunnels appears to be more relevant for *L*/*D* smaller than 2.0, for which a concentration of horizontal stresses occurs in the centre of the pillar. Once again, and for the same reasons mentioned previously, a different behaviour is observed for a *L*/*D* of 0.25, where a maximum concentration is observed adjacent to the 1st Tunnel and a minimum value near the 2nd Tunnel.

In order to highlight the influence of the 2nd Tunnel the changes solely caused by its excavation are depicted in Figure 7. The analysis of the figure confirms that interaction between tunnels mainly occurs for *L*/*D* smaller than 2.0. For higher ratios no significant changes in both vertical and horizontal stresses are noticeable near the 1st Tunnel and all stress variations occur in the proximity of the 2nd Tunnel. In contrast, for ratios smaller than 1.0 there is a significant interaction with the excavation of the 2nd Tunnel increasing significantly the stresses, particularly the horizontal, around the 1st Tunnel. These results also confirm that for a *L/D* of 0.25 the final stress distribution is mainly due to the excavation of the 2nd Tunnel.

The contours of the *MSL* for the two extreme cases of *L/D* considered (0.25 and 4.0) are plotted in Figure 8. The results show an almost symmetrical distribution of the *MSL* in both analyses. However, while for a *L/D* of 0.25 the entire pillar is almost yielded for a *L/D* of 4.0 the vast majority of the pillar remains with a *MSL* similar to the initial value, confirming that in this case both tunnels behave independently. In contrast, for a *L/D* of 0.25 significant interaction occurs with the outer area around the 2nd Tunnel showing higher values of *MSL*.

The results of the *MSL* along the horizontal centreline of the pillar are presented in Figure 9 for the 4 stages modelled in all analyses. After the 1st Stage it is possible to observe that the length of the pillar that yields is nearly the same in all cases and corresponds to about 1.2 m. Naturally, that corresponds to different percentages of *L/D* in the analyses due to the different pillar widths considered. For a *L/D* higher than 2.0 the *MSL* at middle distance of the pillar remains unchanged, suggesting an independent behaviour of the excavations. As mentioned previously, with the installation of the lining the *MSL* decreases in the area adjacent to the 1st Tunnel due to the increase of the horizontal stresses. With the second excavation the inverse behaviour occurs, with the increase of the *MSL* in the vicinity of the 2nd Tunnel. The results also confirm that for an *L/D* smaller than 2.0 a significant interaction between tunnels occurs with a substantial increase of the *MSL* near the centre of the pillar. For *L/D* smaller than 1 almost the entire pillar yields apart from the area adjacent to the 1st Tunnel, which presents smaller values. These are justified by the proximity of both tunnels which results on the transmission by arching of the stresses to the lining of the 1st Tunnel.

Based on the *MSL* results predicted at the vertical centreline of the pillar it is also possible to

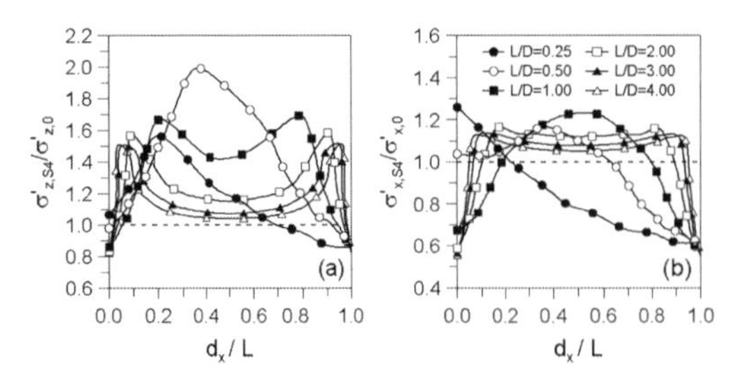

Figure 6. Normalised final stresses along the horizontal centreline of the pillar: (a) vertical stress; (b) horizontal stress.

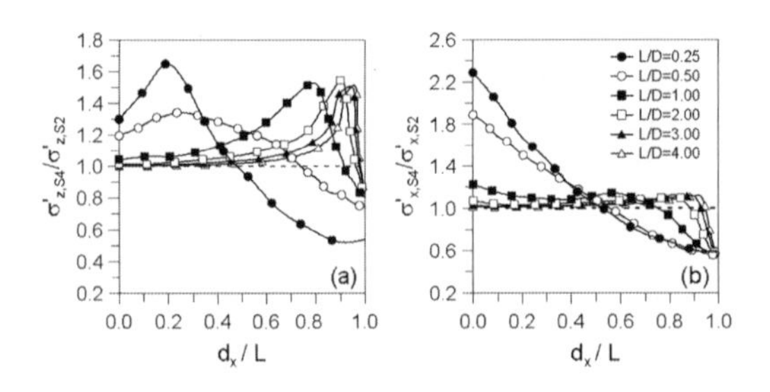

Figure 7. Change on stresses caused by the excavation of the 2nd Tunnel along the horizontal centreline of the pillar: (a) vertical stress; (b) horizontal stress.

confirm that interaction mainly occurs for *L/D* smaller than 2.0 (Figure 10). The zone affected by the excavation extends between 1.5·*D* above and 1.0·*D* below the tunnel axis, although near the ground surface, and particularly after the excavation of the 2nd Tunnel, some fluctuations occur with the analyses with a *L/D* smaller than 1.0 showing a decrease of the *MSL*, while for higher ratios the opposite occurs.

5 INFLUENCE OF THE INITIAL STRESS CONDITIONS

One factor that influence the stress redistribution that occurs during the excavation is the initial stress conditions. In order to assess the importance of this factor additional analyses were carried out by varying the K_0 value, while keeping the other parameters unchanged. In addition to the reference value of 0.6 four new values were adopted, 0.45, 1.0, 1.4 and 1.8. The analyses were conducted for three *L/D* scenarios, the reference value of 1.0 and the extreme cases of 0.25 and 4.0.

The vertical and horizontal stresses induced by the excavation of the 2nd Tunnel in the horizontal centreline of the pillar are plotted in Figure 11. The results show a significant variation of stresses for a *L/D* of 0.25 regardless of the K_0 considered. However, a different behaviour is observed in the vertical stresses of the analyses with a K_0 smaller than 1.0. These show a concentration of stresses near the 1st Tunnel, peaking at a distance of about 0.2*L*, followed by a significant decay that results in a reduction of up to 0.5 in the vertical stresses near the 2nd Tunnel. In contrast, for K_0 values higher or equal to 1.0 the stress variations mainly occur in the centre of the pillar with small influence near both tunnels. The changes in the horizontal stresses show a clearer pattern, increasing with the decrease of the K_0 value. Adjacent to the 1st Tunnel an increase in horizontal stresses of more than 2.0 is predicted for K_0 values smaller than

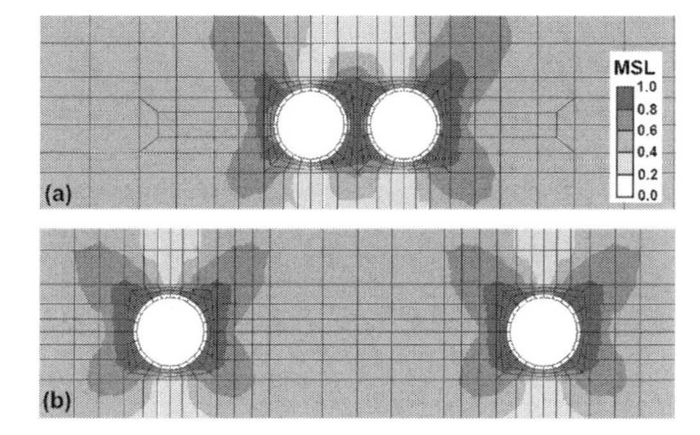

Figure 8. Contours of the *MSL* for the 4th Stage modelled: (a) L/D = 0.25; (b) L/D = 4.00.

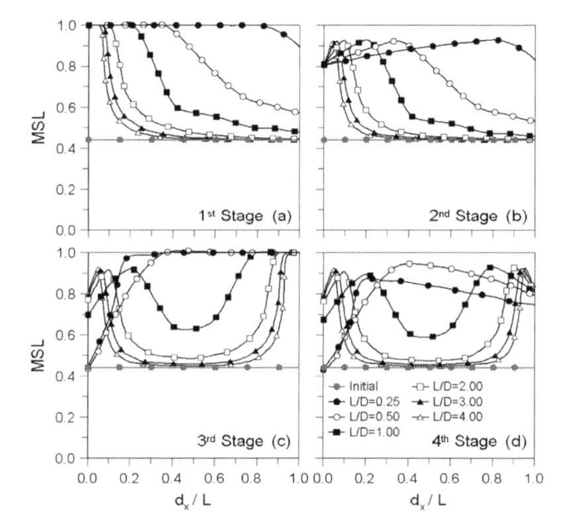

Figure 9. *MSL* along the horizontal centreline of the pillar: (a) 1st Stage; (b) 2nd Stage; (c) 3rd Stage; (d) 4th Stage.

1.0, while adjacent to the 2nd Tunnel a reduction of stresses occurs in all analyses, with a minimum of 0.4 obtained in the analysis with the highest K_0 value of 1.8. For a *L/D* of 1.0 a similar result is observed in the horizontal stresses, although in this case the highest variations obtained for the K_0 values smaller than 1.0 do not surpass 1.4 and only a reduction in stresses in the entire pillar is predicted for K_0 higher than 1.0. In terms of vertical stresses the principal variations occur near the 2nd Tunnel showing that the excavation of this tunnel does not affect significantly the 1st Tunnel, regardless of the K_0 value. However, near the 2nd Tunnel all analyses show a peak concentration of vertical stresses, with a maximum of 1.55 for a K_0 of 1.0, followed by a substantial decay which is more pronounced with the decrease of the K_0, even reaching a reduction of stresses for K_0 values smaller than 1.0.

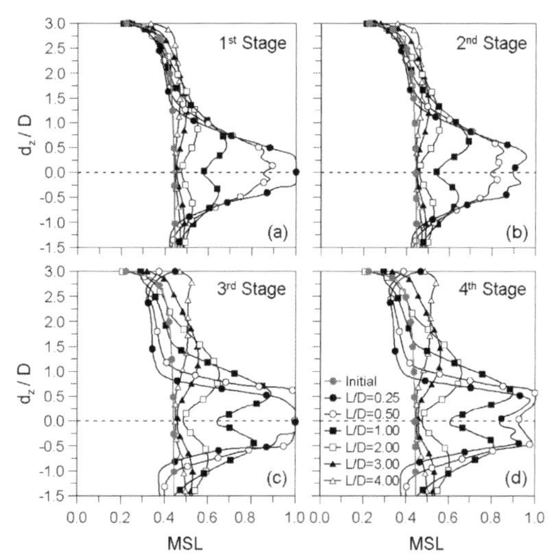

Figure 10. *MSL* along the vertical centreline of the pillar: (a) 1st Stage; (b) 2nd Stage; (c) 3rd Stage; (d) 4th Stage.

Finally, for the analyses with a *L/D* of 4 no visible interaction occurs with both vertical and horizontal stresses near the 1st Tunnel remaining approximately unchanged. In contrast the variations near the 2nd Tunnel are significant and vary depending on the K_0 value adopted. The vertical stresses present a behaviour similar but slightly amplified to that observed in the analyses with a *L/D* of 1.0, with an increase near the 2nd Tunnel, followed by a sudden reduction adjacent to the tunnel. In terms of the horizontal stresses it is possible to visualise that analyses with K_0 higher than 1.0 show a reduction of stresses in the entire pillar while for smaller values an increase followed by a sudden reduction near the 2nd Tunnel is observed.

Since the initial *MSL* is influenced by the K_0 value adopted it was decided to calculate a relative mobilised stress level, R_{MSL}, in order to simplify the comparison of the analyses. The normalised value was determined using Equation 7, where *MSL,0* is the initial mobilised stress value and *MSL,Si* corresponds to the *MSL* determined at stage *i* of the analysis.

$$R_{MSL} = 1 - \frac{1 - MSL,Si}{1 - MSL,0} \tag{7}$$

According with the proposed expression a R_{MSL} of 0 indicates that the *MSL* remains equal to the initial value, while a value of 1.0 corresponds to yielding of the soil. Negative values are also possible and indicate a reduction of the *MSL*.

In Figure 12 the R_{MSL} determined along the horizontal centreline of the pillar for the 2nd ($R_{MSL,S2}$) and 4th stages ($R_{MSL,S4}$) of the analyses with a L/D ratio of 0.25, 1.0 and 4.0 is presented. Naturally, the results obtained for the 2nd Stage are equal in all analyses (note that the difference is the L used in the normalisation of the x-axis) and show that an increase of the K_0 value tends to modify the behaviour of the R_{MSL}. For K_0 values higher or equal than 1.0 an increase of R_{MSL} is observed adjacent to the 1st Tunnel followed by a decay that for the highest values of K_0 considered (1.4 and 1.8) originates a reduction of the stress level.

The exact opposite occurs if the K_0 values are smaller than 1.0, with a reduction of the stress level adjacent to the tunnel followed by an increase. However, despite the K_0 value considered it is possible to observe that the R_{MSL} tends to the initial value for a distance of about $3.0{\cdot}D$, as can be seen in Figure 12c) for a d_x/L of 0.7. With the excavation of the 2nd Tunnel significant variations in the R_{MSL} occur, with the obtained results confirming the observations made about Figure 11. For a L/D of 0.25 the majority of the pillar yields for K_0 higher than 1.0. For smaller values smaller stress levels are mobilised due to the increase of the horizontal stresses. For a L/D of 1.0 the R_{MSL} generally increases in the pillar, and particularly near both tunnels, for the highest K_0 values, where yielding is observed in a small extension. Once again for K_0 smaller than 1.0 a lower R_{MSL} is observed for the same reasons, i.e. increase of horizontal stress. With the increase of the pillar width it is possible to observe that the R_{MSL} near the centre of the pillar tends to reduce considerably, inclusively reaching negative values (reduction of MSL) for the most extreme K_0 considered (i.e. 0.45, 1.4 and 1.8).

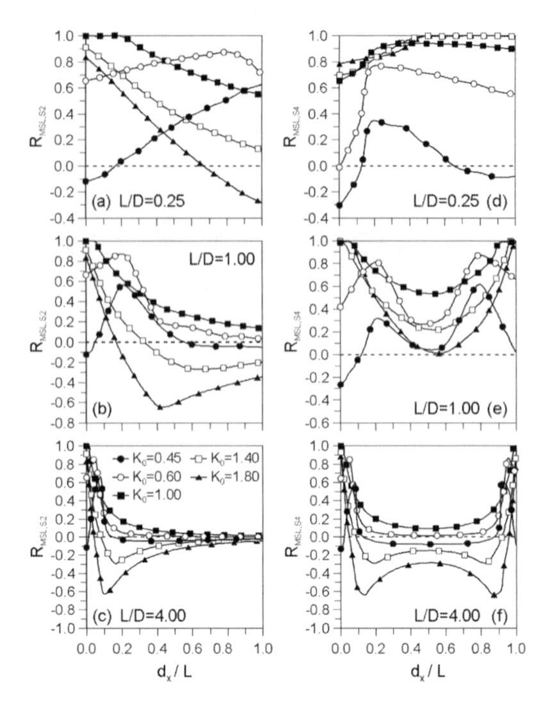

Figure 12. R_{MSL} along the horizontal centreline of the pillar: 2nd Stage (a) L/D = 0.25, (b) L/D = 1.0, (c) L/D = 4.0; 4th Stage (d) L/D = 0.25, (e) L/D = 1.0, (f) L/D = 4.0.

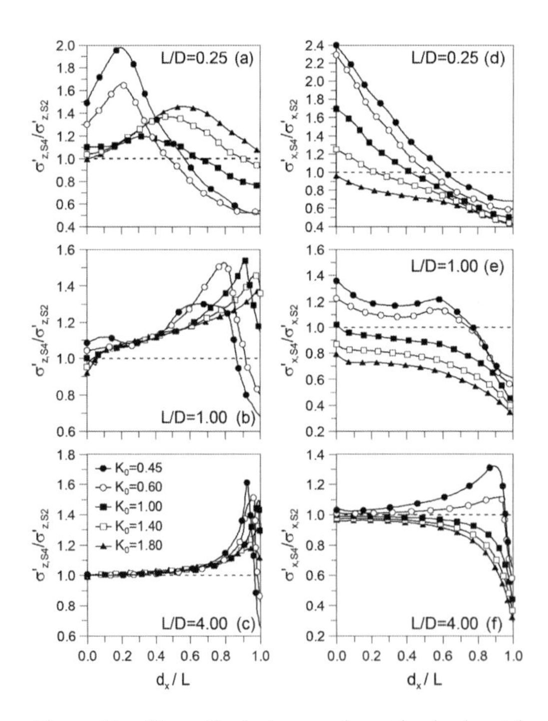

Figure 11. Normalised stresses along the horizontal centreline of the pillar: vertical stress (a) L/D = 0.25, (b) L/D = 1.0, (c) L/D = 4.0; horizontal stress (d) L/D = 0.25, (e) L/D = 1.0, (f) L/D = 4.0.

6 CONCLUSIONS

The congestion of the subsoil in big cities has led to the construction of new tunnels in close proximity with existing ones. In such scenarios it is acknowledged that the width of the pillar between tunnels is of critical importance, but the assessment of the stress redistribution and mobilised stress levels in the pillar has not been thoroughly investigated in the past. In this paper a numerical study is conducted in order to clarify these aspects. Based on the obtained results the following conclusions can be drawn:

- The interaction between tunnels is particularly important for L/D smaller than 2.0, with

the tunnels behaving almost independently for higher ratios.
- The excavation of the 2nd Tunnel originates an increase of stresses in the pillar, higher *MSLs* and even yielding of part of the soil in the pillar.
- For *L*/*D* smaller than 1.0 almost the entire soil in the pillar yielded, apart from the area adjacent to the 1st Tunnel, where, due to arching, the load was directly transferred to its lining.
- The area affected by the excavation extends almost 1.5·*D* above and 1.0·*D* below the tunnels axis.
- K_0 values smaller than 1.0 originate a substantial increase of stresses near the 1st Tunnel and a reduction near the 2nd Tunnel. For higher K_0 values more moderate variations of stresses are observed in the pillar, although with a similar trend. The *MSL* in the pillar also depends on the initial stress conditions, with K_0 smaller than 1.0 presenting smaller values, mainly due to the increase of the horizontal stress.

REFERENCES

Addenbrooke, T. I. & Potts, D. (2001) Twin Tunnel Interaction: Surface and Subsurface Effects. *International Journal of Geomechanics*, **1** (2), pp. 249–271.

Bartlett, J. & Bubbers, B. (1970) Surface movements caused by bored tunnelling. In *Proceedings of the Conference on Subway Construction* sl]: Budapest-Balatonfured, Vol. 539.

Cooper, M., Chapman, D., Rogers, C. & Hansmire, W. (2002) Prediction of settlement in existing tunnel caused by the second of twin tunnels. *Transportation Research Record: Journal of the Transportation Research Board*, (1814), pp. 103–111.

Cording, E. J. & Hansmire, W. (1975) Displacements around soft ground tunnels. In *Proceedings of the 5th Pan American Conference on Soli Mechanics and Foundation Engineering, Buenos Aires, Argentina*. Vol. 4, pp. 571–633.

Divall, S. & Goodey, R. J. (2015) Twin-tunnelling-induced ground movements in clay. *Proceedings of the Institution of Civil Engineers—Geotechnical Engineering*, **168** (3), pp. 247–256.

Elwood, D. E. Y. & Martin, C. D. (2016) Ground response of closely spaced twin tunnels constructed in heavily overconsolidated soils. *Tunnelling and Underground Space Technology*, **51** (Supplement C), pp. 226–237.

Gercek, H. (2005) Interaction between parallel underground openings. In *Proceedings of the The 19th International Mining Congress and Fair of Turkey*, pp. 73–82.

Ghaboussi, J. & Ranken, R. E. (1977) Interaction between two parallel tunnels. *International Journal for Numerical and Analytical Methods in Geomechanics*, **1** (1), pp. 75–103.

Hoyaux, B. & Ladanyi, B. (1970) Gravitational stress field around a tunnel in soft ground. *Canadian Geotechnical Journal*, **7** (1), pp. 54–61.

Janbu, N. (1967) *Settlement calculations based on the tangent modulus concept*. Trondheim, Technical University of Norway. pp. 57.

Kim, S. H., Burd, H. J. & Milligan, G. W. E. (1998) Model testing of closely spaced tunnels in clay. *Géotechnique*, **48** (3), pp. 375–388.

Koungelis, D. & Augarde, C. (2004) Interaction between multiple tunnels in soft ground. *School of Engineering, University of Durham, UK*.

Mair, R. J. & Taylor, R. N. (1997) Bored tunnelling in the urban environment. In *Proceedings of the 14th International Conference on Soil Mechanics and Foundation Engineering, State-of-the-art Report and Theme Lecture, Hamburg*. Balkema, Vol. 4, pp. 2353–2385.

Moller, S. (2006) *Tunnel induced settlements and structural forces in linings*. PhD thesis. University of Stuttgard, Stuttgard.

Nyren, R. (1998) *Field measurements above twin tunnels in London Clay*. PhD thesis. Imperial College London.

Pedro, A. M. G., Cancela, T., Almeida e Sousa, J. & Grazina, J. (2017) Deformations caused by the excavation of twin tunnels. In *Proceedings of the 9th International Symposium on Geotechnical aspects of Underground Construction in Soft Ground, Sao Paulo, Brazil*. pp. 10.

Potts, D. M. & Zdravković, L. (2001) *Finite element analysis in geotechnical engineering: application*. Thomas Telford. London.

Withers, A. D. (2001) *"Surface displacements at three surface reference sites above twin tunnels through the Lambeth Group" Building response to tunnelling*, pp. 735–754.

Numerical Methods in Geotechnical Engineering IX – Cardoso et al. (Eds)
© 2018 Taylor & Francis Group, London, ISBN 978-1-138-33203-4

Numerical analysis of interaction behavior of yielding supports in squeezing ground

A.-L. Hammer & R. Hasanpour
Institute for Tunneling and Construction Management, Ruhr-University Bochum, Germany

C. Hoffmann
DIANA FEA BV, Delft, The Netherlands

M. Thewes
Institute for Tunneling and Construction Management, Ruhr-University Bochum, Germany

ABSTRACT: Yielding elements are installed and embedded in the lining segments in tunneling through squeezing ground. This has been proved to be an effective measure to control convergence, provide stability and allow higher advancing rates. Although it is recognized that these practical measures are highly effective in providing tunnel stability, there is no well-established method to transfer and extend these beneficial effects into the design. This paper presents a series of numerical analyses where different modelling approaches are used, based on FE method. Results are presented and compared to those from the actual measurements. Special spring elements were included in the model to represent the yielding elements. A 2D plain strain approach is compared with a full 3D model, where a detailed construction sequence was established. Interfaces in the lining-rock contact and time evolution of the shotcrete properties were also included in the model.

1 INTRODUCTION

In conventional tunneling through squeezing ground, the use of a shotcrete support system with integrated yielding elements has been proved to be an economical and safe alternative in comparison to the use of a unified shotcrete (system) as a relatively rigid lining. Using yielding elements between a segmented lining allows a controlled deformation at tunnel boundaries while the forces acting on the young shotcrete is adjusted at the same time.

The high ground stresses in squeezing conditions usually exceed the support resistance of the shotcrete lining, therefore, employing yielding elements with controlled deformation can prevent a spalling failure of the applied shotcrete. Besides that, a uniform convergence at the tunnel circumferences can be created as a result.

Despite the fact that the theoretical concept of this type of combined support systems has not been fully recognized yet, the numerical study of the supporting mechanism of shotcrete lining, combined with various types of yielding elements under high ground pressure, is the state-of-the-art,. There is also no standardized procedure regarding computational analysis for integrating yielding elements into numerical calculations, and only a few studies have been published related to this topic. For example, John & Poscher (2004) conducted some numerical calculations at the design stage of the Strenger Tunnel in which yielding elements are simulated using separated beam elements by means of limiting their normal forces and bending moments. Brandtner & Lenz (2017) have used a non-linear load-displacement relationship of structural beam elements in order to simulate the mechanical behavior of the yielding support systems into a computational model. Radoncic (2011) and Barla et al. (2011) used an elastic-perfectly plastic Mohr-Coulomb material model for yielding elements modeling; the stiffness and strength values of the elements were obtained from the load-deformation curve. Further numerical calculations can be found in the studies by John & Poscher (2004), Cantieni (2011) and Likar et al. (2013). However, the modeling of the yielding elements is not described explicitly in these studies.

This paper presents an attempt to implement the load-deformation behavior of yielding elements based on experimental results through the numerical modeling of a tunnel construction using conventional methods through squeezing ground. In this case, the yielding elements combined with the shotcrete lining, the steel ribs, and the rock bolting have been used as the tunnel support system. The Tauern Tunnel, in Austria, was considered for this purpose, with results validation. At first, the comparison of the results from various approaches with the measured data is performed using a simplified 2D model by means of applying spring elements into the 2D model. Once the multilinear load-deformation behavior of the yielding elements is simulated appropriately with respect to the data from the laboratory tests, the same simulation procedure is integrated into a comprehensive 3D model and a detailed 2D model for the modeling of yielding elements at the next stage.

Two numerical models are described in the following sections of this study. Both models are performed with the DIANA software. The first model is based on a plain strain analysis, that is simple to be applied for the studied cases, and, that also needs a low computational time. This makes it appropriate for design purposes and comparative analysis of the different support components. Then, a comprehensive three-dimensional model was employed for a detailed investigation of the stress histories within tunneling and for the effect on tunnel advance rate on magnitude for the convergences. The application of a 3D model would be very useful for detailed investigations, as soon as a detailed knowledge about rock mass properties and support is available.

2 CONVENTIONAL TUNNELING IN SQUEEZING GROUND

2.1 *Squeezing ground*

Squeezing behavior as defined by the International Society of Rock Mechanics (ISRM) is "time-dependent large deformation occurring around a cavity" (Barla 2016). This ground behavior can be observed in rock types with creep behavior, low strength, and high ductility. The ground convergence due to the squeezing behavior may occur during excavation or continue for a long period of time. Rock properties, groundwater table, pore water pressures and ground in-situ stresses are the eomechanical factors influencing the squeezing behavior of rocks. Also dependent on the influence of these geomechanical properties are the magnitude and rate of the convergence, as well as the size of the plastic zones around the tunnel excavated through the squeezing ground.

2.2 *Applying ductile supports in tunneling through squeezing ground*

There are two different assumptions for supporting rock mass in a tunnel construction, including an active and a passive one. Through the active approach, also known as the resistance principle, attempts are made to prevent rock deformation by applying rigid support system. In this method, the installed support system must be able to absorb or resist against the applied rock pressures without damage that is usually unsuccessful, particularly when a shotcrete lining is used in tunneling through squeezing ground with high ground stresses.

Through the passive approach (yielding principle), in contrast, deformations are allowed to occur, leading to a decrease in the amount of redistributed stresses at the tunnel boundaries. For this purpose, a set of over-profiles (gap) is created longitudinally into the main lining system with specified distance between each other. The size of the gaps is designed with respect to the expected tunnel convergence at the tunnel circumference (Schneider & Spiegl 2015). Some types of yielding elements are installed in the gap between the lining segments to keep the unity of the support system and prevent bending moments.

Nowadays, yielding elements, which are integrated between segments of the shotcrete lining, are frequently utilized as a passive support system in conventional tunneling through squeezing and difficult grounds. The method has been applied at several tunneling projects, such as the Strenger Tunnel (Radoncic 2011), Tauern Tunnel (Weidinger & Lauffer 2009), Brenner Base Tunnel (Poisel et al. 2017), and Koralm Tunnel (Schubert et al. 2010). Figure 1 illustrates schematically the concept of the passive supporting, which consists of shotcrete lining segments and yielding elements placed in the gap between these segments. This is followed by an installing radial rock bolting and lattice girders.

The stiffness of yielding elements provides the mobilization of the support resistance during the time the initial deformations are taking place. The prerequisite for this is that the amount of stress applied to the shotcrete lining lies below the shotcrete strength at all times. Therefore, the initial resistance of the yielding elements has to be predesigned with a low initial stiffness, considering that the young shotcrete has a low strength initially. Furthermore, the load-deformation behavior of the yielding elements should be based on the strength development of the shotcrete to make the best possible use of the mobilized support capacity (Wiese 2011, Radoncic 2011).

2.3 *Yielding elements*

The current applied yielding elements in tunneling industry can be divided into two groups: a) steel elements and b) porous elements based on binding materials. Figure 2 shows yielding elements:

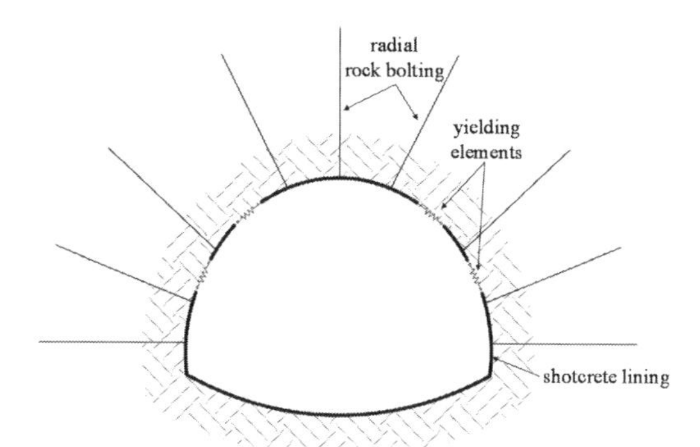

Figure 1. Concept of passive supporting in squeezing ground.

the steel type, including LSC and Wabe; and the porous type, including hiDCon. The difference between various types of yielding elements can be explained in their structure and their response to the applied load.

The LSC yielding element (Dywidag-Systems International) consists of a set of steel tubes that are fixed at the top and bottom by means of two rigid plates. Guide tubes are installed concentrically around the yielding tubes, which limit the outward and inward displacements of the yielding tubes. The yielding tubes are subjected to axial loads by applying it to rigid plates. By loading the yielding tubes, the rotationally symmetrical buckling and the post-buckling process can be observed during deformation of these tubes. The maximum achievable deformation is defined by the length of the steel tubes. (Moritz 1999, 2011).

The beam-shaped hiDCon yielding elements (High Deformable Concrete), which has been developed by Solexperts Company, consist of a high-strength concrete with a porous aggregates matrix, like gypsum or glass foam particles. The required compound of the support system is provided by using special reinforcement levels in form of plates, ring reinforcements and adding steel fibers. The hiDCon yielding elements can be produced for a specific project by making modifications of its geometry, reinforcement and material composition. Under uniaxial compressive loading, the hiDCon elements can absorb a displacement up to 50% of its height showing plastic behavior in the consolidation phase (Anagnostou & Cantieni 2007, Kovári 2009, Stolz & Steiner 2010).

The Wabe yielding elements, developed by Bochumer Eisenhütte Heintzmann, consist of some circular hollow steel profiles that are connected in layers by means of intermediate plates. The cavities between a honeycomb structure and plates provide enough space for the deformation of the support. Then, the flexibility of the yielding element is

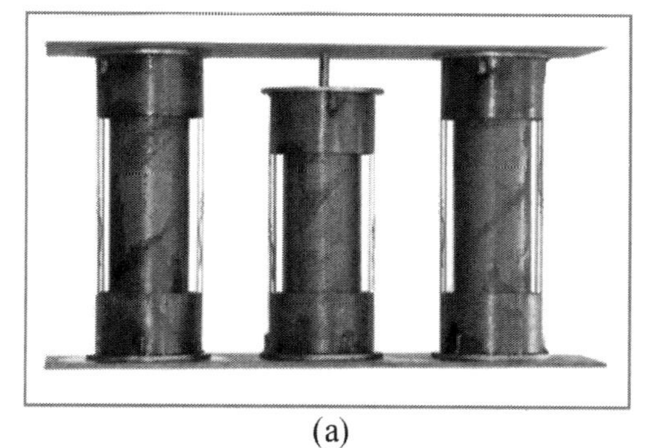

(a)

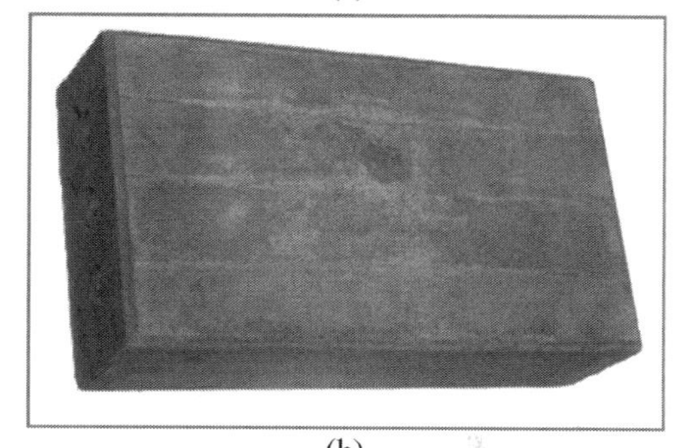

(b)

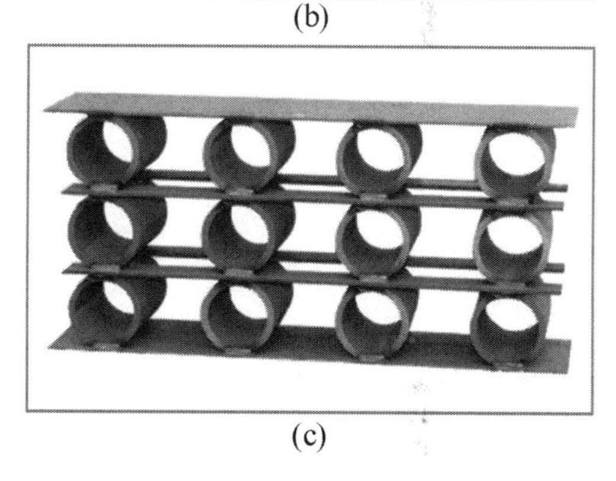

(c)

Figure 2. Various types of the applied yielding elements in tunneling a) LSC b) hiDCon c) Wabe.

entirely related to the sum of the inner diameter of the steel tubes as a geometrical factor, and to the resistance or yielding point of the steel as material characteristics. Depending on the ground behavior, further steel tubes can be inserted into the cavities of the yielding elements to increase its load-bearing capacity, even after partial deformations have taken place. The various load-bearing capacities can be achieved by inserting additional tubes with different diameters and thicknesses (Podjadtke 2009).

Opolony et al. (2011) and Wiese (2011) conducted several experimental tests on common types of yielding supports using a uniaxial compressive test under the same boundary conditions to characterize the system-specific property and the load-deformation behavior of yielding elements. The load-deformation behaviors that resulted from these investigations are used here in numerical simulation of different kinds of yielding elements, performing comparative investigations.

3 NUMERICAL MODELLING

3.1 *General model description*

In this study, a 2D plain strain model and a full 3D model of a tunnel construction in squeezing ground were developed and, relevant results, compared (see Figure 3). Two different numerical models were used due to the fact that only limited information about the characteristics of rock mass exists at the design stage of a tunnel with high overburden. Correspondingly, using simple numerical calculation models in plain strain mode would be more useful for initial prediction of displacements and stresses around the tunnel using rough knowledge on rock mass properties.

Nevertheless, the complex numerical calculation that takes into account the plastic or time-dependent material behavior of rock mass can be applied when a detailed knowledge about the structure of the rock mass is available, such as structural layers and discontinuities. Of course, this information can be obtained with tunnel face mapping along the excavation.

In modeling using the plain stress approach, the construction sequence is considered perpendicular to the direction of the excavation before, during and after installation of the support. During tunneling, the ground is mainly loaded to transverse in the longitudinal direction. The transverse load-bearing effect of the rock mass acting in the considered plane is taken into account for the calculations. The longitudinal load-bearing effect, in which the rock mass is supported lengthwise on the tunnel lining and the rock mass in front of the face, is taken into account by residual load factors. An approximate transfer of the three-dimensional to the two-dimensional stress state is possible by modeling the excavation in the first calculation phase by deactivating the elements within the tunnel cross-section and reducing the primary stresses by using residual forces.

3.2 *3D model approach*

To analyze the interaction between tunnel supports and examine the response of the support system with the adopted yielding elements, the developed full 3D model in DIANA (see Figure 4) was implemented in two steps, as follows:

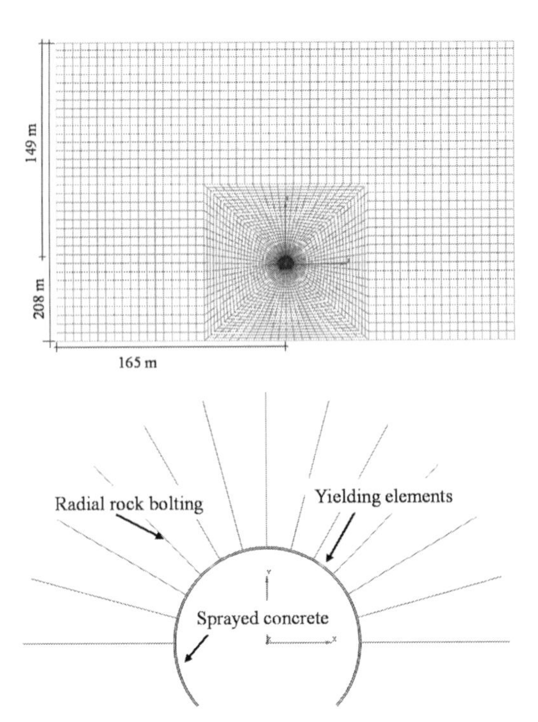

Figure 3. Discretization of the 2D model.

- Block modeling, entering ground properties, definition of ground behavior model, assigning boundary conditions, application of in-situ stress and solving the unexcavated model for damping the unbalanced forces;
- Modelling tunnel advance using step by step excavation, applying tunnel support systems such as steel ribs, rock bolts, and shotcrete by considering their mechanical constitutive models, deploying interface elements for simulation of interactions between various support systems, and modeling the yielding support using special interface elements.

The lining-rock interfaces and time-dependent properties of the shotcrete are also considered in the developed 3D model.

3.3 *Modeling of the rock mass*

Green phyllite with interposed anhydrite was mainly observed in the investigated area of the Tauern Tunnel. During tunneling, the problem of water ingress into the tunnel has not been observed, therefore the influence of groundwater was neglected. The Hoek-Brown failure criterion model was used for modeling the rock mass. The in-situ stress ratio was assumed to be 0.64, based on the information from the studied case. The

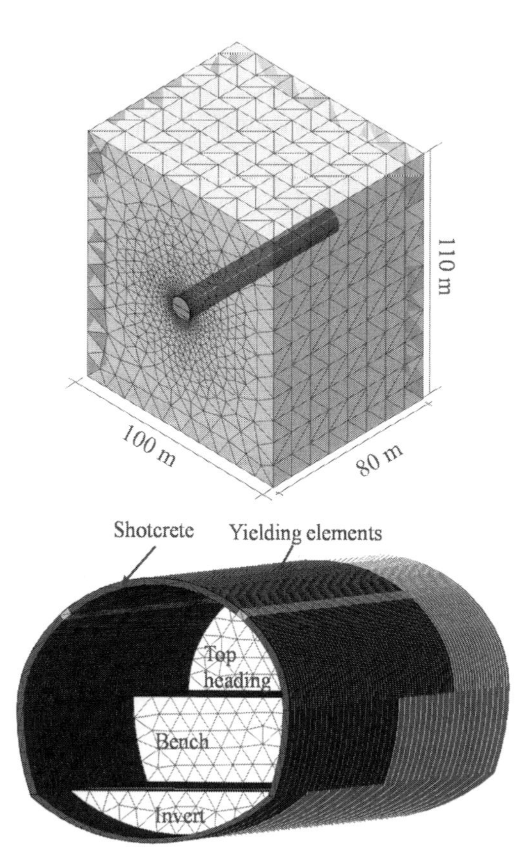

Figure 4. Discretization of the 3D model with excavation steps.

ground parameters used in the numerical investigations are listed in Table 1. Moreover, the contour of the ground stresses in the vertical direction and before the tunnel excavation for the initial state of modeling procedure is shown in Figure 5. This allows damping and redistribution of the ground stresses regarding the applied in-situ stresses observed in the field. According to the Figure 5, the ground stresses surrounding the tunnel were calculated as 24 MPa, which meets the observed one in the Tauern Tunnel for the depth of 905 m.

3.4 *Modeling of the shotcrete*

The shotcrete is simulated via a linear-elastic material model in which the elastic modulus of the shotcrete is gradually increased regarding its time-dependent behavior. In this paper, the increase in elastic modulus was simulated based on the examined data from laboratory tests that have been carried out at the Institute for Tunneling and Construction management at Ruhr-University Bochum. Figure 6 shows the mean values of the

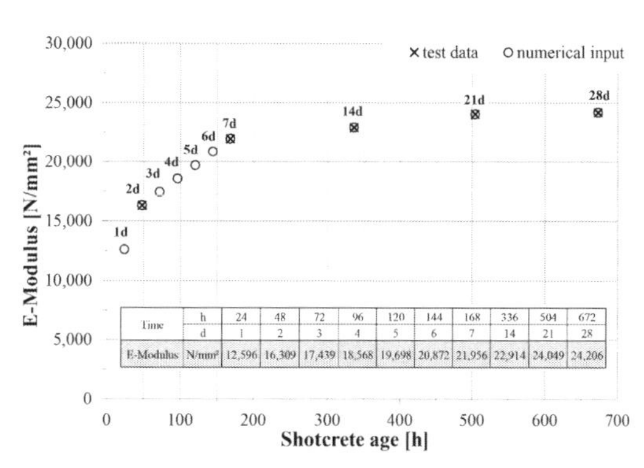

Time	h	24	48	72	96	120	144	168	336	504	672
	d	1	2	3	4	5	6	7	14	21	28
E-Modulus	N/mm²	12,596	16,309	17,439	18,568	19,698	20,872	21,956	22,914	24,049	24,206

Figure 5. Contour of ground stresses in the vertical direction in the block model and surrounding tunnel.

elastic modulus from three examinations at times of 2, 7, 14, 21 and 28 days.

As it can be seen in Figure 6, the increase in the elastic modulus is considerable in the first seven days of applying shotcrete, changing daily. Thereafter, it remains relatively constant for one week, as the increase in elastic modulus no longer changes significantly. The implementation of changing the elastic modulus of shotcrete in the numerical modeling assumes3 m/day of tunnel advance rate, as an average achievement for the reference project, in the construction of the second tube of the Tauern Tunnel.

3.5 *Modeling of the rock bolts and steel ribs*

The rock bolts and the steel ribs are modeled as embedded reinforcements, into the ground and shotcrete, respectively. They are defined in the model regarding their locations, material properties and dimensional parameters.

3.6 *Simulation of interfaces between ground and lining*

The shotcrete integrated by yielding elements represents a composite system behavior during its interaction with the rock mass. A question that

Table 1. Ground parameters used in numerical investigations.

Rock mass parameters	Value	Unit
GSI	40	-
m_i	8	-
Uniaxial compressive strength	40	MPa
Elastic modulus of the intact rock	3.56	GPa
Unit weight	27.5	kN/m^3
In-situ stress	24.0	MPa

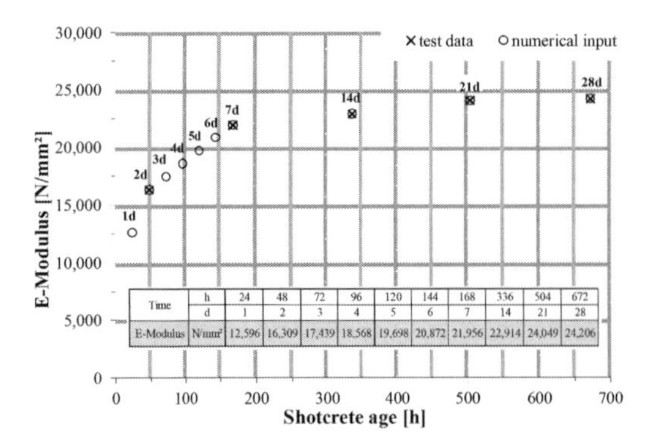

Time	h	24	48	72	96	120	144	168	336	504	672
	d	1	2	3	4	5	6	7	14	21	28
E-Modulus	N/mm²	12,596	16,309	17,439	18,568	19,698	20,872	21,956	22,914	24,049	24,206

Figure 6. Mean values of the measured elastic modulus versus shotcrete ages used in the numerical modeling.

arises is how the tangential lining shortening, which is allowed by yielding elements, affects the composite behavior of the shotcrete and the rock mass at the interaction surfaces, in these particular locations. In this study, this kinematic peculiarity of the overall support system was taken into account by simulating the interaction surfaces between the rock mass, the shotcrete, and the gaps between the yielding elements and the rock mass, by means of two different interface models. By varying the interfacial properties in terms of normal and shear stiffness, it would be possible to analyze either the behavior of the whole composite system or the sliding behavior in the interaction surfaces for the corresponding system of the overall configuration.

3.7 *Modeling of load-deformation behavior of yielding elements*

The interaction between shotcrete and yielding element governs the concept of passive support of a flexible shotcrete lining with integrated yielding elements. When implementing the yielding elements in the computational models, the attempt should be accurately focused on defining the initial rigidity of the yielding elements and, at the same time, simulating the time-dependent strength development of the shotcrete as real as possible. If the initial rigidity of the yielding elements was considered high, this would lead to an early-stage overloading of the shotcrete lining.

In literature, the implementation of yielding elements in the numerical calculations has been studied through various approaches. In this study, the experimental test results, presented by Wiese (2011) and Opolony et al. (2011), are used in a distinct model for simulating the load-deformation behavior of the yielding elements individually. For this purpose, the spring elements, which are fixed between two rigid plates, are proven to be the best simulation method for modeling of the yielding support (see

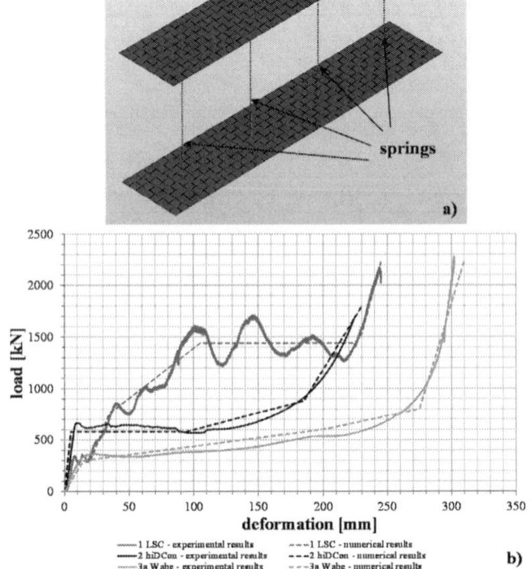

Figure 7. a) Numerical modeling of yielding supports using spring elements, b) The load-deformation curves of three types of yielding support systems obtained from numerical calculations and experimental tests (dashed curves).

Figure 7a). The springs are defined by their normal and shear stiffness, as well as assigning the pair values from the multilinear load-deformation curve obtained from the experimental data as shown in Figure 7b. The experimental tests have been carried out at the Institute for Tunneling and Construction Management (Ruhr-University Bochum).

Figure 7b also depicts the mean load-deformation characteristics of the three tested yielding systems and compares them with the results determined using numerical calculations. In an iterative process, the numerical results are compared to the experimental outcome and the final multilinear behavior is decided to be adjusted in a full 3D model.

3.8 *Modelling of tunnel construction stages*

The simulation of tunnel construction stages and excavation phases has been adjusted regarding information from the Tauern Tunnel. Figure 8 illustrates the excavation and construction stages that are considered in this study for simulation of tunnel advance and excavation process. As it can be seen in Figure 8, at a first stage of tunneling, the 50 meters of the top heading is mined (1 to 50 meters) with one meter

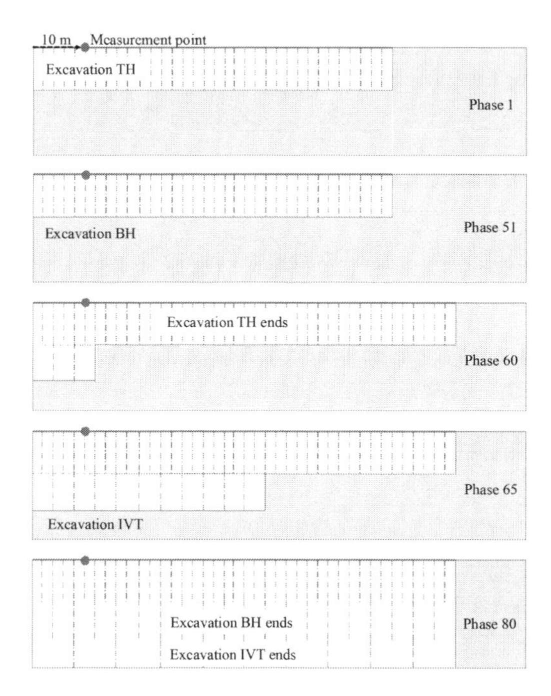

Figure 8. Simulation of the tunnel construction stages and excavation phases adjusted regarding information form Tauern Tunnel.

solving step. The average advance rate is supposed to be 3 m/d. The elastic modulus of the shotcrete lining is gradually increased according to its time-dependent behavior and tunnel advance rate, as explained in Section 2. It was also observed that the radial deformations and redistribution of ground stresses around the tunnel boundaries are fixed after 50 m excavation of the top heading. Hence, it was decided to commence excavating the tunnel bench after the 50th phase. The solving step for the tunnel bench was defined two meters with an excavation rate of 6 m/d.

After 60 meters, it was simulated a completion of the top heading excavation while the excavation of the tunnel bench was still ongoing. The tunnel invert was excavated on phase 66 with solving a 4 meters step. The excavation stages are completed at phase 80, after 60 m excavations of the bench and the invert. After ten meters of the Phase 1 excavation, the influence of a step-by-step tunnel excavation on deformation values and redistribution of the stresses was evaluated with respect to a reference point.

4 RESULTS AND DISCUSSION

4.1 *Comparison of the results with measured data*

To validate numerical modeling, it is attempted here to compare the results from the simulated model

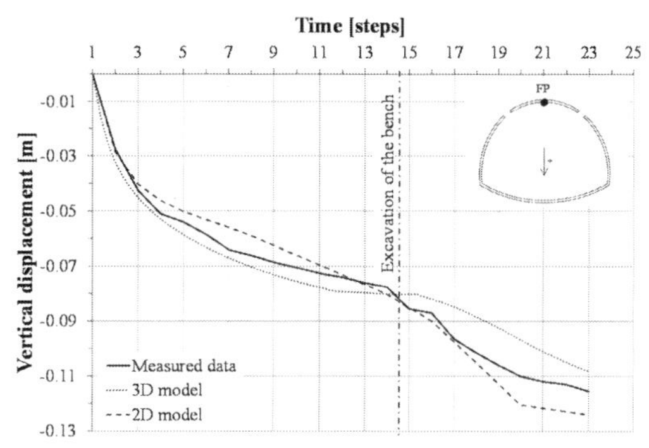

Figure 9. Comparison of the calculated displacements versus tunnel advancing for computational results and measured values.

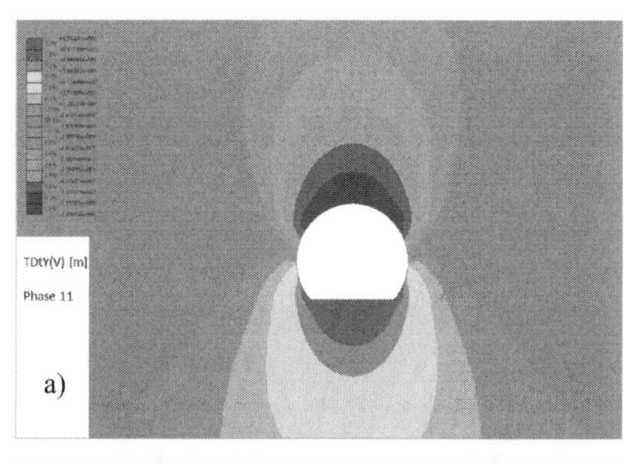

Figure 10. Contour of the displacements in vertical direction for a) 2D model b) 3D model.

with the measured data. The comparison between vertical displacements and measured deformations for the Tauern Tunnel is given in Figure 9, proving that there is a good agreement between the results.

In addition, the displacements as contours are given for both 2D and 3D models. It can be seen that the maximum convergence occurred at the crown in both models, however, there is a little, but acceptable, difference between the calculated results (see Fig. 10).

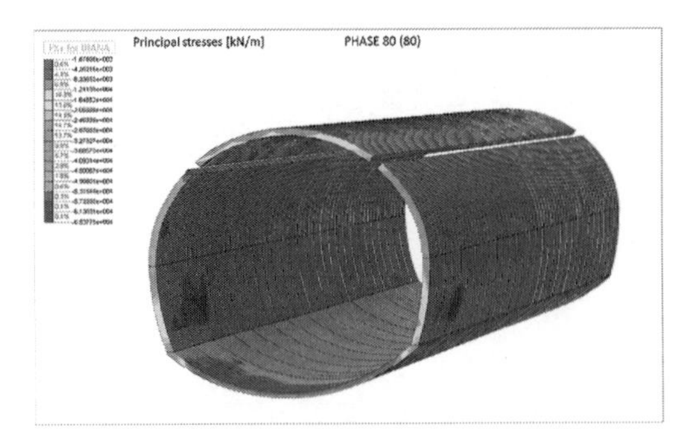

Figure 11. Contour of the mean stresses around the support system and the yielding elements.

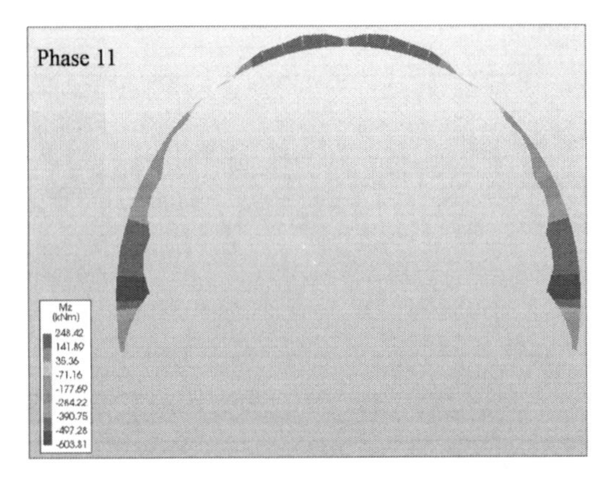

Figure 12. Applied moments (M_z) on the lining system.

The measured values on the applied ground stresses on the tunnel support system have not been reported on behalf of the construction companies at the Tauern Tunnel, comparing them with the findings from computational calculations. The convergences around the tunnel may be considered for the evaluation of the impact of applying yielding elements on allowing the ground to be deformed until a certain amount. The deformation values at the tunnel circumference can be compared with the measured data of the case study. In this study, the calculated displacements using 2D and 3D FEM models, are in a very good agreement with the measured deformations at the Tauern Tunnel.

The redistribution of the stresses around the tunnel after installing yielding supports is given by Figure 11. The applied mean stresses around the tunnel decreased considerably. That proves that the yielding elements were simulated appropriately according to their load-deformation behavior, which allows the ground to be deformed. It also leads to a gradual decrease of the applied ground stresses on the whole support system simultaneously with the occurrence of ground convergences.

The interaction between simulated yielding elements and the shotcrete system can be illustrated by means of representing the numerical results in terms of applied axial forces and the moments on the shotcrete lining (see Figure 12).The maximum applied moments were calculated about 604 kNm at the tunnel sidewalls (Figure 12).

5 CONCLUSIONS

This study presents a series of numerical analyses based on the program DIANA, in which different modeling approaches were used. The attempt in this research was to realistically model the complex interactions between the integrated tunnel support system with the yielding elements for a tunnel construction through squeezing ground. For this purpose, the merits of the developed models, which consider the real behavior of the applied tunnel supports, can be described as follows:

- The explicit description of the time-dependent material behavior of shotcrete, which has not been simulated accurately in the recent studies. This might be due to a lack of enough experimental or measured data for validation and adaptation of the numerical models. Hence, the experimental studies on the time-dependent material behavior of shotcrete are intended to support the consideration of higher-quality material models. In this study, the time evolution of the shotcrete properties was included in the model with respect to the experimental results carried out at Ruhr University in Bochum.
- The real behavior of yielding elements should be applied in the numerical calculation to achieve reliable results. In this study, the special spring elements, which represent the load-deformation behavior of yielding elements appropriately, were included in the developed model between segments of the shotcrete lining using data from experiments.
- The installed steel ribs and rock bolts were modeled regarding their material properties from the studied field.

Results were presented and compared to those from the actual measurements. Moreover, a 2D plain strain approach was compared with a full 3D model where a detailed construction sequence was established. The results also prove that the developed models can be used in the design stage of a conventional tunnel excavation in difficult ground. They can also be used for detailed investigations if an explicit knowledge of ground formations and geological uncertainties are available.

REFERENCES

Anagnostou, G. & Cantieni, L. 2007. Design and analysis of yielding support in squeezing ground. *11th Congress of the ISRM*. Lisbon, Portugal, 09–13. July, pp. 829–832.

Barla, G. 2016. Challenges in the Understanding of TBM Excavation in Squeezing Conditions. *16th ISRM Online Lecture. International Society for Rock Mechanics.* http://www.isrm.net/gca/?id = 1276.

Barla, G., Bonini, M., Semeraro, M. 2011. Analysis of the behavior of a yield-control support system in squeezing rock. In: *Tunnelling and Underground Space Technology* 26 (1): 146–154.

Barla, G. 2001. Tunnelling under squeezing rock conditions. In: D. Kolymbas (ed.): *Tunnelling mechanics, Eurosummer School*. Innsbruck: Logos Verlag: 169–268.

Brandtner, M. & Lenz, G. 2017: Checking the system behavior using a numerical model. *Geomechanics and Tunneling 10 (4)*: 353–365.

Cantieni, L. 2011. Spatial effects in tunneling through squeezing ground. *Ph.D. Thesis*. ETH, Zurich, Switzerland.

John, M., Spöndlin, D., Mattle, B. 2004. Lösung schwieriger Planungsaufgaben für den Strenger Tunnel. *Felsbau* 22 (1): 18–24.

Kovari, K. 2009. Design Methods with Yielding Support in Squeezing and Swelling Rocks. In: Kocsonya P. (ed.): *Safe Tunnelling for the City and for the Environment. ITA-AITES World Tunnel Congress*. Budapest, Hungary, 23–28 May.

Likar, J., Marolt, T., Likar, A. 2013. Adequacy of the yielding elements selection for underground construction in high squeezing grounds. *Proc. of the 12th international conference underground construction*, Prague, Czech Republic.

Moritz, B. 1999. Ductile Support System for Tunnels in Squeezing Rock. *Ph.D. Thesis*. University of Technology, Graz, Austria.

Moritz, B. 2011. Yielding elements—requirements, overview and comparison. In: *Geomechanics and Tunnelling* 4 (3): 221–236.

Opolony, K., Einck H.-B., Thewes, M. 2011. Testing of yielding elements for ductile support. *ITA-AITES World Tunnel Congress*. Helsinki, Finland.

Podjadtke, R. 2009. Entwicklung und Einsatz von stählernen Stauchelementen—Das System WABE im modernen Tunnelbau. In: *Felsbaumagazin* (2): 84–89.

Poisel, A., Weigl, J., Schachinger, T. Vanek, R. & Nipitsch, G. 2017: Semmering Base Tunnel—Excavation of the emergency station in complex ground conditions. In: *Geomechanics and Tunnelling* 5 (10): 458–466.

Radoncic, N. 2011. Tunnel design and prediction of system behaviour in weak ground. *Ph.D. Thesis*. University of Technology, Graz, Austria.

Schneider, E. & Spiegl, M. 2015. Nachgiebiger Ausbau für druckhaftes Gebirge. In: Deutsche Gesellschaft für Geotechnik e.V. (ed.): *Tunnelbau 2015*. Berlin: Ernst & Sohn: 230–256.

Schubert, P., Hölzl, H., Selner, P. & Fasching, F. 2010. Geomechanical knowledge gained from the Paierdorf investigation tunnel in the section through the Lavanttal main fault zone. In: *Geomechanics and Tunnelling* 3 (2): 163–173.

Stolz, M. & Steiner, P. 2010. Der Einsatz von hochdeformierbaren Betonelementen beim Tunnelbau in druckhaften und quellfähigen Gebirgsverhältnissen. *7. Kolloqium, Bauen in Boden und Fels*. Stuttgart, Germany, 26–27 January: 1–5.

Weidinger, F. & Lauffer, H. 2009. The Tauern Tunnel first and second tubes from the contractor's viewpoint. In: *Geomechanics and Tunnelling* 2 (1): 24–32.

Wiese, A.-L. 2011. Vergleichende Untersuchungen von Stauchelementen für den Einsatz in druckhaftem Gebirge. Berlin, Germany. STUVA-Tagung: 148–156.

Numerical Methods in Geotechnical Engineering IX – Cardoso et al. (Eds)
© 2018 Taylor & Francis Group, London, ISBN 978-1-138-33203-4

A practical tool for the preliminary estimation of stability of underground quarries excavated in jointed chalk layers of North France

F. Rafeh
Department of Civil Engineering and Constructions, Issam Fares Faculty of Technology, University of Balamand, Lebanon

H. Mroueh
Laboratory of Civil Engineering and geo-Environment (LGCgE)—Polytech'Lille, University Lille1 Sciences and Technologies, Villeneuve d'Ascq, France

S. Burlon
Institut Français des Sciences et Technologies des Transports, de l'Aménagement, et des Réseaux (IFSTTAR), Marne La Vallée, France

ABSTRACT: The presence of huge surfaces in North France under-lied by unexploited quarries constitutes a risk of damage on both people and constructions. In this paper, a stability analysis of quarries, excavated in jointed chalk substratum of North France by the method of rooms and pillars, is presented. Using geometric data collected from site: width of the pillar W, height of the pillar H, span between pillars L, and the inclination of joints θ, a parametric numerical stability study is conducted. Factors of safety $Fs_{(num)}$ are numerically computed using a classical shear strength reduction technique. A correlated regression technique between $Fs_{(num)}$ at one hand and W, H, L, and θ at the other hand, is integrated for the development of extrapolated formulae factor of safety $Fs_{(ex)}$. These proposed formulae designate a practical tool for the rapid preliminary estimation of the stability state of jointed chalk quarries having known only few geometric data.

1 INTRODUCTION

The presence of hundreds of unexploited underground quarries in North France and its region, regarding their close proximity to urbanized areas, induces serious problems in terms of security of both people and constructions on one hand, and creates a constraint in terms of planning and urban development on the other hand. Hence, it is of high importance to be aware of the circumstances generated by this phenomenon and work on preventing the encountered risks. In North France, and basically in Lille Metropolis, 300 hectares of the above surface are classified under risk while the area of underground quarries is estimated equal to 120 hectares (Ineris, 2007, Ineris, 2012, and Ville de Lille, 2013). Most of these quarries have been excavated in the jointed chalk layers by the method of rooms and pillars (see figure 1). In the domain of underground quarries different studies were performed to anticipate the stability behavior (Martin C.D. et al, 2000, Canbulat I., and coworkers 2000 and 2002, Castellanza R. et al., 2010, and Ferrero A.M. and coworkers, 2010), however, no particular work addressed directly the jointed chalk quarries of North France. In this paper, a stability analysis of these quarries is conducted to provide a better understanding of the safety conditions from a mechanical point of view. A parametric numerical study that covers diverse geometries of room-and-pillar quarries, inspired from site investigations, is performed. A series of 3D nonlinear numerical models of the excavation in chalk layers, with or without joints, is simulated from initial state until failure. Based on the numerical results, a regression analysis is performed and numerical based formulae are extrapolated for the calculation of the pillar resistance R_p and the quarry factor of safety *Fs*. This can be a preliminary risk indicator by which it is possible to estimate the stability state of room-and-pillar quarries having known only few data related to geometry.

2 NUMERICAL MODEL

2.1 *Factor of safety Fs*

Using the 3D nonlinear finite difference code FLAC3D (Itasca Consulting Co., 2009), a series

Figure 1. Room-and-pillar quarry in Lille, North France.

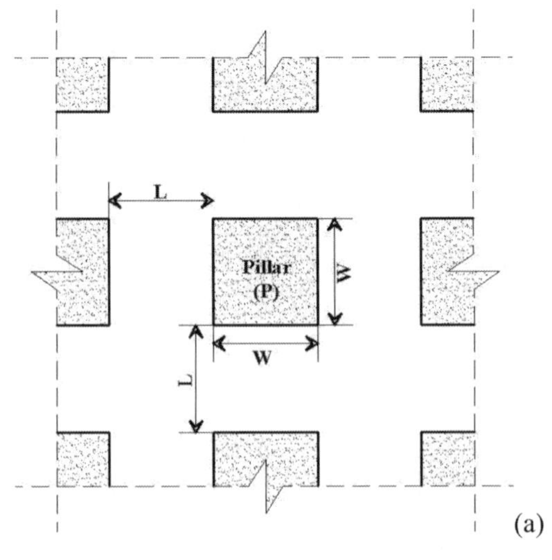

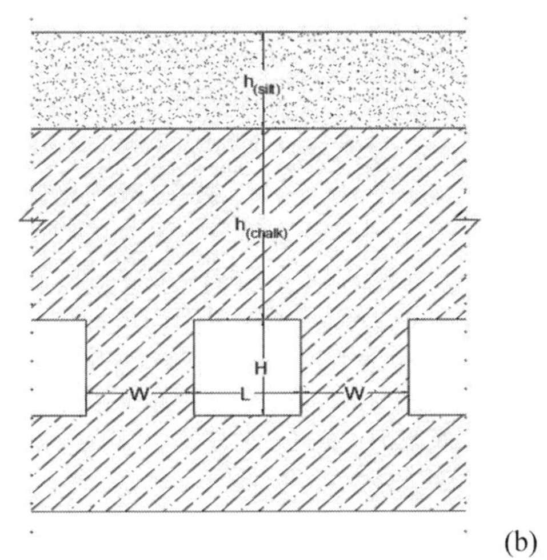

Figure 2. (a) Top view and (b) vertical section of the room-and-pillar quarries.

of excavation models accounting for a range of geometries is simulated. These geometries are inspired from site investigations performed at the underground quarries of North France (Ineris, 2012, and Ville de Lille, 2013). In this work, the room and pillar quarries considered are assumed of symmetric conditions. Each study case represents a multi-pillar system where the pillars are equidistant (same spacing L between adjacent pillars), they have same height (H) and same width (W), and they share same overburden made of a layer of chalk and a layer of silt on the top. For a better illustration see figures 2a and 2b. With the assumption of symmetric geometries, and in order to save simulation time of the three dimensional numerical modelling, half room-pillar systems are considered. Figure 3 illustrates an example of the numerical models showing both the mesh and the assigned boundary conditions. The case presented corresponds to a quarry with pillars of width W = 3 m, height H = 3 m, spacing L = 3 m, overburden height equal to 7 meters (h_{silt} = 2 m plus h_{chalk} = 5 m), and a base of height equal to 5 meters. The mesh considered is cubical with side equal to 0.25 m. Finer meshes are examined and had shown very slight discrepancies in terms of displacements and negligible ones in terms of stability.

In this 3D modelling, boundary conditions are selected of first order along x and y where roller supports are assigned. However, fixed supports were assigned at the bottom. In all models, the height of the base is taken 5 m. Upon trial and error using larger base heights, 5 m, with the rigid nature of the base, seemed sufficient for the current stability study. Site investigations did not show a wide variability in the overburden height. In order to lessen the variables, which is of a non-negligible importance while running a parametric study, the height of the overburden is limited to 10 m: 7 m of chalk and 3 m of silt on top. Consequently, the variables in this study are W, H, and L.

Based on site investigations, the range of these variables considered are listed in table 1. The mechanical properties of both chalk and silt are inspired from the works of Mikolajczak A. (1996) and listed in table 2. The initial stress state is generated by applying gravity forces. For this condition, the value of the initial earth pressure coefficient at rest is close to $\nu/(1-\nu)$ where ν. is Poisson ratio of the chalk.

For each case of study, failure is tested and a factor of safety *Fs* is computed at ultimate conditions. This is based on the conventional shear strength reduction method (Zienkiewicz, 1975, and Griffiths, 1980) where the plastic strength properties of chalk (cohesion c, and friction angle φ) are reduced to a minimum such that no failure is yet attained. In other words, Fs at initial state, denoted Fs_i, is set equal to one ($Fs_i = 1$) where c and φ are not yet reduced. Then, Fs_i is systematically incremented where at each increment new reduced properties are calculated. This is repeated in the numerical process until reaching the highest increment before failure occurs. This highest increment is called Fs_f and at this stage the shear properties are reduced to maximum (c^* and φ^*) just before failure is attained. In this work, Fs_f is called the numerical factor of safety and denoted Fs_{num} (see equation 1).

$$Fs_{num} = \frac{c}{c^*} = \frac{\tan\varphi}{\tan\varphi^*} \tag{1}$$

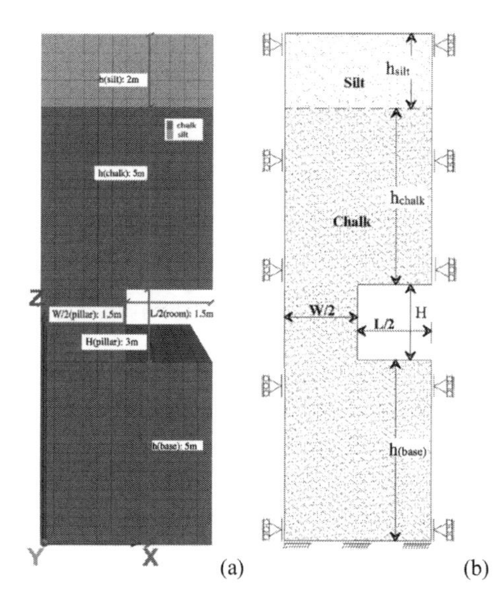

Figure 3. The symmetric quarry model showing (a) the mesh, and (b) the boundary conditions.

Table 1. Summary of geometric charactestics used in the numerical study.

	Width (m)	Height (m)
Pillar	2 (W (8	3 (H (5
Room	2 (L (5	3 (H (5
Overburden Chalk	--	$h_{chalk} = 7$
Overburden Stik	--	$h_{silt} = 3$

2.2 *Resistance of pillar R_p*

In room-and-pillar quarries with a multiple-pillar system, the overall stability depends largely on the stability of the pillar itself. In this case, an analytical factor of safety Fs_{anal} can be defined as the ratio of the resistance of the pillar R_p to the applied loads acting on it q_p (see equation 2).

$$Fs_{anal} = \frac{R_p}{q_p} \tag{2}$$

The applied loads q_p represent the imposed loads at the roof and surface. In this study, no external surface loads are considered, and the pillar is expected to support its own self-weight in addition to the weight of the overburden. Hence, q_p can be evaluated by multiplying the tributary area of the pillar to its density. Room-and-pillar quarries in this study are assumed to have regular sections (see figure 2). So, the applied load q_p can be calculated using equation 3.

$$q_p = (\rho_{ch} g h_{ch} + \rho_{si} g h_{si})(W + L)^2 + (\rho_{ch} g H_{pi})(W)^2 \tag{3}$$

According to data collected from site investigations of underground quarries in Lille and its region, the parameters chosen for this study are listed in tables 1 and 2. The constants ρ_{ch} and ρ_{si} represent the densities of chalk and silt (see table 2 for numerical values), g is the gravity (g = 9.81), h_{ch} and h_{si} are the height of chalk layer and silt layer respectively (see table 2 for numerical values), W, H, and L are the variables of width of the pillar, height of the pillar (where H_{pi} in equation 3 is the same H in equation 4), and the distance between pillars respectively. By replacing these parameters in equation 3, this latter can be simplified into equation 4. The applied load q_p depends on the geometric variables of the pillar, W, H, and L.

$$q_p = 21\left[9.572(W + L)^2 + W^2 H\right] \tag{4}$$

Concerning R_p, it usually depends on the compressive strength of the constituent material of the pillar, and its geometry. In this work, and since a range of pillar geometries is considered, R_p is extrapolated from the numerical results

Table 2. Summary of mechanical properties used in the numerical study.

	Density ρ (kg/m³)	Young Modulus E (MPa)	Poisson Ratio ν(-)	Cohesion c (kpa)	Friction Angle φ(°)	Dilation Angle ψ(°)
Chalk	2100	250	0.3	1000	30	5
Silt	1800	10	0.3	50	30	5

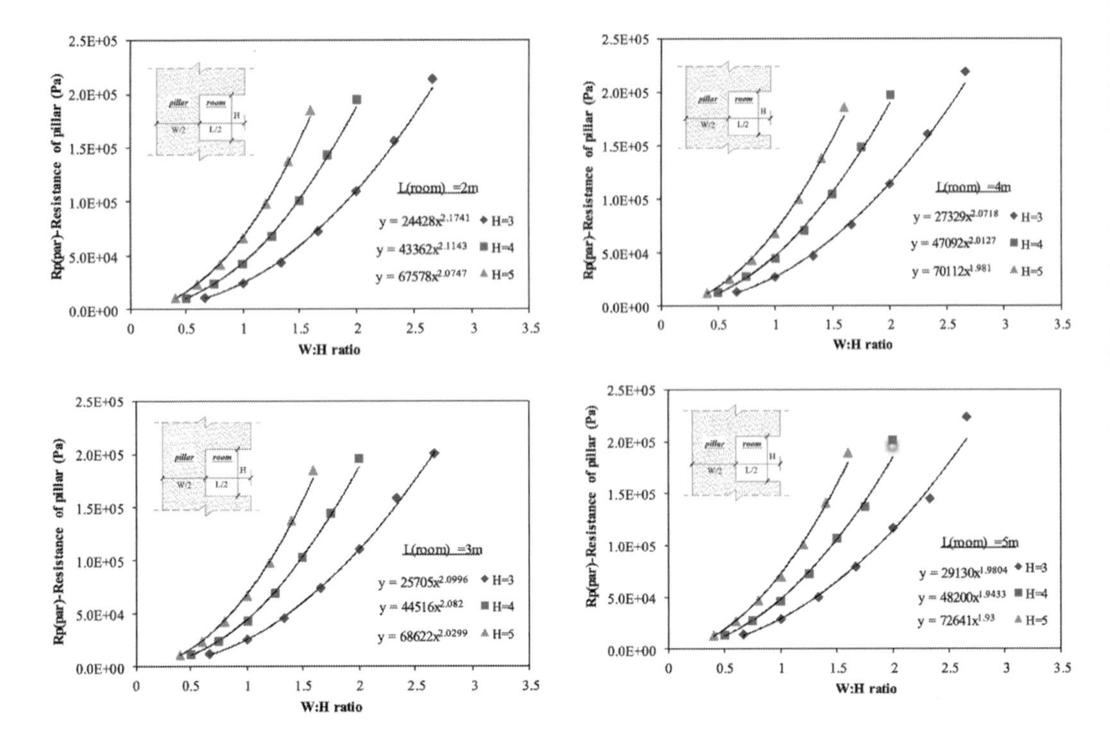

Figure 4. Results of the first regression analysis performed between simulated $R_{p(par)}$ and W:H ratios.

(designated by Fs_{num}) that correspond to the different geometries (designated by W, H, and L); these R_p values are called parametric R_p and denoted $R_{p(par)}$. They are deduced from the relation in equation 5.

$$R_p(par) = FS_{\text{num}} \times q_p \tag{5}$$

3 EXTRAPOLATED FORMULAE

In a quarry of multi-pillar system, the resistance of the pillar is directly related to its geometry: W, H, and L (which is unified in this study; i.e., consideration of equidistant pillars) and the height of the overburden (assuming a constant 10 m total height of the overburden in this study which includes 7 m of chalk and 3 m of silt on top) (see figure 1(a)). A series of excavation models that accounts for the range of geometric variables, notably W, H and L values is simulated. For each model, the safety factor Fs_{num}, based on the conventional c-phi reduction method (Zienkiewicz, 1975, is computed (equation 1). Then, having calculated the pillar pressure q_p by using equation 4, it is possible, and using equation 5, to determine the pillar resistance $R_{p(par)}$. Afterwards, a coupled regression analysis of the obtained results is performed. The coupled regression integrated in developing the proposed formulae represents a series of three dependent regression analyses. First regression is done between $R_{p(par)}$ and W:H, the second is performed between coefficients obtained from the first regression and H:L ratios, and the third is carried out between coefficients obtained from the second regression and L values. Figure 4 illustrates an example of the first regression.

Relating the previous regressions, a nonlinear relation between the obtained values of $R_{p(par)}$ and W:H ratios and L values is developed. This relation denoted as $R_{p(ex)}$, is an extrapolated formula that enables the calculation of the resistance of the pillar in a multi-pillar system in terms of W, H and L (see equation 6).

$$R_{p(ex)} = 2613.2 \frac{(W)^{\alpha}}{(H)^{\beta}.(L)^{\gamma}} \tag{6}$$

Getting back to the relation in equation 1, similar relation between extrapolated values of $R_{p(ex)}$ and $Fs_{(ex)}$ can be written (see equation 7).

$$FS_{(ex)} = \frac{R_{p(ex)}}{q_p} \tag{7}$$

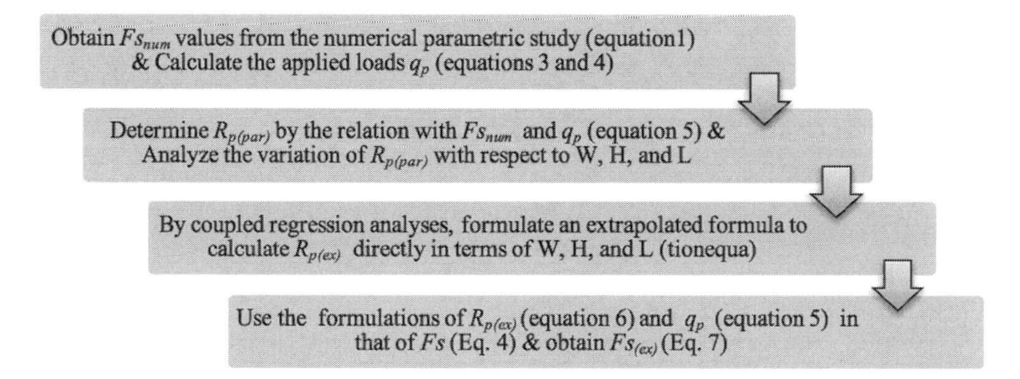

Figure 5. Diagram of the approach proposed for the formulation of $Fs_{(ex)}$.

Substituting $R_{p(ex)}$ of equation 6 in $Fs_{(ex)}$ of equation 7, the extrapolated formula for the calculation of the factor of safety in function of geometric variables is obtained (see equation 8).

$$Fs_{(ex)} = 124.438 \frac{(W)^{\alpha}}{\left[9.572(W+L)^2 + W^2.H\right](H)^{\beta}.(L)^{\gamma}} \tag{8}$$

where,

$$\alpha = 2.4797 - 0.0157H - 0.081L - 0.073 H/L \tag{9}$$

$$\beta = 0.351 - 0.0157H - 0.012L - 0.073 H/L \tag{10}$$

$$\gamma = 0.0572 - 0.069L \tag{11}$$

The regression approach used in these formulations, is explained in a diagram that summarizes the considered steps (see figure 5). Using the formula in equation 7, it is possible to estimate the factor of safety of a room-and-pillar quarry without the need to regenerate and run numerical models at every time.

4 PRESENCE OF JOINTS

At shallow depths and low-stress environments existing fractures may control the potential failure modes and the associated extent of failure. This occurs when the low confining pressures are not capable of blocking the fracturing phenomenon which in turn leads to progressive fracturing and thus causes reduction in the loading capacity of the pillar at one hand, and provides less resistance to sliding at the other hand. In this case, shear failure in the pillar may appear and even dominate the compression failure.

Investigations of the quarries in North France have shown evidence to the presence of natural embedded joints in the chalk substratum as well as mechanical fractures emerged in weak overstressed zones of the pillars basically (see figure 6). Hence, it is quite important to account for the presence of joints or fractures in the stability study of underground cavities.

4.1 *Effect of W:H and L*

A numerical approach similar to the one presented before is conducted for the selected range of study geometries (Table 1) taking into consideration this time the presence of a single joint set at different inclinations with respect to the vertical axis of the pillar. This is modelled using the ubiquitous-joint model built in FLAC3D (Itasca Consulting Co., 2009). The ubiquitous-joint model is an anisotropic plasticity model that includes weak planes of specific orientation embedded in a Mohr-Coulomb solid. In this study, the chalk continuum consists of the chalk matrix and the joints defined as weak planes with reduced shear strength properties (joints cohesion is set equal to 50% that of the intact matrix). Factors of safety are then simulated according to the conventional shear strength reduction technique applied on the ubiquitous-joint model including both the matrix and the joints in an identical manner.

Results corresponding to three selected cases of joint inclinations θ (30°, 45°, & 60°) are plotted and compared with previous results obtained when no joints were defined (see figure 7).

Using a simplified regression approach, three extended extrapolated formulae for the calculation of factor of safety of quarries in jointed chalk $Fs_{(ex\_j\theta)}$ with joint sets inclined at $\theta = 30°$, $\theta = 45°$, and $\theta = 60°$ are developed:

- Joint set inclined at an angle $\theta = 30°$:

$$Fs_{(ex\_j30)} = 1.0394 * Fs_{(ex)} - 0.5771 \tag{12}$$

Figure 6. Fracture at 45° in the pillar of the Cavaignac quarry at Lille, North France, 2011.

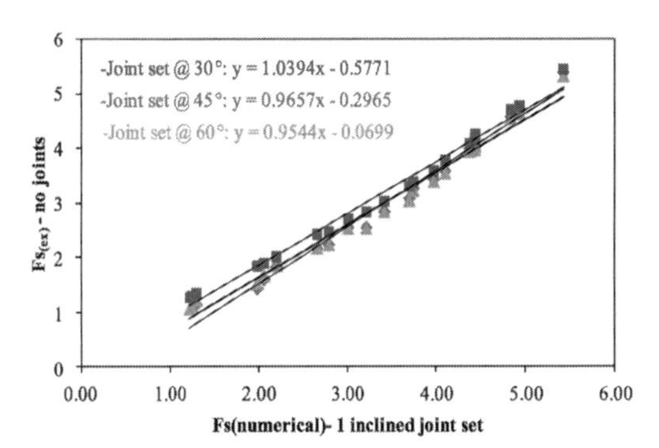

Figure 7. Linear relation between empirical and numerical *Fs* of cases without and with joints, respectively.

- Joint set inclined at an angle $\theta = 45°$:

$$Fs_{(ex\_j45)} = 0.9657 * Fs_{(ex)} - 0.2965 \quad (13)$$

Joint set inclined at an angle $\theta = 60°$:

$$Fs_{(ex\_j60)} = 0.9544 * Fs_{(ex)} - 0.0699 \quad (14)$$

where $Fs_{(ex)}$ is calculated from equation 7.

This permits the rapid calculation of the factor of safety for each case of study having known only the dimensions W, H, and L, in addition to the joint set inclination. These formulae can be used as a practical tool for the preliminary estimation of stability of underground quarries excavated in jointed chalk layers of North France.

5 CONCLUSION

The phenomenon of unexploited underground quarries reveals a serious problem in terms of stability of the underground strata which eventually provokes damage risks on the surfaces lying on top. This imposes security threats against both people and constructions at one hand, and against urban development at the other hand. North France, and particularly Lille and its region are seriously exposed to such risks since a large number of abandoned quarries have been identified. For this reason, the City of Lille has launched the project on risk assessment and control associated to this phenomenon. In terms of this project, this paper comes to present some of the results of the numerical study performed to test the stability of quarries excavated in jointed chalk layers by the method of room-and-pillar. A parametric study accounting for the wide variation in the geometries of investigated quarries is conducted. This provides a better understanding of the effect of such variations, mainly the width to height ratios of pillars (W:H) and the width of void rooms which represents the spacing between pillars (L), on the overall stability of the quarry. Afterwards, a coupled regression analysis is affected and accordingly numerical based extrapolated formulae are developed to facilitate the estimation of the resistance of the pillar in a multi-pillar system R_p and the factor of safety *Fs*. Moreover, the presence of joints which is evident in quarries of Lille is considered in a separate numerical study. Based on the proposed extrapolated formulae for quarries without joints, new formulae are developed for the calculation of factors of safety of jointed quarries. By this, it is now possible to estimate stability state of room-and-pillar quarries of North France and similar regions given only their geometric characteristics (W, H, and L) and the angle of inclination of existing joints.

ACKNOWLEDGEMENT

The authors would like to thank the City of Lille for making us a part of this project. Special thanks goes to Mr. G. Cheppe and Miss G. Berrehouc. Gratification is also honored to Semofi, the research engineering company for geotechnical problems. Special thanks is dedicated to Mr. G.M. Gallet de Saint Aurin for all the support.

REFERENCES

Canbulat I., & Van der Merwe J.N., M. van Zyl, A. Wilkinson, A. Daehnke and J. Ryder (December 2002). Task 6.9.1 The development of techniques to predict and manage the impact of surface subsidence. COALTECH 2020.

Castellanza, R., Nova, R., and Orlandi, G. (2010). Evaluation and remediation of an abandoned gypsum

mine. Journal of geotcchnical and geo-environmental engineering 136.4: 629–639.
Ferrero A.M., Segalini A., & Giani G.P. 2010. Stability analysis of historic under-ground quarries. Elsevier, Computers and Geotechnics37 (2010) 476–486.
Griffths, D. V., 1980. Finite element analyses of walls, footings and slopes. PhD thesis dissertation ed. s.l.:University of Manchester.
Ineris 2007. Mise en sécurité des cavités souterraines d'origine anthropique: Surveillance—Traitement, Guide Technique. Rapport d'Etude, February 2007.
Ineris 2012. Carrières souterraines sur le territoire de Lille; Hellemmes; Lomme. Inspection et avis sur l'état géotechnique des carrières souterraines de craie années 2011 et 2012. Rapport d'Etude INERIS DRS-11–123081–10891 A PROJET 2, March 2012.
Itasca Consulting Co. 2009. FLAC3D—Fast Lagrangian Analysis of Continua in 3 Dimensions, Ver. 4.0 Command Reference Manual. Minneapolis: Itasca.
Martin C.D, & Maybee W.G. 2000. The strength of hard-rock pillars. International Journal of Rock Mechanics & Mining Science 37 2000 1239–1246.
Mikolajczak A., 1996. Modélisation du comportement de craies sous sollicitations simples et complexes. PhD thesis dissertation, University of Lille.
Ville de Lille 2013. D.I.C.R.I.M. Document d'Information Communal sur les Risques Majeurs. Lille, Hellemmes, Lomme Les 9 Risques Majeurs. Document of City of Lille, Service of Urban Risks; Juin 2013.
Zienkiewicz O.C., Humpheson C., & Lewis R.W., 1975. Associated and non-associated visco-plasticity and plasticity in soil mechanics, Géotechnique, 25(4): 671–689.

Numerical Methods in Geotechnical Engineering IX – Cardoso et al. (Eds)
© 2018 Taylor & Francis Group, London, ISBN 978-1-138-33203-4

Numerical analysis of old masonry vaults of the Paris subway tunnels

E. Bourgeois
Université Paris Est, IFSTTAR, Marne la Vallée, France

O. Moreno Regan
Setec tpi, Paris, France

A.S. Colas & P. Chatellier
Université Paris Est, IFSTTAR, Marne la Vallée, France

J.F. Douroux & A. Desbordes
RATP, Fontenay sous Bois, France

ABSTRACT: Most of the Paris subway tunnels comprise a masonry vault built between 1900 and 1960. In order to identify potential cracking areas and to predict the deformations resulting from loads associated to the construction of new infrastructures, RATP and IFSTTAR have developed a constitutive model specifically aimed at representing the main features of the masonry lining. The approach consisted of adapting an existing model combining a homogenization procedure with damage. The differences between the approach adopted and the original model are highlighted. The resulting model makes it possible to account for a global heterogeneous anisotropic elastic behavior including damage. It has been implemented in the finite element code CESAR-LCPC. As a first step, the model has been used to simulate tests, taken from the literature, performed on semicircular masonry vaults.

1 INTRODUCTION

The structural health of the tunnels of the Paris subway is a major concern for the Régie Autonome des Transports Parisiens (RATP), which is in charge of ensuring the continuity of the metro service. Given the current development of the network within the Grand Paris project, it is necessary to predict the effects of new engineering works near existing tunnels. Masonry vaults can withstand displacements in the order of several tens of millimeters without reaching failure, which is not properly accounted for by classical elastoplastic models. This is why RATP and IFSTTAR (French Institute of science and technology for transport, development and networks) have developed a specific constitutive model. Following previous works, the geometrical arrangement of blocks and mortar joints is taken into account by means of a homogenization technique. The orientation of the mortar joints follows the shape of the vault, which implies that the properties of the vault are heterogeneous. In the last place, the quasi-brittle behavior of the masonry components was described by an isotropic damage model. The model has been implemented in the finite element code CESAR-LCPC, and appropriate parameters have been determined on the basis of an experimental program undertaken to characterize the Paris subway tunnels. The aim is to identify potential cracking areas in the tunnel as well as its deformation resulting from a given load case. To validate the approach, the model has been used to simulate tests performed on semicircular masonry vaults, taken from the literature.

2 MECHANICAL MODELLING OF MASONRY

2.1 *Homogenization procedure*

Masonry is a heterogeneous material made of elementary blocks (bricks or stone blocks) and mortar joints. Its behavior has been investigated in numerous scientific publications (Angelillo, 2014). Many approaches have been developed to evaluate the stability of vaults on the basis of the static equilibrium of the structure, by means of limit analysis (Heyman, 1969, 1997), or of the yield design theory (Salençon, 1983), (Delbecq, 1982). Such stability analyses do not provide the deformations of the structure under service or accidental loads.

Among all the techniques to model the deformability masonry structures, the finite element method is frequently used. It makes it possible to discuss the interaction between the masonry vault and the surrounding ground. Masonry itself can be dealt with by different approaches (Lourenço, 1996) $\tilde{\varepsilon}$. Modelling blocks, joints and interfaces separately implies a detailed discretization and a large number of degrees of freedom, which is not very adapted to the analysis of the whole structure. On the other hand, one can adopt a macroscopic modeling in which all components of the masonry are represented as a continuous homogeneous medium "equivalent" to the masonry.

To transform the masonry into an equivalent continuum medium, we consider the geometry of the heterogenous assembly of blocks and joints as periodic, and define a basic cell, on which an auxiliary problem is solved.

The approach proposed by Zucchini and Lourenço (2002, 2004) relies on an engineering approach to get an approximate solution of the auxiliary problem, which proves very efficient from a numerical point of view. The homogenized procedure provides the macroscopic anisotropic elastic tensor of the homogeneous equivalent material.

Two approaches are possible: in the first one, the "macroscopic stress" is converted in loads applied on the boundaries of the basic cell; in the second one, the auxiliary problem is defined in terms of strains applied to the basic cell. For the coupling of the homogenization procedure with damage models in the components, it is preferable to adopt the second "strain-driven" approach. Yet, this approach does not give access (in the case of the approximate solution scheme adopted here) to all the required elastic parameters, which makes it necessary to perform also a stress-driven homogenization procedure. More details can be found in Moreno Regan (2016) and Moreno Regan *et al.* (2017).

2.2 *Isotropic damage model*

Contrary to the more complex approach used by Zucchini & Lourenço (2004), the nonlinear behavior of the masonry components (blocks and joints) is represented, inside each component, by an isotropic damage model. Inside the base cell, the apparent stiffness is equal to the initial stiffness multiplied by a damage coefficient *(1-d)*. The damage coefficient takes the same value at any point of the blocks inside the base cell. Two other damage coefficients are associated with the vertical and horizontal joints. In other words, we introduce, at the macroscopic level, three variables describing the damage state of the components of the masonry.

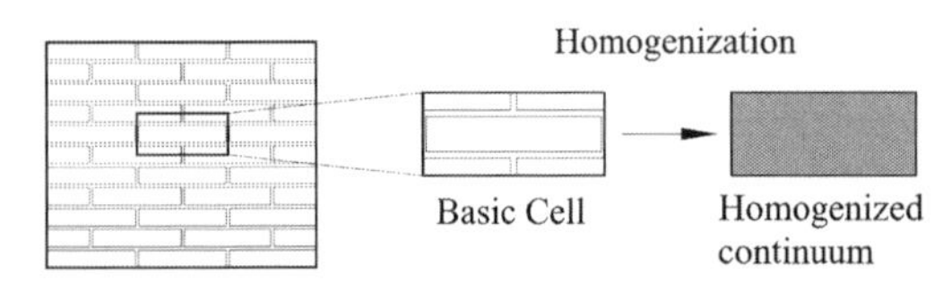

Figure 1. Basic cell of a masonry wall (after Zucchini & Lourenço, 2002).

We adopt the modified version of the Mazars (1986) model proposed by Davenne *et al.* (1989) for concrete. Damage evolution is controlled by an equivalent strain given by:

$$\tilde{\varepsilon} = \gamma\sqrt{\langle\varepsilon_1\rangle_+^2 + \langle\varepsilon_2\rangle_+^2 + \langle\varepsilon_3\rangle_+^2} \tag{1}$$

where ε_i (i = 1,2,3) represent the principal strains, $<\varepsilon_i>_+ = \varepsilon_i$ if $\varepsilon_i \geq 0$ or 0 otherwise. In eq. (1) the coefficient γ is defined by the following expression:

$$\gamma = -\frac{\sqrt{\langle\bar{\sigma}_1\rangle_-^2 + \langle\bar{\sigma}_2\rangle_-^2 + \langle\bar{\sigma}_3\rangle_-^2}}{\langle\bar{\sigma}_1\rangle_-^2 + \langle\bar{\sigma}_2\rangle_-^2 + \langle\bar{\sigma}_3\rangle_-^2} \tag{2}$$

where $\bar{\sigma}_i$ represents the principal effective stress in the direction i; and $<\sigma_i>_-$ is the negative part of the principal effective stress: $<\sigma_i>_- = \sigma_i$ if $\sigma_i \leq 0$ and 0 otherwise. The value of γ lies in the interval [0,1] and is calculated only when at least a one principal effective stress is negative, i.e. compressive.

No evolution of the damage variable occurs if the condition $F \leq 0$ is fulfilled, where $F = \tilde{\varepsilon} - \tilde{\varepsilon}_M$, $\tilde{\varepsilon}_M$ denoting the current threshold, equal to the initial threshold ε_{D0} if it has never been reached, or to the maximum value reached by $\tilde{\varepsilon}$ otherwise. The initial threshold ε_{D0} is the strain at the maximum stress in a uniaxial direct tension test (Mazars, 1986), assuming quasi-linear behavior up to failure.

The damage variable d is a linear combination of two variables d_t and d_c, associated with the tensile and compression stresses (Mazars, 1986):

$$d = \alpha_t d_t + \alpha_c d_c \tag{3}$$

where $\alpha_t + \alpha_c = 1$ (see Mazars (1986) for a detailed definition of α_t and α_c). The evolution of d_c is given by:

$$d_c = 1 - \frac{\varepsilon_{D0}(1 - A_c)}{\tilde{\varepsilon}_M} - \frac{A_c}{\exp\left[B_c(\tilde{\varepsilon}_M - \varepsilon_{D0})\right]} \tag{4}$$

where A_c and B_c are obtained experimentally from the stress-strain curves of a compression test.

For d_t we adopt the expression proposed by La Borderie (2003) to avoid excessive mesh sensitivity problems in the finite element modeling of the softening behavior of the damage law (see Murakami, 2012):

$$d_t = 1 - \frac{\varepsilon_{D0}}{\tilde{\varepsilon}_M} \exp\left[-B_t\left(\tilde{\varepsilon}_M - \varepsilon_{D0}\right)\right] \tag{5}$$

where the parameter B_t depends on a characteristic length l_c, the mode I fracture energy G_{ft} (assumed to be a material property), and the tensile strength of the material f_t:

$$B_t = \frac{l_c f_t}{G_{ft}} \tag{6}$$

We let $l_c = \sqrt{S}$, S being the area of the finite element to which the integration point belongs. This is another difference with the model of Zucchini & Lourenço (2004), where the characteristic length is equal to the dimension of the brick or joint perpendicular to the crack direction, producing mesh-sensitivity.

The proposed model combines a homogenized anisotropic elasticity and isotropic damage models at the component level and makes it possible to reproduce a global anisotropic damage behavior.

2.3 *Introduction of the vault geometry*

In the last place, variations of the directions of the block-mortar bond in the vault are also taken into account. This adds up to computing the orientation of the local axis attached to the masonry at a given point of the vault. We have dealt with this question by assuming that the vault has an elliptical shape and by introducing the appropriate parameters (coordinates of the ellipse center, ellipse semi-axes, and orientation of the major axis), see Moreno Regan *et al.* (2017).

3 NUMERICAL IMPLEMENTATION

3.1 *Global iterative resolution procedure*

The model for masonry vaults was programmed in the finite element solver CESAR-LCPC (Humbert *et al.* 2005), developed by IFSTTAR (cesar.ifsttar.fr).

In the finite element analysis of damage problems, an elastic solution is computed first, with the initial value of the damage parameters. The displacements obtained with the homogenized model are converted into strains at the level of the masonry components, which induce a modification of the damage parameters and of the stiffness of the material: the stress state must be updated to account for this stiffness reduction, which produces out-of-balance nodal forces. The procedure follows a classical Newton-Raphson iterative scheme. The homogenization – damage procedure is carried out at each integration point.

The numerical scheme is summarized in Figure 2. The global damage criterion F_G is simply the maximum value of the damage criterion in all components (brick, joint, joints intersection).

3.2 *Local iterative resolution procedure*

The stresses in each component k (i.e. block and the two mortar joints) depend on the damage variable d_k which depends on the stress and strain state in that component: a local iterative process is therefore necessary. It is performed at the base cell level. For each strain increment at the global level, the stresses and strains in the cell components (bricks, joints and joint intersection) are computed using the damage coefficients of the previous iteration. Damage variables are then updated using the model of paragraph 3.1, from the new stresses. The process is iterated until convergence of the damage variables d_k. The procedure is illustrated by Figure 3.

4 APPLICATION

4.1 *Presentation of the tests taken as reference*

The model was used to simulate experimental tests performed by Krajewski & Hojdys (2015) on semicircular masonry vaults. Two cases are studied here: in case 1, a point load is applied directly to the vault (Figure 4), whereas in case 2, a vault of same geometry is buried and a point load is applied on the surface of the backfill material layer (Figure 5).

The vaults were supported by two reinforced concrete blocks connected by steel rods to maintain the distance between them.

The bricks length, width and thickness were equal to 250 mm, 125 mm and 65 mm, respectively. The thickness of the mortar joints was 13 mm.

(i) Compute incremental macroscopic stresses and strains in each integration point
(ii) Compute damage coefficients in basic cell, see 3.2
(iii) Check global criterion:
 $F_G > 0$?
 YES: damage loading. Proceed to (iv)
 NO: no further damage. Exit
(iv) Compute local damaged homogenized elastic tensor
(v) Update global stiffness matrix
(vi) Update macroscopic stresses

Figure 2. Global iterative algorithm for masonry vaults.

For each integration point:
(i) Compute local macroscopic strains
(ii) Compute stresses and strains in the components
(iii) For each component k :
  Compute equivalent strain, Eq. (1)
  Check damage criterion $F_k>0$?
  YES: damage variable d_k increases: Proceed to (iv)
  NO: no further damage for component k. Exit
(iv) Update damage variables
(v) Convergence?
  YES: Proceed to (vi)
  NO: Proceed to (ii)
(vi) Update damaged homogenized parameters

Figure 3. Local iterative algorithm in the base cell.

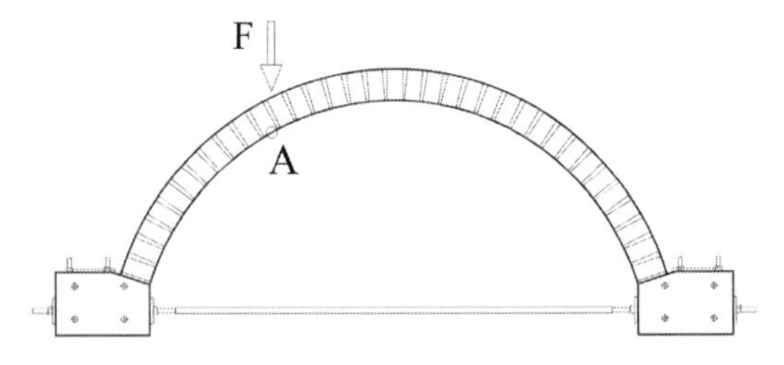

Figure 4. Case 1: vault without backfill material, Krajewski & Hojdys (2015).

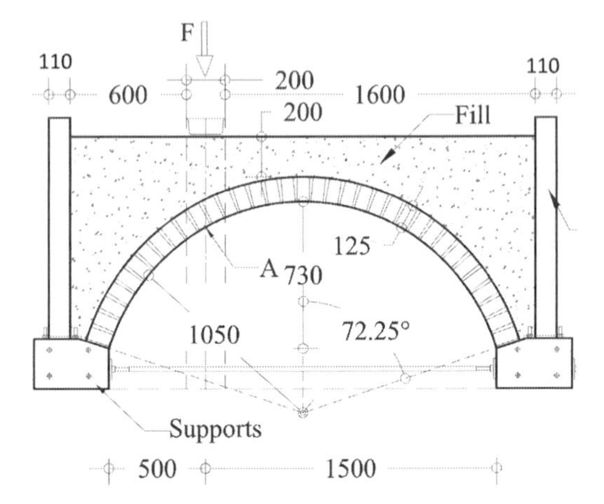

Figure 5. Case 2: buried vault (dimensions in mm) (after Krajewski & Hojdys, 2015).

In the case of the buried vault (figure 5), vertical concrete walls were built to retain the backfill material. Two steel beams connected the walls to prevent their horizontal displacements and rotations.

A set of LVDT sensors was used to monitor the deformations of the vault. In what follows, the analysis is focussed on the radial displacement of the point of the intrados located just below the load (point A of Figures 4 and 5). The mean ellipse (in the sense of paragraph 2.3) is actually a circle of radius 111 cm for both cases.

4.2 *Material parameters*

The simulations presented have been carried out with the parameters given in Table 1.

In the case of the buried vault (case 2), the mechanical behavior of the backfill material is described by a classical elastic-perfectly plastic model, using the Mohr-Coulomb criterion, with $\rho = 300$ kg/m^3, E = 10 MPa, $\nu = 0.3$, c = 4.5 kPa, $\varphi = 37$ degrees, $\psi = 10$ degrees.

4.3 *Vault without fill*

FEM calculations were carried out in plane strain conditions. The mesh used is presented in figure 6. It comprises 1000 nodes and 268 quadratic elements (6-node triangles and 8-node quadrangles).

The boundary conditions set to zero the vertical and horizontal displacements of the concrete blocks. In a first step, the self-weight of the vault is applied to the model.

Then, a vertical force is applied on one single node of the vault extrados and increased progressively, by equal increments of 60 N. Given the softening, quasi-brittle behavior, of the masonry and to simulate a displacement controlled test, we have introduced in the model a vertical spring connected to this node, so that part of the applied force is transmitted to the spring. The load applied to the vault is the difference between the total load applied and the force in the spring.

The results are shown in Figure 7, where u_r is the radial displacement of point A (Figure 4).

The numerical load-displacement curve shows a change in slope for a load of around 3.2 kN. The vault is still able to bear an increase in load up to a maximum value of 3.7 kN. After this value, the vault fails in the simulation: the curve is not smooth, and the displacement increases rapidly under quasi constant load. The numerical results are in reasonably good agreement with the measurements (Figure 7).

Figure 8 presents the results of the test performed by Krajewski and Hojdys (2014, 2015), and compares them with the deformed mesh obtained numerically. The figure also represents the maximum extension strain: zones in black correspond to the areas where the maximum extension strain is greater than 0.5%. The numerical simulation reproduces the appearance of a hinge below the load, then the formation of another hinge, not located exactly at the same place as in

Table 1. Properties of mortar and bricks.

	bricks	mortar
density (kg/m^3)	1700	1700
Young's modulus (MPa)	14000	250
Poisson's ratio (–)	0.2	0.16
Tensile strength f_t (MPa)	0.9	0.08
Fracture energy G_{ft} (Pa.m)	100	46
Damage parameter A_c (–)	1.0	0.5
Damage parameter B_c (–)	560	270
Damage threshold ε_{D0} (–)	9 10^{-5}	2 10^{-4}

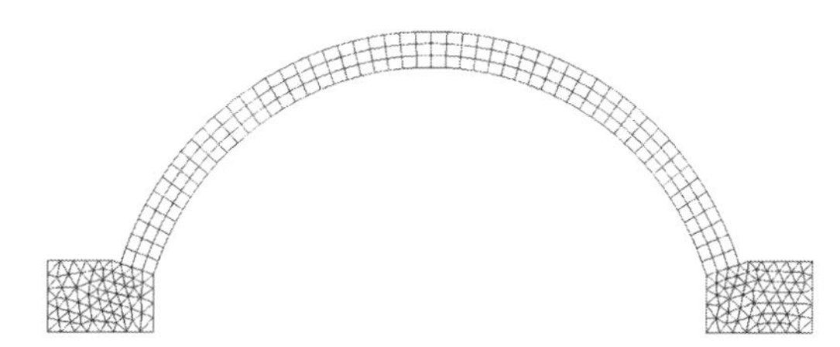

Figure 6. Mesh used for case 1.

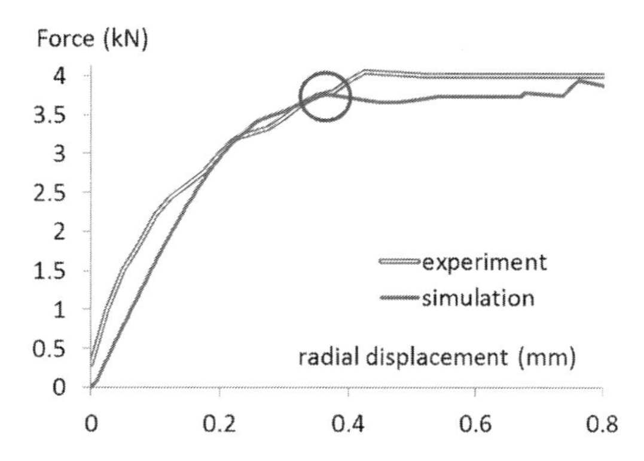

Figure 7. Comparison of load-displacement curves for case 1.

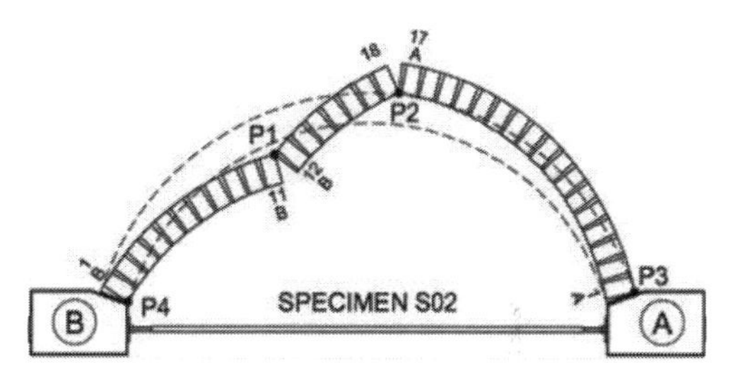

(a) experimental result (after Krajewski & Hojdys, 2015)

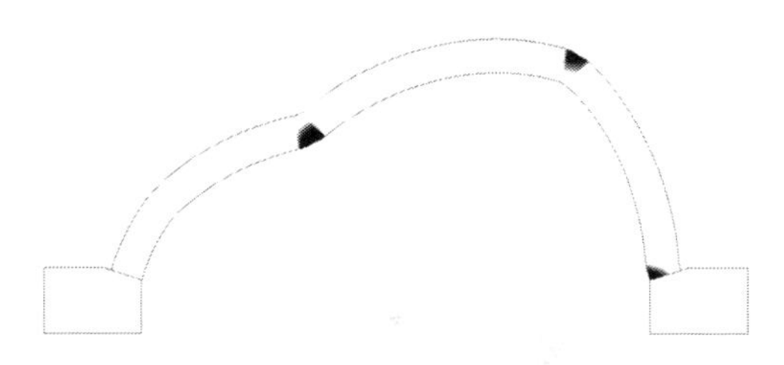

(b) results of the finite element simulation : deformed mesh and contour lines of the maximum extension strain

Figure 8. Deformation of the vault in case 1.

the experiment. However, from a qualitative point of view the model produces satisfactory results.

From a numerical point of view, the convergence of the algorithm is relatively easy until the point when the maximum force applied on the vault is obtained (indicated by a circle in figure 7). Beyond this point, the damage zone initiated at the intrados of the vault spreads over almost its entire thickness and numerical convergence requires large numbers of iterations (several thousands).

4.4 *Buried vault*

FEM calculations were carried out in plane strain conditions. The mesh used is presented in figure 8. It comprises 3000 nodes and 1300 quadratic elements (6-node triangles and 8-node quadrangles).

The calculation was conducted in plane strain conditions in two steps: (1) application of the weight of the filling and vault and (2) application of the load. The contact between the fill and the vault was considered are a perfect bonding, to avoid the introduction of a third type of non-linearity besides the damage model for the vault and the elastoplastic model for the backfill material.

The vertical and horizontal displacements of the concrete blocks are set to zero. In the simulation, the horizontal displacement is set to zero along the retaining walls (although in the experiment the upper edge of the wall moved horizontally by around 0.8 mm). Also the simulation assumes that there is no friction between the wall and the backfill. The vertical force is applied in equal increments of 1500 N.

Results are shown in Figure 10. The overall shape of the curve, the initial stiffness and the progressive reduction in stiffness, and eventually the failure load are reproduced with a good precision by the numerical model.

The numerical load-displacement curve is smoother in both the experiment and the numerical simulations than for case 1, which may be explained by the fact that the load is spread over a larger area of the vault extrados.

Figure 11 presents the results of the test and the deformed mesh obtained numerically, with an

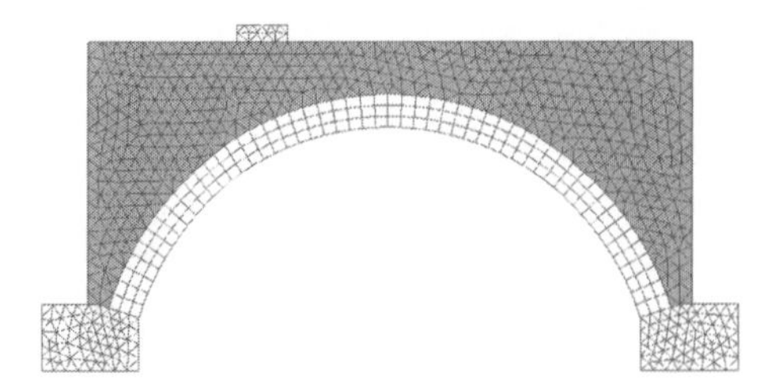

Figure 9. Mesh used for case 2.

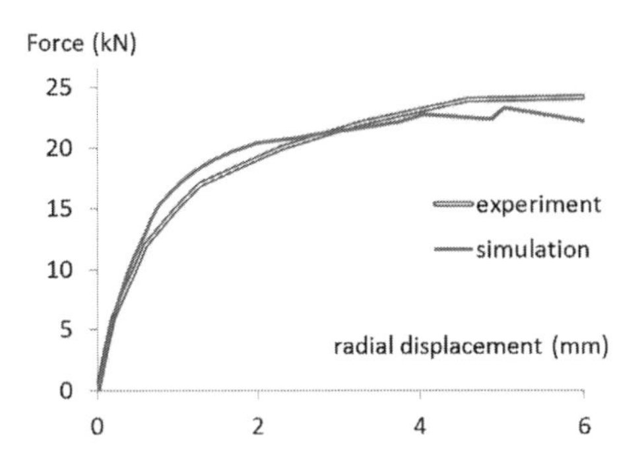

Figure 10. Comparison of load-displacement curves for case 2.

indication of the areas where the extension strain is greater than 1%.

Krajewski and Hojdys (2015) note that in the final state, the vault exhibits four hinges; the numerical model reproduces satisfactorily the four-hinge failure mechanism.

The apparition of hinges in the numerical simulation in case 2, where the loads are spread over a wider area of the vault extrados is very interesting qualitatively, in view of the application of the approach to real masonry tunnels. Also, the fact that the simulation makes it possible to account for realistic displacements of the vault before failure is encouraging.

Note however that, from the numerical point of view, the apparition of damage and of hinges in the vault leads to large numbers of iterations of obtain convergence.

5 DISCUSSION

On the whole, the model seems to reproduce fairly correctly the behavior of the vault, alone or buried, with the same set of parameters for the homogenized-damage model.

In particular, it seems that, in case 2, the conjunction of two non-linearities of different nature, the reduction in stiffness of the masonry and the plastic strains in the backfill, can be dealt with simultaneously by the numerical algorithm, even if

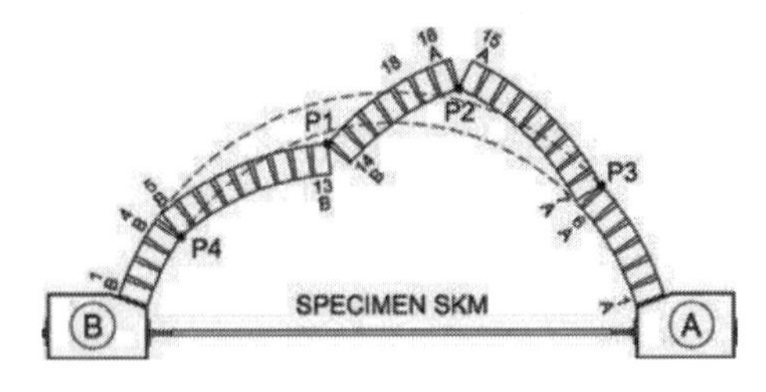

(a) experimental result (after Krajewski & Hojdys, 2015)

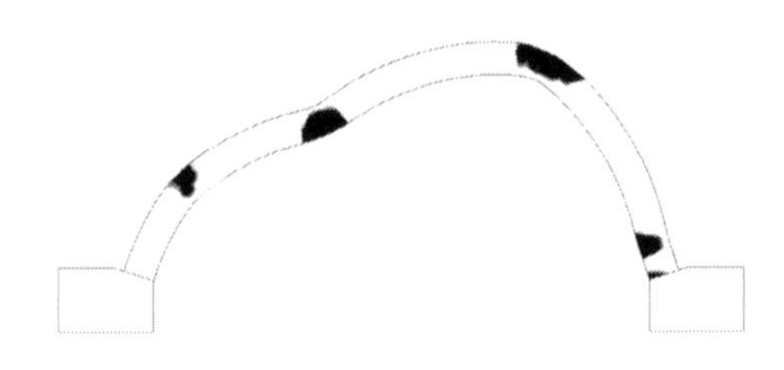

(b) deformed mesh and zones of maximum extension

Figure 11. Deformation of the vault in case 2.

convergence requires large numbers of iterations, and consequently large computing times.

Also, it was checked that the results are not appreciably modified if the load increments are changed.

However, it must be emphasized that the model needs a large number of parameters whose determination is, in some cases, complicated. To cope with this problem, several parametric studies have been carried out to discuss the influence of the various parameters of the model.

5.1 *Influence of the damage model parameters*

The damage model used for the vault involves 8 parameters for the bricks and 8 parameters for the mortar, and their determination is not of equal difficulty. Parameters A_c and B_c, which characterize the evolution of damage in compression, can be relatively easily obtained in compression tests carried out on samples of the materials. Besides, it appears that their numerical values, for the bricks and the mortar, have a small influence on the results. This can be explained by the fact that the overall behavior of the vault (given the geometry and the applied loads) is controlled by the apparition of tension-damaged zones at the intrados of the vault.

The tensile strength f_t is also rather familiar and can be determined experimentally with little ambiguity.

The same holds for the threshold strain, which has a clear influence on the simulation results.

By contrast, the determination of the mode I fracture energy G_{ft} is difficult and this is a serious obstacle to the practical use of the model, because the value adopted for this parameter, especially for the mortar, has a significant influence on the results: for a value of 100 Pa.m in the mortar instead of the reference value of 46 MPa.m, we obtained a maximum force of 4.4 kN (+ 22% with respect to the reference simulation), and a smoother load-displacement curve.

5.2 *Influence of material parameters*

It is also worth noting that, in the case of the buried vault, the friction angle was measured (by Krajewski & Hojdys, 2015) with a good precision. The value of the cohesion is less well known. A parametric study was carried out to assess the influence of this parameter. The results are presented in Figure 12.

5.3 *Influence of boundary conditions*

In the specific situation of case 1, it can also be noted that the conditions of contact between the vault and the concrete blocks are not perfectly well controlled. In the simulations, the vault is perfectly bonded to its supports (concrete blocks), but this is in contradiction with the observed behavior, and it may account for the discrepancy between the deformed shape of the vault in figure 7a and 7b. In other words, the model used for the vault may not be responsible for the difference in the deformed shapes.

5.4 *Other parameters*

Among the other parameters that are likely to have a significant influence on the results of the simulations, it is worth mentioning the contact between the vault extrados and the backfill material, given that the behavior of the buried vault is certainly largely controlled by the conditions in which the load applied at the surface is transferred to the vault. This implies to introduce in the model appropriate elements to represent contact, which also introduces new parameters. At present, we have not investigated this question.

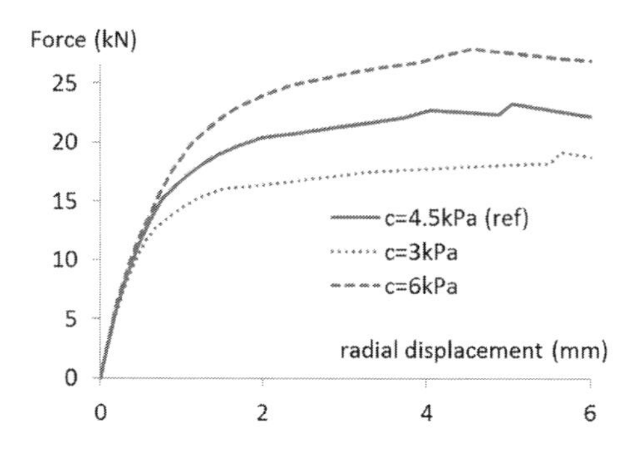

Figure 12. Influence of the backfill cohesion on the load-displacement curve for case 2.

5.5 *Application to the Paris subway tunnels*

The objective is to define alert thresholds for future works nearby existing masonry vaults of the Paris tunnels. The idea is to define these thresholds in terms of maximum admissible values of displacement of the vaults. The comparison between simulations and experiments on well-described brick vaults in the laboratory provides a partial validation of the approach. However, to apply it to actual structures, it is necessary to perform additional investigations on their properties.

In the first place, the geometry of the blocks and joints in the actual vaults is less well-known than in the experimental one. A survey has been carried out by Moreno Regan (2016) on the basis of pictures taken during the construction works (Figure 13), which made it possible to evaluate the typical dimensions of the blocks and of the joints in actual Paris subway tunnels.

In the second place, it was necessary to perform mechanical tests on samples directly taken from the

Figure 13. Photographic analysis of the geometry of the masonry of Paris subway tunnels (Moreno Regan, 2016).

Figure 14. Sample of millstone taken from the masonry (after Moreno Regan, 2016).

subway vaults. A total of more than 170 samples have been extracted, both from the stone blocks (Figure 14) and from the mortar joints, and subjected to various tests aiming at determining their density, porosity, compressive and tensile strength, Young's modulus, and damage parameters.

Some results showed significant discrepancy, which is of particular importance given the sensitivity of the simulation results to the value of some of the parameters. More details can be found in Moreno Regan (2016)

6 CONCLUSION

We have proposed a model, based on that proposed by Zucchini & Lourenço (2002, 2004), that combines a homogenization technique to account for the anisotropy induced by the geometrical arrangement of bricks and joints, and a damage model specifically chosen for our application (simpler than that of the original model).

The preliminary simulations performed to model the experiments by Krajewski and Hojdys (2015) are encouraging from a qualitative point of view, in that they reproduce more or less accurately the failure mechanisms obtained in the laboratory, and also give correct order of magnitude for the displacements of the vault under various types of loading.

However, the determination of the parameters of the model is difficult and, given the influence of some of them on the results, encourage to interpret the results with caution. Further developments include the application of the approach to back-analysis of case histories in which the vaults of the subway have undergone significant deformations, and the extension of the approach to genuinely three-dimensional problems.

ACKNOWLEDGEMENT

This paper is based on the PhD of the second author (Moreno Regan, 2016), carried out between 2013 and 2016 and funded by RATP in the framework of the CIFRE convention 2012–1605, with the support of the ANRT (French National Association for Research and Technology).

REFERENCES

Angelillo, M. 2014. *Mechanics of Masonry Structures*. Springer-Verlag Wien.

Davenne, L., Saouridis, C., & Piau, J. 1989. Un code de calcul pour la prévision du comportement de structures endommageables en béton, en béton armé ou en béton de fibres, *Annales de l'Institut technique du bâtiment et des travaux publics*, 478, 137–156.

Delbecq, J. 1982. *Les ponts en maçonnerie. Constitution et Stabilité.* Ministère des Transports, Direction des routes, Département des Ouvrages d'Art du SETRA.

Heyman, J. 1969. The safety of masonry arches. *International Journal of Mechanical Sciences*, 11(4), 363–385.

Heyman, J. 1997. *The Stone Skeleton*. Cambridge University Press.

Hojdys, L., Kaminski, T., & Krajewski, P. 2013. Experimental and numerical simulation of collapse of masonry arches. *ARCH '13. Proceedings of the 7th International Conference on Arch Bridges. Trogir-Split, Croatia. October 4-6, 2013*, 715–722.

Humbert P, Fezans G, Dubouchet A, Remaud D (2005). CESAR-LCPC: a modeling software package dedicated to civil engineering, Bull. Lab Ponts et Chaussées 256–257: 7–37.

Krajewski, P., & Hojdys, L. 2014. Buried vaults with different types of extrados finishes - experimental tests. In F. P. Chávez (Ed.), *9th International Conference on Structural Analysis of Historical Constructions. Mexico City, Mexico, 14–17 October 2014*.

Krajewski, P., & Hojdys, L. 2015. Experimental studies on buried barrel vaults. *International Journal of Architectural Heritage*, 9(7), 834–843.

La Borderie, C. 2003. *Stratégies et Modèles de Calculs pour les Structures en Béton.* habilitation à diriger les recherches, Université de Pau et des Pays de l'Adour, France.

Lourenço, P., 1996. Computational Strategies for Masonry Structures., Ph.D. dissertation, Technische Universiteit Delf.

Mazars, J. 1986. A description of micro- and macroscale damage of concrete structures. *Engineering Fracture Mechanics*, 25, 729–737.

Moreno Regan, O. 2016. *Study of the behavior of the masonry tunnels of the Paris subway system (in French)*, Ph.D. dissertation, Université Paris-Est.

Moreno Regan, O., Bourgeois, E., Colas, A.S., Chatellier, P., Desbordes, A., Douroux, J.F. 2017. Application of a coupled homogenization-damage model to masonry tunnel vaults, *Computers and Geotechnics,* 83, 132–141.

Moreno Regan, O., Colas, A.S., Chatellier, P., Bourgeois, E., Desbordes, A., Douroux, J.F. 2017. Experimental characterization of the constitutive materials composing an old masonry vaulted tunnel of the Paris subway system, accepted for publication in *International Journal of Architectural Heritage*.

Murakami, S. 2012. Continuum Damage Mechanics. A Continuum Mechanics Approach to the Analysis of Damage and Fracture, Springer Netherlands.

Salençon, J. 1983. *Calcul à la rupture et analyse limite*, Presses de l'ENPC, Paris.

Zucchini, A., & Lourenço, P. 2002. A micro-mechanical model for the homogenisation of masonry. *International Journal of Solids and Structures*, 39, 3233–3255.

Zucchini, A., & Lourenço, P. 2004. A coupled homogenisation-damage model for masonry cracking. *Computers and Structures*, 82(11–12), 917–929.

Numerical Methods in Geotechnical Engineering IX – Cardoso et al. (Eds)
© 2018 Taylor & Francis Group, London, ISBN 978-1-138-33203-4

Investigation of the response of bored tunnels to seismic fault movement

K. Tsiripidou & K. Georgiadis
School of Civil Engineering, Aristotle University of Thessaloniki, Thessaloniki, Greece

ABSTRACT: The response of Twin Circular Bored tunnels (TBM) to seismic fault movement is investigated through three-dimensional finite element analysis. Both normal and reverse fault rupture is considered. The permanent tunnel lining is simulated with separate elastic rings of finite thickness. These rings are connected to each other in the developed finite element models using interface elements with equivalent material properties. The displacement of each elastic ring of the tunnels due to different seismic fault movements is computed and results are presented in terms of the absolute total, vertical and horizontal lining displacements and of the relative shear and normal displacements at the circumferential ring joints. Finally, simplified analyses in which the tunnels are modelled as continuous elastic tubes are also presented. Comparison of the results using the two modelling approaches shows that the conventional simplified assumption of continuous elastic lining cannot accurately reproduce the deformation of the tunnels due to seismic fault rupture.

1 INTRODUCTION

Seismic fault rupture can have devastating consequences on neighboring surface structures or crossing underground structures such as tunnels. Tunnel response during earthquakes has been studied over the last decades mostly through case studies (e.g. Kawakami, 1984, Dean et al. 2006). Past studies on the impacts of Taiwan's severe earthquake in 1999 have shown that tunnels are likely to suffer serious damage when an earthquake occurs especially when they cross a seismic fault (Wang et al. 2001).

Research has been conducted regarding bored tunnel modelling in order to accurately simulate their behaviour during uneven longitudinal loading (Wang et al. 2014), as well as during seismic fault rupture based on case studies (Gregor et al. 2007). Investigations on fault rupture—soil—structure interaction in a uniform soil layer overlying the bedrock level have shown that the existence of rigid underground structures can alter the way that seismic faults propagate to the ground surface (Loli et al. 2012).

In this study, three-dimensional numerical analysis is presented in order to investigate the behaviour of bored tunnel—soil interaction during seismic fault rupture. The phenomenon of seismic fault rupture is considered quasi-static not taking into account any possible effect of the seismic wave propagation during an earthquake. A similar approach was adopted by Loukidis et al. (2009) to study ground movements due to seismic fault rupture. A modelling approach for bored tunnels is presented below, in which the tunnel lining is modelled using volume elements and interface elements are used to model the lining ring joints. Numerical analyses are performed in order to evaluate the tunnel behaviour for normal and reverse fault rupture and several relative fault displacements. The results of the analyses are compared to the corresponding results using the conventional tunnel modelling method widely used in tunnel design, where the structure is modelled as a continuous elastic tube using plate elements.

2 PROBLEM DEFINITION

Twin circular bored tunnels with an 11 m axis-to-axis spacing and 18 m axis depth are modelled. The tunnels have a 5.9 m outer diameter, 0.30 m lining thickness and consist of 1.5 m long segments. They are situated in a 50 m deep sandy clay soil layer underlain by bedrock. Geometrical and loading symmetry are taken advantage of and only half of the problem including one of the tunnels is modelled, as seen in Fig. 1.

Seismic fault rupture at the bedrock level is simulated by applying a prescribed displacement at the left half of the bottom of the model and on the left vertical boundary of the model (Fig. 1). In this way, the model is divided into two parts; the right footwall block part that remains stationary and the left hanging wall block part that moves according

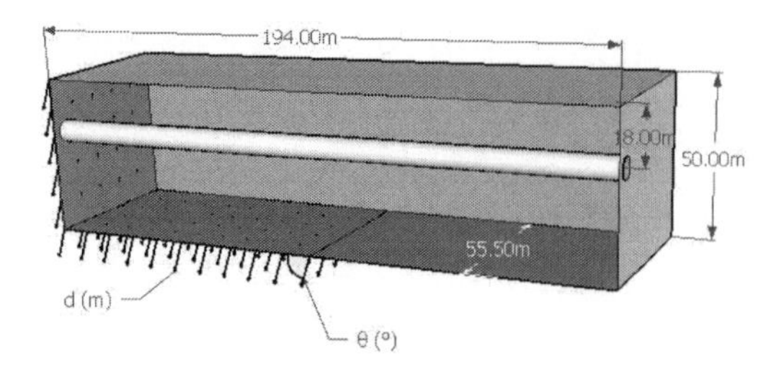

Figure 1. Problem definition: model dimensions and prescribed displacements.

to the prescribed boundary displacement. The two blocks are separated by a zone, the thickness, shape and location of which depends on the way that the seismic fault propagates to the ground surface and is affected by the mechanical behavior of the soil and the presence and the characteristics of the tunnel. Six fault angles were considered, two for normal faults (θ = 50° and 70°), two for vertical slips (θ = 90° and -90°) and two for reverse faults (θ = 110° and 130°). However, the results for only two fault angles are discussed in this paper: one normal fault with θ = 70° and one reverse fault with θ = 110°. Relative sliding at the bedrock level, is applied in increments of prescribed displacement of 12.5, 25 and 50 cm.

3 NUMERICAL MODEL

The numerical analyses were performed with the finite element program Plaxis 3D (Brinkgreve et *al.* 2016). In most analyses, both the soil and the tunnel lining were simulated using 10-noded triangular volume elements. Discrete lining rings were modelled by placing interface elements at the circumferential joints between adjacent rings. This set of analyses is referred to in the following as discrete ring analysis.

Analyses were also performed, according to the conventional assumption that the tunnel lining is a continuous elastic tube. In these analyses 6-noded triangular plate elements were used for the tunnel lining. 12-noded interface elements, which consist of two pairs of nodes, compatible with the 6-noded triangular side of a volume or a plate element, were placed between the lining and the soil, in both sets of analyses. This set of analyses is referred to in the following as elastic tube analysis.

The dimensions of the finite element mesh are shown in Fig. 1. As noted above, the height of the mesh is 50 m. The other two dimensions are: 55.5 m width and 194 m length, and were selected based on preliminary analyses that were also used to determine the appropriate mesh refinement. The finite element mesh for the analyses in which the tunnel lining is modelled with volume elements (discrete ring analyses) consists of approximately 135000 elements and is illustrated in Fig. 2. A similar mesh is used in the conventional analyses, where linear elastic plate elements are used to model the tunnel lining (elastic tube analyses), that consists of slightly fewer finite elements (approximately 110000 elements). It is noted that part of the mesh including soil elements only is not shown in Fig. 2 in order to illustrate the tunnel mesh in the area of the propagated fault. A finer mesh is used in the area through which the fault rupture propagates from the bedrock level up to the ground surface since severe deformations are expected in this area.

3.1 *Modelling of the tunnel lining*

Tunnels constructed using the TBM method are structures made of discrete structural elements. The permanent tunnel lining consists of precast concrete segments, which are modelled in the discrete ring set of analyses as rings of finite thickness (lining). Each ring is connected to its adjacent rings using interface elements as shown in Fig. 3a. These elastoplastic interface elements with equivalent parameters are used to approximate the effect of the circumferential joint interconnecting bolts and shear keys on the tunnel response.

As noted above, analyses were also performed with the conventional elastic tube approach, in which the tunnel lining is modelled with linear elastic plate elements as shown in Fig. 3b. In this way the tunnel is assumed continuous and relative displacements between adjacent rings due to fault rupture cannot be captured. The results of these analyses are compared to the "discrete ring" analyses below, in order to investigate the effect that the discrete nature of the lining has on its behavior during fault movements. In addition to the "conventional" (continuous plate elements) and "discrete ring" (volume elements with interfaces) analyses, preliminary analyses were also conducted in which the tunnel lining was modelled as continuous using volume elements without interfaces in order to investigate the effect of the element type on the results. Very small differences in the computed tunnel displacements were observed between these analyses and the "conventional" analyses, suggesting that the element type has little influence on the tunnel behavior.

4 MATERIALS AND PROPERTIES

4.1 *Tunnel lining*

The thickness of the lining segments is 30 cm, the outer ring diameter is 5.9 meters and the length

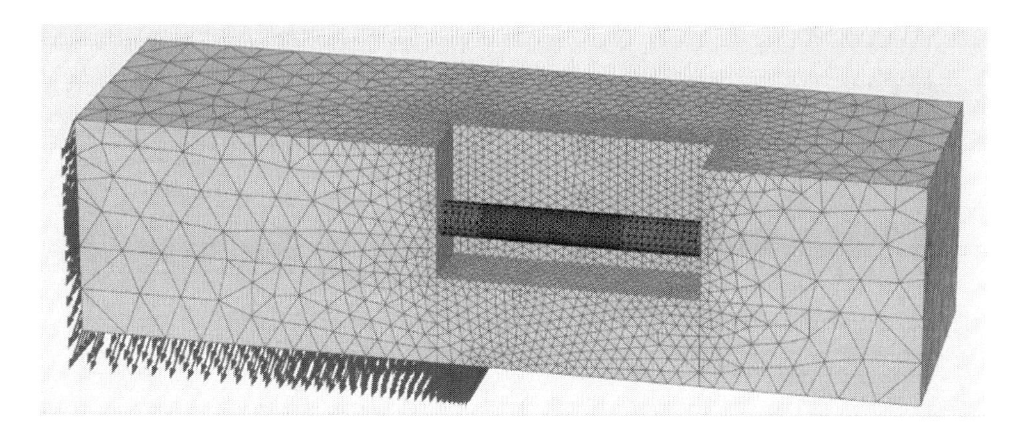

Figure 2. Typical soil and tunnel mesh.

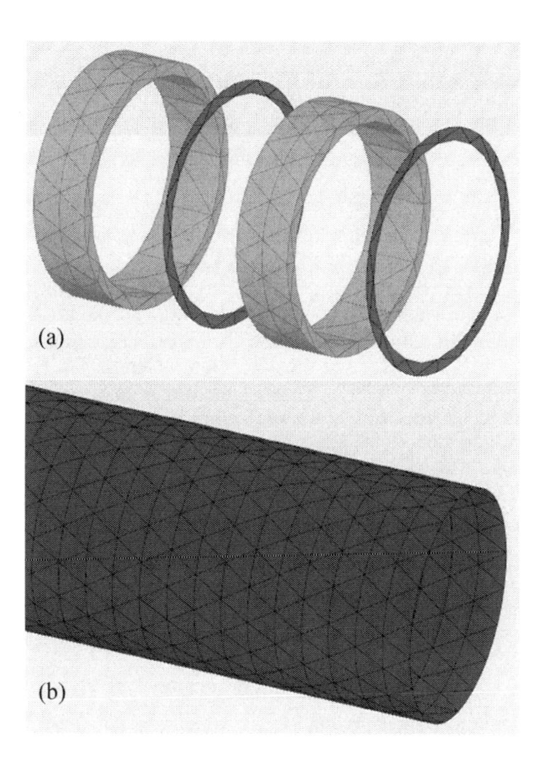

Figure 3. Modelling of the tunnel lining with: (a) discrete rings and interface elements and (b) elastic plate elements.

of each ring is 1.5 meters. The lining is considered to be linear elastic with a Young's modulus of $E = 35{\cdot}10^6$ kPa and a Poisson's ratio of $v = 0.2$.

4.2 *Circumferential joints*

The interconnecting bolts—shear keys systems that link the successive rings are modelled by placing interfaces between the rings in the numerical model. Several equivalent strength parameters of the interfaces were examined in this study, however, only the conservative case of zero tensile and shear strength ($\sigma_{n,t,}$ $\tau = 0$) is presented here. Zero interface strength allows free relative sliding and gap opening between successive rings. Because of this selection of zero strength parameters, the value of the elastic properties of the interfaces, has no effect on their relative sliding and tensile normal behavior. However, the compressive normal behavior that is especially relevant in the case of reverse fault movement, is significantly affected by the choice of Young's modulus of the interface material. This is demonstrated in Fig. 4, which shows the deformation of the tunnel due to reverse fault movement for the cases of (a) zero and (b) $E = 10^6$ kPa interface elasticity modulus. As seen in Fig. 4a, overlapping of the lining rings due to the development of compressive deformation between successive rings, is observed when zero elasticity modulus is assigned to the interfaces between the lining rings. For this reason, all analyses presented in the following paragraphs were performed with $E = 10^6$ kPa (as in the analysis of Fig. 4b). The tunnel deformation for normal fault movement and $E = 10^6$ kPa interface modulus of elasticity is shown in Fig. 5.

4.3 *Soil*

The constitutive model used for the soil mechanical behavior is the Hardening Soil with Small Strain stiffness model. Typical material properties for the Thessaloniki sandy clay used in the analyses are summarized in Table 1. The groundwater table was taken at the ground surface. Both drained and undrained loading conditions were modelled, but only the undrained case is presented here.

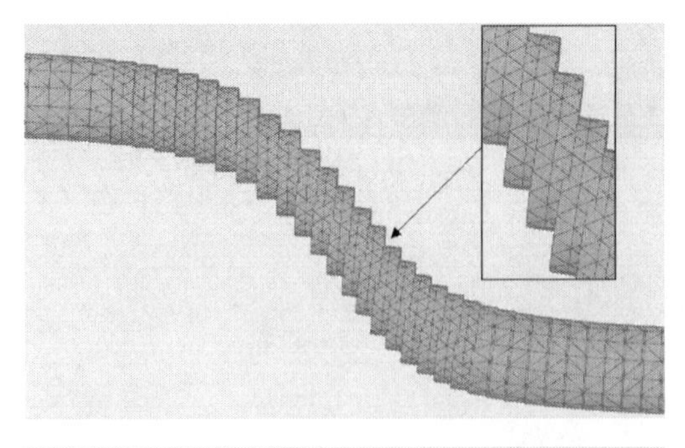

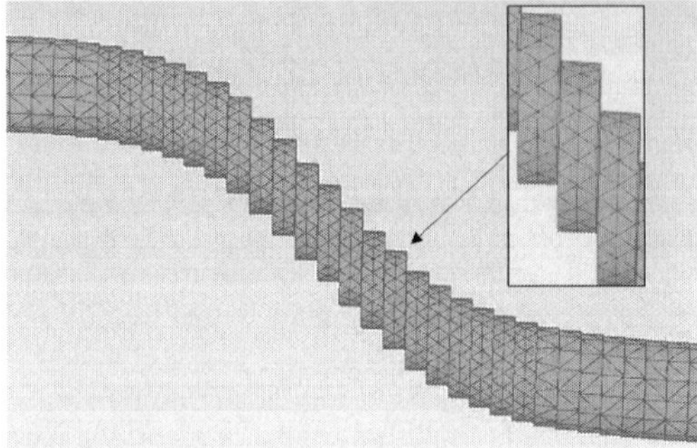

Figure 4. Tunnel deformation due to reverse fault rupture for: (a) E = 0 and (b) E = 10^6 kPa interface modulus of elasticity.

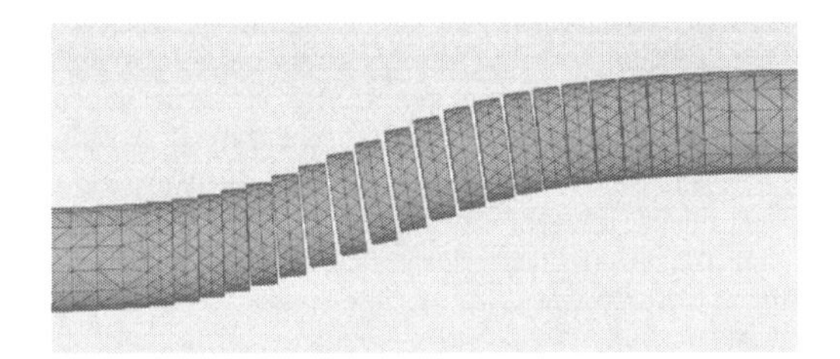

Figure 5. Tunnel deformation due to normal fault rupture.

5 RESULTS

The displacement field at 50 cm fault movement for the case of a $\theta = 70°$ normal fault is presented in Fig. 6. As seen in this figure, the soil mass is separated by the developed shear zone, into two almost rigid blocks, the moving hanging wall block on the left and the stationary footwall block on the right. The tunnel deformation due to the soil movement is examined below in terms of total, vertical and horizontal tunnel lining displacements and shear and normal relative displacements between the individual rings. Due to space limitations only the total and relative displacements at the invert level (z = –20.8 m in Fig. 7) are presented. It is noted, however, that very similar responses were observed at the crown level (z = –15.2 m in Fig. 7).

In order to estimate the relative ring displacements in the analyses with linear elastic plate element tunnel lining, it was assumed that all relative displacements develop at the locations of the circumferential joints between successive rings and that the individual rings are perfectly rigid. Based on this assumption, the relative displacements were calculated as the change in the displacement (either vertical or horizontal) over 1.5 m increments (equal to the individual ring length).

Figures 8 and 9 show the variation of the total, vertical and horizontal displacements with horizontal distance at the tunnel invert level for a $\theta = 110°$ reverse fault and a $\theta = 70°$ normal fault, respectively. The obtained displacement curves for fault movements of d = 12.5 cm, 25 cm and 50 cm are plotted in each case. Both tunnel modelling approaches are examined in these figures. Displacement curves from "discrete ring" analyses, in which the tunnel lining was modelled using volume elements and the joints between the rings were modelled with interface elements, are plotted as continuous curves. The displacement curves obtained from the "conventional" elastic tube analyses are plotted as dashed lines.

It can be observed in Figures 8 and 9 that the conventional elastic tube modelling approach, as expected, cannot reproduce the stepwise variation of total and vertical displacements that the "discrete ring" analyses predict. These displacement steps correspond to vertical slips at the ring joint locations. Steps can also be observed in the horizontal displacement diagram of Figure 9c for the case of normal fault movement. These steps correspond to gap openings between the rings due to the tensile longitudinal loading. In contrast, in the case of reverse fault movement (Figure 8c) in which the tunnel is compressed longitudinally, no relative horizontal displacements take place at

Table 1. Soil parameters.

Parameter	Value	Units
γsat	21	kN/m^3
E50	19	MPa
Eoed	25	MPa
Eur	95	MPa
v′	0.2	–
c′	15	kN/m^2
φ'	26	o
K0,x	0.59	–
K0,y	0.59	–
Go	96.3	MPa
γ0,70	0.0001	–

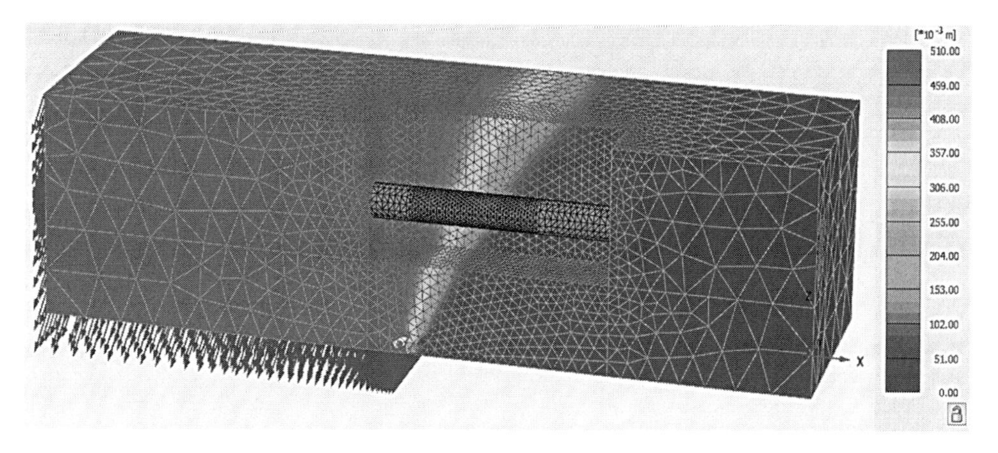

Figure 6. Total displacement field at 50 cm fault movement for 70° normal fault and "discrete ring" tunnel.

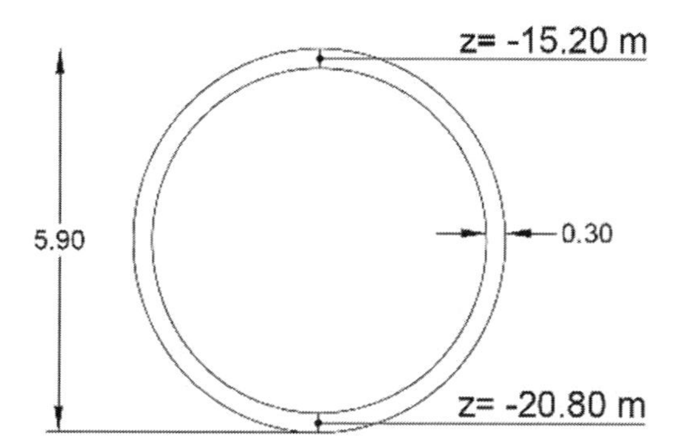

Figure 7. Reference levels of tunnel's section.

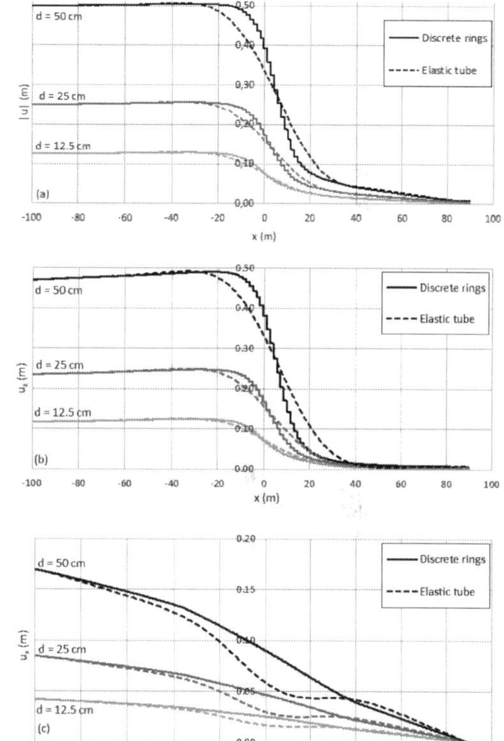

Figure 8. Tunnel displacements for reverse seismic fault rupture (θ = 110°): (a) total displacements, (b) vertical displacements and (c) horizontal displacements.

the ring joints and the longitudinal compressive displacement of the tunnel leads to continuously varying longitudinal strain.

It can also be observed in these figures that in the more realistic "discrete ring" modelling approach the resulting displacement variations are concentrated in the area of the fault zone, leading to large relative displacements between adjacent rings within this zone. In contrast, in the conventional approach the tunnel has a much stiffer response and the displacement variation is distributed in a much larger part of the tunnel.

Comparing the discrete rings analysis results for normal and reverse fault rupture (Figures 8 and 9), it can be observed that while in the case of reverse fault movement the lining rings translate vertically without rotation, as evidenced in the u_z - x plot (Fig. 8b), in the case of normal fault movement, the u_z - x plot (Fig. 9b) indicates rotation of the tunnel lining rings during fault movement. This rotation can be attributed to the gap opening observed in this case. The rotation of the lining rings in this case can be observed more clearly in Fig. 5.

Figures 10 and 11 show the computed relative displacements between the rings for reverse and normal fault rupture, respectively. As also discussed above it can be seen clearly in these figures

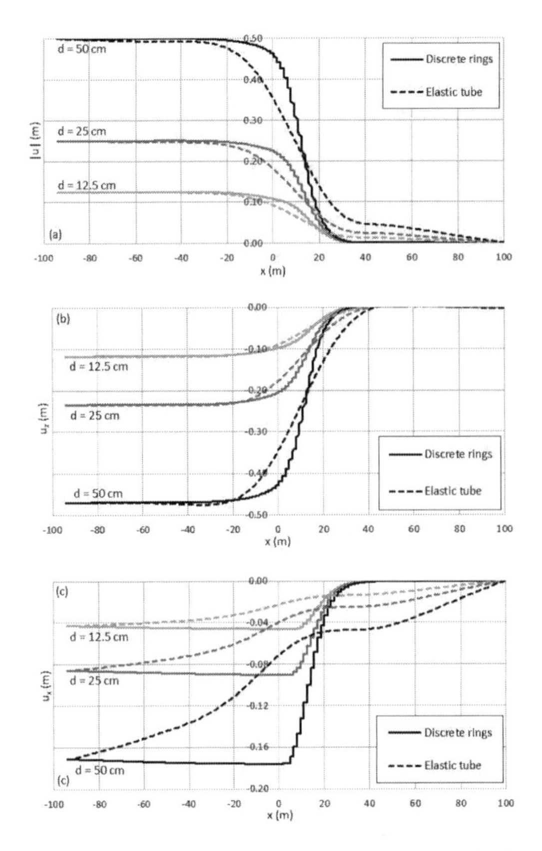

Figure 9. Tunnel displacements for normal seismic fault rupture ($\theta = 70°$): (a) total displacements, (b) vertical displacements and (c) horizontal displacements.

that in the discrete ring analyses results, the relative displacement is concentrated in a relatively small number of circumferential ring joints. This is also evident in Figure 12, which shows the relative displacements at the interfaces, calculated for the discrete ring analyses at 50 cm fault movement.

For fault movement of 50 cm, the discrete ring analyses give a relative displacement at the joints 112% and 96% greater than that computed from the elastic tube analyses for reverse and normal fault movement, respectively.

Comparing the relative displacement curves for d = 25 cm and d = 50 cm, it is also interesting to note that the percentage difference between the calculated relative displacements with the two modelling approaches increases as the fault movement increases. However, the length of the zones of influence does not change. For the discrete ring analyses, it is equal to approximately 23 ring lengths (34.5 m) and 26 ring lengths (39 m) for normal and reverse faults, respectively. For the elastic tube analyses, over 50 ring lengths (75 m) are affected for both normal and reverse faults.

It can also be observed in Figures 10 and 11 that the position at which the maximum relative displacement takes place differs for the normal and reverse fault movements. In the case of normal fault movement ($\theta = 70°$), the maximum relative displacement develops at x = 12.5 m and its value is $|u|_{rel,max} = 3.36$ cm, while in the case of reverse fault movement ($\theta = 110°$), it is observed at x = 4.0 m and its value is $|u|_{rel,max} = 3.76$ cm. This difference can be attributed to the different shear zones that

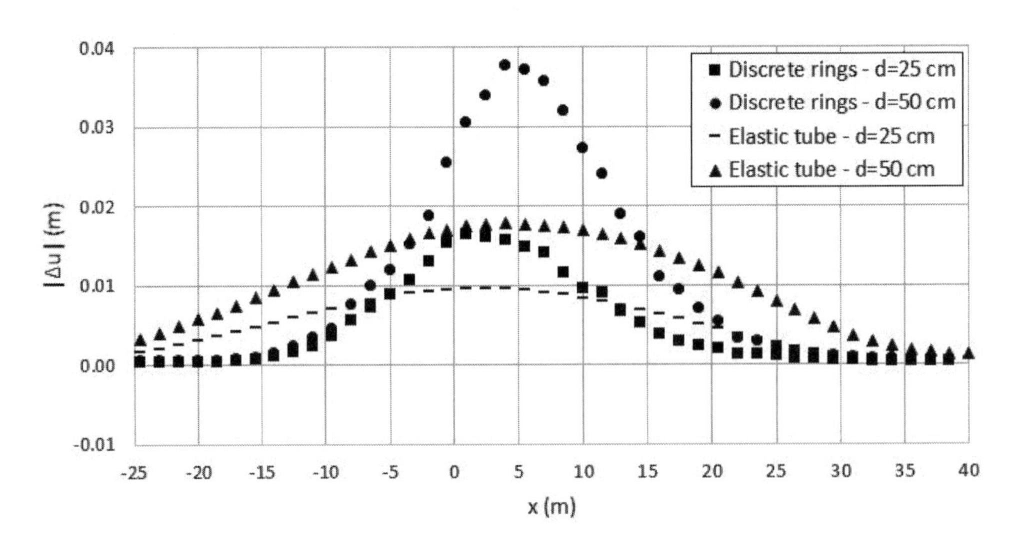

Figure 10. Relative displacements between rings (reverse fault).

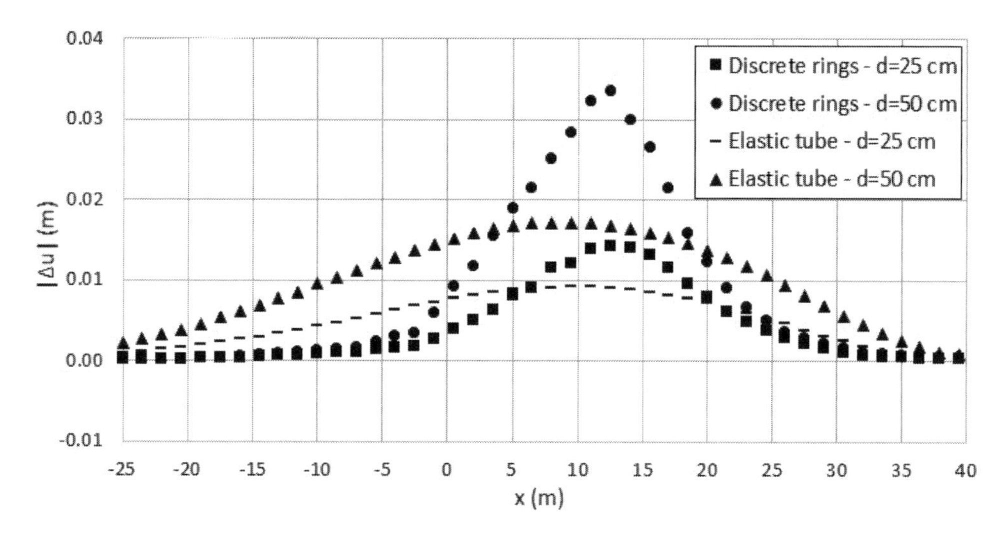

Figure 11. Relative displacements between rings (normal fault).

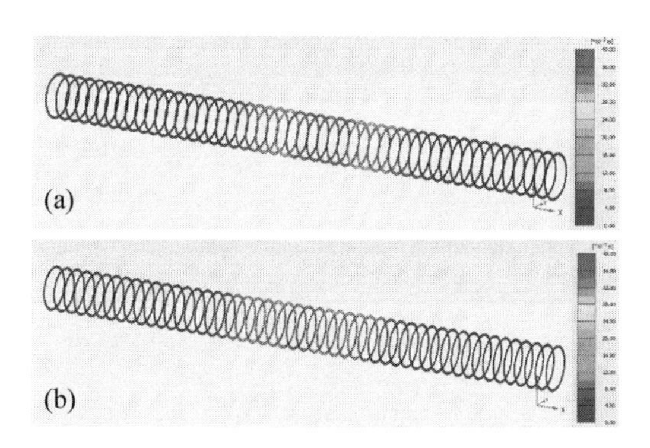

Figure 12. Relative circumferential joint interface displacements for: (a) reverse seismic fault rupture ($\theta = 110°$) and (b) normal seismic fault rupture ($\theta = 70°$).

develop as the fault propagates from the bedrock to the ground surface in the two modes of fault movement. As seen in the same figures, although the position at which the maximum relative displacement takes place also differs in the case of the elastic tube analyses, the computed maximum relative displacements are almost identical.

6 CONCLUSIONS

Finite element analyses of the response of twin bored tunnels to seismic fault rupture were presented. Two modelling approaches were used for the tunnel lining. In the first, the true thickness of the tunnel was modelled using volume elements and the circumferential joints were modelled by placing interface elements between adjacent lining rings. In the second modelling approach the conventional assumption of a continuous tunnel lining was made, as done routinely in design. The tunnel lining was modelled as a continuous elastic tube using plate elements. Fault rupture was simulated be applying a uniform prescribed displacement at part of the bottom and one side of the finite element mesh. Both normal and reverse fault rupture was modelled.

The first modelling approach that uses interface elements to model the circumferential lining joints was shown to be capable of modelling the discrete nature of the tunnel lining. This approach allows the direct calculation of the normal and shear relative joint displacements. These relative displacements are critical in the design of the lining segment interconnecting bolts and shear connectors.

In contrast, the second approximate modelling approach (elastic tube) cannot satisfactorily capture the deformation of the tunnels due to seismic fault rupture. Consequently, the relative lining displacements, which can only be calculated indirectly in this case, are significantly underestimated. It was also observed that this underestimation expressed in terms of the percentage difference between the calculated relative displacements with the two approaches, increases as the fault movement increases.

Finally, it was shown that normal fault movement causes rotation along with the vertical and horizontal translational of the lining rings in the part of the tunnel length that is affected by the fault rupture. Gap opening is also observed in the circumferential joints in this part of the tunnels.

Neither rotation nor gap opening is observed in the case of reverse fault rupture. The maximum relative joint displacements computed for reverse and normal faults are, however, similar.

REFERENCES

Brinkgreve, R.B.J., Kumarswamy, S., Swolfs, W.M. 2015. Plaxis 3D user's manual. Netherlands: Plaxis B.V.

Dean, A., Young, D.J., Kramer, G.E. 2006. Analyses of underground structures crossing an active fault in Coronado, California. In Barták, Hrdina, Romancov & Zlámal (eds), Underground Space—the 4th Dimension on Metropolises (1): 445–450. London: Taylor & Francis Group.

Gregor, T., Garrod, B., Young, D. 2007. The Use and Performance of Precast Concrete Tunnel Linings in Seismic Areas. *Proceedings of the 10th Iaeg Congress*: 679. Nottingham: Geological Society Publishing House.

Kawakami, Hideji 1984. Evaluation of deformation of tunnel structure due to Izu-Oshima-Kinkai earthquake of 1978. *Earthquake engineering and structural dynamics* 12(3): 369–383.

Loli, M., Bransby M.F., Anastasopoulos, I., Gazetas, G. 2012. Interaction of caisson foundations with a seismically rupturing normal fault: centrifuge testing versus numerical simulation. *Géotechnique* 62(1): 29–43.

Loukidis, D., Bouckovalas G.D., Papadimitriou, A.G. 2009. Analysis of fault rupture propagation through uniform soil cover. *Soil Dynamics and Earthquake Engineering* 29(11–12): 1389–1404.

Wang, Z., Wang, L., Li, L., Wang, J. 2014. Failure mechanism of tunnel lining joints and bolts with uneven longitudinal ground settlement. *Tunnelling and Underground Space Technology* 40: 300–308.

Wang, W.L., Wang, T.T., Su, J.J., Lin, C.H., Seng, C.R., Huang, T.H. 2001. Assessment of damage in mountain tunnels due to the Taiwan Chi-Chi Earthquake. *Tunnelling and Underground Space Technology* 16: 133–150.

Numerical Methods in Geotechnical Engineering IX – Cardoso et al. (Eds)
© 2018 Taylor & Francis Group, London, ISBN 978-1-138-33203-4

Numerical study on water-jet cutting technique applied in underground coal mines

W.L. Gong, Y.X. Sun, X. Gao, J.L. Feng & Z.H. Li
State Key Laboratory for Geomechanics and Deep Underground Engineering, China University of Mining and Technology, Beijing, China

L.R. Sousa
State Key Laboratory for Geomechanics and Deep Underground Engineering, China University of Mining and Technology, Beijing, China
Construct, University of Porto, Portugal

G.X. Xie
Municipal Construction Engineering Co. Ltd., Beijing, China

ABSTRACT: The present paper is related to the development of a cutting system based on water jet with directly injected abrasives (DIA-jet). DIA-jet intends to cut rock bolts during mining works. In order to optimize the design of the jet nozzle and the abrasive tank, computational fluid dynamics (CFD) were used to obtain the behavior of the liquid-solid two-phase flow inside the nozzle and abrasive tank where the intense liquid-solid interactions took place. Based on numerical analyses, better understanding of the two-phase flow was achieved and the optimal parameters, including length of the jet nozzle, positions and numbers of the water orifices, as well as their relation to the slurry outlet on the abrasive tank, were obtained. Finally, performance of the water jet cutting system designed based on the optimized parameters from the numerical simulation was experimentally verified.

1 INTRODUCTION

Roof State Key Laboratory for Geomechanics and Deep Underground Engineering, China University of Mining and Technology, Beijing, China caving timely is critical for mitigate dynamic disasters such as coal-gas outbursts during the face development in underground coal mines. For caving the bolt-meshed supported roof, cutting of the bolt or steel mesh will be required. Cutting operation in the burst-prone coal mines could be dangerous, since the sparks or heat produced may trigger a coal-gas outburst. Therefore, the development of a safety cutting technique for the roof caving is very important.

Water jet added with abrasive particles (e.g. silica or garnet), termed as abrasive water jet (AWJ), can be utilized to cut a wide range of materials. It is suitable for cutting operations under inflammable and explosive environment. By accelerating the abrasive particles with a high-speed water jet, cutting ability of the AWJ is much higher than that of the pure water jet (Gong et al., 2005). Thus, AWJ cutting techniques have found wider applications in varied industries.

AWJ is generally classified into two types by article-accelerating mechanisms (Nanduri et al., 1999). One is the abrasive-entrained waterjet (AEWJ) where a system is used to feed abrasive particles, and another is to directly inject to feed abrasives (DIA-jet) where abrasive particles are premixed with water in a high pressure tank (Gong & An, 2001). In the AWJ system, jet nozzle and the abrasive tank are two important component parts. Better understanding of the multiphase flow mechanisms inside them is crucial for the optimal design of the cutting system based on AWJ techniques.

The AEWJ technique has been investigated extensively, for instance, in relation to the entrainment of air on the particle acceleration process (Tazibt et al., 1996); to the energy transfer during the mixing and acceleration process of abrasive particles (Momber, 2001); to the resonant frequency for self-resonated AWJ (Gong et al., 2005); to the internal flow and to the particle movement in the abrasive waterjet nozzle (Long et al., 2017); to the ultra-fine grading (Gong & An, 2001); and to the cutting polycrystalline diamond and meat with bone and multiple layered materials (Axinte et al., 2009; Wang & Shanmugam, 2009; Lee et al., 2012).

DIA-jet system can be made compact in size and volume. However, the system should stop working when feeding the abrasives. This hinders the application of the DIA-jet techniques. In contrast, numbers of the publications on the DIA-jet are relatively small. A recent case reported is application of the DIA-jet in cutting notch slot for blocking the blast-induced vibration in tunnel excavation (Kim & Song, 2015). Investigations on the mechanisms of the multi-phase flow in the DIA-jet system are rarely reported in recent decades.

The present paper is related to the development of a DIA-jet based cutting system to cut the bolt supporting immediate roof in the mining works. In order to optimize parameters and structures of the jet nozzle and abrasive tank, computational fluid dynamics (CFD) was used to obtain the behavior of the multiphase flow inside them which will not be obtained by the experimental methods because of the complexity of the liquid-solid interactions. The experiments were conducted for testing the system performance designed based on the CFD modeling results.

2 CUTTING SYSTEM DEVELOPMENT

The requirements for designing a cutting tool for the burst-prone coal mines involve; 1) no thermal effect and sparks in the cutting process; 2) ability to cut materials with a wide range of hardness; 3) light in weight and small in volume, so as to move easily in the roadway tunnels or mine workings. Under these considerations, a water-jet cutting system was developed based on DIA-jet technique as shown schematically in Figure 1.

It is seen that the high-pressure water pump was powered by a hydraulic motor driven by the emulsion fluid from the pipe line of the hydraulic support system. The hydraulic motor is much smaller than that of the electric motor at same rated power, as well as free from the sparking problem during the working. The system has a working pressure of 30 MPa, flow rate 12 *l*/min, and the rated power 11 kW. Major task for the developed cutting tool is to cut the tray seat mounted on the rebar in the upper corner of the roof. Key components in the developed cutting tool are the jet nozzle and abrasive tank which will be numerically analyzed in the following.

3 PHYSICAL AND NUMERICAL MODELS

3.1 *Abrasive tank*

Configuration for the abrasive tank is shown in Figure 2. It is seen that the cylindrical tank has a cone-shaped bottom where water inlet orifices and slurry outlet are mounted. Pressurized water enters into the tank and is mixed with abrasive particles. The cone-shaped body, indicated with dashed line in the figure, plays a key role for mobilizing the abrasive particles to a quasi-fluid state (slurry). The question that needs an answer is how many water orifices are required and how they are placed on the bottom to fluidize the abrasives and flowing out smoothly. According to axisymmetric nature of the cone-shaped body, the flow is consequently axisymmetric.

Figure 3 shows the studied area where a triangular finite mesh was generated. A triangular

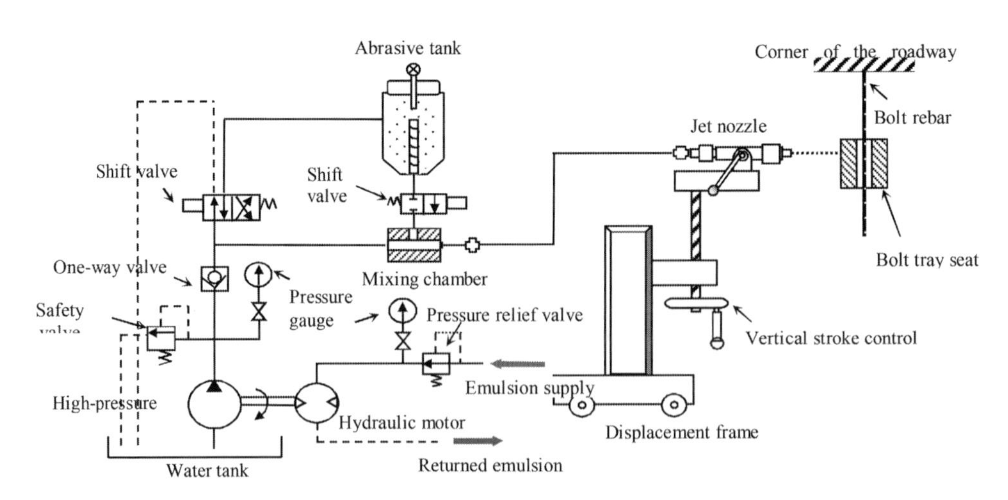

Figure 1. Schematic diagram for the DIA-jet based cutting system used in underground mines.

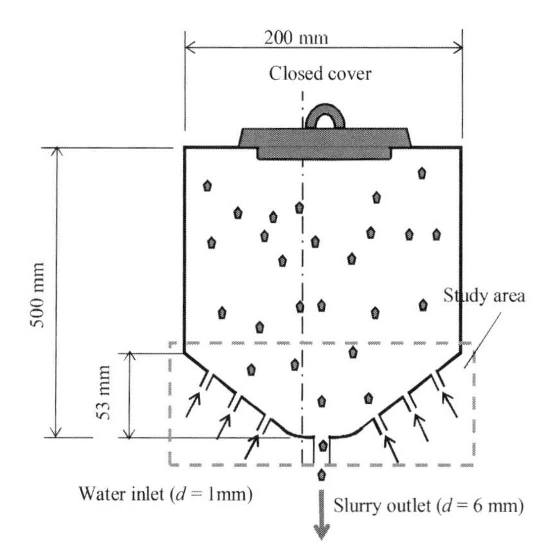

Figure 2. Schematic of the abrasive tank.

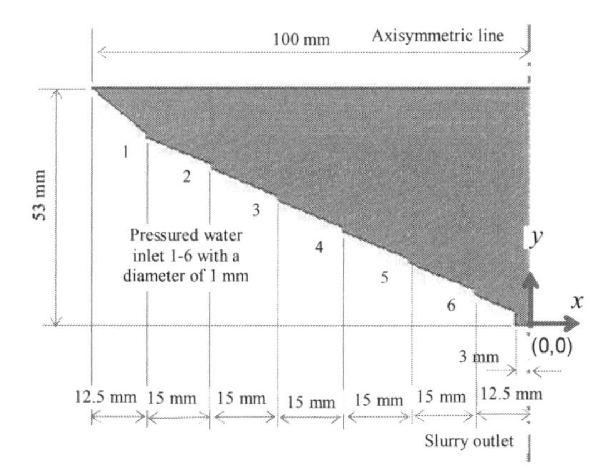

Figure 3. Domain for the study area of the abrasive tank.

structured grid was used to discretize the computational domain. The origin of the coordinate system was placed at end of the slurry outlet and y axis along the axis of symmetry line. The slurry outlet was set as pressure outlet (at atmospheric pressure 101,325 Pa).

3.2 *Jet nozzle*

Figure 4a shows schematically the jet nozzle. It has three segments, i.e., the inlet segment (6 mm in diameter) for connecting the working pipe line, convergence segment for transition from the inlet to the jet orifice, and the collimating segment for accelerating the particles (0.8 mm in diameter). The fluidized abrasives are further mixed in the

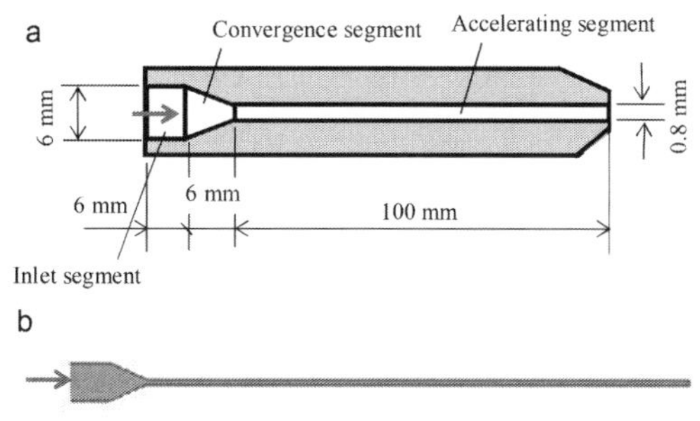

Figure 4. Jet nozzle: (a) scheme of the jet nozzle, and (b) meshed area of the flow field in the nozzle.

chamber with pressurized water and then enter into the jet nozzle. Figure 4b shows the studied area where a triangular finite mesh was generated. The nozzle inlet velocity was set at as 100 m/s and outlet was set at atmospheric pressure 101325 Pa. The question that needs to answer is how long is the collimating segment of the nozzle along which the accelerating particles have the maximum velocity.

3.3 *Numerical models*

For the flow field computations, commercial CFD code FLUENT 6.3 was used. It is a powerful finite-volume-based program package for modelling different flowing problems. The basic idea is to divide the calculation area into a series of non-repetitive control grid volumes. Each control volume is integrated by differential equations to be solved, and then come to a set of discrete equations where the unknowns are the values of the dependent variables at the grid volumes.

In this work, liquid-particle two phase flow was considered and Euler-Euler approach was adopted. The continuous phase is water and no-slip velocity condition was imposed on the wall boundaries. The flow simulation work was conducted under the postulation that the fluid is steady and incompressible. Standard two-phase flow and turbulent governing equations and finite volume CFD methods used in this paper could be seen in the referred publications (Mahmoud et al., 2010; Cebeci et al., 2005; Guan and Tu, 2010).

4 RESULTS AND DISCUSSIONS

4.1 *Abrasive tank*

In the numerical analysis, the second order upwind scheme was used for solving advection terms in the conservation equations for the mass and momentum, and the standard $k-\varepsilon$ models for solving the

Reynolds stresses. QUICK (Quadratic Upwind Interpolation of Convective Kinematics) scheme was employed to discretize the phase volume fraction, and SIMPLE (Semi-Implicit Method for Pressure-Linked Equations) scheme were employed for solving the discretized equations (Mahmoud et al., 2010). For the particle phase, VOF (Volume Of Fraction) model was applied in the computation which is considered suitable for simulating the motions of the particles having a wide distribution in the water phase (Mahmoud et al., 2010).

Six water orifices (numbered from 1–6) were placed on the meridian plane of the cone-shaped body. The water orifice was set as velocity inlet and their velocities were set at 100, 98, 96, 94 and 90 m/s, respectively. The volume fraction of abrasive particles was set as 25% in the tank. At beginning, the abrasive particles were assumed being stationary and distributed along a line. The initial coordinate of the particle group is set in a straight line with the starting point (–10, 10) mm and the end point is (–90, 52) mm. The particles diameter is 0.178 mm and density is 3000 kg/m^3.

Figure 5 shows velocity distributions obtained in the simulation. It is seen that there is a large vortex in the middle of the axisymmetric center line. At the slurry outlet, the velocity of the particles is the highest.

The velocity vector field is presented in Figure 6. The maximum velocities of the outlet is: $u_x = 204$ m/s and $u_y = 215$ m/s. The lateral velocity produces vortices and motion of the vortex makes the abrasive fully mixed with high-pressure water forming a fluidized two-phase flow.

The pressure distribution in the cone-shaped area is seen in Figures 7 and 8. It is seen that the outlet is the zone with the lowest pressures. In the vicinity of the outlet, two very high pressure zones exist, indicating the fact that the slurry will flow out of the orifice driven by the high pressure nearby the outlet area.

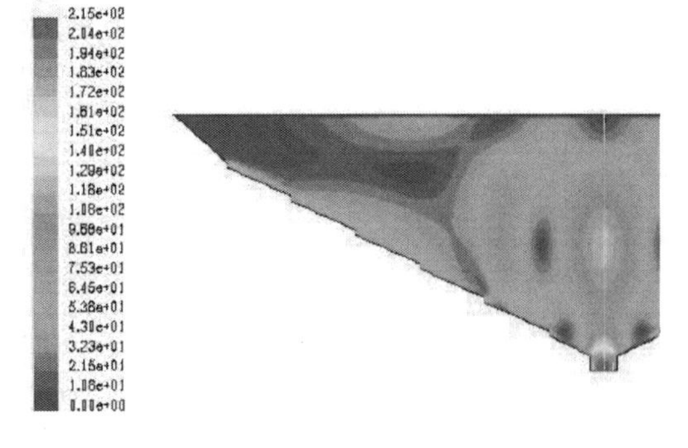

Figure 5. Contour of the velocity distribution in the cone-shaped area of the abrasive tank.

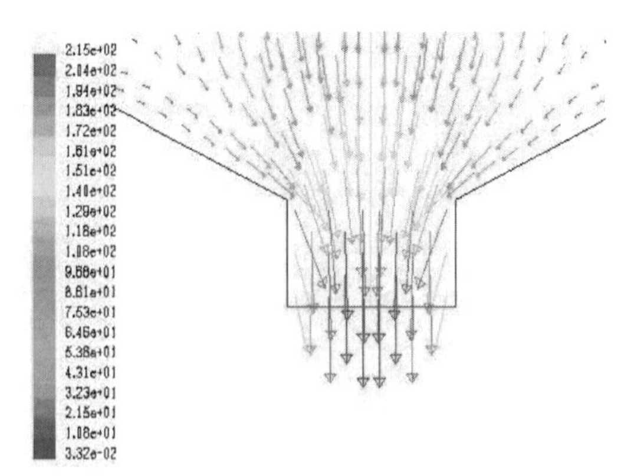

Figure 6. Velocity vectors around the slurry outlet of the tank.

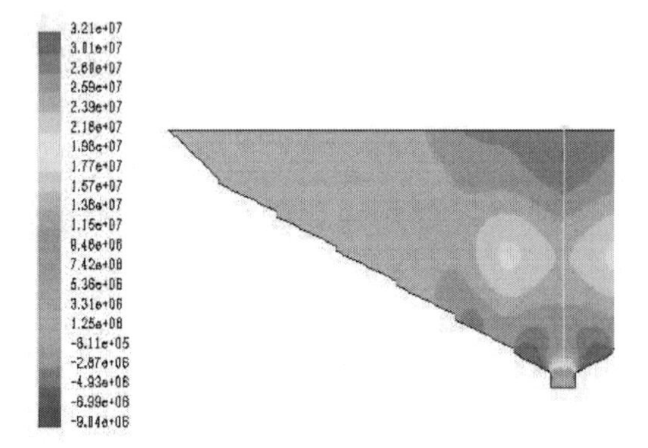

Figure 7. Contour of the pressure distribution in the cone-shaped area of the abrasive tank.

Figure 9 shows the particle trajectories. It is seen that they move from the initial place at the coordinates (–10, 10) mm and (–90, 52) mm to the outlet along large circles, demonstrating the fact that the particles are fully mobilized and flowing into the outlet smoothly.

Figure 10 shows the flow paths for the liquid injected from the water orifices. Compared with Figure 9, it is seen that the particle trajectories are consistent with the flowing paths for the water representing a good fluidization effect.

The velocity distribution along the axisymmetric line of the tank is shown in Figure 11. The maximum outlet velocity is at the slurry outlet whose coordinates are (0, 0) mm as shown in Figure 3.

Based on the numerical analysis, following optimal results for designing the abrasive tank were obtained: 1) when the water orifices 2, 4 and 6 are opened while 1, 3 and 5 are closed, the maximum velocity is situated at the outlet area; 2) when the orifices 1, 3 and 5 are opened, and the others are

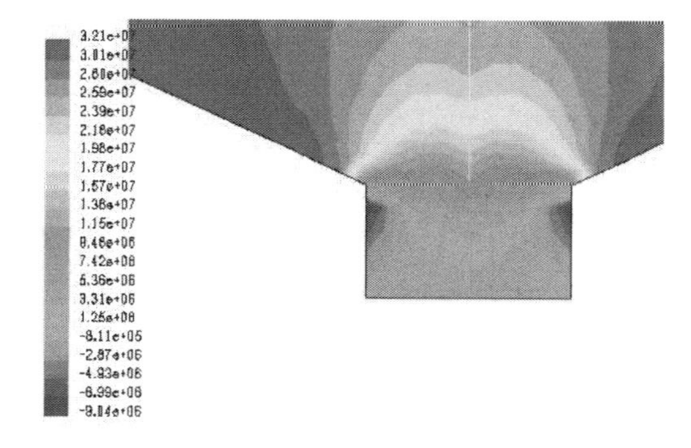

Figure 8. Pressure field around the slurry outlet of the tank.

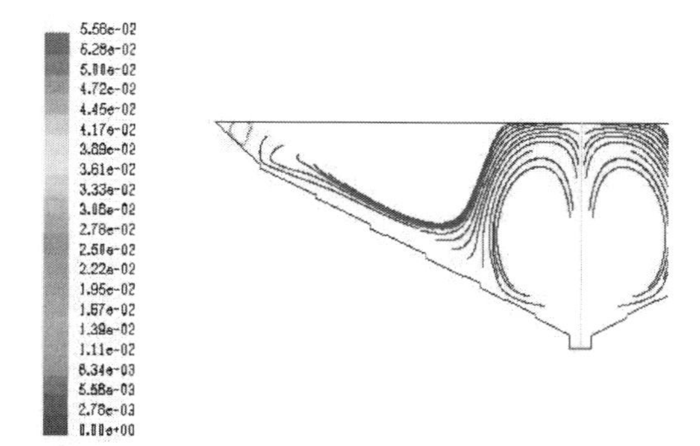

Figure 9. Trajectories for the mobilized particle.

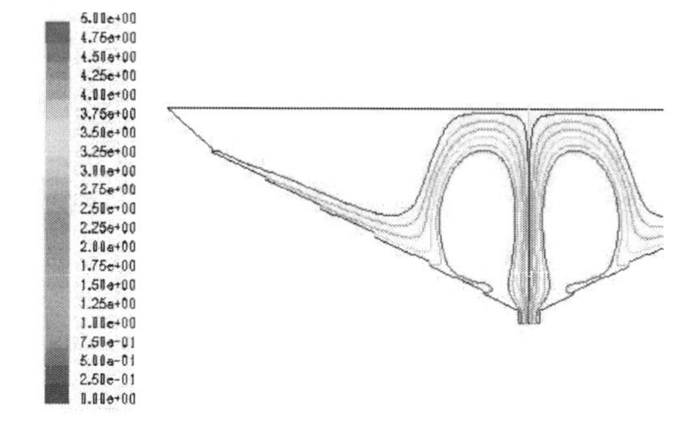

Figure 10. Flow paths of the water.

closed, the maximum velocity is no longer at the outlet; 3) when the orifices 1 and 6 are opened and the rest are closed, the velocity value is not at the outlet, and 4) when the orifices 3, 4 and 5 are opened and the else are closed, the best fluidization effect is achieved (the outlet velocity is the highest).

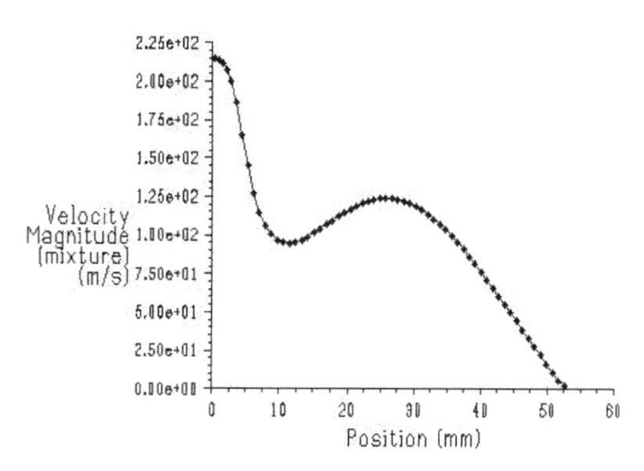

Figure 11. Velocity distribution along the half outlet orifice (also see Figure 3).

4.2 *Jet nozzle*

The key parameter for jet nozzle is the length of the collimating segment and nozzle orifice. In this research, the nozzle orifice diameter was chosen at 0.8 mm. The nozzle length includes a 6 mm long inlet segment and a 6 mm long convergence segment. The task for the numerical analysis is to look for the optimal lengths of the collimating segment along which the abrasive particles could be accelerated to the maximum. In the computation, the inlet velocity was set at 100 m/s. Model implementation of the two phase flow and turbulence for the nozzle is the same as that of the abrasive tank in the previous section.

Figures 12–15 illustrate the particle velocity distribution along the nozzle centerline over 0–30, 30–60, 60–80, and 80–100 mm, respectively, including all the segments. It is seen from that in the inlet segment (0–6 mm), the velocity is almost constant; in the convergence segment (6–12 mm), the velocity increases sharply; and when the particles enter into the collimating segment (12–30 mm), particle velocity increases slowly. The linear increase of the particle velocity with respect to the longitudinal position represent the linear accelerating process (Figure 13). At the position 75 mm, the velocity attains the maximum and then decreases (Figure 14). Finally Figure 15 illustrates the decrease of the velocity process. The general law of the particle accelerating process can be summed up as: rapid acceleration, linear acceleration, acceleration to the maximum and deceleration.

Details of the particle in the different segments in the jet nozzle can be observed on the velocity vector maps. Figure 16–17 show the velocity vector field of the jet nozzle in two intervals, 0–30 and 30–60 mm. A sharp increase of the velocity for the particles in the convergence segment is seen (Figure 16). Linear increase of the velocity is seen in Figure 17.

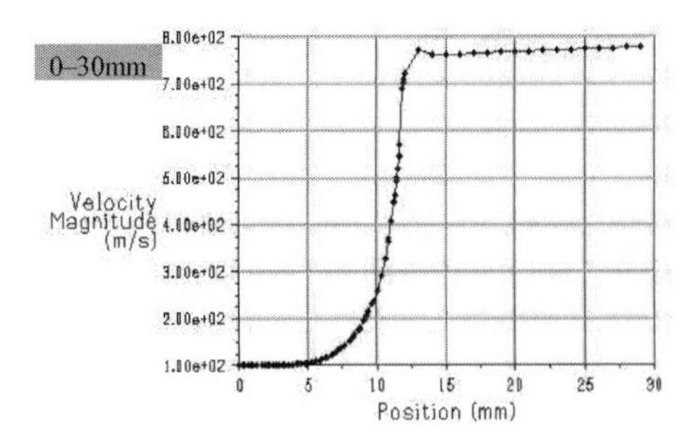

Figure 12. Particle velocity distributions over 0–30 mm including the inlet segment (0–6 mm), the convergence segment (6–12 mm) and part of the collimating segment (12–30 mm).

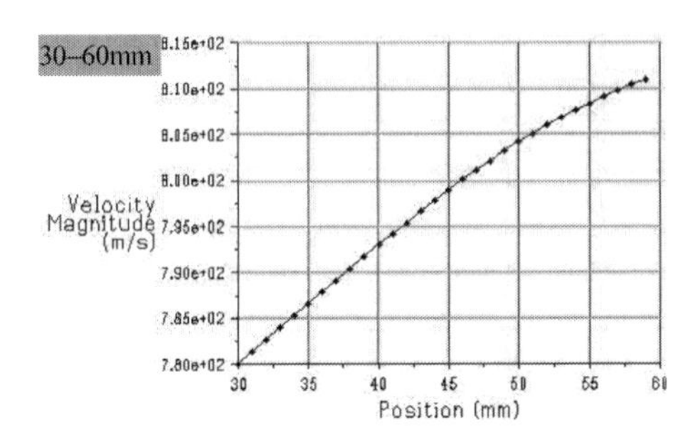

Figure 13. Particle velocity distributions over 30 – 60 mm in the collimating segment.

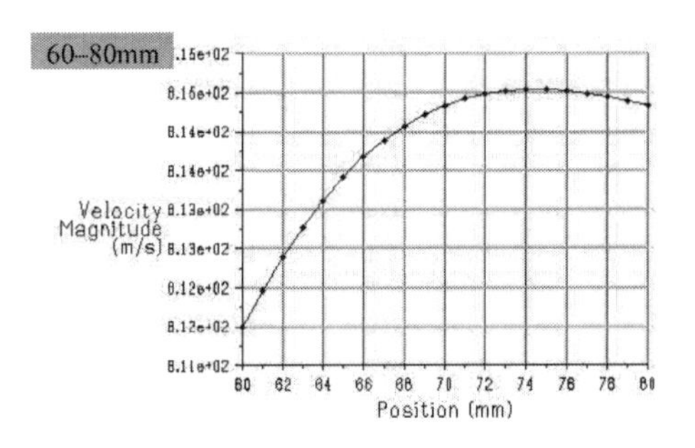

Figure 14. Particle velocity distributions over 60–80 mm in the collimating segment.

One of the principles for designing the abrasive jet cutting nozzle is the selection of the optimal length for the nozzle at which particles can be accelerated to the maximum. Thus, numerical analyses for optimizing the length of the nozzle at a predetermined orifice diameter of 0.8 mm were carried out. The obtained results about the length of the collimating lengths against positions of the maximum velocity for the nozzle are reported in Table 1. It is seen that at position around 72 mm, the maximum particle velocity was attained for the most of the collimating pipe lengths. Therefore, 72 mm length would be the optimal length for the collimating segment.

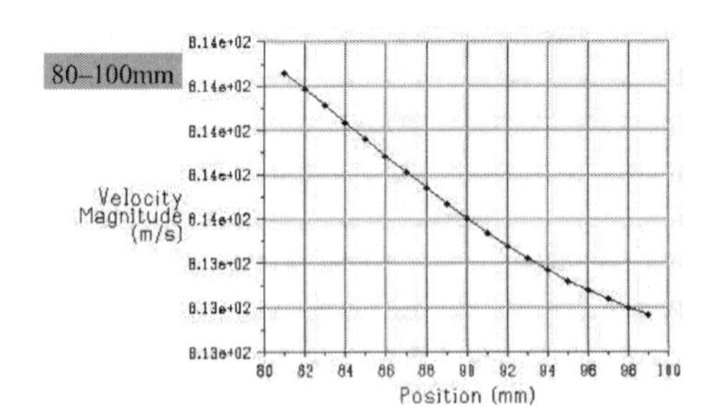

Figure 15. Particle velocity distributions over 80–100 mm in the collimating segment.

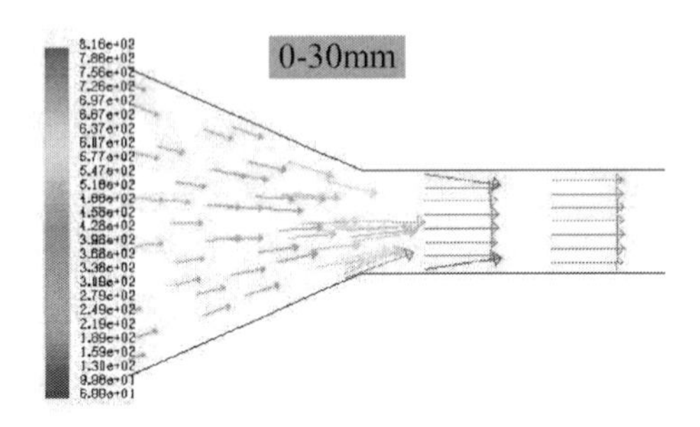

Figure 16. Particle velocity vector map in different over 0–30 mm section in the jet nozzle.

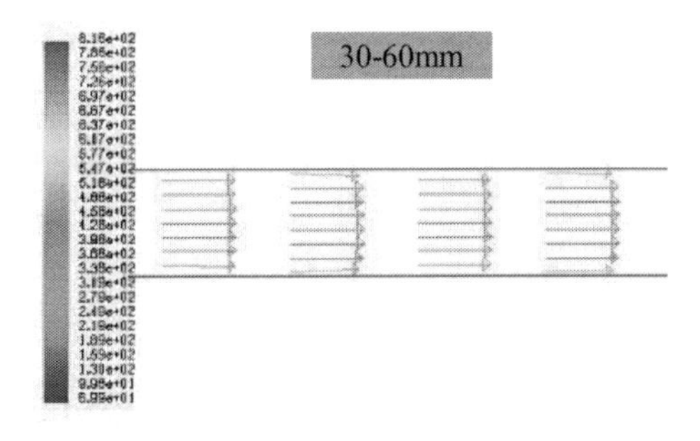

Figure 17. Particle velocity vector map over 30–60 mm in the nozzle.

5 EXPERIMENTAL RESULTS

For validating the system design using the key parameters for the cutting system design obtained in the CFD simulation, materials cutting experiments were conducted at High Pressure Water Jet Laboratory (HPWJL) at China University of Mining and Technology Beijing (CUMTB). The

Table 1. Length of the collimating segment vs. positions where the maximum velocity of the particles is attained (counted from the start point of the collimating segment).

Length of the collimating segment (mm)	70	73	74	75	95	100	140	150
Positions of the maximum velocity (mm)	70	73	72	72	73	75	105	114

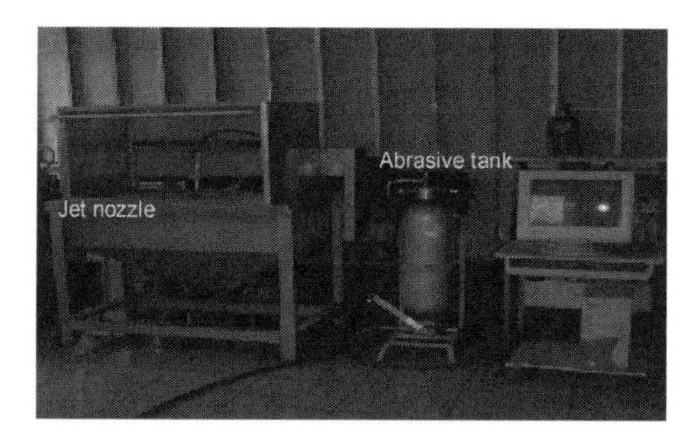

Figure 19. Prototype of the DIA-jet cutting system.

abrasive material is garnet with an average diameter of 0.2 mm. The orifice diameter of the jet nozzle is 0.8 mm and the collimating segment length is 72 mm. Two types of samples were selected. One is granite with UCS equal to 160 MPa, Young's modulus 60 GPa, Poisson's ratio 0.2 and another is bolt tray seat made of # 45 steel with tensile strength of 588 MPa, creep strength of 294 MPa, hardness of 217 HB and impact toughness 294 kJ/m.

Figure 19 shows the prototype of the DIA-jet cutting system and the samples on the test stand. The jet nozzle and abrasive tank are indicated in the picture.

Figure 20a presents the 100 × 100 × 180 mm^3 granite specimen sample from Laizhou, Shandong province, and Figure 20b the bolt tray seat with a cylindrical configuration. Velocity of the jet nozzle for cutting granite is 0.83 mm/s and working pressure for the DIA-jet is 40 MPa. Moving velocity of the jet nozzle for cutting bolt tray seat is 1 mm/s and working pressure for the DIA-jet is 35 MPa.

Figure 21 shows the plot for cutting depth against standoff distance. It is seen that the optimal standoff is 3 mm at which the cutting depth is the greatest. The cutting depths decrease with the increase of the standoff distance. At a standoff larger than 5 mm, the jet loses its cutting ability. At jet pressure of 40 MPa and moving velocity 0.83 mm/s, the cutting depth is 100 mm.

Figure 22 shows the profile for cutting depths vs. moving velocity of the jet nozzle. It is seen that

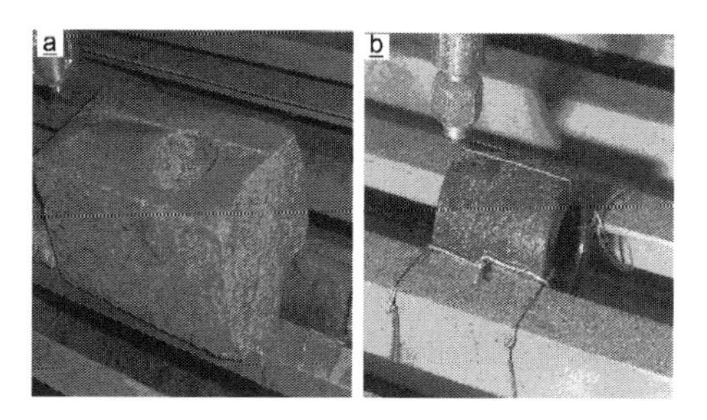

Figure 20. The testing samples: (a) granite; and (b) bolt tray.

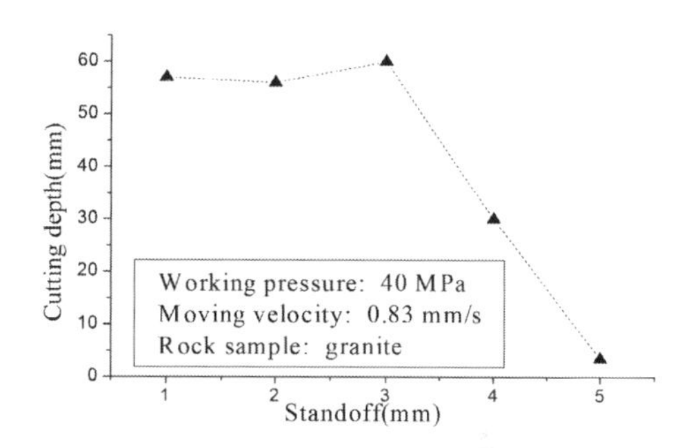

Figure 21. Experimental results for cutting granite: standoff vs. cutting depth.

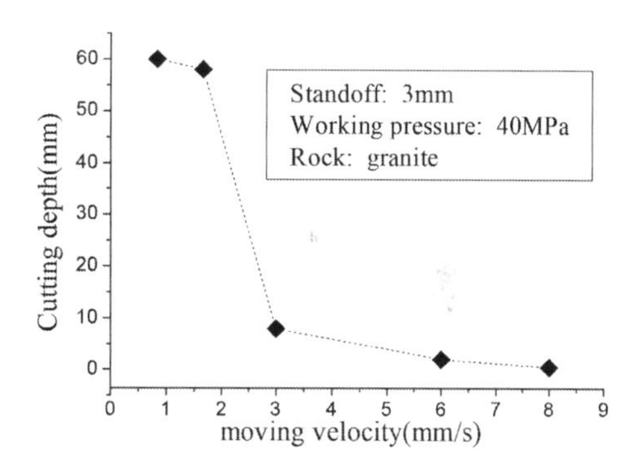

Figure 22. Experimental results for cutting granite: moving velocity vs. cutting depth.

the cutting depths decrease dramatically with the increase of the moving velocity of the jet nozzle. When the moving velocity is larger than 3 mm/s, the DIA-jet is unable to cut the granite sample with a jet pressure of 40 MPa and standoff of 3 mm.

Figure 23 illustrates the relation between the cutting velocity and standoff distance for cutting bolt tray. At a standoff distance of 2 mm, the cutting velocity is the greatest. It takes one minute for cutting a single bolt tray which is 60 mm long and

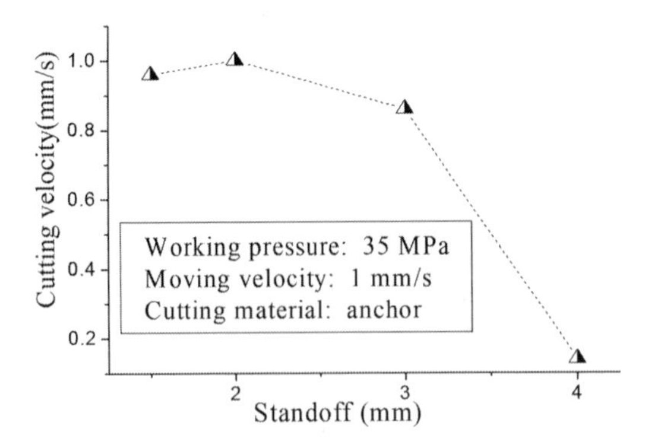

Figure 23. Experimental results for cutting bolt tray seat: standoff vs. moving velocity.

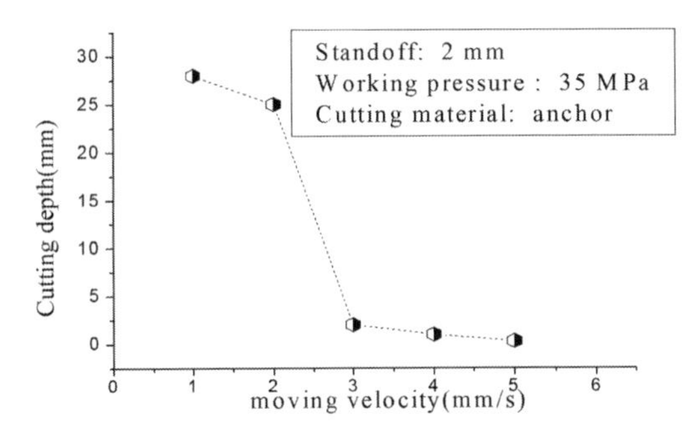

Figure 24. Experimental results for cutting bolt tray seat: nozzle movement velocity vs. cutting depth.

28 mm thick, at working pressure 35 MPa, standoff 2 mm, cutting velocity 1 mm/s.

Figure 24 shows the cutting depths vs. jet nozzle moving velocity. It is seen that the cutting depths decrease dramatically with the increase of the moving velocity of the jet nozzle. When the moving velocity is larger than 3 mm/s, the DIA-jet is unable to cut the bolt tray seat sample.

6 CONCLUSIONS

Based on the CFD simulations conducted in this research, the following optimal parameters were obtained for the design of the DIA-jet cutting system for the burst-prone underground coal mines: 1) the best fluidization effect is achieved by playing three water orifices on the cone-shaped bottom at the positions 3, 4 and 5; 2) the optimal length for the jet nozzle collimating segment is 72 mm. The obtained results for designing the cutting system were experimentally verified.

ACKNOWLEDGEMENT

The Key National Research and Development program of China under Grant No.2016YFC0600901 is gratefully acknowledged.

REFERENCES

Axinte D.A., Srinivasu D.S., Kong M.C. & Butler-Smith P.W. 2009.Abrasive waterjet cutting of polycrystalline diamond: a preliminary investigation. Int J. Mach. Tools Manuf., 49, 797–803.

Cebeci T., Shao J.P., Kafyeke F. & Laurendeau E. 2005. Computation Fluid Dynamics for Engineers. Long Beach, California: Horizons Publishing.

Gong W.L. & An L.Q. 2001. Investigation and application of the high-pressure water jet ultra-fine comminution technology. China Power Technology, 7(3), 35–40.

Gong W.L., An L.Q. & Chi L.L. 2005.Study of Inherent Frequency of Helmholtz Resonator. Proceedings of the 2005 WJTA American Waterjet Conf., August 21–23, 2005, Houston Texas, 2005, 501–509.

Guo B.Y., Langrish T.A.G. & Fletcher D.F. 2001. An assessment of turbulence models applied to the simulation of a two-dimensional submerged jet. App. Mech. Model., 25, 635–653.

Guan H.Y. & Tu J.H. 2010. Computational techniques for multi-phase flows. Oxford, UK: Butterworth-Heinemann Publications.

Kim J.G. & Song J.J. 2015. Abrasive water jet cutting methods for reducing blast-induced ground vibration in tunnel excavation. Int. J. Rock Mech. Min. Sci., 75, 147–158.

Long X.P., Ruan X.F., Liu Q., Chen Z.W., Xue S.X. & Wu Z.Q. 2017.Numerical investigation on the internal flow and the particle movement in the abrasive waterjet nozzle. Power Technol., 314, 635–640.

Lee J.H., Park K.S., Kang M.C., Kang B.S. & Shin B.S. 2012. Experiments and computer simulation analysis of impact behavior of micro-sized abrasive in waterjet cutting of thin multiple materials. Trans. Nonferrous Metals. Soc. China, s3, 864–869.

Mahmoud H., Kriaa W., Mhiri H., Le Palec G. & Bournot P. 2010. A numerical study of a turbulent axiymmetric jet emerging in a xo-flowing stream. Energy Convers. Manag., 51, 2117–2126.

Momber A.W. 2001. Energy transfer during the mixing of air and solid particles into a high-speed waterjet: an impact-force study. Exp. Thermal. Fluid Sci. 25, 31–41.

Kyriakou M., Missirlis D. & Yakinthos K. 2010. Numerical modelling of the vortex breakdown phenomenon on a delta wing with trailing-edge jet-flap. Int. J. Heat Fluid Flow, 31, 1087–1095.

Nanduri M., Taggart D., Kim T., 1999. Cutting efficiency of abrasive water jet nozzles. In: Mohamed Hashish ed. Proceedings of the 10th American Water J. Conf. Vol., Houston, Texas, 1999:217–232.

Tazibt A., Parsy F. & Abriak N. 1996. Theoretical analysis of the particle acceleration process in abrasive water jet cutting. Comput. Mater. Sci., 5, 243–254.

Wang J. & Shanmugam D.K. 2009. Cutting meat with bone using an ultrahigh pressure abrasive waterjet. Meat. Sci., 81, 671–677.

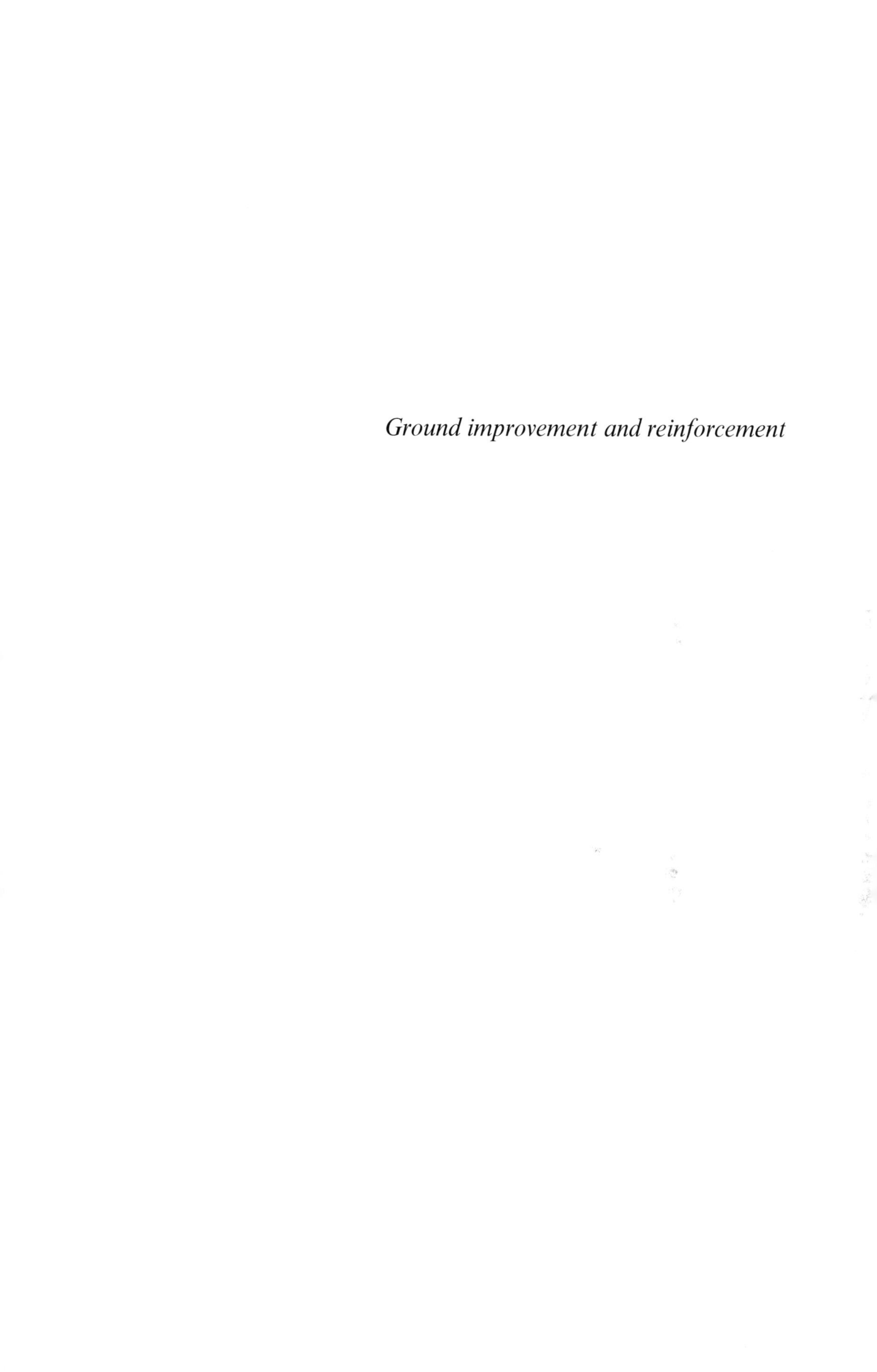

Ground improvement and reinforcement

Numerical Methods in Geotechnical Engineering IX – Cardoso et al. (Eds)
© 2018 Taylor & Francis Group, London, ISBN 978-1-138-33203-4

A new method for finite element modelling of prefabricated vertical drains

A.M. Lester, G.P. Kouretzis & S.W. Sloan
Centre of Excellence for Geotechnical Science and Engineering, University of Newcastle, Callaghan, NSW, Australia

ABSTRACT: Prefabricated Vertical Drains (PVDs) are frequently used in civil engineering projects where there is a need to speed up the consolidation of low permeability soils subjected to structural loads. However, when it comes to finite element modelling, capturing the geometry of individual PVDs is often a tedious exercise, particularly when incorporating their surrounding 'smear zones' of reduced horizontal permeability (k_h) that result from installation. In this paper, a new method is introduced which involves the attachment of a smear radius to 1D drainage elements. This allows for automatic reduction of k_h at mesh integration points that fall within the smear zone, as well as the consideration of non-uniform hydraulic properties. The results of analyses carried out for a unit cell with a single PVD and smear zone show that the new method achieves the same accuracy as the case where the smear zone is modelled as a separate meshed region.

1 INTRODUCTION

The need for prefabricated vertical drains (PVDs) typically arises in projects where large structures are to be built over soft clay soils with high compressibility and low permeability. PVDs have the effect of shortening drainage path lengths within the soil, which results in faster dissipation of excess pore pressures and hence shorter consolidation times. This will often translate into significant time savings for construction programs, and as such PVDs have been very widely used on sites where soft clays are present. In order to quantify the time savings that PVDs will offer, it is necessary to carry out analysis which will often involve the consideration of complicated geometries and loading patterns, such as those associated with the construction of a road embankment for example. Additionally, complex models of soil behaviour may be adopted with the aim of determining settlements accurately. In these situations, it becomes necessary to use a generalised analysis method which will take account of these various complexities, and finite element analysis has been popular in this respect.

The installation of PVDs usually involves pushing them into the ground using a steel mandrel attached to a large excavator, and this process results in a region of disturbed soil surrounding the PVD which is known as a smear zone. The soil in the smear zone is characterised by a reduced horizontal permeability (k_h) compared to the free-field, which works against the flow of pore water that the PVD is supposed to be facilitating. The reduction in k_h is often significant enough that it necessitates the modelling of smear zones in finite element analysis of soils where PVDs have been installed (e.g. Basu & Prezzi 2007, Indraratna & Redana 1997, Indraratna & Redana 1998). Usually, the PVDs and their surrounding smear zones are modelled as they appear after the installation process, and analysis of such a model is termed a 'wished in-place' analysis.

One of the main issues when it comes to modelling PVDs and smear zones in finite element analysis is that their dimensions are often quite small in comparison to the domain of the problem to be analysed. Additionally, there are often very high pore pressure gradients in the vicinity of the PVDs, and this means that a very fine mesh is required to capture the geometry and flow conditions in these regions. Unsuitably thin mesh elements and/or excessive computation times can result from such constraints. To avoid these problems, a number of simplified methods for modelling PVDs and smear zones in finite element analysis have been introduced in the past. One of these is the use of 1D drainage elements, which can be made to represent individual PVDs with material properties different to the surrounding soil, but do not require the introduction of any extra nodes into the mesh. The smear zones are then typically meshed as separate regions surrounding the PVD, with a reduced k_h compared to the undisturbed soil (e.g. Indraratna & Redana 2000, Indraratna et al. 2008, Li & Rowe

2001, Ong et al. 2012, Rowe & Taechakumthorn 2008, Sathananthan 2005).

Whilst it gives users full control over their geometry and material properties, meshing individual smear zones can be a tedious exercise, particularly in real design situations involving multiple PVDs, where optimising the PVD spacing parametrically requires altering the geometry of each smear zone one-by-one. Further complication may be introduced if there is a desire to incorporate variation in k_h over the width of the smear zone (e.g. Basu & Prezzi 2009, Walker & Indraratna 2006, Walker & Indraratna 2007) or if the PVDs become close enough together such that their smear zones start to overlap. In this paper, a new method of modelling smear zones around drainage elements is introduced, which aims to simplify the modelling process from the analyst's perspective. This involves the use of the finite element program to calculate the radius from the drainage element to integration points (IPs) in other surrounding elements. Comparison of these radii with a user-defined smear radius (r_s) then allows the program to determine whether the IPs fall within the smear zone and, if so, reduce their k_h values in accordance with a user-defined function for k_h over the width of the smear zone.

Following a description of the method for defining smear zones attached to drainage elements, a formulation for addressing the issue of overlapping smear zones is introduced. Details of analyses carried out for a benchmark problem involving a single drain (frequently referred to as the unit cell problem) are then presented to demonstrate the capabilities of the new modelling techniques. It is noted that the formulations presented in this paper are for 2D finite element analysis, although it is possible to extend them to 3D analysis.

2 METHOD FOR DEFINING SMEAR ZONES

An outline of the intended finite element procedure for defining a smear zone attached to a drainage element is as follows:

1. The geometry of the soil elements and the drainage element are specified via the nodal coordinates and element topology of the finite element mesh.
2. Material properties are assigned to the soil elements, including their undisturbed k_h values.
3. Material properties are assigned to the drainage element, including its characteristic dimension. In a 2D analysis this would normally be the width of the plane strain or axisymmetric 'drain wall' that the element represents. This is unless the analysis is axisymmetric and the element is located on the axis of symmetry, in which case the characteristic dimension is the diameter of the drain.
4. The drainage element is activated.
5. The smear zone is activated by specifying the drainage element to which it is attached, together with r_s and the k_h function that applies over the width of the smear zone. The identification of IPs that lie within the smear zone then takes place, followed by a reduction in the k_h values corresponding to these IPs. This procedure will only take place once for each smear zone that is specified.
6. The reduced k_h values are stored in an array of integration point properties, which can be called upon when required for assembly of the governing finite element equations.

In this study, it is assumed that drainage elements to which smear zones will be attached are initially aligned vertically with respect to the chosen global coordinate system. This is to say that for a 2D analysis, where the global horizontal and vertical axes are denoted x and y respectively, the drainage elements are parallel to the y-axis right along their length at the time the smear zone is activated. This simplifies the process of identifying IPs that lie within smear zones considerably, and it is expected that this assumption will be sufficient for many analyses involving PVDs. However, it is possible to develop a procedure for identifying IPs in smear zones attached to drainage elements which are initially inclined, and this is currently the subject of further research by the authors.

For vertical 1D drainage elements, determining whether an IP lies within the smear zone is simply a matter of comparing the coordinates of the IP to the nodal coordinates of the drainage element. An IP will lie within the smear zone if both of the following conditions are met:

- The difference between the x-coordinate of the IP and the x-coordinate of one of the drainage element nodes is less than r_s.
- The y coordinate of the IP lies in between the y coordinates of the end nodes of the drainage element.

If an IP is identified as lying within the smear zone of a drainage element, the k_h value corresponding to this IP can then be reduced according to the user specified function for k_h over the width of the smear zone.

An important point to note with regards to the above methodology is that because k_h values are only adjusted at the locations of IPs within the finite element mesh, the accuracy with which the user-defined k_h function is represented depends on the discretisation of the problem. A fine mesh with an ample number of IPs in the smear zone is critical in ensuring that changes in k_h are accurately captured.

3 OVERLAPPING SMEAR ZONES

With the extent of a smear zone defined only by the position of a drainage element and an attached smear radius, it is desirable to have an inbuilt means of accounting for variations in k_h which occur in overlaps between smear zones, should these be present. This saves the user having to manually identify overlapping zones and mesh them separately to the rest of the smear zone. There has been very little research done on determining k_h in overlaps between smear zones of different PVDs. To the authors' knowledge, only two studies (Basu & Prezzi 2009, Walker & Indraratna 2007) have addressed this problem before. Both involved the consideration of overlap between two identical smear zones extending from two adjacent identical PVDs, and both adopted the assumption that k_h in the overlapping zone is constant and equal to k_h at the edge of the overlapping zone. This is depicted in Figure 1, using a linear function for k_h in each of the individual smear zones as per Walker & Indraratna 2007. Here, k_u denotes the horizontal permeability of the undisturbed soil, k_x the constant horizontal permeability in the overlapping zone and k_d the horizontal permeability of the PVD.

It has been argued that the design of PVD systems should be focused on preventing overlaps between smear zones from occurring at all, owing to the detrimental effect that they have on the efficacy of PVDs (e.g. Indraratna et al. 2016). In some situations overlapping smear zones would only occur when the PVDs are in very close proximity, and in this case installation of the PVDs may be impractical from a construction viewpoint. However, estimates of r_s as a function of mandrel radius (r_m) vary widely, with previous laboratory tests (Bergado et al. 1991, Indraratna & Redana 1998, Indraratna et al. 2015, Onoue et al. 1991, Sathananthan & Indraratna 2006, Sharma & Xiao 2000) indicating r_s values of about 1.5–5.5r_m and recent field tests (Indraratna et al. 2014) indicating an r_s of about 6.3r_m. Additionally, the r_s adopted in finite element analysis will often be adjusted depending on the chosen function for k_h within the smear zone, and also the way in which the geometry of the PVD and smear zone is modelled. For example, in a 2D analysis with multiple PVDs, the drainage elements represent drain walls rather than individual drains of circular cross section. The adoption of a large r_s could lead to overlapping smear zones in analyses where PVDs are closely spaced. The effect that this has on k_h, and hence the functionality of the PVDs, needs to be quantified so that designers are then able to choose appropriate spacings for PVDs.

From a computational perspective it is convenient to develop a method which will calculate the value of k_h in an overlapping zone via a subtractive process, considering the contribution of each individual smear zone one at a time. This has a number of advantages which include:

- The smear zone data can be read sequentially from an input file, with the corresponding reductions in k_h applied one smear zone at a time.
- The method can be used to calculate k_h at all IPs, regardless of whether an overlap between smear zones occurs or not.
- Differences in r_s and k_h functions between overlapping smear zones can be easily taken into account. Although this feature is unlikely to be used in many analyses, owing to groups of PVDs usually being assigned the same r_s and k_h function for the smear zone, it is considered useful for the sake of generality.

The subtractive method proposed here for determining k_h at an IP where up to two overlapping smear zones may occur is expressed mathematically as:

$$k_h(r) = k_u - \left[k_u - k_{h,1}(r)\right] - \left[k_u - k_{h,2}(r)\right] \qquad (1)$$

where k_u is the undisturbed horizontal permeability, $k_u - k_{h,1}(r)$ is a term subtracted due to the presence of smear zone 1 and $k_u - k_{h,2}(r)$ is a term subtracted due to the presence of smear zone 2. It can be shown that Equation 1 recovers the thick black line representing $k_h(r)$ in Figure 1, for the case where two smear zones have the same r_s and

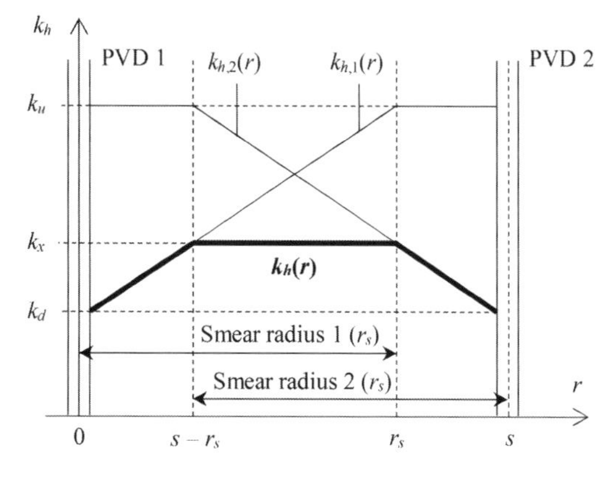

Figure 1. Horizontal permeability (k_h) in overlapping smear zones of two adjacent PVDs.

identical linear k_h functions. In other cases, $k_h(r)$ will be equal to $k_{h,1}(r)$ at the left edge of the overlapping zone, and equal to $k_{h,2}(r)$ at the right edge with intermediate values occurring inside the overlapping zone.

Another desirable extension to Equation 1 would be to account for overlaps where more than two smear zones are involved, and this is also a case that has not been addressed before. However, the continued subtraction of terms of the form $k_u - k_{h,i}(r)$, where i is the smear zone number, could result in unrealistically low permeability values. Currently, there is no experimental data to describe what happens to k_h in an overlap between more than two smear zones, and in the absence of such evidence a means of accounting for this will not be pursued here.

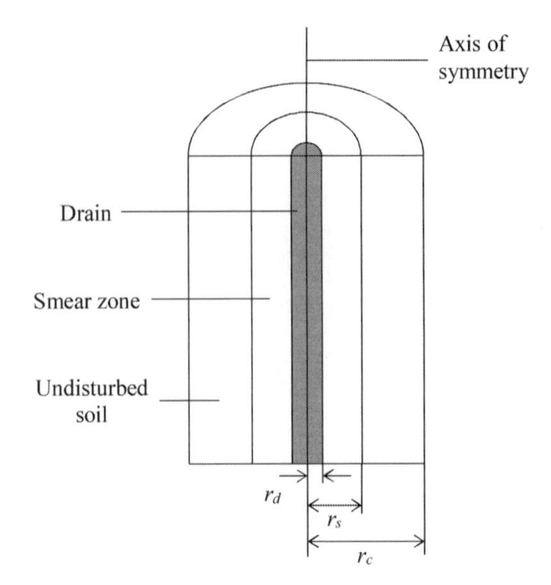

Figure 2. Unit cell model of PVD surrounded by smear zone and undisturbed soil.

4 ANALYSIS OF A BENCHMARK PROBLEM

In order to investigate the performance of the new smear zone modelling techniques, a series of 1D drainage elements were implemented into the finite element program SNAC, which has been developed at the University of Newcastle, Australia over many years. The methods described in Sections 2 and 3 for defining smear zones attached to drainage elements, and taking account of overlap between smear zones, were then also implemented and a series of analyses based on a simple benchmark problem were carried out. This problem is the elastic consolidation of a unit cell consisting of a drain with an assumed circular cross section, surrounded by soil with a low k_h value representing the smear zone, and this in turn is surrounded by undisturbed soil (Figure 2). An approximate (semi-analytical) solution to this problem can be found by combining the vertical consolidation theory of Terzaghi with the radial consolidation theory of Hansbo (1981), assuming a constant k_h in the smear zone.

Three sets of 2D axisymmetric analyses were carried out, with each set corresponding to a different value of the unit cell radius (r_c = 0.53 m, 0.73 m, 1.03 m). The radius of the drain (r_d) and smear zone (r_s) were kept constant for all analyses at 0.03 m and 0.23 m respectively, such that only the width of the undisturbed soil unit was changed. If the unit cell is viewed as representative of a single PVD located within a group of many others, then changing the r_c in this way is representative of an increase in PVD spacing. Therefore, by varying r_c, the effect of PVD spacing on the performance of the new modelling techniques could be investigated. The first analysis in each of the three sets (Analysis 1) was carried out using drainage elements with an attached smear radius, the second (Analysis 2) was carried out using drainage elements with a separate region of the mesh specified for the smear zone, and the third (Analysis 3) was carried out with separate mesh regions specified for both the drain and the smear zone.

The 2D axisymmetric mesh used to perform the analyses with r_c = 1.03 m is shown in Figure 3a. Close ups of the mesh detail are then provided in Figures 3b, 3c and 3d for Analyses 1, 2 and 3 respectively. The mesh consists of 2550 triangular elements, each with six displacement nodes and three pore pressure nodes, and the elements are concentrated in the vicinity of the drain and the top (free draining) boundary of the mesh where high pore pressure gradients occur. 1D drainage elements, each with three displacement nodes and two pore pressure nodes, were added over the left hand edge of the mesh in Analyses 1 and 2. In Analysis 3, extra triangular elements were added over the left hand edge of the mesh in order to represent the drain.

For all analyses, the boundary conditions applied to the mesh were zero horizontal displacement ($u_x = 0$) at the right hand edge and zero vertical displacement ($u_y = 0$) at the base. Additionally, zero pore pressure ($p = 0$) was prescribed at the top of the mesh to allow free drainage at this boundary. A uniform load was applied at the top of the mesh, increasing linearly from 0 to 30 kPa over a time interval of 0.0001 years and then held constant at 30 kPa for 1.5 years. The elastic deformation parameters adopted for the soil and drainage elements were

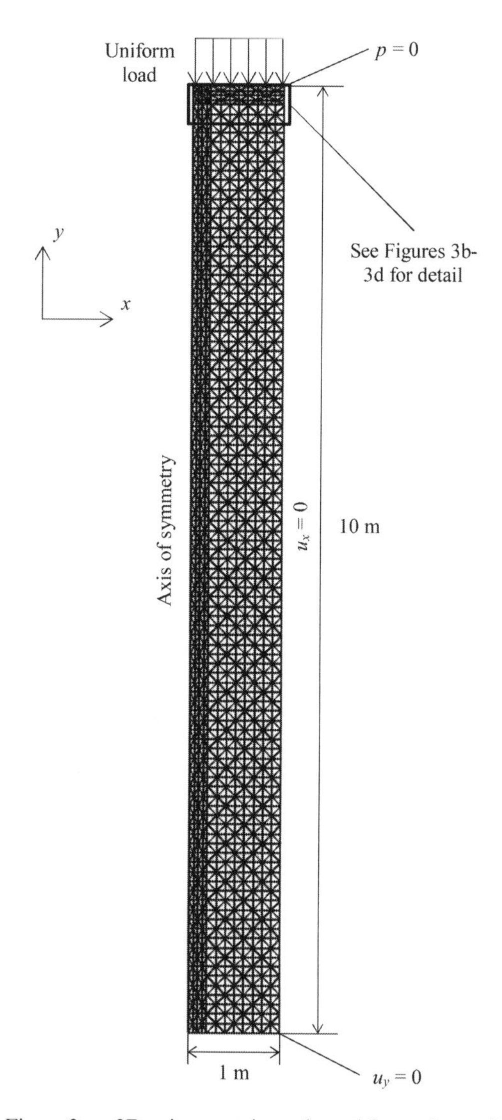

Figure 3a. 2D axisymmetric mesh used for analyses of unit cell with $r_c = 1.03$ m.

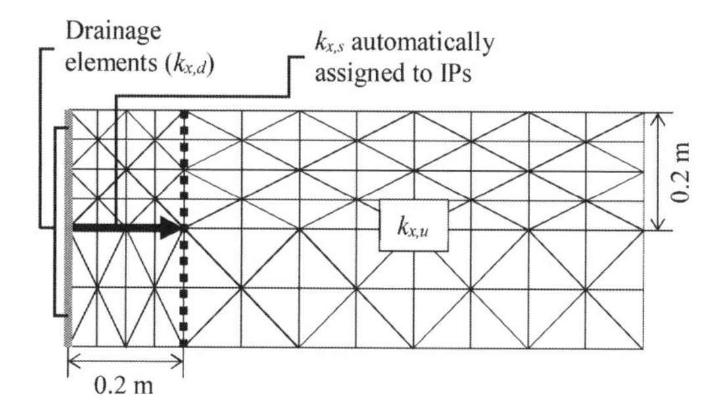

Figure 3b. Close up of mesh for Analysis 1.

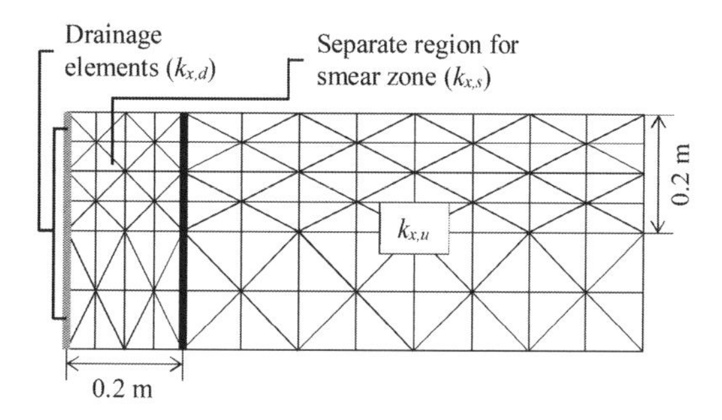

Figure 3c. Close up of mesh for Analysis 2.

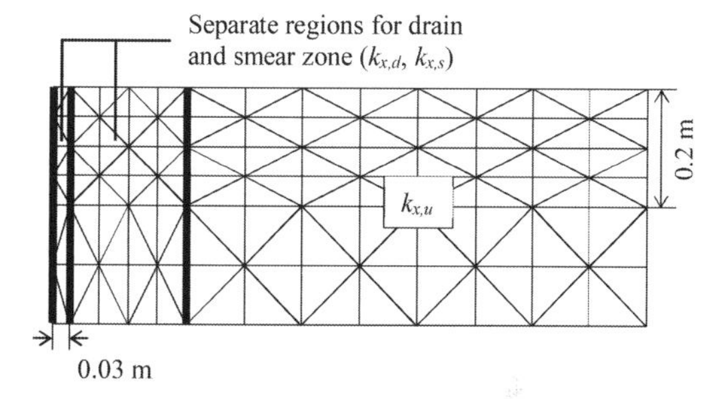

Figure 3d. Close up of mesh for Analysis 3.

an elastic modulus (E) of 6000 kPa and Poisson's ratio (v) of 0.35. The permeability values assigned were $k_{x,u} = k_{y,u} = 0.016$ m/year for the undisturbed soil, $k_{x,s} = 0.005$ m/year and $k_{y,s} = 0.016$ m/year for the smear zone and $k_{x,d} = k_{y,d} = 9 \times 10^3$ m/year for the drain. The high permeability of the drain was accounted for in the semi-analytical solution by assuming zero well resistance.

The governing finite element equations were solved using an automatic time substepping scheme for elastic consolidation (for more details see Abbo 1997, Sloan & Abbo 1999). One coarse time step was used for the 0.0001 year period over which the linearly increasing load was applied, and 150 coarse time steps were used for the 1.5 year period over which the load was held constant. A displacement error tolerance (DTOL) of 10^{-4} m was imposed for each time substep determined by the solution scheme. The results of the analyses are summarised in the settlement vs. time graphs shown in Figures 4a-4c, which are plotted for the node in the top left corner of the mesh. The graphs also include the curves obtained using Hansbo's semi-analytical solution.

It is noted that the settlement vs. time curves for Analyses 1 and 2 in each of Figures 4a-4c are identical, which confirms that the application of a smear radius to a drainage element leads to the same results as creating a separate meshed region

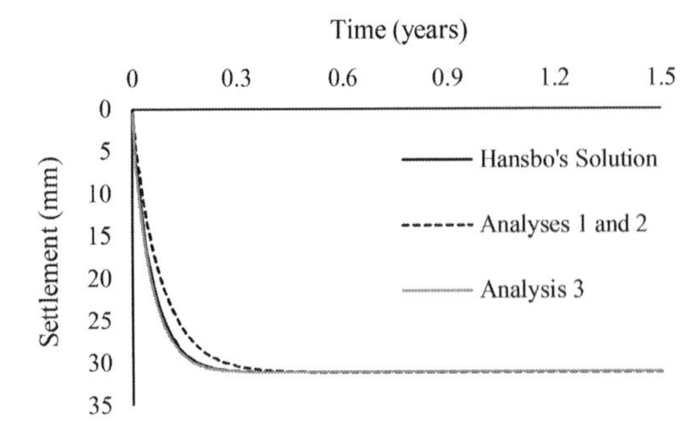

Figure 4a. Settlement vs. time curves for $r_c = 0.53$ m.

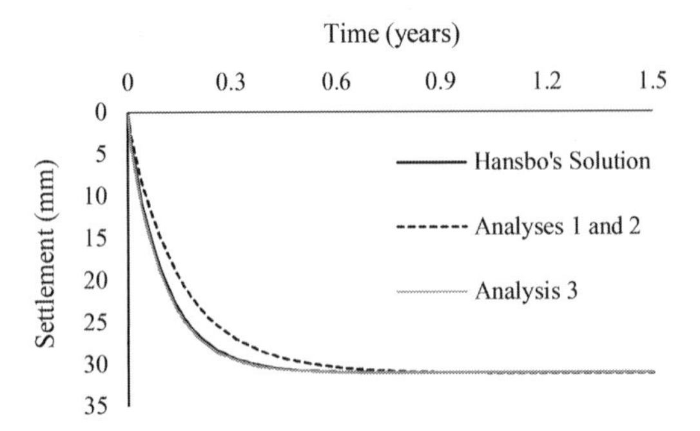

Figure 4b. Settlement vs. time curves for $r_c = 0.73$ m.

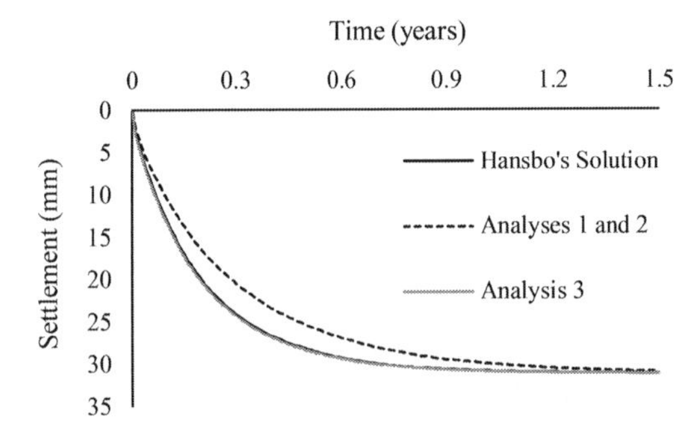

Figure 4c. Settlement vs. time curves for $r_c = 1.03$ m.

for the smear zone. The former method is easier from the analyst's perspective in terms of setting up the model, and so presents a favourable alternative when it comes to finite element analyses of problems involving many PVDs with surrounding smear zones. However, the settlement vs. time curves for Analyses 1 and 2 do not match Hansbo's solution for any value of r_c. The curves for Analysis 3 are much closer, and this demonstrates that most of the difference between Hansbo's solution and Analyses 1 and 2 is due to the approximation of a drain of finite radius using 1D drainage elements. The slight difference between Analysis 3 and Hansbo's solution can then be explained by the so-called 'equal strain' assumption, which was adopted by Hansbo (1981) for the radial consolidation of the unit cell. This assumes that the displacement is the same everywhere over the surface of the unit cell whilst it is subjected to the constant applied load during consolidation. Although this gives a close approximation of reality, it is not strictly accurate.

In a simple analysis such as the unit cell problem, where there is only one drain and smear zone, it is relatively easy to set up an analysis such as Analysis 3 where the drain cross section is meshed with 2D elements. Indeed this should be done where practical, as it will give more accurate results due to the greater precision with which the geometry of the drain is modelled. In analyses with many drains though, this may necessitate the use of very thin elements and these could lead to inaccuracy in the results, or even singularities which could compromise the progression of the analysis. Additionally, meshes that contain such elements would be time consuming for the analyst to set up. Modelling PVDs and smear zones using drainage elements with an attached smear radius provides a method which is easy to set up, and gives results to the same degree of accuracy as creating separate meshed regions for the smear zones. It will not be as accurate as meshing the drain cross section, but in many analyses this will not be practical and in this situation it is considered sufficient to use drainage elements as an approximation to the real geometry of the PVDs.

5 CONCLUSIONS

A numerical method for simplifying finite element modelling of PVDs and their surrounding smear zones has been introduced. The use of 1D drainage elements with an attached smear radius makes setting up finite element analyses with multiple PVDs much simpler from the analyst's perspective, in comparison to specifying smear zones as separate regions in the mesh. Overlapping smear zones, which would otherwise introduce further complications, can also be readily taken into account. Analyses carried out for the unit cell benchmark problem showed that the use of a smear radius gave identical results to the case where the smear zone was specified as a separate region in the mesh. There is a loss of accuracy in the use of drainage elements to approximate the real geometry of the PVDs, but in analyses where many PVDs are involved this is often a necessary sacrifice to ensure that mesh elements do not end up too thin.

ACKNOWLEDGEMENTS

The first author is thankful for the financial support of an Australian Government Research Training Program Scholarship received whilst undertaking the research reported in this paper.

REFERENCES

Abbo, A.J. 1997. *Finite Element Algorithms for Elastoplasticity and Consolidation*. PhD Thesis, University of Newcastle.

Basu, D. & Prezzi, M. 2007. Effect of the smear and transition zones around prefabricated vertical drains installed in a triangular pattern on the rate of soil consolidation. *International Journal of Geomechanics* 7(1): 34–43.

Basu, D. & Prezzi, M. 2009. Design of prefabricated vertical drains considering soil disturbance. *Geosynthetics International* 16(3): 147–157.

Bergado, D.T. *et al.* 1991. Smear effects of vertical drains on soft Bangkok clay. *Journal of Geotechnical Engineering* 117(10): 1509–1530.

Hansbo, S. 1981. Consolidation of fine-grained soils by prefabricated vertical drains. *Proceedings of the 10th International Conference on Soil Mechanics and Foundation Engineering, Stockholm*. Rotterdam: A.A. Balkema. 677–682.

Indraratna, B. & Redana, I.W. 1997. Plane-strain modelling of smear effects associated with vertical drains. *Journal of Geotechnical and Geoenvironmental Engineering* 123(5): 474–478.

Indraratna, B. & Redana, I.W. 1998. Laboratory determination of smear zone due to vertical drain installation. *Journal of Geotechnical and Geoenvironmental Engineering* 124: 180–184.

Indraratna, B. & Redana, I.W. 2000. Numerical modelling of vertical drains with smear and well resistance installed in soft clay. *Canadian Geotechnical Journal* 37: 132–145.

Indraratna, B. et al. 2008. Analytical and numerical modelling of consolidation by vertical drain beneath a circular embankment. *International Journal of Geomechanics* 8(3): 199–206.

Indraratna, B. et al. 2014. Soil disturbance analysis due to vertical drain installation. *Proceedings of the Institution of Civil Engineers: Geotechnical Engineering* 168(3): 236–246.

Indraratna, B. et al. 2015. Characterization of smear zone caused by mandrel action. *Proceedings of the International Foundations Congress and Equipment Expo 2015, San Antonio (Texas)*. Reston (Virginia): American Society of Civil Engineers. 2225–2232.

Indraratna, B. et al. 2016. Predictions using a) Industry standard soil testing and b) Unconventional large diameter specimens: A designer's perspective of the trial embankment at Ballina. *Proceedings of the Embankment and Footing Prediction Symposium 2016, Newcastle*. Newcastle: Centre of Excellence for Geotechnical Science and Engineering. 181–208.

Li, A.L. & Rowe, R.K. 2001. Combined effects of reinforcement and prefabricated vertical drains on embankment performance. *Canadian Geotechnical Journal* 38: 1266–1282.

Ong, C.Y. et al. 2012. Degree of consolidation of clayey deposit with partially penetrating vertical drains. *Geotextiles and Geomembranes* 34: 19–27.

Onoue, A. et al. 1991. Permeability of disturbed zone around vertical drains. *Proceedings of the Geotechnical Engineering Congress (ASCE), Boulder (Colorado)*. New York: American Society of Civil Engineers. 2: 879–890.

Rowe, R.K. & Taechakumthorn, C. 2008. Combined effect of PVDs and reinforcement on embankments over rate-sensitive soils. *Geotextiles and Geomembranes* 26: 239–249.

Sathananthan, I. 2005. *Modelling of Vertical Drains with Smear Installed in Soft Clay*. PhD Thesis, University of Wollongong.

Sathananthan, I. & Indraratna, B. 2006. Laboratory evaluation of smear zone and correlation between permeability and moisture content. *Journal of Geotechnical and Geoenvironmental Engineering* 132(7): 942–945.

Sharma, J. & Xiao, D. 2000. Characterization of a smear zone around vertical drains by large-scale laboratory tests. *Canadian Geotechnical Journal* 37: 1265–1271.

Sloan, S.W. & Abbo, A.J. 1999. Biot consolidation analysis with automatic time stepping and error control. Part 1: Theory and implementation. *International Journal for Numerical and Analytical Methods in Geomechanics* 23(6): 467–492.

Walker, R. & Indraratna, B. 2006. Vertical drain consolidation with parabolic distribution of permeability in smear zone. *Journal of Geotechnical and Geoenvironmental Engineering* 132(7): 937–941.

Walker, R. & Indraratna, B. 2007. Vertical drain consolidation with overlapping smear zones. *Géotechnique* 57(5): 463–467.

Numerical Methods in Geotechnical Engineering IX – Cardoso et al. (Eds)
© 2018 Taylor & Francis Group, London, ISBN 978-1-138-33203-4

Finite element modelling of reinforced road pavements with geogrids

J. Neves
CERIS, CESUR, Instituto Superior Técnico, Universidade de Lisboa, Lisboa, Portugal

M. Gonçalves
Infraestruturas de Portugal, Lisboa, Portugal

ABSTRACT: Finite Element Modelling (FEM) is one of the most common methods used to solve complex problems in civil engineering. The aim of this paper is to demonstrate the use of FEM to analyse the reinforcing influence of geogrids in road pavements. Using ADINA software, a parametric study was carried out, taking into consideration diverse pavement structures, climatic and traffic conditions, and geogrid reinforcement solutions. The paper describes the two-dimensional axisymmetric finite element model developed for unreinforced and reinforced pavement analysis. In general, the research confirmed the adequacy of ADINA for the analysis of reinforced road pavements that use geogrids. Results from the numerical analysis allowed for the following main conclusions: 1) fatigue resistance was improved when the geogrid was applied on top of unbound granular layers; 2) rutting decreased when the geogrid was applied on top of the subgrade soil.

1 INTRODUCTION

Alongside other numerical methods, finite element modelling (FEM) is one of the most common methods used to solve complex problems in civil engineering. This is also the case for road pavements with certain specific conditions, such as advanced structures and materials, layers geometry and interface behaviour, and traffic loading.

The use of geogrids in road pavements could prove to be a means of engineers achieving technical and economical solutions to issue of construction and performance. Indeed, several authors have confirmed that the performance of pavements can be significantly improved when geogrids are used (Barksdale et al. 1989, Zornberg 2011, Zornberg & Gupta 2010). As a layered system, one of the main functions of geogrids in pavements is reinforcement, which mainly derives from lateral restraint, increased bearing capacity and membrane support. Reinforcement using geogrids is a complex phenomenon and depends on several factors, such as the type of pavement, the characteristics of the geogrids and their placing in the pavement (Erickson & Drescher 2001, Mounes et al. 2011, Zornberg et al. 2012).

The design of reinforced pavement requires a somewhat more complex approach in terms of analysis and study thereof. FEM has been extensively used to analyse the presence of geogrids in the pavement structure (Bohagr 2013, Kim & Lee 2013, Lees 2017, Saad et al. 2006, Ouf et al. 2016, Perkins 2011). Several codes have been used to simulate pavement reinforcement. Some of the most common in the literature are: ABAQUS (AlAbdullah & Taresh 2017, Calvarano et al. 2017, Kim & Lee 2013, Huang 2014); PLAXIS (Ahirwar & Mandal 2017, Al-Jumaili 2016, Faheem & Hassan 2014, Ibrahim et al. 2017); FLAC (Benmebarek et al. 2013); ANSYS (Melaku & Qiu 2015); and ADINA (Gonçalves 2015, Saad et al. 2006).

In addition to other modelling techniques, such as Discrete Element Modelling (DEM) (Chen et al. 2013, Horton et al. 2015, McDowell et al. 2006, Qian et al. 2011 and 2015, Tran et al. 2015), the geogrids' properties in terms of stiffness, thickness, and interface frictional behaviour in practical design can be more easily taken into consideration by FEM.

Depending on the boundary conditions (traffic load and pavement geometry), analysis can be performed in two-dimensional (2D) or three-dimensional (3D) finite element models. The 2D models may have certain advantages in cases of routine use for practical design calculations on account of their simplicity and lower computational requirements. Their main disadvantage is in the area of load traffic modelling.

The purpose of this paper is to present a case of the use of FEM to analyse the reinforcement influence of geogrids in pavement design. It demonstrates the use of ADINA software in 2D analysis of reinforced flexible pavements using geogrids in the unbound granular layers. Based on a parametric analysis considering several structures and materials' properties, this paper sets out to confirm the influence of the geogrid on pavement reinforcement in terms of the fatigue cracking (asphalt concrete layers) and rutting resistance (subgrade soil) criteria.

2 FEM ANALYSIS

2.1 *Pavements and materials*

The applicability of FEM-2D was demonstrated for a flexible pavement consisting of upper layers of asphalt concrete (AC) and bottom base and sub-base layers of unbound granular material (UGM) (Fig. 1).

The pavement reinforcement was defined for the three positions of the geogrid in the granular layers as presented in Figure 1: top of the base layer (R1); interface of base and sub-base layers (R2); and bottom of the sub-base layer (R3).

The parametric study considered differing geogrid positions and thickness and mechanical properties (elastic modulus) of the pavement layers. Table 1 shows the range of thicknesses adopted for UGM (t_{UGM}) and AC (t_{AC}) layers, in accordance with the pavement catalogue of Portuguese Road Agency (JAE, 1995). The numerical simulations considered elastic linear behaviour for all pavement materials including the subgrade soil, which is characterized by the elastic modulus (E) and the Poisson's ratio (ν). Table 1 presents the mechanical properties of the base (E_B), sub-base (E_{SB}), and subgrade (E_{SG}) layers. In the case of AC layers, 3000, 4000 and 6000 MPa were taken into consideration in term, indirectly, of the viscoelastic behaviour influenced by temperature. Poisson's ratios of 0.40 and 0.35 were used for the AC and UGM layers, respectively.

Geogrids were modelled as a rigid layer of 2.54×10^{-3} cm thickness (e_G) and the mechanical characteristics of elastic modulus (E_G) and Poisson's ratio (ν_G), as presented in Table 2.

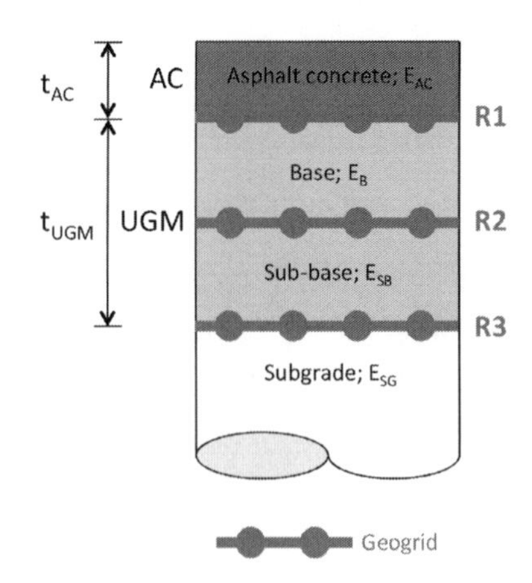

Figure 1. Pavement structure.

Table 1. Pavement layers characteristics.

Subgrade	Unbound granular material			Asphalt concrete
E_{SG} (MPa)	E_{SB} (MPa)	E_B (MPa)	t_{UGM} (cm)	t_{AC} (cm)
30**	60**	120**	30**(15+15)	17**/20/25
			40 (20+20)	17/19/23
60	120	*	20	16/18/22
		240	30 (15+15)	13/15/20
			40 (20+20)	13/15/19

* Non-existent base layer
** Pavement characteristics used in the study of load modelling

Table 2. Geogrid characteristics.

e_G (cm)	E_G (MPa)	ν_G
2.54×10^{-3}	4230	0.35

2.2 *Numerical modelling*

FEM analysis was performed using ADINA software (ADINA 2001) and considering a 2D model (Fig. 2). Figure 2a shows the geometry and the dimensions of the axisymmetric model. The modelled domain of the subgrade was 2.50 m wide and 2.50 m long. The pavement geometry depended on the parametric study under analysis (Table 1). With regard to the boundary conditions, the conventional kinematic conditions were considered: in both vertical boundaries, fixed horizontal movements (x and y axes) were considered, with freedom of vertical movements (z axis); in the lower horizontal boundary, a fixed support with blocking of the movements in all directions was taken into consideration.

The pavement layers, the geogrid material, and the subgrade were modelled with eight-node isoparametric elements. The interface conditions between pavement layers and geogrid corresponded to full friction that represents maximization of the pavement performance. Figure 2b presents an example of the FE mesh used in the numerical analysis.

As is known, the standard axle cannot be modelled directly when a 2D analysis is applied. A circular contact area of 15 cm diameter, 570 kPa tyre pressure and 40 kN load was considered equivalent to the effect of the standard axle load of 80 kN. Gonçalves (2015) has previously performed a study comparing the tensile strain at the bottom of the AC layer (ε_t) and the compression strain at the top of the subgrade soil (ε_c), on the vertical line through the load centre, applying the multi-layers programs BISAR (SHELL) and ADINA. In the case where BISAR was used, two calculations were performed for the load modelling: two circular tyre contact areas of 20 kN load and 10.5 cm diameter (2×20 kN);

Table 3. Validation of load modelling.

Strains ($\times 10^{-6}$)	BISAR		ADINA
	(2×20 kN)	(1×40 kN)	(1×40 kN)
ε_t	185.8	207.7	205.0
ε_c	–556.7	–596.0	–575.0

and one circular contact area of 40 kN load and 15 cm diameter (1×40 kN). Table 3 presents the results for the pavement characteristics identified in Table 1. It can be concluded that there is a reasonable level of agreement between the obtained values for design purposes that validates the load modelling in ADINA.

3 RESULTS AND DISCUSSION

Numerical simulations were performed in both scenarios: unreinforced and geogrid-reinforced pavements. The results from ADINA calculations showed strains and displacements due to loading. In general, a reduction in strains and displacements was achieved through the use of the geogrid.

The most important effect of the reinforcement was observed in terms of horizontal displacements. This effect was more significant close to the geogrid interface. Indeed, the main mechanism involved in the benefit achieved by geogrid reinforcement is expected to be the lateral constraint. Numerical calculations confirmed this confinement, not only in the base layer but also in the subgrade soil. Figure 3 shows the benefit for reinforcement R3.

Figure 4 shows the distribution of horizontal strains in unreinforced pavement (Fig. 4a) and geogrid-reinforced subgrade (R3) (Fig. 4b). Figure 5 presents an example of the vertical strains obtained for the unreinforced pavement (Fig. 5a) and geogrid-reinforced subgrade (R3) (Fig. 5b). In general, the geogrid-reinforcement led to an overall reduction in strains, which was more marked in the case of vertical strains. This is a consequence of the increased pavement bearing capacity due to the effect of the geogrid reinforcement.

One potential outcome of the reduction in strains is an increase in the admissible number of load cycles (N), expressed by the maximum total accumulated number of equivalent standard axle load (ESAL = 80 kN) during the 20-year design period.

Taking into account the design criteria commonly used in flexible pavements—fatigue cracking and rutting resistance—tensile strains (ε_t) and compression strains (ε_c) were each assessed at the bottom of the AC layer and the top of the subgrade soil, on the vertical line through the load centre.

The equations from Shell (1978) pertaining to fatigue resistance and permanent deformation

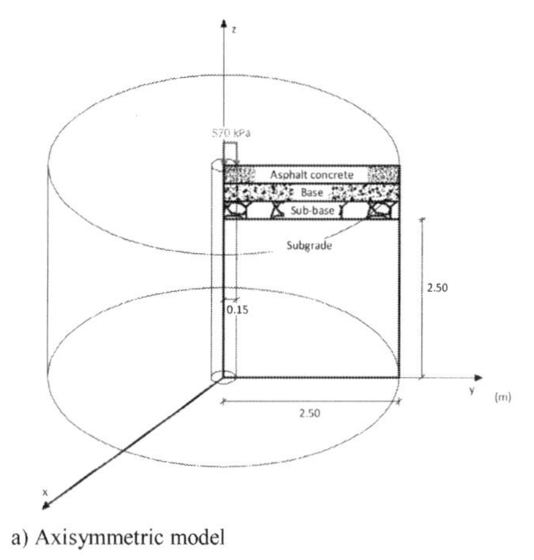

a) Axisymmetric model

b) Finite element mesh

Figure 2. Two-dimensional pavement modelling.

criteria can be used to estimate the total accumulated number of load cycles (N). Considering a 9% volume of asphalt binder, the equations are respectively:

$$\varepsilon_t = 8.78\, E^{-0.36} N^{-0.20} \tag{1}$$

$$\varepsilon_c = 0.018\, N^{-0.25} \tag{2}$$

where ε_t = horizontal tensile strain; ε_c = vertical compression strain; and E = elastic modulus of AC (Pa). Several values for the AC elastic modulus were adopted in the numerical modelling (see section 2.1).

Based on the strains obtained from the numerical analysis and using the aforementioned equations, the admissible numbers of load cycles for the unreinforced pavement ($N_{unreinforced}$) and the reinforced pavement ($N_{reinforced}$) were calculated for all the pavement structures (Table 1). Figure 6 shows the relationship between the N obtained for reinforced

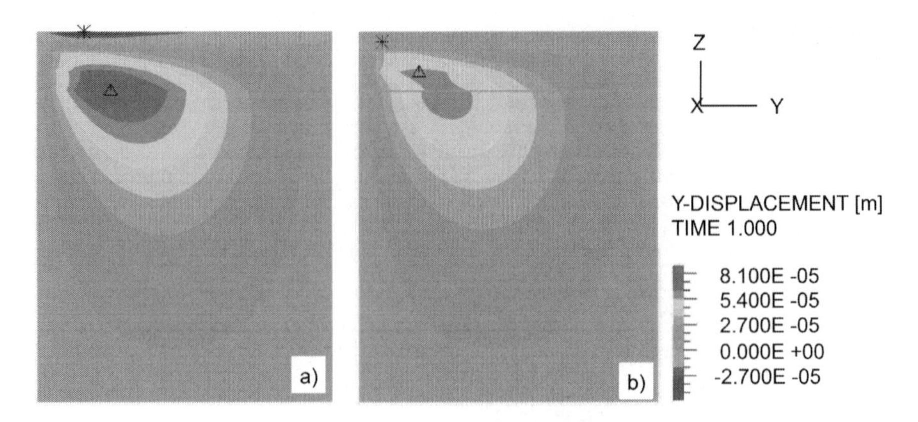

Figure 3. The distribution of horizontal displacements (R3): a) unreinforced pavement; b) reinforced pavement.

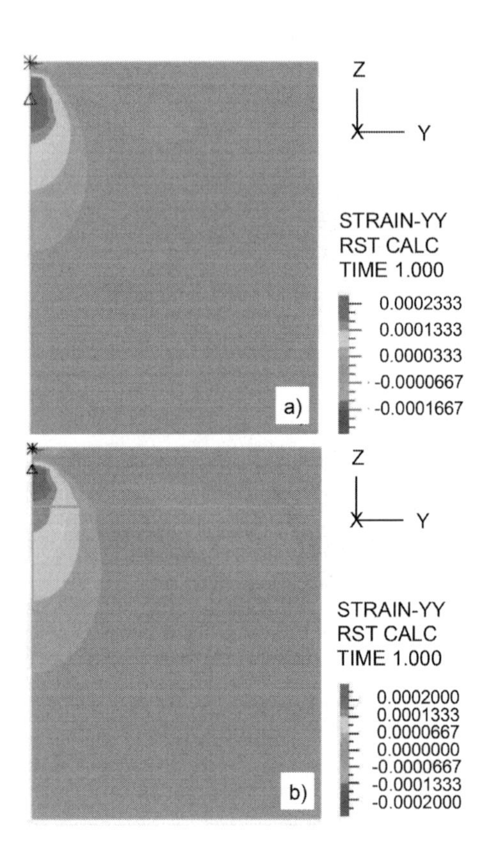

Figure 4. The distribution of horizontal strains (R3): a) unreinforced pavement; b) reinforced pavement.

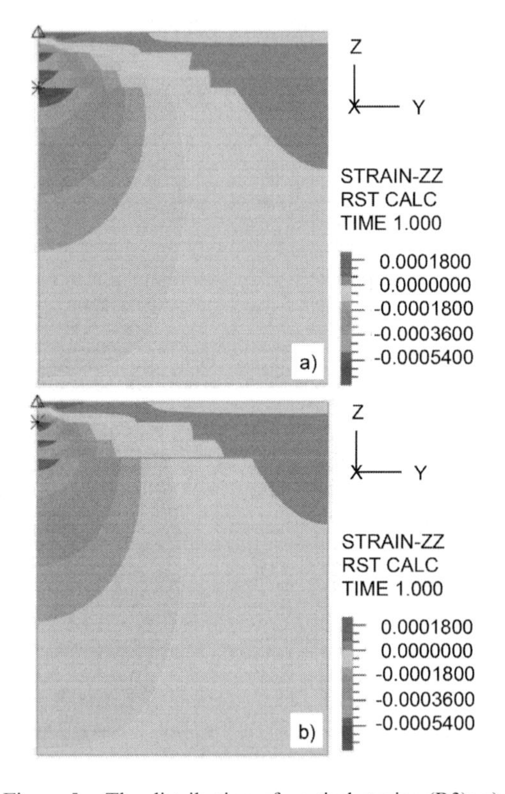

Figure 5. The distribution of vertical strains (R3): a) unreinforced pavement; b) reinforced pavement.

pavement and unreinforced pavement for the fatigue (Fig. 6a) and rutting (Fig. 6b) criteria of pavement with 40 cm of UGM layer (E_{SG} = 30 MPa). Calculations were performed for the three different locations of the geogrid: R1, R2, and R3 (see Fig.1).

Figure 6 allows the following conclusions:

- A general trend of an increase in the number of load cycles for the reinforced pavement was observed, which was more significant for permanent deformation criteria.
- The greatest benefit in terms of the fatigue criterion was obtained for reinforcement on the top of the granular layers (R1). In this case, the increase of the admissible number of load cycles for the reinforced pavement was 20%.

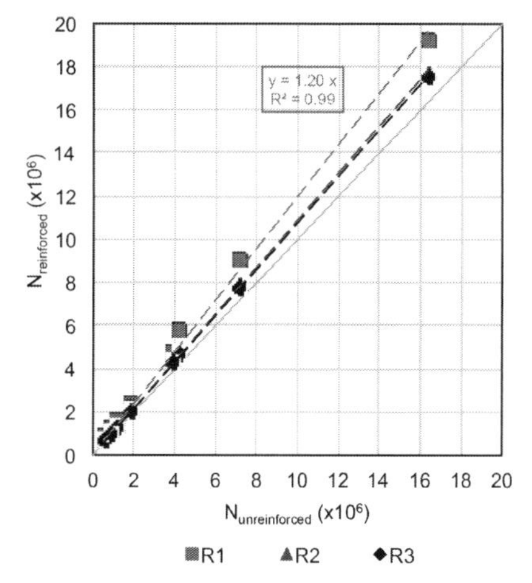

a) Fatigue resistance

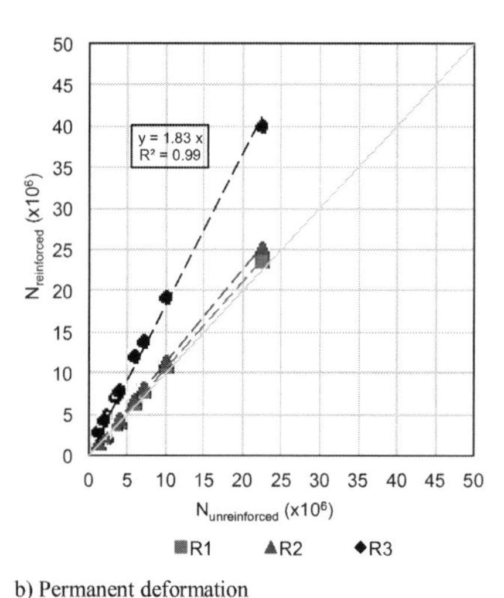

b) Permanent deformation

Figure 6. Influence of the geogrid on the admissible number of load cycles.

- With respect to permanent deformations, the greatest benefit in terms of increasing the number of load cycles was obtained for reinforcement R3 (geogrid placed at the bottom of granular layers). The increase in the admissible number of load cycles for the reinforced pavement was 83%.

It was possible to achieve a global benefit in pavement performance through geogrid reinforcement. The results presented herein showed that the most significant effect for pavement design purposes was related to the rutting criterion. Gonçalves (2015) demonstrated that the effect of geogrid could be more pronounced when the pavement presents a low bearing capacity (weak subgrade, thick granular layer and thin asphalt concrete layer).

4 CONCLUSIONS

The research presented in this paper confirmed the adequacy of FEM for analysis of the reinforcement of road pavements using geogrids.

This paper highlights the main advantages and disadvantages (limitations) in the use of ADINA software for modelling geogrid-reinforced road pavements using two-dimensional finite element models. 2D models can have advantages in the case of routine use for practical design calculations on account of their simplicity and lower computational requirements. The main disadvantage has to do with the load traffic modelling. However, the paper validated the use of a circular tyre contact area equivalent to the effect of the standard axle load of 80 kN.

With regard to the impact of geogrid reinforcement on pavement design, considering the fatigue resistance and permanent deformation criteria, from the results of the use of this numerical tool allowed for the following main conclusions: (1) fatigue resistance was improved when the geogrid was applied on top of the unbound granular layers; (2) rutting decreased when the geogrid was placed on top of the subgrade soil. The effect of geogrids could be more pronounced if the pavement presents a low bearing capacity: weak subgrade, thicker granular layer and thinner asphalt concrete layer. In this case, a maximum increase in the admissible number of load cycles for the reinforced pavement of 20% and 83% was obtained for the fatigue resistance and permanent deformations, respectively.

All these results confirmed that an important general benefit in pavement can be had from an improvement in bearing capacity generated by the use of geogrids. Furthermore, the research demonstrated a reduction in the horizontal displacements due to lateral restraint furthered by the geogrids.

Total friction in the interface of geogrid and pavement material was always considered in FEM analysis. This corresponds to a total interlocking between aggregates and geogrid, that is the maximum reinforcement effect of geogrid. In reality, the reinforcement function of the geogrid may be not so effective. However, in FEM using ADINA other interface conditions can be simulated.

The results presented in this paper were obtained for elastic linear behaviour of pavement materials and subgrade soil. FEM analysis can consider other behaviour of the materials such as elastoplastic models. For those more complex models, identical conclusions will be obtained as demonstrated by Gonçalves (2015).

REFERENCES

ADINA 2001. Finite element computer program, theory and modeling guide. Rep. ARD 01-7, Vol. 1, Adina R&D Inc., Watertown, Mass.

Ahirwar, S.K. & Mandal, J.N. 2017. Finite element analysis of flexible pavement with geogrids. Transportation Geotechnics and Geoecology, TGG 2017, 17–19 May, Saint Petersburg, Russia. Procedia Engineering 189, 411–416.

AlAbdullah, S.F.I. & Taresh, N.S. 2017. Evaluation of soil reinforced with geogrid in subgrade layer using finite element techniques. International Journal of GEOMATE, July, 2017, Vol.13, Issue 35, 174–179. DOI: http://dx.doi.org/10.21660/2017.151216.

Al-Jumaili, M.A.H. 2016. Finite element modelling of asphalt concrete pavement reinforced with geogrid by using 3-D Plaxis Software. International Journal of Materials Chemistry and Physics, Vol. 2, No. 2, 62–70.

Barksdale, R.D., Brown, S.F. & Chan, F. 1989. Potential benefits of geosynthetics in flexible pavement systems. Transportation Research Board. National Cooperative Highway Research Program Report No. 315. Washington DC, USA.

Benmebarek, S., Remadna, M.S. & Lamine, B. 2013. Numerical modeling of reinforced unpaved roads by geogrid. The Online Journal of Science and Technology, 3 (Issue 2), 109–115.

Bohagr, A.A. 2013. Finite element modeling of geosynthetic reinforced pavement subgrades. Master Thesis, Washington State University, USA.

Calvarano, L.S., Leonardi, G. & Palamara, R. 2017. Finite element modelling of unpaved road reinforced with geosynthetics. Transportation Geotechnics and Geoecology, TGG 2017, 17–19 May 2017, Saint Petersburg, Russia. Procedia Engineering 189, 99–104.

Chen, C., McDowell, G.R. & Thom, N.H. 2013. A study of geogrid reinforced ballast using laboratory pullout tests and discrete element modelling. Journal of Geomechanics and Geoengineering, Vol. 8, No. 4, 244–253, DOI: 10.1080/17486025.2013.805253.

Erickson, H. & Drescher, A. 2001. The use of geosynthetics to reinforce low volume roads. Minnesota Department of Transportation.

Faheem, H. & Hassan, A.M. 2014. 2D PLAXIS finite element modeling of asphalt concrete pavement reinforced with geogrid. Journal of Engineering Sciences, Assiut University, Vol. 42, No. 6, 1336–1348.

Gonçalves, M.M. 2015. Influence of the use of geosynthetics in the design of road pavements. Master Thesis, University of Lisbon, Portugal. (in Portuguese)

Horton, M., Oliver, T., Buckley, J. & Jas, H. 2015. Comparative discrete element modelling of geogrid stabilised aggregates and pavement performance. 16th AAPA International Flexible Pavements Conference, Australia, 1–10.

Huang, W-C. 2014. Improvement evaluation of subgrade layer under geogrid-reinforced aggregate layer by finite element method. International Journal of Civil Engineering, Vol. 12, No. 3, Transaction B: Geotechnical Engineering, 204–215.

Ibrahim, E.M., El-Badawy, S.M., Ibrahim, M.H., Gabr, A. & Azam, A. 2017. Effect of geogrid reinforcement on flexible pavements. Innov. Infrastruct. Solut., 2:54. DOI 10.1007/s41062–017–0102–7.

J.A.E. 1995. Pavement Design Manual. Junta Autónoma de Estradas. (in Portuguese)

Kim, M. & Lee, J.H. 2013. Effects of geogrid reinforcement in low volume flexible pavement. Journal of Civil Engineering and Management, S14-S22.

Lees, A. 2017. Simulation of geogrid stabilisation by finite element analysis. Proceedings of the 19th International Conference on Soil Mechanics and Geotechnical Engineering, 1370–1380.

McDowell, G.R, Harireche, O., Konietzky, H., Brown, S.F. & Thom, N.H. 2006. Discrete element modelling of geogrid-reinforced aggregates. Proceedings of the Institution of Civil Engineers, Geotechnical Engineering 159, Issue GEI, 35–48.

Mounes, S.M., Karim, M.R., Mahrez, A. & Khodaii, A. 2011. An overview on the use of geosynthetics in pavement structures, Scientific Research and Essays, Volume 6, 2234–2241.

Melaku, S. & Qiu, H. 2015. Finite element analysis of pavement design using ANSYS Finite Element Code. The Second International Conference on Civil Engineering, Energy and Environment, 64–69.

Ouf, M.S., Mostafa, A.E.A., Fathy, M. & Elgendy, M.F. 2016. Evaluation of stabilized pavement sections using finite element modeling. International Journal of Scientific & Engineering Research, Volume 7, Issue 4, 1749–1756.

Perkins, S.W. 2011. Numerical modeling of geosynthetic reinforced flexible pavements. Montana Department of Transportation.

Qian, Y., Tutumluer, E. & Huang, H. 2011. A validated Discrete element modeling approach for studying geogrid-aggregate reinforcement mechanisms. Geo-Frontiers 2011, ASCE Geo-Institute, March 13–16, Dallas, Texas.

Qian, Y., Mishra, D., Kazmee, H. & Tutumluer, E. 2015. Geogrid-aggregate interlocking mechanism investigated via Discrete Element Modeling. Geosynthetics Conference, Portland, USA.

Saad, B., Mitri, H. & Poorooshsb, H. 2006. 3D FE analysis of flexible pavement with geosynthetic reinforcement. Journal of Transportation Engineering, 402–415.

Shell 1978. Shell Pavement Design Manual. Asphalt Pavements and Overlays for Road Traffic. Shell International Petroleum Comp., UK.

Tran, V.D.H., Meguid, M.A. & Chouinard, L.E. 2015. Three-dimensional analysis of geogrid-reinforced soil using a finite-discrete element framework. International Journal of Geomechanics, 15(4). DOI: 10.1061/(ASCE)GM.1943–5622.0000410.

Zornberg, J.G. 2011. Advances in the use of geosynthetics in pavement design. Geosynthetics India. India Institute of Technology Madras, Chennai, India, Vol. 1, 3–21.

Zornberg, J.G. & Gupta, R. 2010. Geosynthetics in pavements: North American contributions. 9th International Conference on Geosynthetics, Brazil, 379–398.

Zornberg, J.G., Ferreira, J.A.Z. & Roodi, G.H. 2012. Geosynthetic-reinforced unbound base courses: quantification of the reinforcement benefits. University of Texas at Austin.

Numerical Methods in Geotechnical Engineering IX – Cardoso et al. (Eds)
© 2018 Taylor & Francis Group, London, *ISBN 978-1-138-33203-4*

Optimal design of reinforced slopes

Javier Gonzalez-Castejon
LimitState Ltd., Sheffield, UK
University of Glasgow, UK

Colin Smith
Department of Civil and Structural Engineering, University of Sheffield, UK

ABSTRACT: The analysis of mechanically stabilised earthworks using geotextiles or geogrids is typically addressed by means of modified limit equilibrium approaches or the Finite Element Method. However the design of the reinforcement layout and strength is generally specified by the engineer following the recommendations of design guidance or based on experience. Limit analysis and in particular Discontinuity Layout Optimization (DLO) presents an alternative for both the analysis and design of reinforced slopes. The current study presents initial work into developing an automated approach for determining the optimum layout of reinforcement for a given earthwork geometry. In this paper, a modified formulation of DLO termed Reinforcement Strength Optimization (RSO) is presented that is able to find the minimum tensile strength of the reinforcing material required for the stability of the system for a given layout and vertical spacing. This value can then be used as a parameter to guide the engineer in manually developing an optimal reinforcement layout, or as a component part of a fully automatic reinforcement layout optimization procedure.

1 INTRODUCTION

1.1 *Analysis of mechanically reinforced earthworks*

Mechanical reinforcement of earth slopes and walls is a well established practice in geotechnical engineering. Incorporating a geotextile or geogrid provides the soil with an additional tensile strength that will contribute to the overall stability of the system. Among the benefits of using this technique is the construction of steeper slopes and the reduced environmental impact of the solution. A current common analysis approach is the use of a modified limit equilibrium slip circle approach (Bishop 1955), where the action of the reinforcement is represented by an external load acting on the slip circle. Some authors have extended the method in order to account for the compatibility of strains between soil and geotextile (Rowe & Soderman 1985) or to assess the tensile force in each layer (Leshchinsky et al. 2017). The Finite Element Method (FEM) is also used when the problem requires an in depth study involving deformations and more realistic soil-reinforcement interaction in terms of admissible strains and differential displacements of the different parts of the system. An alternative to these two approaches is Discontinuity Layout Optimisation, DLO (Smith & Gilbert 2007), a limit analysis technique based on the theory of perfect plasticity. Clarke et al. 2013 presented a DLO model for soil nails and more recent work has focused on the application to geotextiles (Smith & Tatari 2016).

1.2 *Conventional design approach*

Various design codes/manuals provide guidelines for the design and analysis of reinforced slopes based mainly on a combination of empirical and limit equilibrium approaches. For instance the British Standard (BS8006-1 2010) assumes a two-part wedge mechanism (Woods & Jewel 1990) for the collapse of the unreinforced system and it sets out a series of steps to follow in order to ensure stability by means of adding a geotextile or geogrid.

Application of the method can be found *e.g.* in the UK Highways Agency manual Design Methods for the Reinforcement of Highway Slopes by Reinforced Soil and Soil Nailing Techniques (HA 68/94 1994). The first step is to determine the value of the applied horizontal force (T_{sum}) that should be applied to the unreinforced system to maintain stability. This can be achieved by classic limit equilibrium procedures or taken directly from design charts.

The number of necessary reinforcement layers can be obtained by dividing T_{sum} by the design tensile strength of the reinforcement. The design tensile strength is typically specified by the engineer based on their experience. The embedment length is then obtained by imposing that pull-out resistance is ensured. Finally HA68/94 suggests a variable spacing with depth of the reinforcement layer.

1.3 *Proposed design approach*

The previously mentioned methods all have their advantages and disadvantages. However some common points for the design are the need to specify the available tensile strength alongside the layout and vertical spacing of the layers of geotextile. In the current paper an automatic procedure for finding the minimum required tensile strength for the reinforcement utilising DLO is proposed. The approach is applied to a reinforced slope using horizontal layers of geotextiles. The designer can benefit from the rapidity of the model for ascertaining the minimum tensile strength needed and by performing a parametric study, manually vary the length and positioning of the reinforcements in order to achieve an optimised solution.

2 DISCONTINUITY LAYOUT OPTIMIZATION

2.1 *Introduction*

Discontinuity Layout Optimization, DLO, (Smith & Gilbert 2007) is a computational limit analysis technique that offers an alternative procedure to those based on limit equilibrium or finite element methods to assess the factor of safety of a wide range of geotechnical problems. It provides an automatic general purpose procedure for finding critical upper bound solutions in the form of slip-line failure mechanisms that are generally within a few percent of the true solution. It uses a mathematical optimization formulation in the context of the two main assumptions of Limit Analysis: (i) the soil is a perfectly plastic material; (ii) soil obeys an associated flow rule.

Qualitatively the DLO procedure can be described by the process shown in Figure 1. In contrast to a FEA approach, DLO discretises the domain using nodes and potential slip-lines connecting those nodes rather than elements. The method can be presented in two mathematically equivalent forms. One in terms of displacement parameters (kinematic formulation) and one in terms of force parameters (equilibrium formulation). The former will be used in this paper.

2.2 *DLO kinematic formulation*

The primal kinematic problem formulation for the tranlational plane strain analysis of a quasi-statically loaded, perfectly plastic cohesive-frictional body discretised using m nodal connections (slip-line discontinuities), n nodes and a single load case can be stated as follows:

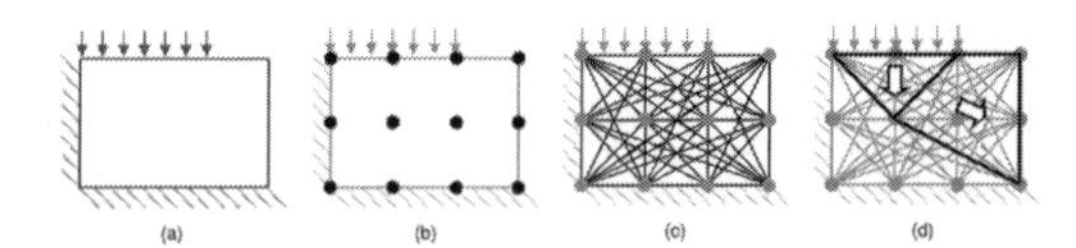

Figure 1. Stages in DLO procedure: (a) Initial problem definition, surcharge applied to block of soil close to vertical cut; (b) discretization of soil block using nodes; (c) interconnection of nodes with potential discontinuities; (d) identification of critical subset of potential discontinuities using optimisation, giving the layout of slip-lines in the critical failure mechanism.

$$\min \lambda \mathbf{f}_L^T \mathbf{d} = -\mathbf{f}_D^T \mathbf{d} + \mathbf{g}^T \mathbf{p} \quad (1)$$

subject to:

$$\mathbf{Bd} = 0 \quad (2)$$

$$\mathbf{Np} - \mathbf{d} = 0 \quad (3)$$

$$\mathbf{f}_L^T \mathbf{d} = 1 \quad (4)$$

$$\mathbf{p} \geq 0 \quad (5)$$

where $\mathbf{f}_D$ and $\mathbf{f}_L$ are vectors containing respectively specified dead and live loads, $\mathbf{d}$ contains displacements along the discontinuities, where $\mathbf{d}^T = \{s_1, n_1, s_2, n_2 ... n_m\}$, and s_i and n_i are the relative shear and normal displacements between blocks at discontinuity i; $\mathbf{g}^T = \{c_1 l_1, c_1 l_1, c_2 l_2, ... c_m l_m\}$, where l_i and c_i are respectively the length and cohesive shear strength of discontinuity i. $\mathbf{B}$ is a suitable $(2n \times 2m)$ compatibility matrix, $\mathbf{N}$ is a suitable $(2m \times 2m)$ flow matrix and $\mathbf{p}$ is a $(2m)$ vector of plastic multipliers. The discontinuity displacements in $\mathbf{d}$ and the plastic multipliers in $\mathbf{p}$ are the LP variables.

In this formulation the linear optimization problem aims to minimise the energy dissipated in the system (equation 1) subject to compatibility constraints at every node (equation 2) and the associated flow rule (equations 3 and 5), while imposing unit work by the live loads (equation 4). The nature of this formulation ensures that in effect $\mathbf{p} = |\mathbf{d}|$ and thus energy dissipation $(\mathbf{g}^T \mathbf{p})$ is always positive. Full details of the formulation may be found in Smith & Gilbert 2007.

3 SOIL REINFORCEMENT IN DLO

Clarke et al. 2013 proposed an extended model implemented in DLO able to represent both geotextiles and soil nail reinforcement. Failure could occur either by bending, tensile or compressive rupture, controlled by three parameters. In the current paper only the parameters relevant to planar flexible horizontal reinforcement are utilised as follows:

- reinforcement is modelled as a one dimensional element with a finite tensile strength, zero compressive strength and negligible flexural rigidity.
- slip-lines can only cross the reinforcement at a node.
- slip can occur independently above and below the reinforcement. The reinforcement may also extend longitudinally and plastically at a node.
- soil reinforcement can freely move relative to the soil.
- the shear strength of the soil/reinforcement interface may be lower than the shear strength of the adjoining soil by a factor α.

Figure 2 shows a section of reinforcement within the soil making distinction between the upper and the lower parts of soil. In order to ensure that slip can take place independently above and below, the horizontal compatibility equations are verified separately for the upper (**Soil u**) and lower (**Soil l**) blocks of soil at each node. Based on the previously stated assumptions the following expressions can be deduced for compatibility at a node lying on the reinforcement:

$$\sum \alpha_i s_i^u - \beta_i n_i^u = d_r \tag{6}$$

$$\sum \alpha_i s_i^l - \beta_i n_i^l = -d_r \tag{7}$$

$$\sum \beta_i s_i^{ul} + \alpha_i n_i^{ul} = 0 \tag{8}$$

where α_i and β_i are respectively horizontal and vertical direction cosines for each discontinuity i, and the subscripts u and l refer to the upper and lower blocks of soil.

Equations (6) and (7) maintain horizontal compatibility, while allowing the reinforcement to undergo plastic extension by an amount d_r as required, resulting in additional plastic work $|d_r|g_r$ where g_r denotes the tensile strength of the reinforcement R.

Equation (8) ensures conventional vertical compatibility (in this model the reinforcement is assumed to be flexible enough not to hinder vertical displacements in line with conventional approaches).

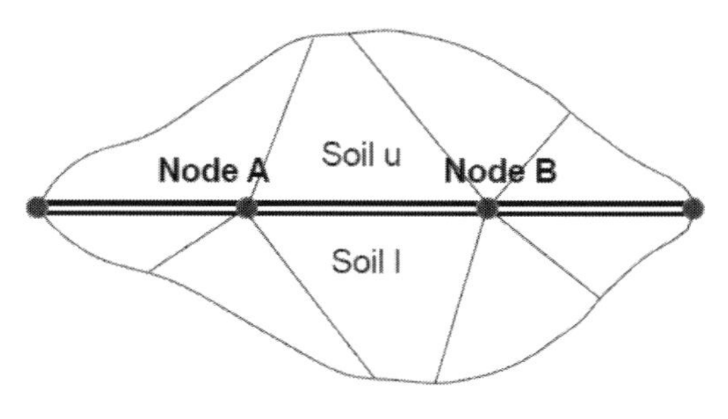

Figure 2. Detail of a soil block with an embedded segment of reinforcement.

4 REINFORCEMENT STRENGTH OPTIMISATION (RSO)

A revised formulation based on the previously described DLO is now provided, able to determine the maximum global reduction factor on the tensile strength that can be applied to a pre-determined extent of reinforcement and that still maintains stability.

This may be formulated as follows:

$$\min \lambda \mathbf{g}_r^T \mathbf{p}_r = -\mathbf{f}_D^T \mathbf{d} + \mathbf{g}_s^T \mathbf{p}_s \tag{9}$$

subject to:

$$\mathbf{Bd} = 0 \tag{10}$$

$$\mathbf{Np} - \mathbf{d} = 0 \tag{11}$$

$$\mathbf{g}_r^T \mathbf{p}_r = 1 \tag{12}$$

$$\mathbf{p_s}, \mathbf{p_r} \geq 0 \tag{13}$$

Where the subscripts s and r represents the soil and the reinforcement respectively, $\mathbf{p} = \{\mathbf{p}_s, \mathbf{p}_r\}^T$ and $\mathbf{f}$ now represents all loading in the system (there is no longer a distinction between live and dead loads). g_s is now equal to cl for a discontinuity in the soil and αcl for a discontinuity running along the edge of the reinforcement.

In this case the linear programming solution determines a uniform factor applied to the reinforcement tensile capacity needed in order to carry a given external load.

5 DESIGN BASED ON HA68/94

The method will now be illustrated using an example based on HA68/94. A design tensile strength of 14.4 kN/m was assumed for the 70 degree slope in Fig. 3. The necessary horizontal force for stability of 113 kN/m can be obtained from the charts given in HA68/94 for the specified geometry and soil properties. This gives a total number of 8 layers that will be irregularly spaced according to the HA68/94 recommendation and with proportional embedment length to the depth of each layer. The final solution can be seen in Fig. 3

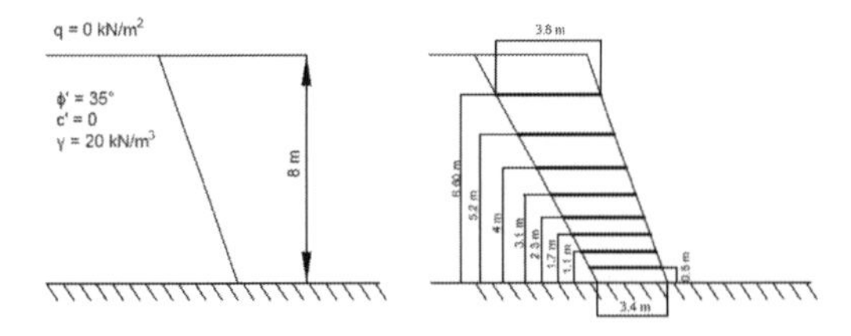

Figure 3. Hand obtained solution extracted from HA 68/94.

6 APPLICATION OF RSO

6.1 *Introduction*

The RSO approach can be utilised to automatically find the necessary tensile strength needed for stability and can be used for the design of the reinforced slopes according to the following procedure:

- Select the vertical spacing of the reinforced system. This could be according to standard design guidance or engineering criteria.
- Create an over-reinforced system with extended lengths of reinforcements *i.e.* the reinforcements should be significantly longer than needed.
- Apply the RSO optimisation. This will give the minimum strength of the reinforcement needed for stability.
- Using this reinforcement strength, manually reduce the reinforcement lengths while maintaining stability.

The last step can be achieved by means of an iterative procedure informed by recommendations in standard guidance. Ideally this phase would be automated by a second optimisation to give an overall optimised solution in terms of the length and strength of reinforcements that can be achieved. In the current paper, the guidelines of the standards are taken as a first guess. This has been found to give close to unity factors of safety. A manual refinement has then been performed to attain a factor of safety equal to the target (1.0 for the present case).

6.2 *Parametric study*

A parametric study was performed based on the example problem in Fig. 3 using the version of DLO implemented in the commercial software LimitState:GEO (LimitState 2017). Three different vertical spacings of reinforcements have been studied: one metre,half a metre and a variable spacing dependent on the depth of each layer.

One of the challenges with comparing a fully general stability analysis such as DLO with Limit Equilibrium approaches is that the latter typically ignore the possibility of local failures. In the context of reinforced soils, this local failure typically occurs at the slope face between the reinforcing layers. To mitigate this, while minimising the effect on the overall analysis, two measures were taken: (i) a small cohesion intercept value of $c' = 1\text{kN/m2}$ has been added to the main soil body, (ii) a thin facing layer with an enhanced cohesion of $c' = 5\text{kN/m2}$ has been modeled all the way down the face of the slope. This aims to represent vegetation, a soil crust or weak facing close to the surface. The layer is thin enough (0.1 m) such that it would have only a minor effect on the results of a conventional limit equilibrium analysis.

6.3 *Comparison*

The parameter chosen for comparison of the various solutions was the normalised volume of reinforcement per unit width, where the volume is defined as the overall length of reinforcement multiplied by the tensile strength of geotextile and divided by the yield strength of the geotextile material. Since the latter is assumed to be the same for all cases it is set to 1.0 to give the normalised value. The measure of volume is a simple approach to determining the cost of each option.

Configuration A is a first estimation using full width reinforcements and then determining the optimal strength for that layout using RSO. This solution gives the lowest necessary strength for an ideal over reinforced system.

Configuration B was performed in the following fashion: the previously obtained tensile strength from Configuration A was used as the selected design value. A hand optimisation was then undertaken using the approach in HA68/94 (with an embedment length proportional to the depth of the layer) as a first guess and then manually shortening the lengths of each layer in the direction of the collapse mechanism until the target factor of safety was reached (1.0 for this case).

Finally Configuration C set the embedment lengths according to the HA 68/94 criteria and then the RSO optimisation was run to obtain the required tensile strength.

A summary of the different strengths for each comparison is given in Table 1. The mecha-

Table 1. Tensile strengths (kN/m²) for the different comparisons and spacings.

Spacing	0.5 m	1 m	variable
A	5.5	13.6	10.0
B	1.3	0.51"	0.42
C	1.15	0.45"	1.45
HA68/94	-	-	14.4

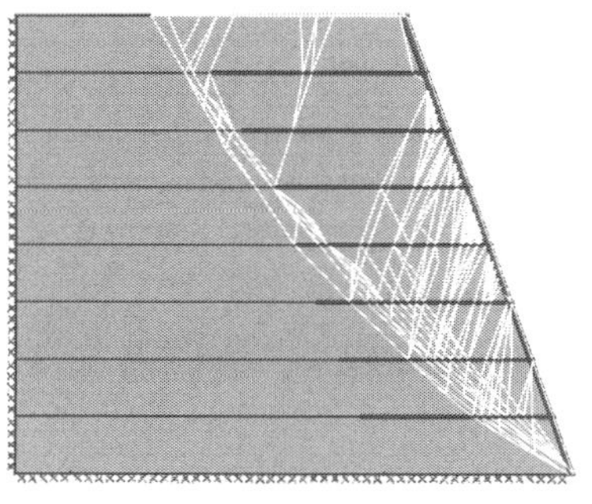

(a) 1m reinforcement spacing

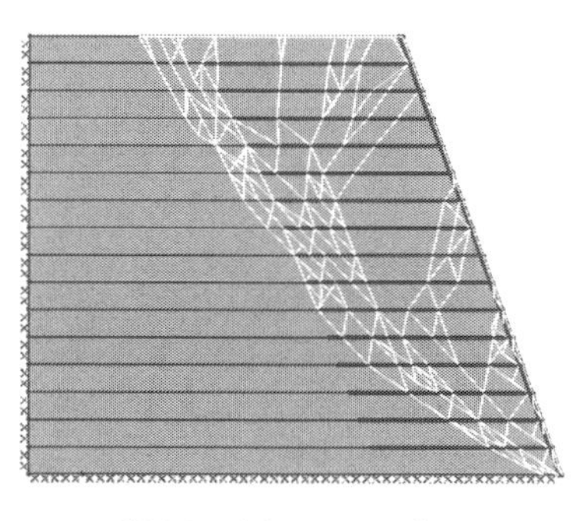

(b) 0.5m reinforcement spacing

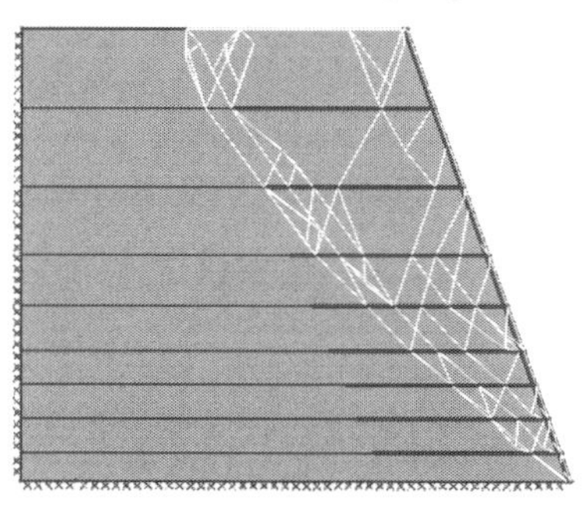

(c) variable reinforcement spacing

Figure 4. Reinforced slope using three vertical spacings: 1 metre, 0.5 metres and a variable spacing using the HA 68/94 guidelines. The failure mechanisms shown correspond to the Configuration C layout and tensile strength (see Table 1) and a factor of safety of 1.0.

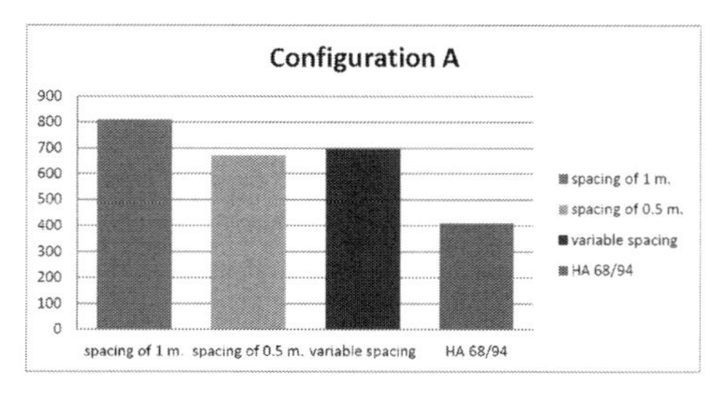

(a) Configuration A: idealised reinforced system with full lengths of reinforcements

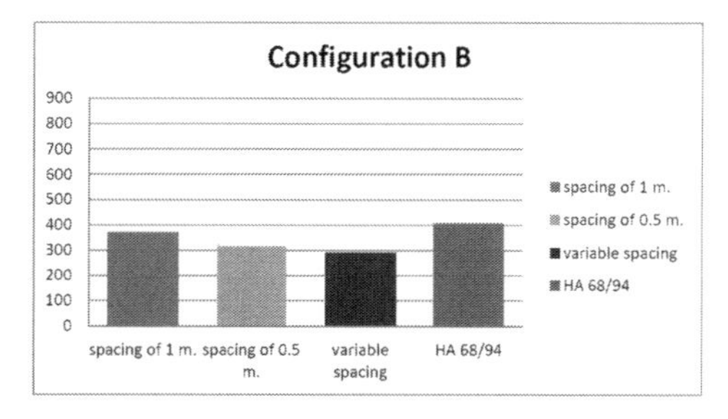

(b) Configuration B: uses the tensile strength obtained in Configuration A and an iterative manual optimisation was made to find the final layout

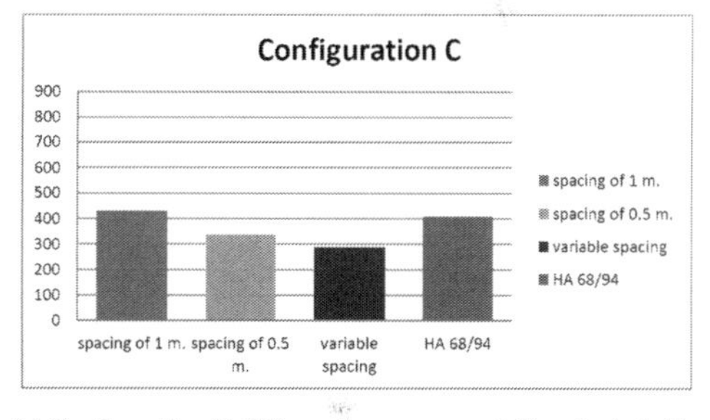

(c) Configuration C: follows the recommendations included in the HA 68/94 for the embedment lengths and finds the tensile strength using RSO for that layout

Figure 5. Comparisons made in terms of volume of reinforcement (Tensile strength × total length).

nisms and normalised volume results for each configuration (A, B and C) are given in Figures 4 and 5 respectively.

7 DISCUSSION

The results of the parametric studies using RSO show that a more optimal solution from the point of view of the total volume of the material can be achieved using RSO, even allowing for the slightly enhanced soil strengths in the RSO models as discussed in Section 6.2.

While the gains in this example are relatively minor, the method should demonstrate significant advantage for more complex scenarios: the HA68/94 approach makes use of the principles of limit equilibrium, *i.e.* assuming a specific form of collapse mechanism and proposing a reinforced solution for that collapse mode, on the other hand the RSO procedure uses a fully general limit analysis approach and therefore all possible collapse mechanisms are examined. For simple slope stability this may not vary significantly from the limit equilibrium mechanism. However for a more complex loading pattern *e.g.* surface loading at or behind the crest of the slope, the RSO procedure will capture any changed failure mode and the most prone to fail mechanism is the one for which the reinforcement strength is selected.

The procedure described in this paper involves a combined automated and manual optimization procedure. There is however scope to automate the manual stage by reformulating the optimization objective function in terms of the overall volume of reinforcement. Research into this approach is ongoing with promising results so far.

8 CONCLUSIONS AND FUTURE WORK

A new automatic approach, termed RSO, for finding the minimum reinforcement tensile strength required for a given configuration of reinforcement has been presented.

Starting with a given vertical spacing, the optimal tensile strength can be calculated by applying the RSO procedure to an over reinforced system (reinforcement lengths extended to a large value). This will generally give the lowest tensile strength needed for the given vertical spacing.

Based on this value, further hand optimisation can be carried out to determine the optimal lengths of reinforcement. Ideally this second manual stage would be replaced by a second automated optimisation procedure that minimises the overall volume of reinforcement by automatically reducing lengths to give a complete solution in terms of strength and layout. This is ongoing research with promising results so far.

ACKNOWLEDGEMENTS

The authors wish to acknowledge the support of the European Commission via the Marie Sklodowska-Curie Innovative Training Networks (ITN-ETN) project TERRE Training Engineers and Researchers to Rethink Geotechnical Engineering for a low carbon future' (H2020-MSCA-ITN-2015–675762).

REFERENCES

Bishop, A. W. (1955). The use of the slip circle in the stability analysis of slopes. *Gotechnique 5*(1), 7–17.

BS8006–1 (2010). *Code of practice for strengthened/ reinforced soils and other fills.* BS 8006–1.

Clarke, S., C. Smith, & M. Gilbert (2013). Modelling discrete soil reinforcement in numerical limit analysis. *Canadian Journal of Civil Engineering 50*(7), 705–715.
dx.doi.org/10.1139/cgj-2012–0387.

HA 68/94 (1994, August). *Design Methods for the Reinforcement of Highway Slopes by Reinforced Soil and Soil Nailing Techniques.* Highways Agency, London, UK, HA. Volume 4 Section 1 Part 4 (HA 68/94).

Leshchinsky, D., B. Leshchinsky, & O. Leshchinsky (2017). Limit state design framework for geosynthetic-reinforced soil structures. *Geotextiles and Geomembranes.* http://dx.doi.org/10.1016/j.geotexmem.2017.08.005.

LimitState (2017, February). *LimitState:GEO Manual Version 3.4.a.* (LimitState Ltd ed.).

Rowe, R. K. & K. L. Soderman (1985). An approximate method for estimating the stability of geotextilereinforced. *Canadian Geotechnical Journal 22*, 392–398.

Smith, C. & M. Gilbert (2007). Application of discontinuity layout optimization. *Royal Society A: Mathematical, Physical and Engineering 463*, 2461–2484. doi:10.1098/rspa.2006.1788.

Smith, C. C. & A. Tatari (2016). Limit analysis of reinforced embankments on soft soil. *Geotextiles and Geomembranes Journal 44*(4), 504–514. doi:10.1016/j.geotexmem.2016.01.008.

Woods, R. & R. Jewel (1990). A computer design method for reinforced soil structures. *Geotextiles and Geomembranes* (9), 233–259.

Numerical Methods in Geotechnical Engineering IX – Cardoso et al. (Eds)
© 2018 Taylor & Francis Group, London, ISBN 978-1-138-33203-4

Simulation of a jet injection into an elastic perfectly-plastic soil using uGIMP

Daniel Ribeiro
CERIS, Instituto Superior Técnico, University of Lisbon, Portugal

João Ribas Maranha
Geotechnics Department, National Laboratory for Civil Engineering (LNEC), Portugal

Rafaela Cardoso
CERIS, Instituto Superior Técnico, University of Lisbon, Portugal

ABSTRACT: Jet Grouting is a ground improvement technique which has been largely used worldwide. Its versatility and advantages justify its popularity, while the difficulties one can find when defining the injection mechanism and the phenomena involved motivate the use of empirical approaches to predict the properties of the treated (or grouted) volume. It is recognized the interest on developing new designing tools coupling the geometry prediction and the displacements induced to the soil by the jet. Some studies have reported promising results when numerical particle methods are utilized. In this study the capabilities and limitations of Uniform Generalized Interpolation Material-Point (uGIMP) Method to handle a water-jet injection into a target soil is analyzed. Simulations were performed for 3 initial jet velocities (50, 100, and 150 m/s) typical of jet grouting injections, where the target soil was a soft clay. Improvements to be considered for future developments are identified and reported.

1 INTRODUCTION

Jet Grouting is a ground improvement technique which has been largely used worldwide. Its versatility and advantages justify its popularity, while the difficulties one can find to define the injection mechanism and the phenomena involved motivate the use of empirical approaches to predict the properties of the treated (or grouted) volume. Moreover, the existing models and methods, as well as the classic theories were found inaccurate and mostly limited to the geometry prediction, which has been motivating emerging studies through particle methods (Ribeiro & Cardoso 2016, for example).

Several studies have been showing significant installation effects (during construction) (Wang et al. 1999, Wong and Poh 2000, Hsiung et al. 2001, Poh and Wong 2001, Wang et al. 2013, Wu et al. 2016), such as displacements induced in the soil and pore-pressure variations, which are usually neglected when designing jet grouting solutions. Indeed, their construction can significantly affect surrounding structures and infrastructures. So far, attempts to quantify these effects through cavity expansion (Wang et al. 2013, Shen et al. 2013, Wu et al. 2016, Wang et al. 2016) were found in Literature. However, as they do not reproduce the phenomena or the physical effects involved, their application still needs data from a careful instrumentation of the site, as the most of the cavity expansion applications to jet grouting are found by reverse engineering.

In this study, some numerical experiments were conducted where a fluid jet injection into a target soil was simulated using a Material–Point Method (MPM) descendant known as Uniform Generalized Interpolation Material–Point (uGIMP) method. The abilities and the limitations of the uGIMP method were analyzed, focusing on the penetration rate and on the displacements produced in the soil.

The Uniform GIMP (uGIMP) method was the one adopted for these simulations because it is one of the simplest, thereby, more efficient GIMP methods, from a computational point of view, and it was found suitable for studying the problem under analysis.

In GIMP methods the stage where the update of the stresses take place in the algorithm is also an important topic which requires some attention. However, an extensive analysis of each updating approach is not provided, as they have been well examined in Literature (Bardenhagen 2002, Nairn 2003, for example).

Broadly speaking, if the stresses (and strains) are updated after updating momentum at the material-points the updating method is termed Update–Stress–Last (USL), and otherwise it is termed Update–Stress–First (USF) when the algorithm updates the stresses (and strains) before updating momentum (Bardenhagen 2002, Wallstedt and Guilkey 2008). Alternatively, one can update stresses and strains at the beginning and at the end (including the re-extrapolation of the nodal momenta) of each computing cycle, and this is known as the USAVG approach (Update–Stress–Averaged) (Nairn 2003). The USAVG stresses' update scheme was the one adopted in this study, as it was shown advantageous through energy conservation analysis conducted by Nairn (2003).

2 UNIFORM GIMP

In the 1990s, Sulsky et al. (1994, 1995) came forward with a particle method with especial interest for time-dependent materials, which was named Material-Point Method (MPM). This method descends from the Particle–in–Cell (PIC) (Harlow and Welch 1965) methods, and is pointed-out as a solid mechanics extension of the fluid-implicit-PIC method (FLIP) (Brackbill 2005).

As in FLIP (Brackbill and Ruppel 1986, Brackbill et al. 1988) this method is a mixed Lagrangian-Eulerian method where the material is discretized in a set of Lagrangian material points (or particles), while a background grid is utilized for solving the momentum equation through standard finite element methods (FEM).

At each time-step, the scattered information of the continuum (e.g., mass, momentum, energy and constitutive properties) carried by the material-points is projected (or mapped) to the grid. No permanent information is stored in the grid, as the information carried by the particles is enough to describe the continuum and its properties.

As an attempt to face some numerical problems commonly found in standard MPM formulations, such as those related to spatial integration errors, Bardenhagen and Kober (2004) proposed the Generalized Interpolation Material-Point (GIMP) Method. Assumed as an extension of the Standard MPM, through the introduction of a general particle characteristic function $\chi_{mp}(\boldsymbol{x})$, centered at the material–point position $\boldsymbol{x}_{mp}$, in the shape function, this method somehow considers the material-points as having extent within the quadrature scheme (Steffen et al. 2008). Furthermore, the shape function φ_{imp} (or interpolation function) and the gradient of the the shape function $\nabla\varphi_{imp}$ are now defined by the convolution of the characteristic function with, respectively, the shape function of the grid and its gradient,

$$\phi_{imp} = \frac{\int_{\Omega} \phi_i(\boldsymbol{x})\chi_{mp}(\boldsymbol{x})\,d\Omega}{\int_{\Omega} \chi_{mp}(\boldsymbol{x})\,d\Omega}, \tag{1}$$

$$\nabla\phi_{imp} = \frac{\int_{\Omega} \nabla\phi_i(\boldsymbol{x})\chi_{mp}(\boldsymbol{x})\,d\Omega}{\int_{\Omega} \chi_{mp}(\boldsymbol{x})\,d\Omega}, \tag{2}$$

where φ_{imp} results on a weighting function with support in adjacent cells and in next nearest neighbor cells (subscripts: i – grid node; mp – material–point).

While in the original MPM the material-points were spatially Dirac delta functions, which means that the volume of material represented by each particle is assumed to exist only in a single point in space, in GIMP Methods material-points are treated as blocks of material. This feature reduces the errors and instabilities often found when particle distributions become disordered, such as the particle grid crossing which was well examined by Bardenhagen and Kober (2004) and Steffen et al. (2008). Adopting weighting functions as shape functions (or interpolation functions), in a similar way to other meshless methods, such as Smoothed Particle Hydrodynamics, is found as the main reason for the successful advance.

Because of its various domains of interest, especially over the engineering disciplines, various GIMP Methods have been proposed. One of the most popular GIMP analysis is the Uniform (or Uintah) GIMP (uGIMP) method.

In a briefly rough explanation, during the analysis, if the uGIMP approach is adopted then the material–point domains are assumed to be stationary (or unchanged) and the particle domains are always aligned with the grid (Wallstedt and Guilkey 2008, Sadeghirad et al. 2011). As a consequence, uGIMP may not be accurate for simulations involving large rotations or important shear deformations: material–point domains are expected to become parallelograms, when shear deformation takes place; large rotations are likely to cause misalignment between material–point domains and the grid (Sadeghirad et al. 2011).

Another inconvenience of uGIMP is the extension instabilities, which are found when the particles become separated to an extent to which their regions of influence no longer overlap. (Sadeghirad et al. 2011).

Moreover, in the case of a continuum penetrating another, such as in the case of a jetting fluid into a solid porous material, for a short domain of analysis (i.e., simulation time), if a perfectly inelastic collision is assumed for the particles, then extension instabilities can be avoided without compromising the quality of the results.

3 NUMERICAL EXPERIMENTS – DETAILS

3.1 *Models for the materials*

3.1.1 *Liquid model*

The liquid model comprehends the governing equations of mass and momentum,

$$\frac{\partial \rho}{\partial t} + \rho \nabla \cdot \boldsymbol{v} = 0, \tag{3}$$

and

$$\rho \frac{D\boldsymbol{v}}{Dt} = \nabla \cdot \sigma + \rho \boldsymbol{b}, \tag{4}$$

respectively (where: ρ is the mass density, $\boldsymbol{v}$ the velocity vector, σ the stress tensor, $\boldsymbol{b}$ is the body force and $\frac{D\boldsymbol{v}}{Dt}$ represents the material derivative of velocity), the equation of state adopted, and other relevant details (e.g., viscous component for stresses calculation).

In the GIMP code used for the simulations provided, NairMPM (Nairn 2003), Tait's equation was the Equation of State adopted.

The Tait's Equation (TE) is a well known semi–empirical equation relating volume change of a material under hydrostatic compression to the applied pressure. Originally proposed for water, it has been also applied to other fluids. The standard TE can be written in its isothermal form as (Macdonald 1966):

$$p = C\,K\left[e^{\left(\frac{1-J}{C}\right)} - 1\right], \tag{5}$$

where K is the bulk modulus of the fluid, C is the universal Tait constant (0.0894) and J is the determinant of the deformation gradient ($J = det\boldsymbol{F}$), also referred to as the *Jacobian* of deformation, which is here defined for volume variations $\left(J = V/V_o\right)$.

3.1.2 *Soil model*

An elastic perfectly-plastic model was chosen for the simulations here conducted. A von-Mises yield criterion was adopted, and the elastic (pre-yield) regime was ruled by a neo-Hookean model.

The neo-Hookean model is an extension of the Hooke's law for the case of large deformations, and it can be expressed as (Nairn 2003):

$$\sigma = \lambda \frac{lnJ}{J}\boldsymbol{I} + \frac{\mu}{J}(\boldsymbol{B} - \boldsymbol{I}), \tag{6}$$

where is J the Jacobian ($J = |\boldsymbol{F}|$; $\boldsymbol{F}$ is the deformation gradient), $\boldsymbol{B}$ is the left Cauchy-Green deformation tensor ($\boldsymbol{B} = \boldsymbol{F}\boldsymbol{F}^T$), λ and μ are, respectively, the first and the second Lamé constants. The first constant can be estimated as:

$$\lambda = \frac{3K\nu}{1+\nu}, \tag{7}$$

where, ν is the Poisson's rate and K is the bulk modulus, while the second constant corresponds to the shear modulus G.

The von-Mises yield criterion can be expressed as (von Mises 1913, Dill 2006):

$$\sqrt{3J_2} - Y = 0, \tag{8}$$

where Y is the yield (shear) stress, and J_2 the second stress invariant which reads

$$J_2 = \frac{1}{2}\boldsymbol{s} : \boldsymbol{s}, \tag{9}$$

and $\boldsymbol{s}$ is the deviator stress tensor.

Because no yield occurs under hydrostatic compression, when considering the von-Mises yield criterion, the volumetric behavior of the pore water is not suitable to contribute for rupture in the simulations performed.

3.2 *Geometry and other inputs*

The problem under analysis is the injection of a fluid jet into a soil box, which is schematically presented in Figure 1.

In geotechnical applications, such as the case of jet grouting injections and vertical water jets, the diameter of the nozzle is usually ranging between 2 and 4 mm. Nevertheless, in previous studies found in Literature for jet injections (Bui et al. 2007, Stefanova et al. 2012, for example), the nozzle diameter was of the order of centimeters. In this study, the nozzle diameter is adopted to be 4 mm.

The simulations were performed for the initial jet velocities of 50, 100 and 150 m/s, which are in agreement to the usual injection properties for the single fluid jet grouting system (EN 2001, Ribeiro 2014, Spagnoli 2013).

The domain of analysis is defined looking for a commitment between computational costs and the requirements of the analysis. The code is running on CPU and for the density of material–points adopted (the voxel of each material–point was of 1 mm^2), the simulations were computationally demanding. Thereby, a box of 200 mm (50 times the nozzle diameter) per 500 mm of soil was adopted as the jetting target. The simulation time is of 5 ms, with an archive time–interval 0.5 ms.

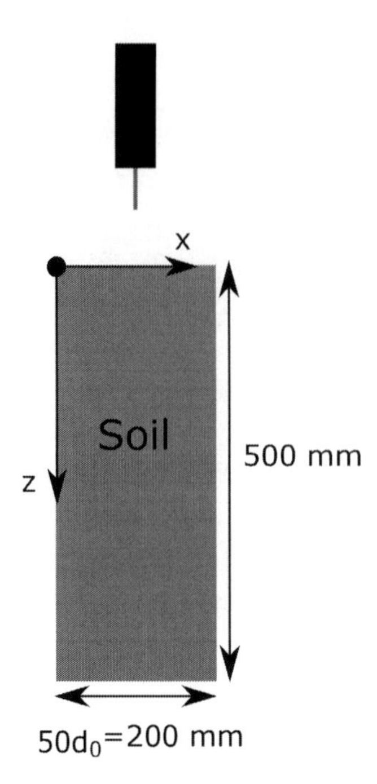

Figure 1. Scheme for the injection simulations.

The soil is adopted to be a soft clay. The properties of the soil are according to typical values found in Bowles (1996) and presented in Table 1 (where K is the drained bulk modulus, G is the shear modulus, ρ is the density of the material, and Y is the yield stress which corresponds to the unconfined compressive strength for the von-Mises criterion).

The soil is under undrained conditions while submitted to a sudden loading due to the hydrodynamic action of the jet. Thereby, if saturated, the behavior of the soil cannot be disregarded of the pore–water effects. However there are not any implemented rules in the code for computing the volumetric behavior of water.

Therefore, the undrained bulk modulus was used as an indirect manner to consider the volumetric behavior of the pore–water. The undrained bulk modulus K_u pertains to the drained bulk modulus K and the fluid bulk modulus K_f (2200 MPa) as follows:

$$K_u = K + \frac{K_f}{n}, \tag{10}$$

where n is the porosity of the material that was adopted to be 0.5, accordingly to reference values from Bowles (1996) and Terzaghi et al. (1996), which corresponds to a K_u of 4484 MPa.

The results of the simulations performed are provided and analyzed, highlighting all the relevant details and limitations, in the following section.

Table 1. Properties of the soft clay.

K (MPa)	84
G (MPa)	10
ρ (g/cm3)	1.8
Y (kPa)	50

4 DISCUSSION OF THE RESULTS

A graphical output of the simulations is provided in Figure 2, for the 3 different velocities (50, 100 and 150 m/s), after 3 ms.

In Figure 2 the cutting mechanism is found as a penetration of the soil rather than its erosion. As the soil is a soft clay, this cutting mechanism is in agreement to Croce et al. (2014) who states that the jet action in clays is expected to be of this type. For this reason, the downstream distance of the jet at a given instant of time will be referred in this study as *penetration depth*.

In the NairnMPM code (Nairn 2003), as much as it was found, there is not a direct way to track the penetration of the jet in the soil. Thereby the penetration depth was graphically obtained for each instant of time. In Figure 3 the penetration depth is shown varying along time for the simulations performed.

The penetration (depth) shows a linear variation along time, allowing the definition of constant penetration rates. The constant penetration rates are due to some of the hypotheses adopted, and which were already assumed by other authors, however not explicitly explained (Ribeiro and Cardoso 2016). These hypotheses are essentially related with the current limitations of the available tools, which are:

- Submerged Jet – When the injections are performed inside a borehole, such as the case of jet grouting, the jet is expected to be submerged. It was attempted to perform a water jet injection where an ambient fluid (at rest) layer was existing before the target soil. The results were very inconclusive as the water jet was literally cutting the water layer at rest. Thereby, these studies were not continued;
- Counterflow – In real jet grouting injections, counterflow is expected and it provides an important attenuation factor for the penetration rate of the jet. The time of analysis is not enough for the jet to become submerged and this artifact cannot be analyzed. However, this would face difficulties very similar to those pointed in the above item;
- Jet Spread – The jet is expected to spread. While spreading the jet is also expected to follow a reduction in its hydrodynamic action.

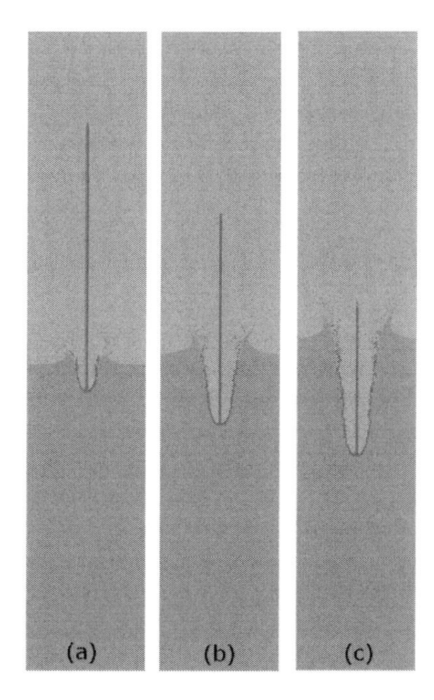

Figure 2. Simulations after 3 ms for velocities of: (a) 50 m/s, (b) 100 m/s and (c) 150 m/s.

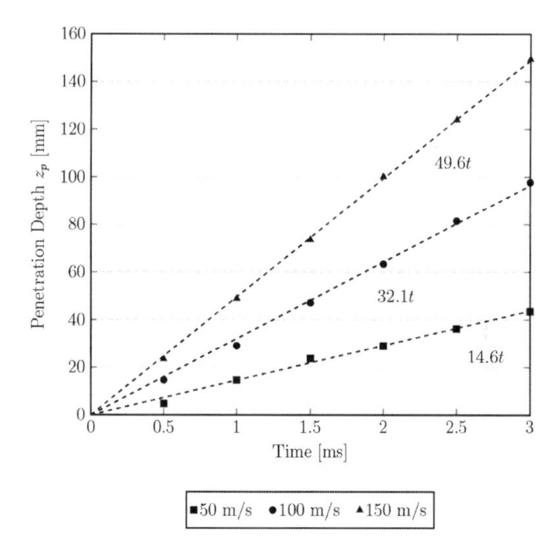

Figure 3. Penetration depth along time for 50, 100 and 150 m/s.

This feature was not possible to consider in the simulations, thereby, the jet was assumed not to spread;

- Soil-Hardening – The hardening of the soil is not expected to be significant near the impact surface, as a sudden stress variation is expected and the rupture of the soil occurs very fast. Nevertheless, far enough from the jet impact, the hardening of the soil maybe relevant as the soil properties will change and it probably will slow down the jet penetration. The soil models adopted cannot consider the hardening of the soil.

It is also important to note that the penetration rates are independent from the bulk modulus adopted, as in the von-Mises criterion the yielding is controlled by shear. Moreover, through this yield criterion, the volumetric behavior of water is assumed to have no effects in the yield of the soil. This feature needs to be analyzed in the future.

Despite of the limitations identified and the simplifying hypotheses adopted, the vertical displacements are consistent. As expected, major displacements are induced by the jet at a velocity of 150 m/s while the minimum stands for 50 m/s. In Figure 4 one can find the profiles obtained from the simulations, at 150 mm of depth (from the target surface, z), as illustrated in Figure 4, for the initial jet velocities of 150 m/s, 100 m/s and 50 m/s, after 3 ms of running simulation.

Analyzing the cross-sectional profiles of vertical displacements, one can find the maximum displacements for each velocity occurring at the jet impact (or jet center line, x = 100 mm) while a slight swelling is observed far enough from that point. When a high loading rate is applied to a saturated soil, the material is expected to exhibit an undrained response where the soil volume is maintained. Thereby, if the soil is placed inside a rigid box, and it is loaded in the middle, under undrained loading conditions, it is expected to observe

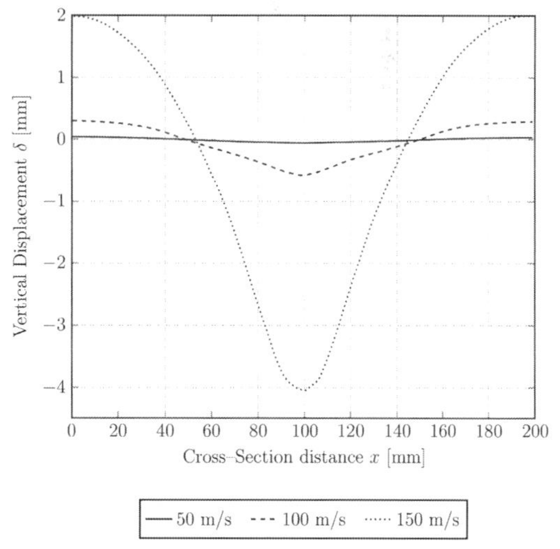

Figure 4. Vertical Displacements' profiles at z = 150 mm for 50, 100 and 150 m/s (jet velocity) after 3 ms.

swelling near the borders of the rigid box. Thus, the results observed in the simulations are realistic.

The gravitational acceleration (9.81 m/s^2) was set cross directional to the jet flow (x axis in Figure 1), from left to right. However, in these simulations the effect of gravity appears to be negligible, as it seems not to have an evident influence in the results and the displacement profiles are nearly symmetrical.

5 CONCLUSIONS

In this study the capacities of uGIMP to handle fluid jet injections into target soils were analyzed. A water jet injection was simulated for a soft clay target, and the results were found promising. Nevertheless, this type of tools are not ready for the current design practice of jet injections throughout geotechnical applications.

Suitable experimental data is absent for validation of the numerical experiments conducted. Although, some limitations were identified and insights for future developments came forward. Thereby, the main conclusions and remarks were selected, which are summarized as follows:

- Accordingly to the results obtained, it is possible to state that under further developments, such has implementing a proper constitutive model for the soil, and developing an adequate approach for accounting with the volumetric behavior of water, these tools will be suitable for reproducing fluid jet injections into target soils;
- The volumetric behavior of water, and hence the pore-water effects such as pore-pressure variations, cannot be explicitly computed, for these kind of problems, through the available constitutive equations. In Literature a model was found for GIMP methods (Homel et al. 2016), but it was not proven to work for the stress levels experienced by the target soil during injection. Furthermore, in this study the undrained bulk modulus was adopted as an indirect manner to consider the influence of the volumetric porewater behavior in the overall volumetric behavior of the soil. It is worth to note that pore–water had no influence in the rupture of the material, as a von–Mises yield criterion was adopted. Furthermore, in a total stress criterion, in undrained conditions, the pore–pressure needs not be known;
- A cutting mechanism due to the jet action was found. Some authors have reported it as an expected output for clayey materials, and it was here found for a soft clay target soil. As the von-Mises yield criterion was adopted, one can conjecture that when the target soil is a clayey material, the jet penetration is essentially controlled by shear;
- The accuracy of the simulations provided cannot be analyzed, as well as the significance of the loading rate effects in the soil, as a result of the jet impact, due to the lack of suitable experimental data for comparison. It is worth to mention that loading rate effects are not explicitly considered in the constitutive models adopted for the soil.

Further developments on these tools will require adequate laboratory experimental data. The main variables to be experimentally measured along time should be the penetration depth (or rate), the displacements induced to the soil and the excess of pore–water pressure caused by the jet action in the target soil.

ACKNOWLEDGMENTS

Acknowledgement is due to Fundação para a Ciência e a Tecnologia (FCT), Portugal for the financial support provided through the PhD Grant [SFRH/BD/86978/2012].

REFERENCES

Bardenhagen, S. (2002). Energy conservation error in the material point method for solid mechanics. *Journal of Computational Physics 180*(1), 383–403.

Bardenhagen, S. & E. Kober (2004). The generalized interpolation material point method. *Computer Modeling in Engineering and Sciences 5*(6), 477–496.

Bowles, J.E. (1996). *Foundation design and analysis* (5th ed.). McGraw-Hill, New York.

Brackbill, J., D. Kothe, & H. Ruppel (1988). Flip: A lowdissipation, particle-in-cell method for fluid flow. *Computer Physics Communications 48*(1), 25–38.

Brackbill, J. & H. Ruppel (1986). Flip: A method for adaptively zoned, particle-in-cell calculations of fluid flows in two dimensions. *Journal of Computational Physics 65*(2), 314–343.

Brackbill, J.U. (2005). Particle methods. *International Journal for Numerical Methods in Fluids 47*(8–9), 693–705.

Bui, H.H., K. Sako, & R. Fukagawa (2007). Numerical simulation of soilwater interaction using smoothed particle hydrodynamics (sph) method. *Journal of Terramechanics 44*(5), 339–346.

Croce, P., A. Flora, & G. Modoni (2014). *Jet grouting: technology, design and control*. Boca Raton: CRC Press, Taylor and Francis Group. ISBN: 9780415526401.

Dill, E.H. (2006). *Continuum mechanics: elasticity, plasticity, viscoelasticity*. CRC press.

EN (2001). 12716:2001. execution of special geotechnical works – jet grouting. *CEN*.

Harlow, F.H. & J.E. Welch (1965). Numerical calculation of time-dependent viscous incompressible flow of fluid with free surface. *The physics of fluids 8*(12), 2182–2189.

Homel, M.A., J.E. Guilkey, & R.M. Brannon (2016). Continuum effective-stress approach for high-rate

plastic deformation of fluid-saturated geomaterials with application to shaped-charge jet penetration. *Acta Mechanica 227*(2), 279–310.
Hsiung, B.-C.B., D. Nash, M. Lings, H. Hsieh, I.H. Wong, & T.Y. Poh (2001). Discussion on the paper: Effects of jet grouting on adjacent ground and structures. *Journal of Geotechnical and Geoenvironmental Engineering 127*(12), 1076–1078.
Macdonald, J.R. (1966, Oct). Some simple isothermal equations of state. *Rev. Mod. Phys. 38*, 669–679.
Nairn, J.A. (2003). Material point method calculations with explicit cracks. *Computer Modeling in Engineering and Sciences 4*(6), 649–664.
Poh, T.Y. & I.H. Wong (2001). A field trial of jet-grouting in marine clay. *Canadian Geotechnical Journal 38*(2), 338–348.
Ribeiro, D. (2014). Study and analysis of some methods used to predict the diameter of columns made by jet grouting in sands. In *5rd Meeting of young geotechnicals*, Covilhã, Portugal, pp. 11–20. Portuguese Geotechnical Society.
Ribeiro, D. & R. Cardoso (2016). A review on models for the prediction of the diameter of jet grouting columns. *European Journal of Environmental and Civil Engineering 0*(0), 1–29.
Sadeghirad, A., R.M. Brannon, & J. Burghardt (2011). A convected particle domain interpolation technique to extend applicability of the material point method for problems involving massive deformations. *International Journal for Numerical Methods in Engineering 86*(12), 1435–1456.
Shen, S.-L., Z.-F. Wang, W.-J. Sun, L.-B. Wang, & S. Horpibulsuk (2013). A field trial of horizontal jet grouting using the composite-pipe method in the soft deposits of shanghai. *Tunnelling and Underground Space Technology 35*, 142–151.
Spagnoli, G. (2013). A review of the energy balance equations of the jet grouting. *International Journal of Geotechnical Engineering 7*(4), 438–442.
Stefanova, B., K. Seitz, J. Bubel, & J. Grabe (2012). Water-soil interaction simulation using smoothed particle hydrodynamics. 6th international conference on scour and erosion. In *6th International conference on scour and erosion*, Paris, France, pp. 695–704.
Steffen, M., R.M. Kirby, & M. Berzins (2008). Analysis and reduction of quadrature errors in the material point method (mpm). *International Journal for Numerical Methods in Engineering 76*(6), 922–948.
Sulsky, D., Z. Chen, & H. Schreyer (1994). A particle method for history-dependent materials. *Computer Methods in Applied Mechanics and Engineering 118*(1), 179–196.
Sulsky, D., S.-J. Zhou, & H.L. Schreyer (1995). Application of a particle-in-cell method to solid mechanics. *Computer Physics Communications 87*(12), 236–252. Particle Simulation Methods.
Terzaghi, K., R.B. Peck, & G. Mesri (1996). *Soil mechanics in engineering practice*. John Wiley & Sons.
von Mises, R. (1913). Mechanik der festen korper im plastischdeformablen zustand. *Nachrichten von der Gesellschaft der Wissenschaften zu Gottingen, Mathematisch-Physikalische Klasse 1913*, 582–592.
Wallstedt, P. & J. Guilkey (2008). An evaluation of explicit time integration schemes for use with the generalized interpolation material point method. *Journal of Computational Physics 227*(22), 9628–9642.
Wang, J., B. Oh, S. Lim, & G. Kumar (1999). Effect of different jet-grouting installations on neighboring structures. *Field Measurements in Geomechanics: Proceedings of the 5th international symposium FMGM99, Singapore*.
Wang, Z.-F., X. Bian,&Y.-Q.Wang (2016). Numerical approach to predict ground displacement caused by installing a horizontal jet grout column. *Marine Georesources & Geotechnology*, 1–8.
Wang, Z.-F., S.-L. Shen, C.-E. Ho, & Y.-H. Kim (2013). Investigation of field-installation effects of horizontal twin-jet grouting in shanghai soft soil deposits. *Canadian Geotechnical Journal 50*(3), 288–297.
Wong, I.H. & T.Y. Poh (2000). Effects of jet grouting on adjacent ground and structures. *Journal of Geotechnical and Geoenvironmental Engineering 126*(3), 247–256.
Wu, Y.D., H.G. Diao, C.W. Ng, J. Liu, & C.C. Zeng (2016). Investigation of ground heave due to jet grouting in soft clay. *Journal of Performance of Constructed Facilities 30*(6), 06016003.

Numerical Methods in Geotechnical Engineering IX – Cardoso et al. (Eds)
© 2018 Taylor & Francis Group, London, ISBN 978-1-138-33203-4

Numerical modelling of limit load increase due to shear band enhancement

K.-F. Seitz & J. Grabe
Institute of Geotechnical Engineering and Construction Management, Hamburg University of Technology, Germany

ABSTRACT: Strengthening of just the shear bands is a patented idea to prevent a critical failure mechanism in order to increase the capacity of a geotechnical system with a minimum of material. Numerical modelling is used to examine this proposed method for limit load increase. Two basic geotechnical systems are analysed using OptumG2 for finite element limit analysis (fela): a shallow foundation with bearing failure and a retaining wall with horizontal wall movement resulting in the active earth wedge. An automated procedure identifies and enhances the shear band using a coupling of OptumG2 with MATLAB. The fela is performed with OptumG2 and run from a MATLAB script, which identifies the shear band based on the simulation result, sets the area of shear band enhancement and reruns the fela. The numerical simulations show that the shear band enhancement leads to a capacity increase for a simple geotechnical system.

1 INTRODUCTION

Strengthening of just the shear bands is a patented idea (Grabe & Pucker 2012) to increase limit loads with a minimum of material. Following the concept of one critical failure mechanism, which determines a system's limit load, the limit load is increased if the critical failure mechanism is prevented. The idea is to strengthen the shear bands, along which soil bodies are moving, when the limit load has been reached. Two well known failure mechanisms have been presented in literature and will be analysed in the present paper: the bearing capacity of a shallow foundation (Prandtl 1920) and the earth wedge occuring when a retaining wall is horizontally displaced (Coulomb 1776). Fig. 1 shows a sketch of the presented concept. When the critical failure mechanism is prohibited through the enhancement of the shear bands, a new critical failure mechanism will evolve leading to an increased limit load: the bearing capacity of a shallow foundation will be increased and the force resulting from active earth pressure will be decreased.

This concept has been presented by Grabe & Seitz 2016 and Seitz & Grabe (2018). Grabe & Seitz (2016) show the results of 1 g model tests of a retaining wall. Seitz & Grabe (2018) present further model tests with an improved method for shear band enhancement. They analyse two geotechnical systems: a shallow foundation and a retaining wall. Despite the fact that the execution of these 1 g model tests is challenging (bearing in mind the: reproducibility of the model test, the disturbance of the soil body when enhancing the shear bands and the enhancement itself), the results of the 1 g model tests have been quite promising and are supporting the hypothesis of limit load increase when the shear bands are enhanced.

The present paper is analysing the concept of limit load increase through shear band enhancement numerically. OptumG2 is used to model

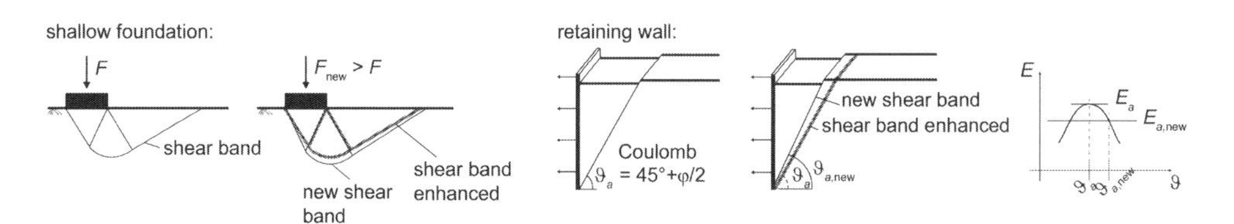

Figure 1. Limit load increase for geotechnical systems due to shear band enhancement: shallow foundation (left) retaining wall (right) (modified from Grabe & Seitz 2016).

the two basic geotechnical systems, which have been presented above. The shear band enhancement is carried out automatically by coupling the OptumG2 (Optum CE 2017) analysis to a MATLAB routine.

2 NUMERICAL MODEL

2.1 *Numerical method*

In order to simulate both boundary value problems (bvp) OptumG2 is used. The analysis type *Limit Analysis* is chosen for the simulation, which is based on finite element limit analysis (fela). Elements of type *lower* and *upper* are used to calculate the respective bounds for limit loads of the system. Adaptive mesh refinement according to the shear dissipation $D_s = (\sigma - \mathbf{m}p)^T(\varepsilon - \mathbf{m}\varepsilon_v)$, where p is the mean stress, ε_v is the volumetric strain and $\mathbf{m} = (1,1,1,0,0,0)^T$, is chosen, compare Krabbenhoft (2016). Shear dissipation is the dissipation of shear strain energy into plastic shearing. A sensitivity analysis is carried out for each bvp in order to determine the influence of the number of adaptivity steps and elements on the limit loads. To evaluate the quality of the limit analysis the maximum error is calculated by

$$Error = \frac{UB - LB}{UB + LB} \cdot 100, \tag{1}$$

where UB and LB stand for the limit loads of upper and lower bound.

2.2 *Material model*

The soil body consists of loose sand, which is modelled as a Mohr Coulomb soil with non-associated flow rule. For material parameters cohesion $c = 0$ kPa, Young's modulus $E = 30$ *Mpa*, friction angle $\phi = 30°$, delation angle $\psi = 1°$, Poisson's ratio $v = 0.25$ and unit weight $\gamma = 18$ kN/m^3 see fig. 2.

2.3 *Shallow foundation*

In the present study a shallow foundation is modelled as a rigid plate, which is vertically loaded with a line load q_u. Fig. 2 shows the FE-model which is symmetric with respect to the y-axis. The half of the shallow foundation measures 0.5 m in width (b) and the contact to the soil surface is modelled rough. The model of the subsoil measures 5 by 5 meters. The limiting load q_u on the shallow foundation is determined through limit analysis.

Fig. 3 shows the limit load results for the shallow foundation calculations with varying number of elements. In course of mesh refinement the number of elements is increased to the target number of elements. The start value is chosen to be half of the target elements. The error between lower and upper bound of the limit load is beneath ± 5% from 5,000 elements on. 10,000 elements are chosen for the following analysis.

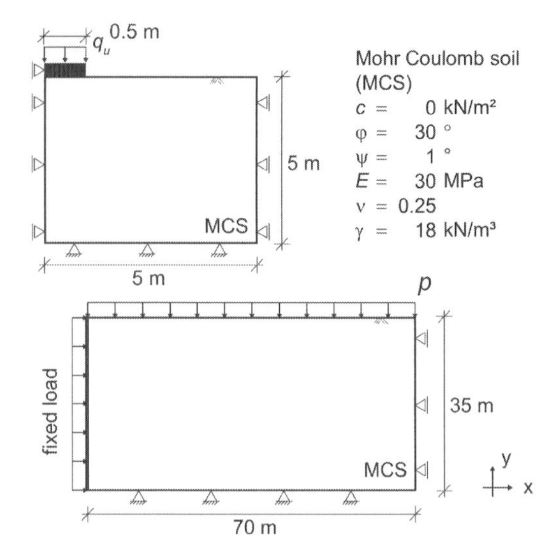

Figure 2. Numerical models for OptumG2 analyses: retaining wall (above), shallow foundation (below) with a Mohr Coulomb Soil.

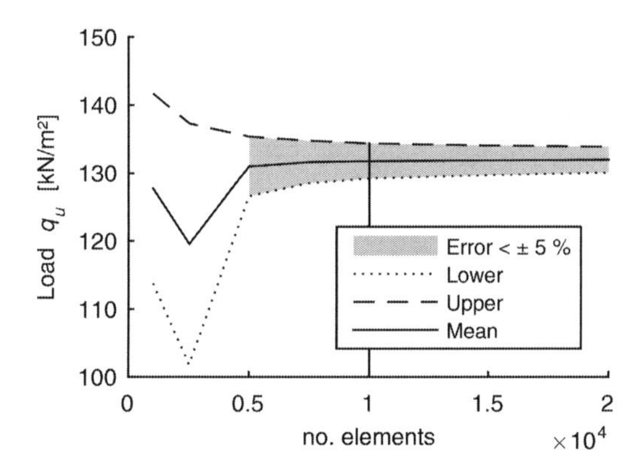

Figure 3. Limit loads for varying number of elements in the shallow foundation bvp. Mesh refinement: 5 adaptivity steps. The grey area shows, for which number of elements the error is less than ± 5%. 10,000 elements are chosen for the following analysis.

The variation of adaptivity steps (fig. 4) shows, that three adaptivity steps yield upper and lower bounds with a maximum error of less than ± 5%. Five adaptivity steps are chosen for the following analysis.

The resulting mesh for 10,000 elements (type *upper*) after five adaptivity steps can be seen in

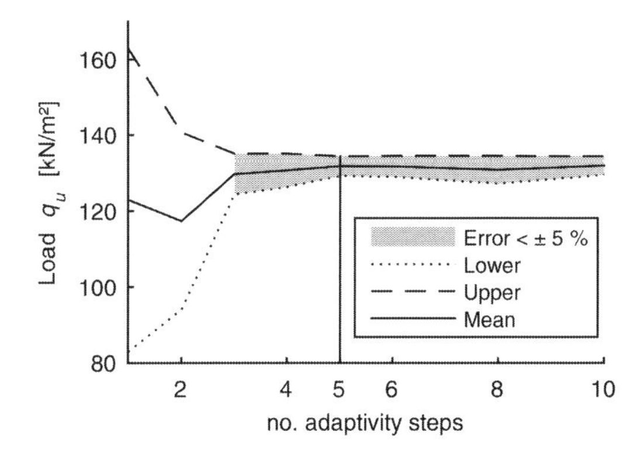

Figure 4. Limit loads for varying number of adaptivity steps in the shallow foundation bvp with 10,000 elements. The grey area shows, for which number of adaptivity steps the error is less than ± 5%. 5 adaptivity steps are chosen for the following analysis.

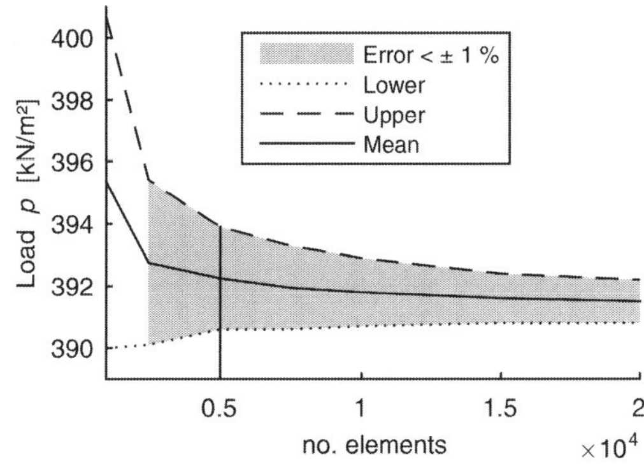

Figure 6. Limit loads p for varying number of elements in the retaining wall model. Mesh refinement: 5 adaptivity steps. The grey area shows, for which number of elements the error is less than ± 1%. 5,000 elements are chosen for the following analysis.

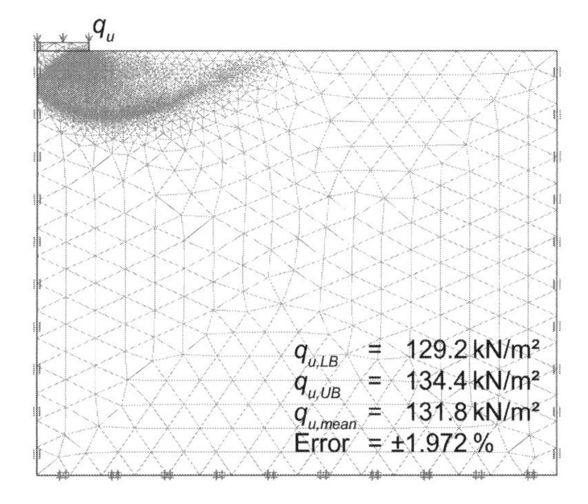

Figure 5. Shallow foundation: mesh discretization for 10,000 elements of type *upper* after 5 adaptivity steps (not true to scale).

Figure 5. The distance between the critical failure mechanism and the model boundaries is sufficiently large in order to fully enable failure. Also any increase in model dimensions has only negligible influence on the limit loads of the system. The upper and lower limit loads for the surcharge on the foundation q_u are 129.2 kN/m^2 and 134.4 kN/m^2. The maximum error of these bounds is 1.97%.

The result is compared to the analytic solution for the classic shallow foundation problem presented in Hjiaj, Lyamin, & Sloan (2005). They determine an expression for approximation of the bearing capacity factor

$$N_\gamma = e^{\frac{1}{6}(\pi+3\pi\tan(\varphi))}\tan^{\frac{2\pi}{5}}(\varphi), \quad (2)$$

which is based on Terzaghi's equation (Terzaghi 1943) for the limiting foundation surcharge q_u:

$$q_u = cN_c + qN_q + 0.5\gamma b N_\gamma. \quad (3)$$

The first two terms for cohesion and surcharge on the soil surface are 0 in the presented bvp. Only the the last term is evaluated with the width of the foundation b, γ and N_γ. The bearing capacity q_u for the presented bvp is 131.6 kN/m^2. The deviation from the mean of lower and upper bound (131.8 kN/m^2) is 0.15%, which is a very good agreement. The analytical solution lies within the calculated bounds for upper and lower limit load.

2.4 *Retaining wall*

The retaining wall is modelled as a rigid plate, which is supported through a line load of 200 kN/m^2 on the left hand side, in order to ensure stability. The reduction factor of the interface between soil and retaining wall is 2/3. The model of the soil measures 70 by 35 meters. The lower boundary is fixed in x- and y-direction and the boundary on the right hand side is fixed in x-direction. The surface of the model is loaded with the line load p. The limit load of p is determined during the limit analysis.

Fig. 6 shows the results for limit load p on the soil surface with varying number of elements. Analogous to the shallow foundation model, the starting value of elements is half of the number of target elements. 5,000 elements are chosen for the following analysis.

The variation of adaptivity steps (fig. 7) shows, that already with two adaptivity steps the error can be reduced to less than ± 1%. However, five adaptivity steps are chosen for the following analysis.

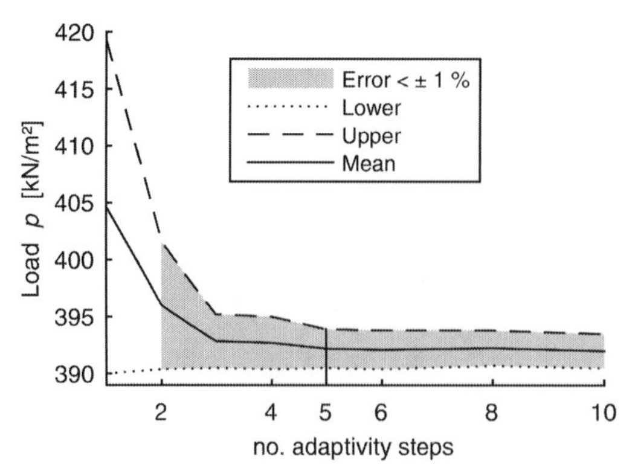

Figure 7. Limit loads p for varying number of adaptivity steps in the retaining wall model. No. elements 10,000. The grey area shows, for which number of adaptivity steps the error is less than ± 1%. 5 adaptivity steps are chosen for the following analysis.

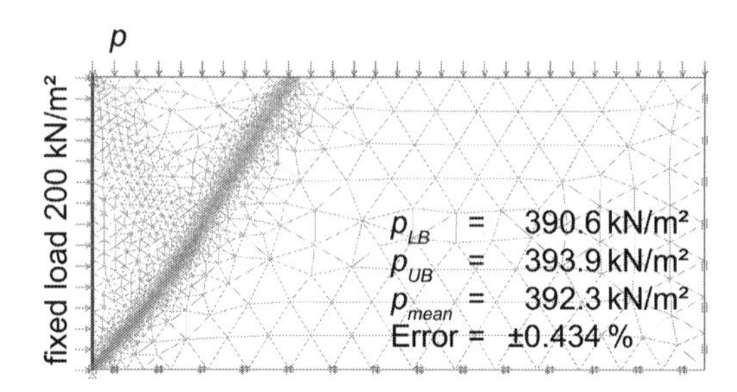

Figure 8. Retaining wall: mesh discretization for 5,000 elements of type *upper* after 5 adaptivity steps.

The resulting mesh for 5,000 elements (type *upper*) after five adaptivity steps is shown in Figure 8. The soil model is sufficiently large in order to fully enable the failure mechanism. The upper and lower bounds for p are 390.6 kN/m^2 and 393.9 kN/m^2, leading to an error of ± 0.43%.

The numerical result is compared to an analytical result obtained from the calculation according to DIN Deutsches Institut f¨ur Normung e.V. (2016). The analytical limit load for p is 399.3 kN/m^2 and therefore 1.8% higher than the mean value of the numerical result. The numerical upper and lower bounds do not bracket the analytical solution, in contrast to the shallow foundation model. However, the result is sufficiently exact for the following analysis.

3 NUMERICAL ENHANCEMENT

3.1 *Method*

The failure mechanism of a geotechnical system can be predicted using limit analysis provided in

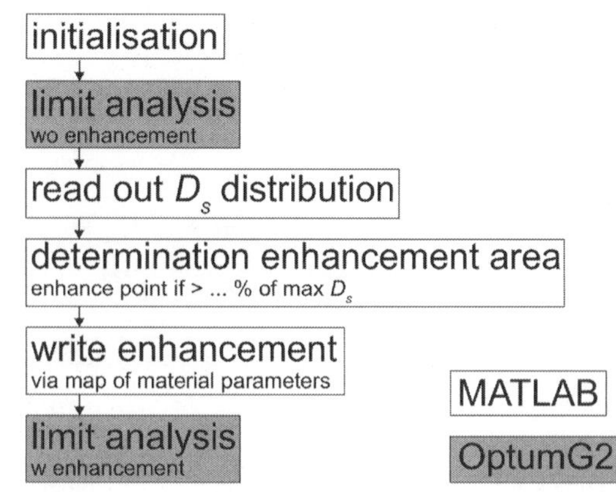

Figure 9. Flowchart of the MATLAB routine for automated shear band enhancement. Procedures with white background are carried out in MATLAB, those with grey background are carried out with OptumG2.

OptumG2. The mechanism consists of soil bodies which are moving along the shear bands. These develop in regions where the shear dissipation D_s is large. The distribution of shear dissipation is calculated during the limit analysis for upper and lower bound using OptumG2. The analysis result is written in a g2x-file, which is encoded in Extensible Markup Language (XML). MATLAB can be used to read out the XMLdocument and as well to execute OptumG2 through DOS command.

An automated procedure to identify and enhance the shear bands is written in a MATLAB routine. Fig. 9 shows the routine flowchart. After initialisation MATLAB sends out a command to execute the calculation of an OptumG2 input file. When the calculation of the limit analysis for upper and lower bound has finished, MATLAB reads out the distribution of shear dissipation. The area of enhancement comprises all calculation points with a shear dissipation D_s, which are bigger than a threshold. The threshold is determined as a percentage of the maximum shear dissipation. Since the shear dissipation distributions vary for upper and lower bounds, the thresholds for each calculation stage vary as well as the corresponding enhancement points. The enhancement points for both stages are taken together in order to determine the over all enhancement area. In those enhancement points the material parameters are modified in order to model enhancement material. A new distribution of material parameters is written to a text file. Points which shall not be enhanced keep their original material parameters. Followingly, a new input file including the updated material distribution is created and the limit analysis with enhancement is invoked trough MATLAB and carried out with OptumG2.

3.2 *Enhancement material*

Using the presented method for automated shear band enhancement, the material model of the soil and the enhancement need to be the same. Therefore, the enhancement is also modelled with the Mohr-Coulomb model. The enhancement is modelled as a soil-cement mixture, which could arise from jet grouting. Based on the investigations of Axtell & Stark (2008) a Young's modulus of E_{enh} = 1000 MPa is chosen. The shearing parameters are derived from the unconfined compressive strength measured in Axtell & Stark (2008) through applying a procedure presented in Wietek (2010), which is based on the fracture relations of Mohr's circle. The friction angle φ_{enh} = 32.6° and the cohesion c_{enh} = 2200 kPa. The unit weight of the enhancement material is $\gamma_{enh} = 23$ kN/m^3.

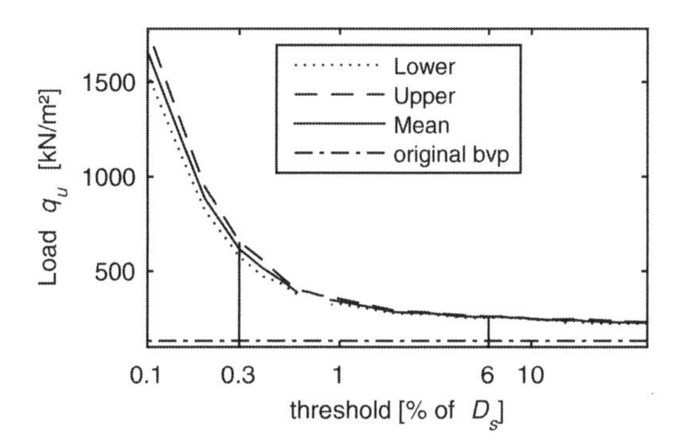

Figure 10. Limit loads of q_u in the shallow foundation model for shear band enhancement depending on the applied threshold. Exemplary chosen results for the enhancement are marked.

4 RESULTS

The automated shear band enhancement is carried out for 31 thresholds for the shallow foundation bvp and the retaining wall bvp: 0.1, 0.2 ... 1, 2, ... 10, 20 ... 40% of the maximum shear dissipation. The higher the threshold (higher percentages of max. shear dissipation) for the enhancement is, the less points will reach this enhancement threshold. Therefore, lower thresholds (lower percentages of max. shear dissipation) lead to a higher increase in the system's limit load.

4.1 *Shallow foundation with enhancement*

The shear band enhancement in the shallow foundation model shows a significant increase of bearing capacity (see fig. 10). The original bearing capacity is exceeded for all tested threshold. The determined increase of bearing capacity ranges between 72 and 1158% of the original bearing capacity.

Exemplary two enhancement areas corresponding to two thresholds are shown in fig. 11 with the original and the enhanced shear dissipation distributions. It can be seen, that the shear dissipation distribution and therefore also the location

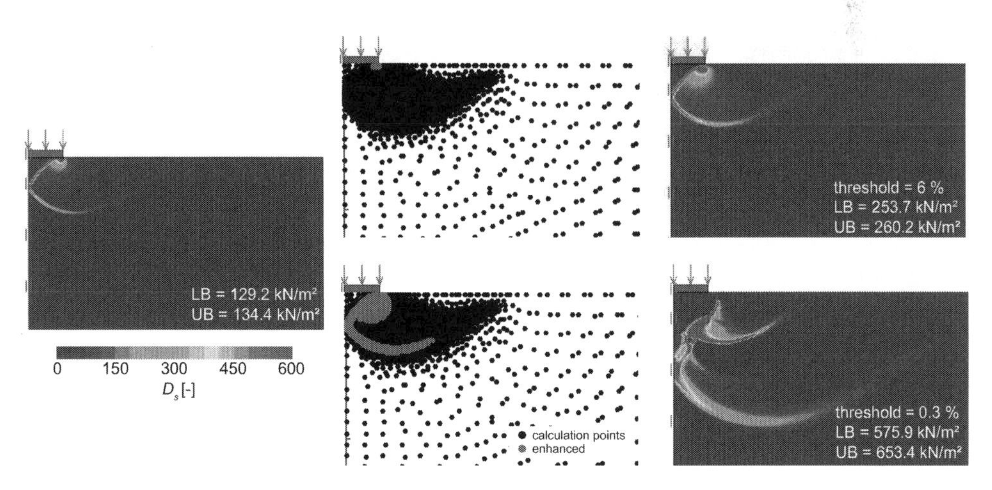

Figure 11. Enhancement results for the shallow foundation model: original bvp (left), enhancement area (middle) and distribution of shear dissipation with enhancement (right).

of the shear band is influenced by shear band enhancement. The enhancement area for a threshold of 6% of the maximum shear dissipation is quite small but has an significant influence on the limit loads. It can also be observed, that shear band enhancement should be located in the area, where the shear bands begin to develop: underneath the foundation edges.

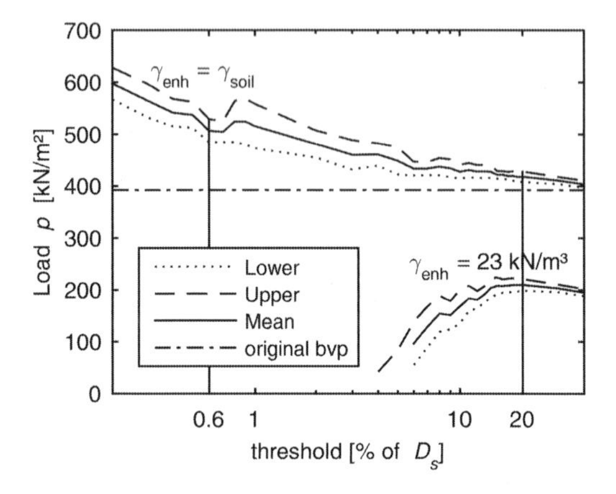

Figure 12. Limit loads of p in the retaining wall model for shear band enhancement depending on the applied threshold. Exemplary chosen results for the enhancement are marked.

4.2 *Retaining wall with enhancement*

The enhancement of shear bands in the retaining wall model leads for low thresholds to an unstable system, see fig. 12. This can be explained through the increase of loads on the retaining wall due to the increase in unit weight when using enhancement material instead of soil material. Therefore, the analysis is carried out with a material which has the same unit weight as the soil but improved shearing parameters as the original enhancement material.

For the enhancement with the unit weight $\gamma = 18$ kN/m^3 it can be seen, that the limit load for p increases. The determined increase of limit load ranges between 3 and 52% of the original limit load.

Fig. 13 shows the enhancement result for two exemplary thresholds. The distribution of shear dissipation for the original bvp (left) as well for the enhanced models (right) are shown. It can be seen, that a larger area of enhancement has an obvious impact on the location of the shear band, whereas a higher threshold has no significant influence. However, the limit loads can be increased in both cases with a factor of 6.5 and 28.8%. As it has been observed for the shallow foundation problem also for the retaining wall the enhancement is located at the beginning of the shear band: lower end of the shear band.

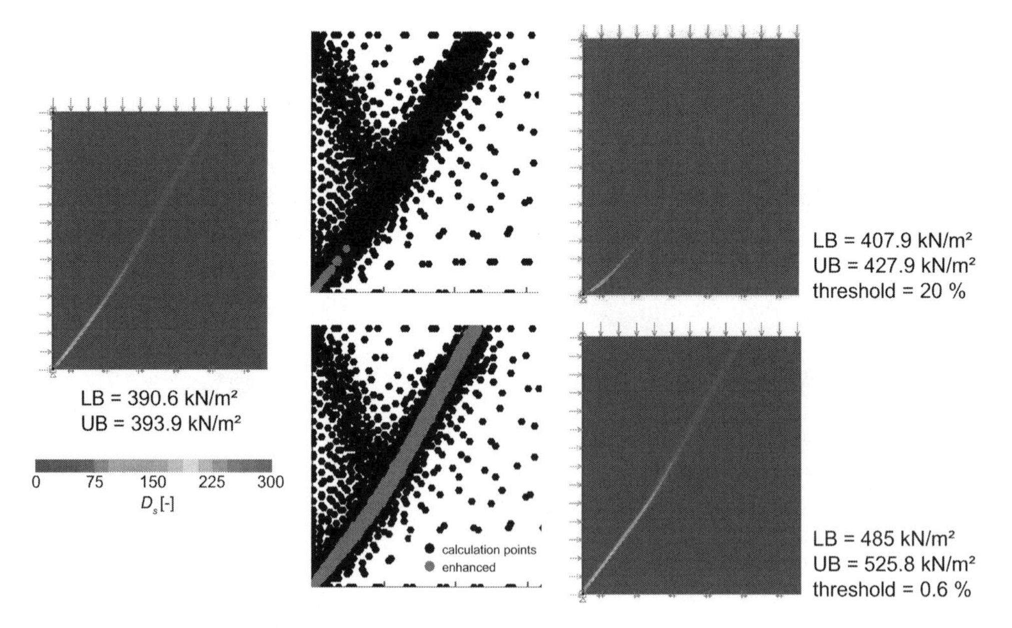

Figure 13. Enhancement results for the retaining wall model: original bvp (left), enhancement area (middle) and distribution of shear dissipation with enhancement (right).

5 CONCLUSION

This paper presents an automated method for the realization of numerical shear band enhancement using OptumG2. The limit loads can be successfully increased using this enhancement technique. Already small amounts of enhancement material, applied efficiently regarding the location, result in an increase of limit load. The increase is especially significant in the case of the shallow foundation bvp. The retaining wall model may become unstable, when applying shear band enhancement. However, choosing a different enhancement material solves this issue and proves, that shear band enhancement is also applicable in such a geotechnical system leading to an limit load increase.

Currently it is difficult to evaluate the quantity of enhancement material due to the unstructured mesh resulting from mesh adaptivity. A structured mesh should be tested in the application of shear band enhancement in order to quantify the amount of enhancement material. Other possible enhancement materials besides the soil-cement mixture arising from jet grouting should be evaluated in future studies to provide suitable enhancement material for the retaining wall model.

ACKNOWLEDGEMENT

The authors thank Kristian Krabbenhoft for ideas on the implementation of automated material enhancement in OptumG2.

This work is funded by the German Research Foundation (DFG) "Increase of bearing capacity for geotechnical constructions through shear zone enhancement" (GR 1024/23-1).

REFERENCES

Axtell, P.J. & T.D. Stark (2008). Increase in shear modulus by soil mix and jet grout methods. *DFI Journal 2*(1), 11–21.

Coulomb, C.A. (1776). *Essai sur une application des re`gles de maximis & minimis a` quelques proble`mes de statique, relatifs à l'architecture.* Paris: De l'Imprimerie Royale.

DIN Deutsches Institut für Normung e.V. (2016). Baugrund—berechnung des erddrucks.

Grabe, J. & T. Pucker (2012). Bodenertü chtigungsverfahren sowie Anordnungen dafür.

Grabe, J. & K.F. Seitz (2016). Optimization of geotechnical structures for states of serviceability and ultimate loads. In A. Zingoni (Ed.), *Insights and Innovations in Structural Engineering, Mechanics and Computation*, pp. 2048–2053. Boca Raton: CRC Press.

Hjiaj, M., A.V. Lyamin, & S.W. Sloan (2005). Numerical limit analysis solutions for the bearing capacity factor n_γ *International Journal of Solids and Structures 42*(5–6), 1681–1704.

Krabbenhoft, K. (2016). *OptumG2: Analysis.* OptumCE. Manual.

Optum CE (2017). OptumG2 (version 2017.02.07).

Prandtl, L. (1920). Über die härte plastischer körper. *Nachrichten von der königlichen Gesellschaft der Wissenschaften zu Göttingen, mathematisch-physikalische Klasse.*

Seitz, K.F. & J. Grabe (2018). 1g-modeling of limit load in—crease due to shear band enhancement—to be published. In *Physical modelling in geotechnics: Proceedings of the 9th International Conference on Physical Modelling in Geotechnics 2018 (ICPMG 2018), London, Great Britain, 17–20 July 2018.*

Terzaghi, K. (1943). *Theoretical soil mechanics.* New York and London: J. Wiley and Sons, inc. and Chapman and Hall, limited.

Wietek, B. (2010). *Stahlfaserbeton: Grundlagen und Praxisan-wendungen* (2 ed.). Wiesbaden: Vieweg + Teubner.

Numerical Methods in Geotechnical Engineering IX – Cardoso et al. (Eds)
© 2018 Taylor & Francis Group, London, ISBN 978-1-138-33203-4

Finite element modelling of rigid inclusion ground improvement

K. Lődör & B. Móczár
BUTE Department of Engineering Geology and Geotechnics, Budapest, Hungary

ABSTRACT: During the past decades, ground improvement has successfully been able to provide competitive and economical technical foundation solutions by increasing the ground bearing capacity, and reducing the settlements. Reinforcement of soft soils by rigid vertical inclusions has been an increasingly used technique over the last few years. Nowadays by using three dimensional numerical modelling, the real behavior of complex foundation systems can be investigated successfully. It is not enough to define the soil properties correctly, creating a perfect model geometry is also important. In this study Plaxis 3D finite element software was used to examine the pile geometry. Beam, embedded beam, or volume pile elements can be used to simulate real pile behavior. Piles are principally used to transfer the loads from a superstructure, through weak, compressible strata onto stronger, less compressible soil. This behavior is the base to find how piles can be modelled in the most perfect way.

1 INTRODUCTION

Ground improvement through vertical rigid piles is an interesting method for foundations on soft, compressible soils. Recently, this process has become widespread in Europe against the traditional pile or pile raft foundations. The behavior of traditional pile foundation concepts are now well known and many researches have dealt with this area of deep foundations. These can be designed by detailed standards and prescribed methods. In contrast, there are still unclear, undeveloped areas in the design of soil improvement methods.

Rigid inclusion contain elements that are slender, often cylindrical shape, mechanically continuous and typically vertical. They are laid out according to a regular mesh pattern, which must be adapted both to the nature and geometry of applied loads and to soil conditions. The rigid inclusion concept implies that inclusion caps are not structurally connected to the supported structure. There is a well-compacted, granular load transfer platform (LTP) between the inclusions and the raft. The LTP has a very important role to ensure the transfer of loads to the ends of the piles and to uniform settlements. A minimum load transfer platform thickness is necessary to allow for appropriate load transfer between inclusions and soils, as well as to limit forces within the supported structure. This thickness, often on the order of 40 to 80 cm, proves essential in deriving an optimal design for the supported structure, particularly with the aim of reducing bending moments in the slabs.

In our research, 3D numerical modelling of rigid inclusion foundation system, the most important question is the pile modelling opportunities. In Plaxis 3D, there are two generally used options for modelling piles, but both has its disadvantages. By the combination of basic elements, so called hybrid elements can be created of which behavior are more favorable.

2 LOAD TRANSFER MECHANISMS

2.1 *Negative skin friction*

Negative skin friction (NSF) is in fact a downward friction imposed on foundation pile shaft as a result of subsoil settlement. Negative skin friction is usually mobilized to ultimate stress limit in most cases except at very close proximity to the neutral plane. It needs only few millimeters of relative displacement between the settling subsoil and the pile shaft surface. The neutral plane is at the depth, where the settlement of pile and soil is the same. Above the neutral plane where the surrounding soil settles more than the pile, negative skin friction develops along the piles. This causes additional load to be transferred to the piles. Axial load-distribution inside a rigid inclusion in case of negative skin friction is shown in Figure 1.

There are many conventional methods to determine the critical height (h_c), which take into account the negative skin friction. The intensity of this negative friction can be calculated based on the following relation:

$$F_N = 2\pi r_p \int_0^{h_c} K tan\delta \sigma_v'\left(z, r_p\right) dz \quad (1)$$

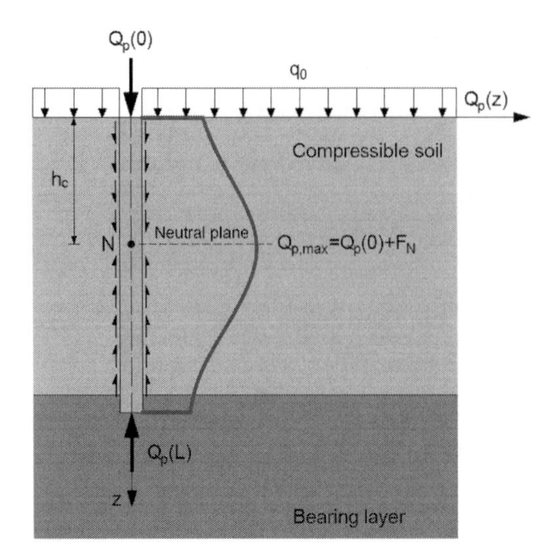

Figure 1. Axial load-distribution inside a rigid inclusion (ASIRI 2012).

Where r_p = inclusion radius; K = earth pressure coefficient; tand = coefficient of soil-pile friction; $s`_v(z,r_p)$ = effective vertical stress at the inclusion contact in the final state, while taking into account the downdrag effect.

Combarieu (1985) went on to define a radial variation law for vertical stress at height z, by introducing the notation of downdrag effect of soft soil around the inclusion. (Fig. 2).

In the presence of several inclusions, the group effect, which increases as the mesh becomes tighter, combines with the downdrag effect. The average vertical stress between the inclusions is expressed, over the height where negative skin friction is applied, as follows:

$$\sigma_v^*(z)=\sigma_1'(z)-\left[\sigma_1'(z)-\sigma_v'(z,r_p)\right]\frac{2Ktan\delta r_p}{(R^2-r^2)\mu\left(\lambda,\frac{R}{r_p}\right)} \quad (2)$$

Where $s`_1(z)$ = effective vertical stress in free field within the ground in the final state, as calculated without taking into account the presence of the inclusions; R = radius of influence; l = downdrag coefficient; m = depends on the downdrag coefficient and the ratio of radius of influence and the inclusion radius.

The critical height is often determined by adapting the hypothesis that negative skin friction only acts if the stress in contact with inclusion is greater than the initial stress.

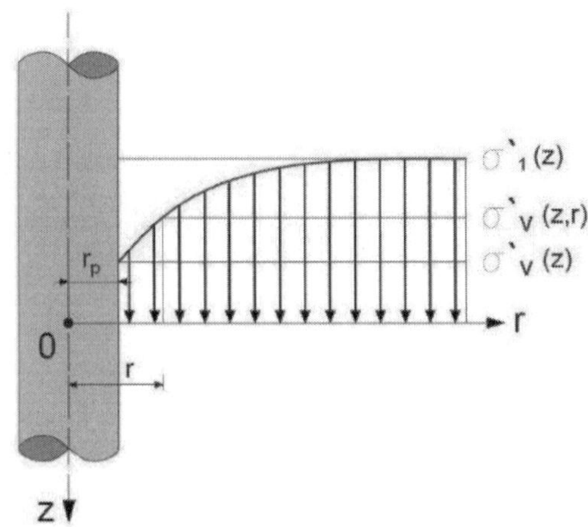

Figure 2. Distribution of stress around the inclusion (ASIRI 2012).

2.2 *Load transfer in the platform*

The columns in rigid inclusion foundation system are not structural elements, which is fundamentally different to the classic piled raft foundation concept. The column heads punch into the load transfer layer and two typical failure modes are possible (Prandl failure diagram and shear cone type failure mode). The LTP is responsible to transmit the structural loads into the inclusion heads and into the soil between the columns. Among the various analytical methods available to evaluate load transfer in the platform, the ASIRI National Project forwards the two following tested methods.

The first method is the fictitious inclusion method, which offers the advantage of providing a homogeneous approach consistent with the negative skin friction evaluation method.

The second method is the diffusion cone method, which entails an approach compliant with the mechanisms exposed during the various experiments and modeling exercises conducted within the scope of the ASIRI National Project.

3 PILE MODELLING OPPORTUNITIES IN PLAXIS 3D

Using Plaxis 3D finite element software, two basic solutions can be used for pile modelling. Depending on the calculation problem and the available parameters there are advantages and disadvantages of these two methods, which are the following:

- Volume pile concept
- Embedded beam concept

3.1 *Volume pile concept*

The volume pile concept which is a well-known and widespread modelling process for modelling of pile foundations corresponds to a full discretization of pile volume using 3D finite elements. This process allows a precise, complex modelling of the pile geometry, which is a very important question for real simulation of pile-soil interaction along the pile shaft. The material behavior of the pile is described by a linear elastic, non-porous model, and the soil-structure contact is modelled with interface elements with zero thickness. Elastic-plastic model is used to describe the behavior of interface for the modelling of soil-structure interaction. The Coulomb criterion is used to distinguish between elastic behavior, where small displacements can occur within the interface, and plastic interface behavior when permanent slip may occur. The elastic zone is limited by value of the maximum shear stress t which can be mobilized at the contact of the pile shaft and the soil. Normal stresses in the soil are also limited by tension cut-off criterion. The most important interface property is the strength reduction factor (R_{inter}). The interface properties are calculated from the soil properties in the associated data set and the strength reduction factor. The structural forces and stresses in volume elements can be queried automatically in Plaxis 3D which is a great advantage when evaluating the results.

3.2 *Embedded beam concept*

The embedded beam approach was introduced by Dadek and Shahrour (2004). The embedded beam is a structural object composed of beam elements that can be placed in arbitrary direction in the sub-soil and that interacts with the sub-soil by means of special interface elements. The beam elements are considered to be linear elastic and are defined by the same material parameters as a regular beam element. The interaction may involve a skin resistance as well as a foot resistance. The skin friction and the tip force are determined by the relative displacements between the soil and the pile. Although embedded beam does not occupy volume, a particular volume around the pile (elastic zone) is assumed in which plastic soil behavior is excluded. The size of this elastic zone is equal to the real pile diameter which can be defined in the parameter settings. This is the reason why embedded beam behaves like a volume pile.

For the skin and the base resistance a failure criterion is used to distinguish between elastic and plastic interface behavior. For elastic behavior relatively small displacement differences occur within the interface, and for plastic behavior permanent slip may occur. For the interface to remain elastic the shear force (t_s) at a particular point is given by:

$$|t_s| < T_{max} \tag{3}$$

There are three possible options to define the maximum skin resistance of the embedded beam which are the following:

- Linear
- Multi-linear
- Layer dependent

The input for the shaft resistance is defined by means of the skin resistance at the pile top ($T_{top,max}$) and bottom ($T_{bot,max}$). Using this approach the total pile bearing capacity (N_{pile}) is given by:

$$N_{pile} = F_{max} + \frac{1}{2} L_{pile} \left(T_{top,max} + T_{bot,max} \right) \tag{4}$$

The base resistance is only mobilized when the pile body moves in the direction of the base.

4 NUMERICAL CALCULATION

In our numerical calculations, a unit cell model was used to perform the parametric study. Unit cell models are well adapted to the design of piles in the central part of the grid and under uniform vertical surface loading. Only one homogeneous, compressible silt layer is defined in the model space. The 10 m long (L) concrete pile with a diameter (D) of 80 cm is located on the soft soil layer and the pile spacing (R) is 2.0 × 2.0 m. The compacted granular load transfer platform of 50 cm thickness is placed over the inclusion head and topped with a raft having a thickness of 30 cm. Vertical prescribed displacements are activated on the raft. The behavior of the different pile concepts are investigated with three different prescribed displacements which are determined as a function of the pile diameter. In this case, a force is obtained as output, which is necessary for the prescribed displacement to occur. For the evaluation of skin and base resistance mobilization, the following prescribed displacements are defined in the calculations:

- $u_z = 0.02 \cdot D = 0.016\ m$
- $u_z = 0.10 \cdot D = 0.080\ m$
- $u_z = 0.20 \cdot D = 0.160\ m$

Based on in-situ pile load testing and the DIN 1054, the following load-settlement curve is used for the determination of prescribed displacements:

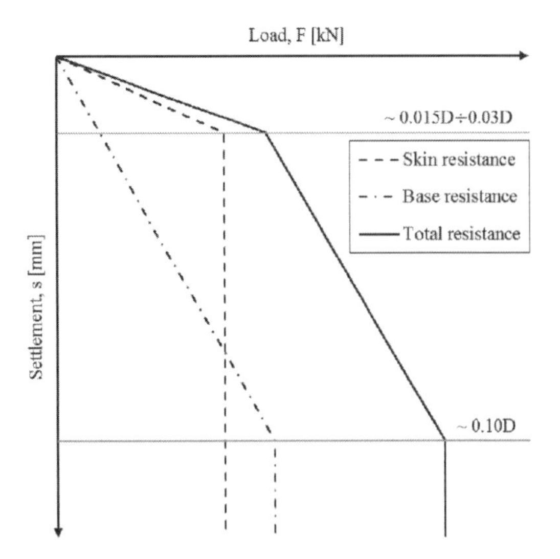

Figure 3. Load-settlement curve (DIN 1054).

Table 1. Material properties.

Properties	Soft silt	LTP	Pile VP	Pile EB
Material model	HSS	HS	EL	EL
g (kN/m³)	19	20	24	5
E (MPa)	–	–	33·10³	
n (–)	0.3	0.3	0.2	
E_{oed}^{ref} (MPa)	5	70	–	
E_{50}^{ref} (MPa)	5	70	–	
E_{ur}^{ref} (MPa)	18	210	–	
m (–)	0.75	0.5	–	
c'_{ref} (kPa)	16.9	1	–	
f' (°)	26.9	40	–	
$g_{0.7}$ (–)	2.5·10⁻⁴	–	–	
G_0^{ref} (MPa)	65	–	–	
R_{int} (–)	0.9	1	–	

The soft silt layer is modelled with HSS (Hardening Soil model with small-strain stiffness) and the LTP is modelled with HS (Hardening Soil) material model. The pile and the raft are modelled with elastic (EL) material model. The properties of the investigated rigid inclusion foundation components are summarized in Table 1.

In this study, especially the deformations of the foundation system (soil and pile) and the stress distribution in the piles are addressed. In the numerical calculations, using Plaxis 3D, 15-noded wedge finite elements are used.

5 RESULTS

5.1 *Volume pile approach*

During the calculations the distribution of the normal forces and the skin resistances and the soil-pile interaction (pile and surrounding soil) are investigated in case of different prescribed displacements. The inclusion is a cylindrical shape CFA (Continuous Flight Auger) pile.

In the first step, the model sensitivity was examined as a function of the mesh coarseness of the pile with 0.02D prescribed displacement. Multiple numerical analyses were performed with very coarse, coarse, medium, fine, very fine mesh and with local refinement mesh in the volume element. The effects of the coarseness are shown in Figure 4 and Figure 5.

The skin resistance could be calculated from the shear stresses (t_1) of the interface element with the following equation:

$$T_{skin} = \tau_1 D\pi \tag{5}$$

Shear stress in interface elements only depends on the earth pressure, coefficient of soil-pile friction, interface reduction factor and the effective normal stress so the skin resistance could be limited using the interface reduction factor; skin resistance however cannot be limited by manually defining a maximum cut-off value.

As it can be seen in Figure 4 and Figure 5, the maximum normal force in the VP increases with

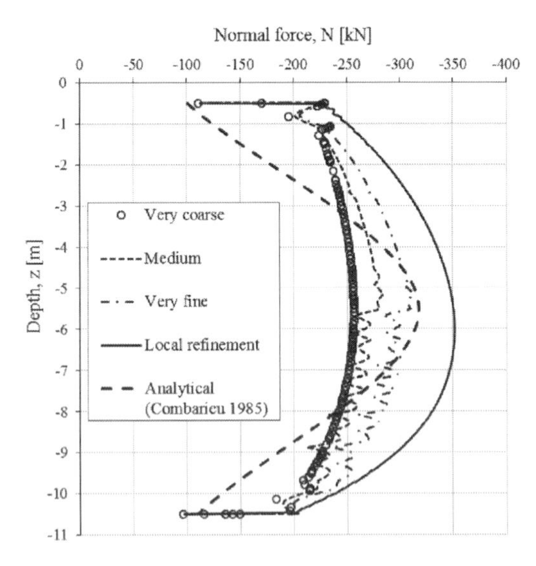

Figure 4. Sensitivity of normal force distribution in VP in case of different mesh coarseness with 0.02D prescribed displacement.

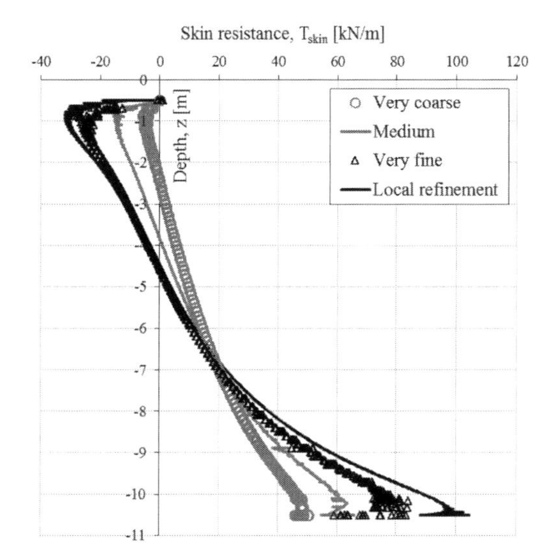

Figure 5. Sensitivity of skin resistance distribution in VP in case of different mesh coarseness with 0.02D prescribed displacement.

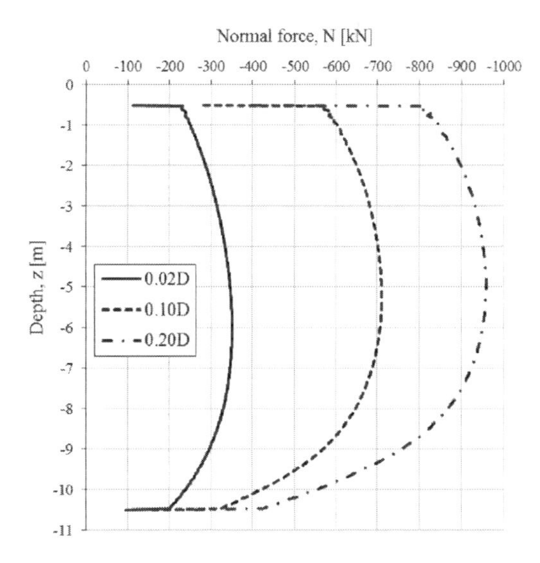

Figure 6. Normal force distribution in VP in case of different prescribed displacements.

mesh refinement. At the same time, the neutral plane moves deeper so the negative skin friction increases, which induces additional normal force. The neutral plane is located at 2.5 m depth if the mesh is very coarse, while with local refinement the neutral plane shifts to 4.57 m depth. In all cases, the neutral plane and the location of the maximum normal force are not at same depth. Using various

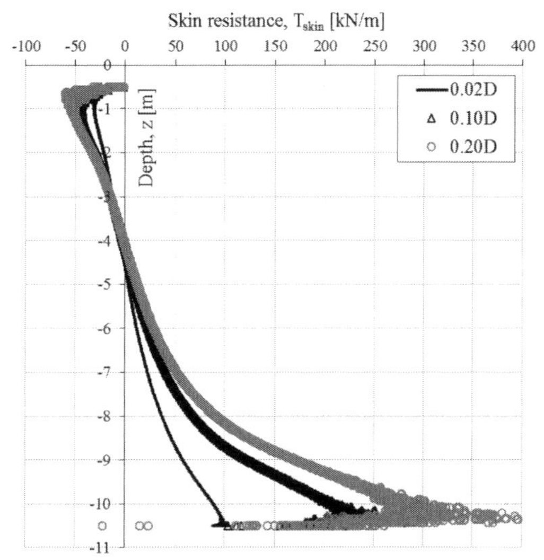

Figure 7. Skin resistance distribution in VP in case of different prescribed displacements.

mesh coarseness, there is no big difference in the forces, that are necessary for the prescribed displacements to occur.

The influence of different prescribed displacements on the stresses are shown in Figure 6 and Figure 7.

Using various (increasing) prescribed displacements the positive skin friction increase while the negative part is almost the same in all cases. Increasing the prescribed displacements the neutral plane moves upwards which is, however, not significant in the calculations.

5.2 *Embedded beam approach*

Assuming embedded beam, three special options were used in the numerical calculations to define the maximum skin resistance. The first option is the layer dependent (LD) option. With this definition, the maximum shear stress of the embedded pile is related to the strength parameters of the soil and the normal stress along the interface. When using the layer dependent option, the embedded interface elements behave similar to normal interface elements as used in the volume pile approach, with the difference that the interaction is modelled along the line element. In addition, an ultimate cut-off value for the skin resistance can be defined (LD-Tmax). Based on the results of in-situ CPT (Cone Penetration Test) the ultimate cut-off value for the skin resistance could be calculated using the following recommendation of Szepesházi (2011):

$$T_{skin} = q_s D\pi = 1.2\mu_s \sqrt{q_c} D\pi \tag{5}$$

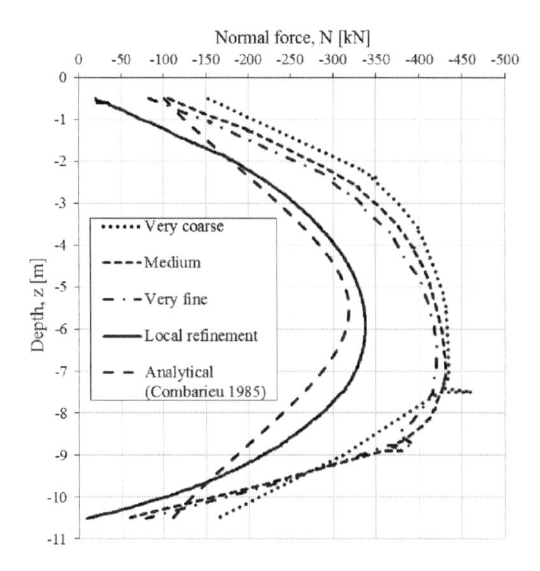

Figure 8. Sensitivity of normal force distribution in EB in case of different mesh coarseness with 0.02D prescribed displacement (layer dependent option of skin resistance).

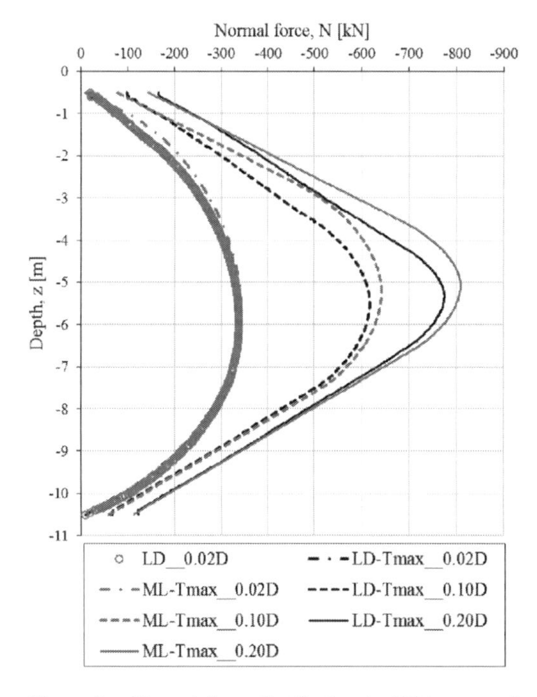

Figure 9. Normal force distribution in EB in case of different prescribed displacements and using different skin resistance options.

where q_s = specific skin resistance; q_c = specific tip resistance; m_s = technological factor (1.0 for CFA

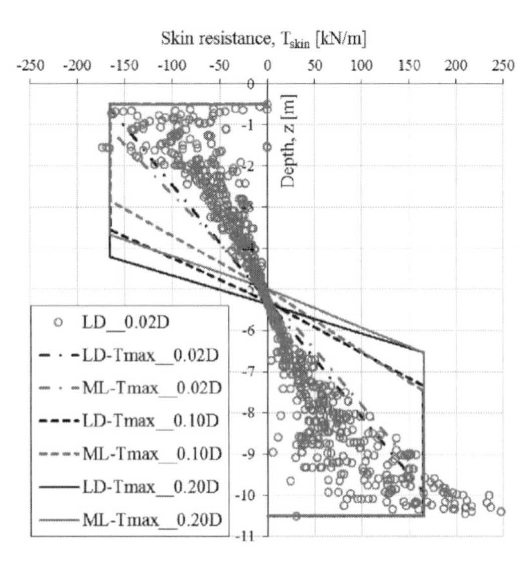

Figure 10. Skin resistance distribution in EB in case of different prescribed displacements and using different skin resistance options.

pile). The second option to set the skin friction of EB is the multi-linear (ML-Tmax) distribution, with which it is possible to define values for the skin friction at certain depths.

The model sensitivity was examined as a function of the mesh coarseness of the pile with 0.02D prescribed displacement. The effect of the mesh coarseness is not significant in the forces, that are necessary for the prescribed displacements to occur.

As it can be seen in Figure 8, the maximum normal force in the EB decreases with mesh refinement, while the location of the neutral plane is almost the same. At the pile ends, the normal force is almost zero; it seems, that because the embedded beam is not connected to other structural elements at the ends, the load cannot be directly transferred to the pile.

To examine the mobilization, 0.10D and 0.20D prescribed displacements were used in the calculations. The influence of different prescribed displacements on the stresses are shown in Figure 9 and Figure 10. As Figure 9 shows, as a result of increasing prescribed displacements, the normal force at the top and the bottom of the embedded beam does not increase significantly, which can be attributed to the connection problem of the EB element. Increasing the prescribed displacement, the mobilization of the shaft is getting bigger, nevertheless the position of the neutral plane is nearly the same even in that cases when diverse options are used to define the skin resistance.

5.3 *Comparison of different methods and the hybrid element*

As seen in the previous results, the behavior of different pile models and the distribution of stresses are dissimilar. The maximum normal forces are almost the same in both cases, whereas the distributions are different due to the differences in the distribution of the skin resistances and the connection problem at the ends of the EB element.

To solve the connection problem in the EB approach, a hybrid pile element (EB+VP) could be

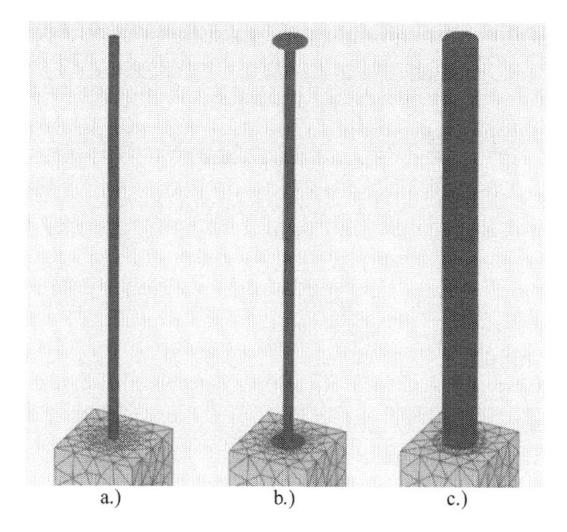

Figure 11. Investigated pile models—Embedded beam (a.); Hybrid pile element (b.); Volume pile (c.).

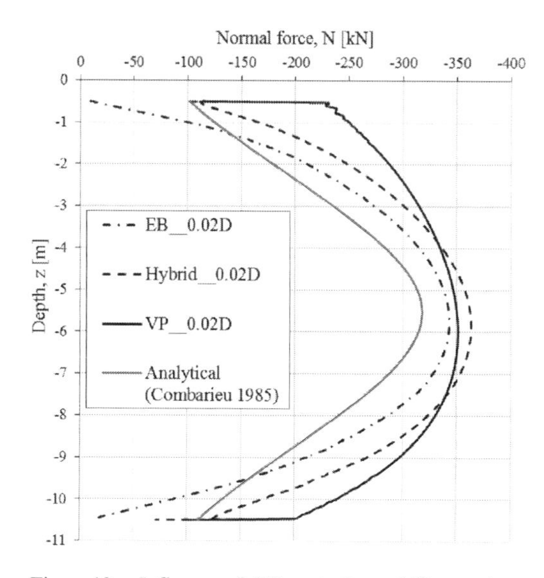

Figure 12. Influence of different pile modelling options on the distribution of normal force.

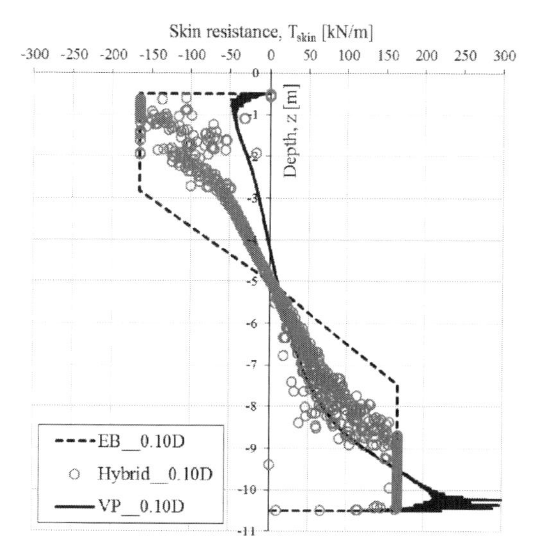

Figure 13. Influence of different pile modelling options on the distribution of skin resistance.

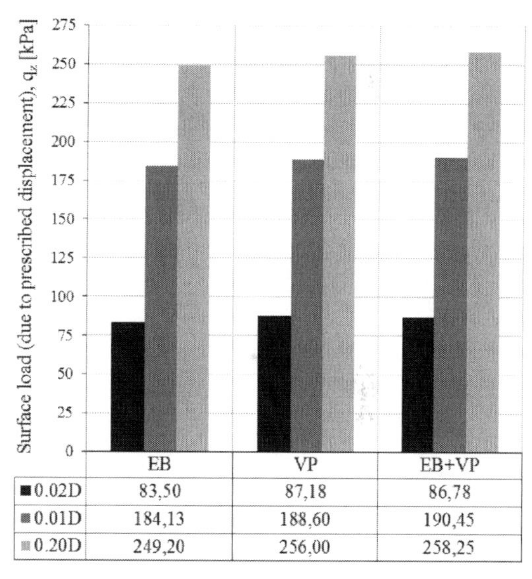

	EB	VP	EB+VP
0.02D	83,50	87,18	86,78
0.01D	184,13	188,60	190,45
0.20D	249,20	256,00	258,25

Figure 14. Sum of the forces that are necessary for the prescribed displacements to occur.

created. In this case, at both ends of the embedded pile, 1 cm thick VP elements are inserted, which provide appropriate stiffness and load transfer surfaces. Investigated pile models are shown on Figure 11.

As Figure 12 shows, the normal force at the ends increases if the pile is modelled with hybrid elements. The distribution of the skin resistance of the hybrid element is getting closer to that in the VP, the positive part is almost the same; the negative skin resistance decreases compared to the EB concept. Using cut-off value for the skin resistance in hybrid element is a great advantage of this approach and maybe gives more realistic pile behavior.

While the displacements are the most important results in the design calculations, the influence of the model geometry of the piles is not significant (Fig. 14). In most cases, the role of the rigid inclusion foundation is to improve the soil properties. One of the most important aspects in the design method of rigid inclusion foundation system is the improvement factor, which is the ratio of the maximum displacement of the raft with and without inclusions. It means that if the improving factor is the question in the numerical calculation, the pile modelling option is not so important.

6 CONCLUSIONS

The parametric study on rigid inclusion foundation technique—especially the pile modelling opportunities—has been performed using 3D FEM via Plaxis 3D to analyze its performance in reducing the settlement and transferring the load. For the evaluation of skin and base resistance mobilization, the prescribed displacements were defined on the top of the raft. Three different types of pile model geometries were examined in the aspect of distribution of normal force, skin resistance and displacement of the complex foundation system.

The calculation results shows that the influence of the mesh coarseness is not significant in the forces, that are necessary for the prescribed displacements to occur, on the other hand, creating finer mesh, the distribution of the normal force and the skin resistance are changed. Using VP, the maximum skin resistance cannot be limited, which is a disadvantage of this modelling approach, while using EB an ultimate cut-off value for the skin resistance can be defined. At the ends of the EB element, the normal force is almost zero; it seems, that because the embedded beam is not connected to other structural elements at the ends, the load cannot be directly transferred to the pile. To solve the connection problem, a hybrid pile element could be created. Since the rigid inclusion foundation is also a soil improvement technique, the most important question is the displacement behavior of the system, which could be reached with each pile modelling options. The results could be refined by back analysis which could help in the design methods (Lődör et al. 2016).

REFERENCES

ASIRI National Project. 2012. Recommendations for the design, construction and control of rigid inclusions ground improvements. Paris: Presses des Ponts.

DIN 1054:2005–01. 2005. Ground – Verification of the safety of earthworks and foundations. Berlin: Beuth Publisher.

Hor, B., Song, M.-J., Jung, M.-H., Song, Y.-H. & Park, Y.-H. 2015. A 3D FEM analysis on the performance of disconnected piled raft foundation. *Japanese Geotechnical Society Special Publication.* vol. 2: 1238–1243.

Jenck, O., Dias, D. & Kastner, R. 2005. Soft ground improvement by vertical rigid piles two-dimensional physical modelling and comparison with current design methods. *Soils and Foundations.* 45(6): 15–30.

Lődör, K., Móczár, B., Mahler, A., Bán, Z. 2016. Back analysis of settlements beneath the foundation of a sugar silo by 3D finite element method. *Plaxis Bulletin.* Issue 39: 12–17.

Marjanovic, M., Vukicevic, M., König, D., Schanz, T. & Schafer, R. 2016. Modeling of laterally loaded piles using embedded beam elements. *4th International Conference, Contemporary achievements in civil engineering, Subotica, Serbia, 22 April 2016.*: 349–358.

Plaxis 3D Anniversary Edition Manual. *Plaxis BV, Delft, 2016*

Szepesházi, R., Scheuring, F., 2016. Practical briefing for pile design and testing – Hungarian practice. *ISSMGE – ETC 3 International Symposium on Design of Piles in Europe, Leuven, Belgium, 28–29 April 2016.*: 199–211.

Tschuchingg, F. 2013. 3D Finite Element Modelling of Deep Foundations Employing an Embedded Pile Foundation. *PhD Dissertation, Graz University of Technology, 2013.* Graz.

Tschuchingg, F., Schweiger, H., F. 2015. The embedded pile concept – Verification of an efficient tool for modelling complex deep foundations. *Computers and Geotechnics.* vol. 63: 244–254.

Tschuchingg, F., Schweiger, H., F. 2013. Comparison of Deep Foundation Systems using 3D Finite Element Analysis Employing Different Modelling Techniques. *Geotechnical Engineering Journal of the SEAGS & AGSSEA.* 44 (3).: 40–46.

Varaksin, S., Hamidi, B., Huybrechts, N. & Denies, N. 2016. Ground Improvement vs. Pile Foundations? *ISSMGE – ETC 3 International Symposium on Design of Piles in Europe, Leuven, Belgium, 28–29 April 2016.*: 1–48.

Numerical Methods in Geotechnical Engineering IX – Cardoso et al. (Eds)
© 2018 Taylor & Francis Group, London, ISBN 978-1-138-33203-4

Stone column-supported embankments on soft soils: Three-dimensional analysis through the finite element method

D.O. Marques & J.L. Borges
CONSTRUCT-GEO, Faculty of Engineering (FEUP), University of Porto, Porto, Portugal

ABSTRACT: In this study, the three-dimensional behavior of a stone column-supported embankment over a soft soil is analyzed using a computer code based on the finite element method. The computer code incorporates the Biot consolidation theory (fully coupled formulation of the flow and equilibrium equations) with soil constitutive relations simulated by the p-q-θ critical state model. Special emphasis is given to the analysis, during and after the construction period, of the excess pore pressures, settlements, horizontal displacements, effective stresses, stress levels, arching effect and stress concentration ratio for each column and its variation in depth. To study the overall effect of using the stone column technique, comparisons between the embankment with stone columns (three-dimensional analysis) and the embankment without stone columns (two-dimensional analysis) are also included. Several overall conclusions are put forward.

1 INTRODUCTION

Geotechnical engineers face several difficulties when designing embankments over soft soils. These difficulties are related to the weak geotechnical properties of the soft soil: (i) its low shear strength significantly limits the embankment height that is possible to construct with adequate safety for short term stability; (ii) high deformability and low permeability provoke high settlements that develop slowly as pore water flows and excess pore pressure dissipates (consolidation).

Among all the techniques to improve geotechnical behaviour of embankments on soft soils, the stone columns method (Figure 1) is the most adequate when the main purpose is, simultaneously, to increase the load carrying capacity and reduce and accelerate settlements.

In this paper, the three-dimensional behaviour of an embankment on soft soils reinforced with stone columns is analysed, during and after the construction period, with a finite element program developed by Borges (1995). The initial version of the program (2-D version) was presented in 1995 and several improvements were posteriorly developed and implemented, particularly a 3-D fully coupled analysis version (Borges, 2004).

Basically, for the present applications, the program uses the following theoretical hypotheses: a) coupled analysis of the flow and equilibrium equations considering the soil constitutive relations formulated in effective stresses (Biot consolidation theory) (Borges, 1995; Borges and Cardoso 2000; Lewis and Schrefler, 1987); this formulation is applied to all stages of the problem, both during the embankment construction and in the post-construction period; b) use of the p-q-θ critical state model (Borges, 1995; Borges and Cardoso, 1998; Lewis and Schrefler, 1987), an associated plastic flow model, to simulate constitutive behaviour of soil.

The accuracy of the finite element program has been assessed in several ground structures involving consolidation by comparing numerical results to field measurements. For instance, Borges (1995) compared results of two geosynthetic-reinforced embankments, one constructed up to failure (Quaresma, 1992) and the other observed until the end of consolidation (Yeo, 1986; Basset, 1986a,b). The accuracy was considered adequate in both cases. Very good agreements of numerical and field results were also observed both in an embankment on soft soils incorporating stone columns (Domingues, 2006) and in a braced excavation in very soft ground (Costa, 2005; Costa et al., 2007).

For three-dimensional application, the program uses two types of the 20-noded brick element. Figure 2a shows the element used in the soft soil (element with 60 displacement degrees of freedom, at the corners and at middle of the sides, and with 8 more excess pore pressure degrees of freedom, at the corners), where consolidation analysis is considered. In the fill and in the columns, it is the 20-noded brick element, with only 60 displacement degrees of freedom (at the corners and at middle of the sides), that is used (Figure 2b).

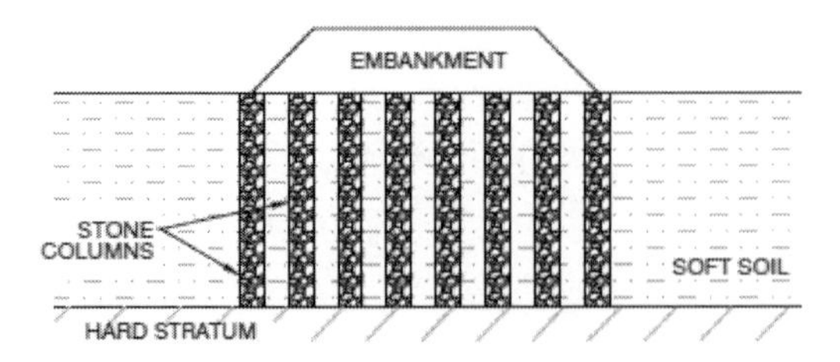

Figure 1. Typical section of an embankment over soft soil reinforced with stone columns.

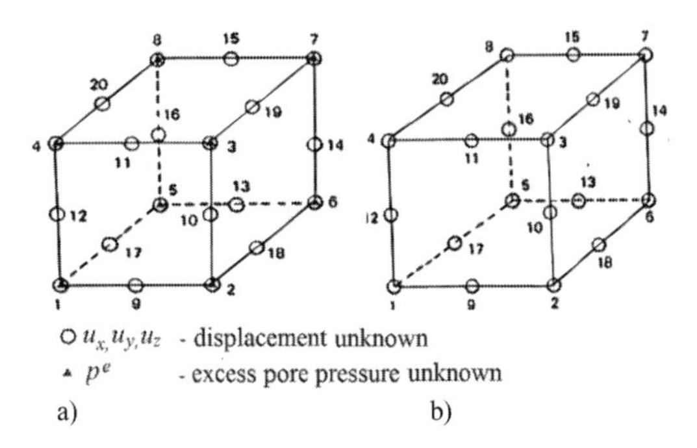

Figure 2. 20-noded brick element: a) with 60 displacement degrees of freedom and 8 excess pore pressure degrees of freedom; b) with 60 displacement degrees of freedom.

Similarly, for two-dimensional analyses, two types of the six-noded triangular element are considered (Figure 3): (i) with 12 displacement degrees of freedom, at the vertices and at middle of the edges (for fill and elements) and (ii) with 3 more excess pore pressure degrees of freedom at the vertices (for soft soil elements).

2 DESCRIPTION OF THE PROBLEM

The problem comprises the construction of a 2.5 m height symmetric embankment, with a 12.5 m crest width, 2/3 (V/H) inclined slopes and very large longitudinal length. The foundation is a 6 m thick saturated clay layer lying on a rigid and impermeable soil, which constitutes the lower boundary. The clay is lightly overconsolidated to 2 m depth and normally consolidated from 2 m to 6 m. The water level is at the ground surface. The diameter of the stone columns is 1.13 m and the spacing between columns is 2.0 m in square grid.

Figure 4 shows the finite element mesh used in the three-dimensional analysis of the embankment incorporating the stone columns. To simplify the mesh, the stone columns are modelled as elements with a square section in plan with the

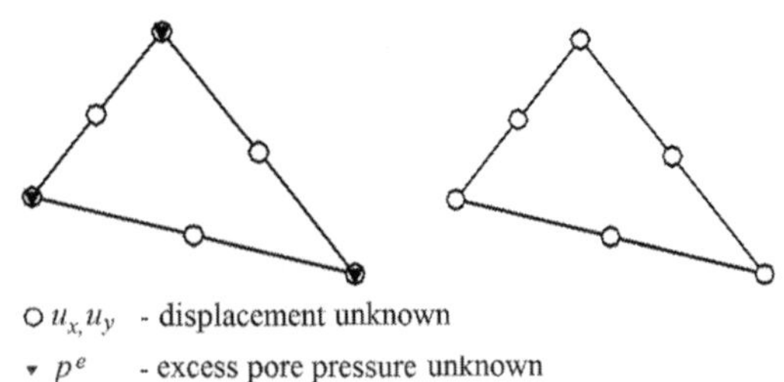

Figure 3. 6-noded triangular element: a) with 12 displacement degrees of freedom and 3 excess pore pressure degrees of freedom; b) with 12 displacement degrees of freedom.

same sectional area, instead of a circular section. To evaluate the impact of this simplification, the central part of the embankment was modelled by three-dimensional unit cells with the same area in plan and equal spacing between columns, one with a square section and the other with a circular section. Results shown that the evolution of the settlements along the plane of symmetry is similar on both cases and the difference in settlements is, at most, 1 cm. Thus, the results were considered sufficiently close and the square section was adopted to model the stone columns in the subsequent calculations.

The displacement boundary conditions were defined considering that the soft clay and the stone columns lay on a hard stratum ($y = 0$ plane, where displacements are set as zero in the three directions, x, y and z). One the other hand, symmetry conditions imply: (i) zero displacement in x-direction for nodes on the $x = 0$ plane; (ii) zero displacement in z-direction for nodes on the $z = 0$ plane, vertical plane containing the columns' centre in a row of columns; (iii) zero displacement in z-direction for nodes on the $z = 1$ m plane, vertical plane equidistant from two rows of columns in x-direction. Assuming that the horizontal displacement can be defined as zero at nodes that are enough distant from the embankment, the plane of $x = 30.0$ m was considered as the lateral boundary with zero displacement in the x-direction.

Regarding drainage boundary conditions, excess pore pressure was set as zero on the ground level (upper drainage surface), i.e. on the $y = 6$ m plane, and on the drainage surfaces defined by the lateral surfaces of the columns, i.e. with y-coordinate varying from 0 to 6 m, z-coordinate from 0 to 0.5 m (which means that centres of the columns are on the $z = 0$ boundary plane) on the following planes: $x = 0$, $x = 0.5$, $x = 1.5$, $x = 2.5$, $x = 3.5$, $x = 4.5$, $x = 5.5$, $x = 6.5$, $x = 7.5$, $x = 8.5$, $x = 9.5$ and $x = 10.5$ m.

The embankment construction was simulated activating the elements that form the fill layers. Five 0.5 m height layers were considered, constructed in 7 days each, without pause periods.

The constitutive relations of the embankment, stone columns and foundation soil were simulated using the p-q-θ critical state model with the parameters indicated in Table 1 (λ, slope of normal consolidation line and critical state line; k, slope of swelling and recompression line; Γ, specific volume of soil on the critical state line at mean normal stress equal to 1 kPa; N, specific volume of normally consolidated soil at mean normal stress equal to 1 kPa).

Table 2 shows other geotechnical properties: γ, unit weight; ν', Poisson's ratio for drained loading; c' and ϕ', cohesion and angle of friction defined in effective terms; kx and ky, coefficients of permeability in x and y directions. Table 3 indicates the variation with depth of the at-rest earth pressure coefficient, K0, and over-consolidation ratio, OCR, in the foundation. The embankment soil was considered with 0.43 for K0 and 1 for OCR. All these parameters were defined considering typical values for this kind of soils.

To study the overall effect of using the stone column technique, the embankment without stone columns was modelled. Numerical results have shown that the same embankment without stone column is not stable. After the 5th week of the embankment construction, i.e., when building the 2.0–2.5 m layer, the overall safety factor reaches the value of 1.0, i.e. the overall failure of the embankment without columns occurs. For this reason, the embankment without stone columns was modelled with 2 m height and the construction of the embankment was completed in an overall time of 28 days (4 weeks).

Figure 5 shows the 2D finite element mesh for the embankment without the stone columns,

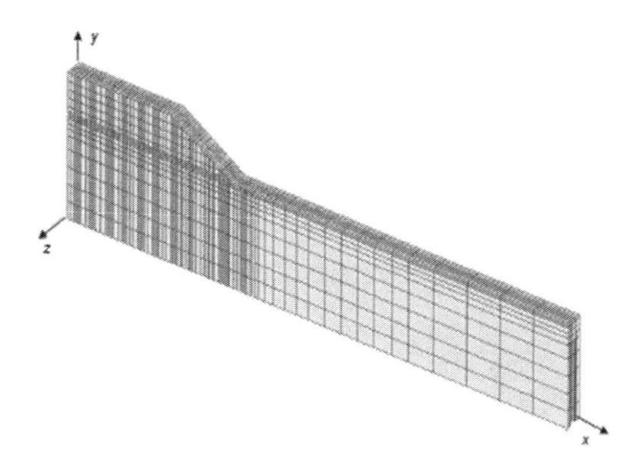

Figure 4. 3D finite element mesh for the problem with stone columns.

Table 1. Parameters of the p-q-θ critical state model.

	λ	k	Γ	N
Soft soil	0.18	0.025	3.05	3.16
Stone columns	0.0038	0.00095	1.914	1.916
Embankment	0.03	0.005	1.8	1.819

Table 2. Geotechnical properties of the foundation, stone columns and embankment soils.

	γ	ν'	ϕ'	$k_x = k_y$
	(kN/m^3)		(°)	(m/s)
Soft soil	16	0.25	26	10^{-9}
Stone columns	22	0.3	40	–
Embankment	20	0.3	35	–

Table 3. At rest earth pressure coefficient, K_0, and over consolidation ratio, OCR, on soft soil.

Depth (m)	K_0	OCR
0–2	0.86–0.56	2.92–1
2–6	0.56	1

problem that can be considered as a plane strain problem, given the very large longitudinal length of the embankment; y-axis is the symmetry line and, with exception of the boundary conditions for excess pore pressure (set as zero only on the upper drainage surface, i.e. at nodes with $y = 6$ m) and the fill height, all the other characteristics of the problem, when compared with the three-dimensional problem, are maintained.

3 ANALYSIS OF THE RESULTS

The construction of the embankment fill provokes variations of effective stress and of pore pressure in the soft soil foundation.

Figure 6 show results of excess pore pressure for the problem without stone columns at the end of construction. Figure 7 shows results of excess pore pressures for the problem with stone columns at the end and after the construction period. These results are shown both on the vertical plane that contains one row of column centers, $z = 0$ plane (on the right side), and on the vertical plane equidistant from two rows of columns, $z = 1$ m plane (on the left side).

Considering the problem without stone columns, at the end of the construction period, the maximum value happens in the central zone and

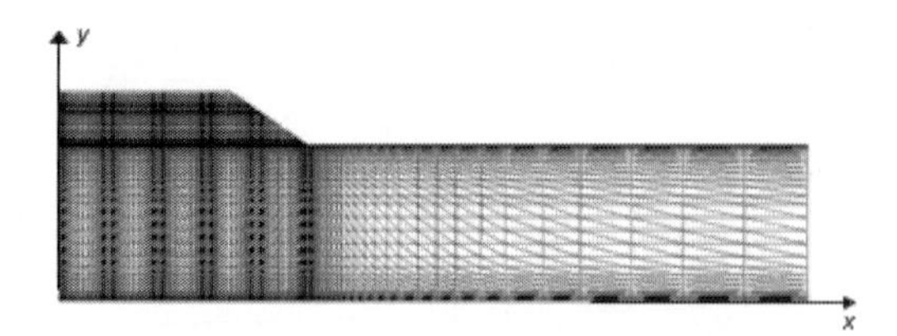

Figure 5. 2D finite element mesh for the problem without stone columns.

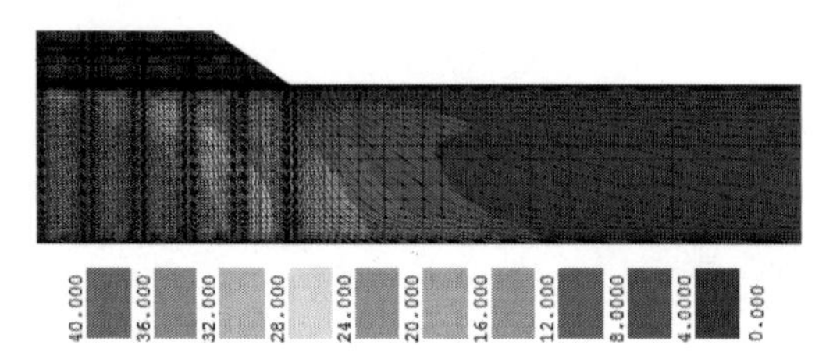

Figure 6. Excess pore pressure (u) for the embankment without stone columns at the end of the construction.

is similar to the vertical stress due to the embankment weight.

The excess pore pressure decreases from the central zone to the periphery, where its value is not significant.

Concerning the embankment with stone columns, the shape of the isovalue curves shows the three-dimensional condition of the problem, with drainage occurring both horizontally and vertically towards the several drainage surfaces (stone columns and upper drainage surface). The results show that stone columns play a key role in the dissipation process of the foundation. The consolidation process begins during the construction period, and at the end of construction the maximum value of excess pore pressure reached 16.8 kPa, which is significantly lower than vertical stress corresponding the embankment weight (50 kPa).

After construction, excess pore pressure continues to dissipate and, at week 7, maximum value of excess pore pressure is 12.9 kPa. Around 105th week (approximately 2 years), variation of excess pore pressure is not observed in the color maps. The comparison with the problem without columns makes clear how the presence of these reinforcing elements, which are simultaneously drains, accelerate the consolidation. As typically occurs in this type of works, the maximum values of excess pore pressure take place in the central zone under the embankment and decreases towards the periphery.

Figures 8 and 9 show the deformed mesh at the end of construction and at the end of consolidation (with the displacement scale increased five times) for both analyses. These results are complemented by the results shown in Figures 10–13, namely: settlements for the problem with columns at the embankment base, for several stages in the soft soil between columns (z = 1) (Figure 10); settlements at the embankment base at the end of construction and at the end of consolidation for both analyses (for the problem with columns results are shown on the line that contains one row of column centres, z = 0, and on the line in the soft soil equidistant from two rows of columns, z = 1 m) (Figure 11); variation in time of settlement at the middle point under the embankment centre on the ground level, i.e. point with x = 0, y = 6 m and z = 1 for both problems (Figure 12); horizontal displacements for the problem with columns at the end of construction and at the end of consolidation, along the interface between columns and soft soil, i.e. along points with x = 0.5 m, x = 4.5 m, x = 8.5 m, x = 10.5 m (and z = 0) (Figure 13).

Figure 9 and 10 show that, for the problem with stone columns, settlements are higher in the central loaded zone and there are upward vertical displacements, although not significant, near the embankment toe. At the end of the construction, the maximum settlement of the soft soil is 11.19 cm and the maximum long-term settlement at the end of consolidation is 14.02 cm (Fig. 10), which means that 80% of the maximum long-term settlement is reached at the end of construction, at week 5.

Two weeks after the construction is finished, the maximum settlement reaches 93.5% of the maximum long-term settlement. As expected, settlements are higher in the soft soil than in column (Fig.11), although the difference is significantly low when compared to the total value of the settlement. In the end of consolidation, the maximum differential settlement at the surface of the soft soil is 15 cm; the maximum differential settlement under the embankment platform is 3.58 cm and on the top of the embankment is approximately 1 cm. Analyzing these results, it should be pointed out the very low value of the differential settlement on the top of the embankment fill. This fact is very significant towards the effects on eventual works constructed on the embankment surface and is justified by load transfer within the embankment fill.

In the problem without stone columns there are important upward vertical displacements near the embankment toe, reaching 7.8 cm in the end of consolidation. The stone columns have the effect of reducing very expressively the consolidation time (from approximately 6000 days to 100 days, as shown in Figure 12).

Figure 13 also shows that horizontal displacements are outwards, and are higher in the columns near the embankment toe, where higher values generally occur between depths of 1.5–2.5 m

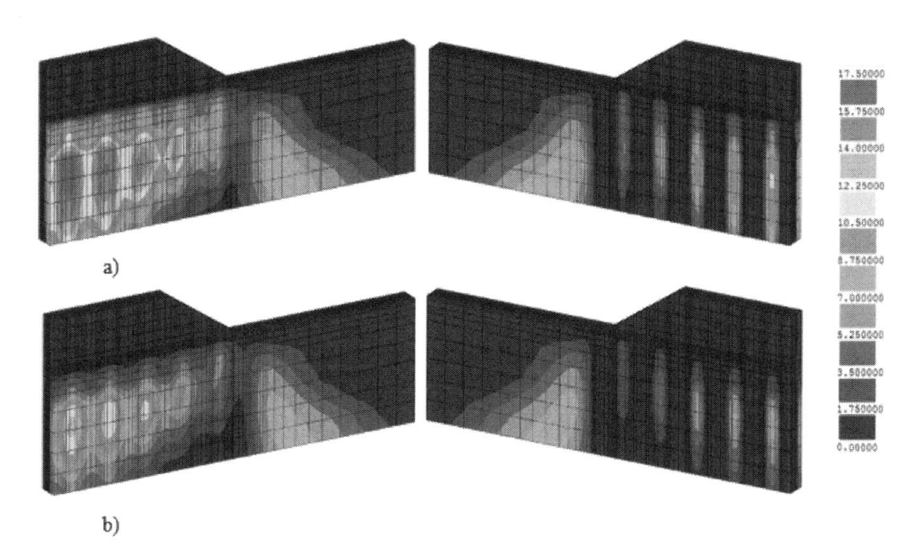

Figure 7. Excess pore pressure (u) for the embankment with stone columns: a) 2.5 m height embankment (end of construction); u_{max} = 16.8 kPa; b) 14 days after construction; u_{max} = 12.9 kPa.

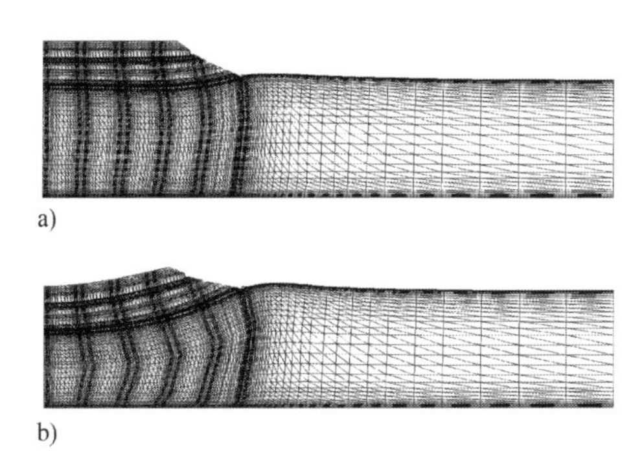

Figure 8. Deformed mesh for the embankment (2 m height) without stone columns: a) end of construction; b) end of consolidation.

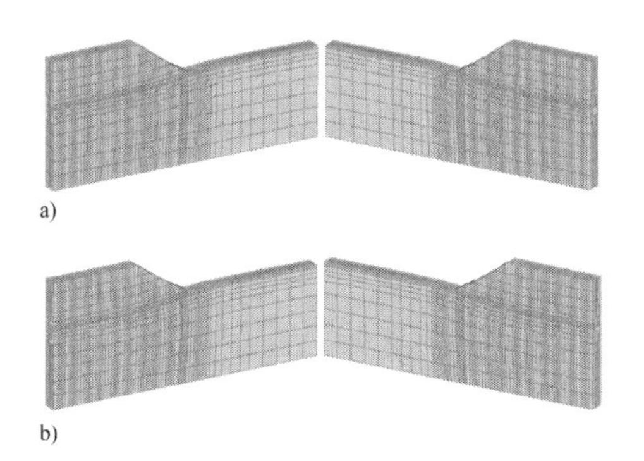

Figure 9. Deformed mesh for the embankment with stone columns: a) end of construction; b) end of consolidation.

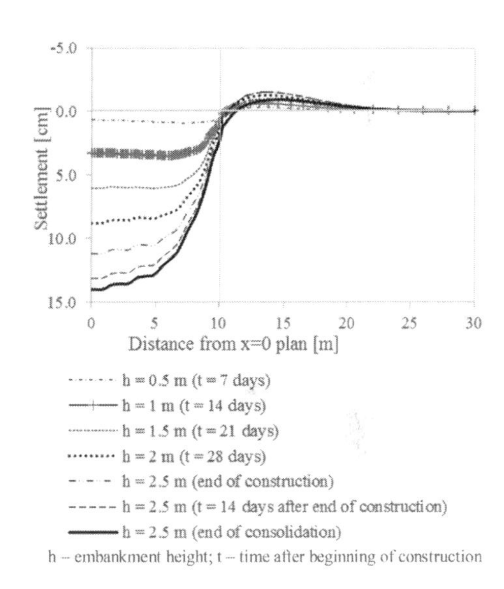

Figure 10. Settlements at embankment base for the embankment with stone columns (z = 1 m).

(maximum value of 4.37 cm). Borges (1995, 2009), explains that the outward horizontal displacements are in consonance with experimental results observed in real works and are associated with shear strains during the consolidation process—which are adequately simulated only by elastoplastic models with closed yielding surfaces, which is the case of the p-q-θ critical state model used in this study. Important horizontal displacements occur until the end of construction and increase slightly until the end of consolidation. The maximum displacement difference registered between the end of construction and end of consolidation is lower than 0.5 cm.

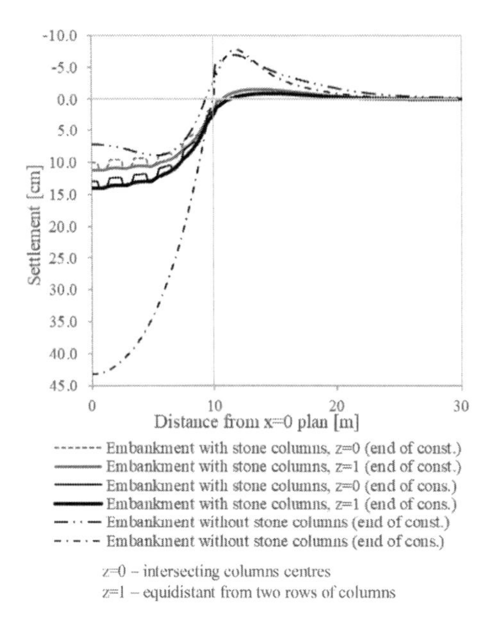

Figure 11. Settlements at the embankment base at the end of construction and the end of consolidation for the embankment with and without columns.

As the dissipation of excess pore pressures takes place, the vertical effective stresses increase (Fig. 14), mainly in the stone columns. The increments of vertical effective stress are very high in the columns and much lower in the soft soil. The columns located closer to the periphery (columns 5 and 6 counted from the centre of the embankment) are under the embankment slope. In these columns the increments of vertical effective stress are lower than in the central columns, presenting instead higher increments in the lower part of the columns.

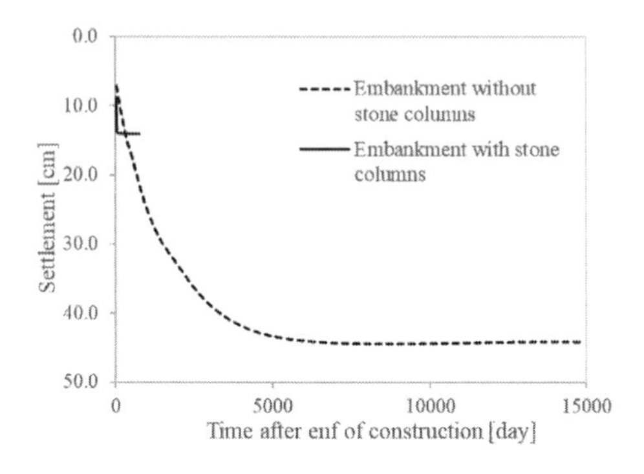

Figure 12. Settlement in time at the middle point under the embankment (point with $x = 0$, $y = 6$ and for 3D case, $z = 1$) for the embankment with and without stone columns.

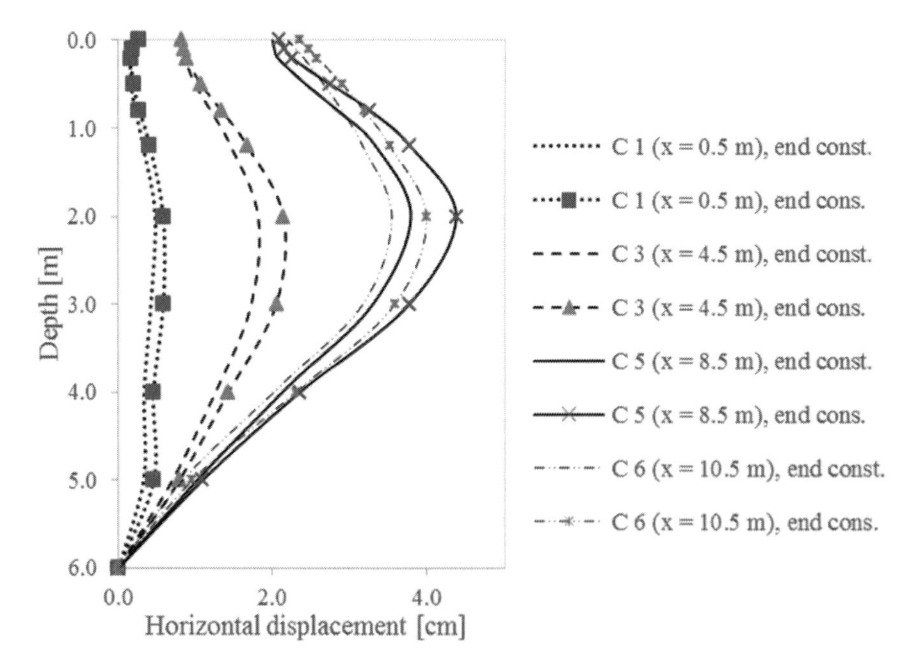

Figure 13. Horizontal displacements at the end of construction and at the end of consolidation along several vertical lines at the interfaces between columns and soft soil for the embankment with stone columns.

Figure 15 shows colour maps of stress level in the ground at the end of construction and at the end of consolidation for the problem with stone columns. Stress level, SL, measures the proximity to the soil critical state and is defined as follows:

$$SL = \frac{q}{pM} \tag{1}$$

where q = deviatoric stress; p = effective mean stress; and M = parameter that defines the slope of the critical state line according with p-q-θ model. In normally consolidated soils, SL varies from zero to 1, the latter being the critical state level. In overconsolidated soils, because of the peak strength behaviour, stress level may be higher than 1. By analysing these figures, several points can be issued. A zone of critical state in the soft soil beneath the embankment base is identified since the construction of the embankment. This fact is related with the increase of deviatoric stress associated to shear strain in that zone. Stress level reaches high values in columns in almost all length. During the post-construction period the stress level increases in the columns and decrease in the soft soil, mainly in the area outward the embankment toe. Due to consolidation, stress level reduces in the soft soil because effective mean stress increases and there are low variations of deviatoric stress.

The variation in depth of the stress concentration factor (SCF = $\Delta\sigma'_c/\Delta\sigma'_s$, where $\Delta\sigma'_c$ is the increment of effective stress in the column and $\Delta\sigma'_s$, in the soft soil), for each column, is shown in Figure 16, at the end of construction and at the end of the consolidation. The SCF for the 6th column counted from the center of the embankment was calculated considering the interior half of the

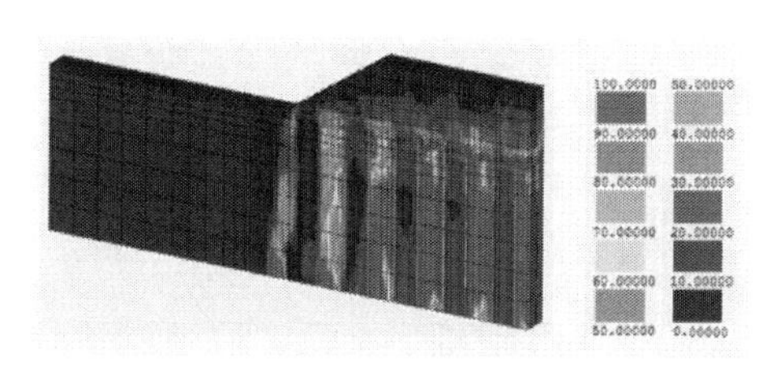

Figure 14. Increments of vertical effective stress (kPa) at the end of construction for the embankment with stone columns.

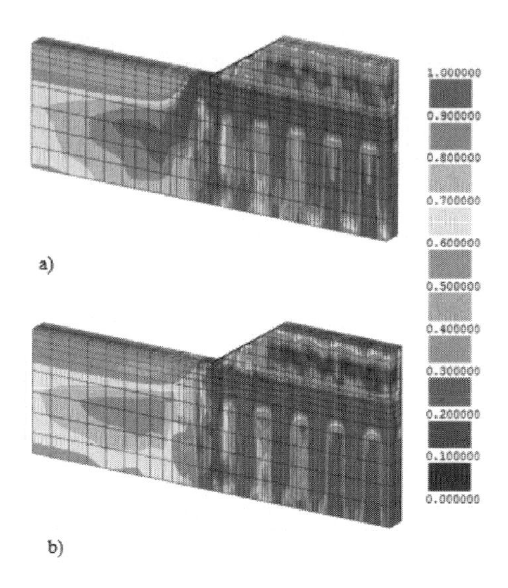

Figure 15. Stress level distribution for the embankment with stone columns: a) end of construction; b) end of consolidation.

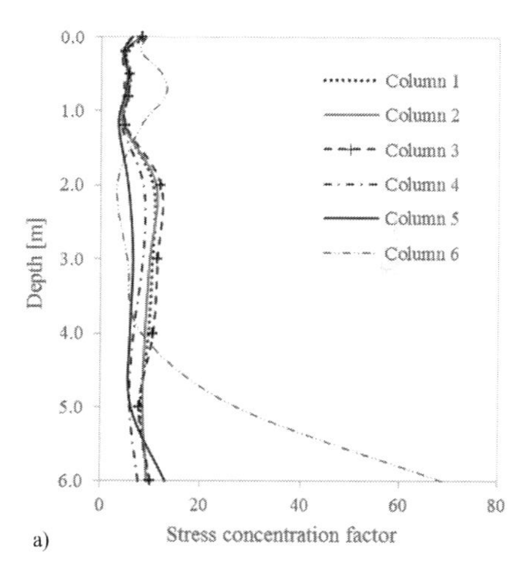

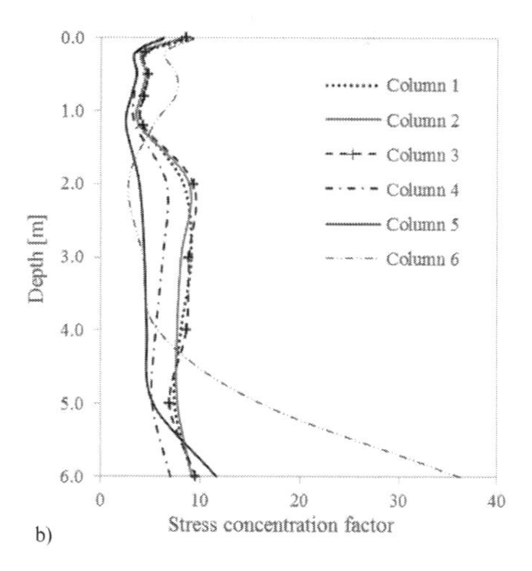

Figure 16. Variation of stress concentration ratio, SCR, for the embankment with stone columns: a) end of construction; b) end of consolidation.

column influence area, since the exterior half is outside the embankment.

These results show that: (i) SCF values at the end of construction are close to those at the end of consolidation, except for column 6, which is in agreement with the conclusions drawn on the evolution of the excess pore pressure and the increments of vertical effective stress; (ii) in general, SCF varies between 5 and 10, which is typical in this type of projects (Sexton et.al, 2004); (iii) the three columns closest to the centre of the embankment have very similar SCF values. The SCF presents a high value at the base of the embankment, which is due to the distribution of charges to the columns by arching effect on the body of the landfill. The SCF reduces from the surface to 1 m depth and increases from 5 to 10 between the depths of 1 m and 2 m, remaining approximately constant at lower depths; (iv) in the peripheral column 6, the SCF evolution profile is very different from the other columns, with a very high value in the base of the column. This is because the soft soil near the base takes low increments of vertical effective stress, which makes that the SCF, with very low value in the denominator, is very high. This effect is also visible, with less expression, in column 5.

4 CONCLUSIONS

A study was conducted to analyse the 3D time-dependent behaviour of an embankment on soft soils reinforced with stone columns. The analyses were performed by a finite element program that incorporates the Biot consolidation theory and constitutive relations simulated by the p-q-θ critical state model, which was also applied on the same embankment without stone columns. The following overall conclusions are pointed out:

- The interaction among the embankment, soft soil and columns provide load transfer to the more rigid material—the stone columns—and stress redistributions, resulting in very significant reduction of settlements both at the base and on the top of the embankment.
- The stone columns also play a key role in the consolidation process of the foundation, enormously reducing the consolidation time, due to drainage which occurs both horizontally and vertically towards the stone columns and upper drainage surface.
- Consequently, most of the settlements and horizontal displacements occur during the construction period, which is very significant towards the planning of eventual works constructed on the embankment.

REFERENCES

Basset, R.H. 1986a. The instrumentation of the trial embankment and of the Tensar SR2 grid. *Proc. Prediction Symposium on a Reinforced Embankment on Soft Ground*, King's College, London.

Basset, R.H. 1986b. Presentation of instrumentation data. *Proc. Prediction Symposium on a Reinforced Embankment on Soft Ground*, King's College, London.

Borges, J.L. 1995. Geosynthetic reinforced embankments on soft soils. Analysis and design. PhD Thesis in Civil Engineering, Faculty of Engineering, University of Porto, Portugal, 1995 (in Portuguese).

Borges, J.L., 2004. Three dimensional analysis of embankments on soft soils incorporating vertical drains by finite element method. Computers and Geotechnics, 31(8): 665–676, Elsevier.

Borges, J.L. 2009. Numerical analysis of embankments on soft soils incorporating vertical drains. In Karstunen & Leoni (ed.), *Geotechnics of Soft Soils: Focus on Ground Improvement: Proc. 2nd Intern. Workshop,* Glasgow, 3–5 September 2008.

Borges, J.L. & Cardoso, A.S. 1998. Numerical simulation of the p-q-_ critical state model in embankments on soft soils. *R. Geotecnia, nº 84, pp. 39–63* (in Portuguese).

Borges, J.L. & Cardoso, A.S. 2000. Numerical simulation of the consolidation processes in embankments on soft soils. *R. Geotecnia, nº 89, pp. 57–75* (in Portuguese).

Costa, P.A. 2005. Braced excavations in soft clayey soils—Behavior analysis including the consolidation effects. MSc Thesis, Faculty of Engineering, University of Porto, Portugal (in Portuguese).

Costa, P.A., Borges, J.L., Fernandes, M.M., 2007. Analysis of a braced excavation in soft soils considering the consolidation effect. Geotechnical and Geological Engineering, 25(6): 617–629.

Domingues, T.S., 2006. Foundation reinforcement with stone columns in embankments on soft soils—Analysis and design. MSc Thesis, Faculty of Engineering, University of Porto, Portugal (in Portuguese).

Lewis, R.W. & Schrefler, B.A., 1987. The Finite Element Method in the Deformation and Consolidation of Porous Media. John Wiley and Sons, Inc., New York.

Quaresma, M.G. 1992. Behaviour and modeling of an embankment over soft soils reinforced by geotextile. PhD Thesis, Universite Joseph Fourier, Grenoble I (in French).

Sexton, B.G., McCabe B.A., Castro J. 2014. Appraising stone column settlement prediction methods using finite element analyses. *Acta Geotechnica*. 9:993–1011.

Yeo, K.C. 1986. Simplified foundation data to predictors. *Proc. of the Prediction Symp. on a Reinforced Embankment on Soft Ground*, King's College, London, 1986.

Numerical Methods in Geotechnical Engineering IX – Cardoso et al. (Eds)
© 2018 Taylor & Francis Group, London, ISBN 978-1-138-33203-4

Three-dimensional parametric study of stone column-supported embankments on soft soils

D.O. Marques & J.L. Borges
CONSTRUCT-GEO, Faculty of Engineering (FEUP), University of Porto, Porto, Portugal

ABSTRACT: Parametric studies of an embankment on soft soils reinforced with stone columns are performed using a computer program based on the finite element method. The computer program incorporates the Biot consolidation theory (fully coupled formulation of the flow and equilibrium equations) with constitutive relations simulated by the p–q–θ critical state model. A 3D numerical approach is used to study the influence of the following parameters: construction time of the embankment; permeability anisotropy of the soft soil; deformability of the stone column. Special emphasis is given to comparisons, during and after the construction period, of the excess pore pressures, settlements, horizontal displacements, stress levels and stress concentration factor. The results indicate that the consolidation of soft soil is significantly influenced by the permeability anisotropy and construction time of the embankment. The deformability of the stone column expressively influences the long term settlements and horizontal displacements.

1 INTRODUCTION

Geotechnical engineers face several difficulties when designing embankments over soft soils. These difficulties are related to the weak geotechnical properties of the soft soil: (i) its low shear strength significantly limits the embankment height that is possible to construct with adequate safety for short-term stability; (ii) high deformability and low permeability provoke high settlements that develop slowly as pore water flows and excess pore pressure dissipates (consolidation).

A variety of techniques can be used to overcame these difficulties: (i) preloading, (ii) use of lightweight materials in the embankment fill, (iii) replacement of the inadequate soil, (iv) reinforcement with geosynthetics (Rowe, 1984; Borges and Cardoso, 2001, 2002), (v) consolidation acceleration using prefabricated vertical drains (Borges, 2004; Shen et al., 2005); (vi) vacuum preloading (Indraratna et al. 2004), (vii) stage construction of the embankment and (viii) the application of pile/column-reinforced foundations, such as stone column (Greenwood, 1970; Hughes et al. 1975; Borges et al. 2009), geosynthetic encased stone columns (Keykhosropur et al. 2012, Almeida et al. 2015), concrete piles, lime or lime-cement columns (Rajasekaran and Rao, 2002), deep soil mixing and jet-grout columns (Borges and Marques, 2011).

Among all the techniques to improve geotechnical behaviour of embankments on soft soils, the stone columns method (Figure 1) is the most adequate when the main purpose is, simultaneously, to increase the overall stability and reduce and accelerate settlements.

In this paper, a three-dimensional numerical study is performed to analyse the influence of three factors on the performance of embankments on soft soils reinforced with stone columns: the construction time of the embankment, the permeability anisotropy of the soft soil and the deformability of the stone column material.

In the analyses, some results of excess pore pressure, settlements, horizontal displacements and stress concentration factor are analysed, so that a better geotechnical interpretation could be accomplished.

All the analyses were performed with a finite element program developed by Borges (1995). The initial version of the program (2-D version) was presented in 1995 and several improvements were posteriorly developed and implemented, particularly a 3-D fully coupled analysis version (Borges, 2004).

Basically, for the present applications, the program uses the following theoretical hypotheses: a) coupled analysis of the flow and equilibrium equations considering the soil constitutive relations formulated in effective stresses (Biot consolidation theory) (Borges, 1995; Borges and Cardoso 2000; Lewis and Schrefler, 1987); this formulation is applied to all stages of the problem, both during the embankment construction and in the post-construction period; b) use of the *p*-*q*-θ critical state

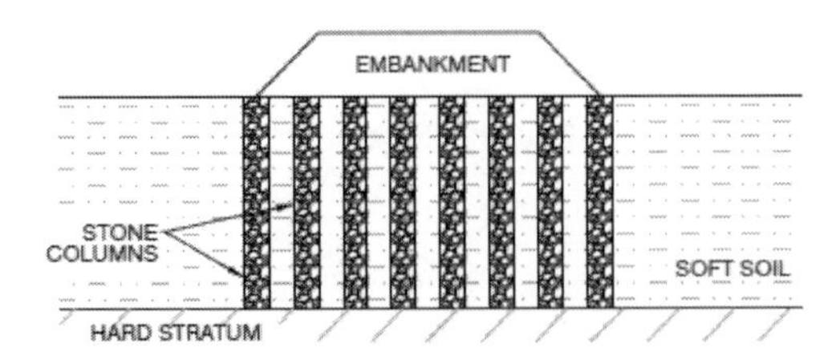

Figure 1. Typical section of an embankment over soft soil reinforced with stone columns.

model (Borges, 1995; Borges and Cardoso, 1998; Lewis and Schrefler, 1987), an associated plastic flow model, to simulate constitutive behaviour of soil.

The accuracy of the finite element program has been assessed in several ground structures involving consolidation by comparing numerical results to field measurements. For instance, Borges (1995) compared results of two geosynthetic-reinforced embankments, one constructed up to failure (Quaresma, 1992) and the other observed until the end of consolidation (Yeo, 1986; Basset, 1986a,b). The accuracy was considered adequate in both cases. Very good agreements of numerical and field results were also observed both in an embankment on soft soils incorporating stone columns (Domingues, 2006) and in a braced excavation in very soft ground (Costa et al., 2007; Costa, 2005).

For three-dimensional application, the program uses two types of the 20-noded brick element. Figure 2a shows the element used in the soft soil (with 60 displacement degrees of freedom, at the corners and at middle of the sides, and with 8 more excess pore pressure degrees of freedom, at the corners), where consolidation analysis is considered. In the fill and in the columns, it is the 20-noded brick element, with only 60 displacement degrees of freedom (at the corners and at middle of the sides), that is used (Figure 2b).

2 DESCRIPTION OF THE PROBLEM

The problem consists of a 2.5 m height symmetric embankment, with a 12.5 m crest width, 2/3 (V/H) inclined slopes and very large longitudinal length. The soft ground is a 6 m thick saturated clay layer lying on a rigid and impermeable soil, which constitutes the lower boundary. The clay is lightly overconsolidated to 2 m depth and normally consolidated from 2 m to 6 m. The water level is at the ground surface. The diameter of the stone columns is 1.13 m and the spacing between columns is 2.0 m in square grid.

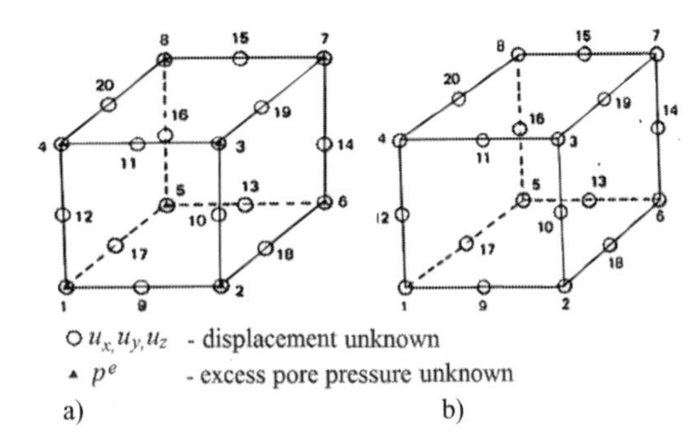

Figure 2. 20-noded brick element: a) with 60 displacement degrees of freedom and 8 excess pore pressure degrees of freedom; b) with 60 displacement degrees of freedom.

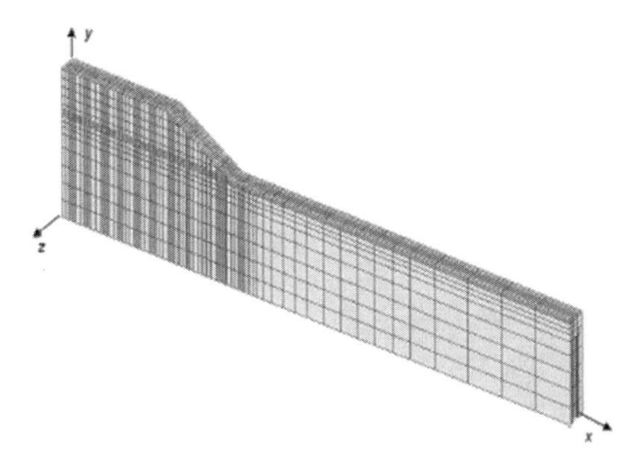

Figure 3. 3D finite element mesh.

Figure 3 shows the finite element mesh used in the three-dimensional analysis of the embankment incorporating the stone columns. To simplify the mesh, the stone columns are modelled as elements with a square section in plan with the same sectional area, instead of a circular section. To evaluate the impact of this simplification, the central part of the embankment was modelled by three-dimensional unit cells with the same area in plan and equal spacing between columns, one with a square section and the other with a circular section. Results shown that the evolution of the settlements along the plane of symmetry is similar on both cases and the difference in settlements is, at most, 1 cm. Thus, the results were considered sufficiently close and the square section was adopted to model the stone columns in the subsequent calculations.

The displacement boundary conditions were defined considering that the soft clay and the stone columns lay on a hard stratum ($y = 0$ plane, where displacements are set as zero in the three directions,

Table 1. Mechanical and hydraulic properties of the materials.

	λ	k	Γ	N	γ(kN/m^3)	ν'	ϕ'(°)	$k_x = k_y$ (m/s)
Soft soil	0.18	0.025	3.05	3.16	16	0.25	26	10^{-9}
Stone columns	0.0038	0.00095	1.914	1.916	22	0.3	40	–
Embankment	0.03	0.005	1.8	1.819	20	0.3	35	–

Table 2. At?rest earth pressure coefficient, K_0, and over consolidation ratio, OCR, on soft soil.

Depth (m)	K_0	OCR
0–2	0.86–0.56	2.92–1
2–6	0.56	1

Table 3. Construction time for cases T0 to T3.

Case	T0	T1	T2	T3
Time of construction (weeks)	5	1	2	10

x, y and z). One the other hand, symmetry conditions imply: (i) zero displacement in x-direction for nodes on the $x = 0$ plane; (ii) zero displacement in z-direction for nodes on the $z = 0$ plane, vertical plane containing the columns' centre in a row of columns; (iii) zero displacement in z-direction for nodes on the $z = 1$ m plane, vertical plane equidistant from two rows of columns in x-direction. Assuming that the horizontal displacement can be defined as zero at nodes that are enough distant from the embankment, the plane of $x = 30.0$ m was considered as the lateral boundary with zero displacement in the x-direction.

Regarding drainage boundary conditions, excess pore pressure was set as zero on the ground level (upper drainage surface), i.e. on the $y = 6$ m plane, and on the drainage surfaces defined by the lateral surfaces of the columns, i.e. with y-coordinate varying from 0 to 6 m, z-coordinate from 0 to 0.5 m (which means that centres of the columns are on the $z = 0$ boundary plane) on the following planes: $x = 0$, $x = 0.5$, $x = 1.5$, $x = 2.5$, $x = 3.5$, $x = 4.5$, $x = 5.5$, $x = 6.5$, $x = 7.5$, $x = 8.5$, $x = 9.5$ and $x = 10.5$ m.

The embankment construction was simulated activating the elements that form the fill layers. Five 0.5 m height layers were considered, constructed in 7 days each, without pause periods.

The constitutive behaviour of the materials is modeled by the p-q-θ critical state model with the parameters indicated in Table 1 (λ, slope of normal consolidation line and critical state line; k, slope of swelling and recompression line; Γ, specific volume of soil on the critical state line at mean normal stress equal to 1 kPa; N, specific volume of normally consolidated soil at mean normal stress equal to 1 kPa). Table 1 also shows other geotechnical properties: γ, unit weight; ν', Poisson's ratio for drained loading; c' and ϕ', cohesion and angle of friction defined in effective terms; kx and ky, coefficients of permeability in x and y directions. Table 2 indicates the variation with depth of the at-rest earth pressure coefficient, K0, and overconsolidation ratio, OCR, in the foundation. The embankment fill was considered with 0.43 for K0 and 1 for OCR. All these parameters were adopted taking into account typical values reported in bibliography for these kind of soils (Lambe & Withman 1969; Borges 1995; Costa 2005; Domingues 2006).

3 ANALYSIS OF RESULTS

3.1 *Influence of the embankment construction time*

In order to analyze the influence of this factor, four cases were considered: T0 - construction time of 5 weeks (the case described in previous section); T1 - construction time of 1 week; T2 - construction time of 2 weeks and T3 - construction time of 10 weeks, as indicated in Table 3.

Figure 4 shows the distribution of excess pore pressure at the end of construction. These results are shown both on the vertical plane that contains one row of column centers, z = 0 plane (on the right side), and on the vertical plane equidistant from two rows of columns, z = 1 m plane (on the left side). As expected, the results show that the longer the construction period, the lower is the maximum value of excess pore pressure in the soft soil at the end of construction. Once the embankment construction occurs in partially drained conditions, the slower the construction rate, the more advanced the dissipation process at the end of construction.

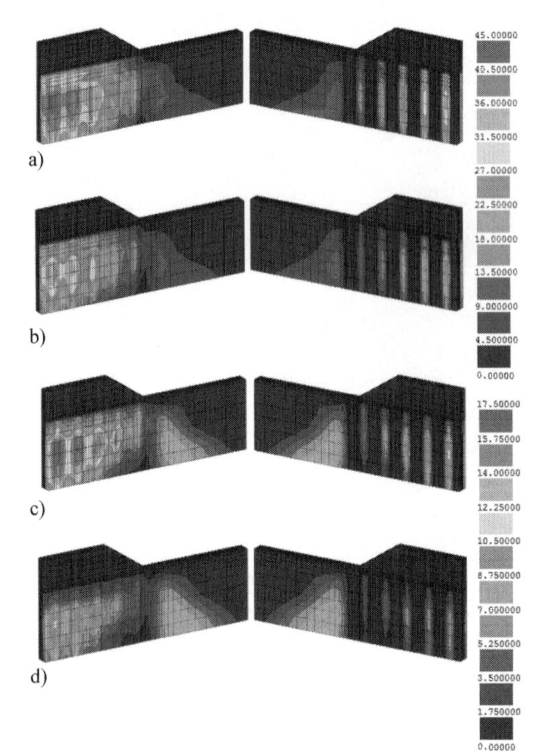

Figure 4. Excess pore pressure (kPa) at the end of the construction for cases: a) T1 (u_{max} = 41.7 kPa); b) T2 (u_{max} = 35.3 kPa); c) T0 (u_{max} = 16.8 kPa) and d) T3 (u_{max} = 9.3 kPa).

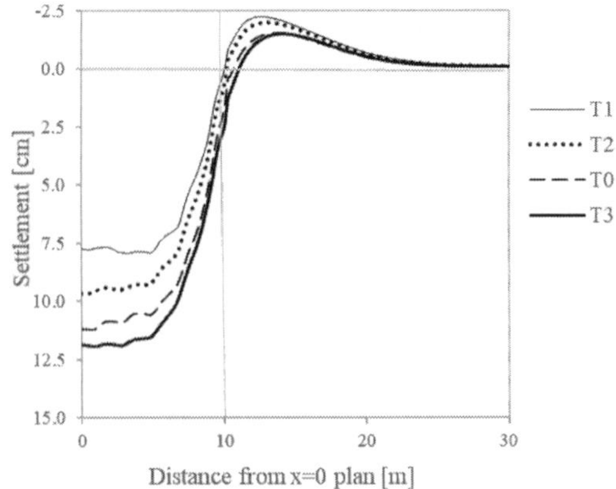

Figure 5. Settlements at the base of the embankment at the end of the construction for cases T0-T3.

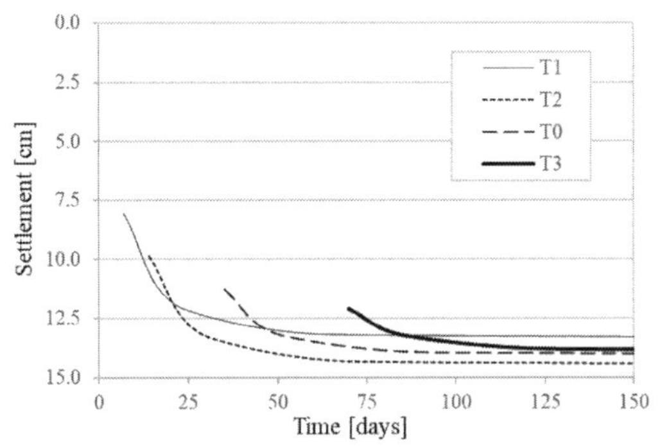

Figure 6. Maximum settlement at the base of the embankment, after construction period, for cases T0-T3.

Table 4. Influence of the construction time.

Case	Maximum settlement (cm)	Settlement at end of construct. (cm)	Maximum horizontal displacement (cm)
T0	14.1	9.87	3.99
T1	13.4	8.07	4.65
T2	14.6	11.28	4.55
T3	13.9	12.10	3.93

Figure 5 shows the settlements at the base of the embankment in the soft soil equidistant from two rows of columns (plane z = 1 m) at the end of construction. These results show that, as expected, maximum settlement increases with the time of construction. As typically occurs in embankments over soft soils, when the loading is fast, the soft soil deforms under practically undrained conditions which entails horizontal outward displacements and vertical upward displacements. With slower construction rates, the distortions caused by loading on the foundation during the construction process are lower.

The evolution in time of the maximum settlement at the embankment base after the construction period is shown in Figure 6. It should be noted that, at the end of consolidation, the maximum settlement is similar for all cases (Table 4). The maximum horizontal displacement at the interface between the column and the soft soil, along vertical line x = 10.5 m, increases slightly with the increase of the construction rate.

3.2 *Influence of the permeability anisotropy of the soft soil*

The influence of the permeability anisotropy is investigated by varying the ratio between the coefficient of horizontal permeability, k_x, and the coefficient of vertical permeability, k_y, as shown

Table 5. Permeability anisotropy of the soft soil.

Case	K0	K1	K2	K3
k_h/k_v	1	2	4	10

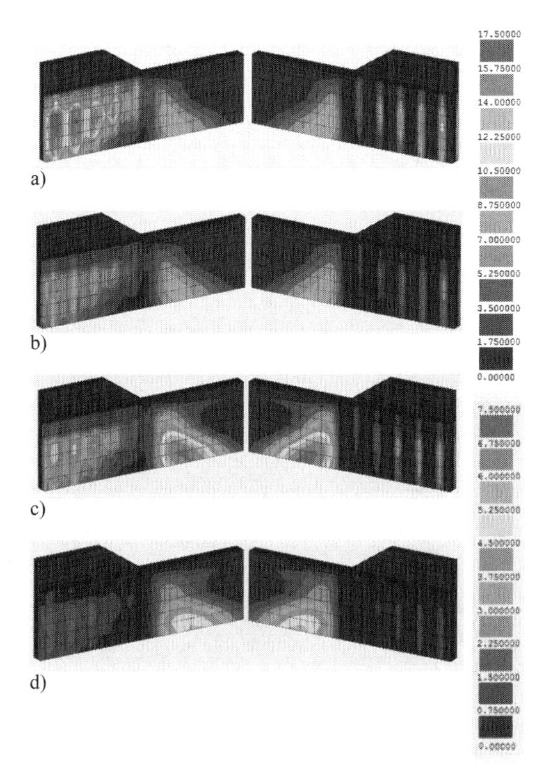

Figure 7. Excess pore pressure (kPa) at the end of the construction for cases: a) K0 (u_{max} = 16.8 kPa); b) K1 (u_{max} = 9.7 kPa); c) K2 (u_{max} = 7.1 kPa) and d) K3 (u_{max} = 5.5 kPa).

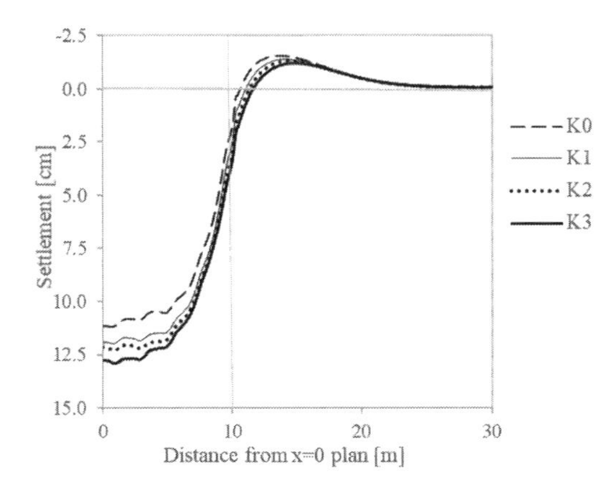

Figure 8. Settlements on the ground surface at the end of the construction for cases K0-K3.

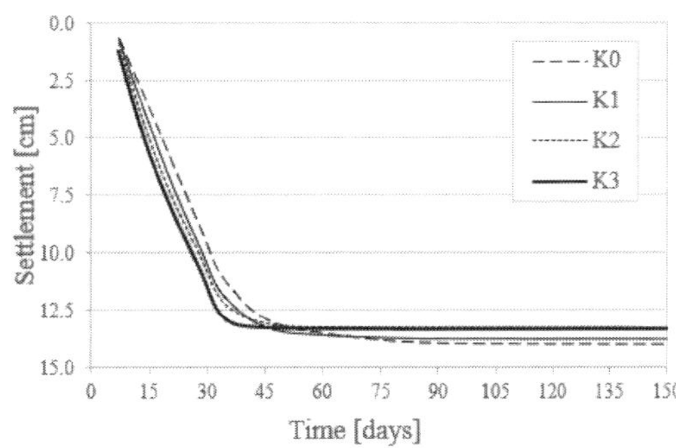

Figure 9. Maximum settlement on the ground surface for cases K0-K3.

Table 6. Influence of the permeability anisotropy.

Case	Maximum settlement (cm)	Maximum horizontal displacement (cm)
K0	14.1	3.99
K1	13.8	3.77
K2	13.3	3.63
K3	13.3	3.56

Table 7. Values of parameter λ of column.

Case	M0	M1	M2	M3
λ_{col}	0.0038	0.023	0.0076	0.0029
$r = \lambda_{col} / (\lambda_{col})_{M0}$	1	6	2	0.75

Table 8. Influence of parameter λ of column.

Case	Maximum settlement (cm)	Maximum horizontal displacement (cm)
M0	14.1	4.00
M1	24.5	7.17
M2	18.2	5.36
M3	11.3	3.11

in Table 5, while all other parameters are kept constant.

Figure 7 shows distributions of the excess pore pressure and Figure 8 shows the radial variation of the settlement at the base of the embankment, both at the end of the construction. The acceleration of the consolidation with permeability anisotropy is clearly observed in Figure 7, with the

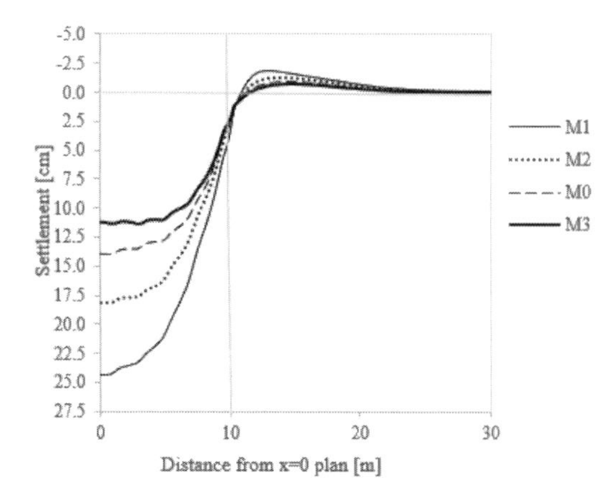

Figure 10. Settlements on the ground surface at the end of the consolidation for cases M0-M3.

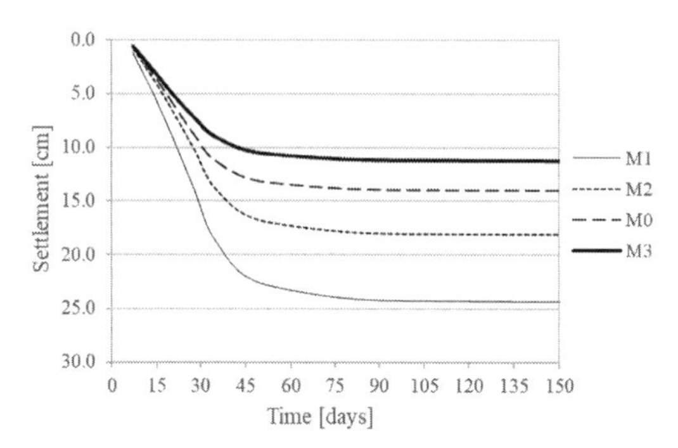

Figure 11. Maximum settlement on the ground surface for cases M0-M3.

highest dissipation of excess pore pressure corresponding to the highest value of k_h/k_v (case K3). As expected, settlement at the end of construction (Figure 8) increases with the increase of anisotropy permeability. In fact, the permeability anisotropy accelerates the consolidation process, and, the higher the permeability of the soil in the horizontal direction, the faster the soil water flows to the drainage surfaces (soft soil surface and columns) and, consequently, consolidation settlements are processed faster. The evolution in time of the maximum settlement at the embankment base is shown in Figure 9. As also expected, at the end of the consolidation the settlements at the base of the embankment show approximately the same magnitude in all the calculations.

Table 6 shows maximum long term settlement and maximum horizontal displacement at the interface between the column and the soft soil, along vertical line $x = 10.5$ m, for all cases.

3.3 *Influence of the deformability of the column material*

The influence of the deformability of the columns material is investigated by varying the parameter that defines the slope of the normal consolidation line of the column material, λ_{col}.

This parameter is related to the traditional compression index, C_c, by the equation $C_c/\ln_{10}$.

Four cases were considered, as indicated in Table 7. M0 is equal to case T0 presented previously.

The radial variation of settlement at the end of consolidation is shown in Figure 10. These results clearly show that reducing the deformability of the column material significantly reduces settlements, as well as vertical upward displacements. Table 8 shows maximum settlement and maximum horizontal displacement at the interface between the column and the soft soil, along vertical line line $x = 10.5$ m, for all cases. These results show that maximum settlement significantly reduces with the stiffness of column, as well as maximum horizontal displacement, as expected.

Variation in time of the maximum settlement at the middle point of the base of the embankment and distributions of the excess pore pressure at the end of construction are shown in Figures 11 and 12,

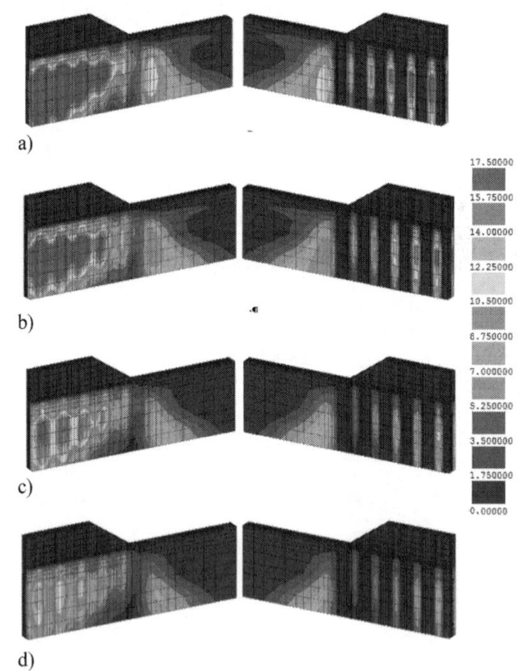

Figure 12. Excess pore pressure (kPa) at the end of the construction for cases: a) M1 (u_{max} = 31.6 kPa); b) M2 (u_{max} = 25.5 kPa); c) M0 (u_{max} = 16.8 kPa) and d) M3 (u_{max} = 13.6 kPa).

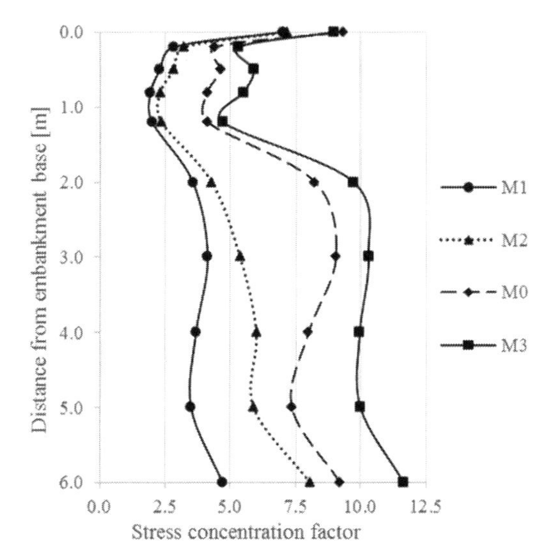

Figure 13. Stress concentration ratio, SCR, evolution in depth:at the end of consolidation, in the central column, for cases M0-M3.

respectively. Results show clearly that the velocity of consolidation increases with the stiffness of column.

Figure 13 shows that the stress concentration factor (SCF = $\Delta\sigma'_c/\Delta\sigma'_s$, where $\Delta\sigma'_c$ is the increment of effective stress in the column and $\Delta\sigma'_s$, in the soft soil) increases with the stiffness of column.

4 CONCLUSIONS

A parametric study was conducted to analyse the influence of the following factors on the three-dimensional time-dependent behaviour of embankments on soft soils reinforced with stone columns: the construction time of the embankment; the permeability anisotropy of the soft soil and the stiffness of column. The analyses were performed by a finite element program that incorporates the Biot consolidation theory and constitutive relations simulated by the p-q-θ critical state model.

The results showed that the stiffness of column significantly influences the behaviour of the embankment. Increasing the stiffness of column reduces settlements and horizontal displacements, accelerates the consolidation and increases the stress concentration factor. The results also showed that the soil permeability anisotropy and the construction time significantly influence the consolidation.

REFERENCES

Almeida. M.S.S., Hosseinpour. I., Riccio. M & Alexiew D. 2015. Behavior of Geotextile-encased granular columns supporting test embankment on soft deposit. *Journal of Geotechnical and Geoenvironmental Engineering*, 141, 04014116, 1–9.

Basset, R.H. 1986a. The instrumentation of the trial embankment and of the Tensar SR2 grid. In *Proc. Prediction Symposium on a Reinforced Embankment on Soft Ground*, King's College, London.

Basset, R.H. 1986b. Presentation of instrumentation data. In *Proc. Prediction Symposium on a Reinforced Embankment on Soft Ground*, King's College, London.

Borges, J.L. 1995. Geosynthetic reinforced embankments on soft soils. Analysis and design. PhD Thesis in Civil Engineering, Faculty of Engineering, University of Porto, Portugal, 1995 (in Portuguese).

Borges, J.L. 2004. Three-dimensional analysis of embankments on soft soils incorporating vertical drains by finite elementmethod. *Comput. Geotech.* 31 (8),665–676.

Borges, J.L. 2009. Numerical analysis of embankments on soft soils incorporating vertical drains. In Karstunen & Leoni (ed.), *Geotechnics of Soft Soils: Focus on Ground Improvement: Proc. 2nd Intern. Workshop, Glasgow, 3–5 September 2008.*

Borges, J.L. & Cardoso, A.S. 1998. Numerical simulation of the p-q-theta critical state model in embankments on soft soils. *R. Geotecnia, nº 84, pp. 39–63* (in Portuguese).

Borges, J.L. & Cardoso, A.S. 2000. Numerical simulation of the consolidation processes in embankments on soft soils. *R. Geotecnia, nº 89, pp. 57–75* (in Portuguese).

Borges, J.L. & Cardoso, A.S. 2001.Structural behaviour and parametric study of reinforced embankments on soft clays. *Comput. Geotech.* 28 (3), 209–233.

Borges, J.L. & Cardoso, A.S. 2002.Overall stability of geosynthetic-reinforced embankments on soft soils. *Geotext. Geomembr.* 20 (6), 395–421.

Borges, J.L, Domingues, T.S. & Cardoso, A.S. 2009. Embankments on Soft Soil Reinforced with Stone Columns: Numerical Analysis and Proposal of a New Design Method. *Geotech. Geol. Eng.* 27:667–679.

Borges, J.L., Marques, D.O, 2011. Geosynthetic-reinforced and jet grout column-supported embankments on soft soils: Numerical analysis and parametric study. Computers and Geotechnics, 38(7): 883–896.

Costa, P.A. 2005. Braced excavations in soft clayey soils— Behavior analysis including the consolidation effects. MSc Thesis, Faculty of Engineering, University of Porto, Portugal (in Portuguese).

Costa, P.A., Borges, J.L., Fernandes, M.M., 2007. Analysis of a braced excavation in soft soils considering the consolidation effect. Geotechnical and Geological Engineering, 25(6): 617–629.

Domingues, T.S., 2006. Foundation reinforcement with stone columns in embankments on soft soils—Analysis and design. MSc Thesis, Faculty of Engineering, University of Porto, Portugal (in Portuguese).

Greenwood D.A. 1970. Mechanical improvement of soils below ground surfaces. In *Proc. Of Ground Engineering Conference*, Institution of Civil Engineers, London: 11–22.

Hughes J.M.O, Withers N.J., Greenwood D.A. 1975. Field trial of reinforcement effect of a stone column in soil. *Geotechnique*, 25(1):31–44

Indraratna, B. Bamunawita, C. and Khabbaz, H. 2004. Numerical modeling of vacuum preloading and field

applications. *Canadian Geotechnical Journal*, 41 (6): 1098–1110.
Keykhosropur, L., Soroush, A., and Imam, R. 2012. 3D numerical analyses of geosynthetic encased stone column. *Geotextiles and Geomembranes*, 35, 61–68.
Lambe T.W. & Whitman R.V. 1969. *Soil mechanics*. New York: Wiley.
Lewis, R.W. & Schrefler, B.A., 1987. The Finite Element Method in the Deformation and Consolidation of Porous Media. John Wiley and Sons, Inc., New York.
Quaresma, M.G. 1992. Behaviour and modeling of an embankment over soft soils reinforced by geotextile. PhD Thesis, Universite Joseph Fourier, Grenoble I (in French).
Rajasekaran G, Rao S. 2002. Permability characteristics of lime treated marine clay. *Ocean Engineering* 29(2): 113–127.
Rowe, R.K. 1984. Reinforced embankments: Analysis and design. *J.Geotech. Eng*. 110 (2), 231–246.
Sexton, B.G., McCabe B.A., Castro J. 2014. Appraising stone column settlement prediction methods using finite element analyses. *Acta Geotechnica*. 9:993–1011.
Shen, S.L., Chai,J.C., Hong,Z.S., Cai,F.X. 2005. Analysis of field performance of embankments on soft clay deposit with and without PVD-improvement. *Geotextiles and Geomembranes*, 23(6),463–485.
Yeo, K.C. 1986. Simplified foundation data to predictors. In *Proc. of the Prediction Symp. on a Reinforced Embankment on Soft Ground*, King's College, London, 1986.

Numerical Methods in Geotechnical Engineering IX – Cardoso et al. (Eds)
© 2018 Taylor & Francis Group, London, ISBN 978-1-138-33203-4

Finite element analysis of performance of bearing reinforcement earth wall

S. Horpibulsuk & P. Witchayaphong
School of Civil Engineering, Center of Innovation in Sustainable Infrastructure Development, Suranaree University of Technology, Thailand

C. Suksiripattanapong
Department of Civil Engineering, Faculty of Engineering and Architecture, Rajamangala University of Technology Isan, Thailand

A. Arulrajah
Department of Civil and Construction Engineering, Swinburne University of Technology, Australia

R. Rachan
Department of Civil Engineering, Mahanakorn University of Technology, Thailand

ABSTRACT: This paper presents a numerical simulation of the bearing reinforcement earth wall by PLAXIS 2D. The bearing reinforcement was regarded as a cost-effective earth reinforcement. The model parameters for the simulation were obtained from the conventional laboratory tests and back analyses from the laboratory pullout tests of the bearing reinforcement. The change in bearing stresses, settlements, lateral movement and tensions in the reinforcements during and after construction is simulated. Overall, the simulated test results are in good agreement with the measured ones. The simulated bearing stress presents a trapezoid distribution shape as generally assumed by the conventional method of examination of the external stability of MSE walls. The simulated settlement is almost uniform due to a high stiffness of the rigid foundation and the bearing reinforcements. The maximum lateral movement occurs at about the mid-height of the wall, resulting in the bi-linear maximum tension plane.

1 INTRODUCTION

The use of inextensible reinforcements to stabilize earth structures has grown rapidly in the past two decades. When applied for retaining walls or steep slopes, they can be laid continuously along the width of the reinforced soil system (grid type) or laid at intervals (strip type). Both grid and strip reinforcements are widely employed around the world, including Thailand and Australia. The construction cost of the mechanically stabilized earth (MSE) wall is mainly dependent upon the transportation of backfill from a suitable borrow pit and the reinforcement type. The backfill is generally granular materials. The transportation of the backfill is thus a fixed cost for a particular construction site. Consequently, the reinforcement becomes the key factor, controlling the construction cost for a particular site.

Horpibulsuk and Niramitkornburee (2010) have introduced a cost-effective earth reinforcement designated as "Bearing reinforcement". It is simply installed, conveniently transported and possesses high pullout and rupture resistances with less steel volume. Figure 1 shows the typical configuration of the bearing reinforcement, which is composed of a longitudinal member and transverse (bearing) members. The longitudinal member is a steel deformed bar and the transverse members are a set of steel equal angles. This reinforcement has been introduced into practice in Thailand since 2008 by the Geoform Co., Ltd. Several earth walls stabilized with the bearing reinforcements were constructed at various parts of Thailand. This reinforcement has been considered to be one of the standard earth reinforcements for the Department of Highways, Thailand. The earth wall stabilized by the bearing reinforcements is designated as "bearing reinforcement earth (BRE) wall".

For a MSE wall design, an examination of external and internal stability is a routine design procedure. The examination of external stability is generally performed using the conventional method (limit equilibrium analysis) assuming that the composite backfill-reinforcement mass behaves as a rigid body (McGown et al., 1998). The external stability of the BRE wall with the vertical spacing of the reinforcement less than 750 mm was successfully examined by the conventional method (Horpibulsuk et al., 2010 and 2011). The internal stability of the BRE wall deals with the rupture and pullout resistances of the reinforcement. The practical equations for estimating pullout resistance of the bearing reinforcement with different transverse members were proposed by Horpibulsuk and Niramitkornburee (2010). The equations were successfully used to design the BRE wall in Thailand. To verify the concept, a full-scale test BRE wall was designed by the limit equilibrium analysis and constructed in the campus of Suranaree University of Technology (Horpibulsuk et al., 2010 and 2011). The performance of the BRE wall was measured and reported. The small lateral movement and settlement were observed. The practical method of designing the BRE found on the hard stratum was introduced. This method has been adopted to design several BRE walls under the Department of Highways, Thailand.

The performance of MSE walls was extensively studied using the full-scale, laboratory model tests and numerical simulation (Bergado et al., 2000; Bergado and Teerawattanasuk, 2007; Park and Tan, 2005; Skinner and Rowe, 2005; Al Hattamleh and Muhunthan, 2006; Hatami and Bathurst, 2005 and 2006; and Abdelouhab et al., 2011). The PLAXIS program has been proved as a powerful and precise tool for predicting the performance of the MSE wall and pullout test results (Bergado et al., 2003; and Khedkar and Mandal, 2007 and 2009). This paper presents a numerical simulation of the performance of the BRE wall during and after construction, which includes settlement, bearing stress, lateral movement, lateral earth pressure and tension force in the reinforcements. The full-scale test results by Horpibulsuk et al. (2011) were taken for this simulation. The simulation was performed using the finite element code (PLAXIS 2D). The bearing reinforcement was modeled as the geotextile with an equivalent friction resistance. The equivalent friction resistance is represented by the soil/reinforcement interface parameter and was obtained from the back analysis of the laboratory

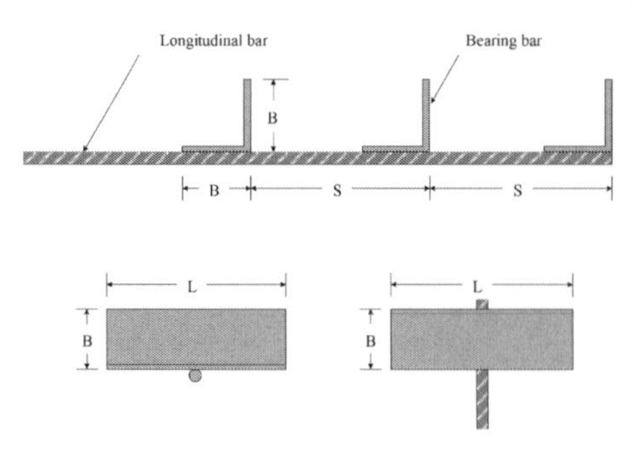

Figure 1. Configuration of the bearing reinforcement (Horpibul-suk and Niramitkornburee, 2010).

pullout test results. The other model parameters were obtained from the conventional laboratory tests. The knowledge gained from this simulation provides a useful information for further analysis and design of the other BRE walls with different wall heights, ground conditions and features of bearing reinforcement.

2 MODEL PARAMETERS

The bearing reinforcement earth wall was modeled as a plane strain problem. The finite element mesh and boundary condition are shown in Figure 2. The finite element mesh involved 15-node triangular elements for the backfill and the foundation. The nodal points at the bottom boundary were fixed in both directions and those on the side boundaries were fixed only in the horizontal direction. The simulation was performed in drained condition because the groundwater was not detected during the study. The model parameters related to the compressibility were obtained from the conventional laboratory test that did not consider the time dependent behavior such as creep. The creep model is not within the scope of this study because this paper aims to simulate the wall behavior with the simple and well-known soil models for practical design.

The weathered crust layer was classified as a silty clay. The water content was 12% and the dry unit weight, γ_d was 17 kN/m^3. The apparent cohesion and the friction angle were determined using drained direct shear tests and equal to c' = 20 kPa and φ' = 26 degrees. An elastic, perfectly plastic Mohr-Coulomb model was used to simulate the behavior of the weathered crust layer. The material properties of the weathered crust layer used for the finite element simulations are also shown in Table 1.

The medium to very dense sand layer was classified as clayey sand, according to the USCS. It consisted of 15–18% gravel, 48–60% sand, 8–10% silt and 16–23% clay. The natural water content was 12–20% and the dry unit weight, $\gamma_{d\max}$ was 17–19 kN/m^3. Based on a drained direct shear test, the strength parameters were c' = 0 and φ' = 37 degrees. This is typical of the residual soil in the SUT campus (Horpibulsuk et al., 2008). An elastic, perfectly plastic Mohr-Coulomb model was used to simulate the behavior of this medium to very dense sand. The material properties used for the finite element simulations are shown in Table 1.

The geotextile elements, which cannot resist the bending moment, were employed to model the bearing reinforcement, even though it is composed of longitudinal and transverse members. This modeling converts the contribution of both the friction

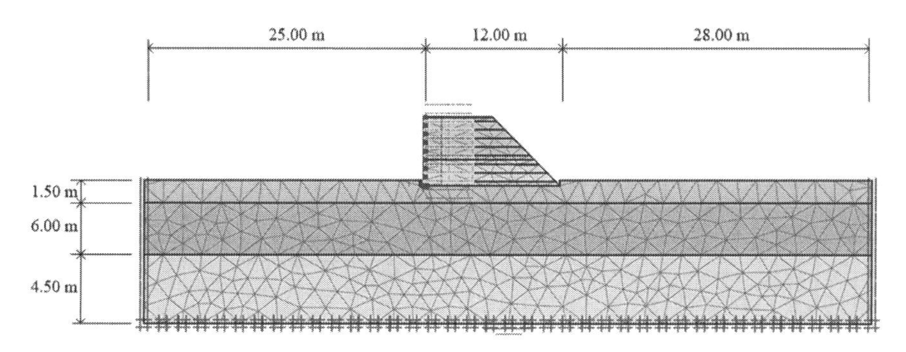

Figure 2. Finite element model of BRE wall.

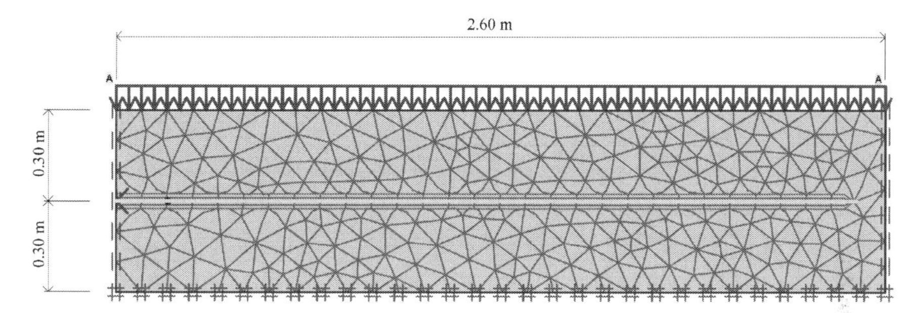

Figure 3. Finite element model for pullout tests.

Table 1. Model parameters for backfill and subsoil.

Item	Backfill soil	Weathered crust	Medium dense sand	Very dense sand
Material model	Mohr-Coulomb	Mohr-Coulomb	Mohr-Coulomb	Mohr-Coulomb
Material type	Drained	Drained	Drained	Drained
γ_{dry}	17 kN/m^3	17 kN/m^3	17.15 kN/m^3	18 kN/m^3
γ_{wet}	18.15 kN/m^3	18 kN/m^3	18.15 kN/m^3	19 kN/m^3
k_x	1 m/day	1 m/day	1 m/day	1 m/day
k_y	1 m/day	1 m/day	1 m/day	1 m/day
E_{ref}	35,000 kN/m^2	1,875 kN/m^2	40,000 kN/m^2	50,000 kN/m^2
ν'	0.33	0.30	0.25	0.25
c'	1	20 kPa	1	1
φ'	40°	26°	35°	38°
Ψ	8°	0°	3°	8°

and bearing resistances to the equivalent friction resistance. The equivalent friction resistance is represented by the interface factor, *R*. The input parameter for this element is an axial stiffness, *AE*, where *A* is the cross-sectional area of longitudinal member and *E* is the modulus of elasticity of the material (steel). The test longitudinal member was 12 mm diameter and 2.6 m length. The axial stiffness of bearing reinforcement used in laboratory model test is shown in Table 2. The width of the transverse member in the laboratory model test was 0.15 m. The soil/bearing reinforcement interface parameter, *R*, was from the back analysis of the laboratory pullout tests by Horpibulsuk and Neramitkornburee (2010) (Figure 3). The elastic perfectly-plastic model was used to simulate the constitutive relation of the interface between soil and bearing reinforcement. There was no evidence of the bending of the transverse members from the retrieved bearing reinforcements, which indicates that the deformation of the transverse members during pullout was in the elastic range with a very

Table 2. Model parameters for bearing reinforcement in laboratory model test.

Type	Modulus of elasticity (GPa)	Axial stiffness, EA (kN/m)
Bearing reinforcement	200	150,796

Table 3. Model parameters for reinforced element structure.

Item	Bearing reinforcement	Facing concrete
Material model	Elastic	Elastic
EA	4.5E+4 kN/m	3.556E+6 kN/m
EI	-	5,808 kNm²/m
W	-	3.36 kN/m/m
v'	-	0.15

small magnitude. It is thus assumed that the transverse members are rigid. Consequently, the pullout displacement and pullout force mobilized insignificantly varies over the length of the reinforcement and the R value is dependent on only the numbers of transverse member, n. As n increases, the R value increases (stiffness increases). The $n = 2$ and 3 were considered to determine the R that are the same as the full-scale BRE wall. The laboratory pullout test was modeled as a plane strain problem. The nodal points at the bottom boundary were fixed in both directions and those on the side boundaries were fixed only in the horizontal direction. The finite element mesh was comprised of 15-nodes triangular elements. The finite element mesh consisted of 558 triangular soil elements not including interface elements. The parameters for bearing reinforcement used in the BRE wall model test are tabulated in Table 3.

The wall face was made of segmental concrete panel, which measured 1.50 x 1.50 x 0.14 m in dimension. The facing panel was modeled as a beam element. The values for the strength parameters and the modulus of elasticity are shown in Table 3. The soil-facing panel interface, R, was taken as 0.9, which is generally used for concrete panels (Bathurst, 1993).

3 FINITE ELEMENT ANALYSES

3.1 *Soil-reinforcement interface coefficient, R*

Figures 4 and 5 show the measured and simulated total pullout force and displacement relationship of the 2.6 m length bearing reinforcements with 2 and 3 transverse members (n = 2 and 3), respectively. The test results within a small displacement of less than 5 mm were used to determine the interface coefficient, R which is consistent with the field wall movement. The small lateral wall movement was observed due to the base restriction effect of the hard stratum (Rowe and Ho, 1997). The interface coefficient, R, was derived by a back analysis varied until the modeled curves coincided with the laboratory curves. The R values of 0.65 and 0.75 provide the best simulation for 2 and 3 transverse members, respectively. These values were used for simulating the field performance of the BRE wall. This method of determining, R is analogous to that suggested by Bergado et al., (2003); and Khedkar and Mandal (2007 and 2009) for hexagonal wire mesh and cellular reinforcement.

3.2 *Bearing stress*

Figure 6 shows the relationship between bearing stress and construction time in both reinforced

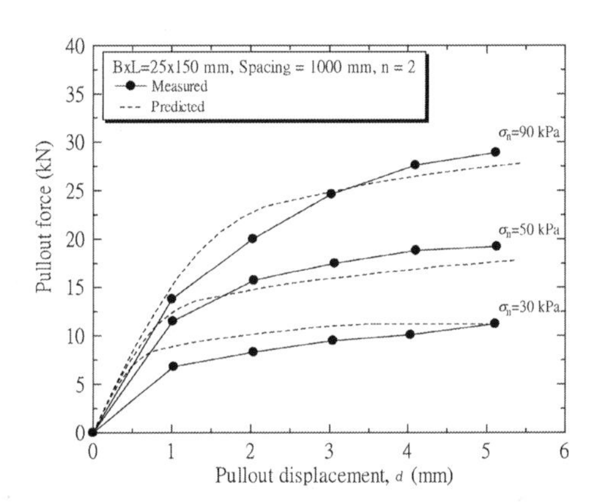

Figure 4. Comparison between the simulated and measured pullout test result of the bearing reinforcement with two transverse members.

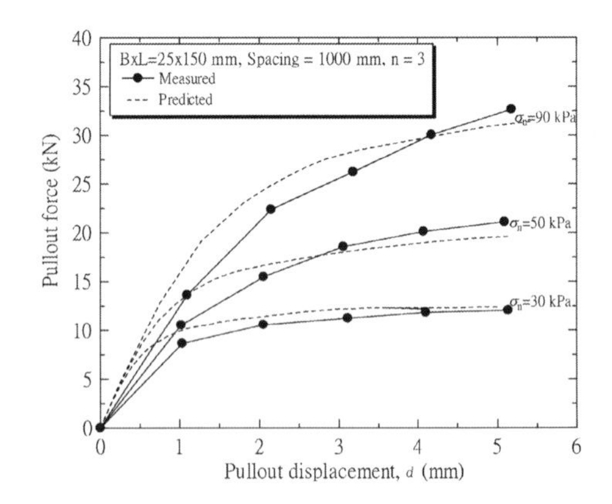

Figure 5. Comparison between the simulated and measured pullout test result of the bearing reinforcement with three transverse members.

(0.5 and 2.4 m from wall facing) and unreinforced (4.5 m from facing) zones. The bearing stresses increased during construction due to the backfill placement. The bearing stress changed insignificantly with time after the completion of construction. The simulated bearing stresses for both front and back are very good in agreement with the measured ones. At 2.4 m from facing, the bearing stresses during 10 days (1st and 2nd loading) of construction are very close to the measured ones but after the 2nd loading, the simulated bearing stress is higher than the measured one, and hence the simulated final bearing stress is lower. The difference between the simulated and measured bearing stress might be due to the non-uniformity of compaction at this particular location; therefore, the earth pressure cell sank into the ground at about 32 kPa vertical pressure (2nd loading). The bearing stress could be again recorded after the 3rd loading that the earth pressure cell located on the hard compacted foundation.

Figure 7 shows the measured and the simulated distribution of bearing stresses at the end of construction from the front to back. The simulated and measured bearing stresses patterns are in good agreement. Within the reinforced zone, the bearing stress distributes approximately in trapezoid shape, which is normally observed for embankments constructed on rigid foundation. The simulated bearing stress in the reinforced zone decreases from the front to back because the BRE wall behaves as a rigid body, retaining the unreinforced backfill. The maximum bearing stress at front is thus due to a eccentric load caused by the lateral thrust from the unreinforced backfill and the vertical load from the weight of segmental panels. The bearing stress insignificantly changes with distance in the unreinforced zone and being equal to that at the end of bearing reinforcement.

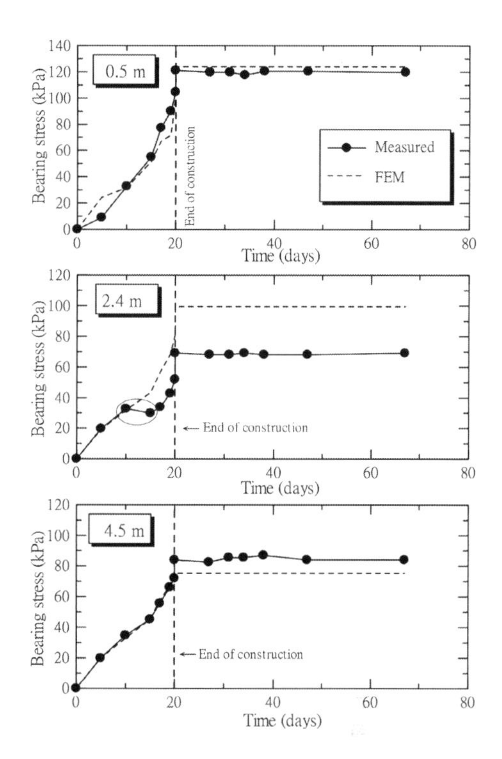

Figure 6. Comparison between the simulated and measured bearing stress change with construction time.

3.3 *Settlement*

The measured and simulated settlements of the BRE wall are illustrated in Figures 8 and 9. The observed data from the four settlement plates at the center of the BRE wall were compared with the simulation. The settlement increased with construction time (Figures 8). Because the wall was founded on the relatively dry and hard stratum, the immediate settlement was dominant (insignificant consolidation settlement). The simulated settlements during construction are very close to the measured ones. The simulated settlements decrease from front that is close to the facing panel (82 mm) to back (77 mm) (*vide* Figure 9). Even though the BRE wall behaves as a rigid block, which causes the large bearing stress at front (due to eccentric load), the settlement is almost uniform due to the contribution of the stiffness of the foundation and the reinforcements. Among the four measuring points, 2 measured data divert from the simulation results: at 0.8 m and 5 m (unreinforced zone) from the facing. The measured settlement at 0.8 m from the facing is slightly higher than the simulated one possibly because the foundation might be disturbed during the foundation excavation for making the leveling pad. The measured settlement in unreinforced zone (5 m from facing) is higher than the simulated one because the stiffness of the foundation in the unreinforced zone is lower than that in the reinforced zone (the foundation in the reinforced zone was compacted before constructing the BRE wall). In this simulation, the same modulus of elasticity, E was applied to both unreinforced and reinforced zones for simplicity. A better simulation could be found if different E is used for the simulation. Overall speaking, the computed settlements in the reinforced zone from the FEM analysis agree reasonably well with the measured ones.

3.4 *Lateral movements*

The simulated and measured lateral movements are compared and shown in Figure 10. The measured lateral movement was the sum of the lateral movements during construction (measured by a theodolite) and after end construction (measured by digital inclinometers). The measured lateral

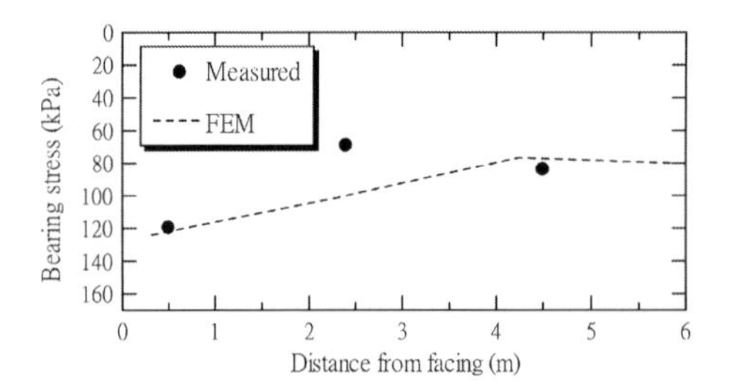

Figure 7. Comparison between the simulated and measured bearing stress distribution.

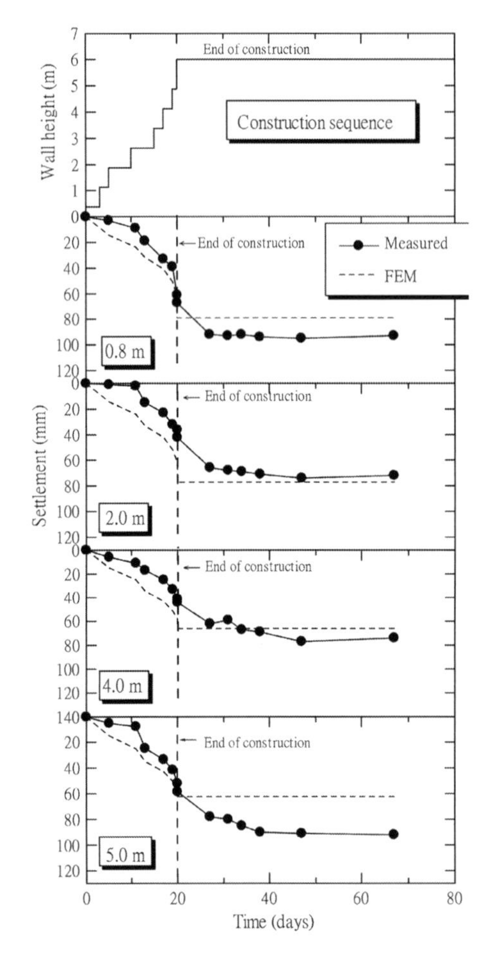

Figure 8. Comparison between the measured and computed settlements change with construction time.

movement is lower than the simulated one because the stiffness of the inclinometer casing prevents the soil lateral movement and the inclinometer casing was installed close to the leveling pad, which obstructs the movement of the inclinometer. However, based on the R values obtained from the back analysis of the laboratory pullout tests,

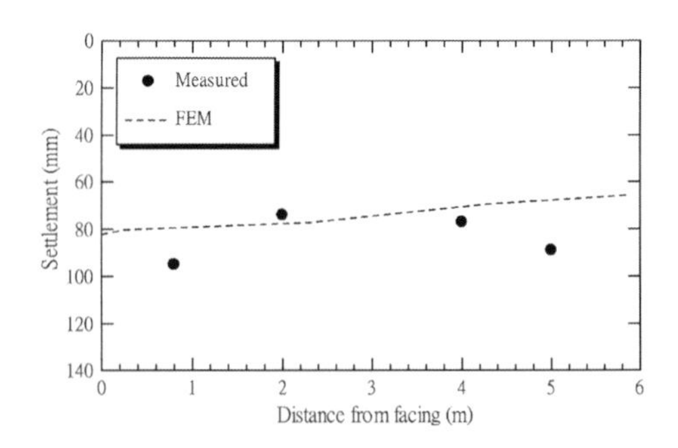

Figure 9. Comparison between the measured and computed settlements.

the patterns of the lateral movement from both the simulation and measurement are almost the same. Lateral movement is caused by the wall settlement and pullout displacement of the reinforcement, which is governed by the R value. The R value also controls the tension in the reinforcement. The lower the R value, the greater the lateral movement and the lower the tension in reinforcement. The R values obtained from the back analysis are considered as suitable for simulating the field performance of the BRE wall because both the simulated lateral movement and the simulated tension in the reinforcement (presented in the following section) are in good agreement with the measured ones.

The simulated maximum lateral movement in the subsoil occurs between 0.5–1.5 m depth below original ground surface corresponding to the weathered crust. The simulated maximum wall movement occurs at about the mid-height with a small magnitude of 23.5 mm. Although the BRE wall rotated about the toe, at a certain stage of deformation process, a crack developed at about the middle of the wall height and the wall started to deform as two rigid panels with a progressive opening of the crack. This finding is in agreement with that by Pinto and Cousens (1996 and 2000).

3.5 *Tensions in the bearing reinforcement*

Figure 11 shows the comparison between the simulated and the measured tension forces in the bearing reinforcements at 14 days after the completion of construction and 10 days after additional surcharge load 20 kPa. The smooth relationships between tension and distance are found for both the measured data and simulation results. The smooth curves (without sharp peaks) from the measured data are because all the strain gauges were attached to the longitudinal members (no strain gauge was on the transverse members) and the stresses in the transverse members might be insignificant or the strains in the steel might be too small.

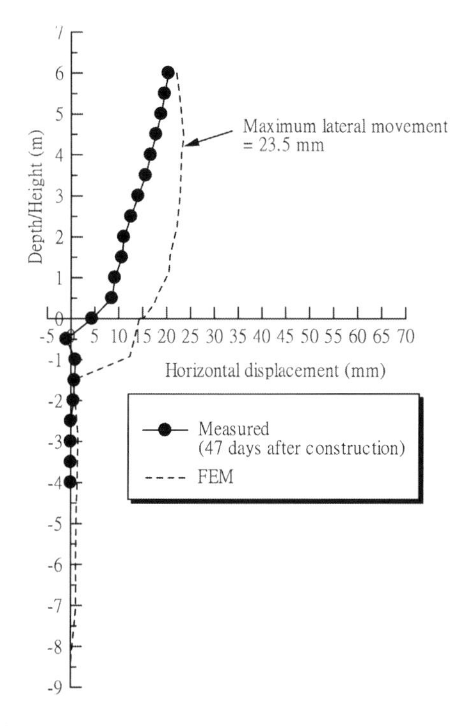

Figure 10. Comparison between the simulated and measured lateral movements.

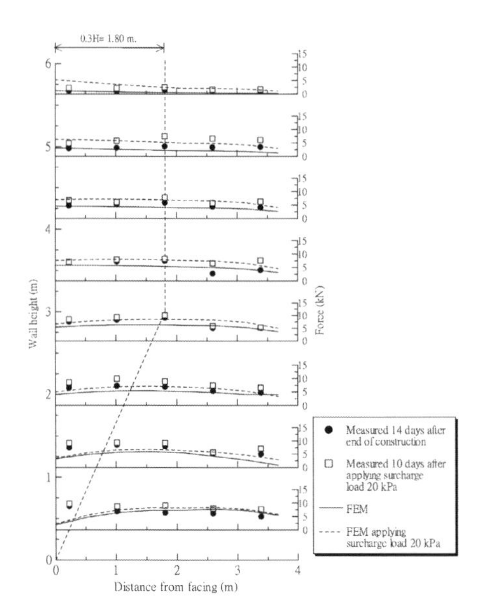

Figure 11. Comparison between the simulated and measured tension forces in the reinforcements.

The possible failure plane recommended by AASHTO (2002) for inextensible reinforcements is also shown in the figure by a dash line. Most of the simulated maximum tension forces lie on the recommended possible failure plane. In practice, the maximum tension (possible failure) plane recommended by AASHTO (2002) can be thus used to examine the internal stability of the BRE wall using the limit equilibrium analysis. This simulated maximum tension pattern is approximate bi-linear and similar to the previous studies for different types of reinforcement (Chai, 1992; Bergado et al., 1995; Alfaro et al., 1997; and Bergado and Teerawattanasuk, 2007). This approximate bi-linear maximum tension plane is caused by the lateral movement of two facing panels at about the mid-height of the wall.

4 CONCLUSIONS

This paper presents a numerical analysis of the bearing reinforcement earth (BRE) wall constructed on the hard stratum by PLAXIS 2D. The goetextile elements, which cannot resist the bending moment, were used to model the bearing reinforcements by converting the contribution of both the friction and bearing resistances to the equivalent friction resistance. This modeling is considered to be applicable and practical for working state (small pullout displacement). The equivalent friction resistance is represented by the interface factor, R, which was determined from the back analysis of the laboratory pullout test. The R values of 0.65 and 0.75 were obtained for the bearing reinforcements with 2 and 3 transverse members, respectively. The BRE wall was modeled under a plane strain condition. Overall, the behavior of the BRE wall is simulated satisfactorily and agreed well with the predictions. The bearing stress distribution is approximately trapezoid shape as generally observed for embankments found on hard stratum. The foundation settlement is almost uniform due to the effect of high stiffness of the foundation and reinforcements. The simulated maximum lateral wall movement occurs at about the mid-height. This results in the approximate bilinear maximum tension (possible failure) plane. This maximum tension (possible failure) plane is very close to that recommended by AASHTO (2002) for inextensible reinforcements. In practice, this recommended maximum tension plane is acceptable to examine the internal stability of the BRE wall. The simulation approach presented was successfully applied to investigate the performance of the BRE wall in Thailand.

ACKNOWLEDGEMENTS

The first author is grateful to the Suranaree University of Technology for financial support, facilities and equipment provide.

REFERENCES

AASHTO, 2002. Standard specifications for highway and bridge, *seventh ed. American Association of State Highway and Transportation Officials*, Washington D.C.

Abdelouhab, A., Dias, D., Freitag, N., 2011. Numerical analysis of the behaviour of mechanically stabilized earth walls reinforced with different types of strips. *Geotextiles and Geomembranes* 29, 116–129.

Alforo, M.C., Hayashi, S., Miura, N., Bergado, D.T., 1997. Deformation of reinforced soil-embankment system on soft clay foundation. *Soils and Foundations* 37 (4), 33–46.

Al Hattamleh, O., Muhunthan, B., 2006. Numerical procedures for deformation calculations in the reinforced soil walls. *Geotextitles and Geomembranes* 24 (1), 52–57.

Bathurst, R.J., 1993. Investigation of footing resistant on stability of large-scale reinforced soil wall tests. *Proceedings of 46th Canadian Geotechnical Conference.*

Bergado, D.T., Chai, J.C., Miura, N., 1995. FE analysis of grid reinforced embankment system on soft Bangkok clay. *Computers and Geotechnics* 17, 447–471.

Bergado, D.T., Chai, J.C., Miura, N., 1996. Prediction of pullout resistance and pullout force-displacement relationship for inextensible grid reinforcements. *Soils and Foundations* 36 (4), 11–22.

Bergado, D.T., Teerawattanasuk, C., 2007. 2D and 3D numerical simulations of reinforced embankments on soft ground. *Geotextiles and Geomembranes* 26 (1), 39–55.

Bergado, D.T., Teerawattanasuk, C., Youwai, S., Voottipruex, P., 2000. FE modeling of hexagonal wire reinforced embankment on soft clay. *Canadian Geotechnical Journal* 37 (6), 1–18.

Bergado, D.T., Youwai, S., Teerawattanasuk, C., Visudmedanukul, P., 2003. The interaction mechanism and behavior of hexagonal wire mesh reinforced embankment with silty sand backfill on soft clay. *Computers and Geotechnics* 30, 517–534.

Chai, J.C., 1992. Interaction between Grid Reinforcement and Cohesive-Frictional Soil and Performance of Reinforced Wall/Embankment on Soft Ground, *D.Eng. Dissertation, Asian Institute of Technology*, Bangkok, Thailand.

Hatami, K., Bathurust, R.J., 2005. Development and verification of numerical model for the analysis of geosynthetic-reinforcement soil segmental wall under working stress condition. *Canadian Geotechnical Journal* 42, 1066–1085.

Hatami, K., Bathurust, R.J., 2006. Numerical model for the analysis of geosynthetic-reinforced soil segmental wall under surcharge loading. *Journal of Geotechnical and Geoenvironmental Engineering* 136 (6), 673–684.

Horpibulsuk, S., Niramitkornburee, A., 2010. Pullout resistance of bearing reinforcement embedded in sand. *Soils and Foundations* 50 (2), 215–226.

Horpibulsuk, S., Kumpala, A., Katkan, W., 2008. A case history on underpinning for a distressed building on hard residual soil underneath non-uniform loose sand. *Soils and Foundations* 48 (2), 267–286.

Horpibulsuk, S., Suksiripattanapong, C., Niramitkornburee, A., 2010. A method of examining internal stability of the bearing reinforcement earth (BRE) wall. *Suranaree Journal of Science and Technology* 17 (1), 1–11.

Horpibulsuk, S., Suksiripattanapong, C., Niramitkornburee, A., Chinkulkijniwat, A., Tangsuttinon, T., 2011. Performance of earth wall stabilized with bearing reinforcements. *Geotextiles and Geomembranes* 29 (5), 514–524.

Hufenus, R., Rueegger, R., Banjac, R., Mayor, P., Springman, S. M., Bronnimann, R., 2006. Full-scale field tests on geosynthetic reinforced unpaved roads on soft subgrade. *Geotextiles and Geomembranes* 24 (1), 21–37.

Khedkar, M.S., Mandal, J.N., 2007. Pullout response study for cellular reinforcement. *Proceedings of Fifth International Symposium on Earth Reinforcement, IS Kyushu '07, November 14–16, 2007,* Fukuoka, Japan, pp. 293–298.

Khedkar, M.S., Mandal, J.N., 2009. Pullout behavior of cellular reinforcements. *Geotextiles and Geomembranes* 27 (4), 262–271.

McGown, A., Andrawes, K.Z., Pradhan, S., Khan, A.J., 1998. Limit state analysis of geosynthetics reinforced soil structures. Keynote lecture. *Proceedings of 6th International Conference on Geosynthetics, March, 25–29, 1998*, Atlanta, GA, USA, pp. 143–179.

Park, T., Tan, S.A., 2005. Enhanced performance of reinforced soil walls by the inclusion of short fiber. *Geotextiles and Geomembranes* 23 (4), 348–361.

Pinto, M.I.M., Cousens, T.W., 1996. Geotextile reinforced brick faced retaining walls. *Geotextiles and Geomembranes* 14 (9), 449–464.

Pinto, M.I.M., Cousens, T.W., 2000. Effect of the foundation quality on a geotextile-reinforced brick-faced soil retaining wall. *Geosynthetics International* 7 (3), 217–242.

Rowe, R.K., Ho, S.K., 1995. Continuous panel reinforced soil walls on rigid foundations. *Journal of Geotechnical and Geoenvironmental Engineering* 123 (10), 912–920.

Skinner, G.D., Rowe, R.K., 2005. Design and behavior of a geosynthetic reinforced retaining wall and bridge abutment on a yielding foundation. *Geotextitles and Geomembranes* 23 (3), 234–260.

Numerical Methods in Geotechnical Engineering IX – Cardoso et al. (Eds)
© 2018 Taylor & Francis Group, London, ISBN 978-1-138-33203-4

Numerical studies on the influence of column reinforcements with soil-binders on railway tracks

A. Paixão, A. Francisco & E. Fortunato
National Laboratory for Civil Engineering (LNEC), Lisbon, Portugal

J.N. Varandas
CEris, ICIST, FCT, Universidade NOVA de Lisboa, Caparica, Portugal

ABSTRACT: The reinforcement of railway track foundations, through injection of binders, without the need of removal of track components (i.e. rails, sleepers and ballast) has many advantages compared to traditional approaches, which normally are more complex, costly and strongly affect train operations. To contribute to a deeper understanding of issues associated with soil-binder column reinforcements, numerical studies were developed to analyse and compare different column layouts and to assess the effect of such reinforcements on the dynamic train-track interaction. To that aim, two three-dimensional numerical programs were used. Aspects of the train-track interaction and values of stresses and displacements on the track and trackbed layers are analysed and discussed. The results suggest that in similar problems of track platforms with uneven track stiffness, it is important to take into account the non-linear elastic behaviour of the ballast to obtain a more accurate three-dimensional load distribution.

1 INTRODUCTION

1.1 *Description of the problem*

Although in railway maintenance works the replacement of the track components, such as the rail, fastenings, sleepers and ballast, can be achieved with automated means with minimal disturbance in train operations, interventions (i.e. reinforcements or renewals) in the track platform are significantly more complex, time-consuming and costly (Selig & Waters, 1994, INNOTRACK, 2008). Although different reinforcement techniques have been used (INNOTRACK, 2008, Indraratna et al., 2011, Li et al., 2016), from soil replacement to reinforcement with geosynthetics, only a few allow interventions without removal of the track elements above the platform, allowing less disruptive process for the railway traffic (Blacklock, 1978, Kouby et al., 2010). In this context, platform reinforcement by injection of binders may be particularly interesting in the rehabilitation of old single freight railway tracks.

1.2 *Proposed alternative approach*

In this paper, the authors study the structural behaviour of the railway track submitted to a reinforcement method that consists of introducing soil-binder columns through injection, by drilling through the ballast layer. This method primarily intends to be an alternative rehabilitation approach of old railway platforms without sub-ballast layer. Some of these structures tend to develop well-known pathologies in the upper part of the platform (Selig & Waters, 1994), thus requiring frequent maintenance works to restore adequate track geometric quality. To assess the viability of this method, numerical studies were developed to evaluate the influence of the introduction of the stiffer zones, caused by the presence of columns, on the train-track system. Initially, a static study was developed using FLAC3D (Itasca, 2015) where displacements and stresses of the railway structure were evaluated. Then, another study was carried using a 3D FEM program to evaluate the dynamic response of the vehicle-track-foundation system. With these two approaches it was possible to obtain more insight into the proposed reinforcement method.

2 STATIC APPROACH USING FLAC3D

In this study, the track modelling in FLAC3D followed the same approach presented in earlier works (Paixão & Fortunato, 2010), where the rails, rail pads, sleepers ballast and subgrade (only soil foundation; no sub-ballast) were modelled using a 8-node hexahedral grid. A specific model was developed for each of the column reinforcement layouts presented in Figure 1. The models were about 4.6 m long (corresponding to 8 sleepers), 6.4 m wide and 3.4 m deep, below the rail, comprising about 16.882 zones and 20.161 grid-points—the actual simulated dimensions were twice the above due to the consideration of a vertical transverse symmetry plane. The soil-binder columns where 1 m long, inserted below the ballast layer, with a simplified geometry as depicted in Figure 2. The track gauge was 1435 mm and comprised 54E1 rails (E = 210 GPa, ν = 0.3), rail pads with a vertical stiffness of 500 kN/m and monoblock sleepers (E = 30 GPa, ν = 0.25), with the dimensions 2.40 $\times$0.26 $\times$0.22 m, spaced 0.6 m. Horizontal displacements were restricted in the vertical boundary planes and vertical displacements restricted in the lower horizontal boundary. A static axle load of 200 kN of a freight train was considered in the analysis, in the positions identified in Figure 1.

Table 1 presents the properties of the geomaterials considered in the numerical models. Because the modelled columns were intended to represent soil-binder inclusions obtained by injection below the ballast layer, they were made of two zones (centre and outer zones) with different mechanical characteristics, associated with different binder contents.

Table 2 presents the calculated peak vertical displacements of the rail under the load and the static track stiffness (as the ratio between the wheel load of 100 kN and the peak vertical displacement of the rail). It was verified that the columns provided a stiffness increment of only about 5 to 11%, with respect to the unreinforced layout (F-O). Moreover,

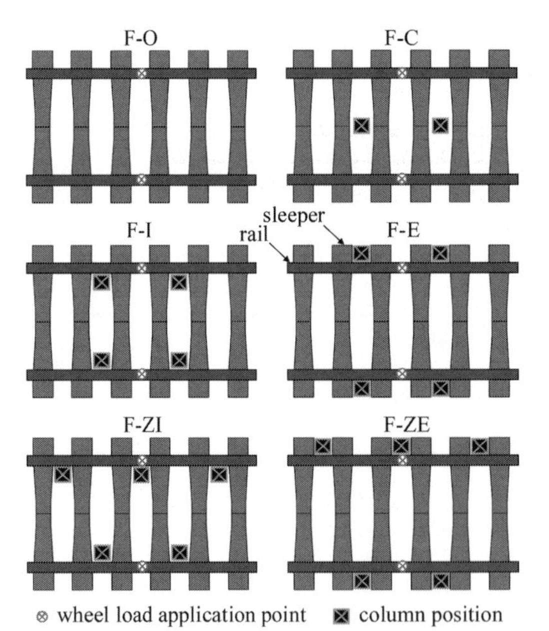

Figure 1. Representation of the tested layouts in FLAC3D.

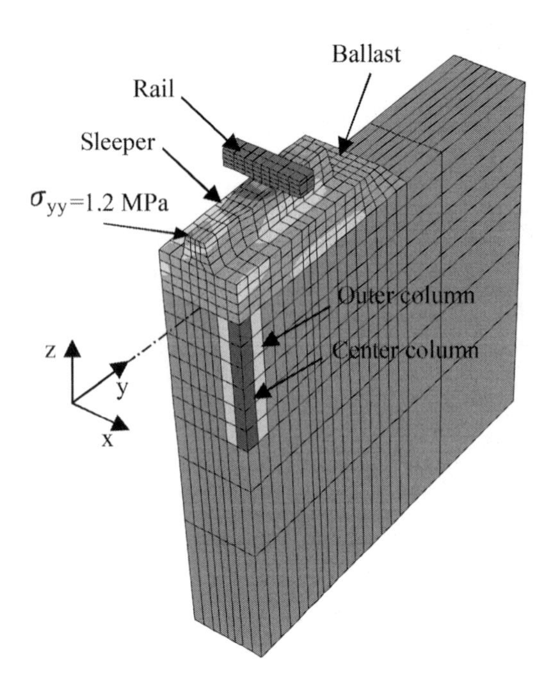

Figure 2. Partial representation of the model F-I, denoting the high transverse stresses in sleeper.

the response of the reinforced structures was somewhat equivalent, regardless of the column layout, although F-I and F-E layouts were more efficient in increasing stiffness than F-ZI and F-ZE. On the other hand, layouts F-ZI and F-ZE may provide a more stable support to the sleepers. Track stiffness unevenness is reported to increase track degradation (Steenbergen, 2013, Grossoni et al., 2015), thus it should be avoided. In the next section, track stiffness unevenness along the line caused by the columns will be addressed in more detail.

It was also found that layout F-C (with columns at the track centre line) caused higher bending moments in the sleepers, which may limit their service life. Peak tensile stresses of 1.2 MPa were obtained in the upper central fibers of the sleepers (Fig. 2), which were at least 70% higher than in other layouts and 3 times greater than in layout F-O. As a results, layout F-C was descarded in subsequent analyses.

3 DYNAMIC TRAIN-TRACK-SUBSTRUCTURE INTERACTION USING PEGASUS 3D FEM

3.1 *General aspects of the 3D FEM program*

Pegasus is a 3D FEM program (Varandas, 2013, Varandas et al., 2016) fully implemented in MATLAB (2013) and was specifically developed

Table 1. Parameters of the geomaterials in FLAC3D models.

Geomaterial	Young Modulus, E_i MPa	Poison's ratio, v (-)	Dimensions m
Ballast	160	0.20	0.3 (thickness)
Foundation	40	0.45	2.8 (height)
Column			
centre	1000	0.25	$0.2 \times 0.2 \times 1.0$
outer	400	0.25	$0.4 \times 0.4 \times 1.0$

Table 2. Peak rail vertical displacements and static stiffness.

	Displacement	Static stiffness
Column layout	mm	kN/mm
F-O	1.24	80.3
F-C	1.18	84.5
F-I	1.12	89.5
F-E	1.12	89.4
F-ZI	1.15	86.6
F-ZE	1.16	86.5

to analyse the dynamic response of the railway tracks in the time domain. Recently, its calculation efficiency has been further improved with a mixed implicit-explicit time integration scheme (Varandas et al., 2017), combining the well-known Newmark implicit scheme and the Zhai (1996) explicit scheme. In Pegasus, the vehicle, the track superstructure and the ballast-subgrade layers form three separate structural systems: the vehicle interacts with the track superstructure by wheel-rail contact forces and the track superstructure with the ballast-subgrade by sleeper-ballast contact forces. The mathematical formulation is represented by the following system of equations:

$$\begin{cases} \boldsymbol{K}_v\boldsymbol{u}_v + \boldsymbol{C}_v\boldsymbol{v}_v + \boldsymbol{M}_v\boldsymbol{a}_v = \boldsymbol{f}_{g,v} + \boldsymbol{f}_{a,w} \\ \boldsymbol{K}_t\boldsymbol{u}_t + \boldsymbol{C}_t\boldsymbol{v}_t + \boldsymbol{M}_t\boldsymbol{a}_t = \boldsymbol{f}_{g,t} - \boldsymbol{f}_{a,w} + \boldsymbol{f}_{a,b} \\ \boldsymbol{K}_s\boldsymbol{u}_s + \boldsymbol{C}_s\boldsymbol{v}_s + \boldsymbol{M}_s\boldsymbol{a}_s = \boldsymbol{f}_{g,s} - \boldsymbol{f}_{a,b} \end{cases} \tag{1}$$

where ***K***, ***C*** and ***M*** are the global stiffness, damping and mass matrices; subscripts v, t and s refer to the vehicle, track and ballast-subgrade systems; ***u***, ***v*** and ***a*** are nodal displacement, velocity and acceleration vectors; and $\boldsymbol{f}_g$, $\boldsymbol{f}_{a,w}$ and $\boldsymbol{f}_{a,b}$ are the gravity load, wheel-rail and sleeper-ballast interaction force vectors, respectively. The damping matrices of the track and ballast-soil system, where determined following the Rayleigh damping approach.

Different vehicle systems can be considered, consisting in an assemblage of rigid bodies with concentred masses connected by dampers and springs. The track superstructure comprises Euler-Bernoulli beam elements representing the rails and sleepers, connected by 3D spring-damper elements, representing the rail pads. The ballast-substructure system is discretized with low-order eight-node solid hexahedral elements. Non-linear elastic constitutive models can be considered to represent the behaviour of the geomaterials. In these studies, an extension of the k - θ model (Brown & Pell, 1967) was considered for the ballast layer. The k - θ model establishes that the resilient modulus, E_r, increases with the first stress invariant (sum of principal stresses), θ. The extension of this model is described in detail in (Varandas, 2013): under tension, a constant value for the resilient modulus, E_{min}, is considered; under compression, the resilient modulus is given by the two expressions presented in Figure 3, whether θ is greater or lower than θ_t.

To achieve a more realistic cylindrical shape of the columns, compared with the previous prismatic configuration used with FLAC3D (Fig. 4), a new function was introduced in Pegasus. The generation of the columns' mesh was parametrised to apply an automatic transformation to the elements' nodes coordinates. The diameter of the centre and outer parts of the columns were 0.2 and 0.4 m, respectively.

3.2 *Description of the models*

Figure 5 depicts the column layouts studied with Pegasus 3D FEM. It is noted that a new layout was introduced, P-IE, to provide a more uniform track support. The superstructure of the models was equivalent to the FLAC3D models; however, it was necessary to significantly increase the length of the models to $50.1 \times 8.4 \times 3.8$ m (corresponding to 87 sleepers) to allow for the train pass by. At the lateral vertical boundaries of the model, local transmitting boundaries, consisting of visco-elastic dampers were placed to absorb impinging

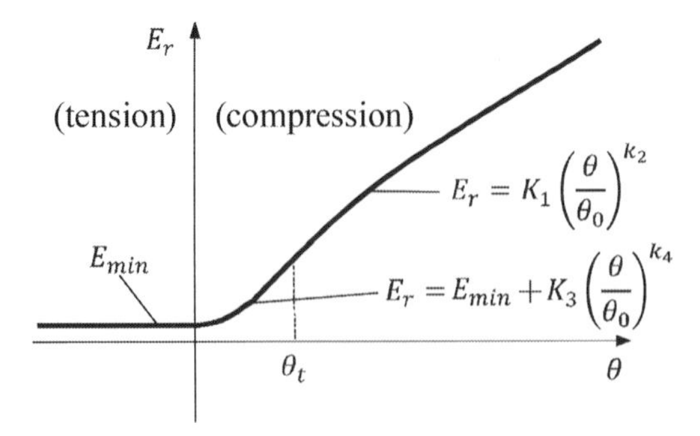

Figure 3. The E_r - θ relationship in the extended k - θ model.

waves (Lysmer & Kuhlemeyer, 1969, Kouroussis et al., 2011).

Apart from the ballast layer, the parameters of the geomaterials in Pegasus were the same as those presented in section 2. The k - θ model was assigned to the ballast, considering E_{min} = 16 MPa and θ_t = 8 kPa (Fig. 3), assuming K_1 = 110 MPa and k_2 = 0.6 (Aursudkij et al., 2009). A Rayleigh damping ratio of 3% was applied for frequencies of 2 and 100 Hz.

To simulate the effect of a 200 kN axle load freight train passing by at 80 km/h, it was considered a 8-dof vehicle comprising a bogie and half car body (Fig. 6) with the parameters presented in Table 3.

In total, the models comprised 249.560 elements and 731.856 degrees-of-freedom. A time step of 6.54×10^{-6} s was considered in the analyses.

3.3 *Dynamic train-track interaction*

Figure 7 shows the calculated wheel-rail interaction forces after applying a 80 Hz low-pass filter. The solid and the dotted lines represent the left and the right wheels, respectively. It is visible the typical excitation of the sleeper spacing causing a variation of less than ± 2 kN about the static wheel load of 100 kN. It is also visible that the effect of the columns on the train-track interaction is practically negligible because the wheel-rail interaction forces are approximately the same as in model P-O.

For a better contrast between the results, Figure 8 shows the differences in vertical displacements of the wheels, $\Delta u_{z,w}$, with reference to model P-O. It is noted that the vertical displacements of both wheels were generally overlapped; P-ZI and P-ZE models were exceptions because their column layouts are not symmetrical (Fig. 5). Nevertheless, in these two cases, the differential

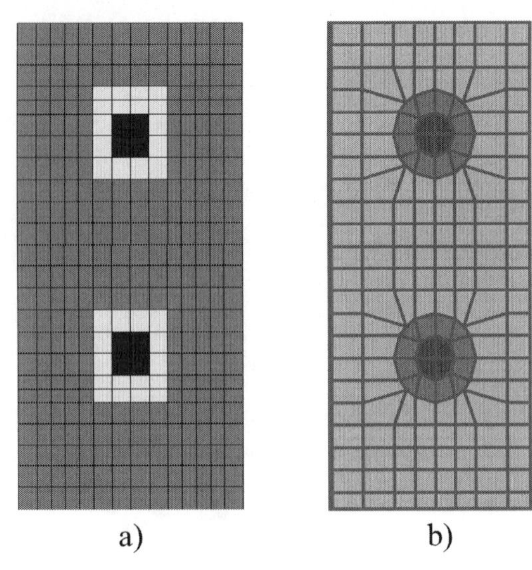

Figure 4. Cross-section of the columns' mesh in FLAC3D (a) and in Pegasus (b) models.

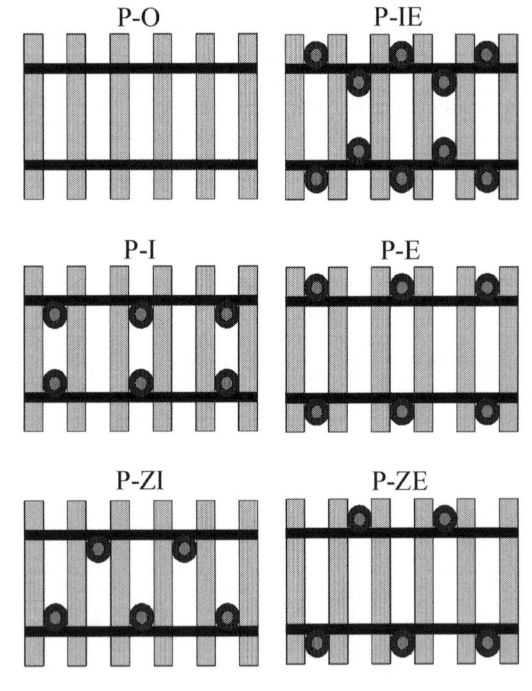

Figure 5. Representation of the tested layouts in Pegasus.

displacement between the two wheels was quite low: less than 0.02 mm. It is also verified that all reinforcement layouts reduced rail displacements ($\Delta u_{z,w} > 0$) (assuming that wheel and rail practically

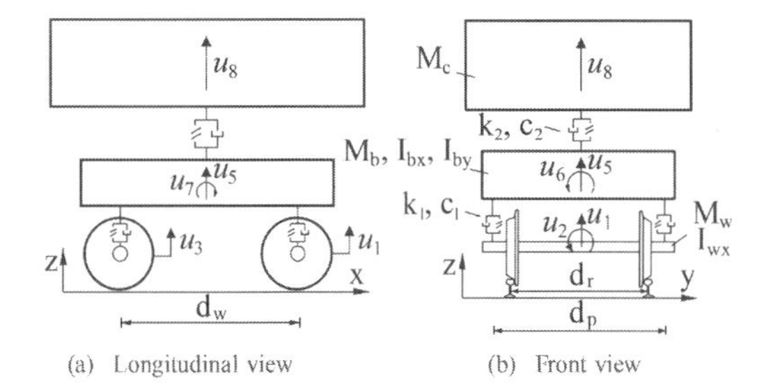

Figure 6. Freight vehicle model (see parameters in Table 3).

Table 3. Characteristics of the freight vehicle model.

Parameter	Units	Value
Car body mass, M_c	t	45.19
Secondary suspension stiffness, k_2	kN/m	6000
Secondary suspension damping, c_2	kN.s/m	120
Bogie mass (without axles), M_b	t	7.91
Inertial bogie (x) direction, I_{bx}	t.m^2	3.39
Inertial bogie (y) direction, I_{by}	t.m^2	4.2
Primary suspension stiffness, k_1	kN/m	1340
Primary suspension damping, c_2	kN.s/m	20
Wheelset mass, M_w	t	5.14
Inertial axle (x) direction, I_{wx}	t.m^2	1.45

have identical displacement). This agrees with the results in section 2 (Table 2); however, Pegasus models were somewhat more deformable, probably because of their greater depth.

The dynamic track stiffness presented in Figure 9 was calculated as the ratio between the wheel-rail interaction force and the wheel displacement along the track. Although there is some variability of this parameter along the track (again, because of to the sleeper spacing—the rail is not continuously supported), there seems to be three levels of track stiffness, depending on the number of columns per sleepers (c/s): i) the lowest level, around 65 kN/mm for P-O, corresponds to c/s = 0; ii) the middle level, around 70 kN/mm for all remaining models, corresponds to c/s = 1; iii) the highest level, about 74 kN/mm for model P-IE, with c/s = 2.

Another relevant aspect to analyse is the effect of the reinforcements in the sleeper-ballast contact forces, because higher sleeper-ballast contact forces increase ballast degradation (Selig & Waters, 1994, Nurmikolu & Kolisoja, 2011). The results presented in Figure 10 regard the differences in peak sleeper-ballast contact forces, again with reference to model P-O (negative values indicate downward increments in sleeper-ballast contact forces). To consider the fact that P-ZI and P-ZE are not symmetrical along the track, the results in Figure 10 are relative only to half sleeper so that results can be comparable between models. The results also do not include the self-weight of the track elements. Regarding these interaction forces, all reinforcement layouts yielded higher contact forces under the sleepers because the track became stiffer (Fig. 9) and more load was being transferred directly to the sleepers under the wheels. Nevertheless, this increase of load was only about 8% for model P-IE and less than 5% for the remaining layouts. Three loading levels were also identified, which further emphasize the influence of the ratio c/s. The lower scattering in the peak contact forces in model P-IE is probably because of to the more uniform support conditions it provides to the ballast layer, as was intended.

3.4 *Response of the ballast-substructure system*

Peak vertical stresses in the ballast layer, $\sigma_{z,peak}$, caused only by the passing vehicle are presented in Figure 11, depicting a horizontal section through the layer, about 0.2 m below the sleepers (z = 0.11 m). To help in the interpretation of the figure, the black lines indicate the position of the sleepers and the central zone of the columns. The colour maps clearly evidence the change in the load distribution due to the concentration of stresses in the areas aligned with the columns. Among the reinforcement layouts, P-IE appears to yield lower peak stresses and a more uniform load distribution

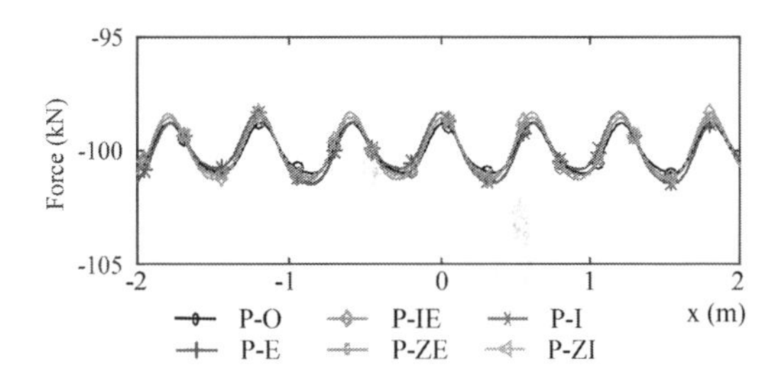

Figure 7. Wheel-rail interaction forces.

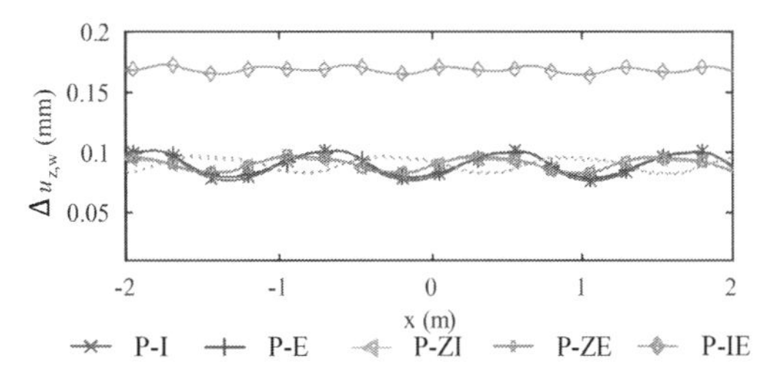

Figure 8. Differences in wheel vertical displacements with regard to model P-O (positive values indicate upward variations).

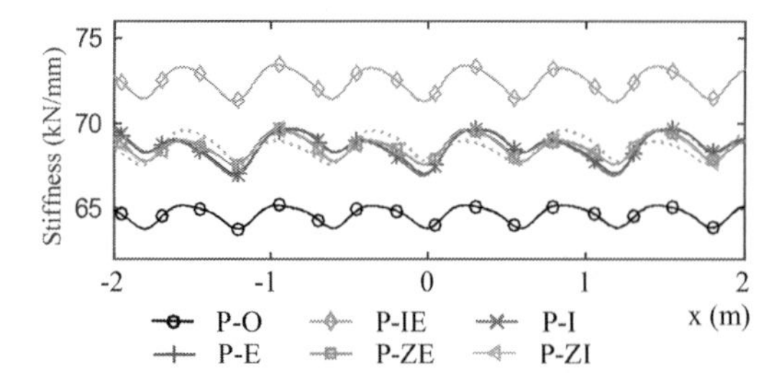

Figure 9. Comparison of the dynamic track stiffness.

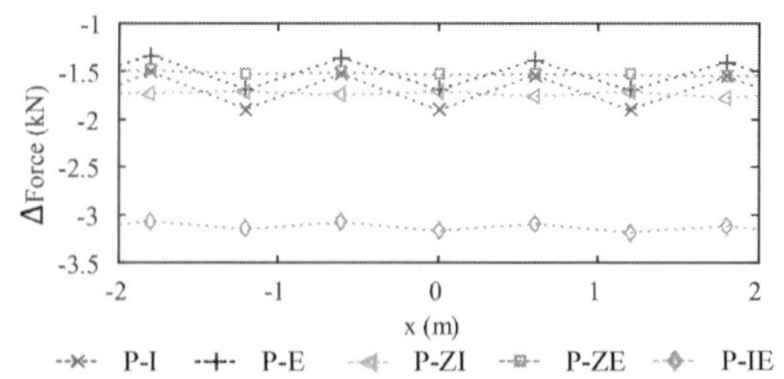

Figure 10. Differences in peak sleeper-ballast contact forces with reference to model P-O (negative values indicate downward contact force increments).

under the rail alignment, which is in agreement with what was mentioned regarding the sleeper-ballast contact forces. Due to the consideration of the k - θ model, this higher variability of stress levels in the ballast layer, compared to model P-O, both in the longitudinal and transverse directions, creates important variations in the peak resilient modulus of the ballast material, as visible in the colour maps in the right column of Figure 11.

The k - θ model evidences a more realistic and direct load transfer from the sleepers to the columns, which also influences the load distribution from the wheels to the sleepers. Without the consideration of the k - θ model, the stress levels in the ballast and in the subgrade could have been underestimated. To illustrate the load transfer mechanism from the sleepers to the columns, Figure 12 shows colour maps of the peak vertical stresses in the ballast-substructure system, as well as the peak ballast resilient modulus after the vehicle has passed by, in vertical sections of the track structures, aligned with the rail, at y = 0.75 m.

Because a more direct load transfer between the sleepers and the deep layers of the platform is achieved with the column reinforcements, the untreated soils between and around the columns undergo slightly lower stress levels. This behaviour is clearly visible in layout P-IE in Figure 12a, denoted by the dark red areas between columns. This stress reduction is mostly important at the top of the platform, where stress levels are usually more severe, in that unwanted long-term behaviours, such as progressive shear failure and accumulated plastic deformation, may be mitigated (Selig & Waters, 1994). To illustrate this effect, Figure 13 shows peak deviatoric stresses resulting from the train passage, in a horizontal section through the top of the platform (z = –0.05 m). The areas that turned from light blue, in layout P-O, to blue and dark blue, in the reinforced layouts, correspond to locations where there has been a beneficial reduction in the deviatoric stress. On the other hand, strong concentrations of stresses, denoted by the dark red colours, were formed at the treated soil-binder columns, as would be expected.

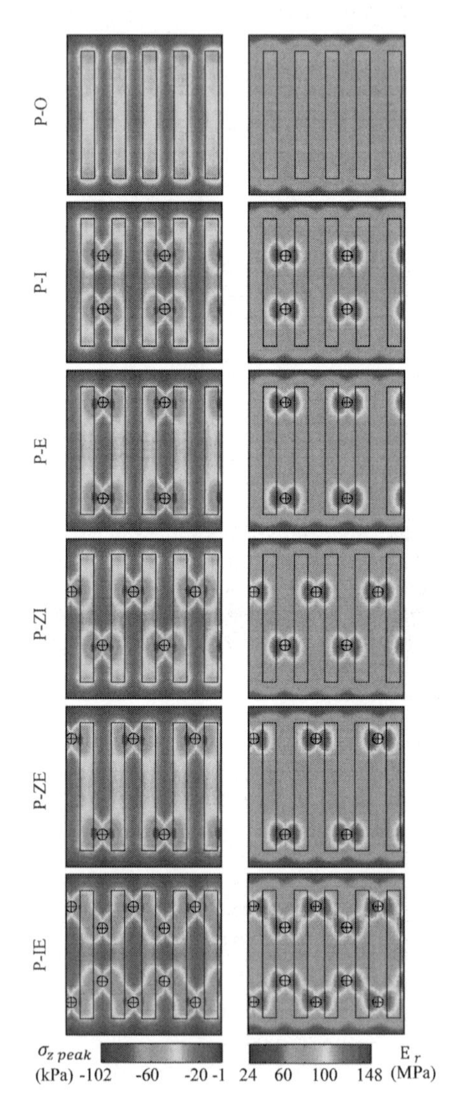

Figure 11. Peak vertical stresses (left) and resilient modulus (right) in the ballast layer at z = 0.11 m.

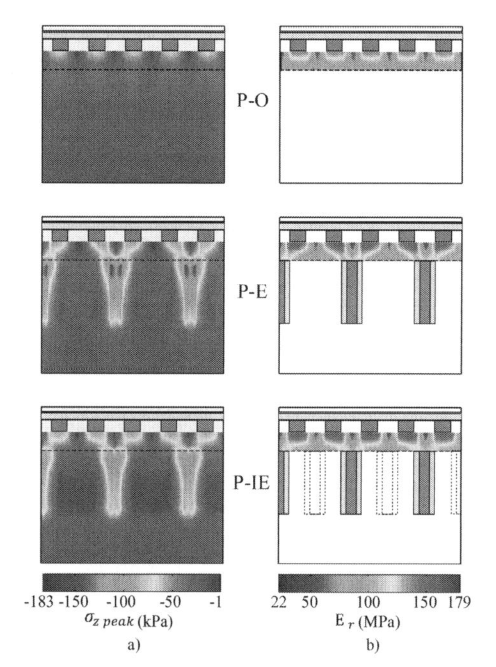

Figure 12. Peak vertical stresses in the ballast-substructure system (a) and $E_{r,peak}$ values in the ballast layer (b) at y = 0.75 m.

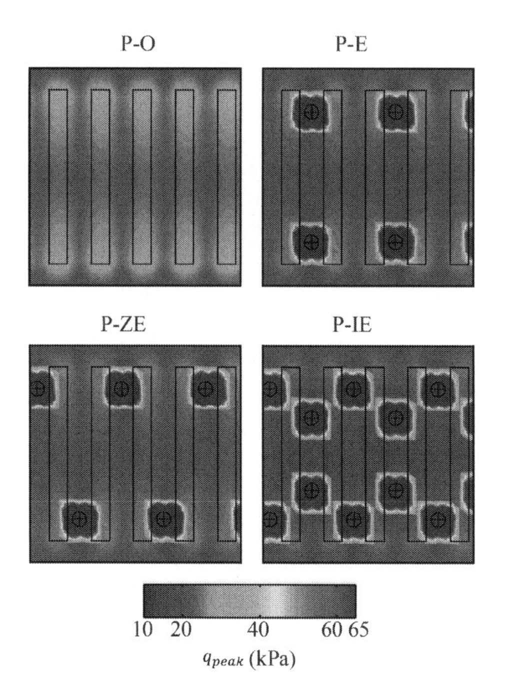

Figure 13. Peak deviatoric stress at the top of the platform (z = -0.05 m).

4 FINAL REMARKS

A study on the influence of soil-binder column reinforcements in the track platform was presented. Firstly, FLAC3D was used to carry out a parametric study on the response of the track under static loading, considering different layouts of the columns under the ballast layer. The results of this study showed that all tested reinforcement layouts yielded more or less comparable results. Moreover, the reinforcement layout with columns aligned with the track centre line should be avoided because it could reduce the service life of sleepers due to higher tension stresses on the concrete monoblock sleepers.

Because the columns were significantly stiffer elements introduced in the subgrade, it was decided to study in more detail the effect of the reinforcements on the train-track interaction resulting from passing freight trains at 80 km/h. To carry out such dynamic analyses, Pegasus 3D FEM program was used, that also allowed to consider the ballast layer as a non-linear elastic material. The results showed that the dynamic track stiffness and the sleeper-ballast contact forces increased with the number of columns per sleeper, as was expected, but the reinforcement layouts had minimal effect on the wheel-rail interaction forces. In general, the reinforced structures caused higher stress concentration in the track platform at the column alignments, but slightly lower stress levels around and between the columns. The results suggested that the consideration of the non-linear elastic behaviour of the ballast layer in similar scenarios of uneven track stiffness along the track is rather important to accurately assess the load distribution between sleepers and the load propagation with depth.

Further studies should be carried out to assess the effect of higher train speeds on the train-track interaction and evaluate its consequences in the long-term behaviour. Column-soil interfaces may also be considered to allow for friction and slip.

ACKNOWLEDGMENTS

This work was carried out under the R&D project *GroutRail*, in collaboration with the company *Mota-Engil Construções, S.A*, and was co-funded by the European Regional Development Fund, through the POCI, in the scope of Portugal2020 and Lisb@2020 programs [POCI-01–0247-FEDER-017978]. Part of the work was conducted in the framework of the TC202 national committee of the Portuguese Geotechnical Society (SPG) "Transportation Geotechnics", in association with the International Society for Soil Mechanics and Geotechnical Engineering (ISSMGE-TC202).

The first author's postdoctoral fellowship [SFRH/BPD/107737/2015] was financially supported by *FCT. Fundação para a Ciência e a Tecnologia*, through POCH, being co-funded by ESF and the national funds of MCTES, Portugal.

REFERENCES

Aursudkij, B., McDowell, G.R. & Collop, A.C. 2009. Cyclic loading of railway ballast under triaxial conditions and in a railway test facility. *Granular Matter,* 11, 391–401.

Blacklock, J.R. 1978. Evaluation of Railroad Lime Slurry Stabilization—Final Report. Washington, D.C.: U.S. Department of Transportation, Federal Railroad Administration.

Brown, S. & Pell, P. 1967. An experimental investigation of the stresses, strains and deflections in layered pavement structure subjected to dynamic loads. *2nd Int. Conf. on Structural Design of Asphalt Pavements.* Michigan, Ann Arbor.

Grossoni, I., Andrade, A.R., Bezin, Y. & University of, H. 2015. Assessing the Role of Longitudinal Variability of Vertical Track Stiffness in the Long-Term Deterioration. *IAVSD 24th International Symposium on Dynamics of Vehicles on Roads and Tracks.* Graz, Austria.

Indraratna, B., Ngo, N.T. & Rujikiatkamjorn, C. 2011. Behavior of geogrid-reinforced ballast under various levels of fouling. *Geotextiles and Geomembranes,* 29, 313.

INNOTRACK 2008. State of the art report on soil improvement methods and experience; INNOTRACK—Innovative Track Systems. Paris: INNOTRACK Consortium.

Itasca 2015. FLAC3D—Fast Lagrangian Analysis of Continua in 3 Dimensions: User Manual. Minneapolis, MN: Itasca Consulting Group.

Kouby, A.L., Bourgeois, E. & Rocher-Lacoste, F. 2010. Subgrade Improvement Method for Existing Railway Lines—an Experimental and Numerical Study. *Electronic Journal of Geotechnical Engineering,* 15, 461–493.

Kouroussis, G., Verlinden, O. & Conti, C. 2011. Finite-Dynamic Model for Infinite Media: Corrected Solution of Viscous Boundary Efficiency. *Journal of Engineering Mechanics,* 137, 509–511.

Li, D., Hyslip, J., Sussmann, T. & Chrismer, S. 2016. *Railway Geotechnics,* Boca Raton, FL, CRC Press.

Lysmer, J. & Kuhlemeyer, R.L. 1969. Finite dynamic model for infinite media. *Journal of the Engineering Mechanics Division,* 95, 859–877.

MATLAB 2013. *MATLAB release R2013b documentation,* Natick, MA, USA, The MathWorks, Inc.

Nurmikolu, A. & Kolisoja, P. 2011. Mechanism and effects of railway ballast degradation. *GEORAIL 2011, International Symposium Railway geotechnical engineering.* Paris: IFSTTAR.

Paixão, A. & Fortunato, E. 2010. Rail track structural analysis using three-dimensional numerical models. *In:* Benz, T. & Nordal, S. (eds.) *7th European Conference on Numerical Methods in Geotechnical Engineering (NUMGE2010).* Trondheim: Taylor & Francis Group.

Selig, E.T. & Waters, J.M. 1994. *Track geotechnology and substructure management,* London, Thomas Telford.

Steenbergen, M.J.M.M. 2013. Physics of railroad degradation: The role of a varying dynamic stiffness and transition radiation processes. *Computers & Structures,* 124, 102–111.

Varandas, J.N. 2013. *Long-term behaviour of railway transitions under dynamic loading.* Ph.D., Faculdade de Ciências e Tecnologia da Universidade Nova de Lisboa.

Varandas, J.N., Hölscher, P. & Silva, M.A.G. 2016. Three-dimensional track-ballast interaction model for the study of a culvert transition. *Soil Dynamics and Earthquake Engineering,* 89, 116–127.

Varandas, J.N., Paixão, A. & Fortunato, E. 2017. A study on the dynamic train-track interaction over cut-fill transitions on buried culverts. *Computers & Structures,* 189, 49–61.

Zhai, W.M. 1996. Two simple fast integration methods for large-scale dynamic problems in engineering. *International Journal for Numerical Methods in Engineering,* 39, 4199–4214.

Numerical Methods in Geotechnical Engineering IX – Cardoso et al. (Eds)
© 2018 Taylor & Francis Group, London, ISBN 978-1-138-33203-4

Finite element analysis of soil-structure interaction in soil anchor pull-out tests

H.J. Seo
Xi'an Jiaotong—Liverpool University, Shaanxi Sheng, China

L. Pelecanos
University of Bath, UK

ABSTRACT: This study presents new data from soil anchor field pull-out tests that were carried out on in-situ ground anchor systems, using strain gauges to evaluate the changes in the variations of axial load and skin friction along the nail during the tests. The results of these field tests provide details about the development of skin friction with induced displacements, thus offering the opportunity to perform load-transfer analyses of the soil anchor. A Finite Element (FE) model based on the load-transfer approach is set up to analyse this problem. The nonlinear load-transfer (t-z) relation requires 3 distinct parameters to be defined, which are related to the initial soil stiffness, stiffness degradation and ultimate strength. These parameters are defined from the field test data. Subsequent FE analyses using these parameters are run to validate the new load-transfer models which are compared with the field test results and exhibit an excellent agreement.

1 INTRODUCTION

Soil anchors are widely used to provide support to earth retaining structures and also assist in the control of displacements of retaining walls. They provide additional resistance to structural movements by mobilising the available pre-stress. The pre-stress is provided from enhancing the shear resistance of soil and the skin friction at the interface between the grout material and the surrounding soil. A critical component of the capacity of soil anchors is the interface shear strength between the grout and the surrounding soil (Tagaya et al., 1988; Powell & Watkins, 1990, Kumar & Kouzel, 2008). Several field tests and laboratory experiments can be carried out to derive the shear resistance between the grout and the soil. A widely used field technique is the pull-out test (Merrifield & Sloan, 2006; White et al., 2008; Seo et al., 2011a, 2017).

The interface resistance of a soil-nail and the soil mass is very complicated due to the combination of shear banding development at the nail-soil interface, hoop stresses around the circumference of the nail and confining pressure development with the depth (Seo et al., 2012, 2014a, b). In cases of a hybrid soil-nail and compression anchor system, the behavior mechanisms are even more complicated (Seo et al., 2010, 2011b) as the response is dictated by both a nail compression and an anchor tension. Extensive experimental and numerical research (Seo et al., 2016) is required to achieve an detailed understanding of the problem.

The study presents an approach to deriving load-transfer curves for soil anchors based on monitored anchor pull-out field tests. A similar technique was followed earlier in distributed fibre optic sensing (Soga et al., 2015, 2017; Kechavarzi et al., 2016, 2019; Di Murro et al., 2016, 2019; Acikgoz et al., 2016, 2017) and was applied to monitoring axially-loaded piles (Ouyang et al., 15; Pelecanos et al., 2016, 2017a,b, 2018) and heat-exchanger foundation piles (Ouyang et al., 2018a,b,c).

This paper described the proposed approach and then shows the field data from a field test with the relevant developed load-transfer curves. These curves are then used as input to a finite-element analysis of the same field test and provide a verification which proves that the load-transfer method is an appropriate technique for modelling the deformation response of soil anchors.

2 FIELD TESTS

2.1 *Definition of the problem*

Field pull-out tests of soil anchors are carries out to evaluate their bearing capacity and acquire some relevant geotechnical design parameters. Specifically, for performance-based design, where the pre-yield behaviour (e.g. magnitude of expected displacements and deformations) is also of interest, apart from the ultimate limit state, such field pull-out tests can define the load-transfer response of soil-anchor systems.

The problem is defined schematically in Figure 1: (a) an anchor pull-out test can be carried out in the field by applying tensile stresses on the strands within the anchor, which are attached at the bottom of the anchor. This effectively results in a force load acting at the bottom of the anchor with an upward direction; (b) in order to model this response numerically, a potential simple method would be to use a finite-element beam-spring (load-transfer) model that uses nonlinear load-transfer (t-z) curves to describe the soil springs. Therefore, the problem subsequently becomes how to define appropriate parameters for the nonlinear soil springs.

2.2 *Pull-out tests*

Such a pull-out field test of an anchor was carried out in Seoul, Korea. The examined PC strands were made of steel with a diameter of 12.7 mm and 3 m long. The ground anchor consisted of two PC strands within plastic pipes, and an anchor body. The maximum yield load of two PC strands was 318 kN. The PC strands were located in a plastic pipe which was filled with oil so that no significant friction was developed between two PC strands and the plastic pipes when the PC strands were pulled. Consequently, the applied pull-out load was transferred directly to the anchor body, and therefore the anchor body could apply a stress on the grout material. To measure this load development within the grout, vibrating wire strain gauges

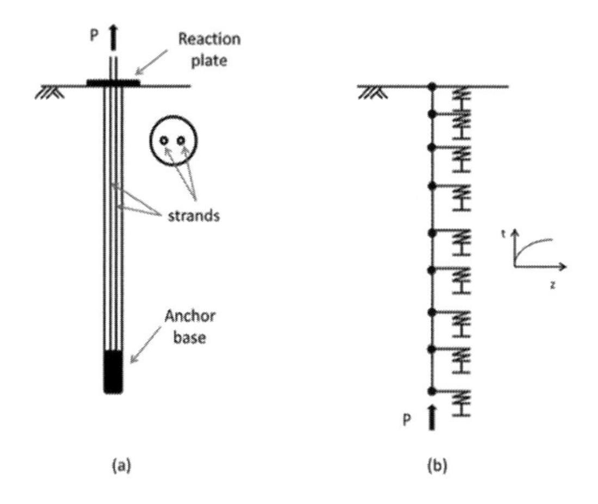

Figure 1. Problem definition: (a) an anchor pull-out field test and (b) a relevant finite element beam-spring model.

(VWSGs) (SG4150 manufactured by GTC cooperation) were deployed using steel wires to fix on an auxiliary steel bar. A total of seven VWSGs were installed along the bar with some discrete intervals.

The anchor pull-out load was increased progressively and the developed strain values from the VWSGs at 4 load steps (40, 70, 100, 130 kN) were recorded, as shown in Figure 2. Figure 2 (a) shows the soil anchor top lop-displacement curve, (b) the profiles of axial force, whereas (c) the profiles of anchor skin friction. Axial force was calculated by multiplying the monitored axial strains by the axial rigidity, EA, (E is the anchor's Young's modulus and A is anchor cross-sectional area). The values of skin friction were calculated by differentiating the values of calculated axial strain with the depth.

It may be observed that the anchor reaches an ultimate load value at about 130 kN. Also, the values of axial force and skin friction increase gradually with the applied load. As anticipated, larger values of both axial force and skin friction occur close to the bottom of the anchor, which is nearer to the point of action of the load.

2.3 *Developed load-transfer curves*

Relevant analysis of the monitored field data offers associated values for the skin friction and the vertical displacements of the soil anchor. Skin friction values are obtained by direct numerical differentiation (such as central difference method) of the axial force values with the depth. Vertical displacement values can be obtained by numerical integration (such as trapezium rule etc.) of the axial strain values, with the depth. Integration of strains offers values of relative displacement. Thus, total displacement can be obtained by adding the integrated values to absolute values from readings at a fixed point. Here, the fixed point was located at the top of the anchor, of which the response is shown in the top load-displacement curve. The development of skin friction with the vertical displacement at different depths along the soil anchor provides the load-displacement (t-z) curves at those depths, which are shown in Figure 3.

3 FINITE ELEMENT ANALYSIS

3.1 *Numerical model*

The field data from the monitored field tests presented before were processed to derive a numerical model of the field test. The computational model is based on a finite-element (FE) beam-spring formulation in which the anchor is simulated with two-noded beam elements with axial displacement degrees-of-freedom, whereas the soil is modelled

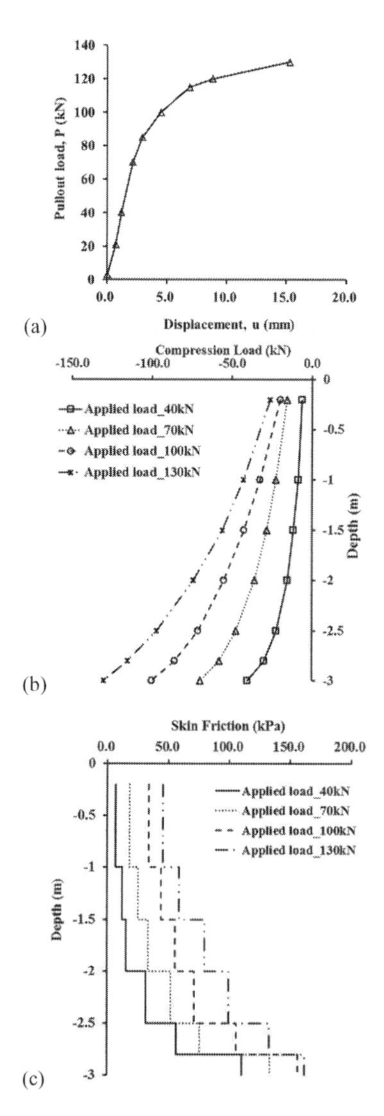

Figure 2. Field test results: (a) anchor top load-displacement curve, (b) profiles of axial anchor force and (c) profiles of anchor skin friction.

with nonlinear springs, again with one degree-of-freedom. The governing FE equation (described by Equation 1) is global equilibrium and the global stiffness matrix comprises of the anchor stiffness matrix, $[K_a]$, and the soil stiffness matrix, $[K_s]$. Applied loads are added at the bottom of the anchor and define the load vector, $\{P\}$, whereas the solution of the equation provides the vector of vertical nodal displacements, $\{u\}$.

$$([K_a]+[K_s])\cdot\{u\}=\{P\} \tag{1}$$

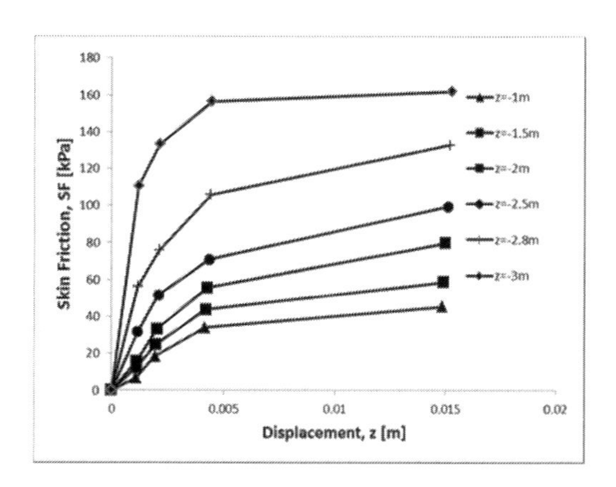

Figure 3. Developed load-transfer curves from the field test at different depths along the anchor length.

Table 1. Parameters of the load-transfer model.

Parameter	k_m	t_m	d	h
Unit	kN/m^3	kPa	-	-
Value	22500	52	1	1

The soil springs adopt a nonlinear load-transfer curve (t-z) defined by a hyperbolic-type relation, according to Equation 2. This equation provides the relation between shaft friction, t, with the local nodal displacement, z. It is a 4-parameter equation where t_m and k_m are related to the maximum value of friction and initial soil stiffness respectively. Lastly, the two dimensionless parameters d and h are related to stiffness degradation and hardening/softening respectively (Pelecanos & Soga, 2017a,b; 2018a,b).

$$t=\frac{k_m\cdot z}{\sqrt[d]{\left[1+\left(\frac{k_m}{t_m}\cdot z\right)^{(h\cdot d)}\right]}} \tag{2}$$

The parameters of the load-transfer model were calibrated based on the field tests presented in Section 2. Table 1 presents the back-calculated parameters of the load-transfer model that were used in the FE analysis.

3.2 *Analysis of field test*

The derived load-transfer curves were used as input in a FE beam-spring analysis of the field pullout test. The results of the FE analysis are shown in Figure 4 and compared with the observed field data. Figure 4 (a) presents the top

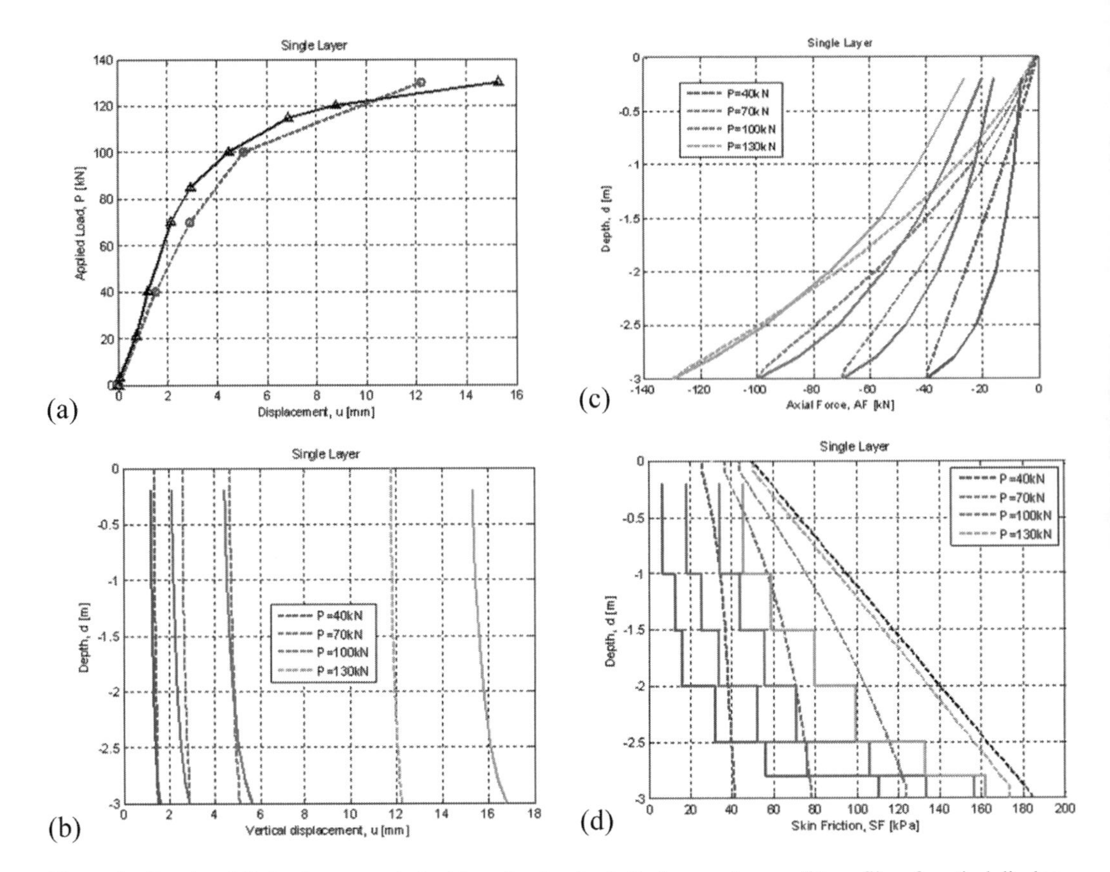

Figure 4. Results of finite element analysis: (a) anchor top load-displacement curve, (b) profiles of vertical displacement, (c) profiles of axial force and (d) profiles of skin friction.

load-displacement curve, (b) the profiles of vertical anchor nodal displacement, (c) the profiles of the axial anchor force whereas (d) the profiles of the anchor skin friction. It may be observed that there is a very good agreement between the field observed data and the computational predictions from the eployed FE model. This may prove that the derived load-transfer curves are suitable for modeling axial displacements, deformations and the ultimate response of such soil anchors.

It is also observed that there is a small mismatch between the numerical predictions of axial anchor force and those monitored during the field test, particularly closer to the top of the anchor. In particular, it is shown that the FE model predicts zero values of axial force at the top boundary. These zero values from the FE model were anticipated, as the boundary condition at the top of the anchor is free and therefore the stress should be equal to zero. It is expected that the non-zero values of axial force from the field data are due to the existence of the stiff steel reaction plate at the top of the anchor, at the ground surface, which has yielded some plate-soil-anchor interaction. Nevertheless, this observed difference however is moderately small and therefore negligible

4 CONCLUSIONS

The interaction behavior between soil anchors and the surrounding soil is a complicated phenomenon and still an unresolved engineering challenge. This is true because the soil-structure interaction is governed by a number of different mechanisms, such as the confining stress with the depth, the development of hoop stresses around the circumference of the nail due to soil dilation and the shear banding behavior at large displacements.

This study proposes a practical method for deriving load-transfer curves for soil anchors under pull-out forces. The approach takes advantage of monitoring data from field pull-out tests to calculate the skin friction and vertical displacement values. The latter two parameters if plotted together may present the load-transfer development.

Monitoring data from a recent field test in Seoul, Korea are analyzed, processed and discussed. Afterwards, relevant load-transfer curves are derived and subsequently appropriate simplified FE analyses are carried out that match reasonably well the observed field data. The latter exercise offers a validation of the suggested approach which proves that this method may be employed to predict the deformation behavior of soil anchors, i.e. some information beyond the ultimate bearing capacity of an anchor.

It is therefore proved that soil anchors follow a load-transfer response. This is the same as axially-loaded foundation piles. It was shown that the developed skin friction is consistent with the vertical displacements. Consequently, it is suggested that the use of a 1D finite element beam-spring formulation with relevant t-z curves is a suitable practical approach to predict the deformation and displacement behaviour of soil anchors and associated soil-structure interaction.

ACKNOWLEDGEMENTS

The Authors would like to acknowledge the assistance of the technicians and other colleagues in the field tests.

REFERENCES

Acikgoz, MS, Pelecanos, L, Giardina, G, Aitken, J & Soga, K 2017, 'Distributed sensing of a masonry vault during nearby piling', Structural Control and Health Monitoring, vol 24, no. 3, p. e1872.

Acikgoz, MS, Pelecanos, L, Giardina, G & Soga, K 2016, 'Field monitoring of piling effects on a nearby masonry vault using distributed sensing.', International Conference of Smart Infrastructure and Construction, ICE Publishing, Cambridge.

Di Murro, V, Pelecanos, L, Soga, K, Kechavarzi, C, Morton, RF & Scibile, L 2016, 'Distributed fibre optic long-term monitoring of concrete-lined tunnel section TT10 at CERN.', International Conference of Smart Infrastructure and Construction, ICE Publishing, Cambridge.

Di Murro, V, Pelecanos, L, Soga, K, Kechavarzi, C, Morton, RF, Scibile, L, 2019, Long-term deformation monitoring of CERN concrete-lined tunnels using distributed fibre-optic sensing., Geotechnical Engineering Journal of the SEAGS & AGSSEA. 50 (1) (Accepted).

Kechavarzi, C, Soga, K, de Battista, N, Pelecanos, L, Elshafie, MZEB & Mair, RJ 2016, Distributed Fibre Optic Strain Sensing for Monitoring Civil Infrastructure, Thomas Telford, London.

Kechavarzi, C., Pelecanos, L., de Battista, N., Soga, K. 2019. Distributed fiber optic sensing for monitoring reinforced concrete piles. Geotechnical Engineering Journal of the SEAGS & AGSSEA. 50 (1) (Accepted)

Kumar, J. and Kouzer, K.M., 2008. Vertical uplift capacity of horizontal anchors using upper bound limit analysis and finite elements. Canadian Geotechnical Journal, 45(5), pp.698–704.

Merifield, R.S. and Sloan, S.W. (2006). "The ultimate pullout capacity of anchors in frictional soils." Canadian Geotechnical Journal, Vol. 43, No. 8, pp. 852–868.

Ouyang, Y, Broadbent, K, Bell, A, Pelecanos, L & Soga, K 2015, 'The use of fibre optic instrumentation to monitor the O-Cell load test on a single working pile in London', Proceedings of the XVI European Conference on Soil Mechanics and Geotechnical Engineering, Edinburgh.

Ouyang, Y., Pelecanos, L., Soga, K. 2018a, Thermo-mechanical response test of a cement grouted column embedded in London Clay, Geotechnique (Under Review).

Ouyang, Y., Pelecanos, L., Soga, K. 2018b, 'Finite element modelling of thermo-mechanical behaviour and thermal soil-structure interaction of a thermo-active cement column buried in London Clay'. 9th European Conference on Numerical Methods in Geotechnical Engineering, Porto, Portugal. (Accepted)

Ouyang, Y., Pelecanos, L., Soga, K. 2018c, 'The field investigation of tension cracks on a slim cement grouted column'. DFI-EFFC International Conference on Deep Foundations and Ground Improvement: Urbanization and Infrastructure Development-Future Challenges, Rome, Italy. (Accepted)

Pelecanos, L, Soga, K, Hardy, S, Blair, A & Carter, K 2016, 'Distributed fibre optic monitoring of tension piles under a basement excavation at the V&A museum in London.', International Conference of Smart Infrastructure and Construction, ICE Publishing, Cambridge.

Pelecanos, L., Skarlatos, D. and Pantazis, G., 2017a. Dam performance and safety in tropical climates—recent developments on field monitoring and computational analysis. 8th International Conference on Structural Engineering and Construction Management (8th ICSECM), Kandy, Sri Lanka.

Pelecanos, L, Soga, K, Chunge, MPM, Ouyang, Y, Kwan, V, Kechavarzi, C & Nicholson, D 2017b, 'Distributed fibre-optic monitoring of an Osterberg-cell pile test in London.', Geotechnique Letters, vol 7, no. 2, pp. 1–9.

Pelecanos, L., Soga, K., Elshafie, M., Nicholas, d. B., Kechavarzi, C., Gue, C.Y., Ouyang, Y. and Seo, H., 2018. Distributed Fibre Optic Sensing of Axially Loaded Bored Piles. Journal of Geotechnical and Geoenvironmental Engineering. 144 (3) 04017122.

Pelecanos, L., & Soga, K. 2017a. Innovative Structural Health Monitoring Of Foundation Piles Using Distributed Fibre-Optic Sensing. 8th International Conference on Structural Engineering and Construction Management 2017, Kandy, Sri Lanka.

Pelecanos, L., & Soga, K. 2017b. The use of distributed fibre-optic strain data to develop finite element models for foundation piles. 6th International Forum on Opto-electronic Sensor-based Monitoring in Geo-engineering, Nanjing, China.

Pelecanos, L & Soga, K 2018a, 'Using distributed strain data to evaluate load-transfer curves for axially loaded piles', Journal of Geotechnical & Geoenvironmental Engineering, ASCE (Under Review).
Pelecanos, L., Soga, K. 2018b, 'Development of load-transfer curves for axially-loaded piles based on inverse analysis of fibre-optic strain data using finite element analysis and optimisation'. 9th European Conference on Numerical Methods in Geotechnical Engineering Porto, Portugal. (Accepted)
Seo, H.J., Kim, H.R., Jeong, N.S. and Lee, I.M., 2010. Behavioral Mechanism of Hybrid Model of Soil-nailing and Compression Anchor. Journal of the Korean Geotechnical Society, 26(7), pp.117–133.
Seo, H.J., Jeong, K.H., Choi, H. and Lee, I.M., 2011a. Pullout resistance increase of soil nailing induced by pressurized grouting. Journal of geotechnical and geoenvironmental engineering, 138(5), pp.604–613.
Seo, H.J., Kim, H.R., Jeong, N.S., Shin, Y.J. and Lee, I.M., 2011b, December. Behavior of hybrid system of soil-nailing and compression anchor. In 14th Asian Regional Conference on Soil Mechanics and Geotechnical Engineering, ARC 2011.
Seo, H.J., Lee, G.H., Park, J.J. and Lee, I.M., 2012. Optimization of Soil-Nailing Designs Considering Three Failure Modes. Journal of the Korean Geotechnical Society, 28(7), pp.5–16.
Seo, H.J. and Lee, I.M., 2014a. Shear Behavior between Ground and Soil-Nailing. Journal of the Korean Geotechnical Society, 30(3), pp.5–16.
Seo, H.J., Lee, I.M. and Lee, S.W., 2014b. Optimization of soil nailing design considering three failure modes. KSCE Journal of Civil Engineering, 18(2), pp.488–496.
Seo, H.J., Choi, H. and Lee, I.M., 2016. Numerical and experimental investigation of pillar reinforcement with pressurized grouting and pre-stress. Tunnelling and Underground Space Technology, 54, pp.135–144.
Seo, H., Pelecanos, L., Kwon, Y.S. and Lee, I.M., 2017. Net load-displacement estimation in soil-nailing pullout tests. Proceedings of the Institution of Civil Engineers-Geotechnical Engineering.
Seo, H. –J., Pelecanos, L., 2017. Load transfer in soil anchors—Finite Element analysis of pull-out tests. 8th International Conference on Structural Engineering and Construction Management (8th ICSECM), Kandy, Sri Lanka, December, 2017.
Soga, K, Kechavarzi, C, Pelecanos, L, de Battista, N, Williamson, M, Gue, CY, Di Murro, V & Elshafie, M 2017, 'Distributed fibre optic strain sensing for monitoring underground structures—Tunnels Case Studies ', in S Pamukcu, L Cheng (eds.), Underground Sensing, 1st edn, Elsevier.
Soga, K, Kwan, V, Pelecanos, L, Rui, Y, Schwamb, T, Seo, H & Wilcock, M 2015, 'The role of distributed sensing in understanding the engineering performance of geotechnical structures', Proceedings of the XVI European Conference on Soil Mechanics and Geotechnical Engineering, Edinburgh.
Tagaya, K., Scott, R.F. and Aboshi, H., 1988. Pullout resistance of buried anchor in sand. Soils and Foundations, 28(3), pp.114–130.
White, D.J., Cheuk, C.Y. and Bolton, M.D., 2008. The uplift resistance of pipes and plate anchors buried in sand. Géotechnique, 58(10), pp.771–779.

Numerical Methods in Geotechnical Engineering IX – Cardoso et al. (Eds)
© 2018 Taylor & Francis Group, London, ISBN 978-1-138-33203-4

The effect of non-linear soil behavior on mixed traffic railway lines

K. Dong & O. Laghrouche
Heriot Watt University, Edinburgh, UK

D.P. Connolly & P.K. Woodward
University of Leeds, Leeds, UK

P. Alves Costa
University of Porto, Porto, Portugal

ABSTRACT: Railway freight services can be added to lines that have previously only be used for passenger services, with the aim of increasing network capacity. Freight trains have larger axle loads and thus can have a negative effect on track longevity, particularly on ballasted lines supported by sub-optimal ground conditions. This is because larger subgrade strains are generated, which can result in non-linear behavior. Therefore it is important to be able to determine the effect of the new rolling stock on track behavior before operation. This is challenging to do because non-linear soil behavior is challenging to simulate. As a solution, this paper presents an equivalent non-linear, thin layer element soil model, coupled to an analytical track model. It is capable of quickly and accurately computing the response of non-linear track behavior. The model is used to investigate the effect of introducing freight wagons on an existing ballasted passenger line with poor ground conditions.

1 INTRODUCTION

Railway operators who wish to tweak network capacity, may add freight services to tracks that have previously only be used for passenger services. If these lines were designed without freight in mind and/or were constructed at a time when compaction techniques were less scientific than today, then freight trains potential could have a detrimental impact.

To investigate and predict the track performance and ground response under various train loads and speeds, a number of modelling techniques have been proposed. The approaches include analytical models (Krylov 1995, Degrande & Lombaert 2001, Takemiya & Bian 2005), semi-analytical models (Sheng et al. 1999, Madshus & Kaynia 2000, Sheng et al. 2003, Kaynia et al. 2000, Thompson 2008, Triepaischajonsak & Thompson 2015). There are also numerical models: 2.5D models (Yang et al. 2003, Alves Costa et al. 2012, Alves Costa et al. 2010) and fully 3D models using finite element (FE) and possibly boundary element (BE) theories (Hall 2003, Kouroussis et al. 2011, Arlaud et al. 2015, Kacimi et al. 2013).

For freight trains, the dominant frequency components of the vibration are within 4–30 Hz (Jones & Block 1996). In order to study the vibrations induced by the freight trains, both dynamic and quasi-static generation mechanism, a track response model combined with transfer functions from sleeper to ground was utilized by (Jones & Block 1996). Another numerical model was proposed for the studies of longitudinal dynamics of the trainset (Belforte et al. 2008). On-site tests can be costly (Jones 1994), meaning theoretical models are often used to examine the track performance and ground response from freight trains.

In modelling the ground vibrations from railways, linear elastic models of the soil are commonly used, because strains are small. Nonetheless, when axle loads increase and/or the train speed gets close to the critical velocity, the track deflections increase and non-linear soil response occurs (Madshus & Kaynia 2000, Alves Costa et al. 2010). To simulate this non-linear behavior, soil stiffness' can be artificially reduced (Madshus & Kaynia 2000, Kaynia et al. 2000). Alternatively, using an automated, equivalent non-linear approach, the shear modulus can be adjusted based on the maximum effective octahedral shear strain in each soil element. Then it can be updated element by element until a tolerance requirement is met (Alves Costa et al. 2010).

This paper therefore provides a robust and efficient semi-analytical approach to model non-linear soil effects. The track is modelled analytically and allows for 1D wave propagation. The soil is modelled using a non-linear equivalent thin-layer method (TLM). The soil stiffness is updated in an iterative manner to simulate the non-linear behavior of the soil with the minimum computational effort.

2 NUMERICAL MODEL DEVELOPMENT

Freight trains carry heavier loads than passenger trains, thus causing elevated strains within the supporting subgrade. Large strains cause non-linear soil behavior, resulting in reduced support stiffness. To model this in a computationally efficient manner, a thin-layer finite element model was developed, and then combined with an equivalent non-linear procedure. To simulate the combined track-soil behavior, the track was coupled to the surface of the soil model.

2.1 *Track model*

Ballasted track was modelled as shown in Figure 1. One dimensional wave propagation was considered in the ballast and an equivalent spring was used to couple the track to the soil using (Dieterman & Metrikine 1996):

$$\begin{bmatrix} a_{11} & a_{12} & 0 \\ a_{21} & a_{22} & a_{23} \\ 0 & a_{32} & a_{33} \end{bmatrix} \begin{Bmatrix} \tilde{u}_r(k_1,\omega) \\ \tilde{u}_s(k_1,\omega) \\ \tilde{u}_{bb}(k_1,\omega) \end{Bmatrix} = \begin{Bmatrix} \tilde{P}(k_1,\omega) \\ 0 \\ 0 \end{Bmatrix} \tag{1}$$

$$a_{11} = EI_r k_1^4 + k_p^* - \omega^2 m_r \tag{2}$$

$$a_{12} = a_{21} = -k_p^* \tag{3}$$

$$a_{22} = k_p^* + \frac{2\omega E_b^* b\alpha}{\tan(\frac{\omega h}{C_p})C_p} - \omega^2 m_s \tag{4}$$

$$a_{23} = a_{32} = \frac{-2\omega E_b^* b\alpha}{\sin(\frac{\omega h}{C_p})C_p} \tag{5}$$

$$a_{33} = \frac{2\omega E_b^* b\alpha}{\tan(\frac{\omega h}{C_p})C_p} + k_{eq} \tag{6}$$

Where EI_r is the bending stiffness of the rail; m_r is the mass of rails per meter; m_s is the equivalent distributed mass of sleepers; k_p^* is the complex stiffness of the railpad; k_{eq} is the equivalent stiffness of the ground; E_b^* is the Young's modulus of the ballast; C_p is the compression wave speed in the ballast; h is the ballast layer height; α is the adimensional parameter, taken as 0.5; b is the half-width of the track.

The ballasted track model included the coupling between the track and the soil, using the ratio between the load and average displacement along the track-soil interface (Steenbergen & Metrikine 2007). It was calculated as:

$$\tilde{k}_{eq}(k_1,\omega) = \frac{2\pi}{\int_{-\infty}^{+\infty} \tilde{u}_{ZZ}^{G}(k_1,k_2,0,\omega)\frac{\sin(k_2 b)^2}{(k_2 b)^2}dk_2} \tag{7}$$

Where u_{zz} is the Green's function of vertical displacement of the ground in the wavenumber-frequency domain, and k_1 and k_2 are the Fourier images of coordinate x and y, respectively. The Green function was computed using the Haskell-Thompson approach (Sheng et al. 1999).

2.2 *Soil model*

The soil was modeled using the Thin-Layer Method (TLM) as illustrated in the Figure 2.

It is worth noting that:

- The thickness of the thin layer quadratic elements were computed as h = wavelength/8 = $\pi/4\ k_{max}$, where k_{max} was the maximum wavenumber defined
- After obtaining the displacement of each node, the strains/stresses were calculated using Equations 8 and 9

$$\{\boldsymbol{\varepsilon}\} = [\mathbf{B}]\{\mathbf{u}\} \tag{8}$$

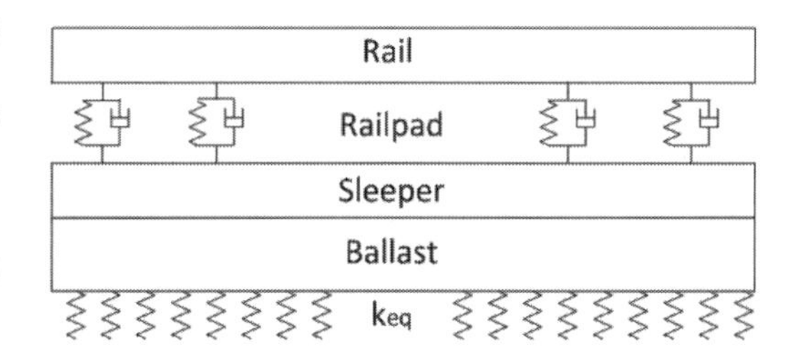

Figure 1. Analytical ballasted track model layout.

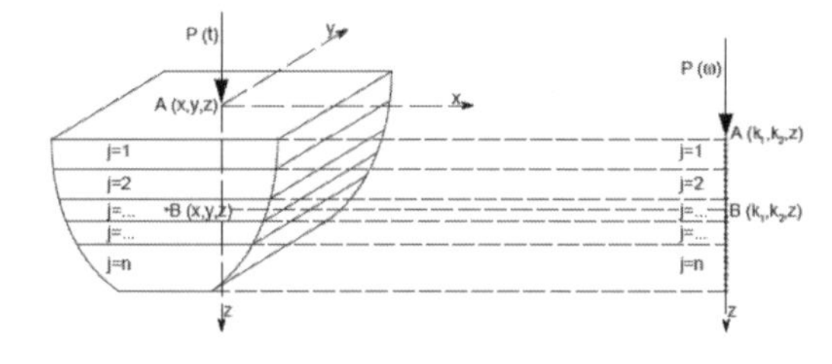

Figure 2. Schematic diagram of Thin-Layer Method modeling process (Alves Costa 2011).

$$\{\boldsymbol{\sigma}\} = [\mathbf{D}]\{\boldsymbol{\varepsilon}\} = [\mathbf{D}][\mathbf{B}]\{\mathbf{u}\} \tag{9}$$

Where $[B] = [B_1\ B_2\ B_3]$ and

$$[B_i] = \begin{bmatrix} ik_1N_i & 0 & 0 \\ 0 & ik_2N_i & 0 \\ 0 & 0 & \frac{\partial N_i}{\partial z} \\ ik_2N_i & ik_1N_i & 0 \\ 0 & \frac{\partial N_i}{\partial z} & ik_2N_i \\ \frac{\partial N_i}{\partial z} & 0 & ik_1N_i \end{bmatrix} \tag{10}$$

$$\begin{aligned} N_1(\xi) &= \frac{1}{2}\xi^2 - \frac{1}{2}\xi \\ N_2(\xi) &= 1 - \xi^2 \\ N_3(\xi) &= \frac{1}{2}\xi^2 + \frac{1}{2}\xi \end{aligned} \tag{11}$$

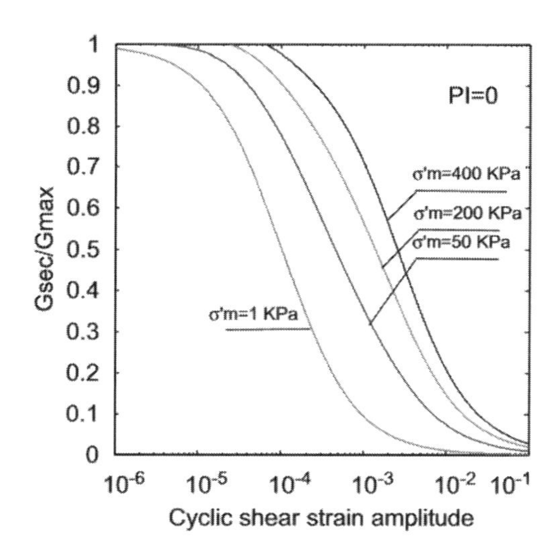

Figure 3. Modulus reduction curves for non-plastic soil (Alves Costa 2010).

2.3 *Equivalent non-linear model*

If low stiffness soil is found on freight lines, it is likely to experience high levels of strain. This can result in soil stiffness degradation, thus increasing the track displacements and causing track deterioration. To simulate this, a non-linear equivalent mod-el, based on an iterative stiffness updating procedure, was used. This meant that each studied case was repeated multiple times until convergence was reached:

1. Assume low/zero strain within all elements
2. Use track-soil model to compute strain time histories and determine the maximum effective octahedral shear strain values for all elements
3. Use stiffness degradation curves (Figure 3), to obtain the new stiffness for all elements
4. Use damping curves, to obtain the new damping properties for all elements
5. Repeat steps 2–4 until the established tolerance is met for all elements (3% used in this case)

3 MODEL VALIDATION

The model contained 3 main components: track, soil and the track-soil coupling mechanism. To ensure these were working correctly, validation was performed using an example outlined in (Chen et al. 2005). In order to validate the TLM model for the ground response, same case was studied and the stresses in the soil compared against the published result.

The train-embankment-ground model contained a Euler beam resting on top of the half-space with a concentrated moving force acting on the beam (Figure 4). The stresses generated by the contact force between the embankment and ground were calculated at 2 m depth below the loading point.

Key embankment and ground properties related to the validation are listed in the Table 1 and Table 2 respectively. The load was a vertical 160 kN point load moving with a speed of 30 m/s.

Figure 5 reveals strong agreement between the model and the benchmark.

4 ANALYSIS AND RESULTS

Simulations were run to determine the effect of adding 25 tonne fright axle loads to a previous passenger-only (17 tonne) ballasted line, with the aim of determining increases in track displacement and soil strain. To do so, the following track properties were assumed: m_r = 120 kg/m, m_s = 490 kg/m, k_p^* = 5 × 10^8 N/m^2, E_b^* = 125 MPa, h = 0.35 m, b = 2.5 m. The soil was modelled as a homogenous half-space using the following properties: density = 2000 kg/m^3, Young's modulus = 25 MPa, Poisson's ratio = 0.35, damping = 0.03. The stiffness degradation profile was the same as that shown previously. Train speed for both the passenger and freight axle loads was 26 m/s.

Figure 6 (left) shows the variation of strain versus depth within the soil. The maximum octahedral strains is located approximately 1 m below the ground surface and decays rapidly with depth.

In comparison, Figure 6 (right) shows maximum strain and the resulting effect on soil stiffness.

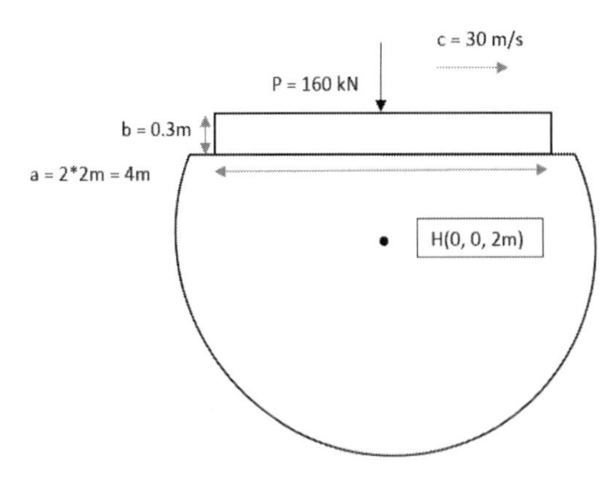

Figure 4. Schematic diagram of Chen et al. 2005 validation model.

Table 1. Properties of the embankment.

Density (kg/m^3)	Young's modulus (MPa)	Width (m)	Height (m)	Mass (kg)	Second moment of area (m^4)
1900	30000	4	0.3	2280	0.009

Table 2. Properties of the ground.

Shear modulus (MPa)	Poisson ratio	Density (kg/m^3)	Secondary wave speed (m/s)
10	0.45	1800	74.54

After the first iteration, the soil drops to 67% of its original stiffness and by the third (and final) iteration, it has reached a value of 59%.

The resulting reduction in stiffness (Young's modulus) with depth is shown in Figure 7 (left). For iteration 1, stiffness is constant with depth, however after strain updating, the subsequent iterations show large variations with depth, and are all lower than the starting value, particularly near the soil surface. For the passenger train, track displacements are 3.7 mm, however for the freight train, the linear value is 5.5 mm displacement, and the non-linear (iteration 3) is 8.4 mm. Therefore, it can be seen that the soil behavior is significantly non-linear, and that traditional linear analysis would greatly underestimate track deflections. This would result in much faster loss of track geometry and require frequent tamping. In addition, it is interesting to note that as the soil stiffness decreases, dynamic effects become more prevalent, with iteration 3 displacements appearing less symmetric than iteration 1.

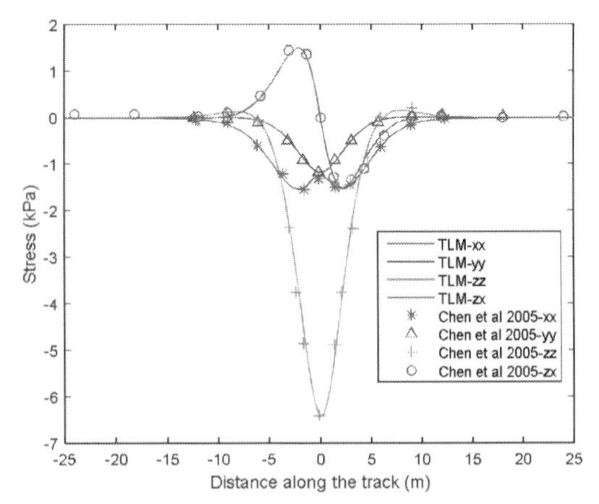

Figure 5. Comparisons of the dynamic stresses of an element with 2 m depth underneath the moving load.

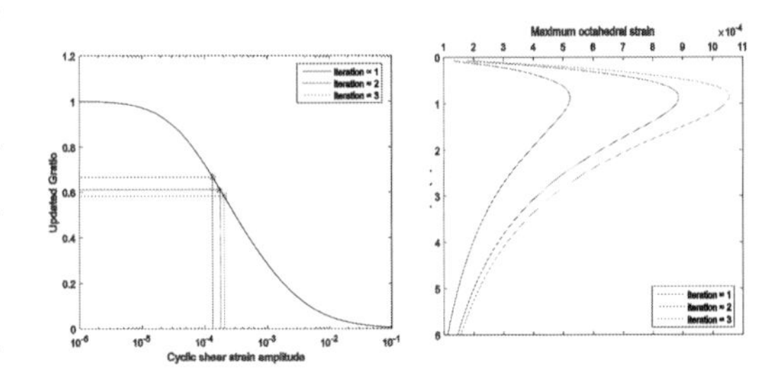

Figure 6. Left: Octahedral strain vs soil depth; Right: Soil stiffness degradation during freight train passage.

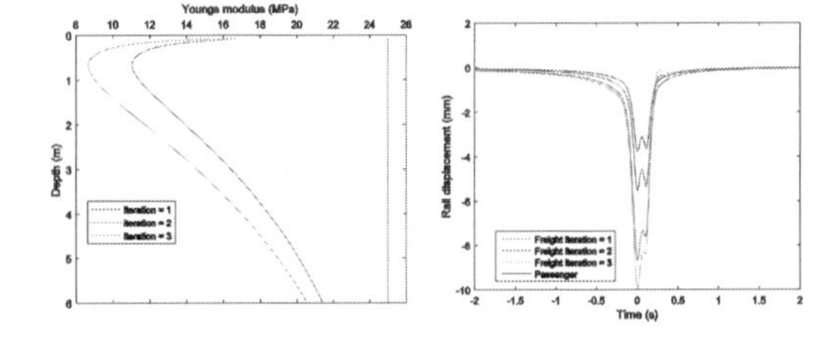

Figure 7. Left: Young's modulus reduction with depth; Right: Track displacements.

5 CONCLUSIONS

There are increased pressures on network operators to run freight trains on ballasted track originally designed for passenger services. These tracks may not have the desired subgrade characteristics for heavy axle loads, possibly giving rise to non-linear soil behavior. To analyse this problem, an equivalent non-linear numerical model was developed, capable of quickly assessing soil stresses and strains, and resulting track displacements. The model was validated against a published benchmark case and then used to compare freight and passenger train response on a low stiffness ballasted line. It was shown that the track displacements have the potential to become high, due to non-linear stiffness reduction and the resulting dynamic amplification.

REFERENCES

Alves Costa, P. Calçada, R. Silva Cardoso, A. & Bodare, A. 2010. Influence of soil non-linearity on the dynamic response of high-speed railway tracks. *Soil Dynamics and Earthquake Engineering*, 30(4), 221–235.

Alves Costa, P. 2011. Vibrações Do Sistema Via-Maciço Induzidas Por Tráfego Ferroviário. Modelação Numérica E Validação Experimental.

Alves Costa, P. Calçada, R. & Silva Cardoso, A. 2012. Track-ground vibrations induced by railway traffic: In-situ measurements and validation of a 2.5D FEM-BEM model. *Soil Dynamics and Earthquake Engineering*, 32(1), 111–128.

Arlaud, E. Costa D'Aguiar, S. & Balmes, E. 2015. Validation of a reduced model of railway track allowing long 3D dynamic calculation of train-track interaction. Computer Methods and Recent Advances in Geomechanics - *Proceedings of the 14th Int. Conference of International Association for Computer Methods and Recent Advances in Geomechanics, IACMAG 2014*, (September), 1193–1198.

Belforte, P. Cheli, F. Diana, G. & Melzi, S. 2008. Numerical and experimental approach for the evaluation of severe longitudinal dynamics of heavy freight trains. *Vehicle System Dynamics*, 46(SUPPL.1), 937–955.

Chen, Y. Wang, C.J. Chen, Y.P. & Zhu, B. 2005. Characteristics of stresses and settlement of ground induced by train. In Environmental vibrations: Prediction, monitoring, mitigation and evaluation: *Proceedings of the International symposium on environmental Vibrations* (pp. 33–42).

Degrande, G. & Lombaert, G. 2001. An efficient formulation of Krylov's prediction model for train induced vibrations based on the dynamic reciprocity theorem. *The Journal of the Acoustical Society of America*, 110(3), 1379–1390.

Dieterman, H.A. & Metrikine, A.V. 1996. The equivalent stiffness of a half-space interacting with a beam. Critical velocities of moving load along the beam. *European Journal of Mechanics A/Solids*, 15, 67–90.

El Kacimi, A. Woodward, P.K. Laghrouche, O. & Medero, G. 2013. Time domain 3D finite element modelling of train-induced vibration at high speed. *Computers and Structures*, 118, 66–73.

Hall, L. 2003. Simulations and analyses of train-induced ground vibrations in finite element models. *Soil Dynamics and Earthquake Engineering*, 23(5), 403–413.

Jones, C.J. 1994. Use of numerical models to determine the effectiveness of anti-vibration systems for railways. *Proceedings of the Institution of Civil Engineers—Transport*, 105(1), 43–51.

Jones, C.J.C. & Block, J.R. 1996. Prediction of ground vibration from freight trains. *Journal of Sound and Vibration*, 193(1), 205–213.

Kaynia, A.M. Madshus, C. & Zackrisson, P. 2000. Ground vibration from high-speed trains: Prediction and countermeasure. *Journal of Geotechnical and Geoenvironmental Engineering*, 126, 531–537.

Kouroussis, G. Gazetas, G. Anastasopoulos, I. Conti, C. & Verlinden, O. 2011. Discrete modelling of vertical track-soil coupling for vehicle-track dynamics. *Soil Dynamics and Earthquake Engineering*, 31(12), 1711–1723.

Krylov, V.V. 1995. Generation of ground vibrations by superfast trains. *Applied Acoustics*, 44(2), 149–164.

Madshus, C. & Kaynia, A.M. 2000. High-Speed Railway Lines on Soft Ground: Dynamic Behaviour At Critical Train Speed. *Journal of Sound and Vibration*, 231(3), 689–701.

Sheng, X. Jones, C.J.C. & Petyt, M. 1999. Ground Vibration Generated By a Harmonic Load Acting on a Railway Track. *Journal of Sound and Vibration*, 225(1), 3–28.

Sheng, X. Jones, C.J.C. & Thompson, D.J. 2003. A comparison of a theoretical model for quasi-statically and dynamically induced environmental vibration from trains with measurements. *Journal of Sound and Vibration*, 267(3), 621–635.

Steenbergen, M.J. & Metrikine, A.V. 2007. The effect of the interface conditions on the dynamic response of a beam on a half-space to a moving load. *European Journal of Mechanics, A/Solids*, 26(1), 33–54.

Takemiya, H. & Bian, X. 2005. Substructure Simulation of Inhomogeneous Track and Layered Ground Dynamic Interaction under Train Passage. *Journal of Engineering Mechanics*, 131(7), 699–711.

Thompson, D. 2008. Railway noise and vibration: mechanisms, modelling and means of control. *Elsevier*.

Triepaischajonsak, N. & Thompson, D.J. 2015. A hybrid modelling approach for predicting ground vibration from trains. *Journal of Sound and Vibration*, 335, 147–173.

Yang, Y.B. Hung, H.H. & Chang, D.W. 2003. Train-induced wave propagation in layered soils using finite/infinite element simulation. *Soil Dynamics and Earthquake Engineering*, 23(4), 263–278.

Offshore geotechnical engineering

Numerical Methods in Geotechnical Engineering IX – Cardoso et al. (Eds)
© 2018 Taylor & Francis Group, London, ISBN 978-1-138-33203-4

The dynamics of an offshore wind turbine using a FE semi-analytical analysis considering the interaction with three soil profiles

Dj. Amar Bouzid
Department of Civil Engineering, Faculty of Technology, University of Saad Dahled, Blida, Algeria

R. Bakhti
Department of Civil Engineering, Faculty of Science and Applied Science, University of Akli Mohaned Oulhadj, Bouira, Algeria

S. Bhattacharya
Department of Civil and Environmental Engineering, Tomas Telford Building, University of Surrey, Surrey, UK

ABSTRACT: Nowadays, Offshore Wind Turbines (OWTs) are placed in harsh environments seeking for a better outcome in energy production. However, the accurate determination of the first natural frequency of an OWT is crucial for the safe design of the OWTs where the resonance phenomenon must be avoided at all costs. Since both the OWT and its supporting monopile are all tubular, the whole system exhibits an axisymmetric geometry under non-axisymmetric loading. This type of problems can be analyzed by a semi-analytical approach commonly called Fourier Series Aided Finite Element Method (FSAFEM). In the present paper, a computer code has been written to estimate natural vibrations of an OWT supported by a monopile embedded in three soil profiles. The profiles considered are those found in the API and DNV offshore guidelines, which are constant, linear and parabolic variation of stiffness with depth. Comparisons with measured natural frequencies of OWTs are performed.

1 INTRODUCTION

In the past few years, wind energy has become more and more competitor of fossil fuel energy. Seeking a better outcome, Offshore Wind Turbines (OWTs) should be installed in remote locations of harsh environments. Consequently, the designer is put in a challenging situation of two levels. Firstly, the nature and the magnitude of loading required for an appropriate design an OWT structure should be determined properly. Secondly, a decreasing as much as possible in the weight of the turbine should be achieved by minimizing the thickness of the tower wall, and therefore, the OWT becomes more and more flexible. As a result, the estimation of the natural frequency of the OWTs is a very important design calculation to avoid resonance and resonance related effects.

The principal source of excitation in the wind turbines is the rotor motion which is the first excitation due to the rotational speed resulting in a frequency equal to 1P, and the second excitation frequency caused by the rotor blades provides a frequency equal to N_bP, where N_b is the number of blades in the wind turbine. As we know, the natural frequency should be kept apart from the excitation frequencies of the dominant forces, and hence, three design approaches can be considered according to the frequency ranges formed before, in between or after $1P$ and N_bP:

1. The OWT design may lead to very flexible structure if the natural frequency lies in the Soft-soft range where it is less than 1P,
2. The design of the OWT may result in a very stiff structure which is economically unfeasible if the natural frequency lies in the stiff-stiff range where it is higher than N_bP.
3. The most appropriate range is the soft-stiff range where the natural frequency is sufficiently far from the other excitations avoiding thus the resonance.

Many strategies have been followed to estimate the natural frequency of an OWT (Jin-Hak et *al.* 2015 and Prendergast et *al.* 2015). The common procedure is to consider first the OWT a s a structure connected to a fixed base and compute its corresponding natural frequency. Then, a relationship which accounts for the finite stiffness of the soil should be considered. Many authors proposed solutions for the fixed base natural frequency. For instance, Belvins (2000), proposed the following formula:

$$f_{FB}=\frac{1}{2\pi}\sqrt{\frac{3(EI)_T}{L^3\left(m_{RNA}+\frac{33}{144}m_T\right)}} \quad (1)$$

For the estimation of the natural frequency taking account of is the foundation flexibility effect, Arany et *al.* (2015, 2016), provided a simple formula:

$$f_n=C_L C_R f_{FB} \quad (2)$$

where C_L and C_R are the foundation flexibility coefficients, given by:

$$C_R=1-\frac{1}{1+a\left(\eta_R-\frac{\eta_{LR}^2}{\eta_L}\right)},\quad C_L=1-\frac{1}{1+b\left(\eta_L-\frac{\eta_{LR}^2}{\eta_R}\right)} \quad (3)$$

With η_L,η_R,η_{LR} are the non-dimensional foundation stiffness parameters given by:

$$\eta_L=\frac{K_L L_T^3}{(EI)_\eta},\eta_R=\frac{K_R L_T}{(EI)_\eta},\eta_{LR}=\frac{K_{LR}L_T^2}{(EI)_\eta} \quad (4)$$

The monopile head stiffnesses $K_L,K_R\,K_{LR}$ are the key elements in these expressions. They account for the soil/monopile interaction and consequently they affect the fixed base frequency to properly assess the first natural frequency of the system. To assess these parameters several methods were used. Indeed, Arany et *al.* (2017) used monopile spring model, whereas Abed et *al.* (2016) and Aissa et *al.* (2017) used the FE Semi-Analytical approach in both homogenous and non-homogeneous soils.

The whole system tower/monopile is axisymmetric since both components are axisymmetric. Consequently, the Finite element analysis using Fourier Series is the best choice to assess properly the natural frequency of the offshore wind tower. In this paper the FSAFEM is used to analyze an OWT founded on a monopile and its surrounding soil. The interface elements are also included in the study, where the extreme states of the soil/monopile interface are considered.

2 AN OWT AND ITS FOUNDATION SOIL AS AN AXISYMMETRIC PROBLEM UNDER NON-AXIMETRIC LOADING

Any civil engineering problem can be considered as an axisymmetric problem if three conditions are fulfilled. The first condition is that the solid geometry must be generated by a rotating plane area around an axis in the same plane, the second is that the properties of the material must be symmetric to the same axis and the third is that the boundary conditions must also be axisymmetric. For most OWTs, all these conditions are satisfied (Figure 1), where, the shape, the properties of the material and the boundary condition are symmetrical with respect to the z-axis. Therefore, ring elements can be used for modelling the OWT structure, and a large cylinder containing both soil and monopile. The study is performed numerically in two-dimensional radial plane and analytically in the circumferential direction. The loading on the structure is divided into a number of harmonics, where the whole response of the structure is computed as a result of superposition of all harmonics considered in the analysis.

2.1 *Description of displacement and applied loading fields*

For non-axisymmetric loading, every field quantity has to be expressed in the Fourier series form where the final outcome is obtained by appending the results for each harmonic term. The displacement components *u*, *v* and *w* are given by:

$$\begin{aligned}
u_r(r,z,\theta)&=\sum_{i=0}^{L}\bar{u}_{ri}\cos i\theta+\sum_{i=1}^{L}\bar{\bar{u}}_{ri}\sin i\theta\\
v_z(r,z,\theta)&=\sum_{i=0}^{L}\bar{v}_{zi}\cos i\theta+\sum_{i=1}^{L}\bar{\bar{v}}_{zi}\sin i\theta\\
w_\theta(r,z,\theta)&=\sum_{i=1}^{L}\bar{w}_{\theta i}\sin i\theta+\sum_{i=0}^{L}\bar{\bar{w}}_{\theta i}\cos i\theta
\end{aligned} \quad (5)$$

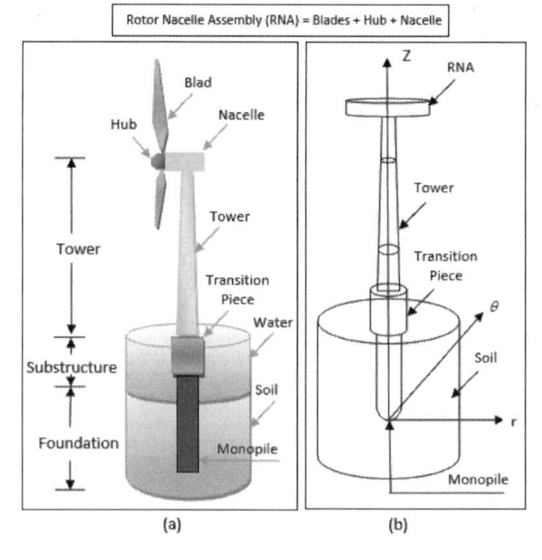

Figure 1. Axisymmetric model for an OWT (a): main components, (b): axisymmetric model.

A Fourier series representation of loads components R, Z and $T$is:

$$R(r,z,\theta)\bar{R}_0+\sum_{i=1}^{\infty}\left(\bar{R}_i(r,z)\cos i\theta+\bar{\bar{R}}_i(r,z)\sin i\theta\right)$$
$$Z(r,z,\theta)=\bar{Z}_0+\sum_{i=1}^{\infty}\left(\bar{Z}_i(r,z)\cos i\theta+\bar{\bar{Z}}_i(r,z)\sin i\theta\right) \quad (6)$$
$$T(r,z,\theta)\dot{T}_0+\sum_{i=1}^{\infty}\left(\bar{T}_i(r,z)\sin i\theta-\bar{\bar{T}}_i(r,z)\cos i\theta\right)$$

where the single barred series describe symmetric amplitudes and double barred describe antisymmetric amplitudes.

2.2 *Volume elements used to model the OWT structure, the soil and the monopile*

For modelling the tower, the monopile or even the surrounding soil, the eight-noded ring element is used. This element is characterized by three degrees of freedom for each node (Fig 2), resulting in 24 degrees of freedom for the whole number of nodes.

The shape functions of the eight-noded ring element written in terms of local coordinates are given by:

For nodes 1, 3, 5 and 7:

$$N_j=\frac{1}{4}(1+\xi\xi_j)(1+\eta\eta_j)(\xi\xi_j+\eta\eta_j-1) \quad (7)$$

At mid-side nodes

$$N_j=\frac{1}{2}(1-\xi^2)(1+\eta\eta_j)\, for\ j=4,8 \quad (8)$$

$$N_j=\frac{1}{2}(1-\eta^2)(1+\xi\xi_j)\, for\ j=2,6 \quad (9)$$

The stiffness matrix associated with the first harmonic fundamentally depends on the strain-displacement matrix $\boldsymbol{B}_v$ as well as on elasticity matrix $\boldsymbol{D}_v$. It can be written as:

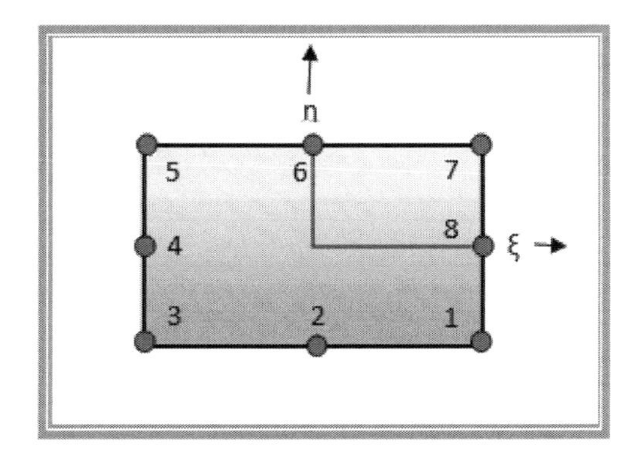

Figure 2. The eight nodes axisymmetric element.

$$\boldsymbol{K}_{vol}=\int_A\int_{-\pi}^{\pi}\boldsymbol{B}_v^T\boldsymbol{D}_v\boldsymbol{B}_v r\,dr\,dz\,d\theta \quad (10)$$

Where A is the cross-section area of the volume finite element. As the matrix $\boldsymbol{B}_v$ is important, it is useful to divide it into sub-matrices. Indeed:

$$\boldsymbol{B}_v=\left[\boldsymbol{B}_1\boldsymbol{B}_2\ldots\ldots\ldots\boldsymbol{B}_N\right] \quad (11)$$

Where $\boldsymbol{B}_j$ is a sub-matrix corresponding to the node j. N is the total number of element nodes. $\boldsymbol{B}_j$ can be written in a more explicit form:

$$\begin{bmatrix} \frac{\partial N_j}{\partial r}\cos\theta & 0 & 0 \\ 0 & \frac{\partial N_i}{\partial z}\cos\theta & 0 \\ \frac{N_j}{r}\cos\theta & 0 & \frac{N_j}{r}\cos\theta \\ \frac{\partial N_j}{\partial z}\cos\theta & \frac{\partial N_j}{\partial r}\cos\theta & 0 \\ 0 & \frac{N_j}{r}\sin\theta & \frac{\partial N_j}{\partial r}\sin\theta \\ -\frac{N_j}{r}\sin\theta & 0 & \left(\frac{\partial N_j}{\partial r}-\frac{N_j}{r}\right)\sin\theta \end{bmatrix} \quad (12)$$

2.3 *Interface element*

Amar Bouzid et *al.* (2004) have formulated joint element for modelling the soil-pile interfaces of an axisymmetric problem under non-axisymmetric loading. This element can be used with eight or nine-nodes axisymmetric volume elements for modelling an OWT with its foundation system (Figure 3).

The shape functions of this element written in terms of local coordinates are given by:

$$N_1=\frac{\xi}{L}\left(\frac{2\xi}{L}-1\right), N_2=\left(1-\frac{4\xi^2}{L^2}\right), N_3=\frac{\xi}{L}\left(\frac{2\xi}{L}+1\right) \quad (12)$$

The stiffness matrix $\boldsymbol{K}_{int}$ of the interface element is given by the equation:

$$\boldsymbol{K}_{int}=\int_A\boldsymbol{B}_i^T\boldsymbol{D}_i\boldsymbol{B}_i\,dA \quad (13)$$

where, $\boldsymbol{D}_i$ is the interface constitutive matrix and A is the cross-section area of the interface element. The thickness of the interface element is

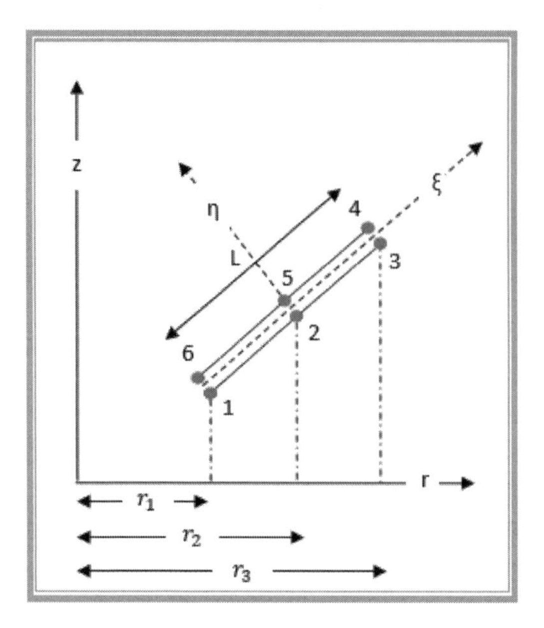

Figure 3. Zero thickness six-noded interface element.

taken to be zero. $\boldsymbol{B}_i$ is the strain-displacement matrix given by:

$$\boldsymbol{B}_i = \begin{bmatrix} -N_1 & 0 & 0 & -N_2 & 0 & 0 & N_1 & 0 & 0 \\ 0 & -N_1 & 0 & 0 & -N_2 & 0 \;\ldots\ldots & 0 & N_1 & 0 \\ 0 & 0 & -N_1 & 0 & 0 & -N_2 & 0 & 0 & N_1 \end{bmatrix} \quad (14)$$

and $\boldsymbol{D}_i$ is the interface constitutive matrix:

$$\boldsymbol{D}_i = \begin{bmatrix} k_s & 0 & 0 \\ 0 & k_n & 0 \\ 0 & 0 & k_s \end{bmatrix} \quad (15)$$

with k_s is the interface shear stiffness and k_n is the interface normal stiffness.

2.4 *Mesh adopted*

Only half of the domain was meshed for the 3D semi-analytical FEM study due to the symmetry of the problem. The mesh used for the study is shown in Fig. 4. The distances from boundaries are chosen large enough to eliminate the boundary effect. Applied boundary conditions to the soil are pinned support at the bottom with no displacements in the horizontal, vertical and circumferential directions $(u_r = v_z = w_\theta = 0)$ and roller support at sides with no movement in the horizontal direction $(u_r = 0)$. The bottom dimension of the mesh is given in terms of monopile length L_p, whereas the lateral dimension is given in function of monopile diameter D_p. All are shown on Figure 4.

3 SOIL PROFILES CONSIDERED IN THIS STUDY

A more general, and often appropriate class of soils where deposits of soils have stiffnesses which increase with depth. This consideration of soil non-homogeneity has proved to be more realistic in many practical cases where the effective stresses increase with depth. The soil modulus is usually taken to have a power law variation with depth as expressed by the equation:

$$E_s(z) = E_{sD}\left(\frac{z}{D_p}\right)^{\alpha} \quad (16)$$

where E_{sD} is the soil modulus at a depth z equal to the monopile diameter D_p, and α is an exponent that varies between zero and one (Figure 5).

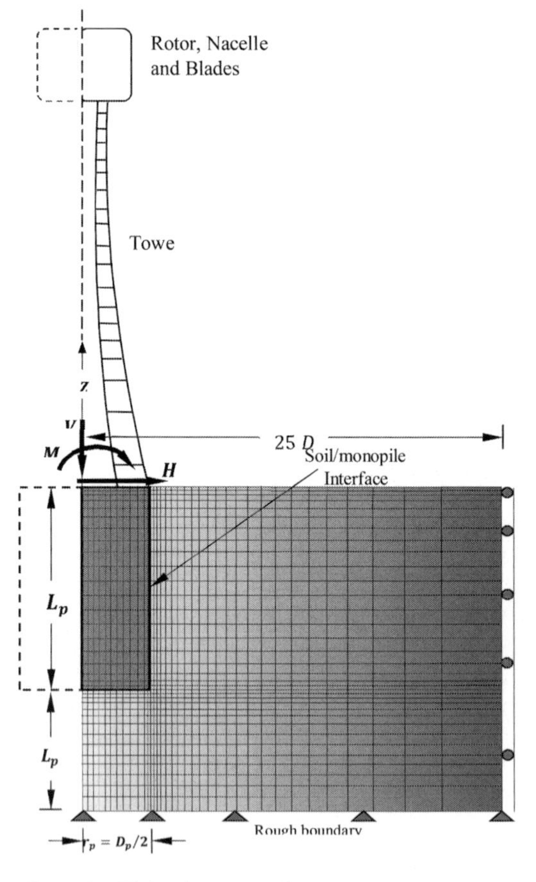

Figure 4. Finite element mesh.

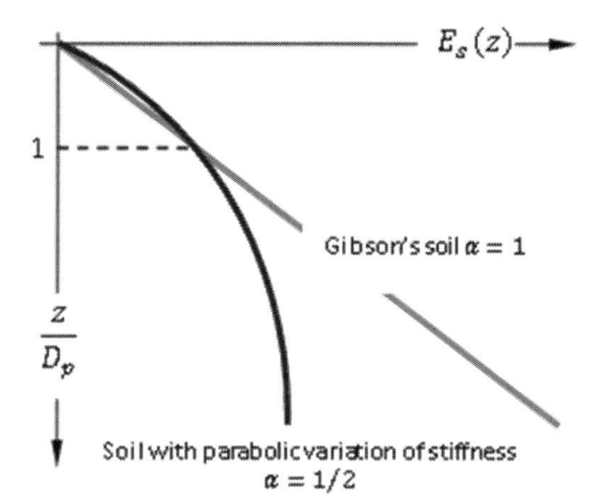

Figure 5. Variation of soil Young's modulus with depth according to a power law pattern.

The pattern of equation (16) is adopted in this paper to study OWTs in three soil profiles.

4 THE ESTIMATION OF THE NATURAL FREQUENCY OF THE OW TOWER

The estimation of the natural frequency of the OWTs by the finite element method depends on both global stiffness matrix $\boldsymbol{K}$ and the global mass matrix $\boldsymbol{M}$. The well-known equation to obtain the eigenvalues is:

$$det\left(\boldsymbol{K} - \omega^2 \boldsymbol{M}\right) = 0. \tag{17}$$

In this expression, $\boldsymbol{K}$ is the global stiffness matrix of the system soil, monopile and tower and $\boldsymbol{M}$ its global mass matrix.

4.1 *The mass matrix*

To evaluate the natural frequency of any vibrating structures two mass matrix options may be adopted; consistent mass matrix or lumped mass matrix. The lumped mass matrix is employed in this paper. It consists in converting the consistent mass matrix into a diagonal matrix.

The consistent mass matrix of an element i can be computed by:

$$\boldsymbol{m}_i = \pi\rho \int_A \boldsymbol{N}_i^T \boldsymbol{N}_i r dr dz \tag{18}$$

where ρ is and N_i. is the shape function matrix of the element i given by:

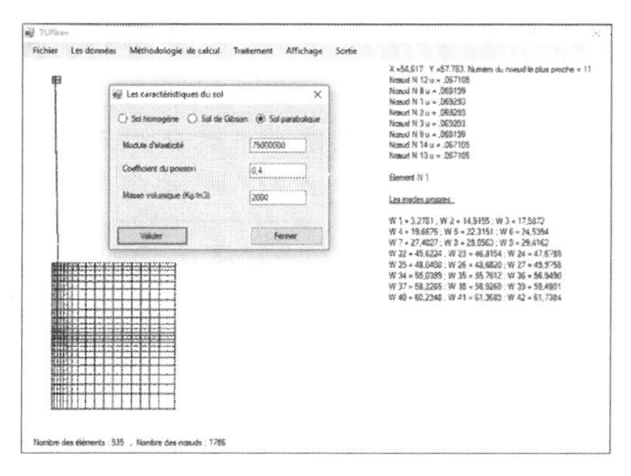

Figure 6. Screenshot of the computer code TURbaxi.

$$N_i = \begin{bmatrix} N_1 & & & \\ & N_1 & & \\ & & N_1 & \underbrace{\cdots\cdots\cdots}_{2,3,4,5,6,7,8} \end{bmatrix} \tag{19}$$

The HRZ lumping (Kaiser 1972) has been chosen to compute the mass matrix of each element. For full details the reader is referred to the aforementioned reference.

5 Description of THE computer code TURbaxi

TURbaxi is a finite element program written in visual studio community 2013 (Fig 4). This computer code uses the finite element method combined with the Fourier series for modelling the OWT structure and its supporting monopile as a soil/structure interaction problem. Computation of both mass matrix and stiffness matrix are based on the eight-noded ring element for modelling the system tower/monopile/soil and the six-noded interface element for modelling soil/monopile interface. The natural frequency of the whole problem is obtained by resolving the equation (17).

Three types of soil profiles and two interface (soil-monopile) states have been implemented. According to the value of α in equation (16), the soil can be homogenous, of Gibson's or having a stiffness which varies with depth in a parabolic pattern. The soil/monopile interface can be either smooth or rough.

6 ANALYSIS OF SOME INSTALLED OWTS

For the full availability of their structural data, five (05) Offshore Wind Turbines have been chosen from five (05) wind farm sites. These are: Lely (UK), Irene Vorrink (Netherlands), Belwind 1 (Belgium),

Table 1. Wind farms name and location with data source.

No.	Wind farm name and location	OWT	Data source
I	Lely offshore wind farm (Netherlands)	Lely A3	Arany et al. (2016), Zaaijer (2002)
II	Irene Vorrink offshore wind farm (Netherlands)	Irene A	Arany et al. (2016), Zaaijer (2002)
III	Belwind 1 offshore wind farm (Belgium)	Belwind 1	Arany et al. (2016), Damgaard et al. (2013)
IV	Gunfleet Sands offshore wind farm (UK)	Gunfleet Sands	Arany et al. (2016), Leblanc et al. (2011)
V	Burbo Bank Offshore wind farm (UK)	Burbo Bank	Arany et al. (2016), Versteijlen et al. (2011)

Table 2. Input parameters of the wind turbines chosen for this study.

Component dimension	I	II	III	IV	V
Tower height (m)	39	48.8	53	60	66
Substructure height (m)	7.1	5.2	37	28.0	22.8
Tower top diameter (m)	1.9	1.7	2.3	3.0	3.0
Tower bottom diameter (m)	3.2	3.5	4.3	5.0	5.0
Tower wall thickness (mm)	12.0	13.0	28	33.0	28.0
Substructure wall thickness (mm)	35	28	70	50	45
Monopile diameter (m)	3.7	3.5	5.0	4.7	4.7
Monopile wall thickness (mm)	35	35	70	65	75.0
Monopile embedded length (m)	21	20	35	38	24
Tower material Young's modulus (GPa)	210	210	210	210	210
Top mass (ton)	32.0	35.7	130.8	234.5	234.5
Soil density (Kg/m^3)	2000	2000	1800	2000	2090
Soil's Young's Modulus (MPa)	72	70	64	81	83
Soil's Poisson's ratio	0.4	0.4	0.4	0.4	0.4

Table 3. Measured and estimated frequencies of each OWT.

OWT	Fixed base	Measured	Interface state	Homogeneous	Gibson	Parabolic
					Computed frequency	
I	0.799	0.735	Smooth	0.742	0.737	0.739
			Rough	0.750	0.746	0.747
II	0.615	0.560	Smooth	0.570	0.565	0.567
			Rough	0.577	0.573	0.574
III	0.426	0.372	Smooth	0.376	0.374	0.374
			Rough	0.381	0.379	0.380
IV	0.376	0.314	Smooth	0.326	0.320	0.322
			Rough	0.333	0.328	0.330
V	0.339	0.292	Smooth	0.299	0295	0.296
			Rough	0.305	0.301	0.303

Gunfleet Sands (UK) and Burbo Bank (UK). Wind farm name and location with the data source of each are presented in Table 1. The Input parameters for each wind turbine are summarized in Table 2.

The natural frequencies estimated by the computer code TURbaxi for each state of interface and for each soil profile are given in Table 3.

By comparing the natural frequencies estimated by TURbaxi and the measured frequencies, three points are worth noticing:

1. As whole, computed natural frequencies are in perfect agreement with those established by measurements.
2. Numerical natural frequencies with smooth interface are much closer to the measured frequencies than those with rough interface for all profiles considered.
3. The Gibson‘s pattern gives the closest frequencies to the measured ones. It confirms its rep-

resentation of sand stiffness as linear variation with depth, since all offshore subsoils were sand.

7 CONCLUSIONS

Since the influence of its value has a direct impact on the cost of an OWT, the natural frequency is a parameter of paramount significance for a safe design. Hence, it is very hard to estimate the natural frequency of an OWT properly by simply using analytical procedures, such as the combination of the fixed base tower solution with other analytical expressions developed to account for the soil/monopile interaction.

Because a wind tower along with its monopile foundation exhibit a problem of axisymmetric geometry subjected to non-axisymmetric loading, the Fourier Series Aided Finite Element Method is perfectly appropriate to analyze this problem under lateral loading and to find the natural frequency of the vibrating tower in a problem where soil stiffness is taken into account. Based on this approach a computer code called TURbaxi has been written and applied in this paper to provide natural frequencies of five chosen OWTs installed in Europe, for which measured natural frequencies were known.

The numerical study encompassed three soil profiles: a soil with a constant stiffness with depth, a soil with a linear increasing of stiffness and a soil with parabolic variation of stiffness with depth. Furthermore, for each soil profile two interface states, namely rough and smooth were considered.

Computed results showed a perfect agreement with those of measured data regardless the interface state, although the smooth results were more accurate. The deviation in the worst case was under 5%. This shows clearly the reliability of the numerical procedure used to assess this crucial parameter which requires only elastic analysis.

REFERENCES

Abed Y., Amar Bouzid Dj., Bhattacharya S. and Aissa M.H. (2016). Static impedance functions for monopiles supporting offshore wind turbines in nonhomogeneous soils-emphasis on soil/monopile interface characteristics. Earthquakes and Structures, 10(5), 1143–1179.

Aissa M.H, Amar Bouzid Dj. and Bhattacharya S. (2017). Monopile head stiffness for serviceability limit state calculations in assessing the natural frequency of offshore wind turbines. International Journal of Geotechnical Engineering, 1–17.

Amar Bouzid Dj., Tiliouine B. and Vermeer, P.A. (2004). Exact formulation of interface stiffness matrix for axisymmetric bodies under non-axisymmetric loading. Computers and Geotechnics, 31(2), 75–87.

Arany L., Bhattacharya S., Adhikari S., Hogan S.J. and Macdonald J.H.G. (2015). An analytical model to predict the natural frequency of offshore wind turbines on three-spring flexible foundations using two different beam models. Soil Dynamics and Earthquake Engineering 74: 40–45

Arany L., Bhattacharya S., Macdonald J.H.G. and Hogan S.J. (2016). Closed form solution of Eigen frequency of monopile supported offshore wind turbines in deeper waters incorporating stiffness of substructure and SSI. Soil Dynamics and Earthquake Engineering 83:18–32

Arany L., Bhattacharya S., Macdonald J.H.G. and Hogan S.J. (2017). Design of monopiles for offshore wind turbines in 10 steps. Soil Dynamics and Earthquake Engineering, 92, 126–152.

Belvins R.D. (2001). Formulas for Frequencies and Mode shapes, Malabar, Florida, USA.

Damgaard M., Ibsen L.B., Andersen L.V. and Andersen J.K.F.. (2013). Cross-wind modal properties of offshore wind turbines identified by full scale testing. J Wind Eng Ind Aerodyn 116:94–108.

Jin-Hak Y., Sun-Bin K., Gil-Lim Y. and Andersen L.V. (2015). Natural frequency of bottom-fixed offshore wind turbines considering pile-soil-interaction with material uncertainties and scouring depth. Wind and Structures, An International Journal, 21(6), 625–639.

Kaiser H.F. (1972). The JK method: a procedure for finding the eigenvectors and eigenvalues of a real symmetric matrix, The Computer Journal, 15(3): 271–273.

Leblanc Thilsted C. and Tarp-Johansen N.J. (2011). Monopiles in sand-stiffness and damping. In: Proceedings of the European wind energy conference.

Prendergast L.J., Gavin K. and Doherty P. (2015). An investigation into the effect of scour on the natural frequency of an Offshore Wind Turbine. Ocean Engineering 101, 1–11.

Versteijlen W.G., Metrikine A.V., Hoving J.S. and de Vries W.E. (2011). Estimation of the vibration decrement of an offshore wind turbine support structure caused by its interaction with soil. In: Proceedings of the EWEA offshore conference.

Zaaijer M.B. (2002). Foundation models for the dynamic response of offshore wind turbines.

Numerical Methods in Geotechnical Engineering IX – Cardoso et al. (Eds)
© 2018 Taylor & Francis Group, London, ISBN 978-1-138-33203-4

Numerical method for evaluation of excess pore pressure build-up at cyclically loaded offshore foundations

Martin Achmus & Jann-Eike Saathoff
Leibniz Universität Hannover, Germany

Klaus Thieken
Achmus + CRP Planungsgesellschaft für Grundbau mbH, Berlin, Germany

ABSTRACT: In particular during storm events, a build-up of excess pore pressures may occur in the soil around cyclically loaded offshore foundations. Such accumulated excess pore pressure reduces the effective stresses in the soil and hence negatively affects the structural integrity. Even though the consideration of this degradation effect on the bearing capacity is commonly demanded by the involved certification or approval bodies, no general applicable and accepted method for the calculative verification currently exists. The paper presents a novel approach which allows for the transfer of the soil behavior obtained in cyclic DSS tests to the bearing capacity of the foundation structure by means of a 3D numerical model. The respected transfer function enables the consideration of site-specific cyclic DSS test results by taking into account the mean stress, the cyclic shear stress amplitude and the number of load cycles at each integration point of the numerical model. Hence, the numerical approach may contribute to the optimisation of common foundation solutions as well as to the verification of innovative foundation structures even in complex soil conditions.

1 INTRODUCTION

Offshore wind turbines (OWTs) will be increasingly used for the production of renewable energy in future. Experienced solutions (e.g. monopile) as well as innovative foundation solutions (e.g. mono-bucket) will be utilized for the support of the OWT structures. Regardless of the actual foundation type, considerable cyclic loading from the offshore environment has to be transferred by the structure to the subsoil. Under undrained or partially drained conditions, the corresponding shear stresses in the soil may lead to a build-up of accumulated excess pore pressures which in turn may cause reductions in shear strength. As the result, the cyclic loading may cause degradation of the bearing capacity which has to be accounted for in the design of cyclically loaded offshore foundations.

Even though the consideration of this degradation effect on the bearing capacity is commonly demanded by the involved certification or approval bodies (e.g. DNVGL 2017, BSH 2015), no general applicable and accepted method for the calculative verification currently exists.

From the theoretical point of view, four different approaches for the verification of the bearing capacity are available. The first and most reliable approach is the execution of a load test on a prototype foundation (or at least on a large-scale model) in same or comparable soil conditions. In fact, this option is ruled out for the practical design due to the immense time and cost effort of such tests in the offshore environment.

The second option is the application of empirical methods such as the p-y method for monopiles. However, empirical approaches are valid only for soil conditions and foundation geometries the method is calibrated on and hence appropriate conservatism is a basic prerequisite in order to enable the applicability for varied soil conditions. Hence, optimisation of common foundation solutions (e.g. monopile) would accompany with site-specific (or cluster-specific) adaption of the empirical approach which in turn would require further field testing or numerical considerations. Empirical approaches are seen only as transfer function from more sophisticated verifications and not as stand-alone approach. Moreover, widely accepted empirical approaches for innovative foundation concepts (e.g. monobucket) are generally missing and are—due to the complex bearing behaviour—hardly to be achieved (see Achmus et al. 2013).

Nowadays, the foundation behavior due to storm events can in principle be simulated by means of 3D numerical models incorporating complex constitutive models that enable the prediction of the soil behavior due to detailed load-time series of the entire storm event (see Tasan 2011, Cuellar 2011, Zachert 2015). However, the main problem in such approaches is the immense calculation effort to simulate an entire storm event and the accumulated calculative error due to simulating thousands of load cycles. At the time being, these methods are not suitable for practical application.

The fourth general approach, which is followed in this paper, combines the advantages of simplified numerical simulations for monotonic loading by means of a 3D numerical model with the high quality results regarding the soil behavior obtained from cyclic soil testing with large number of load cycles. Such approach is for instance applied by Andresen et al. 2011 who combine the information from site-specific cyclic DSS and triaxial tests as input for the numerical simulations. This method is termed "NGI approach". The simulations are finally executed by using the user—defined soil model UDCAM (Jostad et al. 2014) being implemented in the FE-Software Plaxis3D (Brinkgreve et al. 2016).

This paper presents a novel approach which is—due to its simplicity compared to the NGI approach—only valid for the simulation of the cyclic bearing capacity degradation whereas the cyclic deformation accumulation is currently not considered. The approach is solely based on the results of cyclic DSS tests which are seen as most representative for the considered build-up of excess pore pressure. However, the approach can be extended to enable the incorporation of further cyclic soil tests as outlined at the end of the paper.

2 CONCEPT OF THE APPROACH

The following concept is designed to consider the excess pore pressure build-up and hence the degradation in capacity due to cyclic lateral loading during an storm event. As loading input, the minimum and maximum loads acting on the foundation structure (H_{min}, H_{max}) and the number of load cycles N must be given. This means that the real load packages calculated for a certain storm event with varying mean loads and cyclic load amplitudes has to be transferred to a reference load with an equivalent number of cycles $N_{equi} = N$.

The presented approach consists of five main calculation steps to be executed in the numerical simulations. The single calculation steps are listed in the following and explained afterwards.

1. Application of mean load for determination of the reference stress state in the soil (Figure 1 left).
2. Increasing the load by cyclic load amplitude for determination of the cyclic stress amplitude (Figure 1 middle).
3. Calculation of excess pore pressure (for undrained conditions) based on stress amplitude and cyclic DSS test results.
4. Consolidation analysis for estimation of pore pressure dissipation during the storm event (Figure 1 right).
5. Calculation of shear strength degradation from maximum pore pressure and determination of post-cyclic bearing capacity due to the storm event.

2.1 *Calculation phase 1: Mean load*

Firstly, the numerical model is generated based on the site-specific soil conditions and the intended foundation geometry (see section 3). The foundation is subsequently subjected to the mean load $H_{mean} = 0.5 \cdot (H_{max} - H_{min})$ corresponding to the cyclic load conditions under consideration. The effective stress conditions at the integration points in terms of the mean principal effective stress σ'_m (see Equation 1) and the von-Mises stresses σ'_v (see Equation 2) are stored (by means of an appropriate *USER subroutine in the Abaqus program used).

$$\sigma_m' = \left(\sigma_I' + \sigma_{II}' + \sigma_{III}'\right)/3 \tag{1}$$

$$\sigma_v' = \sqrt{\frac{1}{2}\cdot\left(\sigma_I' - \sigma_{II}'\right)^2 + \left(\sigma_{II}' - \sigma_{III}'\right)^2 + \left(\sigma_{III}' - \sigma_I'\right)^2} \tag{2}$$

2.2 *Calculation phase 2: Cyclic amplitude*

The load applied in the numerical model is increased by the cyclic amplitude in order to achieve the stress conditions at the integration points for the maximum load. Subsequently, the cyclic stress ratio CSR is calculated from the stress differences between step 1 and step 2 as stated by Equation 3:

$$CSR = \frac{\sigma_m'^{(2)} - \sigma_m'^{(1)}}{\sigma_v'^{(2)} - \sigma_v'^{(1)}} \tag{3}$$

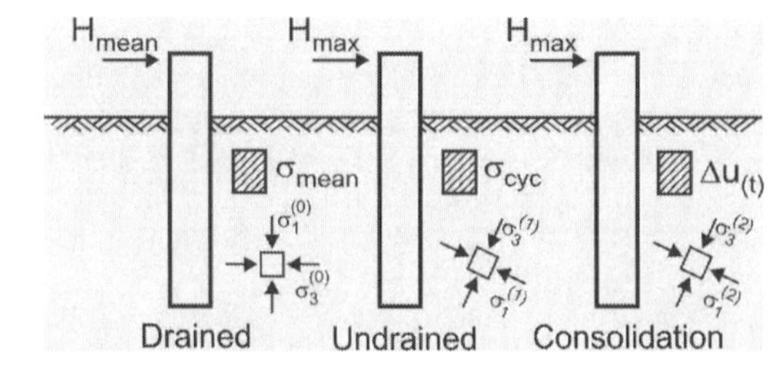

Figure 1. Different steps of the presented approach.

The CSR values are stored by the *USER subroutine for the following calculation step.

2.3 *Calculation phase 3: Excess pore pressure*

The following calculation step aims to calculate the accumulated excess pore pressure Δu under undrained conditions at the integration points corresponding to the applied cyclic load and the number of load cycles. Here, the approach proposed by De Alba et al. (1975) is utilized:

$$\frac{\Delta u}{\sigma'_m} = 0.5 + \frac{1}{\pi} \cdot \arcsin\left(2 \cdot \left(\frac{N}{N_{liq}(CSR)}\right)^{1/\alpha} - 1\right) \quad (4)$$

Herein, the degradation of shear strength $\Delta u/\sigma'_m$ (see Figure 2) depends on the ratio of applied load cycles N to the number of load cycles that yield full liquefaction N_{liq} when applying the CSR as obtained in calculation phase 2. The factor α represents a shape factor for the pore pressure build-up with the number of load cycles. The dependence of N_{liq} on the cyclic stress ratio for a certain soil has to be determined by a series of cyclic DSS tests. Hence, by applying the dependence of N_{liq} on the CSR, Equation 4 combines the behavior of the foundation-soil system with the results from site-specific cyclic element tests.

The excess pore pressures are calculated and stored by the *USER subroutine for each integration point. Figure 2 qualitatively shows the decrease in the effective stresses and the build-up of excess pore pressure as it occurs according to Equation 4.

2.4 *Calculation phase 4: Consolidation*

The fourth calculation step aims to account for the consolidation that takes place during the storm event. For this, the excess pore pressure field from phase 3 corresponding to a load cycle number $N_{equi} = 1$ ($\Delta u_{max}/N_{equi}$) is applied to the numerical model in order to simulate the consolidation process by means of a coupled pore fluid diffusion and stress analysis. The amount of the pore pressure dissipation is ruled by the permeability of the soil. The outcome of this analysis is a decay curve (see Figure 3, left) for each integration point of the numerical model. It is assumed that the accumulated excess pore pressure is generated due to an equivalent amount from each cycle ($\Delta u_{max}/N_{equi}$). The actual accumulated pore pressure Δu_{acc} in each element at the end of the storm is found by distributing the maximum excess pore pressure (Δu_{max}) over the duration of the storm (10800s) and superposing the pore pressure decay curves (exemplarily from black to light grey) for each pore pressure increase by means of an analytical consideration (Figure 3 right). The decay superposition in each integration point is done by means of the *USER subroutine. The number of applied decay curves has to meet with the actual number of equivalent load cycles corresponding to the storm event. The decay curves are placed with certain time duration to each other at which the initial pore pressure for $N_{equi} = 1$ is added to the remaining value from the previous load cycles.

Figure 4 shows the distribution of storm phases during a 35 h storm event as given in the German BSH standard (BSH 2015). A crucial question regards the consolidation time to be considered for the analysis. Taking into account the entire storm duration would not be representative for the actual behavior due to varying storm intensities which will yield in relative larger pore pressure build-up within the peak phase. One possibly for handling these different intensities would be the individual consideration of the storm phases in row. However, as such detailed information is commonly not available in a practical project, a conservative approach would be the application of the entire cyclic load from the storm event only in the 3 h peak phase (cf. Fig. 4).

In the exemplary calculations presented in this paper, a total effective consolidation time of 3h (10800s) is assumed, i.e. after each cycle a consolidation of 10800s/N_{equi}, takes place. For a 35 h storm as given in Fig. 4, this appears as a rather conservative approach.

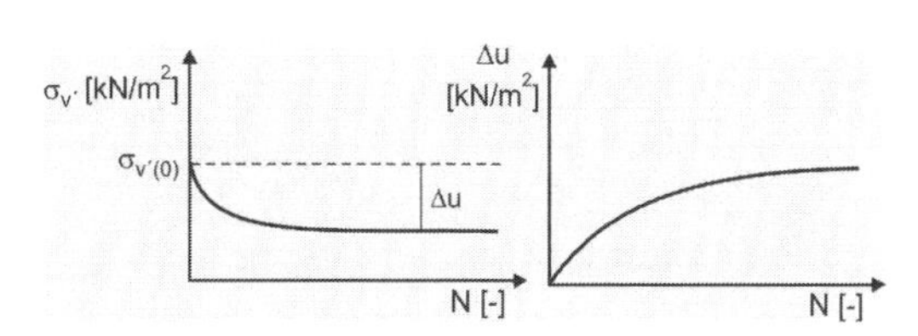

Figure 2. Degradation of effective stresses with number of load cycles.

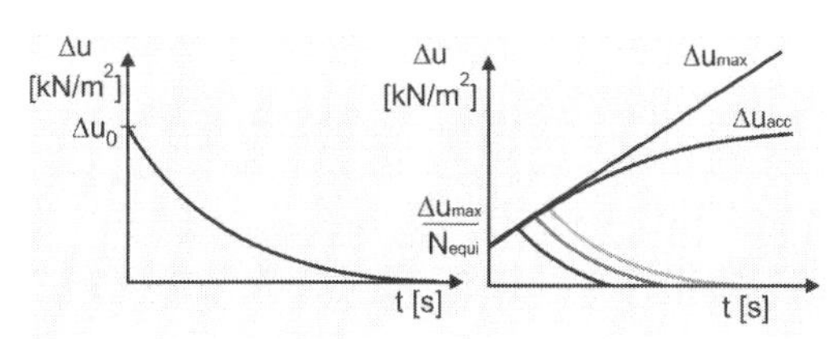

Figure 3. Schematic sketch for a decay curve (left), illustration of the analytical superposition (right).

2.5 *Calculation phase 5: Post-cyclic bearing capacity*

In the fifth step the calculated excess pore pressure for each element is used to derive an equivalent angle of friction respectively:

$$\varphi'_{equi} = \arctan\left[\left(1-\frac{\Delta u}{\sigma'_m}\right)\tan\left(\varphi'_{soil}\right)\right] \tag{5}$$

By applying these friction angles to the related element in the FE model the post-cyclic pile capacity can be determined.

3 NUMERICAL SIMULATION

In the following a monopile with a diameter of 7.6 m and an embedded length of 30 m is investigated under perfect swell load with a maximum load of 15 MN (H_{min} = 0, H_{max} = 15 MN) with a load eccentricity of 30 m (i.e. the corresponding maximum moment with respect to soil surface oscillates between M_{min} = 0 and M_{max} = 450 MNm). The number of load cycles is set to N = 40.

The execution and evaluation of a cyclic DSS test series is beyond the scope of this paper. For the exemplary calculations presented here, α was chosen to 5 and the following function N_{liq} (CSR) is applied:

$$N_{liq} = \frac{CSR^{-6}}{0.04} \tag{6}$$

The described analysis was carried out in the finite element program Abaqus. The three-dimensional numerical model of a monopile consists of approximately 30,000 C3D8(P) elements (ABAQUS 2016). Based on the symmetry only one half is modeled to reduce the computational effort (Figure 5). In preliminary analyses the mesh resolution and the model dimension have been

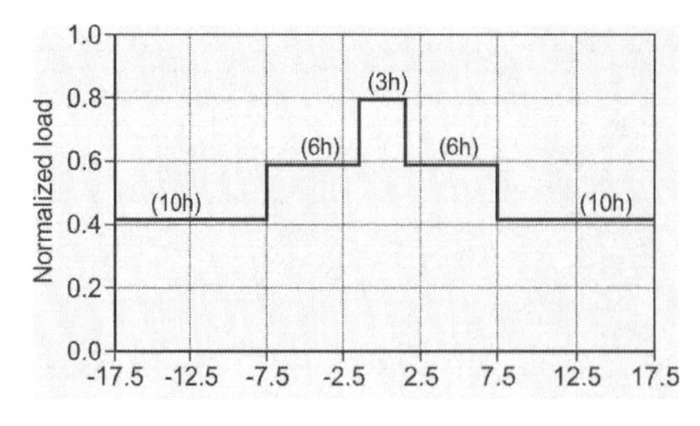

Figure 4. Characterisation of BSH 35h storm (see BSH 2015).

optimized to reach an appropriate computational effort and sufficiently accurate results.

The final model has a width of 12-times the diameter and a depth of 1.5-times the pile length. The model is fixed in all degrees of freedom at the bottom, in normal direction at the periphery and in y-direction at the symmetry plane. The monopile is modeled with a linear-elastic behavior with a Young's modulus E = 2.1 10^8 kN/m^2, a Poisson's ratio ν = 0.27 and a steel unit weight γ_{steel} = 68 kN/m^3. The wall thickness of the monopile varies along the length, and a scour protection on top of the subsoil was considered (Figure 6). The load is applied on a reference point which is connected to the monopile with a coupling constraint.

The soil parameters for dense sand used are shown in Table 1. The initial horizontal earth pressure at rest was calculated according to the common Jaky approach with k_0 = 1 - sin(φ') and the angle of dilatancy with $\psi = \varphi'$-30°. An elasto-plastic material law with Mohr-Coulomb failure criterion and stress-dependent stiffness was used. The stress dependent stiffness modulus, i.e. the oedometric stiffness, is considered with the following equation:

$$E_{oed} = \kappa \cdot 100\,kN/m^2 \cdot \frac{\sigma m}{100\,kN/m^2}^{\lambda} \left[\frac{kN}{m^2}\right] \tag{7}$$

Herein p_{ref} is the atmospheric reference stress, σ_m is the current mean principal stress in the considered soil element and κ and λ are soil dependent stiffness parameters.

For the contact modeling the elasto-plastic master-slave concept between the monopile and the adjusted soil was used in a way that a connection between the soil and the structure is present as well as their relative displacement is possible. The maximum coefficient of friction in the sand-steel interface is set to tan (2/3φ') and linearly mobilized within an elastic slip value of Δuel = 1 mm.

The calculation is executed in several steps. First, the initial conditions are set, in which the horizontal stress is calculated with the relation

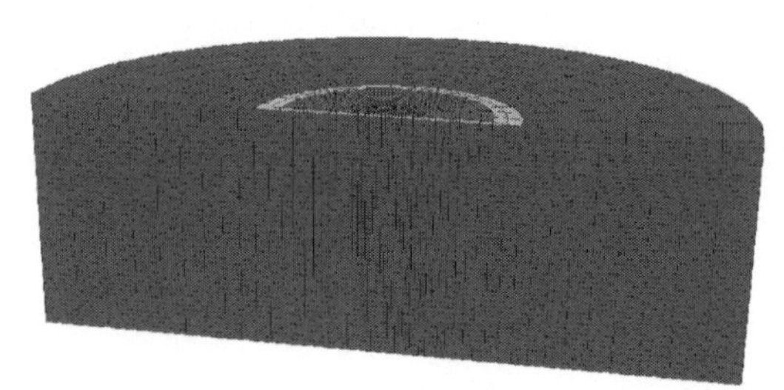

Figure 5. Abaqus model.

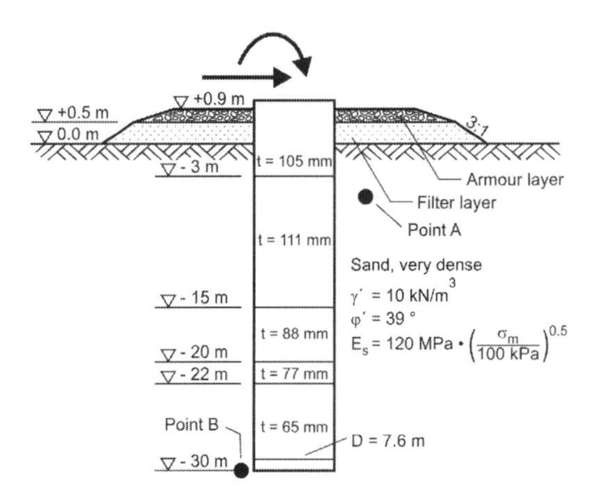

Figure 6. Schematic sketch of the monopile system with parameters used.

of Jaky. Subsequently, the scour protection and afterwards both the monopile and the contact are activated with a wished-in-place method. A permanent vertical load of 4.2 MN is applied in the next step. Afterwards, the mean lateral load (7.5 MN) of the perfect swell load and the related moment with a lever arm of h = 30 m is applied. Eventually, the maximal lateral load (15 MN) is applied in order to calculate the pore pressure with an empirical equation from the stresses which correspond to the entire storm event. As mentioned in the preceding section, the equivalent number of cycles is assumed to N = 40 and the effective consolidation time in the storm event is set to 3 h. The pore pressure is calculated according to Equation 4 at every integration point. The equivalent angle of friction and the excess pore pressure are derived using Equation 5.

For the consolidation analysis, the Abaqus model is extended in order to enable a coupled pore fluid and stress analysis. For the hydraulic consolidation analysis the drained model was converted into a simple linear-elastic coupled model by changing the element type to C3D8P. The boundary conditions were adapted for the additional degree of freedom.

Table 1. Soil parameters for dense sand.

Soil	Armour layer	Filter layer	Sand
Depth from [m]	0.9	0.5	0.0
Depth to [m]	0.5	0.0	45
Cohesion c' [kN/m²]	0.1	0.1	0.1
Unit weight γ' [kN/m³]	9.5	11	10
Friction angle φ'_{ref} [°]	35	32.5	39
Dilatancy angle ψ [°]	5	2.5	9.0
Poisson's ratio ν [1]	0.275	0.27	0.225
Stiffness parameter κ [−]	900	750	1200
Stiffness parameter λ [−]	0.4	0.4	0.5
Permeability k_f [m/s]	10^{-3}	10^{-4}	10^{-5}

4 RESULTS

In the following, the results for the excess pore pressures at two distinct points A and B are presented. These points are shown in the schematic drawing in Fig. 6. Point A is situated near the mudline and point B at the base of the monopile at the plane of symmetry.

The consolidation of excess pore pressures during the storm event has to be considered in order to estimate the reduction in capacity realistically. The decay curves for Points A and B calculated in the consolidation analysis step are depicted in Figure 7. After 40 cycles, a ratio of accumulated excess pore pressure to initial effective vertical stress of 1.0 for point A and of 0.84 for point B was calculated from Equation 4. The consolidation analysis was done for $1/N \cdot \Delta u$. For the case considered here, there is almost no excess pore pressure after 2000s and a large reduction can already be recognized after a few seconds.

To elucidate the effect of the soil permeability, also the decay curves for sand with double the permeability than given in Table 1 are shown in Fig. 7. A considerably faster decay of the excess pore pressures occurs in that case.

Figure 8 (left) shows the distribution of the normalized excess pore pressure, i.e. related to the initial effective stress. The field was calculated by using Equation 4, i.e. the results apply to totally undrained loading without any consolidation. The dark grey colour symbolizes a ratio of 1, which means that the excess pore pressure is equal to the vertical effective stress at the start of the storm. In these regions there is going to be no support for the pile capacity. The related distribution of the absolute excess pore is depicted in Figure 8 (right). It shows the main influenced area in loading direction and the influence at the base of the pile. The excess pore pressure distribution is the same as shown in Fig. 8 (left), but the absolute values were factored with $1/N_{equi}$. This is the stress state at the beginning of the consolidation calculation, which yields the decay curves exemplary shown in Fig. 7.

The calculation of the normalized accumulated excess pore pressure with consideration of the consolidation over the chosen effective consolidation time is shown in Figure 9 for the points A and B. The axis intercept for t = 0 s represents the pore pressure for N = 1. Thereby, the development of the maximum excess pore pressure calculated with Equation 4 is assumed to be equally distributed over the loading event, i.e. each load cycle creates an excess pore pressure of $1/N_{equi} \cdot \Delta u_{max}$. As mentioned above, the duration of the storm was conservatively set to 3h (10800s).

The arrows in Fig. 9 show the increase due to the next cycle, respectively. The dark line represents the upper bound (UB) with the added excess pore pressure for the individual cycle and the grey line the lower bound (LB) after the consolidation of the excess pore pressure of the considered load cycle. For point A (near mudline) a higher normalized accumulated excess pore pressure than for point B can be seen. However, the absolute accumulated excess pore pressure is much higher in point B than in point A, because the initial effective stress at point B is much higher than in point A.

This calculation procedure was done for all integration points in the finite element model,

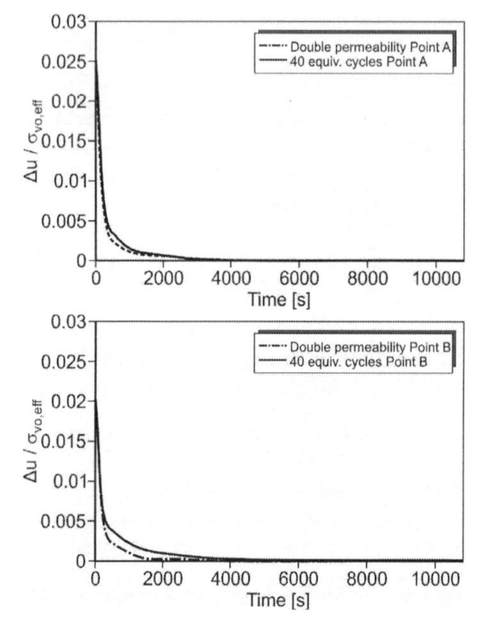

Figure 7. Excess pore pressure decay curves for Point A (top) and for point B (bottom).

applying the individual decay curves derived for these points. The result is the accumulated excess pore pressure field at the end of the storm event is depicted in Figure 10.

Figure 11 shows the accumulation curves for the point A under the assumption that the permeability of the sand soil is doubled with respect to the value given in Table 1, i.e. $k_f = 2\ 10^{-5}$ m/s. The start value for N = 1 is the same, but a higher permeability correlates with a flatter accumulation curve, because more excess pore pressure can dissipate in the same time.

Figure 12 elucidates the effect of the number of equivalent load cycles N_{equi}. The same cyclic load and the same effective consolidation time was applied, but the number of load cycles was once halved (N_{equi} = 20) and once doubled (N_{equi} = 80) with respect to the reference case (N_{equi} = 40). The start value of the curves ($\Delta u/N_{equi}$) has now changed. The value for 20 equivalent cycles is 4 times greater than for 80 cycles. Furthermore, the distance between the LB and UB and the inclination do not change, because the same decay curve is used over the whole 3h (compare Figure 11). The result is that for a higher number of load cycles a smaller accumulated excess pore pressure is obtained. The relative accumulation of pore pressure with respect to the start value is of course higher for a greater number of cycles. However, the final pore pressure is strongly affected by the start value, which is greater for smaller cycle numbers.

Using the resulting excess pore pressure field (Fig. 10) as input, the equivalent angles of friction for each element of the numerical model can be figured out and the post-cyclic capacity can be calculated by loading the monopile system monotonically.

The results in terms of load over pile head displacement are shown in Figure 13. The load-displacement curves for the case without (static) and with consideration of the storm event are depicted. The load displacement curve shows results up to a head rotation of 3° which is the limit given in DNVGL (2017). It can clearly be seen that there is a reduction in capacity and also a decreased stiffness due to the cyclic loading. The smallest reduction in capacity is given for an equivalent number

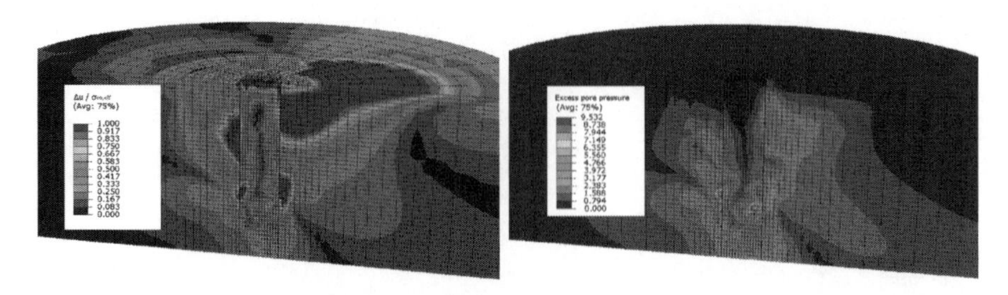

Figure 8. Visualization of $\Delta u/\sigma_{0,eff}$ (left) after 40 undrained cycles and of the excess pore pressure $\Delta u/N_{equi}$ (right).

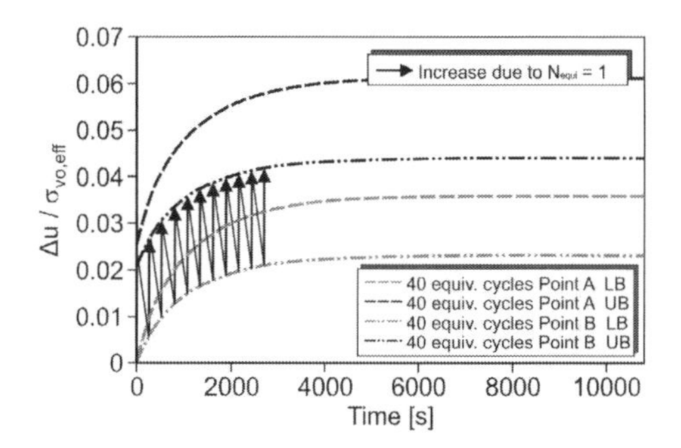

Figure 9. Accumulation of excess pore pressure for points A and B for an effective consolidation time of 3h (10800s).

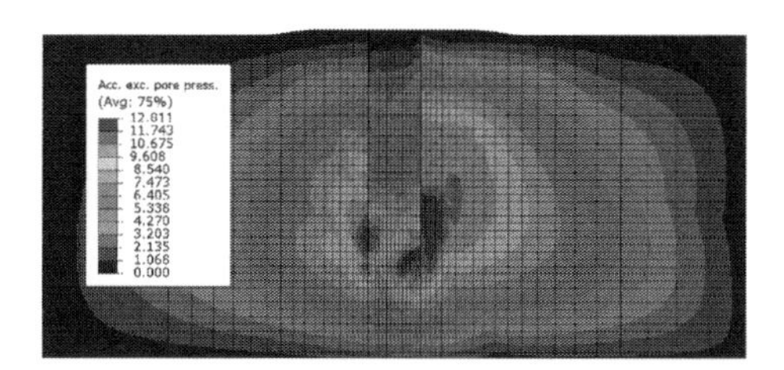

Figure 10. Calculated accumulated excess pore pressures at the end of the storm (N_{equi} = 40, t_{ges} = 10,800 s).

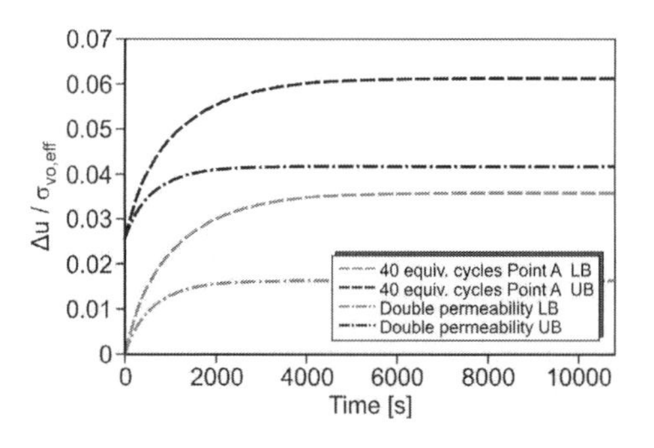

Figure 11. Comparison of accumulation of excess pore pressure for different permeabilities for 40 equivalent cycles.

of cycles of 80 and the greatest for 20 cycles. However, the effect of the number of equivalent load cycles on the degraded capacity is relatively small. For the case considered here, a reduction of capacity due to the storm event amounts to about 11.2% of the monotonic capacity.

A comparison of the bending moments after a displacement of 6 cm (which coincides approximately with the displacement under the maximum load H_{max} = 15 MN) is shown in Figure 14. A degradation of 2% in the moment curve over the depth can be seen.

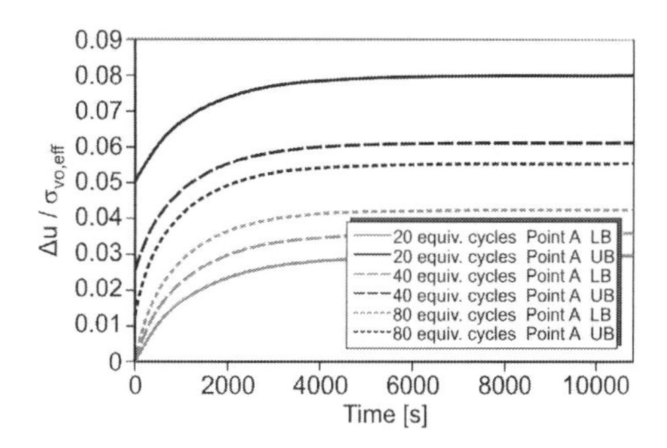

Figure 12. Comparison of accumulation of excess pore pressure for different numbers of equivalent cycles.

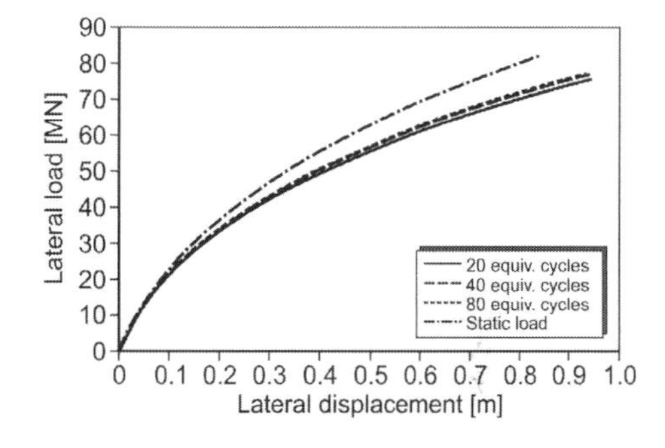

Figure 13. Comparison of load-displacement curves with and without the effect of the storm event.

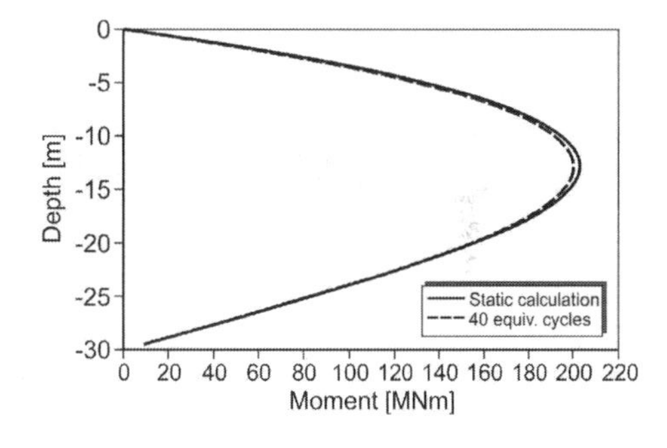

Figure 14. Comparison of bending moment prior to and after the storm event.

5 DISCUSSION

The conceptual framework presented and exemplarily applied here is capable to quantify capacity degradation of monopiles due to storm events. The first results gained are promising. However, several assumptions and simplifications were made and have to be checked and possibly modified. The most important aspects of the method which require further investigation are as follows:

- The general applicability of the approach of De Alba et al. (1975) (Equation 4) in combination with a function N_{liq} (CSR) needs to be checked thoroughly. Anyway, for practical projects comprehensive cyclic DSS tests have to be executed.
- For the example calculation, the effective consolidation time was conservatively estimated. However, no reliable method exists for a more accurate estimation of the consolidation time.
- In the current method, the cyclic load amplitude H_{max} - H_{min} is applied under drained condition which represents a conservative approach due to a less stiff behaviour compared to the actual existing nearly undrained behaviour. Further analyses should investigate the sensitivity regarding the drainage behaviour.
- So far, negative excess pore pressures which might also occur and their favourable effect regarding pile capacity are neglected.

A crucial point in practical design is the determination of a reference load with an equivalent number of load cycles which has the same effect as the real storm load scenario consisting of numerous load packages with varying mean load, cyclic load amplitude and number of cycles. Concepts for the calculation of equivalent load cycle numbers can already be found in literature, see e.g. EAP (2012) or LeBlanc (2010). However, the suitability of such methods for the problem of accumulated excess pore pressures considered here has to be checked.

The presented framework is a valuable tool to estimate the influence of accumulation of excess pore pressures on the capacity of offshore foundations. However, the method needs further development and refinement. It should also be noted that currently it is not validated by laboratory or field tests.

6 CONCLUSION

A concept for the estimation of excess pore pressure build-up was outlined and results for an exemplary application to a monopile system were presented. The method combines the results of cyclic element tests on soil with numerical simulations of manageable effort and complexity. It thus enables a prediction of capacity degradation due to storm events in a comprehensible and manageable way.

In the current state of the model, several simplifications and assumptions were made which require thorough reconsideration in the further development of the method. Also, a validation of the method by comparison with model or field tests is necessary in future. However, the first results gained and presented in this paper are promising and encourage further research on that matter.

REFERENCES

ABAQUS User's Manual 2016.Simulia: Providence, RI, USA.

Achmus, M., Akdag, C.T., Thieken, K. 2013: Load-bearing behavior of suction bucket foundations in sand, Applied Ocean Research 43, 157–165.

Andresen, L., Jostad, H.P. & Andersen, K.H. 2011: Finite Element Analyses applied in Design of Foundations and Anchors for Offshore Structures. *International Journal of Geomechnics*: 417–430.

American Petroleum Institute—API 2014. Recommended Practice 2GEO—Geotechnical and Foundation Design Considerations.

Brinkgreve R.B.J., Engin, E., Swolfs, W.M. 2016. PLAXIS 3D Manual.

BSH, Bundesamt für Seeschifffahrt und Hydrographie 2015: Standard Konstruktion, Mindestanforderungen an die konstruktive Ausführung von Offshore Bauwerken in der ausschließlichen Wirtschaftszone (AWZ).

Cuellar, P. 2011: Pile foundations for offshore wind turbines: numerical and experimental investigations on the behavior under short term and long term cyclic loading. *Dissertation*, Technische Universität Berlin.

De Alba, P., Chan, C.K. & Seed, H.B. 1975. Determination of soil liquefaction characteristics by large scale laboratory tests. EERC REP NO 75–114, University California, Berkely CA.

DNVGL 2017. Offshore Standard: Design of Offshore Wind Turbine Structures DNV-ST-0126. Det Norske Veritas, Norway.

EA-Pfähle (EAP), Empfehlungen des Arbeitskreises Pfähle 2012. German Society for Geotechnics e.V. DGGT, 2nd Edition, Ernst & Sohn, Berlin.

Glasenapp, R. 2016. Das Verhalten von Sand unter zyklischer irregulärer Belastung, *Dissertation*, Technische Universität Berlin.

Jostad, H.P., Grimstad, G., Andersen, K.H., Saue, M., Shin, M. & You, D. 2014. A FE Procedure for Foundation Design of Offshore Structures—applied to study a potential OWT Monopile foundation in Korean Western Sea. *Geotechnical Engineering Journal of ESATGS & AGSSEA* Vol. 45 No. 4: 63–72.

Kluge, K. 2008. Soil liquefaction around Offshore pile foundations-scale model investigation. *Dissertation*, Technische Universität Braunschweig.

LeBlanc, C., Houlsby, G.T. & Byrne, B.W. 2010. Response of stiff piles in sand to long-term cyclic lateral loading. *Geotechnique* Vol.6 0, No. 2: 79–90.

Tasan, H.E. 2011. Zur Dimensionierung der Monopile Gründungen von Offshore Windenergieanlagen. *Dissertation*, Technische Universität Berlin.

Zachert, H. 2015. Zur Gebrauchstauglichkeit von Gründungen für Offshore-Windenergieanlagen. Veröffentlichungen des Instituts für Bodenmechanik und Felsmechanik am Karlsruher Institut für Technologie, Heft 180.

Numerical Methods in Geotechnical Engineering IX – Cardoso et al. (Eds)
© 2018 Taylor & Francis Group, London, ISBN 978-1-138-33203-4

Stiffness of monopile foundations under un- and reloading conditions

Klaus Thieken, Martin Achmus, Jann-Eike Saathoff, Johannes Albiker & Mauricio Terceros
Leibniz University Hannover, Germany

ABSTRACT: The foundation stiffness under operational loads is a crucial point in the design of monopile foundations. As key design issues, structural eigenfrequencies as well as the structural fatigue loading are significantly affected. According to measurement results, the current calculation methods lead to an underestimation of the actual existing foundation stiffness. The reason is most likely that the stiffness is usually figured out from load-displacement curves for virgin monotonic loading, whereas actually the stiffness under un- and reloading conditions is relevant for the operational behavior. Concerning this matter, the proposed paper presents numerical simulations to derive the foundation stiffness under un- and reloading condition. Herein, the HSsmall soil model is applied, that accounts for the strain dependency of soil stiffness. By varying the load level and the cyclic load amplitude, it is demonstrated that the secant foundation stiffness for un- and reloading is only slightly affected by the load level and the magnitude of the un- and reloading amplitude. Hence, the application of constant foundation stiffness in the overall dynamic structural analysis appears reasonable.

1 INTRODUCTION

Monopile foundations are currently a widely used foundation concept in water depths up to 30 m or even 40 m. For such depths and the current generation of wind turbines with rated energy output > 5 MW, pile diameters of 6 m and more are necessary to fulfill the design requirements. A crucial requirement for monopiles with such diameters regards the stiffness under operational loading, which strongly affects the eigenfrequency of the overall structure. For the usual "soft-stiff" design it must be ensured that the eigenfrequency lies sufficiently above the 1P excitation frequency resulting from the rotational frequency of the wind turbine in order to avoid resonance effects and associated high fatigue loads.

It is common practice in the design of monopiles to calculate the bearing behavior with an integrated model, in which the soil is represented by springs with non-linear load-displacement relationships (Figure 1). This approach is called p-y method and recommended by the current offshore guidelines (API 2014, DNV 2013). For piles in sandy soils, based on investigations of Reese et al. (1974) and Murchison et O'Neill (1984) a hyperbolic tangent function is used to describe the p-y relationship. The p_u-value and the initial stiffness of the p-y curve E_{py} determine the p-y curve. These values are depth-dependent and calculated with only the angle of internal friction φ' and the buoyant unit weight γ' of the sandy soil as input parameters. The method was originally developed and calibrated for flexible piles. For the application to large-diameter monopiles a modification of this method is necessary in order to account for the bearing behavior which is more similar to the behavior of a rigid pile.

The p-y curves resemble the behavior of the pile-soil system under virgin monotonic loading. In a fully integrated model, due to constraints regarding the model complexity normally only linear springs can be considered. In many design approaches, the initial stiffness of the p-y curves, i.e. the E_{py}-value (see Fig. 1), is applied. Actually, under operational loading un- and reloading from a certain maximum load level predominates. Therefore, the stiffness under un- and reloading should be used in dynamic calculations, which in general is dependent on the level of the maximum load and the magnitude of the un- and reloading amplitude.

This paper gives a short overview of the p-y approaches currently applied in design practice. The results of these approaches are compared with the results of a numerical simulation model for an exemplary monopile-soil configuration. Herein, secant stiffnesses for both virgin loading and un- and reloading are considered.

The aim of this paper is to show the need for a more sophisticated approach, which incorporates the un- and reloading stiffness into a p-y approach.

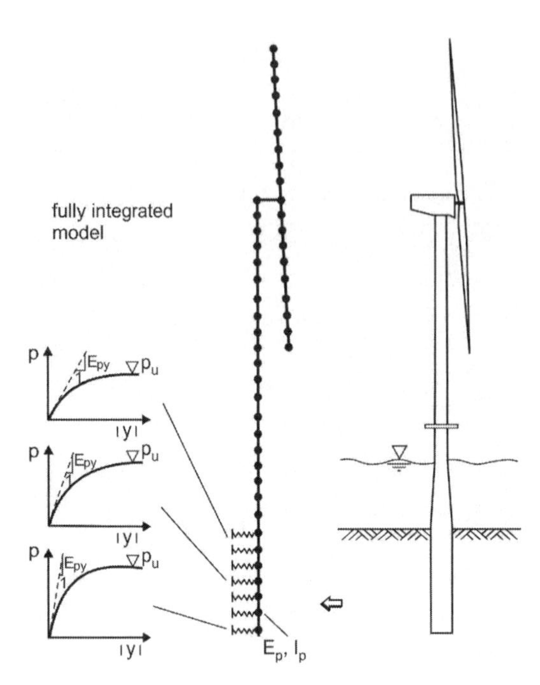

Figure 1. Modelling monopile-soil interaction in a fully integrated model by p-y curves.

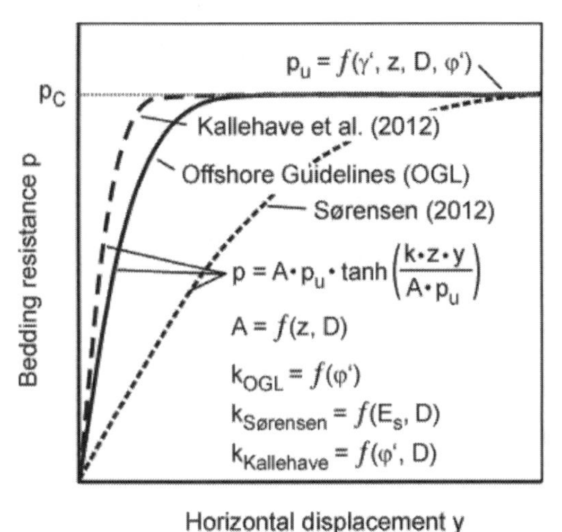

Figure 2. p-y-relation according to Offshore Guidelines and modifications after Kallehave et al. (2012) and Soerensen (2012).

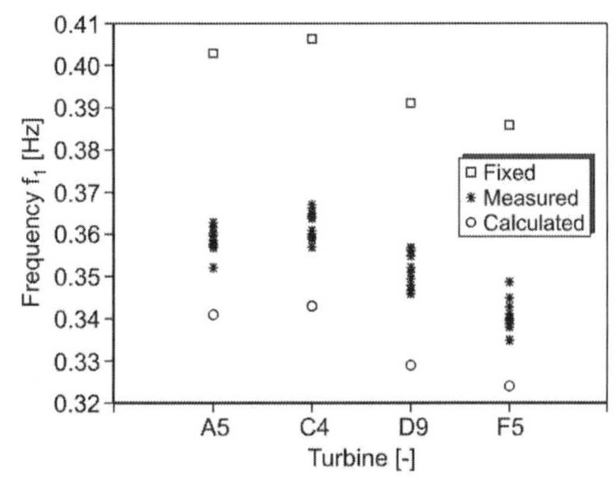

Figure 3. Measured and calculated eigenfrequencies (Damgaard et al., 2014).

2 P-Y APPROACHES FOR MONOPILES IN SANDY SOIL

The original p-y approach stated in the offshore guidelines (OGL) was calibrated with load tests on flexible piles (Reese et al. 1974). Several numerical investigations showed that the deflections for large-diameter monopiles are significantly underestimated for maximum service loads (e.g. Abdel-Rahman et Achmus 2005, Sørensen 2012, Thieken et al. 2015a). Sørensen (2012) proposed a modification of the original p-y curves by applying an initial stiffness value dependent on oedometric modulus of the soil and the diameter of the pile. For large-diameter piles, this adaptation leads to a softened p-y curve (see Fig. 2).

In contrast to these findings, field measurements on operating offshore wind turbines founded on monopiles indicated that the actual eigenfrequency was higher than predicted in the design and thus the foundation stiffnesses for operational dynamic loads were obviously underestimated (Damgaard et al. 2014, see Fig. 3). Even the application of the initial stiffness of the p-y curves in the calculations did not yield sufficient stiffness. Therefore, Kallehave et al. (2012) also proposed a modification of the initial stiffness value in the original p-y method, which this time leads to a hardened p-y curve (Fig. 2).

Improved p-y approaches for monopiles have recently been developed (Thieken et al. 2015b, Thieken 2015, Byrne et al. 2017). However, up to now in most practical projects the approaches of API and of Kallehave et al. were applied in the determination of the eigenfrequency of wind turbine structures founded on monopiles in sand soils. Therefore, the numerical simulation results will be compared here to these approaches.

3 NUMERICAL SIMULATION

3.1 *General*

Numerical simulations of the load-bearing behavior of a monopile with a diameter of 8 m and an

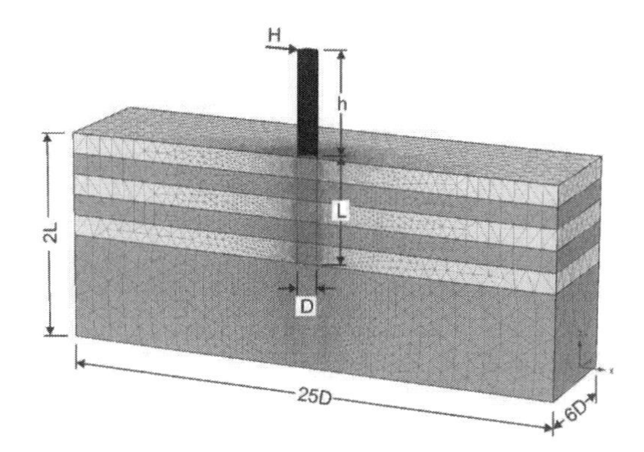

Figure 4. Overview of Plaxis3D model.

embedded length of 32 m (i.e. with an aspect ratio of L/D = 4) were carried out. The wall thickness of the steel pipe pile amounts t = 86.35 mm and was chosen according to the respective recommendation of API (2014). A horizontal load with a certain lever arm h (Fig. 4) with respect to the soil surface was applied. By varying h, different combinations of horizontal and moment loading could be investigated.

The numerical analysis was carried out with the finite element program Plaxis3D (Brinkgreve et al, 2014). Due to the symmetry conditions of the loading and the pile only one half of the system could be modeled. By means of preliminary analyses a sufficient mesh fineness with regard to solution accuracy was found and the model dimensions were chosen such that there is no impact of the boundaries. The final model is discretized with roughly 200,000 elements and has a width of 25D in loading direction, a depth of 2 L and 6D perpendicular to the loading direction (Figure 4).

The monopile is modeled tubular and open-ended. For the steel material, linear-elastic behavior was assumed with a Young's modulus $E = 2.1E^8$ kN/m², a Poisson's ratio $\nu = 0.27$ and a steel unit weight $\gamma_s = 68$ kN/m³.

For modeling the contact behavior between steel and soil, elasto-plastic interface elements are attached to the monopile. The maximum shear stress in the contact surface τ_{max} results from the product of the horizontal stress σ_H and the tangent of the contact friction angle $\delta = 2/3\varphi'$.

In the numerical simulation, first the initial stress state in the soil was calculated with a k_0-procedure, i.e. $\sigma_H = k_0 \cdot \sigma_V$. Then the monopile was installed in a wished-in-place procedure, i.e. effects of the installation procedure on the stress state of the soil were disregarded. A vertical load of 4 MN was then applied to the pile. In the last stage, the lateral and moment loading was applied with alternating virgin loading steps and un- and reloading cycles.

3.2 *Constitutive law for the soil*

For consideration of the non-linear soil behavior at small strains, an advanced material law is necessary. The increased stiffness for small strains can be considered with the hardening soil small strain (HSsmall) model according to Benz (2007). This model takes the strain dependency of soil stiffness into account, which is very important for the topic investigated in this study. This model is an upgraded version of the sophisticated Hardening Soil Model (HSM) according to Schanz (1998), which is an elasto-plastic model with isotropic hardening and also enables the consideration of stress-dependent soil stiffness, for instance.

In the HSsmall soil model reference values for the stiffness moduli are defined, which are calculated based on a stress-dependent power law controlled by an exponent m and a reference stress p_{ref}. Either the maximum principal stress σ_1 or the minimum principal stress σ_3 is used as a stress measure:

$$E_{oed} = E_{oed,ref} \cdot \left(\frac{\sigma_1}{p_{ref}} \right)^m \tag{1}$$

$$E_{50} = E_{50,ref} \cdot \left(\frac{\sigma_3}{p_{ref}} \right)^m \tag{2}$$

$$E_{ur} = E_{ur,ref} \cdot \left(\frac{\sigma_3}{p_{ref}} \right)^m \tag{3}$$

$$G_0 = G_{0,ref} \cdot \left(\frac{\sigma_3}{p_{ref}} \right)^m \tag{4}$$

For the present investigations, the exponent m in equations (1) to (4) is assigned the same value as λ in equation (5). Equations (6) and (7) are applied to calculate E_{50} and E_{ur} in dependency on the oedometric stiffness E_{oed}:

$$E_{50} = \frac{1-\nu-2\nu^2}{1-\nu} \cdot E_{oed} \tag{6}$$

$$E_{ur} = 3 \cdot E_{50} \tag{7}$$

To correctly approximate the aimed trend of the shear modulus and the stiffness moduli over the depth, the pile is embedded in 5 layers of equal thickness and a 6th layer reaches downward to the bottom of the model (see Fig. 4). This had to be done due to the different exponents m. The different stress exponents ($m \neq m_{G0}$, with $m_{G0} = 0.5$)

result in a deviation, which is minimized by choosing a sufficiently small layer thickness.

The dependence of the shear modulus G and thus of the soil stiffness on the magnitude of shear strain is modeled according to the approach of Santos and Correia (2001) (see also Figure 5):

$$\frac{G}{G_0} = \frac{1}{1 + 0.385\frac{\gamma}{\gamma_{0.7}}} \tag{8}$$

Here, the reference shear strain $\gamma_{0.7}$ corresponds to the shear strain at which the soil stiffness is degraded to 72.2% with respect to the maximum value G_0.

For the simulations in this paper, the soil parameters given in Table 1 were applied, which are typical for a dense to very dense sand.

3.3 *Validation of the numerical model*

In order to validate the numerical model the experimental field test of Li et al. (2015) in intermediate scale has been back-calculated. The test pile had an embedded length of L = 2.2 m, a wall thickness of t = 14 mm and a diameter of D = 0.34 m, implying an aspect ratio of about L/D = 6.5, which is more slender, but not totally out of range for a monopile. The soil was composed of homogeneous dense to very dense sand. The horizontal load was applied with a lever arm of h = 0.4 m.

In Figure 6, the field test results are compared to the results of the numerical back-calculation. A quite satisfying agreement for both, the virgin loading curve as well as the un- and reloading cycles can be stated. The hysteresis loop from the numerical simulation is more pronounced than it was found in the experimental tests. However, regarding the average (secant) stiffness under un- and reloading at least fair agreement is obtained.

Table 1. Soil parameters.

Buoyant Unit weight γ	[kN/m³]	10.31
Void ratio e_{init}	[-]	0.6
Cohesion e'_{ref}	[kN/m²]	0.1
Angle of internal friction φ'_{ref}	[°]	40
Angle of dilatancy ψ	[°]	10
Shear strain $\Upsilon_{0,7}$	[-]	0.001
Poisson's ratio (un- and reloading) u'_{ur}	[-]	0.25
Reference stress p_{ref}	[kN/m²]	100
Earth pressure coefficient at rest k_0	[-]	0.357
Stiffness parameter λ (= m)	[-]	0.5
Reference Tangent stiffness for oedometric initial loading $E_{oed,ref}$	[MN/m²]	120

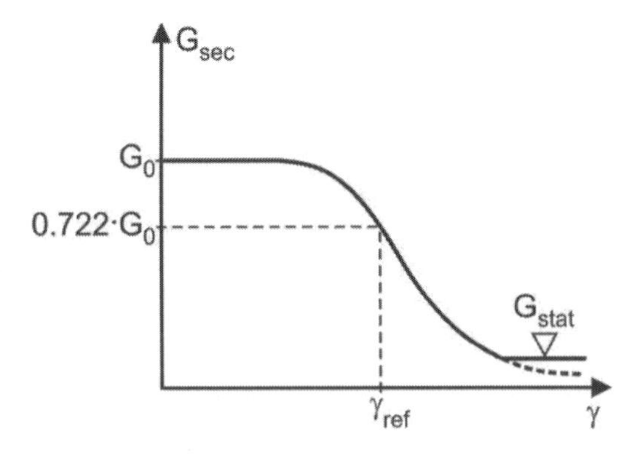

Figure 5. Relation between shear modulus and shear strain.

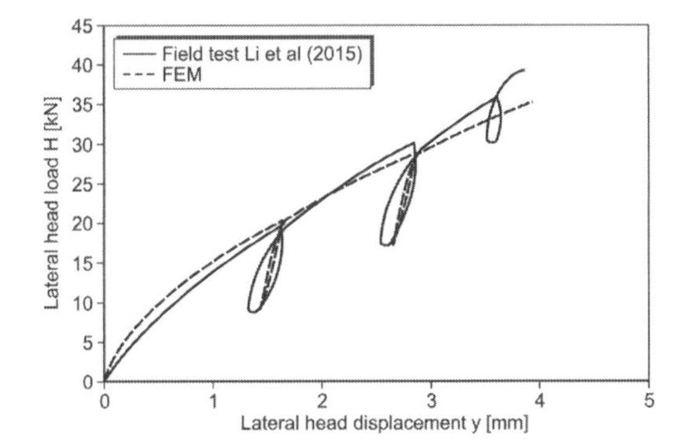

Figure 6. Load-deflection curve from a field test of Li et al. (2015) and from a numerical back-calculation.

Therefore, it can be concluded that the finite element model is capable to reproduce the load-bearing behavior of the test pile realistically.

4 RESULTS FOR A REFERENCE SYSTEM

In the following, detailed calculation results in terms of load-deflection and moment-rotation curves and secant stiffnesses are presented for a reference system with a lever arm h = 50 m. First a simulation with monotonic loading was carried out. The load was increased until a pile head rotation of 0.08° occurred. Subsequently, simulations with alternating loading and un- and reloading cycles were carried out. The magnitude of the unloading amplitudes was varied with 25%, 50%, 75% and 100% (full unloading) of the formerly reached maximum load.

Figure 7 shows the load-deflection and moment-rotation relations for the monotonically

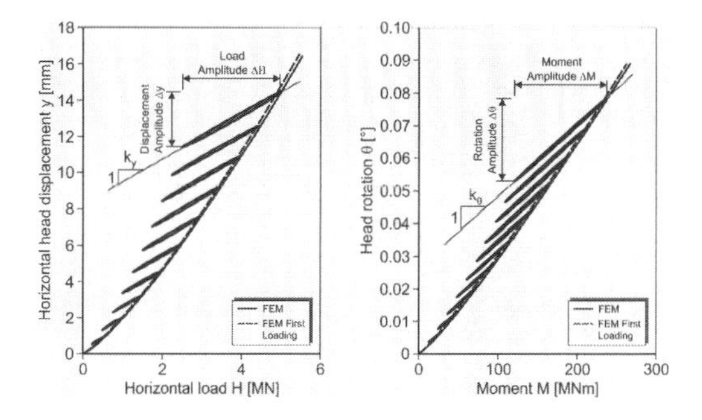

Figure 7. Numerically obtained load-deflection and moment-rotation curves for static and cyclic loading, exemplary for the reference system.

and cyclically loaded reference system exemplarily for an unloading amplitude of 50% of the formerly reached load. The alternating application of virgin loading and un- and reloading cycles is theoretically not equivalent to a purely monotonic application of the load. However, as can be seen in Fig. 7, the influence of intermediate un- and reloading cycles is rather small and can be neglected. Therefore, only the curves achieved with alternating virgin and un- and reloading steps are evaluated in the following.

From the curves given in Fig. 7, the secant stiffnesses $K_{s,y}$ and $K_{s,\theta}$ can be calculated. For virgin loading, the total loads and deformations are used, for un- and reloading, the magnitude of unloading and the respective deformation amplitude are used.

- For virgin monotonic loading:

$$K_{s,y} = \frac{H}{y}, \qquad K_{s,\theta} = \frac{M}{\theta} \tag{9}$$

- For un- and reloading:

$$K_{s,y} = \frac{\Delta H}{\Delta y}, \qquad K_{s,\theta} = \frac{\Delta M}{\Delta \theta} \tag{10}$$

Figure 8 shows the so-defined secant stiffness $K_{s,y}$ for the first loading and different un- and reloading stages with regard to the cyclic loading amplitude ΔH, and in Figure 9 against the deformation amplitude Δy. Obviously, the secant stiffnesses for un- and reloading are significantly larger than the secant stiffnesses for first loading, except for the initial part of very low load levels. It is recognized that the secant stiffness for un- and reloading slightly decreases with the magnitude of the loading amplitude and additionally with increasing degree of unloading. However, these

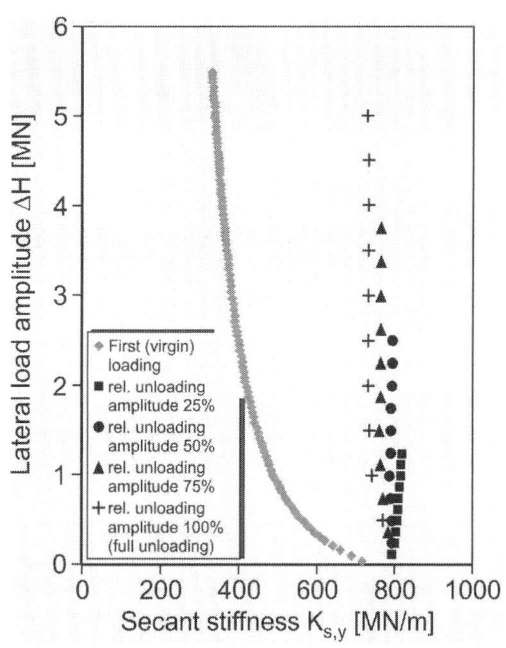

Figure 8. Secant stiffness for un- and reloading against the loading amplitude, for different relative levels of unloading.

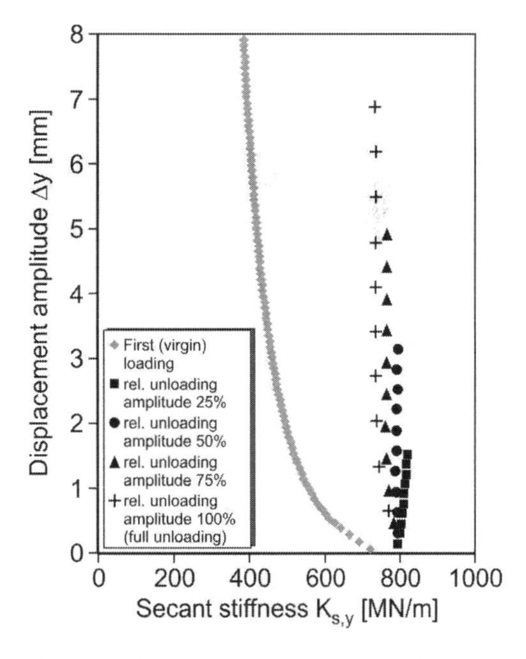

Figure 9. Secant stiffness for un- and reloading against the deformation amplitude, for different relative levels of unloading.

decreasing tendencies are not very pronounced for the regarded reference system with a lever arm of h = 50 m. These statements apply to both, the load-deflection and the moment-rotation relations.

5 VARIATION OF THE LEVER ARM

The results presented in Figs. 8 and 9 are valid for a lever arm of the horizontal load of h = 50 m. In order to investigate the effect of the lever arm on the secant stiffnesses and their load-dependence, numerous calculations with varying lever arms were carried out. In total, 21 load-displacement curves have been calculated each again with 10 un- and reloading steps with magnitude of 25% to 100% of the formerly achieved load level. In Figure 10, the results of the calculations with 50% un- and reloading are shown in terms of the pile head displacement in dependence of horizontal and moment load. It should be noted that also negative moments and thereby negative lever arms were considered in the calculations, since it is possible that such load combinations occur at the embedment point of a monopile (see Dubois et al. 2016).

In the following, first the curves for monotonic loading and then the curves for un- and reloading are presented and analyzed with respect to the secant stiffnesses.

5.1 *Secant stiffnesses for virgin monotonic loading*

Figure 11 shows the load-displacement and load-rotation surfaces for virgin monotonic loading only. The maximum considered load represents a typical ULS load of the investigated monopile system. A zero crossing of the surfaces can be seen in both figures. This crossing does not occur for a zero moment loading, but for a configuration where the negative moment load and the positive lateral load cancel each other out with respect either to head deflection or head rotation.

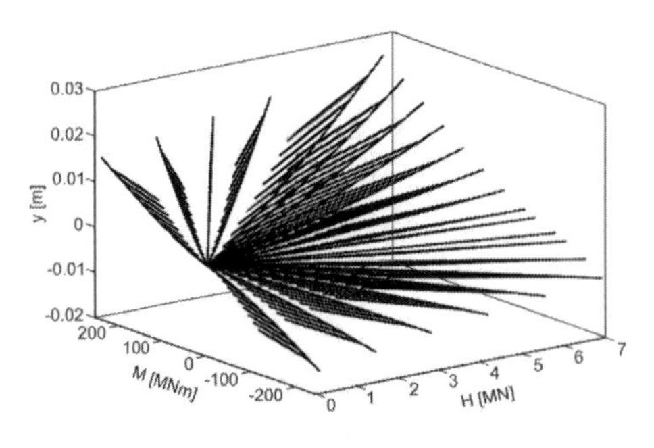

Figure 10. Load-deflection curves for virgin loading and 50% un- and reloading for varying lever arms.

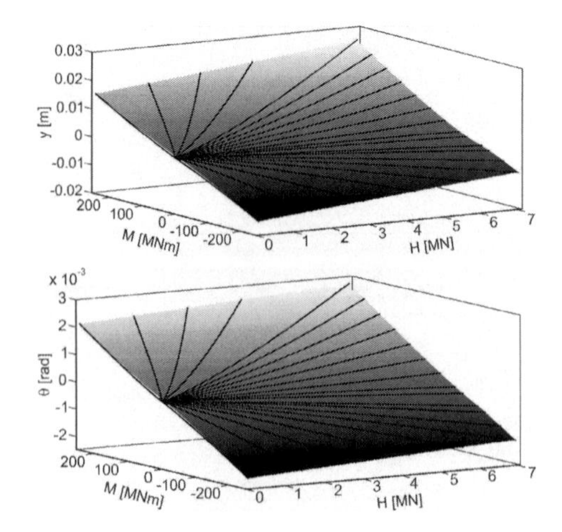

Figure 11. Load-displacement surface (top) and load-rotation surface (bottom) for virgin monotonic loading.

From the curves depicted in Fig. 11, secant stiffnesses $K_{s,y}$ and $K_{s,\theta}$ can be calculated as defined in Eq. 9. Fig. 12 shows these secant stiffnesses resulting of the numerical simulations (black dots). Only the results for positive moments are presented. For H-M combinations with different signs of horizontal load and moment, $\theta = 0$ or $y = 0$ can occur, which leads to infinite stiffness values (cf. Eq. 9). The reason for that is that in the simplified stiffness definitions used here, the coupling between pile head rotation and horizontal load is neglected, i.e. the coupling terms in the stiffness matrix were set to zero.

The shape of all the secant stiffness curves is similar to the shape found for the reference system (see Fig. 8). The stiffness considerably decreases with increasing load, which is the outcome of the non-linearity of the load-deflection and moment-rotation curves.

For comparison, also the secant stiffnesses derived from the static p-y curves of the API (2014) and the Kallehave et al. (2012) approach are shown in Fig. 12. Both approaches predict an almost constant, i.e. load-independent secant stiffness. The Kallehave approach yields significantly higher stiffnesses than the API approach, which of course was the intention of the modification made by Kallehave et al. However, compared to the initial stiffness gained from the numerical simulations, the Kallehave approach either under- or overestimates the stiffness, dependent on the considered lever arm.

5.2 *Secant stiffnesses for un- and reloading*

Figure 13 shows the load-displacement and load-rotation surfaces for the extracted load amplitudes

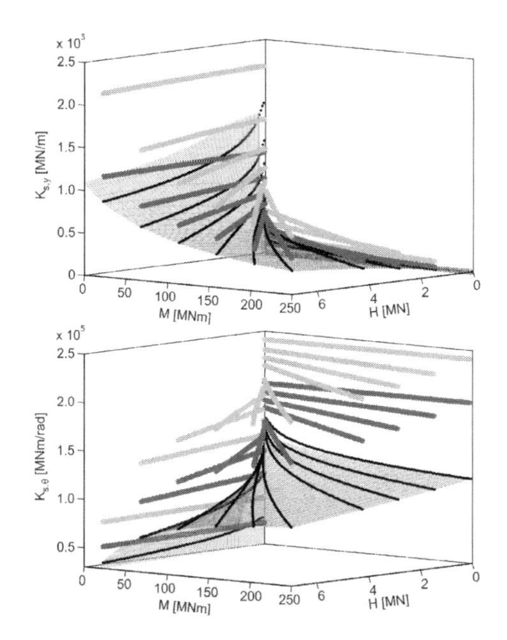

Figure 12. Response stiffness surfaces for monotonic virgin loading (gray surface with black dots) made from load-displacement/rotation surface: lateral stiffness (left) and moment stiffness (right) in comparison with API (dark gray) and Kallehave et al. (light gray).

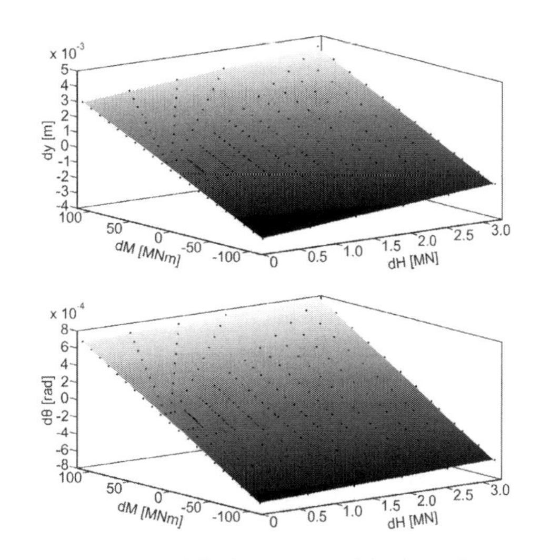

Figure 13. Load-displacement and load-rotation surfaces for 50% un- and reloading.

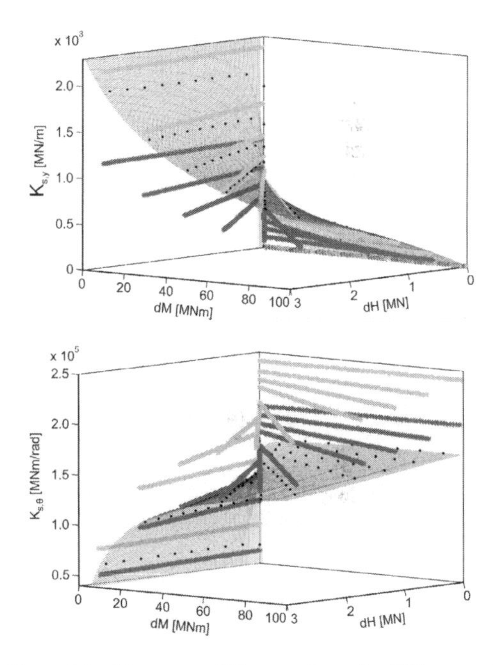

Figure 14. Response stiffness surfaces for 50% un- and reloading (gray surface with black dots) made from load-displacement/rotation surface: lateral stiffness (top) and moment stiffness (bottom) in comparison with API (dark gray) and Kallehave et al. (light gray).

against the deformation amplitudes. The magnitudes of the loads are valid for an un- and reloading level of 50%.

The plane on which all curves for un- and reloading lie is evidently only slightly curved, which indicates that the secant stiffnesses derived from these curves are almost constant.

This is confirmed in Figure 14, where the secant stiffnesses $K_{s,y}$ and $K_{s,\theta}$ for un- and reloading (calculated with Eq. 10) are presented (black dots). For a given lever arm, the secant stiffnesses are almost constant. This result was also obtained for the reference case with h = 50 m (Fig. 8).

Again, the secant stiffnesses from the API approach and the Kallehave approach are also depicted for comparison. Regarding the approximate independence from load level, the results agree with the results from the numerical simulations for un- and reloading. However, the magnitude of the stiffness values does not agree. Both approaches can lead to either over- or underestimation of the stiffness values, dependent on the considered lever arm.

Similar stiffness values as presented in Fig. 14 were obtained for un- and reloading levels of 25%, 75% and 100%. This confirms the findings for the reference system (Fig. 9), that the secant stiffness for un- and reloading is only slightly dependent on the level of the un- and reloading amplitude.

6 CONCLUSIONS

The eigenfrequency of a wind energy converter is a crucial parameter in the design and—for wind turbines with monopile foundations—it is significantly affected by the stiffness of the monopile-soil system. In design practice, the soil around a monopile is modeled by springs with constant stiffnesses in dynamic calculations of the overall system. The spring stiffnesses are usually derived from p-y curves.

In this paper, the API approach (API 2014) and a modified approach of Kallehave et al. (2012) for large-diameter piles are considered. These approaches resemble the pile behavior under virgin monotonic loading, whereas during operation mainly un- and reloading takes place.

For a monopile with a diameter of 8 m embedded over a length of 32 m in dense sand, numerical simulations of the behavior under virgin loading and under un- and reloading with varying amplitudes and arbitrary loading conditions, viz. lever arms, were conducted. The sophisticated HSsmall material law, which is capable to realistically simulate soil behavior also for un- and reloading, was applied. For the considered system, it was found that the un- and reloading stiffness is only slightly dependent on the load level from which the unloading starts and on the amplitude of the un- and reloading cycle. This finding is basically in agreement with results from eigenfrequency measurements on operating offshore wind turbines reported by Kallehave et al. (2012). For a large bandwidth of mean wind speeds, Kallehave et al. obtained no significant change in the eigenfrequencies of the offshore wind energy converters which is not in line with the non-linear foundation stiffness corresponding to virgin loading.

The comparison of the results of the p-y approaches from API and Kallehave et al. with the numerical simulation results showed that neither of these approaches is suitable for a sufficiently accurate prediction of un- and reloading stiffnesses. Depending on the loading conditions, i.e. the combination of moment and horizontal load, both significant under- and overestimation of the stiffness is possible. It is strongly desirable to develop a new calculation approach, which yields realistic system stiffnesses for arbitrary system and loading conditions.

REFERENCES

Abdel-Rahman, K., Achmus, M. 2005. Finite Element Modeling of Horizontally Loaded Monopile Foundations for Offshore Wind Energy Converters in Germany, Proceedings of the 1st International Symposium on Frontiers in Offshore Geotechnics (ISFOG), Australia, pp. 391–396.

American Petroleum Institute - API 2014. Recommended Practice 2GEO - Geotechnical and Foundation Design Considerations.

Benz, T. 2007. Small Strain Stiffness of Soils and its Numerical Consequences. PhD Thesis, University of Stuttgart.

Brinkgreve R.B.J., Engin, E., Swolfs, W.M. 2014. PLAXIS 3D Manual.

Byrne, B.W., McAdam, R.A., Burd, H.J., Houlsby, C.T., Martin, C.M., …, Plummer, M.A.L. 2017. PISA: New design methods for offshore wind turbine monopiles, *Proceedings of the 8th International Conference Offshore Site Investigations and Geotechnics (OSIG)*, London/UK.

Damgaard, M., Bayat, M., Andersen, L.V., Ibsen, L.B. 2014. Assessment of the dynamic behavior of saturated soil subjected to cyclic loading from offshore monopile wind turbine foundations. *Computers and Geotechnics*. Vol 61: pp 116–126.

Det Norske Veritas-DNV, 2013. Offshore Standard DNV-OS-J101, Design of Offshore Wind Turbine Structures.

Dubois, J., Thieken, K., Terceros, M., Schaumann, P., Achmus, M. 2016. Advanced Incorporation of Soil-Structure Interaction into Integrated Load Simulation, Proc. of the 26th International Offshore and Polar Engineering Conference (ISOPE), pp. 754–762.

Kallehave, D., Le Blanc Thilsted, C., Liingaard, M.A. 2012. Modification of the API p-y formulation of initial stiffness in sand. *Proceedings of the 7th International Conference for Offshore Site Investigation and Geotechnics*. London/UK.

Li, W., Igoe, D., Gavin, K. 2015: Field tests to investigate the cyclic response of monopoles in sand. *Proc. Of the Institution of Civil Engineers – Geotechnical Eng.*. Vol. 168, Issue GE5: pp. 407–421.

Murchison, J.R., O'Neill, M.W., 1984. Evaluation of p-y-Relationship in Cohesionless Soils. *Analysis and Design of Pile Foundations*. Editor J.R. Meyer, ASCE, New York, pp. 174–191.

Reese, L.C., Cox, W.R., Koop, F.D. 1974. Analysis of laterally loaded piles in sand. *Proc. Of Offshore Technology Conf.*. No. OTC 2080.

Santos, J.A., Correia, A.G. 2001: Reference threshold shear strain of soil and its application to obtain a unique strain-dependent shear modulus curve for soil. *Proceedings of the 15th International Conference on Soil Mechanics and Geotechnical Engineering*, pp. 267–270.

Schanz, T. 1998. Zur Modellierung des mechanischen Verhaltens von Reibungsmaterialien. *Habilitation*. University of Stuttgart.

Sørensen, S.P.H. 2012. Soil-structure interaction for non-slender, large-diameter offshore monopiles. PhD Thesis, Aalborg University, Denmark, Department of Civil Engineering.

Thieken, K., Achmus, M., Lemke, K., Terceros, M. 2015a. Evaluation of p-y Approaches for Large Diameter Monopiles in Sand, International Journal of Offshore and Polar Engineering 25(2), pp. 134–144

Thieken, K., Achmus, M. and Lemke, K. 2015b. A new static p-y approach for piles with arbitrary dimensions in sand, Geotechnik 38 (4), pp. 267–288.

Thieken, K. 2015. Geotechnical Design Aspects of Foundations for Offshore Wind Energy Converters. PhD Thesis, Leibniz University of Hannover.

Numerical Methods in Geotechnical Engineering IX – Cardoso et al. (Eds)
© 2018 Taylor & Francis Group, London, ISBN 978-1-138-33203-4

3D FE dynamic modelling of offshore wind turbines in sand: Natural frequency evolution in the pre-to after-storm transition

Evangelos Kementzetzidis
Section of Offshore Engineering, Department of Hydraulic Engineering, Delft University of Technology, Delft, The Netherlands

Willem Geert Versteijlen
Siemens Gamesa Renewable Energy, Den Haag, The Netherlands

Axel Nernheim
Siemens Gamesa Renewable Energy, Hamburg, Germany

Federico Pisanò
Section of Geo-Engineering, Department of Geoscience and Engineering, Section of Offshore Engineering, Department of Hydraulic Engineering, Delft University of Technology, Delft, The Netherlands

ABSTRACT: 3D non-linear finite element analyses are proving increasingly beneficial to analyse the foundations of Offshore Wind Turbines (OWTs) in combination with advanced soil modelling. For this purpose, the well-known SANISAND04 bounding surface plasticity model (Dafalias & Manzari 2004) is adopted in this work to incorporate key aspects of critical state soil mechanics into the analysis of monopile foundations in sand. The final 3D soil-foundation-OWT model is exploited to simulate the response of an 8 MW OWT to a long loading history of approximately 2 hours duration. The scope is to investigate/explain the drops in natural frequency observed in the field during storms, as well as its subsequent recovery. The numerical results point out a strong connection between transient frequency drops and pore pressure accumulation, whereas the original OWT natural frequency seems to be restored as a consequence of post-storm re-consolidation.

1 INTRODUCTION

A surge of interest on the dynamic response of OWTs has been recorded in recent years. According to van Kuik et al. 2016 improved insight from advanced 3D simulations could lead to major breakthroughs, including possible pile eigenfrequency fine tuning as a function of soil characteristics and other key variables. As dynamic-sensitive structures, OWTs and their foundations must be designed with special concern for cyclic/dynamic loading conditions.

Multiple factors may affect in reality the dynamics of an OWT during its lifetime, and particularly its first fundamental frequency f_0. Hereafter, the effects of relevant geotechnical aspects on f_0 are investigated, with focus on the operational shifts in eigen-frequency induced by (i) evolution of the pore pressure field around the monopile during loading, and (ii) changes in the local state of soil (e.g. through plastic straining and compaction/dilation) predicted via advanced constitutive modelling. Variations in soil geometry around the foundation, for instance due to scour (Germanische Lloyd 2005), are instead disregarded.

The ultimate goal of this work is to shed new light on the operational evolution of f_0 as related to fundamental hydro-mechanical processes in the soil foundation. The case of a monopile foundation founded in homogeneous medium-dense sand is explicitly considered, in the same modelling framework recently developed by Corciulo et al. 2017, Kementzetzidis et al. 2017.

2 INTEGRATED SOIL-MONOPILE-TURBINE 3D FE MODELLING

A 3D FE model of the whole sand-monopile-OWT system has been built through the OpenSees simulation platform (http://opensees.berkeley.edu; (McKenna 1997)). Its main modelling ingredients include (i) use of an advanced critical state, cyclic sand model, and (ii) dynamic time-domain simulation of the OWT response to an environmental loading history of remarkable duration (≈ 2 hours). To accommodate the second ingredient, the trade-off between accuracy and computational burden has been resolved closer to the latter through a rather coarse FE discretisation of the 3D soil domain. This peculiar aspect of the present work is imposed by the unavoidable long duration of FE analyses aiming to examine the effect of post-storm re-consolidation. Accordingly, the main value of the results being presented lies on the qualitative side, though with the merit of highlighting fundamental aspects of OWT dynamics never tackled so far through 3D time-domain non-linear simulations.

2.1 *Hydro-mechanical FE modelling of saturated low-frequency soil dynamics*

The low-frequency dynamics of the water-saturated soil is described via the *u–p* formulation by Zienkiewicz and coworkers, based on the assumption of negligible soil-fluid relative acceleration (Zienkiewicz et al. 1999).

Spurious checkerboard pore pressure modes near the 'undrained-incompressible limit' are avoided by employing the H1-P1 ssp stabilised elements proposed by (McGann et al. 2015). These 8-node equal order brick elements exploit a non-residual-based stabilisation (Huang et al. 2004) that produces an additional Laplacian term in the pore water mass balance equation. The stabilisation of the pore pressure field is controlled by a numerical parameter α to be set as suggested by (McGann et al. 2015), which can be set as a function of the average element size in the FE mesh and the elastic moduli of the soil skeleton. Importantly, two-phase ssp bricks (stabilised single-point integration hexahedra elements) also feature an enhanced assumed strain field that mitigates both volumetric and shear locking.

Time marching is performed through the well-known Newmark algorithm with parameters $\beta = 0.6$ and $\gamma = (\beta + 1/2)^2/4 = 0.3025$ (Hughes 1987), combined with explicit forward Euler integration of soil constitutive equations at each stress point (Sloan 1987).

It should be noted that a uniform and steady distribution of soil permeability is considered for the sake of simplicity, although in reality it may vary substantially as a function of the evolving void ratio (Shahir et al. 2012).

2.2 *SANISAND04 modelling of cyclic sand behaviour*

Modelling accurately the cyclic hydro-mechanical behaviour of sands plays a major role in the time-domain simulation of dynamic soil-foundation interaction. This study relies on the predictive capability of the SANISAND04 model by (Dafalias & Manzari 2004), available in OpenSees after the implementation developed at the University of Washington (http://opensees.berkeley.edu/wiki/index.php/Manzari _Dafalias_Material); (Ghofrani and Arduino 2017). While readers are referred to the relevant literature available, it is here worth recalling the main features of the SANISAND04 model:

- critical state theory included through the 'state parameter' concept proposed by (Been and Jefferies 1985, Wood et al. 1994).;
- bounding surface formulation with kinematic/rotational hardening;
- transition from compactive to dilative response across the so-called 'phase transformation' surface, evolving in the stress-space as a function of the state parameter;
- phenomenological modelling of post-dilation fabric changes upon load reversals via a fabric-related tensor, with beneficial impact on the prediction of pore pressure build-up under undrained symmetric/two-way cyclic loading.

Despite many successful applications, the SANISAND04 model cannot predict accurately ratcheting phenomena (Niemunis et al. 2005, Corti et al. 2016), vital for a reliable prediction of monopile deformations. This limitation has been recently remedied by (Liu et al. 2017, Liu et al. 2018).

Soil parameters and soil-pile interface properties
A homogenous sand deposit of Toyoura clean sand is considered, with SANISAND04 constitutive parameters listed in Table 1 after (Dafalias & Manzari 2004).

The sharp HM (Hydro-Mechanical) discontinuity at the sand-pile interface is handled by inserting a thin continuum layer of 'degraded' Toyoura sand around the monopile, both along its shaft and under the tip. The weaker interface sand features elastic shear modulus and critical stress ratio 2/3 and 3/4 times lower than in the intact material, respectively.

2.3 *OWT and monopile structures*

The OWT-monopile set-up assumed in this study is representative of the current industry practice and concerns a large 8 MW OWT founded in medium-dense/dense sand. Relevant structural details—courtesy of Siemens Gamesa Renewable Energy (The Hague, Netherlands) – have been all incorporated in the numerical model, although incompletely reported in this paper due to confidentiality issues. In particular, the

Table 1. Toyoura SANISAND04 parameters.

Description	Parameter	Value
Elasticity	G_0 [1]	125
	v	0.3
Critical state	M	1.25
	c	0.712
	λ_c	0.019
	e_0	0.934
	ξ	0.7
Yielding	M	0.01
Hardening	h_0	7.05
	c_h	0.968
	n^b	1.1
Dilatancy	A_0	0.704
	n^d	3.5
Fabric	z_{max}	4
	c_z	600
Density [t/m³]	ρ_{sat}	19.4

left side of Figure 1 illustrates the prototype OWT taken into account, featuring (i) a monopile with diameter D = 8 m, underground length L_{pile} =27 m and average thickness t = 62 mm, (ii) a superstructure with mudline-to-hub distance of approximately 150 m, and (iii) a rotor with blade length L_{blade} in the order of 75 m. The OWT model also includes structural and equipment masses (flanges, transition piece, boat landing and working platforms, etc.), as well as the RNA lumped mass M_{RNA} (Rotor-Nacelle Assembly) at the top with suitable rotational inertia I_M associated with nacelle mass imbalances. Added mass effects due to the surrounding sea water are simplistically introduced in the form of nodal lumped masses evenly distributed along the water depth H_w = 26 m and calculated as twice the water mass in the submerged OWT volume (Newman 1977).

The steel structure above the mudline (wind tower and part of the monopile) is modelled as an elastic beam with variable cross-section, and subdivided into approximately 160 Timoshenko beam elements with consistent (non-diagonal) mass matrix. The underground portion of the tubular monopile is instead modelled as a 3D hollow cylinder, discretised by using 8-node, one-phase ssp bricks $(H1ssp)$ (Figure 1).

A major issue in the dynamic simulation of OWTs concerns the modelling of all sources of energy dissipation (damping). In particular:

- most energy dissipation takes place within the soil domain as plastic/hysteretic damping and wave radiation away from the monopile. Absorbing viscous dampers to prevent spurious reflections are set along the lateral domain boundaries—see also (Corciulo et al. 2017);
- structural damping is introduced based on Eurocode 1 (BS EN 1991). A (Rayleigh) damping ratio $\zeta_{steel} = 0.19\%$ is assigned to all steel cross-sections at the pivot frequencies 0.1 and 80 Hz;
- hydrodynamic damping is incorporated following Leblanc and Tarp-Johansen 2010, where a damping ratio of 0.12% due to wave radiation is obtained for an OWT with $f_0 = 0.3$ Hz, pile diameter of 4.7 m and water depth at 20 m. In the lack of more specific data, a damping ratio of $\zeta_w = 0.12\%$ is assigned to the added water mass nodes (Figure 1);
- aerodynamic damping is not part of the total damping identified later in this study, although it is implicitly included in the wind loading histories applied to the OWT.

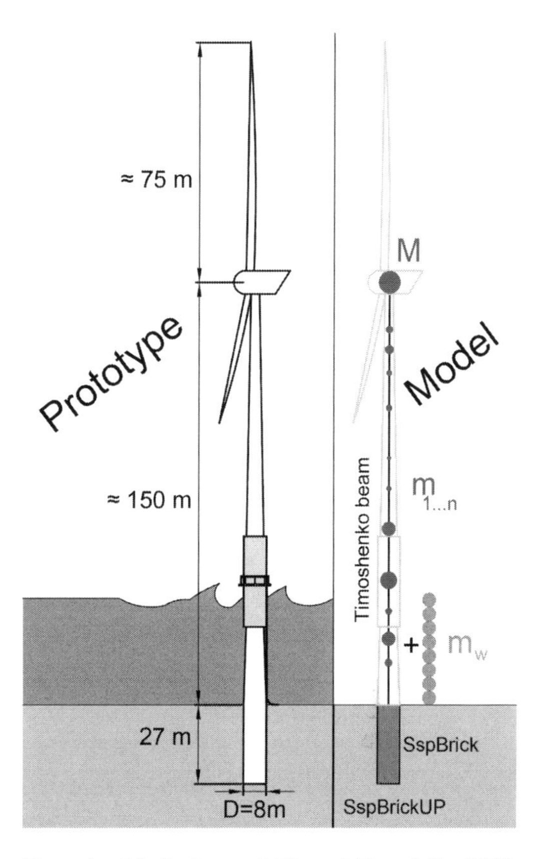

Figure 1. Idealisation and FE modelling of the OWT-foundation-soil system.

[1] Stiffness is described by the dimensionless parameter G_0 in the relevant elastic law.

3 SENSITIVITY TO DISCRETISATION/ SIMULATION PARAMETERS

3.1 *Space/time discretisation*

Under the common assumption of mono-directional lateral loading, only half OWT has been modelled for computational convenience. The accuracy and efficiency of FE results depends strongly on space/time discretisation, i.e. on the FE mesh and time-step size adopted. As mentioned above, efficiency has been privileged here over accuracy to allow for the simulation of long time histories. The chosen domain size and mesh density are illustrated in Figure 2.

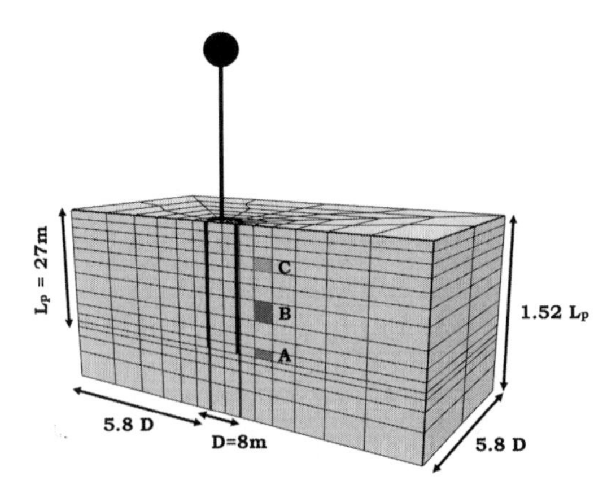

Figure 2. 3D soil mesh formed by ≈950 ssp bricks—A, B and C are the control points considered in the post-processing stage.

To enable 3D non-linear simulations under very long loading histories, special attention must be devoted to discretisation/simulation parameters. The sensitivity of numerical results to the time-step size Δt has been explored along with the sensitivity of the simulated pore pressure field to the stabilisation parameter α (McGann et al. 2015). All adopted Δt values lie within the range examined in the following, whereas α has been finally set—and found satisfactory—according to the indications in McGann et al. 2015 – $\alpha = 3 \times 10^{-5}$. .

3.2 *Sensitivity to time-step size*

Time-step sensitivity analyses have been performed both for short-duration (20 s—Figure 3a, load applied at the hub) and long(er)-duration (660 s—Figure 3b, loads applied as described later in Section 4.1) loading histories. The former has been designed to investigate the effect of a wide range of time-steps, while the latter has been devised to confirm for an 11 minutes simulations the inferences from shorter 20 s tests. It is worth noting that time integration with adaptive time-step size has been included within the global time marching scheme.

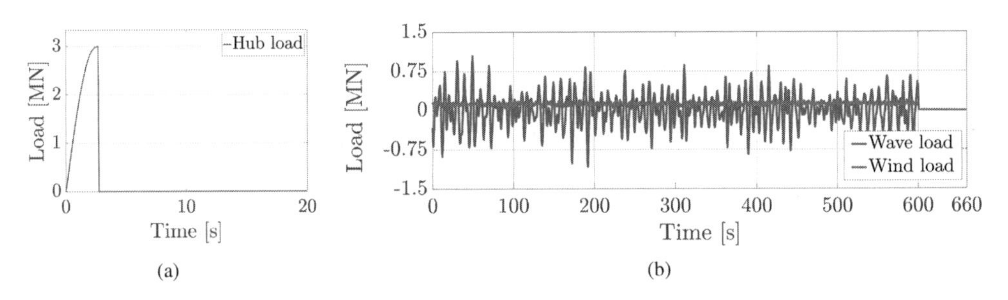

Figure 3. Time-step sensitivity analysis: (a) short and (b) long loading histories.

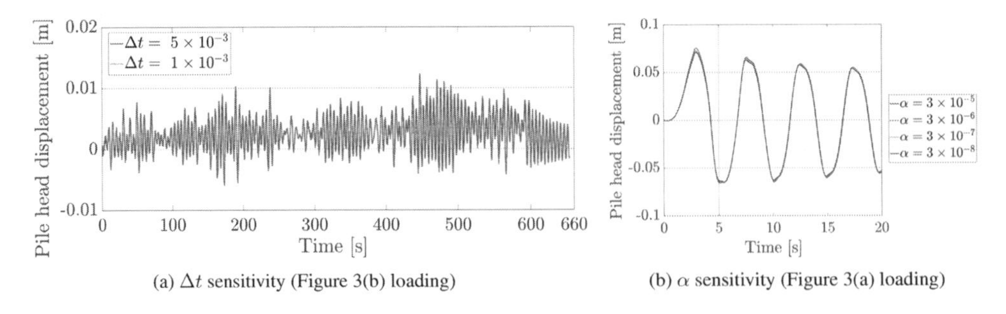

Figure 4. Sensitivity of the monopile head lateral displacement to Δt and α.

Table 2. Explored values of time-step size and pore pressure stabilisation parameter with associated calculation time for the relevant load case. All simulations run on a i7-4790 4.00GHz CPU.

Time-step	$\Delta t = 1\times10^{-2}$ s	$\Delta t = 5\times10^{-3}$ s	$\Delta t = 1\times10^{-3}$ s	
Analysis time [mins]	39 mins	51 mins	215 mins	
Stabilization parameter α	$\alpha = 3\times10^{-5}$	$\alpha = 3\times10^{-6}$	$\alpha = 3\times10^{-7}$	$\alpha = 3\times10^{-8}$
Analysis time [mins]	36 mins	41 mins	47 mins	48 mins

Δt values in the range $[10^{-3};10^{-2}]$ have been considered in short 20 s simulations, and indistinguishable results obtained at the control points A, B and C in Figure 2 – thus not reported for brevity. Then, two different Δt have been extracted from the same range and applied to the longer simulation scenario depicted in Figure 3b. Although different by half an order of magnitude, the tested Δt values produced very similar results, for instance in terms of monopile head lateral displacement (Figure 4a). The impact on the computational burden of different time-step sizes is documented in Table 2, and justifies the adoption of any Δt within the range examined. As such different sections of the analysis were calculated with different Δt sizes to accommodate for the varying demand in accuracy caused by the alternating amplitude of the cyclic loads applied.

3.3 *Sensitivity to the stabilisation parameter*

The sensitivity of numerical results to the pore pressure stabilisation parameter α has been also studied over a range spanning three orders of magnitude, i.e. $\alpha = 3\times10^{-5} - \alpha = 3\times10^{-8}$. The results obtained indicate a very mild influence on the global performance, for instance on the monopile deformation. It is interesting to note that choice of a specific α value also affects the computational efficiency, as shown in Table 2. In agreement with McGann et al. 2015 for the dominant element size in the FE mesh, values in the range $\alpha = 10^{-5} - \alpha = 10^{-6}$ have been considered appropriate for the present application.

4 DYNAMIC OWT PERFORMANCE DURING AND AFTER A STORM

An OWT founded on a monopile embedded in dense Toyoura sand with relative density $D_r = 80\%$ has been considered. In order to promote faster pore pressure dissipation, a relatively high permeability value has been set in the whole soil domain, $k = 10^{-4}$ m/s.

4.1 *Loading scenario*

This work aims to relate transient f_0 drops experienced by an OWT during storms to the evolution of the pore pressure field, including after-storm re-consolidation. In this spirit, an analysis case has been conceived to let the OWT go through different loading stages. Strong, weak loading and load removal phases to allow for consolidation are included. The overall loading scenario (sum of wind and wave loads with limited wind component due to OWT feathering)[2] is illustrated in Figure 5 and features:

1. 150 s of weak loading to estimate the 'small strain' f_0;
2. 1200s of strong storm loading $(v_{wind} > 24m/s)$ to induce transient f_0 drops;
3. 150 s of the same weak loading scenario to explore possible frequency drops caused by storm-induced, pore pressure build-up;
4. 1.7 hours (6000s) of no loads in the domain to allow for excess pore pressure dissipation;
5. 150 s of the same weak loading scenario to observe the expected regain in f_0 due pore-pressure dissipation and void ratio variations (re-consolidation);

The last 150 s of loading have been applied at excess pore pressures entirely dissipated. Therefore, any differences recorded in the response, compared to the initial 150 s of loading, should be related to previous plastic straining and changes in void ratio in the sand.

Load application

The total wave force is distributed along the submerged OWT nodes, accounting for the actual wave height—nodes above the mean sea level are loaded during wave impact to ensure realistic simulation. The OWT blades are significantly pitched out under such storms, only the wind drag along

[2] Load time-history created by manipulating/altering load segments estimated at Siemens Gamesa Renewable Energy for an 8 MW OWT, almost fully feathered, under a strong storm. It must be noted that due to the factorisation, the loads are no more one-to-one related to the 8MW turbine.

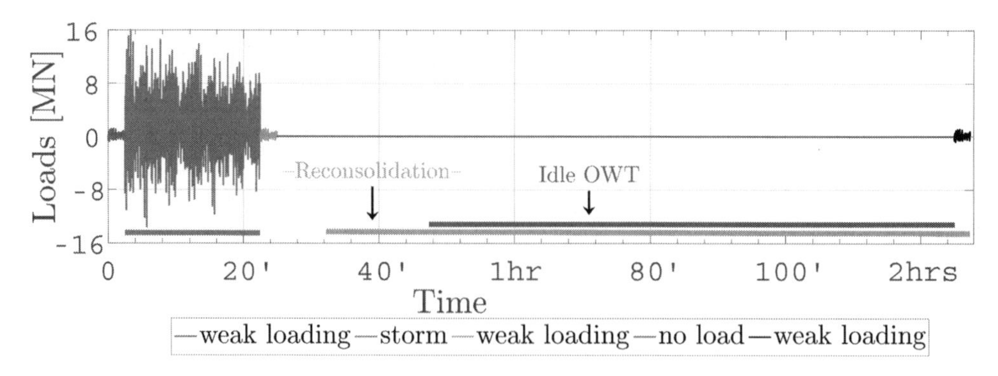

Figure 5. Assumed load time history—sum of wind and wave thrust forces.

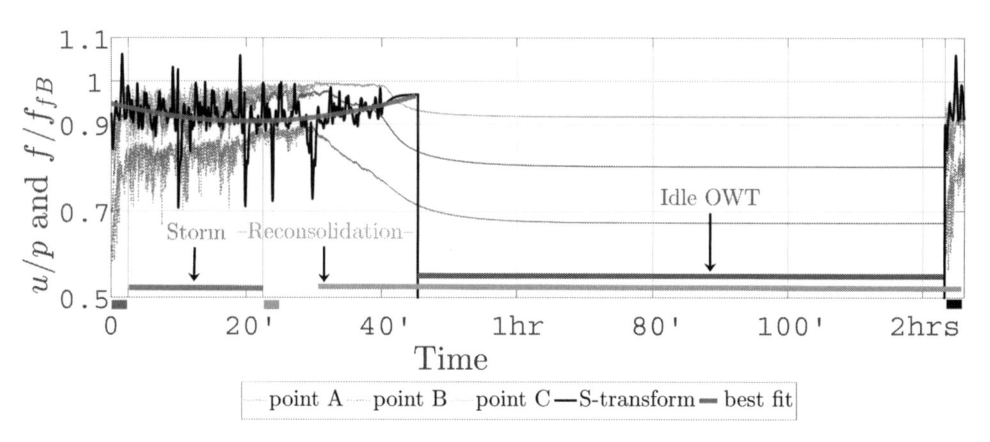

Figure 6. Thick black line: normalised OWT peak frequency; thick red line: best quadratic fit of the peak frequency time evolution; dotted lines: u=p ratios at the control points in Figure 2; thick blue, green and black lines: time range of each sub-stage in the global loading scenario in Figure 5.

the hub and tower is considered and applied to the tower bottom through a pair of equivalent point force and moment.

4.2 *Simulation results*

The evolution of the frequency content in the OWT response has been monitored by applying so-called S(Stockwell)-transformation to the simulated time history of the hub lateral displacement (Stockwell et al. 1996) – see Figure 6. As the S-transform returns the (time-varying) frequency content within a relevant band, the outcropping value associated with the maximum normalised S-amplitude at each time step has been extracted to track f0 drops (black line in Figure 7) with respect to the fixed base natural frequency f_{FB}[3] – the same concept is also used later in Figure 8.

It is evident from Figure 6 that the natural frequency of the OWT drops during the storm, as suggested by the quadratic best-fit on the variable peak frequency extracted from the S-transform. At the same time, an increase in pore pressure—and most importantly in *u/p* (pore pressure-to-total mean pressure) ratio—is observed at all the control points along the embedded length of the foundation. It is comforting to observe that the local minimum of the fitting parabola lies close to onset of load removal: this evidence supports the belief that the recovery of f_0 may start right after the end of a strong loading event. A few abrupt drops of the peak response frequency are also observed, most likely due to temporary (and particularly severe) reductions in soil stiffness and, possibly, interaction with higher vibration modes.

The after-storm *u/p* trends in Figure 6 keep on their increasing branches even right after load removal, with further impact on the operational

[3] The fixed base natural frequency was calculated by performing an eigenvalue analysis on the OWT fixed at the mudline.

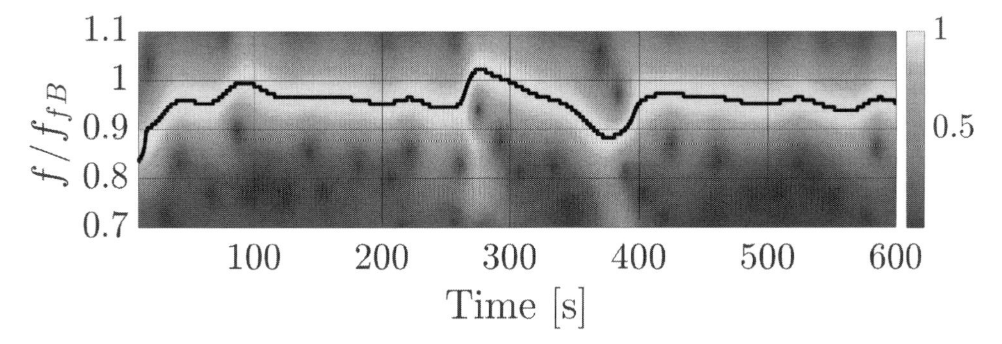

Figure 7. S-transform of the OWT response for the load time history in Figure 3(b). All frequency values are normalized with respect to the fixed-base natural frequency f_{FB} of the OWT. The colorbar indicates the magnitude of all harmonics, the thick black line underlines the evolution of the peak frequency.

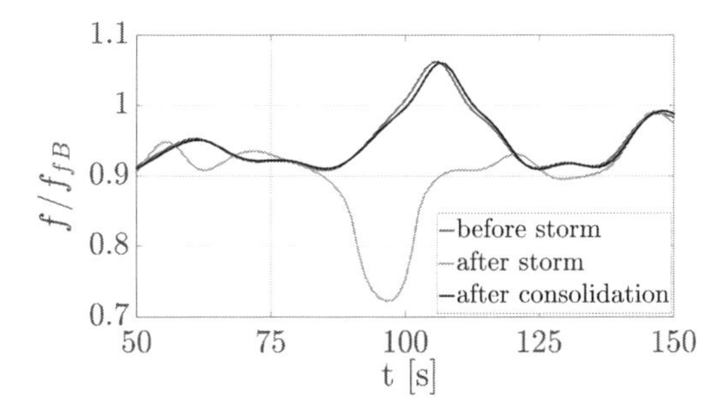

Figure 8. Time evolution of the OWT peak frequency (normalised with respect to f_{FB}) in correspondence of the three weak loading events over time (see Figures 5–6). The last 100 of the 150 seconds aredisplayed as the effect of the previously applied loads, for the the after storm case (green line), is significant.

stiffness of the sand and cantilever-like free vibration of the OWT. The gradual decay of the free vibration amplitude allows for the dominance of the re-consolidation process, first starting at deeper soil locations (Figure 6). It can be seen that f_0 tends prominently to its pre-storm range as re-consolidation starts occurring at the deepest control point A. It can also be observed that, as soon as the excess pore pressure at the shallowest point C is dissipated, the natural frequency of the OWT appears as fully restored. This should be attributed to the low effective confinement of shallow sand layers, more prone to pore pressure build-up and loss of shear stiffness/strength under cyclic loading. In these conditions, the upper portion of the sand deposit cannot contrast effectively the lateral loading, with immediate and apparent effect on the global foundation stiffness.

Finally, it should be noted in Figure 8 that the pre-storm and after-consolidation responses of the OWT are practically coincident. This supports the conclusion that, even during strong storm events, monopiles in (medium-dense) sand experience only temporary losses in lateral stiffness, eventually remedied by excess pore pressure dissipation and re-consolidation. However, this might not be the case, for instance, in fine-grained materials, in which cyclic loading does not only induce pore pressure build-up but also mechanical destructuration (Seidalinov and Taiebat 2014).

5 CONCLUSIONS

A long-lasting time-domain analysis including a 20 minutes storm event was performed for an 8MW OWT supported by a monopile in sand. A state-of-the-art plasticity model was employed to simulate the hydro-mechanical cyclic soil behaviour, with specific ability to describe the response of sands under a wide range of void ratio and effective confinement. A model disregarding void ratio effects would have not fully allowed to obtain the results presented in this study. Based on the evolution of the OWT dynamics from pre-storm to post-consolidation stages, it is concluded that the monopile stiffness degradation induced by even strong storm is not expected to be permanent. This inference confirms the observations from previous fields measurements, for instance from those reported by (Kallehave et al. 2015).

REFERENCES

Been, K. & M. Jefferies (1985). A state parameter for sands. *Géotechnique 35*(2), 99–112.

BS EN (1991). 1–4: 2005 eurocode 1: Actions on structures—general actions—wind actions.

Corciulo, S., O. Zanoli, & F. Pisanò (2017). Transient response of offshore wind turbines on monopiles in sand: role of cyclic hydro–mechanical soil behaviour. *Computers and Geotechnics 83*, 221–238.

Corti, R., A. Diambra, D. M. Wood, D. E. Escribano, & D. F. Nash (2016). Memory surface hardening model for granular soils under repeated loading conditions. *Journal of Engineering Mechanics,* 04016102.

Dafalias, Y. F. & M. T. Manzari (2004). Simple plasticity sand model accounting for fabric change effects. *Journal of Engineering mechanics 130*(6), 622–634.

Germanische Lloyd (2005). Guideline for the certification of offshore wind turbines.

Ghofrani, A. & P. Arduino (2017). Prediction of LEAP centrifuge test results using a pressure-dependent bounding surface constitutive model. *Soil Dynamics and Earthquake Engineering.*

Huang, M., Z. Q. Yue, L. G. Tham, & O. C. Zienkiewicz (2004, September). On the stable finite element procedures for dynamic problems of saturated porous media. *International Journal for Numerical Methods in Engineering 61*(9), 1421–1450.

Hughes, T. J. R. (1987). *The Finite Element Method: linear static and dynamic finite element analysis.* Prentice-Hall.

Kallehave, D., C. Thilsted, & A. T. Diaz (2015). Observed variations of monopile foundation stiffness. In *The 3rd Internationl symposium on Frontiers in offshore Geotechnics*, pp. 717–722. CRC Press LLC.

Kementzetzidis, E., S. Corciulo, W. G. Versteijlen, & F. Pisanò (2017). Geotechnical aspects of offshore wind turbine dynamics from 3d non-linear soil-structure simulations. *Soil Dynamics and Earthquake Engineering submitted for publication.*

Leblanc, C. & N. J. Tarp-Johansen (2010). Monopiles in sand. stiffness and damping.

Liu, H., J. A. Abell, A. Diambra, & F. Pisanò (2018). A three-surface plasticity model capturing cyclic sand ratcheting. *Géotechnique submitted for publication.*

Liu, H., F. Zygounas, A. Diambra, & F. Pisan`o (2017). Enhanced plasticity modelling of high-cyclic ratcheting and pore pressure accumulation in sands. In *In Proceedings of the Xth European conference on Numerical Methods in Geotechnical Engineering (NUMGE 2017), Porto, Portugal, submitted for publication.*

McGann, C. R., P. Arduino, & P. Mackenzie-Helnwein (2015). A stabilized single-point finite element formulation for three-dimensional dynamic analysis of saturated soils. *Computers and Geotechnics 66*, 126–141.

McKenna, F. T. (1997). *Object-oriented finite element programming: frameworks for analysis, algorithms and parallel computing.* Ph. D. thesis, University of California, Berkeley.

Newman, J. N. (1977). Marine hydrodynamics. MIT press.

Niemunis, A., T. Wichtmann, & T. Triantafyllidis (2005). A high-cycle accumulation model for sand. *Computers and geotechnics 32*(4), 245–263.

Seidalinov, G. & M. Taiebat (2014). Bounding surface saniclay plasticity model for cyclic clay behavior. *International Journal for Numerical and Analytical Methods in Geomechanics 38*(7), 702–724.

Shahir, H., A. Pak, M. Taiebat, & B. Jeremić (2012). Evaluation of variation of permeability in liquefiable soil under earthquake loading. *Computers and Geotechnics 40*, 74–88.

Sloan, S. W. (1987). Substepping schemes for the numerical integration of elastoplastic stress–strain relations. *International Journal for Numerical Methods in Engineering 24*(5), 893–911.

Stockwell, R. G., L. Mansinha, & R. Lowe (1996). Localization of the complex spectrum: the s transform. *IEEE transactions on signal processing 44*(4), 998–1001.

van Kuik, G. A. M., J. Peinke, R. Nijssen, D. J. Lekou, J. Mann, J. N. Sørensen, C. Ferreira, J. W. van Wingerden, D. Schlipf, P. Gebraad, et al. (2016). Long-term research challenges in wind energy–a research agenda by the European Academy of *Wind Energy. Wind Energy Science 1*, 1–39.

Wood, D. M., K. Belkheir, & D. F. Liu (1994). Strain softening and state parameter for sand modelling. *Géotechnique 44*(2), 335–339.

Zienkiewicz, O. C., A. H. C. Chan, M. Pastor, B. A. Schrefler, & T. Shiomi (1999). *Computational geomechanics.* Wiley Chichester.

Numerical Methods in Geotechnical Engineering IX – Cardoso et al. (Eds)
© 2018 Taylor & Francis Group, London, ISBN 978-1-138-33203-4

Multiscale investigations on the failure mechanism of submarine sand slopes with coupled CFD-DEM

M. Kanitz & J. Grabe
Institute of Geotechnical Engineering and Construction Management, Hamburg University of Technology, Hamburg, Germany

ABSTRACT: In terms of the construction of gravity offshore foundation systems, the short- and longterm stability of submarine sand slopes is of crucial importance. The heavy structures cannot be placed directly on the sea ground. The weak topsoil has to be excavated to place the structure on more stable ground, which results in a pit. A steep but stable inclination of its slopes would thereby meet both economic and ecologic interests as the disturbed area is reduced. Different hydrodynamic disturbances as tidal streaming or waves are hereby influencing the stability of the submarine slopes. The changes in pore water pressure due to the hydrodynamic disturbances and the interaction of the soil grains with the surrounding pore water significantly influences the stability of the slope and its failure mechanism. To take into account these effects, the numerical simulations are carried out with a multiscale approach, namely a coupling of the Computational Fluid Dynamics (CFD) and the Discrete Element Method (DEM). In this method, the soil particles are modelled on a microscale level, tracking velocity and position in a lagrangian way. The fluid phase is modelled at a macroscale level using CFD. The numerical modelling aims on investigating the effect of tidal streaming on submarine sand slopes with different packing densities to take into account the influence of dilatancy and contractancy.

1 INTRODUCTION

The placement of heavy structures on the marine topsoil gains more and more importance in the industry due to the growing sector of offshore wind energy and the use of the ocean for marine infrastructure, e.g. immersed tunnel systems. These structures have to be placed on stable ground to avoid settlements, that can neither be predicted nor controlled. Therefore, the weak marine topsoil has to be excavated, resulting in a pit, which is naturally dependent of the slope angle realized during the excavation. A steep but stable slope thereby meets both economic and ecologic interests as the disturbance of the marine fauna and flora is reduced and less material has to be removed. The excavation process of sand through suction or grab dredging is already investigated regarding the process improvement (Entenmann and Boley 2001), (van Rhee 1998). In these investigations, hydrodynamic disturbances of tidal streaming or wave loading are not included. Their effect on the stability of the subaqueous slopes is still not fully quantified. Peire et al. (2009) describes the dredging of the foundation pits for gravity base foundations in the North Sea. The excavation depth was about 7 m below the surrounding seabed with slope angles of 1:8 to 1:5. These inclinations were at least stable for the construction period. As these inclinations are quite flat and the slope angle significantly smaller than the critical angle, more research is needed, to investigate the stability of submarine slopes under hydrodynamic disturbances with steeper inclinations. The investigation of the tidal currents plays a major role regarding the mechanisms of erosion and rearrangement of the sediments. Its effect is more of an indirect nature as the erosional and depositional processes caused by the tidal currents can lead to an undercutting or overloading of the submarine slope (Evans 1995).

To investigate these effects, a numerical method is used, that includes the interaction forces between the pore water and the free water as well as the interaction forces between the pore water and the soil skeleton. In this context, the Discrete Element Method (DEM) combined with the Computational Fluid Dynamics (CFD) is able to exchange phase specific information for the determination of interaction forces. Thus, it is possible to investigate flow induced processes and the influence of soil movement on pore water pressure changes and vice versa. In this contribution, the coupled CFD-DEM method will be introduced and first numerical investigations on the stability of submarine slopes due to tidal disturbances are presented. The calculations are carried out using the open source software package CFDEMcoupling®, which combines the discrete element code LIGGGHTS® with CFD solvers based on OpenFOAM®.

2 CFD-DEM METHOD

The coupled CFD-DEM approach combines the CFD and the DEM to solve particle-fluid flow problems and was firstly proposed by Tsuji et al. (1993). Using this method, the dynamics of the fluid are calculated as a continuum with the CFD, while the motion of the particles is determined in a Lagrangian way with the DEM. So, the coupled CFD-DEM allows to investigate dynamic information of a process like particle-fluid interaction forces and the trajectories of individual particles. The knowledge about these information gives the possibility to gain inside-knowledge of multiphase processes. The following sections gives a short introduction in the coupled CFD-DEM method.

2.1 *Discrete Element Method*

DEM is a Lagrangian method to track the particle position and movement in an explicit way and was originally developed by Cundall and Strack (1979). The governing equations to determine the translational and rotational movement of a particle i with mass m, Newton's second law of motion is used:

$$m_i \frac{\partial v_i}{\partial t} = \sum_j \mathbf{F}_{ij}^c + \mathbf{F}_i^f + \mathbf{F}_i^g, \tag{1}$$

$$I_i \frac{\partial \omega_i}{\partial t} = \sum_j \mathbf{M}_{ij}^c, \tag{2}$$

where v_i and ω_i describe the translational and angular velocity of the particle i and I_i is the momentum of inertia. $\mathbf{F}_{ij}^c$ and $\mathbf{M}_{ij}^c$ represent the contact force and the torque acting on particle i by particle j or the walls. $\mathbf{F}_i^f$ and $\mathbf{F}_i^g$ are the particle-fluid interaction force and the gravitational force. To calculate the contact force, the soft sphere approach is used, proposed by Cundall and Hart (1992). In this method, the particles overlap slightly when they are in contact. The elastic, plastic and frictional forces are then determined by the spatial overlap. The contact traction distribution over the contact area is composed by a tangential and a normal contact plane. The accurate description of the contact traction distribution is complex as it is related to the shape of the particle, the material properties and the motion state of the particle. In general, there exist a variety of models to calculate the contact forces. The simplest model is a linear spring-dashpot model firstly proposed by Cundall and Strack (1979), see Fig. 1. The spring is used for the elastic deformation while the dashpot accounts for the viscous dissipation. The normal force tending to repulse the particles can be calculated from the spatial overlap $\delta\mathbf{x}_p$ through

$$\mathbf{F}_n = k_n \delta\mathbf{x}_p - c_n \delta\mathbf{v}_{p,n}, \tag{3}$$

where $\delta\mathbf{v}_{p,n}$ is the relative normal velocity. The tangential force can be determined similarly through

$$\mathbf{F}_t = k_t \delta\mathbf{x}_p - c_t \delta\mathbf{v}_{p,t}. \tag{4}$$

The coefficients k_n and k_t are the spring stiffness in normal and tangential direction and consequently c_n and c_t the damping coefficient. The numerical simulations in this work are carried out using the more sophisticated simplified Hertz-Mindlin contact model implemented by Langston et al. (1994). The computationally expensive original Hertz-Mindlin contact model includes a nonlinear relationshop between the normal force and the normal displacement (Hertz 1881) and a force-displacement relationship depending on the whole loading history and the rate of change of normal and tangential displacement (R. W. Mindlin, H. Deresiewicz 1953). Langston et al. (1994) adopted a more intuitiv version with a direct force-displacement relation for the tangential force. Additionally, interparticle forces are causing a rotation, as the forces are not acting at the mass centre of the particle. The generated torque consists of two components of a tangential and asymmetrical normal traction distribution and is called rolling friction torque. It can be solved by using rolling friction moddels. In the presented numerical simulations,

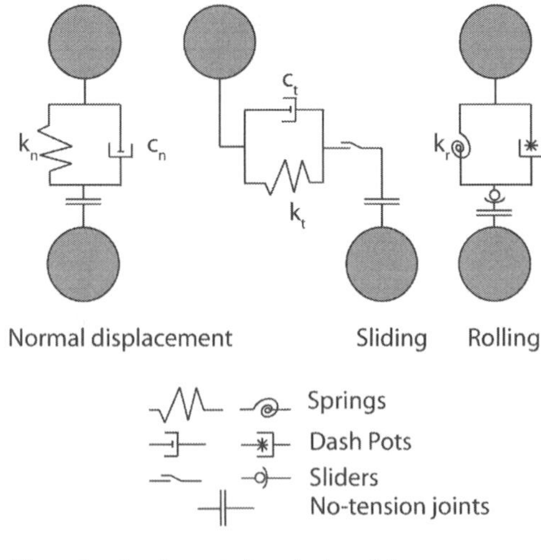

Figure 1. Implemented contact model.

the rolling friction model introduced by Iwashita (1998) is used. In addition to the spring and the dash pot used to model normal and tangential displacement, a supplementary set of an elastic spring, a dashpot and a slider is located at each contact point, see Fig. 1. The spring and the dashpot work as a resistance against the relative rotation from rolling and the slider starts working when the torque is exceeding the rolling resistance.

2.2 *Computational Fluid Dynamics*

The basis for the numerical investigations with CFD is formed by the Navier-Stokes equations. In this contribution, the Finite Volume Method (FVM) is applied, dividing the computational domain into cells, for which the locally averaged Navier-Stokes equations are solved. The governing equations are formed by the conservation of mass (Eqn. 5) and momentum (Eqn. 6) in terms of locally averaged variables (Anderson and Jackson 1967) through

$$\frac{\partial}{\partial t}\rho_f\alpha_f + \nabla\cdot(\rho_f\alpha_f\mathbf{u}_f) = 0, \tag{5}$$

$$\begin{aligned}\frac{\partial}{\partial t}(\rho_f\alpha_f\mathbf{u}_f) + \nabla\cdot(\rho_f\alpha_f\mathbf{u}_f\mathbf{u}_f) \\ = -\alpha_f\nabla p + \nabla\cdot(\alpha_f\tau_f) + \alpha_f\rho_f\mathbf{g} + R_{pf},\end{aligned} \tag{6}$$

where $\mathbf{u}_f$ and p describe the fluid velocity and pressure. τ_f and ρ_f are the fluid viscous stress tensor and the density of the fluid. There are two major differences compared to the classic locally averaged Navier-Stokes equations. In Eqn. 5 and Eqn. 6 the influence of the solid phase is already included in the conservation equations. The factor α_f is the porosity, so the occupied volume of fluid in a cell with the presence of particles. R_{pf} is the term for momentum exchange between the fluid and the particles. It describes the particle-fluid interaction forces and is permanently exchanged between the CFD- and the DEM-solver. R_{pf} is determined through the following equation

$$R_{pf} = K_{pf}(\mathbf{u}_f - \mathbf{v}), \tag{7}$$

with the interaction force coefficient K_{pf}. It is defined as the superposition of the particle-fluid interaction forces.

2.3 *Coupling scheme*

Regarding the coupling of the two methods described in the former sections, it is important to keep in mind that the CFD solves its governing equations on a cell basis and the DEM on the individual particle level. Hence, the scale level between these two methods is crucial. Generally, there are two approaches to treat this multiphase problem in CFD-DEM: the resolved and the unresolved CFD-DEM. In the resolved CFD-DEM the fluid cells are significantly smaller than the particle, see Fig. 2 a). This method allows a quite accurate determination of the fluid flow around the particles, but it is computationally expensive. Recently, only systems with some hundreds of particles are reasonable to compute with the resolved method. The unresolved method thereby allows to investigate problems with some millions of particles as the CFD cell are notably larger than the particles, see Fig. 2 b). Nevertheless, using this method, the fluid flow around the particles can not be solved explicitly and force models have to be used.

2.4 *Particle-fluid interaction forces*

Using the unresolved method in CFD-DEM, the particle-fluid interaction forces are determined using force models, that are dependent of the porosity α_f in the fluid cells. The fluid surrounding the particles generates various particle-fluid interaction forces, that can be of interest depending on the investigated process. In CFD-DEM, different particle-fluid forces as the particle-fluid drag force, pressure gradient force, unsteady forces like the virtual mass force, the Basset history force and lift forces are already implemented (Li et al. 1999), (Xiong et al. 2005).

In the presented simulation, the drag force is calculated following the correlation of (Koch and Hill 2001) and (Koch and Sangani 1999). It is based on results obtained from numerical simulations of fluid flow trough a collection of randomly dispersed particles in a bubbling fluidised bed using Lattice-Boltzmann equations. Additionnally, the pressure gradient force and the viscous forces are included in the presented numerical simulations. The pressure gradient force is derived by (Anderson

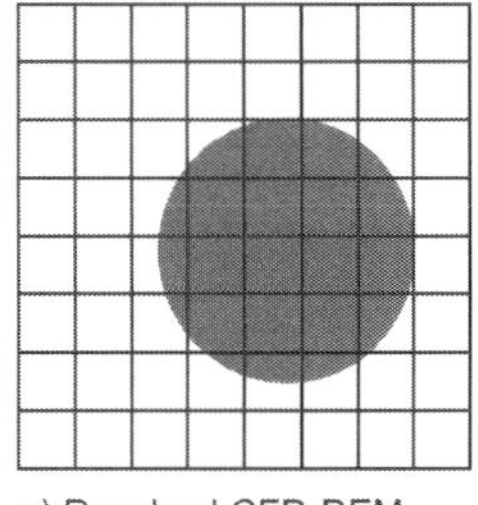

a) Resolved CFD-DEM

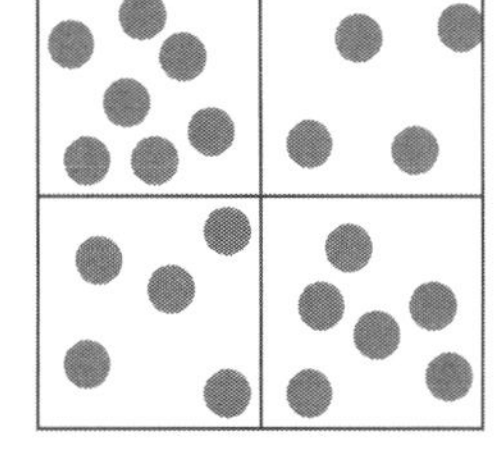

b) Unresolved CFD-DEM

Figure 2. Treatment of the scaling between cell-based CFD and particle-based DEM after Hager (2014).

Table 1. Particle-Fluid interaction forces.

Forces	Correlations
Drag force	$F_D = F_0(\alpha_f) + F_1(\alpha_f)Re_p^2$ for $Re_p < 20$
	$F_D = F_0(\alpha_f) + F_3(\alpha_f)Re_p^2$ for $Re_p > 20$
	$F_0(\alpha_f) = \frac{1+3(\alpha_f/2)^{1/2}+(135/64)\alpha_f \ln \alpha_f + 16.14\alpha_f}{1+0.681\alpha_f - 8.48\alpha_f^2 + 8.16\alpha_f^3}$
	for $\alpha_f < 0.4$
	$F_0(\alpha_f) = 10\alpha_f/(1-\alpha_f)^3$ for $\alpha_f > 0.4$
	$F_1(\alpha_f) = 0.110 + 5.10 \cdot 10^{-4} \exp^{11.6\alpha_f}$
	$F_3(\alpha_f) = 0.0673 + 0.212\alpha_f + 0.0232(1-\alpha_f)^5$
Pressure gradient force	$F_p = -V_p \partial p / \partial \mathbf{x} = -V_p(\rho_f \mathbf{g} + \rho_f \mathbf{u}_f \partial \mathbf{u}_f / \partial \mathbf{x})$

and Jackson 1967). It describes the phenomenon, that an immersed particle in a flow field with slowly diverging streamlines experiences a force resulting from the additional pressure gradient. The pressure gradient force model of (Anderson and Jackson 1967) includes the buoyancy force due to gravity and the effects resulting from the accelerated pressure gradients in the fluid. The used force models with the determination of the particle-fluid interaction forces are shown in Tab. 1. The viscous force is described by the derivation of the viscous stress tensor τ_f in Eqn. 6.

3 NUMERICAL MODEL

In order to simulate the influence of tidal streaming on submarine slopes, the open source software package CFDEMCOUPLING® is used for the numerical simulation. The geometry of the numerical model can be seen in Fig. 3. The generation of the slope is initalized in a pure DEM calculation. In the DEM, microscopic properties of the particles are needed to characterize the behaviour of the used material. In the numerical simulations presented in this contribution, a coarse-grained sand, the so-called Hamburger Sand, is used. Different tests were carried out experimentally und numerically to determine and calibrate the coefficients for internal friction and rolling friction and the coefficient for the loss of energy, due to a particle collision. The parameters used can be found in Tab. 2. The particle size distribution is shown in Fig. 4. A polydisperse particle packing is used in the numerical model that represents the particle size distribution.

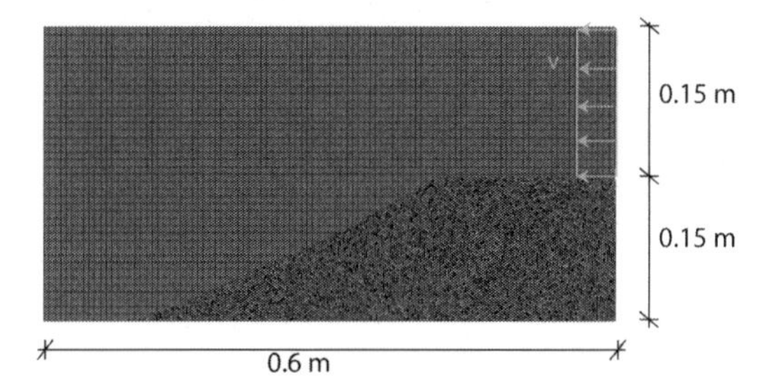

Figure 3. Geometry of the numerical model.

Table 2. Material parameters for Hamburger Sand.

Maximum and minimum void ratio	
n_{max}	0.446
n_{min}	0.342
Coefficient of friction	
μ	0.918
Coefficient of rolling friction	
Particle-Particle: μ_R	0.057
Particle-Wall: μ_R	0.44
Young's modulus	
E (N/mm²)	3157.5

A growing diameter algorithm is used to initiate different packing densities. Hence, a loosely packed seabed (void ratio n_{max} = 0.44) and a densely packed seabed (void ratio n_{min} = 0.37) can be investigated. The dimensions of the model in x- and z-direction can be seen in Fig. 3, in y-direction, a width of approximately 10 particle diameters and a periodic boundary in both DEM and

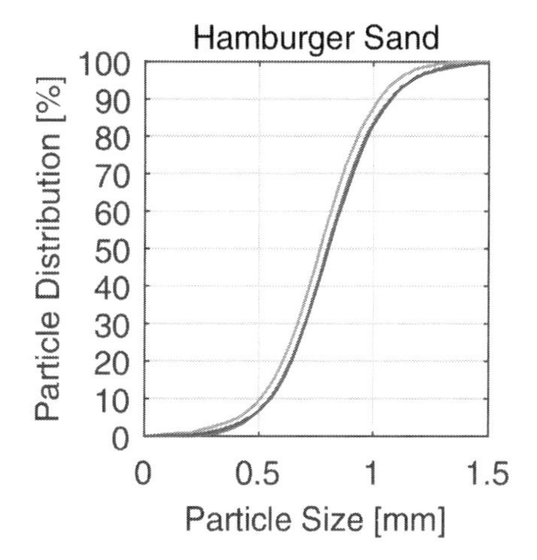

Figure 4. Particle size distribution of Hamburger Sand.

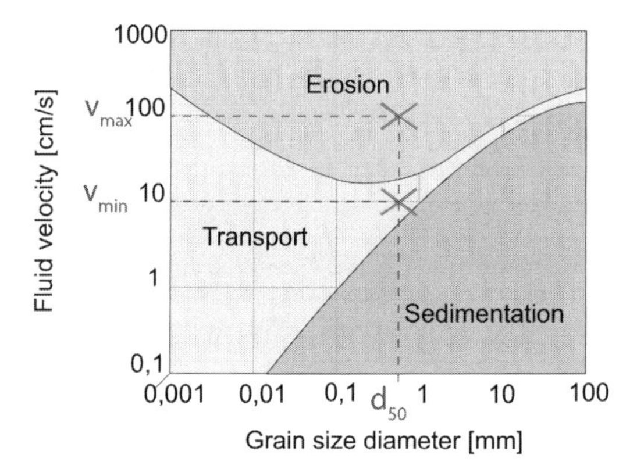

Figure 5. Hjülström diagramm for sedimentation, erosion and transport after Hjulström (1935).

CFD is used to reduce the computational costs. The investigated slope angle is $\beta = 20°$. To investigate the process of transport, erosion and collapse, different streaming velocities are investigated. The minimum velocity is v_{min} = 0.1 m/s and the highest velocity is v_{max} = 1.0 m/s. The d_{50} of the investigated Hambuger Sand is 0.7 mm, hence, according to the Hjülström diagramm, transport would take place at v_{min} and erosion at v_{max}, see Fig. 5. The inlet velocity is constant in order to investigate the evolution of the velocity profile and the boundary layer at the slope surface.

4 RESULTS

Due to high computational costs, the simulation time was limited to t = 1.18 seconds. Fig. 6 shows the start of the tidal streaming and the initiated disturbances in a loosely packed slope (top) and a densely packed slope (bottom). Both the fluid phase and the solid phase are displayed so the fluid flow and the particle motion can be analyzed. As it can be seen, there is literally no motion in the slope at v_{min} = 0.1 m/s, only some few particles on the inclined slope surface are moving. At v_{max}, erosional processes are remarkable. The shape of the slope starts to change. Between the loosely and densely packed slope, the difference in particle motion is minimal. The boundary of motion in the slope is somewhat steeper around the edge of the slope.

After t = 0.58 seconds, a distinct boundary layer is created at the slope surface. It is more pronounced at a higher velocity, see Fig. 7 on the right side. At a lower velocity, the boundary layer is significantly thinner. Concerning the erosional processes, at high velocity, the particles over the whole slope surface are in motion. The shape of the initial slope has remarkably changed for both

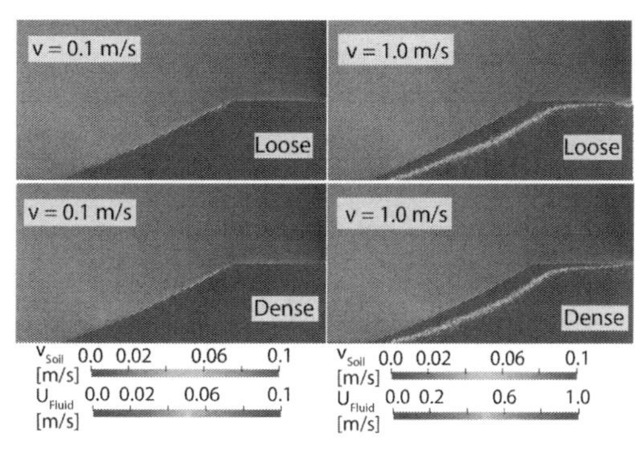

Figure 6. Fluid flow and particle movement at t = 0.025 seconds for loosely and densely packed sand at v_{min} and v_{max}.

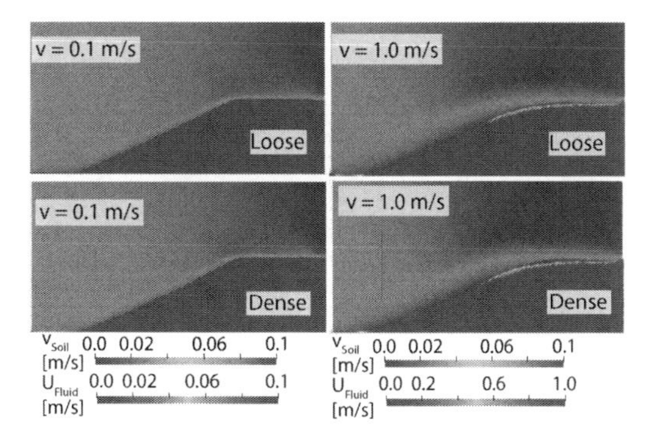

Figure 7. Fluid flow and particle movement at t = 0.58 seconds for loosely and densely packed sand at v_{min} and v_{max}.

loosely and densely packed slopes and the edge of the surface is rounded. It can be observed, that the edge of the slope gets steeper in the densely packed slope. This could be due to the decreased pore volume between the grains. The flow velocity in the pores is higher due to the decreased space between the particles and carries away a larger amount of particles. Regarding the slow streaming velocity v_{min}, the shape of the slope has not changed and there are no intense transport mechanism visible neither for a loosely nor a densely packed slope.

At the end of the simulation at t = 1.18 seconds, the boundary layer has increased notably in the case of the high stream velocity v_{max}, see Fig. 8. In the case of the low velocity, the change is not as striking. At v_{max}, a turbulent vortex is created towards the slope toe. As can be observed in Fig. 8 right, a shaded area (where the flow has no influence) has developed. The particles are hence eroded on the shoulder of the slope and are transported towards the edge. Once they reach the shaded area and the inclination of the slope is stable, the movement of the particles stops. Thus, more and more particles are tranported to the slope edge resulting in a bump. This bump is then further increasing the shaded area on the left side of the slope. At this time step, there is no more difference between the densely and the loosely packed slope. The flow through the pore space and the resulting transport mechanisms have loosened the densely packed soil bed resulting in the similar behaviour of the slopes. In summary, the packing density plays a minor role concerning the stability or the transport mechanisms of submarine slopes at tidal disturbances. Concerning the comparison with empirical relations, there is a good agreement with the Hjülström diagram as erosion takes place at v_{max} = 1.0 m/s.

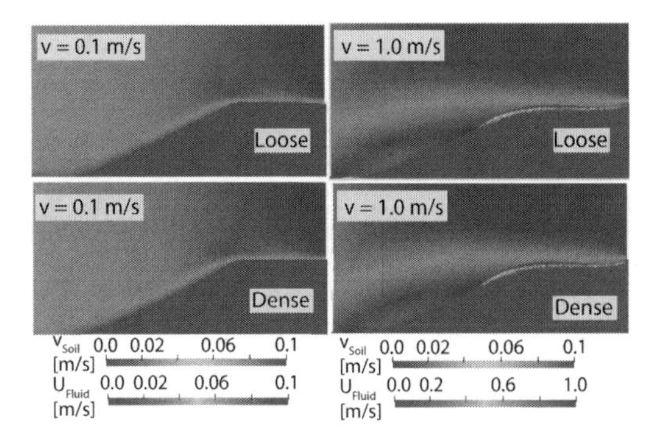

Figure 8. Fluid flow and particle movement at t = 1.18 seconds for loosely and densly packed sand at v_{min} and v_{max}.

To further investigate the transport mechanism at a low velocity, the particle movement at the slope surface is displayed more detailed in Fig. 9. As it can be seen, the particles are in motion at the beginning of the simulation. The motion is even more pronounced in a densely packed slope with maximum particle velocities of 0.01 m/s. The built-up of the boundary layer with advancing time results in a decreased velocity of the current near the slope surface, so the particles begin to slow down, which can already be identified after t = 0.187 s. In the densely packed slope, there is almost a layer of particles moving with a velocity of 0.002 m/s, in the loosely packed slope, there are just some few particles moving. After t = 0.375 the motion of the particles almost completely stopped in both cases. Hence, in summary, the motion of particles takes places at the beginning of the simulation, when the boundary layer is thin enough and until the rearrangement of the particles is finished.

As the built-up of the boundary layer plays a major role concerning the particle motion, the evolution of the velocity profile is investigated in detail. In Fig. 10 the evolution of the velocity profile over time and height is displayed. The profile at the last time step is in darker color than the former time steps. The velocity profile was plotted at x = 0.45 m, so it is just befor the slope edge and includes the stream velocity in the particle bed. The sediment layer is displayed in the figure and represents the height of the slope at the last time step, so the change of the slope height due to transport mechanisms is considered. The decrease of the stream velocity in both low and high stream velocity due to the particle presence can be identified by a change of the gradient in the profile. At v = 0.1 m/s, the flow velocity in the particle bed has decreased over time while the flow velocity in the free water phase has increased. The same tendency can be observed at a velocity of v = 1.0 m/s at the right side of Fig. 10. Though, the velocity in the soil bed seems to vary. An increase of the velocity in the soil bed could be explained by the ongoing rearrangement of the particles at the high velocity. With a decrease of particles in that region, the influence on the flow velocity is decreased as well. The shape of the velocity profile is logarithmic for both velocities until a height of z = 0.045 m for v_{min} = 0.1 m/s and z = 0.06 m for v_{max} = 1.0 m/s followed by a parabolic shape. The variation of the velocity profile is more extensive regarding v = 1.0 m/s. So, at this point, it is not clear, whether the displayed profile is the final one. Hence, the simulation time has to be increased to further investigate the evolution.

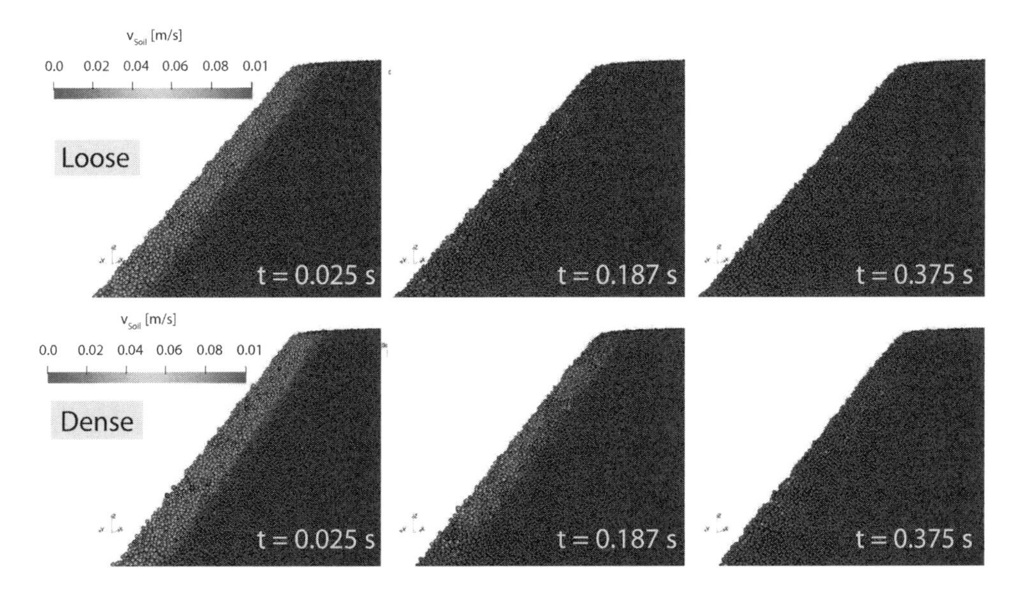

Figure 9. Evolution of particle motion at v_{min} = 0.1 m/s in loosely and densely packed slope.

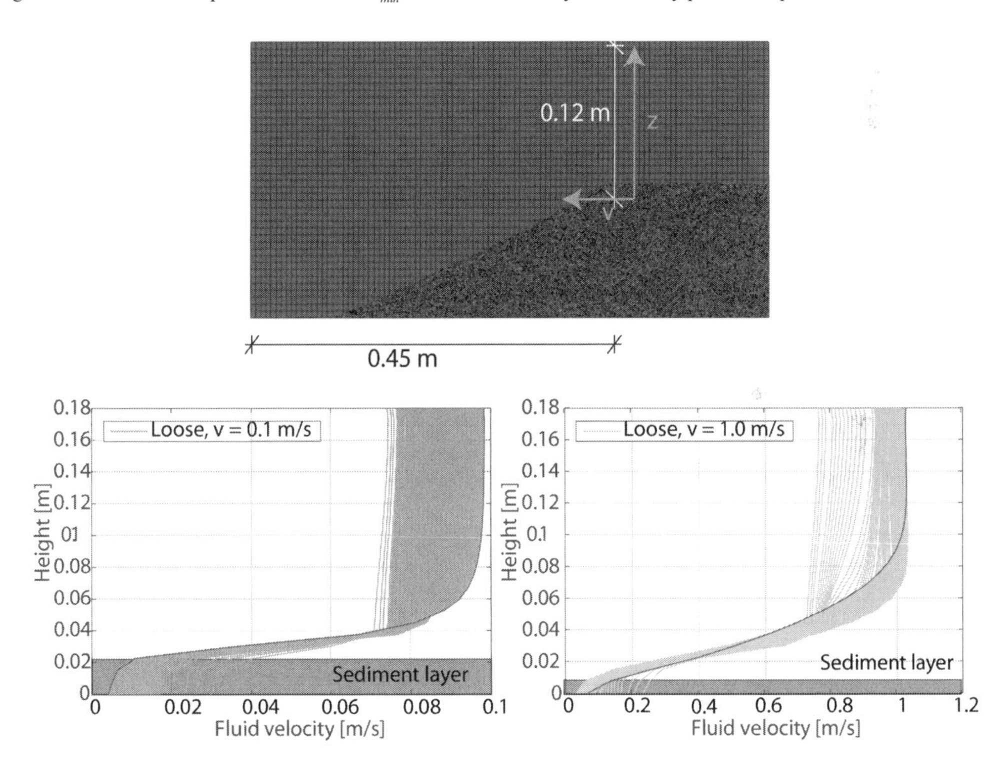

Figure 10. Evolution of the velocity profile over time and height at v = 1.0 m/s and v = 0.1 m/s in a densely packed slope.

5 CONCLUSION

The presented contribution shows early stage results of the investigation of tidal disturbances on the submarine slope stability. Different velocities and different packing densities of the seabed were investigated in small scale numerical simulations with a coupled CFD-DEM approach. The flow field shows the characteristic logarthmic distribution and evolution of a boundary layer in the vicinity of the particle bed. It could be observed, that the packing density plays a minor role regarding erosional processes. Once, the soil body has been loosened due to the flow through the pores, the behaviour of the slopes at different initial packing densities is nearly identical. The evolution of the boundary layer has a major impact on the erosional processes and influences the final shape of the slope. The numerical simulations with CFDEMcoupling® show already promising results concerning the erosion and the initiation of motion of the particles, but it still has to be validated as the comparison with the Hjülström diagram signifies only a qualitive one. For this reason, small scale physical models will be carried out in order to compare the experimental results with the numerical ones. The physical models have to be kept small due to the high computational costs of the simulation. Hence, the numerical simulation containing around a million of particles can take several days.

AKNOWLEDGEMENT

The authors thank the German Research Foundation (DFG) for funding this project under the project number GR 1024/22-1 and Christoph Goniva and Alice Hager from DCS Computing for the support in the calibration of the material parameters and in the numerical simulations.

REFERENCES

Anderson, T.B. & R. Jackson (1967). A fluid mechanical description of fluidized beds. *Industrial and Engineering Chemistry Fundamentals 6*(4), 527–538.

Cundall, P.A. & R.D. Hart (1992). Numerical Modelling of Discontinua. *Engineering Computations 9*(2), 101–113.

Cundall, P.A. & O.D.L. Strack (1979). A discrete numerical model for granular assemblies. *Géotechnique* (29), 47–65.

Entenmann,W. & C. Boley (2001). Abbau von Ton und Sand unterhalb des Grundwasserspiegels; aktuelle geotechnische und hydrogeologische Aspekte, dargestellt an Fallbeispielen aus Niedersachsen. *Zeitschrift für Angewandte Geologie 47*(1).

Evans, N.C. (1995). Stability of submarine slopes. *Geo Report* (47).

Hager, A. (2014). *CFD-DEM on Mulitple Scales - An Extensive Investigation of Particle-Fluid Interactions*. Dissertation, Johannes Kepler Universität Linz, Linz.

Hertz, H. (1881). Über die Berührung fester elastischer Körper. *Journal für die reine und angewandte Mathematik 92*, 156–171.

Hjulström, F. (1935). Studies on the morphological activity of rivers as illustrated by the river Fyris. *Les Études rhodaniennes 11*(2), 255–256.

Iwashita, K. (1998). Rolling resistance at contacts in simulation of shear band development by DEM. *Journal of Engineering Mechanics* (124), 285–292.

Koch, D.L. & R.J. Hill (2001). Inertial effects in suspension and porous-media flows. *Annual Review of Fluid Mechanics 33*(1), 619–647.

Koch, D.L. & A.S. Sangani (1999). Particle pressure and marginal stability limits for a homogeneous monodisperse gas-fluidized bed: Kinetic theory and numerical simulations. *Journal of Fluid Mechanics 400*, 229–263.

Langston, P.A., U. Tüzün, & D.M. Heyes (1994). Continuous potential discrete particle simulations of stress and velocity fields in hoppers: Transition from fluid to granular flow. *Chemical Engineering Science 49*(8), 1259–1275.

Li, Y., J. Zhang, & L.-S. Fan (1999). Numerical simulation of gas–liquid–solid fluidization systems using a combined CFD-VOF-DPM method: Bubble wake behavior. *Chemical Engineering Science 54*(21), 5101–5107.

Peire, K., H. Nonneman, & E. Bosschem (2009). Gravity base foundations for the Thornton Bank offshore wind farm. *Terra et Aqua* (115), 19–29.

R.W. Mindlin, H. Deresiewicz (1953). Elastic spheres in contact under varying oblique forces. *Journal of Applied Mechanics 20*, 327–344.

Tsuji, Y., T. Kawaguchi, & T. Tanaka (1993). Discrete particle simulation of two-dimensional fluidized bed. *Powder Technology 77*(1), 79–87.

van Rhee, C. (1998). *The Breaching of Sand Investigated in Large-Scale Model Tests*. Virginia: American Society of Civil Engineers.

Xiong, Y., M. Zhang, & Z. Yuan (2005). Three-dimensional numerical simulation method for gas–solid injector. *Powder Technology 160*(3), 180–189.

Numerical Methods in Geotechnical Engineering IX – Cardoso et al. (Eds)
© 2018 Taylor & Francis Group, London, ISBN 978-1-138-33203-4

One-dimensional finite element analysis of the soil plug in open-ended piles under axial load

T.M. Joseph, H.J. Burd & G.T. Houlsby
Department of Engineering Science, University of Oxford, UK

P. Taylor
Atkins Ltd., Warrington, UK

ABSTRACT: In the offshore industry, open-ended steel piles are primarily designed using one of three pile design methods: the API (2011), the ICP (2005) and the UWA (2005). The soil plug, inside the pile, is considered differently in each of the methods, producing different pile lengths due to the varying assumptions adopted. A one-dimensional finite element method has been used to analyse the behaviour of the resistive components of the soil-pile-plug system. This adopts the principle of virtual work to solve the coupled differential equations enabling the contribution of each component to be isolated. Using this method, a more realistic axial pile response can be modelled which directly influences the behaviour of the soil plug. Predictions from the method have been compared to the results of double walled pile tests which show the mobilisation of the external shaft friction from the top downwards, and the internal friction from the base upwards.

1 INTRODUCTION

Many different methods are used to design offshore open-ended piles (OEP). Few of these consider the contribution of the soil plug directly and none of these methods model its behaviour accurately. Investigation of the behaviour of the soil plug has shown that the plug is mobilised from the base upwards. The compression of instrumented piles has been used to establish empirical pile design methods. However, in design, a rigid pile is often assumed, thereby simplifying the process. Improved modelling which captures the different responses of the soil, pile and plug, and the interaction between them, can only be achieved using a numerical approach, where a mathematical procedure is used to solve the equations of equilibrium that governs the soil-pile-plug response. A one-dimensional finite element (FE) procedure of this sort is presented in this paper.

The FE procedure has been used in combination with three pile design methods: the API (2011) method, the ICP (2005) method, and the UWA (2005) method, to produce a FE variant of each. A case study is presented which examines the predicted response from each FE variant and compares these to published pile test data.

2 FINITE ELEMENT ANALYSIS

2.1 *Open ended pile model*

Consider an OEP as shown in Figure 1 with an axially applied load F_t. In this system, the elastic pile is assumed to be in equilibrium as the load applied at the pile head is balanced by the summation of the external and internal shear stresses, $\tau_e(z)$ and $\tau_i(z)$, respectively, plus the end-bearing resistance $q_{p,b}$ on the base of the pile annulus. The action of τ_i is supported by the end-bearing resistance across the base of the plug, $q_{pl,b}$. Each of these components has separate stiffnesses which are mobilised at different rates to support the pile. These stiffnesses and the relative displacement across the interfaces dictate the mobilisation of the interface shear stresses, and the end bearing resistance. In this model, zero thickness interface layers are assumed.

2.2 *Strong and weak forms of the equations of equilibrium*

For the resistive components for the OEP, the axial stress, σ_p, that acts on the cross-sectional pile area, A_p, satisfies the following strong (differential equation) form (compression positive):

$$\frac{d\sigma_p}{dz}A_p + \tau_e P_e + \tau_i P_i = 0 \quad (1)$$

where τ_e, τ_i = shear stress in the external and internal interfaces, respectively; and P_e, P_i = external and internal perimeter of the pile, respectively. The plug satisfies the separate equilibrium equation (compression positive):

$$\frac{d\sigma_{pl}}{dz}A_{pl} - \tau_i P_i = 0 \quad (2)$$

where, σ_{pl} = the axial stress in the plug and A_{pl} = the cross-sectional area of the plug.

Equations (1) and (2) are converted to the weak (virtual work) form by multiplying by an arbitrary set of virtual displacements, δw, and integrating over the pile length (L) and plug length ($L—L_{pl}$) as appropriate. The application of the boundary conditions at the limits of the pile and plug is then used to obtain an expression which balances the virtual work associated with the applied load and the sum of the internal virtual work terms, giving:

$$\begin{aligned}\delta w_t(F_t) = {} & \delta w_{p,b}F_{p,b} - \int_0^L \sigma_p A_p\left(\frac{d(\delta w_p)}{dz}\right)dz \\ & +\delta w_{pl,b}F_{pl,b} - \int_{L-L_{pl}}^{L} \sigma_{pl}A_{pl}\left(\frac{d(\delta w_{pl})}{dz}\right)dz \\ & +\int_0^L \delta w_s \tau_s P_e dz + \int_{L-L_{pl}}^{L}(\delta w_p - \delta w_{pl})P_i\tau_i dz \\ & +\int_0^L(\delta w_p - \delta w_s)P_e\tau_e dz\end{aligned} \quad (3)$$

where δw_t = pile head virtual displacement; $\delta w_{p,b}$ = pile base virtual displacement; δw_p = virtual displacement along the pile shaft; $\delta w_{pl,b}$ = plug base virtual displacement; δw_{pl} = virtual displacement along the soil plug; δw_s = external soil virtual displacement; and τ_s = shear stress in the soil. Equation (3) may be expressed as:

$$0 = \delta W_E + \delta W_I \quad (4)$$

where δW_E and δW_I are expressions for the external and internal virtual work, respectively.

The external virtual work is associated with the applied load at the top of the pile:

$$\delta W_E = -(\delta w_t F_t) \quad (5)$$

The internal virtual work is associated with the internal stresses in the pile, the plug and the interfaces. This is given by:

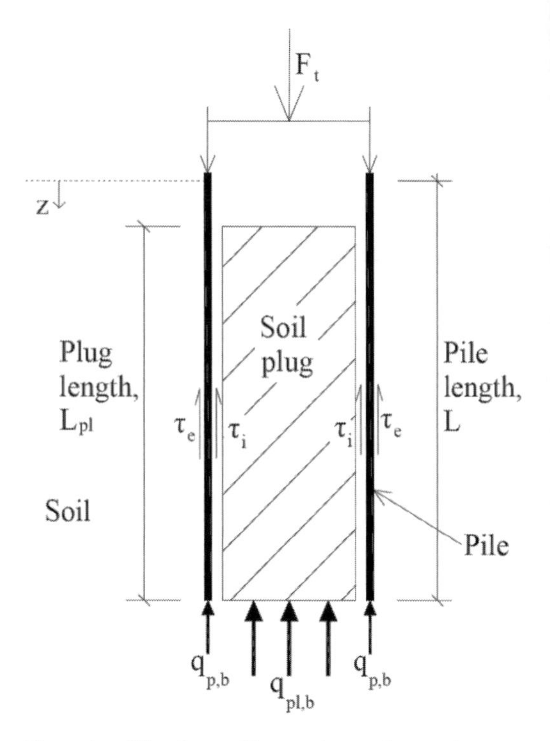

Figure 1. Pile, plug and base resistance interactions.

$$\begin{aligned}\delta W_I = {} & \delta w_{p,b}F_{p,b} - \int_0^L \sigma_p A_p\left(\frac{d(\delta w_p)}{dz}\right)dz \\ & +\delta w_{pl,b}F_{pl,b} - \int_{L-L_{pl}}^{L} \sigma_{pl}A_{pl}\left(\frac{d(\delta w_{pl})}{dz}\right)dz \\ & +\int_0^L \delta w_s \tau_s P_e dz + \int_{L-L_{pl}}^{L}(\delta w_p - \delta w_{pl})P_i\tau_i dz \\ & +\int_0^L(\delta w_p - \delta w_s)P_e\tau_e dz\end{aligned} \quad (6)$$

3 FINITE ELEMENT PROCEDURE

The virtual work equations specified above are used to determine a set of discretized 1D finite element equations using the standard Galerkin approach. The OEP is discretised as a series of finite elements along its entire length. The pile, plug and soil are modelled as sets of 2-noded line elements. The pile-soil and pile-plug interfaces are modelled as 4-noded, zero-thickness, interface elements; these interface elements are used to connect the pile elements to the (internal) plug elements and the (external) soil elements. The end-bearing of the pile and plug are modelled as lumped elements. Figure 2 indicates diagrammatically the general form of the finite element model.

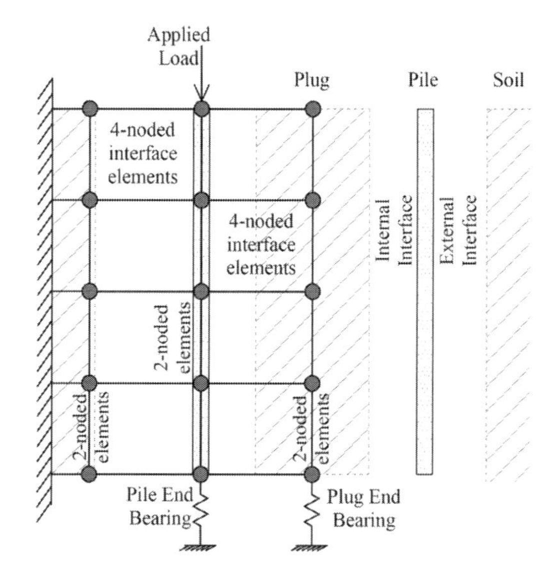

Figure 2. Simple discretised model of pile-plug-soil, interfaces and end bearing components.

The axial displacements within each element are related to the nodal displacements via Lagrangian shape functions The displacements and strains determined using this approach are:

$$w = \mathbf{N}\boldsymbol{w}_e \quad \text{and} \quad \frac{dw}{dz} = \mathbf{B}\boldsymbol{w}_e \tag{7}$$

where $\boldsymbol{N}$ is the shape function matrix; $\boldsymbol{w}_e$ is the vector of element nodal displacements; and $\boldsymbol{B}$ is a matrix formed from the derivative of the element shape function, $\boldsymbol{N}$.

3.1 *Pile base*

The pile base is a lumped element and therefore adopts the value of an equivalent point load:

$$f_{p,b} = F_{p,b} \tag{8}$$

3.2 *Pile*

The virtual work term for the pile is:

$$\int_0^L \sigma_p A_p \left(\frac{d(\delta w_p)}{dz} \right) dz \tag{9}$$

Adopting the Lagrangian shape functions and their derivatives, and on the basis of the Galerkin approach, the virtual work associated with the pile is approximated by:

$$\sum_{pile\ elements} \delta \boldsymbol{w}_e^{\boldsymbol{T}} \int_0^L \boldsymbol{B}^{\boldsymbol{T}} \sigma_p A_p .dz \tag{10}$$

The internal force within each pile element is determined from:

$$f_{element,p} = \int_{elements} \boldsymbol{B}^{\boldsymbol{T}} \sigma_p A_p .dz \tag{11}$$

3.3 *Plug base*

The plug base is also a lumped element and adopts a point load value as:

$$f_{pl,b} = F_{pl,b} \tag{12}$$

3.4 *Plug*

The virtual work expression for the plug is:

$$\int_{L-L_{pl}}^L \sigma_{pl} A_{pl} \left(\frac{d(\delta w_{pl})}{dz} \right) dz \tag{13}$$

Converting to the Galerkin form gives:

$$\sum_{plug\ elements} \delta \boldsymbol{w}_e^{\boldsymbol{T}} \int_{L-L_{pl}}^L \boldsymbol{B}^{\boldsymbol{T}} \sigma_{pl} A_{pl} .dz \tag{14}$$

The internal force within each of the plug elements is then determined from:

$$f_{element,pl} = \int_{elements} \boldsymbol{B}^{\boldsymbol{T}} \sigma_{pl} A_{pl} .dz \tag{15}$$

3.5 *Soil*

The virtual work expression is:

$$\int_0^L \delta w_s \tau_s P_e dz \tag{16}$$

This gives the Galerkin form as:

$$\sum_{soil\ elements} \delta \boldsymbol{w}_e^{\boldsymbol{T}} \int_0^L \boldsymbol{N}^{\boldsymbol{T}} \tau_s P_e .dz$$

The internal force within each of the soil elements is determined from:

$$f_{element,s} = \int_{elements} \boldsymbol{N}^{\boldsymbol{T}} \tau_s P_e .dz \tag{17}$$

3.6 *Internal interface*

Adopting the Langrangian shape functions, the virtual work (Galerkin form) of the 4-noded internal interface elements is:

$$\sum_{int-int\ elements} \delta w_e^T \int_{L-L_{pl}}^{L} N^T \tau_i P_i .dz \tag{18}$$

The internal force within each 4-noded internal interface element is:

$$f_{element,int-int} = \int_{elements} N^T \tau_i P_i .dz \tag{19}$$

3.7 *External interface*

Similarly, adopting the Langrangian shape functions, the virtual work (Galerkin form) of the 4-noded external interface elements is:

$$\sum_{ext-int\ elements} \delta w_e^T \int_{0}^{L} N^T \tau_i P_i .dz \tag{20}$$

The internal force within each 4-noded external interface element is:

$$f_{element,\ ext-int} = \int_{elements} N^T \tau_i P_i .dz \tag{21}$$

3.8 *Assembly*

The global internal force vector (F_{INT}) is obtained by assembling the individual forces of the nodes for each element and the base node for the pile and the plug. A similar assembly process is performed for the virtual displacement vector (δw). This process is performed by using an assembly function [°] which populates the internal forces vector, displacements vector and the stiffness matrix employing an appropriate degree of freedom mapping:

$$\begin{aligned} F_{INT} = \mathbb{A}\left[f_{p,b} \right] - \mathbb{A}\left[f_{element,p} \right] \\ + \mathbb{A}\left[f_{pl,b} \right] - \mathbb{A}\left[f_{element,pl} \right] \\ + \mathbb{A}\left[f_{element,s} \right] + \mathbb{A}\left[f_{element,int-int} \right] \\ + \mathbb{A}\left[f_{element,ext-int} \right] \end{aligned} \tag{22}$$

$$\begin{aligned} \delta w = \mathbb{A}\left[\delta w_{p,b} \right] - \mathbb{A}\left[\delta w_p \right] + \mathbb{A}\left[\delta w_{pl,b} \right] - \mathbb{A}\left[\delta w_{pl} \right] \\ + \mathbb{A}\left[\delta w_s \right] + \mathbb{A}\left[\delta w_{i,i} \right] + \mathbb{A}\left[\delta w_{e,i} \right] \end{aligned} \tag{23}$$

The internal work is now written as:

$$\delta W_I = \delta w^T F_{INT} \tag{24}$$

The external virtual work is assembled in the same manner, giving:

$$\delta W_E = -\delta w^T \mathbf{F}_{EXT} \tag{25}$$

From Equation (4):

$$0 = \delta w^T \left(\mathbf{F}_{INT} - \mathbf{F}_{EXT} \right) \tag{26}$$

As the virtual displacements are arbitrary on the basis of the first lemma of variational calculus:

$$G = 0 = \left(\mathbf{F}_{INT} - \mathbf{F}_{EXT} \right) \tag{27}$$

The vector equation, $G = 0$, is solved using the modified Newton-Raphson method.

4 CASE STUDY

The selected case study consists of a double wall pile test that was performed in Pigeon Creek, Indiana, USA. The soil comprised predominantly gravelly sand to a depth of 13 m. CPT data available for the site is presented in Figure 3. The test pile was 8.24 m long with dimensions as shown in Figure 4. This was driven to the required depth and loaded incrementally to failure. The separation of the pile walls and the installation of strain gauges over their lengths allowed the differential response of both pile walls to be measured. The results of these tests are published in Salgado et al. (2002) and Paik et al. (2003). Using this published

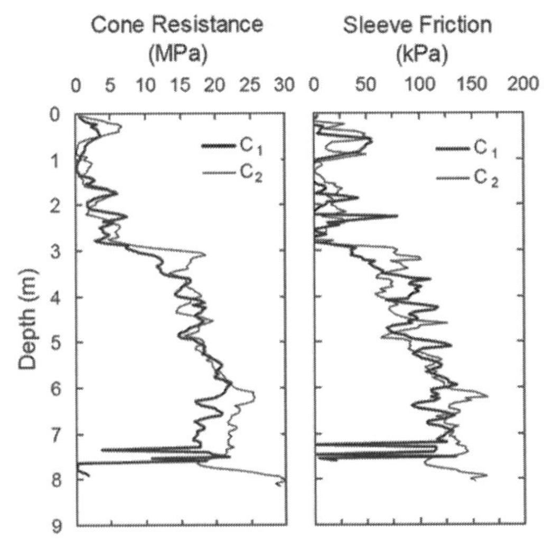

Figure 3. q_c and f_s measurements (Salgado et al., 2002).

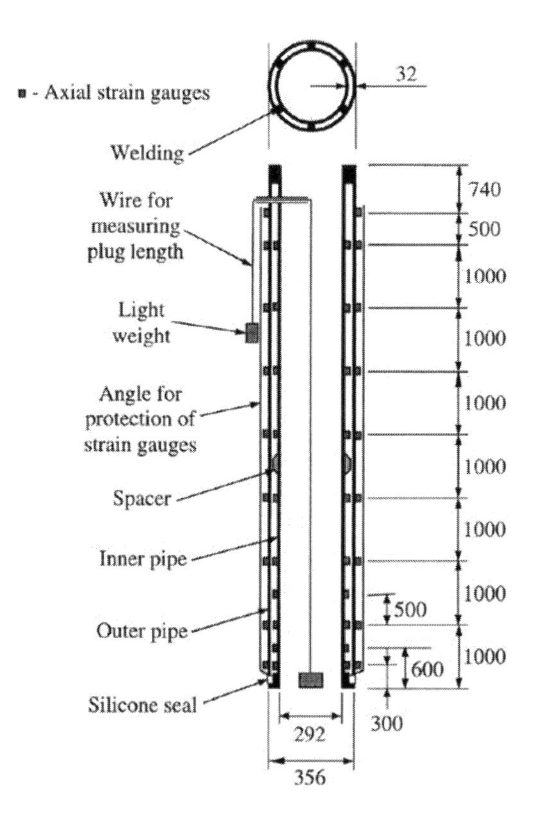

Figure 4. Double-walled open-ended test pile (Salgado et al., 2002).

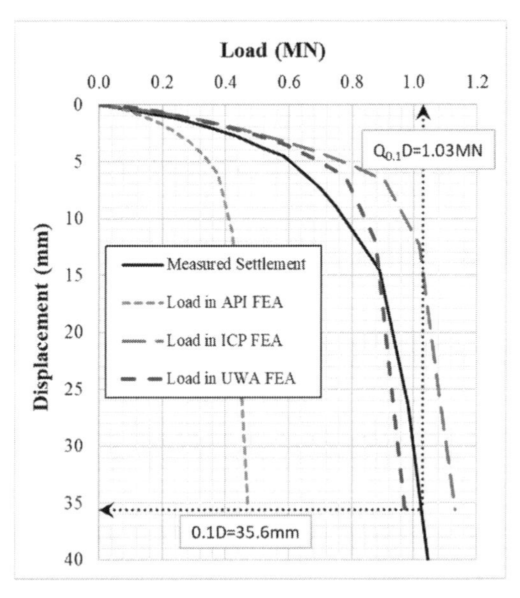

Figure 5. Load displacement responses using FE variants of the API, ICP and UWA design methods.

data, a comparison is made of these results and those estimated by the FE procedure adopting the input parameters of the design methods to obtain their FE variants.

The ultimate external shaft friction, τ_{ult}, and the ultimate end bearing resistance, q_{ult}, are input to the FE procedure from the design methods. By using the design methods directly, without the FE procedure, each predicts a plugged pile, therefore q_{ult} is applied over the area of both the annulus and the pile plug.

In addition, the Q-z and t-z soil reaction curves, taken directly from the API, are specified as constitutive models in the finite element analysis and employed for the ICP and UWA methods. The t-z soil reaction curves are applied to both the external and internal sides of the pile with τ_{ult} applied as the limiting resistance on both sides.

The computed pile head load-displacement responses are presented in Figure 5. Here it is shown that the estimated capacity from the API-FEA significantly underpredicts the capacity at a pile head displacement of $0.1D$, where D is the external pile diameter, with the ratio of calculated to measured capacity, $Q_c/Q_m = 0.45$. The ICP-FEA and UWA-FEA methods both estimate the capacity to within 10% of the measured value.

Using the FE model, the mobilised base capacity, which includes the internal shaft and annular end bearing capacity on the pile, can be extracted under the action of loads applied at the pile head. This has been done for each of the three design methods and presented in Figure 6 to Figure 8. The dotted lines represent the FE estimated values and the solid lines are those measured. These figures demonstrate the ability of the FE model to mobilise the soil plug from the base upwards; a characteristic not currently possible with the existing design methods.

From Figure 6 the API-FEA does not mobilise the full measured internal loads, as the overall capacity is lowest, therefore a smaller load is attributed to each component. The ICP-FEA (Figure 7) and the UWA-FEA (Figure 8) both demonstrate an improved but under-estimation of the development of the base capacity although this is not fully developed to match the measured loads.

The external shaft friction developed along the pile can be extracted from the FE model and compared to the load measured along the external pile in the pile tests. These comparisons are presented in Figure 9 to Figure 11. During the installation, some of the strain gauges below 4 m and 5 m were damaged resulting in missing data.

For a pile head load of 1130 kN, the resulting shaft friction measured along the pile ranged from

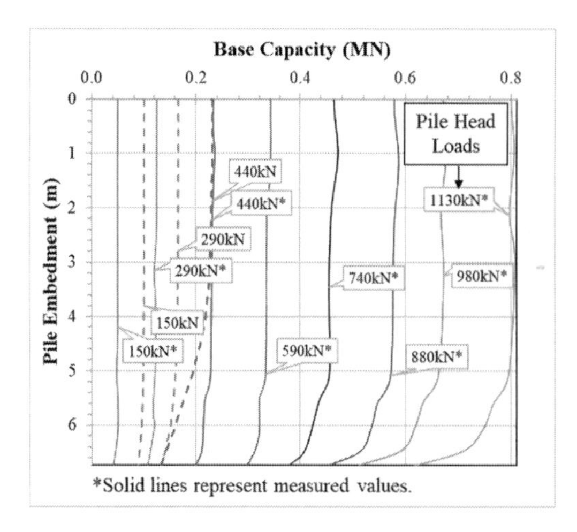

Figure 6. Comparison of measured base capacity to estimates using the API-FEA procedure with axial pile head loads.

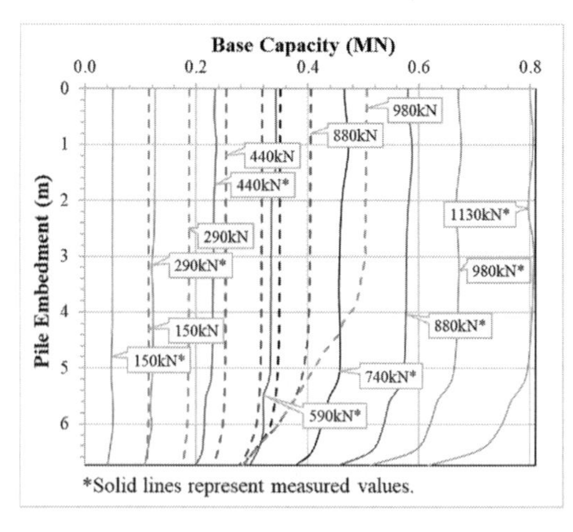

Figure 8. Comparison of measured base capacity to estimates using the UWA-FEA procedure with axial pile head loads.

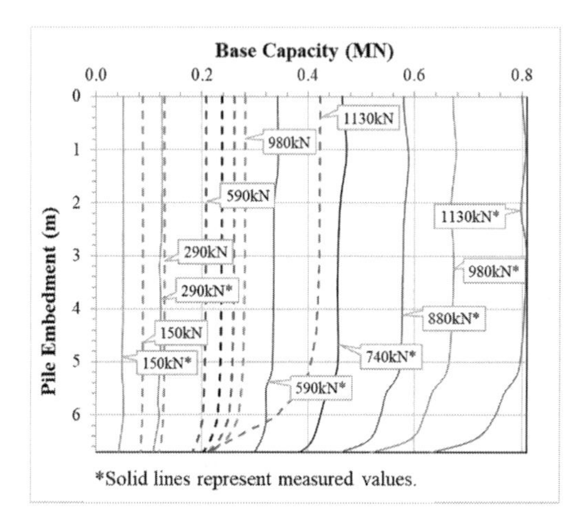

Figure 7. Comparison of measured base capacity to estimates using the ICP-FEA procedure with axial pile head loads.

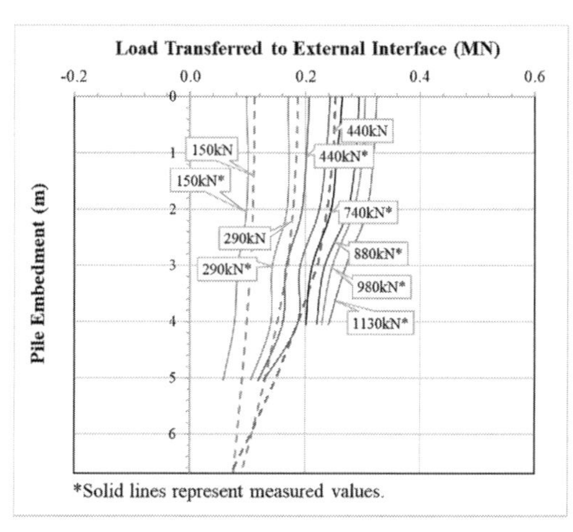

Figure 9. Comparison of measured load distribution along external pile to estimates of external shaft capacity using the API-FEA procedure with axial pile head loads.

approximately 320 kN to 240 kN in the pile segments. The API-FEA procedure (Figure 9) shows that the full load measured in the pile was not mobilised as the maximum FEA value estimated was 440 kN. The ICP-FEA procedure (Figure 10) estimates a maximum of 500 kN in the top segment of the pile. The UWA-FEA (Figure 11) estimates a similar load of approximately 500 kN in the top segment, however this is achieved under a smaller pile head load. This comparison demonstrates that, with the assumptions adopted, whilst ICP and UWA provide good estimates of total pile capacity (Figure 5), none of the design methods predict the relative contributions from the base and shaft identified by the FEA.

Figure 12 to Figure 14 compares the mobilisation of the end bearing capacity. The API-FEA comparison shows that the base capacity of both the annulus and the plug is underpredicted; the total base capacity is approximately 40% of that measured. The ICP-FEA improves the accuracy of model with an estimation of total end bearing of about 70% of that measured. The total base capacity from the UWA-FEA method is the best approximation with 75% of the measured capacity estimated. In these plots however, the distribu-

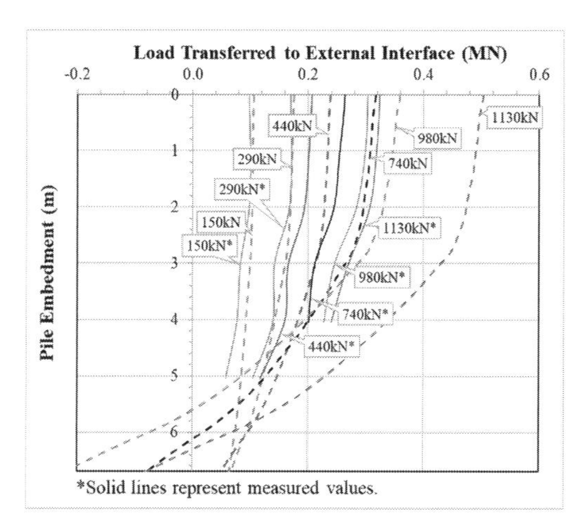

Figure 10. Comparison of measured load distribution along external pile to estimates of external shaft capacity using the ICP-FEA procedure with axial pile head loads.

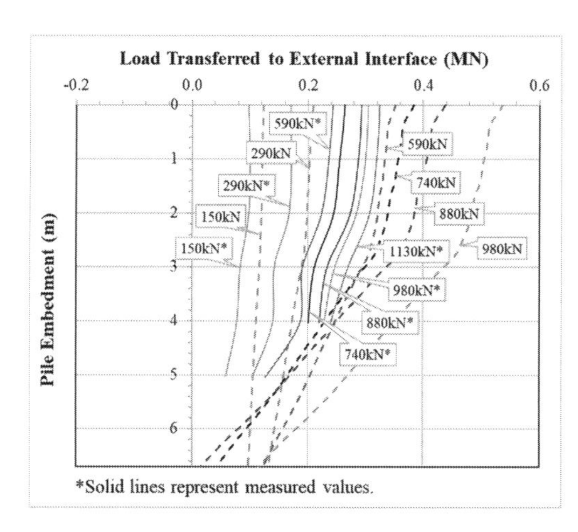

Figure 11. Comparison of measured load distribution along external pile to estimates of external shaft capacity using the UWA-FEA procedure with axial pile head loads.

tion of base capacity between the annulus and soil plug, from the FE analysis, is somewhat different to that measured.

5 DISCUSSION

The equation of equilibrium used to represent a loaded open-ended pile has been solved using a finite element procedure. This method has been unique in that the soil-pile-plug model has been incorporated. The use of the Galerkin method with Lagrangian shape functions is used to repre-

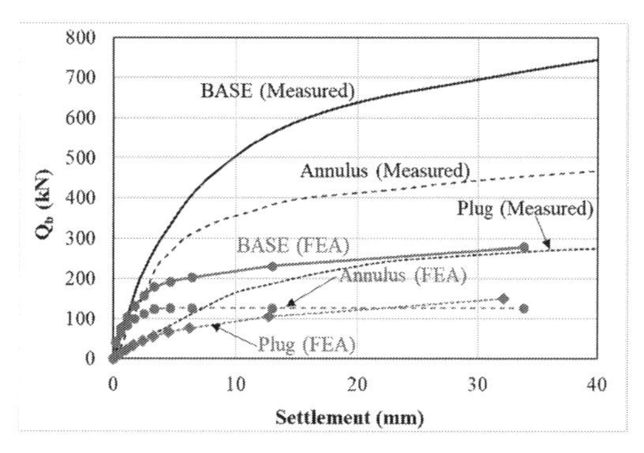

Figure 12. Comparison of measured base capacity and API-FEA estimates vs pile head settlement.

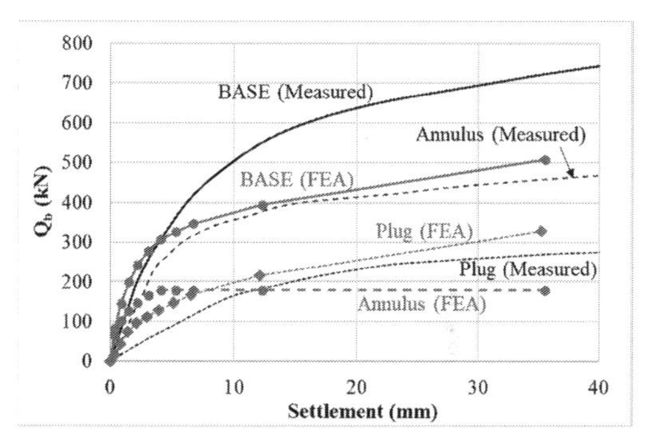

Figure 13. Comparison of measured base capacity and ICP-FEA estimates vs pile head settlement.

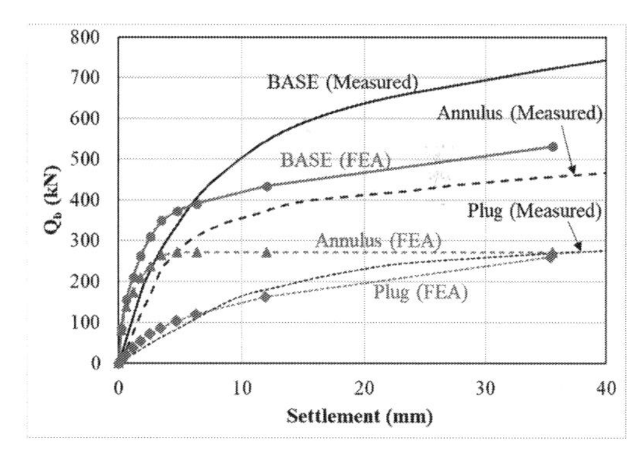

Figure 14. Comparison of measured base capacity and UWA-FEA estimates vs pile head settlement.

sent the various components of the model. Three existing pile design methods were used in the FE model to estimate the capacities and response of these components.

The design methods have derived their capacities based on detailed studies of existing pile test data. These relate the ultimate compressive capacity of the pile to detailed soil test results and derived behaviours of the interface and end bearing. These were deduced for general use by the industry with a rigid pile assumption. The true behaviour of the pile under loading however, is to compress and modelling this response in design is usually an unsophisticated process combining simple soil reaction curves and an elastic pile in a Winkler-type analysis. The finite element model presented here, however, enables a forensic examination of the responses of the individual pile components.

Using the geotechnical engineering database, such as Yang et al. (2015), the ICP and the UWA pile design methods have been shown to provide improved estimates of pile capacity. This tendency was also found in the case study presented in Figure 5 where these methods estimate the measured pile capacity within 10% accuracy.

The end bearing behaviour is difficult to predict. Each of the pile design methods presented adopts a different approach to estimate this behaviour since each considers the response of the pile tip, plug base and internal shaft friction differently. In each method, if the pile is considered 'plugged', the base resistance is applied over the full base area. In the case study, the measured resistance at the base of the pile was approximately 14 MPa which is similar to the measured q_c values. As shown in Figure 6 to Figure 8, with small loads applied, the pile tip provides the only contribution to end bearing. As more load is applied, this quantity rapidly increases demonstrating the high stiffness of this component.

The mobilisation of the plug's capacity was clearly observed in the case study due to the double walled pile configuration. As shown in Figure 6 to Figure 8, as the load increases, the plug capacity is gradually mobilised. This behaviour was replicated in the FE model. Essentially, the plug conveys load from the pile to the soil beneath the base of the plug. The interaction between the pile and the internal plug will therefore influence the end bearing capacity. If the efficiency of load transfer is considered perfect, all the load that is resisted by the internal plug-pile interface is supported by the end bearing of the plug. Also, as demonstrated by Randolph et al. (1991) and other researchers, for drained sandy soils, the compression from the base of a confined soil plug will cause lateral expansion thereby increasing the internal interface stress from the pile tip, attenuating upwards the pile shaft.

The stiffness and capacity of the plug's base is difficult to determine due to the parameters involved in the mobilisation. Again considering sandy soils, these parameters include: the interface friction angle, δ_{cv}, between the pile and plug, accounting for the interface roughness between the two materials; the effective stress in the soil, σ'_{v0}, which depends on the density and length of the plug; a factor that considers the dilatant behaviour of sand when subjected to shearing, K_0; and the confinement stress of the plug, which is dependent on the internal diameter of the pile, D_i. This confinement stress increases with a reducing D_i, and influences the end bearing resistance. Therefore, in sands:

$$q_{pl,b} = f\left(\delta_{cv}, \sigma'_{v0}, K_0, \frac{1}{D_i}\right) \tag{28}$$

Due to the number of factors included in mobilisation, the plug stiffness is much lower than that of the pile tip. Current estimates of plug capacity, such as Lehane et al. (2015) suggest 0.2 to 0.3 of the measured q_c.

The mobilised resistance of the external shaft friction is shown in Figure 9 to Figure 11. The UWA demonstrates the best match of the measured to estimated data followed by the ICP and then the API method. However, both the ICP and UWA methods appear to significantly over-estimate the external shaft capacity.

In general, the pile design methods have been shown to underestimate the end bearing capacity. Figure 14, which highlights the base response of the UWA method, shows that the use of the ultimate values coupled with the API's Q-z soil reaction curves has the best comparison to the values measured. Further work is to be done to improve on this trend.

It is therefore observed from the use of the FE model that the design methods do not readily predict the base and shaft capacity contributions in open-ended piles.

6 CONCLUSIONS

The numerical procedure presented herein has been able to demonstrate that by using finite elements, a critical analysis can be performed on the behaviour of the individual components of an open-ended pile. A case study is presented which compares published results from physical testing using a double walled pile, to those estimated by a simple one-dimensional FE procedure, adopting the ultimate design parameters from three pile design methods, coupled with the API's axial soil reaction curves. The procedure has shown that with the ICP and UWA parameters adopted, the overall capacity is well estimated, however the mobilised distribution of load across each of the resistive components is not well matched. In addition, the true response and capacity of the soil plug is not well captured by the design methods considered. Further work is therefore necessary to ensure that the mobilised stiffness of the individual components required to support axial loads are accurately considered in design. One approach to achieving this is by further improvement of the soil-pile-plug model using a FE procedure.

ACKNOWLEDGEMENTS

This work was supported by grant EP/L016303/1 for Cranfield University and the University of Oxford, Centre for Doctoral Training in Renewable Energy Marine Structures—REMS (http://www.rems-cdt.ac.uk/) from the UK Engineering and Physical Sciences Research Council (EPSRC). The authors also acknowledge the contribution of Atkins (c/o Andrew Benson and Dr David French) for the support of this research work.

REFERENCES

American Petroleum Institute Recommended Practice 2GEO, Geotechnical and Foundation Design Considerations, API RP 2GEO, Apr 2011.

Burd, H. 2015. Finite Element Analysis of Axially Loaded Pile. Internal document.

Khennane, A. 2013. Introduction to finite element analysis using MATLAB and Abaqus". Boca Raton: CRC Press.

Jardine, R., Chow, F., Overy, R. & Standing, J. 2005. *ICP design methods for driven piles in sands and clays*. London: Thomas Telford.

Lehane, B.M. & Gavin, K.G. 2001. Base resistance of jacked pipe piles in sand. *J. Geotech. Geoenviron. Eng.* 128(3): 198–205.

Lehane, B.M. & Randolph, M.F. 2002. Evaluation of a Minimum Base Resistance for Driven Pipe Piles in Siliceous Sands. *J. Geotech. Geoenviron. Eng.* 127(6): 473–480.

Lehane, B.M., Schneider, J.A. & Xu, X. 2005. The UWA-05 method for prediction of axial capacity of driven piles in sand. In: Gourvenec S and Cassidy M (eds.) 2005. *Proc. Int. Symp. Frontiers in Offshore Geotechnics, ISFOG 2005.* London: Taylor Francis, 683–689.

Lehane, B.M., Schneider, J.A. & Xu, X. 2005. CPT based design of driven piles in sand for offshore structures. Report Number: 05345. University of Western Australia: Geomechanics Group Publication.

Joseph, T.M., Houlsby, G.T., Burd, H. & Taylor, P. 2017. Finite Element Analysis of Soil Plug Behaviour Within Open-Ended Piles. SUT OSIG, London, Sept 2017.

Paik, K., Salgado, R., Lee, J. & Kim, B. 2003. Behaviour of open-and closed-ended piles driven into sands. *J. Geotech. Geoenviron. Eng.* 129(4): 296–306.

Randolph, M.F., Leong, E.C. & Houlsby, G.T. 1991. One-dimensional analysis of soil plugs in pipe piles. *Geotechnique* 41(4): 587–598.

Salgado, R., Lee, J., Kim, K. & Paik, K. 2002. Load tests on pipe piles for development of CPT based design method. Final Report FHWA/IN/JTRP-2002/4 for the Indiana Department of Transportation and the US Department of Transportation Federal Highway Administration.

Smith, I.M. & Griffiths, D.V. 2004. *Programming the finite element method.* Chichester: John Wiley & Sons.

Wood, D.M. 2004. Geotechnical modelling. Applied Geotechnics Vol. 1. Abingdon: Spon Press.

Yang, Z., Jardine, R., Guo, W. & Chow, F. 2015. *A Comprehensive Database of Tests on Axially Loaded Piles Driven in Sand.* London: Academic Press.

Numerical Methods in Geotechnical Engineering IX – Cardoso et al. (Eds)
© 2018 Taylor & Francis Group, London, ISBN 978-1-138-33203-4

Spudcan installation and post installation behaviour in soft clay: The press-replace method

Wang Ze-Zhou & Goh Siang Huat
Department of Civil and Environmental Engineering, National University of Singapore, Singapore

ABSTRACT: The life cycle of a jack-up spudcan foundation consists of installation, operations and extraction. A proper simulation of the continuous installation process to capture the stress state of the soil can help provide a better understanding of spudcan-soil interaction response during the subsequent operational stage. This paper presents the application of the Press-Replace Method (PRM) to simulate (i) the continuous installation of a jack-up spudcan foundation into a single layer seabed using an undrained effective stress analysis and (ii) its post-installation consolidation behaviour during the operational stage. The simulated penetration resistance and excess pore pressures generated during the installation stage, and their subsequent pore pressure dissipation during the operational stage, are validated through comparisons with published data. In general, the PRM is able to replicate experimental observations quite well. A significant advantage of this technique is that it allows a better-integrated analysis of the spudcan-soil interaction mechanism incorporating both the undrained response during the installation process and the consolidation response during the operational phase. With the PRM, the use of an additional computer script to map the stress state at the end of installation stage to a separate file for consolidation analysis, which may be necessary for some techniques, is no longer needed. Therefore, a dual stage analysis can be effectively simplified to a one-step analysis with minimum user intervention.

1 INTRODUCTION

Spudcans are commonly used as foundations for offshore jack-up rigs. As shown on Figure 1, a spudcan, which may vary in size and dimension, is typically cylindrical in elevation view. The entire life cycle of a spudcan mainly consists of three stages, namely (i) installation, which refers to the jacking of a spudcan into the seabed, (ii) operation, during which a spudcan is subjected to working loads, and (iii) extraction, which refers to the removal of a spudcan from the seabed. During installation, the continuous pushing of a spudcan into the seabed leads to an altered state of soil, which eventually creates a high confining stress around the spudcan, which in turn enhances the bearing capacity of the soil. In addition, the time-dependent dissipation of the excess pore pressures generated during the installation stage can affect the subsequent operation and eventual extraction of the spudcan. Therefore, the penetration process during the installation stage should be properly modelled so as to capture more accurately the excess pore pressures generated in the vicinity of the spudcan, which then allows its performance during the subsequent operation and extraction stages to be better evaluated. However, such effects cannot be explicitly simulated by the conventional wish-in-place approach.

With the development of computer technology, the incorporation of spudcan installation effects can be achieved using finite element techniques. However, the severe element deformations and distortions created by the continuous pushing of a spudcan into the seabed cannot be handled by conventional Lagrangian-based finite-element package (Tho et al., 2012). Therefore, several modified finite element methods have been proposed to study such large-deformation problems. The updated Lagrangian formulation was adopted by Yi et al. (2012) to study the cone penetration process that allows for consolidation effects. Remeshing and Interpolation Technique with Small Strain (RITSS), which is a form of the Arbitrary Lagrangian Eulerian (ALE) algorithm, was developed by Hu & Randolph (1998) and has been applied to cone penetration studies (Lu et al., 2004;) and spudcan installation analysis (Mehryar et al., 2002; Hossain et al., 2005; Ragni et al., 2016). The explicit Eulerian finite element technique was adopted by Tho et al. (2012) and Yi et al. (2012) to simulate the spudcan installation process.

In this paper, the adaptation of a recently developed finite element technique known as the Press-Replace Method (PRM) (Andersen et al., 2004; Engin, 2013) to analyse spudcan installation and post installation behaviour is presented.

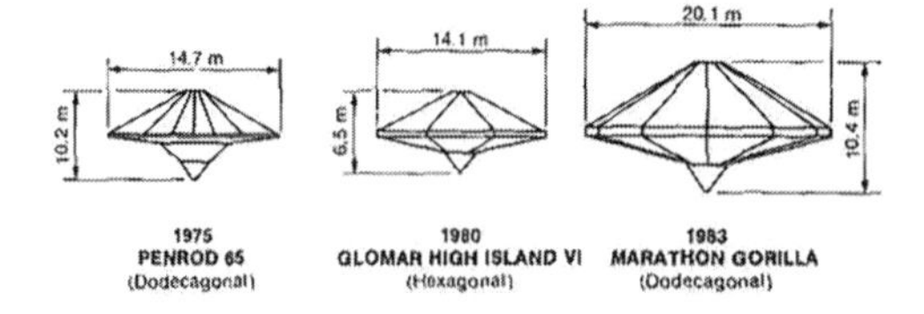

Figure 1. Typical spudcan dimension.

The computed penetration resistance, the excess pore pressure data, as well as the undrained shear strength data during both the undrained installation and post-installation consolidation phases will be extracted and compared with published numerical and centrifuge results. In general, the PRM gives highly promising results. Advantages of the PRM over other existing techniques will also be highlighted in this paper.

2 PRESS-REPLACE METHOD (PRM)

The PRM is a simplified approach to simulate problems associated with continuous penetration of an object into a continuum. This method was first used for simulating the load-controlled penetration of a suction anchor in clay (Andersen et al., 2004) and was further used to simulate pile and cone penetration processes (Engin et al., 2012; Paniagua et al., 2014; Tehrani et al., 2016). As its name suggests, the PRM consists of two phases, namely the press phase and the replace phase. The entire installation process is broken down into numerous press-replace paired phases. The press phase is to simulate the advancement of the penetrating object. A Dirichlet boundary condition (displacement) is prescribed at every press phase to simulate the penetration. During every press phase, only a portion of the total penetration depth is achieved. The replace phase is performed to replace the soil body that was displaced by the penetrating object with the material of the penetrating object. This replace phase results in the modification of the global stiffness matrix and boundary conditions, which then forms the initial condition for the next press phase. These two phases are repeated until the desired penetration depth is achieved.

The PRM formulates the global stiffness matrix based on the undeformed geometry of the model. This means that the updated global stiffness matrix only includes the part of the soil body that is replaced by the material of the penetrating object. Therefore, compensation in the form of straining inside and near the zone that is replaced by the penetration object is needed. However, the elastic strain energy to be compensated is expected to be small due to the much higher stiffness of the penetrating object compared to that of the soil. Hence, it is not unreasonable to assume that total spent energy is mostly dissipated by plastic deformation, and as such the errors introduced by neglecting this elastic straining is not expected to be significant (Engin, 2013; Tehrani et al., 2016).

3 FINITE ELEMENT MODEL

3.1 *Choice of soil model*

For a normally consolidated or lightly over-consolidated soft clay, the Drucker-Prager and Mohr-Coulomb models overestimate the undrained shear strength due to errors in the effective stress path. Yi et al. (2012) reported that the Modified Cam Clay model yields better agreement with centrifuge experimental data than the Drucker-Prager and Mohr-Coulomb models. Hence, the Modified Cam Clay model is also adopted in this study to model the Malaysian Kaolin clay used in the centrifuge spudcan experiments performed by Purwana et al. (2005). Table 1 summarises the Modified Cam Clay parameters of this soil.

3.2 *Model setup*

The commercial software Plaxis 2D AE (Brinkgreve et al, 2014) is used in the current study. The finite element model is created following the 100-g centrifuge experiment configuration described by Purwana (2006). As shown in Figure 2, a 50 m × 50 m prototype soil domain is created. In the centrifuge experiment, a 5 mm layer of free water was placed above the soil domain to ensure soil saturation. This corresponds to 5 m of free water under prototype conditions, the effect of which was included in the finite element model.

An axisymetric spudcan is modelled. As indicated in Figure 2, the entire penetration process is

Table 1. Modified cam clay parameters of Malaysia.

Parameter	Value
Slope of critical state line, M	0.9
Slope of isotropic normal compression line, λ	0.244
Slope of isotropic swelling line, κ	0.053
Specific volume of soil at p' = 1 kPa, N	3.22
Effective unit weight of soil, γ' (kN/m^3)	6
Effective Poisson's ratio of soil, ν'	0.33
Coefficient of lateral earth pressure at rest, K_0	0.6
Coefficient of permeability, k (m/s)	2.0×10^{-8}

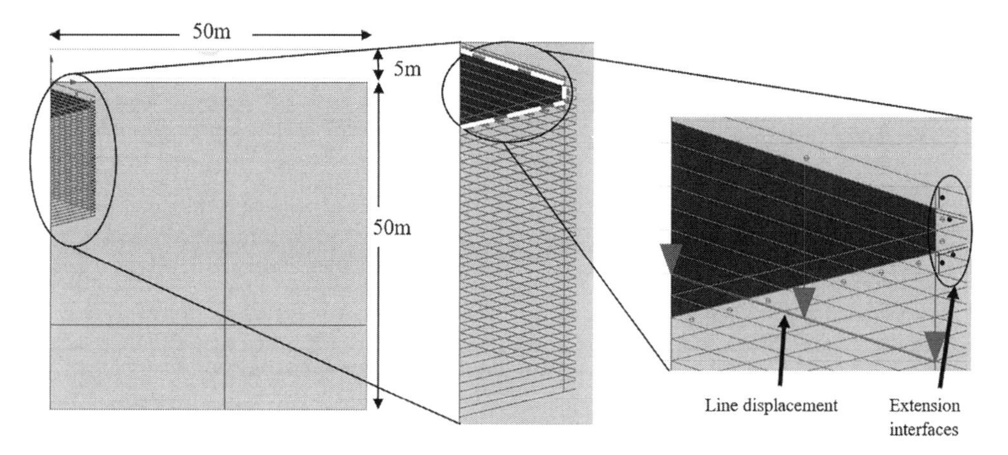

Figure 2. Model set up and graphic illustration of the PRM procedure.

decomposed into numerous press-replace paired phases. According to Engin (2013), a single press phase should be no larger than 1/10 of the diameter of the penetrating object. In the current study, a 1 m step size is chosen and this is achieved by imposing a Dirichlet boundary condition on the top face of the spudcan, as indicated by the downward arrow in Figure 2. An additional parametric study (not shown) has indicated that this step size is sufficient to achieve a converged analysis. There are in total 4,870 15-node triangular elements in this 50 m × 50 m soil domain. A mesh convergence study has also been performed which indicated that the current mesh size is sufficient. A smooth spudcan soil interface is simulated in the present study. Extended interfaces are placed at the tip corner of the spudcan so as to minimize numerical stress oscillations. The detailed dimensions of the spudcan used in the current study are shown on Figure 3.

Compared to the application of the PRM in earlier studies involving pile installation and cone penetration, there is one major modification made in the current finite element model of spudcan installation. During the replace phases of pile installation and cone penetration, only soil elements below the penetrating objects were replaced by the material of the penetrating object. However, duirng spudcan installation, replacement of materials for elements above and below the spudcan were necessary in order to simulate the penetration process of the finite thickness object. In the conventional PRM analysis, a non-porous concrete material was typically assigned to the increasing length of the penetrating object in the soil. However, in the case of spudcan penetration, the authors observed that unreasonable values of excess pore pressures developed in the elements

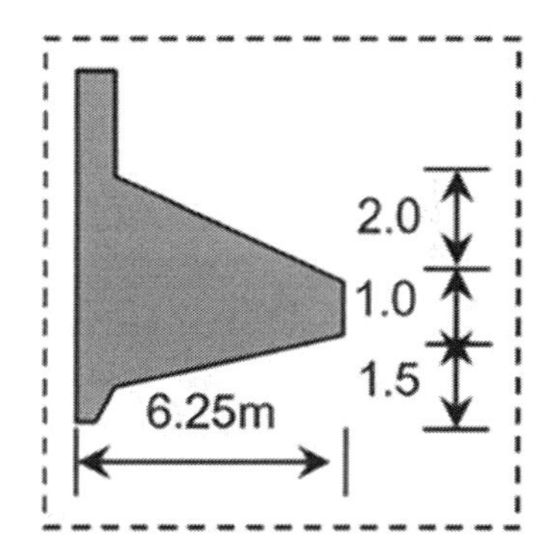

Figure 3. Dimension of spudcan used in the analysis.

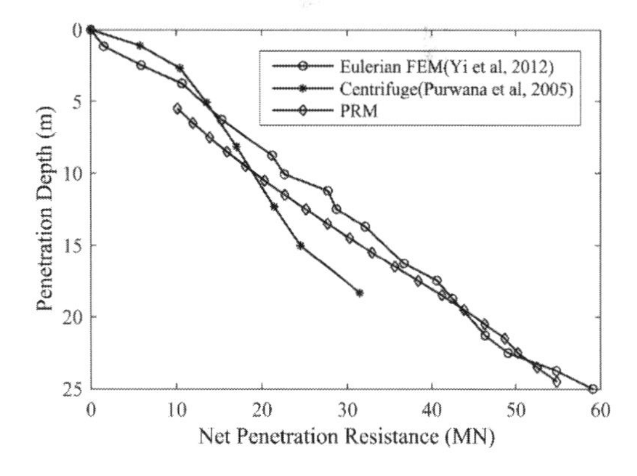

Figure 4. Comparison of net penetration resistance.

above the spudcan when the non-porous concrete material was replaced back by the soil materials upon passage of the spudcan. It was found that

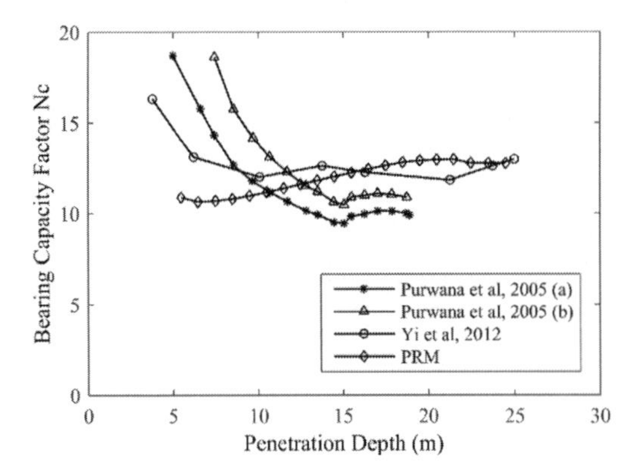

Figure 5. Bearing capacity factor against penetration depth.

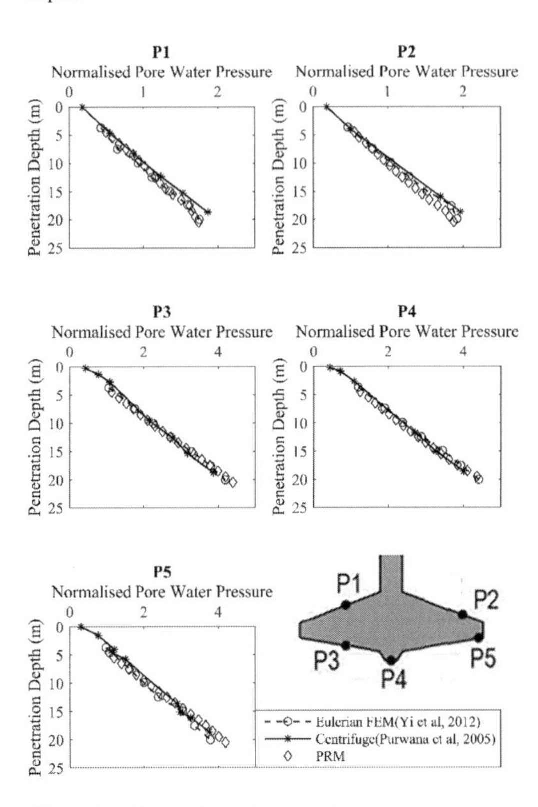

Figure 6. Comparison of pore water pressure.

this problem could be avoided by assigning the spudcan material to be the same as the surrounding soil, but with very stiff, impermeable plate elements defined along the spudcan periphery so that it still behaves like a rigid body. This is indicated by the thick dashed line in Figure 2. The authors have performed additional checks to verify that the strains within the spudcan elements are much smaller than those of the soil outside.

4 SPUDCAN INSTALLATION

4.1 *Penetration resistance*

The net spudcan penetration resistance with depth computed by the PRM is presented in Figure 4. Net penetration resistance refers to the force output minus the self-weight of the spudcan. The penetration depth is with respect to the lowest point of the spudcan at the end of each press phase. On the same figure, the numerical simulation results using Eulerian FEM (Yi et al., 2012) and centrifuge experimental results (Purwana et al., 2005) are also shown. It is seen that the PRM simulation yields a generally linear relationship between the net penetration resistance and penetration depth, which tends to agree quite well with Yi et al.'s Eulerian results at greater depths. However, the PRM under-predicts the net penetration resistance at shallow depths. In the PRM, the penetration process starts from the state where the spudcan is just fully embedded into the seabed. There are two implications related to this setting. First, the PRM can only predict penetration resistance at depths larger than the thickness of the spudcan, as can be seen from Figure 4 in which the PRM curve starts at 5.5 m below ground level. Second, it follows that any changes in the state of the soil for the first 5.5 m cannot be captured by the PRM. This may in part account for the lower penetration resistances predicted by the PRM at shallower penetration depths. Nevertheless, the net penetration resistance from the PRM at greater depths, which is usually the more critical condition, agrees quite well with the results from Eulerian FEM.

It is observed that, at greater depths, both the PRM and the Eulerian FEM predict larger penetration resistances than those measured in the centrifuge experiment. The lower resistance measured in the experiment was likely due to strength degradation (Hossain & Randolph, 2003; Erbrich, 2005), in which a trapped wedge of weak soil was observed below the spudcan during the penetration process. This soil mass was subjected to very large strains, thus becoming fully remolded and possessing a lower shear strength. Both the Eulerian FEM and the PRM do not capture this degradation.

Figure 5 plots the bearing capacity factor N_c derived from the current PRM analysis, together with other published data. N_c is calculated based on an undrained shear strength profile of 1.47z, wherein z refers to the depth beneath the soil surface. This undrained shear strength profile is calculated using the Modified Cam Clay parameters and Wroth's relation (Wroth, 1984). The PRM predicts a converged Nc of 12.7, which agrees well with the value obtained by Yi et al. (2012), which is approximately 12.5.

4.2 *Pore water pressure*

The pore pressures computed during the PRM installation process are compared with the centrifuge experimental results (Purwana et al., 2005). Figure 6 shows the locations of the pore pressure transducers in the centrifuge experiment. There are in total five pore pressure transducers, labeled as P1, P2, P3, P4 and P5. P1 and P2 are on the top face of the spudcan. P3, P4 and P5 are on the bottom face of the spudcan. Figure 6 also plots the pore pressure results from the PRM, the Eulerian FEM and the centrifuge experiment at all five transducer locations. The normalised pore water pressure represents the measured pore water pressure normalised by the spudcan diameter and unit weight of water.

There is good agreement between the PRM pore pressure results and those obtained from the Eulerian FEM and centrifuge experiments at P3, P4 and P5. However, the PRM slightly under-predicts the pore water pressures at P1 and P2, compared with the results from Eulerian FEM and centrifuge experiment. This is most likely due to PRM's inability to model the full flow of soil above the spudcan, which consequently predicts a less significant remolding of the soil in the zone of P1 and P2.

Figures 7 and 8 plot the excess pore pressure contours developed during the penetration process. The comparisons between the PRM (on the left) and Eulerian FEM (on the right) contours are made at 10 m and 20 m penetration depth respectively. Good agreement between the PRM and the Eulerian FEM are observed in these two figures. In addition, the PRM pore water pressure contours (on the left) are plotted and compared with Eulerian FEM (on the right) results in Figures 9 and 10. Generally good agreement between the PRM and Eulerian FEM results are also observed below the spudcan. At 10 m penetration depth, the PRM 50 kPa contour line is at almost the same depth as that predicted by Eulerian FEM. However, at 20 m penetration depth, some discrepancies are observed in the plotted contours (e.g. 150 kPa) above the spudcan. This indicates that the PRM yields a lower pore water pressure above the spudcan, which is consistent with what was observed in Figure 6. The implication of this observation will be discussed later.

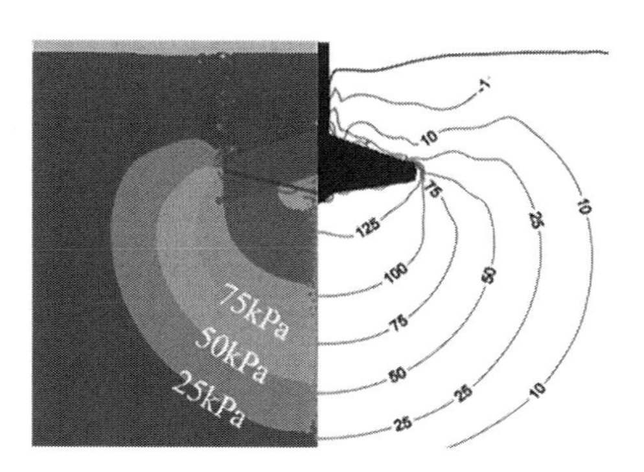

Figure 7. Excess pore pressure contour at 10m penetration depth (Left: PRM Right: Eulerian FEM).

5 CONSOLIDATION

5.1 *Pore water pressure*

During the operational stage, the spudcan is subjected to working loads. Depending on the operational duration, the degree of dissipation of the excess pore water pressures generated during the installation stage is expected to be different. Purwana et al. (2005) reported that suction pressures below the spudcan, which depends on the dissipation of the excess pore water pressure generated during the installation stage, significantly affects the extraction force of the spudcan. In the centrifuge experiment (Purwana et al, 2005), the service load was 24.3MN. Therefore, in the PRM analysis, an unloading phase was introduced after the spudcan installation analysis to reduce the

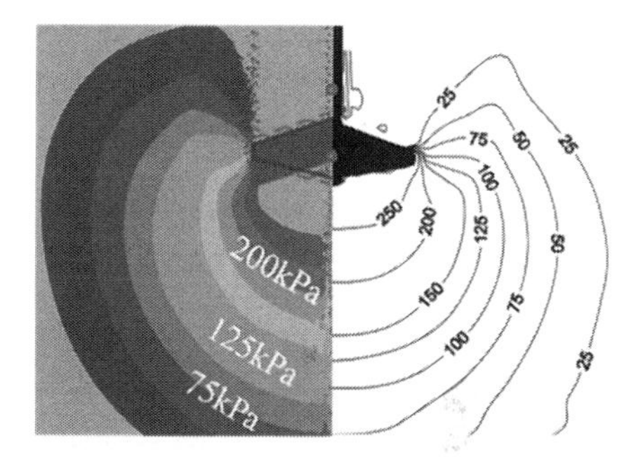

Figure 8. Excess pore pressure contour at 20m penetration depth (Left: PRM Right: Eulerian FEM).

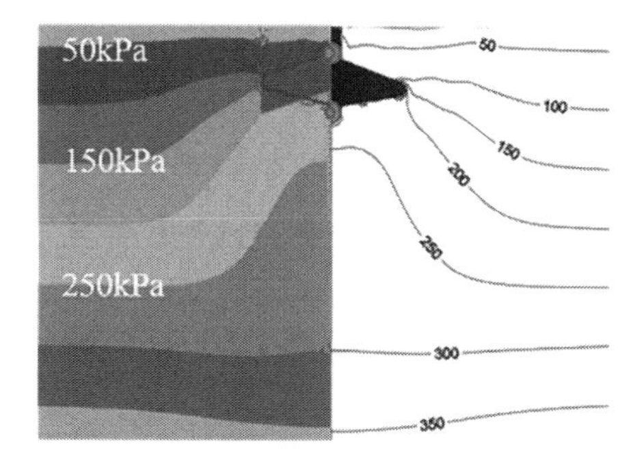

Figure 9. Pore pressure contour at 10m penetration depth (Left: PRM Right: Eulerian FEM).

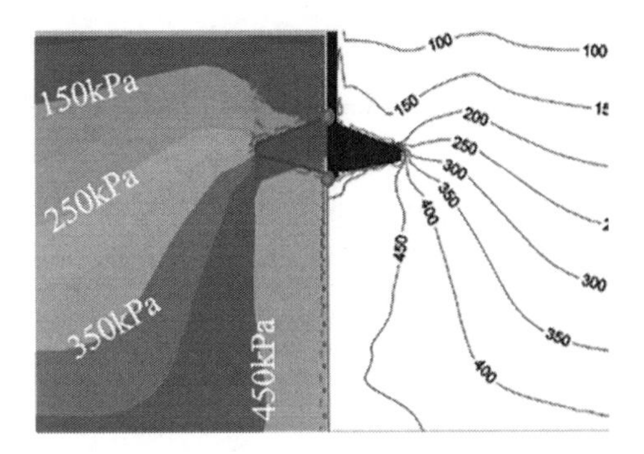

Figure 10. Pore pressure contour at 20m penetration depth (Left: PRM Right: Eulerian FEM).

spudcan load to 24.3 MN. This service load was maintained during the subsequent consolidation stage and pore water pressure changes were monitored at the locations of the five transducers.

Figure 11 plots the PRM results showing the dissipation of excess pore water pressures with time. Comparisons with the centrifuge data reported by Purwana et al. (2005) are made. Except for P1 and P2 (which were located on the top face of the spudcan), the agreement between the PRM and centrifuge test results are generally quite favourable. Two observations are highlighted.

First, the PRM predicts a slightly higher rate of dissipation of pore water pressure as compared to the centrifuge results. The authors believe this could be due to some differences in the soil permeability between the centrifuge experiment and the present numerical analysis. Figure 12 shows further comparison with results reported by Zhou et al. (2009), who adopted the same permeability as that used in the current study. This figure compares the change in average excess pore water pressure at the base of the spudcan at different times during the consolidation phase. It is observed that both PRM and Zhou et al. (2009) report a higher rate of dissipation as compared to the centrifuge results. The reasonable agreement between the excess pore pressure results of the PRM and Zhou et al. (2009) suggests that the discrepancies with the centrifuge results could be due to inhomogeneity in the soil permeability in the experimental test sample.

The second observation is that, compared to the experimental results, the PRM yields lower excess pore water pressures at the beginning and end of the consolidation phase at P1 and P2, which are on the top face of the spudcan. The initial excess pore water at P1 and P2 is lower because the PRM does not account for soil remolding and strength loss on the top face of the spudcan. The gradients of the P1 and P2 pore pressure histories indicate that PRM produces a slightly higher rate of dissipation, which is consistent with the results of P3, P4 and

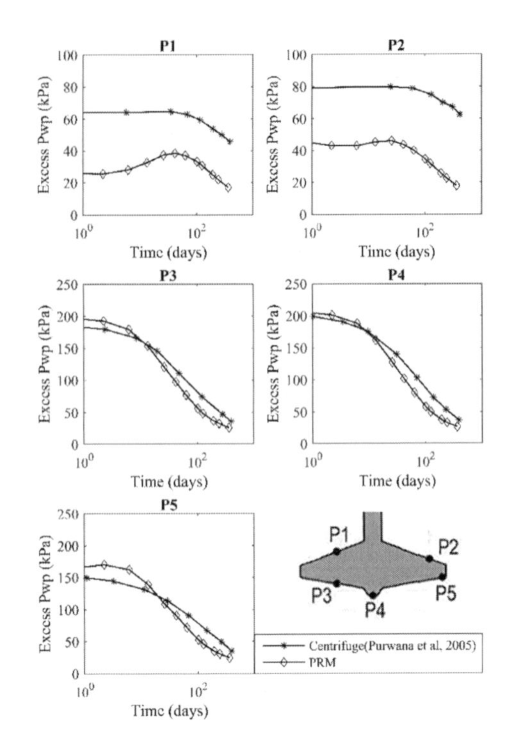

Figure 11. Dissipation of excess pore water pressure with time.

P5 on the underside of the spudcan. The implication of this observation is that the PRM predicts a higher effective stress above the spudcan, which will lead to a higher computed resistance during the extraction stage. However, both Purwana et al. (2005) and Zhou et al. (2009) reported that the contribution of the top resistance is relatively small compared to the suction contribution below the spudcan.

In addition, Figures 13 and 14 plot pore water pressure contours after 35 days and 90 days respectively. The left half of each figure shows the PRM results, while the right half shows the numerical results reported by Yi et al. (2014). The two sets of results are in generally good agreement. Although there are some discrepancies in the pore water pressure immediately above the spudcan, the PRM pore pressure contour some distance above the spudcan agrees well with Yi et al. (2014). This indicates that although the PRM cannot account for changes in the soil strength due to backflow and remolding above the spudcan, the effect of this limitation diminishes quite rapidly with distance away from the top face of the spudcan. Therefore, the discrepancies in pore water pressure at P1 and P2 are not expected to have a significant impact on the total resistance above the spudcan.

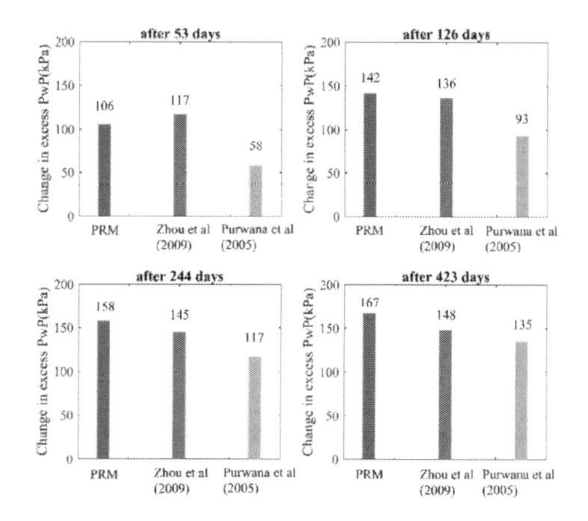

Figure 12. Comparison of dissipation of excess pore water pressure with other numerical results.

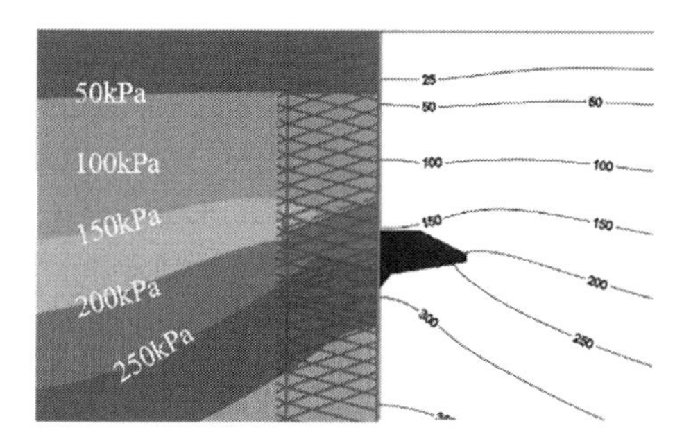

Figure 13. Pore pressure contour after 35 days of consolidation (Left: PRM Right: Yi et al., 2014).

5.2 *Undrained shear strength*

The dissipation of excess pore water pressure during the consolidation phase will eventually result in an increase in undrained shear strength. The computed stress states (mean effective stress, overconsolidation ratio) from the PRM analysis, together with the Modified Cam Clay parameters, may be used to derive the undrained shear strength following Wroth (1984). The strength improvement ratio, defined as the ratio of the long-term undrained shear strength to the in-situ undrained shear strength, are computed using the PRM results and compared with those reported by Yi et al. (2013). In the current study, the in-situ undrained shear strength is derived as 1.47*z, wherein z refers to the depth beneath the soil surface. Figures 15 and 16 show the contours of the computed strength improvement ratio, for both the PRM and Yi et al. (2014), after 180 days and 5 years of consolidation respectively.

It is observed that consolidation eventually creates a zone of soil with improved undrained shear strength below the spudcan. The size and the shape of the improved zone is similar to the excess pore water pressure contours presented in Figures 7 and 8. This is not surprising because the improved undrained shear strength is the result of the dissipation of excess pore water pressure. It is also observed that, compared to the Yi et al. (2014), the PRM appears to predict a lower ratio further away from the spudcan. For a strength improvement ratio of 1.1, the contour derived from by Yi et al. (2014) extends over a larger zone compared to that from the PRM. Nonetheless, there is good agreement between the PRM and results of Yi et al. (2014) on the size and the shape of the improved zone directly below the spudcan. This is important because an accurate estimation of the undrained shear strength

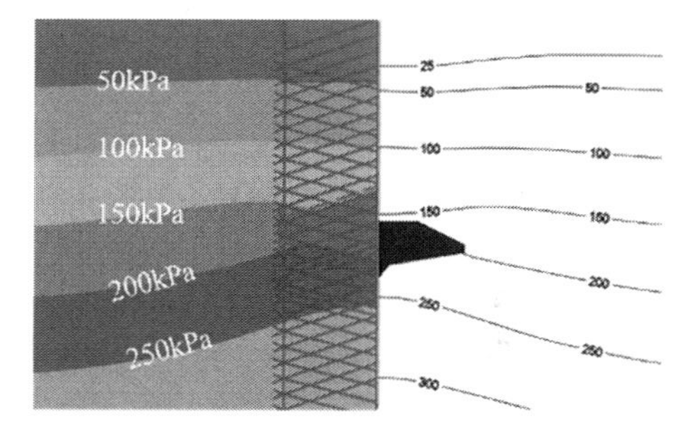

Figure 14. Pore water pressure after 90 days of consolidation (Left: PRM Right: Yi et al., 2014).

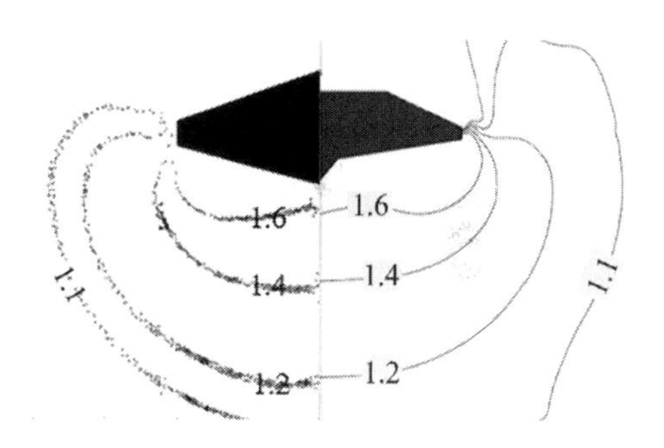

Figure 15. Strength improvement ratio after 180 days of consolidation (Left: PRM Right: Yi et al., 2014).

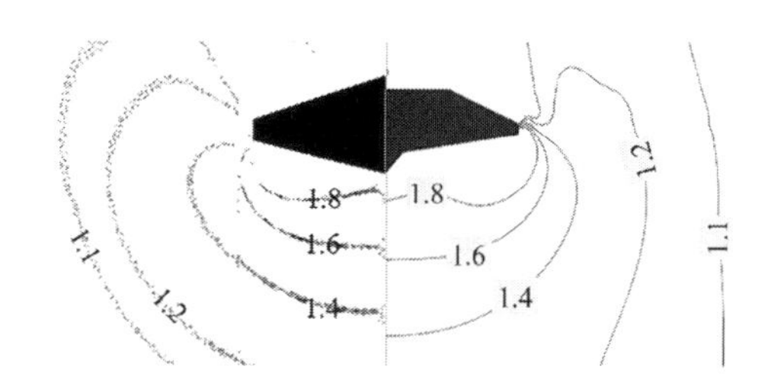

Figure 16. Strength improvement ratio after 5 years of consolidation (Left: PRM Right: Yi et al., 2014).

directly below the spudcan is necessary for a proper analysis on spudcan punching behaviour.

6 LIMITATIONS AND CONCLUSIONS

There are limitations associated with the PRM. The PRM is not able to model the flow of the soil below and around the peripheral zone of the spudcan during installation. Therefore, the PRM may not be useful when deformation information is need. For example, the PRM may not produce reasonable results of punching behaviour.

In conclusion, the results from this study show that the PRM is a promising numerical modelling method to simulate spudcan installation. The PRM can predict penetration resistance and excess pore water pressure that agrees reasonably well with centrifuge results and results of other numerical techniques. A key contribution of the present study is that the PRM can effectively bridge the installation and consolidation analysis for modelling both the short-term and long-term response of the spudcan. A single working file is used to perform the whole analysis, therefore eliminating potential information loss that may occur through mapping errors when a separate subroutine or script is adopted to bridge these two stages. The transition from the installation to the consolidation analysis is seen to be effective because the rate of dissipation of excess pore water pressure predicted by the PRM agrees reasonably well with both centrifuge measurements and other published results.

REFERENCES

Andersen, K.H., Andresen, L., Jostad, H.P. and Clukey, E.C., 2004, January. Effect of skirt-tip geometry on set-up outside suction anchors in soft clay. In *ASME 2004 23rd International Conference on Offshore Mechanics and Arctic Engineering*(pp. 1035–1044). American Society of Mechanical Engineers.

Erbrich, C.T., 2005, October. Australian frontiers–spudcans on the edge. In *Proc., 1st Int. Symp. on Frontiers in Offshore Geotechnics* (pp. 49–74). London: Taylor & Francis Group.

Engin, H.K., 2013. Modelling pile installation effects: a numerical approach.

Engin, H.K., Brinkgreve, R.B.J. and Tol, A.F., 2015. Simplified numerical modelling of pile penetration–the press-replace technique. *International Journal for Numerical and Analytical Methods in Geomechanics, 39*(15), pp.1713–1734.

Hu, Y. and Randolph, M.F., 1998. A practical numerical approach for large deformation problems in soil. *International Journal for Numerical and Analytical Methods in Geomechanics, 22*(5), pp.327–350.

Hossain, M.S., Hu, Y. and Randolph, M.F., 2003, January. Spudcan foundation penetration into uniform clay. In *The Thirteenth International Offshore and Polar Engineering Conference*. International Society of Offshore and Polar Engineers.

Hossain, M.S., Hu, Y., Randolph, M.F. and White, D.J., 2005. Limiting cavity depth for spudcan foundations penetrating clay. *Géotechnique, 55*(9), pp.679–690.

Lu, Q., Randolph, M.F., Hu, Y. and Bugarski, I.C., 2004. A numerical study of cone penetration in clay. *Géotechnique, 54*(4), pp.257–267.

Mehryar, Z., Hu, Y. and Randolph, M.F., 2002, January. Penetration analysis of spudcan foundations in NC clay. In *The Twelfth International Offshore and Polar Engineering Conference*. International Society of Offshore and Polar Engineers.

Brinkgreve, R.B.J., Engin, E. and Swolfs, W.M., 2014. PLAXIS 2D AE manual. *AA Balkema.*

Purwana, O.A., Leung, C.F., Chow, Y.K. and Foo, K.S., 2005. Influence of base suction on extraction of jack-up spudcans. *Géotechnique, 55*(10), pp.741–753.

PURWANA, O.A., 2007. Centrifuge model study on spudcan extraction in soft clay (Doctoral dissertation).

Paniagua, P., Nordal, S. and Engin, H.K., 2014. Back calculation of CPT tests in silt by the Press-Replace technique.

Ragni, R., Wang, D., Mašín, D., Bienen, B., Cassidy, M.J. and Stanier, S.A., 2016. Numerical modelling of the effects of consolidation on jack-up spudcan penetration. *Computers and Geotechnics, 78*, pp.25–37.

Tho, K.K., Leung, C.F., Chow, Y.K. and Swaddiwudhipong, S., 2010. Eulerian finite-element technique for analysis of jack-up spudcan penetration. *International journal of geomechanics, 12*(1), pp.64–73.

Tehrani, F.S., Nguyen, P., Brinkgreve, R.B. and van Tol, A.F., 2016. Comparison of Press-Replace Method and Material Point Method for analysis of jacked piles. Computers and *Geotechnics, 78*, pp.38–53.

Wroth, C.P., 1984. The interpretation of in situ soil tests. *Geotechnique, 34*(4), pp.449–489.

Young, A.G., Remmes, B.D. and Meyer, B.J., 1984. Foundation performance of offshore jack-up drilling rigs. *Journal of Geotechnical Engineering, 110*(7), pp.841–859.

Yi, J.T., Lee, F.H., Goh, S.H., Zhang, X.Y. and Wu, J.F., 2012. Eulerian finite element analysis of excess pore pressure generated by spudcan installation into soft clay. *Computers and Geotechnics, 42*, pp.157–170.

Yi, J.T., Goh, S.H., Lee, F.H. and Randolph, M.F., 2012. A numerical study of cone penetration in fine-grained soils allowing for consolidation effects. *Géotechnique, 62*(8), p.707.

Yi, J.T., Goh, S.H. and Lee, F.H., 2013. Effect of sand compaction pile installation on strength of soft clay. *Géotechnique, 63*(12), p.1029.

Yi, J.T., Zhao, B., Li, Y.P., Yang, Y., Lee, F.H., Goh, S.H., Zhang, X.Y. and Wu, J.F., 2014. Post-installation pore-pressure changes around spudcan and long-term spudcan behaviour in soft clay. *Computers and Geotechnics, 56*, pp.133–147.

Zhou, X.X., Chow, Y.K. and Leung, C.F., 2009. Numerical modelling of extraction of spudcans. *Géotechnique, 59*(1), pp.29–39.

Numerical Methods in Geotechnical Engineering IX – Cardoso et al. (Eds)
© 2018 Taylor & Francis Group, London, ISBN 978-1-138-33203-4

Behaviour of laterally loaded pile

S. Ahayan
ArGEnCo Departement, University of Liège, Belgium
Institut de Recherche en Génie Civil et Mécanique (GeM), École Centrale Nantes, France

B. Cerfontaine & F. Collin
ArGEnCo Departement, University of Liège, Belgium

P. Kotronis
Institut de Recherche en Génie Civil et Mécanique (GeM), École Centrale Nantes, France

ABSTRACT: Monopiles have been by far the most common support structure for offshore turbines. They have always been an appropriate solution for complex site conditions.

Design of monopiles is usually based on the use of nonlinear p-y curve. These curves are primordially based on the use of simple constitutive models for soil, such as Tresca criterion. Since soil behavior is highly non-linear and very complex, fundamental features of soil such as anisotropy, creep, or destructuration have to be taken into consideration. Accordingly, it is necessary to consider more complex soil behavior constitutive models. This work aims to explain, via FEM simulations, the influence of different constitutive laws for soil, on the laterally loaded pile responses. Tresca, Mohr-Coulomb criteria and Modified Cam-Clay model have been compared and their effect on p-y curve is analyzed.

1 INTRODUCTION

Offshore wind energy applications have great potential towards achieving European Unions energy target for 2020. Accordingly, in the last decade, there is an exponential increase in the offshore wind energy industry. Offshore Wind Turbines (OWT) substructure is an important concern. This is due to the huge construction cost of the foundations and substructures, which counts for 25 to 30% of the total construction cost (EWEA 2015). Larger wind parks are increasingly developed away from the coastline, at increasing water depths. Current projects reach a distance of 150 *Km* offshore and at a water depth up to 45 *m*. The scale of online wind farms follows an upward trend since the early 2000s, and is correlated to the rising size and production capacity of individual windmills. In the last 25 years, the energy production of one single OWT has been increased twentyfold, while rotor diameter has been grown from 35 m up to more than 150 *m* (Schaumann, Lochte-Holtgreven, & Steppeler 2011). The continuous increase of OWT dimensions attests the rapid growth of the offshore wind market.

The type and design of the foundation solutions for OWT projects depend on the size of the turbine, water depth, and local seabed conditions. According to those constraints, the challenge is to develop the most cost-effective solution for the support structure and onsite installation process (Arshad & OKelly 2013). The support structure of OWT comprises the substructure on the top of the foundation system. On one hand, the substructure is chosen based on the mean sea level. On the other hand, the foundation type is bonded to the OWT dimensions, transfer of applied loads to the soil, and ground geotechnical properties.

In a shallow and moderate water depth, between 10 *m* and 70 *m*, the use is made of bottom-mounted substructures that are firmly fixed to the seabed by underwater foundations. Three main layouts are commonly used for shallow OWT applications: monopod, multipod, and jacket structure. Additionally, there are three practical options for the foundation system, comprising the gravity base foundations (GBF), piled foundations, and the skirt or bucket foundations, also known as suction caissons foundations.

Based on the database of The European Wind Energy Association (EWEA) for the year 2016, it shows that 88% of the installed substructure foundation systems in Europe used monopiles, whereas the remaining 12% corresponded to jacket structures. In this research, the focus is made on offshore wind turbine based on single piles, which are designed to resist lateral loading, such as wind, wave and rotor reactions.

P-y curve is one of the most common methods that has been widely used (API (RP2A-WSD 2000), DNV (Veritas 2004), ect). P-y curves represent the relationship between the lateral load (p) and lateral displacement (y) at a certain point of a pile. Several numerical techniques, such as finite element and finite difference methods, are employed to predict the p-y curve for laterally loaded pile design (Byrne, McAdam, Burd, Houlsby, Martin, Zdravkovi, Taborda, Potts, Jardine, & Sideri 2015), (Zdravkovi´c, Taborda, Potts, Jardine, Sideri, Schroeder, Byrne, McAdam, Burd, Houlsby, et al. 2015)).

The work of Matlock (Matlock 1970) is widely used by the offshore pile industry. Matlocks recommendations are primordially based on the use of simple constitutive models for soil, meaning elastic perfectly plastic model such as Tresca or Mohr- Coulomb criteria. Since the behaviour of the clay soil is highly nonlinear and very complex, it is necessary to consider advanced constitutive models (Ahayan, Yin, Kotronis, & Collin 2016). This research aims to explain, via FEM simulations, the influence of using different constitutive laws of the soil on the responses of laterally loaded piles. Those laws vary from the simplest model, such as Tresca, to the critical state law of Cam-Clay model (Roscoe & Burland 1968). The effect of each model on the prediction of p-y curve is presented in the following sections.

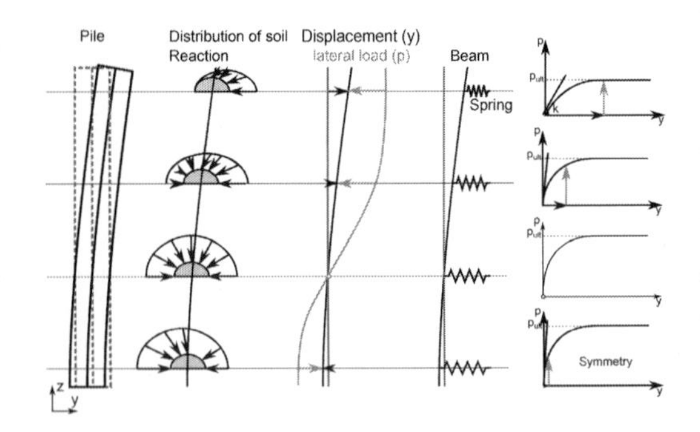

Figure 1. The p-y curve method.

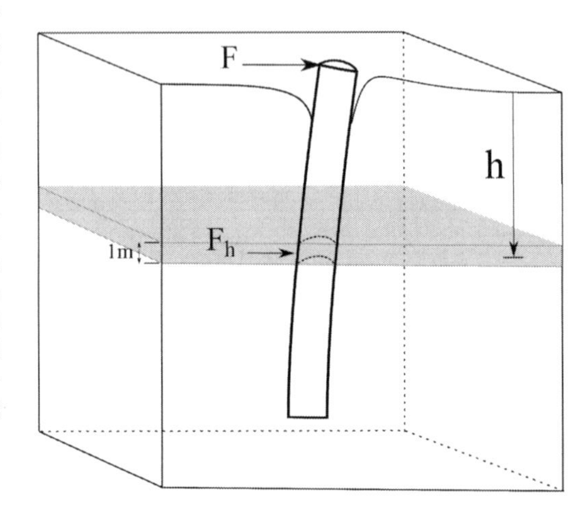

Figure 2. From 3D to 2D representation of laterally loaded pile problem.

2 FINITE ELEMENT MODEL

The finite element model presented in this section allows analysing pile behaviour under lateral monotonic loading. The main objective is to assess the influence of soil behaviour on the p-y curve.

The p-y curve method considers the pile as a beam, while the soil is considered as a series of uncoupled non-linear springs (Reese, Cox, & Koop 1974). Each of those springs are defined by a p-y curve, as shown in Figure 1.

The non-linear finite element code LAGAMINE, developed at the University of Liege (Collin 2003), is selected as the main numerical platform for this study.

2.1 *2D simulation of laterally loaded pile*

In the present work, the 3D problem of laterally loaded pile is simplified to a 2D problem to determine the characteristic p-y curve by numerical modeling. This simplification takes place by considering a slice of one meter-wide soil at a given depth and by assuming plane strain conditions, as illustrated in Figure 2. Similar assumption can be found in (Zhang, Ye, Noda, Nakano, & Nakai 2007).

2.2 *Geometry and boundary conditions*

Representative dimensions are set for an offshore monopile of 2 m diameter. Figure 4 shows the model dimensions and the considered mesh. Since the pile is assumed to be submitted to a symmetrical lateral loads, only half of the pile is simulated.

2.3 *Pile loading*

Monopile, as an offshore substructure, is designed to resist mainly to cyclic lateral loads, coming from wind, wave and rotor solicitations. However, the most critical influence of the cyclic loads occurs during the first cycles. Therefore, the investigation has been conducted for the first cyclic loads on the offshore monopile.

In the following, the analysis is carried out by applying a lateral displacement equivalent to 10% of the pile diameter to ensure that the lateral bearing capacity of the soil is reached. Lateral displacement is unloaded afterwards to reach its initial position, and applied again on the negative direction with the same value (see Figure 3). Based on this displacement, a large deformation has to be considered, in addition to gapping between the pile and the soil.

2.4 *FE mesh*

The pile section is considered as a rigid body, and the soil matrix is discretized using 3040 quadrangular (8 nodes) elements.

As a consequent of lateral displacement forces, a gap is created between the soil and the pile under lateral loading, an interface element is considered to reproduce the lost contact between both materials as shown in Figure 5.

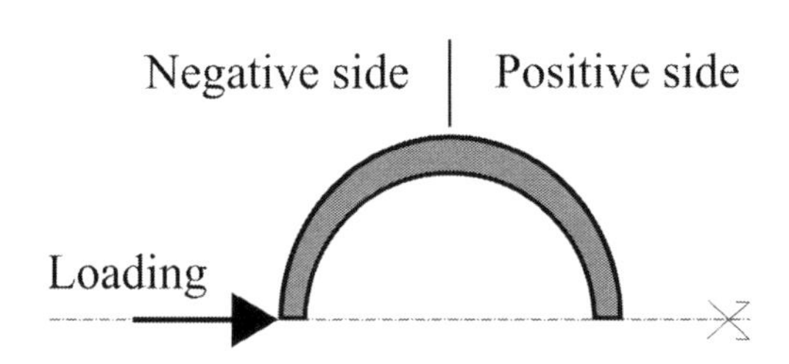

Figure 3. Loading pile configuration.

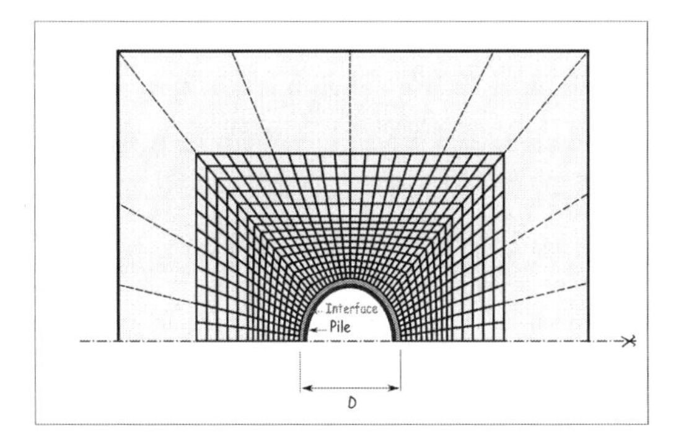

Figure 4. 2D finite element model.

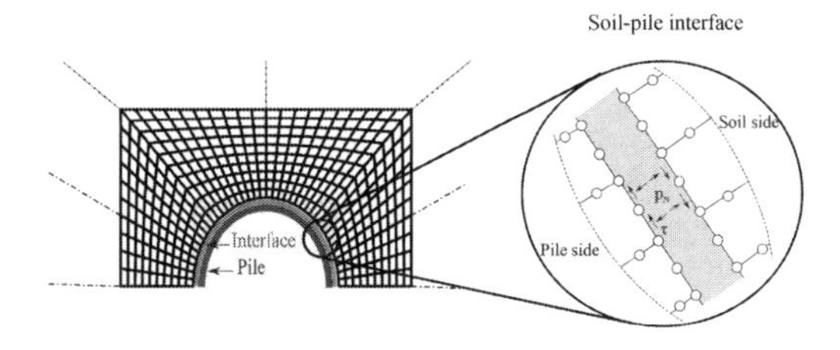

Figure 5. Interface finite element.

The interface element, as described in previous literature (Cerfontaine, Dieudonn´e, Radu, Collin, & Charlier 2015) and(Cerfontaine, Collin, & Charlier 2015), belongs to the zero-thickness family:

The probable zone of contact respects the "ideal contact constraint states":

$$g_N \geq 0, p_N \geq 0, p_N . g_N = 0 \tag{1}$$

where p_N is the normal pressure, and g_N is the gap between two sides of the interface. This ideal constraint establishes that contact holds if the gap is equal to zero (closed), giving rise to a normal contact pressure. Whereas, if the gap is positive (open), it indicates a non-contact pressure. The shear behavior of the interface is described similarly, in which maximum shear resistance along the interface is ruled by Mohr-Coulomb criterion, and according to:

$$\tau_{max} = \sigma_n \tan \phi \tag{2}$$

where φ is the steel-soil friction angle, which is equal to two thirds the shear angle of the soil.

If the shear stress is lower than τ_{max} soil and pile are considered "stuck", and the relative tangential displacement g_T is null. If the maximal shear stress is reached, both sides of interface encounter a relative displacement.

2.5 *p-y curve from 2D finite element method*

The soil reaction curve is extracted from the 2D simulation from post-processing routines. As illustrated in Figure 1, computed total stress (normal pressure and shear stress) in the interface elements is integrated along the pile circumference to determine the local lateral load (p), which is related to the local pile displacement (y).

2.6 *Constitutive models*

In order to analyze the influence of the constitutive model, with respect to simplified model such as Tresca criterion, two additional constitutive laws have been considered. The Mohr-Coulomb (MC) criterion is firstly used as a reference model for soils, secondly the Modified Cam-Clay model (MCC) is also used in the modeling. The MCC model is the first critical state model that describes the behavior of soft soils such as clay. Assuming isotropic behavior, the MCC model is defined by a yield criterion represented, in $p-q$ plane, by an ellipse oriented in line with the p axis as shown in Figure 6.

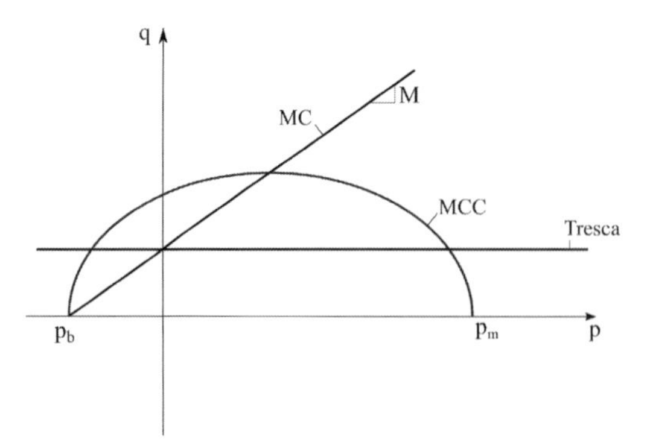

Figure 6. MCC, MC and Tresca yield surface in $p-q$ plane.

Table 1. Soil parameters.

	v	κ	E (MPa)	λ	φ	c (kPa)	ocr
Tresca	0.25	-	2	-	0	10	-
MC	0.25	-	2	-	32	10	-
MCC	0.25	0.04	-	0.26	32	10	1.2

The model is described by the following equation in generalized space:

$$f(\sigma_d, p, p_m) = \frac{3}{2}\sigma_d : \sigma_d + M^2(p + p_b)(p - p_m) = 0 \quad (3)$$

σ_d is the shear stress tensor $\left(\sigma_{d_{ij}} = \sigma_{ij} - \delta_{ij}p\right)$, and p is the mean stress. M is the slope of the critical state line in $p-q$ plane. This parameter is calculated directly from the shear angle of soil φ, as follow:

$$M = \frac{6sin\phi}{3 - sin\phi} \quad (4)$$

p_m represents the preconsolidation pressure of soil. It controls the size of the yield surface, due to the following isotropic hardening rule:

$$\dot{p}_m = \frac{1+e}{\lambda - \kappa} p_m \dot{\varepsilon}_v^p \quad (5)$$

where e is the current void ratio, κ and λ are the compression and swelling slope of the isotropic consolidation line in $e - \ln p$ respectively.

The set of soil properties, presented in Table 1, is considered for this analysis. The soil is assumed as a slightly overconslidated clay (ocr = 1.2). In-situ stresses are estimated with an assumption of a unit weight of 15 $kN.m^{-}3$, and the isotropic initial conditions are respected. The chosen soil is a natural and very soft clay, representative of some marine clays such as the clay of Gulf of Mexico and the shallow clay in the North Sea (Bjerrum 1973).

3 RESULTS ANALYSIS

The aim of this section is to study the influence of the interface between the soil and the pile, in addition to the soil behavior on p-y curves of laterally loaded pile.

3.1 *Soil-pile interaction*

Before analyzing the behavior of different soil models, the p-y curve has been explained using one type of soil behavior model. The p-y curve is predicted by using MCC model at different depth levels, illustrated at different loading levels.

As shown in Figure 8 and 7, p-y curves are illustrated at different depths, with the distribution of the total contact pressure $p_T = p_N + \tau$ around the pile in different loading levels. Figure 8 shows the p-y curve at a depth of 5 m, while Figure 7 shoes the p-y curve at a depth of 15 m.

It has been observed that the evolution of the p-y curve at 5 m of depth is going through different phases as explained below:

- phase 1 (portion AB): Initially, at point A, the stresses are uniform and normal to the pile section, assuming that the foundation is vertical, and installed without inducing any initial friction. The pile loading induces a redistribution of the total pressure around the pile, shear stresses appear, and the normal pressure increase in the positive side of the pile section, and remains positive and non-null around the pile until point B. In addition, the p-y curve has a linear form.
- phase 2 (portion BC): p-y curve has a non-linear form during this phase. This non linearity is related to the non-linearity of soil, and on the other hand to the non-linearity of soil-pile interface behavior. Therefore, during loading, the total pressure around the pile evolves until it reaches a null value. Thus, a gap between soil and pile appears, which is considered as an irreversible deformation, and then allows non-linear behavior of the structure. When the maximum displacement is applied (Point C), the gap reaches all the negative side, while the total pressure remains non-null in the positive side and reaches its maximum value in front of the pile section.
- phase 3 (portion CD): The unloading phase is started. In the positive side, the soil reaction (p)

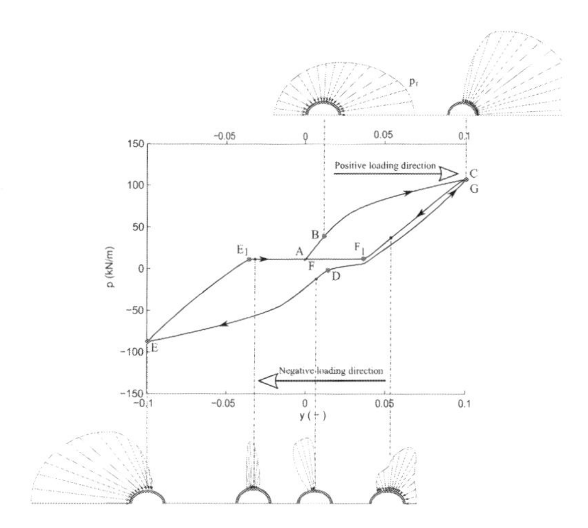

Figure 7. p-y curve and total pressure contact at 5 m of depth.

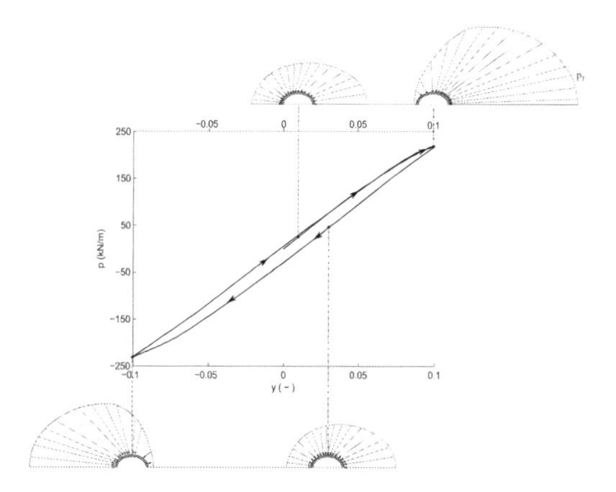

Figure 8. p-y curve and total pressure contact at 15 m of depth.

decreases with unloading, until a gap appears. In the negative side, the contact pressure increases.

- phase 4 (portion DE): first, the pile is loaded until its initial position, and then loaded in the negative direction up to 10% of pile diameter. During this phase, the contact pressure is increasing in the negative side, and decreasing in the positive side, until its reaches its maximum value at point E. The gap increases during this phase in the positive side.
- phase 5 (portion EF): During this phase, a horizontal displacement is applied in the positive direction until its initial position. The contact pressure decreased in the negative side, and increases in the positive side. Then, the gap evolves in the negative and positive sides. Due to the gap behavior, the soil reaction (p) reaches its minimum value in the portion (E1F), where the two structures are in contact just in the head of the pile section.
- phase 6 (portion FG): This section can be subdivided into two portions, the first portion (FF1, which is similar to the portion E1F, and characterized by the gap in the negative and the positive sides, and the second portion (F1G), when the gapdisappears in the positive side. The segment F1G has the same form of the portion CD, during the unloading.

At 15 meters of depth, the distribution of contact pressure remains non-null around the pile during loading and unloading phases. The resulted p-y curve has continuous form, without any slop break.

As shown in Figures 8 and 7, the p-y curve has entirely different form compared to the extracted curve in 5 meters of depth. The resulted p-y curves have different evolutions due to two main reasons. The first reason is related to the depth of the analyzed surface, while the second reason is due to the soil-pile interface behavior resulting from the lateral loading on the pile. However, the consequence of the soil-pile interface behavior highly differs from one depth to another.For instance, at the depth of 5 meters, lateral loading shifts the soil together with the pile. This shift causes a redistribution of the contact pressure around the pile and creates a gap between the soil and the pile. Thus, a null value is obtained for the total contact pressure as a result of that gap. However, the same loading does not induce a gap at a depth of 15 meters, it only redistributes the contact pressure. From this analysis, it is concluded that the evolution of the lateral capacity (p) at a certain depth is entirely related to the soil behavior.

3.2 *Influence of the constitutive model*

The main objective of this section is to assess the influence of soil behavior on the p-y curve. For this reason, three soil models, Tresca, Mohr-Coulomb (MC) and Modified Cam-Clay model (MCC) have been chosen for the modelling, the mesh and boundary conditions remaining the same as described in the section 2. The considered soil parameters for each constitutive law are given in Table 1. The extracted p-y curves are illustrated in Figure 9 at the same depth of 5 meters. At this depth, it is observed that the p-y curve behaves similarly, in terms of pile interface behavior, for the different soil models as shown in the Figure 9. Pile interface behavior is characterized by the gap formation between the two materials (soil and pile). However, the impact of different soil

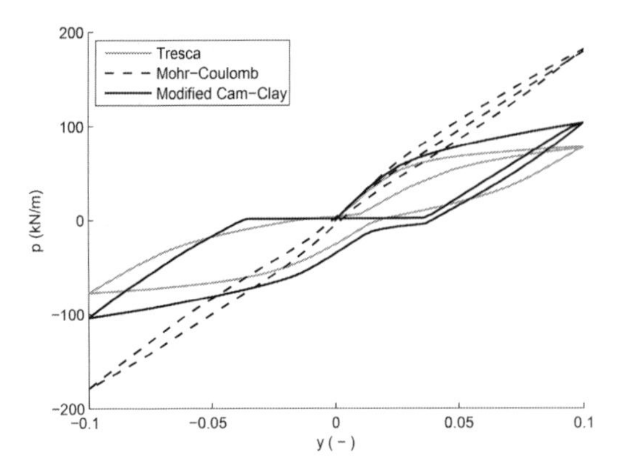

Figure 9. p-y curves at 5m of depth.

behaviors appears in the mobilization of the pile lateral capacity.

It has been observed that Mohr-Coulomb model has the highest mobilization effect of contact pressures. MC induces a higher value of lateral capacity compared to the MCC and Tersca models. This difference is explained by the elastic domain predicted by each model. For instance, MC model has a larger elastic domain compared to Tresca criterion as shown in Figure 5. With the MCC, the soil is considered normally consolidated, which means that the soil reached the plasticity state once loading path. High elastic domains correspond to high values of lateral capacity.

It has been observed from the same Figure that MCC model predicts higher value of lateral capacity compared to Tresca prediction. On one hand, Tresca model is an elastic perfectly-plastic model. While on the other hand, MCC predicts the hardening and softening states of the soil. For slightly overconsolidating soil (with ocr = 1.2), hardening phase occurs rapidly while elastic limit increases constantly with increasing the loads.

Figure 10 shows different soil zones where the plasticity is reached in the case of Tresca, MC and MCC simulations at 10%D of horizontal displacement in the positive side. For the three simulations, the soil plasticity evolves during loading, but not with the same extent: the simulated soil with MC model remains elastic around the pile except at the pile head where plastic strains appears. In the case of MCC model, the plasticity is reached around the pile once the horizontal displacement is applied, because of the assumption of slightly overconsolidating soil. While the soil is in plasticity just in the positive side in the case of Tresca model. These observations support the interpretations that it has been done from Figure 9.

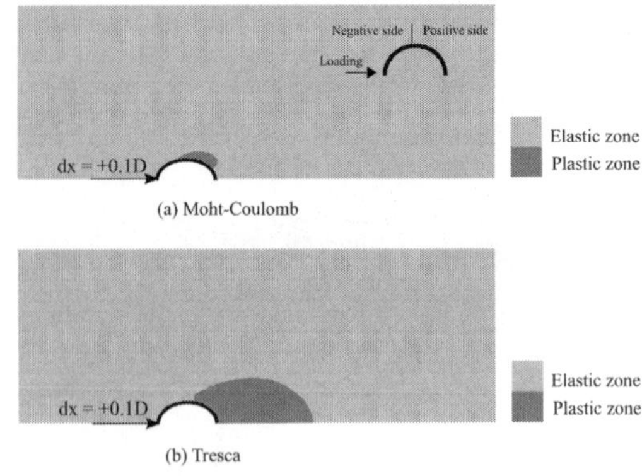

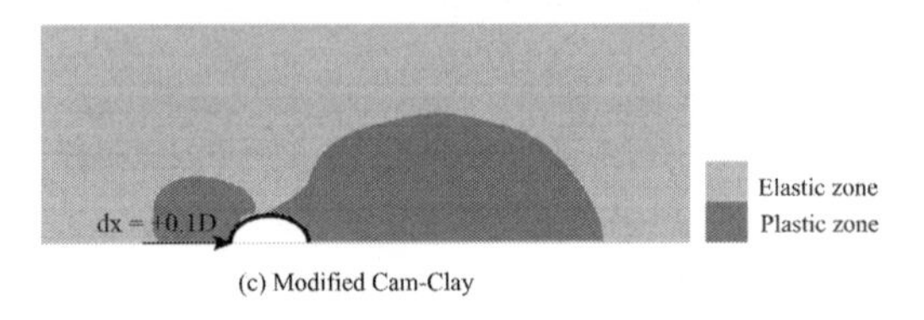

Figure 10. Plasticity zones for 10%D of lateral displacement.

4 CONCLUSIONS

The laterally loaded pile has been investigated in this article, in which Tresca, Mohr-Coulomb, and Modified Cam-Clay models have been applied during the investigation. Those soil models are considered to be the most widely used in laterally loaded pile modeling. Moreover, the interface impact was outlined, and the gapping phenomenon was explained at different depths of the pile.

The extracted p-y curves have shown critical differences related to the soil behavior. It has been observed that Tresca criterion underestimates the soil capacity compared to Mohr-Coulomb and Modified Cam-Clay models, and Mohr-Coulomb criterion predicts high value of lateral capacity compared to other models. This remarks have been related directly to the evolution of soil plasticity during loading. While MC and Tresca models are elastic perfectly plastic models, MCC model predicts some soil hardening. Moreover, MC model allows a large elastic domain compared to Tresca criterion.

The analyzed results are based on 2D simulation with the assumption of plane deformation. However, it should be noted that laterally loaded pile is actually a 3D problem. The assumption of plane strain remains only relevant for certain depth, which is considered to be one limitation in this research. In fact, two soil mechanisms may exist in the soil resistance for laterally loaded pile as explained in (Reese, Cox, & Koop 1974). In the upper part, soil fails in a conical wedge that extends to the soil surface. The soil failure mechanism transits into a localized plane mechanism at a certain depth. Therefore, 3D simulations are recommended to be provided within further research. Moreover, a complete cyclic load is recommended to be calculated to investigate the interface behavior in later stages of cyclic load.

REFERENCES

Ahayan, S., Z. Yin, P. Kotronis, & F. Collin (2016). L'effet de l'écrouissage rotationnel sur le comportement des sols argileux. In *34émes Rencontres de l'AUGC*. University of Liège.

Arshad, M. & B.C. OKelly (2013). Offshore wind-turbine structures: a review. *Proceedings of the Institution of Civil Engineers-Energy 166*(4), 139–152.

Bjerrum, L. (1973). Problems of soil mechanics and construction on soft clays and structurally unstable soils (collapsible, expansive and others). In *Proc. of 8th Int. Conf. on SMFE*, Volume 3, pp. 111–159.

Byrne, B., R. McAdam, H. Burd, G. Houlsby, C. Martin, L. Zdravkovi, D. Taborda, D. Potts, R. Jardine, & M. Sideri (2015). New design methods for large diameter piles under lateral loading for offshore wind applications. In *3rd International Symposium on Frontiers in Offshore Geotechnics (ISFOG 2015), Oslo, Norway, June*, pp. 10–12.

Cerfontaine, B., F. Collin, & R. Charlier (2015). Numerical modelling of transient cyclic vertical loading of suction caissons in sand. *Géotechnique 65*(12).

Cerfontaine, B., A.-C. Dieudonné, J.-P. Radu, F. Collin, & R. Charlier (2015). 3d zero-thickness coupled interface finite element: formulation and application. *Computers and Geotechnics 69*, 124–140.

EWEA, E. (2015). The european offshore wind industry key trends and statistics 1st half 2015. *European Wind Energy Association (EWEA) 1*, 8.

EWEA, E. (2017). The european offshore wind industry-key trends and statistics 2016. *European Wind Energy Association (EWEA) 1*, 8.

Matlock, H. (1970). Correlations for design of laterally loaded piles in soft clay. *Offshore Technology in Civil Engineerings Hall of Fame Papers from the Early Years*, 77–94.

Randolph, M.F. & G. Houlsby (1984). The limiting pressure on a circular pile loaded laterally in cohesive soil. *Geotechnique 34*(4), 613–623.

Reese, L.C.,W.R. Cox, & F.D. Koop (1974). Analysis of laterally loaded piles in sand. *Offshore Technology in Civil Engineering Hall of Fame Papers from the Early Years*, 95–105.

Roscoe, K. & J. Burland (1968). On the generalized stress-strain behaviour of wet clay.

RP2A-WSD, A. (2000). Recommended practice for planning, designing and constructing fixed offshore platforms–working stress design–. In *Twenty-*.

Schaumann, P., S. Lochte-Holtgreven, & S. Steppeler (2011). Special fatigue aspects in support structures of offshore wind turbines. *Materialwissenschaft und Werkstofftechnik 42*(12), 1075–1081.

Veritas, D.N. (2004). Design of offshore wind turbine structures. *Offshore Standard DNV-OS-J101 6*, 2004.

Zdravković, L., D. Taborda, D. Potts, R. Jardine, M. Sideri, F. Schroeder, B. Byrne, R. McAdam, H. Burd, G. Houlsby, et al. (2015). Numerical modelling of large diameter piles under lateral loading for offshore wind applications. In *Proceeding 3rd International Symposium on Frontiers in Offshore Geotechnics. Norway:[sn]*.

Zhang, F., B. Ye, T. Noda, M. Nakano, & K. Nakai (2007). Explanation of cyclic mobility of soils: Approach by stressinduced anisotropy. *Soils and Foundations 47*(4), 635–648.

Numerical Methods in Geotechnical Engineering IX – Cardoso et al. (Eds)
© 2018 Taylor & Francis Group, London, ISBN 978-1-138-33203-4

Modelling of the lateral loading of bucket foundations in sand using hydro-mechanical interface elements

B. Cerfontaine
School of Science and Engineering, University of Dundee, UK

R. Charlier & F. Collin
Urban and Environmental Engineering, University of Liege, Belgium

ABSTRACT: Behaviour of offshore foundations such as piles or bucket foundations is highly dependent on interface properties. Shear mobilisation along their boundaries is crucial, not only under tension but also lateral loading. Gap opening plays an important role to develop the suction effect under bucket foundations.

Offshore foundation behaviour is often computed in purely drained (long-term) or undrained (short-term) conditions. However, dissipation of over-/under- water pressures may dissipate fast especially in sand. Subsequently in some cases the true behaviour is not purely drained nor purely undrained, but partially drained.

We developed a hydro-mechanically coupled finite element of interface able to reproduce loss of contact, friction mobilisation (Coulomb criterion), sliding, water flow along the interface and through it. Hydro-mechanical couplings arise from the definition of an effective stress, the dependence of longitudinal flow and storage to gap opening.

A bucket foundation is modelled upon lateral loading to illustrate the capabilities of the finite elements as well as the effects of couplings on the overall caisson behaviour.

1 INTRODUCTION

The exponential development of offshore renewable energy devices, exploiting wind, waves or tides, has increased the demand of innovative geotechnical engineering solutions to replace classical but sometimes costly offshore monopiles. Indeed as projects are planned in further and deeper waters, jacket superstructure or floating devices become competitive. Suction caissons or skirted foundations represent an interesting solution for wind turbine foundations, where the moment applied at the top is transferred to ground mainly via a push/pull loading (Houlsby and Byrne 2005, Houlsby et al. 2006). The partially drained behaviour due to the generated suction effect was proven to increase the strength of these foundations against transient uplift (Achmus et al. 2013, Cerfontaine et al. 2015).

The soil-caisson interface behaviour plays a crucial role in the global response of offshore foundations in general and suction caisson in particular, as reported in (Kourkoulis, Lekkakus, Gelagoti, & Kaynia 2014, Cerfontaine, Collin, & Charlier 2015). Shear mobilisation and gap opening both strongly affect the soil structure interaction. However interface simulation in partially drained conditions requires the formulation of special finite element, encompassing hydro-mechanical couplings.

Modelling of the mechanical interface problem with finite elements is not new and has been widely studied (Wriggers 2006) since the early work of Goodman (Goodman et al. 1968). Several approach were developed based on the discretisation method, the enforcement of the contact constraint or the interface constitutive law. Basically the mechanical problem relates normal and relative displacement along the interface to normal and shear stress.

In many geotechnical problems, the interface is a preferential path for fluid flow. These hydro-mechanical couplings have been devoted less attention than the purely mechanical problem over recent years (Guiducci et al. 2002, Jha and Juanes 2014, Cerfontaine et al. 2015). The interface introduces a discontinuity in the field of fluid pressures due to the longitudinal and transversal fluid flows. It is then a non-classical fluid boundary conditions applied on each side of the solid porous bodies, since it is not pressure imposed nor flux imposed.

In this work we investigate the role of the interface in the response of a suction caisson subjected to an imposed lateral displacement. Both purely drained and partially drained simulations are carried out to illustrate the increase of strength that could be mobilised in these conditions. P-y curves related to the rigid body motion of the caisson are also derived for more practical and simplified applications.

2 STATEMENT OF THE PROBLEM

In this section we introduce the methodology used to simulate the hydro-mechanically coupled interface behaviour by the finite element method. The discretisation procedure and different couplings are defined. The constitutive law of the soil material is also defined.

2.1 *Interface finite elements*

The hydro-mechanically coupled finite element of interface is based on the zero-thickness approach (Charlier and Cescotto 1988). Boundaries of two solid porous bodies that are likely to contact are discretised using special boundary elements, having no thickness. For instance in Figure 1, the two solids Ω^i (i = 1,2) are in contact along their common boundary Γ_c. If external load tends to push the solids closer, normal load increase and the surface area may increase also. If the loads tend to separate them, normal stresses release and the contact area decreases (gap creation). If a relative displacement is generated, shear stress are likely to be created. Therefore solving the mechanical contact problem requires 1) detecting the evolution of the contact zone, 2) defining a constitutive law to describe the normal/shear behaviour of the interface, 3) discretising the problem in finite elements. A detailed description of the method could be found in (Cerfontaine et al. 2015).

One of the most important variables in contact problem is the gap function g_N, which is a measure of the distance between the two sides of the interface. It is theoretically positive if there is no contact and null otherwise. There are many different

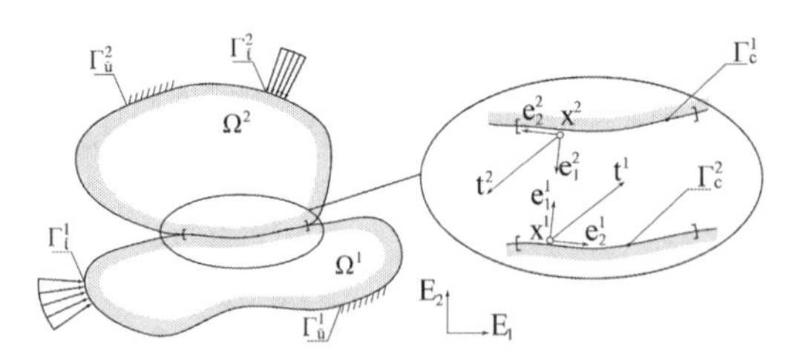

Figure 1. Definition of the contact problem between two solids Ω^1 and Ω^2.

ways of calculating this function (Wriggers 2006), depending on the reference system used (projection on one side or the other) or the place where it is computed (nodes, integration point). Our formulation considers that one of the two sides is given more importance (mortar/non mortar approach) (Belgacem et al. 1998). The gap function is defined at each integration point of the segments corresponding to this side with respect to the local system of coordinates (e_1, e_2, e_3).

Most of solids in contact are not able to overlap. Therefore contact between both sides generates normal stress (p_N). The ideal normal contact conditions may be summarised by the Hertz-Signorini-Moreau condition (Wriggers 2006)

$$g_N \geq 0, \qquad p_N \geq 0, \qquad g_N\, p_N = 0. \tag{1}$$

Its implementation requires a Lagrange multiplier formulation to ensure there is no interpenetration of two solids in contact. In this formulation we considered a penalty method to regularise this criterion (Charlier and Cescotto 1988). We assume that a slight interpenetration ($g_N < 0$) is possible such that

$$\dot{p}_N = -K_N\, \dot{g}_N, \tag{2}$$

Where K_N is a penalty coefficient that should be appropriately defined. It should be high enough to avoid spurious a too high non-physical interpenetration of the solids but sufficiently low to avoid bad conditioning of the stiffness matrix.

Shear stresses (τ_1, τ_2) result from relative displacement between both sides of the interface. Their evolution is similarly defined with respect to variations of the relative displacement variations $\dot{g}_{T1}$ and $\dot{g}_{T2}$ such that

$$\dot{\tau}_{1,2} = K_T\, \dot{g}_{T1,2}. \tag{3}$$

The penalty coefficient related to shear stress K_T may be related to physical properties of the interface. In addition, the maximum shear stress is defined according to an elastic perfectly plastic Coulomb criterion without dilatancy such that

$$f\left(p_N, \tau_1, \tau_2\right) \equiv \sqrt{\tau_1^2 + \tau_2^2} - \mu p_N = 0. \tag{4}$$

The fluid flow problem within the interface is summarised in Figure 2 which is a cross-section of a 3D problem. In this case the interface models a gap between the two porous media Ω_i (i = 1,2). This gap creates a new volume Ω^3 in which a fluid flow takes place. This also generates two boundary conditions Γ_q^i (i = 1, 2) between Ω^3 and the

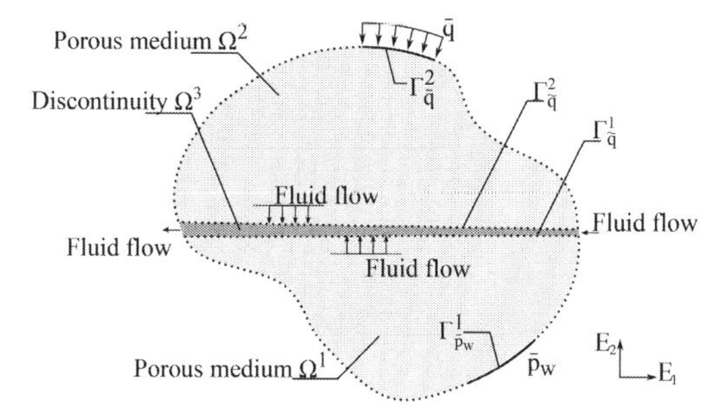

Figure 2. Definition of the flow problem corss-section of the 3D case in a plane perpendicular to the interface.

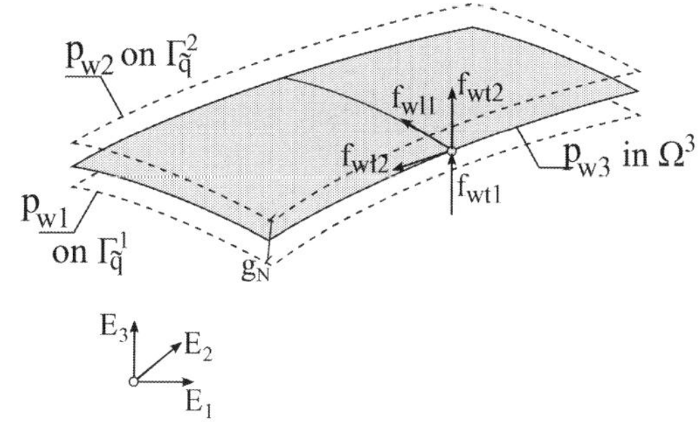

Figure 3. Definition of transversal and longitudinal fluid flows.

porous solid bodies Ω^i. Fluid exchange takes place between inside the interface and the porous media.

The problem may be idealised as depicted in Figure 3. The field of pore water pressure is already discretised on both sides of the interface $\Gamma^i_{\bar{q}}$. It is assumed in the following that the pore water pressure is homogeneous over the width of the gap, though it varies in both longitudinal directions. Therefore an additional field of pressure p_{w3} is discretised inside Ω^3. Additional nodes with a single degree of freedom are necessary to discretise this field.

At each integration point four fluxes must be computed as represented in Figure 3. Two of them are parallel to the interface (f_{wl1} and f_{wl2}) in both principal directions (e_2, e_3). They are assumed to obey the Darcy's law such that,

$$f_{wl(i-1)} = -\frac{k_l}{\mu_w}\left(\nabla_{e_i} p_{w3} + \rho_w g \nabla_{e_i} z\right)\rho_w \quad \text{for} \quad i = 2,3 \tag{5}$$

Where ∇_{e_i} is the gradient in the direction $\mathbf{e}_i$, μ_w is the dynamic viscosity of the fluid, g the acceleration of gravity, ρ_w is the density of the fluid and k_l is the permeability.

Two other fluxes are perpendicular to the interface (f_{wl1} and f_{wl2}). They represent the exchange between the the porous media and the interface. They depend on the drop of pressure between each side of the interface and inside it such that

$$f_{wt1} = \rho_w T_w \left(p_{w1} - p_{w3}\right), \tag{6}$$

$$f_{wt2} = \rho_w T_w \left(p_{w3} - p_{w2}\right), \tag{7}$$

Where p_w denotes a water pressure, T_w is a transversal conductivity and ρ_w is water density.

The formulation induces three hydro-mechanical couplings. The normal pressure acting on each side of the interface is decomposed into an effective normal pressure (and a pore water pressure, according to the Terzaghi's definition

$$p_N = p'_N + p_w. \tag{8}$$

The effective normal pressure is used in the mechanical part of the model (contact detection and Coulomb criterion).

The void created when two side saturated bodies lose contact introduces a storage term in the fluid mass balance equation. In this work the variation of the total mass of fluid M_f stored in Ω^3 comes mainly from the opening/closingof the gap if the fluid is assumed incompressible, namely

$$d\dot{M}_f = \rho_w \dot{g}_N \, d\varphi_{\bar{q}}, \tag{9}$$

Where $d\varphi_{\bar{q}}$ is an infinitesimal part of the interface area.

Finally longitudinal permeability used in the Darcy's law depends on the side of the gap g_N according to a cubic law such that

$$k_l = \begin{cases} \dfrac{(D_0)^2}{12} & \text{if} \quad g_N \leq 0 \\ \dfrac{(D_0 + g_N)^2}{12} & \text{otherwise.} \end{cases} \tag{10}$$

If contact exists ($g_N \leq 0$), the gap is negative and permeability should be equal equal to zero. However from a physical point of view, there could be a path for fluid flow through asperities of a rough surface (residual opening). Moreover from the numerical point of view, a null permeability may lead to a badly conditioned problem. Therefore a very low residual opening D_0 is added. Hence, the permeability is computed according to (Olsson & Barton 2001, Guiducci, Pellegrino, Radu, Collin, & Charlier 2002).

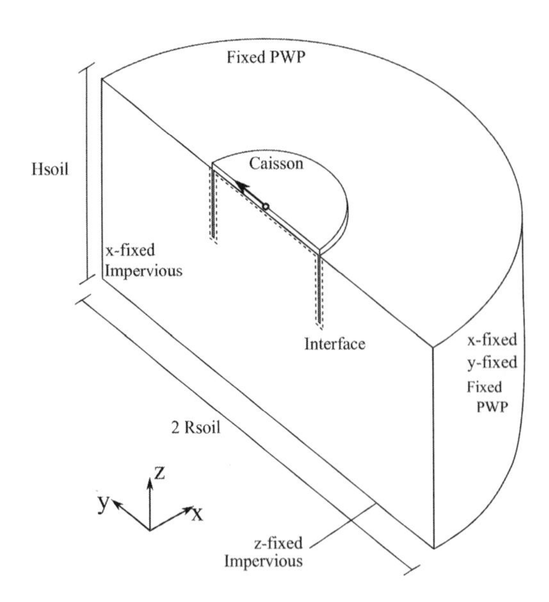

Figure 4. Global sketch of the problem geometry.

Table 1. Geometrical parameters: R_{int} inner radius, R_{out} outer radius, L length, t_{skirt} thickness of the skirt, t_{lid} thickness of the lid.

R_{int}	R_{out}	L	t_{skirt}	t_{lid}
3.8 m	3.9 m	4 m	0.1 m	0.4 m

2.2 *Numerical model*

A steel suction caisson is assumed embedded in an homogeneous sandy soil layer as shown in Figure 4. Interface elements are set up all along the boundary between the soil and the caisson. The radius of the soil layer is equal to 24 m and its depth to 12 m. The mesh comprises 16058 nodes and 14190 eight-node elements. The caisson whose parameters are provided in Table 1, is composed of a rigid lid at the top and a vertical skirt. It is 7.8 m diameter and its length is equal to 4 m. The lateral loading consists on an imposed lateral displacement at the top centre of the caisson. The installation is not modelled and the caisson is supposed wished in place. The initial stresses within the surrounding soil are defined using a K_0 coefficient equal to 1. A confinement corresponding to 1 m of soil is applied at the soil surface.

The mechanical behaviour of the caisson is assumed purely elastic. It is also deemed impervious. The behaviour of the soil is modelled through the elastoplastic internal friction model PLASOL, implemented in the finite element code LAGAMINE (Barnichon 1998). The law is based on friction angle hardening from the initial φ_i to the final φ_f friction angle, as a function of the deviatoric plastic deformation ϵ^p_{eq} according to

Table 2. Material parameters: E Young modulus, ν Poisson's ratio, K_N,K_T penalty coefficients, μ friction coefficient, T_w transversal conductivity.

Soil				
E [MPa]	ν[-]	φ_i[°]	φ_f[°]	$B\varphi$[-]
2E2	0.3	5	30	0.005
Caisson				
E [MPa]	ν[-]			
2E5	0.3			
Interface				
K_N/K_T [N/m^3]		μ [-]	T_w [m.Pa^{-1}.s^{-1}]	
1E10		0.50	1.E-8	

$$\phi = \phi_i + \left(\phi_f - \phi_i\right)\frac{\epsilon^p_{eq}}{B_\phi + \epsilon^p_{eq}}, \tag{11}$$

where $B_{<1>\varphi</1>}$ is a material parameter. The effective weight of the soil is equal to 10.56 kN/m^3 while its permeability is equal to 1.E-4 m/s. The soil caisson friction coefficient is equal to 0.5. All material parameters are defined in Table 2.

3 RESULTS

In this section we analyse the lateral loading of the caisson previously defined in drained and partially drained conditions. In the first case the pore water pressure within the soil are not allowed to vary. In the second case, pore water pressure are generated and dissipated during loading, leading to partially drained conditions (not totally drained and not totally undrained). The global load displacement relation is explained in the light of the interface behaviour.

3.1 *Drained simulation*

The soil in front of the caisson is progressively loaded due to the imposed lateral displacement. As a consequence it starts plastifying from the surface, where the strength is the lowest, due to the low confinement. Therefore the soil lateral movement islarger close to the surface, as reported in Figure 5. A wedge of highly loaded soil is formed in front of the caisson. Subsequently it tends to rotate, as described in Figure 6.

The lateral loading induces an increase in normal stress within the front interface. Therefore maximum shear stress available is increased and the soil around is pushed down due to the rotation movement. On the contrary at the back of the caisson, there is a discontinuity between vertical/lateral displacement of the soil and caisson. This is due to the creation

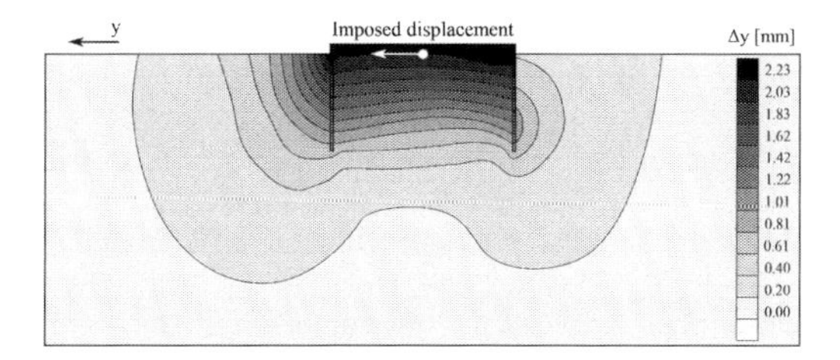

Figure 5. Cross section of the lateral displacement $\Delta y[mm]$ in the central vertical plane, drained simulation after a lateral imposed displacement of 2.5 mm.

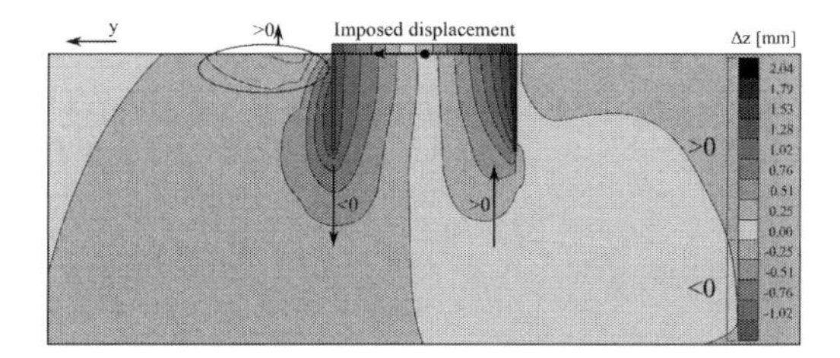

Figure 6. Cross section of the vertical displacement $\Delta z[mm]$ in the central vertical plane, drained simulation after a lateral imposed displacement of 2.5 mm.

of a gap between them. In this zone, no shear stress can be generated. This gap may be important if the caisson must face an uplift load since it creates a preferential path to drain fluid flow from inside it.

Inside the caisson, the soil seems plugged and accompanies the caisson movement. However the slight discontinuity of vertical displacement between soil and caisson at the top may indicate a very small gap. In conclusion the caisson movement may be summarised as the combination of three rigid movements: 1) a lateral displacement, 2) a slight vertical movement and 3) a rotation with respect to the top centre.

Cross-sections of the normal effective and shear stress along the interface are useful to understand how the caisson behaves. They are represented radially in a 2D plot as sketched in Figure 7. Results are normalised and depicted with respect to the initial position of the caisson. If we consider each radial direction, we can define

- an application point: the intersection between radial direction and trace of the caisson;
- a magnitude: distance between the application point and the trace of results;
- a sense: positive (outside the trace of the caisson), negative (inside).

Cross-sections of the normal effective stress at four different depths is provided in Figure 8 after 2.5 mm of imposed lateral displacement. This figure confirms the creation of a gap which is larger is size at the top than at the bottom of the caisson.

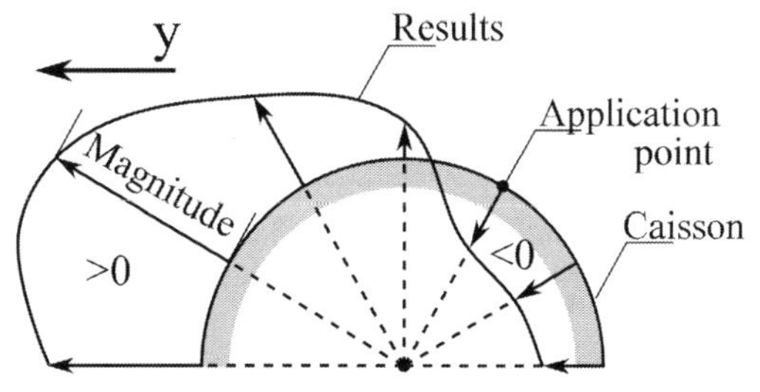

Figure 7. Normalised radial distribution of results within interface.

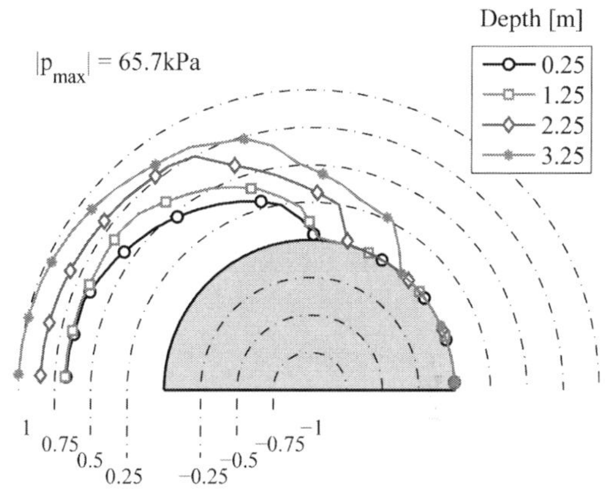

Figure 8. Distribution of normal effective in horizontal cross-sections at different depths, drained simulation.

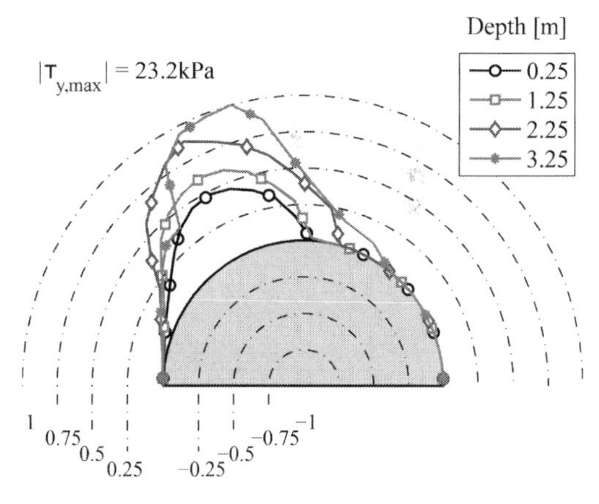

Figure 9. Distribution of shear stress projected in the y direction in horizontal cross-sections at different depths, drained simulation.

It also exhibits that the main reaction component to the loading is obviously the increase in normal load. This load increases the initial stress within the soil by approximatively 50%.

Cross-section of the projection of the shear stress in the y direction are provided in Figure 9.

This underlines that friction plays a non-negligible role in the resistance to lateral loading. This value at the front is equal to zero since the tangent to the caisson is perpendicular to the loading direction. It increases progressively with the direction of the radial direction. It is obviously null at the back of the caisson since there is a gap.

The total horizontal (ΔF_{tot}) load versus lateral displacement (Δ_y) is represented in Figure 10. It presents a classical shape with a clear change of slope after around 2 mm of imposed displacement (Achmus and Thieken 2010, Bienen et al. 2012). Before the slope breakage, initial stresses at the back of the caisson are released and stresses in front increase. This is similar to an elastic unloading. After the slope breakage, a gap is formed and there is no more unloading. The global tangent stiffness corresponds only to the loading of the soil in front of the caisson.

Rotation of the caisson generates a relative vertical displacement between the caisson and the soil, giving birth to non-zero vertical shear stresses inside and outside the caisson. The integration of the stresses over the surface is summarised into ΔF_{in} and ΔF_{out} variables, as reported in Figure 10 and sketched in Figure 11.

The shear components have different magnitudes and signs. Direction of shear forces is directed upwards at the front of the caisson and downwards behind it due to the rotation. However the magnitude of shear stress is different in absolute value between inside and outside the caisson, due to the gap opening (outside at the back). Maximum friction is reached in both cases since a plateau is attained.

In addition, shear is mobilised at the base of the caisson. It may lead to a global failure if the shear capacity is reached. However it is not in this case since the caisson has a large diameter. Finally the pore water pressure variation is null since drained conditions are considered.

3.2 *Partially drained simulation*

During the partially drained simulation the rate of imposed lateral displacement is 0.05 mm/s. The total horizontal load is compared with the drained result in Figure 12. During the first part of the loading, before the slope breakage, both responses are almost identical. However they diverge afterwards due to the late mobilisation of a pore water pressure effect. Subsequently the capacity of the caisson is increased if larger displacement is allowed.

Indeed suction effects within interface oppose to gap creation. This could be observed in Figure 13. Negative variations of pore water pressure are

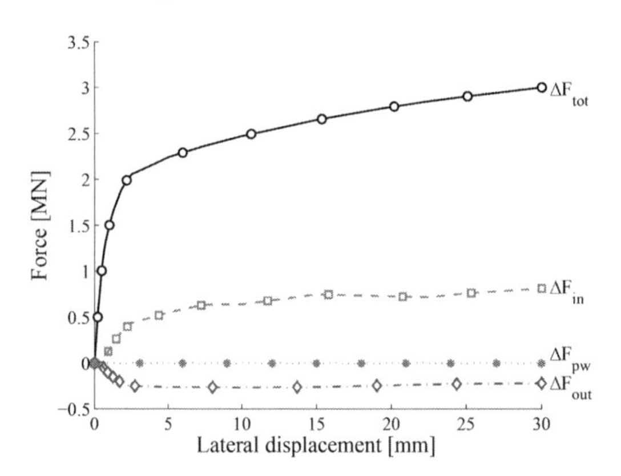

Figure 10. Variation of total horizontal load ΔF_{tot}, PWP under the lid ΔF_{pw} and shear forces $\Delta F_{in}/\Delta F_{out}$, drained simulation.

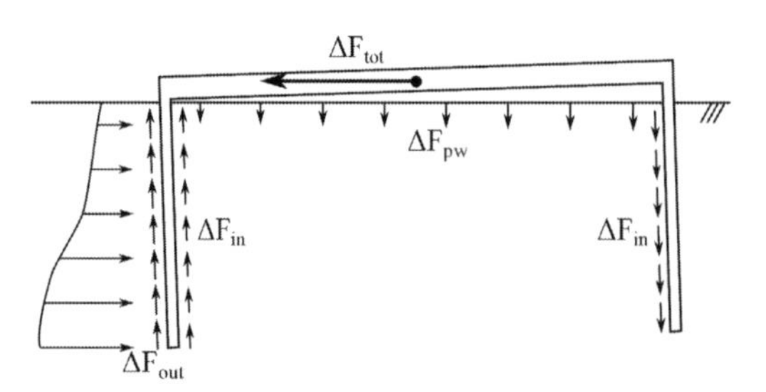

Figure 11. Definition of different components.

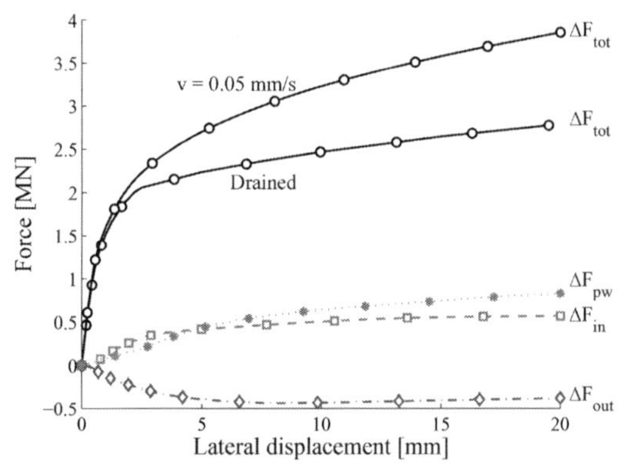

Figure 12. Variation of total horizontal load ΔF_{tot}, PWP under the lid ΔF_{pw} and shear forces $\Delta F_{in}/\Delta F_{out}$, partially drained simulation.

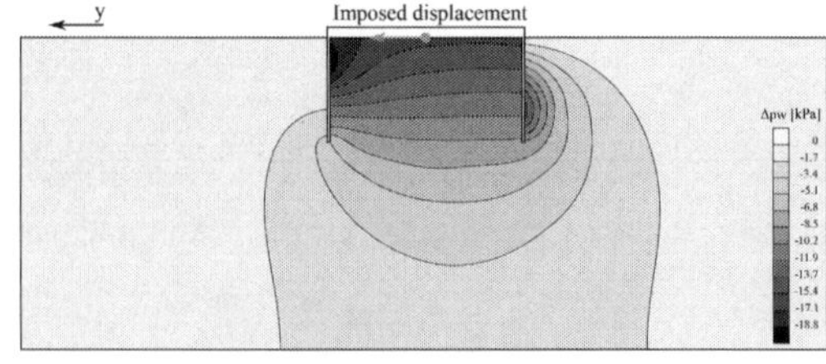

Figure 13. Cross section of pore water pressure variations Δp_w [kPa] in the central plane, drained simulation after 20mm, partially drained simulation.

generated inside at the front close to the top due to the complex movement. It is also negative right under the lid, restraining the rotation movement. The integration of pore water pressure variations under the lid are gathered into ΔF_{pw}, presented in Figure 12. Finally negative variations are even surprisingly noted at the back outside of the caisson, close to the bottom. They also oppose gap creation, despite these underpressures are more easily dissipated. However this effect may suddenly disappear as the gap opening increases vertical permeability within the interface. Therefore if the gap opens up to the surface where water pressure is fixed, the negative overpressure is expected to suddenly disappear.

The distribution of normal effective stress p'_N in horizontal cross-sections is depicted in Figure 14. The partially drained conditions do not increase significantly the maximum normal effective stress in front of the caisson. However as previouly mentioned, it reduces the gap opening at the back (zones where $p'_N = 0$). This is evident by comparison with Figure 8.

The distribution of pore water pressure outside of the caisson follows an opposite pattern. It is almost null at the front where there is no gap and pore water pressures generated by soil volumetric reduction are fast dissipated. However it is negative at the back where gap tends to open and creates this suction effect.

The drained and partially drained p-y curves are represented in Figure 33 for comparison. Each p-y curve includes different components acting on each side of the caisson: normal effective pressure, horizontal shear stress and pore water pressure. The curves close to the surface (depth = 0.25 m) have more or less an identical shape. All others partially drained curves exhibit an increase in force for an identical displacement. The distribution of effective stress around the shaft is not really modified as depicted in Figure 34a. Only the gap is less open since the normal pressure is not equal to zero behind the caisson (depth = 3.25 m). Therefore it could be reasonably stated that the difference between p-y curves is due to the pore pressure distribution.

P-y curves are often defined for piles (Reese and Van Impe 2010). They represent lateral load p mobilised within the soil per meter depth, opposing a given displacement y. Similarly the normal and tangential stresses as well as pore water pressure can be integrated along the boundaries of the caisson. However the modelled caisson is far from a pile: it has a length to diameter ratio of 0.5. Firstly the caisson should be modelled like a rigid body rather than a bending pile. Secondly, the p-y curves result from the difference between inside and outside reactions from the soil onto the caisson, in the y direction. The y displacement is the displacement of the mean fibre of the caisson, at each depth.

The force versus displacement is provided at four depths in Figure 16 in drained and partially drained conditions. This figure confirms two previous observations: 1) load mobilised increases with depth, 2) the larger displacement at the top of the caisson indicates a rotation. The curve shapes are also similar to the global load displacement relation. Indeed the slope breakage corresponds to the gap opening. It appears for a larger displacement in drained conditions as depth increases. However transition becomes smoother as partially drained conditions are considered. The effect of partial drainage is to shift the p-y curves up, mobilising a

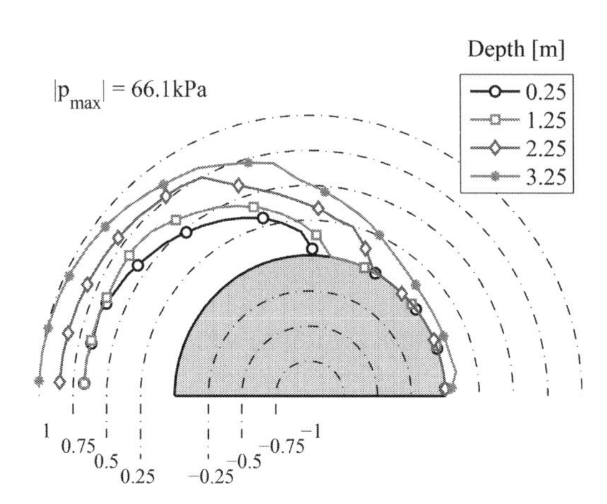

Figure 14. Distribution of normal effective in horizontal cross-sections at different depths, drained simulation after a lateral imposed displacement of 2.5 mm.

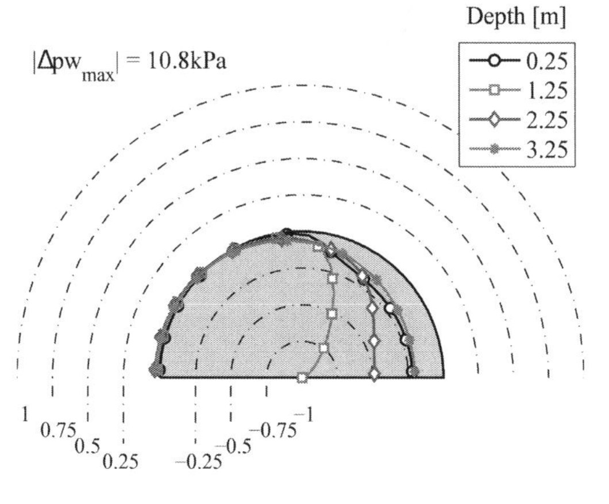

Figure 15. Distribution of pore water pressure out side of the caisson in horizontal cross-sections at different depths, drained simulation after 2.5 mm.

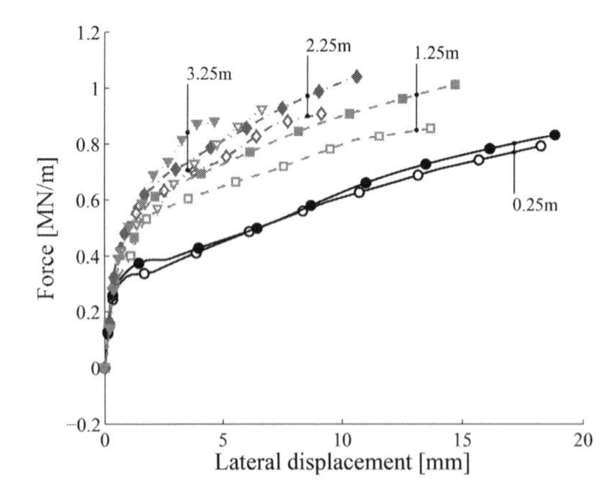

Figure 16. Comparison of p-y curves at different depths in drained (open markers) and partially drained (filled markers) conditions.

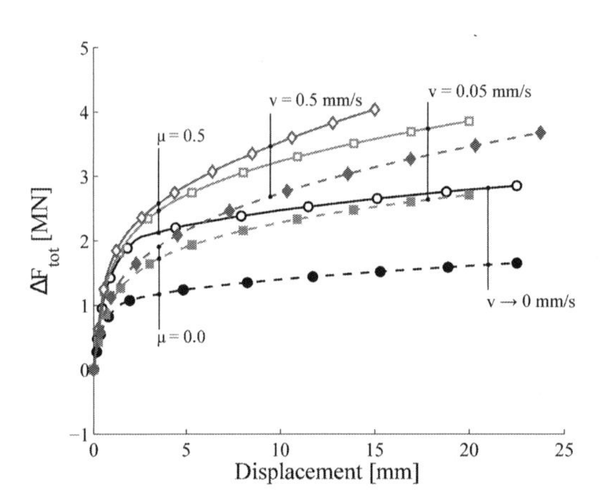

Figure 17. Variation of total horizontal load ΔF_{tot} for different horizontal loading rate (v) and friction coefficients (μ).

high load for a given displacement. This is mainly due to the pore water pressure effect and does not appear close to the surface where these pore pressure are fast dissipated.

The influence of the loading rate and friction coefficient and depicted in Figure 17. The partially drained effect seems to decrease at higher loading rate. Indeed increasing the loading rate from 0.05 mm/s to 0.5 mm/s only slightly raises the total load mobilised.

It could be assumed that friction plays a minor role in the strength development towards lateral loading, contrary to normal reaction of the soil. However results prove that friction is crucial. In the drained case, The mobilised load drops to half of the value if there is no shearing allowed to develop along the shaft. This effect is still important in partially drained conditions. This drop is probably due to the lose of horizontal shearing along the shaft. In addition the rotation movement is not balanced by friction any more, resulting in a different load distribution.

4 CONCLUSIONS

Throughout this work, we have proven that interface behaviour is crucial in the modelling of lateral loading of suction caisson foundation. Opening of gap and shear mobilisation both strongly influence the strength, stiffness and shape of the load-displacement curve. The partially drained behaviour, induced by generation and dissipation of negative pore water pressures is shown to increase the total horizontal resistance for a given displacement. It also opposes to gap opening. This is particularly important since vertical gap may strongly affect the vertical behaviour by introducing preferential path to dissipate inner negative fluid pressures as reported in (Cerfontaine et al. 2015, Cerfontaine et al. 2015).

For all of these reasons the development of hydro-mechanically coupled finite elements of interface is necessary. They must take into account gap opening, shear mobilisation, longitudinal and transversal fluid flows. In this work the zero-thickness approach is coupled with a three-node formulation to discretise the field of pressure inside the interface. Hydro-mechanical couplings arise from the definition of an effective mechanical contact pressure, the dependence of the longitudinal permeability on the gap opening and the definition of a storage term in the mass balance equation as the gap opens.

P-y curves were also defined similarly to piles. They relate the mobilisation of shear along the shaft and normal pressure to the lateral displacement of the caisson. They are obtained from variables available in the interface elements. Such p-y curves could be used to model the caisson as a rigid body connected to non-linear soil springs.

ACKNOWLEDGEMENT

This work was partly supported by a FRIA FRS-FNRS scholarship. The first author has received funding from the European Unions Horizon 2020 research and innovation programme under the Marie Sklodowska-Curie grant agreement No 753156.

REFERENCES

Achmus, M., C.T. Akdag, & K. Thieken (2013). Load-bearing behavior of suction bucket foundations in sand. *Applied Ocean Research 43*, 157–165.

Achmus, M. & K. Thieken (2010). On the behavior of piles in non-cohesive soil under combined horizontal and vertical loading. *Acta Geotechnica 5*(3), 199–210.
Barnichon, J. (1998). *Finite element modelling in structural and petroleum geology*. Ph. D. thesis, University of Liege, Belgium.
Belgacem, F., P. Hild, & P. Laborde (1998). The mortar finite element method for contact problems. *Mathematical and Computer Modelling 28*(4–8), 263–271.
Bienen, B., J. Dührkop, J. Grabe, M. Randolph, & D. White (2012). Response of Piles with Wings to Monotonic and Cyclic Lateral Loading in Sand. *Journal of Geotechnical and Geoenvironmental Engineering 138*(3), 364–375.
Cerfontaine, B., F. Collin, & R. Charlier (2015). Numerical modelling of transient cyclic vertical loading of suction caissons in sand. *Géotechnique 65*(12).
Cerfontaine, B., A. Dieudonné, J. Radu, F. Collin, & R. Charlier (2015). 3D zero-thickness coupled interface finite element: Formulation and application. *Computers and Geotechnics 69*, 124–140.
Charlier, R. & S. Cescotto (1988). Modélisation du phénomène de contact unilatéral avec frottement dans un contexte de grandes déformations. *Journal de Mécanique Théorique et Appliqueé 7*(Suppl. 1).
Goodman, R., R. Taylor, & T. Brekke (1968). A model for the mechanics of jointed rock. *Journal of the Soil Mechanics and Foundations Division 94*.
Guiducci, C., A. Pellegrino, J.-P. Radu, F. Collin, & R. Charlier (2002). Numerical modeling of hydromechanical fracture behavior. In *Numerical models in Geomechanics*, pp. 293–299.
Houlsby, G. & B. Byrne (2005). Design procedures for installation of suction caissons in sand. In *Proceedings of ICE, Geotechnical Engineering*, Number July, pp. 135–144.
Houlsby, G., R. Kelly, J. Huxtable, & B. Byrne (2006). Field trials of suction caissons in sand for offshore wind turbine foundations. *Géotechnique 56*(1), 3–10.
Jha, B. & R. Juanes (2014). Coupledmultiphase flow and poromechanics: A computationalmodel of pore pressure effects on fault slip and earthquake triggering. *Water Resources Research 50*, 3776–3808.
Kourkoulis, R., P. Lekkakus, F. Gelagoti, & A. Kaynia (2014). Suction caisson foundations for offshore wind turbines subjected to wave and earthquake loading: effect of soil? foundation interface. *Géotechnique 64*(3), 171–185.
Olsson, R. & N. Barton (2001). An improved model for hydromechanical coupling during shearing of rock joints. *International Journal of Rock Mechanics and Mining Sciences 38*(3), 317–329.
Reese, L. & W. Van Impe (2010). *Single piles and pile groups under lateral loading* (Second ed.). CRC Press.
Wriggers, P. (2006). *Computational contact mechanics* (Second ed.). Wiley: Chichester.

Numerical Methods in Geotechnical Engineering IX – Cardoso et al. (Eds)
© 2018 Taylor & Francis Group, London, ISBN 978-1-138-33203-4

Effect of scour on the behavior of a combined loaded monopile in sand

Q. Li, L.J. Prendergast, A. Askarinejad & K. Gavin
Faculty of Civil Engineering and Geosciences, Delft University of Technology, Delft, The Netherlands

ABSTRACT: Pile foundations used for offshore wind structures are subjected to large lateral loading from wind and waves while in service as well as significant vertical loading from the top structure. Erosion of soil from around these structures, termed scour, poses a significant problem for the structural stability. In order to better understand the performance of piles facing scour problems, the effect of local scour on the behavior of monopiles installed in sand under combined lateral and vertical loading has been investigated using the Finite Element Method (FEM) using PLAXIS in this paper. The simulation results showed that vertical loading can decrease pile lateral displacement and improve the lateral capacity of piles in the absence of scour and under scour. The increase of scour depth will largely reduce lateral capacity of piles.

1 INTRODUCTION

Offshore wind energy is being generated at a tremendous pace, with the EU capacity predicted to be 150 GW by 2030. In addition to reducing carbon dioxide emissions by 315 million tons, this growth would satisfy 14% of the EU electricity demand (Zervos & Kjaer, 2006). Together with the EU, Governments of other countries with major economies such as the US, South Korea and China have earmarked significant investment, to the tune of $38 billion USD, for offshore wind and other renewable sources of energy (Green & Vasilakos, 2011). Owing to its economy, simple manufacture and installation procedures, monopiles account for approximately 75% of offshore wind foundations (Gavin et al., 2011). Vertical loading transferred from the self-weight of the structure and lateral loading due to wind and wave actions are imposed on piles. However, in view of the complexity involved in analyzing the piles under combined loading, the current practice tends to ignore the interaction effects in combined loaded piles. Instead, these are broadly analyzed independently, i.e. for vertical loading to determine their bearing capacity and settlement and for the lateral loading to determine their flexural behavior (Karthigeyan et al. 2006 and Anagnostopoulos & Georgiadis 1993).

Cylindrical structures such as monopiles are prone to scour, which induces loss of soil support around the piles, reducing the lateral capacity and changing the structural stiffness and natural frequency (Prendergast et al. 2018, Prendergast et al. 2015 and Sørensen & Ibsen 2013). This can pose problems for the superstructure through the generation of excessive fatigue stress as well as operational issues with the turbine. Therefore, effects of scour must be considered during the analysis and design of combined loaded pile foundations unprotected against scour.

The lateral resistance of piles under combined loading has been numerically studied by a number of researchers. With an increase in the vertical load, Karthigeyan et al. (2006), Achmus & Thieken (2010b) and Taheri et al. (2015) observed an increase in the lateral capacity while Madhav & Sarma (1982) and Meera et al. (2007) observed a decrease. Moreover, Klein & Karavaev (1979) and Karthigeyan et al. (2007) obtained a result of both increase and decrease depending on the pile and soil properties. Furthermore, Trochanis et al. (1991) and Abdel-Rahman & Achmus (2006) found the effect of vertical load on the lateral capacity of piles to be negligible, which implies that the combined action can be ignored in design. The findings of previous studies on the influence of vertical load on the lateral capacity of a pile subjected to combined loading is summarized in Table 1. Considering the contradictory conclusions discussed above, there is recognized ambiguity in the results of previous works describing whether vertical load combined with lateral load increases or decreases the lateral resistance properties of piles.

Offshore monopiles can fail due to severe scour caused by currents and waves. Because of the formation of the scour hole around the pile, the depth of embedment of the pile reduces, and consequently there is a reduction in load carrying capacity of piles (Kishore et al., 2009). Experiments have determined that local scour depth (d_s) in sandy soils equates to 1.3 times pile diameter (D)

Table 1. Summary of the effect of vertical load on the lateral response of piles using FEM.

Literature	Effect of vertical load on the lateral response of pile	Soil type
Klein & Karavaev (1979)	Increased capacity	Dense soil
	Decreased capacity	Weak soil
Madhav & Sarma (1982)	Decreased capacity and bending moment	Clay
Trochanis et al. (1991)	Unaffected	Multilayer soil
Abdel-Rahman & Achmus (2006)	Unaffected	Sand
Karthigeyan et al. (2006, 2007)	Increased capacity	Sand
Meera et al. (2007)	Decreased capacity and bending moment	Loose sand
Achmus & Thieken (2010a, b)	Both increased and decreased capacity	Sandy soil
Taheri et al. (2015)	Increased capacity	Silty sand

with a mean of 0.7 (Sumer et al., 1992). In other words, the maximum scour depth is about 2 times the pile diameter (*i.e.*, $ds/D = 2$). Several investigators numerically studied the scour depth and scour pattern around piles in cohesive and cohesionless soils. Lin et al. (2010) modified lateral load–displacement (p–y) curve for a pile in sand and input into the computer software, LPILE Plus V 5.0. It is indicated that by considering the stress history effect on the behavior of piles in sand under scour, ignoring the stress history could result in a conservative estimate. Achmus & Thieken (2010a) developed a 3D FE model using the finite element program ABAQUS to study the lateral deformation response of monopile foundations with scour under monotonic and one-way cyclic loading. With this model, a case study on a planned wind turbine in Taiwan Strait was analyzed and the economic considerations of different design options were discussed. Mostafa (2012) investigated the effect of local and global scour on the behavior of laterally loaded piles installed in different soil conditions using the software program PLAXIS. Various parameters were analyzed such as soil type, scour depth, scour hole dimension, pile material, magnitude of lateral load and load eccentricity. The results showed that scour has a significant impact on piles installed in sand and a less significant impact on piles installed in clay, and global scour has a significant impact on pile lateral displacement and bending stresses. The effect of scour is more significant if piles are subjected to large lateral loads due to the nonlinear response of the pile-soil system. Effect of scour of stiff clayey soils on piles is more pronounced than that of soft clayey soils. Based on the FLAC 3D, Li et al. (2013) calibrated the numerical model of a single pile in soft marine clay against field test data without scour and analyzed several key factors of scour, such as the depth, width and slope of the scour hole and the diameter and head fixity of the pile. The relationships of the ultimate lateral capacity of a single pile with depth, width and slope angle of the scour hole were obtained. The numerical results show that the scour depth had more significant influence on the pile lateral capacity than the scour width. In addition, the pile with a free head was more sensitive to scour than the pile with a fixed head condition.

From the literature, it is evident that not much work has been carried out on the combination of scour with combined loading on piles. Hence, in this investigation, numerical studies were carried out to explore the effect of scour on the lateral capacity of a monopile under combined loading conditions. Different parameters were considered such as vertical load magnitude and scour depth. In all cases considered, the pile was embedded in homogeneous sand with a unit weight of $\gamma = 20$ kN/m^3.

2 MODEL FEATURES

The finite element method is used to simulate the behavior of a vertical pile under combined vertical and lateral loading under the effect of scour. The computations were carried out using the finite element program system PLAXIS (Brinkgreve et al., 2015). A three dimensional model of the pile-soil system was generated in Figure 1.

Figure 2 shows schematic of local scour and the simplification in the modelling. *D* denotes the pile diameter, W_t denotes the top width of scour hole,

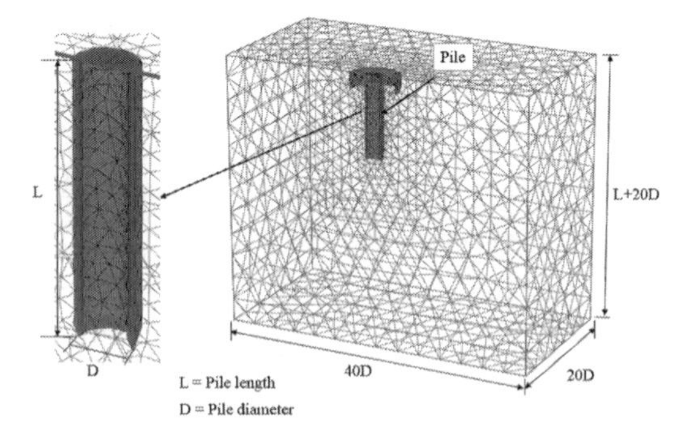

Figure 1. Typical mesh for three-dimensional finite element analyses.

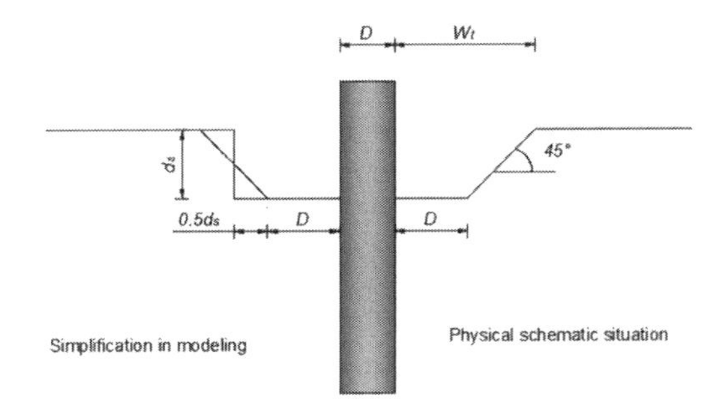

Figure 2. Local scour hole and simplification in the modelling.

Table 2. Sand characteristics used in the analysis.

Material	Sand
Initial Elastic Modules/E [MPa]	20
γ_{sat} [kN/m^3]	20
Poisson Ratio/v	0.25
Friction angle/φ'	37.5°
Dilation Angle/ψ	7.5°
E_{inc} [MPa/m]	0.5
Cohesion/c [kPa]	0
Soil Model	Mohr-Coulomb (Drained)

ds denotes scour depth. Scour hole depths equivalent to 0.5 and 1.5 times the pile diameter were considered (i.e., $ds/D = 0.5, 1.5$).

Local scour represents the case of scour hole occurring in the direct vicinity of a pile which results in a localised reduction in effective stress. Normally the local scour hole is conical in shape with a trapezoidal cross section. For simplify, the scour hole was modelled as a foundation pit with circular shape cross section, assuming the scour hole base extends around the pile at a distance of *D*, see the schematic in Figure 2. This was necessary as the implementation of a conical shaped hole to represent the scour hole resulted in numerical instabilities related to the slope. According to Askarinejad et al. (2017) the effective zone of influence around a lateral loaded pile is within one pile diameter, thus it was considered that outside the 1*D* region from the pile outer surface the small change of the scour shape will have negligible influence on the pile load behaviour. Li et al. (2013) indicated that when the scour hole bottom width is larger than 1*D*, the influence of the slope angle on the pile lateral displacement is negligible. The scour hole side slope angle was assumed to be 45° in the physical situation, therefore the modelling of the scour slope was simplified by adding an additional length of scour hole base of 0.5*ds* in the analysis and maintaining the side slope angle at 90°. At a scour depth of more than 1*D*, the impact of scour hole shape on the effective soil pressure diminishes (Zaaijer and Van der Tempel, 2004). To account for scour around the pile, before the application of a horizontal load, soil elements located inside the scour hole to be modelled were removed.

The sand properties considered in these analyses are reported in Table 2. For the steel pipe pile, the diameter is 1.8 m, the length is 9 m, the wall thickness is 25 mm, the elastic modulus (*E*) is 210 GPa and the unit weight (.) is 78.5 kN/m^3. The pile was modelled as a non-yielding elastic continuum medium and the soil was modelled as a linearly elastic-perfectly plastic material with the Mohr-Coulomb failure criteria.

The response of the piles under pure lateral load was analysed in the first instance. With regard to studying the response of piles under combined loads, the influence of vertical loads of $0.4V_{ult}$ and $0.8V_{ult}$ were considered. In this context. V_{ult} is determined as the vertical load which causes a vertical displacement equating to $0.1D$, obtained through analysis of a single pile subjected to a pure vertical load. The analysis in the lateral direction was performed using load control and the lateral displacement developed at various lateral load magnitudes could be evaluated.

The FE calculations were executed in several phases. Firstly, the initial stress state in the system due to the self-weight of the soil was generated using soil elements only. Subsequently, the pile was generated and 'wished in place', i.e. the installation of the pile was not modelled. The soil elements in the scour hole were then removed. The various load stages were specified in the model, more details are provided in the following section.

3 ANALYSIS AND RESULTS

In the simulation of influence of combined vertical and lateral load, the vertical load is applied prior to lateral load.

3.1 *Pile lateral load-displacement curve*

The ultimate vertical load (V_{ult}) capacity of a single pile was evaluated a priori in a separate numerical analysis, whereby loading was incrementally applied to the pile until a vertical displacement of $0.1D$ = 180 mm was mobilized. V_{ult} was found to be 4.4 MN. Once a datum ultimate capacity was obtained, the response of piles under combined loading were subsequently analysed separately with the vertical load applied in separate cases as follows: $v = 0$, $v = 0.4V_{ult}$ and $v = 0.8V_{ult}$.

Figure 3 shows the pile lateral load-displacement curve under no scour (presented in dotted line), local scour of 0.5D (presented in full solid line) and local scour of 1.5D (presented in dashed line). The pile under a vertical load of $0.4V_{ult}$ is marked with squares and the pile under a vertical load of $0.8V_{ult}$ is marked with circles. From Figure 3 it can be seen that under the same lateral load pile under higher vertical load has less lateral displacement, for both scour and no scour conditions. For the pile under same vertical load and lateral load, the pile with larger scour depth will have larger lateral displacement. With the increase of lateral load, the lateral displacement of pile under $ds = 0$, $v = 0$ and $ds = 0.5D$, $v = 0.8V_{ult}$ are really close to each other, which means, in some case the deleterious effect of scour can be compensated by the reinforcing effect of vertical load.

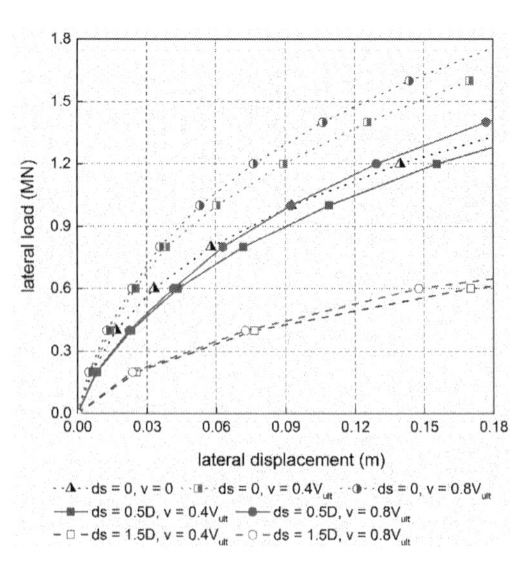

Figure 3. Pile lateral load-displacement curve.

3.2 *Influence of vertical load on pile lateral capacity*

Figure 4 presents how the ultimate lateral load capacity varies with vertical loading under no scour and local scour. The ultimate lateral load capacity (L_{ult}) is defined as the load corresponding to a lateral displacement equating to 10% of the pile diameter ($0.1D = 180$ mm). The results indicate that the larger the vertical load is, the higher the ultimate lateral load capacity of the pile. For example, under no scour condition, with vertical load equating to 40% and 80% of the ultimate vertical load (V_{ult}), the pile lateral ultimate capacity (L_{ult}) increased 23% and 33% respectively as compared to the case with no vertical load. A similar trend of vertical load increasing the pile lateral capacity under different scour depth conditions of local scour can also be observed from Figure 4 and the details of improvement is presented in Table 3.

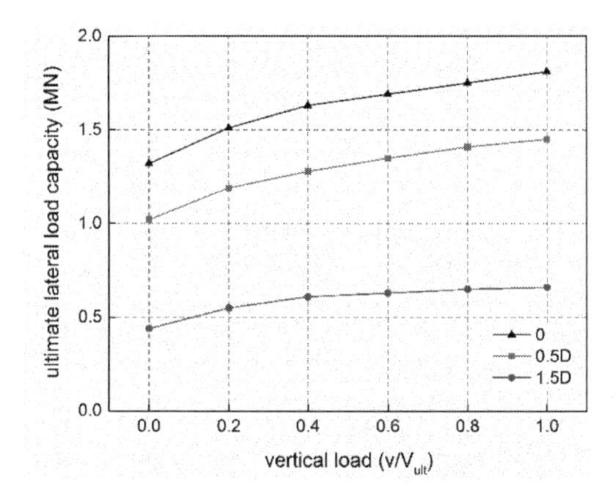

Figure 4. Relationship of pile ultimate lateral load capacity with vertical load level.

Table 3 shows the improvement in pile ultimate lateral load capacity under increasing vertical load, considering no scour and local scour at various scour depths. From the data presented, it is noteworthy that the lateral load capacity increases considerably with increasing vertical load. The vertical load has a higher influence on pile lateral ultimate capacity for deeper scour compared with shallow scour depths.

Table 3. Improvement in pile ultimate lateral load capacity (L_{ult}) with increasing applied vertical load.

	Scour type and depth		
		Local scour	
Vertical load	No scour	0.5D	1.5D
$0.4 V_{ult}$	0.23	0.25	0.39
$0.8 V_{ult}$	0.33	0.38	0.48

3.3 *Influence of scour depth on pile lateral capacity*

Figure 5 shows the ultimate lateral load capacity of a single pile under local scour varying with the normalized scour depth. Ultimate lateral load capacity was found to be 1.76 MN when scour was neglected. The ultimate lateral pile capacity under local scour depth equating to D and $2D$ were found to be 0.95 MN and 0.42 MN respectively, which are approximately

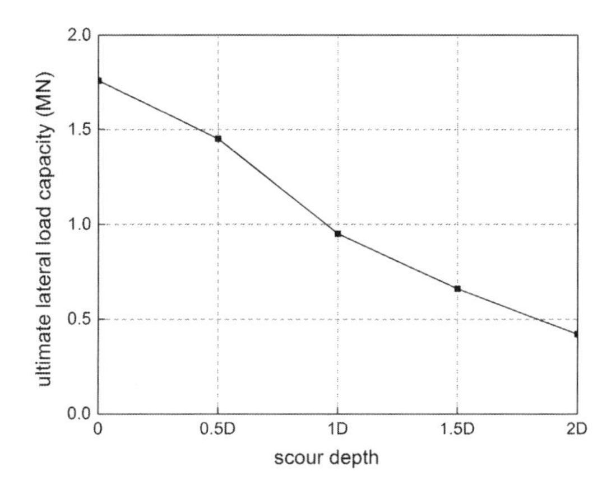

Figure 5. Pile ultimate lateral load capacity with normalized scour depth. (Vertical load = $0.8V_{ult}$).

54% and 24% of the ultimate capacity when there is no scour. This indicates that, when the scour depth reaches *2D*, the pile ultimate lateral load capacity is reduced by approximately 75%. It is also shown that the ultimate lateral load capacity of the single pile decrease significantly and almost linearly with the scour depth in current research.

4 CONCLUSIONS

The effect of scour on the response of combined loaded monopiles in marine sand is an important subject for the safety of offshore structures. A three dimensional numerical analysis was conducted to investigate this effect. The following conclusions can be drawn from the numerical results:

If the vertical load is applied prior to lateral load, the presence of vertical load will decrease pile lateral displacement and increase the pile lateral capacity. For example, under no scour condition, under vertical load equating to $0.4V_{ult}$ and $0.8V_{ult}$ respectively, the pile lateral ultimate capacity increases by 23% and 33% compared with no vertical load when there is zero scour. At the scour depth of 1.5D, the pile lateral ultimate capacity increases 39% and 48% corresponding to vertical load of $0.4V_{ult}$ and $0.8V_{ult}$ respectively. The vertical load shows a higher effectiveness in improving the pile lateral ultimate capacity under deep scour depth compared with shallow scour depth.

The pile ultimate lateral load capacity decrease almost linearly with the increase of the scour depth. The percentage reduction in the pile lateral load capacity is approximately 50% when the scour depth reaches 1D and 76% when the scour depth reaches 2D.

REFERENCES

Abdel-Rahman, K. & Achmus, M. 2006. Numerical modeling of the combined axial and lateral loading of vertical piles. 6th European Conference on Numerical Methods in Geotechnical Engineering, Graz, Austria. 575–581.

Achmus, M. & Thieken, K. 2010a. Behavior of piles under combined lateral and axial loading. Second International Symposium on Frontiers in Offshore Geotechnics (ISFOG) II, Perth, Austrilia. 465–470.

Achmus, M. & Thieken, K. 2010b. On the behavior of piles in non-cohesive soil under combined horizontal and vertical loading. *Acta Geotechnica,* 5(3), 199–210.

Anagnostopoulos, C. & Georgiadis, M. 1993. Interaction of axial and lateral pile responses. *Journal of Geotechnical Engineering,* 119, 793–798.

Askarinejad, A., Sitanggang, A.P.B. & Schenkeveld, F. 2017. Effect of pore fluid on the behavior of laterally loaded offshore piles modelled in centrifuge. 19th International Conference on Soil Mechanics and Geotechnical Engineering (ICSMGE 2017), Seoul, Korea. 897–900.

Brinkgreve, R., Kumarswamy, S. & Swolfs, W. 2015. Plaxis 3D Anniversary Edition Manual. *Plaxis bv. The Netherlands. Delft.*

Gavin, K., Igoe, D. & Doherty, P. 2011. Piles for offshore wind turbines: a state of the art review. *Proceedings of the ICE - Geotechnical Engineering,* 164(4), 245–256.

Green, R. & Vasilakos, N. 2011. The economics of offshore wind. *Energy Policy,* 39, 496–502.

Karthigeyan, S., Ramakrishna, V. & Rajagopal, K. 2006. Influence of vertical load on the lateral response of piles in sand. *Computers and Geotechnics,* 33(2), 121–131.

Karthigeyan, S., Ramakrishna, V. & Rajagopal, K. 2007. Numerical investigation of the effect of vertical load on the lateral response of piles. *Journal of Geotechnical and Geoenvironmental Engineering,* 133(5), 512–521.

Kishore, Y.N., Rao, S.N. & Mani, J. 2009. The behavior of laterally loaded piles subjected to scour in marine environment. *KSCE Journal of Civil Engineering,* 13(6), 403–408.

Klein, G. & Karavaev, V. 1979. Design of reinforced-concrete piles for vertical and horizontal loading. *Soil Mechanics and Foundation Engineering,* 16(6), 321–324.

Li, F., Han, J. & Lin, C. 2013. Effect of scour on the behavior of laterally loaded single piles in marine clay. *Marine Georesources & Geotechnology,* 31(3), 271–289.

Lin, C., Bennett, C., Han, J. & Parsons, R.L. 2010. Scour effects on the response of laterally loaded piles considering stress history of sand. *Computers and Geotechnics,* 37(7), 1008–1014.

Madhav, M. & Sarma, C. 1982. Analysis of Axially and Laterally Loaded Long Piles, Proceedings of 2nd International Conference on offshore Pilings, Austin, Texas. 577–596.

Meera, R., Shanker, K. & Basudhar, P. 2007. Flexural response of piles under liquefied soil conditions. *Geotechnical and Geological Engineering,* 25(4), 409–422.

Mostafa, Y.E. 2012. Effect of local and global scour on lateral response of single piles in different soil conditions. 4(6), 297–306.

Prendergast, L.J., Reale, C. & Gavin, K. 2018. Probabilistic examination of the change in eigenfrequencies of an offshore wind turbine under progressive scour incorporating soil spatial variability. *Marine Structures,* 57, 87–104.

Prendergast, L.J., Gavin, K. & Doherty, P. 2015. An investigation into the effect of scour on the natural frequency of an offshore wind turbine. *Ocean Engineering,* 101, 1–11.

Sørensen, S.P.H. & Ibsen, L.B. 2013. Assessment of foundation design for offshore monopiles unprotected against scour. *Ocean Engineering,* 63, 17–25.

Sumer, B.M., Fredsøe, J. & Christiansen, N. 1992. Scour around vertical pile in waves. *Journal of waterway, port, coastal, and ocean engineering,* 118(1), 15–31.

Taheri, O., Moayed, R.Z. & Nozari, M. 2015. Lateral Soil-Pile Stiffness Subjected to Vertical and Lateral Loading. *Journal of Geotechnical and Transportation Engineering,* 1(2), 30–37.

Trochanis, A.M., Bielak, J. & Christiano, P. 1991. Three-dimensional nonlinear study of piles. *Journal of Geotechnical Engineering,* 117(3), 429–447.

Zaaijer, M. & Van Der Tempel, J. 2004. Scour protection: a necessity or a waste of money. Proceedings of the 43 IEA Topixal Expert Meeting. 43–51.

Zervos, A. & Kjaer, C. 2006. Pure Power. Wind Energy Scenarios up to 2030.

Numerical Methods in Geotechnical Engineering IX – Cardoso et al. (Eds)
© 2018 Taylor & Francis Group, London, ISBN 978-1-138-33203-4

Nonlinear finite-element analysis of soil-pipe interaction for laterally-loaded buried offshore pipelines

H.E.M. Mallikarachchi
University of Cambridge, UK
University of California, Berkeley, USA

L. Pelecanos
University of Bath, UK
Formerly: University of Cambridge, UK

K. Soga
University of California, Berkeley, USA
Formerly: University of Cambridge, UK

ABSTRACT: Secure transportation of oil and gas over large distances means that steel pipes need to pass through different regions and conditions in order to reach onshore processing plants. The long distance, on which the pipelines lay, implies that the pipe material is subjected to a variety of weather conditions and associated temperatures. In an event of submarine landslide, the pipe may be subjected to large lateral force due to moving ground. Different parameters affect lateral movements and uplift of the pipes and their effects need to be addressed carefully. This paper presents a numerical study using nonlinear coupled hydro-mechanical finite element analysis aiming to investigate the effects of loading rate, hydraulic boundary conditions and soil strength on the soil-structure interaction involved in lateral movements of pipelines.

1 INTRODUCTION

The need for secure transportation of oil and gas over large distances by steel pipes which pass through different regions and conditions in order to reach onshore oil and gas processing plants. The long distance, on which the pipelines lay, implies that the pipe material is subjected to a variety of weather conditions and associated temperatures. In an event of submarine landslide, the pipe may be subjected to large lateral force due to moving ground. For that reason, there has been a particular interest into the behaviour of offshore buried pipelines and especially the phenomenon of lateral and upheaval buckling. Different parameters affect lateral movements and uplift of the pipes and their effects need to be addressed carefully.

Pipe-soil interaction involves complicated mechanisms and therefore has been studied widely (O'Rourke, 2010), using analytical (Klar et al., 2005), numerical (Cheong, 2006; Vazouras et al., 2010), laboratory experimental (Oliveira et al., 2005) and field monitoring approaches. Large-scale field monitoring studies using advanced and innovative fibre-optic technology (Soga et al., 2015, 2017; Kechavarzi et al., 2016, 2019; Di Murro et al., 2016, 2019; Ouyang et al., 2015, 2018; Pelecanos et al., 2016, 2017b, 2018a; Pelecanos & Soga, 2017a, b; Seo et al., 2017a,b; Acikgoz et al., 2016, 2017) were conducted by Vorster et al. (2006a, b) and highlighted the importance of accurately identifying localised stress and strain concentrations which may threaten the structural integrity of the pipe system.

Advanced computational analyses have been applied using a series of two-dimensional (2D) (Zhang et al., 2015) and three-dimensional (3D) finite element (FE) analysis with complicated geometries (Cheong et al., 2011), static (Robert et al., 2016) and dynamic (Karamitros et al., 2007; 2011) conditions or subjected to extreme earth movements such as earthquake faulting (Vazouras et al., 2012, 2015). Special attention was paid on the effect of the trench geometry (Kouretzis et al., 2013), the interaction of pipes with surrounding structures, partially embedded pipelines (Ansari et al., 2014), partially-saturated soil behaviour (Robert et al., 2016a, b, c). Of particular importance is the response of pipes due to uplift and lateral and upheaval buckling (Trautmann & O'Rourke, 1985; Trautmann et al., 1985; Yimsiri et al., 2011, Jung et al., 2013, Soga & Pelecanos, 2014; Soga et al., 2015, 2016). The most fundamental work on the lateral pipe-soil interaction is perhaps the one provided by Trautmann & O'Rourke (1985) who investigate a wide range of soil properties and pipe geometries due to lateral buckling movements. However, since soil is saturated and there are hydro-mechanical interactions one needs to investigate the effects of saturation and consolidation, therefore different loading rates and hydraulic boundary conditions need to be investigated.

The aim of this study is to explore soil-pipe interaction in off-shore conditions considering hydro-mechanical soil response. The investigation is particularly focusing on offshore buried pipelines subjected to lateral and upheaval buckling. The study presents FE simulations about soil-pipe interaction models in order to investigate how several physical and numerical parameters affect the response of the soil-pipe system and in particular the loads experienced by the pipe. New numerical FE analyses of soil-pipe interaction for buried pipes under lateral loading are performed and two cases of shallow and deep backfill under various conditions are considered.

The influence of several parameters is investigated, such as pipe depth, soil strength parameters, loading rate and drainage boundary conditions. All the cases considered in this study were compared with the results of Trautmann & O'Rourke (1985), i.e. in terms of the non-dimensional horizontal bearing capacity factor, N_q, dependent on the ratio of pipe diameter to depth, and were found to be in a very good agreement.

2 NUMERICAL MODEL

2.1 *Finite element mesh*

The problem under study is a pipeline buried into a sand backfill of trapezoidal shape surrounded by sandstone. Two scenarios with different geometries are considered: Scenario (A) with a shallow backfill and Scenario (B) with a deep backfill, shown in Figure 1 (a) and (b) respectively. The steel pipe and has a thickness of 17.5 mm and a diameter of 0.610 m. The ratio of depth to diameter for the two scenarios A and B considered is H/D = 2.2 and 5.4 respectively.

The aim of this work is to examine the behaviour of buried offshore pipes under lateral loading. Four cases are considered as listed in Table 1 and these are related to the different assumptions regarding the soil behaviour, the nature of the applied load and the imposed boundary conditions. Each case considers three values for the angle of shearing resistance of the backfill, ⊠ = 30°, 37.5° and 45°. Finally, each case is applied on both scenarios (A) and (B) described above, i.e. the shallow and the deep backfill respectively.

The software Abaqus/Implicit was used and the employed FE mesh is shown in Figure 1 (a) and (b) for the shallow and deep backfill scenarios respectively. The lateral boundaries are placed 10 m away from the problem of interest (the backfill) whereas the bottom boundary is placed at -10 m, i.e. 5 m below the backfill. The elements used are second-order iso-parametric 8-noded quadrilateral with full integration of CPE8 type for Case 1. Effective stress analysis is employed for Cases 2–4 and therefore, 8-noded quadrilateral elements with pore pressure of CPE8P type are used.

The interaction between the pipe and the soil of the backfill is treated with a GAP relation, i.e. using ABAQUS's contact algorithms. Both normal (hard contact option) and tangential (penalty option with friction coefficient = 0.24) interaction properties were specified. This allows the soil and the pipe elements to separate when tensile stresses develop at their interface.

2.2 *Constitutive models & material properties*

Both FE meshes consist of three domains, the subgrade, the backfill and the pipe, which have different material properties and adopted constitutive models.

The subgrade domain consists of sandstone and it is modelled as an elastic material. The backfill domain consists of sand material and it is modelled with a Mohr-Coulomb failure criterion (with parameters cohesion, c, angle of shearing resistance, φ and angle of dilation, ψ). Finally, the pipe domain consists of steel material and it is modelled as an elastic material as the anticipated strain values are well within the elastic range. The solution algorithm employed was the Modified Newton-Raphson which performs iterations to integrate the plasticity equations, as is usually the case with nonlinear soil-structure interaction problems

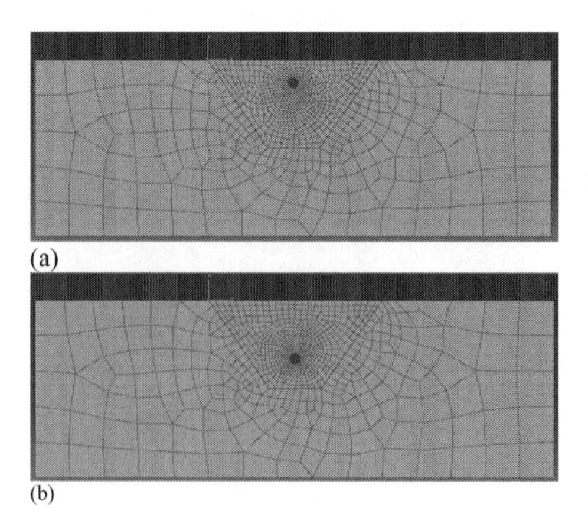

Figure 1. FE mesh for the (a) shallow and (b) deep embedment cases considered.

(Pelecanos et al., 2015, 2018b). The assigned material properties are listed in Table 2.

2.3 *Stages of analysis*

The analyses carried out for Case 1 (dry case) were plane strain (Static, general). For Cases 2–4, coupled pore fluid/stress (Soils, transient consolidation) was specified for the backfill and the subgrade domains which were considered as consolidating materials.

The FE analysis consisted of three steps:

- Step 1: Initial stress conditions

In this step, the following boundary conditions (BC) were applied: full fixity on the bottom boundary, horizontal fixity on the lateral boundaries, full fixity on the inner and outer nodes of the pipe and on the side of the backfill which interacts with the pipe.

Additionally, for the coupled pore fluid/stress analyses (Cases 2–4), the pore water pressure was specified to be zero at the top boundary, so that the water table is placed at ground level.

Predefined fields specified in this step were effective initial stresses on the whole domain. Additionally, for the coupled pore fluid/stress analyses (Cases 2–4), the initial hydrostatic pore water pressure was specified along with uniform values for the void ratio and saturation according to Table 2, for the consolidating materials.

- Step 2: Geostatic state

In this step, the same boundary conditions and predefined fields specified in Step 1 were propagated. Additionally, gravity was applied as a load on the whole model with a value of 9.81 m/s^2. This was performed so that equilibrium is achieved between the specified boundary conditions and predefined fields and the material properties.

- Step 3: Lateral loading of the pipe

In the last step, the pipe was moved laterally. This was achieved by specifying imposed displacements on the outer and inner nodes of the pipe. Two loading rates were considered, for case 2, the loading rate was $V = 2 \cdot 10^{-5}$ m/s, whereas for cases 3 and 4, $V = 10^2$ m/s. The value of the imposed displacement was 0.03 m and therefore the time step for this step was specified according to the required loading rate, i.e. 1500s and 0.0003s respectively. For case 1 which is dry, time-dependent behaviour of the soil is not modelled, and therefore the loading rate does not have any effect.

In this step, the full fixity of the inner and outer nodes of the pipe and on the backfill surface which is in contact with the pipe (as described before in step 1) was deactivated to accommodate the imposed displacements. Also, the soil-pipe contact interaction, as described before, was activated too to allow the formation of gap between the soil and the pipe.

Table 1. Cases considered in this study.

Case	1	1	2	3	4
Saturation	Dry	Fully saturated			
Top hydraulic BC	Dry	Drained			Undrained
Loading	Dry	Slow		Fast	

Table 2. Material properties used in the analysis.

Layer Property	Part Material Unit	Subgrade Sandstone	Backfill Sand	Pipe Steel
Density, ρ	kg/m^3	-	-	7870
Dry unit weight, γ	kg/m^3	16.4	16.4	-
Bulk unit weight, γ	kg/m^3	20.02	20.02	-
Poisson ratio, ν	-	0.25	0.3	0.2
Elastic modulus, E	kPa	20000000	50000	206000000
Shearing angle, φ	deg	-	[30, 37.5, 45]	-
Dilation angle, ψ	deg	-	[1, 5, 16.3]	-
Cohesion, c	kPa	-	0.1	-
Permeability, k	m/s	0.001	0.001	-
Void ratio, e	-	0.585	0.585	-
Saturation, S	-	1	1	-
Specific gravity, G_s	-	2.65	2.65	-

3 FINITE ELEMENT ANALYSIS RESULTS

3.1 *Horizontal bearing capacity factor*

Figure 2 shows the results of the finite element analysis. This figure shows the horizontal bearing capacity factor, N_q, with respect to the pipe embedment H/D and value of shearing angle, φ. The horizontal bearing capacity factor, N_q, is defined as:

$$N_q = \frac{F}{\gamma HD} \tag{1}$$

Where, F, is the maximum value of the horizontal reaction force, γ is the unit weight of the soil, H is the embedment depth and D is the diameter of the pipe.

The following conclusions may be drawn from Figure 2:

- For all the cases considered larger values of angle of shearing resistance, φ, resulted in larger values of the applied force on the pipe and hence higher values of the horizontal bearing capacity factor, Nq. This is in agreement with the observations of Trautmann & O'Rourke (1985).
- For both scenarios of shallow and deep backfill, cases 1 and 2 (dry and slow loading drained) yield very similar values for the horizontal bearing capacity factor. However, their values are larger than those from Trautmann & O'Rourke (1985). This may be attributed to the fact that Trautmann & O'Rourke (1985) considered a uniform homogeneous soil layer, whereas in this study the backfill is surrounded by a much stiffer subgrade which imposes higher forces on the pipe.
- The fast loading drained and undrained analyses (cases 3 & 4) showed very high values of the bearing capacity factor. This was expected as on soil which is prone to dilate, undrained loading results in higher effective stress and stiffer response. This is in agreement to the suggestions of Cheong (2006).
- The difference between the undrained (case 4) and fast loading drained (case 3) is more pronounced for the shallow backfill scenario, whereas this difference seems to be small for the deep backfill scenario. This may be due to the fact that the top boundary which in the undrained case serves as an impermeable boundary is too far from the pipe in the deep backfill scenario, and therefore its effect is small. In contrast, in the shallow backfill scenario, higher pore water pressures develop in the undrained case (A4) and therefore higher forces are acting on the pipe.
- For the dry and slow drained cases (case 1 and 2), larger H/D ratios result in higher values of the bearing capacity factor. This is in agreement with Trautmann & O'Rourke (1985).
- However, this is not the case for the fast loading drained and undrained cases (case 3 and 4) where the shallow backfill (small H/D) resulted in higher values of the bearing capacity factor. This may be due to the fact that the pipe is very close to the surface and it is affected by the hydraulic boundary condition at the top.

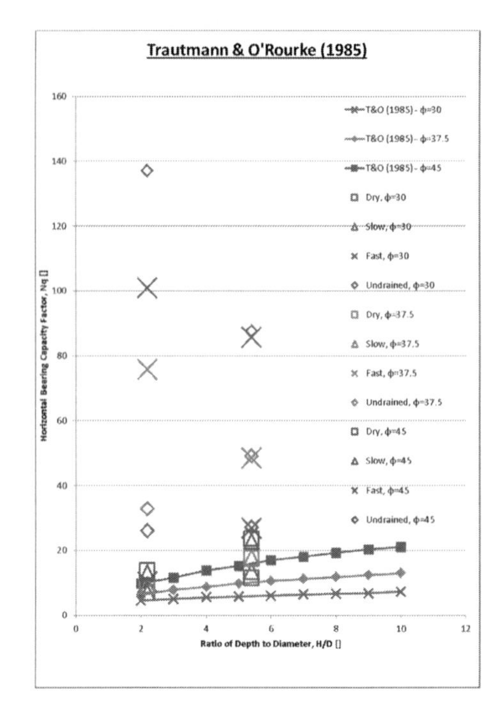

Figure 2. Horizontal bearing capacity factor, N_q.

3.2 *Force-displacement curves*

Figure 3 shows representative relevant force-displacement curves for the case of deep embedment. It is observed that, as expected, larger values of the shearing angle, φ, yield larger values of the resultant force, F. This is because the surrounding soil is stronger and therefore provides more resistance to the lateral movement of the pipe.

It is also shown that for all three values of shearing angle, φ, considered, the fast and undrained cases tend to result in the largest values of total resultant force. In addition, the dry case seems to results in slightly smaller values of the resultant force, whereas the fast loading case provides the smallest value of the resultant force. This was expected, as the fast (and undrained) case results in an somehow undrained response in which case the sudden increase in stress is taken by the pore water which seems to provide additional resistance to the lateral load. In contrast, the slow loading case provide a smaller resultant force which is due to the dissipation of the excess pore water pressures, as the loading is "partially drained", since the slow application of the load allows some time for some of the water pressures to dissipate.

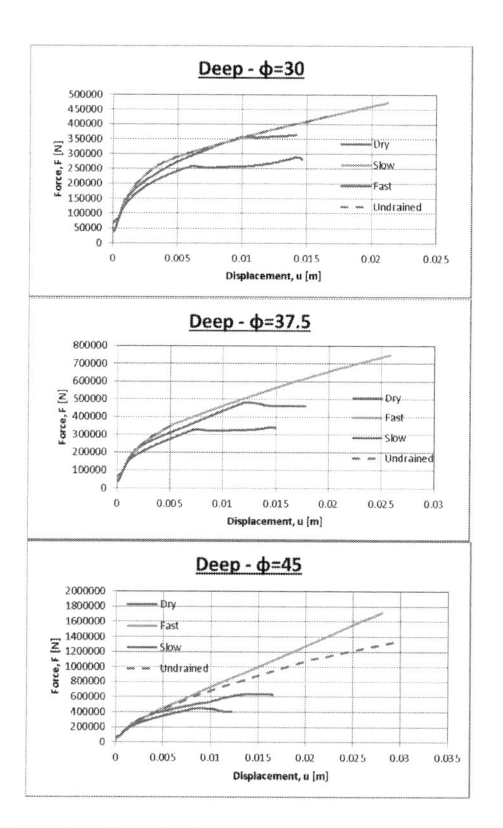

Figure 3. Force-displacement curves.

4 CONCLUSIONS

This paper presents a numerical study related to pipe-soil interaction for offshore buried pipelines. It aims to investigate the effects of loading rate and drainage conditions for buried pipelines. Nonlinear coupled hydro-mechanical finite element analysis is employed which considers a range of different soil properties and pipe embedment values.

The findings of this work may be summarised as follows: (a) Loading rates under undrained or fast drained conditions result in higher horizontal reaction forces, whereas dry or slow drained conditions result in smaller values of reaction forces; (b) Higher values of soil strength properties (angle of shearing resistance, c) and pipe embedment result in larger resultant forces.

ACKNOWLEDGEMENTS

This research was conducted at the University of Cambridge, Department of Engineering and was funded by Tokyo Gas Co. Ltd. Their contribution and support is gratefully acknowledged.

REFERENCES

Acikgoz, MS, Pelecanos, L, Giardina, G, Aitken, J & Soga, K 2017, 'Distributed sensing of a masonry vault during nearby piling', Structural Control and Health Monitoring, vol 24, no. 3, p. e1872.

Acikgoz, MS, Pelecanos, L, Giardina, G & Soga, K 2016, 'Field monitoring of piling effects on a nearby masonry vault using distributed sensing.', International Conference of Smart Infrastructure and Construction, ICE Publishing, Cambridge.

Ansari, Y., Kouretzis, G.P., Sheng, D. 2014, An effective stress analysis of partially embedded offshore pipelines: Vertical penetration and axial walking. Computers and Geotechnics, 58, 69–80.

Cheong, T.P., Soga, K. and Robert, D.J. 2011, "3-D FE Analyses of Buried Pipeline with Elbows subjected to Lateral Loading", Journal of Geotechnical and Geoenvironmental Engineering, American Society of Civil Engineers, Vol. 137, No. 10, pp. 939–948

Cheong, T.P. 2006, Numerical modeling of soil-pipeline interaction. PhD thesis. University of Cambridge.

Di Murro, V, Pelecanos, L, Soga, K, Kechavarzi, C, Morton, RF & Scibile, L 2016, 'Distributed fibre optic long-term monitoring of concrete-lined tunnel section TT10 at CERN.', International Conference of Smart Infrastructure and Construction, ICE Publishing, Cambridge.

Di Murro, V, Pelecanos, L, Soga, K, Kechavarzi, C, Morton, RF, Scibile, L, 2019, Long-term deformation monitoring of CERN concrete-lined tunnels using distributed fibre-optic sensing., Geotechnical Engineering Journal of the SEAGS & AGSSEA. 50 (1) (Accepted).

Jung, J.K., O'Rourke, T.D. and Olson, N.A., 2013. Uplift soil–pipe interaction in granular soil. Canadian Geotechnical Journal, 50(7), pp.744–753.

Karamitros, D.K., Bouckovalas, G.D. and Kouretzis, G.P., 2007. Stress analysis of buried steel pipelines at strike-slip fault crossings. Soil Dynamics and Earthquake Engineering, 27(3), pp.200–211.

Karamitros, D.K., Bouckovalas, G.D., Kouretzis, G.P. and Gkesouli, V., 2011. An analytical method for strength verification of buried steel pipelines at normal fault crossings. Soil Dynamics and Earthquake Engineering, 31(11), pp.1452–1464.

Kechavarzi, C, Soga, K, de Battista, N, Pelecanos, L, Elshafie, MZEB & Mair, RJ 2016, Distributed Fibre Optic Strain Sensing for Monitoring Civil Infrastructure, Thomas Telford, London.

Kechavarzi, C., Pelecanos, L., de Battista, N., Soga, K. 2019. Distributed fiber optic sensing for monitoring reinforced concrete piles. Geotechnical Engineering Journal of the SEAGS & AGSSEA. 50 (1) (Accepted)

Klar, A., Vorster, T.E.B., Soga, K. and Mair, R.J. 2005, "Soil-pipe-tunnel interaction:Comparison between Winkler and elastic continuum solutions," Geotechnique, Vol. 55, No. 6, pp. 461–466

Kouretzis, G.P. Sheng, D, Sloan, S W. 2013, Sand-pipeline-trench lateral interaction effects for shallow buried pipelines. Computers and Geotechnics 54, 53–59.

Oliveira, J.R.M.S., Almeida, M.S.S., Almeida, M.C.F., Borges, R.G. (2010) Physical Modelling of Lateral Clay-Pipe Interaction. Journal of Geotechnical and Geoenvironmental Engineering, 136 (7) 950–956.

O'Rourke, T.D., 2010. Geohazards and large, geographically distributed systems. Géotechnique, 60(7), pp.505–543.
Ouyang, Y, Broadbent, K, Bell, A, Pelecanos, L & Soga, K 2015, 'The use of fibre optic instrumentation to monitor the O-Cell load test on a single working pile in London', Proceedings of the XVI European Conference on Soil Mechanics and Geotechnical Engineering, Edinburgh.
Ouyang, Y., Pelecanos, L., Soga, K. 2018, 'The field investigation of tension cracks on a slim cement grouted column'. DFI-EFFC International Conference on Deep Foundations and Ground Improvement: Urbanization and Infrastructure Development-Future Challenges, Rome, Italy. (Accepted)
Pelecanos, L., Kontoe, S. and Zdravković, L., 2015. A case study on the seismic performance of earth dams. Géotechnique, 65(11), pp.923–935.
Pelecanos, L, Soga, K, Hardy, S, Blair, A & Carter, K 2016, 'Distributed fibre optic monitoring of tension piles under a basement excavation at the V&A museum in London.', International Conference of Smart Infrastructure and Construction, ICE Publishing, Cambridge.
Pelecanos, L., Skarlatos, D. and Pantazis, G., 2017a. Dam performance and safety in tropical climates – recent developments on field monitoring and computational analysis. 8th International Conference on Structural Engineering and Construction Management (8th ICSECM), Kandy, Sri Lanka.
Pelecanos, L, Soga, K, Chunge, MPM, Ouyang, Y, Kwan, V, Kechavarzi, C & Nicholson, D 2017b, 'Distributed fibre-optic monitoring of an Osterberg-cell pile test in London.', Geotechnique Letters, vol 7, no. 2, pp. 1–9.
Pelecanos, L., Soga, K., Elshafie, M., Nicholas, d. B., Kechavarzi, C., Gue, C.Y., Ouyang, Y. and Seo, H., 2018a. Distributed Fibre Optic Sensing of Axially Loaded Bored Piles. Journal of Geotechnical and Geoenvironmental Engineering. 144 (3) 04017122.
Pelecanos, L., Kontoe, S., Zdravković, L., 2018b. Nonlinear analysis of nonlinear dynamic earth dam-reservoir interaction during earthquakes. Journal of Earthquake Engineering. (In Press)
Pelecanos, L., & Soga, K. 2017a. Innovative Structural Health Monitoring Of Foundation Piles Using Distributed Fibre-Optic Sensing. 8th International Conference on Structural Engineering and Construction Management 2017, Kandy, Sri Lanka.
Pelecanos, L., & Soga, K. 2017b. The use of distributed fibre-optic strain data to develop finite element models for foundation piles. 6th International Forum on Opto-electronic Sensor-based Monitoring in Geo-engineering, Nanjing, China.
Robert, D.J., Soga, K. and O'Rourke, T.D., 2016a. Pipelines subjected to fault movement in dry and unsaturated soils. International Journal of Geomechanics, 16(5), p.C4016001.
Robert, D. and Soga, K., 2016b. Behaviour of pipes subjected to fault movements in unsaturated soils using spring analysis. In AP-UNSAT 2015 (pp. 753–758). CRC Press.
Robert, D.J., Soga, K., O'Rourke, T.D. and Sakanoue, T., 2016c. Lateral load-displacement behavior of pipelines in unsaturated sands. Journal of Geotechnical and Geoenvironmental Engineering, 142(11), p.04016060.
Seo, H, Pelecanos, L, Kwon, Y-S & Lee, I-M 2017a, 'Net load-displacement estimation in soil-nailing pullout tests' Proceedings of the Institution of Civil Engineers - Geotechnical engineering, 170 (6) 534–547.
Seo, H & Pelecanos, L 2017b, 'Load Transfer In Soil Anchors – Finite Element Analysis Of Pull-Out Tests'. 8th International Conference on Structural Engineering and Construction Management 2017, Kandy, Sri Lanka.
Soga, K, Kwan, V, Pelecanos, L, Rui, Y, Schwamb, T, Seo, H & Wilcock, M 2015, 'The role of distributed sensing in understanding the engineering performance of geotechnical structures', Proceedings of the XVI European Conference on Soil Mechanics and Geotechnical Engineering, Edinburgh.
Soga, K, Kechavarzi, C, Pelecanos, L, de Battista, N, Williamson, M, Gue, CY, Di Murro, V & Elshafie, M 2017, 'Distributed fibre optic strain sensing for monitoring underground structures - Tunnels Case Studies ', in S Pamukcu, L Cheng (eds.), Underground Sensing, 1st edn, Elsevier.
Trautmann, C.H. and O'Rourke, T.D., 1985. Lateral force-displacement response of buried pipe. Journal of Geotechnical Engineering, 111(9), pp.1077–1092.
Trautmann, C.H., O'Rourke, T.D. and Kulhawy, F.H., 1985. Uplift force-displacement response of buried pipe. Journal of Geotechnical Engineering, 111(9), pp.1061–1076.
Vazouras, P., Karamanos, S.A. and Dakoulas, P., 2010. Finite element analysis of buried steel pipelines under strike-slip fault displacements. Soil Dynamics and Earthquake Engineering, 30(11), pp.1361–1376.
Vazouras, P., Karamanos, S.A. and Dakoulas, P., 2012. Mechanical behavior of buried steel pipes crossing active strike-slip faults. Soil Dynamics and Earthquake Engineering, 41, pp.164–180.
Vazouras, P., Dakoulas, P. and Karamanos, S.A., 2015. Pipe–soil interaction and pipeline performance under strike–slip fault movements. Soil Dynamics and Earthquake Engineering, 72, pp.48–65.
Vorster, T.E.B., Klar, A., Soga, K. and Mair, R.J. (2005): "Estimating the effects of tunneling on existing pipelines," Journal of Geotechnical and Geoenvironmental Engineering, ASCE, Vol. 131, No. 11, pp. 1399–1410
Vorster, T.E.B., Soga, K., Mair, R.J., Bennett, P.J., Klar, A. and Choy, C.K., 2006a. The use of fibre optic sensors to monitor pipeline response to tunnelling. In GeoCongress 2006: Geotechnical Engineering in the Information Technology Age (pp. 1–6).
Vorster, T.E.B., Mair, R.J., Soga, K., Klar, A. and Bennett, P.J., 2006b, July. Using BOTDR fibre optic sensors to monitor pipeline behaviour during tunnelling. In Third European Workshop on Structural Health Monitoring, Granada.
Yimsiri, S., Soga, K., Yoshizaki, K., Dasari, G.R. and O'Rourke, T.D. 2004, "Lateral and upward soil-pipeline interactions in sands for deep embedment conditions," Journal of Geotechnical and Geoenvironmental Engineering, American Society of Civil Engineers, Vol. 130, No. 8, pp. 830–842
Zhang, X., Sheng, D., Kouretzis, G.P., Krabbenhoft, K., Sloan S.W. 2015, Numerical investigation of the cylinder movement in granular matter. Physical Review E 91 022204, 1–9.

Numerical Methods in Geotechnical Engineering IX – Cardoso et al. (Eds)
© 2018 Taylor & Francis Group, London, ISBN 978-1-138-33203-4

Development and validation of a numerically derived scheme to assess the cyclic performance of offshore monopile foundations

J. Albiker & M. Achmus
Institute for Geotechnical Engineering, Leibniz University of Hannover, Germany

ABSTRACT: Large diameter monopiles are an established foundation type for Offshore Wind Energy Converters (OWECs), although the knowledge about the behavior of these piles under cyclic loading in particular is still not comprehensive. The so-called Stiffness Degradation Method (SDM) is a numerical scheme that enables the prediction of deformation accumulation of a pile-soil system under cyclic one-way loading, using finite element simulations in connection with results of cyclic laboratory tests. Using this method, comprehensive parametric studies were carried out in order to identify the main factors affecting the cyclic pile performance. From the results, a practicable scheme could be derived that enables the estimation of the degree of cyclic increase of the pile head deformation by means of an analytical approach that captures the main influencing parameters. Herein, first a measure of pile-soil system stiffness is determined. Then, dependent on this system stiffness measure and the relative load level, cyclic accumulation parameters can be read off from diagrams. The results of field tests on steel pipe piles conducted on an intermediate scale could be regarded for an evaluation of this scheme, which shows that the proposed novel way to assess the cyclic pile deformation behavior yields plausible results.

1 INTRODUCTION

A large number of offshore wind farms are planned in north European offshore reservoirs. In water depths up to 30 m or even 40 m, the monopile often is the most favorable foundation type. The diameters of these single steel pipe piles (as sketched in Figure 1, left), that are primarily designed for bearing cyclic horizontally acting wind and wave loads, vary between 4.0 m and about 8.0 m, whereby the tendency goes towards increasing diameters.

Under cyclic loading, piles exhibit an accumulation of permanent head deflections and rotations. Since wind energy converters are relatively sensitive to permanent deformations, in particular, tilting, it is very important to estimate these as accurately as possible. A number of approaches have been developed recently (see e.g. Achmus et tom Wörden 2012), but none of these have yet been approved.

The 'stiffness degradation method' (SDM) is a numerical procedure, that was developed at the Institute for Geotechnical Engineering at the Leibniz University of Hannover (Kuo 2008). This scheme allows a prediction of the deformation behavior of cyclic horizontally loaded piles using a series of 3D finite element simulations in connection with results of cyclic laboratory tests.

For the present paper, this method was used to carry out parametric studies in order to investigate the influence of various parameters (viz. pile geometry, pile stiffness, loading type and soil density) on the cyclic pile performance. Through these studies, the main factors affecting the behavior of cyclic laterally loaded piles were identified and an easy-to-handle novel scheme to assess the cyclic pile performance under consideration of these factors was developed. This scheme is applied to assess the behavior of medium scale field tests documented by Li et al. (2015) and the results are comparatively evaluated.

2 NUMERICAL MODELING TECHNIQUE FOR CYCLIC HORIZONTAL LOADING

The SDM is described in detail in Kuo (2008) and in other publications, e.g. Achmus et al. (2009) and Kuo et al. (2012), and thus here the method shall be outlined only briefly. A 3-dimensional finite element model like the one shown in Figure 1, right, is used to calculate the pile deformation behavior. The soil behavior under static load is modeled by an elastoplastic material law with Mohr-Coulomb failure criterion and stress-dependent stiffness. The stiffness modulus, i.e. the oedometric stiffness determined under constrained lateral strain, is defined after Ohde (1939) as follows:

$$E_s = \kappa \sigma_{at} \left(\frac{\sigma_m}{\sigma_{at}} \right)^{\lambda} \quad (1)$$

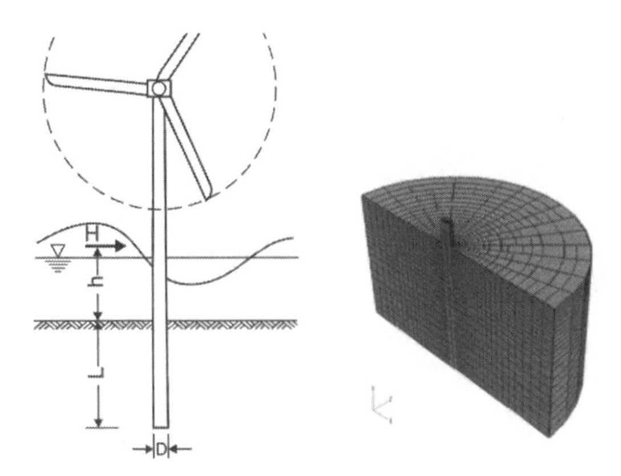

Figure 1. Sketch of a monopile with parameter designation (left), and 3-dimensional finite element mesh used for the numerical calculations (right).

Herein σ_{at} = 100 kN/m² is a reference (atmospheric) stress, σ_m is the current mean principal stress in the considered soil element and κ and λ are soil stiffness parameters.

For describing the increase of the plastic strain of soil with the number of load cycles in element tests (cyclic triaxial tests) an approach by Huurman (1996) is used. The decrease of the (secant) stiffness modulus of the soil with the number of cycles can be approximated by the following equation:

$$\frac{E_{S,N}}{E_{S,1}} = N^{-b_1 X^{b_2}} \tag{2}$$

Herein N is the number of cycles, b_1 and b_2 are regression parameters and $X = \sigma_{1,cyc}/\sigma_{1,f}$ is the cyclic stress ratio, whereby $\sigma_{1,cyc}$ is the maximum of the principal stress in a cycle and $\sigma_{1,f}$ is the main principal stress at failure (in a static test). A cyclic stress ratio is derived for each element of the finite element model from the consideration of the initial stress state and the stress state after applying the horizontal load. With the reduced stiffness values obtained from equation (2) the system's behavior under the lateral load is calculated again. The result represents the increased pile deformation after the considered number of cycles. An illustration of the procedure is given in Figure 2.

For cyclic one-way loading and drained conditions the SDM allows the evaluation of the pile deformation behavior under consideration of the site specific soil conditions as well as the loading conditions and the number of cycles. The application requires the definition of six material parameters accounting for the soil behavior under static loading, for which comprehensive experiences exist, and two parameters b_1 and b_2 describing the stiffness degradation under cyclic loading.

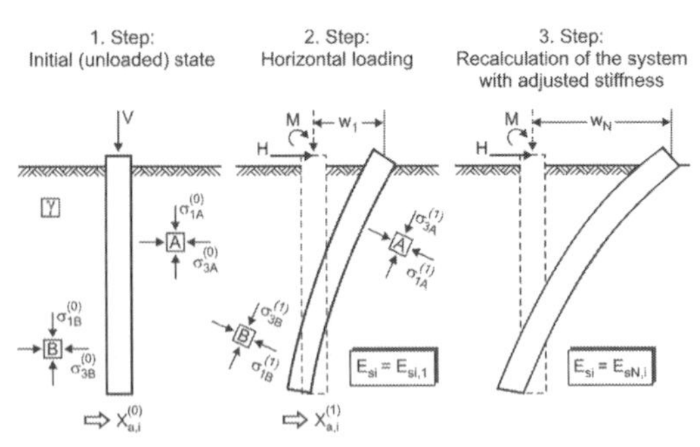

Figure 2. Calculation steps of the SDM.

The method has been validated by the back-calculation of various series of model tests in medium dense as well as in dense sand (see Albiker, 2016). From these back-calculations, it could be concluded that for comparable relative densities of the soil also consistent sets of values for b_1 and b_2 have to be chosen. Value sets for medium dense and dense sand were determined by comparison of calculations and experimental measurements, and the ranges of values were also verified by regression values derived from cyclic triaxial tests (Albiker, 2016).

3 DEVELOPMENT OF A CYCLIC PREDICTION SCHEME

3.1 *Parametric studies*

The SDM was used to conduct comprehensive parametric studies with the goal to identify the principal factors affecting the cyclic pile deformation behavior. In this turn, piles with varying bending stiffness E_pI_p, slenderness ratio L/D and load eccentricity h/L in medium dense and dense sand were investigated. Simulations were carried out with different load levels between H/H_{ult} = 0.125 and 0.5, whereby the ultimate loads H_{ult} were determined by evaluating equilibrium conditions assuming that the spatial passive earth pressures above and below a point of rotation are fully mobilized along the embedded pile length. This procedure is easy to apply and also implemented in several software tools (e.g. IGtH-Pile, Terceros et al., 2015).

The behavior of a pile under cyclic lateral loading can be described by accumulation curves (see Fig. 3), which indicate the relative increase of the pile head deflection y or rotation θ (with respect to the corresponding value under static load) with the number of cycles N. Such curves can usually be well approximated by power law functions as outlined in equation (3) after Little et Briaud (1988) and equation (4) after Le Blanc (2009), under

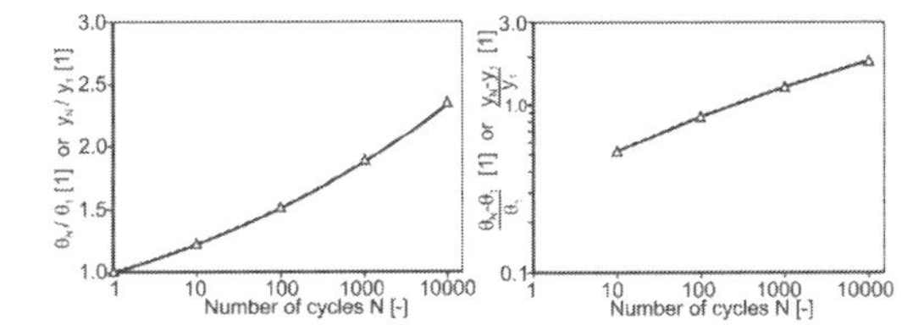

Figure 3. Exemplary cyclic deformation accumulation lines.

adjustment of the accumulation parameters m and T_b:

$$\frac{\theta_N}{\theta_1} = N^m \tag{3}$$

$$\frac{\theta_N - \theta_1}{\theta_1} = T_b T_c N^\alpha \tag{4}$$

Here, θ_1 is the pile head rotation under static loading. The parameter T_c in equation (4) equals 1.0 for one-way loading with full unloading (as considered here), and the parameter α varies only slightly for different pile-soil systems. Consequently, in the following only the parameter T_b is considered.

In the given equations, m and T_b describe the accumulation of pile head rotation. The same approaches could be used to describe the cyclic increase in pile head deflection, but then in general – except for the case of a rigid pile –different m- and T_b-values would apply.

3.2 *Results and deduction of a design proposal*

Figure 4 shows the SDM calculation results for two exemplary systems. Evidently, the stiff pile which behaves almost rigid has a worse cyclic performance (m = 0.136) than the flexible pile (m = 0.094, note that m here describes the accumulation of pile head displacement). The parametric study confirmed that in fact the intensity of bending of the (statically) loaded pile is an indicator for the prediction of the degree of cyclic increase of deformation. For stiff piles, that do not exhibit a pronounced curvature in the deflection line, soil deformation and respective cyclic stiffness degradation occurs along the whole pile length, whereas for flexible piles this applies only to the upper regions close to the embedment point. More vividly, this finding can be explained when considering a very flexible, firmly clamped pile. Here, the clamping of the pile toe will inhibit a continuous cyclic increase of deformation.

For a first orientation, the κ_s-criterion after Poulos et Hull (1989) can be regarded to assess the pile-soil system stiffness:

$$\kappa_s = \frac{E_{soil} L^4}{E_{pile} I_{pile}} \tag{5}$$

where E_{soil} is the Young's modulus of the soil, L the pile length, and $E_{pile} I_{pile}$ is the pile bending stiffness. For $\kappa_s < 4.8$, the pile behavior is considered as perfectly stiff and for $\kappa_s > 388$ as fully flexible.

When examining the results of the parametric studies, firstly it became obvious that systems, that are geometrically similar and have the same value of κ_s but have different absolute dimensions and thus different scaling, show a comparable cyclic deformation accumulation when loaded at comparable relative load levels. Thus, κ_s is applicable in arbitrary scales and it seems that a scaling effect has not to be considered.

Furthermore, it was observed that with increased pile bending stiffness $E_p I_p$ (as already stated) and with increased load level the cyclic deformation accumulation is also increased. This was observed in the progress of the accumulation curves (as exemplary shown in Figure 4 for the case of varied $E_p I_p$) and, consequently, also in the derived values of the accumulation parameters of equations (3) and (4).

However, the effects of pile slenderness ratio L/D and load eccentricity h/L are not captured adequately in κ_s. This became evident when simulations with systems with identical κ_s-values but differing L/D and h/L ratios showed differing cyclic accumulation behavior. A reduced slenderness causes higher spatial soil resistance, while a load eccentricity caused by load application over a lever arm leads to an additional moment load with respect to the embedment point. Both circumstances effectuate a more pronounced pile bending and thus a reduced cyclic deformation accumulation, as long as the piles are not perfectly rigid.

It was thus intended to quantify the characteristic of the (static) pile bending as an alternative indicator for the system stiffness. The envisioned procedure is illustrated and explained in the following:

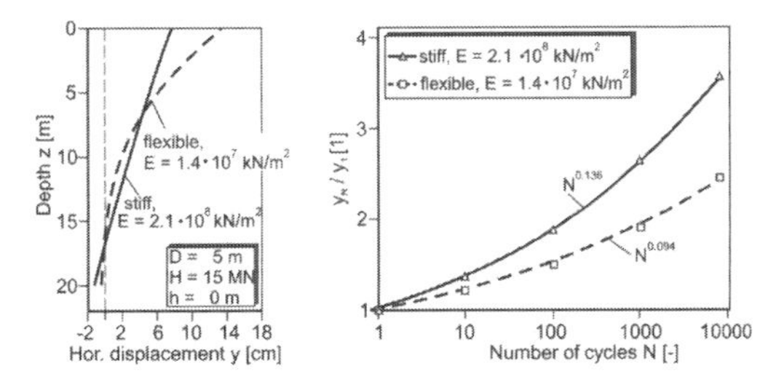

Figure 4. Cyclic deformation accumulation of a stiff and a flexible pile system; left: static deflection lines, right: accumulation of pile head displacement.

Figure 5 shows the construction of pile deflection curves in normalized form. From these curves, the inclination at the top, marked with the designator S_o, can be read off. With this definition, $S_o = 1.0$ applies to rigid piles, and S_o decreases towards 0 with increasing pile flexibility.

Figure 6 shows the calculated values of the accumulation parameters m and T_b depicted over the determined S_o values for all evaluated systems. Clear tendencies are observed that can be well approximated by regression curves. Therefore, the behavior of a monopile under cyclic loading can be predicted when the system stiffness parameter S_o is known.

To determine the parameter S_o, the static pile deflection line of a system must be evaluated. In a subsequent step, a simple equation for the estimation of S_o was sought. It was found that the parameter κ_s after Poulos et Hull (1989) captures basic influencing factors on the system stiffness quite well, though not all relevant factors are regarded. Thus, from the results of the numerical parametric studies a basic mathematic relation between κ_s and S_o was disposed which is valid for the case of a 'reference' slenderness ratio of L/D = 5 and purely horizontally loaded systems with h/L = 0. Following, the effects of varying slenderness ratios and varying load eccentricities, respectively, on S_o, were evaluated in separate examinations, and finally the basic relation between κ_s and S_o was enhanced by two additional terms that cover the influences of varying L/D and h/L values, respectively. The resulting equation (6) enables the determination of S_o in dependency of all investigated system boundary conditions:

$$S_o = \frac{1}{\left(0.066 - 0.0465\frac{H}{H_{ult}}\right)\kappa_s^{0.75}\left(\frac{L}{D}\right)^{-0.5}\left[1+0.9\left(\frac{h}{L}\right)^{0.35}\right]+1} \quad (6)$$

Concluding, the following simple procedure to assess the cyclic accumulation of pile head deflections or rotations is facilitated: S_o, the top inclination of the normalized (static) pile deflection line, can be calculated after equation (6), and

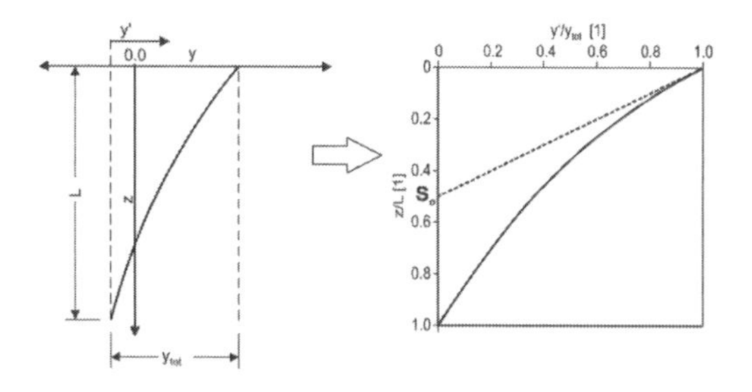

Figure 5. Determination of the system stiffness measure S_o from the pile deflection line under static loading.

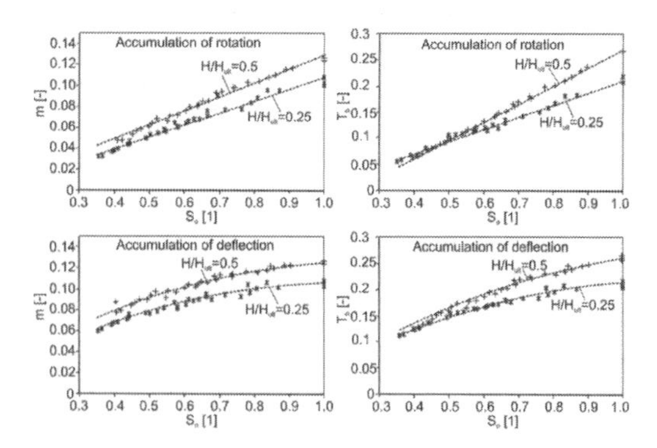

Figure 6. Accumulation parameters m and T_b (exemplary for medium dense sand) in dependency of the system stiffness measure S_o, for load levels $H/H_{ult} = 0.25$ and 0.5.

subsequently, in dependency on S_o and the load level H/H_{ult}, values for the empiric parameters m, T_b and α of the presented power law functions, indicating the degree of cyclic increase in pile deformation, can be read off from diagrams that summarize regression curves like those shown in Figure 6. Such diagrams were developed for the development of pile head rotation as well as deflection and for systems embedded in medium dense and in dense sand. The full information, including all soil parameters used, is documented in Albiker (2016). This novel proposal to assess the cyclic pile deformation accumulation of monopiles is therewith applicable for a substantial range of boundary conditions without the need to carry out further numerical simulations.

4 VALIDATION ON FIELD TESTS

The developed scheme is not yet validated on measurements on real scale monopiles, as such data are not available to date. However, recently Li et al. (2015) presented the results of field tests on driven steel pipe piles in intermediate scale, which shall be regarded for an evaluation.

According to the authors, the field tests were carried out specially with the goal to investigate the behavior of relatively stiff monopiles under cyclic lateral loading; as they saw a lack of available experimental data of such pile-soil systems. Two identical steel pipe piles with diameters of D = 0,34 m, wall thicknesses of d = 14 mm and embedded lengths of L = 2.2 m were driven into unsaturated sand soil. The horizontal load was applied under a lever arm of h = 0.4 m by a hydraulic loading system. The soil is described as over-consolidated dense, fine sand with a relative density close to 100%. Results of cone penetration tests indicated a relatively homogeneous density distribution across depth with a cone tip resistance value of 10 MPa at the top, and linearly increasing values up to about 25 MPa in a depth of 10 m. It is mentioned that the density conditions at the test site in Blessington, Co. Wicklow, Ireland would resemble those found in North Sea wind farm resorts.

In order to determine ranges to be applied for the cyclic loads, initially static tests were conducted. A pile head displacement of 0.05D was defined to be the limit state, and the corresponding ultimate load necessary to provoke this displacement was determined to H = 110 kN. The subsequently chosen load amplitudes for the cyclic tests were taken as a portion of this ultimate load. On each of the two cyclically loaded piles PC1 and PC2, three different load packages of varying load levels were applied consecutively, whereby the cycle numbers of the packages were not constant. Table 1 summarizes the loading characteristics of the two cyclic tests.

In the course of the evaluation of the results, Li et al. (2015) first present cyclic load-displacement lines. Subsequently, they transform these lines into curves depicting the accumulated head displacement and rotation, respectively, over the number of cycles. These curves, in turn, are approximated by the power law function after Little et Briaud (1988, equation (3)) under best possible adjustment of the empiric parameter m. Figure 7 shows the deformation accumulation relations with inclusion of the approximations for pile PC1 (displacement) and pile PC2 (displacement and rotation). For the approximations, Li et al. (2015) took the y_1 values for all load amplitudes from an initial static test under equal boundary conditions. They found a value of m = 0.085 to fit best the accumulation of displacement, and a value of m = 0.06 accordingly for the rotation.

Analyzing these results, first of all it is noted that the m value for the cyclic increase of deflection is slightly higher than the value describing the increase of rotation. This finding is consistent with

Table 1. Load characteristics of the cyclic tests of Li et al. (2015); H_{ult} determined by a deform criterion.

Pile no.	Load series	H_{max} kN	H_{max}/H_{ult} 1	N(tot) –	N(range) –
PC1	1	22	0.2	970	1–970
	2	40–60	0.36–0.55	4007	971–4977
	3	70–100	0.64–0.91	40	4978–5017
PC2	1	33	0.3	1054	1–1054
	2	44	0.4	1305	1055–2359
	3	77	0.7	814	2360–3173

the trendlines of the numerically derived diagrams shown in Figure 6. These diagrams show nearly equal m and T_b values for rotation and deflection for perfectly rigid systems with $S_o = 1.0$. However, with decreasing S_o values and thus increasing flexibility, the values for m and T_b for rotation are decreasing linearly and the intensity of decrease is more pronounced than for deflection.

To calculate the top inclination S_o of the statically loaded pile after equation (6) for the test piles of Li et al., firstly the value of κ_s after equation (5) has to be determined. The Young's modulus of the soil can be back calculated in dependency of the Poisson ratio ν (here $\nu = 0.25$ is assumed) from the stiffness modulus E_s, which in turn can be calculated after equation (1). A numerical back-calculation of an initial static test given in Li et al. (2015) revealed a good agreement of the load-deflection lines with values for the stiffness dependent parameters of κ between 900 and 1100 and $\lambda = 0.5$, while for σ_m the effective overburden pressure in a depth of 0.73 m and thus 1/3 of the embedded pile length L was assigned. For the scheme presented in Chapter 3

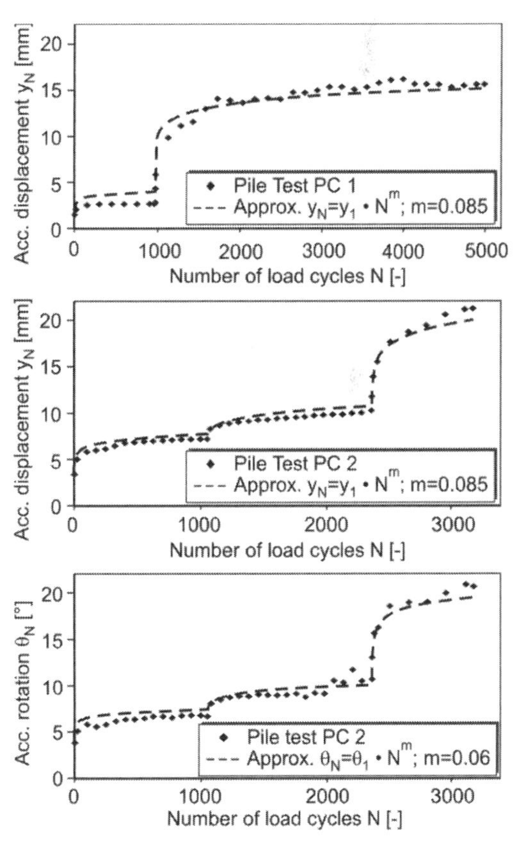

Figure 7. Measured and predicted pile head responses at the pile embedment point after Li et al. (2015).

σ_m generally was determined in the same way. The pile bending stiffness E_pI_p could be calculated with the elastic modulus of steel of 2.1E8 kN/m^2 and the given values of diameter D and wall thickness d. The resulting values for κ_s were determined to $\kappa_s = 16.7$ when assuming $\kappa = 900$, and $\kappa_s = 20.5$ when assuming $\kappa = 1100$. Based on experience, values in this range for the parameter after Poulos et Hull (1989) indicate a relatively stiff, but not rigid pile deformation behavior, as it is typical for monopiles.

To calculate S_o, furthermore the load level has to be determined. The value of $H_{ult} = 110$ kN given by Li et al. is affirmed by application of the above described approach of evaluating equilibrium conditions regarding the spatial passive earth pressure components acting on the incrementally loaded pile, when assuming a friction angle of $\varphi = 45°$ for the sand soil. The latter value of φ was also applied for the mentioned back-calculation of the static test of Li et al. (2015) and leads to a good fit. In Table 1 the resulting relative load levels H/H_{ult} are listed.

The geometric parameters of the test piles, finally, amount to L/D = 6.47 and h/L = 0.18.

Introducing these input parameters into equation (6), values for S_o between 0.77 and 0.82 are calculated, depending on the not clearly defined value for the soil stiffness dependent parameter κ and the value for the load level H/H_{ult}. With these values, from the diagrams in Figure 8 corresponding ranges for the parameter m after Little et Briaud (1988) can be read off. Here, the curves for $H/H_{ult} = 0.25$ and 0.5 are taken for boundaries, as the essential part of loadings lies in this range and investigations indicated that for relative load levels beyond 0.5 the m-values tend to get nearly constant (Albiker 2016).

However, the soil parameters given by Li et al. and also the stiffness and strength parameters that had to be used for the mentioned back-calculation of their static field test indicate a density state that is rather very dense, related to the parameter configuration applied for the numerical deduction of the diagrams in Figure 8 for dense sand. Assuming a reasonable linear extrapolation of the read off m-values from dense to very dense sand, values would result in about m = 0.055 to 0.075 (rotation) and m = 0.06 to 0.085 (deflection). Comparing these values to the m-values determined by Li et al. (2015), a good agreement is obtained (see Table 2).

Table 2. Values for m after Li et al. (2015) and determined by use of the numerically derived scheme.

Source of determination	m (rotation) –	m (deflection) –
Field tests of Li et al. (2015)	0.060	0.085
Num. derived scheme	0.055–0.075	0.060–0.085

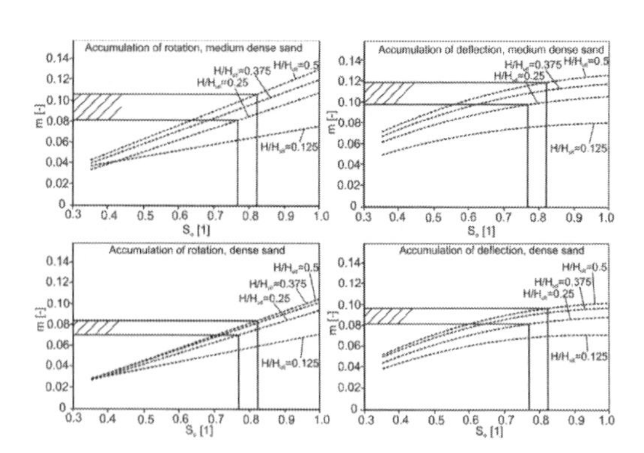

Figure 8. Determination of values for the accumulation parameter m.

5 CONCLUSIONS

The performance of cyclic horizontally loaded piles was investigated numerically by carrying out comprehensive parametric studies under a variation of the system boundary conditions. The principal factors determining the cyclic increase in pile deformations could be identified. The intensity of bending of the normalized deflection line of the (statically) loaded pile was found to be a crucial indicator for the prediction of the cyclic increase of pile deformations. Based on this finding, an easy-to-handle design method could be derived that enables the assessment of the cyclic pile performance in dependency of the system boundary conditions without the need to carry out sophisticated numerical simulations.

The results of field tests on steel pipe piles conducted on an intermediate scale presented by Li et al. (2015) could be regarded for an evaluation of the developed scheme. The results show that the proposed novel way to assess the cyclic pile deformation behavior yields plausible results, both qualitatively and quantitatively. Although the test piles were not real scale offshore piles, the principal boundary conditions regarding pile geometry and stiffness as well as the soil conditions were close to reality and thus this replication can be considered as a successful step of validation.

ACKNOWLEDGMENT

The presented study was partly carried out in the scope of the research project "Ventus efficiens" funded by the *Niedersachsen Vorab* of the "VolkswagenStiftung". The authors sincerely acknowledge the support of this project by the Ministry for Science and Culture of Lower Saxony.

REFERENCES

Achmus, M. & tom Wörden, F. 2012. Geotechnische Aspekte der Bemessung der Gründungskonstruktionen von Offshore-Windenergieanlagen. *Bauingenieur* (87), pp. 331–341

Albiker, J. 2016: Untersuchungen zum Tragverhalten zyklisch lateral belasteter Pfähle in nichtbindigen Böden. Ph.D. Thesis, Leibniz Universität Hannover, Heft 77

Achmus, M., Kuo, Y.-S. & Abdel-Rahman, K. 2009. Behavior of monopile foundations under cyclic lateral load. *Computers & Geotechnics* 36 (2009), pp. 725–735.

American Petroleum Institute - API 2000. Recommended Practice for Planning, Designing and Constructing Fixed Offshore Platforms – Working Stress Design. *RP 2A-WSD. Version 12/2000, errata and supplement 10/2007.*

Kuo, Y.-S., Achmus, M. & Abdel-Rahman, K. 2012. Minimum embedded length of cyclic horizontally loaded monopiles. *Journal of Geotechnical and Geoenvironmental Engineering*, Vol. 138, No. 3, March 1, 2012, pp. 357–363.

Huurman, M. 1996. Development of traffic induced permanent strains in concrete block pavements. *Heron*, Vol. 41, No.1, pp. 29–52.

Kuo, Y.-S. 2008. On the behavior of large-diameter piles under cyclic lateral load. Ph.D. thesis, Leibniz Universität Hannover, Hannover, Heft 65.

LeBlanc, C. 2009. Design of Offshore Wind Turbine Support Structures. Ph.D. thesis, Aalborg University, Denmark, DCE Thesis No. 18.

Li, W., Igoe, D., Gavin, K. 2015: Field tests to investigate the cyclic response of monopiles in sand. *Proc. of the Institution of Civil Engineers - Geotechnical Engineering*, Volume 168, Issue GE5, pp. 407–421

Little, R.L., Briaud, J.-L. 1988. Full scale cyclic lateral load tests on six single piles in sand. *Miscellaneous paper GL-88-27, Geotechnical Division*, Texas A&M University

Ohde, J. 1939. Zur Theorie der Druckverteilung im Baugrund. *Der Bauingenieur* (20), pp. 451–459.

Poulos, H.G. & Hull, T. 1989. The role of analytical geomechanics in foundation engineering. *Found. Eng.: Current princ. and pract.*, ASCE, Reston, 2, pp. 1578–1606.

Terceros, M., Schmoor, K., Thieken, K. 2015: IGtH-Pile Software & Calculation Examples. http://www.igth.uni-hannover.de/downloads

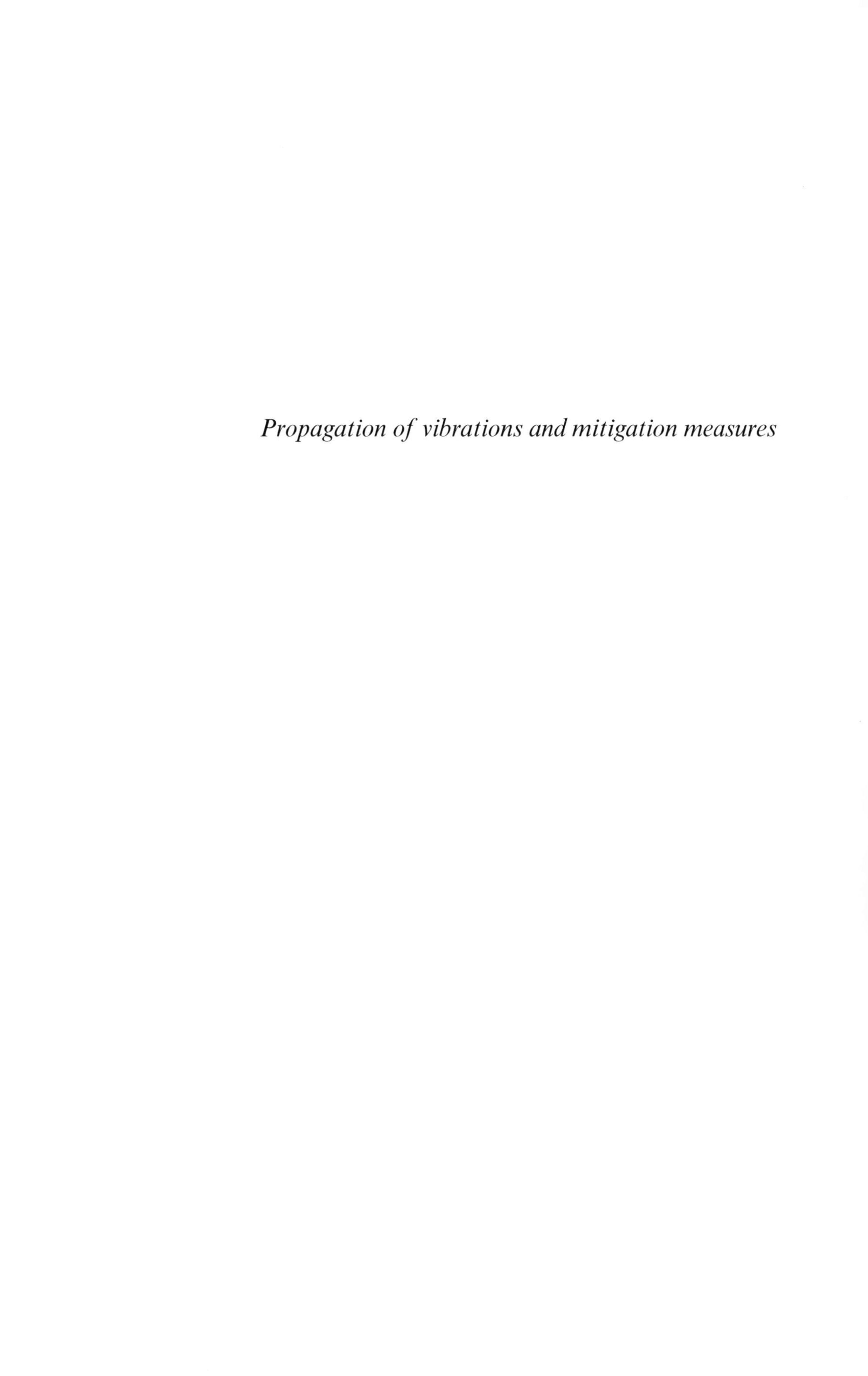

Propagation of vibrations and mitigation measures

Numerical Methods in Geotechnical Engineering IX – Cardoso et al. (Eds)
© 2018 Taylor & Francis Group, London, ISBN 978-1-138-33203-4

Dynamic soil excitation from railway tunnels

P. Bucinskas & L.V. Andersen
Department of Engineering, Aarhus University, Aarhus, Denmark

ABSTRACT: Underground railway lines generate a large amount of ground-borne vibrations, which has negative effects to the inhabitants and structures nearby. However, due to the large number of variables affecting the generation and propagation of vibrations, numerically predicting these effects is a complex task. When accounting for all factors, the numerical models become complicated to produce and take long time to compute. Alternatively, severely simplified models do not provide satisfactory precision. The aim of the paper is identification of the most important factors when modelling the propagation of vibrations from railway tunnels. Identifying the most important parameters allows creation of computational models that account for the most significant phenomena, while reducing the computational complexity as much as possible. For this task a number of different embedded railway tunnel models are created. To account for soil–tunnel interactions a semi-analytical soil model is used, which is based on the transfer matrix method. The tunnel structure is modelled using finite elements which are coupled to the semi-analytical soil solution, thus achieving a fully coupled structure–soil model.

1 INTRODUCTION

Underground metro lines are an essential part large urban centres, as they allow fast and efficient transportation for a large number of passengers. Therefore, new lines are constantly being constructed and old ones upgraded across the world. However, rail lines underneath a densely built city environment can lead to serious problems caused by ground borne vibrations. Hence, numerical models capable of predicting these effects are needed. This becomes especially important in the early project design phases, where the uncertainty regarding the soil conditions and specifications of the structure are high. Thus, computationally efficient models are required that can be used to evaluate larger number of cases.

Over the years, several models have been proposed to analyse railway tunnels. The most commonly used approaches are the finite element (FE) and the boundary element (BE) methods. These methods offer great flexibility to model different conditions and structure geometries, they can also be combined to create coupled FE-BE solutions. An example of coupled FE-BE model can be found in the work by Andersen & Jones (2006). However, these approaches can be extremely time consuming.

To reduce the computation times, so-called 2.5-dimensional models are often used, in which Fourier transformation is carried out in the spatial direction along the tunnel. François *et al.* (2010) as well as Rieckh *et al.* (2012) introduced 2.5-dimensional BE tunnel models in horizontally layered soil. The work assumed a straight tunnel with infinite extent in one direction. Thus, the soil response can be split between separate wavenumbers, allowing efficient parallel computations.

Alternatively, modelling the soil using a semi-analytical approach can also greatly reduce the computational times. The pipe-in-pipe model by Forrest & Hunt (2006) uses a full-space Green's function to model the soil behaviour. The obtained solution is computationally efficient and reproduces the most important phenomena in the system. However, for shallow cut-and-cover type tunnels, the full-space assumption might not be sufficient.

This work uses a Green's function calculated for a half-space as described by Andersen & Clausen (2008), this way taking into account the effect of soil surface. The semi-analytical solution is then coupled with FE, which models the tunnel structure, similar to the models of railways and surface structures previously proposed by the authors (Persson *et al.* 2017; Bucinskas *et al.* 2016; Bucinskas & Andersen 2017).

The modelling approach is described in detail in Section 2. Using the described approach, a number of tunnel models are created and presented in Section 3. These models have varying degree of complexity to investigate the most important modelling parameters of the system. Results obtained are then presented in Section 4. Finally, the conclusions are given in Section 5.

2 MODELLING APPROACH

2.1 *Semi-analytical soil model*

A semi-analytical model based on the Green's function is used to model the soil. The soil stratum is assumed to be homogeneous and visco-elastic with hysteric damping. Further, the surface and the interfaces between layers are perfectly horizontal. To assemble multiple soil layers and obtain the Green's function, the transfer matrix method is used, as originally proposed by Thompson (1950) and Haskell (1953).

An analytical expression for the Green's function is only obtained after transforming the solution into frequency and wavenumber domain over the two horizontal directions, while keeping the vertical coordinates in spatial domain. Due to the approach used, it is only possible to apply external forces on the interfaces between two layers or on the soil surface. Therefore, additional interfaces need to be created for every depth where the structure and soil interact. The Green's function is then calculated for all combinations of created interfaces. After obtaining the Green's function, a unit load is applied at the origin of the coordinate system and the corresponding displacements calculated for a range of wavenumbers. A so-called bell-shaped unit load is created using a two-dimensional Gaussian distribution. This shape of unit load is advantageous as it approaches zero for high wavenumbers, compared to a point force which provides a constant Fourier coefficient in the whole wavenumber domain. This results in a solution where the response of the system at high wavenumbers is also approaching zero, thus acting as a filter. Two bell-shaped loads applied on nearby nodes and the overlap of the loaded areas are shown in Figure 1. It should be noted that the loads can only be distributed over the horizontal coordinates, even when a horizontal load is applied.

Further, double inverse Fourier transformation is applied to transform the solution into frequency-spatial domain. The domain itself is discretized into a number of structure-soil interaction (SSI) nodes, where the response is of interest. To obtain the flexibility matrix, a unit load is placed on a single node and the resulting response is investigated at all nodes. After repeating the process for all SSI nodes a flexibility matrix of the system is obtained. Finally, inverting the flexibility matrix the dynamic stiffness matrix $\mathbf{D}_s$ of the soil is obtained.

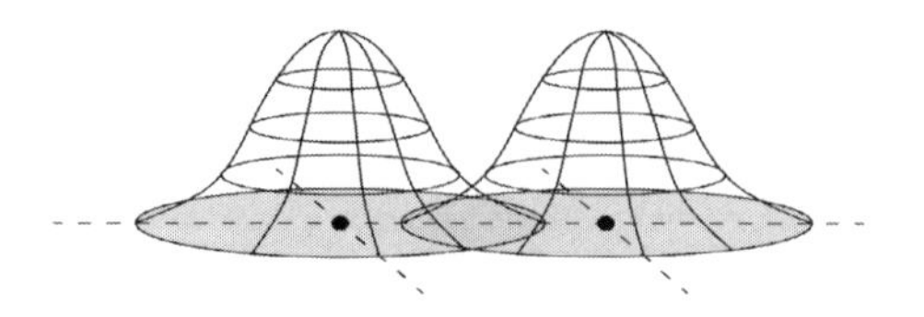

Figure 1. Two bell-shaped loads applied on nearby SSI nodes.

2.2 *Sub-structuring approach to model cavities*

The semi-analytical soil model offers an efficient approach when modelling the soil behaviour. However, a tunnel is a rather complex structure that among other things involves excavation of a substantial amount of material from the continuous stratum. That could significantly influence the overall system behaviour. Unfortunately, directly modelling cavities inside the soil is not possible with the used semi-analytical soil model.

Therefore, the soil–tunnel system is divided into sub-structures. The continuous soil stratum is modelled using the semi-analytical approach, while the volume of the cavity inside the soil is discretized using solid FEs with the material properties of the soil. The cavity inside the soil can then be recreated by subtracting the FE substructure from the continuous stratum. Further, additional shell FEs are added to the exterior of the cavity to model the tunnel lining. The approach is illustrated in Figure 2.

The dynamic stiffness matrix of the system becomes:

$$D_{full} = \begin{bmatrix} D_s^{11} & D_s^{12} \\ D_s^{21} & D_s^{22} \end{bmatrix} - \begin{bmatrix} D_c^{11} & D_c^{12} \\ D_c^{21} & D_c^{22} \end{bmatrix} + \begin{bmatrix} D_l & 0 \\ 0 & 0 \end{bmatrix}, \quad (1)$$

where $\mathbf{D}_s$ and $\mathbf{D}_l$ is the dynamic stiffness matrices of the cavity and tunnel lining respectively. The superscript 11 denotes nodes on the outer boundary of the tunnel, 22 are nodes inside the tunnel, and 21/12 are the coupling terms. The dynamic stiffness matrices of the FE parts are calculated from standard stiffness, **K**, damping, **C**, and mass, **M**, matrices. The dynamic stiffness matrix for the tunnel lining includes linear viscous damping:

$$D_l = K_l - \omega^2 M_l + i\omega C_l, \quad (2)$$

where ω is the circular frequency and i is the unit imaginary number. A hysteric damping model is used for the soil, where damping is independent of the frequency. Therefore, the dynamic stiffness matrix for the cavity becomes:

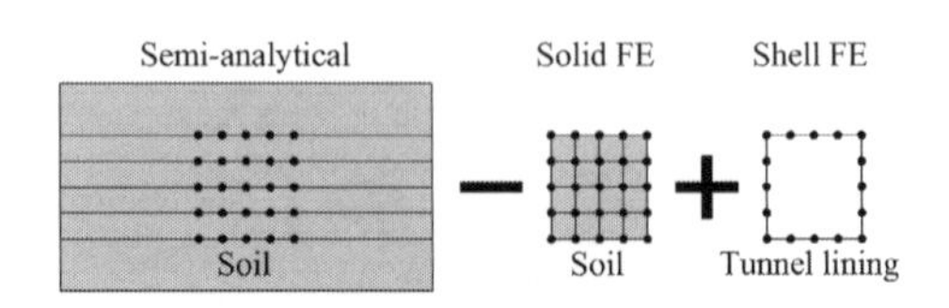

Figure 2. Sub-structuring approach to model a tunnel.

$$D_c = K_c - \omega^2 M_c + iC_c. \quad (3)$$

To properly couple both approaches, additional considerations are needed. Firstly, the SSI nodes inside the stratum are placed in the exact same positions as the nodes in the FE sub-model. Further, the bell-shaped load radiuses are directly related to the element size of the FE, such that the area loaded by the SSI nodes in an interface would be equal to the interaction area between the two sub-models. Eight-node solid FEs are used to model the soil inside the cavity and four-node Mindlin shell elements are utilized to model the tunnel lining. Practice showed that elements with linear interpolation work better with the semi-analytical soil model, compared to elements with quadratic interpolation, as all nodes inside an equal-sided element distribute the stiffness evenly, without additional weighting depending on the position of the node inside the element. This way the obtained dynamic stiffness matrices better represent the results from the semi-analytical soil model.

The sub-structuring approach encounters some problems when modelling cavities inside the soil, without the additional stiffness from the tunnel lining. Most likely the problems arise due to the differences in the two modelling approaches. The two approaches are especially different when dealing with horizontal forces. As already mentioned, the semi-analytical model distributes the loads over horizontal interfaces, thus leading to somewhat different behaviour in vertical and horizontal directions, while in the FE method all directions are handled in the same manner. Further, the convergence between the two approaches also differs—poor discretization tends to underestimate the stiffness in the semi-analytical approach, while it is generally overestimated in the FE method. To an extent these problems can be avoided by properly discretizing the modelled structure.

Further, the damping values obtained in the stratum are often very close to the damping of the FE cavity part. Thus, when subtracting the two sub-structures, the damping can become extremely small or even negative for direct terms. That leads to unphysical behaviour of the system. To avoid these problems, the damping term $\mathbf{C}_c$ in Equation 3 was not used in the computations. The accuracy of the obtained results is somewhat reduced. However, it is assumed to be insignificant, especially when investigating the excitation on the soil surface.

3 ANALYSED CASE

A cut-and-cover type tunnel is investigated in this paper. Construction of this type of tunnel involves digging a trench in which the structure is constructed and then covered back with soil. Cut-and-cover tunnels are easy to construct and are commonly encountered in the urban environment. Further, due to the relatively shallow depth, the excitation of the tunnel can lead to high vibrations on the soil surface.

The modelled tunnel structure has a square profile, with the width and the height both equal to 4.9 m. The tunnel lining is 0.5 m thick and constructed from concrete (material properties used are given in Table 1). The total length of the modelled structure is 49.7 m to minimize the effects of the ends of the structure. It is embedded by 3.1 m from the soil surface to the tunnel crown or 8.0 m from the soil surface to the tunnel floor. The structure is excited by a unit vertical force applied on the tunnel floor in the centre of the structure.

Analysis is performed within the frequency range 1–50 Hz. To ensure a sufficient number of elements to model wave propagation accurately, all FEs are created with a mesh size of 0.7 m, which ensures at least nine elements per S-wavelength at 50 Hz. Four different cases are tested:

- Full model—the tunnel is modelled as described in Section 2. The system is divided into three substructures: the continuous soil stratum modelled with the semi-analytical soil model, cavity discretized into solid FEs and tunnel lining modelled with shell FEs. The unit force is equally divided over 4 nodes, loading a $0.7 \times 0.7\ m^2$ square in the middle of the tunnel.
- Tunnel lining—the full model is simplified by removing the solid FEs that model the cavity. In this case only the tunnel lining is modelled with shell FEs and coupled to the semi-analytical stratum solution.
- Beam—the tunnel-lining model is further simplified by replacing the shell elements with three-dimensional beam elements. In this case a single beam is placed along the centre line of the tunnel, with the corresponding cross-sectional properties. The excitation force is placed on a single node at the centre of the beam.
- Only soil—reference case. No structure is coupled to the semi-analytical soil model. The load is distributed the same way as in the full case.

The soil is modelled as a half-space of drained stiff sand, with a P-wave speed of 580 m/s and an S-wave speed of 310 m/s.

Table 1. Material properties.

Material	Young's modulus (MPa)	Poisson's ratio (-)	Mass density (kg/m3)	Damping ratio (-)
Sand	250	0.30	2000	0.045
Concrete	30000	0.15	2400	0.040

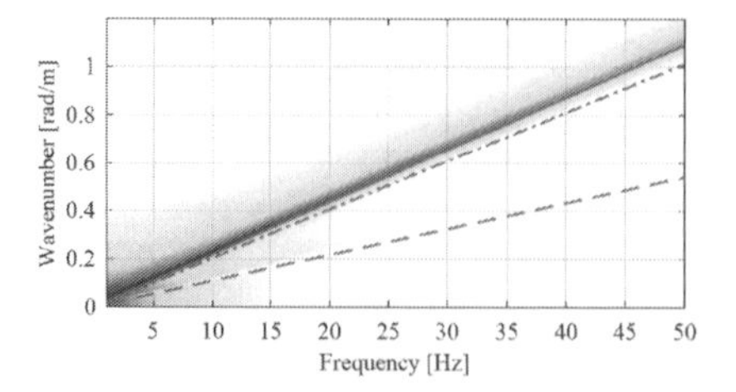

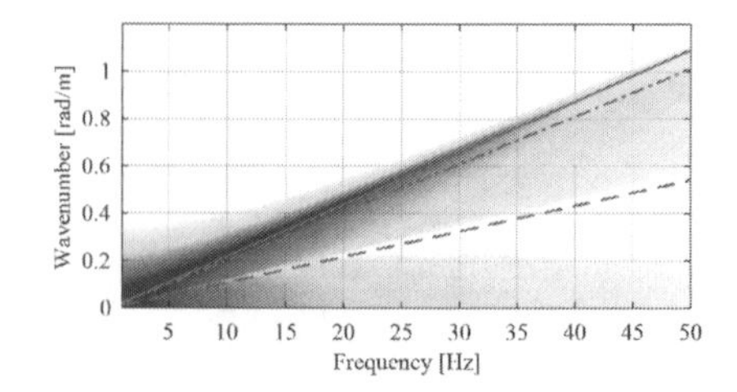

Figure 3. Dispersion diagrams for vertical soil surface displacement. On the left—from a vertical load applied at the soil surface, on the right— from a vertical load applied at 8.0 m depth. Lines indicate P-waves (– –), S-waves (– – –) and R-waves (—). Darker colors indicate higher response. Colors in both plots are not on the same scale.

Two dispersion diagrams for these soil conditions are shown in Figure 3. The dispersion diagram on the left shows the vertical soil response from a vertical load applied on the soil surface, while the dispersion diagram on the right is the soil surface response from a vertical load applied at 8.0 m depth, i.e. where the tunnel floor will be placed.

Comparing the two diagrams it can be seen that in both cases the response is dominated by Rayleigh waves (R-waves). Thus, a strong response of the ground surface can be seen at combinations of the wavenumber and frequency corresponding to the dispersion line of the R-wave. However, when the force is embedded, the response is spread out through a wider wavenumber range.

4 RESULTS

4.1 *Soil surface response over frequencies*

The response of the soil is investigated at four points on the soil surface. The most complex model, i.e. the full tunnel model, is used as the reference for the correct system behaviour, investigating how different simplifications affect the soil response. Figure 4 shows the soil surface response at four observation points. The positions of the points can be seen in Figures 5 and 6.

Figure 4a shows the absolute vertical displacements on the soil surface directly above the excitation point inside the tunnel (the positions of observation nodes can be also seen in Figures 5 and 6). It can be seen that introducing a tunnel structure inside the soil significantly influences the response, when comparing with the only soil case.

Further, modelling the tunnel with or without the cavity does not influence the results significantly. Especially, the responses observed in the different models are similar at higher frequencies. However, the differences between the simple beam model and more complex models that model the tunnel lining with shell elements are quite significant. The beam model predicts a distinct peek at around 28 Hz, which is not predicted in the other cases. This could lead to an overestimation of the response in the middle frequency range, when the simplified tunnel model is utilized.

Further, Figure 4b shows the soil surface response 12.5 m away from the tunnel perpendicularly. The overall behaviour is very similar to that in Figure 4a, with both the full model and the tunnel-lining model showing very similar overall behaviour. In this case, the beam model shows better agreement in very low (up to 5 Hz) and high (above 40 Hz) frequency ranges. However, at middle frequencies the differences are still large. This trend continues with distance increased up to 25.0 m away from the tunnel, as shown in Figure 4c. The simple beam model shows even better agreement with the more complex models. However, the middle frequency range is still represented poorly.

Figure 4d shows the vertical soil surface displacements 20.0 m away from the excitation point along the direction of the tunnel. It is evident that in this direction the tunnel model is very important in the frequency range above 5 Hz, when compared with the only-soil case. Once again, the beam model has a significant peak at around 28 Hz, with poor results in the middle frequency range.

When comparing all four observation points, it can be seen that both the full and lining-only models provide very similar results for all cases. The beam model is not able to reproduce the same behaviour, especially at medium frequency ranges, indicating that modelling the tunnel as a single beam is an oversimplification of the system. However, excluding the cavity inside the tunnel does not have a significant influence. This could be used to simplify the computational model.

4.2 *Displacement field of surrounding soil*

To investigate the effects of the cavity inside the soil, full and just tunnel lining cases are investigated in more detail. The soil displacements are observed on a 30 × 30 × 18 m^3 box is placed around the tunnel structure and the response calculated at two frequencies: 10 Hz and 50 Hz.

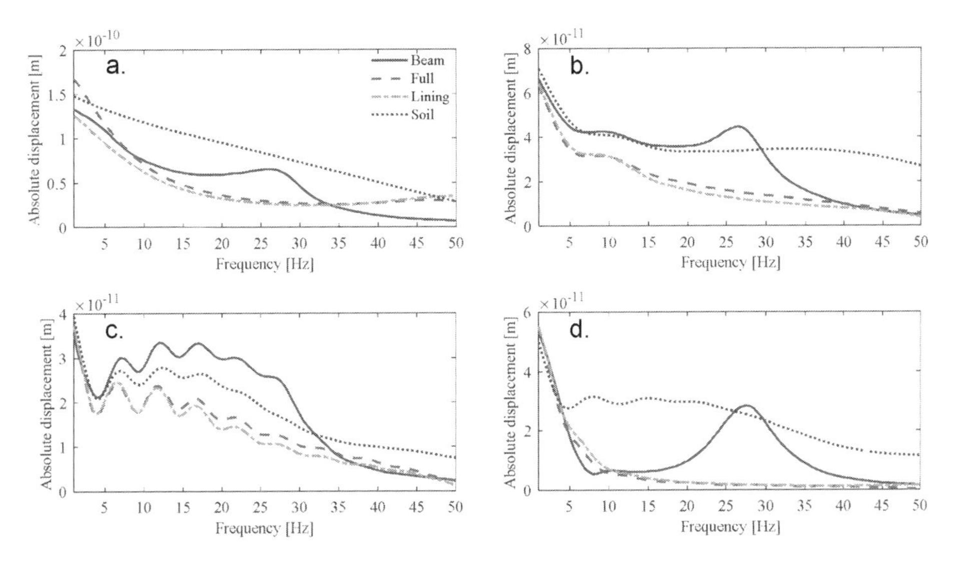

Figure 4. Absolute vertical soil surface displacements at selected observation points: a—directly above the excitation point in the tunnel, b—12.5 m away orthogonally to the tunnel, c—25.0 m away orthogonally to the tunnel, d—20.0 m along the centre line of the tunnel. The point positions are illustrated in Figures 5 and 6.

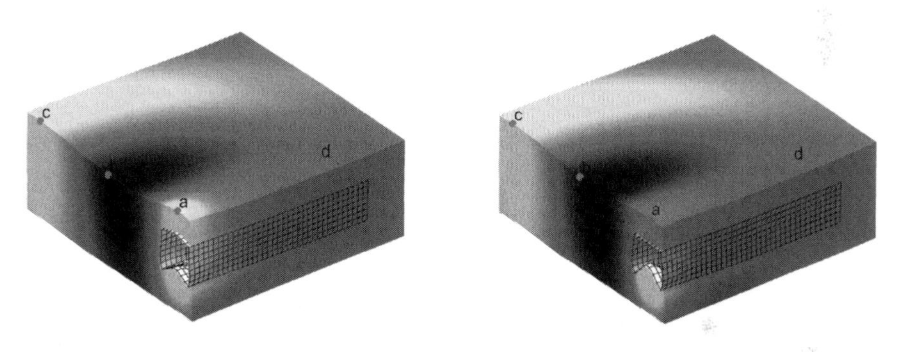

Figure 5. Field response at 10 Hz. On the left—full model; on the right—model with only tunnel lining present. Dark and light shades indicate negative and positive displacement in the vertical direction. Nodes show the positions of observation points used in Figure 4.

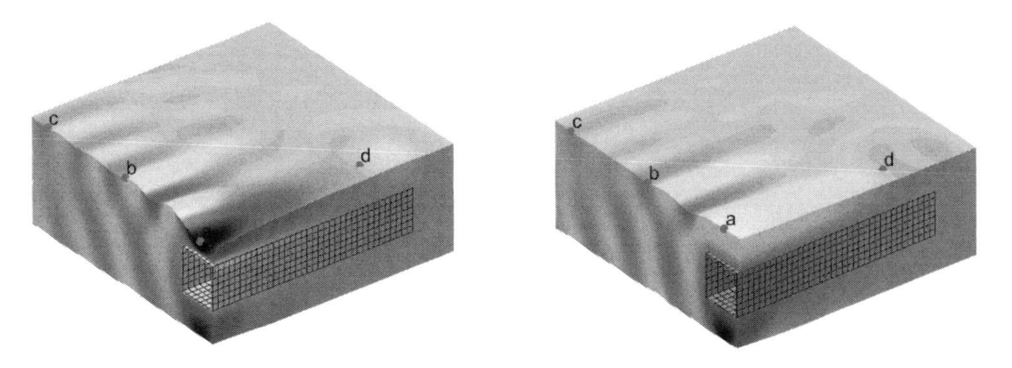

Figure 6. Field response at 50 Hz. On the left—full model; on the right—model with only tunnel lining present. Dark and light shades indicate negative and positive displacement in the vertical direction. Nodes show the positions of observation points used in Figure 4.

Figure 5 shows the absolute vertical soil displacements at 10 Hz. The displacement fields for the investigated cases are almost identical, except the zone around the tunnel excitation point. Here the full model predicts somewhat higher overall displacements, while the zone with large displacements is more localized within the tunnel-lining-only model. Further, the waves are quickly attenuated by the tunnel structure and mostly propagate away from the structure. Similar behaviour is seen in Figure 6, where the displacement field is illustrated at 50 Hz. Once again, the two approaches provide very similar displacement fields further away from the zone around the excitation point. In this case the soil displacements above the tunnel have completely opposite phase angles. This illustrates that the effects from the cavity are present in the nearfield and quickly disappear further away from the structure.

5 DISCUSSION AND CONCLUSION

A computational model to simulate the behaviour of a tunnel structure and the surrounding soil was presented. The proposed method combines a semi-analytical soil model with the FE method which is used to model the cavity inside the tunnel as well as the tunnel lining. A computationally efficient system that can be used to analyse large numbers of cases is obtained. Due to the differences in the two modelling approaches, i.e. the semi-analytical method and the FE method, some difficulties can arise when modelling cavities without the supporting structure. Therefore, further investigation of the proposed methodology is needed. However, adding the supporting structure (in this case the tunnel lining) stabilizes the system and provides satisfactory results.

Using the described methodology, a cut-and-cover type tunnel was modelled. Several different models of the tunnel structure were investigated, with varying degree of simplifications, focus of the analysis being the vertical response of the soil surface. It was determined that modelling the cavity inside the soil is not essential, as the model with only the tunnel lining provides similar results with significantly lower computational effort. However, when considering the displacements close to the excitation point, or when a bigger cavity is present in the soil, the effects from the cavity might become more important. Further, a very simple model that models the whole tunnel as a single beam did not provide satisfactory results. Especially, it overestimates the response in the middle frequency range around 25 Hz. Thus, the beam model provides poorer results than a model with no tunnel at all, i.e. with soil only.

ACKNOWLEDGMENT

The research was carried out in the framework of the project "Urban Tranquility" under the Interreg V programme. The authors of this work gratefully acknowledge the European Regional Development Fund for the financial support.

REFERENCES

Andersen, L. and Clausen, J., 2008. Impedance of surface footings on layered ground. *Computers & Structures*, 86(1), 72–87.

Andersen, L. and Jones, C.J.C., 2006. Coupled boundary and finite element analysis of vibration from railway tunnels—a comparison of two-and three-dimensional models. *Journal of Sound and Vibration*, 293(3), 611–625.

Bucinskas, P., Andersen, L.V., Persson, K., 2016. Numerical modelling of ground vibration caused by elevated high-speed railway lines considering structure-soil-structure interaction, in: Proceedings of the INTER-NOISE 2016—45th International Congress and Exposition on Noise Control Engineering: Towards a Quieter Future.

Bucinskas, P., Andersen, L.V., 2017. Semi-analytical approach to modelling the dynamic behaviour of soil excited by embedded foundations. *Procedia Engineering*, 199, 2621–2626.

Forrest, J.A. and Hunt, H.E.M., 2006. A three-dimensional tunnel model for calculation of train-induced ground vibration. *Journal of Sound and Vibration*, 294(4), 678–705.

François, S., Schevenels, M., Galvín, P., Lombaert, G., Degrande, G., 2010. A 2.5D coupled FE-BE methodology for the dynamic interaction between longitudinally invariant structures and a layered halfspace. *Comput. Methods Appl. Mech. Eng.* 199, 1536–1548.

Haskell, N.A., 1953. The dispersion of surface waves on multilayered media. *Bulletin of the seismological Society of America*, 43(1), pp.17–34.

Persson, P., Andersen, L.V., Persson, K., Bucinskas, P., 2017. Effect of structural design on traffic-induced building vibrations. *Procedia Engineering*, 199, 2711–2716.

Rieckh, G., Kreuzer, W., Waubke, H., Balazs, P., 2012. A 2.5D-Fourier-BEM model for vibrations in a tunnel running through layered anisotropic soil. *Eng. Anal. Bound. Elem.* 36, 960–967.

Thomson, W.T., 1950. Transmission of elastic waves through a stratified solid medium. *Journal of applied Physics*, 21(2), 89–93.

Numerical Methods in Geotechnical Engineering IX – Cardoso et al. (Eds)
© 2018 Taylor & Francis Group, London, ISBN 978-1-138-33203-4

Efficient finite-element analysis of the influence of structural modifications on traffic-induced building vibrations

P. Persson
Department of Construction Sciences, Lund University, Sweden

L.V. Andersen
Department of Engineering, Aarhus University, Denmark

ABSTRACT: Population growth and urbanization results in densified urban areas, where new buildings are being built closer to existing vibration sources such as road-, tram- and rail traffic. In addition, new transportation systems are constructed closer to existing buildings. High vibration levels should be considered in planning urban environment and densification of cities. Vibrations can be annoying for humans but also disturbing for sensitive equipment in, for example, hospitals. In determining the risk for excessive vibration levels, the distance between the source and the receiver, the ground properties, and the type and size of the buildings are governing factors. This paper presents a study of the influence of various parameters of a building's structural design on its vibration levels caused by external ground surface loads. Especially, variations of the slab thickness and the construction material are considered. The parametric studies were conducted by employing an efficient numerical methodology described in the paper. It uses model reduction of a three-dimensional finite element ground model, the reduced soil model being coupled to a building structure of beam and shell finite elements.

1 INTRODUCTION

The densification occurring today in the urban development of cities increases the risk of having disturbing building vibrations. An example of an increasingly disturbing source is faster trains, and an example of a more sensitive receiver is the more advanced equipment housed in hospitals and research facilities. The distance between source and receiver is nowadays shortened due to densification of urban areas, e.g. having buildings closer to railways. Moreover, due to economic and environmental reasons, the building industry tends to build with lighter structural elements and thus use less material, for example, using wooden structures and long-span hollow-core concrete slabs. Analysis of traffic-induced building vibrations is complicated since the problem involves a vast amount of unknown factors, the problem can, in general, roughly be divided into different parts as illustrated in Figure 1: the external source; the medium; and the receiver. The parameters and properties of each region affect the vibration levels occurring in a building. As an example, the building itself can amplify the ground surface vibration levels by more than a factor of two due to resonance in the building structure (FRA 2012). For extensive reading on soil dynamics, wave propagation and vibration-reduction measures—see, for example, (Thompson 2009, Yang & Hung 2009).

1.1 *Vibration mitigation*

In case of predicting or measuring undesirably high vibration levels in a building, vibration-reduction measures can be applied, for example: elastic mats (Sol-Sánchez et al. 2015), soil stabilization—at the source (Andersen & Nielsen 2005) or at the receiver (Persson et al. 2016b)—and wave barriers (Andersen & Nielsen 2005, Persson et al. 2014, Persson et al. 2016a). Researchers have also developed and investigated measures to be applied to the building structure. One such concept is base isolation in terms of mounting a building on springs, see for example (Talbot 2001). Other researchers have evaluated the mitigation effects of employing a thicker lower slab in a building, a so-called blocking floor (Zhao et al. 2010, Clot 2017), or a thicker slab-on-grade (Amick et al. 2004, Xiong et al. 2007), or slitting the slab into isolated islands (Amick et al. 2004, Amick et al. 2005), or supporting the slab with piles (Amick et al. 2005). Clot et al. (2017) also studied the effect of piles, as well as the influence of varying the cross-sectional area of load-bearing columns. The effect of the cross-sectional area of columns was also studied

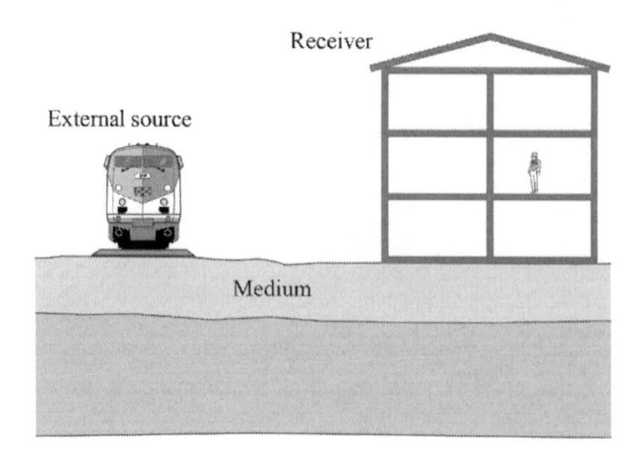

Figure 1. Simplified illustration of the components in a vibration propagation problem in the built environment.

in (Villot 2011). Andersen (2014) investigated differences in vibrational slab response obtained by using heavy-weight material (concrete) or light-weight material (wood). Further, buildings with or without a cellar were compared. Also Persson et al. (2017), compared a heavy-weight building to a light-weight one. They studied the slab vibrational response solely at frequencies near the first resonance frequency of the soil for varying slab thicknesses. From the aforementioned studies it can be concluded that it is possible to mitigate the slab response by modifying the structural design.

1.2 *The present study*

The aim of the investigation presented here is to study, by using an efficient computational methodology (see Section 2), to which extent the design of the structure can be used in reducing externally excited vibrations entering the building via the ground. The objective is to investigate the effect of various geometrical and material parameters of the structural building elements on the vibrational response on the slabs. The paper is a continuation of a conference abstract (Persson 2017), presented at NSCM30 in Lyngby, Denmark, October 2017 by the first author of the present paper.

2 COMPUTATIONAL MODEL

The length scale of local variations in the soil such as granularity is small compared with the wavelengths occurring in the applied frequency range. Thus, homogeneous materials were used in the numerical model. The present study focused on vibrations stemming from traffic. Here the shear strain magnitudes are small and the assumption of linear elasticity is applicable. The finite-element (FE) formulation of a dynamic system can be realized by using the equation of motion:

$$\mathbf{M}\ddot{\mathbf{u}} + \mathbf{C}\dot{\mathbf{u}} + \mathbf{K}\mathbf{u} = \mathbf{f} \tag{1}$$

where $\mathbf{M}$ is the mass matrix, $\mathbf{C}$ the damping matrix, $\mathbf{K}$ the stiffness matrix, $\mathbf{f}$ the load vector and $\mathbf{u}$ being the displacement vector. In steady-state analysis, the displacements and the load can be expressed as complex harmonic functions:

$$\mathbf{f} = \hat{\mathbf{f}}\mathrm{e}^{\mathrm{i}\omega t}, \quad \mathbf{u} = \hat{\mathbf{u}}\mathrm{e}^{\mathrm{i}\omega t} \tag{2}$$

where $\hat{\mathbf{f}}$ and $\hat{\mathbf{u}}$ are the complex amplitudes, i is the complex unit number, and ω is the angular frequency. Inserting Equation 2 into Equation 1 results in the following equation:

$$\mathbf{D}(\omega)\hat{\mathbf{u}} = \hat{\mathbf{f}} \tag{3}$$

where $\mathbf{D}(\omega)$ is the dynamic stiffness matrix:

$$\mathbf{D}(\omega) = -\omega^2\mathbf{M} + \mathrm{i}\omega\mathbf{C} + \mathbf{K} \tag{4}$$

In order to avoid polluting reflections from the artificial boundaries in the FE model, absorbing boundaries, introduced by Lysmer & Kuhlemeyer (1969), are used. To account for material damping, the rate-independent (hysteretic) structural damping matrix with the loss factor, η, is used. See Section 3 for values applied to the various materials.

To ensure that the FE analyses provide results of adequate accuracy, the influence of various parameters such as the aspect ratio of the finite elements, the size of the model geometry and reflection of waves at artificial boundaries have been studied. Quadratic interpolation and at least six nodes per wavelength are employed.

The applied computational methodology essentially consists of two parts: first, model reduction of an FE model of the ground; second, parametric studies using the reduced ground model assembled with an FE building model.

2.1 *Reduced ground model*

In the first part, a large three-dimensional (3D) FE model of the ground is reduced by eliminating degrees-of-freedom (d.o.f.) through a four-step approach: (i) create a full 3D FE model; (ii) set up multiple load cases where the retained d.o.f. are loaded, one at the time; (iii) using Abaqus for steady-state multiple load-case analysis (frequency domain), the displacement results are saved for the retained d.o.f.; (iv) the results are then exported to Matlab where the dynamic compliance matrix is established—it is then being inverted to obtain a reduced dynamic stiffness matrix for each of the frequencies considered. The d.o.f. to be used for loading and for coupling the structure to the ground model are chosen as retained d.o.f.

Making use of the frequency-dependent dynamic stiffness matrix, $\hat{\mathbf{D}}(\omega)$, and the complex displacement amplitudes, $\hat{\mathbf{u}}$, the governing equation of motion with unit loading, $\mathbf{I}$, for each frequency and in each of the retained d.o.f. can be written as:

$$\begin{bmatrix} \mathbf{D}_{rr}(\omega) & \mathbf{D}_{re}(\omega) \\ \mathbf{D}_{er}(\omega) & \mathbf{D}_{ee}(\omega) \end{bmatrix} \begin{bmatrix} \hat{\mathbf{u}}_r \\ \hat{\mathbf{u}}_e \end{bmatrix} = \begin{bmatrix} \mathbf{I} \\ \mathbf{0} \end{bmatrix} \quad (5)$$

where the index r denotes the retained d.o.f. and index e denotes the eliminated ones, $\mathbf{0}$ being a null matrix since the eliminated d.o.f. are unloaded. Denoting the reduced dynamic stiffness matrix as $\mathbf{D}^*(\omega)$ we can write Equation 5 as $\mathbf{D}^*(\omega)\hat{\mathbf{u}}_r = \mathbf{I}$ and hence the calculated amplitudes of the complex displacements give the reduced compliance matrix, $\mathbf{C}^* = (\mathbf{D}^*)^{-1}$, which is then inverted to obtain the sought reduced dynamic stiffness matrix for the layered ground, $\mathbf{D}^*(\omega)$.

The ground was modeled with 3D 20-node solid elements using reduced integration. Given the materials and frequency range of interest described in Section 3, the mesh convergence study resulted in an element size of $1.25 \times 1.25 \times 1.25$ m^3 for the soil layer and $1.25 \times 1.25 \times 20$ m^3 for the bedrock generating about 1.4 million d.o.f. (prior to the model reduction). The dimensions of the layered ground were 10 m deep soil layer on top of a 40 m deep bedrock, the radius of the cylinder-shaped geometry was 70 m, see Figure 2. The number of retained d.o.f. was selected to 1446 (one to be used for loading and the rest for coupling with the building) and hence the reduced dynamic stiffness matrix becomes a full 1446×1446 matrix for each individual frequency.

2.2 *Building model and coupling to ground model*

An FE model of a building structure is established in Abaqus/Standard using structural finite elements. Thereafter the system matrices are exported to MATLAB and the dynamic stiffness matrix for each frequency is calculated. The dynamic stiffness matrices of the building structure are then assembled, by sharing d.o.f. with the reduced dynamic stiffness matrices (one for each frequency) of the ground model. The complex displacement amplitudes were obtained by using Matlab.

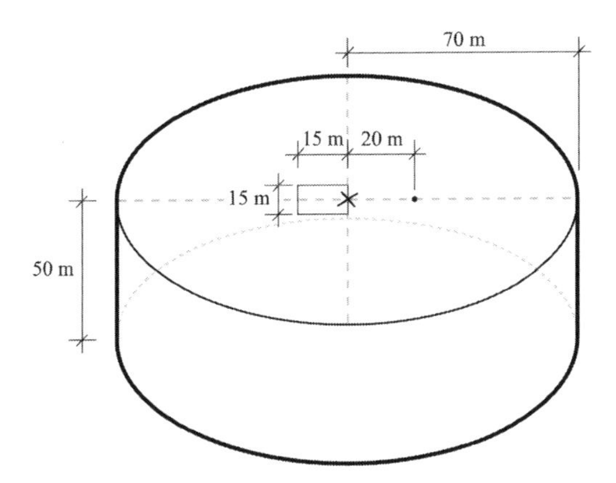

Figure 2. Dimensions of the ground model as well as the slab-on-soil and the position of the excitation source.

The columns were modeled with 2-node linear beam elements, the slabs (excluding the one coupled to the soil) were modeled with 4-node linear shell elements using reduced integration. The slab-on-soil was meshed with 20-node solid 3D element using reduced integration. The element sizes were 0.2 m for the beam elements, 0.25×0.25 m^2 for the shells and $1.25 \times 1.25 \times 0.2$ m^3 for the solid slab-on-soil. The total amount of d.o.f. for the building structure is about 30,000. The columns are fully tied to the slabs. Full coupling between the slab-on-soil and the ground surface was assumed.

3 PARAMETRIC STUDY

To exemplify the use of the developed methodology, a parametric study was performed. The methodology was used to analyze vibrations in a three-story building (representing a typical small residential house in Sweden) originating from a ground surface load.

3.1 *Example case*

The ground in the example case consists of a 10 m deep rather stiff soil layer resting on top of bedrock, see Figure 3. In the figure, the mass density (ρ), the Young's modulus (E), Poisson's ratio (ν) and the loss factor (η) for the ground materials are shown; conditions common in the south of Sweden. The dimensions of the building are shown in Figure 4. The building structure consists of columns and slabs. The material properties used for the concrete as well as the wood are provided in Table 1 and their dimensions are given in Table 2. The structural parts were designed according to the European building code (Eurocode). Two different parameters were varied, one at the time, while the

Table 1. Material properties used for concrete and wood, respectively: mass density (ρ), loss factor (η), Young's modulus (E), Poisson's ratio (v), shear modulus (G). Note that concrete is modeled as isotropic, and wood as orthotropic.

Property	Concrete	Wood
ρ [kg/m^3]	2500	500
η [-]	4%	6%
E [MPa]	32 000	
E_1 [MPa]		8500
E_2 [MPa]		350
E_3 [MPa]		350
v [-]	0.2	
v_{12} [-]		0.2
v_{13} [-]		0.2
v_{23} [-]		0.3
G_{12} [MPa]		700
G_{13} [MPa]		700
G_{23} [MPa]		50

Table 2. Dimensions of the structural parts of the building. The wooden slabs are made of cross-laminated timber (CLT), 7 layers of 40 mm each.

Building part	Concrete	Wood
Slab thickness [mm]	300	280 (7 × 40)
Beam [mm^2]	400 × 200	360 × 115
Column [mm^2]	200 × 200	160 × 160

others were kept constant, the varied parameters being:

- slab thickness,
- light-weight wooden structure versus heavy-weight concrete structure.

The building slabs are subjected to ground vibrations stemming from a vertical harmonic unit point load applied 20 m from the long-side of the building, see Figure 2. The frequency range was set to 5–50 Hz. The load was applied in steps of 1 Hz. Results of the parametric study are given in this section. The root mean square (RMS) value of the magnitudes of the complex vertical velocities for all nodes of the slab in question is used as a measure of the vibration level.

3.2 *Light-weight vs heavy-weight building structure*

In Figure 5, the vibrational responses in the slabs at the ground surface as well as the first and second floors are shown for the wooden structure (left) and for the concrete structure (right). Two similar trends are seen in the left and the right plot, respectively.

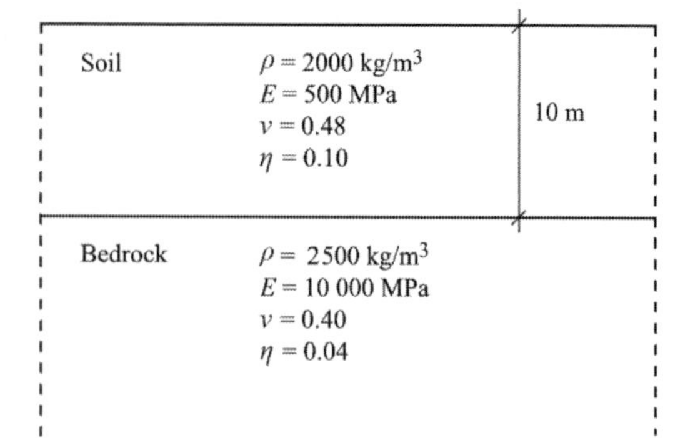

Figure 3. Parameters of the ground.

Firstly, the vibration levels in the building are higher for the low frequencies than for the higher ones. Secondly, the levels of vibration in the slabs are higher than on the ground surface for the lower frequencies, and especially for frequencies were the building has a vibration mode—visible in the frequency spectrum as peaks. As seen in the figure, the response is higher in the heavier concrete building than in the lighter wooden building. This might be explained by two reasons: firstly, the excited vibration modes of the wooden building were of higher order than those excited in the concrete building—thus they require more energy to oscillate; secondly, higher damping was applied to the wooden structure than to the concrete structure.

3.3 *Slab thickness*

The effect of varying the slab thickness in the concrete building on the vibration levels at the first floor (left) and the second floor (right) can be seen in Figure 6, and for the wooden building in Figure 7. As seen in Table 3, concerning the concrete building, the influence being modest

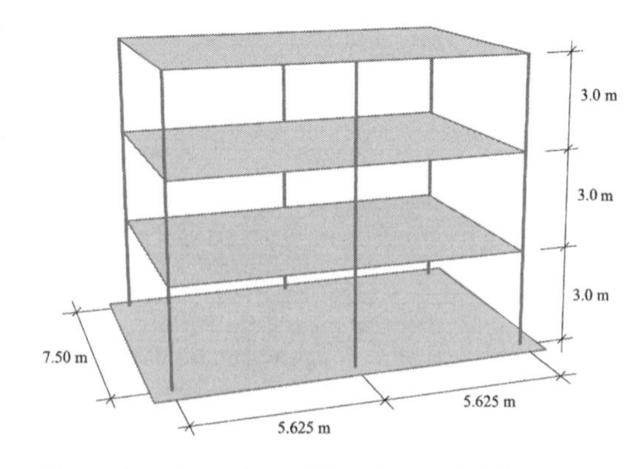

Figure 4. Dimensions of the reference building.

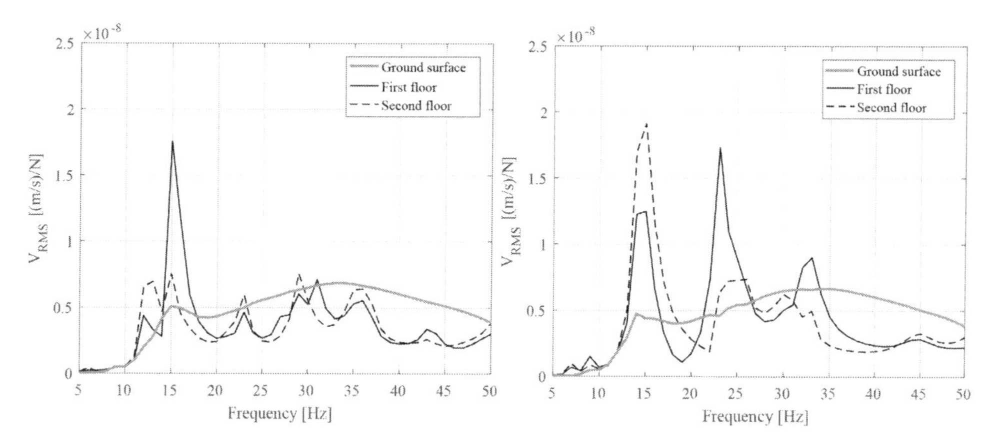

Figure 5. The vertical velocity evaluated for the ground surface and the first and second floors, respectively. Left: light-weight wooden structure. Right: heavy-weight concrete structure.

considering the RMS value over the whole frequency range—the vibration level was only decreased by 10–15% by increasing the thickness from 200 to 400 mm. In this case, it is simply not a realistic vibration-reduction measure to increase the slab thickness. It should be noted, however, that for single frequencies the difference in response can be appreciable due to vibration modes being shifted in frequency. The same conclusions are drawn for the lighter wooden building (cf. Figure 7 and Table 4) when varying the slab thickness from 210 mm to 350 mm, namely that no clear differences in response are seen—the vibration level of the first floor was even increased when the slab thickness was increased, whilst the opposite is seen for the second floor.

3.4 *Coinciding resonance frequencies*

In general, all peaks in the various frequency spectra coincide with eigenfrequencies of the building structure. The eigenfrequencies were calculated by solving the eigenvalue problem in Abaqus for the building with a fixed base foundation. Coinciding resonance frequencies of the ground and of the slabs should be avoided. As an example, at the first

Table 3. RMS values of slab vibrations for the first and second floor, respectively, for the heavy concrete building. The related frequency spectra are shown in Figure 6.

	First floor	Second floor
Slab thickness [mm]	10^{-9} (m/s)/N	10^{-9} (m/s)/N
200	6.15	6.50
300	5.57	5.56
400	5.53	5.46

Table 4. RMS values of slab vibrations for the first and second floor, respectively, for the light wooden building. The related frequency spectra are shown in Figure 7.

	First floor	Second floor
Slab thickness [mm]	10^{-9} (m/s)/N	10^{-9} (m/s)/N
210 (7 × 30)	4.48	4.74
280 (7 × 40)	4.64	3.88
350 (7 × 50)	4.77	3.86

two eigenfrequencies, peaks in the frequency spectra (cf. Figures 6 and 7) can be seen for both the first and the second floor. In Table 5, the resonance frequencies for both the light-weight and the heavy-weight building are listed for the two first modes of major contribution to the slab vibrations for the reference thickness, in the format of one value corresponds to a peak in the frequency spectra and the other value to the calculated eigenfrequency.

3.5 *Computational efficiency*

For the example case used in this study (see Section 3.1), the computational time for the 46 frequency steps in a steady-state analysis took 129 s when using the reduced ground model in the aforementioned described methodology to be solved on two Intel Xeon E5-2650v3 of 2.3 GHz 10-core processors with 62 GB of RAM available. On the same set-up it took 44 390 s with the non-reduced ground model to be solved by using Abaqus/Standard. Hence, the computational time was reduced 344 times. Moreover, by using the developed methodology, the parametric studies are now possible to be solved on a regular desktop computer, e.g. Intel Core i5 3.4 GHz processor with 4 GB available RAM.

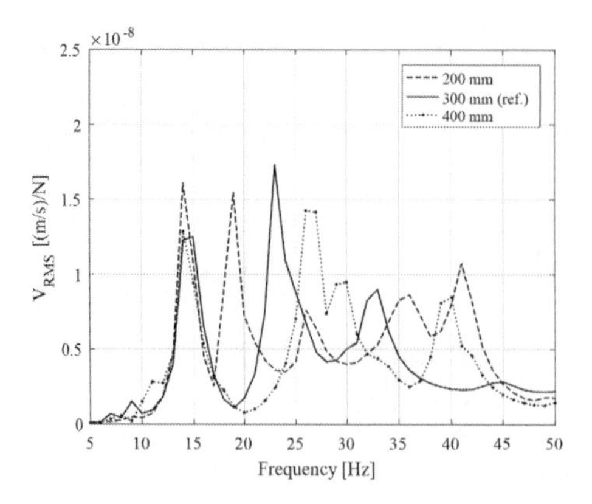

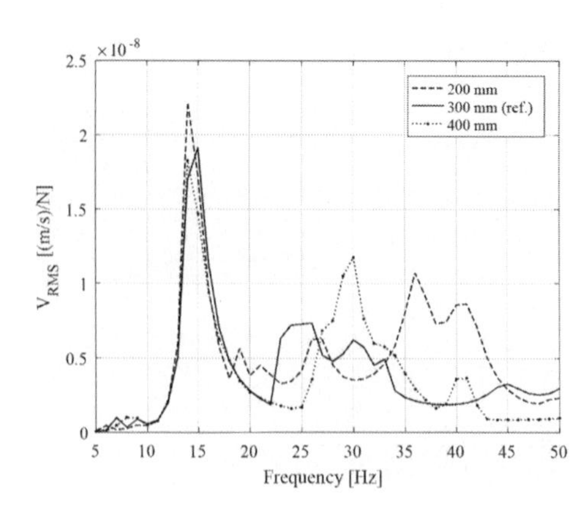

Figure 6. The vertical velocity evaluated for the heavy concrete structure with different slab thicknesses. Left: first floor. Right: second floor.

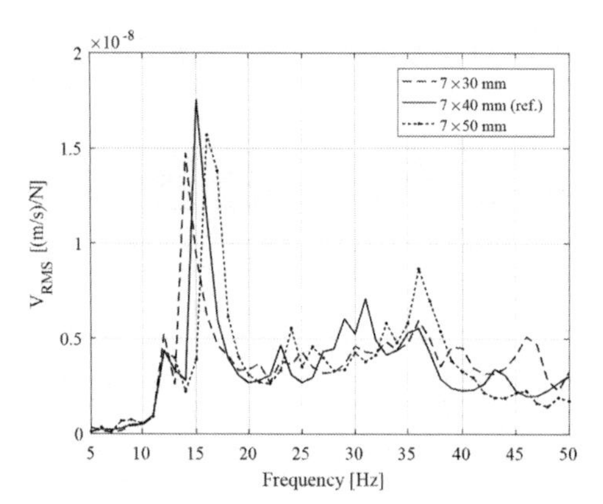

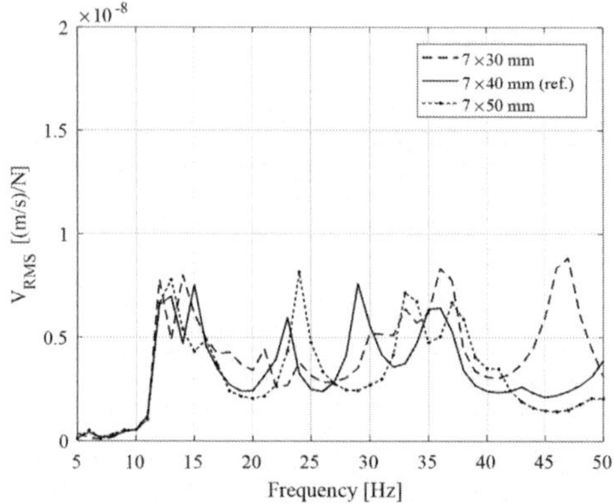

Figure 7. The vertical velocity evaluated for the light wooden structure with different slab thicknesses. Left: first floor. Right: second floor.

However, the reduction of the soil model is a computational heavy process. It took about 5 hours per frequency to reduce the full ground model (approximately 1.4 million d.o.f.) into 1446 × 1446 matrices on two Intel Xeon E5-2650v3 of 2.3 GHz 10-core processors with 62 GB of RAM available. The reduction of the ground model only needs to be performed once, then it can be stored to be used for latter computations, for example, a different project within the same geographical area, or at an area with the same geotechnical conditions. Several reduced ground models can be stored to form a library containing the most common ground types. They would then be available for steady-state analysis on a regular desktop computer. It should be noted that the reduction process can be parallelized, i.e. the reduction for different frequencies can be submitted to different CPUs.

Table 5. Resonance frequencies of the slabs with reference thickness (first and second floor). Format (*yy*/*zz*): (*yy*) peak in frequency spectra, and (*zz*) calculated eigenfrequency. Note the 1 Hz resolution for the steady-state analysis.

	Resonance frequencies (Hz)	
Mode	Light-weight	Heavy-weight
1	12 / 12.4	15 / 15.9
2	15 / 15.1	23 / 23.2

4 CONCLUDING REMARKS

The paper presents a numerical study using an efficient methodology for investigating the effects of structural modifications to a building structure exposed to ground vibrations. The methodology was proven to be efficient by reducing the computational time by 344 times and making it possible to run the numerical analyzes on a regular desktop computer.

Two different parameters were varied in order to investigate differences in the vibrational response of the slabs in the building structure. It was found, for example, that varying the slab thickness was of low importance in affecting the vibrational response—it should be pointed out, for the example case used, that the response could be amplified when increasing the slab thickness. It was also shown that a light-weight wooden building may experience lower slab response when subjected to ground surface excitation than a heavier concrete building would. It should be noted that the eigenfrequencies of the building were lower for the wooden building than for the concrete building, which is explained by weaker columns (which function as slab supports) in the wood building.

In broad strokes, all peaks in the frequency spectra coincide with eigenfrequencies of the building structure. Hence, coinciding resonance frequencies of the ground and of the building modes with large amplitudes in the slabs should be avoided.

To further enhance the computational efficiency of the presented methodology, the coupling of the reduced ground model and the building structure may be realized by coupling forces (instead of displacements). In this way, inversion of compliance matrices is avoided.

As the results from the parametric study shows, it can be difficult to draw any general conclusions in terms of guidelines for the structural design of a building in minimizing traffic-induced vibrations. This may be possible, however, if the model and material parameters can be treated in a stochastic manner in order to quantify the uncertainties. This would be possible using a reduced model of the ground combined with a simple structural model as the one proposed in the paper.

ACKNOWLEDGEMENTS

The research was carried out in the framework of the project "Urban Tranquility" under the Interreg V program. The authors gratefully acknowledge the European Regional Development Fund for the financial support. Moreover, the former Master's dissertation students Rickard Torndahl and Tobias Svensson (Svensson & Torndahl 2017) are gratefully acknowledged for their contributions to the research presented in the paper. Prof. Kent Persson at the Department of Construction Sciences, Lund University, is gratefully acknowledged for his guidance in conducting the presented research.

REFERENCES

Amick, H., Xu, T. & Gendreau M. 2004. The role of buildings and slab-on-grade in the suppression of low-amplitude ambient ground vibrations. *In: Proc of 11th ICSDEE & 3rd ICEGE, Berkeley, CA, 7–9 January*.

Amick, H., Wongprasert, N., Montgomery, J., Haswell, P. & Lynch D. 2005. An experimental study of vibration attenuation performance of several on-grade slab configurations. *In: Proc. SPIE vol. 5933, Bellingham, WA*.

Andersen, L.V. & Nielsen S.R.K. 2005. Reduction of ground vibration by means of barriers or soil improvement along a railway track. *Soil Dynamics and Earthquake Engineering* 25(7): 701–16.

Andersen, L.V. 2014. Influence of dynamic soil–structure interaction on building response to ground vibration. *In: M.A Hicks, R.B.J. Brinkgreve, A. Rohe (Eds.), Numerical Methods in Geotechnical Engineering Vol. 2, CRC Press/Balkema, London, 2014, 1087–1092*.

Clot, A., Arcos, R. & Romeu. J. 2017. Efficient Three-Dimensional Building-Soil Model for the Prediction of Ground-Borne Vibrations in Buildings. *Journal of Structural Engineering* 143(9) 04017098, ASCE.

FRA (Federal Railroad Administration). 2012. *High-speed ground transportation noise and vibration impact assessment*. U.S. Department of Transportation.

Lysmer, J., & Kuhlemeyer, R.L. 1969. Finite dynamic model for infinite media. Journal of Engineering Mechanics Division, ASCE, 95: 859–77.

Persson, P. 2017. Analysis of the effect of structural modifications on traffic-induced building vibrations. *In: Proc of 30th Nordic Seminar on Computational Mechanics NSCM-30, J Høgsberg. N.L. Pedersen (Eds.), 25–27 October 2017*.

Persson, P., Persson, K. & Sandberg, G. 2014. Reduction in ground vibrations by using shaped landscapes. *Soil Dynamics and Earthquake Engineering* 60: 31–43.

Persson, P., Persson, K. & Sandberg, G. 2016a. Numerical study of reduction in ground vibrations by using barriers. *Engineering Structures* 115: 18–27.

Persson, P., Persson, K. & Sandberg G. 2016b. Numerical study on reducing building vibrations by foundation improvement. *Engineering Structures* 124: 361–375.

Persson, P., Andersen, L.V., Persson, K. & Bucinskas, P. 2017. Effect of structural design on traffic-induced building vibrations. *Procedia Engineering* 199: 2711–2716.

Sol-Sánchez, M., Moreno-Navarro, F., Rubio-Gámez, M.C. 2015. The use of elastic elements in railway tracks: A state of the art review. *Construction and Building Materials* 75: 293–305

Svensson, T. & Torndahl R. 2017. *Methodology for analysis of traffic-induced building vibrations*. Report TVSM-5224, Department of Construction Sciences, Lund University (Master's dissertation).

Talbot JP. 2001. *On the performance of base-isolated buildings: a generic model* (PhD thesis).

Thompson, D. 2009. *Railway noise and vibration: mechanisms, modelling, and means of control*. Elsevier.

Villot, M., Ropars, P., Jean, P., Bongini, E. & Poisson, F. 2011. Modeling the influence of structural modifications on the response of a building to railway vibration. *Noise Control Eng. J.* 59(6): 641–651.
Xiong, B., Amick, H. & Gendreau, M. 2007. The effect of buildings on ground vibration propagation. *In: Proc of Noise-Con 2007, Reno, NV, 22–24 October*.
Yang, Y.B. & Hung, H.H. 2009. *Wave propagation for train-induced vibrations*. World scientific publishing Co. Pte. Ltd.
Zhao, N., Sanayei, M., Moore, J.A., Zapfe, J.A. & Hines, E.M. 2010. Mitigation of train-induced floor vibrations in multi-story buildings using a blocking floor. *In: Structures congress 2010*: 1–11.

Numerical Methods in Geotechnical Engineering IX – Cardoso et al. (Eds)
© 2018 Taylor & Francis Group, London, ISBN 978-1-138-33203-4

Assessment of measures to mitigate traffic induced vibrations by means of advanced validated 3D-Finite Element Analyses

T. Meier
BAUGRUND DRESDEN Ingenieurgesellschaft mbH, Dresden, Germany

F. Walther
Landesamt für Straßenbau und Verkehr Sachsen, Germany

ABSTRACT: As part of environmental considerations vibrations caused by traffic infrastructure are of in-creasing importance. In this context there was for example an EU research and development project called RIVAS (Railway Induced Vibration Abatement Solutions). Together with Keller Holding, Offenbach, Germany as one of the members of this project, BAUGRUND DRESDEN together with the Geotechnical Institute of the Technical University of Dresden conducted large-scale field tests and numerical investigations, to firstly demonstrate the applicability and effectiveness of an installation method developed by Keller in conjunction with available damping materials and to validate numerical models, such that they can be used in practice for feasibility studies and design purposes. These FE based analyses are described in the paper at hand together with an application example where the vibrations induced by train traffic are modelled by means of actually moving loads.

1 INTRODUCTION

Keil and Walther 2016 describe the increasing necessity of vibration mitigation measures along traffic infrastructure, e.g. high speed railway tracks. This can be achieved by the installation of vertical inclusions in the ground. Depending on the material used for this purpose there are two main mechanisms involved in the reduction of ground vibrations behind such a submerged protection wall:

1. Reflection of arriving mechanical waves due to a change in stiffness at the interface soil—damping wall
2. Energy dissipation due to the damping capabilities of the installed materials

Different kinds of materials have already been tested for the purpose of vibration mitigation, e.g. concrete, mixtures of soil and binding materials mixed in place (e.g. SoilCrete®) and plates made from different foams (e.g. Regupol® vibration 450, Sylomer® SR 18/25). Thus depending on the in situ subsoil conditions the geotechnical engineer can choose from a variety of materials and installation techniques. In this context finite element analyses (FEA) as described below can be used as a valuable tool for the investigation of the effectives of such measures or to assess materials and geometries.

In the following section field tests, performed in Dresden, Germany are described in short. The next section treats the numerical modelling followed by an application example. Eventually, conclusions are drawn and an outlook is given.

2 LARGE SCALE FIELD TESTS

2.1 *Field and laboratory investigations*

In the field cone penetration tests (CPT) trial pits were carried out and representative soil samples were taken for a comprehensive laboratory testing program, including shear box tests, triaxial tests, oedometric compression tests together with standard index tests for the determination of loosest and densest packing, grain density and shape and angle of repose tests. The results of these tests served the calibration of the parameters of the applied hypoplastic constitutive model.

2.2 *Site description*

The large-scale in situ tests were conducted in the sand pit "Sandgrube Heller – Radeburger Straße" of SBU Sandwerke Dresden GmbH situated in the Dresdner Heide. The underground consists mainly of poorly graded medium sands (SP). The groundwater table is located approx. 70 m underneath the ground surface. Based on the CPT in conjunction with three different interpretation methods (Schmertmann 1976, Karlsruhe Interpretation Method "KIM" Cudmani and Osinev 2001 and German standard DIN 4094-1) the upper approx. 2.5 m below the ground surface show a loose to medium dense packing (ID ≈ 21% to 40%) increasing with depth. Below roughly 7.5 m the sand is in dense state (density in-dex ID ≈ 70%).

2.3 *Installation and test setup*

The different vibration isolators ("screening walls") were installed by driving a wedge box developed by Keller into the ground using a conventional Müller vibrator (Fig. 1). The same vibrator is then used during the tests together with a circular steel plate on the surface of the ground to generate the vibrations. The vibrations in different distances from the vibrator were measured with the aid of geophones (Fig. 2). Three setups were installed with different damping materials (cf. above) and a series of tests with different frequencies was carried out. The test program together with the complete results are not within the scope of this study. For details, please refer to e.g. Keil and Walther 2016. The short description here illustrates the efforts undertaken to obtain a validated numerical model for practical design purposes.

3 NUMERICAL MODELLING AND VALIDATION

The mechanical behavior of the non-cohesive granular material (sand) was allowed for by means of a hypoplastic constitutive model with intergranular strains (so-called "small strain extension") (von Wolffersdorff 1996, Niemunis & Herle 1997). It incorporates density (void ratio) and the most recent deformation history as state variables and allows

Figure 1. Installation of screening walls.

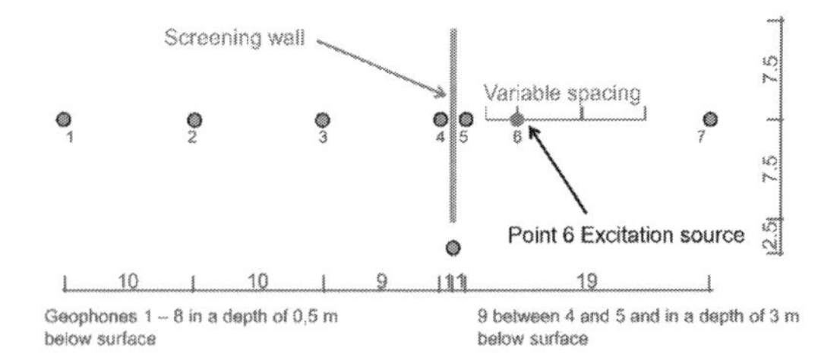

Figure 2. Test setup [Keil and Walther 2016].

for small strain effects, i.e. changes in material stiffness due to changes in the direction of deformation and subsequent shear strain dependent shear stiffness due to further monotonic shearing. Stiffness also de-pends on density as does the shear strength of the material. The constitutive equation also allows for hysteretic damping. The description of the model parameters and of a practical calibration process are described briefly in Meier 2014 and in more detail in Meier 2009.

The FE model used for the back analyses of the field tests is described in Keil & Walther 2016.

A comparison between the measurements and FEA, which is seen as a validation of both the applied constitutive model and the FE approach using Plaxis® is exemplarily shown in Figure 3.

The curve for the "re-calibrated" FEA was obtained after the loosening of the sand adjacent to the damping wall due to the installation process as found by means of pre- and post-installation CPTs was taken into account. But even without this fine tuning, the calculation results agree well with the measurements, i.e. it has been shown that the hypoplastic constitutive law in conjunction with a suitable FE model can be seen as a validated tool for the numerical investigation of wave propagation problems in geotechnical practice.

4 APPLICATION EXAMPLE

4.1 *Introduction*

The following simple application example shall demonstrate

- that the basic method summarized above is also suited for more complex situations and
- how the most recent advances in finite element modelling regarding moving loads can be applied productively in practice.

Up until now, moving loads had to be modeled by means of discrete point loads along the rails. The loads to be applied in each calculation step had to be calculated e.g. with an extra 2D framework program. This procedure, which for

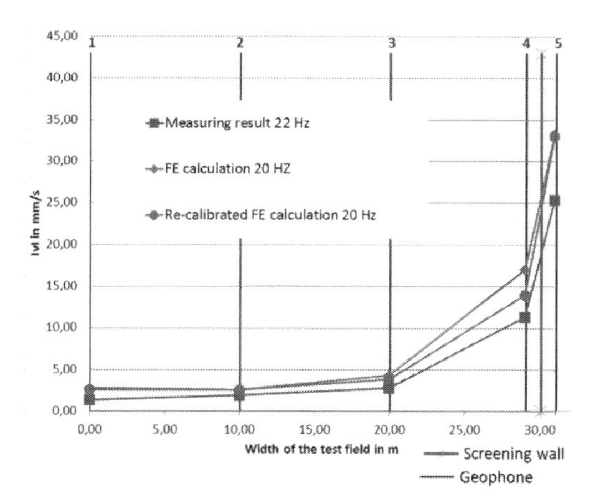

Figure 3. Exemplary comparison between measurements and numerical simulations (test field 1 – isolator material concrete).

practical purposes is very laborious is described e.g. in Meier et al. 2014 or in a simplified form in Sayeed & Shahin 2016.

This shortcoming has been overcome with the latest versions of the FE software Plaxis® 3D, which allows for dynamic point loads actually moving through the model in accordance with given constant velocity vector, a feature, which originates from former activities of the Plaxis Development Com-munity (PDC) and was initiated by the authors. De-tails about the implementation can be found in Gala-vi & Brinkgreve 2014.

4.2 *Description of the model*

A straight single railway track system with a length of 191 m was modelled allowing for sleepers (B 70) with elastics rail pads (15 × 15 × 5 cm³ with an equivalent spring stiffness 100 MN/m) in a spacing of 60 cm and the rails modelled as elastic beams (bending stiffness according to UIC 60).

As a loading regime the wheel loads of a X-2000 high speed train (HST) were considered (cf. Sayeed & Shahin 2016). To save computational costs only two of the actual three cars were considered here.

For the simple demonstration here one observation point approximately 10 m away from the track was defined on the ground surface to assess the effect of the 0.5 m thick and 5 m deep vibration isolation wall between the track and the building. The damping material was assumed to be incompressible and linear-elastic with a Young's modulus according to the Reguplo® damping material of E = 450 kPa, which is very soft compared to the medium dense to dense surrounding sandy soil (Fig. 4).

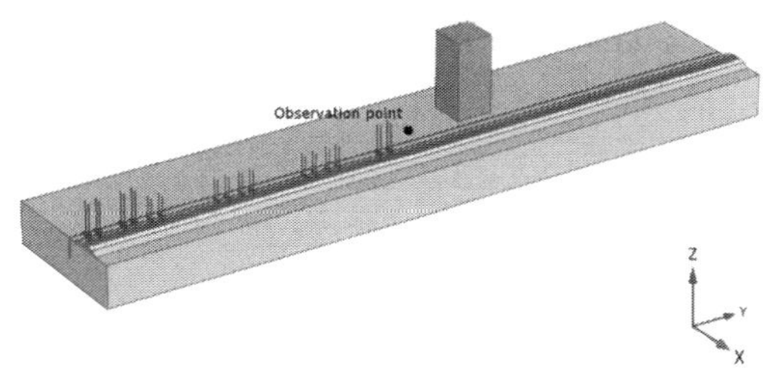

Figure 4. 3D-FE model of a single ballasted railway track together with X-2000 high speed train.

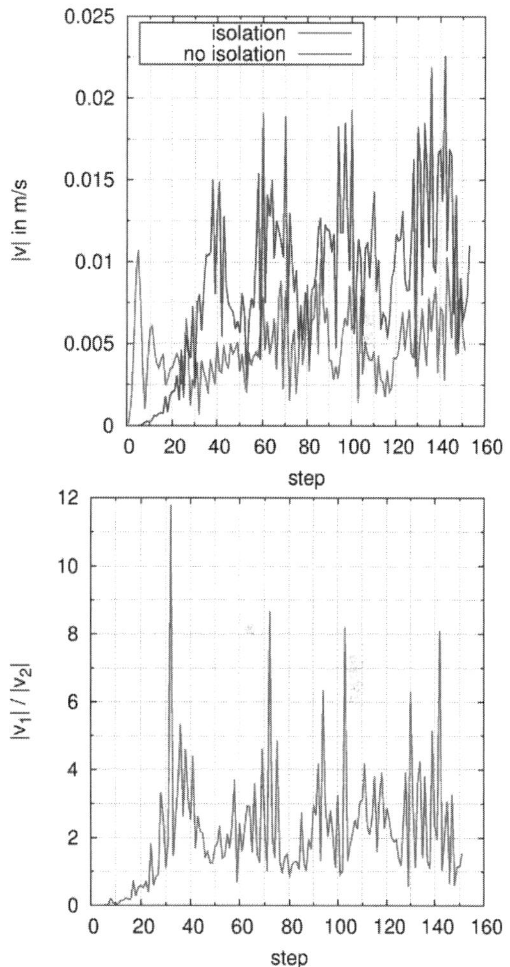

Figure 5. Absolute value of oscillation velocity in the observation point versus time.

The vertical boundaries of the model, which are fixed in their normal direction allow for viscous damping, such that the surface waves produced by the train moving with a velocity of 252 km/h are dissipated as far as possible. The constant time step of the dynamic analyses was chosen in accordance with the Courant-Friedrichs-Lewy condition (cf. e.g. Galavi & Brinkgreve 2014).

4.3 *Results*

The upper diagram in Figure 5 depicts the absolute value of the oscillation velocity vector of the observation point versus time for the two comparative calculations with and without the damping wall. In the diagram below the ratio of these two courses is shown as an indicator of the effectiveness of the investigated vibration mitigation measure.

As expected, the vibrations in the observation point behind the damping wall are significantly reduced by a factor of approximately 2.

5 SUMMARY AND OUTLOOK

The paper at hand describes large scale field tests carried out to validate finite element models in con-junction with a hypoplastic constitutive model for the investigation of wave propagation problems in terms of geotechnical vibration mitigation measures.

A simple application example demonstrates how such a powerful numerical tool, now allowing for actually moving loads, can be used for practical de-sign purposes of such measures.

Due to the low level of shear strain amplitudes induced in the ground by transient loading from traffic infrastructures a less costly linear-elastic approach seems appropriate. First comparative FEA with the subsoil modelled by means of the linear-elastic constitutive model suggest, that such a simplified approach can also be productive, provided suitable Rayleigh damping and stiffness distributions with depth are allowed for.

REFERENCES

Keil, J. & Walther, F. 2016. Vertical vibration mitigation along rail way lines. In *18. International Conference on Railway Engineering and Management*, Copenhagen.

Sayeed, M.A. & Shahin, M.A. 2016. Three-dimensional numerical modelling of ballasted railway track foundations for high-speed trains with special reference to critical speed. In *Transportation Geotechnics* 6(2016) 55–65

Meier, T.; Shahraki, M.; Sadaghiani, M.R.S.; Witt, K.J. 2014: 3D Modelling of Train Induced Moving Loads on an Embankment, *PLAXIS Bulletin 36*, Issue Autum 2014, p. 10–15

Meier, T. 2014. Micro-piles under dynamic horizontal excitation: Field tests and Numerical Modeling, *Proceedings 8th European Conference on Numerical Methods in Geotechnical Engineering (NUMGE 2014)*, Delft, p. 181–186

Galavi, V. & Brinkgreve, R.B.J. 2014. Finite Element modelling of geotechnical structures subjected to moving loads. In Hicks, Brinkgreve & Rohe (Eds) *Numerical Methods in Geotechnical Engineering*, Taylor & Francis Group, London, 2014.

Meier, T. 2009. Application of hypoplastic and viscohypoplastic constitutive models for geotechnical problems. PhD thesis, *Veröffentlichung des Instituts für Bodenmechanik und Felsmechanik der Universtität Fridericiana in Karlsruhe*, Heft 171

Cudmani, R.& Osinov, V.A. 2001. The cavity expansion problem for the interpretation of cone penetration and presiometer tests. *Canadian Geotechnical Journal* (38): 622–638.

Herle, I. & Gudehus, G. 1999. Determination of parameters of hypoplastic constitutive model from properties of grain assemblies. *Mech. Cohes.-Frict. Mater* 1999,4(5):461–486.

Niemunis. A. & Herle, I. 1997. Hypoplastic model for cohesionless soils with elastic strain range. *Mech. Cohes.-Frict. Mater* 4(2):279–299.

Wolffersdorff, P. v. 1996. A hypoplastic relation for granular materials with a predefined limit state surface . *Mech. Cohes.-Frict. Mater* 1:251–271.

Schmertmann, J.H. 1976. An updated correlation between relative density, d_r, and Fugrotype electric cone bearing, q_c. *Technical Report Contract Report DACW 39-76 M6646*, Waterway Experimental Station, U.S.A., WES, Vicksburg, MS.

Numerical Methods in Geotechnical Engineering IX – Cardoso et al. (Eds)
© 2018 Taylor & Francis Group, London, ISBN 978-1-138-33203-4

Experimental validation of a 3D FEM numerical model for railway vibrations

J. Fernández Ruiz
University of La Coruña, La Coruña, Spain

P. Alves Costa
University of Porto, Porto, Portugal

ABSTRACT: In this paper a three dimensional numerical model formulated in time/space domain is applied for railway vibrations. The model is developed in code PLAXIS which is used to simulate the wave propagation in the ground whilst the train loads are calculated with a 2D train-track interaction model. The main objective is to show that commercial programs such as PLAXIS are reliable tools to model railway vibrations. Furthermore, since the wave propagation model is formulated in the time domain and PLAXIS software is able to model excess pore pressure and nonlinear soil behaviour, it could also be a useful tool to consider these two effects, which may be relevant in some cases. A real case located in Portugal is presented, being used for the experimental validation. The numerical results show an acceptable agreement with real measurements. Therefore, the proposed numerical approach can be used as a reliable prediction tool based on PLAXIS software.

1 INTRODUCTION

Recently, the metropolitan population has strongly grown, generating large urban areas, causing the need for more effective and environmentally friendly transport systems. As response to these needs, rail transport systems have been imposing on other. However, rail transport systems can generate a relevant environmental impact in terms of vibration and noise. In fact, the impact of vibrations on working and living environments is considered as an important problem of advanced societies (Smith et al., 2013).

Scientists have made an important effort to develop predictive mathematical models ranging from empirical/semi-empirical models (Bahrekazemi, 2004; Madshus et al., 1996; With et al, 2006) to complex numerical models (Andersen & Nielsen, 2005; Galvín & Domínguez, 2007; Mohammadi & Karabalis, 1995; Sheng et al., 2003). Three-dimensional numerical models developed on time domain are capable of modelling complex geometries and material behaviours, as for instance nonlinear responses. Nevertheless, they concern a very high computational effort (Kouroussis et al., 2013; Connolly et al., 2014).

In contrast, the 2.5D models have been presented as a good tool for predicting railway vibrations with lower computational effort. The main important assumptions of these models are that the geometry and the characteristics (track and soil stiffness, etc.) in the longitudinal direction of the track are invariant and it is subjected to a domain transformation by a Fourier transform, only requiring the cross-section discretization. This technique has been adopted in several studies (Alves Costa et al., 2010a, 2010b; Francois et al., 2010; Muller et al., 2008; Sheng et al., 2005; Yang et al., 2003; Yang & Hung, 2008). Moreover, these models achieve the solution on the frequency domain, so the inclusion of nonlinear behaviour of material (such as soils) is not possible. Actually, 3D models developed in the time domain can overcome all these situations, allowing a better simulation of real behaviour of railway systems.

The main limitation of the models formulated in the time/space domain is the high computational effort added to the fact that the train load cannot be considered from $-\bot$ to $+\bot$, i.e., the domain must be truncated. Examples of 3D numerical models formulated in the time domain to study railway vibrations can be seen in (Connolly et al., 2013; Hall, 2003; Kettil et al., 2008; Kouroussis et al., 2011, Kouroussis et al., 2013) among others.

In this paper, a 3D numerical model formulated in the time domain, through the PLAXIS software, is experimentally validated using an uncoupled numerical approach: i) firstly the dynamic train-track interaction is computed with a 2D model; ii) the resulting loads on each sleeper are applied in the 3D numerical model where wave propagation in the ground is simulated. The main objective of this paper is to show that commercial programs such as PLAXIS are reliable tools to model railway vibrations and if the proposed methodology could be applied to similar problems. This aspect is relevant for engineering practitioners, since almost all times they don't have accesses to complex numerical approaches, which application is usually limited to academia. Furthermore, since the wave propagation model is formulated in the time domain and PLAXIS software is able to model excess pore pressure and nonlinear soil behaviour, it could also be a useful tool to consider these two effects, which could be important in some specific cases. Also, the fact of using a 3D model experimentally validated would imply the possibility of modelling inhomogeneous geometries in the longitudinal direction of the track, as discussed in previous paragraphs.

2 NUMERICAL APPROACH

Numerical procedure is based on an uncoupled scheme: i) firstly the dynamic train-track interaction is computed with a 2D model; ii) the resulting loads on each sleeper are applied in the 3D numerical model formulated in the time domain (PLAXIS software) where wave propagation in the ground is simulated. It should be noted that in order to simulate the train-track interaction problem all masses of train (wheel, bogie and car body), rail and sleeper have been taking into account. Note that only the vertical movement is considered and that only rail unevenness has been taken into account as dynamic excitation source. A detailed explanation of these models can be consulted in Fernández et al. (2017) and in PLAXIS (2011).

3 THE CASE STUDY OF CARREGADO

3.1 *Generalities*

In order to validate the proposed approach, an experimental test site has been selected, near the Carregado city (Portugal). The experimental activities developed in this case study have been already presented in detail in some studies published by the authors, so, the reader is advised to consult Alves Costa et al. (2012) and Fernández et al. (2017) for a detailed description of the experiments. In this paper, only a brief description of the main aspects used in this research is performed.

It should be noted that the vibrations were measured in several points of free-field surface at distances between 3.5 and 45 meters from the track. The train was Alfa Pendular with a speed of 164 km/h.

3.2 *Geodynamic properties of the ground*

The main soil properties are showed in the Table 1. In what concerns to numerical modelling of soil damping, a Rayleigh model has been adopted. Due to Rayleigh damping formulation is frequency-dependent, the α and β coefficients have been calculated in order to obtain an almost "constant damping" in the range of frequencies of interest or main expected frequencies, that in this case are from 10 to up 50 Hz. Figure 1 shows the damping curve considered for all soil layers.

3.3 *Mechanical properties of the track and train characteristics*

Table 2 shows the properties of track elements (sleeper, ballast and subballast) which have been considered according to Alves Costa et al. (2012). The sleeper properties are well defined while ballast and subballast properties were obtained

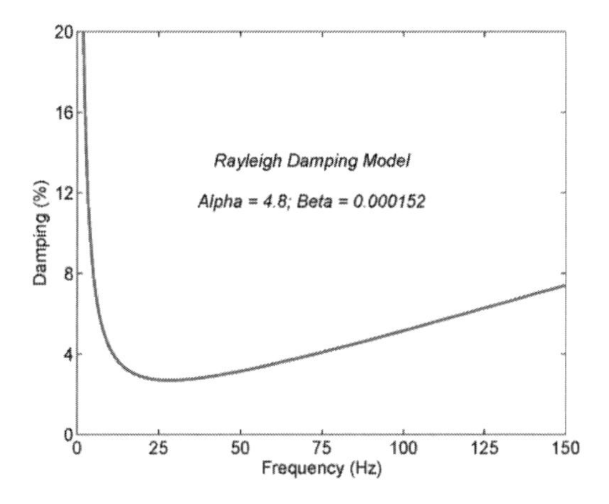

Figure 1. Rayleigh damping for ground.

Table 1. Elastic properties of ground.

	Thickness (m)	Density (kg/m³)	E (kN/m²)	ν	ξ (%)
Layer 1	1.5	1900	110.8×10^3	0.48	3
Layer 2	1	1900	95.8×10^3	0.49	3
Layer 3	1	1900	163.7×10^3	0.49	3
Layer 4	1	1900	119.5×10^3	0.49	3
Layer 5	1	1900	145.4×10^3	0.49	3
Layer 6	1	1900	226.6×10^3	0.49	3
Layer 7	5.5	1900	339.0×10^3	0.48	3
Layer 8	18	1900	539.6×10^3	0.47	3

through the inversion of receptance tests. Detailed information about this procedure is provided in Alves Costa et al. (2012). The stiffness of railpad is 600 kN/mm, the rail is UIC-60 type and the space between sleepers is 0.6 metres.

Figure 2 shows the Rayleigh damping for sleeper, ballast and subballast, where the α and β coefficients were selected in order to obtain a "constant damping" in the range of frequencies of interest or main expected frequencies (from 10 Hz to up 50 Hz).

To perform a reliable train-track interaction analysis, the track unevenness was measured for a length of 100 m before and after the section of the track that was monitored during the passage of trains. The track unevenness profile and the power spectral density of the rail in the wavelengths range and the main properties of Alfa-Pendular can be seen in Alves Costa et al. (2012) and in Fernández et al. (2017).

3.4 *Description of the 3D numerical model*

3.4.1 *Generalities*

A relevant aspect in the modelling of moving loads through 3D FEM is the consideration of a finite distance where train loads are applied. In order to minimize the computational effort and taking into account the compromise between accuracy and computational efficiency, a length of 30 meters for the implementation of railway loads has been considered. This option was supported from previous theoretical tests which can be consulted in Fernández et al. (2017).

3.4.2 *3D numerical model form the simulation of Carregado case study*

The 3D model can see in Figure 3 and in Figure 4 where a region of dimensions 30×45 meters (the free-field measurements of the dynamic response induced by the traffic were performed for distances to the track up to 45 meters) has been considered in the numerical model where the finite element mesh is denser than the rest of the model. This leads to a lower computational effort without important loss of numerical accuracy. As it can be seen in Figure 3, among the refined region of the mesh, the model was extended in order to allow a reasonable reproduction of the travelling waves.

Table 2. Elastic properties of track elements.

	Density (kg/m^3)	E (kN/m^2)	ν	ξ (%)
Sleeper	2500	30×10^6	0.20	1
Ballast	1600	97×10^3	0.12	6
Subballast	1900	212×10^3	0.20	4

Therefore, the global dimensions of the models are: 65×70x30 m.

Figure 4 shows a schematic cross-section of the finite element mesh, where the details of the railway track are highlighted. It is worth explaining that the 3D numerical model for studying wave propagation does not include rail and railpad. Actually, the time histories of the computed loads in each sleeper, which are calculated through the 2D vehicle-track dynamic model, are then directly imposed in the 3D model developed in PLAXIS. This aspect is relevant since PLAXIS does not present adequate features to simulate the moving vehicle along the track.

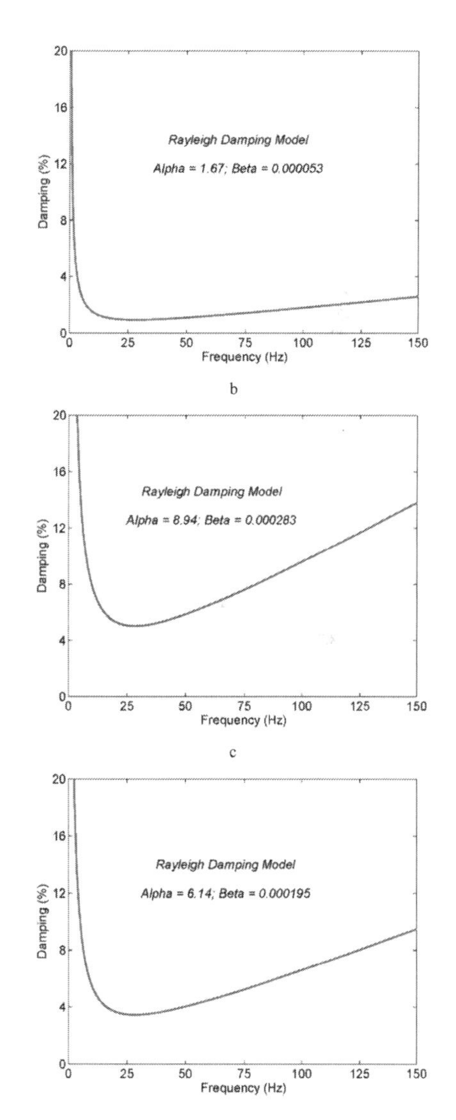

Figure 2. Rayleigh damping: (a) sleeper; (b) ballast; (c) subballast.

a

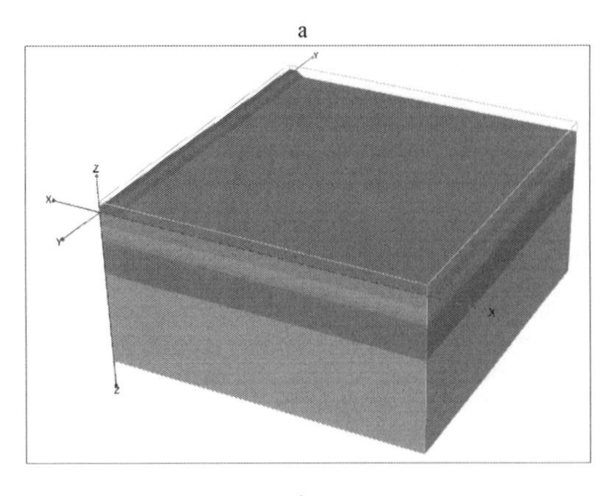

b

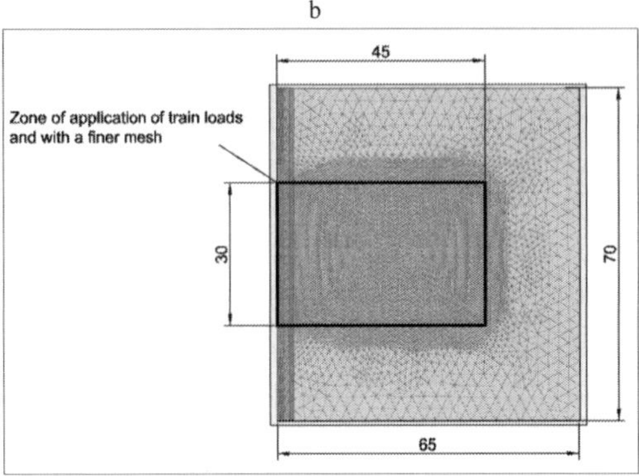

Figure 3. 3D numerical model: 3D view mode (a); plan (b, in metres).

a

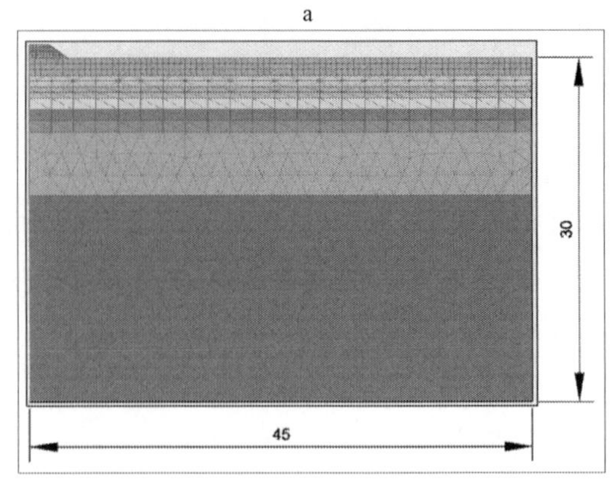

b

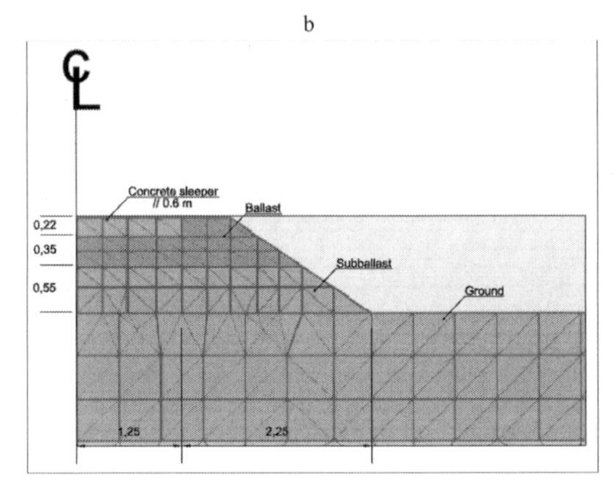

Figure 4. Finite element mesh: schematic general section (a, in metres); track (b, in metres).

The time step is chosen to ensure that a wave front does not move during a single step a distance larger than the minimum dimension of an element ($l_{e,min}$). Due to the stiffest element used in the 3D numerical model corresponds with sleeper, it determines the critical time step. This is calculated as $l_{e,min}/c_p$, that in this case is equal to 6.25×10^{-5} s.

4 RESULTS AND DISCUSSION

The time record of the vertical velocity is shown in Figure 5, where numerical prediction is overlapped with the experimental measurements. It is possible to see that the amplitude of the vertical velocity decreases with increasing distance while the duration of the event increases with increasing distance, as expected. The numerical responses show a reasonable agreement with the real measured but it is not equal at all points, emphasizing that in those closest to the track the numerical response is overestimated, while the points located at a distance greater than 22.5 meters the numerical response is quite good, except for the point at 30 meters. One possible explanation for this fact could be produced by local changes in the soil stiffness which have not been considered in the numerical model or by heterogeneities in the soil or even by some numerical amplification in a frequency range between 30–40 Hz. This can be seen in figure 6d, where a numerical amplified response can be highlighted around 30 Hz. This amplification is not noticed in real measurements.

A better comparison between numerical response and experimental results is provided in the frequency domain (Figure 6). From a general point of view, the numerical responses follow the main trends of the experimental data. Actually, the differences are generally lower than 10 dB. Exception is made for frequencies around 30 Hz to 50 Hz, where the differences are higher.

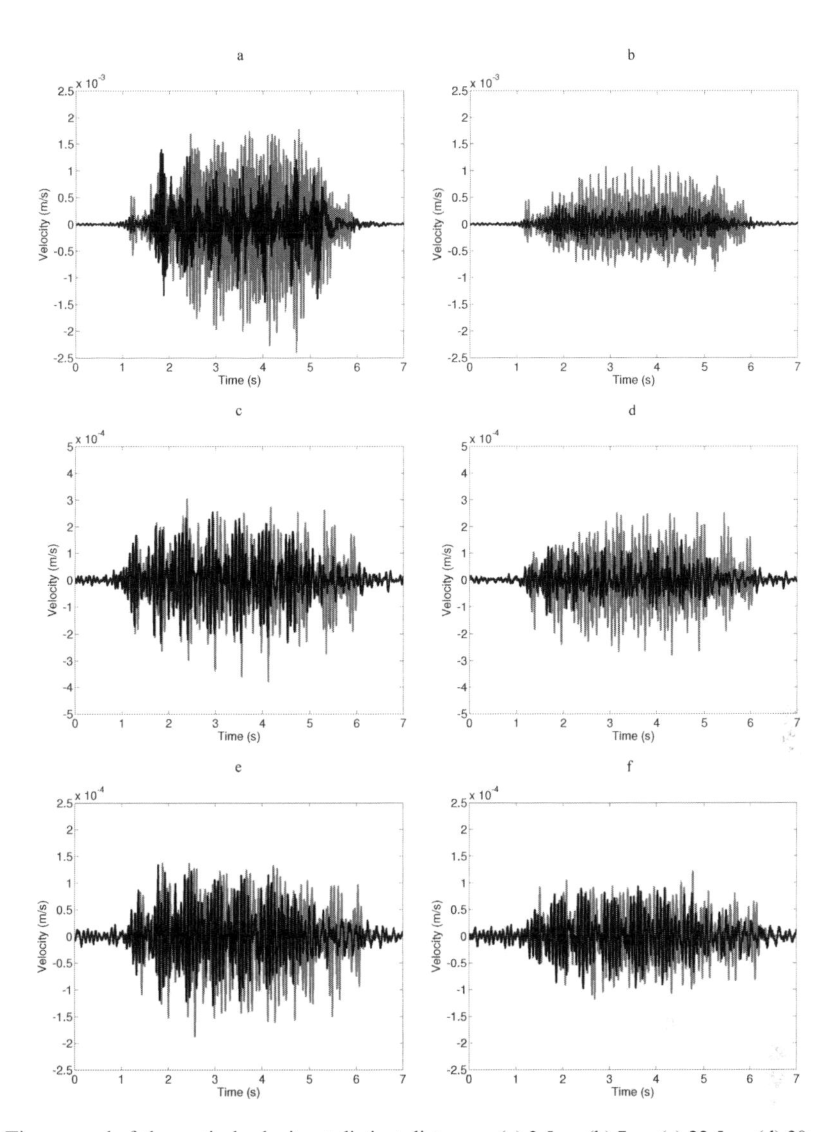

Figure 5. Time record of the vertical velocity at distinct distances: (a) 3.5 m; (b) 7 m; (c) 22.5 m; (d) 30 m; (e) 37.5 m; (f) 45 m (black line-measured; grey line-computed). Alfa-Pendular v = 164 km/h.

A possible reason could be that the geometrical damping provided by the ground was not included in the train-track interaction model, giving rise to a higher amplification of the dynamic train-track interaction loads or the uncertainty regarding the unevenness profile.

However, outside this frequency range the differences are not too relevant, being at closest points to the track the biggest differences in frequencies up to 5 Hz while the points located further from the track (from 22.5 meters) the differences are more important in frequencies above 80 Hz. One possible explanation for it is the damping model. Actually, to Rayleigh damping model is frequency-dependent, which gives rise to a strong attenuation of the vibration amplitude in the higher frequency range.

In frequencies above 70–80 Hz the possible causes of the differences, besides of the mentioned Rayleigh damping, and which are more important from 22.5 meters of track, may be the consideration of the train-track-ground dynamic interaction model, in which different excitation mechanisms such as wheel flats or welds (among others) have not been taken into account. Usually, these excitation sources generate high frequency vibrations. Also, possible heterogeneities in soil stiffness could involve some differences, having more influence at

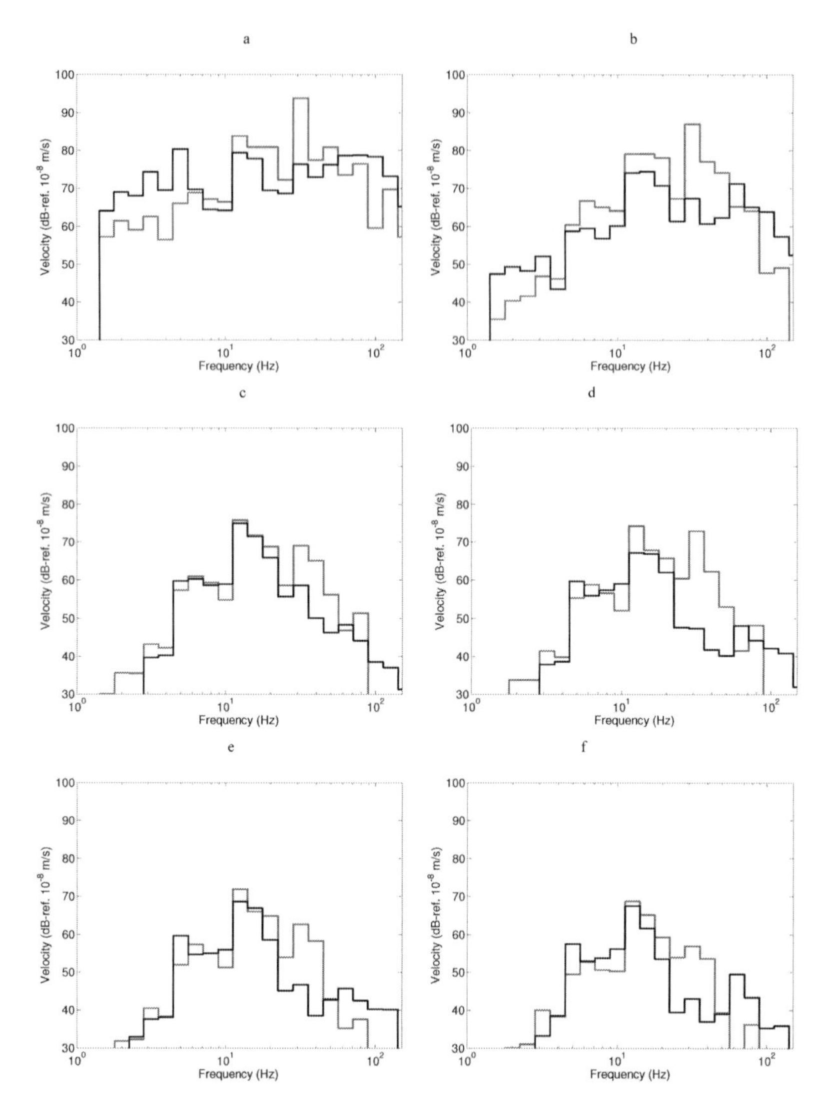

Figure 6. One-third octave spectra of the vertical velocity at distinct distances: (a) 3.5 m; (b) 7 m; (c) 22.5 m; (d) 30 m; (e) 37.5 m; (f) 45 m (black line-measured; grey line-computed). Alfa-Pendular v = 164 km/h.

high than at low frequencies. The fact of finding more differences in the high frequency range at the points located further away from the track seems to be reasonable due to the response at these points is basically controlled by the dynamic excitation, which has only been considered as generated by rail unevenness, not taking into account other possible sources.

Nevertheless, given the large number of modelling uncertainties, the numerical results show a quite good agreement with the real measurements, especially at low frequencies, finding more differences at medium-high frequencies.

5 CONCLUSIONS

This paper focuses on an experimental validation of a 3D FEM model formulated in the time domain through PLAXIS software to study ground-vibrations induced by railway traffic, finding an acceptable agreement between numerical results and real data. In this way, PLAXIS software with the proposed methodology can constitute a reliable tool to study railway vibrations. This aspect is relevant for engineering practitioners, since almost all times they don't have accesses to complex numerical approaches, which application is usually limited to academia Even so, some differences may be attributed to different causes:

- Limitations of the numerical model, such as the fact of considering a finite distance into 3D numerical model where the train loads are applied.
- A frequency-dependent damping model (Rayleigh damping) involves a loss of accuracy in some frequency ranges.
- The only source of dynamic excitation was the rail unevenness, i.e., neglecting wheel flats, welds, etc. Usually, these mechanisms generate high frequency vibrations, being a possible cause for the differences found in the frequency range around 70–80 Hz in the points located further from the track (from 22.5 meters).
- The ground is not a perfect homogeneous medium; therefore its stiffness and damping could vary locally.

Finally, it is noticeable that the computing time has been too high (0.1 hours per second for solving train-track interaction and 36 hours per second for simulating wave propagation in a computer with processor: Intel (R) Core (TM) i7-2600 K CPU @ 3.4 GHz) but the proposed numerical approach may be a good tool to simulate railway vibrations in some specific cases such as soft soils, non-linear soil behaviour, inhomogeneous geometries in the longitudinal direction of the track and transition zones.

ACKNOWLEDGMENTS

The authors also wish to acknowledge IP-Infraestruturas de Portugal, the company responsible for the management of the Portuguese road and railway networks, for all the information provided about Carregado test site and for all support given during the experimental tests.

REFERENCES

Alves Costa, P., Calçada, R., & Silva Cardoso, A. (2012). Track-ground vibrations induced by railway traffic: in-situ measurements and validation of a 2.5D FEM–BEM model. *Soil Dynamics and Earthquake Engineering*; 32:111–128.

Alves Costa, P., Calçada, R., Couto Marques, J., & Silva Cardoso, A. (2010a). A 2.5D finite element model for simulation of unbounded domains under dynamic loading. In: Benz T, Nordal S editors. *In: 7th European conference on numerical methods in geotechnical engineering*. Trondheim. p. 782–790.

Alves Costa, P., Calçada, R., Silva Cardoso, A., & Bodare, (2010b). A. Influence of soil non- linearity on the dynamic response of high-speed railway tracks. *Soil Dynamics and Earthquake Engineering*; 30:221–235.

Andersen, L., Nielsen, S.R.K. (2005). Reduction of ground vibration by means of barrier or soil improvement along a railway track. *Soil Dynamics and Earthquake Engineering*; 25:701–716.

Bahrekazemi, M. (2004). Train-induced ground vibration and its prediction. PhD Thesis, Royal Institute of Technology.

Connolly, D., Giannopoulos, A., & Forde, M.C. (2013). Numerical modelling of ground borne vibrations from high speed rail lines on embankments. *Soil Dynamics and Earthquake Engineering*; 46:13–19.

Connolly, D., Kouroussis, G., Woodward, P.K., Giannopoulos, A., Verlinden, O. & Forde, M.C. (2014). Scoping prediction of re-radiated ground-borne noise and vibration near high speed rail lines with variable soils. *Soil Dynamics and Earthquake Engineering*; 66:78–88.

Francois, S., Schevenels, M., Galvín, P., Lombaert, G., & Degrande, G. (2010). A 2.5D coupled FE–BE methodology for the dynamic interaction between longitudinally invariant structures and a layered half-space. *Computer Methods in Applied Mechanics and Engineering*; 199:1536–1548.

Fernández Ruiz, J., Alves Costa, P., Calçada, R., Medina, L., & Colaço, A. (2017). Study of ground vibrations induced by railway traffic in a 3D FEM model formulated in the time domain: experimental validation. *Structure and Infraestructure Engineering*; 13 (5): 652–664.

Galvín, P., & Domínguez, J. (2007). High-speed train-induced ground motion and interaction with structures. *Journal of Sound and Vibration*; 307: 755–777.

Hall, L. (2003). Simulations and analyses of train-induced ground vibrations in finite element models. *Soil Dynamics and Earthquake Engineering*; 23:403–413.

Kettil, P., Lenhof, B., Runesson, K., & Wiberg, N.E. (2008). Coupled simulation of wave propagation and water flow in soil induced by high-speed trains. *International Journal for Numerical and Analytical Methods in Geomechanics*; 32:1311–1319.

Kouroussis, G., Van Parys, L., Conti, C. & Verlinden, O. (2013). Prediction of ground vibrations induced by urban railway traffic: an analysis of the coupling assumptions between vehicle, track, soil and buildings. *International Journal of Acoustics and Vibration*; 18(4):163–172.

Kouroussis, G., Verlinden, O., & Conti, C. (2011). Free field vibrations caused by high-speed lines: Measurement and time domain simulation. *Soil Dynamics and Earthquake Engineering*; 31:692–707.

Madshus, C., Bessason, B., & Harvik, L. (1996). Prediction model for low frequency vibration from high speed railways on soft ground. *Journal of Sound and Vibration*; 193(1):195–203.

Mohammadi, M., & Karabalis, D.L. (1995). Dynamic 3-D soil-railway track interaction by BEM-FEM. *Earthquake Engineering and Structural Dynamics*; 24: 1177–1193.

Muller, K., & Grundmann, H., Lenz, S. (2008). Nonlinear interaction between a moving vehicle and a plate elastically mounted on a tunnel. *Journal of Sound and Vibration*; 310:558–586.

PLAXIS. (2011). Reference and scientific manual 3D. Delft (Netherlands).

Sheng, X., Jones, C., & Thompson, D. (2003). A comparison of a theoretical model for quasi-statically and dynamically induced environmental vibration from trains with measurements. *Journal of Sound and Vibration*; 267:621–635.

Sheng, X., Jones C., & Thompson, D. (2005). Responses of infinite periodic structures to moving or stationary harmonic loads. *Journal of Sound and Vibration*; 282:125–149.

Smith, M.G., Croy I., Ögren, M., & Persson Waye, K. (2013). On the Influence of Freight Trains on Humans: A Laboratory Investigation of the Impact of Nocturnal Low Frequency Vibration and Noise on Sleep and Heart Rate. *PLoS ONE*. 8(2): p. e55829.

With, C., Bahrekazemi, M., & Bodare, A. (2006). Validation of an empirical model for prediction of train-induced ground vibrations. *Soil Dynamics and Earthquake Engineering*; 26:983–990.

Yang, Y., & Hung, H. (2008). Soil vibrations caused by underground moving trains. *Journal of Geotechnical and Geoenvironmental Engineering*; 134:1633–1644.

Yang, Y., Hung, H., & Chang, D. (2003). Train-induced wave propagation in layered soils using finite/infinite element simulation. *Soil Dynamics and Earthquake Engineering*; 23:263–278.

Numerical Methods in Geotechnical Engineering IX – Cardoso et al. (Eds)
© 2018 Taylor & Francis Group, London, ISBN 978-1-138-33203-4

Probabilistic assessment of ground-vibration transfer in layered soil

L.V. Andersen & P. Bucinskas
Department of Engineering, Aarhus University, Aarhus, Denmark

P. Persson
Department of Construction Sciences, Lund University, Lund, Sweden

ABSTRACT: Risk assessment of ground vibration in early stages of design and construction rely on limited data, and usually the geological and geotechnical site characterization does not involve direct measurements of dynamic soil properties. However, combining a probabilistic approach and empirical knowledge with a semi-analytical model of ground vibration, the paper proposes a method for evaluating ground vibration based on information obtained from borehole logs. The computational model assumes horizontal stratification of soil with homogeneous material in each layer. The properties of each deposit and the layer depths are treated as stochastic variables, and correlation between the properties of each layer is accounted for. Three example cases are studied in which the transmission loss with distance from a vertical harmonic load has been determined. The proposed method is computationally efficient and may be used for preliminary assessment of the risk associated with spreading of ground vibrations from construction work or traffic.

1 INTRODUCTION

When construction work is carried out, the vibration levels are often monitored in the proximity of the construction site as a precaution from damage to nearby buildings. However, it may be difficult and expensive to place geophones in all relevant positions, and sometimes the vibration may propagate over longer distances than anticipated. Especially, weak soil layers can serve as wave guides in certain frequency ranges. If the source, e.g. pile driving, provides strong input in the same frequency range, and if this further collides with resonance frequencies of neighbouring buildings, strong vibration and ground borne noise occur in these buildings. This may annoy inhabitants, and it may possibly lead to structural damage. Similar problems may occur when heavy road traffic passes, for example, road humps or uneven pavements—or when railways pass through built-up areas.

Assessment of risk associated with ground vibration should preferably be done in the early stages of design, prior to the construction. Here, the quality and quantity of information regarding soil properties are usually low. However, judgement carried out in this phase may have higher potential for cost savings and optimal solutions than better qualified but expensive decisions and solutions performed at a later stage. Hence, the idea proposed in this paper is to use information that may be available to a contractor or consultant on the basis of typical geotechnical site investigations. This could, for example, be borehole logs performed to obtain information about the stratification and soil types in the area.

Uncertainties related to soil properties and stratification can be accounted for by stochastic analysis. In this regard, Manolis (2002) presented a thorough review concerning soil dynamics. More recently, Kreuzer et al. (2011) proposed a formulation in horizontal wavenumber domain. The layered ground was modelled, treating the stiffness of each layer as a stochastic field. While this may provide a good representation of the uncertain soil properties, it will also lead to long computation times. Fast assessment of ground vibration requires a simpler model. Vahdatirad et al. (2014) employed a stochastic process for the variation of soil properties with depth in a model of a monopile foundation for an offshore wind turbine. They assumed that the soil properties were independent of horizontal position. A further simplification was made by Vahdatirad et al. (2015) and by Damgaard et al. (2015), who studied surface footings on layered ground, considering the properties of each soil layer as stochastic variables. This approach is taken in the present work.

Section 2 presents the computational model used to calculate wave propagation in the strata. Next, Section 3 describes the methodology that has been applied for stochastic analysis. Numerical examples (denoted as cases) of three different strata with varying complexity are given in Section 4. Finally, Section 5 provides an overall discussion and conclusion.

2 GROUND VIBRATION MODEL

The present analysis of wave propagation in layered soil assumes that all layers are horizontal and that each layer consists of a homogeneous, isotropic and linear viscoelastic material. The ground surface is horizontal and free of traction except for a circular area on which a harmonically varying vertical load is applied. The steady state response of the ground can then be determined by the following approach:

1. The transfer matrices for all soil layers are found in the horizontal wavenumber–frequency domain. The matrices account for primary (P) and secondary (S) waves moving up and down in the layer.
2. Multiplication of the layer transfer matrices leads to a single transfer matrix for the entire strata.
3. Based on the boundary conditions at the top and bottom of the stratum, the Green's function for ground surface displacement is determined in the horizontal wavenumber–frequency domain for a load distributed over a circular area.
4. Double inverse Fourier transformation over the horizontal directions provides the Green's function in space–frequency domain.
5. Applying a vertical load of unit magnitude, the displacement response is evaluated at a number of positions on the ground surface.

Any number of soil layers can be included, and the model allows analysis of strata over a homogeneous half-space or fixed at the base. There is practically no limitation regarding the depth of the strata for the frequency range relevant to ground vibration.

The load can be distributed uniformly over a circular area—or it may be applied on a rigid, circular, massless plate tied to the ground surface. The latter case is computationally more time consuming, since the soil–plate interface is discretized into a number of points. The total force on the plate is then divided into distributed "bell-shaped" loads centred around the interaction points. The magnitudes of these loads must comply with the assumed rigid-body motion.

A FORTRAN code has been developed for efficient analysis of the wave propagation in the layered soil. See Section 4 for the computation times of the different example cases. The model is based on the layer transfer matrices originally proposed by Thomson (1950) and Haskell (1953). In order to counteract numerical problems with singularities of the transfer matrix arising due to combinations of high wavenumbers and deep layers, the method proposed by Wang (1999) has been used. Further details on the semi-analytical ground-vibration model can be found in the work by Andersen and Clausen (2008).

3 STOCHASTIC ANALYSIS OF SOIL STRATA

The paper aims at investigating the uncertainty of ground vibration levels (measured as transmission loss over distance) due to uncertainties of the soil properties, including the stratification. A MATLAB code has been developed for this purpose, allowing simple user definition of the soil stratification based on a library of soil types such as sand, clay, or till. The basic idea is to allow the user to provide input in terms of soil type and overall stiffness (soft, medium or hard) as well as the saturation level (dry, medium, soaked). These additional inputs are used to define a relative increase or decrease of the wave speeds and the mass density of the materials.

The random variables of the model for each layer are the shear modulus G, Poisson's ratio ν, the mass density ρ, and the loss factor η, as well as the layer depth, h. For a homogeneous half-space, h is infinite and therefore not treated as a random variable.

Two probability distributions can be chosen for each of the parameters: 1) the Lognormal distribution (default); 2) the Normal distribution. It is noted that the Normal distribution may lead to negative values of the parameters, which is obviously not possible in reality (except for auxetics, which are not relevant here). Hence, the Normal distribution is truncated at zero and the probabilities associated with positive outcomes are scaled to provide a cumulative probability of 1 as the parameter goes towards infinity. As an exception, if the Normal distribution is assumed for the layer thickness, h, the probability associated with negative layer thickness is treated as the probability that the layer completely disappears. This is deemed realistic, since the stratification may itself be uncertain. The presence of all layers in all positions within a region representative for a wave-propagation problem is not guaranteed.

Based on the mean values and standard deviations of the populations, realizations of the random variables follow a four-step procedure:

1. Five quasi-random numbers are generated based a Sobol sequence (Bratley and Fox, 1988).

2. The five quasi-random numbers are converted into five statistically independent standard Normal distributed random numbers.
3. Four of the random numbers will be used to generate values of G, ν, ρ, and η. These numbers are stored in the column vector x and turned into correlated standard Normal distributed random numbers, $\mathbf{y}$, defined as $\mathbf{y} = \mathbf{Lx}$, $\mathbf{LL}^T = \mathbf{C}$, where $\mathbf{C}$ is the covariance matrix and $\mathbf{L}$ is a lower triangular matrix found by Cholesky factorization.
4. The statistically correlated standard Normal distributed random numbers are converted into random variables following the assumed probability distributions for G, ν, ρ, η, and h.

In the present analysis, unless otherwise stated, all parameters (including the layer thicknesses) are assumed Lognormal distributed. The following coefficients of variation are assumed: $CV_G = 0.20$, $CV_\nu = 0.1$, $CV_\rho = 0.05$, $CV_\eta = 0.2$, and $CV_h = 0.1$. Further, the covariance matrix

$$C = \begin{array}{c} \\ \end{array}\begin{matrix} G & \nu & \rho & \eta \end{matrix} \\ C=\begin{bmatrix} 1.0 & 0.1 & 0.5 & -0.5 \\ 0.1 & 1.0 & 0.2 & 0.0 \\ 0.5 & 0.5 & 1.0 & -0.2 \\ -0.5 & 0.0 & -0.2 & 1.0 \end{bmatrix}\begin{matrix} G \\ \nu \\ \rho \\ \eta \end{matrix} \tag{1}$$

has been employed. The correlations indicated by Equation 1 are based on the overall assumption that a higher relative mass density of the soil will lead to higher G and ρ and to a slightly more dilatational response, i.e. a slightly higher ν. The loss factor will, on the other hand, decrease when the material becomes denser and stiffer. In any case, the layer thickness is independent of the other parameters.

As an alternative, the generation of correlated random variables could be based on Young's modulus, E, or the bulk modulus, K, instead of G. However, using a combination of, for example, G and K (or the phase velocities c_P and c_S of P and S waves) is impractical, since it must be ensured that $K > G$ (or $c_P > c_S$). This can only be guaranteed by defining a high correlation coefficient, which will, on the other hand, deteriorate the Cholesky factorization of the covariance matrix in accordance with Equation 1. On the other hand, using one of the moduli (here the shear modulus G) together with Poisson's ratio, ν, ensures a sound material behaviour.

4 EXAMPLE ANALYSES

In the following subsections, three example cases are studied. Firstly, a single layer over a half-space is considered (Case 1). Secondly, a deposit of normal consolidated clay is analysed, accounting for increase of the stiffness with increasing depth (Case 2). Finally, a multi-layered half-space with a thin peat layer below the top layer is considered (Case 3).

In all cases, the properties and stratification of the soil are presented in a table and a figure. Only the mean values, μ_h, μ_G, etc., are listed in the tables. The standard deviations are determined from the CVs given in the previous section, just above Equation 1. Subsequently, the magnitudes of vertical surface displacements obtained in the horizontal wavenumber–frequency domain are presented, including dispersion lines for the P, S, and Rayleigh (R) waves in the respective materials, based on the mean values of the properties. The results are given in the space–frequency domain for the steady state surface displacements to a vertical load applied at 25 Hz.

Finally, for each case a crude Monte Carlo simulation has been carried out with 50 realizations of the parameters, based on a Sobol sequence as described in the previous section. The results are given in terms of histograms normalized as cumulative distribution functions (CDFs) for the parameters and diagrams with the transmission loss (TL) in decibels as function of the excitation frequency, f:

$$TL_j(f) = 20\log\left(\frac{|V_0(f)|}{|V_j(f)|}\right), \quad j = 1,2,3. \tag{2}$$

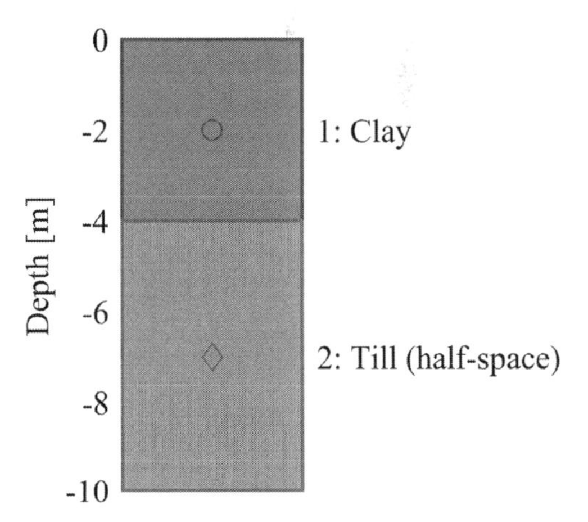

Figure 1. Case 1—stratification of the ground (with mean value of layer thickness). The colours/symbols are reused in the plots of the dispersion lines for P, S, and R waves (Figure 2) and the cumulative distribution functions (Figure 5).

Here V_0 is the vertical displacement amplitude at a reference point placed 2 m from the centre of the source, whereas V_j, $j = 1, 2, 3$, are the vertical displacement amplitude obtained at three points placed 8, 25 and 50 m from the source centre, respectively. The frequency range 0–160 Hz is considered, since it is relevant for ground-borne vibration and noise generated by traffic and construction work.

4.1 *Single clay layer over till half-space*

The material properties and stratification in Case 1 are presented in Table 1 and Figure 1, and results are shown in Figures 2–5. The calculations were done on a dual-socket workstation with 64 GB RAM and two Intel Xeon E5-2620 v3 hexacore CPUs running at 2.40 GHz. The total CPU time was about 1.3 s per frequency in calculations where the ground surface displacements were determined at 20305 observation points. Effective parallelization was achieved. Using all 24 threads, wall-clock time was only 76 ms per frequency. In the calculations performed as part of the Monte Carlo simulation, the space domain solution has only been evaluated at the reference point and three observation points. Here, the total CPU time was about 120 ms per frequency, and wall-clock time was less than 10 ms per frequency.

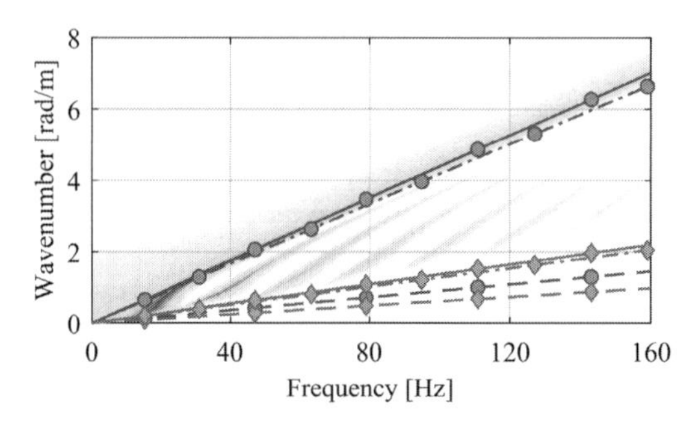

Figure 2. Case 1—dispersion with mean values of properties: Lines for R waves (—), S waves (– · – ·) and P waves (– –) in the different layers (for colours/symbols, refer to Figure 1); magnitudes from computational model with shades indicating the magnitude of vertical ground surface response to the vertical source (dark = high level, pale = low level).

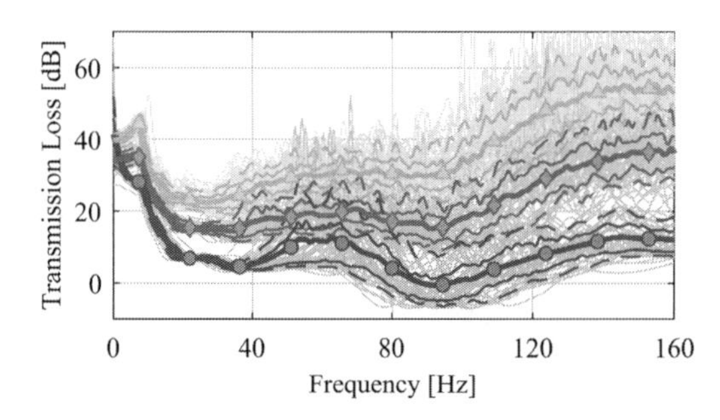

Figure 3. Case 1—transmission loss at the three points with positions and colours/symbols as shown in Figure 4. The lines indicate: Mean values (—); 25% and 75% quantiles (—); 10% and 90% quantiles (– –); individual realizations (—).

Figure 2 shows that the R wave of the till half-space dominates the surface wave propagation at frequencies below the fundamental frequency of the top layer (about 10 Hz). At very low frequencies, the top layer is much thinner than the R wavelength and basically follows the motion of the interface between the top layer and the underlying half-space. In the frequency range from about 10 Hz to 30 Hz, a transition into R waves propagating in the top layer can be seen, and with increasing frequency more wave modes cut on.

Example results of the steady state ground surface displacements at 25 Hz are shown in Figure 4. Since the excitation frequency is above the cut-on frequency, a pattern of concentric circular wave fronts can be observed with a radial distance corresponding to the Rayleigh wavelength of the clay (about 5.7 m).

Histograms showing the CDFs for 50 realizations of the properties are shown in Figure 5. The

Table 1. Case 1—properties of soil layers.

Soil type	μ_h [m]	μ_G [MPa]	μ_ν [-]	μ_ρ [kg/m³]	μ_η [%]
Clay	4.0	35	0.475	1550	3.0
Till	⊥	500	0.350	2100	2.0

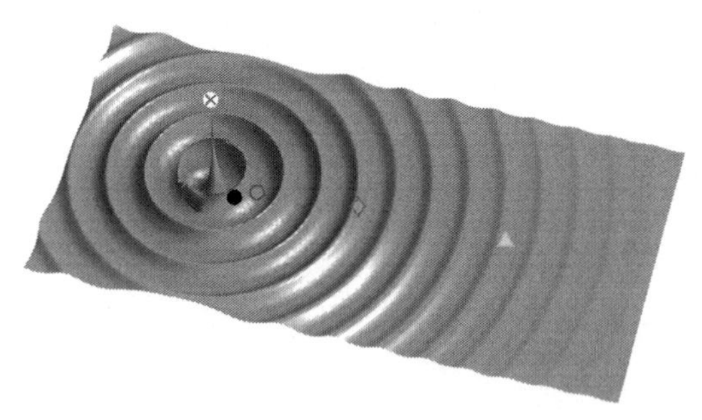

Figure 4. Case 1—surface displacement in phase with a vertical load applied at 25 Hz in the point indicated with a white bullet and a cross. The black point is 2 m from the source and serves as reference point for transmission loss. The other points are placed 8, 25 and 50 m from the source. The colours/symbols are used in Figure 3 to identify the position of points.

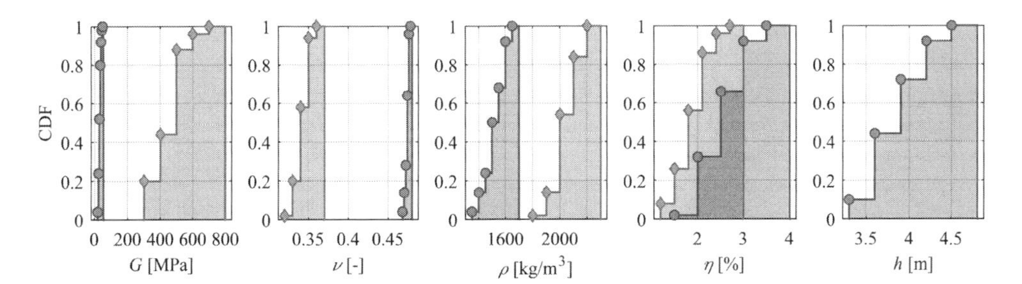

Figure 5. Case 1—histograms normalized as CDFs for the material properties and the thickness of each layer. The colours/symbols refer to the layers as shown in Figure 1 Figure 1. The histograms are obtained from Sobol samples with 50 realizations of each parameter.

results in terms of TL (cf. Equation 2) are given in Figure 3. It is recalled that reference point 0 is placed 2 m from the centre of the source, whereas observation points 1, 2 and 3 are placed 8, 25 and 50 m from the source as indicated by Figure 4. As a general observation, the TL is positive and it increases with distance from the source. At very low frequencies, TL_1 (i.e. the TL for the point 8 m away from the source) is up to 50 dB, which is actually higher than the TLs observed 25 and 50 m from the source, i.e. TL_2 and TL_3. However, with increasing frequency, TL_2 and TL_3 become larger, reaching a local peak right below the cut-on frequency of the clay layer (at 10 Hz). Beyond the cut-on frequency, TL_1 decreases to about 10 dB. For comparison, TL_2 and TL_3 are 15–35 dB and 25–50 dB, respectively. For all positions, a clear increase of the TL occurs in the range 100–140 Hz.

At frequencies near 100 Hz, negative IL occurs in many realizations for the point at 8 m distance, i.e. the vibration level is higher than that at the reference point 2 m from the source. This indicates positive interference of waves travelling along the surface and being reflected from the layer interface at about 4 m depth. However, at 50–70 Hz relatively large values are recorded for TL_1. Actually, TL_1 may be higher than TL_2 and TL_3. Thus, the 90% quantile of TL_1 is higher than the 90% quantile of TL_2. This can be explained by negative interference of waves at observation point 2. Hence, in the first example (Case 1) with a single layer of relatively soft soil on top of a stiffer half-space, the vertical response to a vertical load does not necessarily diminish with distance for all distances and frequencies.

4.2 *Layered half-space of normal consolidated clay*

The material properties and stratification in Case 2 are presented in Table 2 and Figure 6. As indicated by the histograms in Figure 10, a significant

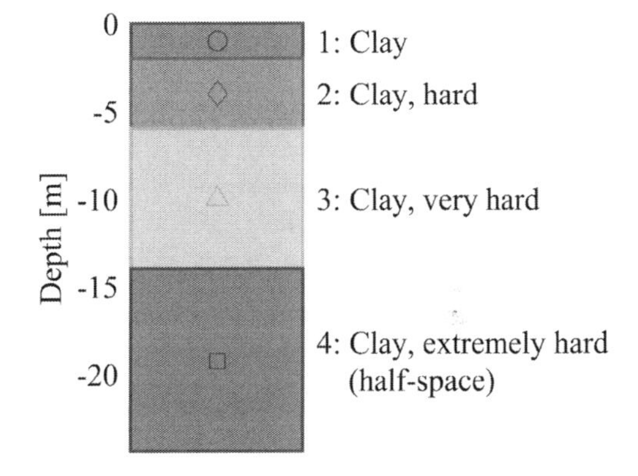

Figure 6. Case 2—stratification of the ground (with mean values of layer thicknesses). The colours/symbols are reused in the plots of the dispersion lines for P, S, and R waves (Figure 7) and the cumulative distribution functions (Figure 10).

overlap exists between the CDFs for the shear moduli of consecutive layers. Hence, the stiffness of layer 1 can, in some realizations, be higher than the stiffness of layer 2, even though layer 2 lies deeper. This may not be realistic behaviour for normal consolidated clay. An improved stochastic model would not only include correlation between the various properties of a single layer as given by Equation 2; it would also account for correlation between the properties of different layers belonging to the same formation—for example in terms of a correlation length.

The total CPU time was about 3 s per frequency when the displacements were calculated at 20305 points on the ground surface. This was reduced to about 330 ms per frequency in the Monte Carlo simulation where only the reference point and three observation points on the surface were considered. Using parallel computing, the wall-clock times were about 150 ms and 20 ms with and without evaluation of the ground surface displacements, respectively. Comparing these numbers to the computation times in Case 1, it can be seen that adding more layers to the ground model does not imply large increase of the calculation time.

The results for Case 2 are shown in Figures 7–10. Figure 7 indicates that the dispersion of waves in the soil profile consisting of clay with mean stiffness increasing over depth is simpler than that in Case 1. The primary part of the energy is contained in the surface wave which tends towards a Rayleigh wave in the top layer of soft clay as the frequency increases. For reference, the length of the surface wave is about 3 m at 50 Hz. Hence, beyond this frequency, the layers of very hard and extremely hard clay (see Table 2 and Figure 6) have nearly no influence on the motion of the ground surface. Even at lower frequencies, the response of the ground surface is largely dominated by the properties of the top layer. Not surprisingly the surface displacements at 25 Hz, shown in Figure 9, are similar to those in Case 1 (cf. Figure 4). In both cases, regular, concentric, circular, equidistant wave fronts are formed. The surface wavelength is about 5.7 m for this frequency.

Comparing Figure 10 to Figure 5, the stiffnesses (i.e. the shear moduli) of the various clay layers in Case 2 are all quite similar, whereas a strong mismatch occurs between the top clay layer and the underlying till half-space in Case 1. This explains why a cut-on frequency is not clearly present in Case 2, contrary to the situation in Case 1. Whereas a significant drop in the TL is observed at 10 Hz in Case 1, only small variations occur in the low-frequency range, below 20 Hz, in Case 2. Similar to Case 2, the TL for observation point 1 has a minimum around 100 Hz (cf. Figure 8). However, TL_1 does not become negative in as many of the realizations as it was observed for Case 1. On the other hand, the high TL obtained in Case 1 for frequencies in the range 50–70 Hz is not present for Case 2. Generally, TL_1 is smaller than TL_2 which is again smaller than TL_3.

The individual realizations of the TL in Case 2 lie closer compared to Case 1. This can be observed by comparison of Figures 3 and 8. An explanation may be that four relatively similar layers are present in Case 2, whereas two distinct layers are present in Case 1. This leads to a smaller possibility of having extreme events in Case 2. One layer may have a stiffness, for example, much higher than the mean value, but this will be counteracted by other layers having a stiffness below average for the same realization of the parameters. By contrast, the response in Case 1 is dominated by the properties of the topmost layer—primarily its shear modulus and thickness.

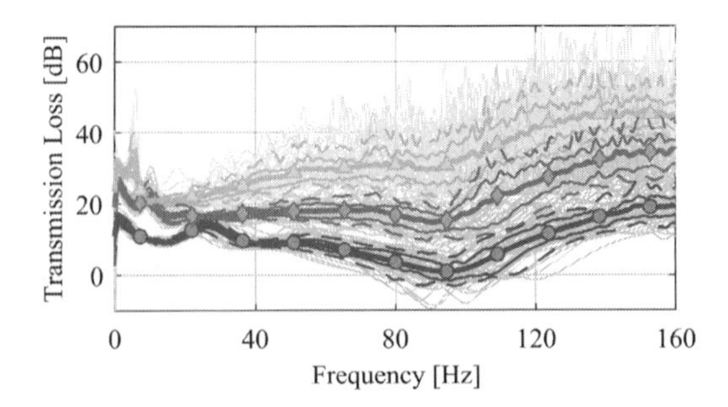

Figure 8. Case 2—transmission loss at the three points with positions and colours/symbols as shown in Figure 9. The lines indicate: Mean values (—); 25% and 75% quantiles (—); 10% and 90% quantiles (– –); individual realizations (—).

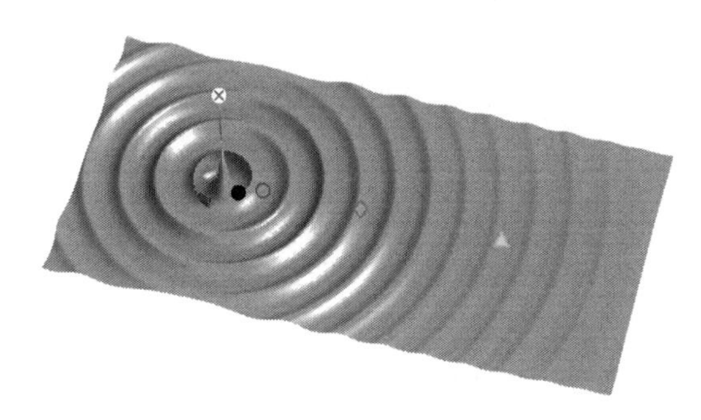

Figure 9. Case 2—surface displacement in phase with vertical load applied at 25 Hz in the point indicated with a white bullet and a cross. The black point is 2 m from the source and serves as reference point for transmission loss. The other points are placed 8, 25 and 50 m from the source. The colours/symbols are used in Figure 8 to identify the position of points.

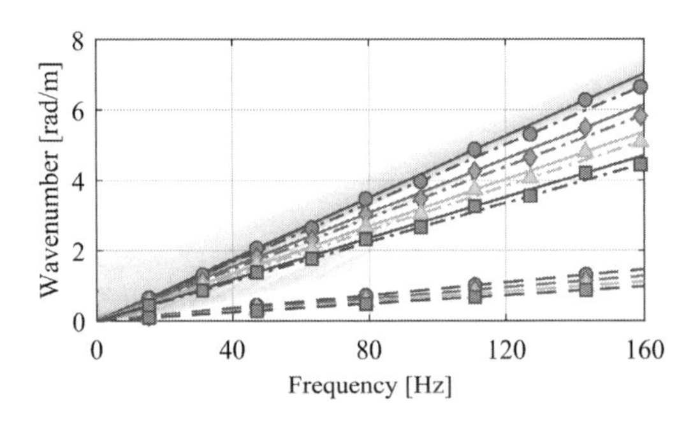

Figure 7. Case 2—dispersion with mean values of properties: Lines for R waves (—), S waves (– · – ·) and P waves (– –) in the different layers (for colours/symbols, refer to Figure 6); magnitudes from computational model with shades indicating the magnitude of vertical ground surface response to the vertical source (dark = high level, pale = low level).

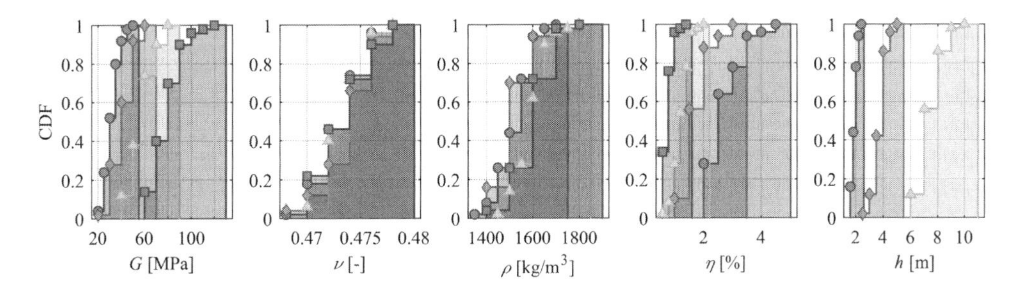

Figure 10. Case 2—histograms normalized as CDFs for the material properties and the thickness of each layer. The colours/symbols refer to the layers as shown in Figure 6. The histograms are obtained from Sobol samples with 50 realizations of each parameter.

Table 2. Case 2—properties of soil layers.

Soil type	μ_h [m]	μ_G [MPa]	μ_ν [–]	μ_ρ [kg/m³]	μ_η [%]
Clay	2.0	35.00	0.475	1550	3.000
Clay, hard	4.0	46.67	0.475	1586	2.000
Clay, very hard	8.0	62.22	0.475	1614	1.333
Clay, extremely hard (half-space)	$\perp$	82.96	0.475	1647	0.889

Table 3. Case 3—properties of soil layers.

Soil type	μ_h [m]	μ_G [MPa]	μ_ν [–]	μ_ρ [kg/m³]	μ_η [%]
Clay, soaked	4.0	40.42	0.4942	1720	3.00
Peat, soaked	1.0	3.464	0.4912	1380	5.00
Till	6.0	500	0.3500	2100	2.00
Limestone	5.0	4300	0.3500	2100	2.00
Limestone, hard	7.0	5733	0.3500	2143	1.33
Bedrock (half-space)	$\perp$	12700	0.3250	2500	1.00

In any case, similarly to Case 1, a systematic increase of the TL occurs within the frequency range 100–140 Hz for all observation points. However, the maximum TL recorded at observation point 3, 50 m from the source, is approximately 45 dB. This is about 5–10 dB smaller than observed in Case 1.

4.3 *Multi-layered ground with a thin peat layer*

As a final example, the material properties and stratification in Case 3 are presented in Table 3 and Figure 11, and results are shown in Figures 12–15. Total CPU times were 5.9 s and 600 ms per frequency for calculations with/without evaluation of the space-domain solution at the additional 20301 points used to plot the ground-surface displacements. Wall-clock times of 280 ms and 32 ms, respectively, were obtained. Comparing this to the previous cases, it can be observed that the calculation time scales linearly with the number of layers.

Figure 12 shows that the dispersion lines for S and R waves in the peat layer lie significantly above the equivalent lines for the clay layer. Even though the peat layer is only 1 m thick and lies below 4 m of clay, it has a strong influence on the ground surface displacements. Especially, at 10–60 Hz the surface wave on the layered half-space is slower than the R wave on the clay. Thus, in this frequency range it can be observed in Figure 12 that the maximum response occurs at wavenumbers slightly higher than the wavenumbers associated with the R wave on the clay. Beginning around 10 Hz, wave propagation bound in the top layer cuts on. Gradually, the surface displacement shifts towards R wave propagation in the top layer of clay. Generally, dispersion in Case 3 is more complex than observed in Cases 1 and 2 with several different propagating modes occurring in the considered frequency range.

Figure 14 shows a peculiar wave pattern at 25 Hz. Obviously, the wave fronts are still circular and concentric. However, as a result of wave interference, the shape of the waves is highly irregular. Therefore, the phase angle of the ground-surface displacement does not vary linearly with the distance away from the source. On the other hand, inspection of Figures 4 and 9 indicates that such linear variation of the phase angle with increasing distance from the source occurs in Cases 1 and 2. It is noted that more regular wave patterns occur at other frequencies.

Figure 13 shows the transmission losses calculated from the 50 realizations of the soil properties and layer depths for which histograms are provided in Figure 15. It can be seen that Poisson's ratio for the soaked clay and peat layers are significantly higher and with less variation than Poisson's ratio for the deeper lying layers. However, it is noted that even small changes in Poisson's ratio close to one

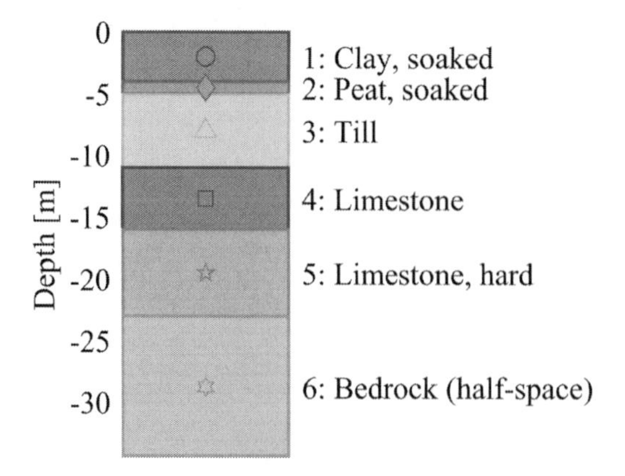

Figure 11. Case 3—stratification of the ground (with mean values of layer thicknesses). The colours/symbols are reused in the plots of the dispersion lines for P, S, and R waves (Figure 12) and the cumulative distribution functions (Figure 15).

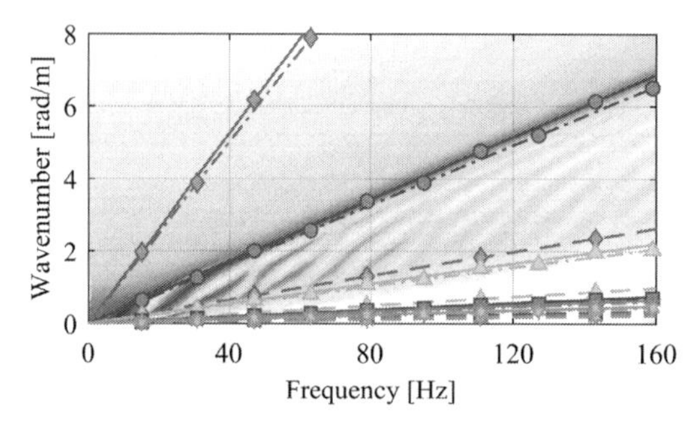

Figure 12. Case 3—dispersion with mean values of properties: Lines for R waves (—), S waves (- - - -) and P waves (– –) in the different layers (for colours/symbols, refer to Figure 6); magnitudes from computational model with shades indicating the magnitude of vertical ground surface response to the vertical source (dark = high level, pale = low level).

half leads to large differences in the P wave speed. Further it is observed that there is a distinct difference between the shear moduli of the three topmost layers (i.e. clay, peat and till) compared to the underlying layers of limestone and bedrock.

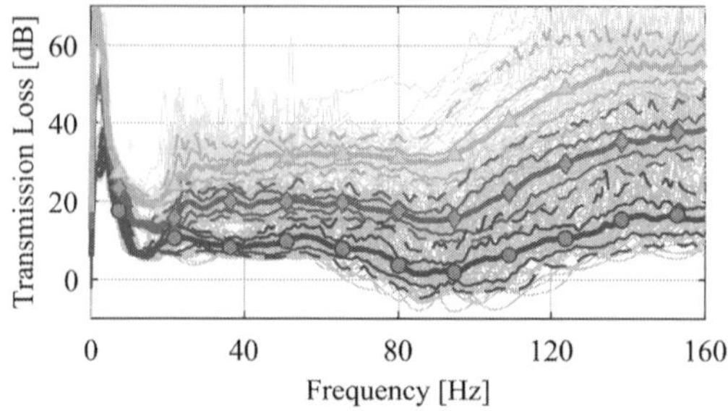

Figure 13. Case 3—transmission loss at the three points with positions and colours/symbols as shown in Figure 14. The lines indicate: Mean values (—); 25% and 75% quantiles (—); 10% and 90% quantiles (– –); individual realizations (—).

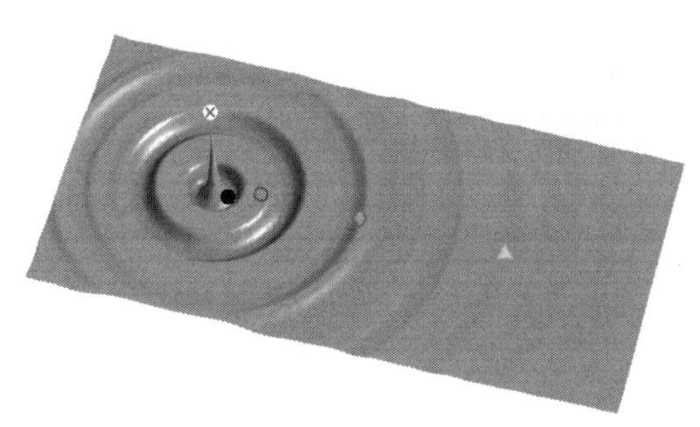

Figure 14. Case 3—surface displacement in phase with vertical load applied at 25 Hz in the point indicated with a white bullet and a cross. The black point is 2 m from the source and serves as reference point for transmission loss. The other points are placed 8, 25 and 50 m from the source. The colours/symbols are used in Figure 13 to identify the position of points.

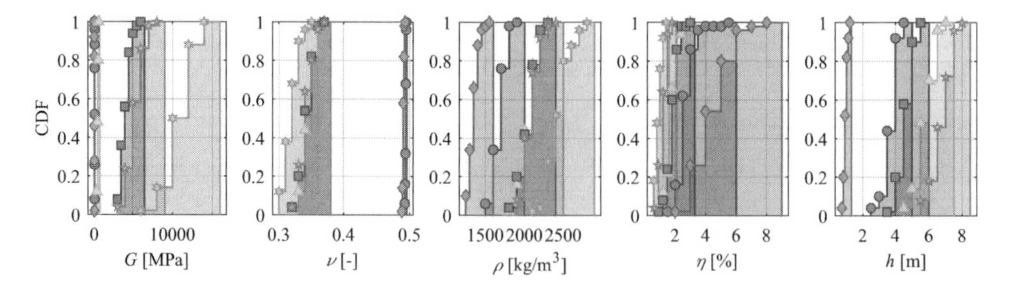

Figure 15. Case 3—histograms normalized as CDFs for the material properties and the thickness of each layer. The colours/symbols refer to the layers as shown in Figure 11. The histograms are obtained from Sobol samples with 50 realizations of each parameter.

Finally, the TL at all three observation points has a high peak (up to about 35, 50 and 65 dB, respectively) followed by a significant drop in the frequency range 5–10 Hz (down to about 10, 15 and 20 dB, respectively). This behaviour is associated with wave propagation in the clay and peat layers. At frequencies above 20 Hz, the TLs in Case 3 become more similar to those observed in Cases 1 and 2. Again, an increase of the TL can be seen in the range from 100 to 140 Hz.

5 DISCUSSION AND CONCLUSION

An efficient method for analysis of the transmission loss (TL) associated with ground vibration in layered soil has been presented, focusing on vertical loads on the ground surface. CPU times are in the order of milliseconds per frequency, which allows stochastic analyses considering the thickness and material properties of the soil layers and as random variables.

Three cases were studied: a single clay layer over a stiffer half-space, normal consolidated clay with stiffness increasing over depth, and a complex soil profile with a thin, soft peat layer. In all cases, 10%, 25%, 75% and 90% quantiles for the TL were determined by Monte Carlo simulation with 50 realizations using Sobol sampling. It has been found that the mean and quantile values of the TL change little by increasing the number of realizations.

In most cases, and for most frequencies in the range 0–160 Hz, the TLs increase with distance from the source. Above 100 Hz, the TLs increase with an increase of the frequency, but below 100 Hz no common trend has been observed. At low frequencies (0–20 Hz) the TLs are strongly influenced by the stratification of the soil but associated with relatively low uncertainty. Contrarily, at higher frequencies (100–160 Hz) the TLs are less influenced by the stratification but associated with high uncertainty. Excluding frequencies below 10 Hz, mean values of the TLs at 8, 25 and 50 m distance from the vertical load were found to be approximately 5–10 dB, 15–35 dB and 25–50 dB, with some variation from case to case. Differences between the 25% and 75% quantiles were about 5 dB for frequencies up to 100 Hz and about 10 dB in the range 100–160 Hz.

The proposed method may be extended to other vibration problems in the built environment, e.g. the influence of structural modifications to a building exposed to external vibration sources (Andersen, 2014; Persson et al., 2017). Further, if vibration levels are predicted to be too high in the early design phase of a construction project, vibration-reduction measures could be installed. Examples of such measures are: elastic mats (Sol-Sánchez et al., 2015), soil stabilization underneath a railway track or barriers along the track (Andersen and Nielsen, 2005), shaped ground surface topology (Persson et al., 2014), or wave impeding blocks (Andersen et al., 2017). These vibration-mitigation measures can be related to high costs that are desirable to avoid. Being able to present results in a statistical manner is highly beneficial in discussions with contractors regarding the location of new buildings, roads or rail lines, or if a vibration-reduction measure should be installed. In doing so, stakeholders can relate a quantified risk of disturbing vibrations to a cost. For example, a 25% risk of having vibration levels above a given threshold can be related to the cost of an appropriate countermeasure.

Finally, the proposed method can be utilized for estimation of the area within which it is necessary to monitor ground vibration during construction without the necessity of dynamic soil testing, instead relying on information from standard borehole logs.

ACKNOWLEDGEMENT

The authors gratefully acknowledge the European Regional Development Fund for financial support via the research project "Urban Tranquility" under the Interreg V programme.

REFERENCES

Andersen, L., Clausen, J., 2008. Impedance of surface footings on layered ground. Comput. Struct. 86, 72–87.

Andersen, L., Nielsen, S.R.K., 2005. Reduction of ground vibration by means of barriers or soil improvement along a railway track. Soil Dyn. Earthq. Eng. 25, 701–716.

Andersen, L.V., 2014. Influence of dynamic soil-structure interaction on building response to ground vibration, in: Numerical Methods in Geotechnical Engineering—Proceedings of the 8th European Conference on Numerical Methods in Geotechnical Engineering, NUMGE 2014.

Andersen, L.V., Peplow, A., Bucinskas, P., Persson, P., Persson, K., 2017. Variation in models for simple dynamic structure-soil-structure interaction problems, in: Procedia Engineering. pp. 2306–2311.

Bratley, P., Fox, B.L., 1988. ALGORITHM 659: implementing Sobol's quasirandom sequence generator. ACM Trans. Math. Softw. 14, 88–100.

Damgaard, M., Andersen, L.V., Ibsen, L.B., Toft, H.S., Sørensen, J.D., 2015. A probabilistic analysis of the dynamic response of monopile foundations: Soil variability and its consequences. Probabilistic Eng. Mech. 41, 46–59.

Haskell, N.A., 1953. The dispersion of surface waves on multilayered media. Bull. Seismol. Soc. Am. 43, 17–34.
Kreuzer, W., Waubke, H., Rieckh, G., Balazs, P., 2011. A 3D model to simulate vibrations in a layered medium with stochastic material parameters. J. Comput. Acoust.
Manolis, G.D., 2002. Stochastic soil dynamics. Soil Dyn. Earthq. Eng. 22, 3–15.
Persson, P., Andersen, L.V., Persson, K., Bucinskas, P., 2017. Effect of structural design on traffic-induced building vibrations, in: Procedia Engineering. pp. 2711–2716.
Persson, P., Persson, K., Sandberg, G., 2014. Reduction in ground vibrations by using shaped landscapes. Soil Dyn. Earthq. Eng. 60, 31–43.
Sol-Sánchez, M., Moreno-Navarro, F., Rubio-Gámez, M.C., 2015. The use of elastic elements in railway tracks: A state of the art review. Constr. Build. Mater.
Thomson, W.T., 1950. Transmission of elastic waves through a stratified solid medium. J. Appl. Phys. 21, 89–93.
Vahdatirad, M.J., Andersen, L.V., Ibsen, L.B., Sørensen, J.D., 2014. Stochastic dynamic stiffness of a surface footing for offshore wind turbines: Implementing a subset simulation method to estimate rare events. Soil Dyn. Earthq. Eng. 65.
Vahdatirad, M.J., Bayat, M., Andersen, L.V., Ibsen, L.B., 2015. Probabilistic finite element stiffness of a laterally loaded monopile based on an improved asymptotic sampling method. J. Civ. Eng. Manag. 21.
Wang, R., 1999. A simple orthonormalization method for stable and efficient computation of Green's functions. Bull. Seismol. Soc. Am. 89, 733–741.

Numerical Methods in Geotechnical Engineering IX – Cardoso et al. (Eds)
© 2018 Taylor & Francis Group, London, ISBN 978-1-138-33203-4

Determining the railway critical speed by using static FEM calculations

J. Estaire & I. Crespo-Chacón
Laboratorio de Geotecnia, CEDEX, Madrid, Spain

ABSTRACT: Trains may encounter a critical speed at which the soil may experience great deformations due to large dynamic amplification of the vertical motion during train passage. The demands of increasing train speeds and/or crossing soft soils have brought this problem to engineering practice. In this paper, we develop a procedure based on the "beam model" to estimate the critical speed of a rail/embankment/ground system. Inter alia, the value of the track stiffness is required as an input. Direct measurements are, however, not always available. In these cases, we take advantage of static FEM calculations to obtain this parameter. Soil non-linear behavior is considered. The developed procedure is validated by using real measurements from Ledsgard, and data from several tests performed in the CEDEX Track Box to simulate the pass-by of trains at different speeds. A set of theoretical cases are also analyzed to check the consistency of the method.

1 INTRODUCTION

Trains may theoretically encounter a critical speed when reaching the surface wave velocity of the ground. If the train speed approaches the wave propagation speed, the soil may experience great deformations that could compromise the safety and stability of the infrastructure. The train speed at which the dynamic response of the railway track and its surrounding ground is intensely amplified is called "critical speed". This dynamic amplification is due to large vibrations that appear in the system because of a "resonance-like" condition. This phenomenon was theoretically shown by Kenney (1954), Frýba (1973), Werkle & Waas (1987), De Barros & Luco (1994) and Krylov (1995).

Development of high-speed train lines is growing throughout Europe and North America. Demands on high train speeds and short travel times call for straight lines, which make, in many occasions, the crossing of soft soil zones unavoidable. Peat, organic clays and soft marine clays may have a shear wave velocity as low as 50 m/s. Problems with high-speed lines, and even traditional tracks, reaching the critical speed can therefore be expected to an increasing amount.

In this context, Dieterman & Metrikine (1997) and Wolfert et al. (1997) compiled observations by railway companies in Germany, Switzerland, Netherlands and Great Britain of substantial increases in the vertical movements in the track as the train speed approaches the surface wave velocity in the ground. Unfortunately, few data from observations, analyses and validated predictions have been reported so far in the literature, and the importance of the problem is sometimes underestimated.

This article shows how to take advantage of static calculations by using the two-dimensional (2D) finite element method (FEM) to obtain an accurate value for the critical speed of a railway track.

Section 2 provides a summary of the theoretical basics to estimate the critical speed of a rail/embankment/ground system on the base of a model that assumes the railway track as a beam on an elastic foundation ("beam model" hereafter). Section 3 shows the procedure to calculate the critical speed by using simple static calculations with PLAXIS 2D. Section 4 is devoted to validating the previous method by using (i) real measurements taken in the well-known case of Ledsgard (Sweden) (Madshus & Kaynia 2000), and (ii) data from several tests that our research team conducted in the CEDEX Track Box facility (CTB) to simulate the pass-by of trains travelling at different speeds (from 0 to 400 km/h). Conclusions are given in Section 5. Limitations of the proposed method and further developments are described in Section 6.

2 BASICS: THE BEAM MODEL APPROACH

The "beam model" assumes that the railway track can be replaced by an infinite long beam, with bending stiffness EI, resting on a viscoelastic half space, with foundation coefficient k (spring constant per unit length). The elastic foundation is assumed to be of the Winkler type, that's to say one with the foundation reaction per unit length, $p(x)$, directly proportional to the beam deflection, $z(x)$, by the factor k. Let us consider that this system is loaded by a wheel load Q according to Figure 1.

The solution of the problem of a constant point load (Q) moving at a constant speed v along this system (from $-\infty$ to ∞) is presented in this section (see Frýba 1973, Krylov 2001 and Estaire et al. 2018, for details).

The vertical rail deflection at the position x and the time t, $z(x,t)$, is described by Equation 1.

$$EI\frac{\partial^4 z(x,t)}{\partial x^4}+m\frac{\partial^2 z(x,t)}{\partial t^2}+2m\omega_N D\frac{\partial z(x,t)}{\partial t}+ +kz(x,t)=Q\delta(x-vt) \quad (1)$$

Where ω_N is the undamped natural frequency of the vibration, also known as "track-on-ballast resonance frequency"; D is the ratio between actual damping and critical damping (damping ratio); and m is the effective track mass per unit length, whose exact meaning is discussed later.

Solution of Equation 1 for $Q = 0$ and $D = 0$ leads to the frequency-dependent phase velocity of track waves propagating in the system, v_{tw} (see Equation 2, where h is the wavenumber):

$$v_{tw}=\sqrt{\frac{k+h^4 EI}{h^2 m}} \quad (2)$$

The velocity v_{tw} has a minimum at $h = (k/EI)^{1/4}$, referred as "track critical velocity" (v_{cr}, Equation 3):

$$v_{cr}=\sqrt[4]{\frac{4kEI}{m^2}} \quad (3)$$

The analysis of the static case leads to the relationship between k and the track stiffness, K (Equation 4). The value of K can be calculated by using Equation 4, where z_0 is the vertical rail deflection in the position in which the static load is being applied.

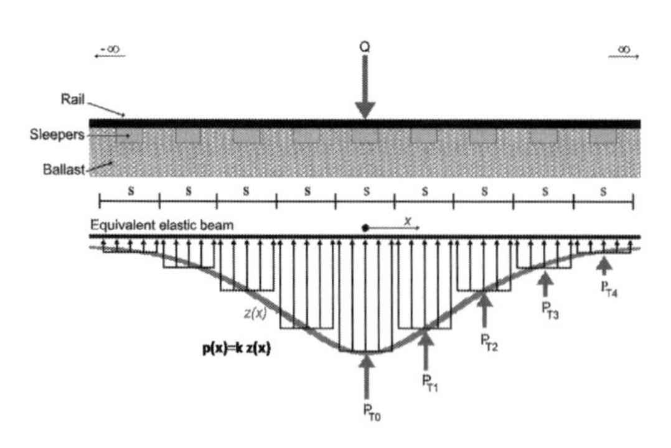

Figure 1. Infinite beam on elastic foundation model.

$$k=\sqrt[3]{\frac{K^4}{64EI}},\ \text{being}\ K=\frac{Q}{z_0} \quad (4)$$

The boundary and initial conditions for finding the general solution of Equation 1 are: at infinite distance to the right and to the left of force Q, the deflection, the slope of the deflection line, the bending moment and the shear force are zero for every time.

The solution of Equation 1 provides the dependence of the rail deflection on the load speed. The relation between these two parameters is shown in Figure 2 for different values of D. From this figure, it can be stated that the parameter v_{cr} is in fact a real critical speed, in the sense that when the velocity of the load approaches that speed, the deflection amplification cannot be neglected. The lower the damping ratio, the higher the dynamic amplification factor.

3 THE PROCEDURE TO DETERMINE THE RAILWAY CRITICAL SPEED

The procedure mainly consists on calculating the critical speed by using Equation 5, which is obtained from Equations 3 and 4.

$$v_{cr}=\sqrt[6]{\frac{K^2 EI}{m^3}} \quad (5)$$

Therefore, the calculation of the critical speed of a railway track requires an accurate determination of EI, m and K.

3.1 *Beam bending stiffness (EI)*

In this approach, as suggested by Fortin (1983), the rail is the only element that provides the bending stiffness of the beam.

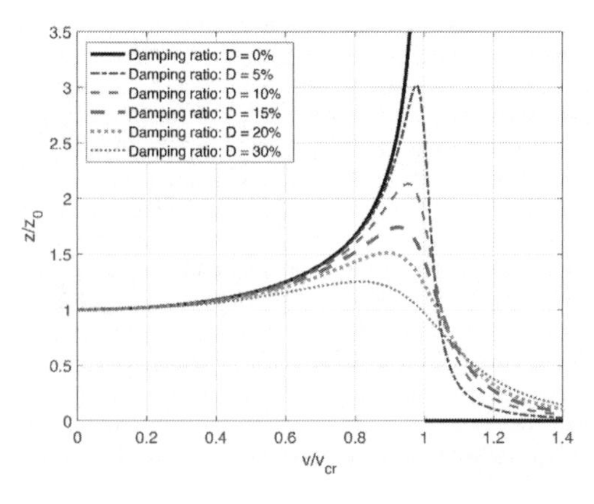

Figure 2. Normalized displacement vs. dimensionless train velocity for different values of the damping ratio.

In the cases analyzed in this paper, the 60 E1 type rail was used so $EI = EI_{rail} = 6415.5$ kN·m^2.

3.2 *Effective mass per unit length (m)*

The effective mass per unit length is the sum of the three following magnitudes[1], as suggested by Fortin (1983) and corroborated in this work with real data:

- Rail mass per unit length ($m_{rail,ef}$).
- Sleeper mass per unit length ($m_{slpr,ef}$): is the ratio of half of mass sleeper (m_{slpr}) to the distance between the center of contiguous sleepers (s).
- Ballast mass per unit length ($m_{balt,ef}$): it is the part of the ballast layer directly loaded, which can be calculated with Equation 6, where B_{balt} = half width of the ballast layer that is directly loaded; H_{balt} = ballast layer height; and d_{balt} = ballast density. The sleeper width in railroad systems throughout Europe is quite similar, so in this work we assume the one used in Spain: $B_{balt} = 1.15$ m.

$$m_{balt,ef} = B_{balt} \cdot H_{balt} \cdot d_{balt} \tag{6}$$

3.3 *Track stiffness (K)*

The track stiffness is the rate between a static point load and the vertical rail deflection in the position in which the load is applied, as shown in Equation 4.

There are two possible cases:

a. When real measurements of wheel load and its corresponding rail deflection are available, the track stiffness can be directly obtained.
b. When real data are not available (e.g., during the design phase of railway tracks), the rail deflection can be obtained by using FEM calculations considering only static loading.

In this context, the main objective of the FEM calculations is to obtain the deflection due to a wheel load in static conditions, in order to calculate the track stiffness.

Although the phenomenon under study is three-dimensional, a two-dimensional model is less time consuming and easier to implement. For this purpose, we have used the PLAXIS-2D v.2015software and worked under plain deformation conditions.

The constitutive material model used in the calculations is the so-called "Hardening Soil model with Small-Strain stiffness" (HSSmall hereafter).

[1] Note that calculations are made considering only one rail.

Limiting states of stress are described by means of the cohesion (c'), the friction angle (φ') and the dilatancy angle (ψ).

Soil stiffness is accurately described by:

- Using three input stiffnesses: the triaxial loading stiffness (E_{50}), the triaxial unloading stiffness (E_{ur}) and the oedometer loading stiffness (E_{oed}).
- Taking into account stress-dependency of stiffness moduli.
- Accounting for the increased stiffness of soils at small strains: i.e., at low strain levels most soils exhibit a higher stiffness than at engineering strain levels, and this stiffness varies non-linearly with strain. This behavior is described by an additional strain-history parameter and two additional material parameters: the small-strain shear modulus (G_0) and the strain level at which the secant shear modulus has reduced to about 72.2% of the small-strain shear modulus ($\gamma_{0.7}$). The non-linear deformational behavior is given by the relation between the normalized shear modulus (G/G_0) and the strain (γ). An example of this relationship is given in Figure 3. This is a very important feature to be able to model the railway track behavior under the loads applied by a moving train.

The steps followed in the calculations are:

Step 1 - Initial conditions: only the natural ground is modelled, and the in-situ stresses are calculated by the K_0 method.

Step 2 - Building the railway structure: the mesh elements corresponding to the embankment and track bed layers are activated.

Step 3 - Activation of the wheel load: the load is modelled as a uniform pressure applied on the whole sleeper length. The value of this load is equal to the reaction in the sleeper underneath the load. In this way, the role of the rail is taken into account even though the rail is not directly included in the mesh.

It is worth noting that when the wheel load is applied on the rail, the load is transferred by the rail not only to the sleeper underneath the load, but also to the neighboring sleepers. The reaction of the sleeper underneath the load (R_{s0}) can be easily calculated by using Equation 7 (Estaire et al. 2018). It can be seen that the reaction of the sleeper underneath the load depends on the track stiffness. An example for R_{s0} vs. K is plotted in Figure 4.

$$R_{s0} = Q\left(1 - \cos\left(\frac{s}{4}\sqrt[3]{\frac{K}{EI}}\right)\exp\left(-\frac{s}{4}\sqrt[3]{\frac{K}{EI}}\right)\right) \tag{7}$$

For usual high-speed lines, $K \approx 100$ kN/mm and $R_{s0}/Q \approx 40\%$.

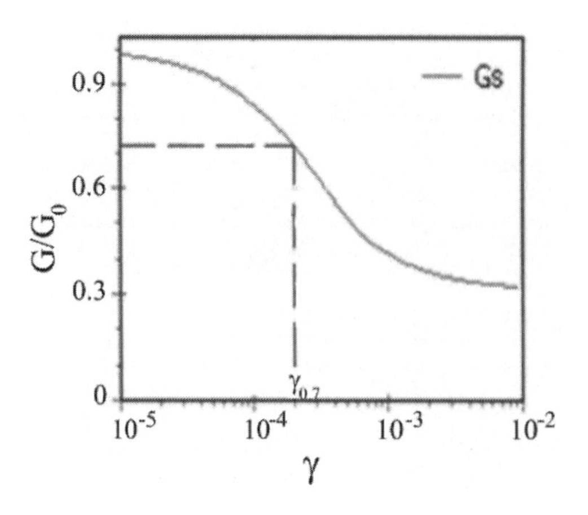

Figure 3. Normalized modulus reduction curve for typical ballast (taken from PLAXIS). *Gs* is the secant shear stiffness.

In Equation 7, the values of s and EI are a-priori known. The Q value is supposed to be equal to typical wheel loads. Values of both K and R_{s0} are unknown so we use an iterative procedure to calculate them.

An initial value for K is assumed to calculate R_{s0}. With this R_{s0} value, a first FEM calculation is done whose output will be the rail deflection in the middle of the sleeper (z_0). With the aid of Equation 4, a new value of K is calculated. The new K value is used to repeat the previous steps. Iterations are performed until convergence. Convergence is guaranteed by the fact that R_{s0} is an increasing function of K, as shown in Figure 4. For an experienced user, convergence is usually got in less than five iterations.

Note that, in any step, for K higher than the real track stiffness, the sleeper reaction determined with Equation 7 and the corresponding deflection obtained with FEM would be overestimated. On the contrary, for K lower than the real track stiffness, the sleeper reaction and deflection would be underestimated. In both cases, K is recalculated and the new K would be smaller (larger) than the previous one if the initial K were overestimated (underestimated).

4 VALIDATION OF THE PROCEDURE

To check the validity of the developed procedure (Section 3), we have used it in two real cases in which the value of the critical speed is provided by other authors and/or methods: Ledsgard (Sweden) and the CEDEX Track Box (Spain). For the sake of completeness, a set of theoretical cases has been analyzed to check the consistency of the method.

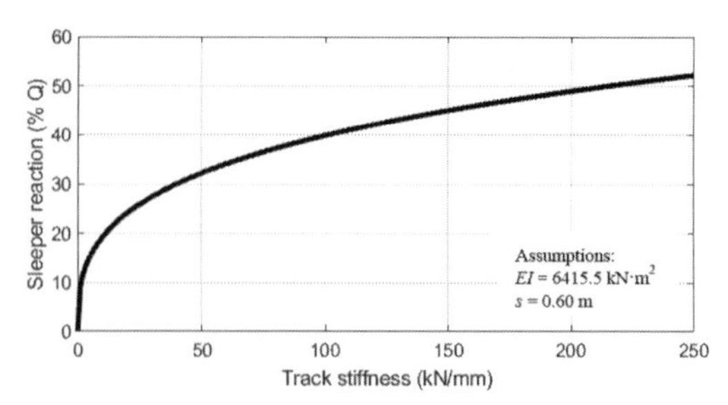

Figure 4. Sleeper reaction vs. track stiffness.

4.1 *The Ledsgard case*

In 1997 a high-speed passenger train service was opened on an existing track in Sweden. Shortly after starting the service, excessive vibrations of the railway embankment were detected at several soft soil locations during train passage at speeds of around 200 km/h. As an immediate action, the train upper speed was limited and the Swedish Railway Administration started to perform research to diagnose the problem, quantify its extent and find solutions.

An X-2000 passenger train was used for the tests. Its wheel load is about 87.5 kN for the locomotive and 60 kN for the rest of the cars. Madshus & Kaynia (2000, 2001) described the measurements, their analysis and results. The cross section of the test site is plotted in Figure 5.

Figure 6 summarizes the downward and upward displacement peaks of the track from all recorded train passages. Results from numerical simulations by Madshus and Kaynia (2000) are also shown. A thorough analysis of the recorded data reveals that the displacement pattern can be decomposed into two fields: (i) a quasi-static displacement field whose pattern and amplitude do not change with train speed (note that it contains only downward motions); and (ii) a dynamic displacement field that depends on the train speed and has equal upward and downward displacement amplitudes.

For this test site, Madshus & Kaynia (2000) concluded that the dynamic amplification of the track response above 70 km/h increases rapidly for increased train speed. 202 km/h was the highest train velocity reached during the tests; but their numerical simulations showed that the amplitude of the dynamic displacement component was expected to continue to rise further, up to train speeds of 235 km/h. For speeds higher than 235 km/h, the amplitude would decrease. The speed at which the maximum dynamic response appeared (235 km/h) was considered as the critical speed for this site. The good agreement between the measured and calculated data provides confidence in the reliability of their physical model.

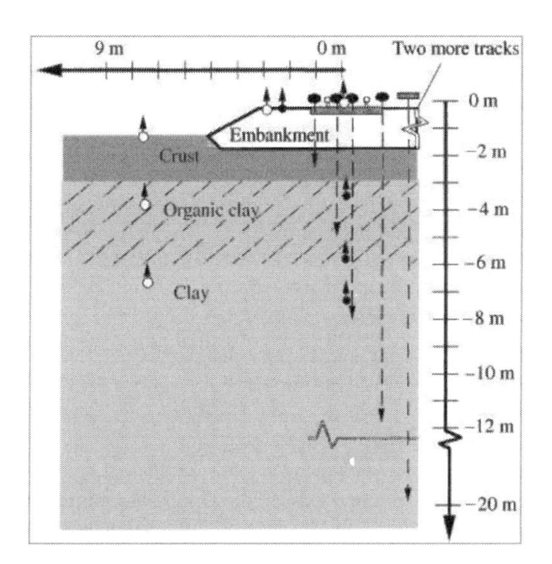

Figure 5. Cross section of Ledsgard site (from Madshus & Kaynia, 2000).

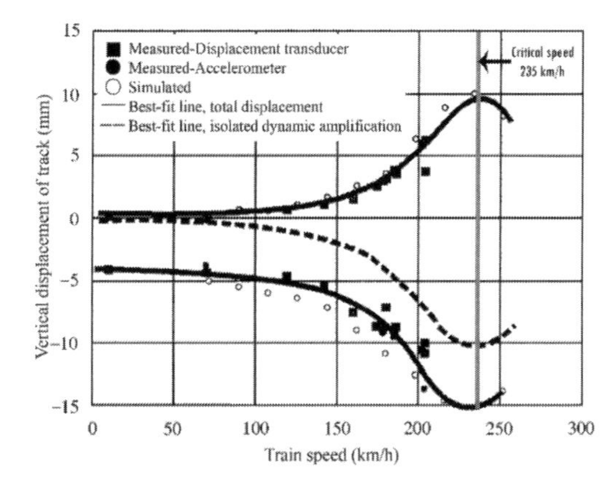

Figure 6. Peak vertical displacement of track vs. train speed: data and simulations (adapted from Madshus & Kaynia, 2000).

4.1.1 *Critical speed inferred from the normalized displacement—velocity relationship*

According to Figure 2, the measured data shown in Figure 6 can provide an estimation for the critical speed. If the vertical deflection is normalized to the static or quasi-static value, the critical speed will be that one which best fits the normalized data (both deflection and train speed) to curves in Figure 2. In this way, not only the critical velocity is inferred but also the damping ratio is estimated.

For these data, Figure 7 provides a critical speed ≈ 215 km/h, which is 91% the value reported by Madshus & Kaynia; and damping ratio ≈ 5%.

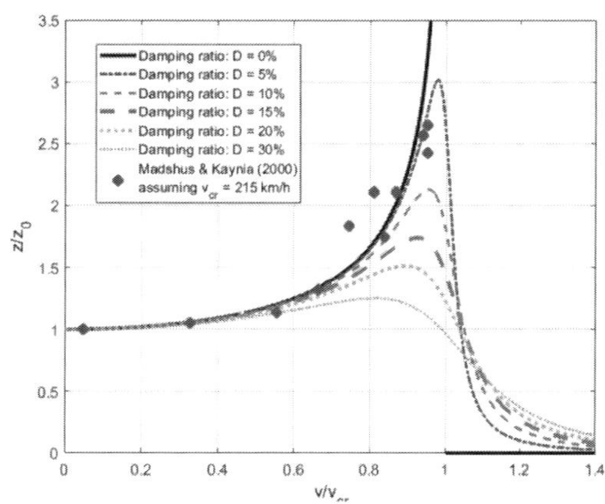

Figure 7. Normalized displacement vs. dimensionless train velocity for the Ledsgard data together with theoretical curves.

4.1.2 *Critical speed inferred from the procedure*

As the wheel load and the quasi-static rail deflection are known for Ledsgard, the track stiffness can be directly calculated (Equation 4) and the critical speed can be estimated by using the input data in Table 1 and Equation 5. Results for both the locomotive and a typical wagon of an X-2000 train are also shown in Table 1.

If rail deflections were not available, the procedure described in Section 3.3 (based on 2D FEM calculations by PLAXIS) may be used.

The geometry and geotechnical properties considered in the 2D FEM models of Ledsgard imitate those of the real site. The geometry has been taken from Figure 5. The phreatic level has been situated at the ground surface. The values of the geotechnical parameters of the layers are listed in Table 2. Some of these parameters were introduced in Section 3.3, except for: unit weight (γ_{ap}), shear wave velocity (Vs), Poisson's ratio (v) and Young's modulus (E_0). The rest of the geotechnical parameters were calculated using Equations 8 to 12, which are mainly based on the shear wave velocity of the different materials.

$$E_0 = 2V_s^2\gamma_{ap}(1+\upsilon) \tag{8}$$

$$E_{50} = E_0/10 \tag{9}$$

$$E_{ur} = 3E_{50} \tag{10}$$

$$E_{oed} = E_{50}(1-\upsilon)/(1-\upsilon-2\upsilon^2) \tag{11}$$

$$G_0 = E_0/(2(1+\upsilon)) \tag{12}$$

The mesh used in the FEM calculations by PLAXIS is plotted in Figure 8. The mesh is formed

Table 1. Input data to calculate the critical speed for both the Ledsgard and CTB cases, and results obtained by using the developed procedure. Results inferred from the normalized displacement—velocity relationship are also shown.

Input data			
	Ledsgard Case		CEDEX tests
EI (kN·m^2)	6415.5		6415.5
$m_{rail,ef}$ (kg/m)	60		60
s (m)	0.67		0.60
m_{slpr} (kg)	325		325
B_{balt} (m)	1.15		1.15
H_{balt} (m)	1.20		0.35
d_{balt} (kg/m^3)	1800		1600
	Locomotive	Wagon	
Q (kN)	87.5	60	79.5
z_0 (mm)	5.0	4.0	0.93
Results for masses per unit length			
	Ledsgard Case		CEDEX tests
$m_{slpr,ef}$ (kN·s^2/m^2)	0.24		0.27
$m_{balt,ef}$ (kN·s^2/m^2)	2.48		0.64
m (kN·s^2/m^2)	2.79		0.97
Results for track stiffness and critical speed from measured deflections (z_0)			
	Ledsgard Case		CEDEX tests
	Locomotive	Wagon	
K (kN/mm)	17.5	15	85.5
v_{cr} (km/h)	241	229	692
Results for track stiffness and critical speed from FEM calculations (i.e., assuming unknown z_0)			
	Ledsgard Case		CEDEX tests
	Locomotive	Wagon	
z_0 (mm)	5.1	3.3	1,01
K (kN/mm)	17	18	78.9
v_{cr} (km/h)	239	243	674
Results for critical speed from the normalized displacement—velocity relationship			
	Ledsgard Case		CEDEX tests
V_{cr} (km/h)	215		660

by 5603 triangular elements of 15 nodes, while the total number of nodes is 45715.

Figure 9 shows the results from FEM calculations by PLAXIS for both the locomotive and a typical wagon of an X-2000 train. The results for the track stiffness and the critical speed are listed in Table 1. Note that these values are very similar to those inferred from real rail deflections. This provides confidence in the reliability of 2D FEM calculations for studying this kind of problems.

The critical speed inferred from the developed procedure for Ledsgard is in the range 229–243 km/h. These values are in good agreement with the one reported by Madhus & Kaynia (2001): 235 km/h. They differ less than 8 km/h (3%).

4.2 *CEDEX Track Box tests*

CEDEX Track Box (CTB) is a facility mainly aimed at testing, at 1:1 scale, complete railway track sections of conventional and high-speed lines for both passenger and freight trains. A complete description of CTB, its operation modes, main uses and advantages are reported by Estaire et al. (2017).

The railway track response is monitored by pressure cells, displacement transducers, geophones and accelerometers. A cross section of the CTB facility is shown in Figure 10.

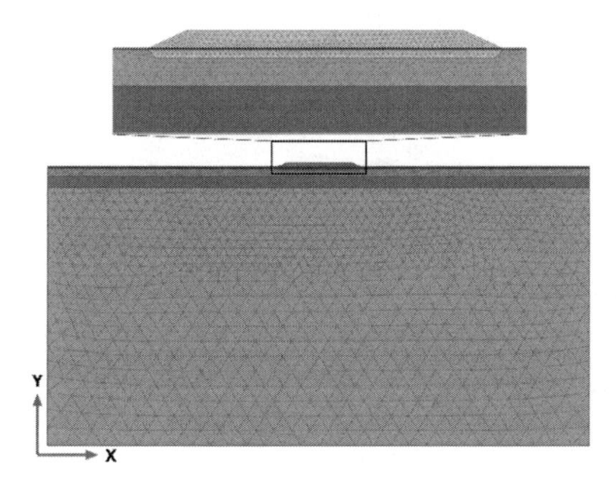

Figure 8. Mesh for the FEM model of the Ledsgard site.

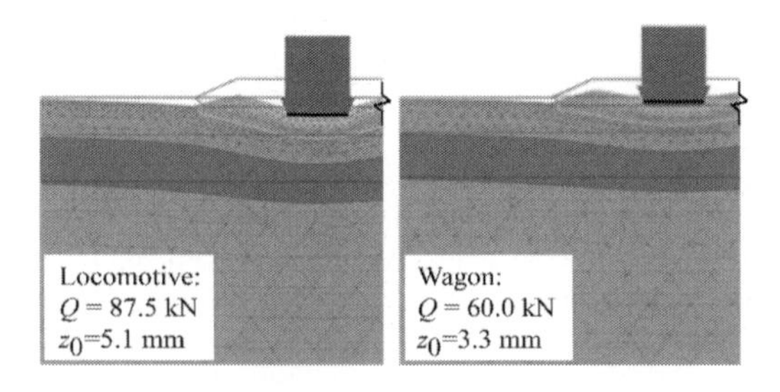

Figure 9. FEM results for the locomotive (left) and a wagon (right) loads at Ledsgard site [Deformed mesh scaled x 300].

Table 2. Values of the geotechnical parameters used in FEM calculations for Ledsgard.

Layer	γ_{ap} (kN/m³)	c' (kN/m²)	φ' (°)	ψ (°)	V_s (m/s)	ν	$\gamma_{0.7}$
Embnkmt	18	20	35	0	250	0.25	$2{\cdot}10^{-4}$
Crust	15	60	0	0	60	0.40	$10{\cdot}10^{-4}$
Org. clay	15	30	0	0	45	0.45	$20{\cdot}10^{-4}$
Clay	16	60	0	0	60	0.40	$20{\cdot}10^{-4}$

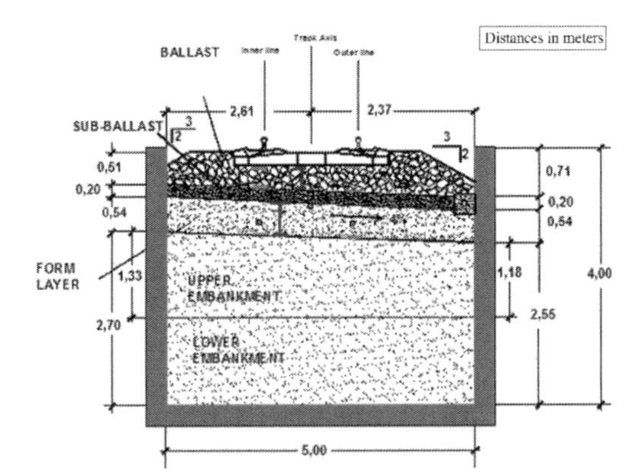

Figure 10. Cross section of the CTB facility.

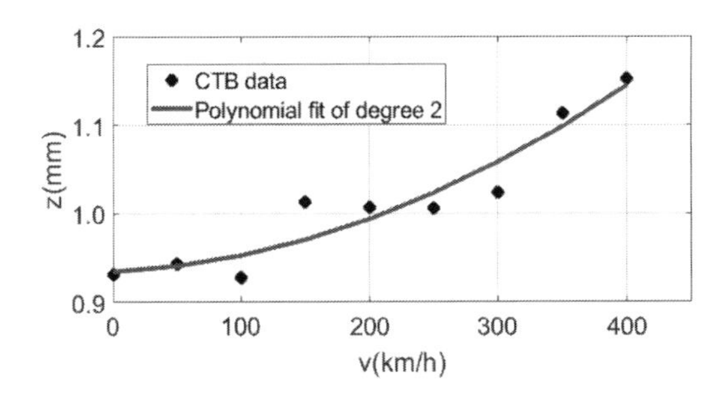

Figure 11. Rail deflection vs. train speed measured at CTB.

In October 2015some tests were performed in CTB to analyze the effect of train speed in the global response of the track. With this aim, a special Siemens S-100 train with a uniform wheel load of 79.5 kN was modelled, instead of the real wheel loads (which are in the range 68.7–85.4 kN). The modelled train speed reached up to 400 km/h. The measured rail deflections are plotted in Figure 11. The deflection can be considered to rise continuously with increased train speed, although between 150 and 300 km/h the measured deflections are almost equal. As no problems were detected in the railway infrastructure after the completion of the tests, the critical speed of CTB should be quite higher than 400 km/h.

4.2.1 *Critical speed inferred from the normalized displacement—velocity relationship*

According to Figure 12, the best fit of the normalized data (both deflection and train speed) to theoretical curves is reached for critical speed around 660 km/h. In this case, the region covered by data is not enough to discriminate between different damping ratios.

4.2.2 *Critical speed inferred from the procedure*

The same method as in Section 4.1.2 was used. Figure 13 shows the CTB geometry implemented in FEM calculations and the deformed mesh obtained, that is formed by 2745 triangular elements of 15 nodes (total of 22343 nodes). Table 3 lists the geotechnical parameters of the layers present in CTB. The rest of geotechnical parameters were calculated using Equations 8 to 12. The input data and results are given in Table 1. The critical speed inferred for CTB from FEM calculations

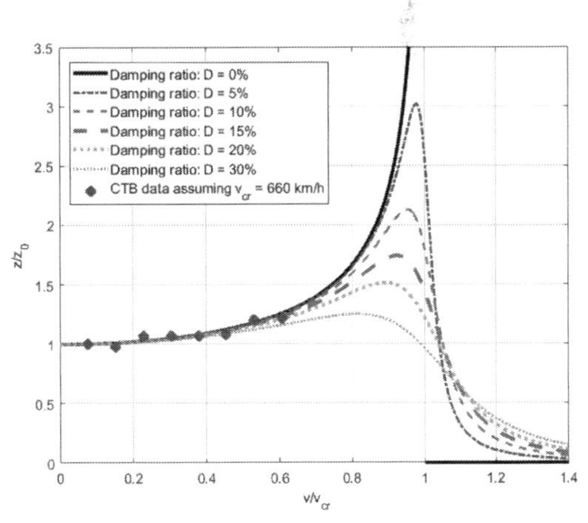

Figure 12. Normalized displacement vs. dimensionless train velocity for the CTB tests together with theoretical curves.

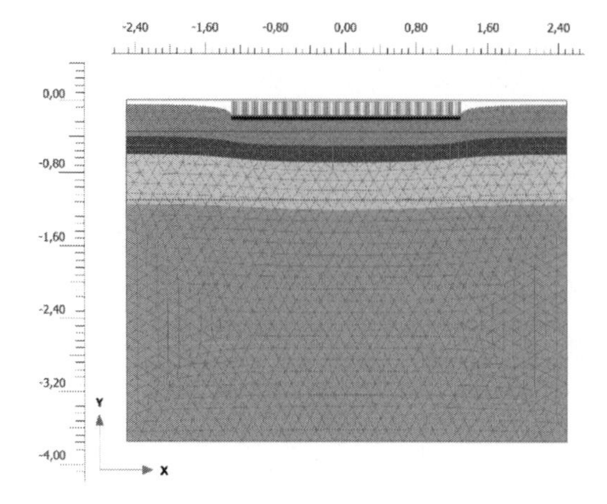

Figure 13. FEM results for CTB test [Deformed mesh scaled x 300].

is 674 km/h, only 3% lower than the value obtained by using the measured static deflection (692 km/h).

4.3 *Consistency tests*

In this section, the developed procedure is applied to a set of hypothetical cases in order to check the

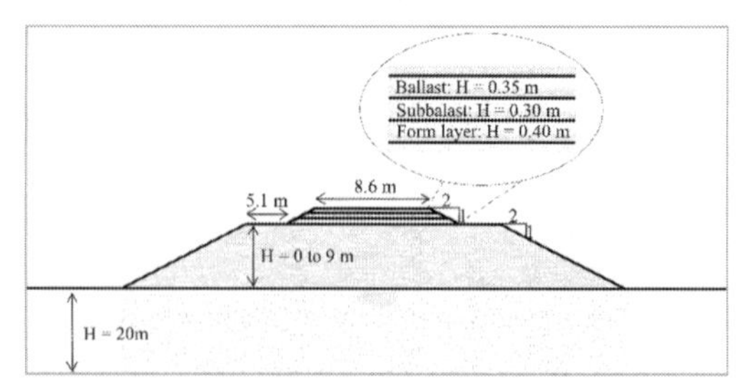

Figure 14. Schematic view for the cross-section of the railway track chosen for applying the consistency test.

consistency of the method as well as to study how the critical velocity depends on (i) the embankment height and (ii) the geotechnical properties of both the embankment and the underlying soil.

For this aim, the simple but typical geometry shown in Figure 14 was assumed for the railway track. Combinations of four different materials for the embankment (E2, E3, E4 and E5) and three for the foundation ground (N4, N5 and N6) were simulated. The geotechnical parameters assumed in all the hypothetical cases are listed in Table 4. The rest of geotechnical parameters were calculated by using Equations 8 to 12. The chosen values for the embankment height were: 0 / 1.5 / 3.9 / 4.6 / 6.1 / 9

Table 3. Values of geotechnical parameters used in FEM calculations for CTB.

Layer	γ_{ap} (kN/m^3)	c' (kN/m^2)	φ' (°)	ψ (°)	V_s (m/s)	ν	$\gamma_{0.7}$
Ballast	16	20	50	20	350	0.20	$3.5\cdot10^{-5}$
Subballast	23	10	40	10	300	0.30	$3.5\cdot10^{-5}$
Form Layer	22	20	35	5	250	0.30	$16.0\cdot10^{-5}$
Embakmt.	20	20	30	0	200	0.30	$22.0\cdot10^{-5}$

Table 4. Values of geotechnical parameters used in FEM calculations for hypothetical cases.

Layer	γ_{ap} (kN/m^3)	c' (kN/m^2)	φ' (°)	ψ (°)	V_s (m/s)	ν	$\gamma_{0.7}$
Ballast	16	20	50	20	350	0.25	$3.5\cdot10^{-5}$
Subballast	23	50	45	15	400	0.25	$3.5\cdot10^{-5}$
Form Layer	22	25	35	5	350	0.30	$16.0\cdot10^{-5}$
Embankments (E2-E3-E4-E5)							
E2	20	20	30	0	233	0.30	$22.0\cdot10^{-5}$
E3	20	20	30	0	250	0.30	$22.0\cdot10^{-5}$
E4	20	10	35	5	305	0.30	$22.0\cdot10^{-5}$
E5	20	20	40	10	366	0.30	$22.0\cdot10^{-5}$
Foundation grounds (N4-N5-N6)							
N4	20	1	28	0	152	0.35	$50.0\cdot10^{-5}$
N5	20	5	30	0	183	0.35	$50.0\cdot10^{-5}$
N6	20	7	32	0	213	0.35	$50.0\cdot10^{-5}$

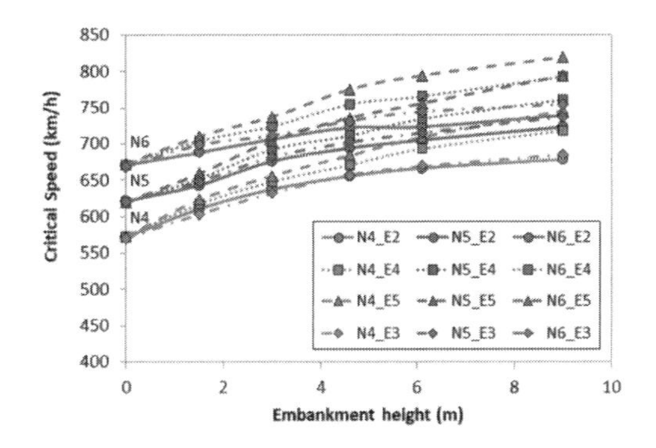

Figure 15. Critical speeds vs. embankment height for different types of embankments and foundation grounds.

.0 m. Results from FEM calculations are shown in Figure 15.

The following conclusions can be inferred from Figure 15:

- The critical velocity increases with the embankment height.
- The stiffer the foundation ground, the lower the increase of critical velocity with embankment height.
- For constant embankment height, the critical speed diminishes when the ground stiffness decreases.
- For a given foundation ground, the increase in critical velocity with embankment height is greater for stiffer embankments.

5 CONCLUSIONS

The beam model has been proved to be a simple but powerful tool to estimate the critical speed of railway tracks. It only needs three features of the track: (i) rail bending stiffness; (ii) effective mass per unit length; and (iii) track stiffness. First, as Fortin (1983) suggested, we have considered that only the rail provides the bending stiffness of the beam and that the only elements that contribute to the effective track mass are the rail, the sleeper, and a specific region of the ballast layer. It is worth noting that the dependency of the beam model on the effective or dynamic track mass limits credibility on its results because the real value of this parameter is hard to determine. Finally, 2D FEM static calculations have been demonstrated to give a reliable value for track stiffness whenever direct measurements for wheel load and rail deflection are not available.

When the normalized displacement—velocity relationships are used to obtain critical speeds, the results are only about 15–30 km/h lower than those inferred from the developed procedure, regardless of the absolute value of critical speed. The developed procedure provides critical speeds in agreement with those reported by other authors and/or methods (they differ by a factor less than 5%). In addition, the results obtained with the procedure for a set of hypothetical cases are consistent. This proves its validity to estimate the critical speed of a railway track.

6 FURTHER DEVELOPMENTS

The train-induced ground vibration is mostly governed by the interaction between the effect of the moving loads in the underneath ground and the waves generated in it, that depends on the train speed and the propagating wave velocity.

The critical speed can be calculated by using two different approaches: the one based on the "beam model" described in this paper; and another one that must be highlighted, which is based on the dispersion curve of surface waves, named as "SASW model".

While the "beam model" consists on a bending wave problem, the "SASW model" considers the critical speed to be the wave velocity of a surface wave with a frequency around the carriage passage frequency.

The comparison between the two methods seems to indicate that the "SASW dispersion curve model" is more appropriate to derive the critical velocity since the ground mechanical parameters involved in the model (mainly the shear wave velocity) better represent the physical phenomenon under study: the propagation of waves in the ground due to a perturbation in the free surface. Furthermore, this method is quite easy to implement and it does not need any further interpretation (Estaire et al., 2018).

REFERENCES

Debarros, F.C.P. & Luco, J.E. 1994. Response of a layered viscoelastic half-space to a moving point load. *Wave Motion* 19: 189–210.

Dieterman, H.A. & Metrikine, A.V. 1997. Steady state displacements of a beam on an elastic half-space due to uniformly moving constant load. *Eur. J. Mech., A/Solids* 16: 295–306.

Estaire, J., Pardo de Santayana, F. & Cuéllar, V. 2017. CEDEX Track Box as an experimental tool to test railway tracks at 1:1 scale. In *Proc. 19th International Conference on Soil Mechanics and Geotechnical Engineering*. Seoul.

Estaire, J., Crespo-Chacón, I. & Santana M. 2018. Determination of the critical speed of railway tracks based on static FEM calculations. In *Proceedings of the 6th European Conference on Computational Mechanics (Solids, Structures and Coupled Problems)*. Glasgow. In preparation.

Estaire, J., Pita, M. & Santana M. 2018. Determination of the critical speed of railway tracks based on the Spectral Analysis of Surface. In preparation.

Fortin, J.P. 1983. Dynamic track deformation. *French Railway Review*, Vol. 1,1: 3–12.

Frýba, L. 1973. *Vibrations of solids and structures under moving loads*. Noordhoof Int. Publ.. Groningen, ISBN 90 01.

Kenney JR, J.T. 1954. Steady state vibrations of beam on elastic foundation for moving load. *J. Appl. Mech. 2*: 359–364.

Krylov, V.V. 1995. Generation of ground vibrations by superfast trains. *Appl. Acoust.* 44: 149–164.

Krylov, V.V. 2001. Generation of ground vibration boom by high-speed trains. *Noise and vibration from High-Speed Trains*. Thomas Telford Publishing: 251–283.

Madshus, C. & Kaynia, A.M. 2000. High-speed railway lines on soft ground: dynamic behaviour at critical train speed. *J. Sound Vibration* 231(3): 689–701.

Madshus, C. & Kaynia, A.M. 2001. High-speed trains on soft ground: track-embankment-soil response and vibration generation. In *Noise and Vibration from High-Speed Trains*, Ed. V.V. Krylov, Thomas Telford Pub., London: 315–346.

Werke, H. & Waas, G. 1987. Analysis of ground motion caused by air pressure waves. *Soil Dyn. Earthq. Eng.* 6: 194–202.

Wolfert, A.R.M., Dieterman, H.A. & Metrikine, A.V. 1997. Passing through the "elastic wave barrier" by a load moving along a waveguide. *J. Sound Vibration* 203: 597–606.

Zimmermann, H. 1888. *Die Berechnung des Eisenbahnoberbaus*. Berlín: Verlag W. Ernst and Sohn.

Numerical Methods in Geotechnical Engineering IX – Cardoso et al. (Eds)
© 2018 Taylor & Francis Group, London, ISBN 978-1-138-33203-4

Stress path evolution in the ground due to railway traffic. Comparison between ballasted and ballastless track systems

A. Ramos & A.G. Correia
School of Engineering—University of Minho, Guimarães, Portugal

R. Calçada & P. Alves Costa
Faculty of Engineering—University of Porto Porto, Portugal

ABSTRACT: In this paper, a numerical study of the stress state evolution in the ground induced by the railway traffic is presented. The numerical models were developed considering two types of railway tracks: ballasted and ballastless tracks. Both structures were modelled in order to evaluate the differences obtained in terms of stresses under two different loading conditions: conventional passenger train (Alfa Pendular) and a freight train. The influence of the train speed and track irregularities were also studied. The numerical models were developed considering the 2.5D formulation based on FEM-PML approach. These results are a contribution to a better understanding the dynamic response of different track systems under distinct traffic scenarios and also to evaluate the comparative performance of the two systems.

1 INTRODUCTION

1.1 *Ballasted track versus ballastless track*

The behavior of the railway tracks is related directly to the performance of the subgrade, namely the stresses induced by the passage of trains. The performance of the railtrack depends on the type of the structure (ballasted or ballastless) and also on the geometry (thickness of the layers), characteristics of the materials (Young modulus, Poisson's ratio, damping ratio, etc.) since these factors will influence the stress distribution.

The ballasted and slab tracks present some differences in terms of materials and elements used but also regarding the track life cost, maintenance, environmental aspects, characteristics of the project, etc. The ballasted track is composed of the superstructure (rail, fastening system and sleepers) and by the infrastructure (ballast, sub-ballast and subgrade). The ballastless track is constituted by rail, railpad, reinforced concrete, support layer, and subgrade. Regarding the slab track, the term ballastless refers to a system where there was a replacement of the ballast by another element as asphalt or reinforced concrete. Firstly, in some systems, the sleepers were kept but posteriorly, this element disappeared. and the rails were embedded into the slab (Gautier, 2015). The ballastless track has been used, mostly in Asia (South Korea, Japan, and China) and some European Countries. The gradual change of the ballasted to the ballastless track is occurring due to the necessity of increasing the train speed, increasing of the load per axle, decrease the maintenance costs and intensify the mixed passenger-freight lines (Robertson et al., 2015).

1.2 *Stress path in the subgrade*

The performance of the subgrade depends on the type of the structure—ballasted and ballastless tracks—and has a significant importance in the global behavior of the rail track. In fact, a few decades ago, the superstructure was the major concern of the railway engineer. However, in the past years, some researchers are also focused on the performance of the substructure and on its influence on the cost of the track maintenance (Selig and Waters, 1994).

The performances of track foundations are also dependent on other factors as the complex stress conditions and rotation of the principal stresses caused by the moving loads. The shear stresses change from positive to negative as wheel passes a specific fixed location and this phenomenon influences the permanent deformation (Chan, 1990). In fact, the work presented in this article is intrinsically related to study of the permanent deformation induced by the trainload. However, in order to determine correctly the permanent deformation, it is necessary to further understand the stresses and the stress paths evolution. This is the main objective of this paper, being the study of permanent deformations treated in another paper (Ramos et al., 2018). Thus, in this work, it will be presented the stress paths obtained in the subgrade of the ballasted and ballastless tracks considering two types of loading (that includes a passenger train—Alfa Pendular—and a freight train), a range of train's speed and also tracks with different unevenness profiles. The stress path in the subgrade will be analyzed in the *p-q* space where *p* is the mean normal stress and *q* is the deviatoric stress, both defined by the following expressions:

$$p = \frac{\sigma_1 + \sigma_2 + \sigma_3}{3} \tag{1}$$

$$q = \sqrt{\frac{1}{2}} * \sqrt{(\sigma_1 - \sigma_3)^2 + (\sigma_2 - \sigma_3)^2 + (\sigma_3 - \sigma_1)^2} \tag{2}$$

where σ_1, σ_2 and σ_3 are the principal stresses.

2 NUMERICAL MODEL

The modelling of the train-track problem requires a sub-structured numerical model, which is divided into two sub-models that include the vehicle and the track-ground system. The dynamic interaction is guaranteed considering the compatibility conditions of the displacements and the equilibrium of forces that there are developed between the two domains due to the train passage.

2.1 *Modelling of the track*

The loading of a train is a tridimensional problem. The purely bidimensional models cannot represent correctly the real behavior of the structure and the ground since only consider the rotation of the principal stresses in the longitudinal plane and the analysis is limited to points on the symmetry's plane of the track (Colaço et al., 2015). Furthermore, the bidimensional models do not take into account the radiation damping since the ground geometry is not correctly attended. In order to capture the tridimensional characteristics, the 2.5D approach was used since presents several advantages when compared to conventional 3D approach. The 3D FEM is a versatile and a powerful tool (allows to use different constitutive models for each layer, complex geometries, and non-linear dynamic analysis) but it is very time-consuming. Furthermore, the definition of the boundary conditions in FEM is also difficult since it is necessary to define a finite length for the track under analysis. The 3D finite element method present advantages in case of studies related to singularities of the track as transition areas.

The 2.5D model present, clearly, advantages in the performance of some parametric studies since the method combines the good calculation processing with the consideration of the tridimensional characteristics of the problem (Yang and Hung, 2001, Alves Costa et al., 2011). The 2.5D formulation is supported, in this case, by the FEM-PML approach. This formulation demands a discretization of the cross-section and implies two essential conditions: the geometry of the cross-section needs to be invariant. The cross-section is discretized by finite elements and the analysis is carried out in the frequency-wavenumber domain, instead of the space-time domain.

The method is performed considering the use of the Fourier transform into the variables time and space (in the longitudinal direction). In the end, the response of the system is presented in the time and spatial domain as shows Figure 1.

This approach combines a numerical technique for the treatment of the cross-section and it is similar to the bidimensional finite elements method except the fact that each node has 3 degrees of freedom and variables are transformed to other domain. The equilibrium of each finite element in the ZY plane is obtained through the following expressions:

$$\left(\iint_{ZY} B^T(-k_1) D B(k_1) dydz - \varpi^2 \iint_{ZY} N^T \rho N dydyz \right) u_n(k_1, \varpi) = p_n(k_1, \varpi) \tag{3}$$

where *B* is the derivative matrix of the shape functions, *N* is the shape functions matrix, *D* is the constitutive matrix, ρ is the density and p_n and u_n are the nodal forces and displacements, respectively. The stiffness and mass matrix are defined by the expressions 4 and 5, respectively:

$$[K] = \iint_{ZY} B^T(-k_1) D B(k_1) dydz \tag{4}$$

$$[M] = \iint_{ZY} N^T \rho N dydz \tag{5}$$

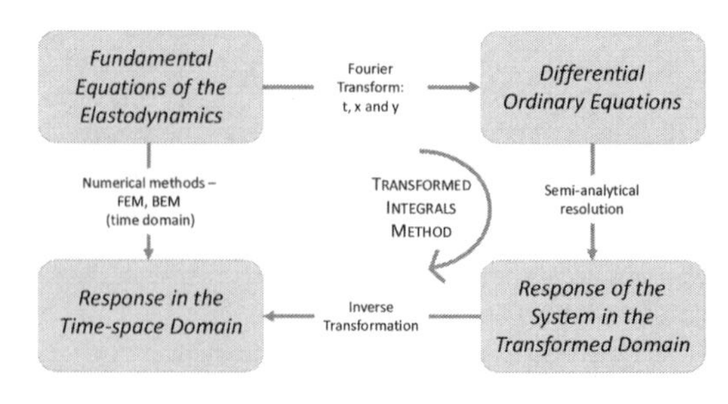

Figure 1. Transformed integrals method—adapted from Alves Costa (2011).

In the direction x (longitudinal), the derivatives are determined analytically since this direction is transformed into the wavenumber domain.

The treatment of the artificial boundaries will be performed considering the PML approach (Perfectly Matched Layers), which is a local method and also presents high acuity. This method is easy to implement and is based on the placement of an external layer to the domain of interest as shown in Figure 2.

This layer is discretized by PML elements that assure the energy absorption without wave reflection in the contact between both domains. It is important to highlight that these characteristics are assured by the stretching of the coordinates of the nodes in the PML domain to the complex domain, fulfilling the requirement of continuity of the analytical solution. Hence, there is a change in the geometric referential of the PML's domain to the complex domain. These changes are only applied in the y and z direction since the direction x is transformed into the wavenumber domain as mentioned previously (Lopes et al., 2014).

2.2 *Modelling of the vehicle*

According to the European standard ISO14837–1 (2005), there are 10 types of excitation mechanisms that include the quasi-static excitation and dynamic excitations. The last one occurs, for example, due to geometric irregularities of the track and the wheels. The quasi-static mechanism is only induced by the movement of the static weight of the train. The dynamic character is related to the temporal change of the stress and deformation experimented in a fixed point of the domain. Otherwise, the generation of the dynamic mechanism is more complex and it is induced by the vehicle-track dynamic interaction that implies the generation of inertia forces and accelerations on the vehicle. Considering this mechanism due to the geometric imperfections in the track, it is necessary to take into account the compatibility of the displacements between the vehicle and the track (assuming the linearity of the system) through the following expression:

$$u_{wheel,i}(t) = u_{rail}(x = ct + a_i, t) + \Delta u\left(t + \frac{a_i}{t}\right), \quad \forall i = 1 \ldots n \tag{6}$$

where c is the train speed, Δu is the geometric irregularity of the track identify by the wheel i in the temporal instant t and a_i is the geometric position of the wheel i in the instant $t=0$.

From this point, the transformation of the irregularity profile of the track into the wavenumber domain is required. The dynamic interaction forces in the frequency domain are determined considering the following expression:

$$N(\Omega)_n = \left([k_v]^{-1}_{nxn} + [A]_{nxn}\right)^{-1} \Delta u_n(\Omega) \tag{7}$$

where N is the vector of the interaction forces associated to each axle of the train and the irregularity with a certain wavelength ($\lambda = 2\pi c/\Omega$). The matrixes [k_v] and [A] represent the dynamic response of train and track due to the dynamic force generated by the irregularity, respectively. These matrixes are complex and are explained in the work developed by Alves Costa (2011), Alves Costa et al. (2011), Alves Costa et al. (2012), Sheng et al. (2003) and Zhai and Cai (1997).

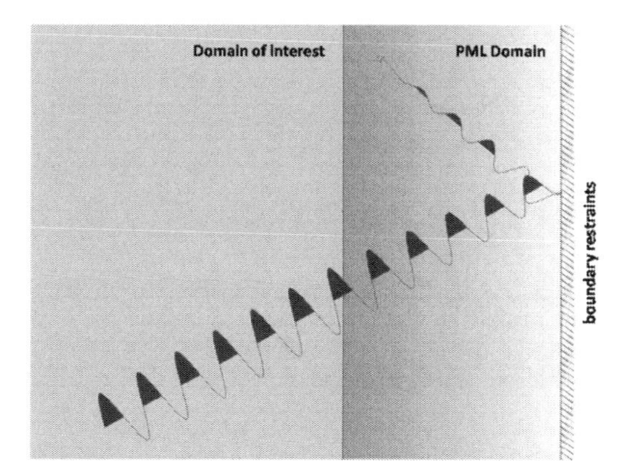

Figure 2. Wave attenuation in the PML (Lopes et al., 2014).

3 CASE STUDY

This case study compares the stress path induced in the subgrade on the ballasted and ballastless tracks. In the ballasted track, the structure is composed of the rail, railpad, ballast, sub-ballast and subgrade. In the case of the ballastless track, a typical cross-section of the Rheda system was considered, as well as its properties. The structure is composed of the rail, railpad, concrete slab, support layer designated as hydraulically bonded layer (HBL), frost protection layer (FPL) and foundation soil. The models of the structures are presented in Figure 3, as well as the identification of the elements. The cross-section of the ballasted track was defined considering the geometric properties of the North railway line of Portugal at km 41 + 600 and are in accordance with the project of this rail track. Some of the mechanical properties of the track

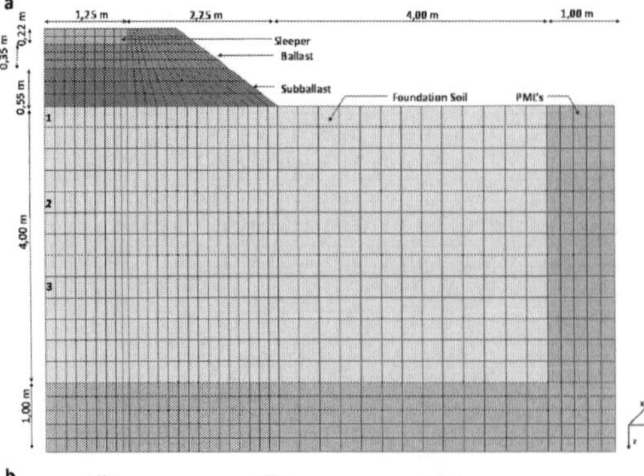

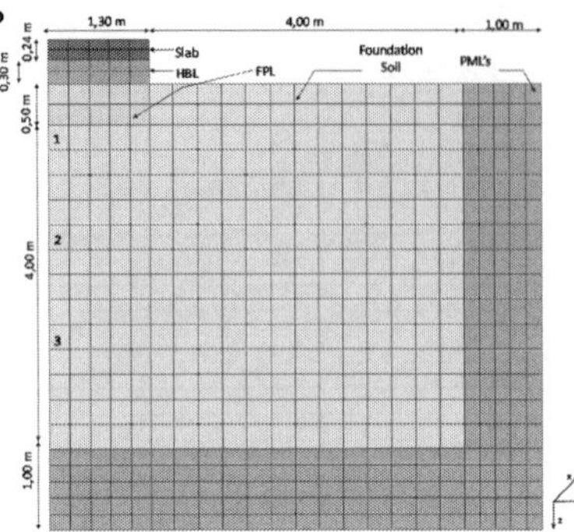

Figure 3. Representative model of the ballasted track (a) and ballastless track (b) (Ramos et al., 2018).

adopted in this study were calibrated through the receptance test, namely the Young modulus of the ballast and sub-ballast, as well as the density and damping coefficient and also the Poisson's ratio of the ballast layer. The values of the foundation soil are in accordance to the bibliography of a dry sand. Regarding the Young modulus of the ballast, the obtained values are slightly lower (97 MPa) when compared to the values described in the bibliography (110 MPa-150 MPa). The value of the density is also lower but is coherent to the value of the Young modulus (the lower the compactness material the greater its deformability). This fact is due to type of modelling (2.5D instead of 3D) since the sleepers were modelled as a continuous and as an orthotropic element (Alves Costa et al., 2010): in the longitudinal plane, it is assumed a Young modulus equal to the ballast since this is the material between the sleepers; in the cross-section, the properties of the sleepers are used. The modelling takes into account the symmetry conditions of the model in order to reduce the time and effort of the calculation. The characteristics of the materials and layers are described in Table 1 and Table 2. The subgrade was modelled considering the characteristics of a dry sand. All the materials were modelled with linear elastic models and were

Table 1. Characteristics of the rails and railpads.

Elements	Characteristics
Rails (UIC 60)	EI = 6110 kN.m^2; m = 60.445 kg/m
Railpads (ballasted track)	k = 600 kN/mm; c = 22.5 kNs/m
Railpads (ballastless track)	k = 40 kN/mm; c = 8 kNs/m

Table 2. Characteristics of the sleepers, ballast, sub-ballast, concrete slab, HBL, FPL and Foundation soil.

Elements	E (MPa)	ν	ξ	ρ (kg/m^3)
Sleepers (ballasted track)	30000	0.20	0.010	1833.3
Ballast	97	0.12	0.061	1591.0
Sub-ballast	212	0.30	0.054	1913.0
Concrete slab	34000	0.20	0.030	2500.0
HBL	10000	0.20	0.030	2500.0
FPL	120	0.20	0.030	2500.0
Foundation soil	120	0.30	0.030	2040.0

characterized by the Young modulus (E), Poisson's ratio (ν) and density (ρ).

Regarding the loading, two trains were considered: Alfa Pendular and a freight train. The Alfa Pendular is composed of 6 car bodies and presents a symmetrical configuration. The average value of the axle load is 135 kN. The freight train was modelled considering the description of the train 5 used for fatigue assessment of railway structures described in Annex D of the EN1991–2 (2003). The average value of the axle load is 225 kN.

In the first analysis, it was only taken into account the static load distribution per axle (quasi-static mechanism). Posteriorly, some analysis considering the interaction vehicle-track were performed.

The evolution of the stresses will be presented in terms effective stresses (initial and increments) in the *p-q* space. The failure line was defined considering the Mohr-Coulomb yielding criteria. In this case, it was assumed a friction angle equal to 30°, a cohesion equal to zero and a k_0 equal to one (vertical stresses equal to horizontal stresses) in order to simplify the reading of the results. The yielding criteria is merely indicative since the analysis is carried out in elastic conditions.

3.1 *Quasi-static mechanism*

The quasi-static mechanism does not take into account the train-track interaction. In Figure 4, it

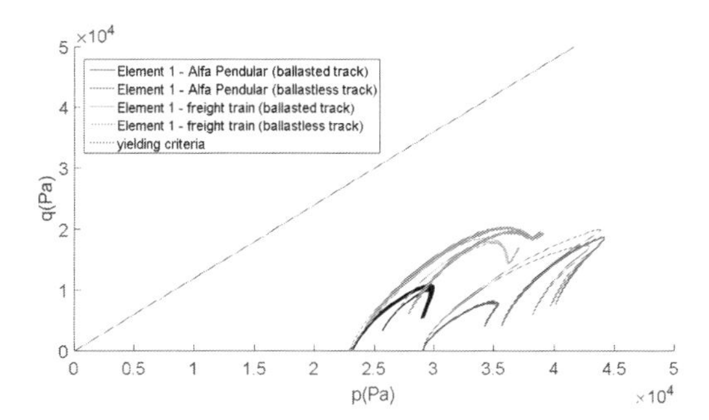

Figure 4. Stress path induced by the Alfa Pendular and a freight train at the same speed (80 km/h) considering the element 1 of each structure.

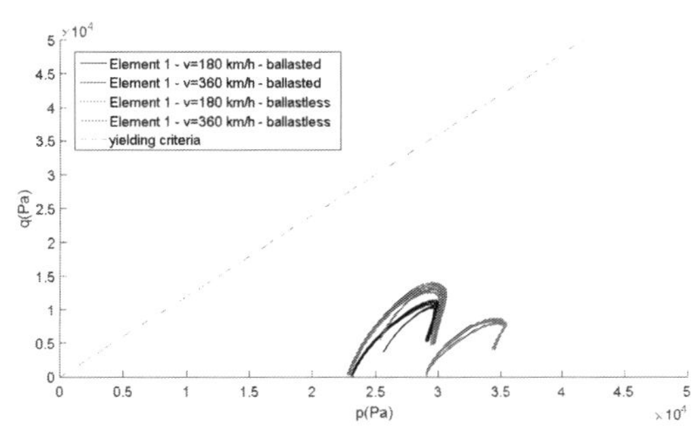

Figure 5. Influence of the train's speed in the ballasted and ballastless track (Alfa Pendular train).

is presented the evolution of the stress levels considering the same train speed (80 km/h) but different train geometry and axle loads: Alfa Pendular and a freight train. Analyzing Figure 4, the main difference between the ballasted and ballastless track is related to the initial state stress. Regarding the loading of the Alfa Pendular train, there is a slight difference in the stress paths (in terms of amplitude), superior in the case of the ballasted track. The freight train induces almost the same amplitude in the subgrade for both structures but the shape of the diagram in *p-q* space is different. In fact, the values of the maximum stresses (σ_z, σ_x, and τ_{xz}) are similar. However, in the ballasted track, the unloading between the passages of the axles is higher. This phenomenon is more significant in the elements near the surface as the element 1, represented in Figure 4. As expected, the freight train induces higher stresses in the subgrade.

Posteriorly, it was investigated the influence of the train's speed (considering only the passage of the Alfa Pendular train) in the stress path of the subgrade. Five speeds were analyzed: 80 km/h, 144 km/h, 180 km/h, 360 km/h and 540 km/h. The train's speed equal to 540 km/h is very high but not unrealistic and it is close to the critical speed since this is a homogenous scenario, which means that the stress path pattern observable in lower speeds (80, 144, 180 and 360 km/h) disappears (there is an amplification of the dynamic effects).

Evaluating the stress path induced in the subgrade, the shape, and amplitude of stresses until 180 km/h is almost equal in both structures and the stress paths are almost overlaid. Between 180 km/h and 360 km/h, it is possible to identify an increment on the amplitude of the stress level, mainly in the ballasted track, as shows Figure 5.

In Table 3 and Table 4 are analyzed the maximum value of the deviatoric and mean stress. The ratio described in both tables is defined by the relation $d_{q;máx}/d_{p;máx}$ and allows evaluating the distortion levels, for each train's speed.

Table 3. Ratio of the maximum stresses in the element 1 of the ballasted track.

Ballasted track				
v (km/h)	80	144	180	360
$d_{q;máx}$ (kPa)	10,10	11.50	11,47	14,14
$d_{p;máx}$ (kPa)	6,80	6.90	6,90	7,48
Ratio	1,49	1.67	1,66	1,89

Table 4. Ratio of the maximum stresses in the element 1 of the ballastless track.

Ballastless track				
v (km/h)	80	144	180	360
$d_{q;máx}$ (kPa)	8,32	8.63	8.36	8,56
$d_{p;máx}$ (kPa)	6,31	6.31	6.32	6,35
Ratio	1,32	1.37	1,32	1,35

Table 3 and Table 4 show that the *Ratio* is higher in the ballasted track. In fact, in the ballastless track, the distortion levels are almost constants. The differences between the two types of structures increase with the increase of the train speed, which means that the ballastless track is less sensitive to this factor. These results are related to the stiffness of each structure and the distribution of the loads and degradation of the stresses.

3.2 *Dynamic mechanism*

The dynamic excitation is due to the train-track dynamic interaction that implies the generation of accelerations and inertia forces on the vehicles. The main causes are the geometric irregularities of the wheel and/or on the track, discontinuities founded on the track and variation of the stiffness in the longitudinal direction. In this case, it was only analyzed the dynamic mechanism due to the existence of track irregularities. The interaction model used, represented in Figure 6, shows the path to obtain the train-track dynamic interaction forces.

The process is based on sub-structured models and can be used uncoupled after the determination of the dynamic interaction forces. The method presents some simplifications, namely the fact that the problem is linear (which means that there isn't any loss of contact between the wheel and the rail). The analysis will be focused on the vertical dynamic interaction forces.

The irregularities are defined by a certain number of harmonics described by a sinusoidal function:

$$y_i(x) = A_i e^{ikix} \tag{8}$$

where A_i is the amplitude of the irregularity and k_i is the wavenumber.

In this particular case, an irregularity profile considering 40 harmonic functions was considered. The irregularities' profile is represented in Figure 7. In this analysis, a certain range of frequency was defined: 2 Hz to 77 Hz, considering a train speed equal to 40 m/s.

Regarding the irregularities, the amplitude was defined according to FRA (Federal Railroad Administration), through the power spectral density:

$$S_{rzz}(k_1) = \frac{10^{-7} A k_3^2 (k_1^2 + k_2^2)}{k_1^4 (k_1^2 + k_3^2)} \tag{9}$$

where A is a parameter function of the geometric quality of the railway track (A = 37.505), k_2 and k_3 are constants (0.1464 rad/m and 0.8168 rad/m, respectively) and k_1 is the wavenumber. This study allows to understand and compare the stress path due to quasi-static (f = 0 Hz) and dynamic mechanism. In this case, the parameter A of the

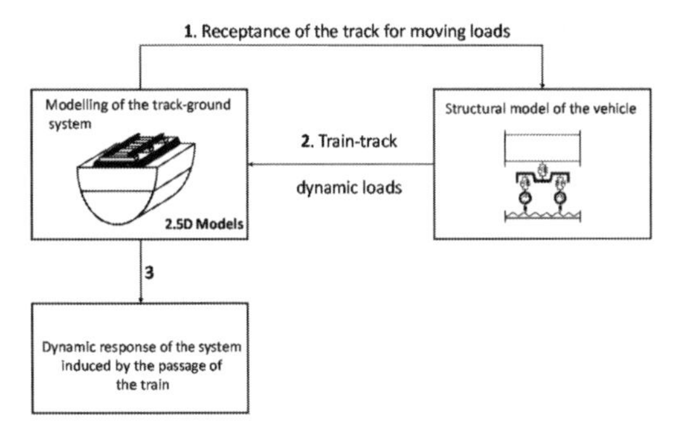

Figure 6. Flowchart representative of the sub-structure models (adapted from Alves Costa (2011).

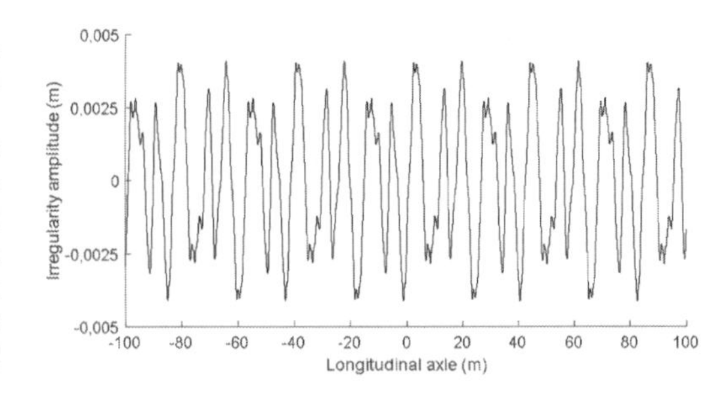

Figure 7. Irregularity's profile.

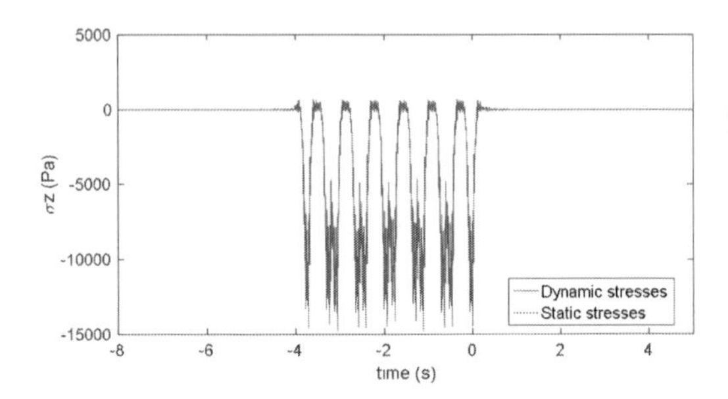

Figure 8. Vertical stresses—comparison between the quasi-static and dynamic mechanism (ballastless track).

expression (9) corresponds to the class 6 of the geometric quality of the track.

In Figure 8, it is presented the vertical stress due to the dynamic mechanism and its comparison to the quasi-static mechanism in the ballastless track (v = 40 m/s). From this Figure, it clear that the dynamic mechanism induces a significant increment in terms of vertical stresses (30%).

In further works, it will be studied the increase dynamic coefficient in the ballasted and ballastless tracks considering the same trains' speed. It is expected that the differences between the ballasted and ballastless tracks increase with the frequencies. It is also expected that the ballasted track show higher sensitivity to the existence of irregularities on track.

4 CONCLUSIONS

This article aims to compare the performance of the ballasted and ballastless tracks considering the stress path induced in the subgrade by the passage of the two types of trains. Hence, this analysis is crucial in the evaluation of the long-term performance, namely the permanent deformation of the track since it is the first step to understand the evolution of the stress level in the railway structures. Thus, two types of structures were modelled: the ballasted and slab track (Rheda system). The influence of the train loading (namely the geometry), the train's speed and the irregularities on the rail were also analyzed.

Regarding the type of train, the freight train induces, as expected higher stresses. This analysis helps to validate the model. The amplitude of the stresses does not vary according to the type of train (the values are similar for both structures). The main difference between the ballasted and slab tracks is related to the identification of the axles in the subgrade in case of the ballasted track, influencing the shape of the stress path. Furthermore, the initial mean stress is higher in the ballastless track, which is an important factor in the permanent deformation analysis as demonstrated by Ramos et al., 2018).

In the parametric analysis (influence of the train's speed), the ballasted track shows higher susceptibility because of its stiffness. The ratios $d_{q;máx}/d_{p;máx}$ are higher in this structure for each speed and the increase of the ratio with the train's speed is more notorious. When the train speed is equal to the critical speed, there is an amplification of the dynamic effects on both structures.

The dynamic mechanism modelled through the consideration of irregularities on track shows that there is a significant increase in the stress level when compared to the quasi-static mechanism and this fact should not be ignored the parametric analysis in both structures.

REFERENCES

Alves Costa, P. (2011) Vibrações do Sistema Via-maciço Induzidas por Tráfego Ferroviário. Modelação Numérica e Validação Experimental. Faculdade de Engenharia da Universidade do Porto, Porto, Portugal.

Alves Costa, P., Calçada, R. & Silva Cardoso, A. (2011) Track-ground vibrations induced by railway traffic: In-situ measurements and validation of a 2.5D FEM-BEM model Soil Dynamics and Earthquake Engineering, 32, 111–128.

Alves Costa, P., Calçada, R. & Silva Cardoso, A. (2012) Influence of train dynamic modelling strategy on the prediction of track-ground vibrations induced by railway traffic. Proceedings of the Institution of Mechanical Engineering, Part F: Journal of Rail and Rapid Transit 226, 434–450.

Alves Costa, P., Calçada, R., Silva Cardoso, A. & Bodare, A. (2010) Influence of soil non-linearity on the dynamic response of high-speed railway tracks. Soil Dynamics and Earthquake Engineering, 30, 221–235.

Chan, A.H. (1990) Permanent deformation resistance of granular layers in pavements. Dept. of Civ. Engr. University of Nottingham,England.

Colaço, A., Alves Costa, P. & Lopes, P. (2015) Análise Numérica da Alteração do Estado de Tensão Geomecânico Induzida pelo Tráfego Ferroviário. Revista Internacional de Métodos Numéricos para Cálculo y Diseño en Ingeniería, 31, 120–131.

En1991–2 (2003) Eurocode 1: Actions on structures—Part 2: Traffic loads on bridges. Annex D—Basis for the fatigue assessment of railway structures. European Committee for Standardization (CEN). Brussels.

Gautier, P.-E. (2015) Slab track: Review of existing systems and optimization potentials including very high speed. Construction and Building Materials, 92, 9–15.

Iso14837–1 (2005) Mechanical vibration—Ground-borne noise and vibration arising from rail systems—Part 1: General guidance.

Lopes, P., Alves Costa, P., Ferraz, M., Calçada, R. & Silva Cardoso, A. (2014) Numerical modeling of vibrations induced by railway traffic in tunnels: From the source to the nearby buildings. Soil Dynamics and Earthquake Engineering, Volumes 61–62, 269–285.

Ramos, A.L., Correia, A.G., Calçada, R. & Costa, P.A. (2018) Influence of permanent deformations of sub-structure on ballasted and ballastless tracks performance. Proceedings of 7th Transport Research Arena TRA 2018, April 16–19. Vienna, Austria.

Robertson, I., Masson, C., Sedran, T., Barresi, F., Caillau, J., Keseljevic, C. & Vanzenberg, J.M. (2015) Advantages of a new ballastless trackform. Construction and Building Materials, 92, 16–22.

Selig, E.T. & Waters, J.M. (1994) Track Geotechnology and Substructure Management, Thomas Telford Services Ltd., London.

Sheng, X., Jones, C.J.C. & Thompson, D.J.A. (2003) Comparison of a theoretical model for quasi-statically and dynamically induced environmental vibration from trains and with measurements Journal of Sound and Vibration, 267, 621–635.

Yang, Y.B. & Hung, H.H. (2001) A 2.5D finite/infinite element approach for modelling viscoelastic body subjected to moving loads. International Journal for NumericalMethods in Engineering 51, 1317–1336.

Zhai, W. & Cai, Z. (1997) Dynamic interaction between lumped mass vehicle and a discretely supported continuous rail track Computers & Structures, 63, 987–997.

Numerical Methods in Geotechnical Engineering IX – Cardoso et al. (Eds)
© 2018 Taylor & Francis Group, London, ISBN 978-1-138-33203-4

Geotechnical challenges in very high speed railway tracks. The numerical modelling of critical speed issues

A. Colaço & P. Alves Costa
Construct-FEUP, University of Porto, Porto, Portugal

ABSTRACT: Recent railway tracks are being designed for traffic speed larger than 350 km/h. This amazing increase of the traffic speed leads to new geotechnical challenges, since the train speed can achieve values close to the wave propagation velocity in the ground, giving rise to the so called "critical speed problem". The present paper approaches this problem focusing on geotechnical engineering. The departing point is the theoretical formulation of the critical speed problem of a moving load on the surface of an elastic solid. From the usage of 2.5D detailed models it was possible to understand the influence of the embankment and of slab track properties on the critical speed.

1 INTRODUCTION

The construction of high-railway networks across the world has brought new problems in terms of railway geotechnics (among others), namely the significant amplification of train-track vibrations at high speeds (Krylov 1995, Madshaus et al. 2004), which can compromise the safety and stability of the infrastructure. Actually, the problem of dynamic amplification of the response of elastic solids due to the increase of the speed of moving loads has been a research topic for more than a century (Lamb 1904, Kenney 1954, Dieterman & Metrikine 1997). However, if in the beginning of the last century the problem has only theoretical interest, this is not true anymore. The increase of the train speed linked with the demand of crossing of alluvionar soft soil can bring this problem to engineering practice, as already observed in the famous case study of Ledsgard, where the train speed approaches the critical speed of the track-ground system (Madshus and Kaynia 2000, Hall 2003, Alves Costa, Calçada et al. 2010). By definition, the critical speed is the velocity of the moving non-oscillating load that conducts to the higher amplification of the dynamic response, i.e., the speed of the load for which a resonant like effect is observed. Therefore, the critical speed is fully dominated by the properties of wave propagation in the embankment-ground system and of the bending wave propagation in the track, as previously discussed by Dieterman and Metrikine (1996 & 1997) and Sheng et al. (2004), among others.

Due to the practical interest of the problem, considerable research effort has been allocated during the latter decade, where several analytical and numerical approaches have been proposed for the assessment of the critical speed (Alves Costa et al. 2010, Woodward et al. 2013, Hall 2013, Sheng et al. 2004, Connolly et al. 2013, among others). So the physical phenomena concerned to the critical speed assessment is well established from the theoretical point of view, being available numerical tools that can deal with the problem. However, several aspects need to be clarified when the focus of the problem is transferred from practice, namely: i) there is a demand of simplified methods, that can be computed in an easier and faster way, for the assessment of the critical speed; ii) there is a demand of a better discernment of how the geotechnical conditions, including the embankment properties, can affect the problem. By other words, a bridge should be developed between physical concepts (in the light of wave propagation theory) and practical geotechnical railway engineering. The present paper aims to present a contribution for reaching these objectives

In the following sections the critical speed problem is revisited from the background theory of wave propagation in a halfspace. In spite of the intensive usage of advanced numerical modelling strategies for the assessment of critical speed, namely the 2.5D FEM-BEM approach, a simplified methodology, previously present by the authors in Alves Costa et al. (2015), is also used. After these two general sections, a case study is presented, where the influence of embankment properties in the critical speed is analysed. From the results obtained, it was shown that the critical speed of the system is conditioned by the layering of the ground as well as by the mechanical properties of the track-ground system. The stiffness and height of embankment also affect the critical speed value.

2 CRITICAL SPEED: PROBLEM OUTLINE

The solution for the problem of dynamic amplification of response in an elastic solid due to traffic loads has been found several years ago. However, due to the didactic value of the problem it is here briefly revisited for two distinct geotechnical conditions, as depicted in Figure 1.

It is assumed a unitary vertical load, spread over an area of 2ax2a = 2 × 2 m². The loading moves along x direction with a speed c and being it center for t = 0 s at the referential origin. In the wavenumber-frequency domain (where k1 is the Fourier image of x and ω is the Fourier image of time), it is given by:

$$p_z(k_1, y, 0, \omega) = \begin{cases} \dfrac{1}{2a}\dfrac{\sin(k_1 a)}{k_1 a}\delta(\omega - k_1 c) \, | \, y \, | < a \\ 0 \end{cases} \quad (1)$$

Which means that expression (1) only has physical meaning when $\omega = k_1 c$ corresponding to a line in the frequency-wavenumber space.

Starting with the scenario 1, the adimensional vertical displacements of the ground surface are depicted in Figure 2 for distinct values of the load speed. The load speed is presented in the adimentional form, being $M = c/C_S$.

From the analysis of the figure, it is quite obvious that when the load speed is lower than the Rayleigh wave velocity in the ground ($C_R = 0.934C_S$), the pattern of deformation of the ground surface is quite similar to the quasi-static deformation, i.e., there is not propagation of ground waves. However, when the load speed achieves this limit the waves propagate in the ground giving rise to the Mach cone, which is quite evident in Figure 2d. As expected, for this case the critical speed is coincident with the Rayleigh wave velocity in the ground. This effect can be deeper understand taking into

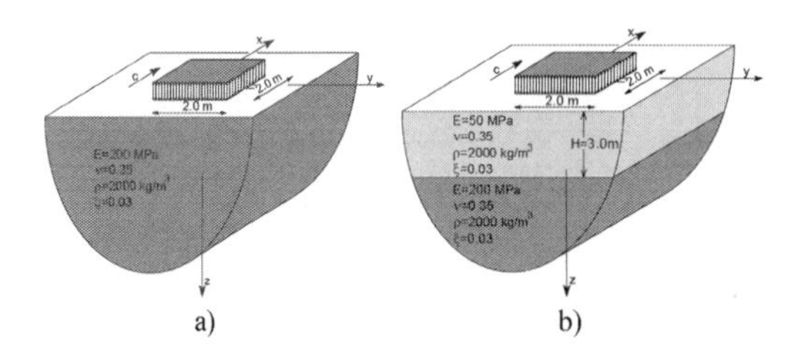

Figure 1. General characteristics of the ground considered in parametric study: a) Ground 1; b) Ground 2.

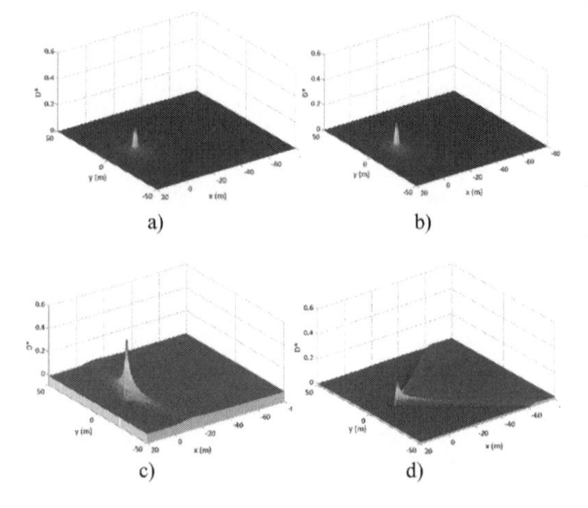

Figure 2. Vertical displacements of the Ground 1 surface, at t = 0 s, for distinct values of the load speeds: a) M = 0,5; b) M = 0,7; c) M = 0,934; d) M = 1,5 (M = c/CS, where C_S is the S wave velocity).

account the wave propagation theory and the k_1-ω chart depicted in Figure 3a. The blue dashed lines correspond to the loading lines (by the imposition of the direct function of eq. (1)) and the red line corresponds to the P-SV dispersion relationship. Since the ground is homogeneous the P-SV waves are non-dispersive, which means that only one P-SV mode occurs in correspondence with the Rayleigh wave velocity (Andersen 2002). The R wave velocity delimits the regions of propagation and no-propagation of waves.

However, if the problem is quite simple for a homogenous half-space, where the P-SV waves are non-dispersive, the same does not happen when the ground layering is taken into account. Figure 3b shows the P-SV dispersion relationship for the scenario 2. Loading lines are overlapped in the same figure (M is the ratio between the load speed and the C_S of the upper layer). The first mode line demits the propagation region from the non-propagation region. So, the critical speed should be somewhat between the Rayleigh and Shear wave velocities of the upper layer.

From the results presented above, it was clear that the phenomena of critical speed of a halfspace can be analyzed taking into account the wave propagation theory. For the assessment of the critical speed it is not mandatory the computation of the amplification factors, which solution is expensive from the computational point of view, since the critical value can be found by a criterious analysis of the interaction effect between the P-SV dispersion relationship and the loading dispersion lines. Anyway, it should be noticed that for a practical point of view the interest of this solution gets lost,

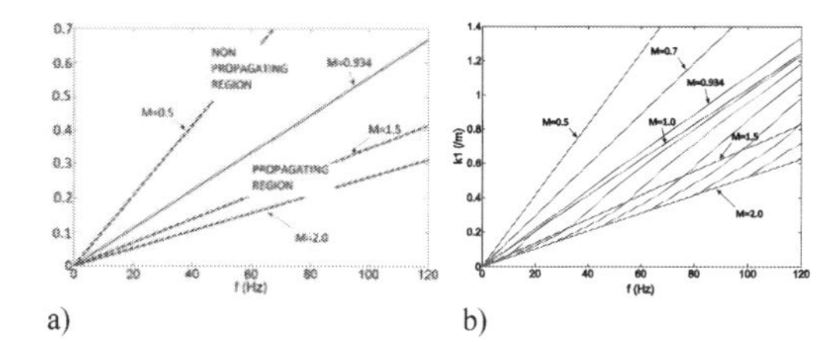

Figure 3. P-SV dispersion relationship in the referential f-k_1: a) Ground 1; b) Ground 2.

because the coupling of the track with the ground gives rise to a change on the dispersion relationship of the system, as exposed in Alves Costa et al. (2015).

3 ASSESSEMENT OF CRITICAL SPEED

3.1 *Direct approach*

The coupling of the track to the ground gives rise to a change on the characteristics of wave propagation through the ground, mainly on the direction of track development. This means that the P-SV dispersion relationship of system cannot be evaluated by a simple procedure as in the previous case. Thus, in the direct approach, the critical speed is estimated by the assessment of the load speed that gives rise to the largest amplification of the track displacements. Therefore, a model needs to be run for different moving load speeds in order to establish a displacement amplification curve that is then used to select the maximum value. Since analytical models and closed-form solutions are restricted in terms of geometry and model properties, a dynamic numerical approach is demanded to obtain the track displacements due to the traffic load.

In this sense, several distinct numerical approaches can be followed, being the Finite Elements Method (FEM), the Finite Differences Method (FDM), the Method of Fundamental Solutions (MFS) and the Boundary Elements Method (BEM) among the most popular. Regarding solution methods, time domain analysis (explicit or implicit solution) or frequency domain analysis are the most common methods that can be applied. It should be stressed that frequency domain analysis is confined to the solution of linear problems since it is based in the overlapping of effects. On other hand, transportation infrastructures, such as roads or railways, can be faced as infinitive and invariant structures. In such cases, the 3D wave propagation solution can be obtained through the combination of distinct plane waves that propagate along the structure development direction. Therefore it is possible to apply a spatial Fourier transformation along that direction and to determine the 3D displacement field as a continuous integral of simpler bi-dimensional solutions, as

$$u^{3D} = \frac{1}{2\pi}\int_{-\infty}^{+\infty} u^{2.5D}(k_1)e^{ik_1(x-x_0)}dk_1 \tag{2}$$

where k_1 is the longitudinal wavenumber.

This approach, usually called as 2.5D, can be extended to different numerical techniques such as FEM, BEM or MFS, or to the combination between them.

The advantage of this method resides in the fact that only the cross-section needs to be discretized without losing the 3D character of the problem. As matter of fact, from the computational point of view, the method is quite efficient since a small system of equations is solved several times (for different wavenumber/frequency) instead of solving an equation system with millions of degrees of freedom (as it is usual in 3D problems). In the present scenario it is adopted a coupled FEM-BEM approach, where the irregular geometry (track-embankment) is simulated by the FEM and the ground is simulated by the BEM. For the computation of the BEM matrices, Green's functions of layered halfspaces are assumed. Following FEM approach for the coupling of both methods, the motion of the domain is described by:

$$\begin{aligned}&(K_1^{global} + ik_1K_2^{global} + k_1^2K_3^{global} + k_1^4K_4^{global} \\ &-\omega^2 M^{global} + K_5^{global}(k_1,\omega))u_n(k_1,\omega) = p_n(k_1,\omega)\end{aligned} \tag{3}$$

where K_1 ^{global} to K_4 ^{global} are the stiffness matrices of the domain described by finite elements, M^{global} is the corresponding mass matrix, k_1 is the Fourier image of the coordinate x, ω is the frequency, u_n is the vector of the nodal displacements, p_n is the vector of the external forces, and, finally K_5 ^{global} is the matrix that collects the impedance terms of the layered ground (computed by the BEM). Solving this system of equations, the displacements on the transformed domain are obtained, which are changed to the space/time domain through a double inverse Fourier transform. Detailed information about the mathematical procedure that can be adopted for the derivation of the matrices can be found in previous works of the present authors (Alves Costa et al. 2010 and Alves Costa et al. 2012).

3.2 *Simplified approach*

As already shown, it is possible to establish a clear boundary, in the wavenumber-frequency domain, between P-SV wave's propagation and non-propagation regions for a layered ground. Actually, this concept was used in a previous section to assess the ground critical speed in the k_1-ω domain. Nevertheless, since it is not easy to establish an analytical procedure to compute the P-SV dispersion relationships for the track-ground system, Alves Costa et al. (2015) proposed an approximated method based on the interaction between the P-SV dispersion relationship of the ground and the properties of propagation of bending waves along the track (in free condition).

The application of the method is quite simple and the computational effort does not represent more than a few seconds. Here, an example is used to better explain the procedure:

i. Consider the track-ground scenario depicted in Figure 4a;
ii. Consider a simplified scenario where the embankment (and remaining granular layers) are extended toward infinite in the cross direction (see Figure 4b)
iii. Compute the P-SV dispersion relationship of the ground-embankment system (black line in Figure 5.
iv. Compute the bending wave dispersion relationship of the track (red line in Figure 5)

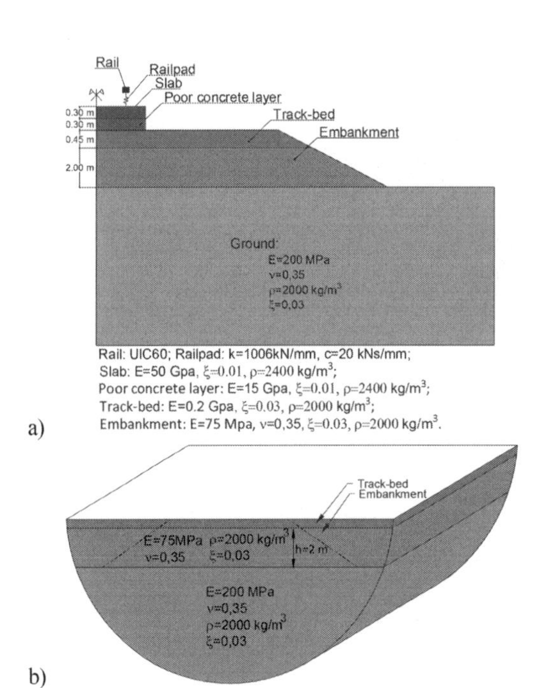

Figure 4. a) Track-ground system considered for 2.5D FEM-BEM model; b) Cross-section assume for expedite methodology.

The critical speed is given by the inverse of the slope of the line that passes in the origin and in the intersection point of the curves above mentioned.

The red regions in the diagram depicted in Figure 5 correspond to the higher energy participation. As can be seen, the red regions only develop bellow the dashed line that defines the critical speed.

Details about how to compute the ground and track dispersion relationships can be found in Alves Costa et al. (2015) and in Mezher et al. (2016).

4 CASE STUDY: EMBANKMENT INFLUENCE ON THE CRITICAL SPEED

4.1 *Geotechnical description*

For this example, it is assumed a ground with increase of stiffness over depth. Therefore, a constant value of the shear wave velocity of 150 m/s is assume in the first 3 m, followed by a linear trend of increase over depth that achieves the value of 475 m/s at 20 m depth. The soil volumetric mass is assumed equal to 1900 kg/m³ and values of 0.49 and 0.03 are adopted for Poisson and damping ratio (hysteretic damping), respectively.

4.2 *Track-embankment configurations*

The track corresponds to a continuous reinforced concrete slab where the rails are fixed through railpads. It is assumed a cross-section of 2.8 x

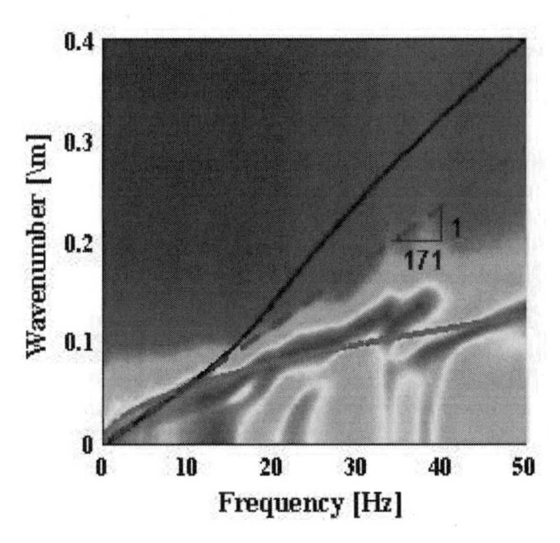

Figure 5. Contour plot of the vertical displacement in the frequency-wavenumber domain with dispersion curves and critical loading line (simplified methodology).

0.3 m, with E = 34 GPa. A value of 3400 kg/m was assumed for the continuous slab mass.

Regarding embankment geometrical and mechanical properties, different configurations were studied taking into account the general properties depicted in Figure 6 and Table 1. As shown in Figure 6, a layered embankment is assumed, where the shallow depth is stiffer than the deeper.

4.3 *Results and discussion*

This section shows the results from analyses presented in Table 1. For each one of the scenarios, the direct and the simplified methods were applied in the assessment of the critical speed.

4.3.1 *Analysis A1*

Starting with the analysis of the case A1, Figure 7 shows the track displacements due to the passage of a single unitary load at different speeds, ranging from the subcritical to the supercritical conditions.

As can be seen, for the lower speeds (Figure 7a and Figure 7b), the system response is characterized by evanescent condition, i.e., the track deformation pattern is similar to that one observed when the track is loaded in static conditions. In spite of that, when the load speed changes from 10 m/s to 70 m/s, slight amplification of track displacement can be identified.

The increase of the load speed for 140 m/s (Figure 7c) did not induce an appreciable change in the deformation pattern. Indeed, just a slight uplift of the track around the loaded region can be noticed when the results are compared with the previous figures. A similar pattern of deformation is also observed when the load speed increases to 160 m/s (Figure 7d). Despite that, it is notorious that higher amplification of the maximum track displacement occurred. However, the deformation pattern still being dominated by evanescent conditions, i.e., it is not possible to detect the energy radiation after the load passage. Indeed, if radiation of energy started to occur it was expected to

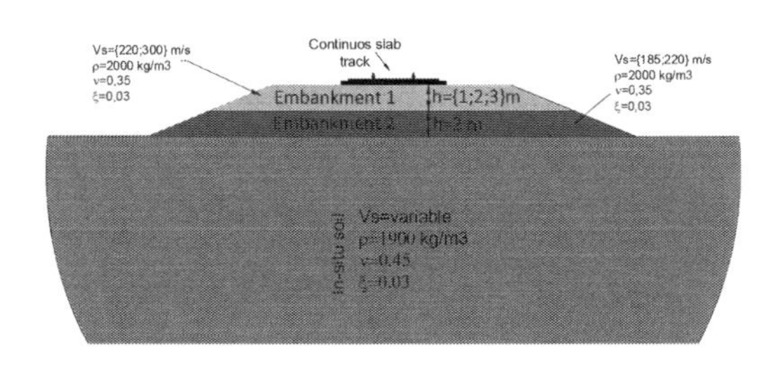

Figure 6. General configuration of the case study—track on embankment.

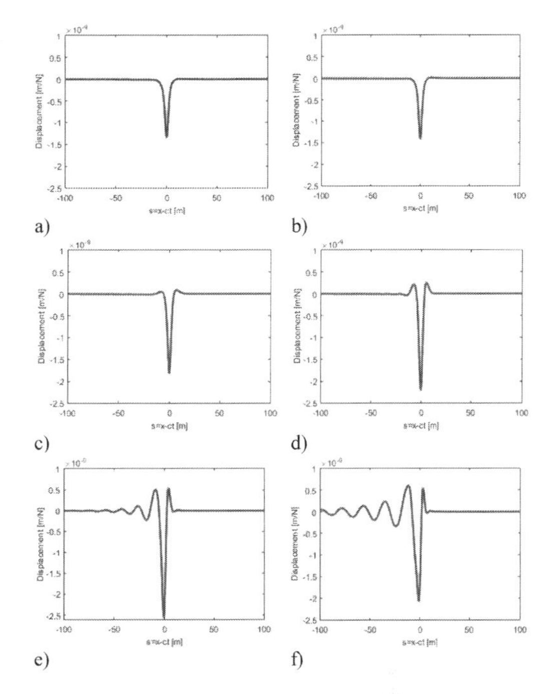

Figure 7. Analysis A1 - Track displacements due to the passage of a single load at speeds of a) 10 m/s; b) 70 m/s; c) 140 m/s; d) 160 m/s; e) 176 m/s; f) 195 m/s.

detect a tail wave traveling behind the load (due to Doppler effect). As matter of fact, this behavior is detected in Figure 7e, i.e., for the load speed of 176 m/s. In this situation, the displacement amplification reaches the maximum value and the behavior pattern changes from evanescent to propagating, i.e., the critical speed is crossed. For speeds larger than the critical value, the displacement amplification starts to decrease and an oscillating tail that moves with the load is detected.

The considerations expressed above are also validated by the analysis of the system dispersion relationships, which are depicted in Figure 8 (the P-SV dispersion relationships of the ground were computed assuming the extension of the embankment toward infinite in the cross direction). As can be seen, for the speeds of 10 m/s up to 140 m/s, the loading lines (dashed lines) develop in the evanescent region. The loading line corresponding to c = 160 m/s crosses the 1st P-SV mode but above the red curve that corresponds to bending wave dispersion relationship of the track. Therefore, the presence of the track prevents the wave propagation. The remaining two loading lines develop in the propagation region and correspond to critical and supercritical conditions.

Table 1. Summary of the different parametric studies performed.

Analysis	$V_{S,Embankment1}$ (m/s)	$V_{S,Embankment2}$ (m/s)	$h_{Embankment1}$ (m)
A1	220	185	1
A2	220	185	2
A3	220	185	3
A4	300	220	1
A5	300	220	2
A6	300	220	3

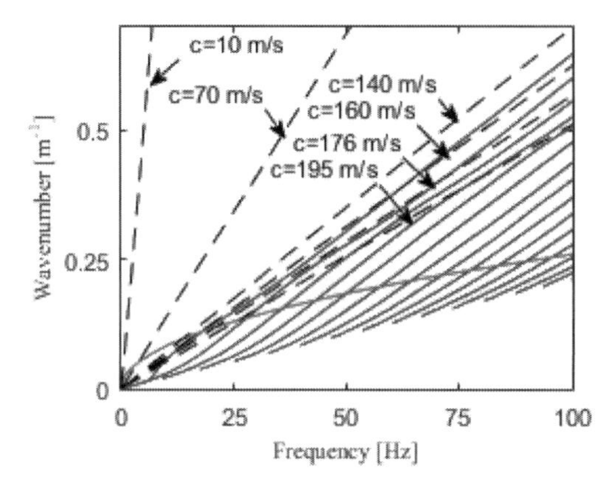

Figure 8. Analysis A1 – P-SV dispersion relationships.

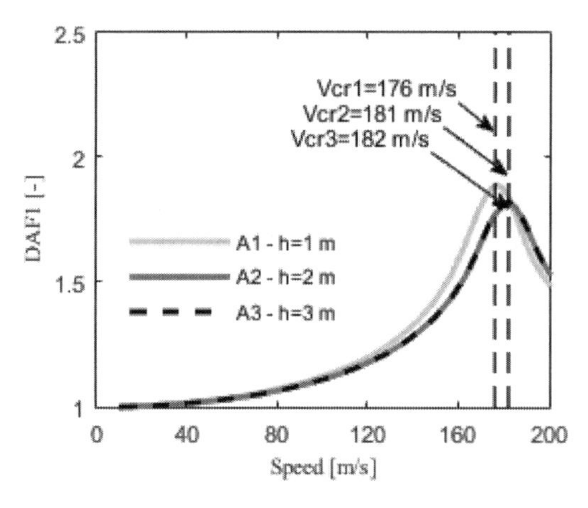

Figure 9. Influence of the embankment height.

4.3.2 *Influence of the embankment height*

Previous research studies performed by the authors show that embankment height plays a relevant role in terms of the dynamic behavior of the railway track when there is a contrast between the geodynamic properties of the embankment and the remaining domain (Alves Costa et al 2015, Kouroussis et al. 2016, Mezher et al. 2016).

To investigate the effect of the embankment1 height in the present case study, scenarios A1, A2 and A3 are considered (A1 – 1 m height; A2 – 2 m height; A3 – 3 m height). In all analyses, the shear wave velocities of the embankment1 and embankment2 are assumed equal to 220 m/s and 185 m/s, respectively. It should be noticed that the stiffness of the embankment1 (and of the embankment2) is larger than the stiffness of the superficial natural soil bellow it.

Figure 9 shows the evolution of DAF (vertical displacement dynamic amplification factor) with the load speed when a single load is attended. The observation of this figure allows concluding that the increase of the embankment1 height gives rise to slight larger critical speed. However, just an insignificant change of the critical speed is observed when increasing embankment1 height from 2 m to 3 m. This is justified by the limited change on the P-SV dispersion relationships that are induced by an increase of the embankment1 height from 2 m to 3 m (as despite in Figure 10a-c).

Results shown in the figure above allow concluding that the increase of the embankment1 height gives rise to negligible differences (for engineering purposes) on the maximum track displacement amplification. Actually, for the present case study, the influence of the embankment1 height on the critical speed is also limited, since the value of the critical speed is controlled by the properties of the embankment2 and of the natural soil, which presents a shear wave velocity of 150 m/s in the first 3 m depth.

As described in the section above, the simplified approach allows an expedite estimation of the critical speed based on the analysis of the dispersion relationships of the system, without the need of computing the dynamic amplification factor curves. Although the accuracy of the method is lower than the direct method, the advantage is notorious for scoping and parametric analysis since the problem can be analyzed in few seconds instead of taking several hours of computational effort. Moreover, the differences between the critical speeds assessed by both approaches are usually small, i.e., with an error usually lower than 5% being the simplified approach conservative (giving lower values for the critical speed when compared with the direct approach).

The assessment of the critical speed for scenarios A1 up to A3 is depicted in Figure 10a-c. Since the same track type is selected for all analyses, the difference on the results is only due to the change of the P-SV dispersion relationships of the embankment-ground system.

Despite the remarks highlighted above, a relevant conclusion can be drawn from this analysis: the critical speed can be larger than the lower shear wave velocity in the ground, since P-SV waves (Rayleigh waves) are dispersive for a layered ground. Therefore, design criteria should not be based on the prescription of a minimum shear wave velocity in the ground, since the critical speed depends on the thickness of the layers and on the contrast of stiffness between them (and also of the track stiffness/mass).

4.3.3 *Influence of the embankment stiffness*

Scenarios A4, A5 and A6 are similar to analyses A1, A2 and A3 with respect to geometry, but the embankment1 and embankment2 have higher stiffness. In these cases, values of 300 m/s and 220 m/s are assumed for the shear wave velocity in the embankment1 and embankment2, respectively. Figure 11 shows the track displacement amplification curves, due to a single load passage, for the scenarios A4, A5 and A6.

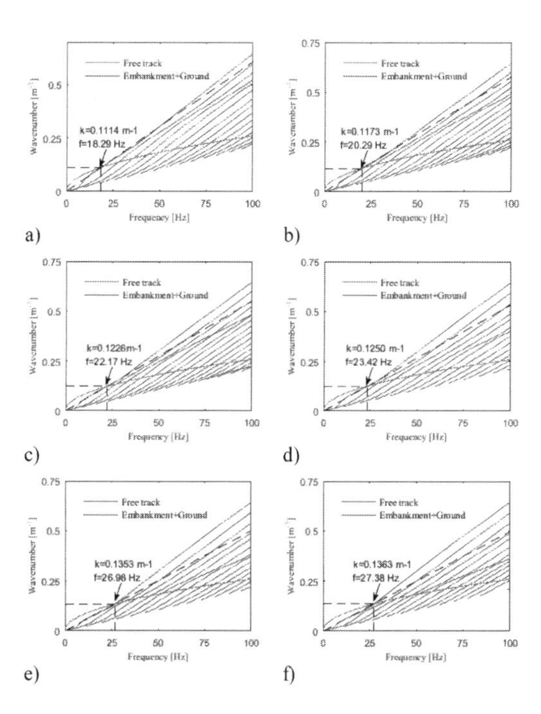

Figure 10. Assessment of critical speed by the simplified approach: a) A1; b) A2; c) A3; d) A4; e) A5; f) A6.

Table 2. Simplified versus detailed analysis.

Analysis	Critical speed (m/s) Direct approach	Simplified approach	Error (%)
A1	176	164	–6.0
A2	181	173	–4.4
A3	182	181	–0.5
A4	195	188	–3.6
A5	204	199	–2.4
A6	206	201	–2.4

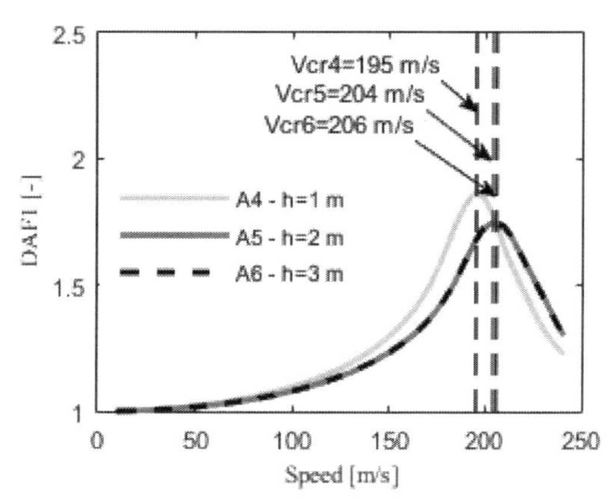

Figure 11. Influence of the embankment height.

Comparing Figure 11 with Figure 9, the increase of the embankment1 and embankment2 stiffness is responsible for an increase of around 10% - 12% in critical speed despite the significant increase of the embankment1 shear wave velocity (from 220 m/s to 300 m/s). Actually, as stressed above, since the natural soil is softer than the embankment1 just a moderate increase of the critical speed is expected due to the improvement of the embankment1 properties. Similar conclusions can be performed analyzing the dispersion curves depicted in Figure 10d-f.

4.3.4 *Summary of results*

Table 2 summarizes the results obtained through the simplified and direct approaches. Comparing the results obtained by the different methodologies it can be seen that the simplified approach is always giving lower values of the critical speed.

In spite of the errors presented by the simplified approach, the method is quite attractive for scoping analyses. The computational effort corresponds to only few seconds and allows investigating which is the influence of several parameters in the track-embankment-ground critical speed.

5 CONCLUSIONS

The studies presented in the present paper were developed assuming linear elastic behavior of the system and two distinct methodologies were presented and applied: i) direct approach; ii) simplified approach. In relation to the last one, based on the interpretation of the wave propagation characteristics, it should be highlighted its ability for parametric and scoping analyses, since the critical velocity of the track-embankment-ground system can be estimated in few seconds.

From the range of studies presented along the paper, it was shown that the critical speed of the system is conditioned by the layering of the ground as well as by the mechanical properties of the track-ground system. The stiffness and height of embankment also affect the critical speed value. From the scenarios analyzed it was shown that increasing stiffness and embankment height gives rise to larger critical speed values. It should be highlighted that in the cases under analysis the stiffness of the embankment was always assumed larger than the stiffness of the ground bellow it.

Despite of the insights obtained from an elastic analysis, it is recognized that the ground strains amplifications that occurs when train speed becomes closer to the critical speed are not compatibles with linear elastic response, demanding non-linear modelling techniques. However, in order to overcome the computational demand usually associated to non-linear modelling, a linear equivalent model, combined with a 2.5D approach, can be considered. On the other hand, a challenge for the future is development of simplified methods that will allow including those effects on the assessment of the critical speed.

ACKNOWLEDGEMENTS

This work was financially supported by: Project POCI-01–0145-FEDER-007457 - CONSTRUCT—Institute of R&D in Structures and Construction funded by FEDER funds through COMPETE2020 - Programa Operacional Competitividade e Internacionalização (POCI) – and by national funds through FCT—Fundação para a Ciência e a Tecnologia; Project PTDC/ECMCOM/1364/2014 and Scholarship SFRH/BD/101044/2014.

The authors are also sincerely grateful to European Commission for the financial sponsorship of H2020 MARIE SKŁODOWSKA-CURIE RISE Project, Grant No. 691135 "RISEN: Rail Infrastructure Systems Engineering Network".

REFERENCES

Alves Costa, P., R. Calçada and A. Silva Cardoso. 2012. Ballast mats for the reduction of railway traffic vibrations. Numerical study. *Soil Dynamics and Earthquake Engineering* **42**(0): 137–150.

Alves Costa, P., R. Calçada, A. Silva Cardoso and A. Bodare. 2010. Influence of soil non-linearity on the dynamic response of high-speed railway tracks. *Soil Dynamics and Earthquake Engineering* **30**(4): 221–235.

Andersen, L. 2002. Wave propagation in infinite structures and media. *PhD thesis*, Aalborg University.

Connolly, D., A. Giannopoulos and M. C. Forde 2013. Numerical modelling of ground vibrations from high speed rail lines on embankments. *Soil Dynamics and Earthquake Engineering* **46**: 13–19.

Costa, P. A., A. Colaço, R. Calçada and A. S. Cardoso 2015. Critical speed of railway tracks. Detailed and simplified approaches. *Transportation Geotechnics* **2**: 30–46.

Dieterman, H. A. and A. Metrikine. 1996. The equivalent stiffness of a half-space interacting with a beam. Critical velocities of a moving load along the beam. *European Journal of Mechanics A/Solids* **15**(1): 67–90.

Dieterman, H. A. and A. Metrikine. 1997. Steady-state displacements of a beam on an elastic half-space due to a uniformly moving constant load. *European Journal of Mechanics A/Solids* **16**(2): 295–306.

Hall, L. 2003. Simulations and analyses of train-induced ground vibrations in finite element models. *Soil Dynamics and Earthquake Engineering* **23**: 403–413.

Kenney, J. T. 1954. Steady-state vibrations of beam on elastic foundation for a moving load. *Journal of Applied Mechanics* **76**: 359–364.

Kouroussis, G., D. P. Connolly, B. Olivier, O. Laghrouche and P. A. Costa. 2016. Railway cuttings and embankments: Experimental and numerical studies of ground vibration. *Science of the Total Environment* **557–558**: 110–122.

Krylov, V. 1995. Generation of ground vibrations by superfast trains. *Applied Acoustics* **44**: 149–164.

Lamb, H. 1904. On the propagation of tremors over the surface of an elastic solid. *Philosophical Transaction of the Royal Society* **203**(Serie A): 1–42.

Madshus, C. and M. Kaynia. 2000. High-speed railway lines on soft ground: dynamic behaviour at critical train speed. *Journal of Sound and Vibration* **231**(3): 689–701.

Madshus, C., S. Lacasse, A. Kaynia and L. Harvik. 2004. Geodynamic challenges in high speed railway projects. *GeoTrans 2004 - Geotechnical Engineering For Transportation Projects*. ASCE. Los Angeles: 192–215.

Mezher, S. B., D. P. Connolly, P. K. Woodward, O. Laghrouche, J. Pombo and P. A. Costa. 2016. Railway critical velocity—Analytical prediction and analysis. *Transportation Geotechnics* **6**: 84–96.

Sheng, X., C. Jones and D. Thompson. 2004. A theoretical study on the influence of the track on train-induced ground vibration. *Journal of Sound and Vibration* **272**: 909–936.

Woodward, P., O. Laghrouche and A. El-Kacimi. 2013. The development and mitigation of ground mach cones for high speed railways. *ICOVP 2013 - International Conference on Vibration Problems*. Z. Dimitrovová, J. Rocha de Almeida and G. Gonçalves. Lisbon.

Numerical Methods in Geotechnical Engineering IX – Cardoso et al. (Eds)
© 2018 Taylor & Francis Group, London, ISBN 978-1-138-33203-4

Numerical modelling of vibration mitigation due to subway railway traffic

P. Lopes
Construct-FEUP, University of Porto, Porto, Portugal
ISEP, Polytechnic Institute of Porto, Porto, Portugal

P. Alves Costa, A. Silva Cardoso & R. Calçada
Construct-FEUP, University of Porto, Porto, Portugal

J. Fernández
University of La Coruña, Campus de Elviña, La Coruña, Spain

ABSTRACT: In the present paper a numerical study is presented in order to evaluate the efficiency of different countermeasures that can be adopted to mitigate vibrations inside buildings due to railway traffic in tunnels. The numerical model is substructured and is composed by three autonomous modules to simulate the main parts of the problem: i) generation of vibrations (train-track interaction); ii) propagation of vibrations (track-tunnel-ground system); iii) reception of vibrations (building coupled to the ground). A comparative analysis is carried out on the potential benefits of improving the geometrical quality of the track and of the installation of resilient elements under the railway slab, being evaluated the vibrations inside of a building due to railway traffic in a shallow tunnel located in Madrid. Finally, some considerations are made regarding an engineering solution that could involve the combination of several types of mitigation measures, taking advantage of the potentialities of the model.

1 INTRODUCTION

The growing trend of the world population and its concentration in urban areas requires the development of efficient and more sustainable transportation systems, bringing new demands to the technical and scientific communities. According to ONU, it is estimated that in 2050 about 66% of the world population lives in urban centres. Subway railway networks correspond to the most efficient mass transportation system in highly populated areas. Despite the advantages provided by subway railway systems, there are also some drawbacks, namely related to vibration and re-radiated noise inside buildings due to traffic, which annoy inhabitants and can promote health problems for long term exposure.

Having in mind the concerns expressed above, technical and scientific communities have allocated considerable effort on attempts to achieve a better understanding of the problem as well as to develop prediction tools that can be used to mitigate this kind of situation. From the previous research, it is possible to establish an understanding of this complex problem (Lai et al., 2005): i) the dynamic interaction between vehicle and track is the source of vibration; ii) vibrations are transmitted from the track to the tunnel and, posteriorly, to the ground; iii) the energy propagated as elastic waves on the ground reaches the foundations of the structures, impinging the building and giving rise to perceptible vibrations and re-radiated noise inside dwellings.

Different vibration prediction approaches have been presented during the latter years, ranging from scoping and empirical rules (Connolly et al., 2014) to advanced numerical methods (Gupta et al., 2007; Yang & Hung, 2008; Lopes et al., 2014b). Although empirical models have the advantage of being derived from experimental results, their application to complex scenarios, where design of mitigation measures is required, is difficult and sometimes even impossible. Alternatively, when the problem's geometry is simple, analytical models, such PiP, developed at U. Cambridge (Hussein & Hunt, 2007), can be applied. However, if the geometry is not so simple, Clouteau et al. (2005) proposed the usage of periodic models based on FEM-BEM in order to reduce the computational effort. Alternatively, if the domain can be assumed as invariant along the tunnel development direction, 2.5D approach can be applied (Alves Costa et al., 2012b; Lopes et al., 2014b).

As highlighted above, a comprehensive modelling approach should also attend to the mechanism of vibration generation and to reception of vibrations inside buildings. Due to the complexity of the problem, the sub-structuring approach is a rational methodology for dealing with this kind of problem. Regarding to the simulation of the source, it is usual to assume that vibrations are generated due to unevenness of the track, being the vehicle simulated through a simple multi-body approach where rigid masses, which represent the main masses of the vehicle, are connected by spring-dashpots to attend to the train's suspensions. A bit more complex is the simulation of the nearby buildings. The 3D FEM is the most suitable method for dealing with complex structures of buildings and the soil-structure interaction (SSI) problem can be rigorously solved by a 3D FEM-BEM approach, where the BEM is adopted for the modelling of the ground. Nevertheless, in an attempt to minimize the computational effort required as well as the complexity of the problem, Lopes et al. (2014a) showed that lumped parameter models can be used to represent the dynamic ground behavior without a considerable loss of accuracy.

From the description presented above, it is clear that the main theoretical bases for the formulation of the problem are established. Comprehensive models for the prediction of vibrations induced by railway traffic in tunnels are desirable and constitute valuable tools for the design of new infrastructures or for the efficiency evaluation of mitigation countermeasures. The present paper aims to contribute to this last aspect, presenting a numerical study developed in order to evaluate the efficiency of different countermeasures that can be adopted to mitigate vibrations inside buildings due to railway traffic in tunnels. The numerical approach followed was previously proposed by Lopes et al.(2014a, 2014b). The selected case study corresponds to a shallow railway tunnel in Madrid which crosses the ground beneath a building. This case study results from a practical problem: the several complaints of the building's inhabitants due to the excessive vibrations inside the building induced by the railway traffic in the tunnel.

One of the great advantages of a comprehensive model including the source, the propagation medium and also the receiver (building) is the possibility to study the influence of countermeasures on the dynamic response of buildings, and to assess the best solutions that can be applied. In the present study a comparative analysis is carried out on the potential benefits that can be achieved by improving the geometrical quality of the track and by the installation of resilient elements beneath the railway slab.

Regarding the paper organization, a brief description of the numerical approach is firstly presented, being followed by the presentation of the case study. Then, the particular aspects of the numerical model are described and both mitigation measures are studied. Finally, the main conclusions are highlighted.

2 BRIEF DESCRIPTION OF THE NUMERICAL APPROACH

The numerical model used in the present study was previously presented by Lopes et al. (2014a, 2014b). For this reason, only a brief description of the model is here presented, being the reader invited to consult previous work for a deeper understanding of the followed approach.

In order to reach high numerical performance, a sub-structuring approach was followed, being the global model composed of three sub-models. Each sub-model is dedicated to one of the main parts of the global problem: i) generation; ii) propagation; iii) reception. This classification can also be established taking into account the sub-domain that is simulated:

i. Modelling of train-track interaction (generation)

The dynamic mechanism results from the generation of inertial forces on the train due to the train-track interaction. These inertial effects can have different sources as, for instance, the track unevenness. The assessment of the dynamic train-track interaction loads requires the solution of an interaction problem between both domains, where the dynamics of the train must be taken into account. Here, the train is simulated through a multi-body approach, where the main masses of the train are simulated as rigid bodies interconnected by spring-dashpot elements to represent the suspensions (Alves Costa et al., 2012a). Train-track dynamic interaction loads are obtained by a compliance formulation developed on the frequency domain, where the source of excitation is given by the track unevenness.

ii. Modelling of track-tunnel-ground system (propagation)

The solution of the 3D wave propagation through the track-tunnel-ground system is obtained by a 2.5D FEM-PML approach, where the equilibrium equations are formulated on the wavenumber-frequency domain. Since the FEM is not suitable to deal with unbounded domains, the discretized region is boxed by perfectly matched layers, also formulated on the 2.5D domain, that avoid the spurious reflection of the waves that reach artificial boundaries, which results from the limitation of the interest domain (François et al., 2012; Lopes et al., 2014b).

This numerical model is used to obtain the transfer functions from the rail to other points of the system, as well as to assess the impedance of the track that is used on the generation modulus for the solution of the train-track dynamic interaction problem.

iii. Modelling of buildings and soil-structure interaction

The most suitable numerical approach for simulation of 3D dynamic behavior of buildings corresponds to the finite element method (FEM). However, soil-structure interaction (SSI) is a relevant aspect that must be properly attended to achieve accurate predictions of vibrations inside buildings due to the railway traffic in the tunnel. A precise approach to deal with the SSI is by the 3D FEM-BEM coupling, where the BEM capabilities to simulate the dynamic behaviour of the ground are notorious. On the other hand, alternative and simpler methods can also be explored such as the lumped parameter models. Lopes et al. (2014a) showed that the solution obtained considering a lumped parameter approach for SSI can be very similar to the one obtained using a detailed 3D BEM approach, but the former is much simpler to implement and can be easily introduced into a commercial finite element code. This aspect is relevant as it facilitates its transfer from academia to engineering practice.

Since it is assumed that the presence of the building does not affect the vibration generation source (Coulier et al., 2014), the vibration fields at the free-field, obtained from the application of the modules mentioned above, are used as excitation source to the structure, which in turn is coupled to the ground. Studies performed by Lopes et al. (2014a) show that it is generally acceptable to neglect the presence of the tunnel on the assessment of the building footings impedance. This simplification allows assuming the ground as a half-space without any disturbance due to the tunnel cavity.

3 CASE STUDY DESCRIPTION

The selected case study corresponds to an ancient shallow tunnel that belongs to a stretch of the railway network in Madrid. The tunnel crosses the ground just beneath an existing building. Figure 1 shows the geometry of the cross-section of the problem as well as the soil properties. These properties were obtained from the studies carried out by Melis (2011) on the geotechnical characterization of the ground for the construction of recent

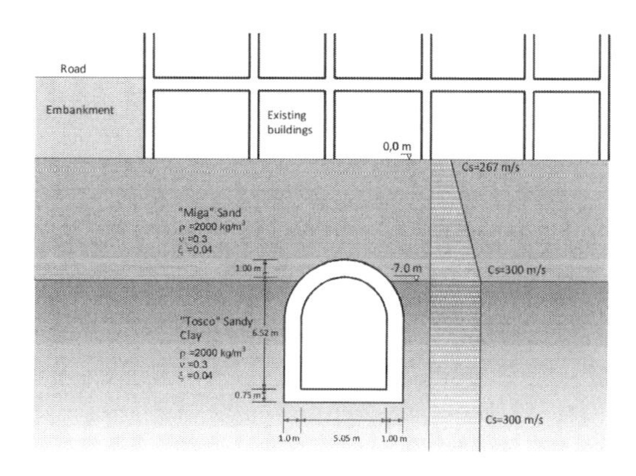

Figure 1. Schematic representation of the cross-section of the tunnel and of the dynamic properties of the soil.

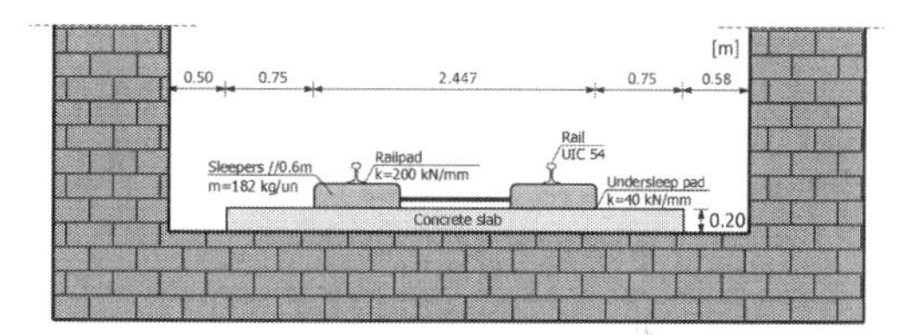

Figure 2. Detail of railway track.

tunnels nearby this case study. As can be seen, the tunnel is quite shallow being the distance between the tunnel's roof and the building of about only 5.5 m. It should be noted that the building was constructed after the excavation of the tunnel.

The liner of the tunnel is in stone masonry, as it was common in old tunnels. A Young modulus of 5 GPa and a Poisson ration of 0.2 were estimated for the homogeneous equivalent elastic dynamic properties of the material.

The railway track is type STEDEF, as depicted in Figure 2, with bi-block sleepers spaced 0.6 m in the longitudinal direction, and rails are type UIC54. The properties of the elements that constitute the track are also depicted in the figure. As can be seen, the concrete slab is resting directly on the tunnel invert, without inclusion of any resilient element.

The building above the tunnel was constructed in the mid 50's of the last century. Figure 3 shows the building façade. The structure is made of concrete and there is one buried floor and eight elevated floors, being the plant of the regular floors depicted in Figure 4. The properties of the building's structural elements are indicated in Table 1.

In addition to the dead weight of the structural elements indicated in Table 1, a load of 450 kg/m^2 distributed on the slabs surface was considered, in order to take into account non-structural masses of the building.

Figure 3. Picture of the building in study.

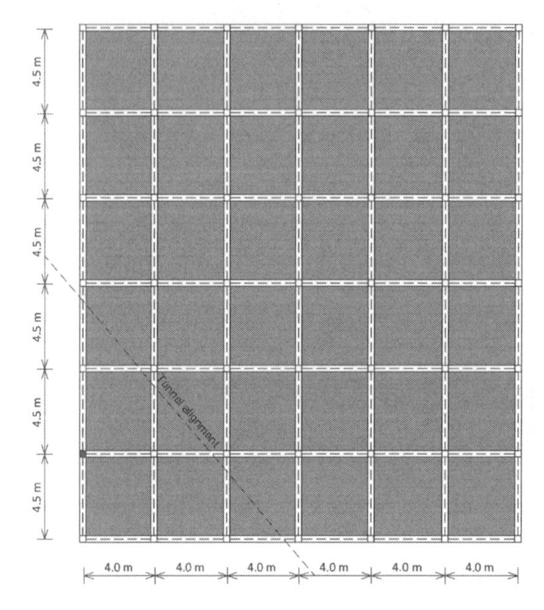

Figure 4. Structural plant of the building.

Table 1. Properties of the structural elements of the building.

Element	Properties E(GPa),	ν,	ρ (kg/m³)	Dimensions
Slabs	30,	0.2,	2500	thickness: 0.25 m
Beams	30,	0.2,	2500	0.30×0.60 m²
Columns	30,	0.2,	2500	0.35×0.35 m²

Regarding the foundations, the building is founded on shallow footings with an area of 2.75×2.75 m².

Regarding rolling stock, the passage of train type 446 of RENFE (double compositions) is considered. The unsprung mass of the vehicle is about 1500 kg per wheelset, and for the present study, the passage at a speed of 14.25 m/s (51.3 km/h) was considered.

It was assumed that the rail unevenness amplitudes can be described by a PSD (power spectral density) function with the following equation (Braun e Hellenbroich, 1991):

$$S(k_x) = S(k_{x,0})\left(\frac{k_x}{k_{x,0}}\right)^{-w} \tag{1}$$

where $k_{x,0} = 1$ rad/m, $S(k_{x,0})$ is a constant that reflects the geometrical quality of the track unevenness and w is a constant that generally assumes a value between 3.0 and 4.0.

For the present case study, the variables take the following values: $S(k_{x,0}) = 1 \times 10^{-6}$ m³/rad and $w = 3$. Since the unevenness profile of the track wasn't measured, the previous values result from an optimization procedure developed in order to obtain an unevenness profile compatible with the measured vertical velocity of the rail due to the train passage (measurement performed by CEDEX (2003)).

4 MODEL DESCRIPTION

As previously mentioned, a sub-structuring approach is followed, adopting different models and modelling techniques for the distinct subdomains. Concerning the track-tunnel-ground system, the dynamic response is assessed by a 2.5D FEM-PML approach (Lopes et al., 2014b).

Regarding the track, an Euler-Bernoulli beam was adopted to simulate the rail, and spring-dashpot elements were used to simulate the railpads and undersleeper pads. Sleepers were simulated as a uniformly distributed mass (303 kg/m) and the resilient elements of the track were also considered uniformly distributed, assuming the following properties: 333 kN/mm/m and 67 kN/mm/m for the railpad and undersleeper pad, respectively. Figure 5 depicts the 2.5D FEM-PML mesh adopted in the present study.

For the simulation of the train-track interaction, the two more relevant excitation mechanisms were considered: i) quasi-static mechanism; ii) dynamic mechanism. The former comprises the movement of the static loads. For the assessment of the latter, Alves Costa et al. (2012a), among others, shown

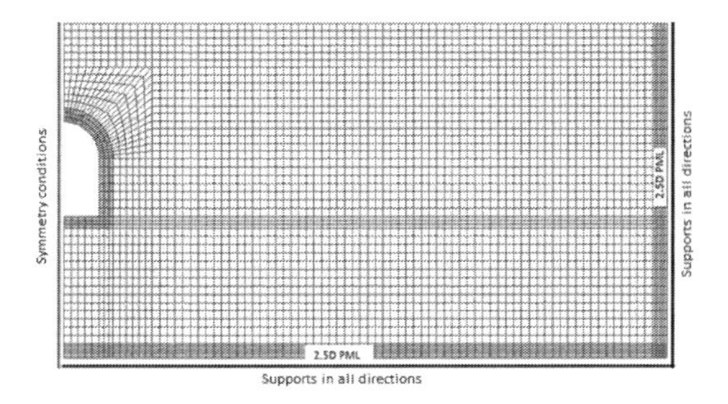

Figure 5. 2.5D FEM-PML mesh.

that consideration of simplified models where only the unsprung masses of the train are taken into account is a reasonable approach (details on the modelling strategy can be found in (Alves Costa et al., 2012a)).

In what concerns to the building, a simplified model based on 3D FEM was constructed, where the main structural elements were considered, namely the ones mentioned in Table 1. Hence, only vertical dynamics is analyzed and since the nearby buildings also have one buried floor, it was assumed that the ground surface corresponds to the level of the footings (simulated as rigid bodies). This simplification is acceptable for the intended modelling. Figure 6 depicts the adopted finite element mesh.

The SSI effects, which are quite relevant for achieving an accurate assessment of the building response due to the train passage (Lopes et al., 2014a), are considered using a lumped parameter approach in order to represent the contribution of the ground. A detailed description of the particular aspects of the modelling strategy can be found on previous works of the authors, namely on (Lopes et al., 2014a, 2014b).

Finally, Rayleigh damping approach was followed, being the α and β parameters selected in order to obtain a damping ratio around 1% for the frequency range between 5 and 80 Hz.

5 CASE STUDY: NUMERICAL MODELLING OF VIBRATION MITIGATION

5.1 *General considerations*

The numerical approach followed allows an efficient simulation of the vibrations induced by traffic in tunnels from the source to the receiver (building), constituting a valuable tool for the efficiency evaluation of mitigation countermeasures.

Therefore, in the present section a comparative analysis, applied to the selected case study, is carried out on the potential benefits of improving the geometrical quality of the track and of the installation of resilient elements under the railway slab.

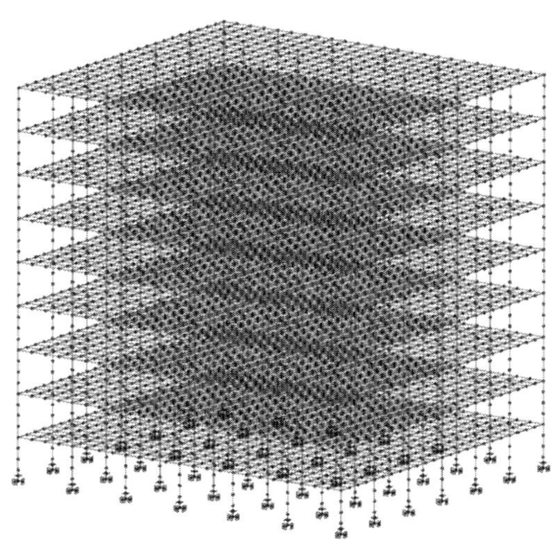

Figure 6. 3D finite elements mesh adopted for the simulation of the building.

5.2 *Improving the geometrical quality of the track*

One of the most efficient vibration mitigation techniques is a rigorous control of the geometrical quality of the track. This type of intervention allows minimizing the train-track dynamic interaction loads, reducing the vibrations that propagate through the ground and that cause discomfort in the nearby buildings. It should be mentioned that the geometrical quality of the track of the present case study was clearly deficient, as reported in Fernandez (2014), justifying a study on the potential benefit of improving it. Thus, a new unevenness profile was generated, but now assuming the parameter $S(k_{x,0}) = 5 \times 10^{-7}$ m^3/rad (half of the previously considered) in the PSD function given by equation (1).

Figure 7 and Figure 8 compare the vertical velocity of the rail for both track conditions previously mentioned in time and frequency domains, respectively. As expected, there is a very significant decrease in the amplitude of the response with the improvement of the geometrical quality of the track. However, the detailed analysis of Figure 8 shows that for frequencies up to about 20 Hz, the vertical velocity is practically independent of the track conditions. The justification for this effect is simple: the lowest frequency range is almost governed by the quasi-static mechanism, thus independent of the larger or lower amplitude of the track unevenness.

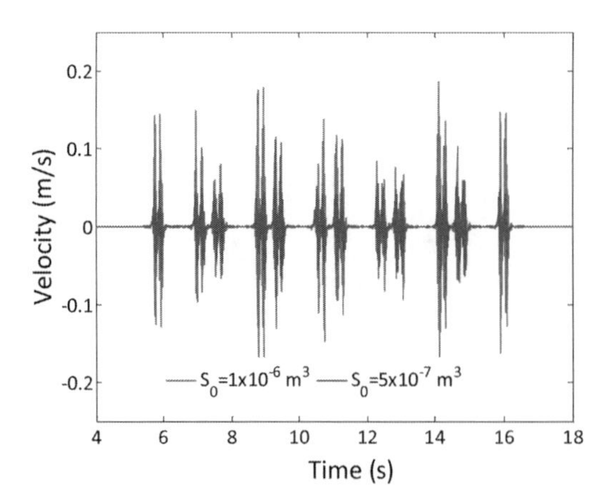

Figure 7. Vertical velocity of the rail: time history.

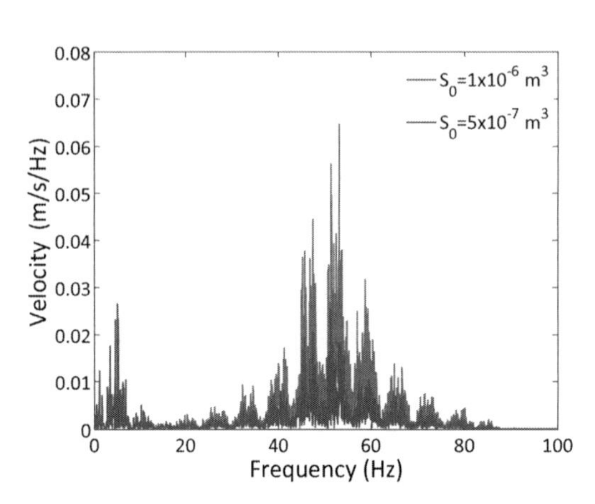

Figure 8. Vertical velocity of the rail: frequency content.

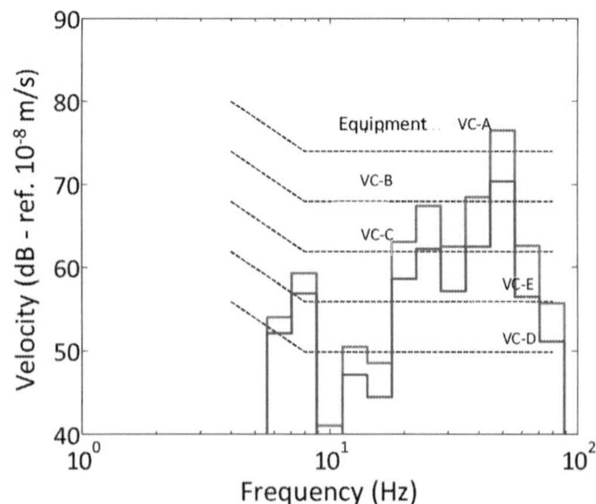

Figure 9. One-third octave spectrum of the vertical velocity in the 5th building floor for different geometrical quality levels of the track (red line-$S_0 = 1 \times 10^{-6}$ m^3/rad;blue line-$S_0 = 5 \times 10^{-7}$ m^3/rad).

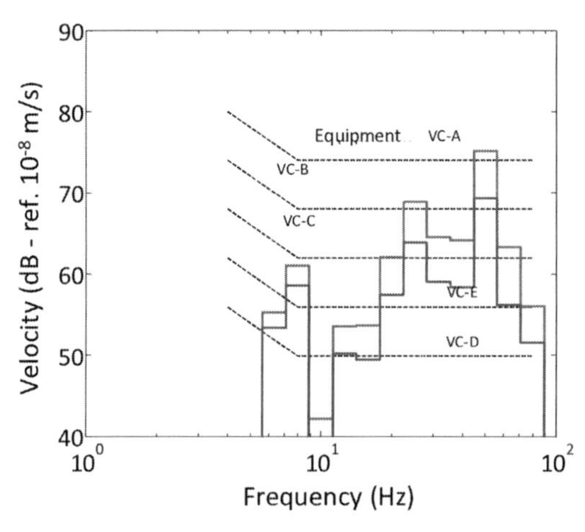

Figure 10. One-third octave spectrum of the vertical velocity in the 7th building floor for different geometrical quality levels of the track (red line-$S_0 = 1 \times 10^{-6}$ m^3/rad;blue line-$S_0 = 5 \times 10^{-7}$ m^3/rad).

The dynamic response of the building is evaluated at observation points on the 5th and 7th floors, located directly above the tunnel. Since one of the main descriptors of the dynamic response is the frequency content, Figure 9 and Figure 10 show the one-third octave spectrum of the vertical velocity at the observation points, for both situations under analysis. Limit curves proposed by Gordon (1999) for vibration-sensitive equipment are superimposed in the figures.

As expected, the improvement of the geometrical quality of the track results in a considerable decrease of the vertical dynamic response at the slab points.

From Figures 9 and 10, it is possible to see that there is a significant gain with the improvement of the geometrical quality of the track, which reaches more than 6 dB in the most relevant frequency range. The gain is lower in the frequency range below 15 Hz, what ends up having little engineering significance, since the energy content associated to those frequency bands is relatively low when compared to energy identified in the frequency range between 30 Hz and 60 Hz.

From the presented analysis, the good geometrical quality of the track is a determinant parameter to achieve a satisfactory performance of the system, especially in what concerns to the vibrations inside buildings near the railway infrastructures.

5.3 *Introduction of a floating slab system*

The introduction of resilient elements on the track is a mitigation measure with high practical potential, since the mitigation occurs at the source, without need of intervention in infrastructures in the vicinity of the railway. In fact, the presence of those elements introduces a cut-off frequency on the track, with the correspondent attenuation of vibrations at the higher frequency range. However, this attenuation in the higher frequency content is accompanied by an increase of the vibration levels in the lower frequency range, which are quite relevant for the vibration analysis in the buildings (Lopes et al., 2014b). Taking advantage of the potentialities of the model, this solution is analysed for the case study presented.

The results presented in Figure 9 and Figure 10 show that the higher frequency content of the vertical velocity of the building slabs is confined between 35 Hz and 60 Hz. Considering the frequency range of interest, the floating slab system will only be efficient if it allows the introduction into the system of a cut-off frequency not exceeding 35 Hz, preferably lower. Hence, an analysis is carried out on the potential benefit of installing a resilient mat under the railway slab that leads to a cut-on frequency of 20 Hz, i.e. introducing a cut-off frequency of about 28 Hz.

The inclusion of the resilient mat gives rise to a considerable amplification of the dynamic response of the rail, as shown in Figure 11, where the vertical velocity of the rail is presented for the floating slab solution and for the reference case. It should be noted that the peak values of the vertical velocity practically double with the softening of the track provided by the mat beneath the slab.

More relevant than the previous result is the one depicted in Figure 12, where the insertion loss curve of the vertical velocity is illustrated at a point located in the tunnel invert (reference section).

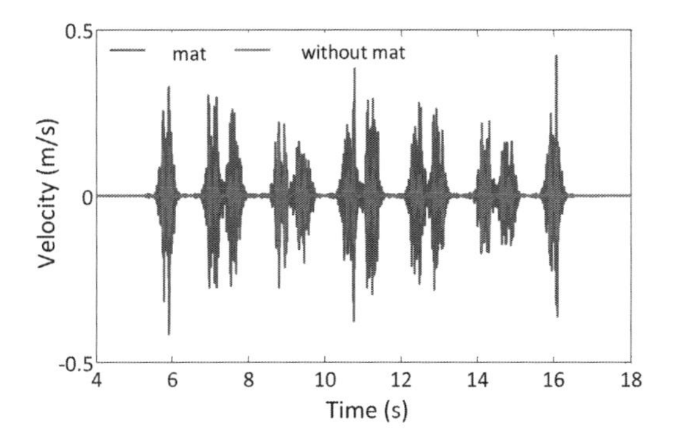

Figure 11. Time history of the vertical velocity of the rail with and without resilient mat.

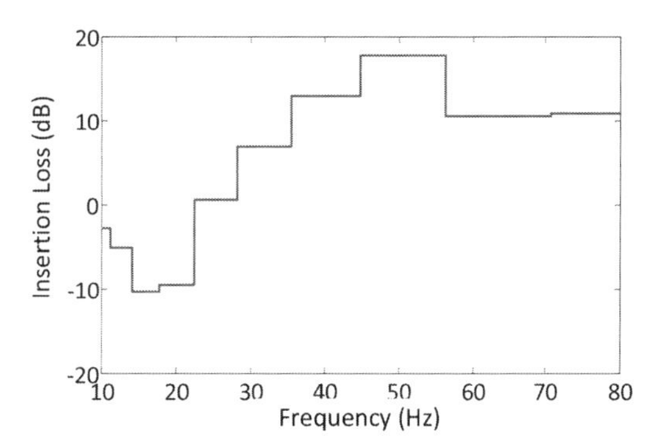

Figure 12. Insertion loss curve at the central point of the invert.

The IL (insertion loss) reflects the efficiency of the mitigation measure, since it corresponds to a measure of the reduction of vibration transmitted from the track to the tunnel and, consequently, to the ground. As can be seen, the introduction of the floating slab system allows a very significant reduction, about 18 dB, in the most relevant frequency range, which is around 50 Hz. This benefit is accompanied by an increase of the frequency content in the band around the cut-on frequency of the isolation system.

Considering the results illustrated in Figure 12 it is expected that the introduction of the floating slab system will result in an increase of the building slabs vibration levels for frequencies close to 20 Hz, being this effect accompanied by a reduction of the dynamic response in the higher frequency range. However, the magnitude of such effects is very dependent not only on the incident wave field that impinges the building footings, but also on the dynamic properties of the building itself (Lopes et al., 2014b).

Figure 13 and Figure 14 show the one-third octave spectrum of the vertical velocity in the 5th and 7th building floors for the floating slab solution and for the reference case.

Analysing the frequency content of the building slabs response, it can be seen that the installation of the floating slab system gives rise to a change in the dominant frequency range. Indeed, if previously the preponderant range was focused on 50 Hz, it is now transferred to near 20 Hz. However, even with the increase in vertical velocity around 20 Hz, there is an appreciable reduction in maximum vibration levels.

From the previous results, the engineering solution could pass through an improvement on the geometrical quality of the track, affecting mainly the higher lengths that are in the origin of vibrations in the lower frequency range, up to 25 Hz,

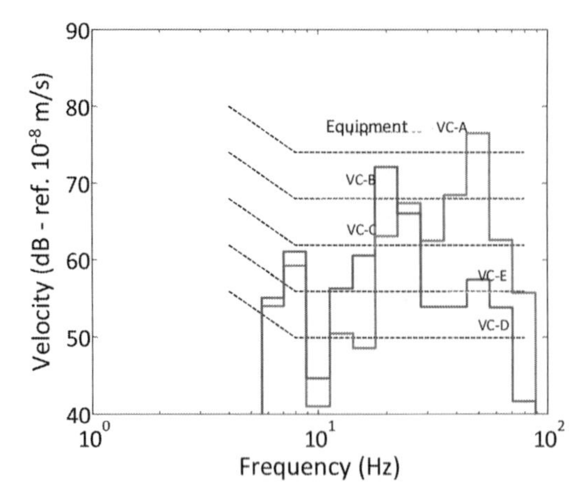

Figure 13. One-third octave spectrum of the vertical velocity in the 5th building floor for the reference case and floating slab solution (red line—without mat; blue line—with resilient mat).

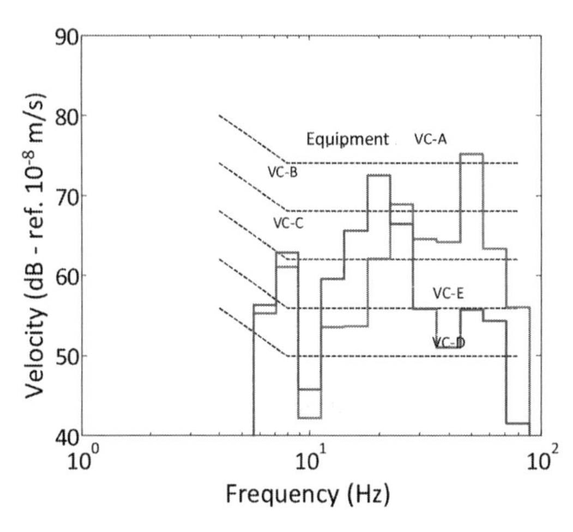

Figure 14. One-third octave spectrum of the vertical velocity in the 7th building floor for the reference case and floating slab solution (red line—without mat; blue line—with resilient mat).

being the solution accompanied by the installation of a floating slab system, as the proposed one, which would allow a significant reduction of the vibration levels in the higher frequency range. It should be noted that this is one of the advantages of having a comprehensive model for the prediction of vibrations, as the presented, which allows the development of studies evaluating the gains associated with synergistic effects resulting from the combination of several types of mitigation measures.

6 CONCLUSIONS

A numerical study to evaluate the efficiency of different countermeasures that can be adopted to mitigate vibrations inside buildings due to railway traffic in tunnels was presented. The followed numerical approach corresponds to a balance between accuracy and complexity. By that reason, different modelling approaches are adopted as function of the specificities of each subdomain, namely: i) a multi-body model for the simulation of the train; ii) a 2.5D FEM-PML approach for the simulation of the track-tunnel-ground system; iii) the 3D FEM for the simulation of the nearby building. Trying to reduce the complexity of the modelling, a lumped-parameter approach was followed to include the soil-structure interaction behavior of the building. This simplification has a good performance with a large reduction of complexity of the modelling strategy.

The case study presented corresponds to a shallow railway tunnel in Madrid and results from a practical problem, the large number of complaints of the inhabitants of a building adjacent to the railway tunnel due to excessive vibrations induced by the traffic in the tunnel.

The comparative analysis on the potential benefits of improving the geometrical quality of the track and of the installation of resilient elements under the railway slab, enabled to draw the following conclusions.

The geometrical quality of the track is a preponderant factor in the vibration levels perceived inside the buildings, being determinant to achieve a satisfactory performance of the system.

Regarding the floating slab system solution, the presence of the resilient mat allows reaching high level efficiency on the attenuation of vibrations inside buildings in the higher frequency range, above the cut-off frequency of the isolation system. However, it should not be forgotten that this type of solution implies amplification of the vibration levels at the lower frequency range (around the cut-on frequency) which are quite relevant for the vibration analysis in the buildings. Therefore, it is possible to conclude that the design of floating slab systems for the mitigation of vibrations induced by railway traffic in tunnels should be performed with care. Attending to the complexity of the problem, the advantage of a comprehensive model for the prediction of vibrations, as the presented, is confirmed.

Taking advantage of the potentialities of the model, an engineering solution that involve the combination of both mitigation measures is proposed and discussed.

From the developed study, the proposed numerical approach, based on a sub-structuring strategy, proves to be an interesting framework for the prediction of vibrations in urban environment due to subway railway traffic and for the design of mitigation measures based on a deep understanding of the problem, allowing obtaining a holistic picture of it.

ACKNOWLEDGEMENTS

This work was financially supported by: Project POCI-01–0145-FEDER-007457 - CONSTRUCT—Institute of R&D In Structures and Construction funded by FEDER funds through COMPETE2020 - Programa Operacional Competitividade e Internacionalização (POCI) – and by national funds through FCT—Fundação para a Ciência e a Tecnologia; Project PTDC/ECMCOM/1364/2014 and Scholarship SFRH/BD/69290/2010.

REFERENCES

Alves Costa, P., R. Calçada and A. Cardoso (2012a). "Influence of train dynamic modelling strategy on the prediction of track-ground vibrations induced by railway traffic." Proceedings of the Institution of Mechanical Engineers, Part F: Journal of Rail and Rapid Transit 226(4): 434–450.

Alves Costa, P., R. Calçada and A. Silva Cardoso (2012b). "Track–ground vibrations induced by railway traffic: In-situ measurements and validation of a 2.5D FEM-BEM model." Soil Dynamics and Earthquake Engineering 32(1): 111–128.

Braun, H. and T. Hellenbroich (1991). "Messergebnisse von strassenunebenheiten." VDI Berichte 877: 47–80.

CEDEX (2003). Estudio de vibraciones inducidas por la explotación ferroviaria en la Cuesta de San Vicente nº36 y en el túnel del campo del moro en el pasillo verde ferroviario de Madrid. Madrid, CEDEX. CEDEX: 82–501–7–006.

Clouteau, D., M. Arnst, T. Al-Hussaini and G. Degrande (2005). "Free field vibrations due to dynamic loading on a tunnel embedded in a stratified medium." Journal of Sound and Vibration 283(1–2): 173–199.

Connolly, D.P., G. Kouroussis, A. Giannopoulos, O. Verlinden, P.K. Woodward and M.C. Forde (2014). "Assessment of railway vibrations using an efficient scoping model." Soil Dynamics and Earthquake Engineering 58: 37–47.

Coulier, P., G. Lombaert and G. Degrande (2014). "The influence of source–receiver interaction on the numerical prediction of railway induced vibrations." Journal of Sound and Vibration 333(12): 2520–2538.

Fernandez, J. (2014). Estudio numérico de vibraciones provocadas por el tráfico ferroviario en túneles en el dominio del tiempo: análisis geotécnico, validación experimental y propuesta de soluciones. PhD, Universidad de Coruna.

François, S., M. Schevenels, G. Lombaert and G. Degrande (2012). "A two-and-a-half-dimensional displacement-based PML for elastodynamic wave propagation." International Journal for Numerical Methods in Engineering 90(7): 819–837.

Gordon, C. (1999). "Generic vibration criteria for vibration-sensitive equipment." Proc. SPIE 3786: 22–39.

Gupta, S., M. Hussein, G. Degrande, H. Hunt and D. Clouteau (2007). "A comparison of two numerical models for the prediction of vibrations from underground railway traffic." Soil Dynamics and Earthquake Engineering 27(7): 608–624.

Hussein, M. and H. Hunt (2007). "A numerical model for calculating vibration from a railway tunnel embedded in a full-space." Journal of Sound and Vibration 305(3): 401–431.

Lai, C., A. Callerio, E. Faccioli, V. Morelli and P. Romani (2005). "Prediction of railway-induced ground vibrations in tunnels " Journal of Vibration and Acoustics 127(5): 503–514.

Lopes, P., P. Alves Costa, R. Calçada and A. Silva Cardoso (2014a). "Influence of soil stiffness on vibrations inside buildings due to railway traffic: numerical study." Computers & Geotechnics 61: 277–291.

Lopes, P., P. Alves Costa, M. Ferraz, R. Calçada and A. Silva Cardoso (2014b). "Numerical modeling of vibrations induced by railway traffic in tunnels: From the source to the nearby buildings." Soil Dynamics and Earthquake Engineering 61–62: 269–285.

Melis, M. (2011). Apuntes de introducción al Proyecto y Construcción de Túneles y Metros en suelos y rocas blandas o muy rotas. C. de and E.T.S. d. I. d. C. ferrocarriles, Canales y Puertos. Madrid, Universidad Politécnica de Madrid.

Yang, Y. and H. Hung (2008). "Soil Vibrations Caused by Underground Moving Trains." Journal of Geotechnical and Geoenvironmental Engineering 134(11): 1633–1644.

Numerical Methods in Geotechnical Engineering IX – Cardoso et al. (Eds)
© 2018 Taylor & Francis Group, London, ISBN 978-1-138-33203-4

Mitigation of vibration induced by railway traffic through soil buried inclusions: A numerical study

A. Castanheira-Pinto & P. Alves Costa
Construct-FEUP, University of Porto, Portugal

L. Godinho & P. Amado-Mendes
ISISE, Department of Civil Engineering, University of Coimbra, Coimbra, Portugal

ABSTRACT: The topic of vibrations induced by rail traffic has received a special attention from the scientific community over the last decades. Nowadays, such thematic is faced as a public health concern. Thus, procedures to minimize the discomfort caused by an efficient railway network have to be proposed. In this paper a preliminary numerical study is presented about the mitigation of vibrations based on the introduction of a set of inclusions in the ground, parallel to the railway track. Analyses in space-frequency, and in frequency-wavenumber domains allowed concluding that the ground wave-propagation is strongly affected by the inclusions' presence. Based on that numerical study, theoretical simple equations were derived and they are here proposed in order to support the engineering design of this kind of mitigation measure.

1 INTRODUCTION

Accordingly to the United Nations, in 2050 urban world population will be twice of the population living in rural areas (United Nations, 2015). Such fact brings a new set of challenges, namely, the construction and development of cities' mass transportation systems with emphasis for efficient railway systems. For buildings in the vicinity of those facilities, a major discomfort can be felt at each train pass (Connolly et al., 2016). Thus, ground-borne vibrations induced by human activities, such as construction or traffic, have become a relevant concern of modern societies. Although structural integrity usually isn't affected by such excitation, the long term continuous exposition of population to vibrations is nowadays faced as a public health problem (Croy et al., 2013, Smith et al., 2013). In this scope, transportation infrastructures in urbanized regions should receive special attention, namely in the case of railways, since there is a global demand for increasing the railway infrastructure capacity. The increase of railway infrastructure capacity will demand efficient solutions for vibration mitigation in order to achieve the societal acceptance of this goal. Actually, the topic of mitigation of vibrations induced by railway traffic has received a considerable attention by the technical and scientific communities, where different solutions have been proposed and analyzed. Usually, the distinct solutions are grouped depending on their location: i) at the source, as for instance the introduction of resilient elements at track level, (Alves Costa et al., 2012, Bongini et al., 2011); ii) at the receiver, through the change of the receiver (building) dynamic behavior (Talbot and Hunt, 2009, Talbot, 2016). Alternatively to these two solutions, mitigation measures can be applied at the propagation path, trying by that way to minimize, or at least to change, the energy that impinges the facilities being protected.

Despite of the recent advances in the topic, the exploration of the potential given by phononic materials, a specific class that can be framed within the so-called metamaterials, has not been extensively analyzed in the context of vibration mitigation. Actually, the propagation of almost all types of waves such as ultrasound, acoustic, elastic, and even electromagnetic and thermal, in specific classes of periodic structures known as phononic or metamaterials, has drawn the interest of a large number of scientists and engineers (Hussein et al., 2014). The present paper aims to give a contribution for the development of vibration shielding solutions based in the introduction of periodic arrays of inclusions parallel to railway infrastructures. The paper starts with a deep analysis of the phenomena, followed by a step-by-step approach, where the physical aspects are analyzed in the space-frequency and in the wavenumber-frequency domains.

2 BRIEF DESCRIPTION OF THE NUMERICAL APPROACH

Transportation infrastructures, such as roads and railways tracks, can be often assumed as infinite and longitudinally invariant structures. In such conditions, if the assumption of linear response can be faced as reasonable, it is possible to achieve the 3D wave propagation solution by a 2.5D approach, where the equilibrium equations are formulated in the wavenumber-frequency domain. This approach takes hand of the Fourier transformation regarding the longitudinal development direction and, by that reason, only the cross-section needs to be discretized. Therefore, in such conditions, the 3D response is given by:

$$u^{3D}(x,y,z,\omega)=\frac{1}{2\pi}\int_{-\infty}^{+\infty}u^{2.5D}(k_1,y,z,\omega)\mathrm{e}^{-\mathrm{i}(x-x_0)k_1}\mathrm{d}k_1 \tag{1}$$

where k_1 and ω are the Fourier images of x coordinate and time, respectively.

The displacement field in the 2.5D domain can be assessed by using a numerical approach. In the present study a 2.5D finite element strategy is followed, since it allows dealing easily with complex geometries. A PML technique is applied to avoid the spurious wave reflection in the artificial boundaries (Lopes et al., 2013). Following the 2.5D FEM-PML approach and after the assemblage of the equations of each individual element, the equilibrium condition is established by the following equation:

$$\left\{\left[K^g_{FEM}\right]+\left[K^g_{PML}\right]-\omega^2\left(\left[K^g_{FEM}\right]+\left[K^g_{PML}\right]\right\}\right)u_n=p_n \tag{2}$$

where ω is the excitation frequency, u_n is the vector of nodal displacements in the transformed domain, p_n is the vector of external nodal loads in the transformed domain. The matrices $\left[K^g_{FEM}\right]$ and $\left[K^g_{PML}\right]$ are the global stiffness matrices of the FEM domain and of the PML domain, respectively, while $\left[M^g_{FEM}\right]$ and $\left[M^g_{PML}\right]$ are the corresponding global mass matrices.

The advantage of this procedure, when compared with a traditional 3D FEM approach, resides in the considerable computational effort saving. Instead of solving a huge system of equations, with millions of degrees of freedom, a small system of equations (since discretization along the development direction is avoided) is solved several times for a range of frequencies and wave numbers.

A detailed description of the numerical procedure can be found in following references: (Lopes et al., 2014, Amado-Mendes et al., 2015).

3 DYNAMIC ANALYSIS OF A SINGLE INCLUSION SYSTEM

3.1 *Wave propagation in the presence of a single stiffer inclusion*

Mitigation measures may either reduce the vibration content induced if adopted in the source, decrease the excitation transmitted to infrastructure when considered at the receiver or even destroy the wave propagation pattern once adopted in the propagation path (Van Hoorickx et al., 2017). The present paper aims to study buried inclusions as a solution for disrupting the wave field induced by railway traffic. The effect induced by the presence of a single stiffer inclusion introduced in the ground was already studied by (Coulier et al., 2013, Barbosa et al., 2015), where multiple behaviors for different frequency ranges were identified.

For the present study, a cross-section adopting a single horizontal inclusion, with the correspondent diameter 0.6m and buried at a depth of 0.4m, was modelled. A harmonic load was applied 10 meters away from the inclusion, as can be observed in Figure 1.

Regarding the mechanical properties of the model, a stiffer inclusion was considered with a shear velocity five times larger than the surrounding soil. Table 1 presents all the relevant properties for each material.

Initially, a direct comparison between the homogeneous scenario and the case with one inclusion was carried out, being plotted the vertical response obtained per frequency for both cases in Figure 2, for a quarter of the section.

A similar behavior between the homogenous case and the case with the inclusion is observed for the 25Hz results. Since the Rayleigh wavelength generated for such frequency is substantially larger than the inclusion's diameter, its presence does not disturb the wave field. However, when the magnitude of the generated wavelength and the diameter of the inclusion start to be similar, an attenuated area emerges. Such fact is due to the wave-guiding behavior promoted by the stiffer inclusion, which was already explained by (Coulier et al., 2013). Additionally, a dependency between the shadow region angle and the excitation frequency is observed.

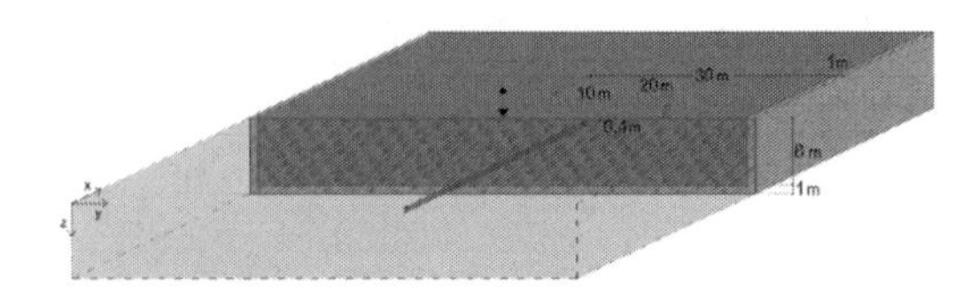

Figure 1. Cross section geometry.

Table 1. Mechanical properties used in the model.

Material	Mass [kg/m^3]	Young Modulus [MPa]	ν[–]	ζ[–]
Soil	1700	116	0.33	0.001
Inclusion	2700	4416	0.2	0.001

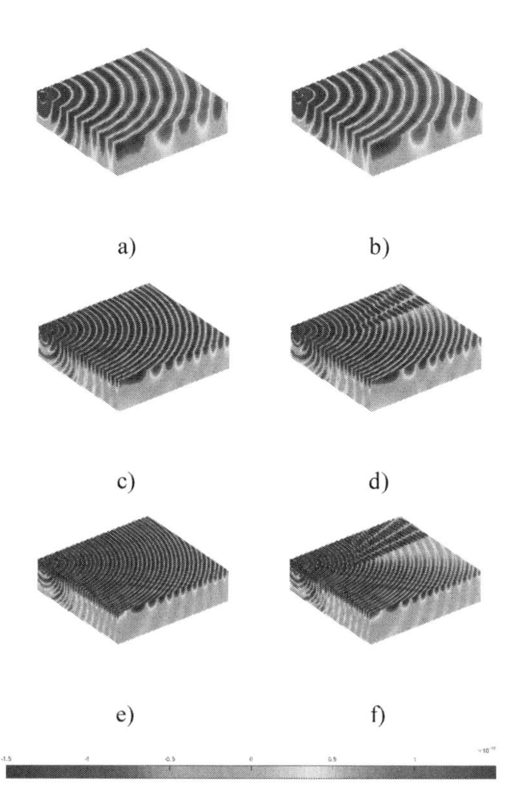

Figure 2. Vertical displacement for homogeneous media: a) 25Hz; c) 50Hz; e) 75Hz; Vertical displacement for a medium with one stiffer inclusion: b) 25Hz; d) 50Hz; f) 75Hz.

Such relationship is expressed by the following equation.

$$\theta = \sin^{-1}\left(\sqrt[4]{\frac{M\omega^2}{EI}}\frac{C_R}{\omega}\right) \tag{3}$$

where M represents the mass of the inclusion, EI the bending stiffness, C_R the Rayleigh wave propagation velocity, ω the excitation frequency and θ the open angle.

In order to clearly express the attenuation provided by the stiffer inclusion, the vertical displacement insertion loss was computed (see equation 4) and plotted in Figure 3:

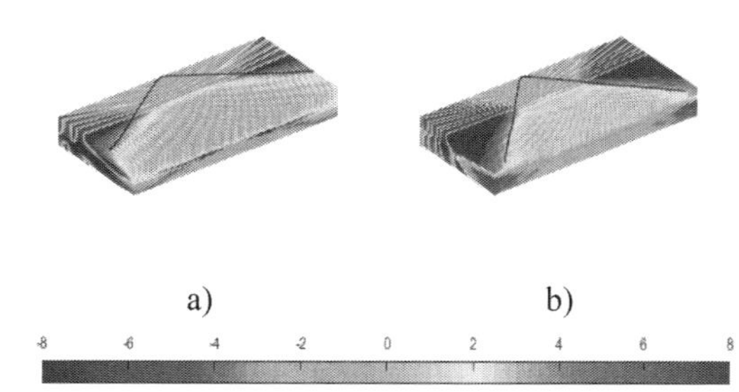

Figure 3. Vertical displacement insertion loss [dB] for single inclusion scenario and for frequencies of: a) 50Hz; b) 75Hz.

$$IL(dB) = 20\log_{10}\left(\frac{u^{ref}(x,y,z,\omega)}{u_1(x,y,z,\omega)}\right) \tag{4}$$

where u^{ref} represents the displacement results obtained for the homogenous scenario and u_1 the results for the variant cases.

The limits between mitigated and unmitigated areas were analytically determined, making use of equation (3), and superimposed to the results.

Positive insertion losses are achieved when the wave's incident angle is such that a guiding effect is activated. A direct comparison between 50Hz (Figure 3 a) and 75Hz (Figure 3 b) results shows a clear dependency between the excitation frequency and the guiding angle. If a Euler-Bernoulli beam can be assumed as being representative of the inclusion's behavior, the dispersion relationship is expressed by the following equation:

$$K_1 = \sqrt[4]{\frac{M\omega^2}{EI}} \tag{5}$$

where K_1 is the wavenumber from which the inclusion's bending mode is triggered and consequently the wave-guiding effect is observed.

To deeper understand the mechanical behavior induced by the inclusions, a wavenumber-frequency analysis was conducted for one point located 20 meters away from the loading point (red rectangle in Figure 1), and presented in Figure 4. Only the propagated region was plotted since wavenumbers greater than Rayleigh wavenumber give rise to evanescent waves. Superimposed, the dispersion relationship for a beam in a free field condition is depicted (equation (5)). The vertical axis corresponds to dimensionless wavenumber, given by:

$$K1 = k_1 * \frac{Cs}{\omega} \tag{6}$$

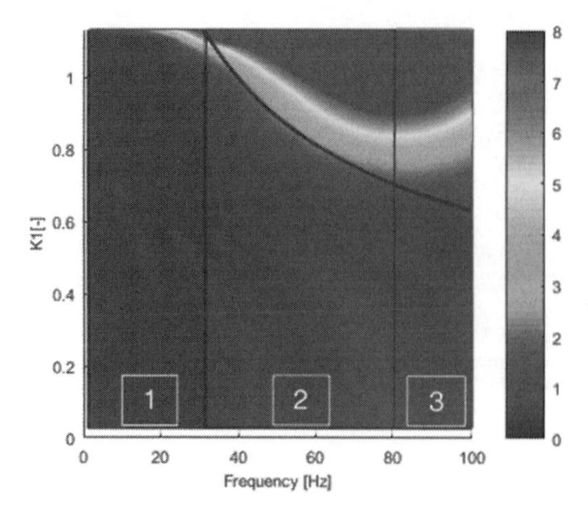

Figure 4. Frequency-wavenumber representation of the computed insertion loss (dB) for the alignment point 20m away from the load (single inclusion).

where k_1 represents the longitudinal wavenumber, C_S the shear wave velocity and $K1$ the dimensionless wavenumber.

Three distinct behaviors are observed, corresponding to the three zones marked in Figure 4. The first zone comprises frequencies among $[0;\omega_c]$, where ω_c makes reference to the frequency from which the inclusion starts to be felt. Associated with this frequency range are wavelengths between [4;∞] meters. Since the magnitude of the generated wavelength is a few orders larger than the inclusion's diameter, the impact of the mitigation countermeasure is almost negligible.

Frequencies in the range $[\omega_c;80\text{ Hz}]$ define the second zone and the expression to determine the critical frequency, is given by:

$$\omega_c = C_R\sqrt{\frac{M}{EI}} \tag{7}$$

The middle frequency range, i.e. zone 2, can be characterized by the activation of the inclusion's bending mode (equation (5)) which is responsible for the guiding effect and inevitably positive insertion losses. When $K1$ (equation (6)) is greater than K_1 (equation (5)), the wave propagation is hindered since the natural bending mode of the inclusion is activated, and, therefore, the wave starts to be guided along the inclusion.

Lastly, a decrease in the insertion loss is observed in the third zone. This is explained by the fact that, for frequencies larger than 80 Hz, the wavelength starts to become smaller than the inclusion's diameter (and also than the inclusion depth), allowing the energy to pass over the top. Thus, a loss of efficiency is perceived when the excitation frequency becomes greater than 80Hz.

3.2 *Influence of cross-section symmetry*

As previously mentioned, the mitigation measure discussed here is particularly interesting when sensitive areas near the track need to be protected. Being so, if some mitigation is needed on both sides of the track a symmetry condition is achieved. Thus, a new symmetric inclusion was added to the previous case, and a frequency-wavenumber analysis was carried out for the same point as before, being the results shown in Figure 5, with the dispersion relationship superimposed (solid black line).

A disturbance in the pattern is observed when comparing with the preceding scenario. This effect is the result of a well-known phenomenon in building acoustics, the cavity resonance (Van Hoorickx et al., 2017). Such effect occurs when standing waves are developed between the two inclusions, leading to a constructive and destructive interference in the generated wave field. When the wavelength developed among the two inclusions is a multiple of half their distance, standing waves appear. Decomposing a surface wave in transversal and longitudinal direction, $k_1 = k_R \sin\theta$ (Figure 6), the following expression, which allows to determine the longitudinal wavenumber for which the cavity resonance effect is triggered, is achieved:

$$k_1 = k_R\sqrt{1-\left(\frac{nC_R}{2df}\right)^2} \tag{8}$$

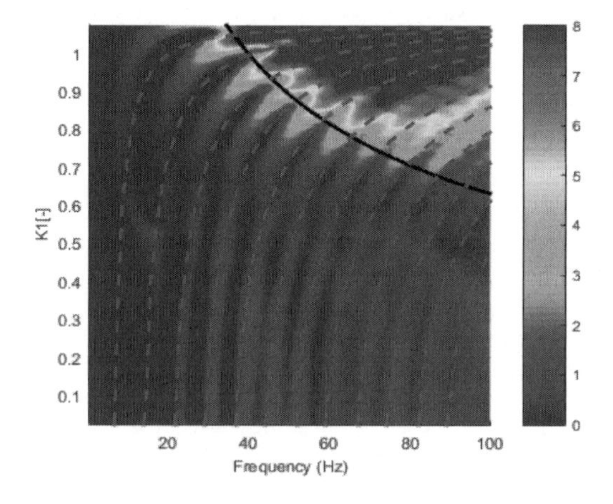

Figure 5. Insertion loss (dB) for the symmetric case with two inclusions in frequency-wavenumber domain, for the alignment 20m away from the load.

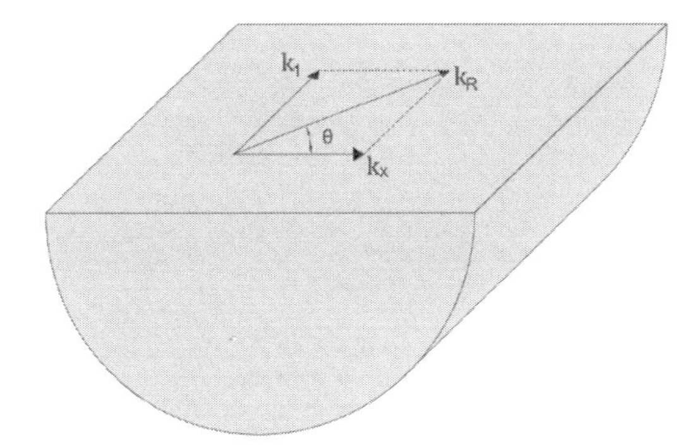

Figure 6. Decomposition of a surface wave.

where k_R represents the Rayleigh wavenumber, f the excitation frequency, d the distance among inclusions and n an integer number which represents the multiple of half wavelength generated between inclusions.

As it can be perceived from equation (8), for a single n there will be a wavenumber per frequency responsible for inducing the appearance of standing waves between inclusions. Superimposed in Figure 5, there are the cavity modes (green dashed lines), determined by the equation (8) while adopting only even numbers for the value of n.

4 DYNAMIC ANALYSIS OF A COMPOSED SYSTEM

4.1 *A row of inclusions*

A meta-structure behavior can be achieved when multiple inclusions are adopted. For the purpose of understanding the benefit induced by a meta-structure solution, a new cross section was modelled, with three inclusions, as can be seen in Figure 7.

The analysis follows a similar strategy to the one used in the previous section. Thus, Figure 8 presents the computed vertical displacement insertion loss.

Comparing the results between the current scenario with the previous one for a frequency of

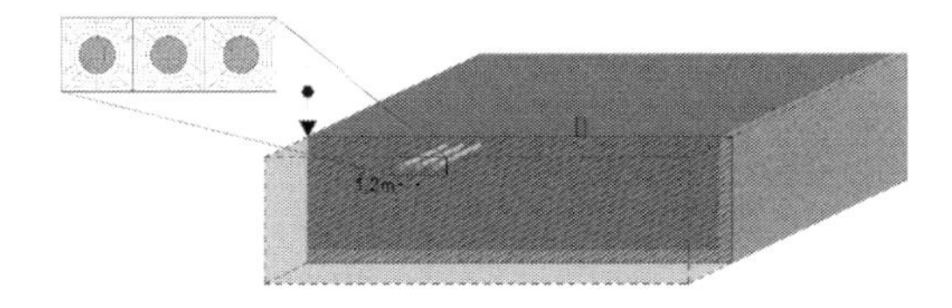

Figure 7. Cross-section with three parallel inclusions with a detail of the generated finite element mesh.

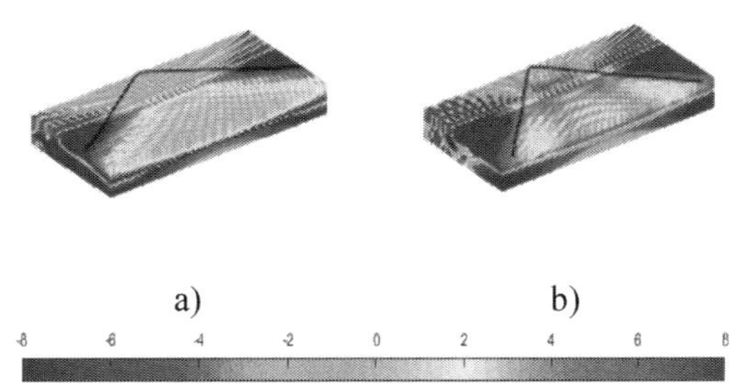

Figure 8. Vertical displacement insertion loss [dB] for the three inclusion scenario and for frequencies of: a) 50Hz; b) 75Hz.

50Hz, a substantial difference is noticed, apart from the attenuation magnitude, justified by the increased number of inclusions. However, a different conclusion is drawn when the comparison is focused in the 75Hz frequency results. Contrary to the results expressed in Figure 3 b), where no attenuation was observed inside the guiding cone, in this case, a positive insertion loss emerges. Such effect is related to the group interaction between the three inclusions.

To understand how the effect is triggered, and as it was done in the previous section, an analysis in frequency-wavenumber was performed, for a point 20 meters away from the load application point, and being presented in Figure 9.

Comparing Figure 4 with the results expressed in Figure 9, some similarities can be identified, namely the first two zones, where the first makes reference to generated wavelengths orders higher than inclusion's diameter and the second is characterized by

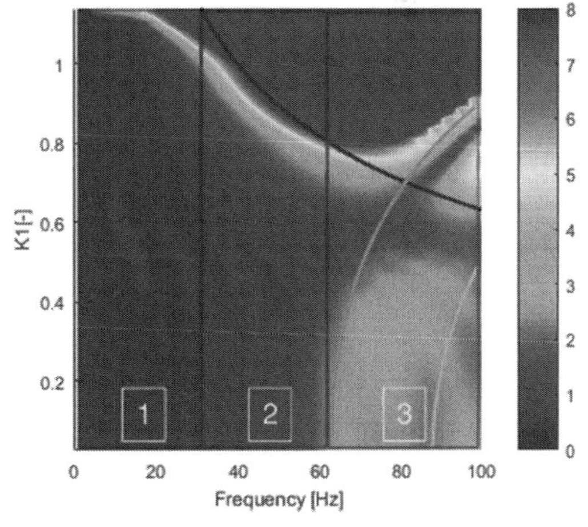

Figure 9. Insertion loss in the frequency-wavenumber domain (dB) for the alignment point 20m away from the load (multiple inclusions scenario).

the activation of the wave-guiding effect. However, differences are observed in the third zone, where a positive insertion loss arises even for plane-strain conditions, $k_1 = 0$. Such gain in the insertion loss is directly related to a group effect, which is a well-known effect in acoustic field, designated by "sonic crystal effect". Superimposed in the Figure 9 there are two curves which delimited the group effect. Those limiting curves can be obtained using the following equations:

$$f(\theta) = \frac{C_R}{2d\cos(\theta)} \tag{9}$$

$$f(\theta) = \frac{C_R}{\sqrt{2}d\cos(\theta)} \tag{10}$$

where $f(\theta)$ represents the frequency range in which the group interaction effect will be induced, according to the wave's incident angle θ.

A mitigation design can be achieved making use of equations (9) and (10) in order to design the solution to a specific frequency range. If a broader frequency range needs to be mitigated, a different distance between consecutive inclusions must be considered. On the other hand, more inclusions have to be added when the magnitude of the attenuation achieved by a given number of inclusions needs to be increased, maintaining the center-to-center distance constant.

4.2 *Influence of the inclusions' group orientation: a column of inclusions*

As the previous section adopted a set of inclusions parallel to the surface, in this section it is assessed the influence of the solution's group orientation. Thus, a new scenario was modelled assuming a set of inclusions aligned in the vertical direction. An illustrative scheme of cross-section geometry is presented in Figure 10, with the vertical displacement insertion loss plotted in Figure 11 (for frequencies of 50Hz and 75 Hz).

As can be seen in Figure 11, the resulting attenuation pattern, adopting a column of inclusions, is

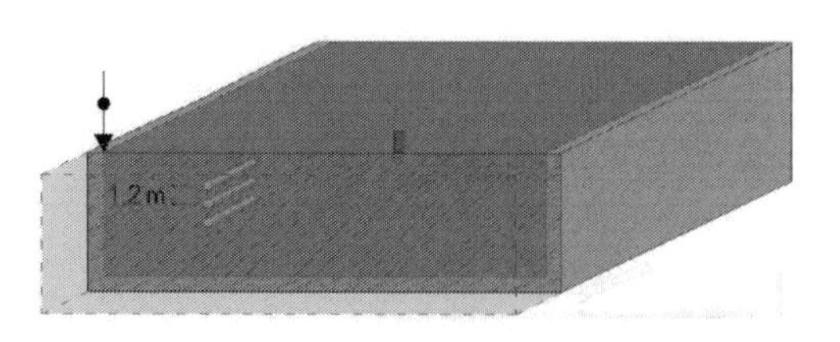

Figure 10. Cross section schematic view with a column of three inclusions.

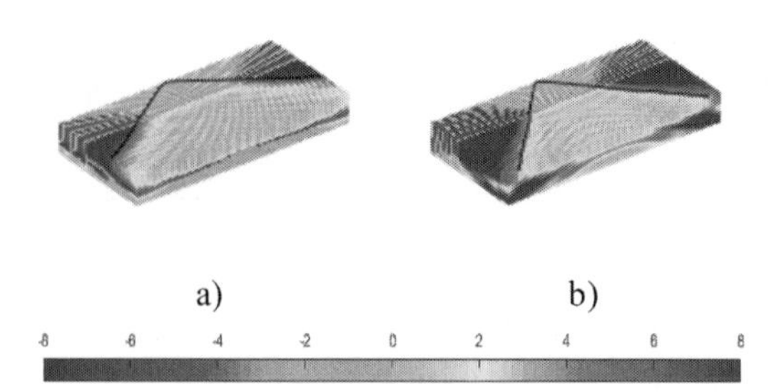

Figure 11. Vertical displacement insertion loss [dB] for a vertical set of three inclusions scenario: a) 50Hz; b)75Hz.

very similar with the single inclusion case, being also possible to identify the wave-guiding effect, previously discussed.

Contrarily, a group interaction isn't noticed, leaving the area between the guiding cones unaffected. The absence of a group interaction could be easily explained making use of Rayleigh's wave front geometry. Figure 12 illustrates the wave field resulting from a harmonic excitation on the surface for both scenarios, a row and a column of inclusions.

Taking into account the illustration in Figure 12, it is possible to see that Rayleigh waves have practically a vertical wave front. On the row of inclusions case (Figure 12 a), since multiple waves hit the group of inclusions in sequence, a group interaction can be achieved. On the other hand, in the

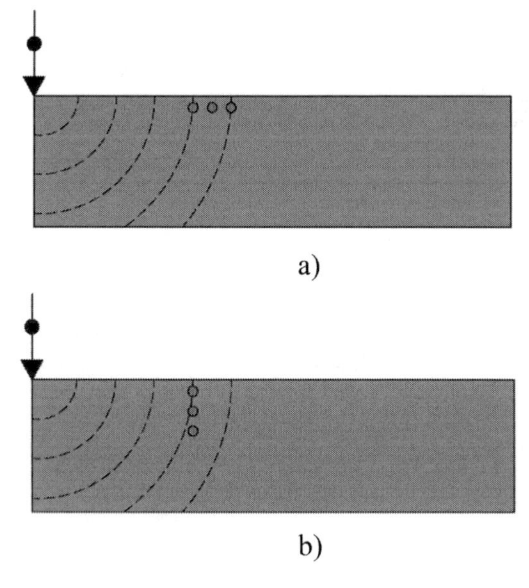

Figure 12. Schematic representation of the wave field induced by a harmonic load for: a) a row of inclusions; b) a column of inclusions.

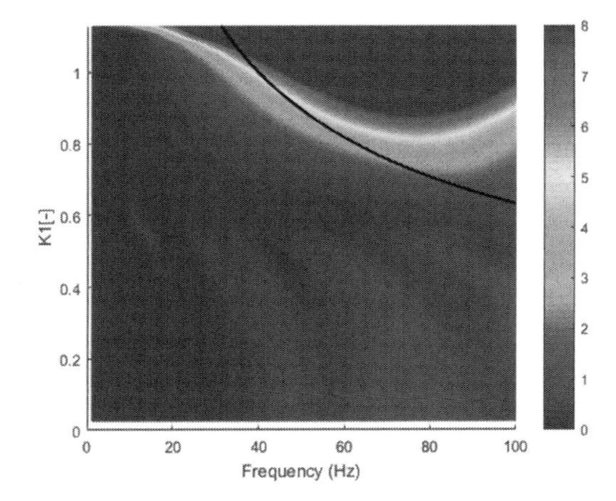

Figure 13. Insertion loss in the frequency-wavenumber domain (dB) for the alignment point 20m away from the load (column of three inclusions scenario).

case of a column of inclusions, all inclusions are hit at the same time.

Thus, all the inclusions will be synchronized not being able to develop constructive and destructive interferences in the wave field generated. A wavenumber-frequency analysis was, also, carried out with the insertion loss of the vertical displacement, for a point located 20 meters away from the load, being presented in Figure 13.

The attenuation behavior produced is very similar to the one induced by adopting an isolated inclusion, which can be seen by comparing Figure 4 and Figure 13. Therefore, the same number of inclusions can induce different attenuation behaviors just by considering different placement and orientations of the inclusions' layouts.

5 CONCLUSIONS

This paper presents a numerical study of a potential mitigation measure for shielding ground-borne vibrations. A FEM-PML formulation model based, in a 2.5D approach, was used. Firstly, the mechanical behavior originated by a single inclusion was studied. Within the several behaviors identified, for different ranges of frequencies, the wave-guiding phenomenon must be highlighted. Such effect, is activated when the generated transversal wavenumber becomes larger than the one determined by the bending wave dispersion relationship for an inclusion in a free-field condition, for a specific frequency. This leads to the formation of a cone which defines the attenuated areas. An assessment of the influence produced by cross-section symmetry allowed to identify a disturbed pattern in the wavenumber-frequency analysis. Such pattern is related to the emerging of standing waves among inclusions, known as cavity resonances, which disturb the wavefield generated. In order to assess the group behavior, a new cross-section was considered, where three inclusions parallel to the surface were modelled. Apart from the previously identified behaviors, a new one arises, being responsible for inducing attenuation within the guiding cones. A wavenumber-frequency analysis was carried out, where the emerging of a new area was detected in comparison with the single inclusion case, which is related to interaction among inclusions. This group effect, is a well-known phenomenon previously identified in the acoustic field, known as the "sonic-crystal" effect. Theoretical expressions were proposed in order to determine the specific range of frequencies in which the effect will be triggered. Finally, a new scenario was constructed, making use of the same number of inclusions as the previous case, but now being oriented in a column alignment. The results produced by a column of inclusions resembles to the initial case, of a single buried inclusion. Since the Rayleigh wave hits all inclusions at the same time, the group effect isn't activated. Such effect shows the importance related with the adopted layout.

ACKNOWLEDGEMENTS

This work was financially supported by: Project POCI-01-0145-FEDER-007457 -CONSTRUCT - Institute of R&D in Structures and Construction and Project POCI-01-0145-FEDER-007633, both funded by FEDER funds through COMPETE2020 - Programa Operacional Competitividade e Internacionalização (POCI) – and by national funds through FCT - Fundação para a Ciência e a Tecnologia; Project PTDC/ECMCOM/1364/2014. The authors are also sincerely grateful to the European Commission for the financial sponsorship of H2020 MARIE SKŁODOWSKA-CURIE RISE Project, Grant No. 691135 "RISEN: Rail Infrastructure Systems Engineering Network".

REFERENCES

Alves Costa, P., Calçada, R. & Silva Cardoso, A. 2012. Ballast mats for the reduction of railway traffic vibrations. Numerical study. Soil Dynamics and Earthquake Engineering, 42, 137–150.

Amado-Mendes, P., Alves Costa, P., Godinho, L.M.C. & Lopes, P. 2015. 2.5D MFS–FEM model for the prediction of vibrations due to underground railway traffic. Engineering Structures, 104, 141–154.

Barbosa, J., Alves Costa, P. & Calçada, R. 2015. Abatement of railway induced vibrations: Numerical

comparison of trench solutions. Engineering Analysis with Boundary Elements, 55, 122–139.
Bongini, E., Lombaert, G., François, S. & Degrande, G. 2011. A parametric study of the impact of mitigation measures on ground borne vibration due to railway traffic. In: DE ROECK, G., DEGRANDE, G., LOMBAERT, G. & MULLER, G. (eds.) EURODYN 2011. Leuven.
Connolly, D.P., Marecki, G.P., Kouroussis, G., Thalassinakis, I. & Woodward, P.K. 2016. The growth of railway ground vibration problems—A review. Science of The Total Environment, 568, 1276–1282.
Coulier, P., François, S., Degrande, G. & Lombaert, G. 2013. Subgrade stiffening next to the track as a wave impeding barrier for railway induced vibrations. Soil Dynamics and Earthquake Engineering, 48, 119–131.
Croy, I., Smith, M.G. & Waye, K. 2013. Effects of train noise and vibration on human heart rate during sleep: an experimental study. BMJ Open, doi:10.1136/bmjopen-2013-002655.
Hussein, M., Leamy, M. & Ruzzene, M. 2014. Dynamics of Phononic Materials and Structures: Historical Origins, Recent Progress, and Future Outlook. Applied Mechanics Reviews, 66, 040802.
Lopes, P., Alves Costa, P., Calçada, R. & Silva Cardoso, A. 2013. Numerical Modeling of Vibrations Induced in Tunnels: A 2.5D FEM-PML Approach. In: XIA, H. & CALÇADA, R. (eds.) Traffic Induced Environmental Vibrations and Controls: Theory and Application. Nova.
Lopes, P., Alves Costa, P., Calçada, R. & Silva Cardoso, A. 2014. Influence of soil stiffness on vibrations inside buildings due to railway traffic: numerical study. Computers & Geotechnics, 61, 277–291.
Smith, M.G., Croy, I., Ögren, M. & Persson Waye, K. 2013. On the Influence of Freight Trains on Humans: A Laboratory Investigation of the Impact of Nocturnal Low Frequency Vibration and Noise on Sleep and Heart Rate. PLoS ONE, 8, e55829.
Talbot, J. 2016. Base-isolated buildings: Towards performancebased design. Proceedings of the Institution of Civil Engineers: Structures and Buildings, 169, 574–582.
Talbot, J. & Hunt, H. 2009. On the Performance of Base-isolated Buildings and Isolation of Buildings from Rail-tunnel Vibration: a Review. In: GIBBS, B., GOODCHILD, J., HOPKINS, C. & OLDHAM, D. (eds.) Collected Papers in Building Acoustics: Sound Transmission. Multi-science.
United Nations, D.O.E. a. S.A., Population Division 2015. World urbanization prospects—highlights.
Van Hoorickx, C., Schevenels, M. & Lombaert, G. 2017. Double wall barriers for the reduction of ground vibration transmission. Soil Dynamics and Earthquake Engineering, 97, 1–13.

Numerical Methods in Geotechnical Engineering IX – Cardoso et al. (Eds)
© 2018 Taylor & Francis Group, London, ISBN 978-1-138-33203-4

Author index